# MEDICAL
# AND VETERINARY
# ENTOMOLOGY

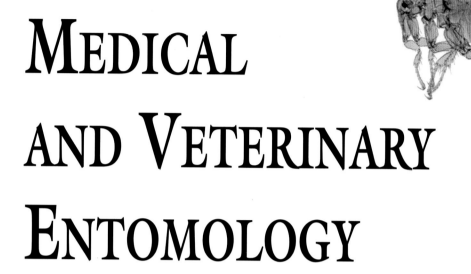

# MEDICAL
# AND VETERINARY
# ENTOMOLOGY

Edited by

**GARY MULLEN**

*Department of Entomology and Plant Pathology*
*Auburn University*
*Auburn, Alabama*

**LANCE DURDEN**

*Institute of Arthropodology and Parasitology*
*Georgia Southern University*
*Statesboro, Georgia*

**ACADEMIC PRESS**
An imprint of Elsevier Science

*Amsterdam    Boston    London    New York    Oxford    Paris    San Diego    San Francisco*
*Singapore    Sydney    Tokyo*

Academic Press
*An Elsevier Science Imprint.*
525 B Street, Suite 1900, San Diego, California 92101-4495, USA
http://www.academicpress.com

Academic Press
32 Jamestown Road, London NW1 7BY, UK
http://www.academicpress.com

Library of Congress Catalog Card Number: 2001097271

International Standard Book Number: 0-12-510451-0

PRINTED IN CHINA
02  03  04  05  06  07  DN  9  8  7  6  5  4  3  2  1

# CONTENTS

# CONTRIBUTORS

*Numbers in parentheses indicate the pages on which the authors' contributions appear.*

**Peter H. Adler** (185), Department of Entomology, Clemson University, Clemson, South Carolina 29634–0365, USA

**Roger D. Akre** (383), Deceased, formerly Department of Entomology, Washington State University, Pullman, Washington 99164–6382, USA

**Richard J. Brenner** (29), USDA-ARS, Office of Technology Transfer, Beltsville, Maryland 20705; formerly USDA-ARS, Center for Medical, Agricultural and Veterinary Entomology, Gainesville, Florida 32604, USA

**E. Paul Catts** (317), Deceased, formerly Department of Entomology, Washington State University, Pullman, Washington 99164–6382, USA

**Lance A. Durden** (1, 45, 103), Georgia Southern University, Institute of Arthropod and Parasitology, Statesboro, Georgia 30460, USA

**Woodbridge A. Foster** (203), Department of Entomology, The Ohio State University, Columbus, Ohio 43210–1220, USA

**Reid R. Gerhardt** (127), Department of Entomology and Plant Pathology, University of Tennessee, Knoxville, Tennessee 37901–1071, USA

**Raj K. Gupta** (147), Headquarters, USAMRC, Fort Detrick, Frederick, Maryland 21702–5012, USA

**Robert D. Hall** (127), Department of Entomology, University of Missouri, Columbia, Missouri 65211, USA

**William L. Krinsky** (67, 87, 303), Department Epidemiology and Public Health, Yale Medical School, New Haven, Connecticut 06520–8034, USA

**Robert S. Lane** (517), Division of Insect Biology, University of California, Berkeley, California 94720, USA

**John E. Lloyd** (349), Department of Entomology, University of Wyoming, Laramie, Wyoming 82071, USA

**John W. McCreadie** (185), Department of Biology, University of South Alabama, Mobile, Alabama 36688, USA

**Roger D. Moon** (279), Department of Entomology, University of Minnesota, St. Paul, Minnesota 55108, USA

**Gary R. Mullen** (1, 163, 317, 425, 427, 449), Department of Entomology and Plant Pathology, Auburn University, Auburn, Alabama 36849, USA

**Bradley A. Mullens** (263), Department of Entomology, University of California, Riverside, California 92521, USA

**William L. Nicholson** (517), Viral and Rickettsial Zoonoses Branch, Centers for Disease Control & Prevention, Atlanta, Georgia 30333, USA

**Barry M. OConnor** (449), University of Michigan, Museum of Zoology, University of Michigan, Ann Arbor, Michigan 48109, USA

**Hal D. Reed** (383), Biology Department, Oral Roberts University, Tulsa, Oklahoma 74171, USA

**William K. Reisen** (15), Arbovirus Field Station, University of California, Bakersfield, California 93312, USA

**Louis C. Rutledge** (147), Mill Valley, California 94941–3420, USA

**Daniel E. Sonenshine** (517), Department of Biological Sciences, Old Dominion University, Norfolk, Virginia 23529–0266, USA

**Scott A. Stockwell** (411), U. S. Army, 25th Medical Detachment (Entomology), Fort Hood, Texas 76544, USA

**Robert Traub** (103), Deceased, formerly Bethesda, Maryland, USA

**Edward D. Walker** (203), Department of Entomology, Michigan State University, East Lansing, Michigan 48824, USA

# PREFACE

This book is the result of the recognized need for an up-to-date, entomologically oriented textbook for teaching courses in medical and veterinary entomology at the college and university level. It was initiated in 1993 by Section D (Medical and Veterinary Entomology) of the Entomological Society of America, following surveys that were conducted to identify the scope and format of such a book that would best meet the needs for classroom instruction. The clear sentiment was to produce a comprehensive textbook covering both medical and veterinary entomology, recognizing the close relationship between these two disciplines. The individuals polled indicated a strong preference for organizing the chapters taxonomically according to the insect and related-arthropod groups involved, similar to that followed in W. B. Herms's original Medical and Veterinary Entomology, published in 1915. That classic work, and the seven editions that followed, served as the standard textbook for medical and veterinary entomology students for more than sixty years, until its last printing in 1979. It is hoped that this book will help to fill the void that has persisted these past two decades in promoting the teaching of medical and veterinary entomology as an important part of college and university curricula.

With the exception of the first two chapters (Introduction and Epidemiology of Vector-Borne Diseases), the chapters are similarly organized with the following major subheadings: Taxonomy, Morphology, Life History, Behavior and Ecology, Public Health Importance, Veterinary Importance, Prevention and Control, and References and Further Reading. This format should be helpful to the reader in locating specific information and to teachers who wish to assign only select portions of a chapter for their students to read. It is also for this reason

that the discussions of Public Health Importance and Veterinary Importance are addressed separately. This should maximize the flexibility in using this as a textbook for courses in either medical or veterinary entomology, or for courses in which the two disciplines are combined.

Literature citations generally have been kept to a minimum throughout the text to make the book as readable and student-friendly as possible. The authors were asked to follow a guideline of providing 45–50 references for each chapter, selecting only those that are particularly pertinent or serve as recommended follow-up sources for additional information relating to the chapter subject. Exceptions were made for some of the longer chapters (e.g., Mosquitoes, Mites, and Ticks) in which the number of references and suggested readings significantly exceeds the guideline, reflecting the breadth and importance of these particular arthropod groups.

In addition to students in the traditional sense, this book is targeted at a much broader audience, including: (1) entomologists in general; (2) specialists in other disciplines who have an interest in arthropods of medical or veterinary importance; (3) physicians, nurses, health officials, and others in the medical field who encounter insects, spiders, scorpions, ticks, and other arthropods in the course of their professional activities; (4) personnel in a variety of public health- and community-related programs (e.g., mosquito control, tick control) and professional pest control operators who wish to have an informative, readable reference source on their shelf; (5) military personnel who have responsibility for dealing with a diversity of entomologically related problems; (6) and veterinarians, wildlife personnel, zoological park officials, and other individuals in animal health-related fields who invariably encounter insects and related arthropods in their respective lines of work.

Traditionally, medical and veterinary entomology has included not only insects but also certain arachnid groups that can present problems for humans and other animals. This book follows that tradition. However, unlike most other books in this subject area, a greater amount of attention is given to these eight-legged relatives of insects in this work. Fully one-quarter of the text is devoted to arachnids, with separate chapters on scorpions, solpugids, spiders, mites, and ticks. In fact, the chapters on mites and ticks (note: ticks are simply a subgroup of mites) represent the first and third largest chapters in the book, respectively, reflecting their diversity and medical-veterinary importance. Mosquitoes are the subject of the second largest chapter, as the most important group of insects from a medical and public health perspective.

In providing appropriate balance to the insect groups discussed, the reader may be surprised to see separate chapters on the Coleoptera (beetles) and Lepidoptera (moths and butterflies). These two groups are not the first that come to mind when one thinks of medical and veterinary entomology and typically are treated only briefly, if at all, in most medical-veterinary books. However, as the space devoted to these groups indicates, the beetles and moths and the problems they cause deserve more attention than they are generally accorded.

We wish to acknowledge the loss of three contributors to this book who passed away after submitting, or contributing to, the initial and one or more revised drafts of their chapters. They are Roger D. Akre and E. Paul Catts of the Department of Entomology, Washington State University, and the late Robert Traub of Bethesda, Maryland. In recognition of their contributions, the respective chapters appropriately bear their names as authors, albeit posthumously. This book is dedicated to the memory of these distinguished entomologists.

*Gary R. Mullen*
*Lance A. Durden*

# ACKNOWLEDGMENTS

Many people have helped in various ways with the preparation of this textbook and we are grateful to all of them. Foremost, we must thank all of the chapter authors who wrote their assigned chapters and patiently accommodated our editorial requests, comments, guidelines, and also responded to comments furnished by outside reviewers. Obviously, without the commitment and dedication of the chapter authors, this project could not have come to fruition.

Rebecca L. Nims (Auburn University, Alabama and Georgia Southern University, Statesboro) was contributing art editor and expertly captured the majority of the figures as digital images and then digitally labeled and improved many of them; she also prepared original figure 23.2. Margo A. Duncan (Gainesville, Florida) was commissioned to prepare original or composite pen and ink drawings for the following figures: 2.1, 2.3, 4.6, 4.8, 7.3, 18.1A, and 18.4. We also wish to thank the following individuals who contributed original artwork in the form of line drawings, with the respective figure numbers indicated in parentheses following their institutional affiliations: the late Dr. E. Paul Catts (Washington State University, Pullman; 16.1–16.9, 16.11, 16.13–16.15, 16.17, 16.20, 16.23, 16.26, 16.31, 16.32, 16.35, 16.36, 16.39), Dr. Woodbridge A. Foster (The Ohio State University, Columbus; 12.14), Susan J. M. Hope (Mebane, North Carolina; 13.1), Takumasa Kondo (Auburn University, Alabama; 8.9), Dr. William L. Krinsky (Yale Medical School, New Haven, Connecticut; 5.6), Dr. Roger D. Moon (University of Minnesota, St. Paul; 14.2), Dr. Bradley A. Mullens (University of California, Riverside; 13.3), Dr. Blair J. Sampson (USDA-ARS, Small Fruit Research Center, Poplarville, Mississippi; 19.5), and Dr. Lawrence W. Zettler (Illinois College, Jacksonville; 11.2).

We are grateful to the following persons for providing original photographs, slides, and other illustrations reproduced in this book, with the respective figure numbers indicated in parentheses: Dr. W. V. Adams, Jr. (Louisiana State University, Baton Rouge; 13.11), Dr. Peter H. Adler (Clemson University, South Carolina; 11.1, 11.5), the late Dr. Roger D. Akre (Washington State University, Pullman; 19.9, 19.14, 19.15, 19.24); Dr. Hans Bänziger (Chaing Mai University, Thailand; 18.18–18.23), Dr. Yehuda Braverman (Kimron Veterinary Institute, Israel; 10.19), Dr. Alberto B. Broce (Kansas State University, Manhattan; 14.7), Dr. Corrie Brown (Foreign Animal Disease Diagnostic Laboratory, Plum Island, New York and University of Georgia, Athens; 10.16), Dr. Jerry F. Butler (University of Florida, Gainesville; 18.13, 19.13, 23.25, 23.40, 23.43), Dr. James D. Castner (University of Florida, Gainesville; 3.3–3.11), Dr. Ronald D. Cave (Escuela Agricola Panamericana, Zamorano, Honduras; 16.25), Valerie J. Cervenka, (University of Minnesota, St. Paul; 14.22), Dr. Jack Kelly Clark (University of California, Davis; 24.16), Dr. George H. D'Andrea (Alabama State Veterinary Diagnostic Laboratory, Auburn; 10.14), Dr. Debbie R. Folkerts (Auburn University, Alabama; 21.1), Dr. Woodbridge A. Foster (The Ohio State University, Columbus; 12.14, 12.17, 12.19, 12.25, 12.27, 12.31), Dr. Ellis G. Greiner (University of Florida, Gainesville; 11.8), Dr. Martin Hall (The Natural History Museum, London; 16.37), Dr. Robert G. Hancock (Cumberland College, Williamsburg, Kentucky; 12.23), Dr. Carl C. Hansen (National Museum of Natural History, Washington, D.C.; 18.12), Dr. Elton J. Hansens (Asheville, North Carolina; 4.12, 4.13, 7.14, 13.6, 14.12, 19.12, 19.30, 23.11), Dr. Charles M. Hendrix,

(Auburn University, Alabama; 23.48, 23.60), Dr. Nancy C. Hinkle (University of California, Riverside and University of Georgia, Athens; 4.14, 7.7–7.10), Paul M. Horton (Clemson University, South Carolina; 19.28), Lacy L. Hyche (Auburn University, Alabama; 18.15), Dr. Robert J. Keiffer (University of California, Hopland; 24.15), Takumasa Kondo (Auburn University, Alabama; 19.23), Peter J. Landolt (USDA-ARS, Wapato, Washington), Dr. Robert S. Lane (University of California, Berkeley; 24.13, 24.19), Lloyd L. Lauerman (Alabama State Veterinary Diagnostic Laboratory, Auburn; 10.8–10.11), Dr. Kriangkrai Lerthudsnee (Mahidol University, Bangkok, Thailand; 12.20), Dr. John E. Lloyd (University of Wyoming, Laramie; 4.15, 17.2, 17.6, 17.10–17.12), Dr. John M. MacDonald (Auburn University, Alabama; 23.45, 23.46, 23.54), the late Dr. Sturgis McKeever (Georgia Southern University, Statesboro; 13.4, 13.13, 16.18, 18.6, 18.8, 18.11, 18.14, 18.17, 22.13, 22.17), Dr. Hendrik J. Meyer (North Dakota State University, Fargo; 14.18, 14.21, 14.23), Dr. Roger D. Moon (University of Minnesota, St. Paul; 14.19), Dr. Gary R. Mullen (Auburn University, Alabama; 5.14, 6.9, 10.4, 10.5, 10.12, 14.20, 16.19, 18.7, 19.8, 19.19–19.22, 19.29, 22.12, 22.19), Dr. Bradley A. Mullens (University of California, Riverside; 10.2, 13.7), Dr. Harold D. Newson (Michigan State University, East Lansing; 12.34, 12.36), Dr. Yoshiro Ohara (Tohoku University School of Medicine, Sendai, Japan; 13.10), Dr. Jonathan D. Patterson (Michigan State University, East Lansing; 12.35), the late Dr. L. L. Pechuman (Cornell University, Ithaca, New York; 13.5), Dr. Hal C. Reed (Oral Roberts University, Tulsa, Oklahoma; 19.18, 19.31), Ross Ritter (Potter Valley, California; 24.14), Mary Elizabeth Rogers (Waukegan, Illinois; 10.17), Dr. William S. Romoser (The Ohio State University, Columbus; 12.20), Dr. Justin O. Schmidt (Carl Hayden Bee Research Laboratory, Tucson, Arizona; 19.4, 19.27), Dr. Philip J. Scholl (USDA-ARS, Lincoln, Nebraska; 16.34), Dr. Scott A. Stockwell (US Army, 25th Medical Detachment, Fort Hood, Texas; 20.1, 20.7, 20.8), Dr. Daniel R. Suiter (University of Georgia, Griffin; 3.2), Dr. Robert B. Tesh (University of Texax Medical Branch, Galveston; 5.4), P. Kirk Visscher (University of California, Riverside; 10.6), Dr. Laurel L. Walters (Lieen-Follican Research, Bishop, California; 9.7), and Dr. Ralph E. Williams (Purdue University, West Lafayette, Indiana; 14.17).

In addition, we express our appreciation to these individuals who assisted in the preparation of figures or in other ways provided illustrations from which the figures that appear in this book were selected: Dr. Anne-Marie Callcott (USDA-APHIS, Gulfport, Mississippi), Dr. Randy Davidson (Southeastern Cooperative Wildlife Disease Study, University of Georgia, Athens), Dr. Harold J. Harlon (Crownsville, Maryland), Dr. Sidney Kunz (U.S. Livestock Insects Laboratory, USDA/ARS, Kerrville, Texas), Dr. Peter Landolt (USDA-ARS, Wapato, Washington), Dr. Donald G. Manley (Pee Dee Research and Education Center, Florence, South Carolina), Dr. Leonard E. Munstermann (Yale University, New Haven, Connecticut), Dr. Edward T. Schmidtmann (USDA-ARS, Arthropod-Borne Animal Diseases Laboratory, Laramie, Wyoming) and Dr. Bryce Walton (Gettysburg, Pennsylvania). Additional figure and illustration credits are provided throughout the book.

We are especially grateful to the following individuals who served as external reviewers or provided advice for one or more of the chapters or chapter sections: Dr. John R. Anderson (Connecticut Agricultural Experiment Station, New Haven), Renee Anderson (Auburn University, Alabama), Dr. Arthur G. Appel (Auburn University, Alabama), the late Dr. Ross H. Arnett (formerly of Gainesville, Florida), Matt Aubuchon (Auburn University, Alabama), Dr. Hans Bänziger (Chaing Mai University, Thailand), Dr. Alberto B. Broce (Kansas State University, Manhattan), Dr. Richard L. Brown (Mississippi State University, Mississippi), Dr. John Burger (University of New Hampshire, Durham), Dr. John B. Campbell (University of Nebraska, North Platte), Dr. Bruce M. Christensen (University of Wisconsin, Madison), Dr. G. B. Edwards (Florida State Collection of Arthropods, Gainesville), Dr. Richard G. Endris (Schering Plough, Union, New Jersey), Dr. Marc Epstein (U.S. National Museum of Natural History, Smithsonian Institution, Washington D.C.), Dr. Richard D. Fell (Virginia Polytechnic and State University, Blacksburg), Dr. Howard Frank (University of Florida, Gainesville), Dr. Reid R. Gerhardt (University of Tennessee, Knoxville), Dr. M. Lee Goff (University of Hawaii at Manoa, and Chaminade University, Honolulu), Dr. Ellis C. Greiner (University of Florida, Gainesville), Dr. William L. Grogan (Salisbury State University, Maryland), Dr. Duane J. Gubler (Centers for Disease Control and Prevention, Fort Collins, Colorado), the late Dr. William L. Hardy (formerly at the University of California, Berkeley), Dr. Cluff E. Hopla (University of Oklahoma, Norman), Dr. James A. House (USDA-APHIS, Foreign Animal Disease Diagnostic Laboratory, Plum Island, New York), Dr. Lawrence J. Hribar (Florida Keys Mosquito Control District, Marathon), Dr. Gregg J. Hunt (USDA/ARS, Arthropod-Borne Animal Diseases Laboratory, Laramie, Wyoming), Lacy L. Hyche (Auburn University, Alabama), Dr. James E. Keirans (Georgia Southern University, Statesboro), Dr. Robert R. Killick-Kendrick (Imperial College at Silwood Park, Ascot, England), Dr. Katherine M. Kocan (Oklahoma State University, Stillwater), Dr. Elliot S. Krafsur (Iowa State University, Ames), Dr. Daniel C. Kurtak (Chewelah, Washington), Dr. Phillip G. Lawyer

(Uniformed Services University of Health Sciences, Bethesda, Maryland), Dr. Robert E. Lewis (Ames, Iowa), Dr. John E. Lloyd (University of Wyoming, Laramie), Dr. Timothy J. Lysyk (Agriculture Canada, Lethbridge, Alberta), Dr. Adrian G. Marshall (University of Aberdeen, Scotland), Dr. Robert Minckley (University of Utah, Salt Lake City), Dr. Roger D. Moon (University of Minnesota, St. Paul), Dr. Charles D. Morris (Winter Haven, Florida), Dr. Bradley A. Mullens (University of California, Riverside), Dr. Roy A. Norton (State University of New York, Syracuse), Dr. Richard S. Patterson (Gainesville, Florida), Michelle Perdue (Auburn University, Alabama), Dr. Peter V. Perkins (Gainesville, Florida), Dr. Robert V. Peterson (Monte L. Bean Life Sciences Museum, Brigham Young University, Provo, Utah), Dr. Oscar J. Pung (Georgia Southern University, Statesboro), Dr. Sarah E. Randolph (University of Oxford, England), Dr. William C. Reeves, Sr. (University of California, Berkeley), Dr. William K. Reisen (Arbovirus Field Station, University of California, Bakersfield), Dr. Richard G. Robbins (Defense Pest Management Information Analysis Center, Forest Glen Section, Walter Reed Army Institute of Research, Washington D.C.), Dr. The Honorable Dame Miriam Rothschild (Ashton Wold, England), Dr. Michael J. Rust (University of California, Riverside), Dr. Raymond E. Ryckman (Redlands, California), Dr. Justin O. Schmidt (Carl Hayden Bee Research Center, Tucson, Arizona), Dr. Edward T. Schmidtmann (USDA-ARS, Arthropod-Borne Animal Diseases Laboratory, Laramie, Wyoming), Dr. Christopher J. Schofield (London School of Tropical Medicine & Hygiene, United Kingdom), Dr. Philip J. Scholl (USDA-ARS, Lincoln, Nebraska), Dr. Craig T. Sheppard (University of Georgia, Coastal Experiment Station, Tifton), Dr. Louis N. Sorkin (American Museum of Natural History, New York, New York), Dr. Scott Stewart (NIH/NAIAD, Hamilton, Montana), Dr. Christine A. Sundermann (Auburn University, Alabama), Dr. Sabina F. Swift (University of Hawaii at Manoa, Honolulu), Dr. Pete D. Teel (Texas A & M University, College Station), Dr. Robert B. Tesh (University of Texas Medical Branch, Galveston), Dr. Stephen Torr (Natural Resources Institute, Chatham Maritime, England), the late Dr. Robert Traub (formerly of Bethesda, Maryland), Dr. Michael J. Turell (U.S. Army Medical Research Institute of Infectious Diseases, Fort Detrick, Frederick, Maryland), Dr. William J. Turner (Washington State University, Pullman), Dr. S. Bradleigh Vinson (Texas A & M University, College Station), Dr. Laurel L. Walters (Lieen-Follican Research, Bishop, California), Amelia Williams (Auburn University, Alabama), Dr. Mark L. Wilson (University of Michigan, Ann Arbor), Dr. James C. Wright (Auburn University, Alabama), Dr. Russell E. Wright (Oklahoma State University, Stillwater), and Dr. David G. Young (Gainesville, Florida).

We are indebted to the following individuals for their assistance in conducting library searches or otherwise providing the authors with relevant literature: Dr. Harold J. Harlan (Crownsville, Maryland), Elizabeth Mason (North Dakota State University, Fargo), Dr. Richard G. Robbins (Defense Pest Management Information Analysis Center, Forest Glen Section, Walter Reed Army Institute of Research, Washington D.C.), the late Dr. Robert Traub (formerly of Bethesda, Maryland) and Anne R. Viera (University of Tennessee, Knoxville).

We also would like to thank the staff with whom we have worked at Academic Press/Elsevier Science in San Diego, especially Dr. Charles R. Crumly, Monique Larson, Molly Wofford, and Christine Vogelei.

# 1

# INTRODUCTION

LANCE A. DURDEN AND GARY R. MULLEN

*Medical entomology* is the study of insects, insect-borne diseases, and other associated problems that affect humans and public health. *Veterinary entomology* is similarly the study of insects and insect-related problems that affect domestic animals, particularly livestock and companion animals (dogs, cats, horses, caged birds, etc.). In addition, veterinary entomology includes insect-associated problems affecting captive animals in zoological parks and wildlife in general. *Medical-veterinary entomology* combines these two disciplines.

Traditionally the fields of medical and veterinary entomology have included health-related problems involving arachnids (particularly mites, ticks, spiders, and scorpions). This broad approach encompassing insects and arachnids is followed in this text. Alternatively, the study of health-related problems involving arachnids is called *medical-veterinary arachnology* or, if just mites and ticks are considered, *medical-veterinary acarology.*

Historically, both medical and veterinary entomology have played major roles in the development of human civilization and animal husbandry. Outbreaks of insect-borne diseases of humans have profoundly influenced human history; these include such diseases as yellow fever, plague, louse-borne typhus, malaria, African trypanosomiasis, Chagas disease, and lymphatic filariasis. Likewise, livestock scourges such as bovine babesiosis, bovine theileriosis, scabies, pediculosis, and botfly infestations, all of which are caused or transmitted by arthropods, have greatly influenced animal production and husbandry practices. Arthropod-related disorders continue to cause significant health problems to humans, domestic animals, and wildlife. At the same time, new strains of known pathogens, as well as previously unrecognized disease agents transmitted by arthropods, are causing newly recognized diseases (e.g., Lyme disease and human granulocytic ehrlichiosis) and the resurgence of diseases that had been suppressed for many years (e.g., malaria). In fact, emerging and resurging *arthropod-borne diseases* are recognized as a growing health concern by public health and veterinary officials (Wilson and Spielman 1994, Walker *et al.* 1996, Gubler 1998, Winch 1998, Gratz 1999).

## GENERAL ENTOMOLOGY

Basic concepts of entomology, such as morphology, taxonomy and systematics, developmental biology, and ecology, provide important background information for medical and veterinary entomologists. General entomology books which the reader will find helpful in this regard include Borror *et al.* (1989), Gullan and Cranston (1994), Gillot (1995), Elzinga (1997), Chapman (1998), and Romoser and Stoffolano (1998). References that provide a more taxonomic or biodiversity-oriented approach to general entomology include works by Arnett (1993), Richards and Davies (1994), Bosik (1997), and Daly *et al.* (1998). General insect morphology is detailed in Snodgrass (1993), whereas a useful glossary of general entomology is Torre-Bueno (1962). Texts on *urban entomology*, the study of insect pests in houses, buildings, and urban areas, which also has relevance to medical-veterinary entomology, have been prepared by Ebeling (1975), Hickin (1985), Mallis (1997), and Robinson (1996). General texts on *acarology* include works by Krantz (1978), Woolley (1987), Evans (1992) and Walter and Proctor (1999).

## MEDICAL-VETERINARY ENTOMOLOGY LITERATURE

Textbooks or monographs pertaining to medical entomology, veterinary entomology, or the combined discipline of medical-veterinary entomology are listed under these headings at the end of this chapter. Most of these publications emphasize arthropod *morphology*, *biology*, *systematics*, and *disease relationships*, whereas some of the more recent texts, such as Beaty and Marquardt (1996) and Crampton *et al.* (1997), emphasize *molecular aspects* of medical-veterinary entomology. Other works are helpful regarding *common names* of arthropods of medical-veterinary importance (Pittaway 1992), *surveillance techniques* (Bram 1978), *control measures* (Drummond *et al.* 1988), or *ectoparasites* (Andrews 1977, Marshall 1981, Kim 1985, Uilenberg 1994, Barnard and Durden

1999). Publications that devote substantial sections to arthropods associated with wildlife and the pathogens they transmit include Davis and Anderson (1971), Davidson *et al.* (1981), Fowler (1986) and Davidson and Nettles (1997).

Several journals and periodicals are devoted primarily to medical and/or veterinary entomology. These include the *Journal of Medical Entomology*, published by the Entomological Society of America (Lanham, MD); *Medical and Veterinary Entomology*, published by the Royal Entomological Society of London (UK); *Journal of Vector Ecology*, published by the *Society of Vector Ecologists* (Corona, CA); *Vector Borne and Zoonotic Diseases*, published by Mary Ann Liebert, Inc., Larchmont, New York; and *Review of Medical and Veterinary Entomology*, published by CAB International (Wallingford, UK). Journals specializing in parasitology, tropical medicine, or wildlife diseases that also include articles on medical-veterinary entomology include *Parasitology*, published by the British Society for Parasitology; *Journal of Parasitology*, published by the American Society of Parasitologists (Lawrence, KS); *Parasite-Journal de la Société Française de Parasitologie*, published by PRINCEPS Editions (Paris, France); *Advances in Disease Vector Research*, published by Springer-Verlag (New York); *Bulletin of the World Health Organization*, published by the World Health Organization (Geneva, Switzerland); *Journal of Wildlife Diseases*, published by the Wildlife Disease Association (Lawrence, KS); *Emerging Infectious Diseases*, published by the Centers for Disease Control and Prevention (Atlanta, GA); the *American Journal of Tropical Medicine and Hygiene*, published by the American Society of Tropical Medicine and Hygiene (Northbrook, IL); and *Memorias Do Instituto Oswaldo Cruz;* published by the Instituto Oswaldo Cruz (Rio de Janeiro, Brazil). Various Internet Web sites pertaining to medical-veterinary entomology can also be accessed for useful information.

## HISTORY OF MEDICAL-VETERINARY ENTOMOLOGY

Problems caused by biting and annoying arthropods and the pathogens they transmit have been the subject of writers since antiquity (Service 1978). Homer (mid-8th century BC), Aristophanes (ca. 448–380 BC), Aristotle (384–322 BC), Plautus (ca. 254–184 BC), Columella (5 BC to AD 65), and Pliny (AD 23–79) all wrote about the nuisance caused by flies, mosquitoes, lice, and/or bedbugs. However, the study of modern medical-veterinary entomology is usually recognized as beginning in the late 19th century, when blood-sucking arthropods were first proven to be vectors of human and animal pathogens.

Englishman Patrick Manson (1844–1922) was the first to demonstrate pathogen transmission by a blood-feeding arthropod. Working in China in 1877, he showed that the mosquito *Culex pipiens fatigans* is a vector of *Wuchereria bancrofti,* the causative agent of Bancroftian filariasis. Following this landmark discovery, the role of various blood-feeding arthropods in transmitting pathogens was recognized in relatively rapid succession.

In 1891, Americans Theobald Smith (1859–1934) and F. L. Kilbourne (1858–1936) implicated the cattle tick, *Boophilus annulatus,* as a vector of *Babesia bigemina,* the causative agent of Texas cattle fever (bovine babesiosis/piroplasmosis). This paved the way for a highly successful *B. annulatus*-eradication program in the United States directed by the US Department of Agriculture. The eradication of this tick resulted in the projected goal: the elimination of indigenous cases of Texas cattle fever throughout the southern United States.

In 1898, Englishman Sir Ronald Ross (1857–1932), working in India, demonstrated the role of mosquitoes as vectors of avian malarial parasites from diseased to healthy sparrows. Also in 1898, the cyclical development of malarial parasites in anopheline mosquitoes was described by Italian Giovani Grassi (1854–1925). In the same year, Frenchman Paul Louis Simond (1858–1947), working in Pakistan (then part of India), showed that fleas are vectors of the bacterium that causes plague.

In 1848, American physician Josiah Nott (1804–1873) of Mobile, AL, had published circumstantial evidence that led him to believe that mosquitoes were involved in the transmission of yellow fever virus to humans. In 1881, Cuban-born Scottish physician Carlos Finlay (1833–1915) presented persuasive evidence for his theory that what we know today as the mosquito *Aedes aegypti* was the vector of this virus. However, it was not until 1900 that American Walter Reed (1851–1902) led the US Yellow Fever Commission at Havanna, Cuba, which proved *A. aegypti* to be the principal vector of yellow fever virus.

In 1903, Englishman David Bruce (1855–1931) demonstrated the ability of the tsetse fly *Glossina palpalis* to transmit, during blood-feeding, the trypanosomes that cause African trypanosomiasis.

Other important discoveries continued well into the 20th century. In 1906, American Howard Taylor Ricketts (1871–1910) proved that the Rocky Mountain wood tick, *Dermacentor andersoni,* is a vector of *Rickettsia rickettsii,* the causative agent of Rocky Mountain spotted fever. In 1907, F. P. Mackie (1875–1944) showed that human body lice are vectors of *Borrelia recurrentis,* the spirochete that causes louse-borne (epidemic) relapsing fever. In 1908, Brazilian Carlos Chagas (1879–1934) demonstrated transmission of the agent that causes American trypanosomiasis, later named Chagas disease in his honor, by the cone-nose bug *Panstrongylus megistus.*

In 1909, Frenchman Charles Nicolle (1866–1936), working in Tunis, showed that human body lice are vectors of *Rickettsia prowazekii,* the agent of louse-borne (epidemic) typhus.

These important discoveries, as well as others of historical relevance to medical-veterinary entomology, are discussed in more detail in the references listed at the end of this chapter. Because of the chronology of many major discoveries relevant to this topic in the 50-year period starting in 1877, this time has been called the "golden age of medical-veterinary entomology" (Philip and Rozeboom 1973).

## IDENTIFICATION AND SYSTEMATICS OF ARTHROPODS OF MEDICAL-VETERINARY IMPORTANCE

Table I provides a list of the eight orders of insects and four orders of arachnids that are of particular interest to medical-veterinary entomologists. Accurate identification of these arthropods is an important first step in determining the types of problems they can cause and, subsequently, in implementing control programs.

Although taxonomy and identification are discussed in more detail with respect to arthropod groups treated in the chapters that follow, some publications provide a broader perspective on the classification, taxonomy,

TABLE I

**Principle Orders of Insects and Arachnids of Medical-Veterinary Interest**

| Order | Common names |
|---|---|
| Class Insecta | |
| Order Blattaria | Cockroaches |
| Order Phthiraptera | Lice |
| Order Hemiptera | True bugs: bedbugs, kissing bugs, assassin bugs |
| Order Coleoptera | Beetles |
| Order Siphonaptera | Fleas |
| Order Diptera | Flies: mosquitoes, black flies, no-see-ums, horse flies, deer flies, sand flies, tsetse flies, house flies, stable flies, horn flies, bot flies, blow flies, flesh flies, louse flies, keds, etc. |
| Order Lepidoptera | Moths and butterflies |
| Order Hymenoptera | Wasps, hornets, velvet ants, ants, bees |
| Class Arachnida | |
| Order Scorpionida | Scorpions |
| Order Solpugida | Solpugids, sun spiders, camel spiders, barrel spiders |
| Order Acari | Mites, ticks |
| Order Araneae | Spiders |

and/or identification of a range of arthropods of medical-veterinary importance. These include two works published by the US Centers for Disease Control and Prevention (1979, 1994), as well as citations by Service (1988), Hopla *et al.* (1994), Lago and Goddard (1994), and Davis (1995). Also, some medical-veterinary entomology books are very taxonomically oriented, with emphasis on identification, e.g., Baker *et al.* (1956), Smith (1973), Lane and Crosskey (1993), Walker (1995) and Baker (1999).

# TYPES OF PROBLEMS CAUSED BY ARTHROPODS

## ANNOYANCE

Irrespective of their role as blood-feeders (hematophages), parasites, or vectors of pathogens, certain arthropods cause severe annoyance to humans or other animals because of their biting behavior. These include lice, bedbugs, fleas, deer flies, horse flies, tsetse flies, stable flies, mosquitoes, black flies, biting midges, sand flies, chiggers, and ticks. Some, however, do not bite but instead are annoying because of their abundance, small size, or habit of flying into or around the eyes, ears, and nose. Nonbiting arthropods that cause annoyance include the house fly, chironomid midges, and eye gnats. Large populations of household or filth-associated arthropods, such as houseflies and cockroaches, can also be annoying. Nuisance arthropods are commonly problems for humans at outdoor recreational areas, including parks, lakes, and beaches.

## ENVENOMATION

Members of several groups of arthropods can inject venom when they bite or sting. Most notable are bees, wasps, ants, spiders, and scorpions. Others, such as blister beetles and certain caterpillars, produce toxins that can cause problems when they are touched or ingested. Envenomation by these arthropods is discussed in more detail in the respective chapters that follow.

In general, envenomation results in medical or veterinary conditions ranging from mild itching to intense debilitating pain or even to life-threatening encounters due to allergic reactions. Envenomation sites on the skin usually appear as reddened, painful, more or less circular lesions surrounding the bite, sting, or point of venom contact. These areas may become raised and can persist for several days, often causing inflammation of adjacent tissues. Caterpillars that cause envenomation typically secrete toxins from specialized setae that

penetrate the skin, causing contact dermatitis. Blisters can also develop at arthropod envenomation sites on contact of the skin with blister beetles (family Meloidae), false blister beetles (family Oedemeridae), and certain rove beetles (family Staphylinidae) which secrete toxins in their body fluids. If meloid beetles are accidentally ingested with fodder by livestock, the resulting systemic reaction can be life threatening.

## ALLERGIC REACTIONS

A relatively wide spectrum of allergic reactions can occur in humans or animals exposed to certain arthropods. Many of the species involved also cause envenomation by biting or stinging, with the allergic reaction resulting from an overresponsive host immune system. Bites or stings from arthropods such as lice, bedbugs, fleas, bees, ants, wasps, mosquitoes, and chiggers all can result in allergic host reactions. Contact allergies can occur when certain beetles or caterpillars touch the skin. Respiratory allergies can result from inhaling allergenic air-borne particles from cockroaches, fleas, or other arthropods. The recirculation of air by modern air-handling systems in buildings tends to exacerbate inhalation of insect allergens.

Humans and animals usually react to repeated exposure to bites or stings from the same or antigenically related arthropods in two possible ways, depending on the nature of the antigen or venom inoculated and the sensitivity of the host: (1) desensitization to the bites or stings with repeated exposure and (2) allergic reactions which, in extreme cases, can develop into life-threatening anaphylactic shock. However, a distinct five-stage sequence of reactions typically occurs in most humans when they are repeatedly bitten or stung by the same, or related, species of arthropod over time. Stage 1 involves no skin reaction but leads to development of hypersensitivity. Stage 2 is a delayed-hypersensitivity reaction. Stage 3 is an immediate-sensitivity reaction followed by a delayed-hypersensitivity reaction. Stage 4 is immediate reaction only, whereas Stage 5 again involves no reaction (i.e., the victim becomes desensitized). These changes reflect the changing host immune response to prolonged and frequent exposure to the same arthropod or to cross-reactive allergens or venoms.

## INVASION OF HOST TISSUES

Some arthropods invade the body tissues of their host. Various degrees of invasion occur, ranging from subcutaneous infestations to invasion of organs such as the lungs and intestine. Invasion of tissues allows arthropods to exploit different host niches and usually involves the immature stages of parasitic arthropods.

The invasion of host tissues by fly larvae, called *myiasis,* is the most widespread form of host invasion by arthropods. Larvae of many myiasis-causing flies move extensively through the host tissues. As they mature, they select characteristic host sites (e.g., stomach, throat, nasal passages, or various subdermal sites) in which to complete the parasitic phase of their development.

Certain mites also invade the skin or associated hair follicles and dermal glands. Others infest nasal passages, lungs, and air sacs or stomach, intestines, and other parts of the alimentary tract of their hosts. Examples include scabies mites, follicle mites, nasal mites, lung mites, and a variety of other mites that infest both domestic and wild birds and mammals.

## ARTHROPOD-BORNE DISEASES

Table II lists the principle groups of insects and arachnids involved in arthropod-borne diseases and the associated types of pathogens. Among the wide variety of arthropods that transmit pathogens to humans and other animals, mosquitoes are the most important, followed by ticks. Viruses and bacteria (including rickettsiae) are the most diverse groups of pathogens transmitted by arthropods, followed by protozoa and filarial nematodes.

All of the viruses listed in Table II are arthropod-borne viruses, usually referred to as *arboviruses,* indicating that they are typically transmitted by insects or other arthropod hosts. The study of arboviruses is termed *arbovirology.* These and related terms are discussed in more detail in Chapter 2, on the epidemiology of vector-borne diseases.

Pathogens are transmitted by arthropods in two basic ways, either biologically or mechanically. In biological transmission, pathogens undergo development or reproduction in the arthropod host. Examples of diseases that involve biological transmission are malaria, African trypanosomiasis, Chagas disease, leishmaniasis, and lymphatic filariasis. In mechanical transmission, pathogens are transmitted by arthropods via contaminated appendages (usually mouthparts) or regurgitation of an infectious blood meal. Examples of diseases that involve mechanical transmission are equine infectious anemia and myxomatosis. Biological transmission is by far the more common and efficient mechanism for pathogen maintenance and transmission.

A wide range of life-cycle patterns and degrees of host associations is characterized by arthropod vectors. Some ectoparasites, such as sucking lice, remain on their host for life. Others, such as mosquitoes and most biting flies, have a more fleeting association with the host, with some being associated with it only during the brief acts of host

TABLE II

**Examples of Arthropod-Borne Diseases of Medical-Veterinary Importance**

| Arthropod vectors | Diseases grouped by causative agents |
| --- | --- |
| Mosquitoes | Viruses: yellow fever, dengue, Rift Valley fever, myxomatosis; eastern equine encephalomyelitis, western equine encephalomyelitis, Venezuelan equine encephalomyelitis, St. Louis encephalitis, LaCrosse encephalitis, Japanese encephalitis, Murray Valley encephalitis, Chikungunya fever, O'nyong nyong fever, Ross River fever, West Nile fever. Protozoans: malaria. Filarial nematodes: Wuchererian filariasis, Bancroftian filariasis, dog heartworm |
| Black flies | Filarial nematodes: human onchocerciasis (river blindness), bovine onchocerciasis |
| Biting midges | Viruses: bluetongue disease, epizootic hemorrhagic disease, African horse sickness, leucocytozoonosis, Oropouche fever. Filarial nematodes: equine onchocerciasis, mansonellosis |
| Sand flies | Viruses: sand fly fever, vesicular stomatitis. Bacteria: Oroya fever (Veruga Peruana). Protozoans: leishmaniasis |
| Horse flies and deer flies | Viruses: equine infectious anemia, hog cholera. Bacteria: tularemia. Protozoans: surra (livestock trypanosomiasis). Filarial nematodes: loiasis, elaeophorosis |
| Tsetse flies | Protozoans: African trypanosomiasis, nagana |
| Triatomine bugs | Protozoans: American trypanosomiasis (Chagas disease) |
| Lice | Viruses: swine pox. Bacteria: epidemic typhus, trench fever, louse-borne relapsing fever |
| Fleas | Viruses: myxomatosis. Bacteria: plague, murine (endemic) typhus, tularemia |
| Ticks | Viruses: tick-borne encephalitis, Powassan encephalitis, Colorado tick fever, Crimean-Congo hemorrhagic fever, African swine fever. Bacteria: Lyme disease, Rocky Mountain spotted fever, Boutonneuse fever, tick-borne ehrlichiosis, Q fever, heartwater fever (cowdriosis), anaplasmosis, tick-borne relapsing fever, avian spirochetosis, theileriosis (East Coast fever), bovine dermatophilosus. Protozoans: babesiosis |
| Mites | Bacteria: tsutsugamushi (scrub typhus), rickettsialpox |

*Note:* For more Comprehensive coverage, see the individual chapters devoted to each arthropod group.

location and blood-feeding. Between these two extremes is a wide range of host associations exhibited by different arthropod groups.

Literature references on vector-borne diseases, together with their epidemiology and ecology, are provided under Arthropod-Borne Diseases at the end of this chapter.

## FOOD CONTAMINANTS

Many arthropods can contaminate or spoil food materials. In addition to causing direct damage to food resources, arthropods or their parts (e.g., setae, scales, shed cuticles, or body fragments) may be accidentally ingested. This can lead to toxic or allergic reactions, gastrointestinal myiasis, and other disorders.

Insects such as the house fly may alight on food and regurgitate pathogen-contaminated fluids prior to, or during, feeding. While feeding they also may defecate, contaminating the food with potential pathogens. Because the alimentary tract of arthropods may harbor pathogenic microorganisms, subsequent consumption of the contaminated food can lead to the transmission of these pathogens to humans or other animals. Similarly, the integument of household pests such as flies and cockroaches (particularly their legs and tarsi) can serve as a contact source of pathogens which may be readily transferred to food items. Some of these arthropods previously may have visited fecal matter, garbage heaps, animal secretions, or other potential sources of pathogens, thereby further contributing to health risks.

Additional information on insects and other arthropods that can contaminate food is provided by Olsen *et al.* (1996) and in reviews by Terbush (1972), Hughes (1976), and Gorham (1975, 1991a,b).

## FEAR OF ARTHROPODS

Some people detest arthropods, or infestation by them, to such a degree that they suffer from *entomophobia*, the fear of insects; *arachnophobia*, the fear of spiders and other arachnids; or *acarophobia*, the fear of mites (including ticks). Showing concern or disapproval towards the presence of potentially injurious arthropods is probably a prudent and healthy reaction, but phobic behaviors reflect an unusually severe psychological response. Such persons exhibit more-than-normal fear when they encounter an arthropod, often resorting to excessive or obsessive measures to control the problem (e.g., overtreatment of themselves or their homes with insecticides and other chemical compounds).

## DELUSORY PARASITOSIS

A relatively common psychological state occurs in which an individual mistakenly believes that he or she is being bitten by, or infested with, parasites. This is called *delusory parasitosis*, also referred to as *delusional parasitosis* or *delusions of parasitosis*. This condition is distinct from simply a fear, or phobia, of insects or other arthropods and represents a more deeply rooted psychological problem. This delusory condition is most frequently experienced by middle-aged or elderly persons, particularly women, and is one of the most difficult situations for entomologists to approach.

Remarkable behavioral traits are sometimes attributed to the parasites by victims. These include descriptions of tiny animals jumping into the eyes when a room is entered or when a lamp is switched on. Some victims have failing eyesight; others may have real symptoms from other conditions such as psoriasis that they may attribute to the imagined parasites. Victims become convinced that the parasites are real, and they often consult a succession of physicians in a futile attempt to secure a diagnosis and satisfactory treatment to resolve the problem. Patients typically produce skin scrapings or samples of such materials as vacuumed debris from carpets, draperies, and window sills, which they believe contain the illusive parasites.

Victims of delusory parasitosis often turn to extension entomologists or medical entomologists as a last resort out of frustration with being unable to resolve their condition through family physicians, allergists, and other medical specialists. Because patients are convinced that arthropods are present, they are usually reluctant to seek counseling or other psychiatric help. Dealing with cases of delusory parasitosis requires careful examination of submitted specimens, tact, and professional discretion on the part of the entomologist. Additional information on delusory parasitosis is provided by Driscoll *et al.* (1993), Koblenzer (1993), Kushon *et al.* (1993), Poorbaugh (1993), Webb (1993a,b), Goddard (1995), and Hinkle (2000).

## TOXINS AND VENOMS

Many arthropods of medical-veterinary importance produce toxins. Notable among these are scorpions, spiders, bees, wasps, ants, and velvet ants; certain beetles (e.g., blister beetles, some rove beetles, and darkling beetles);

and caterpillars, cocoons, and adults of various moths. Additionally, antigenic components in saliva released during blood-feeding by arthropods (e.g., certain fleas, ticks, mosquitoes, and chiggers) cause local or systemic reactions in their hosts.

Toxins produced by arthropods represent a wide range of chemical substances from simple inorganic or organic compounds to complex alkaloids and heterocyclic compounds. The term *venom* refers to toxins that are injected into animal tissues via specialized structures such as stings, chelicerae (fangs), and spines. Venoms are often complex mixtures of toxins and various pharmacologically active compounds that facilitate the spread and effectiveness of the toxic components. They commonly include amines (e.g., histamine, catecholamines, serotonin), peptides, polypeptides (e.g., kinins), specific proteins, and enzymes (e.g., phospholipase, hyaluronidase, esterases) that vary significantly among different arthropod taxa. Depending on what types of cells or tissues they affect, toxins and venoms can be characterized, for example, as neurotoxins, cytotoxins, or hemotoxins. Frequently they cause such symptoms as pain, itching, swelling, redness, hemorrhaging, or blisters, the severity of which is largely dependent on the particular types and amounts of toxin involved.

Further information on arthropod toxins and venoms is provided by Beard (1960), Roth and Eisner (1962), Bücherl and Buckley (1971), Bettini (1978), Schmidt (1982), Tu (1984), and Meier and White (1995).

## HOST DEFENSES

Humans and other animals have developed elaborate means to defend themselves against infestation by arthropods and infection by pathogens they may transmit. Both behavioral and immunological responses are used to resist infestation by arthropods. Behavioral defenses include evasive, offensive, or defensive action against biting flies such as mosquitoes, black flies, ceratopogonids, stable flies, and horse flies. Grooming and preening by animals (e.g., biting, scratching, or licking) are defensive behaviors used to reduce or prevent infestations by ectoparasites and other potentially harmful arthropods. Host immunological defenses against arthropods vary with different arthropods and with respect to previous exposure to the same or antigenically related taxa. Details concerning such host immune responses are beyond the scope of this book, but some general trends are noteworthy. Repeated feeding attempts by the same or antigenically cross-reactive arthropods often lead to fewer arthropods being able to feed successfully, reduced engorgement weights, greater mortality, and

decreased fecundity of female arthropods. Widespread arthropod mortality rarely results. For more information concerning the types of host immune responses and cell types involved against various ectoparasites, see Wikel (1996b) and other works listed at the end of this chapter.

Many blood-feeding arthropods partially or completely counteract the host immune response by inoculating *immunomodulators* or *immunosuppressive compounds* into the bite site. In fact, a wide range of pharmacologically active compounds is known to be released at the bite site by various arthropods (Ribeiro 1995). These compounds range from *anticoagulants* to prevent the blood from clotting, local *analgesics* to reduce host pain, *apyrase* to prevent platelet aggregation and promote capillary location, and various enzymes and other factors for promoting blood or tissue digestion. Some of these compounds are perceived by the host as antigens and may elicit an immune response, whereas others can cause localized or systemic toxic responses and itching.

## FORENSIC ENTOMOLOGY

Forensic entomology is the study of arthropods, especially insects, associated with crimes and other aspects of the courts and judicial system. Forensic entomology usually involves the identification of insects and other arthropods associated with human remains as an aid to determining the time and place of death.

Time of death can often be ascertained based on the ambient temperature and other weather conditions over the preceding days at the crime site and by correlating this information with the developmental rates of key arthropod species present on, or in, the corpse. These arthropods are typically fly larvae, some of which are important primary and secondary decomposers of animal remains. By knowing developmental times and related information for decomposer species at different temperatures, it often is possible to quite accurately estimate the time of death.

The location where a crime took place, if different from the discovery site, also sometimes can be determined based on the presence of unique arthropods with known distributions that do not include the area where the body was found. Similarly, examination of carefully collected insect evidence can aid in solving other crimes (e.g., the origin of drug shipments and sources of vehicles and other accessories used in crimes) in which there is arthropod evidence involving taxa with characteristic geographical distributions.

Further details on the science of forensic entomology are provided by Vincent *et al.* (1985), Smith (1986),

Erzinclioglu (1989), Catts and Haskell (1990), Catts and Goff (1992), and Goff (2000).

## MINOR ARTHROPOD PROBLEMS OF MEDICAL-VETERINARY INTEREST

In addition to arthropod groups detailed in the chapters that follow, a few arthropods in other groups may have minor, incidental, or occasional significance to human and animal health. These include springtails (order Collembola), bark lice (order Pscoptera), walking sticks (order Phasmida), mayflies (order Ephemeroptera), earwigs (order Dermaptera), thrips (order Thysanoptera), caddisflies (order Trichoptera), centipedes (class Chilopoda), and millipedes (class Diplopoda).

On rare occasions, *springtails* have been recorded infesting human skin (Scott *et al.* 1962, Scott 1966). Similarly, some *bark lice* (psocids) are known to cause allergies or dermatitis in humans (Li and Li 1995, Baz and Monserrat 1999). Certain adult *mayflies* and *caddisflies* can cause inhalational allergies, especially when they emerge in large numbers from lakes, rivers, or streams (Seshadri 1955).

In addition to various hymenopterans, arachnids, and other venomous arthropods detailed in the following chapters, a few miscellaneous arthropods produce venoms that can cause medical-veterinary problems. These include *walking sticks* (stick insects) and *millipedes,* some of which utilize venomous defensive secretions or sprays. Defensive sprays of certain walking sticks can cause conjunctivitis (Stewart 1937), whereas defensive sprays of some millipedes contain hydrochloric acid that can chemically burn the skin and can cause long-term skin discoloration (Radford 1975). *Centipedes,* especially some of the larger tropical species, can cause envenomation when they "bite" with their poison claws (maxillipeds), which are equipped with poison ducts and glands (Remington 1950).

*Thrips,* which have tubular mouthparts adapted for sucking plant fluids, occasionally pierce the skin and have been known to imbibe blood (Williams 1921, Hood 1927, Bailey 1936, Arnaud 1970). On rare occasions, *earwigs* also have been recorded as imbibing blood (Bishopp 1961). Bishopp further noted that earwigs have been known to pierce human skin with their pair of caudal pincers (cerci) and may stay attached for an extended period.

Some miscellaneous arthropods inhabit the feathers of birds or the fur of mammals. The exact nutritional requirements of some of these arthropods remain unknown; most of them, however, do not appear to be true ectoparasites. Representatives of two of the three suborders of earwigs (suborders Arixeniina and Hemimerina) live in mammal fur. The Arixeniina are associated with Old World bats, whereas the Hemimerina are found on African cricetomyine rodents (Nakata and Maa 1974). These earwigs may feed on skin secretions or sloughed cells, but their effect on the health of their hosts is poorly understood. Other occasional inhabitants of host pelage, such as various beetles, cheyletid mites, and pseudoscorpions, are predators of ectoparasites and are therefore beneficial to their hosts (Durden 1987).

A few arthropods that are not mentioned in the following chapters can occasionally serve as *intermediate hosts* of parasites that adversely affect domestic and wild animals. These include certain *springtails* and *psocids* (bark lice) as intermediate hosts of tapeworms (Baz and Monserrat 1999).

## REFERENCES AND FURTHER READING

### GENERAL ENTOMOLOGY
Arnett, R. H., Jr. (1993). American insects: a handbook of the insects of America north of Mexico. Sandhill Crane Press, Gainesville, FL.
Borror, D. J., Triplehorn, C. A., and Johnson, N. F. 1989. An introduction to the study of insects (6th ed.). Saunders College Pub., Philadelphia.
Bosik, J. J. (Chairman). (1997). Common names of insects and related organisms. Entomological Society of America, Lanham, Maryland.
Chapman, R. F. (1998). The insects: structure and function (4th ed.). Cambridge Univ. Press, Cambridge, UK.
Daly, H. V., Doyen, J. T., and Purcell III, A. H. (1998). Introduction to insect biology and diversity (2nd ed.). Oxford Univ. Press, London.
Ebeling, W. (1975). Urban entomology. University of California Press, Berkeley.
Elzinga, R. J. (1997). Fundamentals of entomology (4th ed.). Prentice Hall, Upper Saddle River, NJ.
Gillott, C. (1995). Entomology (2nd ed.). Plenum, New York.
Gullan, P. J., and Cranston, P. S. (1994). The insects: an outline of entomology. Chapman & Hall, London.
Mallis, A. (1997). Handbook of pest control: the behavior, life history and control of household pests (8th ed., edited by D. Moreland). Mallis Handbook & Technical Training Co., Cleveland.
Richards, O. W., and Davies, R. G. (1994). Imm's general textbook of entomology (10th ed.). Vol. 1. Structure, physiology and development. Vol. 2. Classification and biology. Chapman & Hall, London.
Robinson, W. H. (1996). Urban entomology: insect and mite pests in the human environment. Chapman & Hall, London.
Romoser, W. S., and Stoffolano, Jr., J. G. (1998). The science of entomology (4th ed.). WCB/McGraw Hill, Boston.
Snodgrass, R. (1993). Principles of insect morphology. Cornell Univ. Press, Ithaca, NY.
de la Torre-Bueno, J. R. (1962). A glossary of entomology. Brooklyn Entomological Society, Brooklyn.

### GENERAL ACAROLOGY
Evans, G. O. (1992). Principles of acarology. CAB International, Wallingford, UK.

Krantz, G. W. (1978). A manual of acarology (2nd ed.). Oregon State University Bookstores, Corvallis.

Walter, D. E., and Proctor, H. E. (1999). Mites: ecology, evolution and behaviour. CABI Publishing, Wallingford, U.K.

Woolley, T. A. (1987). Acarology: mites and human welfare. Wiley, New York.

## MEDICAL-VETERINARY ENTOMOLOGY

Baker, E. W., Evans, T. M., Gould, D. J.,. Hull, W. B, and Keegan, H. L. (1956). A manual of parasitic mites of medical or economic importance. National Pest Control Assoc., New York.

Baker, J. R., Apperson, C. S., Stringham, S. M., Waldvogel, M. G., and Watson, D. W. (eds.). (2000). Insect and other pests of man and animals. Some important, common, and potential pests in the Southeastern United States (2nd ed.). North Carolina State University, Raleigh.

Beaty, B. J., and Marquardt, W. C. (eds.). (1996). The biology of disease vectors. Univ. Press of Colorado, Niwot.

Crampton, J. M., Beard, C. B., and Louis, C. (eds.). (1997). Molecular biology of insect disease vectors: a methods manual. Chapman & Hall, London.

Eldridge, B. F., and Edman, J. D. (eds.) (2000). Medical entomology: a textbook on public health and veterinary problems caused by arthropods. Kluwer Academic, Dordrecht/Norwell, MA.

Harwood, R. F., and James, M. T. (1979). Entomology in human and animal health (7th ed.). Macmillan Co., New York.

Hickin, N. E. (1985). Pest animals in buildings: a world review. G. Godwin, London.

Jobling, B. (1987). Anatomical drawings of biting flies. British Museum (Natural History) and Wellcome Trust, London.

Kettle, D. S. (1995). Medical and veterinary entomology (2nd ed.). CAB International, Wallingford, UK.

Kim, K. C. (ed.). (1985). Coevolution of parasitic arthropods and mammals. Wiley, NY.

Lehane, M. (1991). Biology of blood-sucking insects. Chapman & Hall, London.

Marcondes, C. B. (2001). Entomologia médica e veterinaria. Editora Atheneu, São Paulo.

Mario Vargas, V. (2001). Los acaros en la salud humana y animal. University of Costa Rica, San José.

Marquart, W. C., Demaree, R. S., and Grieve, R. B. (1999). Parasitology and vector biology (2nd ed.). Academic Press, San Diego.

Marshall, A. G. (1981). Ecology of ectoparasitic insects. Academic Press, London.

Nutting, W. B. (ed.). (1994). Mammalian diseases and arachnids. Vol. 1. Pathogen biology and clinical management, 277 pp. Vol. 2. Medico-veterinary, laboratory, and wildlife diseases, and control, 280 pp. CRC Press, Boca Raton, FL.

Parish, L. C., Nutting, W. B., and Schwartzman, R. M. (eds.). (1983). Cutaneous infestations in man and animals. Praeger, New York.

Patton, W. S. (1931). Insects, ticks, mites and venomous animals of medical and veterinary importance. Part II: public health. Liverpool Univ. Press, Liverpool.

Patton, W. S., and Evans, A. M. (1929). Insects, ticks, mites and venomous animals of medical and veterinary importance. Part I: medical. Liverpool Univ. Press, Liverpool.

Pittaway, A. R. (1992). Arthropods of medical and veterinary importance: a checklist of preferred names and allied terms. Univ. of Arizona Press, Tucson.

Walker, A. R. (1995). Arthropods of humans and domestic animals: a guide to preliminary identification. Chapman & Hall, New York.

## MEDICAL ENTOMOLOGY

Alexander, J. O. (1984). Arthropods and human skin. Springer-Verlag, Berlin.

Andrews, M. L. A. (1977). The life that lives on man. Taplinger Pub., New York.

Burgess, N. H. R., and Cowan, G. L. O. (1993). A colour atlas of medical entomology. Chapman & Hall, London.

Busvine, J. M. (1980). Insects and hygiene: the biology and control of insect pests of medical and domestic importance (3rd ed.). Chapman & Hall, London.

Daniel, M., Stramova, H., Absolonova, V., Dedicova, D., Lhotova, H., Maskova, L., and Petras, P. (1992). Arthropods in a hospital and their potential significance in the epidemiology of hospital infections. *Folia Parasitol.* **39**, 159–170.

Furman, D. P., and Catts, E. P. (1982). Manual of medical entomology. Cambridge Univ. Press, Cambridge, UK.

Goddard, J. (1998). Arthropods and medicine. *J. Agromed.* **5**, 55–83.

Goddard, J. (2000). Physician's guide to arthropods of medical importance (3rd ed.). CRC Press, Boca Raton, FL.

Gordon, R. M., and Lavoipierre, M. M. J. (1962). Entomology for students of medicine. Blackwell Sci., Oxford.

Gratz, N. G. (1999). Emerging and resurging vector-borne diseases. *Annu. Rev. Entomol.* **44**, 51–75.

Gubler, D. J. (1998). Resurgent vector-borne diseases as a global health problem. *Emerg. Infect. Dis.* **4**, 442–450.

Harwood, R. F., and James, M. T. (1979). Entomology in human and animal health (7th ed.). Macmillan Co., New York.

Herms, W. B. (1961). Medical entomology (5th ed.). Macmillan Co., New York.

Horsfall, W. R. (1962). Medical entomology: arthropods and human disease. Ronald Press, New York.

James, M. T., and Harwood, R. F. (1969). Herms's medical entomology (6th ed.). Macmillan Co., New York.

Lane, R. P., and Crosskey, R. W. (eds.). (1993). Medical insects and arachnids. Chapman & Hall, London.

Leclercq, M. (1969). Entomological parasitology: the relations between entomology and the medical sciences. Pergamon, Oxford.

Marples, M. J. (1965). The ecology of the human skin. Charles C. Thomas, Springfield, IL.

Matheson, R. (1950). Medical entomology. Comstock Pub. Co., Ithaca, NY.

McClelland, G. A. H. (1992). Medical entomology: an ecological perspective (12th ed.). University of California Press, Davis.

Orkin, M., and Maibach, H. I. (eds.). (1985). Cutaneous infestations and insect bites. Dekker, New York.

Peters, W. (1992). A colour atlas of arthropods in clinical medicine. Wolfe Pub. Ltd., London.

Riley, W. A., and Johannsen, O. A. (1938). Medical entomology: a survey of insects and allied forms which affect the health of man and animals (2nd ed.). McGraw-Hill, New York.

Service, M. W. (1980). A guide to medical entomology. Macmillan Co., London.

Service, M. W. (2000). Medical entomology for students (2nd ed.). Cambridge Univ. Press, Cambridge, UK.

Smith, K. G. V. (ed.). (1973). Insects and other arthropods of medical importance. British Museum (Natural History), London.

Walker, D. H., Barbour, A. G., Oliver, J. H., Jr., Lane, R. S., Dumler, J. S., Dennis, D. T., Persing, D. H., Azad, A. F., and McSweegan, E. (1996). Emerging bacterial zoonotic and vector-borne diseases. *J. Am. Med. Assoc.* **275**, 463–469.

Wilson, M. E., and Spielman, A. (eds.). (1994). Vector-borne terrestrial diseases, pp. 123–224, *In*: Disease in evolution: global changes and emergence of infectious diseases. Proceedings of a conference. Woods Hole, Massachusetts, November 7–10, 1993. *Ann. N. Y. Acad. Sci.* **740**, 1–503.

Winch, P. (1998). Social and cultural responses to emerging vector-borne diseases. *J. Vector Ecol.* **23**, 47–53.

## VETERINARY ENTOMOLOGY

Axtell, R. C., and Arends, J. J. (1990). Ecology and management of arthropod pests of poultry. *Annu. Rev. Entomol.* **35**, 101–126.

Barker, B. (1999). Livestock entomology lab manual (2nd ed.). Kendall/Hunt, Dubuque, IA.

Barnard, S. M., and Durden, L. A. (1999). A veterinary guide to the parasites of reptiles. Vol. 2. Arthropods (excluding mites). Krieger, Melbourne, FL.

Bay, D. E., and Harris, R. L. (1988). Introduction to veterinary entomology. Stonefly Pub., Bryan, TX.

Bowman, D. D. (1999). Georgi's parasitology for veterinarians (7th ed.). Saunders, Philadelphia.

Bram, R. A. (1978). Surveillance and collection of arthropods of veterinary importance. Department of Agriculture, Animal and Plant Health Inspection Service, Washington, DC. US Department of Agriculture, Agriculture Handbook No. 518.

Drummond, R. O., George, J. E., and Kunz, S. E. (1988). Control of arthropod pests of livestock: a review of technology. CRC Press, Boca Raton, FL.

Flynn, R. J. (1973). Parasites of laboratory animals. Iowa State University Press, Ames.

Foil, L. D., and Foil, C. S. (1990). Arthropod pests of horses. *Comp. Cont. Educ. Pract. Vet.* **12**, 723–731.

Georgi, J. R. (1990). Parasitology for veterinarians (5th ed.). Saunders, Philadelphia.

Jones, C. J., and DiPietro, J. A. (1996). Biology and control of arthropod parasites of horses. *Comp. Cont. Educ. Pract. Vet.* **18**, 551–558.

Lancaster, J. L., and Meisch, M. V. (1986). Arthropods in livestock and poultry production. Halstead, New York.

Soulsby, E. J. L. (1982). Helminths, arthropods and protozoa of domesticated animals (7th ed.). Ballière, Tindall & Cassell, London.

Uilenberg, G. (Coordinator). (1994). Ectoparasites of animals and control methods. *Rev. Sci. Tech. Off. Int. Epizoot.* **13**, 979–1387.

Wall, R., and Shearer, D. (1997). Veterinary entomology. Chapman & Hall, London.

Wall, R., and Shearer, D. (2000). Veterinary ectoparasites: biology, pathology & control (2nd ed.). Blackwell Science Ltd., Abingdon, U.K.

Williams, R. E., Hall, R. D., Broce, A. B., and Scholl, P. J. (eds.). (1985). Livestock entomology. Wiley, New York.

## WILDLIFE ENTOMOLOGY

Davidson, W. R., Hayes, F. A., Nettles, V. F., and Kellogg, F. E. (1981). Diseases and parasites of white-tailed deer. Miscellaneous publication No. 7. Tall Timbers Research Station, Tallahassee, FL.

Davidson, W. R., and Nettles, V. F. (1997). Field manual of wildlife diseases in the southeastern United States (2nd ed.). Southeastern Cooperative Wildlife Disease Study, Athens, GA.

Davis, J. W., and Anderson, R. C. (eds.) (1971). Parasitic diseases of wild mammals. Iowa State University Press, Ames.

Fowler, M. E. (ed.). (1986). Zoo and wild animal medicine (2nd ed.). Saunders, Philadelphia.

Fowler, M. E., and Miller, R. E. (eds) (1999). Zoo and wild animal medicine: current therapy (4th ed.). Saunders, Philadelphia.

Samuel, W. M., Pybus, M. J., and Kocan, A. A. (eds.) (2001). Parasitic diseases of wild mammals (2nd ed.). Iowa State University Press, Ames.

## HISTORY OF MEDICAL-VETERINARY ENTOMOLOGY

Anon. (1996). History of CDC [CDC's 50th anniversary]. *Morb. Mortal. Wkly. Rep.* **45**, 526–530.

Augustin, G. (1909). History of yellow fever. Searcy & Pfaff, New Orleans.

Bayne-Jones, S. (1964). Preventive medicine in World War II. Vol. 7. Communicable diseases, arthropod-borne diseases other than malaria. Office of the Surgeon General, Department of the Army, Washington, DC.

Bean, W. B. (1982). Walter Reed—a biography. Univ. Press of Virginia, Charlottesville.

Bockarie, M. J., Gbakima, A. A., and Barnish, G. (1999). It all began with Ronald Ross: 100 years of malaria research and control in Sierra Leone (1899–1999). *Ann. Trop. Med. Parasitol.* **93**, 213–224.

Busvine, J. R. (1976). Insects, hygiene and history. Athlone Press, London.

Busvine, J. R. (1993). Disease transmission by insects: its discovery and 90 years of effort to prevent it. Springer-Verlag, Berlin.

Calisher, C. H. (1996). From mouse to sequence and back to mouse: peregrinations of an arbovirologist. *J. Vector Ecol.* **21**, 192–200.

Cartwright, F. F. (1972). Disease and history; the influence of disease in shaping the great events in history. Hart-Davis, London.

Chernin, E. (1983). Sir Patrick Manson: an annotated bibliography and a note on a collected set of his writings. *Rev. Infect. Dis.* **5**, 353–386.

Chernin, E. (1987). A unique tribute to Theobald Smith, 1915. *Rev. Infect. Dis.* **9**, 625–635.

Collins, W. E. (1976). Fifty years of parasitology: some fulgent personalities in arthropodology. *J. Parasitol.* **62**, 504–509.

Cook, G. C. (1992). From the Greenwich hulks to Old St. Pancras: a history of tropical disease in London. Athlone Press, London.

Cox, F. E. G. (ed.). (1996). The Wellcome Trust illustrated history of tropical diseases. Trustees of the Wellcome Trust, London.

Cushing, E. C. (1957). History of entomology in World War II. Smithsonian Inst., Washington, DC.

Delaporte, F. (1991). The history of yellow fever: an essay on the birth of tropical medicine. MIT Press.

Desowitz, R. S. (1991). The malaria capers: more tales of parasites and people, research and reality. Norton, New York.

Desowitz, R. S. (1997). Who gave Pinta to the Santa Maria? Tracing the devastating spread of lethal tropical diseases. Norton, New York.

Dolman, C. E. (1982). Theobald Smith (1859–1934), pioneer American microbiologist. *Perspect. Biol. Med.* **25**, 417–427.

Eldridge, B. F. (1992). Patrick Manson and the discovery age of vector biology. *J. Am. Mosq. Control Assoc.* **8**, 215–222.

Ellis, J. H. (1992). Yellow fever and public health in the New South. Univ. of Kentucky Press, Lexington.

Geong, H.-G. (2001). Carving a niche for medical entomology: a quest for professional identity in American entomology. Part I: Understanding arthropods as disease spreaders. Part II: Insect control as a sanitary science. *Am. Entomol.* **47**, 236–243.

Gillett, J. D. (1979). Vitamin C, yellow fever and plague: the near misses. *Antenna, Bull. R. Entomol. Soc. Lond.* **3**, 64–70.

Gillett, J. D. (1985). Medical entomology, past, present and future: a personal view. *Antenna, Bull. R. Entomol. Soc. Lond.* **9**, 63–70.

Gorgas, M. D., and Hendrick, B. J. (1924). William Crawford Gorgas: his life and work. Garden City Pub. Co., Garden City, NY.

Harwood, R. F., and James, M. T. (1979). Historical review. *In*: R. F. Harwood (ed.), Entomology in human and animal health (7th ed.). pp. 3–10. Macmillan Co., New York.

Horsman, R. (1987). Josiah Nott of Mobile: southerner, physician, and racial theorist. Louisiana State Univ. Press, Baton Rouge.

Laurence, B. R. (1989). The discovery of insect-borne disease. *Biologist* **36**, 65–71.

Lewis, J. D. (1984). Reminiscences of medical entomology in the last fifty years. *Antenna, Bull. R. Entomol. Soc. Lond.* **8**, 117–122.

Lockwood, J. A. (1987). Entomological warfare: history of the use of insects as weapons of war. *Am. Entomol.* **33**, 76–82.

Miller, G. L. (1997). Historical natural history: insects and the Civil War. *Am. Entomol.* **43**, 227–245.

Nye, E. R., and Gibson, M. E. (1997). Ronald Ross: malariologist and polymath—a biography. Macmillan Co./St. Martin's, New York.

Oldstone, M. B. A. (1998). Viruses, plagues, and history. Oxford Univ. Press, New York.

Peterson, R. K. D. (1995). Insects, disease, and military history. *Am. Entomol.* **41**, 147–160.

Philip, C. B. (1948). Tsutsugamushi disease (scrub typhus) in World War II. *J. Parasitol.* **34**, 169–191.

Philip, C. B., and Rozeboom, L. E. (1973). Medico-veterinary entomology: a generation of progress. *In*: History of entomology. (R. F. Smith, T. E. Mittler, and C. N. Smith, eds.). pp. 333–360. Entomological Society of America, College Park, MD.

Schmidt, C. H., and Fluno, J. A. (1973). Brief history of medical and veterinary entomology in the USDA. *J. Wash. Acad. Sci.* **63**, 54–60.

Service, M. W. (1978). A short history of medical entomology. *J. Med. Entomol.* **14**, 603–626.

Sosa, O., Jr. (1989). Carlos J. Finlay and yellow fever: a discovery. *Bull. Entomol. Soc. Am.* **35**, 23–25.

Woodward, T. E. (1973). A historical account of the rickettsial diseases with a discussion of unsolved problems. *J. Infect. Dis.* **127**, 583–594.

Young, M. D. (1966). Scientific exploration and achievement in the field of malaria. *J. Parasitol.* **52**, 2–8.

Zinsser, H. (1935). Rats, lice and history. Little, Brown & Co., Boston.

Zinsser, H. (1936). Biographical memoir of Theobald Smith, 1859–1934. *Natl. Acad. Sci. Biogr. Mem.* **17**, 261–303 (reprinted in *Rev. Infect. Dis.* **9**, 636–654).

## IDENTIFICATION AND SYSTEMATICS OF TAXA OF MEDICAL-VETERINARY IMPORTANCE

Baker, A. S. (1999). Mites and ticks of domestic animals. An identification guide and information source. The Stationery Office, London.

Centers for Disease Control and Prevention. (1979). Introduction to arthropods of public health importance. Centers for Disease Control and Prevention, Atlanta.

Centers for Disease Control and Prevention. (1994). Pictorial keys: arthropods, reptiles, birds and mammals of public health importance. Centers for Disease Control and Prevention, Atlanta.

Davis, G. M. (1995). Systematics and public health. *Bioscience* **45**, 705–714.

Hopla, C. E., Durden, L. A., and Keirans, J. E. (1994). Ectoparasites and classification. *Rev. Sci. Tech. Off. Int. Epizoot.* **13**, 985–1017.

Lago, P. K., and Goddard, J. (1994). Identification of medically important arthropods. *Lab. Med.* **25**, 298–305.

Service, M. W. (ed.). (1988). Biosystematics of haematophagous insects. Oxford Univ. Press, London.

## TYPES OF PROBLEMS CAUSED BY ARTHROPODS

### Annoyance

Burns, D. A. (1987). The investigation and management of arthropod bite reactions acquired in the home. *Clin. Exp. Dermatol.* **12**, 114–120.

Frazier, C. D. (1973). Biting insects. *Arch. Dermatol.* **107**, 400–402.

Newson, H. D. (1977). Arthropod problems in recreation areas. *Annu. Rev. Entomol.* **22**, 333–353.

### Envenomation

Camazine, S. (1988). Hymenopteran stings: reactions, mechanisms and medical treatment. *Bull. Entomol. Soc. Am.* **34**, 17–21.

Cloudsley-Thompson, J. (1995). On being bitten and stung. *Antenna, Bull. R. Entomol. Soc. Lond.* **19**, 177–180.

Goddard, J. (1994). Direct injury from arthropods. *Lab. Med.* **25**, 365–371.

Keegan, H. L. (1969). Some medical problems from direct injury by arthropods. *Int. Pathol.* **10**, 35–45.

Nichol, J. (1989). Bites and stings: the world of venomous animals. Facts on File, New York.

Papp, C. S., and Swan, L. A. (1983). A guide to biting and stinging insects and other arthropods (2nd ed.). Entomography Pub., Sacramento, CA.

### Allergic Reactions

Bellas, T. E. (1982). Insects as a cause of inhalant allergies: a bibliography (2nd ed.). CSIRO Aust. Div. Entomol. Rep. No. 25. CSIRO, Canberra.

Feinberg, A. R., Feinberg, S. M., and Benaim-Pinto, C. (1956). Asthma and rhinitis from insect allergens. *J. Allergy* **27**, 436–444.

Feingold, B. F., Benjamini, E., and Michaeli, D. (1968). The allergic responses to insect bites. *Annu. Rev. Entomol.* **13**, 137–158.

Frazier, C. A., and Brown, F. K. (1980). Insects and allergy and what to do about them. Univ. of Oklahoma Press, Norman.

Henson, E. B. (1966). Aquatic insects as inhalant allergens: a review of American literature. *Ohio J. Sci.* **66**, 529–532.

Heyworth, M. F. (1999). Importance of insects in asthma. *J. Med. Entomol.* **36**, 131–132.

Levine, M. I., and Lockley, R. F. (eds.) (1981). Monograph on insect allergy. Typecraft, Pittsburgh.

Musken, H., Franz, J. T., Fernandez-Caldas, E., Maranon, F., Masuch, G., and Bergmann, K. C. (1998a). Psocoptera spp. (dust lice): a new source of indoor allergies in Germany. *J. Allergy Clin. Immunol.* **101**, 121.

Musken, H., Franz, J. T., Fernandez-Caldas, E., Masuch, G., Maranon, F., and Bergmann, K. C. (1998b). Psocoptera (dust lice): new indoor allergens? *Allergologie* **21**, 381–382.

Schulman, S. (1967). Allergic responses to insects. *Annu. Rev. Entomol.* **12**, 323–346.

Wirtz, R. A. (1980). Occupational allergies to arthropods—documentation and prevention. *Bull. Entomol. Soc. Am.* **26**, 356–360.

Wirtz, R. A. (1984). Allergic and toxic reactions to non-stinging arthropods. *Annu. Rev. Entomol.* **29**, 47–69.

### Arthropod-Borne Diseases

Beck, J. W., and Davies, J. E. (1981). Medical parasitology. Mosby, St. Louis.

Benenson, A. S. (ed.) (1995). Control of communicable diseases manual (16th ed.). American Public Health Association, Washington, DC.

Beran, G. W., and Steele, J. (1994). Handbook of zoonoses: section A. Bacterial, rickettsial, chlamydial, and mycotic (2nd ed.). CRC Press, Boca Raton.

Busvine, J. R. (1979). Arthropod vectors of disease. The Institute of Biology's studies in biology No. 55. Edward Arnold, London.

Cook, G. C. (ed.). (1996). Manson's tropical diseases (20th ed.). Saunders, Orlando.

Dye, C. (1992). The analysis of parasite transmission by bloodsucking insects. *Annu. Rev. Entomol.* **37**, 1–19.

Ewald, P. W. (1983). Host-parasite relations, vectors and the evolution of disease severity. *Annu. Rev. Ecol. Syst.* **14**, 465–485.

Faust, E. C., Beaver, P. C., and Jung, R. C. (1975). Animal agents and vectors of human disease (4th ed.). Lea & Febiger, Philadelphia.

Goddard, J. (1999). Infectious diseases and arthropods. Humana Press, Clifton, NJ.

Horsfall, F. L., and Tamm, I. (eds.) (1965). Viral and rickettsial infections of man. Lippincott, Philadelphia.

Hubbert, W. T., McCullough, W. F., and Schnurrenberger, P. R. (eds.) (1975). Diseases transmitted from animals to man (6th ed.). Thomas, Springfield, IL.

Jeffrey, H. C., Leach, R. M., and Cowan, G. O. (1991). Atlas of medical helminthology and protozoology. Churchill Livingstone, New York.

McHugh, C. P. (1994). Arthropods: vectors of disease agents. *Lab. Med.* **25**, 429–437.

Mills, J. N., Childs, J. E., Ksiazek, T. G., Peters, C. J., and Velleca, W. M. (1995). Methods for trapping and sampling small mammals for virologic testing. Centers for Disease Control and Prevention, Atlanta.

Monath, T. P. (ed.). (1988). The arboviruses: epidemiology and ecology. Vol. 1. General principles. Vol. 2. African horse sickness to dengue. Vol. 3. Eastern equine encephalomyelitis to O'nyong virus disease. Vol. 4. Oropouche fever to Venezuelan equine encephalomyelitis. Vol. 5. Vesicular stomatitis to yellow fever. CRC Press, Boca Raton, FL.

Moore, C. G., McLean, R. G., Mitchell, C. J., Nasci, R. S., Calisher, C. H., Marfin, A. A., Moore, P. S., and Gubler, D. J. (1993). Guidelines for arbovirus surveillance programs in the United States. Centers for Disease Control and Prevention, Fort Collins.

Service, M. W. (1986). Blood-sucking insects: vectors of disease. Arnold, London.

Service, M. W. (ed.) (1989). Demography and vector-borne diseases. CRC Press, Boca Raton, FL.

Service, M. W. (ed.) (2001). The encyclopedia of arthropod-transmitted infections of man and domesticated animals. CABI Publishing, Wallingford, UK.

Snow, K. R. (1974). Insects and disease. Wiley, New York.

Strickland, G. T. (ed) (1991). Hunter's tropical medicine (7th ed.). Saunders, Philadelphia.

Theiler, M., and Downs, W. G. (1973). The arthropod-borne viruses of vertebrates: an account of the Rockefeller Foundation virus program, 1951–1970. Yale Univ. Press, New Haven.

World Health Organization. (1989). Geographical distribution of arthropod-borne diseases and their principal vectors. WHO/VBC/89.967, Geneva.

World Health Organization. (1995). Vector control for malaria and other mosquito-borne diseases. WHO Tech. Rep. Ser. No. 857.

## FOOD CONTAMINANTS

Gorham, J. R. (1975). Filth in foods: implications for health. *J. Milk Food Technol.* **38**, 409–418.

Gorham, J. R. (1991a). Insect and mite pests in food: an illustrated key. Vols. 1 and 2. USDA Agricultural Handbook No. 655.

Gorham, J. R. (ed.) (1991b). Ecology and management of food-industry pests. FDA Tech. Bull. No. 4. AOAC International, Arlington, VA.

Hughes, A. M. (1976). The mites of stored food and houses. Ministry of Agriculture, Fisheries and Food, Her Majesty's Stationery Office, London.

Olsen, A. R., Sidebottom, T. H., and Knight, S. A. (eds.) (1996). Fundamentals of microanalytical entomology: a practical guide to detecting and identifying filth in foods. CRC Press, Boca Raton, FL.

Terbush, L. E. (1972). The medical significance of mites of stored food. *FDA By-Lines* **3**, 57–70.

## PHOBIAS AND DELUSORY PARASITOSIS

Beerman, H., and Nutting, W. B. (1984). Arachnid-related phobias: symbiphobia, preventions, and treatments. *In*: Mammalian diseases and arachnids, Vol. 2. Medico-veterinary, laboratory, and wildlife diseases, and control (W. B. Nutting (ed.),. pp. 103–112. CRC Press, Boca Raton, FL.

Driscoll, M. S., Rothe, M. J., Grant-Kels, J. M., and Hale, M. S. (1993). Delusional parasitosis—a dermatological, psychiatric, and pharmacological approach. *J. Am. Acad. Dermatol.* **29**, 1023–1033.

Goddard, J. (1995). Analysis of 11 cases of delusions of parasitosis reported to the Mississippi Department of Health. *South. Med. J.* **88**, 837–839.

Hinkle, N. C. (2000). Delusory parasitosis. *Am. Entomol.* **46**, 17–25.

Koblenzer, C. S. (1993). The clinical presentation, diagnosis and treatment of delusions of parasitosis—a dermatologic perspective. *Bull. Soc. Vector Ecol.* **18**, 6–10.

Kushon, D. J., Helz, J. W., Williams, J. M., Lau, K. M. K., Pinto, L., and St. Aubin, F. E. (1993). Delusions of parasitosis: a survey of entomologists from a psychiatric perspective. *Bull. Soc. Vector Ecol.* **18**, 11–15.

Poorbaugh, J. H. (1993). Cryptic arthropod infestations: separating fact from fiction. *Bull. Soc. Vector Ecol.* **18**, 3–5.

Webb, J. P., Jr. (1993a). Delusions of parasitosis: a symposium—coordination among entomologists, dermatologists, and psychiatrists. *Bull. Soc. Vector Ecol.* **18**, 1–2.

Webb, J. P., Jr. (1993b). Case histories of individuals with delusions of parasitosis in southern California and a proposed protocol for initiating effective medical assistance. *Bull. Soc. Vector Ecol.* **18**, 16–25.

## TOXINS AND VENOMS

Beard, R. L. (1960). Insect toxins and venoms. *Annu. Rev. Entomol.* **8**, 1–18.

Bettini, S. (ed.) (1978). Arthropod venoms. Springer-Verlag, Berlin.

Bücherl, W., and Buckley, E. E. (1971). Venomous animals and their venoms. Vol. 3. Venomous invertebrates. Academic Press, San Diego.

Meier, J., and White, J. (eds.) (1995). Handbook of clinical toxicology of animal venoms and poisons. CRC Press, Boca Raton, FL.

Roth, L. M., and Eisner, T. (1962). Chemical defenses of arthropods. *Annu. Rev. Entomol.* **7**, 107–136.

Schmidt, J. O. (1982). Biochemistry of insect venoms. *Annu. Rev. Entomol.* **27**, 339–368.

Tu, A. T. (ed.) (1984). Handbook of natural toxins. Vol. 2. Insect poisons, allergens, and other invertebrate venoms. Marcel Dekker, New York.

## HOST DEFENSES

Barriga, O. O. (1981). Immune reactions to arthropods. *In*: The immunology of parasitic infections: a handbook for physicians, veterinarians, and biologists (O. O. Barriga (ed.). pp. 283–317. University Park Press, Baltimore.

Nelson, W. A., Bell, J. F., Clifford, C. M., and Keirans, J. E. (1977). Interaction of ectoparasites and their hosts. *J. Med. Entomol.* **13**, 389–428.

Nelson, W. A., Keirans, J. E., Bell, J. F., and Clifford, C. M. (1975). Host-ectoparasite relationships. *J. Med. Entomol.* **12**, 143–166.

Ribeiro, J. M. C. (1995). Blood-feeding arthropods: live syringes or invertebrate pharmacologists? *Infect. Agents Dis.*, 143–152.

Wikel, S. K. (1982). Immune responses to arthropods and their products. *Annu. Rev. Entomol.* **27**, 21–48.

Wikel, S. K. (1996a). Host immunology to ticks. *Annu. Rev. Entomol.* **41**, 1–22.

Wikel, S. K. (ed.) (1996b). The immunology of host-ectoparasitic arthropod relationships. CAB International, Wallingford, UK.

Wikel, S. K. (1999). Modulation of the host immune system by ectoparasitic arthropods. *BioScience* **49**, 311–320.

## FORENSIC ENTOMOLOGY

Byrd, J. H., and Castner, J. L. (2000). Forensic entomology: utility of arthropods in legal investigations. CRC Press, Boca Raton.

Catts, E. P., and Goff, M. L. (1992). Forensic entomology in criminal investigations. *Annu. Rev. Entomol.* **37**, 253–272.

Catts, E. P., and Haskell, N. H. (1990). Entomology and death: a procedural guide. Joyce's Print Shop, Clemson, SC.

Erzinclioglu, Y. Z. (1989). Entomology and the forensic scientist: how insects can solve crimes. *J. Biol. Educ.* **23**, 300–302.

Goff, M. L. (2000). A fly for the prosecution: how insect evidence helps solve crimes. Harvard Univ. Press, Cambridge, MA.

Smith, K. G. V. (1986). A manual of forensic entomology. British Museum (Natural History), London.

Vincent, C., Kevan, D. K. McE., Leclerq, M., and Meek, C. L. (1985). A bibliography of forensic entomology. *J. Med. Entomol.* **22**, 212–219.

## MINOR ARTHROPOD PROBLEMS OF MEDICAL-VETERINARY INTEREST

Arnaud, P. H., Jr. (1970). Thrips "biting" man. *Pan.-Pac. Entomol.* **46**, 76.

Bailey, S. F. (1936). Thrips attacking man. *Can. Entomol.* **68**, 95–98.

Baz, A., and Monserrat, J. (1999). Distribution of domestic Psocoptera in Madrid apartments. *Med. Vet. Entomol.* **13**, 259–264.

Bishopp, F. C. (1961). Injury to man by earwigs. *Proc. Entomol. Soc. Wash.* **63**, 114.

Durden, L. A. (1987). Predator-prey interactions between ectoparasites. *Parasitol. Today* **3**, 306–308.

Dziezye, J. (1992). Insect defensive spray-induced keratitis in a dog. *J. Am. Vet. Assoc.* **200**, 1969.

Eisner, T. (1965). Defensive spray of a phasmid insect. *Science* **148**, 966–968.

Hatch, R. L., Lamsens, S. D., and Perchalski, J. E. (1993). Chemical conjunctivitis caused by the spray of *A. buprestoides*, two-striped walking stick. *J. Fla. Med. Assoc.* **80**, 758–759.

Hood, J. D. (1927). A blood-sucking thrips. *Entomologist* **60**, 201.

Li, D.-N., and Li, J.-C. (1995). [Report on human dermatitis caused by *Liposcelis divinatorius*] (in Chinese). *Chin. J. Parasitol. Parasit. Dis.* **13**, 283.

Nakata, S., and Maa, T. C. (1974). A review of the parasitic earwigs (Dermaptera: Arixeniina; Hemimerina). *Pac. Insects* **16**, 307–374.

Radford, A. J. (1975). Millipede burns in man. *Trop. Geogr. Med.* **27**, 279–287.

Remington, C. L. (1950). The bite and habits of a giant centipede (*Scolopendra subspinipes*) in the Philippine islands. *Am. J. Trop. Med.* **30**, 453–455.

Scott, H. G. (1966). Insect pests. Part I. Springtails. *Modern Maintenance Manage.* **18**, 19–21.

Scott, H. G., Wiseman, J. S., and Stojanovich, C. J. (1962). Collembola infesting man. *Ann. Entomol. Soc. Am.* **55**, 428–430.

Seshadri, A. R. (1955). An extraordinary outbreak of caddis-flies (Trichoptera) in the Meltrudam township area of Salem district, South India. *S. Indian J. Entomol.* **3**, 337–340.

Stewart, M. A. (1937). Phasmid injury to the human eye. *Can. Entomol.* **49**: 84–86.

Williams, C. B. (1921). A blood-sucking thrips. *Entomologist* **54**, 163–164.

# 2

# EPIDEMIOLOGY OF VECTOR-BORNE DISEASES

WILLIAM K. REISEN

*Medical/veterinary* entomologists play a pivotal role in understanding the epidemiology of vector-borne diseases and are a key component in programs that research, monitor, and control vector-borne parasites. Successful management is dynamic and requires constant diligence and research for successful program maintenance. Medical entomology especially comes to the forefront in public health during periods of war, famine, or natural disasters which disrupt control programs, displace populations, and increase exposure to vectors. Historically, the large-scale movement of nonimmune populations (such as military troops) into areas endemic for vector-borne disease has had devastating effects. Rapid circumglobal transport and commerce continually introduce parasites of humans and domestic animals and their vectors into new geographical areas, placing previously unexposed populations at risk of infection.

Although methods of investigation vary considerably among the vast array of vector-borne parasites, basic ecological concepts unify the pattern of information necessary to understand the epidemiology of vector-borne disease. Information progresses from discovery of the parasite as the causative agent of the disease, to identification of its mode of transmission among vectors and vertebrate hosts, to monitoring, forecasting, and control. During the discovery period, the pattern of cases in time and space and the identification of the causative agent indicate that an arthropod may be responsible for transmission. The definitive incrimination of the vector requires a combination of field and laboratory investigation that determines vector abundance in time and space, host selection patterns, field infection rates, and vector competence. Although short-term studies rapidly may determine the mode(s) of transmission, establishing transmission cycles and interseasonal maintenance mechanisms typically requires years of careful, often frustrating, ecological investigation and laboratory experimentation. Effective surveillance and control programs are best implemented after maintenance, amplification, and epidemic transmission patterns have been described. Unfortunately, discovery rarely progresses in the orderly fashion outlined above. Frequently, monitoring and management of cases progresses more rapidly than the discovery of the pathogen or the mode(s) of transmission. This chapter provides an introduction to concepts needed to understand the epidemiology of vector-borne diseases. Epidemiology developed as a science through the investigation of outbreaks of infectious diseases. As a modern discipline, *epidemiology* deals with the natural history and spread of diseases within human and animal populations. Vector-borne diseases consist minimally of a triad that includes an arthropod vector, a vertebrate host, and a parasite. The spread of pathogens by arthropods is especially complex, because in addition to interactions between the vertebrate host and the parasite, an arthropod is required for transmission of the parasite to uninfected hosts. Environmental factors such as temperature and rainfall impact these processes by affecting the rate of parasite maturation within the arthropod host, as well as arthropod abundance in time and space.

A *vector* is an arthropod responsible for the transmission of parasites among vertebrate hosts. Vectors transmit parasites, not diseases. *Disease* is the response of the host to invasion by or infection with a parasite. A *parasite* is any organism, including viruses, bacteria, protozoa, and helminths, that is dependent upon the host for its survival. Parasites may or may not cause disease. When a parasite injures its host and causes disease, it is referred to as a *pathogen* or *disease agent*. A *vector-borne disease*, therefore, is an illness caused by a pathogen that is transmitted by an arthropod. *Facultative parasites* have both free-living and parasitic forms, whereas *obligate parasites* are totally dependent upon their host(s) to provide their requirements for life. *Ectoparasites* live on or outside the host, whereas *endoparasites* live inside the host. When interacting with their hosts, ectoparasites produce an *infestation* that typically remains topical or peripheral, whereas endoparasites produce an *infection* upon invasion of host tissues. The occurrence and severity of disease depends upon the host-parasite interaction after infection. A host carrying a parasite is *infected*, whereas an infected host capable of transmitting a parasite is *infective*. A host capable of parasite maintenance without clinical symptoms is a *carrier*.

A complete understanding of the epidemiology of arthropod-borne disease requires knowledge of the ecology, physiology, immunology, and genetics of parasite, arthropod, and vertebrate host populations and how they interact in their environment. The degree of contact between the vertebrate host and vector ranges from intermittent (e.g., mosquitoes) to intimate (e.g., sucking lice). Frequently the host provides the vector not only with food, in the form of blood or other tissues, but also a habitat, or place in which to live. Blood feeding by the vector typically brings parasite, vector, and vertebrate host

together in time and space and ultimately is responsible for the transmission of parasites from infected to susceptible vertebrate hosts. A vector usually must take at least two blood meals during its life-time to transmit a parasite: the first to acquire the infection and the second to transmit it. Blood meals are taken to provide the arthropod with nutrients necessary for metabolism, metamorphosis, and/or reproduction. The *gonotrophic cycle* is the reproductive cycle of blood feeding, blood meal digestion, egg maturation, and oviposition. *Parous* females have completed one or more gonotrophic cycles and have a greater probability of being infected with parasites than females feeding for the first time (i.e., *nulliparous* females). Unlike parasites that are transmitted directly from host to host, parasites transmitted by arthropods generally have replaced free-living or environmentally resistant stages with those that can multiply and develop within the arthropod and be transmitted during the blood-feeding process.

## COMPONENTS OF TRANSMISSION CYCLES

The components of a *transmission cycle* of an arthropod-borne disease are (1) a vertebrate host which develops a level of infection with the parasite that is infectious to a vector, (2) an arthropod host or vector that acquires the parasite from the infected host and is capable of transmission, and (3) one or more vertebrate hosts that are susceptible to infection with the parasite after being fed upon by the vector (Fig. 2.1). Vector-borne parasites have evolved

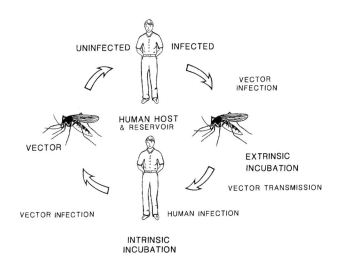

**FIGURE 2.1** Components of the transmission cycle of an anthroponosis such as malaria or louse-borne typhus. (Original by Margo Duncan)

mechanisms for tolerating high constant body temperatures and evading the complex immune systems of the vertebrate hosts as well as for tolerating variable body temperatures and avoiding the very different defensive mechanisms of the arthropod vectors. Asexual parasites, such as viruses and bacteria, employ essentially the same life form to infect both vertebrate and arthropod hosts, whereas more highly evolved heterosexual parasites, such as protozoa and helminths, have different life stages in their vertebrate and arthropod hosts. Some asexual parasites, such as the plague bacillus, intermittently may bypass the arthropod host and be transmitted directly from one vertebrate host to another.

Among sexually reproducing parasites, the host in which gametocyte union occurs is called the *definitive host,* whereas the host in which asexual reproduction occurs is called the *intermediate host.* Vertebrates or arthropods can serve as either definitive or intermediate hosts, depending upon the life cycle of the parasite. For example, humans are the definitive host for the filarial worm, *Wuchereria bancrofti,* because adult male and female worms mate within the human lymphatic system, whereas the mosquito vector, *Culex quinquefasciatus,* is the intermediate host where development occurs without reproduction. In contrast, humans are the intermediate host of the *Plasmodium* protozoan that causes malaria, because only asexual reproduction occurs in the human host; gametocytes produced in the human host unite only in the gut of the definitive mosquito host.

A disease is the response of the host to infection with the parasite and can occur in either vertebrate or arthropod hosts. *Immunity* includes all properties of the host that confer resistance to infection and play an important role in determining host suitability and the extent of disease or illness. Some species or individuals within species populations have *natural immunity* and are refractory to infection. Natural immunity does not require that the host have previous contact with the parasite, but it may be age dependent. For example, humans do not become infected with avian malaria parasites, even though infective *Culex* mosquito vectors feed frequently on humans. Conversely, mosquitoes do not become infected with the measles or poliomyelitis viruses that infect humans, even though these viruses undoubtedly are ingested by mosquitoes blood feeding on viremic hosts.

Individuals within populations become infected with parasites, recover, and in the process actively acquire immunity. This *acquired immunity* to the parasite ranges from transient to lifelong and may provide partial to complete permanent protection. A partial immune response may permit continued infection but may reduce the severity of disease, whereas complete protection results in a cure and usually prevents immediate reinfection.

Acquired immunity may be humeral and result in the rapid formation of antibodies, or it may be cellular and result in the activation of T cells and macrophages. *Antibodies* consist of five classes of proteins called *immunoglobulins* that have specific functions in host immunity. *Immunoglobulin G* (IgG) is most common, comprising over 85% of the immunoglobulins present in the sera of normal individuals. The IgGs are relatively small proteins and typically develop to high concentrations several weeks after infection; they may persist at detectable and protective levels for years. In contrast, IgMs are large macroglobulins that appear shortly after infection but decay rapidly. For the laboratory diagnosis of many diseases, serum samples typically are tested during periods of acute illness and convalescence, 2 to 4 weeks later. A fourfold increase in parasite-specific IgG antibody concentration in these paired sera provides diagnostic serological evidence of infection. The presence of elevated concentrations of IgM presumptively implies current or recent infection. *T cells* and *macrophages* are several classes of cells that are responsible for the recognition and elimination of parasites. In long-lived vertebrate hosts, acquired immunity may decline over time, eventually allowing reinfection.

Clinically, the host response to infection ranges from inapparent or asymptomatic to mildly symptomatic to acute. Generally it is beneficial for the parasite if the host tolerates infection and permits parasite reproduction and/or development without becoming severely ill and dying before infecting additional vectors.

## THE VERTEBRATE HOST

One or more *primary vertebrate hosts* are essential for the maintenance of parasite transmission, whereas *secondary* or *incidental hosts* are not essential to maintain transmission but may contribute to parasite amplification. *Amplification* refers to the general increase in the number of parasites present in a given area. An *amplifying host* increases the number of parasites and therefore the number of infected vectors. Amplifying hosts typically do not remain infected for long periods of time and may develop disease. A *reservoir host* supports parasite development, remains infected for long periods, and serves as a source of vector infection, but it usually does not develop acute disease.

Attributes of a primary vertebrate host include accessibility, susceptibility, and transmissibility.

**Accessibility.**   The vertebrate host must be abundant and fed upon frequently by vectors. Host seasonality, diel activity, and habitat selection determine availability in time and space to host-seeking vectors. For example, the avian hosts of eastern equine encephalomyelitis (EEE) virus generally begin nesting in swamps coincidentally with the emergence of the first spring generation of the mosquito vector, *Culiseta melanura*, thereby bringing EEE virus, susceptible avian hosts, and mosquitoes together in time and space. Diel activity patterns also may be critical. For example, larvae (microfilariae) of *W. bancrofti* move to the peripheral circulatory system of the human host during specific hours of the night that coincide with the biting rhythm of the mosquito vector, *Cx. quinquefasciatus*. Historically, epidemics of vector-borne diseases have been associated with increases in human accessibility to vectors during wars, natural disasters, environmental changes, or human migrations.

**Susceptibility.**    Once exposed, a primary host must be susceptible to infection and permit the development and reproduction of the parasite. *Dead-end hosts* either do not support a level of infection sufficient to infect vectors or become extremely ill and die before the parasite can complete development, enter the peripheral circulatory system or other tissues, and infect additional vectors. Ideal reservoir hosts permit parasites to survive in the peripheral circulatory system (or other suitable tissues) in sufficient numbers for sufficiently long time periods to be an effective source for vector infection. Asexual parasites, such as viruses and bacteria, typically produce intensive infections that produce large numbers of infectious organisms for relatively short periods during which the host either succumbs to infection or develops protective immunity. In the case of EEE virus, for example, 1 ml of blood from an infected bird may contain as many as $10^{10}$ virus particles during both day and night for a 2- to 5-day period; birds that survive such infections typically develop long-lasting, protective immunity. In contrast, highly evolved parasites produce comparatively few individuals during a longer period. *W. bancrofti*, for example, maintains comparatively few microfilaria in the bloodstream (usually <10 microfilaria per cubic millimeter of blood), which circulate most abundantly in the peripheral blood during periods of the day when the mosquito vectors blood feed. However, because both the worms and the human host are long-lived, transmission is enhanced by repeated exposure rather than by an intense parasite presentation over a period of a few days. Infection with >100 microfilaria per female mosquito may prove fatal for the vector; therefore, in this case, limiting the number of parasites that infect the vector may increase the probability of transmission.

**Transmissibility.**    Suitable numbers of susceptible vertebrate hosts must be available to become infected and thereby maintain the parasite. Transmission rates typically decrease concurrently with a reduction in the number of susceptible (i.e., nonimmune) individuals remaining in the host population. The *epidemic threshold* refers to the number of susceptible individuals required for epidemic transmission to occur, whereas the *endemic threshold* refers to the number of susceptible individuals required for parasite persistence. These numerical thresholds vary depending on the immunology and dynamics of infection in the host population. Therefore, suitable hosts must be abundant and either not develop lasting immunity or have a relatively rapid reproductive rate, ensuring the rapid recruitment of susceptibles into the population. In the case of malaria, for example, the parasite elicits an immune response that rarely is completely protective, and the host remains susceptible to reinfection. In contrast, encephalitis virus infections of passerine birds typically produce lifelong protection, but bird life expectancy is short and the population replacement rate is rapid, ensuring the constant renewal of susceptible hosts.

## THE ARTHROPOD VECTOR

Literally, a *vector* is a "carrier" of a parasite from one host to another. An effective vector generally exhibits characteristics that complement those listed above for the vertebrate hosts and include host selection, infection, and transmission.

**Host selection.**    A suitable vector must be abundant and feed frequently upon infective vertebrate hosts during periods when stages of the parasite are circulating in the peripheral blood or other tissues accessible to the vector. Host-seeking or biting activity during the wrong time or at the wrong place on the wrong host will reduce contact with infective hosts and reduce the efficiency of transmission. Patterns of host selection determine the types of parasites to which vectors are exposed. *Anthropophagic* vectors feed selectively on humans and are important in the transmission of human parasites. Anthropophagic vectors which readily enter houses to feed on humans or to rest on the interior surfaces are termed *endophilic* (literally, "inside loving"). Vectors which rarely enter houses are termed *exophilic* (i.e., "outside loving"). *Zoophagic* vectors feed primarily on vertebrates other than humans. *Mammalophagic* vectors blood feed primarily on mammals and are important in the maintenance of mammalian parasites. In contrast, *ornithophagic* vectors feed primarily on avian hosts and are important in the maintenance of avian parasites. There is a distinction between vectors attracted to a host and those which successfully blood feed on the host. Mammalophagic vectors therefore represent a subset of those *mammalophilic* vectors that are attracted to mammalian hosts.

**Infection.**    The vector must be susceptible to infection and survive long enough for the parasite to complete multiplication and/or development. Not all arthropods that ingest parasites support parasite maturation, dissemination, and transmission. For example, the mosquito *Cx. quinquefasciatus* occasionally becomes infected with western equine encephalomyelitis (WEE) virus; however, because this virus rarely escapes the midgut, this species rarely transmits WEE virus. Some arthropods are susceptible to infection under laboratory conditions, but in nature they seldom feed on infected vertebrate hosts and/or survive long enough to allow parasite development. The *transmission rate* is the number of new infections per unit of time and is dependent upon the rate of parasite development to the infective stage and the frequency of blood feeding by the vector. Because many arthropod vectors are poikilothermic and contact their homeothermic vertebrate hosts intermittently, parasite transmission rates frequently are dependent upon ambient temperature. Therefore, transmission rates for many parasites are more rapid at tropical than at temperate latitudes, and at temperate latitudes they progress most rapidly during summer. The frequency of host contact and, therefore, the transmission rate also depend upon the life history of the vector. For example, epidemics of malaria in the tropics transmitted by a mosquito that feeds at 2-day intervals progress faster than epidemics of Lyme disease at temperate latitudes, where the spirochetes are transmitted to humans principally by the nymphal stage of a hard tick vector that may have one generation and one blood meal per life stage per year.

**Transmission.** Once infected, the vector must exhibit a high probability of refeeding on one or more susceptible hosts to ensure the transmission of the parasite. Diversion of vectors to nonsusceptible or dead-end hosts dampens transmission effectiveness. The term *zooprophylaxis* (literally, "animal protection") arose to describe the diversion of *Anopheles* infected with human malaria parasites from humans to cattle, a dead-end host for the parasites. With zooprophylaxis the dead-end host typically exhibits natural immunity, in which host tissues are unacceptable to parasites and do not permit growth or reproduction. Alternatively, transmission to a dead-end host may result in serious illness, because the host-parasite relationship has not coevolved to the point of tolerance by the dead-end host. WEE virus, for example, can cause serious illness in humans, which are considered to be a dead-end host because they rarely produce a viremia sufficient to infect mosquitoes.

# MODES OF TRANSMISSION

The transmission of parasites by vectors may be vertical or horizontal. *Vertical transmission* is the passage of parasites directly to subsequent life stages or generations within vector populations. *Horizontal transmission* describes the passage of parasites between vector and vertebrate hosts.

## VERTICAL TRANSMISSION

Three types of vertical transmission are possible within vector populations: transstadial, transgenerational, and venereal transmission.

*Transstadial transmission* is the sequential passage of parasites acquired during one life stage or stadium through the molt to the next stage(s) or stadium. Transstadial transmission is essential for the survival of parasites transmitted by mites and hard ticks that blood feed once during each life stage and die after oviposition. Lyme disease spirochetes, for example, that are acquired by larval ticks must be passed transstadially to the nymphal stage before transmission to vertebrates.

*Transgenerational transmission* is defined as the vertical passage of parasites by an infected parent to its offspring. Some parasites may be maintained transgenerationally for multiple generations, whereas others require horizontal transmission for amplification. Transgenerational transmission normally occurs *transovarially* (through the ovary) after the parasites infect the ovarian germinal tissue. In this situation most of the progeny are infected. Other parasites do not actually infect the ovary and, although they are passed on to their progeny, transmission is truly transovarial. This situation is usually less efficient and only a small percentage of the progeny are infected. Transgenerational transmission in vectors such as mosquitoes also must include transstadial transmission, because the immature life stages do not blood feed.

*Venereal transmission* is the passage of parasites between male and female vectors and is relatively rare. Venereal transmission usually is limited to transovarially infected males who infect females during insemination, which, in turn, infect their progeny during fertilization.

La Crosse virus (Fig. 2.2) is an example of a vertically maintained parasite where the arthropod host serves as the reservoir. This virus is maintained by transgenerational transmission within clones of infected *Aedes triseriatus* mosquitoes and is amplified by horizontal transmission among squirrels and chipmunks. Because this temperate mosquito rarely has more than two generations per year, La Crosse virus spends long periods in infected vectors and relatively short periods in infected vertebrate hosts. Females infected vertically or horizontally transmit their infection transovarially to first-instar larvae. These larvae transmit the virus transstadially through the four larval stadia and the pupal stage to the adults. These transgenerationally infected females then take a blood meal and oviposit infected eggs, often in the same tree hole from which they emerged. Some blood meal

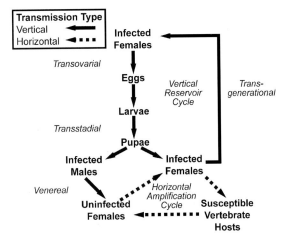

**FIGURE 2.2**   Modes of transmission of a vertically maintained parasite, La Crosse encephalitis virus.

hosts become viremic and amplify the number of infected *Ae. triseriatus* females by horizontal transmission. Venereal transmission of the virus from transgenerationally infected males to uninfected females has been demonstrated in the laboratory and may serve to establish new clones of infected females in nature.

## HORIZONTAL TRANSMISSION

Horizontal transmission is essential for the maintenance of almost all vector-borne parasites and is accomplished by either anterior (biting) or posterior (defecation) routes. *Anterior-station transmission* occurs when parasites are liberated from the mouthparts or salivary glands during blood feeding (e.g., malaria parasites, encephalitis viruses, filarial worms). *Posterior-station* (or *stercorarian*) *transmission* occurs when parasites remain within the gut and are transmitted via contaminated feces. The trypanosome that causes Chagas disease, for example, develops to the infective stage within the hindgut and is discharged onto the host skin when the triatomid vector defecates during feeding. Irritation resulting from salivary proteins introduced into the host during feeding causes the host to scratch the bite and rub the parasite into the wound. Louse-borne relapsing fever and typhus fever rickettsia also employ posterior-station modes of transmission.

There are four types of horizontal transmission, depending upon the role of the arthropod in the life cycle of the parasite: mechanical, multiplicative, developmental, and cyclodevelopmental.

*Mechanical transmission* occurs when the parasite is transmitted among vertebrate hosts without amplification or development within the vector, usually by contaminated mouthparts. Arthropods that are associated intimately with their vertebrate hosts and feed at frequent intervals have a greater probability of transmitting parasites mechanically. The role of the arthropod is essentially an extension of contact transmission between vertebrate hosts. Eye gnats, for example, have rasping, sponging mouthparts and feed repeatedly at the mucous membranes of a variety of vertebrate hosts, making them an effective mechanical vector of the bacteria which cause conjunctivitis or "pink eye." Mechanical transmission also may be accomplished by contaminated mouthparts if the vector is interrupted while blood feeding and then immediately refeeds on a second host in an attempt to complete the blood meal.

*Multiplicative* (or *propagative*) *transmission* occurs when the parasite multiplies asexually within the vector and is transmitted only after a suitable incubation period is completed. In this case, the parasite does not undergo metamorphosis and the form transmitted is indistinguishable from the form ingested with the blood meal. St. Louis encephalitis (SLE) virus, for example, is not transmitted until the virus replicates within and passes through the midgut, is disseminated throughout the hemocoel, and enters and replicates within the salivary glands. However, the form of the virus does not change throughout this process.

*Developmental transmission* occurs when the parasite develops and metamorphoses, but does not multiply, within the vector. Microfilariae of *W. bancrofti*, for example, are ingested with the blood meal, penetrate the mosquito gut, move to the flight muscles, where they molt twice, and then move to the mouthparts, where they remain until they are deposited during blood feeding. These filarial worms do not reproduce asexually within the mosquito vector; i.e., the number of worms available for transmission is always equal to or less than the number ingested.

*Cyclodevelopmental transmission* occurs when the parasite metamorphoses and reproduces asexually within the arthropod vector. In the life cycle of the malaria parasite, for example, gametocytes that are ingested with the blood meal unite within the mosquito gut and then change to an invasive form that penetrates the gut and forms an asexually reproducing stage on the outside of the gut wall. Following asexual reproduction, this stage ruptures and liberates infective forms that move to the salivary glands, from where they are transmitted during the next blood meal.

The *extrinsic incubation period* is the time interval between vector infection and parasite transmission when the parasite is away from the vertebrate host. The *intrinsic incubation period* is the time from infection to the onset of symptoms in the vertebrate host. Repeated lag periods of consistent duration between clusters of new cases at the onset of epidemics were first noticed by early epidemiologists who coined the term extrinsic incubation. These intervals actually represent the combined duration of extrinsic and intrinsic incubation periods.

The duration of the extrinsic incubation period is typically temperature dependent. The rate of parasite development normally increases as a linear degree—day function of ambient temperature between upper and lower thresholds. After being ingested by the mosquito vector, WEE virus, for example, must enter and multiply in cells of the midgut, escape the gut, be disseminated throughout the hemocoel, and then infect the salivary glands, after which the virus may be transmitted by bite. Under hot summer conditions, this process may be completed within 4 days, and the vector mosquito, *Cx. tarsalis*, is capable of transmitting the virus during the next blood meal. In contrast, under cooler spring conditions transmission may be delayed until the third blood meal. Some parasites may increase the frequency of vector blood feeding and thereby enhance transmission. The plague bacillus, for example, remains within and eventually blocks the gut of the most efficient flea vector, *Xenopsylla cheopis*. Regurgitation occurs during blood feeding, causing vector starvation and, therefore, transmission at progressively more closely spaced intervals before the vector succumbs to starvation.

## TRANSMISSION CYCLES

Transmission cycles vary considerably depending upon their complexity and the role of humans as hosts for the parasite. A vector-borne *anthroponosis* is a disease resulting from a parasite that normally infects only humans and one or more anthropophagic vectors (Fig. 2.1). Malaria, some forms of filariasis, and louse-borne typhus are examples of anthroponoses with transmission cycles that involve humans and host-specific vectors. Humans serve as reservoir hosts for these parasites, which may persist for years as chronic infections. Vectors of anthroponoses selectively blood feed upon humans and are associated with domestic or peridomestic environments. Widespread transmission of an anthroponosis with an increase in the number of diagnosed human cases during a specified period of time is called an *epidemic*. When human cases reappear consistently in time and space, transmission is said to be *endemic*.

*Zoonoses* are diseases of animals that occasionally infect humans. Likewise, *ornithonoses* are diseases of wild birds that are transmitted occasionally to humans. In most vector-borne zoonoses, humans are not an essential component of the transmission cycle, but rather become infected when bitten by a vector that fed previously on an infected animal host. Although humans frequently become ill, they rarely circulate sufficient numbers of parasites to infect vectors and thus are termed *dead-end hosts*. The *enzootic transmission cycle* is the basic, or primary, animal cycle (literally "in animals"). When levels of enzootic transmission escalate, transmission may become *epizootic* (an outbreak of disease among animals). Transmission from the enzootic cycle to dead-end hosts is called *tangential transmission* (i.e., at a tangent from the basic transmission cycle). Often different vectors are responsible for enzootic, epizootic, and tangential transmission. *Bridge vectors* transmit parasites tangentially between different enzootic and dead-end host species. Human involvement in zoonoses may depend on the establishment of a secondary *amplification cycle* among vertebrate hosts inhabiting the peridomestic environment.

WEE virus is a zoonosis that exemplifies primary and secondary transmission cycles and tangential transmission to man and equines (Fig. 2.3). In California, WEE virus amplification occurs in a primary enzootic transmission cycle that consists of several species of passerine birds and *Cx. tarsalis* mosquitoes. In addition to birds, *Cx. tarsalis* blood feed on a variety of mammals, including rabbits. Rabbits, especially jackrabbits, develop sufficient viremia to infect some *Cx. tarsalis* and *Ae. melanimon* mosquitoes, thereby initiating a secondary zoonotic transmission cycle. WEE virus activity in the secondary *Aedes*–rabbit cycle usually has been detected after

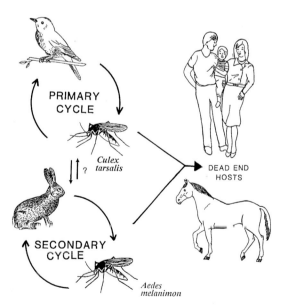

**FIGURE 2.3** Components of the transmission cycles of a zoonosis such as western equine encephalomyelitis (WEE) virus. (Original by Margo Duncan)

amplification in the primary *Cx. tarsalis*–bird cycle. Both *Cx. tarsalis* and *Ae. melanimon* transmit the virus tangentially to humans and equines, which are dead-end hosts for the virus.

## INTERSEASONAL MAINTENANCE

An important aspect of the ecology of vector-borne parasites is the mechanism(s) by which they persist between transmission seasons or outbreaks. Parasite transmission typically is most efficient when weather conditions are suitable for vector activity and population growth. In temperate latitudes, overwintering of parasites becomes problematic when vertebrate or arthropod hosts either enter a winter dormancy or migrate. Similar problems face tropical parasites when transmission is interrupted by prolonged dry or wet seasons. The apparent seasonality that is characteristic of most vector-borne parasites may be due to either the periodic amplification of a constantly present parasite or to the consistent reintroduction of parasites following focal extinction.

Mechanisms of parasite maintenance during periods of unfavorable weather include the following:

**Continued transmission by vectors.** During periods of unfavorable weather, vectors may remain active and continue to transmit parasites, although transmission rates may be slowed by cold temperature or low vector abundance. In temperate latitudes with cold winters, transmission may continue at a slow rate, because the frequency of blood feeding and rate of parasite maturation in the vector is diminished. In tropical latitudes, widespread transmission may be terminated during extended dry seasons that reduce vector abundance and survival. In both instances, transmission may be restricted spatially and involve only a small portion of the vertebrate host population. Human infections during adverse periods usually are highly clumped and may be restricted to members of the same household.

**Infected vectors.** Many vectors enter a state of dormancy as non–blood-feeding immatures or adults. Vertically infected vectors typically remain infected for life and therefore may maintain parasites during periods when horizontal transmission is interrupted. California encephalitis virus, for example, is maintained during winter and drought periods within the transovarially infected eggs of its vector, *Ae. melanimon*. Infected eggs of this floodwater mosquito may remain dormant and infected for up to several years and are able to withstand winter cold, summer heat, and extended dry periods. Inundation of eggs during spring or summer produces broods of adult mosquitoes that are infected at emergence.

Similarly, vectors that inhabit the nests of migratory hosts such as cliff swallows often remain alive and infected for extended periods until their hosts return.

**Infected vertebrate hosts.** Parasite maintenance may be accomplished by infected reservoir hosts that either continue to produce stages infective for vectors or harbor inactive stages of the parasite and then relapse or recrudesce during the season when vectors are blood feeding. Adult filarial worms, for example, continue to produce microfilariae throughout their lifetime, regardless of the population dynamics or seasonality of the mosquito vector. In contrast, some Korean strains of *vivax* malaria overwinter as dormant stages in the liver of the human host and then relapse in spring, concurrent with the termination of diapause by the mosquito vector(s).

Alternatively, parasites may become regionally extinct during unfavorable weather periods and then are reintroduced from distant refugia. Two possible mechanisms may allow the reintroduction of parasites:

**Migratory vertebrate hosts.** Many bird species overwinter in the tropics and return to temperate or subarctic breeding sites each spring, potentially bringing with them infections acquired at tropical or southern latitudes. It also is possible that the stress of long flights and ensuing reproduction triggers relapses of chronic infections. In addition, many large herbivores migrate annually between summer (or wet) and winter (or dry) pastures, bringing with them an array of parasites. Rapid long-range human or commercial transportation is another possible mode for vector and parasite introduction. The seasonal transport of agricultural products and the movements of migratory agricultural workers may result in the appearance of seasonality.

**Weather fronts.** Infected vectors may be carried long distances by prevailing weather fronts. Consistent weather patterns, such as the sweep of the southeastern monsoon from the Indian Ocean across the Indian subcontinent, may passively transport infected vectors over hundreds of kilometers. The onset of WEE virus activity in the north central United States and Canada has been attributed to the dispersal of infected mosquitoes by storm fronts.

## VECTOR INCRIMINATION

To understand the epidemiology of vector-borne disease, it is essential to establish which arthropod(s) is/are the primary vector(s) responsible for parasite transmission. Partial or incomplete vector incrimination has resulted in the misdirection of control efforts at arthropod species that do not play a substantial role in either enzootic maintenance or epidemic transmission. Vector incrimination combines field and laboratory data that measure field infection rates, vector competence, and vectorial capacity.

**Infection rates.** The collection of infected arthropods in nature is an important first step in identifying potential vectors, because it indicates that the candidate species feeds on vertebrate hosts carrying the parasite. Infection data may be expressed as a percentage at one point in time or an *infection prevalence* (i.e., number of vectors infected/number examined × 100). The more commonly employed *infection rate* refers to *infection incidence* and includes change over a specified time period. When the infection prevalence is low and arthropods are tested in groups or pools, data are referred to as a *minimum infection rate* (number of pools of vectors positive/total specimens tested/unit of time × 100 or 1000). Minimum infection rates are relative values with ranges delineated by pool size. For example, minimum infection rates of vectors tested in pools consisting of 50 individuals each must range from 0 to 20 per 1000 females tested.

It is important to distinguish between infected hosts harboring a parasite and infective hosts capable of transmission. In developmental and cyclodevelopmental vectors, the infective stages may be distinguished by location in the vector, morphology, or biochemical properties. Distinguishing infective from noninfective vectors is difficult, if not impossible, with viral or bacterial infections, because the parasite form does not change. The ability to transmit may be implied by testing selected body parts, such as the cephalothorax, salivary glands, or head. With some tick pathogens, however, parasite movement to the mouthparts does not occur until several hours after attachment. As mentioned previously, the transmission rate is the number of new infections per time period. When standardized per unit of population size, the transmission rate may be expressed as an incidence. The annual parasite incidence is the number of new infections per year per 1000 population.

The *entomological inoculation rate* is the number of potentially infective bites per unit of time. This frequently is determined from the human or host biting rate and the proportion of vectors that are infective and is calculated as bites per human per time period × infectivity prevalence.

*Vector competence* is defined as the susceptibility of an arthropod species to infection with a parasite and its ability to transmit this acquired infection. Vector competence is determined quantitatively by feeding the candidate arthropod vector on a vertebrate host circulating the infective stage of the parasite, incubating the blood-fed arthropod under suitable ambient conditions, refeeding the arthropod on a noninfected susceptible vertebrate

host, and then examining this host to determine if it became infected. Because it often is difficult to maintain natural vertebrate hosts in the laboratory and control the concentration of parasites in the peripheral circulatory system, laboratory hosts or artificial feeding systems frequently are used to expose the vector to the parasites. *Susceptibility to infection* may be expressed as the percentage of arthropods that became infected among those blood feeding. When the arthropod is fed on a range of parasite concentrations, susceptibility may be expressed as the *median infectious dose* required to infect 50% of blood-fed arthropods. The ability to transmit may be expressed either as the percentage of feeding females that transmitted or the percentage of hosts that became infected.

Failure of a blood-fed arthropod to become infected with or transmit a parasite may be attributed to the presence of one or more barriers to infection. For parasites transmitted by bite, the arthropod midgut provides the most important barrier. Often parasites will grow in a nonvector species if they are inoculated into the hemocoel, thereby by-passing this *gut barrier*. After penetrating and escaping from the midgut, the parasite then must multiply and/or mature and be disseminated to the salivary glands or mouthparts. Arthropod cellular or humeral immunity may clear the infection at this point, creating a *dissemination barrier*. Even after dissemination to the salivary glands, the parasite may not be able to infect or be transmitted from the salivary glands due to the presence of *salivary gland infection* or *salivary gland escape barriers*, respectively.

For parasites transmitted at the posterior station, vector competence may be expressed as the percentage of infected vectors passing infective stages of the parasite in their feces.

The concept of *vectorial capacity* summarizes quantitatively the basic ecological attributes of the vector relative to parasite transmission. Although developed for mosquito vectors of malaria parasites and most easily applied to anthroponoses, the model provides a framework to conceptualize how the ecological components of the transmission cycle of many vector-borne parasites interact.

*Vectorial capacity* is expressed by the formula:

$$C = ma^2(P^n)/(-\ln P),$$

where $C$ is the vectorial capacity as new infections per infection per day, $ma$ is the bites per human per day, $a$ is the human biting habit, $P$ is the probability of daily survival, and $n$ is the extrinsic incubation period (in days).

The *biting rate (ma)* frequently is estimated by collecting vectors as they attempt to blood feed and is expressed as bites per human per day or night (e.g., 10 mosquitoes per human per night). The human biting habit (*a*) combines vector feeding frequency and host selection. The *feeding frequency* is the length of time between blood meals and frequently is expressed as the inverse of the length of the gonotrophic cycle. Host selection patterns are determined by testing blood-fed vectors to determine what percentage fed on humans or the primary reservoir. Therefore, if the blood feeding frequency is 2 days and if 50% of host-seeking vectors feed on humans, $a = (1/2 \text{ days}) \times (0.5) = 0.25$. In this example, $ma^2 = 10 \text{ bites/human/night} \times 0.25 = 2.5$; *a* is repeated because infected vectors must refeed to transmit.

The probability of the vector surviving through the extrinsic incubation period of the parasite, $P^n$, requires information on the probability of vector survival ($P$) and the duration of the extrinsic incubation period ($n$). $P$ is estimated either vertically, by age-grading the vector population, or horizontally, by marking cohorts and monitoring their death rate over time. In Diptera, $P$ may be estimated vertically from the parity rate (proportion of parous females per number examined). In practice, $P = (\text{parity rate})^{1/g}$, where $g$ is the length of the gonotrophic cycle. The extrinsic incubation period may be estimated from ambient temperature from data gathered during vector competence experiments by testing the time from infection to transmission for infected vectors incubated at different temperatures. Continuing our example, if $P = 0.8$ and $n = 10$ days, then the duration of infective life is $P^n/(-\ln P) = 0.8^{10}/(-\ln \times 0.8) = 0.48$. Therefore $C = 2.5 \times 0.48$, or 1.2 parasite transmissions per infective host per day.

## SURVEILLANCE

The number of cases of most vector-borne diseases typically varies over both time and space. Information on the number of cases can be gathered from morbidity and mortality records maintained by state or national governmental agencies for the human population. *Morbidity data* are records of illness, whereas mortality data are records of the cause of death. These data vary greatly in their quality and timeliness, depending upon the accuracy of determining the cause of illness or death and the rapidity of reporting. In the United States, the occurrence of confirmed cases of many vector-borne diseases, including yellow fever, plague, malaria, and encephalitis, must by law be reported to municipal health authorities. However, infections with many arthropod-borne parasites, including Lyme disease and the mosquito-borne encephalitides, frequently are asymptomatic or present variable clinical symptoms and therefore remain largely undiagnosed and underreported. The frequency of case detection and accuracy of reporting systems are dependent on the type of surveillance employed and the ability of the medical or veterinary community to recognize suggestive symptoms and request appropriate confirmatory laboratory tests. In addition, some laboratory tests vary in their specificity and sensitivity, thus complicating the interpretation of laboratory results. Cases may be classified as suspect or *presumptive*, based on the physician's clinical diagnosis, or *confirmed*, based on a diagnostic rise in specific antibodies or the direct observation (or isolation) of the parasite from the case. Surveillance for clinical cases may be active or passive.

*Active surveillance* involves *active case detection* in which health workers visit communities and seek out and test suspect cases. In malaria control programs, for example, a field worker visits every household biweekly or monthly and collects blood films from all persons with a current or recent fever. Fever patients are treated with antimalarial drugs presumptively, and these suspected cases are confirmed by detection of malaria parasites in a blood smear. Confirmed cases are revisited and additional medication administered, if necessary. This surveillance provides population infection rates regardless of case classification criteria.

Most surveillance programs rely on *passive surveillance*, which utilizes *passive case detection* to identify clinical human or veterinary cases. In this system, individuals seeking medical attention at primary health care organizations, such as physicians' offices, hospitals, and clinics, are diagnosed by an attending physician who requests appropriate confirmatory laboratory tests. However, because many arthropod-borne diseases present a variety of nonspecific symptoms (e.g., headache, fever, general malaise, arthralgia), cases frequently may be missed or not specifically diagnosed. In mosquito-borne viral infections the patient often spontaneously recovers, and cases frequently are listed under fevers of unknown origin or aseptic (or viral) meningitis without a specific diagnosis. In a passive case-detection system, it is the responsibility of the attending physician to request laboratory confirmation of suspect clinical cases and then to notify the regional public health epidemiologist that a case of a vector-borne disease has been documented.

The reporting system for clinical cases of vector-borne diseases must be evaluated carefully when interpreting surveillance data. This evaluation should take into account the disease, its frequency of producing clinically recognizable symptoms, the sensitivity and specificity of confirmatory laboratory tests, and the type and extent of the reporting system. Usually programs that focus on the surveillance of a specific disease and employ active case detection provide the most reliable epidemiological information. In contrast, broad-based community health care systems that rely on passive case detection typically produce the least reliable information, especially for relatively rare vector-borne diseases with nonspecific symptoms.

Diseases that are always present or reappear consistently at a similar level during a specific transmission season are classified as *endemic*. The number of cases in a population is expressed as incidence or prevalence. *Population* is defined as the number of individuals at risk from infection in a given geographical area at a given time. *Incidence* is the number of new cases per unit of population per unit of time. Incidence data are derived from two or more successive samples spaced over time. *Prevalence* is the frequency of both old and new infections among members of a population. Prevalence typically is determined by a single point in time estimate and frequently is expressed as the percentage of the population tested that was found to have been infected.

The level of parasite endemicity in a population may be graded as *hypoendemic* (low), *mesoendemic* (medium), or *hyperendemic* (high), depending upon the incidence of infection and/or the immune status of the population. In malaria surveys, for example, the percentage of children with palpable spleens and the annual parasite incidence are used to characterize the level of endemicity. In endemic disease, the percentage of individuals with sera positive for IgG-class antibodies typically increases as a linear function of age or residence history, whereas in hypoendemic disease with intermittent transmission, this function is disjunct, with certain age groups expressing elevated positivity rates. The occurrence of an extraordinarily large number of human infections or cases is termed an *epidemic*. Health agencies, such as the World Health Organization, typically monitor incidence data to establish criteria necessary to classify the level of endemicity and to decide when an epidemic is under way. A geographically widespread epidemic on a continental scale is called a *pandemic*.

*Serological surveys* (or *serosurveys*) are a useful epidemiological tool for determining the cumulative infection experience of a population with one or more parasite- and host-related factors affecting the efficiency or risk of transmission, and reinfection rates. When coupled with morbidity data, serosurveys provide information on the ratio of apparent to inapparent infections. *Random sampling* during serosurveys representatively collects data on the entire population and may provide ecological information retrospectively by analysis of data collected concurrently with each serum sample. This information may assign risk factors for infection, such as sex, occupation, and residence history, or it may help in ascertaining age-related differences in susceptibility to disease. *Stratified sampling* is not random and targets a specific cohort or subpopulation. Although stratified samples may have greater sensitivity in detecting rare or contiguously distributed parasites, the data are not readily extrapolated

to infection or disease trends in the entire population. Repeated serological testing of the same individuals within a population can determine the time and place of infection by determining when individuals first become *seropositive*, i.e., serologically positive with circulating antibodies against a specific parasite. This change from seronegative to seropositive is called a *seroconversion*.

Forecasting the risk of human infection usually is accomplished by monitoring environmental factors, vector abundance, the level of transmission within the primary and/or amplification cycles, and the numbers of human or domestic animal cases. As a general rule, the accuracy of forecasting is related inversely to the time and distance of the predictive parameter from the detection of human cases. Surveillance activities typically include the time series monitoring of environmental conditions, vector abundance, enzootic transmission rates, and clinical cases.

**Environmental conditions.** Unusually wet or warm weather may indicate favorable conditions for vector activity or population increases, concurrently increasing the risk of parasite transmission. Parameters frequently monitored include temperature, rainfall, snow pack (predictive of vernal flooding), and agricultural irrigation schedules.

**Vector abundance.** Standardized sampling at fixed sites and time intervals can be used to compare temporal and spatial changes in vector abundance that are useful in detecting an increased risk of parasite transmission. Extraordinary increases in vector abundance and survival may forecast accurately increased enzootic transmission and, to a lesser extent, epidemics.

**Enzootic transmission rates.** Monitoring the level of parasite infection in vector or vertebrate populations provides direct evidence that the parasite is present and being actively transmitted (Fig. 2.4). The level of transmission usually is directly predictive of the risk of human or domestic animal involvement. Enzootic transmission activity may be monitored by vector infection rates, vertebrate-host infection rates, sentinel seroconversion rates, and clinical cases.

**Vector infection rates.** Sampling vectors and testing them for parasites determines the level of infection in the vector population (Fig. 2.4, C and D). When vectors are tested individually, prevalence data are expressed as percentages; e.g., 10 females infected per 50 tested is a 20% infection rate. When combined with abundance estimates, infection rates also may be expressed as infected vectors per sampling unit per time interval; 100 bites per human per night × 0.2 infection rate = 20 infective bites per human per night. These data provide an index of the transmission rate. When infection rates are low and vector populations large, vectors usually are tested in lots or pools. It is statistically advantageous to keep the pool size constant and thus keep the chance of detecting

**FIGURE 2.4** Mosquito-borne encephalitis surveillance in southern California. (A) Coop with 10 sentinel chickens; (B) Taking blood sample from chicken; (C) Hanging mosquito trap on permanent standard (components from left to right are trap motor and fan assembly with collecting carton, dry-ice bait in a Styrofoam container, and battery); (D) Sorting mosquito collections by species to estimate relative abundance.

infection the same. Because there may be more than one infected vector per pool, infection rates are expressed as a minimum infection rate = positive pools/total individuals tested × 100 or 1000.

**Vertebrate-host infection rates.** Introduced zoonoses, such as sylvatic plague in North American rodents, frequently produce elevated mortality that may be used to monitor epizootics of these parasites over time and space. In contrast, endemic zoonoses rarely result in vertebrate host mortality. Testing reservoir or amplifying hosts for infection is necessary to monitor the level of enzootic parasite transmission. Stratified sampling for these parasites (directly by parasite isolation or indirectly by seroprevalence) usually focuses on the young of the year to determine ongoing transmission. For example, examining nestling birds for viremia can provide information on the level of enzootic encephalitis virus transmission.

Monitoring the incidence of newly infected individuals in a population over time is necessary to detect increased transmission activity. Because many parasites are difficult to detect or are present only for a limited time period, sampling frequently emphasizes the monitoring of

seropositivity. Monitoring the IgM antibody, which rises rapidly after infection and decays relatively quickly, can indicate the level of recent infection, whereas monitoring the IgG antibody documents the population's historical experience with the parasite. Sampling, marking, releasing, recapturing, and resampling wild animals is most useful in providing information on the time and place of infection in free-roaming animal populations.

**Sentinel seroconversion rates.** Sentinels typically are animals that can be monitored over time to quantify the prevalence of a parasite. Trapping wild animals or birds is labor intensive, and determining seroprevalence may provide little information on the time and place of infection, especially if the host has a large home range. To circumvent this problem, caged or tethered natural hosts or suitable domestic animals of known infection history are placed in sensitive habitat and repeatedly bled to detect infection. A suitable sentinel should be fed upon frequently by the primary vector species, be easy to diagnose when infected, be unable to infect additional vectors (i.e., not serve as an amplifying host), not succumb to infection, and be inexpensive to maintain and easy to bleed or otherwise sample for infection. Chickens, for example, are useful sentinels in mosquito-borne encephalitis virus surveillance programs (Fig. 2.4, A and B). Flocks of seronegative chickens are placed at farmhouses and then bled weekly or biweekly to determine seroconversions to viruses such as WEE or SLE. Because the chickens are confined and the date of seroconversion known, the time and place of infection is determined, while the number seroconverting estimates the intensity of transmission.

**Clinical cases.**   Detecting infection among domestic animals may be an important indication that an epizootic transmission is under way and that the risk of human infection has become elevated. Domestic animals often are more exposed to vectors than are humans and thus provide a more sensitive indication of parasite transmission. Clinical human cases in rural areas in close association with primary transmission cycles may be predictive of future epidemic transmission in urban settings.

Vector-borne diseases frequently affect only a small percentage of the human population, and therefore vector control remains the intervention method of choice. Control programs attempt to maintain vector abundance below thresholds necessary for the transmission of parasites to humans or domestic animals. When these programs fail, personal protection by repellents or insecticide-impregnated clothing, bed nets, or curtains is often the only recourse. Vaccination may be a viable alternative method of control for specific vector-borne diseases, if the vaccine imparts lasting immunity as in the case of yellow fever virus. However, many parasites, such as malaria, have evolved to the point where infection elicits a weak immune response that provides only short-term and marginal protection. The need for continued revaccination at short intervals severely limits their

global usefulness, especially in developing countries. Although breakthroughs in chemotherapy have been useful in case management, it remains the mandate of the medical/veterinary entomologist to devise strategies which combine epidemiological and ecological information to effectively reduce or eliminate the risk of vector-borne diseases.

## REFERENCES AND FURTHER READING

Beaglehole, R., Bonita, R., and Kjellstrom, T. (1993). Basic epidemiology. World Health Organization, Geneva.

Bruce-Chwatt, L. J. (1980). Essential malariology. William Heinemann Medical Books Ltd., London.

Garrett-Jones, C. (1970). Problems of epidemiological entomology as applied to malariology. *Misc. Publ. Entomol. Soc. Am.* 7, 168–178.

Gregg, M. B. (1988). Epidemiological principles applied to arbovirus diseases. *In:* The arboviruses: epidemiology and ecology. Vol. 1 (T. P. Monath, ed.),. pp. 292–309. CRC Press, Boca Raton, Florida.

Herms, W. B. (1961). Medical entomology (5th ed.). Macmillan Co., New York.

Herms, W. B., and James, M. T. (1995). How arthropods cause and carry disease. *In:* A dictionary of epidemiology (3rd ed., J. M. Last, J. H. Abromson, *et al.,* eds.). pp. 15–26. Oxford University Press, London.

Jawetz, E., Melnick, J. L., and Adleberg, E. A. (1972). Host-parasite relationships. *In:* Review of medical microbiology (10th ed., E. Jawetz, J. L. Melnick, and E. A. Adelberg, eds.). pp. 128–135. Lange Medical Publications, Los Altos, California.

Macdonald, G. (1957). The epidemiology and control of malaria. Oxford University Press, London.

Moore, C. G., McLean, R. G., Mitchell, C. J., Nasci, R. S., Tsai, T. F., Calisher, C. H., Marfin, A. A., Moore, P. S., and Gubler, D. J. (1993). Guidelines for arbovirus surveillance programs in the United States. US Department of Health and Human Services, Centers for Disease Control and Prevention, Division of Vector Borne Infectious Diseases, Ft. Collins, Colorado.

Rice, P. L., and Pratt, H. D. (1992). Epidemiology and control of vectorborne diseases. US Department of Health, Education, and Welfare, Public Health Service Pub. No. (HMS0) 72-8245.

# COCKROACHES (*Blattaria*)

RICHARD J. BRENNER

*Cockroaches are* among the oldest and most primitive of insects. They evolved about 350 million years ago during the Silurian Period, diverging together with the mantids from an ancestral stock that also gave rise to termites (Boudreaux 1979). Cockroaches are recognized as the order Blattaria. Although the majority of species are feral and not directly associated with people, a few species have evolved in proximity to human habitations, where they have adapted to indoor environments. Their omnivorous feeding behavior, facilitated by their unspecialized chewing mouthparts, has contributed to a close physical relationship between cockroach populations and humans, with resultant chronic exposure of humans to these pests.

The presence of some species in the home (e.g., German and brownbanded cockroaches) often is an indicator of poor sanitation or substandard housekeeping.

Although they are primarily nuisance pests, their presence can have important health implications. Cockroaches are generalists that feed on virtually any organic substance grown, manufactured, stored, excreted, or discarded by humans. Consequently, food supplies are at risk of contamination by pathogens associated with cockroaches. Because species that infest structures typically have high reproductive rates, humans commonly are exposed to high levels of potentially allergenic proteins associated with cockroaches, which can lead to significant respiratory ailments. Cockroaches also can serve as intermediate hosts of parasites that debilitate domestic animals.

## TAXONOMY

There are about 4000 species of cockroaches worldwide. About 70 species occur in the United States, 24 of which have been introduced from other parts of the world. According to Atkinson *et al.* (1991), 17 of these species are pests of varying degrees. There are five cockroach families, three of which include most of the pest species: Blattidae, Blattellidae, and Blaberidae. Species in the Cryptocercidae are unusual in that they have gut symbionts similar to those found in termites, and they live in family groups in decaying logs. Members of the Polyphagidae include those dwelling in arid regions, where they are capable of moving rapidly through sand. Species in these two families are rarely pests. The family Blattidae includes relatively large cockroaches that are the most common peridomestic pests throughout much of the world. Blattellid cockroaches range in length from less than 25 mm (e.g., *Supella* and *Blattella*) to 35–40 mm (e.g., *Periplaneta* and *Parcoblatta* spp.). *Parcoblatta* species are feral, occasionally invading homes but seldom reproducing indoors. Blaberid cockroaches range greatly in size and include some of the more unusual species, such as the Cuban cockroach, which is green as an adult, and the Surinam cockroach, which is parthenogenetic in North America. Nearly all of the blaberids that occur in the United States are restricted to subtropical regions and have minor medical or veterinary significance. Taxonomic keys for adults are provided by McKittrick (1964), Cornwell (1968), Roth (1985), and Helfer (1987). A pictorial key for identifying the egg cases of common cockroaches is provided by Scott and Borom (1964).

## MORPHOLOGY

Cockroaches have retained their basic ancestral form. The Blattaria are distinguished from other insect orders by morphological characters associated with wing size and venation, biting/chewing mouthparts, and prominent cerci. They differ from other orthopteroid insects by having hind femora which are not enlarged, cerci typically with eight or more segments, a body that is dorsoventrally flattened and generally ovoid, and a head that is largely concealed from above by a relatively large pronotum.

A common indicator of cockroach infestations is their egg cases, or *oothecae* (singular *ootheca*), purse-shaped capsules that typically contain 5–40 embryos (Fig. 3.1). Coloration ranges from light brown to chestnut brown, depending on the degree of sclerotization. A keel that runs the anterior length of the ootheca permits transport of water and air to the developing embryos. Each embryo is contained in a separate compartment that may or may not be obvious externally. In some species (e.g., German and brownbanded cockroaches) lateral, anterior-to-posterior indentations denote the individual developing embryos. Others have only weak lateral indentations (e.g., brown and smokybrown cockroaches), and still others have no lateral indentations but differ in their symmetry (e.g., Oriental, American, and Australian cockroaches).

The mouthparts of cockroach nymphs and adults are characterized by strongly toothed mandibles for biting and chewing. Maxillary and labial palps are well developed, with five and three segments, respectively. Antennae are long and whiplike, originate directly below the middle of the compound eyes, and consist of numerous small segments. The arrangement of three ocelli near the

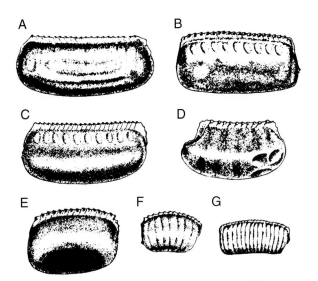

**FIGURE 3.1** Cockroach oothecae (egg cases). A, Australian cockroach (*Periplaneta australasiae*); B, Brown cockroach (*P. brunnea*); C, Smokybrown cockroach (*P. fuliginosa*); D, Oriental cockroach (*Blatta orientalis*); E, American cockroach (*P. americana*); F, Brownbanded cockroach (*Supella longipalpa*); G, German cockroach (*Blattella germanica*). (Courtesy of the US Public Health Service)

antennal sockets is variable: they are well developed in winged species (*macropterous*) but rudimentary or lacking in species with reduced wings (*brachypterous*) or those lacking wings altogether (*apterous*).

Adults generally have two pairs of wings that are folded fanwise at rest. The front wings, called *tegmina* (singular *tegmen*), are typically hardened and translucent, with well-defined veins. The hind wings are membranous and larger. In some species, such as the wood cockroaches (e.g., *Parcoblatta* species), females are brachypterous and incapable of flight, whereas males are macropterous. Other species, such as the Florida woods cockroach (*Eurycotis floridana*), have only vestigial wing buds as adult males and females. In cockroaches, all three pairs of legs are well developed, with large coxae and slender, long segments that aid in the rapid running that is characteristic of these insects. Each femur has two longitudinal keels that typically are armed with spines. The tibiae are often heavily spined and are used for defense against predators. Each tarsus consists of five segments with a pair of claws and may bear a padlike *arolium* that aids in walking on smooth surfaces. Ventral pads, or *pulvilli*, are present on tarsomeres 1–4. A pair of caudal *cerci* have small ventral hairs that are sensitive to vibrations caused by low-frequency sound and air movement; their stimulation initiates an escape response.

The posterior end of the abdomen of some nymphs and all males bears a pair of *styli* (singular *stylus*) between the cerci, arising from the sternum of the ninth abdominal segment. In winged species, the styli may be apparent only when viewed ventrally. The structure of the styli serves to distinguish males from females. Generally the males also can be recognized by their more slender bodies, with laterally tapered and dorsally flattened external genitalia (*terminalia*). The terminalia of the more robust females are notably broader than in males and bear a conspicuous *subgenital plate* that is rounded or keel-like when viewed ventrally. Associated with this plate is a relatively large *genital chamber* (genital pouch) in which the ootheca develops. For a more detailed description of cockroach genitalia, see McKittrick (1964) or Cornwell (1968). Nymphal stages are similar in appearance to adults, but they lack wings, have incompletely developed genitalia, and may vary markedly in color from the adult.

## LIFE HISTORY

Cockroaches are *paurometabolous* insects. The immature cockroaches generally are similar in appearance to the adults except for their undeveloped sexual organs and lack of fully developed wings (Fig. 3.2). Reproduction in cockroaches is typically sexual, although parthenogenesis is reported in a few species. Comparative life history data for some of the more common cockroach pests are provided in Table I.

In cockroaches, *embryogenesis* and *oviposition* occur in one of three ways. Most species are *oviparous*, including all *Periplaneta* species and the Oriental and brownbanded cockroaches. Eggs of oviparous species are protected inside a thick-walled, impermeable ootheca which is deposited soon after it is formed. Embryonic development occurs external to the female. The German cockroach is oviparous, but the female carries the ootheca protruding from the genital chamber until just hours before hatching occurs. The ootheca is softer than in *Periplaneta* species, allowing uptake of water and nutrients from the genital pouch. A few cockroaches, such as *Blaberus* species and the Surinam cockroach, are *pseudo-ovoviviparous*, in that females produce an ootheca which is extruded, rotated, and then retracted into the genital pouch. The eggs are incubated internally until hatching. The only known *pseudo-viviparous* species is *Diploptera punctata*, a pest species in Hawaii; the embryos hatch while still in the genital pouch. Embryogenesis takes 1–8 weeks, depending on the species.

The number of *nymphal instars* varies from 5 to 13, depending on the species, nutritional sources, and microclimate. Development of pestiferous species through the nymphal stadia requires from 6–7 weeks for German cockroaches to well over a year for *Periplaneta* species and other larger cockroaches. Typically, the nymphs exhibit strong aggregation tendencies, governed largely by *aggregation pheromones*. These pheromones act as locomotory inhibitors; when cockroaches perceive the pheromone they become relatively stationary. Studies of various species have shown that development to the adult stage is quicker when nymphs are reared in groups rather

**FIGURE 3.2** Developmental stages of cockroaches, represented by *Periplaneta brunnea*. Left to right: first, second, third, and fourth nymphal instars; adult female, adult male. (Courtesy of Daniel R. Suiter)

TABLE I
Life Histories of Selected Common Species of Cockroaches, Showing the High Degree of Variability
Within Species Due to Environmental Temperatures and Nutritional Availability

| Cockroach | No. eggs/oothecae | No. nymphal instars | Developmental time (days) | Embryonic development |
|-----------|-------------------|---------------------|---------------------------|----------------------|
| German | 30–40 | 5–7 | 103 | Internal/extruded |
| Asian | 37.5 | 5–7 | 52–80 | Internal/extruded |
| Brownbanded | 16–18 | 6–8 | 90–276 | External |
| American | 6–28 | 7–13 | 168–700 | External |
| Smokybrown | 20 | 9–12 | 160–716 | External |
| Australian | 24 | — | 238–405 | External |
| Oriental | 18 | 7–10 | 206–800 | External |
| Surinam | 26 | 8–10 | 127–184 | Internal |

than in isolation. However, aggregation does have a biological cost; those reared in groups typically are smaller in size, and cannibalism may occur. Longevity of cockroaches varies from several weeks to over a year.

## BEHAVIOR AND ECOLOGY

Mating in cockroaches generally is preceded by courtship behavior facilitated by sex pheromones. In some species a blend of volatile compounds is produced by virgin females to attract and orient males (e.g., *Periplaneta* species and the brownbanded cockroach). In the German cockroach, the sex pheromone is a blend of nonvolatile and volatile cuticular components that elicits courtship by males following palpation of the female's integument by the male's antennae. Once courtship is initiated in the male, he turns away from the female and raises his wings to expose dorsal *tergal glands;* the female feeds on pheromones from these glands as the male grasps her genitalia with his pair of caudal claspers. Most species copulate in an end-to-end position. During the hour or so that follows, a *spermatophore* is formed and passed from the male into the *genital chamber* of the female. Only about 20% of females mate again after the first gonotrophic cycle.

Cockroaches can be categorized ecologically as domestic, peridomestic, or feral. *Domestic* species live almost exclusively indoors and are largely dependent on humans for resources (food, water, and harborage) for survival. They rarely are able to maintain themselves outdoors. Although this group contains the smallest number of species, it presents the greatest concern to human health. Domestic species include the German and brownbanded cockroaches. *Peridomestic* species are those which survive in or around human habitation. Although they do not require humans for their survival, they are adept at exploiting the amenities of civilization. This group is represented by American, Australian, brown, and smokybrown cockroaches (all *Periplaneta* species), the oriental cockroach, and the Florida woods cockroach. *Feral* species are those in which survival is independent of humans. This group includes more than 95% of all species in the world. Only a few occur indoors as occasional and inadvertent invaders that typically do not survive in a domestic environment. They are of little or no medical importance.

Cockroach behavior and survival are strongly influenced by their need for food, water, and safe harborage from potential predators and detrimental microclimates. They are *omnivorous* and will consume virtually any organic matter, including fresh and processed foods, stored products, and even book bindings and pastes on stamps and wallpaper when more typical foodstuffs are not available. Cockroaches have the same general problems with water balance as do other terrestrial arthropods. Their relatively small size results in a high surface area to volume ratio and a high risk of losing water through respiration, oral and anal routes, or the cuticle. Temperature, air flow, relative humidity, and availability of liquid water greatly affect water regulation.

As a result of these physiological considerations, physical constraints of the environment usually determine habitat preferences of cockroaches in and around structures. Oriental and American cockroaches, for example, require high moisture and occur in damp terrestrial environments such as septic tanks and municipal sewer systems. Brown, smokybrown, and Florida woods cockroaches occur in a wider range of habitats associated with trees, wood and leaf piles, wall voids, and foundation blocks of buildings. Brownbanded cockroaches are more tolerant of drier conditions and commonly occur in kitchens, pantries, and bedrooms. German cockroaches occupy harborages near food and water. Consequently, they are found primarily in kitchens and pantries, and secondarily in bathrooms, when their populations are high. In mixed populations of German and brownbanded cockroaches, the German cockroach tends to outcompete the brownbanded cockroach within 9 months.

Cockroaches are adept crawlers and are capable of rapid movement even across windows and ceilings. Flight ability varies with species. Some are incapable of flight except for crude, downward gliding used as an escape behavior. Others are weak fliers, occasionally seen flying indoors when disturbed. Still others are relatively strong fliers that are particularly active at sunset, when they may be attracted indoors by lights and brightly lit surfaces. Attraction to light is especially common in the Asian, Surinam, and Cuban cockroaches and in many of the wood cockroaches (*Parcoblatta* species).

Pestiferous cockroaches that occur indoors are typically nocturnal and tend to avoid lighted areas. This enables them to increase their numbers and become established in structures before human occupants even become aware of their presence.

## COMMON COCKROACH SPECIES

The following 11 species of cockroaches are commonly encountered by homeowners in the United States and are the ones most frequently brought to the attention of medical entomologists.

### Oriental cockroach *(Blatta orientalis)*

This peridomestic cockroach (Fig. 3.3) is believed to have originated in northern Africa and from there spread to Europe and western Asia, South America, and North America. It is a relatively lethargic species that prefers cooler temperatures than does the German cockroach and is primarily a concern in temperate regions of the world. Adults are black and 25–33 mm long. Males are winged but do not fly, and females are

**FIGURE 3.4** American cockroach (*Periplaneta americana*), female. (Courtesy of the University of Florida/IFAS)

brachypterous. Their tarsi lack aroliar pads, precluding this cockroach from climbing on smooth vertical surfaces. Oothecae are 8–10 mm long, each typically containing 16 eggs. Also commonly known as *waterbug*, this species is usually associated with damp or wet conditions, such as those found in decaying wood, heavy ground cover (e.g., ivy), water meter boxes, and the lower levels of structures. It infests garbage chutes of apartment complexes, sometimes reaching upper floors. Development is slow compared to that of most other species, requiring about a year depending on temperature conditions. Adults may live for many months. Mobility is fairly restricted, making control easier than for most other species. This species is rarely seen during the daytime.

### American cockroach *(Periplaneta americana)*

The American cockroach (Fig. 3.4) is a large species with adults 34–53 mm in length. It is reddish brown, with substantial variation in light and dark patterns on the pronotum. Adults are winged and capable of flight. Nymphs typically complete development in 13–14 months while undergoing 13 molts. Adults live an average of 15 months, but longevity may exceed 2 years. Females drop or glue their oothecae (8 mm long) to substrates within a few hours or days of formation. Each ootheca has 12–16 embryos. A female generally produces 6–14 egg cases during her life (mean of 9)

The American cockroach is perhaps the most cosmopolitan peridomestic pest species. Together with other closely related *Periplaneta* species, *P. americana* is believed to have spread from tropical Africa to North America and the Caribbean on ships engaged in slave trading. Today this species infests most of the lower latitudes of both hemispheres and extends significantly into the more temperate regions of the world.

**FIGURE 3.3** Oriental cockroach (*Blattella orientalis*), female. (Courtesy of the University of Florida/IFAS)

The habitats of this species are quite variable. American cockroaches infest landfills, municipal sewage systems, storm drainage systems, septic tanks, crawl spaces beneath buildings, attics, tree holes, canopies of palm trees, voids in walls, ships, electronic equipment, caves, and mines. Studies conducted in Arizona indicated movement by a number of individuals several hundred meters through sewer systems and into neighboring homes. This species often can be seen at night on roofs and in air stacks or vents of sewage systems, through which they enter homes and commercial buildings. Entrance also is gained to homes through laundry vent pipes and unscreened or unfiltered attic ventilation systems. This cockroach is known to move from crawl spaces of hospitals via pipe chases into operating theaters, patients' rooms, storage facilities, and food preparation areas. Consequently, the potential of this cockroach for disseminating pathogenic microorganisms can be a significant concern for health care personnel.

### Australian cockroach *(Periplaneta australasiae)*

Adult body coloration is similar to that of the American cockroach, but with paler lateral markings on the upper edges of the tegmina (Fig. 3.5). The pronotum is ringed with similar coloration. Adults are slightly smaller than American cockroaches, measuring 32–35 mm in length. Females mature in about 1 year and typically live for another 4–6 months. A female can produce 20–30 oothecae during her lifetime; the ootheca is about 11 mm long and contains about 24 embryos. Embryonic development requires about 40 days. Nymphs are strikingly mottled, distinguishing them from nymphs of other *Periplaneta* species.

This peridomestic species requires somewhat warmer temperatures than the American cockroach and does not occur in temperate areas other than in greenhouses and

**FIGURE 3.5**   Australian cockroach *(Periplaneta australasiae)*, female. (Courtesy of the University of Florida/IFAS)

**FIGURE 3.6**   Brown cockroach *(Periplaneta brunnea)*, female. (Courtesy of the University of Florida/IFAS)

other pseudotropical environs. In the United States, outdoor populations are well established in Florida and along the coastal areas of Louisiana, Mississippi, Alabama, and Georgia. It commonly is found in environments similar to those inhabited by the smokybrown cockroach. In situations where both species occur (e.g., tree holes, attics), the Australian cockroach tends to displace the smokybrown. It can be a serious pest in greenhouses and other tropical environments in more temperate latitudes, where it can cause feeding damage to plants, notably seedlings.

### Brown cockroach *(Periplaneta brunnea)*

The brown cockroach (Fig. 3.6) is smaller than the American cockroach (33–38 mm), and its pronotal markings are more muted. The most apparent diagnostic characteristic for separating these two species is the shape of the last segment of the cercus; in the brown cockroach, the length is about equal to the width, whereas in the American cockroach the length is about 3 times the width. The ootheca of the brown cockroach usually is larger (7–13 vs 8 mm) and contains more embryos (24 vs 16). The brown cockroach affixes its oothecae to substrates using salivary secretions. They give the ootheca a grayish hue not typical of other *Periplaneta* species that attach their oothecae with salivary secretions. This species is more subtropical than the American cockroach, occurring throughout the southeastern United States, where it infests homes and outbuildings. It is less frequently associated with sewage than is the American cockroach. Because of its similar appearance to the American cockroach, it is often misidentified and may be more widely distributed than is commonly recognized. In Florida, *P. brunnea* is commonly found in canopies of palm trees and attics. It also readily infests various natural cavities and those in human-associated structures. The oothecae can be useful in differentiating species infesting buildings. Most

**FIGURE 3.7** Smokybrown cockroach (*Periplaneta fuliginosa*), female. (Courtesy of the University of Florida/IFAS)

**FIGURE 3.8** Florida woods cockroach (*Eurycotis floridana*), female. (Courtesy of the University of Florida/IFAS)

cockroach oothecae persist in the environment after the nymphs have emerged and provide a history of infestation.

### Smokybrown cockroach (*Periplaneta fuliginosa*)

The smokybrown cockroach (Fig. 3.7) has become a major peridomestic pest throughout the southern United States, including southern California, and extends as far north as the Midwestern states. It can be differentiated from the American cockroach by its slightly smaller size (25–33 mm) and uniform dark coloration. Although developmental times are quite variable, individuals mature in about 10 months. Adults may live for more than a year. Females produce several oothecae, which are 10–11 mm in length with 20 embryos, at 11-day intervals.

Primary foci for this peridomestic species in the southeastern United States are tree holes, canopies of palm trees, loose mulches such as pine straw or pine bark, and firewood piles. Within structures, *P. fuliginosa* seeks the ecological equivalent of tree holes—areas characterized as dark, warm, protective, and moist, with little air flow and near food resources. These include the soffits (eves) of underventilated attics, behind wall panels, the interstices of block walls, false ceilings, pantries, and storage areas. From these harborages, individuals forage for food and water, generally returning to the same refugia. Mark-release-recapture studies using baited live traps have shown that the median distance traveled between successive recaptures is less than 1 m but that some adults may forage at distances of more than 30 m.

### Florida woods cockroach (*Eurycotis floridana*)

This cockroach is restricted to a relatively small area of the United States along the Gulf of Mexico from eastern Louisiana to southeastern Georgia. It is mentioned here only because of its defensive capabilities. It is a large, dark-reddish brown to black cockroach (Fig. 3.8), 30–40 mm long. Although small wing pads are evident, adults are apterous and are relatively slow moving. Oothecae are 13–16 mm long and contain about 22 embryos. *E. floridana* occurs in firewood piles, mulches, tree holes, attics, wall voids, and outbuildings. Last-instar nymphs and adults, if alarmed, can spray a noxious mix of aliphatic compounds that are both odoriferous and caustic. If this is sprayed into the eyes or onto soft tissues, a temporary burning sensation is experienced. Domestic dogs and cats quickly learn to avoid this species. Among its common names are the *Florida cockroach,* the *Florida woods roach,* the *Florida stinkroach,* and *palmettobug.* The last term also is commonly used for other *Periplaneta* species.

### Brownbanded cockroach (*Supella longipalpa*)

Like the German cockroach, this domestic species probably originated in tropical Africa, where it occurs both indoors and outdoors. In North America and Europe it is confined almost exclusively to indoor environments of heated structures. In warm climates, infestations occur particularly in apartments without air conditioning and in business establishments with relatively high ambient temperatures, such as pet stores and animal-care facilities. Adults are similar in size to the those of the German cockroach (13–14.5 mm long) but lack pronotal stripes. Adults have two dark bands of horizontal stripes on the wings (Fig. 3.9), whereas nymphs have two prominent bands running across the mesonotum and first abdominal segment. The brownbanded cockroach derives its name from these bands.

**FIGURE 3.9**    Brownbanded cockroach (*Supella longipalpa*), female. (Courtesy of the University of Florida/IFAS)

**FIGURE 3.10**    German cockroach (*Blattella germanica*), female. (Courtesy of the University of Florida/IFAS)

Populations tend to occur in the nonfood areas of homes, such as bedrooms, living rooms, and closets. Male brownbanded cockroaches occasionally fly and are attracted to lights. Members of this species seek harborage higher within rooms than do German cockroaches. The ootheca is small, only 5 mm long, with an average of 18 embryos and an incubation time of 35–80 days. Females deposit their oothecae by affixing them to furniture, in closets, on or behind picture frames, and in bedding. Transporting *S. longipalpa* with furniture to new locales is common. Although this species occurs with other cockroaches in homes, the German cockroach often outcompetes it within a few months.

### German cockroach (*Blattella germanica*)

This cockroach also is known as the *steamfly* in Great Britain. It is believed to have originated in northern or eastern Africa, or Asia, and has spread from there via commerce. The German cockroach is considered to be the most important domestic pest species throughout the developed world. Adults are about 16 mm long, with two dark, longitudinal bands on the pronotum (Fig. 3.10). It requires warm (optimally 30–33°C), moist conditions near adequate food resources. It primarily inhabits kitchens and pantries, with secondary foci in bathrooms, bedrooms, and other living spaces in heavily infested structures. Although this species is nocturnal, like most other cockroaches, some individuals may be seen moving about on walls and in cupboards during the daylight hours where infestations are heavy. Their wing musculature is vestigial, making them unable to fly except for short, gliding, downward movements. *B. germanica* does not readily move between buildings; however, it does occur in garbage collection containers and outbuildings near heavily infested structures.

The German cockroach has a high reproductive potential. Females produce an ootheca (6–9 mm) containing about 30 embryos within 7–10 days after molting to the adult, or about 2–3 days after mating. The female carries the egg case until a few hours before hatching of the nymphs, preventing access of any oothecal parasitoids or predators. Oothecae are produced at intervals of 20–25 days, with a female producing 4–8 oothecae during her lifetime. Nymphs complete their development in 7–12 weeks.

This species is the main cockroach pest in most households and apartment complexes. Control is difficult, in part because of their movement between apartments through plumbing chases in shared or adjacent walls. Researchers studying over 1,000 apartments in Florida concluded that the median number of cockroaches per apartment was >13,000. This high biotic potential makes this species a major nuisance, as well as a pest with implications for human health.

### Asian cockroach (*Blattella asahinai*)

The Asian cockroach is closely related to the German cockroach, from which it is difficult to distinguish morphologically. In fact, Asian and German cockroaches are capable of hybridizing and producing fertile offspring, which further complicates their identifications. Techniques have been developed to differentiate these two species and their hybrids based on cuticular hydrocarbons in the waxy layer of the integument.

Despite their morphological similarity, *B. asahinai* differs from *B. germanica* in several aspects of its behavior and ecology. It is both a feral and a peridomestic species. Nymphs of the Asian cockroach commonly occur, sometimes in large numbers, in leaf litter and in areas of rich ground cover or well-maintained lawns. Unlike the German cockroach, the adults fly readily and are most active beginning at sunset, when they fly to light-colored walls or brightly lit areas. This behavior can make invasion a nightly occurrence in homes near heavily infested areas. Flight does not occur when temperatures at sunset are below 21°C.

Like those of the German cockroach, Asian cockroach females carry their oothecae until shortly before they are ready to hatch. The ootheca is similar in size and contains the same number of embryos as does that of the German cockroach (38–44). Nymphs are smaller than their *B. germanica* counterparts and are somewhat paler in appearance. Development from egg to adult requires about 65 days, with females producing up to six oothecae during their life span. Adults also are slightly smaller than those of *B. germanica* (average of 13 vs 16 mm).

The Asian cockroach was first described in 1981 from specimens collected in sugar-cane fields on the Japanese island of Okinawa. When it was first discovered in the United States in 1986, the Asian cockroach was found only locally in three counties in Florida, from Tampa to Lakeland; populations already had become established, with densities as high as 250,000 per hectare. By 1993, this species had spread to at least 30 Florida counties and had infested citrus groves throughout the central part of the state. It feeds on succulent early growth of citrus nursery stock, tassels of sweet corn, strawberries, cabbage, tomatoes, and other agricultural products, although there has been no evidence of significant economic damage.

Infestations of apartments by *B. asahinai* have become common in central Florida. This cockroach also has become an increasing problem in warehouses, department stores, hotels, fast-food establishments, automobile dealerships, and other businesses with hours of operation that extend beyond dusk.

### Surinam cockroach *(Pycnoscelus surinamensis)*

This species is believed to have originated in the Indo-Malayan region. It commonly occurs in the southeastern United States from North Carolina to Texas. The adults are fairly stout, 18–25 mm in length, with shiny brown wings and a black body (Fig. 3.11). Nymphs characteristically have shiny black anterior abdominal segments, whereas the posterior segments are dull black and roughened. In North America this species is unusual in that it is parthenogenetic, producing only female offspring; elsewhere both males and females are found. The ootheca is 12–15 mm long, is poorly sclerotized, and contains about 26 embryos. Oothecae are retained inside the genital chamber, from which the nymphs emerge in about 35 days. Females produce an average of three oothecae and live about 10 months in the laboratory. This cockroach commonly burrows into compost piles and the thatch of lawns. Transfer

**FIGURE 3.11** · Surinam cockroach (*Pycnoscelus surinamensis*). (Courtesy of the University of Florida/IFAS)

of fresh mulch into the home for potting plants can result in household infestations. Adult females fly and are attracted to light. They are most likely to be noticed by homeowners at night when they fly into brightly lit television screens. This species commonly is transported in commercial mulch to more temperate areas of the United States, where it has been known to infest greenhouses, indoor plantings in shopping malls, and zoos.

### Cuban cockroach *(Panchlora nivea)*

This medium-sized cockroach (22–24 mm in length) is unusual in that the adults are pale green. The nymphs are dark brown and are found in leaf litter and decaying wood piles. Adults are strong fliers and are attracted to lights. *Panchlora nivea* is believed to be native to the Caribbean basin, Mexico, Central America, and northern South America. In the United States it occurs commonly in Florida and coastal Louisiana and Texas. This cockroach often is seen in the evening, resting on windows and glass patio doors, apparently drawn to the brightness of indoor lighting.

## PUBLIC HEALTH IMPORTANCE

Cockroaches infesting human dwellings and workplaces represent a more intimate and chronic association than do most other pests of medical/veterinary importance. High populations of any cockroach species may adversely affect human health in several ways. These include contamination of food with their excrement, dissemination of pathogens, induced allergies, psychological stress, and bites. Although documentation of bites is limited, there are reports of cockroaches feeding on fingernails, eyelashes, skin calluses of hands and feet, and food residues about the faces of sleeping humans, causing blisters and small wounds (Roth and Willis 1957, 1960). There are other accounts of bites around the mouths of infants in heavily infested homes and even in hospitals. American and Australian cockroaches are the most often implicated species. Bites by the Oriental cockroach have resulted in inflammation of the skin, degeneration of epithelial cells, and subsequent necrosis of the involved tissues.

While many individuals develop a tolerance for cockroach infestations, others may experience psychological stress. The level of stress tends to be proportional to the size of the cockroaches and the magnitude of the infestation. An aversion to cockroaches may be so strong that some people become irrational in their behavior, imagining a severe infestation even when there is none. This illusion of abundant cockroaches has caused some families to move out of their homes. High cockroach populations also produce a characteristic odor that can be unpleasant or even nauseating to some people. Foodstuffs may become contaminated with the excrement of cockroaches, which, on subsequent ingestion, may cause vomiting and diarrhea.

The presence of cockroaches in homes does not necessarily imply poor housekeeping. Peridomestic species such as the American and the Oriental cockroach commonly infest municipal sewage systems or septic tanks and may move into homes through sewage lines. Any of the *Periplaneta* species may develop high outdoor populations, inducing individuals to seek less crowded environments. At such times, they often enter homes through attic vents, breaches in construction joints, or through crawl spaces. This tends to occur in early fall. While they are active at night the smokybrown cockroach, Asian cockroach, and feral wood roaches (*Parcoblatta* species) often find their way into even the best-kept homes. Adults frequently alight on doors illuminated by entrance lights, or on window screens of lighted rooms. Entrance is gained once the door is opened or by squeezing past window-screen frames.

Poor housekeeping and unsanitary conditions contribute significantly to cockroach infestations. The German cockroach and, to a lesser degree, the brownbanded cockroach are the principal bane of apartment dwellers. Their survival is enhanced by crowded living quarters, associated clutter, and the accumulated organic debris associated with food preparation. Construction practices used to build apartment complexes (e.g., common wiring ducts, sewage lines, and refuse areas) can contribute to the spread of cockroaches in multiunit dwellings.

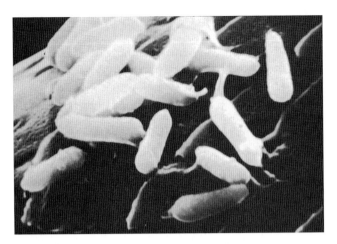

**FIGURE 3.12**    Bacteria adhering to tarsus of German cockroach (*Blattella germanica*). (From Gazivoda and Fish 1985)

## PATHOGENIC AGENTS

The significance of cockroaches in public health remains controversial despite the logical assumption that they play a role in transmitting pathogenic agents. Given that cockroaches are so closely associated with humans and poor sanitation, the potential for acquiring and mechanically transmitting disease agents is very real. They are capable of transmitting microorganisms (Fig. 3.12) and other disease agents indirectly by contaminating foods or food preparation surfaces.

Table II lists pathogenic organisms that have been isolated from cockroaches in domestic or peridomestic environments. At least 32 species of bacteria in 16 genera are represented. These include such pathogens as *Bacillus subtilis,* a causative agent of conjunctivitis; *Escherichia coli* and 9 strains of *Salmonella,* causative agents of diarrhea, gastroenteritis, and food poisoning; *Salmonella typhi,* the causative agent of typhoid; and 4 *Proteus* species, which commonly infect wounds. These isolations primarily have involved American, German, and Oriental cockroaches. Cockroaches also have been found harboring the eggs of 7 helminth species, at least 17 fungal species, 3 protozoan species, and 2 strains of poliomyelitic virus (Brenner *et al,.* 1987; Koehler *et al.,* 1990, Brenner 1995). Researchers in Costa Rica have shown that Australian, American, and Madeira cockroaches become infected with the protozoan *Toxoplasma gondii* after eating feces of infected cats. This suggests the possibility of cockroach involvement in the maintenance and dissemination of this parasite, which causes toxoplasmosis in humans, cats, and other animals.

Although many pathogens have been recovered from natural populations of cockroaches, this does not necessarily mean that cockroaches serve as their vectors. Isolation of pathogens from cockroaches simply may be

## TABLE II
### Bacteria Pathogenic to Humans That Have Been Isolated from Field-Collected Cockroaches

| Pathogen | Associated disease | Cockroach species |
| --- | --- | --- |
| *Alcaligenes faecalis* | Gastroenteritis, secondary infections, urinary tract infections | *Blatta orientalis* *Periplaneta americana* |
| *Bacillus subtilis* | Conjunctivitis, food poisoning | *Blaberus craniifer,* *Blatta orientalis,* *P. americana* |
| *B. cereus* | Food posioning | *Blaberus craniifer* |
| *Campylobacter jejuni* | Enteritis | *Blatta orientalis,* *P. americana* |
| *Clostridium perfringens* | Gas gangrene, food poisoning | Cockroaches |
| *C. novii* | Gas gangrene | *B. orientalis* |
| *C. perfringens* | Food posioning gas gangrene | *B. orientalis* |
| *Enterobacter aerogenes* | Bacteremia | *Blattella germanica,* *P. americana* |
| *Escherichia coli* | Diarrhea, wound infection | *Blatta orientalis,* *Blatella germanica,* *P. americana* |
| *Klebsiella pneumoniae* | Pneumonia, urinary tract infections | Cockroaches |
| *Mycobacterium leprae* | Leprosy | *B. germanica,* *P. americana,* *P. australasiae* |
| *Nocardia spp.* | Actinomycetoma | *P. americana* |
| *Proteus morganii* | Wound infection | *P. americana* |
| *P. rettgeri* | Wound infection | *P. americana* |
| *P. vulgaris* | Wound infection | *Blaberus craniifer,* *Blatta orientalis,* *P. americana* |
| *P. mirabilis* | gastroenteritis, wound infection | *P. americana* |
| *Pseudomonas aeruginosa* | Respiratory infections, gastroenteritis | *Blaberus craniifer,* *Blatta orientalis,* *Blattella germanica,* *P. americana* |
| *Salmonella bredeny* | Food poisoning, gastroenteritis | *P. americana* |
| *S. newport* | Food poisoning, gastroenteritis | *P. americana* |
| *S. oranienburg* | Food poisoning, gastroenteritis | *P. americana* |
| *S. panama* | Food poisoning, gastroenteritis | *P. americana* |
| *S. paratyphi-B* | Food poisoning, gastroenteritis | *P. americana* |
| *S. pyogenes* | Pneumonia | *Blatta orientalis* |
| *S. typhi* | Typhoid | *B. orientalis* |
| *S. typhimurium* | Food poisoning, gastroenteritis | *Blattella germanica,* *Nauphoeta cinerea* |
| *S. bovis-morbificans* | Food poisoning, gastroenteritis | *P. americana* |
| *S. bareilly* | Food poisoning, gastroenteritis | *P. americana* |
| *Serratia marcescens* | Food poisoning | *Blatta orientalis,* *Blattella germanica,* *P. americana* |
| *Shigella dysenteriae* | Dysentery | *B. germanica* |
| *Staphylococcus aureus* | Wound infection, skin infection, infection of internal organs | *Blaberus craniifer,* *Blatta orientalis,* *Blattella germanica* |
| *Streptococcus faecalis* | Pneumonia | *Blatta orientalis,* *Blattella germanica,* *P. americana* |
| *Vibrio spp.* | Not applicable | *Blatta orientalis* |
| *Yersinia pestis* | Plague | *B. orientalis* |

indicative of the natural microbial fauna and flora in our domestic environment. Under certain circumstances, however, cockroaches have the potential for serving as secondary vectors of agents that normally are transmitted by other means. Anecdotal accounts associating diseases in humans with the occurrence of cockroaches and microbes lend some credence to the hypothesis that these pests can serve as vectors. Burgess (1982) reported the isolation from German cockroaches of a serotype of *S. dysenteriae* that was responsible for an outbreak of dysentery in Northern Ireland. Mackerras and Mackerras (1948) isolated *S. bovis-morbificans* and *S. typhimurium* from cockroaches captured in a hospital ward where gastroenteritis, attributed to the former organism, was common. In subsequent experimental studies, *Salmonella* organisms remained viable in the feces of cockroaches for as long as 40 days postinfection (Mackerras and Mackerras 1949). Some of the most compelling circumstantial evidence suggesting that cockroaches may be vectors was noted in a correlation between cases of infectious hepatitis and cockroach control at a housing project during 1956–1962 in southern California (Tarshis 1962). The study area involved more than 580 apartments and 2800 persons; 95% of the apartments had German cockroaches and a lesser infestation of brownbanded and Oriental cockroaches. After pest control measures were initiated, the incidence of endemic infectious hepatitis decreased for 1 year. When treatments were discontinued during the following year because the insecticide was offensive to apartment dwellers, the cockroach population increased, accompanied by a corresponding increase in the incidence of hepatitis. Effective control measures were applied for the following 2 years, and cockroach populations and cases of infectious hepatitis dropped dramatically while hepatitis rates remained high in nearby housing projects where no pest control measures were conducted.

## Intermediate Hosts

Cockroaches can serve as intermediate hosts for animal parasites (Table III). Roth and Willis (1960) published an extensive list of biotic associations between cockroaches and parasitic organisms that potentially infest humans. The eggs of seven species of *helminths* have been found naturally associated with cockroaches. These include hookworms (*Ancylostoma duodenale* and *Necator americanus*), giant human roundworm (*Ascaris lumbricoides*), other *Ascaris* species, pinworm (*Enterobius vermicularis*), tapeworms (*Hymenolepis* species), and the whipworm *Trichuris trichuria*. Development of these helminths in cockroaches has not been observed. These relationships probably represent incidental associations with the omnivorous feeding behavior of cockroaches.

However, cockroaches may serve as potential reservoirs and possible vectors through mechanical transfer in areas where a high incidence of these pathogens in humans is accompanied by substantial cockroach infestations. Human infestations by *spirurid nematodes* associated with cockroaches are known only for the *cattle gullet worm* (*Gongylonema pulchrum*) in the United States, Europe, Asia, and Africa and for the *stomach worm Abbreviata caucasia* in Africa, Israel, Colombia, and Chile. Human cases involving these parasites are rare and cause no pathology.

## Cockroach Allergies

It is only in recent years that the importance of cockroach allergies has been recognized. Allergic reactions result after initial sensitization to antigens following inhalation, ingestion, dermal abrasion, or injection. Allergens produced by cockroaches are rapidly being recognized as one of the more significant indoor allergens of modernized societies. Among asthmatics, about half are allergic to cockroaches. This rate is exceeded only by allergies to house-dust mites. Sensitivity to cockroaches also affects about 10% of nonallergic individuals, suggesting a subclinical level of allergy.

Symptoms exhibited by persons allergic to cockroaches are similar to those described by Wirtz (1980), who reported on occupational allergies in entomologists. They include sneezing and a runny nose, skin reactions, and eye irritation in about two-thirds of the cases. In the more severe cases, individuals may experience difficulty breathing or, even more alarming, anaphylactic shock following exposure to cockroaches. Such allergic reactions can be life-threatening (Brenner *et al.,* 1991).

In recent years, research has focused on determining the specific components of cockroaches that cause allergy. Laboratory technicians exhibit strong allergies to cast skins and excrement of German cockroaches, whereas most patients seen at allergy clinics react primarily to cast skins and whole-body extracts of German cockroaches. Once an individual has become hypersensitized, he or she may experience severe respiratory distress simply by entering a room where cockroaches are held.

Several proteins that can cause human allergies have been identified in the German cockroach. Different exposure histories are likely to result in allergies to different proteins. Cast skins, excrement, and partially consumed food of cockroaches, in addition to living cockroaches, all produce allergenic proteins. Some are extremely persistent and can survive boiling water, ultraviolet light, and harsh pH changes, remaining allergenically potent for decades. Traditionally, whole-body extracts have been used to screen for allergens in skin tests and in bronchial challenges for diagnosing cockroach allergies

**TABLE III**
Cockroaches as Intermediate Hosts of Parasites of Veterinary Importance

| Phylum and parasite | Scientific name | Definitive host | Cockroach intermediate host |
|---|---|---|---|
| Acanthocephala (thorny-headed worms) | | | |
| | *Moniliformis moniliformis* | Rat, mice, dog, cat (primates) | Oriental, German |
| | *M. dubius* | Rat | American, Smokybrown, German |
| | *Prosthenorchis elegans* *P. spirula* | Captive primates | German, Madiera, others |
| Pentastomida (tongue worms) | | | |
| | *Raillietiella hemidactyli* | Reptiles | American |
| Nematoda (round worms) | | | |
| Esophageal and gastrointestinal worm | *Abbreviata caucasica* | Primates (humans) | German |
| Stomach worm | *Cyrnea colini* | Prairie chicken, turkey, bobwhite, quail | German, American |
| Esophagus worm | *Gongylonema neoplasticum* | Rodents, rabbit | Oriental, American |
| Gullet worm | *G. pulchrum* | Cattle (humans) | German |
| Stomach worm | *Mastophorus muris* | Rodents, cat | American, Madiera |
| Eye worm | *Oxyspirura mansoni* | Chicken, turkey | Surinam |
| Eye worm | *O. parvorum* | Chicken, turkey | Surinam |
| Esophageal worm | *Physaloptera rara* | Dog, cat, raccoon, coyote, wolf, fox | German |
| Esophageal worm | *P. praeputialis* | Dog, cat, coyote, fox | German |
| Roundworms | *Protospirura bonnei* *P. muricola* | Monkeys | German, brownbanded |
| Stomach worm | *Spirura rytipleurites* | Cat, rat | Oriental |
| Stomach worm | *Tetrameres americana* | Chicken, bobwhite, ruffed grouse | German |
| Stomach worm | *T. fissipina* | Ducks, geese waterfowl (also chicken, turkey, pigeon, quail) | Unspecified multiple species |

*Note.* Rare definitive hosts are listed in parentheses.

(Fig. 3.13). However, use of more specific antigens that become aerosolized in cockroach-infested homes may be more appropriate, as this is likely to be the sensitizing material. Studies with laboratory colonies have shown that a population of several thousand German cockroaches produced several micrograms of aerosolized proteins in 48 hr. Consequently, the presence of cockroaches may have profound respiratory implications for asthmatic occupants of infested structures. For a general discussion on aerosolized arthropod allergens, see Solomon and Mathews (1988).

Development of an allergy to one insect species can result in broad cross-reactivity to other arthropods, including shrimp, lobster, crab and crawfish, sowbugs (isopods), and house-dust mites. Chronic indoor exposure to cockroach allergens, therefore, may have significant and widespread effects on human health.

# VETERINARY IMPORTANCE

Cockroaches serve as intermediate hosts for a number of parasitic worms of animals (Table III). Most of these relationships are of no economic importance. The majority of the parasites are *nematodes* in the order Spirurida, all members of which use arthropods as intermediate hosts. Species infesting dogs and cats, among other hosts, attach to the mucosa of the gastrointestinal tract, where erosion of tissue may occur at the points of attachment.

**FIGURE 3.13** Apparatus for conducting allergen tests using cockroaches. (Courtesy of R. J. Brenner, USDA/ARS)

Although serious damage seldom occurs, anemia and slow growth may result. Several cockroach-associated nematodes occur in Europe and North America. The *esophageal worms Physaloptera rara* and *P. praeputialis* are the most widespread species in the United States. They develop in the German cockroach, field crickets, and several species of beetles.

Poultry also are parasitized by nematodes which undergo development in cockroaches. The Surinam cockroach is the intermediate host for the *poultry eye worms Oxyspirura mansoni* and *O. parvorum*. Both occur in many parts of the world. In the United States, their distribution is limited to Florida and Louisiana. The German cockroach has been incriminated as the intermediate host for chicken and turkey parasites, including the *stomach worms Tetrameres americana, T. fissispina*, and *Cyrnea colini; C. colini* also develops in the American cockroach. *C. colini* apparently causes no significant damage to poultry, but *Oxyspirura* species can cause pathology ranging from mild conjunctivitis to severe ophthalmia with seriously impaired vision. *T. fissispina* can cause severe damage to the proventriculus of infested birds.

Several nematode parasites of rats and cattle utilize cockroaches as intermediate hosts (Table III). These include *G. neoplasticum* and *Mastophorus muris* in rodents. Both genera occur widely in the United States, where they cause no known pathological problems. The *gullet worm* of cattle, *G. pulchrum*, has been shown experimentally to undergo development in the German cockroach, although the usual arthropod hosts are coprophagous beetles.

Exotic zoo animals also can become infested with parasitic nematodes for which cockroaches serve as possible intermediate hosts. *Protospirura bonnei* and *P. muricola*, for example, have been found in cockroaches collected in cages of monkeys. In a case of "wasting disease" in a colony of common marmosets, more than 50% of German and brownbanded cockroaches captured in the animal room in which they were housed contained the coiled larvae of *Trichospirura leptostoma* in muscle cells (Beglinger *et al.*, 1988).

*Acanthocephalans* (thorny-headed worms) commonly infest primates in zoos and research facilities. *Prosthenorchis elegans* and *P. spirula* occur naturally in South and Central America. Their natural intermediate hosts are unknown. In captivity, primates become infected after eating any of several cockroach species in which the intermediate stages of the parasite have completed development. Heavily infested primates frequently die within a few days. The proboscis of acanthocephalan adults commonly penetrates the intestines of the primate host, causing secondary infections, perforation of the gut wall, and peritonitis.

One *pentastomid* (tongue worm), *Raillietiella hemidactyli*, develops in cockroaches and reptilian hosts. In Singapore, infested geckos are a common occurrence in houses where heavy infestations of *R. hemidactyli* larvae have been found in American cockroaches. Remnants of cockroaches are found commonly in the guts of these lizards.

For additional information on the veterinary importance of cockroaches, see Chitwood and Chitwood (1950), Roth and Willis (1957), Levine (1968), and Noble and Noble (1976).

## PREVENTION AND CONTROL

Traditionally, cockroaches have been controlled using a variety of toxic chemicals applied as residual pesticides to harborage sites or areas frequented by foraging individuals (see Ebling, 1975 and Rust *et. al.*, 1995). Most materials are neurotoxins that disrupt the nervous system, causing locomotory and respiratory failure. These include organophosphates, carbamates, botanicals such as pyrethrins, and pyrethroids. Formulations include wettable powders, emulsifiable concentrates, crack-and-crevice aerosols, dusts, and baits. Several other materials with different modes of action also are currently in use. When ingested, boric acid (delivered as a fine powder or a dilute solution) damages the gut epithelium of cockroaches and kills them by interfering with nutrient absorption. Inorganic silica dust is absorptive, reducing cuticular lipids and causing desiccation. Active ingredients with other modes of action, such as hydramethylnon and sulfluramid, are metabolic inhibitors which disrupt the conversion of food to energy.

The use of baits containing many of the active ingredients mentioned above have been used extensively to control cockroaches. These baits are used indoors in the form of child-resistant bait stations to reduce human exposure. Other bait formulations of gels or pastes are used in crack-and-crevice treatments, making them inaccessible to children and pets. Scatter baits are commonly used outdoors to treat mulches and other landscaping materials that harbor cockroaches.

*Insect growth regulators* (IGRs) can be used to prevent cockroaches from reaching maturity. Two commonly used IGRs are juvenile hormone analogs and chitin synthesis inhibitors. *Juvenile hormone analogs* regulate morphological maturation and reproductive processes. They are highly specific to arthropods, have very low mammalian toxicity, and are effective at exceptionally low rates of application. Such compounds include hydroprene and fenoxycarb. *Chitin synthesis inhibitors* prevent normal formation of chitin during molting. These compounds cause many of the affected nymphs to die during the molting process. Males that survive to the adult stage often have reduced life expectancies, whereas females tend to abort their oothecae.

*Integrated pest management,* which incorporates various control techniques, has contributed significantly to successful control of cockroaches. This approach uses nontoxic agents, such as sticky traps, vacuum devices, diatomaceous earth, or silica-gel repellents and desiccants, and manipulation of harborage sites to reduce or prevent infestations. Desiccants and dusts should be used only in geographic areas or situations with relatively low humidity; high humidity causes these materials to clump and lose their effectiveness. Building designs and construction techniques can significantly influence cockroach survival. By manipulating microclimates in discrete areas of structures frequented by cockroaches, homes and other buildings can be rendered less hospitable to pest species while at the same time greatly reducing aerosolized allergens. Nontoxic repellents can be used to deny access of cockroaches to specific areas.

Biological control of cockroaches has drawn increased attention in recent years. Among the natural agents that have been investigated are parasitic wasps, nematodes, and sporulating fungi. Females of the eulophid wasp *Aprostocetus hagenowii* and the evaniid wasp *Comperia merceti* deposit their eggs in the oothecae of certain peridomestic cockroaches. Major shortcomings in utilizing these wasps are difficulties involved in their mass production and the fact that they do not completely eliminate cockroach infestations. However, *A. hagenowii* has been shown to reduce populations of the peridomestic *Periplaneta* species following inundative or augmentative releases of this wasp. *C. merceti* parasitizes oothecae of the brownbanded cockroach and is the only known parasitoid of a domestic species. The use of *parasitic nematodes* (e.g., *Steinernema carpocapsae*) and several *fungal pathogens* that have been isolated from cockroaches has not yet proved to be effective as a practical management tool. Another drawback to their use is the allergenic nature of several components of nematodes and many sporulating fungi that can become airborne and, upon inhalation, cause asthmatic responses in humans.

Models have been developed for predicting population foci of peridomestic cockroaches based on physical characteristics of residential properties (Smith *et al.,* 1995). However, the use of such models is limited by the scope of the data base used in its development and the complexity of the model itself. The use of traps to detect foci of cockroaches and the analysis of trap counts to determine cockroach abundance and distributional patterns can be helpful in assessing the extent of infestations and monitoring the effectiveness of control programs (Brenner and Pierce 1991).

# REFERENCES AND FURTHER READING

Atkinson, T. H., Koehler, P. G., and Patterson, R. S. (1991). Catalog and atlas of the cockroaches (Dictyoptera) of North America north of Mexico. *Misc. Publ. Entomol. Soc. Am.* **78,** 1–86.

Beglinger, R., Illgen, B., Pfister, R., and Heider, K.. (1988). The parasite *Trichospirura leptostoma* associated with wasting disease in a colony of common marmosets, *Callithrix jacchus. Folia Primatol.* **51,** 45–51.

Boudreaux, H. B. (1979). Arthropod phylogeny with special reference to insects. Wiley, New York

Brenner, R. J. (1988). Focality and mobility of some peridomestic cockroaches in Florida. *Ann. Entomol. Soc. Am.* **81,** 581–592.

Brenner, R. J. (1991). Asian cockroaches: implications to the food industry and complexities of management strategies. *In:* Ecology and management of food-industry pests(J. R. Gorhamed.). pp.121–130US Food and Drug Administration Technical Bulletin No. 4..

Brenner, R. J. (1995). Economics and medical importance of German cockroaches. *In:* Understanding and controlling the german cockroach(M. K. Rust, J. M. Owens, and D. A. Reierson, eds.). Oxford Univ. Press, London.

Brenner, R. J., Barnes, K. C., Helm, R. M., and Williams, L. W. (1991). Modernized society and allergies to arthropods: risks and challenges to entomologists. *Am. Entomol.* **37:** 143–155.

Brenner, R. J., Koehler, P. G., and Patterson, R. S. (1987). *Infections in medicine.* **4,** 349–355, 358, 359, 393.

Brenner, R. J., and Pierce, R. R. (1991). Seasonality of peridomestic cockroaches: mobility, winter reduction, and impact of traps and baits (Orthoptera: Blattaria). *J. Econ. Entomol.* **84,** 1735–1745.

Brenner, R. J., Patterson, R. S., and Koehler, P. G. (1988). Ecology, behavior, and distribution of *Blattella asahinai* (Orthoptera: Blattellidae) in central Florida. *Ann. Entomol. Soc. Am.* **81,** 432–436.

Burgess, N. R. (1982). Biological features of cockroaches and their sanitary importance. *In:* The modern defensive approach of cockroach control(D. Bajomi and G. Erdos, eds.). pp. 45–50.International Symposium, Bologna.

Chitwood, B. G., and Chitwood, M. B. (1950). An introduction to nematology. Section I. Anatomy, p. 213 Baltimore. Monumental.

Cornwell, P. B. (1968). The cockroach. Vol. 1. Hutchinson, London.

Ebling, W. (1975). Urban entomology. Univ. California, Berkeley, CA. pp. viii & 695.

Gazivoda, P., and Fish, D. (1985). Scanning electron microscope demonstration of bacteria on tarsi of *Blattella germanica*. J. N. Y. *Entomol. Soc.* **93**, 1064–1067.

Helfer, J. R. (1987). How to know the grasshoppers, cockroaches and their allies. Brown, Dubuque, IA.

Koehler, P. G., Patterson, R. S., and Brenner, R. J. (1990). Cockroaches. *In:* Mallis handbook of pest control (7th ed., K. Story, ed.). Franzak and Foster, Cleveland. p. 100–175.

Levine, N. D. (1968). Nematode parasites of domestic animals and of man. Burgess, Minneapolis.

Mackerras, I. M., and Mackerras, M. J. (1949). An epidemic of infantile gastroenteritis in Queensland caused by *Salmonella bovis-morbicans* (Basenau). *J. Hyg.* **47**, 166–181.

Mackerras, M. J., and Mackerras, I. M. (1948). *Salmonella* infections in Australian cockroaches. Aust. *J. Sci.* **10:** 115.

Massicot, J. G., and Cohen, S. G. (1986). Epidemiologic and socio-economic aspects of allergic diseases. *J. Allergy Clin. Immunol.* **78**, 954–958.

McKittrick, F. A. (1964). Evolutionary studies of cockroaches. Cornell University Agriculture Experiment Station. Memoir 389. New York State College of Agriculture, Ithaca.

Noble, E. R., and Noble, G. A. (1976). Parasitology (4th ed.). Lea & Feibiger, Philadelphia.

Peterson, R. K. D., and Shurdut, B. A. (1999). Human health risk from cockroaches and cockroach management: a risk analysis approach. *Am. Entomol.* **45**, 142–148.

Roth, L. M. (1985). A taxonomic revision of the genus *Blattella* Caudell (Dictyoptera, Blattaria: Blattellidae). *Entomol. Scand. Suppl.* **22**, 221 pp.

Roth, L. M., and Willis, E. R.. (1957). The medical and veterinary importance of cockroaches. *Smithsonian Misc. Coll.* **134**, 1–147.

Roth, L. M., and Willis, E. R. (1960). The biotic associations of cockroaches. *Smithsonian Misc. Coll.* **141**, 470 pp.

Rust, M. K., Owens, J. M., and Reierson, D. A. (eds.) (1995). Understanding and controlling the German cockroach. Oxford University Press, New York.

Scott, H. G., and Borom, M. R. (1964). Cockroaches: key to egg cases of common domestic species. US Department of Health, Education and Welfare, Public Health Service, Communicable Disease Center, Atlanta. 18 pp.

Smith, L. M., II, Appel, A. G., Mack, T. P., Keever, G. J., and Benson, E. P. (1995). Model for estimating relative abundance of *Periplaneta fuliginosa* (Dictyoptera: Blattidae) by using house and landscape characteristics. *J. Econ. Entomol.* **88**, 307–319.

Solomon, W. R., and Mathews, K. P. (1988). Aerobiology and inhalant allergens. *In:* Allergy principles and practice (3rd ed., E. Middleton, Jr., C. E. Reed, E. F. Ellis, N. F. Adkinson, Jr., and J. W. Yunginger eds.). Mosby, Washington, DC. pp. 312–372.

Tarshis, I. B. (1962). The cockroach—a new suspect in the spread of infectious hepatitis. *Am. J. Trop. Med. Hyg.* **11**, 705–711.

Wirtz, R. A. (1980). Occupational allergies to arthropods—documentation and prevention. *Bull. Entomol. Soc. Am.* **26**, 356–360.

# 4

# LICE (*Phthiraptera*)

LANCE A. DURDEN

*Lice are* a menace to humans, pets, and livestock, not only because of their blood-feeding or chewing habits, but also because of their ability to transmit pathogens. The human body louse has been indirectly responsible for influencing human history through its ability to transmit the causative agent of epidemic typhus. However, most of the 3200 known species of lice are ectoparasites of wild birds or mammals and have no known medical or veterinary importance.

The order Phthiraptera is divided into two main taxonomic groups: the Anoplura (sucking lice) and Mallophaga (chewing or biting lice). All members of the Anoplura are obligate, hematophagous ectoparasites of placental mammals, whereas the more diverse Mallophaga include species that are obligate associates of birds, marsupials, and placental mammals. Although certain chewing lice imbibe blood, most species ingest host feathers, fur, skin, or skin products. Because of the different feeding strategies of the two groups, the blood-feeding Anoplura are far more important than the Mallophaga in transmitting pathogens to their hosts.

## TAXONOMY

Major taxonomic syntheses for the sucking lice include a series of eight volumes by Ferris (1919–1935) that remains the most comprehensive treatment of this group on a worldwide basis. Ferris (1951) updated much of his earlier work in a shorter overview of the group. Kim *et al.* (1986) have compiled an authoritative manual and identification guide for the sucking lice of North America. Durden and Musser (1994a) provide a taxonomic checklist for the sucking lice of the world, with host records and geographical distribution for each species.

The chewing lice are taxonomically less well known than are the sucking lice, and few authoritative identification guides are available. These include a synopsis of the lice associated with laboratory animals (Kim *et al.* 1973), guides to the lice of domestic animals (Tuff 1977, Price and Graham 1997), and an identification guide to the lice of sub-Saharan Africa (Ledger 1980). These publications provide information on both sucking lice and chewing lice. Checklists of the Mallophaga of the world (Hopkins and Clay 1952) and of North America (Emerson 1972) are useful taxonomic references for this group.

Because of the relatively high degree of host specificity exhibited by both chewing and sucking lice, several host–parasite checklists have been prepared. These include a detailed list of both anopluran and mallophagan lice associated with mammals (Hopkins 1949), a host–parasite list for North American Mallophaga (Emerson 1972), a world host–parasite list for the chewing lice of mammals (Emerson and Price 1981), and a host–parasite checklist for the Anoplura of the world (Durden and Musser 1994b).

About 550 species of sucking lice have been described (Durden and Musser 1994a), all of which parasitize placental mammals; these lice are currently assigned to 50 genera and 15 families. About 2650 valid species of Mallophaga have been described; most of these are associated with birds, but about 400 (ca. 15%) parasitize mammals. The Mallophaga can be divided into 3 suborders (Table I), 11 families, and 205 genera.

The Mallophaga are divided into the following three groups (suborders of most authors): Amblycera (seven families, ca. 76 genera, and ca. 850 species), Ischnocera (three families, ca. 130 genera, and ca. 1800 species), and Rhyncophthirina (one family, 1 genus, and 3 species) (Figs. 4.1 and 4.5). However, there has been disagreement regarding the taxonomic rank of these three groups and their relationships to the Anoplura. Many current classifications treat the Phthiraptera as an order and assign suborder (or superfamily) rank to each of the Anoplura, Amblycera, Ischnocera, and Rhyncophthirina. Other classifications treat the Anoplura and Mallophaga as separate orders. Unfortunately, recent phylogenetic analyses of lice based on cladistic principles have produced contradictory results and have failed to resolve this issue. Regardless of current taxonomic interpretations, it is widely agreed that both sucking and chewing lice originated from a common nonparasitic ancestral group closely related to the order Psocoptera (book lice and bark lice). These two groups diverged in the late Jurassic or early Cretaceous Period, 100–150 million years ago.

Sucking lice of medical importance are assigned to two families, the Pediculidae and Pthiridae, whereas sucking lice of veterinary importance are assigned to five families: the Haematopinidae, Hoplopleuridae, Linognathidae, Pedicinidae, and Polyplacidae (Table II). Only one species of chewing louse, the dog biting louse, in the family Trichodectidae, has public health importance. Mallophaga of veterinary significance are typically placed in five families: the Boopiidae, Gyropidae, Menoponidae, Philopteridae, and Trichodectidae (Table I).

**TABLE I**

**Classification and Hosts of Chewing Lice (Mallophaga) of Medical and Veterinary Importance**

| Louse | Host |
| --- | --- |
| Suborder Amblycera | |
| Family Boopiidae | |
| *Heterodoxus spiniger* | Dog, other carnivores |
| Family Gyropidae | |
| Slender guineapig louse, *Gliricola porcelli* | Guinea pig |
| Oval guineapig louse, *Gyropus ovalis* | Guinea pig |
| Family Menoponidae | |
| Chicken body louse, *Menacanthus stramineus* | Domestic fowl |
| Shaft louse, *Menopon gallinae* | Domestic fowl |
| Goose body louse, *Trinoton anserinum* | Goose |
| Large duck louse, *T. querquedulae* | Duck |
| Suborder Ischnocera | |
| Family Philopteridae | |
| Slender goose louse, *Anaticola anseris* | Goose |
| Slender duck louse, *A. crassicornis* | Duck |
| Large turkey louse, *Chelopistes meleagridis* | Turkey |
| Chicken head louse, *Cuclotogaster heterographus* | Domestic fowl |
| Fluff louse, *Goniocotes gallinae* | Domestic fowl |
| Brown chicken louse, *Goniodes dissimilis* | Chicken |
| Large chicken louse, *G. gigas* | Domestic fowl |
| Wing louse, *Lipeurus caponis* | Domestic fowl |
| Slender turkey louse, *Oxylipeurus polytrapezius* | Turkey |
| Family Trichodectidae: | |
| Cattle biting louse, *Bovicola bovis* | Cattle |
| Goat biting louse, *B. caprae* | Goat |
| Angora goat biting louse, *B. crassipes* | Goat |
| Horse biting louse, *B. equi* | Horse |
| *B. limbata* | Goat |
| Donkey biting louse, *B. ocellata* | Donkey |
| Sheep biting louse, *B. ovis* | Sheep |
| Cat biting louse, *Felicola subrostrata* | Cat |
| Dog biting louse, *Trichodectes canis* | Dog, other canids |
| Suborder Rhyncophthirina | |
| Family Haematomyzidae: | |
| Elephant louse, *Haematomyzus elephantis* | Elephant |

# MORPHOLOGY

Lice are small (0.4–10 mm in the adult stage), wingless, dorso-ventrally flattened insects. The elongate abdomen possesses sclerotized dorsal, ventral, and/or lateral plates in many lice (Fig. 4.2); these provide some rigidity to the abdomen when it is distended by a blood meal or other food source. In adult lice the abdomen has 11 segments and terminates in genitalia and associated sclerotized plates. In females, the genitalia are accompanied by finger-like *gonopods*, which serve to guide, manipulate, and glue eggs onto host hair or feathers. The abdomen is adorned with numerous setae in most lice. Immature lice closely resemble adults but are smaller, have fewer setae,

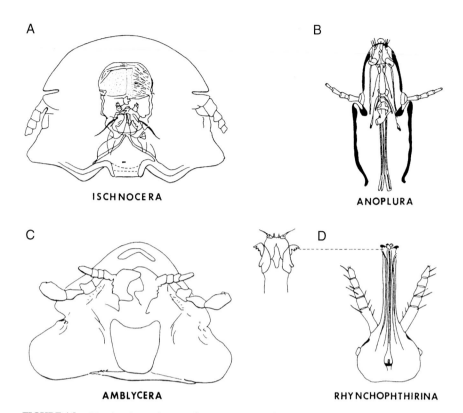

A    **ISCHNOCERA**

B    **ANOPLURA**

C    **AMBLYCERA**

D    **RHYNCHOPHTHIRINA**

**FIGURE 4.1**    Head and mouthparts of representatives of each of the four principal groups of lice. (A) Ishnocera; (B) Anoplura; (C) Amblycera; (D) Rhynchopthirina. (A, from Price and Graham, 1997; B and D, from Ferris, 1931; C, from Bedford, 1932)

and lack genitalia. After each nymphal molt, the abdomen is beset with progressively more setae, and the overall size of the louse increases.

The male genitalia in lice (Fig. 4.3) are relatively large and conspicuous, sometimes occupying almost half the length of the abdomen. The terminal, extrusable, sclerotized *pseudopenis* (*aedeagus*) is supported anteriorly by a *basal apodeme*. Laterally, it is bordered by a pair of chitinized *parameres*. Two or four *testes* are connected to the *vas deferens*, which coalesces posteriorly to form the *vesicula seminalis*. In the female, the *vagina* leads to a large *uterus*, to which several *ovarioles* supporting eggs in various stages of development are connected by the *oviducts*. Two or more large *accessory glands*, which secrete materials to coat the eggs, and a single *spermatheca*, in which sperm is stored, are situated posteriorly in the abdomen. Except for the human body louse, all lice cement their eggs, called *nits*, onto the hair or feathers of their host. Eggs are usually subcylindrical, with rounded ends and a terminal cap, the *operculum* (Fig. 4.4). On the top of the operculum is a patch of holes or areas with thin cuticle, called *micropyles*, through which the developing embryo respires. Most of the egg is heavily chitinized, which helps to protect the embryo from mechanical damage and desiccation. A suture of thin cuticle encircles the base of

the operculum. At the time of hatching, the first-instar nymph emerges from the egg by cracking this suture and pushing off the operculum.

In chewing lice, the head is broader than the thorax (Fig. 4.5). Amblyceran chewing lice have four-segmented antennae and have retained the maxillary palps characteristic of their psocopteran-like ancestor. However, ischnoceran chewing lice have three to five antennal segments and lack maxillary palps. In the Amblycera, the antennae are concealed in lateral grooves, whereas in the Ischnocera and Rhyncophthirina, the antennae are free from the head (Figs. 4.1 and 4.5).

There is a gradation in the specialization of the mouthparts and of the internal skeleton of the head, or *tentorium*, from the psocopteran-like ancestor of the lice through the Amblycera, Ischnocera, Rhyncophthirina, and Anoplura. Although mallophagan lice all possess chewing mouthparts (Fig. 1.6), the components and mechanics of these mouthparts differ for each group. In the Amblycera, the opposable mandibles move in a vertical plane, or perpendicular to the ventral surface of the head, whereas in the Ischnocera they move more or less horizontally. In contrast, the Rhyncophthirina possess tiny mandibles that are situated at the tip of an elongated rostrum (Figs. 4.1D and 4.5F). Through extreme

TABLE II

**Classification and Hosts of Sucking Lice (Anoplura) of Medical and Veterinary Importance**

| Louse | Host |
|---|---|
| Family Echinophthiriidae | |
| *Echinophthirius horridus* | Harbor seals |
| Family Haematopinidae | |
| Horse sucking louse, *Haematopinus asini* | Horse, donkey |
| Shortnosed cattle louse, *H. eurysternus* | Cattle |
| Cattle tail louse, *H. quadripertusus* | Cattle |
| Hog louse, *H. suis* | Swine |
| Buffalo louse, *H. tuberculatus* | Asiatic buffalo, cattle |
| Family Hoplopleuridae | |
| *Hoplopleura captiosa* | House mouse |
| Tropical rat louse, *H. pacifica* | Domestic rat |
| Family Linognathidae | |
| *Linognathus africanus* | Goat, sheep |
| Sheep face louse, *L. ovillus* | Sheep |
| Sheep foot louse, *L. pedalis* | Sheep |
| Dog sucking louse, *L. setosus* | Dog, other canids |
| Goat sucking louse, *L. stenopsis* | Goat |
| Longnosed cattle louse, *L. vituli* | Cattle |
| Little blue cattle louse, *Solenopotes capillatus* | Cattle |
| Family Pedicinidae | |
| *Pedicinus* spp. | Old World primates |
| Family Pediculidae | |
| Head louse, *Pediculus humanus capitis* | Human |
| Body louse, *P. humanus humanus* | Human |
| Family Polyplacidae | |
| Rabbit louse, *Haemodipsus ventricosus* | Domestic rabbit |
| Mouse louse, *Polyplax serrata* | House mouse |
| Spined rat louse, *P. spinulosa* | Domestic rat |
| Family Pthiridae | |
| Crab louse, *Pthirus pubis* | Human |

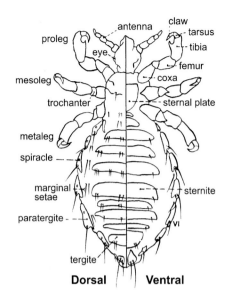

**FIGURE 4.2**  A generalized sucking louse (Anoplura), showing dorsal (left) and ventral (right) morphology. (From Ignoffo, 1959.)

(Fig. 4.8). At rest, the mouthparts are withdrawn into the head and are protected by the snoutlike *haustellum*, representing the highly modified *labrum*. The haustellum is armed with tiny recurved teeth which hook into the host skin during feeding. The *stylets,* consisting of a serrated *labium,* the *hypopharynx,* and two *maxillae,* then puncture a small blood vessel (Fig. 4.8). The hypopharynx is a hollow tube through which saliva (containing anticoagulants and enzymes) is secreted. The maxillae oppose each other and are curved to form a food canal through which host blood is imbibed (Fig. 4.9).

In sucking lice, all three thoracic segments are fused and appear as one segment. In most species, the legs terminate in highly specialized claws for grasping the host

modifications, members of the chewing louse genus *Trochilocoetes* (parasites of humming birds) have evolved mouthparts that can function as sucking organs.

The thorax in chewing lice usually appears as two, and occasionally three, segments. Chewing lice possess one or two simple claws on each leg; species that parasitize highly mobile hosts, especially birds, typically have two claws.

In sucking lice (Figs. 4.2 and 4.7) the head is slender and narrower than the thorax. Anoplura have three- to five-segmented antennae and lack maxillary palps. Eyes are reduced or absent in most sucking lice but are well developed in the medically important genera *Pediculus* and *Pthirus* (Fig. 4.7A, B), and *ocular points,* or eyeless projections posterior to the antennae, are characteristic of sucking lice in the genus *Haematopinus* (Fig. 4.7E).

As indicated by their name, anopluran mouthparts function as sucking devices during blood feeding

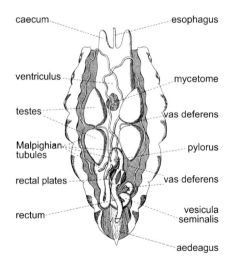

**FIGURE 4.3**  Internal abdominal anatomy of a male human body louse (*Pediculus humanus humanus*). (From Ferris, 1951.)

insects, strong *cibarial* and *esophageal muscles* produce a sucking action during blood feeding. The *esophagus* leads to a spacious *midgut* composed primarily of the *ventriculus*. The posterior region of the midgut is narrow and forms a connection between the ventriculus and the hindgut. Ventrally, *mycetomes* containing symbiotic microorganisms connect to the ventriculus.

## LIFE HISTORY

Lice are *hemimetabolous* insects. Following the egg stage, there are three nymphal instars, the last of which molts to an adult. Although there is wide variation between species, the egg stage typically lasts for 4–15 days and each nymphal instar for 3–8 days; adults live for up to 35 days. Under optimal conditions many species of lice can complete 10–12 generations per year, but this is rarely achieved in nature. Host grooming, resistance, molting and feather loss, hibernation, and hormonal changes, as well as predators (especially insectivorous birds on large ungulates), parasites and parasitoids, and unfavorable weather conditions can reduce the number of louse generations.

Fecundities for fertilized female lice vary from 0.2 to 10 eggs per day. Males are unknown in some parthenogenetic species, whereas they typically constitute less than 5% of the adult population in the cattle biting louse and less than 1% in the horse biting louse.

## BEHAVIOR AND ECOLOGY

Blood from the host is essential for the successful development and survival of all sucking lice. Anoplura are vessel feeders, or *solenophages,* that imbibe blood through a hollow dorsal stylet derived from the hypopharynx (Fig. 4.9). Contraction of powerful cibarial and pharyngeal muscles create a sucking reaction for imbibing blood.

Chewing lice feed by the biting or scraping action of the mandibles. Bird-infesting chewing lice typically use their mandibles to sever small pieces of feather, which drop onto the labrum and are then forced into the mouth. Chewing lice which infest mammals use their mandibles in a similar manner to feed on host fur. Many chewing lice that infest birds and mammals can also feed on other integumental products, such as skin debris and secretions. Some species of chewing lice are obligate, or more frequently facultative, hematophages. Even those species of chewing lice that imbibe blood scrape the host integument until it bleeds. The rhyncophthirinan *Haematomyzus elephantis,* which parasitizes both African and Asian elephants, feeds in this manner.

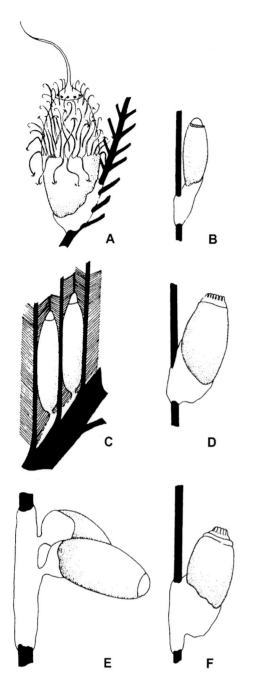

**FIGURE 4.4** Eggs (nits) of representative lice. (A) Chicken body louse, *Menacanthus stramineus;* (B) Oval guineapig louse, *Gyropus ovalis;* (C) Pigeon louse, *Columbicola columbae;* (D) Cattle biting louse, *Bovicola bovis;* (E) Elephant louse, *Haematomyzus elephantis;* (F) Human head louse, *Pediculus humanus capitis.* (From Marshall, 1981).

pelage. These *tibio-tarsal claws* consist of a curved tarsal element which opposes a tibial spur (Fig. 4.10) to enclose a space that typically corresponds to the diameter of the host hair.

The *internal anatomy* of lice (Fig. 4.3) is best known for the human body louse. As in most hematophagous

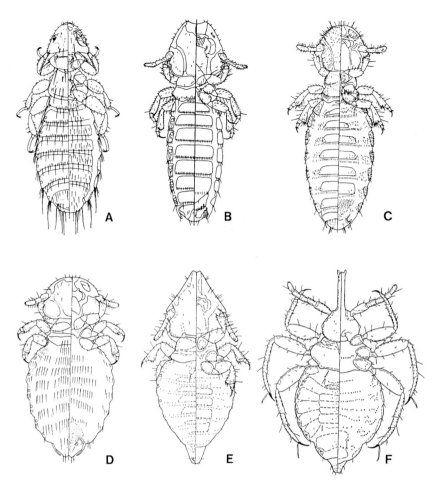

**FIGURE 4.5**  Chewing lice (Mallophaga) of veterinary importance, showing dorsal morphology (left) and ventral morphology (right) in each case. Not drawn to scale. (A) *Heterodoxus spiniger*, male, from carnivores; (B) *Tricholipeurus parallelus,* female, from New World deer; (C) Sheep-biting louse (*Bovicola bovis*), female; (D) Dog-biting louse (*Trichodectes canis*), female; (E) Cat-biting louse (*Felicola subrostrata*), male; (F) Elephant louse (*Haematomyzus elephantis*), male. (A–E, from Emerson and Price, 1975; F, from Werneck, 1950)

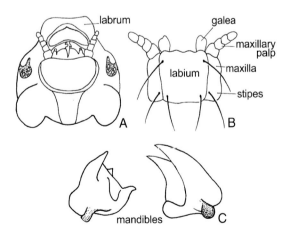

**FIGURE 4.6**  Generalized mouthparts of an amblyceran chewing louse (Mallophaga). (A) Ventral view of head; (B) labium and associated structures; (C) mandibles. (Drawn by Margo Duncan)

*Symbionts* are thought to be present in all lice that imbibe blood. Symbionts in the mycetomes (Fig. 4.3) aid in blood meal digestion, and lice deprived of them die after a few days; female lice lacking symbionts also become sterile. In female human body lice, some symbionts migrate to the ovary, where they are transferred transovarially to the next generation of lice.

Lice in general exhibit *host specificity*, some to such a degree that they parasitize only one species of host. The hog louse, slender guineapig louse, large turkey louse, and several additional species listed in Tables I and II all are typical parasites of a single host species.

Host specificity is broader in some lice. Some lice of veterinary importance parasitize two or more closely related hosts. Examples include the three species which parasitize domestic dogs: *Linognathus setosus, Trichodectes*

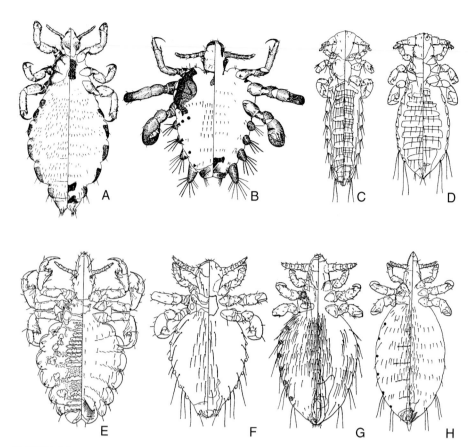

**FIGURE 4.7** Sucking lice (Anoplura) of medical and veterinary importance, showing dorsal morphology (left) and ventral morphology (right) in each case. Not drawn to scale. (A) Human body louse (*Pediculus humanus humanus*), female; (B) Human crab louse (*Phthirus pubis*), female; (C) Flying squirrel louse (*Neohaematopinus sciuropteri*), male; (D) Spined rat louse (*Polyplax spinulosa*), male; (E) Hog louse (*Haematopinus suis*), female; (F) Little blue cattle louse (*Solenoptes capillatus*), male; (G) Dog sucking louse (*Linognathus setosus*), male; (H) Longnosed cattle louse (*L. vituli*), female. (From Ferris, 1923–1935.)

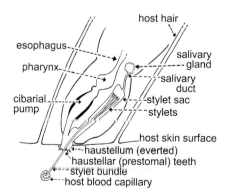

**FIGURE 4.8** Head region of a sucking louse (Anoplura) feeding on a host, showing components of mouthparts and associated internal structures. (Original by Margo Duncan.)

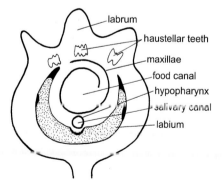

**FIGURE 4.9** Cross-section through the mouthparts of a sucking louse (Anoplura). (Original by Margo Duncan.)

*canis,* and *Heterodoxus spiniger.* These lice also parasitize foxes, wolves, coyotes, and occasionally other carnivores. Similarly, the horse sucking louse (*Haematopinus asini*), parasitizes horses, donkeys, asses, mules, and zebras, whereas *L. africanus* parasitizes both sheep and goats. At least six species of chewing lice are found on domestic fowl, all of them parasitizing chickens, but some also feeding on turkeys, guinea fowl, pea fowl, or pheasants (Table I). Lice found on atypical hosts are termed *stragglers.*

Some sucking lice, such as the three taxa that parasitize humans, the sheep foot louse, and the sheep face louse, are not only host specific, but also infest specific body areas, from which they can spread in severe infestations. Many chewing lice, particularly species that parasitize birds, also exhibit both *host specificity* and *site specificity;* examples include several species that are found on domestic fowl, and species confined to turkeys, geese, and ducks (Table I). Lice inhabiting different body regions on the same host typically have evolved morphological adaptations in response to specific attributes of the host site. These include characteristics such as morphological differences of the pelage, thickness of the skin, availability of blood vessels, and grooming or preening activities of the host. Site specificity in chewing lice is most prevalent in the more sedentary, specialized Ischnocera than in the mostly mobile, morphologically unspecialized Amblycera. For example, on many bird hosts, round-bodied ischnocerans with large heads and mandibles are predominately found on the head and neck. Elongate

forms with narrow heads and small mandibles tend to inhabit the wing feathers, whereas morphologically intermediate forms occur on the back and other parts of the body.

Some chewing lice inhabit highly specialized host sites. These include members of the amblyceran genus *Piagetiella,* which are found inside the oral pouches of pelicans, and members of several amblyceran genera, including *Actornithophilus* and *Colpocephalum,* which live inside feather quills. Several bird species are parasitized by 5 or more different species of site-specific chewing lice, and up to 12 species may be found on the neotropical bird *Crypturellus soui* (a tinamou).

Site specificity is less well documented for sucking lice. However, domestic cattle may be parasitized by as many as five anopluran species, each predominating on particular parts of the body. Similarly, some Old World squirrels and rats can support up to six species of sucking lice.

Because of the importance of maintaining a permanent or close association with the host, lice have evolved specialized *host-attachment mechanisms* to resist grooming activities of the host. The robust tibio-tarsal claws of sucking lice (Fig. 4.10) are very important in securing them to their hosts. Various arrangements of hooks and spines, especially on the heads of lice that parasitize arboreal or flying hosts, such as squirrels and birds, also aid in host attachment. Mandibles are important attachment appendages in ischnoceran and rhyncophthirinan chewing lice. In some species of *Bovicola,* a notch in the first antennal segment encircles a host hair to facilitate attachment.

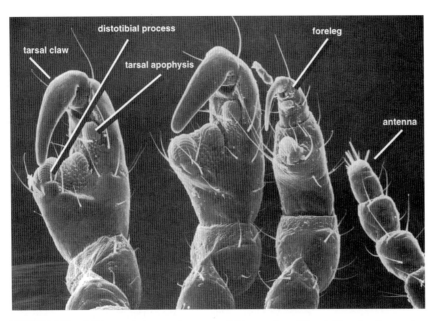

**FIGURE 4.10**     Tibio-tarsal claws and antenna of *Linognathus africanis* (Anoplura): scanning electron micrograph. (From Price and Graham, 1997.)

A few lice even possess *ctenidia* ("combs") that are convergently similar in morphology to those characteristic of many fleas. They occur most notably among lice that parasitize coarse-furred, arboreal, or flying hosts. Additionally, chewing lice that parasitize arboreal or flying hosts often have larger, more robust claws than do their counterparts that parasitize terrestrial hosts.

Because of their reliance on host availability, lice are subjected to special problems with respect to their long-term survival. All sucking lice are obligate blood-feeders; even a few hours away from the host can prove fatal to some species. Some chewing lice also are hematophages and similarly cannot survive prolonged periods off the host. However, many chewing lice, particularly those that subsist on feathers, fur, or other skin products, can survive for several days away from the host. For example, the cattle biting louse can survive for up to 11 days (this species will feed on host skin scrapings), and *Menacanthus* spp. of poultry can survive for up to 3 days off the host. Off-host survival is generally greater at low temperatures and high humidities. At 26°C and 65% relative humidity (RH), 4% of human body lice die within 24 hr, 20% within 40 hr, and 84% within 48 hr. At 75% RH, a small proportion of sheep foot lice survives for 17 days at 2°C, whereas most die within 7 days at 22°C. Recently fed lice generally survive longer than unfed lice away from the host. Although most lice are morphologically adapted for host attachment and are disadvantaged when dislodged, the generalist nature of some amblyceran chewing lice better equips them for locating another host by crawling across the substrate. Amblycerans are more likely than other lice to be encountered away from the host, accounting for observations of these lice on bird eggs or in unoccupied nests and roosts.

*Host grooming* is an important cause of louse mortality. Laboratory mice infested by the mouse louse, for example, usually limit their louse populations to 10 or fewer individuals per mouse by regular grooming. Prevention of self-grooming or mutual grooming by impaired preening action of the teeth or limbs of such mice can result in heavy infestations of more than 100 lice. Similarly, impaired preening due to beak injuries in birds can result in tremendous increases of louse populations. Biting, scratching, and licking also reduce louse populations on several domestic animals.

Whereas most species of lice on small and medium-sized mammals exhibit only minor seasonal differences in population levels, some lice associated with larger animals show clear seasonal trends. Some of these population changes have been attributed to host molting, fur density and length, hormone levels in the blood meal, or climatological factors such as intense summer heat, sunlight, or desiccation. On domestic ungulates in temperate regions, louse populations typically peak during the winter or early spring and decline during the summer. An exception to this trend is the cattle tail louse, whose populations peak during the summer.

Another important aspect of louse behavior is the mode of transfer between hosts. Direct host contact appears to be the primary mechanism for louse exchange. Transfer of lice from an infested mother to her offspring during suckling (in mammals) or during nest sharing (in birds and mammals) is an important mode of transfer. Several species of lice that parasitize livestock transfer during suckling, including the sheep face louse and the sheep biting louse, both of which move from infested ewes to their lambs at this time. Lice can also transfer during other forms of physical contact between hosts, such as mating or fighting. Transfer of lice between hosts also can occur between hosts that are not in contact. The sheep foot louse, for example, can survive for several days off the host and reach a new host by crawling across pasture land. Nests of birds and mammals can act as foci for louse transfer, but these are infrequent sites of transfer.

Dispersal of some lice occurs via *phoresy*, in which they temporarily attach to other arthropods and are carried from one host to another (Fig. 4.11). During phoresy, most lice attach to larger, more mobile blood-feeding arthropods, usually a fly, such as a hippoboscid or muscoid. Phoresy is particularly common among ischnoceran chewing lice. Movement of the mouthparts in a horizontal plane better facilitates their attachment to a fly than in the amblycerans, in which mouthparts move in a vertical plane. Phoresy is rare among sucking lice. This is

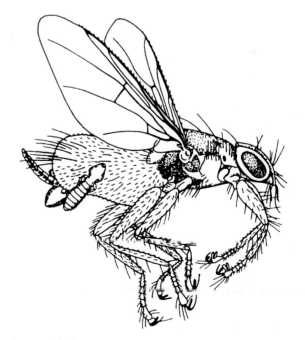

**FIGURE 4.11** Two ischnoceran chewing lice (Mallophaga) phoretic on a hippoboscid fly, attached by their mandibles to the posterior abdomen. (From Rothschild and Clay, 1952).

probably because attachment to the fly is achieved by the less efficient mechanism of grasping with the tarsal claws.

Mating in lice occurs on the host. It is initiated by the male pushing his body beneath that of the female and curling the tip of his abdomen upward. In the human body louse, the male and female assume a vertical orientation along a hair shaft, with the female supporting the weight of the male as he grasps her with his anterior claws. Most lice appear to exhibit similar orientation behavior during mating. Notable exceptions include the crab louse of humans, in which both sexes continue to clasp with their claws a host hair, rather than each other, during mating; and the hog louse, in which the male strokes the head of the female during copulation. Some male ischnoceran chewing lice possess modified hook-like antennal segments, with which they grasp the female during copulation.

*Oviposition* behavior by female lice involves crawling to the base of a host hair or feather and cementing one egg at a time close to the skin surface. Two pairs of finger-like *gonopods* direct the egg into a precise location and orientation as a cement substance is secreted around the egg and hair base. Optimal temperature requirements for developing louse embryos inside eggs are very narrow, usually within a fraction of a degree, such as may occur on a precise area on the host body. For this reason, female lice typically oviposit preferentially on an area of the host that meets these requirements.

## LICE OF MEDICAL INTEREST

Three taxa of sucking lice parasitize humans throughout the world: the *body louse, head louse,* and *crab louse* (*pubic louse*). All are specific ectoparasites of humans; rarely, dogs or other companion animals may have temporary, self-limiting infestations.

Human head and body lice are closely related and can interbreed to produce fertile offspring in the laboratory. For this reason, they generally are recognized as separate subspecies of *Pediculus humanus,* as in this chapter. Nevertheless, they rarely interbreed in nature, which has prompted some epidemiologists to treat them as separate species, *P. humanus* (body louse) and *P. capitis* (head louse).

**Human body louse** (*Pediculus humanus humanus*)
The human body louse (Figs. 4.7A and 4.12) or *cootie* was once an almost ubiquitous companion of humans. Today it is less common, especially in developed nations. Body lice persist as a significant problem in less developed nations in parts of Africa, Asia, and Central and South America. This is significant because *P.h. humanus* is the only louse of humans that is known

**FIGURE 4.12**    Human body lice (*Pediculus humanus humanus*) feeding on a human. (Courtesy of Elton J. Hansens.)

to naturally transmit pathogens. The large-scale reduction in body louse infestations worldwide has led to a concomitant decrease in the prevalence of human louse-borne diseases. However, situations that result in human overcrowding and unsanitary conditions (e.g., wars, famines, and natural disasters) can lead to a resurgence of body louse infestations, often accompanied by one or more louse-borne diseases.

Adult human body lice (Figs. 4.7A and 4.12) are 2.3–3.6 mm long. Under optimal conditions their populations can multiply dramatically if unchecked; e.g., if clothes of infested individuals are not changed and washed in hot water at regular intervals. In unusually severe infestations, populations of more than 30,000 body lice on one person have been recorded. Body lice typically infest articles of clothing and crawl onto the body only to feed. Females lay an average of four or five eggs per day, and these typically hatch after 8 days. Unique among lice, females oviposit not on hair, but on clothing (Fig. 4.13), especially along seams

**FIGURE 4.13**    Eggs of the human body louse (*Pediculus humanus humanus*) attached to clothing. (Courtesy of Elton J. Hansens.)

and creases. Each nymphal instar lasts for 3–5 days, and adults can live for up to 30 days.

Biting by body lice often causes intense irritation, with each bite site typically developing into a small red papule with a tiny central clot. The bites usually itch for several days but occasionally for a week or more. Persons exposed to numerous bites over long periods often become desensitized and show little or no reaction to subsequent bites. Persons with chronic body louse infestations may develop a generalized skin thickening and discoloration called *Vagabond disease* or *Hobo disease,* names depicting a lifestyle that can promote infestation by body lice. Several additional symptoms may accompany chronic infestations. These include lymphadenopathy (swollen lymph nodes), edema, increased body temperature often accompanied by fever, a diffuse rash, headache, joint pain, and muscle stiffness.

Some people develop allergies to body lice. Occasionally, patients experience a generalized dermatitis in response to one bite or small numbers of bites. A form of asthmatic bronchitis has similarly been recorded in response to allergy to louse infestations. Secondary infections such as impetigo or blood poisoning can also result from body louse infestations.

Body lice tend to leave persons with elevated body temperatures and may crawl across the substrate to infest a nearby person. This has epidemiological significance because high body temperatures of lousy persons often result from fever caused by infection with louse-borne pathogens.

### Human head louse *(Pediculus humanus capitis)*

The human head louse is virtually indistinguishable from the human body louse on the basis of morphological characters and its life cycle. Unless a series of specimens is available for analysis it is often impossible to separate the two subspecies. Generally, adult head lice are slightly smaller (2.1–3.3 mm in length) than body lice.

As indicated by their name, human head lice typically infest the scalp and head region, rather than other areas of the body infested by body lice. Females attach their eggs to the base of individual hairs. As the hair grows, the eggs become further displaced from the scalp. An indication of how long a patient has been infested can be gleaned by measuring the farthest distance of eggs from the scalp and comparing this to the growth rate of hair.

Today, head lice are far more frequently encountered than body lice, especially in developed countries. Transmission occurs by person-to-person contact and via shared objects such as combs, brushes, headphones, and caps. School-age children are at high risk because they are often more likely to share such items. Some school districts in the United States and Britain have infestation prevalences approaching 50% in students. It has been estimated that 6–12 million people, principally children, are infested with head lice annually in the United States. Some ethnic groups, such as persons of African origin, have coarser head hairs and are less prone to head louse infestations. The reason for this is simply that the tibio-tarsal claws of these lice cannot efficiently grip the thicker hairs.

Although head lice are not known to transmit pathogens, heavy infestations can cause severe irritation. As is the case with human body lice, the resultant scratching often leads to secondary infections such as impetigo, pyoderma, or blood poisoning. Severe head louse infestations occasionally result in the formation of scabby crusts beneath which the lice tend to aggregate. Enlarged lymph nodes in the neck region may accompany such infestations.

### Human crab louse *(Pthirus pubis)*

The crab louse, or *pubic louse,* is a medium-sized (1.1–1.8 mm long), squat louse (Fig. 4.7B), with robust tibio-tarsal claws used for grasping thick hairs, especially those in the pubic region. It also may infest coarse hairs on other parts of the body, such as the eyebrows, eyelashes, chest hairs, beards, moustaches, and armpits. This louse typically transfers between human partners during sexual intercourse and other intimate contact; in France, crab lice are described as "papillons d'amour" (butterflies of love). Transfer via infested bed linen or toilet seats can also occur. This is uncommon, however, because crab lice can survive for only a few hours off the host.

Female crab lice lay an average of three eggs per day. Eggs hatch after 7–8 days; the three nymphal instars together last for 13–17 days. Under optimal conditions the generation time is 20–25 days. The intense itching caused by these lice is often accompanied by purplish lesions at bite sites and by small blood spots from squashed lice or louse feces on underwear. Crab lice are widely distributed and relatively common throughout the world. They are not known to transmit any pathogens. One epidemiological study, however, revealed a positive relationship between infection with hepatitis B virus and crab louse infestation.

## LICE OF VETERINARY INTEREST

A wide variety of lice infests domestic, livestock, and laboratory animals (Tables I and II). Many hosts, particularly small rodents, often support few if any lice, whereas large hosts such as livestock animals, including poultry, may be parasitized by very large numbers of lice. For example, fewer than 10 mouse lice (*Polyplax serrata*) on a house mouse are a typical burden, but more than a million lice may be present on extremely heavily infested sheep, cattle, horses, or other large animals.

## LICE OF CATTLE

Lice are a major problem in cattle operations worldwide. Domestic cattle are parasitized by six species of lice: three species of *Haematopinus*, one of *Linognathus*, one of *Solenopotes*, and one of *Bovicola*. Domestic Asiatic buffalo are typically parasitized by *H. tuberculatus* (Tables I and II).

Females of the cosmopolitan *cattle biting louse* (*Bovicola bovis*) lay an average of 0.7 eggs per day, which hatch 7–10 days later. Each nymphal instar lasts 5–6 days, and adult longevity can be as long as 10 weeks. Preferred host sites for this louse are the base of the tail, the shoulders, and the top line of the back, but lice may also populate the pollard in severe infestations.

The *longnosed cattle louse* (*L. vituli*) (Fig. 4.7H) also is a worldwide pest. Females deposit about one egg per day, and the life cycle is completed in about 21 days. This louse is most common on calves and dairy stock; it rarely occurs in large numbers on mature cattle. Preferred infestation sites are the dewlap and shoulders; declining spring populations are often confined to the shoulders.

The *little blue cattle louse* (*Solenopotes capillatus*) (Fig. 4.7F) also has a worldwide distribution. Females lay one or two eggs per day; oviposition typically causes the hairs on which eggs are laid to bend. Eggs hatch after about 10 days, and adulthood is reached about 11 days later. Clusters of *S. capillatus* typically occur on the muzzle, dewlap, and neck of mature cattle. Aggregations of this louse may surround the eyes in severely infested animals, giving a spectacled appearance to the host.

The cosmopolitan *shortnosed cattle louse* (*H. eurysternus*) is the largest louse found on North American cattle; adults measure 3.5–4.7 mm in length. Females lay one to four eggs per day for about 2 weeks, nymphs reach adulthood in about 14 days, and adult longevity is 10–15 days. This louse is more common on mature cattle than on young animals. Preferred infestation sites are the top of the neck, the dewlap, and brisket. However, in severe infestations, the entire region from the base of the horns to the base of the tail can be infested. In North America, *H. eurysternus* is most prevalent in the Great Plains and Rocky Mountain regions.

The *cattle tail louse* (*H. quadripertusus*) parasitizes cattle in the warmer regions of the world. It was inadvertently introduced into the United States, where it now occurs in the Gulf Coast states. Females of this louse oviposit on the tail hairs, which become matted with eggs in severe infestations. Infested tail heads may be shed under these circumstances. Eggs hatch after 9–25 days, depending on the season. Under optimal conditions, the entire life cycle can be as short as 25 days. Nymphs migrate over the host body surface, but adults are typically confined to the tail head. Unlike other cattle lice, *H. quadripertusus* is most abundant during the summer.

Except for *H. quadripertusus*, cattle lice increase in numbers during the winter and early spring in temperate regions. During summer, lice persist on 1–2% of the members of a herd; these chronically infested animals typically reinfest other herd members during the winter. Bulls and older cows often serve as reservoirs of lice. Bulls have longer, thicker hair and massive shoulders and neck that compromise self-grooming. During summer, a small number of lice can survive on the cooler ear tips, where lethal temperatures are rarely reached.

## LICE OF OTHER LIVESTOCK ANIMALS

Horses, donkeys, hogs, goats, and sheep are parasitized by one or more species of louse (Tables I and II). Except for hogs, all of these animals are parasitized by both sucking lice and chewing lice. The *horse biting louse* (*B. equi*) is the most important louse of equids worldwide. Females of this louse oviposit on fine hairs, avoiding the coarse hairs of the mane and tail. This louse typically infests the side of the neck, the flanks, and tail base but can infest most of the body (except the mane, tail, ears, and lower legs) in severe infestations. Long-haired horse breeds are more prone to infestation by *B. equi*.

Domestic swine are parasitized by the *hog louse* (*H. suis*) (Fig. 4.7E). This is a large species in which adult females measure ca. 5 mm in length. Hog lice usually frequent skin folds of the head (especially the ears), neck, shoulders, and flanks of swine. Female hog lice lay an average of 3.6 eggs per day. These are deposited singly on hairs along the lower parts of the body, in skin folds on the neck, and on and in the ears. Eggs typically hatch 13–15 days later; each nymphal instar lasts 4–6 days. Adult hog louse longevity can be up to 40 days, and 6–15 generations can be completed per year, depending on environmental conditions.

Domestic sheep and goats are parasitized by several species of sucking lice and chewing lice (Tables I and II). One of these, *L. africanus*, parasitizes both hosts. Lice of sheep and goats, especially chewing lice, are economically important wherever these livestock animals are farmed, but especially in Australia, New Zealand, and the United States. Females of the *sheep biting louse* (*B. ovis*) lay one or two eggs per day and can live for up to 30 days; each nymphal instar lasts 5–9 days. *B. ovis* mainly infests the back and upper parts of the body but may populate the entire body in severe infestations. This louse causes intense irritation, and infested sheep typically rub against fences and trees, tearing the fleece and greatly reducing its value. Sucking louse infestations of sheep rarely cause major economic problems.

## LICE OF CATS AND DOGS

Domestic cats are parasitized by one species of chewing louse, whereas dogs are parasitized by two species of chewing lice and one species of sucking louse. All four species seem to be distributed worldwide, but none is a common associate of healthy cats or dogs in North America.

The *cat biting louse* (*Felicola subrostrata*) (Fig. 4.5E) parasitizes both domestic and wild cats. It may occur almost anywhere on the body.

Both the *dog biting louse* (*T. canis*) (Fig. 4.5D) and the *dog sucking louse* (*L. setosus*) (Fig. 4.7G) parasitize dogs and closely related wild canids. For example, *T. canis* also parasitizes coyotes, foxes, and wolves. A second species of chewing louse of dogs is *Heterodoxus spiniger* (Fig. 4.5A), which evolved in Australasia from marsupial-infesting lice and apparently switched to dingo hosts. It now parasitizes various canids and other carnivores throughout the world. *T. canis* usually infests the head, neck, and tail region of dogs, where it attaches to the base of a hair using its claws or mandibles. *L. setosus* occurs primarily on the head and neck and may be especially common beneath collars. *H. spiniger* can typically be found anywhere on its host.

## LICE OF LABORATORY ANIMALS

The principal species of lice that parasitize laboratory mammals have been described by Kim *et al.* (1973). These lice also parasitize feral populations of their respective hosts.

The house mouse (*Mus musculus*) is often parasitized by the *mouse louse* (*P. serrata*). Populations of this louse are typically low, with 10 or fewer lice per infested mouse, unless self-grooming or mutual host grooming is compromised. Eggs of this louse typically hatch 7 days after oviposition. Together the three nymphal instars last only 6 days under optimal conditions, which can result in a generation time as short as 13 days.

Domestic rats are often parasitized by the *spined rat louse* (*P. spinulosa*) (Fig. 4.7D) and the *tropical rat louse* (*Hoplopleura pacifica*). Common hosts include the black rat (*Rattus rattus*) and the Norway rat (*R. norvegicus*). The spined rat louse parasitizes these hosts throughout the world, whereas the tropical rat louse is confined to tropical, subtropical, or warm temperate regions, including the southern United States.

Laboratory rabbits are parasitized by the *rabbit louse* (*Haemodipsus ventricosis*). This louse originated in Europe but has accompanied its host and been introduced throughout the world.

## LICE OF POULTRY AND OTHER BIRDS

At least nine species of chewing lice commonly infest poultry (Table I) in various parts of the world. Individual birds can be parasitized by multiple species, each of which often occupies a preferred host site.

The *chicken body louse* (*Menacanthus stramineus*) (Fig. 4.14) is the most common and destructive louse of domestic chickens. It has a worldwide distribution and often reaches pest proportions. Adults measure 3–3.5 mm in length. Females lay one or two eggs per day, cementing them in clusters at the bases of feathers, especially around the vent. Eggs typically hatch after 4–5 days. Each nymphal instar lasts about 3 days, and the generation time typically is 13–14 days. These lice are most abundant on the sparsely feathered vent, breast, and thigh regions.

Several other chewing lice are pests of poultry more or less throughout the world (Table I). Adults of the *shaft louse* (*Menopon gallinae*) measure ca. 2 mm in length, and females deposit eggs singly at the base of the shaft on thigh and breast feathers. Eggs of the *wing louse* (*Lipeurus caponis*) hatch 4–7 days after the female has cemented them to the base of a feather. Nymphal stages of this species each last 5–18 days; generation time typically is 18–27 days, and females can live up to 36 days. Females of the *chicken head louse* (*Cuclotogaster heterographus*) attach their eggs to the bases of downy feathers. Eggs hatch after 5–7 days, each nymphal instar lasts 6–14 days, and average generation time is 35 days.

**FIGURE 4.14** Chicken body lice (*Menacanthus stramineus*) on a chicken. (Courtesy of Nancy C. Hinkle.)

Poultry lice typically transfer to new birds by direct host contact. However, because most species can survive for several hours or days off the host, they also can infest new hosts during transportation in inadequately disinfected cages or vehicles.

# PUBLIC HEALTH IMPORTANCE

Three important pathogens are transmitted to humans by body lice. These are the agents of *epidemic typhus, trench fever,* and *louse-borne relapsing fever.* Today, the prevalence and importance of all three of these louse-borne diseases are low compared to times when human body lice were an integral part of human life. However, trench fever has emerged as an opportunistic disease of immunocompromised individuals, including persons who are positive for human immunodeficiency virus (HIV).

## EPIDEMIC TYPHUS

Epidemic typhus is a rickettsial disease caused by infection with *Rickettsia prowazekii.* It is also known as *louse-borne fever, jail fever,* and *exanthematic typhus.* The disease persists in several parts of the world, most notably in Burundi, Democratic Republic of Congo, Ethiopia, Nigeria, Rwanda, and areas of northeastern and central Africa, Russia, Central and South America, and northern China. Epidemic typhus is largely a disease of cool climates, including higher elevations in the tropics. It thrives in conditions of widespread body louse infestations, overcrowding, and poor sanitary conditions. Epidemic typhus apparently was absent from the New World until the 1500s, when the Spanish introduced the disease. One resulting epidemic in 1576–1577 killed 2 million Indians in the Mexican highlands alone.

The *vector* of *R. prowazekii* is the human body louse. Lice become infected when they feed on a person with circulating *R. prowazekii* in the blood. Infective rickettsiae invade cells that line the louse gut and multiply there, eventually causing the cells to rupture. Liberated rickettsiae either reinvade gut cells or are voided in the louse feces. Other louse tissues typically do not become infected. Because salivary glands and ovaries are not invaded, anterior-station and transovarial transmission do not occur. Infection of susceptible humans occurs via louse feces (posterior-station transmission) when infectious rickettsiae are scratched into the skin in response to louse bites. *R. prowazekii* can remain viable in dried louse feces for 60 days. Infection by inhalation of dried louse feces or by crushed lice are less frequent means of contracting the disease.

Transmission of *R. prowazekii* by body lice was first demonstrated by Charles Nicolle, working at the Institut Pasteur in Tunis in 1909. During these studies, Nicolle accidentally became infected with epidemic typhus, from which he fortunately recovered. He was awarded the Nobel prize in 1928 for his groundbreaking work on typhus. Several other typhus workers also were infected with *R. prowazekii* during laboratory experiments. The American researcher Howard T. Ricketts, working in Mexico, and Czech scientist Stanislaus von Prowazek, working in Europe, both died from their infections and were recognized posthumously when the etiologic agent was named.

Infection with *R. prowazekii* is ultimately fatal to body lice as progressively more and more infected gut cells are ruptured. Infective rickettsiae are first excreted in louse feces 3–5 days after the infective blood meal. Lice usually succumb to infection 7–14 days after the infectious blood meal, although some may survive to 20 days.

The disease caused by infection with *R. prowazekii* and transmitted by body lice is called *classic epidemic typhus* because it was the first form of the disease to be recognized. Disease onset occurs relatively soon after infection by a body louse in classic epidemic typhus. Symptoms generally appear after an incubation period of 10–14 days. Abrupt onset of fever, accompanied by malaise, muscle and head aches, cough, and general weakness, usually occurs at this time. A blotchy rash spreads from the abdomen to the chest and then often across most of the body, typically within 4–7 days following the initial symptoms. The rash rarely spreads to the face, palms, and soles, and then only in severe cases. Headache, rash, prostration, and delirium intensify as the infection progresses. Coma and very low blood pressure often signal fatal cases. A case fatality rate of 10–20% is characteristic of most untreated epidemics, although figures approaching 50% have been recorded. Diagnosis of epidemic typhus involves the demonstration of positive serology, usually by microimmunofluorescence. DNA primers specific to *R. prowazekii* can also be amplified by polymerase chain reaction from infected persons or lice. One-time antibiotic treatment, especially with doxycycline, tetracycline, or chloramphenicol, usually results in rapid and complete recovery. Vaccines are available but are not considered to be sufficiently effective for widespread use.

Persons that recover from epidemic typhus typically harbor *R. prowazekii* in lymph nodes or other tissues for months or years. This enables the pathogen to again invade other body tissues to cause disease seemingly at any time. This form of the disease is called *recrudescent typhus* or *Brill–Zinsser disease.* The latter name recognizes two pioneers in the study of epidemic typhus: Nathan Brill, who first recognized and described recrudescent typhus in 1910, and Hans Zinsser, who demonstrated in 1934 that it is a form of epidemic typhus. Zinsser's (1935) book *Rats, Lice, and History* is a pioneering account of the study of epidemic typhus in general.

Recrudescent typhus was widespread during the 19th and early 20th centuries in some of the larger cities along the east coast of the United States (e.g., Boston, New York, and Philadelphia). At that time, immigrants from regions that were rampant with epidemic typhus, such as eastern Europe, presented with Brill–Zinsser disease after being infected initially in their country of origin. Some of these patients experienced relapses more than 30 years after their initial exposure, with no overt signs of infection with *R. prowazekii* between the two disease episodes. Because infestation with body lice was still a relatively common occurrence during that period, the lice further disseminated the infection to other humans, causing local outbreaks. The last outbreak of epidemic typhus in North America occurred in Philadelphia in 1877. Today, even recrudescent typhus is a rare occurrence in North America. However, this form of typhus is still common in parts of Africa, Asia, South America, and, occasionally, in eastern Europe.

The *southern flying squirrel* (*Glaucomys volans*) has been identified as a reservoir of *R. prowazekii* in the United States, where it has been found to be infected in Virginia during vertebrate serosurveys for Rocky Mountain spotted fever. Since the initial isolations from flying squirrels in 1963, *R. prowazekii* has been recorded in flying squirrels and their ectoparasites in several states, especially eastern and southern states. Peak seroprevalence (about 90%) in the squirrels occurs during late autumn and winter, when fleas and sucking lice are also most abundant on these hosts. Although several ectoparasites can imbibe *R. prowazekii* when feeding on infected flying squirrels, only the sucking louse *Neohaematopinus sciuropteri* (Fig. 4.7C) is known to maintain the infection and transmit the pathogen to uninfected squirrels.

Several cases of human infection have been documented in which the patients recalled having contact with flying squirrels, especially during the winter months when these rodents commonly occupy attics of houses. To distinguish this form of the disease from classic and recrudescent typhus, it is called *sporadic epidemic typhus* or *sylvatic epidemic typhus*. Many details, such as the prevalence and mode of human infection, remain unresolved. Because the louse *N. sciuropteri* does not feed on humans, it is speculated that human disease occurs when infectious, aerosolized particles of infected louse feces are inhaled from attics or other sites occupied by infected flying squirrels.

Except for flying squirrels in North America, humans are the only proven *reservoirs* of *R. prowazekii*. Widespread reports published in the 1950s to 1970s that various species of ticks and livestock animals harbored *R. prowazekii* have since been disproved.

Historically, epidemic typhus has been the most widespread and devastating of the louse-borne diseases. Zinsser (1935) and Snyder (1966) have documented the history of this disease and highlighted how major epidemics have influenced human history. For example, the great outbreak of disease at Athens in 430 BC, which significantly influenced the course of Greek history, appears to have been caused by epidemic typhus. Napoleon's vast army of 1812 was defeated more by epidemic typhus than by opposing Russian forces. Soon thereafter (ca. 1816–1819), 700,000 cases of epidemic typhus occurred in Ireland. Combined with the potato famine of that period, this encouraged many people to emigrate to North America; some of these people carried infected lice or latent infections with them. During World War II, several military operations in North Africa and the Mediterranean region were hampered by outbreaks of epidemic typhus. One epidemic in Naples in 1943 resulted in over 1400 cases and 200 deaths. This outbreak is particularly noteworthy because it was the first epidemic of the disease to be interrupted by human intervention through widespread application of the insecticide dichlorodiphenyltrichloroethane (DDT) to louse-infested persons.

Today, epidemic typhus is much less of a health threat than it once was. This is largely because few people, especially in developed countries, are currently infested by body lice. Higher sanitary standards, less overcrowding, regular laundering and frequent changes of clothes, effective pesticides, and medical advances have contributed to the demise of this disease. Nevertheless, epidemic typhus has the potential to re-emerge. This is evidenced by the largest outbreak of epidemic typhus since World War II that affected about half a million people living in refugee camps in Burundi in 1997–1998. Similarly, more than 5600 cases were recorded in China during 1999. Additional information about epidemic typhus is provided by the Pan American Health Organization/World Health Organization (1973), McDade (1987), and Azad (1988).

## LOUSE-BORNE RELAPSING FEVER

Also known as *epidemic relapsing fever*, this disease is caused by the spirochete bacterium *Borrelia recurrentis*. This pathogen is transmitted to humans by the human body louse, as first demonstrated by Sergent and Foley in 1910. Clinical symptoms include the sudden onset of fever, headache, muscle ache, anorexia, dizziness, nausea, coughing, and vomiting. Thrombocytopenia (a decrease in blood platelets) also can occur and cause bleeding, which may initially be confused for a symptom of a hemorrhagic fever. Episodes of fever last 2–12 days (average, 4 days), typically followed by periods of 2–8 days (average, 4 days) without fever, with two to five relapses being usual. As the disease progresses, the liver and spleen enlarge rapidly, leading to abdominal discomfort and labored, painful breathing as the lungs and diaphragm are compressed. At this stage, most patients remain quietly prostrate with a glazed expression, often shivering and taking shallow breaths. Mortality rates for untreated outbreaks range from 5 to 40%. Antibiotic treatment is with penicillin or tetracycline. Humans are the sole known reservoir of *B. recurrentis.*

Body lice become infected when they feed on an infected person with circulating spirochetes. Most of the spirochetes perish when they reach the louse gut, but a few survive to penetrate the gut wall, where they multiply to massive populations in the louse hemolymph, nerves, and muscle tissue. Spirochetes do not invade the salivary glands or ovarian tissues and are not voided in louse feces. Therefore, transmission to humans occurs only when infected lice are crushed during scratching, which allows the spirochetes in infectious hemolymph to invade the body through abrasions and other skin lesions. However, *B. recurrentis* is also capable of penetrating intact skin. As with *R. prowazekii* infections, body lice are killed as a result of infection with *B. recurrentis.*

An intriguing history of human epidemics of louse-borne relapsing fever is provided by Bryceson *et al.* (1970). Hippocrates described an epidemic of "caucus," or "ardent fever," in Thasos, Greece, which can clearly be identified by its clinical symptoms as this malady. During 1727–1729, an outbreak in England killed all inhabitants of many villages. During the present century, an epidemic that spread from eastern Europe into Russia during 1919–1923 resulted in 13 million cases and 5 million deaths. Millions also were infected during an epidemic that swept across North Africa in the 1920s. Several major epidemics subsequently have occurred in Africa, with up to 100,000 fatalities being recorded for some of them. During and immediately after World War II, more than a million persons were infected in Europe alone.

The only current epidemic of louse-borne relapsing fever is in Ethiopia, where 1000–5000 cases are reported annually, accounting for ca. 95% of the world's recorded infections. Other smaller foci occur intermittently in other regions, such as Burundi, Rwanda, Sudan, Uganda, People's Republic of China, the Balkans, Central America, and the Peruvian Andes. Resurgence of this disease under conditions of warfare or famine is an ominous possibility. Additional information on louse-borne relapsing fever is provided by Bryceson *et al.* (1970).

## TRENCH FEVER

Also known as *five-day fever* and *wolhynia*, trench fever is caused by infection with the bacterium *Bartonella* (formerly *Rochalimaea*) *quintana*. Like the two preceding diseases, the agent is transmitted by the human body louse. Human infections range from asymptomatic through mild to severe, although fatal cases are rare. Clinical symptoms are nonspecific and include headache, muscle aches, fever, and nausea. The disease can be cyclic, with several relapses often occurring. Previously infected persons often maintain a cryptic infection which can cause relapses years later, with the potential for spread to other persons if they are infested with body lice. Effective antibiotic treatment of patients involves administering drugs such as doxycycline or tetracycline.

Lice become infected with *B. quintana* after feeding on the blood of an infected person. The pathogen multiplies in the lumen of the louse midgut and in the cuticular margins of the midgut epithelial cells. Viable rickettsiae are voided in louse feces, and transmission to humans occurs by the posterior-station route when louse bites are scratched. *B. quintana* can remain infective in dried louse feces for several months, contributing to aerosol transmission as an alternative route of transmission. Transovarial transmission does not occur in the louse vector. Infection is not detrimental to lice and does not affect their longevity.

Trench fever was first recognized as a clinical entity in 1916 as an infection of European troops engaging in trench warfare during World War I. At that time, more than 200,000 cases were recorded in British troops alone. Between the two world wars, trench fever declined in importance but re-emerged in epidemic proportions in troops stationed in Europe during World War II. Because of the presence of asymptomatic human infections, the current distribution of trench fever is difficult to determine. However, since World War II, infections have been recorded in several European and African nations, Japan, the People's Republic of China, Mexico, Bolivia, and Canada.

Until recently, *B. quintana* was considered to be transmitted solely by body lice. However, several homeless or immunocompromised people, including HIV-positive individuals, particularly in North America and Europe, have presented with opportunistic *B. quintana* infections. This is manifested not as trench fever but as vascular tissue lesions, liver pathology, chronically swollen lymph nodes, and inflammation of the lining of the heart. Because some of these patients were not infested by body lice, an alternate mode of pathogen transmission may have been involved.

TABLE III
Pathogens Transmitted by Lice

| Disease | Disease agent | Vector | Host(s) | Geographic distribution |
|---|---|---|---|---|
| **Viral** | | | | |
| Swinepox | Pox virus | *Haematopinus suis* | Hog | Widespread |
| **Bacterial** | | | | |
| Bovine anaplasmosis | *Anaplasma* spp. | Cattle-sucking lice | Cattle | Global |
| Epidemic typhus | *Rickettsia prowazekii* | *Pediculus humanus humanus* | Human | Global |
| Sporadic epidemic typhus | *R. prowazekii* | Flying squirrel lice | Flying squirrel, human | North America |
| Louse-borne relapsing fever | *Borrelia recurrentis* | *P. h. humanus* | Human | Global |
| Salmonellosis | *Salmonella enteriditis* | *P. h. humanus* | Human | Eurasia |
| Trench fever | *Bartonella quintana* | *P. h. humanus* | Human | Global |
| Tularemia | *Francisella tularensis* | Rodent and lagomorph lice | Rodents, rabbits | Global |
| Rodent brucellosis | *Brucella brucei* | *Hoplopleura acanthopus* | Voles | Northern Hemisphere |
| Murine hemobartonellosis | *Haemobartonella muris* | *Polyplax spinulosa* | Domestic rats | Global |
| Murine eperythrozoonosis | *Eperythrozoon coccoides* | *Polyplax serrata* | Domestic mouse | Global |
| **Fungal** | | | | |
| Bovine dermatomycosis (ringworm) | *Trichophyton verrucosum* | Cattle lice | Cattle | Global |
| **Helminthic** | | | | |
| Seal heartworm | *Dipetalonema spirocauda* | *Echinophthirius horridus* | Harbor seal | Northern Hemisphere |
| Avian filariasis | *Eulimdana* spp. | Bird chewing lice | Charadriiform birds | Widespread |
| Avian filariasis | *Pelecitus fulicaeatrae* | *Pseudomenopon pilosum* | Aquatic birds | Holarctic region |
| Avian filariasis | *Sarconema eurycerca* | *Trinoton anserinum* | Geese, swans | Holarctic region |
| Double-pored tapeworm[a] | *Dipylidium caninum* | *Trichodectes canis* | Dog, human | Global |

[a] Lice are intermediate hosts, not vectors, of this tapeworm.

## HUMAN LICE AS INTERMEDIATE HOSTS OF TAPEWORMS

Occasionally humans become infested with the *double-pored tapeworm* (*Dipylidium caninum*). Although carnivores are the normal definitive hosts for this parasite, humans can be infested if they accidentally ingest dog biting lice (*T. canis*), which serve as intermediate hosts. Although this would appear to be an unlikely event, infants, especially babies playing on carpets or other areas frequented by a family dog, may touch an infested louse with sticky fingers which may then be put into their mouth, thus initiating an infestation.

## VETERINARY IMPORTANCE

Several chewing lice and sucking lice parasitize domestic animals (Tables I, II). Although louse populations are usually low on these hosts, lice can sometimes multiply to extremely high numbers, particularly on very young, old, or sick animals. Often this is because hosts are unable to effectively groom themselves or they are immunocompromised. Except for the possibility of pathogen transmission, small numbers of lice typically cause little harm to the host. However, large numbers of lice can be debilitating by causing anemia, dermatitis, allergic responses, hair or feather loss, and other disorders. Lice also induce intensive host grooming, which can lead to the formation of hair balls in the stomach, especially in cats and calves.

A few pathogens are known to be transmitted to domestic animals by lice (Table III). The most important of these are the viral agent of *swinepox* and the bacterial agents of *murine haemobartonellosis* and *murine eperythrozoonosis*, all of which are widely distributed. In addition to those listed in Table III, several pathogens have been detected in various species of lice, but there is no current evidence that lice are vectors of these organisms.

### LICE OF LIVESTOCK

Although louse populations of a few hundred individuals commonly occur on healthy livestock, sometimes these numbers can reach into the thousands or, rarely, to more than a million per animal. It is under the latter conditions that detrimental effects to the host occur. These include restlessness, pruritus, anemia, low weight gain, low milk yield, dermatitis, hide or fleece damage, skin crusting or scabbing, and lameness. Large louse populations on domestic stock typically develop on juvenile, senile, sick, nutritionally deprived, or immunocompromised hosts.

Sucking louse infestations of cattle, such as those caused by the *shortnosed cattle louse* (*Haematopinus eurysternus*), the *cattle tail louse* (*H. quadripertusus*), and the *longnosed cattle louse* (*Linognathus vituli*) (Fig. 4.7H), can cause serious damage to the host. This can be manifested as frequent rubbing of infested areas, hair loss, scab formation, slow recovery from disease or trauma, and low weight gain. Younger animals are typically more severely affected than older cattle. Mixed infestations of both chewing and sucking lice on cattle, or of both lice and nematodes, can affect weight gains more severely than single infestations. In single or mixed infestations, weight gains are typically lower in stressed cattle and those on low-nutrition diets. Sometimes, cattle sucking lice cause severe anemia, abortions, or even death. Irritation can be caused by small numbers of lice in sensitive cattle and usually results in frequent rubbing and subsequent hide damage. This rubbing also damages livestock facilities. Severely infested cattle often have patches of bare skin and a greasy appearance which results from crushing lice and their feces during rubbing. Under laboratory or confined conditions, at least three pathogens can be transmitted by cattle sucking lice, i.e., the causative agents of *bovine anaplasmosis, dermatomycosis* (*ringworm*) (Table III) and, rarely, *theileriosis*. The importance of cattle lice in transmitting any of these pathogens in nature is unknown but presumed to be low.

Lice of *horses* and other *equids* typically do not greatly debilitate their hosts except when they are present in large numbers. Pruritus, hair loss, and coat deterioration may occur in severely infested animals. Horses with severe louse infestations are nervous and irritable; they typically rub against objects, kicking and stamping. Hair can be rubbed from the neck, shoulders, flanks, and tail base, resulting in an unkempt appearance that may affect the value of the horse. No pathogens are known to be transmitted by equid lice.

Hog lice can imbibe significant volumes of blood from hogs, especially piglets, which often have larger infestations than adult pigs. Hog-louse feeding sites often cause intense irritation, leading their hosts to rub vigorously against objects, which can result in hair loss and reddened or crusty skin lesions. *Haematopinus suis* is a vector of the virus that causes *swinepox* (Table III), a serious and potentially fatal disease characterized by large pockmark lesions, mainly on the belly of infected animals. Some studies have implicated this louse as a vector of *Eperythrozoon suis* and *E. parvum*, causative agents of swine *eperythrozoonosis*, and of *African swine fever* virus. However, transmission of these pathogens by lice appears to be rare, if it occurs at all, in nature.

All species of lice that parasitize sheep and goats (Tables I and II) can cause debilitation, even when present in relatively small numbers, because of the potential

**FIGURE 4.15**    Fleece damage (wool slippage) in a sheep, caused by severe infestation with *Linognathus africanus* (Anoplura). (Courtesy of John E. Lloyd)

damage which they can cause to fleece and wool (Fig. 4.15). Some sheep develop hypersensitivity to the *sheep biting louse* (*Bovicola ovis*) (Fig. 4.16). This louse causes most sheep fleece devaluation worldwide and is the major cause of *cockle*, an economically disfiguring condition of sheep fleece that is particularly prevalent in New Zealand. Any increase in skin lesions or body rubbing in response to lice generally devalues wool or mohair. Different breeds of sheep and goats exhibit contrasting levels of resistance or tolerance to infestation by lice.

## LICE OF CATS AND DOGS

Louse infestations of cats and dogs are most noticeable on sick or senile hosts. Under these conditions, louse populations can increase dramatically. Severe infestations of any of the four species involved usually cause host restlessness, scratching, skin inflammation, a ruffled or matted coat, and hair loss.

The *dog biting louse* (*T. canis*) is an intermediate host of the *double-pored tapeworm* (*D. caninum*) (Table III). Lice become infected when they ingest viable *D. caninum* eggs from dried host feces. The tapeworm develops into a cysticercoid stage in the louse, where it remains quiescent unless the louse is ingested by a dog, usually during grooming. In the dog gut, the cysticercoid is liberated and metamorphoses into an adult tapeworm. The *dog sucking louse* (*L. setosus*) has been shown to harbor immatures of the filarial nematode *Dipetalonema reconditum*, which parasitizes dogs, but whether or not these lice are efficient vectors remains unknown.

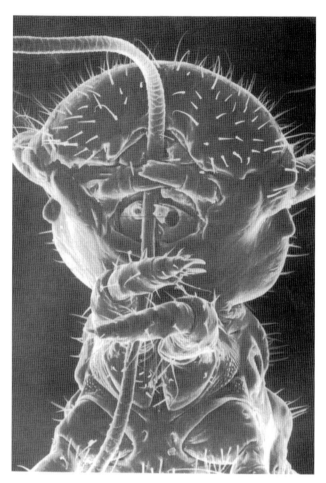

**FIGURE 4.16** Sheep-biting louse (*Bovicola ovis*), showing prothorax and head with mandibles characteristically grasping a host hair: scanning electron micrograph. (From Price and Graham, 1997)

## LICE OF LABORATORY ANIMALS

Some lice that parasitize laboratory animals initiate serious health problems by causing pruritus, skin lesions, scab formation, anemia, and hair loss. Others are vectors of pathogens that can cause severe problems in animal colonies (Table III).

The *mouse louse* (*P. serrata*) is a vector of the bacterium *Eperythrozoon coccoides,* which causes *murine eperythrozoonosis,* a potentially lethal infection of mice that occurs worldwide. Infection of this blood parasite in mice can either be inapparent or result in severe anemia. Transmission of this pathogen in louse-infested mouse colonies is usually rapid. The *spined rat louse* (*P. spinulosa*) is a vector of the bacterium *Haemobartonella muris,* which causes *murine haemobartonellosis* (Table III), another potentially fatal blood infection that can cause severe anemia in laboratory rats.

Laboratory and wild guinea pigs are parasitized by two species of chewing lice, the *slender guineapig louse*

(*Gliricola porcelli*) and the *oval guineapig louse* (*Gyropus ovalis*). Small numbers of these lice cause no noticeable harm, whereas large populations can cause host unthriftiness, scratching (especially behind the ears), hair loss, and a ruffled coat.

Large infestations of the *rabbit louse* (*Haemodipsus ventricosis*) can cause severe itching and scratching, which results in the host rubbing against its cage, often resulting in hair loss. Young rabbits are more adversely affected than are adults and may experience retarded growth as a consequence of infestation by *H. ventricosis.* The rabbit louse is also a vector of the causative agent of *tularemia* among wild rabbit populations (Table III).

### LICE OF POULTRY AND OTHER BIRDS

Although louse populations may be very large on domestic fowl, including domestic chickens, turkeys, guinea fowl, pea fowl, and pheasants, no pathogens are known to be transmitted by these lice. Large populations often occur on birds with damaged beaks whose grooming ability is significantly impaired. The *chicken body louse* (*Menacanthus stramineus*) (Fig. 4.14) often causes significant skin irritation and reddening through its persistent feeding. Occasionally the skin or soft quills bleed from their gnawing and scraping action, with the lice readily imbibing the resultant blood. The *shaft louse* (*Menopon gallinae*) also causes significant losses to the poultry industry, including deaths of young birds with heavy infestations. Large infestations of chicken body lice, shaft lice, and other poultry lice may be injurious to the host by causing feather loss, lameness, low weight gains, inferior laying capacity, or even death.

The vast majority of chewing lice are parasites of wild or peridomestic birds. Several of these lice are suspected vectors of avian pathogens. Some chewing lice of aquatic birds, including geese and swans, are vectors of filarial nematodes (Table III). Pet parrots, parakeets, budgerigars, and other birds also are subject to infestation by chewing lice, which is usually noticed only by the associated host scratching and by ruffled or lost feathers. Large populations of these lice can debilitate their hosts. Ranch birds, such as ostriches, emus, and rheas, are prone to similar adverse effects caused by their associated chewing lice.

## PREVENTION AND CONTROL

Several techniques have been used in attempts to rid humans and animals of lice and louse-borne diseases. Preventing physical contact between lousy persons or animals and the items they contact, as well as various chemical, hormonal, and biological control mechanisms, comprise the current arsenal of techniques. Chemicals used to kill lice are called *pediculicides.*

Clothes of persons with body lice should be changed frequently, preferably daily, and washed in very hot, soapy water to kill lice and nits. Washing associated bed linen in this manner is also advisable. Infested people should also receive a concurrent whole-body treatment with a pediculicide. Overcrowded and unsanitary conditions should be avoided whenever possible during outbreaks of human body lice and louse-borne diseases because it is under these situations that both can thrive.

*Crab lice* can often be avoided by refraining from multiple sexual partners and changing or laundering bed linen slept on by infested persons. Pediculicides should be applied to the pubic area and to any other infested body regions.

To reduce the spread of *head lice,* the sharing of combs, hats, earphones, and blankets, especially by children, should be discouraged. Often, parents of children with head lice are notified to keep youngsters away from school or other gatherings until the infestation has been eliminated. If the parents are also infested, this can further involve ridding the entire family of lice to prevent reinfestations. Various pediculicidal shampoos, lotions, and gels are widely available for controlling head lice. These treatments typically kill all nymphal and adult lice, but only a small proportion of viable louse eggs. Therefore, treatments should be repeated at weekly intervals for 2–4 weeks in order to kill any recently hatched lice. Hatched or dead *nits* which remain glued to hair may be unsightly or embarrassing, and these can be removed with a fine-toothed louse comb. Louse combs have been used, in various forms, since antiquity to remove head lice (Mumcuoglu 1996). A wide range of pediculicides is commercially available. Although its use is now banned in many developed countries, the organochlorine DDT is widely used, especially in less developed countries, for controlling human and animal lice. Several alternative pediculicides, such as lindane, chlorpyrifos, diazinon, malathion, permethrin, or pyrethrins, are currently used throughout the world. Pediculicides can be used in powders, fogs, or sprays to treat furniture or premises for lice. Several general parasiticides show promise as pediculicides. Avermectins such as abamectin, doramectin, and ivermectin can kill human body lice and livestock lice. Prescribed doses of these compounds can be administered orally, by injection, or as topical applications of powders, dusts, and pour-ons. However, many of these compounds have not yet been approved for use on humans. The development of novel control agents for lice is a constant process because resistance to various pediculicides has developed in lice in many parts of the world (Burgess 1995, Mumcuoglu 1996).

Lice of livestock can be controlled by both husbandry practices and chemical intervention. Providing a high-energy diet, especially to cattle, can be an effective louse control strategy. If possible, it is important to keep animals in uncrowded conditions and to spot-treat or quarantine any infested individuals until they have been successfully deloused. Various formulations and applications of pediculicides are typically used to control lice on livestock. Insecticidal dusts, powders, sprays, dips, ear tags, tail tags, resin strips, gut boluses, collars, pour-ons, lotions, and injections are widely used products. Infested animals should be treated twice weekly for 2–4 weeks. Insecticidal *dust bags* or *back rubbers* can be used as self-dosing rubbing stations for cattle and other livestock. Because louse populations on livestock are typically greater during the winter months, pediculicides are usually best applied to them in the late fall. Fall systemic treatments of cattle for both lice and bots are often administered. Shearing wool from sheep removes up to 80% of the lice present on infested animals.

Pets, laboratory animals, and poultry can be treated for lice in several ways. Pets such as dogs and cats can be dipped or bathed with a pediculicidal lotion or shampoo. Various oral or topically applied insecticides used for controlling fleas on pets also are efficacious against lice. Similarly, flea combs also remove lice from pets. Poultry and laboratory animals can be treated with pediculicidal dusts or sprays. Although host treatment is most efficacious, bedding materials and cages can also be treated. Insecticidal feed additives are also available. Insecticide-impregnated resin strips can be added to cages of poultry or laboratory animals to control lice. The bacterium *Bacillus thuringiensis* and the nematodes *Steinernema carpocapsae* and *S. glaseri,* which are effective *biological control agents* against numerous arthropods, can also be used to kill livestock lice. Some *juvenile hormone analogs* and *insect growth regulators* such as diflubenzuron have similarly shown promise as pediculicides.

With respect to louse-borne diseases, *vaccines* have been developed only against epidemic typhus, and none is completely safe or currently approved for widespread use. The live attenuated E-strain vaccine has been administered to humans, particularly in certain African nations, in attempts to quell epidemic typhus outbreaks. However, this vaccine actually caused disease in some patients and did not always prevent subsequent infection.

## REFERENCES AND FURTHER READING

Askew, R. R. (1973). Parasitic insects (2nd ed.). Heinemann, London.
Azad, A. F. (1988). Relationship of vector biology and epidemiology of louse- and flea-borne rickettsioses. *In:* Biology of rickettsial

diseases. Vol. 1(D. H. Walker, ed.). pp. 51–61.CRC Press, Boca Raton, Florida.

Bedford, G. A. H. (1932). Trichodectidae (Mallophaga) found on South African Carnivora. *Parasitology* **24**, 350–364.

Bryceson, A. D. M., Parry, E. H. O., Perine, P. L., Warrell, D. A., Vukotich, D., and Leithead, C. S. (1970). Louse-borne relapsing fever. A clinical and laboratory study of 62 cases in Ethiopia and a reconsideration of the literature. *Q. J. Med.* **39**, 129–170.

Burgess, I. F. (1995). Human lice and their management. *Adv. Parasitol.* **36**, 271–342.

Busvine, J. R. (1969). Lice (4th ed.). British Museum (Natural History), London. Economic Ser. No. 2A.

Butler, J. F. (1985). Lice affecting livestock. *In:* Livestock entomology (R. E. Williams, R. D. Hall, A. B. Broce, and P. J. Scholl, eds.). pp. 101–127.Wiley, New York.

Clay, T. (1938). New species of Mallophaga from *Afroparvo congensis* Chapin. Am. Mus. Novitates No. 1008.

Durden, L. A. (2001). Lice (Phthiraptera). *In:* Parasitic diseases of wild mammals (W. M. Samuel, M. J. Pybus, and A. A. Kocan, eds.). pp. 3–17. Iowa State University Press, Ames.

Durden, L. A., and Musser, G. G. (1994a). The sucking lice (Insecta, Anoplura) of the world: a taxonomic checklist with records of mammalian hosts and geographical distributions. *Bull. Am. Mus. Nat. Hist.* **218**, 1–90.

Durden, L. A., and Musser, G. G. (1994b). The mammalian hosts of the sucking lice (Insecta, Anoplura) of the world: a host-parasite checklist. *Bull. Soc. Vector Ecol.* **19**, 130–168.

Emerson, K. C. (1972). Checklist of the Mallophaga of North America (north of Mexico). Vols. 1–4. Desert Test Center, Dugway, Utah.

Emerson, K. C., and Price, R. D. (1975). Mallophaga of Venezuelan mammals. *Brigham Young Univ. Sci. Bull. Biol. Ser. Ser.* **20(3)**, 1–77.

Emerson, K. C., and Price, R. D. (1981). A host-parasite list of the Mallophaga of mammals. *Misc. Publ. Entomol. Soc. Am.* **12**, 1–72.

Emerson, K. C., and Price, R. D. (1985). Evolution of Mallophaga and mammals. *In:* Coevolution of parasitic arthropods and mammals (K. C. Kim, ed.). pp. 233–255.Wiley, New York.

Ferris, G. F. (1919–1935). Contributions toward a monograph of the sucking lice. Parts 1–8. *Stanford Univ. Publ. Univ. Ser. Biol. Sci.* **2**, 1–634.

Ferris, G. F. (1931). The louse of elephants *Haematomyzus elephantis* (Mallophaga: Haematomyzidae). *Parasitology* **23**, 112–127.

Ferris, G. F. (1951). The sucking lice. *Mem. Pac. Coast Entomol. Soc.* **1**, 1–320.

Gratz, N. G. (1997). Human lice: their prevalence, control and resistance to insecticides. A review 1985–1997. World Health Organization/CTD/WHOPES/97.8, Geneva.

Hopkins, G. H. E. (1949). The host-associations of the lice of mammals. *Proc. Zool. Soc. Lond.* **119**, 387–604.

Hopkins, G. H. E., and Clay, T. (1952). A checklist of the genera and species of Mallophaga. British Museum (Natural History), London.

Ignoffo, C. M. (1959). Keys and notes to the Anoplura of Minnesota. *Am. Midl. Nat.* **61**, 470–479.

Kim, K. C. (1985). Evolution and host associations of Anoplura. *In:* Coevolution of parasitic arthropods and mammals (K. C. Kim, ed). pp. 197–231. Wiley, New York.

Kim, K. C., Emerson, K. C., and Price, R. D. (1973). Lice. *In:* Parasites of laboratory animals (R. J. Flynn, ed.). pp. 376–397.Iowa State University Press, Ames.

Kim, K. C., and Ludwig, H. W. (1978). The family classification of the Anoplura. *Syst. Entomol.* **3**, 249–284.

Kim, K. C., Pratt, H. D., and Stojanovich, C. J. (1986). The sucking lice of North America. An illustrated manual for identification. Pennsylvania State University Press, University Park.

Lancaster, J. L., Jr., and Meisch, M. V. (1986). Lice, *In:* Arthropods in livestock and poultry production. pp. 321–345.Ellis Horwood, Chichester.

Ledger, J. A. (1980). The arthropod parasites of vertebrates in Africa south of the Sahara. Vol. 4. Phthiraptera (Insecta). South African Institute of Medical Research, Johannesburg.

Marshall, A. G. (1981). The ecology of ectoparasitic insects. Academic press, London.

Matthyse, J. G. (1946). Cattle lice, their biology and control. Cornell University Agricultural Experiment Station, Ithaca. Bulletin 832.

McDade, J. E. (1987). Flying squirrels and their ectoparasites: disseminators of epidemic typhus. *Parasitol. Today* **3**, 85–87.

Mumcuoglu, K. Y. (1996). Control of human lice (Anoplura: Pediculidae) infestations: past and present. *Am. Entomol.* **42**, 175–178.

Orkin, M., and Maibach, H. I. (eds.) (1985). Cutaneous infestations and insect bites. Dermatology, Vol. 4. Dekker, New York.

Orkin, M., Maibach, H. I., Parish, L. C., and Schwartzman, R. M. (eds.) (1977). Scabies and Pediculosis. Lippincott, Philadelphia.

Pan American Health Organization/World Health Organization (1973). Proceedings of the International Symposium on the Control of Lice and Louse-borne Diseases. Pan American Sanitary Bureau, Scientific publication No. 263. Pan American Health Organization, Washington, DC.

Price, M. A., and Graham, O. H. (1997). Chewing and sucking lice as parasites of mammals and birds. Technical Bulletin No. 1849, US Department of Agriculture.

Rothschild, M., and Clay, T. (1952). Fleas, flukes and cuckoos. A study of bird parasites. Collins, London.

Snyder, J. C. (1966). Typhus fever rickettsiae, *In:* Viral and rickettsial infections of man (4th ed., F. L. Horsfall and I. Tamm, eds.) pp. 105–114. J. B. Lippincott, Philadelphia.

Townsend, L., and Scharko, P. (1999). Lice infestation in beef cattle. *Comp. Cont. Educ. Pract. Vet.* **21**(Suppl.), S119–S123.

Tuff, D. W. (1977). A key to the lice of man and domestic animals. *Tex. J. Sci.* **28**, 145–159.

Van der Stichele, R. H., Dezeure, E. M., and Bogaert, M. G. (1995). Systematic review of clinical efficacy of topical treatments for head lice. *Br. Med. J.* **311**, 604–608.

Watson, D. W., Lloyd, J. E., and Kumar, R. (1997). Density and distribution of cattle lice (Phthiraptera: Haematopinidae, Linognathidae, Trichodectidae) on six steers. *Vet. Parasitol.* **69**, 283–296.

Werneck, F. L. (1950). Os malofagos de mammiferos. Parte II. Ischnocera (continuacao de Trichodectidae) e Rhyncophthirina. Instituto Oswaldo Cruz, Rio de Janeiro.

Zinsser, H. (1935). Rats, lice, and history. Bantam, New York.

# 5

# TRUE BUGS (*Hemiptera*)

WILLIAM L. KRINSKY

*The order* Hemiptera includes all of the insects known as true bugs. Hemipterans are characterized as soft-bodied insects with piercing and sucking mouthparts and, usually, two pairs of wings. The order traditionally is divided into two major divisions: the Heteroptera and the Homoptera. The name Hemiptera (literally, "half-wings") is derived from the members of the Heteroptera ("different wings"), most of which have fore wings called *hemelytra*. They are composed of a thickened basal portion, the *corium* and *clavus,* and a somewhat transparent or filmy distal portion, the *membrane,* hence the idea of a half-wing (Fig. 5.1). The hind wings are completely membranous. The difference in texture between the fore and hind wings in the heteropterans gives this group its name. By comparison, the Homoptera ("same wings") have two pairs of wings that are very similar in character, both being membranous. The wings of homopterans often are held rooflike over the back of the body, whereas the wings of the heteropterans typically are held flat against the dorsum.

The true bugs, with about 90,000 species worldwide, constitute the largest exopterygote order of insects.

The North American fauna has about 16,000 species of hemipterans, about two-thirds of which are homopterans.

The piercing/sucking mouthparts of almost all true bugs enable these insects to feed on a diversity of fluids. The homopterans feed exclusively on plant juices. Common examples of these insects are aphids, scale insects, psyllids, leafhoppers, treehoppers, and cicadas. All of these are terrestrial. The heteropterans include phytophagous, predaceous, and hematophagous species. Common heteropterans include seed bugs, mirid plant bugs, stink bugs, assassin bugs, water striders, backswimmers, water boatmen, and giant water bugs, as well as the medically important kissing bugs and bed bugs.

Various homopterans and some predaceous and phytophagous heteropterans are known to bite humans. Predaceous or phytophagous bugs in at least 20 families of Hemiptera have been reported as occasionally biting or annoying humans by probing with their mouthparts (Table I). Published reports of these bites have been reviewed by Myers (1929), Usinger (1934), Ryckman (1979), Ryckman and Bentley (1979), and Alexander (1984).

Homopteran species known to cause occasional irritation or pain are leafhoppers (Cicadellidae), treehoppers (Membracidae), spittle bugs (Cercopidae), planthoppers (Fulgoroidea) and cicadas (Cicadidae). Unlike the generally painless bites by hematophagous species, predaceous

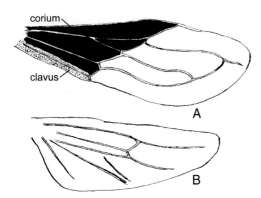

**FIGURE 5.1** Wings of typical triatomine bug (*Triatoma rubrofasciata*). A, Fore wing (hemelytron); B, Hind wing. (Redrawn and modified from Lent and Wygodzinsky, 1979)

and phytophagous bug bites often cause pain or a burning sensation, presumably the result of enzymes and other substances in the saliva that normally digest insect or plant materials. Most of these bites cause only transient discomfort associated with toxic reactions to foreign proteins and the localized erythema and edema that may result.

Common terrestrial heteropterans known to probe human skin are the *wheel bug* (*Arilus cristatus*) and other assassin bugs (*Reduvius personatus, Sinea diadema, Melanolestes picipes*), the *two-spotted corsairs* (*Rasahus*

*biguttatus* and *R. thoracicus*), and certain anthocorids (*Anthocoris musculus, Lyctocoris campestris,* and *Orius insidiosus*). There are fewer reports of nabid, lygaeid, mirid, tingid, and rhopalid bugs biting people. Humans most often are bitten by predaceous species when they enter habitats in which active predation is occurring or when the predatory species are attracted to house lights and enter dwellings. Some of the larger aquatic predaceous hemipterans can stab with their mouthparts, causing pain similar to that of a wasp sting. Species that most often bite in this way are belostomatids (*giant water bugs,* sometimes called "toe-biters") and notonectids (*backswimmers*). The bite of an assassin bug (*Holotrichius innesi*) found in the Sinai and Negev deserts of Israel is considered more neurotoxic and hemotoxic than the bite of venomous snakes in that region (Caras 1974).

The painless bites of the blood-sucking species pose the greatest threat to the health and well-being of humans and other animals because of the pathogens that are often transmitted and the blood loss associated with their feeding. The hematophagous heteropteran species of major medical and veterinary importance are the kissing bugs (triatomines) and the bed bugs (cimicids). These ectoparasitic insects are obligate blood feeders, requiring blood for growth and reproduction. The kissing bugs are vectors of the causative agent of Chagas disease, a significant medical problem in Central

**TABLE I**
**Nonhematophagous Hemiptera That Occasionally Bite Humans**

| Hemiptera | Common names | Locations of published cases |
|---|---|---|
| Homoptera | | |
| Cicadellidae | Leafhoppers | United States (California, Texas); Trinidad; England; North Africa; India; China; Japan; Philippines |
| Cercopidae | Spittle bugs | India |
| Membracidae | Treehoppers | United States (eastern) |
| Heteroptera | | |
| Anthocoridae | Minute pirate bugs | North America; Panama; Brazil; England; Czechoslovakia; Sudan; South Africa |
| Enicocephalidae | Gnat bugs | India |
| Lygaeidae | Seed bugs | Hawaii; Brazil; North Africa; Kuwait; India |
| Miridae | Leaf or plant bugs | North America; Brazil; Europe; Sudan |
| Nabidae | Damsel bugs | United States; Brazil |
| Pyrrhocoridae | Red bugs or stainers | Brazil; North Africa |
| Reduviidae | Assassin bugs | North America; Brazil; North Africa; Israel; India; Philippines |
| Rhopalidae | Scentless plant bugs | North America |
| Belostomatidae | Giant water bugs | North America |
| Notonectidae | Backswimmers | North America |

and South America. Bed bugs are not known to play a role in the transmission of any human disease agents. However, their bites may cause considerable discomfort, and their continued feeding may result in significant blood loss from a host. Some bed bugs feed primarily on non-human hosts, such as bats and swallows. The polyctenids, *bat bugs* that feed exclusively on bat blood, have never been associated with any medical or veterinary problems.

## KISSING BUGS (REDUVIIDAE)

The *kissing bugs* are so named because most of them are nocturnal species which feed on humans, often biting the faces of their sleeping victims. Another common name for them is *conenoses,* referring to the shape of the anterior part of the head (Fig. 5.2). Various common names in South America and where they are used locally

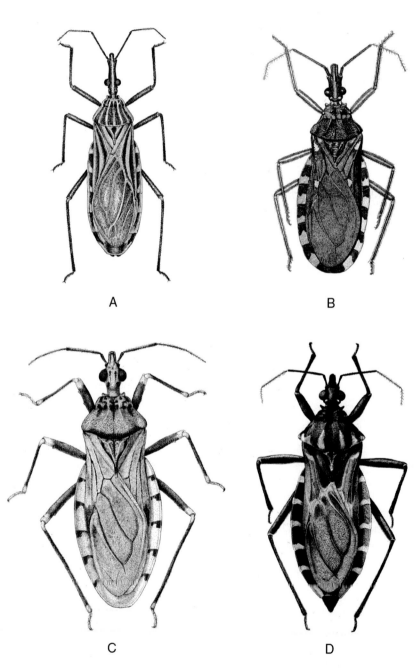

A     B

C     D

**FIGURE 5.2** Triatomine species. (A) *Rhodnius prolixus;* (B) *Triatoma infestans;* (C) *Panstrongylus geniculatus;* (D) *P. megistus.* (Courtesy of the American Museum of Natural History.)

include *barbeiro, bicudo,* or *chupão* (Brazil); *vinchuca* (Bolivia, Uruguay, Paraguay, Chile, Argentina); *bush chinch* (Belize); *chipo* or *pito* (Colombia, Venezuela); *chinchorro* (Ecuador); *chirimacho* (Peru); and *iquipito* or *chupon* (Venezuela) (Schofield *et al.* 1987). They are all members of the subfamily Triatominae in the family Reduviidae.

Lent and Wygodzinsky (1979) wrote an excellent monograph on kissing bugs that includes a survey of the external structures, descriptions of triatomine species, and notes on the vector importance of each species. Triatomine biosystematics, including an assessment of the evolutionary history of the subfamily, were reviewed by Schofield (1988). Biology, taxonomy, public health importance, and control were reviewed by Schofield *et al.* (1987), Schofield and Dolling (1993), and Schofield (1994).

## TAXONOMY

Members of the heteropteran family Reduviidae are commonly called *assassin bugs* because most species attack and feed on other insects. There are 22 subfamilies in the Reduviidae, including the Triatominae, or kissing bugs. Keys for identification of triatomine species are given by Lent and Wygodzinsky (1979).

The Triatominae is divided into 5 tribes and 14 genera; 106 species are known only from the New World, 5 species (*Linshcosteus*) are found only in India, and 7 species (*Triatoma*) are known only from Southeast Asia. The only species found in Africa is *Triatoma rubrofasciata;* it is found throughout the tropics, presumably having spread worldwide via ships.

The New World triatomine species occur from just south of the Great Lakes region of the United States to southern Argentina, with all but a few species concentrated in subtropical and tropical regions. The latter areas are considered the likely places of origin for the subfamily. All triatomines have the potential to transmit *Trypanosoma cruzi,* the etiologic agent of Chagas disease. Of the 119 described triatomine species, about half have been shown to be vectors, and about a dozen of these are considered vectors of major epidemiological importance (Table II).

## MORPHOLOGY

Triatomines range in length from 5 to 45 mm, with the majority of species falling in the range of 20–28 mm. Most species are black or dark brown, often with contrasting patterns of yellow, orange, or red,

notably on the *connexivum* (the prominent abdominal margin at the junction of the dorsal and ventral plates) (Fig. 5.2).

The head of an adult triatomine is constricted posteriorly to form a distinct neck behind the paired ocelli. Prominent hemispherical compound eyes are situated just in front of the ocelli. The region in front of the eyes is cylindrical to conical, hence the name "cone-nosed" bugs. The antennae are filiform and four-segmented. The beak, or *rostrum,* is three-segmented and is formed by the labium, which encloses the stylet-like mouthparts. These stylets are modified portions of the maxillae and mandibles that lie within a dorsal channel of the rostrum and are grooved to form a food canal and a salivary canal. When the bug is not feeding, the straight rostrum is held under, and nearly parallel to, the head (Fig. 5.3). In many nontriatomine reduviids, the rostrum is curved and strongly sclerotized.

The dorsal portions of the thorax include a *collar,* or neck, a somewhat triangular pronotum, and a scutellum. The undersurface of the prothorax (prosternum) has a *stridulatory groove* that has fine transverse sculpturing. When the tip of the rostrum is moved anteriorly to posteriorly in this groove, sound is produced, the function of which is mainly defensive.

The fore wings, or *hemelytra,* have a leathery basal portion (*corium* and base of the *clavus*) and an apical membranous portion (apical clavus and membrane) typical of most heteropterans. The *membrane* is dusky in most species, but it may be spotted or darkened only along the wing veins. The wing veins of the membrane form two elongated closed cells. The hind wings are completely membranous (Fig. 5.1). The hind wings are rarely absent but may be greatly shortened in some species. The relatively slender legs are used for walking. In addition to paired simple claws on each tarsus that allow the bug to crawl over rough surfaces, many species have a spongy structure, the *fossula,* at the apex of the tibia on one or more pairs of legs. The fossulae have adhesive setae on their surfaces that enable the bugs to climb on smooth surfaces, such as leaves and glass.

The triatomine abdomen is 11-segmented, often pointed or lobed in the female, but smoothly rounded in the male. In many species, around the periphery of the abdomen, both dorsally and ventrally, are segmental plates (*connexival plates*) connected to the abdominal segments by intersegmental membranes. These membranes allow for expansion of the abdomen during engorgement. The membranes in different species are folded on themselves in various ways, allowing the plates and membranes to expand in accordion fashion during feeding.

TABLE II
Major Triatomine Vectors of *Trypanosoma cruzi* and Their Geographic Distribution

| Species | Geographic range |
| --- | --- |
| *Rhodnius prolixus* | Southern Mexico, Guatemala, El Salvador, Honduras, Nicaragua, Costa Rica, <u>Colombia</u>, <u>Venezuela</u> |
| *Triatoma infestans* | Peru, Bolivia, Brazil (from Mato Grosso across to northeastern Goiás and Paraiba, south to Rio Grande do Sul), Paraguay, Argentina, Uruguay, and Chile |
| *Triatoma dimidiata* | Mexico south to Ecuador and Peru |
| *Triatoma pallidipennis* | Mexico |
| *Triatoma phyllosoma* | Mexico |
| *Rhodnius pallescens* | <u>Panama</u>, Colombia |
| *Triatoma maculata* | Colombia, <u>Venezuela</u>, Netherlands Antilles, Guyana, Suriname |
| *Triatoma brasiliensis* | northeastern Brazil |
| *Panstrongylus herreri* | northern Peru |
| *Panstrongylus megistus* | <u>Brazil (especially coastal)</u>, Paraguay, Argentina, Uruguay |
| *Triatoma guasayana* | Bolivia, Paraguay, <u>Argentina</u> |
| *Triatoma sordida* | Bolivia, <u>Brazil</u>, Paraguay, Uruguay, <u>Argentina</u> |

*Note.* The two species with the widest geographical distribution are listed first, followed by species arranged generally by their distribution from north to south. In cases where a species is not considered a major vector over the entire range given, countries where the triatomine species is an important vector are underlined (Lent and Wygodzinsky, 1979; Schofield, 1988).

## LIFE HISTORY

As in all Hemiptera, triatomines undergo *hemimetabolous* development. After the egg stage, development occurs through five nymphal instars. Nymphs are distinguished from adults by their smaller eyes, the lack of ocelli and wings, and the presence of thoracic lobes where wings will develop. Both sexes of adults and all nymphal instars require blood for their survival and development.

Female bugs are ready to mate 1 to 3 days after the final molt. Mating involves transfer of a spermatophore from the aedeagus while the male is positioned dorsolateral to the female with his claspers grasping the end of the female's abdomen from below. Copulation lasts from

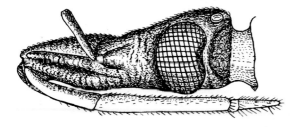

**FIGURE 5.3**    Lateral view of head of *Triatoma dimidiata*. (From Lent and Wygodzinsky, 1979, courtesy of the American Museum of Natural History.)

about 5 to 15 min. Although both sexes usually have had at least one blood meal before mating, unfed males also will mate with fed females.

*Oviposition* by females begins 10 to 30 days after copulation. Each female typically deposits only one or two eggs daily, producing a total of 10 to 30 eggs between blood meals. Depending on the species, a single female may produce up to 1000 eggs in her lifetime, but about 200 is average. Virgin, fed females may lay small numbers of infertile eggs. Each oval egg is about 2–2.5 × 1 mm. The eggs may be white or pink. Most species deposit eggs singly, but some females lay eggs in small clusters or masses. Different species lay eggs freely or glue them to a substrate. Gluing eggs to the substrate is seen in at least two species of *Triatoma* and many species of *Rhodnius, Psammolestes, Cavernicola,* and *Parabelminus.* In those species that glue their eggs to the substrate, the eggs may be single or in clusters. Eggs of some species turn pink or red before hatching 10 to 37 days after oviposition, depending on temperature.

The newly emerged *nymphs* are pink and will take a blood meal 48 to 72 hr after the eggs hatch. The nymphs must engorge fully in order to molt (Fig. 5.4), often requiring more than one blood meal during all but the first instar. The entire life cycle from egg to adult may be as short as 3 to 4 months but more commonly takes 1 to 2 years. The variable developmental times within and

**FIGURE 5.4**  Triatomine nymphs engorging on human foot. (Courtesy of R. B. Tesh)

between species are related to many factors, including environmental temperature, humidity, host availability, host species, feeding intervals, and the length of nymphal diapause.

## BEHAVIOR AND ECOLOGY

The New World triatomines are found in stable, sheltered habitats that are used by reptiles, birds, and a wide variety of mammals for their nests, roosts, or burrows. The kissing bugs can be divided into three general habitat groups: sylvatic, peridomestic, and domestic. *Sylvatic* forms inhabit nests and burrows, as well as a wide array of natural hiding places such as caves, rock piles, fallen logs, tree holes, hollow trees, palm fronds, bromeliads, and other epiphytes. These habitats attract amphibians, lizards (e.g., iguanas), opossums, rodents (e.g., porcupines), armadillos, sloths, bats, and other mammals, upon which the triatomines feed. The *peridomestic* species utilize domestic animals as hosts by living in chicken coops and other bird enclosures, stables, corrals, and rabbit and guinea pig houses. Because *Triatoma infestans* infests the latter, as well as wild guinea pig habitats, this species may have entered the domestic habitat thousands of years ago when people in South America began breeding guinea pigs for use as food. The *domestic* (domiciliary) species, exemplified by *T. infestans,* have colonized human habitations, where they depend on human or domestic animal blood as their source of nourishment. The domestic triatomine species are almost exclusively associated with humans and their pets and are often carried from one region to another in vehicles or concealed in household materials.

Many of the so-called peridomestic species, as well as a few domestic ones, have maintained sylvatic adaptations and may migrate from wild hosts to domestic animals and humans, depending upon the availability of suitable habitats and hosts. Peridomestic species sometimes fly to the lights of houses and thereby are attracted at night to feed on sleeping humans. Passive transport of certain species to human dwellings may occur when palm fronds containing attached triatomine eggs are used as roofing material. This is commonly the case with *Rhodnius prolixus* and other avian-feeding species that cement their eggs to the leaves in and around arboreal birds' nests. The significance of birds in dispersing triatomines is not known, although eggs and young nymphs of *R. prolixus* have been found among the feathers on storks, and *T. sordida* nymphs have been found in the plumage of sparrows.

In whatever habitat triatomines are found, they tend to be secretive, hiding in cracks and crevices of natural and artificial materials (e.g., debris of nests and burrows, rock crevices, and piles of vegetation); in building materials such as wood, shingles, thatch, and palm fronds; and in human dwellings in cracks in the walls, behind pictures or other wall-hangings, in bedding and mattresses, furniture, boxes, suitcases, piles of papers or clothes, and other accumulated materials that provide shelter during the day. Shaded crevices that provide extensive bodily contact with a rough, dry surface are preferred. Nymphs of many species are camouflaged by dirt and debris with which they cover themselves.

Most species of triatomines are nocturnal and actively seek blood from diurnal hosts that are resting or sleeping at night. In some cases, bugs will feed in daylight, typically on hosts that are nocturnal. Kissing bugs can survive for months without a blood meal, making them well adapted to nest habitats in which hosts may be present only intermittently with long intervals in between. When hosts are available, bugs commonly feed every 4 to 9 days. Individual species show definite *host preferences* and may favor bats, birds, armadillos, wood rats, or humans. Those favoring the latter species in South America are most important in the epidemiology of Chagas disease.

As in other hematophagous arthropods, feeding behavior is initiated by a combination of physical and chemical factors. Heat alone stimulates *R. prolixus* to probe, the heat receptors being located on the antennae. Carbon dioxide, which induces feeding responses in various hematophagous arthropods, causes increased activity in triatomines and may alert them to the presence of a host. The possible role of aggregation pheromones in attracting bugs to a host is unclear, but a pheromone in the feces of nymphal and adult *T. infestans* and in nymphal *R. prolixus* attracts unfed nymphs. Soon after feeding, these species defecate on or by the host, so that such a pheromone might attract other bugs to a source of blood.

The *probing* response begins when the rostrum is swung forward. The third segment is flexed upward so that optimal contact with a host occurs when the bug is at the side and just below a host. The serrated mandibular stylets are used to cut through the epidermis of the host, then anchor the mouthparts while the maxillary stylets probe for a blood vessel. When a vessel is penetrated, the left maxillary stylet slides posteriorly on the right stylet, disengaging the two stylets so that the left folds outward from the food canal. The purpose of this action is not known. It may allow a larger opening for ingestion of blood cells, or it may be a mechanism for holding the capillary lumen open (Lehane 1991).

The amount of blood ingested depends on the duration of feeding. This, in turn, is governed by the presence of chemicals in the blood of the host that stimulate the onset of feeding and by stretch receptors in the abdomen of the bug that stimulate cessation. Known phagostimulants of triatomines include various nucleotides and phosphate derivatives of nucleic acids. The salivary glands contain an anticoagulin that presumably helps maintain the flow of blood during feeding.

The time required to engorge fully varies from 3 to 30 min. During feeding, the abdomen becomes visibly distended. Adult bugs may imbibe blood equivalent to about 3 times their body weight, while nymphs may imbibe 6 to 12 times their unfed weight. Blood meals are stored in the anterior, widened portion of the midgut before the blood is passed to the narrower, posterior portion where digestion occurs. After engorging, the bug removes the rostrum from the host and, in most species, defecates on or near the host before crawling away to seek shelter. The interval between feeding and defecation is a major factor in determining the effectiveness of a species as a vector of *Trypanosoma cruzi*. Schofield (1979) reviewed the behavior of triatomines, with particular attention to their role in trypanosome transmission.

## PUBLIC HEALTH IMPORTANCE

Triatomine species that are efficient vectors tend to cause little or no pain when they feed. The bugs stealthily approach their sleeping hosts and engorge without causing much, if any, awareness (Fig. 5.5). However, immediate and delayed skin reactions to bites of *Triatoma infestans* and *Dipetalogaster maxima* have been observed.

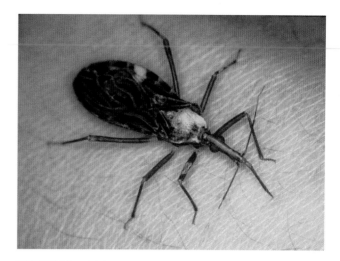

**FIGURE 5.5**  Adult triatomine feeding.

These reactions were not clearly correlated with previous exposure to bugs. Pruritic skin reactions following triatomine bites tend to enhance transmission of *Trypanosoma cruzi* by stimulating the bitten individual to scratch infective feces into the bite wound.

Some individuals react to triatomine feeding with mild *hypersensitivity reactions* such as pruritus, edema, and erythema. These reactions occur most often in response to triatomine species that are not efficient vectors of *T. cruzi* to humans. Within the latter group of species are members of the *Triatoma protracta* complex. These triatomines fly to light and have been known to invade homes situated within natural wood rat habitats. In a small number of cases, individuals have developed severe systemic reactions, including anaphylaxis, following bug bites. Immunotherapy involving multiple injections of *T. protracta* salivary gland extract has been successful in ameliorating the effects of the bite (Marshall and Street 1982).

## Chagas Disease (American Trypanosomiasis)

In 1907, while on an antimalarial campaign in Minas Gerais, Brazil, Carlos Chagas was introduced to the blood-sucking triatomines (*barbeiros*). He found what is now known to be *Trypanosoma cruzi* in the hindguts of several bugs; within 2 years he recognized this same flagellate protozoan in domestic animals and in a sick 2-year-old girl. Chagas disease and its epidemiology was thus first discovered in reverse fashion from that of most diseases. In this case the vector was found first, the nonhuman vertebrate hosts of the parasite second, and the human pathology last. The first report of the disease by Chagas was published in 1909, only

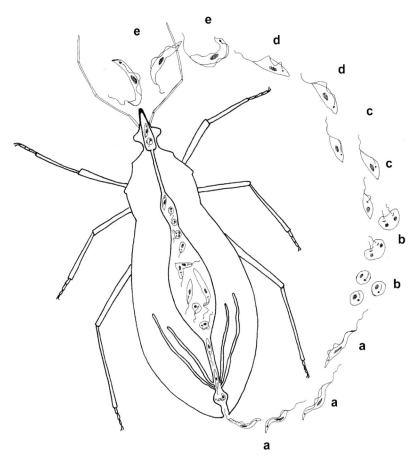

**FIGURE 5.6**    Life cycle of *Trypanosoma cruzi* in a triatomine and vertebrate host (a) metacyclic forms, (b) amastigotes (in vertebrate host), (c) epimastigotes (in vertebrate host), (d) trypomastigotes (in vertebrate host), (e) bloodstream forms ingested by triatomine. (Courtesy of W. L. Krinsky)

20 months after Chagas became aware of the existence of blood-sucking bugs. Chagas disease became known as American trypanosomiasis to differentiate it from African trypanosomiasis (African sleeping sickness), the disease caused by trypanosomes transmitted by tsetse flies in Africa.

Several bibliographies on the vast literature on Chagas disease and its epidemiology have been published, including Olivier *et al.* (1972) and Ryckman and Zackrison (1987). An excellent review of the etiologic agent and its biological associations is provided by Hoare (1972).

Triatomine species that are important *vectors* of *T. cruzi* are listed with their geographic ranges in Table II. *Triatoma infestans* is probably most often responsible for transmission of the trypanosome to humans because of this species' colonization of human dwellings over a wide geographic range in South America. *R. prolixus,* another important vector, is found from southern Mexico through Central America. *Panstrongylus megistus* is generally considered a major vector in the humid coastal regions of eastern Brazil.

The basic features of the life cycle of *Trypanosoma cruzi* are shown in Fig. 5.6. Broad and slender trypanosomes (*trypomastigotes*) circulating in the blood of an infected vertebrate host (Fig. 5.7) are imbibed by the triatomine during feeding. In the proventriculus of the bug, the broad trypomastigotes change into *sphaeromastigotes* and slender *epimastigotes.* The latter multiply by binary fission in the midgut, and as early as the 5th or 6th day after feeding occur in tremendous numbers that carpet the walls of the rectum. As early as the 7th or 8th day after feeding, these epimastigotes become infective *metacyclic forms* (trypomastigotes) that pass out of the bug in the feces and Malpighian tubule secretions. Although alternative *T. cruzi* developmental schemes have been proposed, the life cycle generally is thought to be limited to the gut of the bug, and infective forms occur only in the hindgut and rectum.

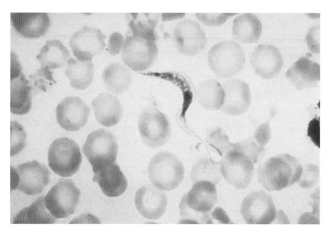

**FIGURE 5.7** *Trypanosoma* species in blood. (US Armed Forces Institute of Pathology, AFIP No. 74-5195)

Transmission to another vertebrate host occurs by this *posterior-station* (stercorarian) route. The trypanosomes infect the vertebrate when infective feces are rubbed into the bite site or other breaks in the skin. Transmission probably occurs most often when infective feces come in contact with the mucosal membranes of the nose or mouth or the conjunctivae of the eyes. Infected triatomines of those species that defecate while engorging or soon after feeding, while still on the host, are the most likely vectors of *T. cruzi*.

The entire development of *T. cruzi* in the lumen of the triatomine gut takes about 6 to 15 days or longer, depending on the ambient temperature and developmental stage of the bug (6 to 7 days in first-instar nymphs and 10 to 15 days in older nymphs and adults). Once a nymph or adult is infected, it is infective for life. Trypanosomes do not pass via the eggs or spermatophore to the next generation. Cannibalism and coprophagy have both been observed in laboratory colonies of triatomines and have been suggested as possible modes of transmission from bug to bug. However, the minimal infection rate among uninfected bugs housed with infected ones indicates that such transmission is not of much consequence under natural conditions. Coprophagy, however, may provide a means for transferring symbionts that are essential for development of some triatomine species (Schofield 1979).

Besides contamination of the skin or mucosal tissues, an alternative mode of transmission from bug to vertebrate is ingestion of a whole infected bug. This is probably common in the case of wild and domestic vertebrate animals (Miles 1983) and is thought to be the most common mode of transmission to wood rats in North America. The possibility of additional trophic levels in such transmission has been demonstrated by the infection of rodents and dogs following their ingestion of house flies that previously had ingested feces of infected triatomines (Hoare 1972). Carnivores may become infected by feeding on infected prey (Miles 1983). Human infection has occurred after accidental contact with freshly squashed triatomine bugs. Trypanosomes in the hindguts of dead triatomines may maintain their infectivity for up to 30 days. The "fecal rain" from triatomine-infested ceilings that falls on inhabitants of some tropical houses may be another source of infection (Miles 1983). Although inadvertent contamination with infective triatomine feces is probably the most common route of human infection, some cultural practices involve deliberate contacts with triatomines. These include the eating of *Triatoma picturata* ("chinche de compostela") for their supposed aphrodisiac properties in Nayarit, Mexico, and the rubbing of feces of *T. barberi* ("chinche voladora") onto the skin to cure warts on children in Oaxaca, Mexico. Both of these activities expose individuals to a high risk of trypanosome infection (Salazar-Schettino 1983).

Other modes of transmission from person to person include infection via blood transfusion and, less commonly, transplacental infection. Although infection of children via the breast milk of infected mothers is rare, the risk of infection during nursing is significantly increased when bleeding of the nipples occurs.

Arthropods other than triatomines have been infected with *Trypanosoma cruzi* in laboratory and field studies. Both the common bed bug (*Cimex lectularius*) and the African argasid tick *Ornithodoros moubata* have been infected by feeding on infected hosts and have maintained infective metatrypanosomes in their guts following normal cyclical development of the parasite. However, these arthropods are not known to have any role in natural transmission cycles.

Many individuals who become infected with *T. cruzi* do not develop symptoms early in the course of infection. These subclinical cases may or may not develop chronic disease. In a small number of cases, especially in children, an acute clinical form of Chagas disease occurs following initial infection. Significant mortality (5−15%) occurs among those showing acute disease.

**Acute Chagas disease**  The acute form of Chagas disease begins with an area of erythematous and indurated skin, called a *chagoma*, at the site of parasite entry. If the infective material is rubbed into the eye, periorbital edema called *Romaña's sign* appears (Fig. 5.8). This swelling, which may be accompanied by regional lymph node enlargements, may last for 2−6 weeks. The tissue changes result from intracellular development of amastigote trypanosomes in subcutaneous tissue and muscles. The amastigotes multiply and transform into

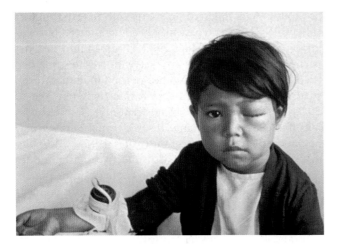

**FIGURE 5.8** Romaña's sign in boy undergoing xenodiagnosis, the feeding of laboratory-reared triatomine bugs on a patient as means of detecting infection with trypanosomes. (Courtesy of US Public Health Service)

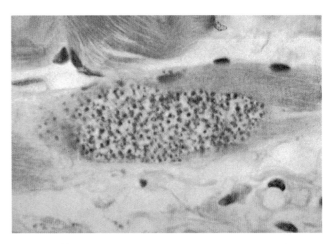

**FIGURE 5.9** *Trypanosoma cruzi* amastigotes developing in heart muscle. (From Peters and Gillies 1981)

trypomastigotes that enter the bloodstream. Other signs of acute Chagas disease include fever, general enlargement of lymph nodes, enlarged liver and spleen, and skin rashes. Complications of the acute phase that may result in death include myocarditis and meningoencephalitis. Most persons with the acute disease survive and enter the *indeterminate phase* of the disease, a stage in which the person appears healthy but still has the potential to develop serious chronic disease.

The indeterminate phase begins when antibodies to *T. cruzi* become detectable by serological testing, and trypanosomes, if detectable at all, may only be demonstrated by special methods, such as culturing blood or xenodiagnosis. The indeterminate phase may last indefinitely without further signs or symptoms of disease; in 30–40% of cases, however, chronic Chagas disease develops within the next few years or as many as 20 years after the initial infection.

**Chronic Chagas disease**   This is most often characterized by cardiac symptoms including palpitations, dizziness, chest pain, and sometimes fainting. The causes of these symptoms are various forms of arrhythmias which may lead to sudden death or persist for several years. The underlying pathology for the cardiac abnormalities is the development of amastigote trypanosomes in the cardiac muscles, accompanied by degeneration of cardiac muscle fibers (Fig. 5.9), followed by fibrosis. The second most often seen type of chronic Chagas disease that occurs south of the Amazon involves enlargement of the esophagus or, less often, the colon. These conditions are known as megasyndromes. Enlargement of these organs is accompanied by gastrointestinal discomfort, including pain

on eating and prolonged constipation in the case of megacolon. The often mammoth enlargement of the esophagus and colon is the result of pathologic destruction of myenteric ganglion cells, so that autonomic parasympathetic innervation (Auerbach's plexus) is greatly diminished. Individual patients may develop chronic cardiac disease, megaesophagus, and megacolon, or only one or two of these syndromes. The most common outcome in patients with chronic Chagas disease is damage to the heart muscle (cardiomyopathy) and conduction fibers; this damage leads to various forms of heart block and, in South America, congestive heart failure. Several years of suffering with cardiac symptoms may precede cardiac failure, which may, after as long as a few more years, result in death.

The restlessness, agitation, irritability, insomnia, and various other vague discomforts experienced by chronic Chagas disease patients have led medical historians to speculate on the basis of Charles Darwin's writings that he may have suffered from this disease. Furthermore, Darwin's palpitations and chest pains were brought on by emotional rather than physical stress, a phenomenon noted in Chagas' first patient when she was a middle-aged woman. Evidence that Darwin may have become infected with Chagas disease trypanosomes while he was in Argentina comes from his own description of being attacked by "the *Benchuca*, a species of *Reduvius*, the great black bug of the Pampas" (Voyage of the H.M.S. Beagle, March 26, 1835). It is impossible to make a definitive diagnosis of Darwin's illness in the absence of pathologic material. The fact that Darwin had some of the same complaints before his voyage on the *Beagle* further complicates the speculation.

In suspected acute Chagas disease, direct examination of anticoagulated blood, buffy-coat preparations, or concentrated serum may reveal living trypanosomes. Blood culture and fixed blood smears are useful for confirmation of the infection. The most sensitive procedure for recovering trypanosomes from both acute and chronic Chagas patients is *xenodiagnosis* (Fig. 5.8). This involves feeding uninfected, laboratory-reared triatomines on a patient, holding the bugs in the laboratory for about 30 days, and then dissecting the hindguts of the bugs to look for trypanosomes. This procedure uses the bugs as living culture chambers. It takes advantage of the natural transmission cycle in which even very small numbers of trypanosomes ingested by a triatomine multiply in great numbers in the alimentary tract. Xenodiagnosis has been most successful when triatomine species and geographic strains from the area in which a person has been infected are used. As in all parasitic diseases, a careful history of travel or activities that may have led to infection is essential for differential diagnosis of the disease.

Diagnosis of chronic Chagas disease requires demonstration of *T. cruzi*–specific antibodies in a patient who has the characteristic cardiac dysfunction and/or megasyndromes. Positive xenodiagnosis and antibody testing alone may indicate only that an individual has been exposed to the parasite and is in the indeterminate phase.

An estimated 16–18 million people in Central and South America are infected with *T. cruzi*, with as many as 500,000 new cases occurring each year. More than half the population of some rural villages is antibody positive. Historically, the nations most affected by infection and disease are Brazil, Argentina, Chile, Bolivia, Paraguay, and Venezuela (Fig. 5.10). The type of disease observed varies from country to country. Cardiomyopathy and megasyndromes are common in Brazil, but cardiac disease alone is common in Venezuela. Cardiac abnormalities are present, but less prevalent, in Colombia and Panama. Cardiac disease is even less common among seropositive people in other parts of Central America and Mexico, where cardiac problems, if they occur, present later in life. Although unproven, differences in clinical presentations may be related to known strain differences in *T. cruzi* isolates.

The social and economic burden caused by Chagas disease is primarily associated with morbidity rather than mortality. Chronically infected individuals often suffer for decades from weakness and fatigue that interfere with their productive enjoyment of life.

Chagas disease affects mostly the poorest people in the population. Typically the incidence of infection is directly associated with poor housing construction and proximity to domestic animal quarters or sylvatic habitats. Substandard houses, such as rough-walled huts made of mud and sticks or adobe mud bricks, often roofed with thatch, provide abundant cracks and crevices in which triatomines can hide during the day and crawl out at night to feed on sleeping people and domestic animals. *R. prolixus,*

naturally occurring on living palms, is especially abundant in palm roofs, whereas *P. megistus,* a species which is naturally found in hollow trees, favors the interstices of timber-framed mud houses. Thousands of triatomines may inhabit individual houses, causing each inhabitant to be bitten by dozens of bugs each night. *Triatoma infestans,* the most important vector of *Trypanosoma cruzi* in southern South America, also may occur inside houses with plastered walls and tiled roofs. The ability of triatomines to fly has even led to their intrusion into luxury high-rise buildings.

More than 100 species of mammals have been found infected with *T. cruzi.* Within domestic settings, humans, dogs, cats, and mice often are involved. Cats may become infected following ingestion of mice, whereas all domestic animals have the potential to become infected following ingestion of triatomines. In peridomestic cycles, chickens are excellent sources of blood for the triatomines but are not susceptible to infection with *T. cruzi.* In sylvatic cycles, opossums and various rodents are often important reservoir hosts that are readily fed upon by bugs. The broad host range of the parasite and the large number of potential triatomine vectors over a vast geographic area contribute to the complexities of the natural cycles of *T. cruzi* in any given region.

In general, the greatest transmission to humans occurs in regions where domestic triatomines are abundant. Not only are these species adapted to survive in close proximity to man, but also most have feeding patterns that cause them to defecate very soon after engorgement while still on or near the host. The lack of truly domestic species of triatomines and the relatively long delay between engorgement and defecation in triatomine species found north of Mexico have been cited as major reasons why there have been only four cases of Chagas disease acquired from triatomines in the United States.

Triatomine species repeatedly found to harbor natural infections with *T. cruzi* in the United States are *Triatoma sanguisuga* in Pennsylvania, Ohio, Maryland, south to Florida, and west into Arizona; *T. lecticularia* in Pennsylvania, Illinois, Maryland, south to Florida, and west into California; *T. gerstaeckeri* in Texas and New Mexico; *T. protracta* in Texas, north to Colorado, and southwest to California; *T. rubida* in Texas and west to California; and *T. recurva* in Arizona. All of these species except *T. recurva* are commonly associated with wood rats (*Neotoma* spp.), which are also naturally infected with *Trypanosoma cruzi.* Other potential reservoirs of infection for *Triatoma sanguisuga* and *T. lecticularia* are raccoons, armadillos, and opossums. The low incidence of human Chagas disease in North America is attributed to the relatively low percentages of infected triatomine bugs (6%) and vertebrate hosts (15%), the sylvatic behavior of the bugs, and the time delay between feeding and

**FIGURE 5.10**   Geographic distribution of human infection with *Trypanosoma cruzi*. (Courtesy of World Health Organization, Vector Biology and Control, 1989.)

defecation (Wood 1951, Lent and Wygodzinsky 1979, Neva 1996).

In the United States, triatomine-associated Chagas disease is much less likely than transfusion-acquired infection. The more than 1 million natives of Latin America now living in the United States are estimated to have an infection rate as high as 10%. Consequently, careful screening of blood donors is essential to prevent chronically infected, asymptomatic individuals from donating blood.

## Other Human Parasites Associated with Kissing Bugs

*Trypanosoma rangeli,* a nonpathogenic trypanosome found in Central and South America, is also transmitted by triatomines. *R. prolixus* is the chief vector. *T. rangeli* is morphologically and serologically distinguishable from *T. cruzi* and, unlike the latter, may be transmitted via the saliva of the bug. It is also found naturally in a wide array of mammals, including monkeys, dogs, opossums, anteaters, raccoons, and humans.

As with all hematophagous arthropods, incidental infections with various blood-borne pathogens can occur when these invertebrates feed on parasitemic hosts. Although recent interest in hepatitis B virus has led to the suggestion that triatomines might at times disseminate this virus, epidemiological evidence is lacking. Experimental feeding by fifth-instar nymphs of *Triatoma infestans* on asymptomatic human immunodeficiency virus (HIV)-infected patients has demonstrated that HIV can survive in the bugs 3 to 7 days after engorgement, but the bugs do not transmit the virus.

## VETERINARY IMPORTANCE

Triatomines transmit *Trypanosoma cruzi* to a variety of domestic and wild animals, including opossums, armadillos, rodents, carnivores, and monkeys. Depending on the strain of the trypanosome, the species and age of the host infected, and other poorly understood factors, the infection can lead to disease. Myocarditis and megaesophagus, similar to the conditions seen in humans, have been observed in dogs. *Canine trypanosomiasis* is of veterinary importance in Central and South America, and many cases have been recognized in southern Texas. Clinical indications of infection in dogs include dyspnea and ascites. There is little evidence that nonhuman wild animal hosts that serve as natural reservoirs of *T. cruzi* develop any pathology.

*T. rangeli* occurs within the distribution of *R. prolixus,* its chief vector. The public health importance of *T. rangeli* in veterinary medicine lies in differentiating this common parasite from the pathogenic *T. cruzi. T. conorhini* is a nonpathogenic parasite of rats transmitted by *Triatoma rubrofasciata* in many tropical regions of the Old and New World. It appears to have a tropicopolitan distribution identical to *T. rubrofasciata,* a domiciliary species of the tropics and subtropics. *Trypanosoma conorhini* is spread by posterior-station transmission (Hoare 1972).

Heavy infestations of triatomines in poultry houses in Central America may cause chronic blood loss in chickens. Even though no avian pathogens are involved, the impact of constant blood feeding may result in significant morbidity and, in the case of young birds, mortality.

## PREVENTION AND CONTROL

The goal of any prevention or control program is to reduce contact between humans and kissing bugs in order to prevent discomfort from bites and the more serious problem of Chagas disease. Control of Chagas disease involves the use of insecticides, improving housing conditions, and treating blood used for transfusions to kill the trypanosome. Residual *insecticides* sprayed on houses or applied to walls in paints are effective in controlling triatomines. However, long-term control requires careful surveillance and selective applications of insecticides. Some simple surveillance techniques, such as placing pieces of colored paper on the inside walls of houses, have been very successful in assessing the presence of triatomines. Their *fecal patterns* on the paper provide an indicator of triatomine activity. Houses treated with insecticides subsequently may be colonized by peridomestic or sylvatic species.

Defecation by *Triatoma infestans* and *R. prolixus* soon after feeding is dependent upon full engorgement by the bugs which, in turn, appears to be related to the density of the bug population in a given habitat; the greatest chance of engorgement is at low-density populations. Presumably, high bug densities with constant host resources lead to smaller blood meal sizes and a slower rate of defecation. Therefore, the chance of inhabitants becoming infected with *Trypanosoma cruzi* tends to be greatest in newly colonized houses where the bug population is rising or in houses being repopulated after vector control has been instituted.

Long-term control of triatomine bugs is best achieved in houses in which rough walls have been covered with plaster, thatch roofs have been replaced with tin or tile, and mud floors have been replaced with concrete. Such changes in construction, as well as the removal of wall hangings, firewood, and accumulated debris or vegetation, which serve as hiding places for triatomines, help to reduce the size of bug populations.

The drugs available for the treatment of acute Chagas disease, such as nifurtimox and benznidazole, are reasonably effective in preventing the development of chronic disease. However, they are associated with a high frequency of side effects, and neither is known to affect the course of chronic Chagas disease once it develops. Gentian violet effectively decontaminates donor blood, but routine screening and treatment of the blood supply is necessary to prevent transmission of *T. cruzi* via blood donors in endemic areas.

Various *biological control* entities, including juvenile hormone mimics, predatory arthropods, and parasitic wasps (e.g., scelionid *Telenomus fariai*) have been studied for the control of triatomines. However, no biological control method that is generally effective has been found for widespread use in Central and South America.

# BED BUGS (CIMICIDAE)

The family Cimicidae includes species known by several common names, including *bed bugs, bat bugs,* and *swallow bugs.* All species in this family are wingless, obligate hematophagous ectoparasites. Their medical and veterinary importance relates primarily to the loss of blood and discomfort caused by their feeding on vertebrate hosts. The monograph on the Cimicidae by Usinger (1966) is still the most comprehensive and best work on the ecology, morphology, reproductive biology, systematics, and taxonomy of the group.

Over 50 common names have been given to bed bugs in different countries. Some of these are: *mahogany-flat* (Baltimore), *heavy dragoon* (Oxford), *red coat* (New York), *wall louse* (*Wandlaus, Wegluis,* and *Wanze* [German]), *Wägglus* (Swedish), *Vaeggelus* (Danish), *Piq-seq* (Chinese), *Chinche* (Old Spanish), *Chinga* (Gallic), *Nachtkrabbler* ("night crawler," German), *Tapetenflunder* ("wallpaper flounder," German), *Punaise* ("stinker," French), *Perceveja* ("pursuer," Portuguese), *Lude* (Finnish), *Plostice* ("flat," Czechoslovakian), *klop* (Russian), *bug* ("ghost, goblin," British), *Buk* (Arabic), *Fusfus* (Syrian), *Pishpesh* (Hebrew), *Ekukulan* (Douala-Bantu), *Kunguni* (Swahili), *Uddamsa* ("biter," Sanskrit), *Rep* (Vietnamese), *Nankinmusi* ("Nanking bug," Japanese), and *Tokozirami* ("bed louse," Japanese). These and other names were reviewed by Usinger (1966).

## TAXONOMY

The family Cimicidae is divided into six subfamilies with 23 genera and 91 described species. The family is related to the predaceous family Anthocoridae, which includes species that feed on insects and mites and occasionally bite humans and other warm-blooded vertebrates. A related family, the Polyctenidae, includes species that are all ectoparasitic on bats and, like some cimicids, are also commonly called bat bugs.

The cimicids include 12 genera with species associated with bats and 9 genera with species associated with birds. In addition, some species in the genus *Cimex* are found on bats and others on birds. Three species are considered

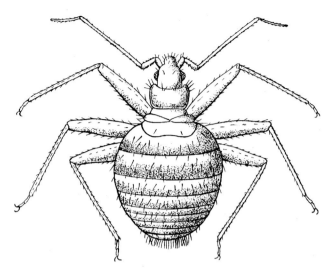

**FIGURE 5.11**   Cimicid bug, *Leptocimex boueti.* (From Brumpt, 1922.)

ectoparasites of humans. *Leptocimex boueti* (Fig. 5.11), a member of the subfamily Cacodminae, occurs on bats and people in West Africa. The other two are members of the subfamily Cimicinae, the bed bugs. *C. hemipterus* (Fig. 5.12) is parasitic on humans and chickens in the Old and New World tropics; *C. lectularius* (Figs. 5.12 and 5.13) is a cosmopolitan species associated primarily with humans, bats, and chickens.

Both *Cimex* species that feed on humans originated in the Old World. The origin of *C. hemipterus* is uncertain; however, there is evidence that *C. lectularius* originated in the Middle East, probably being associated with bats and humans living in caves. *C. lectularius* apparently spread into Europe during historic times, being recorded from Greece by 400 BC, from Italy by AD 77, and from Germany for the first time in the 11th century. The bed bug was known in France in the 13th century and is recorded as occurring in England in 1583. Therefore, the wide dissemination of *C. lectularius* throughout the world probably did not begin until after the 16th century.

## MORPHOLOGY

The most striking feature of cimicids is their dorsoventral flattening. Adults of the oval, mahogany-colored *Cimex* species generally range in length from about 5.5 to 7.0 mm, with abdomens 2.5 to 3.0 mm wide (Fig. 5.13). The females are larger than the males. The bat bug *L. boueti* differs from *C. lectularius* and *C. hemipterus* in having a very narrow pronotum, only slightly wider than the head, and very long legs. It is a smaller species, the

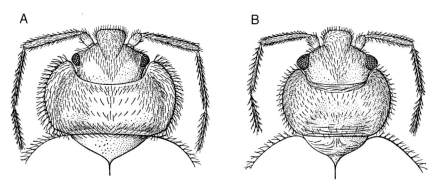

**FIGURE 5.12**    Head and prothorax of adult bed bugs. (A) *Cimex lectularius;* (B) *C. hemipterus.* (From Smart 1943, courtesy of the British Natural History Museum.)

total body length being 2.8 mm in males and 4.0 mm in females (Fig. 5.11).

The cimicid head is small and cylindrical, with two knoblike, multifaceted eyes. Ocelli are not present. The antennae are four-segmented and inserted between the eye and the clypeus. The labium is three-segmented and, as in the triatomines, dorsally encloses the maxillary and mandibular stylets; they in turn enclose a relatively large dorsal food canal and a very small ventral salivary canal. The labium has two sensory lobes at its tip. When the bug is not feeding, the *rostrum*, or beak, composed of the labium and associated mouthparts, is bent below the head, with the tip extending to the middle of the prosternum.

The thorax consists of a narrow canoe-shaped pronotum, a mesonotum which is covered dorso-laterally by reduced fore wings called *hemelytral pads*, and a metanotum hidden below the latter. Nymphs do not have hemelytral pads. *C. hemipterus* can be distinguished from *C. lectularius* by the former's narrower pronotum (Fig. 5.12). The hemelytral pads are oval in *Cimex* species and are reduced to small elevated ridges in *Leptocimex*. Hind wings are never present. The legs are slender, with two-segmented tarsi in the nymphs and three-segmented tarsi in the adults.

The abdomen is 11-segmented and capable of tremendous expansion during blood feeding. In nymphs, membranous areas on the entire ventral surface and on the first

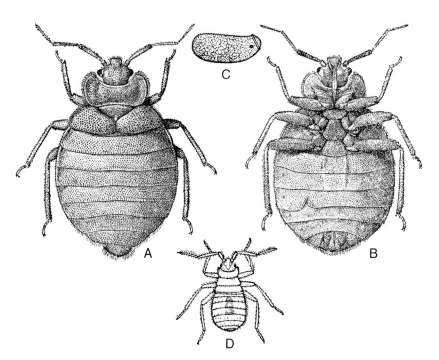

**FIGURE 5.13**    Human bed bug (*Cimex lectularius*), adults. (A) Male, dorsal view; (B) female, ventral view; (C) egg; (D) nymph. (From Busvine, 1966.)

and second and part of the third abdominal terga enable expansion of the abdomen while feeding. In adults, the intersegmental membranes are wide, and the middle of the ventral side of the second to fifth segments of the abdomen is likewise membranous. Female *Cimex* adults are readily distinguished from males by the presence of an indentation on the hind margin of the fifth abdominal sternite (Fig. 5.13). This narrow cleft, called the *paragenital sinus,* is surrounded by bristles and is the point at which the male inserts his aedeagus to intra-abdominally inseminate the female. No paragenital sinus occurs in *L. boueti.*

## LIFE HISTORY

Mating occurs with the male bug straddling the female's back at an oblique angle. In this position the tip of his abdomen is strongly curved against the right side of the venter of the female, where the paragenital sinus is located. The male inseminates the female by injecting sperm into the sinus. This form of *traumatic insemination* that involves introduction of the sperm into an extragenital site occurs in many species of the superfamily Cimicoidea. Specialized structures for reception of the sperm, variously called the *spermalege, organ of Ribaga,* and *organ of Berlese,* are present in these species. Copulation usually lasts from one to several minutes but may take up to half an hour. Females that have been inseminated retain permanent scars that are visible in the integument. The sperm pass from the spermalege into the hemocoel, from which they enter paired outpouchings of the walls of the oviducts called *sperm conceptacles.* From there the sperm travel within the walls of the oviducts via an intraepithelial network of tubular canals called *spermodes* to the bases of the ovarioles.

Mated females usually feed to repletion and then begin to lay eggs 3 to 6 days later. Oviposition lasts for about 6 days, during which 6 to 10 eggs are deposited. Depending on ambient temperature and relative humidity, female bugs may feed every 3 to 4 days. Eggs are laid continuously, with the mean number of eggs per week typically varying from 3 to 8. Some females have been observed to lay as many as 12 eggs in one day and up to 540 eggs in their lifetime. A female is capable of producing viable eggs for 5 to 7 weeks after feeding and mating. After that time, an increasing number of eggs are sterile.

The eggs are elongate/oval, about 1 mm long, and pearly white. They are laid singly and are coated with a transparent cement that causes them to adhere to various surfaces. The eggs are usually deposited in groups or clusters. Hatching usually takes place in 4 to 12 days, depending on the temperature. There are five *nymphal stages,* each lasting 2.5 to 10 days. The temperature threshold for development is about 15°C, with optimal development at 30°C. Humidity, except at the extremes, has little or no effect on development. The total developmental time from egg to adult for *C. lectularius* varies from 24 days (at 30°C) to 128 days (at 18°C), and for *C. hemipterus* it ranges from 25 days (at 30°C) to 265 days (at 18°C).

The nymphs are pale straw-colored before they feed but bear a resemblance to red berries after they have fed. Feeding generally occurs within 24 hr after hatching or molting. At low temperatures, nymphs may survive for 5 to 6 months without feeding, whereas adults can survive even longer. This makes them efficient nest parasites which are able to survive long periods when a host is absent. Nymphs feed at least once during each instar. Engorgement usually takes about 3 min for first-instar nymphs and 10 to 15 min for older nymphs and adults. After fully engorging, nymphs are 2.5–6 times heavier than unfed nymphs, and the adults are 1.5–2 times heavier than unfed individuals. As in triatomines, liquid fecal matter is excreted soon after feeding. Half the weight of the entire blood meal is lost within the first 5 hr after feeding.

## BEHAVIOR AND ECOLOGY

Cimicids are similar to triatomines in their choices of hiding places, the nature of the substrates selected, and their feeding patterns. They hide in cracks and crevices in human and animal habitations and in nests, caves, and tree holes in natural settings. They prefer rough, dry substrates which allow maximum contact of the bugs with the surface. Their attraction to such harborages between feedings often results in large aggregations. In domestic situations, bed bugs prefer hiding in wood and paper accumulations rather than in materials made of stone, plaster, metal, or textiles. Both *C. lectularius* and *C. hemipterus* may infest mattresses, box springs, and upholstered furniture. Other common sites of infestations include public facilities such as theaters and office waiting rooms and upholstered seats in buses. Cimicids will crawl into the narrowest crevices, such as those formed behind loose wallpaper, pictures, or electrical switch or socket plates. The harborages and infested premises are often stained with conspicuous fecal spots that range in color from white to yellow to brown to reddish-brown to black. Areas infested by *C. lectularius* may be identified by a characteristic sweet odor.

Bugs leave their hiding places primarily to feed. They are negatively phototactic, tending to feed mostly in darkness or subdued light when the temperature is above 10°C. Warmth and carbon dioxide, as in many other hematophagous arthropods, appear to be major factors in attracting bed bugs to a host. A temperature differential of only 1 to 2°C is sufficient to induce probing (Lehane 1991).

When the bug has located a host, it approaches with its antennae outstretched and its beak directed downward at a 90° angle. It grabs the host with the tarsal claws of the front legs. After contact is made, the antennae are pulled backwards, and the entire bug makes rocking, pushing movements as the vertically directed stylets are embedded in the skin. As the stylets penetrate the skin, the labium becomes more and more bent at the skin surface. The mandibles, which have retrorse teeth at their tips, move in and out of the skin in alternating fashion, producing a passage for the maxillae. The bundle of feeding stylets probes actively within the skin until a blood vessel of suitable size is penetrated. Only the maxillae, and possibly only the right maxilla, actually enter the lumen of the vessel. The salivary secretion contains an anticoagulin which prevents the blood from clotting. After engorging, the bug withdraws its stylets; sometimes this requires considerable effort owing to the teeth on the mandibles. Once the stylets are again encased in the straightened labial sheath, the beak is folded back under the head. The bugs are quick to retreat when disturbed at the beginning or end of feeding; however, while the stylets are fully embedded in the skin, the bug is unable to withdraw its mouthparts even if handled or rotated.

The two species of *Cimex* most commonly associated with humans have been dispersed over wide areas of the globe, with *C. lectularius* most often being found in temperate regions and *C. hemipterus* in the tropics. These species are carried concealed in luggage, furniture, and all manner of packing materials. They have been transported on land vehicles, ships, and planes.

## PUBLIC HEALTH IMPORTANCE

Usinger (1966) listed 27 human pathogens, including viruses, bacteria, protozoa, and helminths, that have been shown to survive for varying lengths of time in *C. lectularius* and *C. hemipterus*. However, there is little or no evidence to incriminate bed bugs as vectors of these or any other disease agents.

Recent attempts to explain transmission of hepatitis B virus and, to a lesser extent, HIV in otherwise unexplained situations have focused on the possibility of cimicid transmission. Hepatitis B antigens survive in cimicid tissues and feces under laboratory conditions; however, attempts to transmit the virus from infected bugs to chimpanzees failed. These results and those from transmission studies with mosquitoes suggest that it is unlikely that hepatitis B transmission occurs via either infective feces or interrupted feedings.

Although transmission studies with bed bugs and HIV indicate that these insects may harbor the virus for up to 8 days, replication of the virus does not occur and the virus is not present in cimicid feces. These observations, together with failed attempts to experimentally transmit HIV by interrupted feedings, suggest that cimicids are neither biological nor mechanical vectors of HIV.

Despite the fact that cimicids do not play a significant role as vectors of human pathogens, bed bugs are medically important because they cause unpleasant bite reactions and significant blood loss in people living in dwellings that are chronically infested. The actual feeding by bed bugs generally does not produce any pain. If interrupted, a bug will often bite again close to the previous site, thereby creating a linear array of punctures that is characteristic of cimicid bites. People are most often bitten on the limbs, trunk, and face.

*Sensitivity reactions* to bed bug bites are the result of substances injected during feeding. These reactions may be localized cutaneous responses, or they may be generalized and systemic. The most common local reactions are wheals similar to uncomplicated mosquito bites or, in some individuals, large fluid-filled bullae. Erythema is not a common response but may occur as a result of multiple feedings that cause extensive hemorrhaging under the skin. Individual reactions to cimicid bites vary from no response to severe immediate or delayed sensitivity reactions, including *anaphylaxis*. In most cases, swelling and itching associated with the bites can be relieved by application of ice and use of an oral antihistamine. Chronic bed bug bites are sometimes misdiagnosed as allergic dermatitis or other skin disorders. Accurate clinical assessment often requires careful epidemiological evaluation of a patient's living quarters.

People living with chronic infestations of bed bugs often are subject to nightly attacks, resulting in a marked loss of blood and associated iron deficiency. Children who are marginally nourished are especially vulnerable to developing anemia and other medical problems as a result of such chronic blood loss. Individuals subjected to continued feeding by bed bugs may also develop extreme irritability that results from restless nights and chronic sleep deprivation. If the source of the disturbance goes undetected, the emotional stress caused by such infestations may be misdiagnosed as a neurosis.

### Other Cimicids Which Occasionally Attack Humans

In addition to the three cimicid species directly associated with human habitations, there are several species that occasionally feed on people. These include *swallow bugs* of the genus *Oeciacus;* the *bat bugs C. pilosellus* (New World) and *C. pipistrelli* (Europe); and the *bird bugs,* such as the

Mexican chicken bug *Haematosiphon inodorus* and *Cimexopsis nyctalis* from the nests of chimney swifts. Human bites by these species generally occur only in the vicinity of the nesting or roosting sites of their natural hosts.

The swallow bugs that occur in mud nests of swallows include two species: *Oeciacus hirundinis* in Eurasia south to Morocco and *O. vicarius* in North America south to Durango in Mexico. Both species are members of the subfamily Cimicinae and will bite people who disturb infested bird nests. Swallows may be heavily infested with the bugs, and nestlings often die as a result of blood loss. The eggs of *Oeciacus* species are attached to the outer surfaces of swallow mud nests, often being so abundant that they can be seen from a distance. There is some evidence that *Oeciacus* species are carried as nymphs by the birds from nest to nest.

An arbovirus has been isolated from the cliff swallow bug *O. vicarius*, nestling cliff swallows, and house sparrows in eastern Colorado. This is an *alphavirus*, part of the western equine encephalitis complex called Fort Morgan virus. It is not known to cause pathology in its avian hosts or in humans. Occurrence of this virus in bugs and birds suggests that viruses can overwinter in swallow bugs that occupy nests left vacant by their migrating hosts.

## VETERINARY IMPORTANCE

Cimicids can be significant pests in commercial *poultry* production. Cimicids attacking domestic poultry include *Cimex lectularius* in North America, Europe, and the former Soviet Union; *H. inodorus* in Central America; and *Ornithocoris toledoi* in Brazil.

Raised slats and wood shavings in nest boxes in broiler breeder houses provide harborage for the bugs. Indications of cimicid infestations include fecal spots on eggs, nest boxes (Fig. 5.14), and wooden supports, skin lesions on the breasts and legs of birds, reduced egg production, and increased consumption of feed. Chicken bugs are not known to transmit any avian pathogens. However, chickens and other fowl raised in poultry houses heavily infested with chicken bugs are irritable and often anemic. Morbidity in such cases may be high, and young birds may succumb from blood loss.

Two species of nonpathogenic trypanosomes which undergo development in cimicids have been isolated from bats in North America. *Trypanosoma hedricki* and *T. myoti*, both closely related to *T. cruzi*, have been found in big brown bats and little brown bats in southern Ontario, Canada. Developmental stages infective to bats form in the rectum of *C. brevis* and *C. lectularius*, which suggests that these trypanosomes are transmitted by the

**FIGURE 5.14**    Fecal spots, indicative of cimicid activity, along seams of nesting boxes of laying hens in a poultry house heavily infested with *Cimex lectularius*. (Photo by G. R. Mullen)

bugs via the posterior-station route. Because bats also are known hosts for *T. cruzi*, the differentiation of other bat trypanosomes and the elucidation of their transmission are important. Furthermore, because of the similarities of the life cycles and transmission of these nonpathogenic trypanosomes to those of *T. cruzi*, they could be suitable candidates for developing laboratory models of the Chagas disease pathogen.

## PREVENTION AND CONTROL

Measures to prevent cimicid infestations should begin with household sanitation. Removing accumulations of paper and wood trash eliminates hiding places and harborages for the bugs. However, once an infestation occurs, eliminating cimicids requires thorough fumigation with residual insecticides that must be sprayed on surfaces over which the bugs crawl to reach their hosts. Organophosphates have proved most useful when compared with chlorinated hydrocarbons, carbamates, or pyrethrins. For temporary control, such as is needed by a traveler occupying an infested room for one or a few nights, any of various insecticides supplied in aerosol cans can be used to thoroughly spray bed frames, mattresses, and box springs.

Control of cimicids in premises in which people are bothered by the bites of bird bugs or bat bugs requires identification and removal of the source of the bugs. Such sources include bats roosting in attics or eaves, bird nests on window ledges or air conditioners, and birds roosting in chimneys. Removal of the nonhuman vertebrate hosts must be accompanied by use of an insecticide, or the hungry bugs will seek human blood more aggressively in the absence of their natural hosts.

Various arthropods are natural predators of cimicids. These include the masked bed bug hunter *Reduvius personatus,* other hemipterans, ants, pseudoscorpions, and spiders. None of these, however, has been effectively used for controlling bed bugs.

# REFERENCES AND FURTHER READING

Alexander, J. O. (1984). Arthropods and human skin. Springer-Verlag, Berlin. 422 pp.

Asin, S. N., and Catalá, S. S. (1991). Are dead *Triatoma infestans* a competent vector of *Trypanosoma cruzi? Mem. I. Oswaldo Cruz* **86**, 301–305.

Axtell, R. C., and Arends, J. J. (1990). Ecology and management of arthropod pests of poultry. *Annu. Rev. Entomol.* **35**, 101–126.

Bower, S. M., and Woo, P. T. K. (1981). Development of *Trypanosoma (Schizotrypanum) hedricki* in *Cimex brevis* (Hemiptera: Cimicidae). *Can. J. Zool.* **59**, 546–554.

Brenner, R. F., and de la M. Stoka, A. (eds.) (1987). Chagas' disease vectors. 3 vols. CRC Press, Boca Raton, Florida.

Brumpt, E. (1922). *Précis de parasitologie.* Masson & Co., Paris. 1216 pp & IV pls.

Busvine, J. R. (1966). Insects and Hygiene. Methuen & Co. Ltd., London.

Calisher, C. H., Monath, T. P., Muth, D. J., Lazuick, J. S., Trent, D. W., Francy, D. B., Kemp, G. E., and Chandler, F. W. (1980). Characterization of Fort Morgan virus, an alphavirus of the western equine encephalitis virus complex in an unusual ecosystem. *Am. J. Trop. Med. Hyg.* **29**, 1428–1440.

Caras, R. (1974). Venomous animals of the world. Prentice Hall International, Englewood Cliffs, NJ.

Costa, C. H. N., Costa, M. T., Weber, J. N., Gilks, G. F., Castro, C., and Marsden, P. D. (1981). Skin reactions to bug bites as a result of xenodiagnosis. *Trans. R. Soc. Trop. Med. Hyg.* **75**, 405–408.

Foil, L. D., and Issel, C. J. (1991). Transmission of retroviruses by arthropods. *Annu. Rev. Entomol.* **36**: 355–381.

Ghauri, M. S. K. (1973). Hemiptera (bugs). *In:* Insects and other arthropods of medical importance (K. G. V. Smith, ed.). . pp. 373–393. British Museum (Natural History), London

Gooding, R. H. (1972). Digestive processes of haematophagous insects. I. A literature review. *Quaest. Entomol.* **8**, 5–60.

Hoare, C. A. (1972). The trypanosomes of mammals. Blackwell Sci., Oxford.

Jupp, P. G., Purcell, R. H., Phillips, J. M., Shapiro, M., and Gerin, J. L. (1991). Attempts to transmit hepatitis B virus to chimpanzees by arthropods. *S. Afr. Med. J.* **79**, 320–322.

Kirk, M. L., and Schofield, C. J. (1987). Density-dependent timing of defaecation by *Rhodnius prolixus,* and its implications for the transmission of *Trypanosoma cruzi. Trans. R. Soc. Trop. Med. Hyg.* **81**, 348–349.

Lacey, L. A., D'Alessandro, A., and Barreto, M. (1989). Evaluation of a chlorpyrifos-based paint for the control of *Triatominae. Bull. Soc. Vector Ecol.* **14**, 81–86.

Lehane, M. J. (1991). Biology of blood-sucking insects. Harper Collins, London.

Lent, H., and Wygodzinsky, P. (1979). Revision of the Triatominae (Hemiptera, Reduviidae), and their significance as vectors of Chagas' disease. *Bull. Am. Mus. Nat. Hist.* **163**, 123–520.

Lewinsohn, R. (1979). Carlos Chagas (1879–1934): the discovery of *Trypanosoma cruzi* and of American trypanosomiasis (foot-notes to the history of Chagas's disease). *Trans. R. Soc. Trop. Med. Hyg.* **73**, 513–523.

Lowenstein, W. A., Romaña, C. A., Ben Fadel, F., Pays, J. F., Veron, M., and Rouzioux, C. (1992). Survie du virus de l'immunodéficience humaine (VIH-1) chez *Triatoma infestans* (Klug 1834). *Bull. Soc. Pathol. Exot.* **85**, 310–316.

Marshall, N. A., and Street, D. H. (1982). Allergy to *Triatoma protracta* (Heteroptera: Reduviidae). I. Etiology, antigen preparation, diagnosis and immunotherapy. *J. Med. Entomol.* **19**, 248–252.

Miles, M. A. (1983). The epidemiology of South American trypanosomiasis—biochemical and immunological approaches and their relevance to control. *Trans. R. Soc. Trop. Med. Hyg.* **77**, 5–23.

Minter, D. M., Minter-Goedbloed, E., and de C. Marshall, T. F. (1978). Comparative xenodiagnosis with three triatomine species of different hosts with natural and experimental chronic infections with *Trypanosoma (Schizotrypanum) cruzi. Trans. R. Soc. Trop. Med. Hyg.* **72**, 84–91.

Myers, J. G. (1929). Facultative blood-sucking in phytophagous Hemiptera. *Parasitology* **21**, 472–480.

Neva, F. A. (1996). American trypanosomiasis (Chagas' disease). *In:* Cecil textbook of medicine (20th ed., J. C. Bennett and F. Plum, eds.). pp. 1899–1903. Saunders, Philadelphia.

Olivier, M. C., Olivier, L. J., and Segal, D. B. (eds.) (1972). A bibliography on Chagas' disease (1909–1969). Index-catalogue of medical and veterinary zoology. Special publication No. 2. US Govt. Printing Office, Washington, DC.

Olsen, P. F., Shoemaker, J., Turner, H. F., and Hays, K. L. (1964). The incidence of *Trypanosoma cruzi* (Chagas) in wild vectors and reservoirs in east central Alabama. *J. Parasitol.* **50**, 599–603.

Peters, W., and Gillies, H. M. (1981). A colour atlas of tropical medicine and parasitology, 2nd ed. Year Book Med. Publ., New York.

Ryckman, R. E. (1962). Biosystematics and hosts of the *Triatoma protracta* complex in North America (Hemiptera: Reduviidae) (Rodentia: Cricetidae). *U. Calif. Publ. Entomol.* **27**, 93–240.

Ryckman, R. E. (1979). Host reactions to bug bites (Hemiptera, Homoptera): a literature review and annotated bibliography, Part I, Part II (with Bentley, D. G.). *Calif. Vector Views* **26**, 1–49.

Ryckman, R. E. (1985). Dermatological reactions to the bites of four species of Triatominae (Hemiptera: Reduviidae) and *Cimex lectularius* L. (Hemiptera: Cimicidae). *Bull. Soc. Vector Ecol.* **10**, 122–125.

Ryckman, R. E., and Bentley, D. G. (1979). Host reactions to bug bites (Hemiptera, Homoptera): a literature review and annotated bibliography, Part II. *Calif. Vector Views* **26**, 25–49.

Ryckman, R. E., Bentley, D. G., and Archbold, E. F. (1981). The Cimicidae of the Americas and Oceanic Islands, a checklist and bibliography. *Bull. Soc. Vector Ecol.* **6**, 93–142.

Ryckman, R. E., and Zackrison, J. L. (1987). Bibliography to Chagas' disease, the Triatominae and Triatominae-borne trypanosomes of South America (Hemiptera: Reduviidae: Triatominae). *Bull. Soc. Vector Ecol.* **12**, 1–464.

Salazar-Schettino, P. M. (1983). Customs which predispose to Chagas' disease and cysticercosis in Mexico. *Am. J. Trop. Med. Hyg.* **32**, 1179–1180.

Schofield, C. J. (1979). The behaviour of Triatominae (Hemiptera: Reduviidae): a review. *Bull. Entomol. Res.* **69**, 363–379.

Schofield, C. J. (1980). Density regulation of domestic populations of *Triatoma infestans* in Brazil. *Trans. R. Soc. Trop. Med. Hyg.* **74**, 761–769.

Schofield, C. J. (1988). Biosystematics of the Triatominae. *In:* Biosystematics of Haematophagous insects (M. W. Service, ed.). pp. 285–312. Clarendon, Oxford.

Schofield, C. J. (1994). Triatominae: biology and control. Eurocommunica, West Sussex.

Schofield, C. J., and Dolling, W. R. (1993). Bedbugs and kissing bugs. *In:* Medical insects and arachnids (R. P. Lane and R. W. Crosskey, eds.). pp. 483–516. Chapman & Hall, London.

Schofield, C. J., Minter, D. M., and Tonn, R. J. (1987). XIV. The triatomine bugs—biology and control. *In:* Vector Control Series: triatomine bugs—training and information guide. World Health Organization Vector Biology and Control Division 87.941, Paris.

Schofield, C. J., and White, G. B. (1984). Engineering against insect-borne diseases in the domestic environment/house design and domestic vectors of disease. *Trans. R. Soc. Trop. Med. Hyg.* **78,** 285–292.

Smart, J. (1943). A handbook for the identification of insects of medical importance. British Museum (Natural History), London.

Solari, A., Venegas, J., Gonzalez, E., and Vasquez, C. (1991). Detection and classification of *Trypanosoma cruzi* by DNA hybridization with nonradioactive probes. *J. Protozool.* **38,** 559–565.

Theis, J. H. (1990). Latin American immigrants—blood donation and *Trypanosoma cruzi* transmission. *Am. Heart J.* **120,** 1483.

Trumper, E. V., and Gorla, D. E. (1991). Density-dependent timing of defecation by *Triatoma infestans. Trans. R. Soc. Trop. Med. Hyg.* **85,** 800–802.

Usinger, R. L. (1934). Blood sucking among phytophagous Hemiptera. *Can. Entomol.* **66,** 97–100.

Usinger, R. L. (1944). The Triatominae of North and Central America and the West Indies and their public health significance. Public Health Bulletin No. 288. US Govt. Printing Office, Washington, DC.

Usinger, R. L. (1966). Monograph of Cimicidae (Hemiptera-Heteroptera). Thomas Say Foundation, Vol. 7. Entomological Society of America, College Park, MD.

Welch, K. A. (1990). First distributional records of *Cimexopsis nyctalis* List (Hemiptera: Cimicidae) in Connecticut. *Proc. Entomol. Soc. Wash.* **92,** 811.

Woo, P. T. K. (1991). Mammalian trypanosomiasis and piscine crytobiosis in Canada and the United States. *Bull. Soc. Vector Ecol.* **16,** 25–42.

Wood, S. F. (1951). Importance of feeding and defecation times of insect vectors in transmission of Chagas' disease. *J. Econ. Entomol.* **44,** 52–54.

Zapata, M. T. G., Schofield, C. J., and Marsden, P. D. (1985). A simple method to detect the presence of live triatomine bugs in houses sprayed with residual insecticides. *Trans. R. Soc. Trop. Med. Hyg.* **79,** 558–559.

Zeledón, R., Alvarado, R., and Jirón, L. F. (1977). Observations on the feeding and defecation patterns of three triatomine species (Hemiptera: Reduviidae). *Acta Trop.* **34,** 65–77.

Zeledón, R., and Vargas, L. G. (1984). The role of dirt floors and of firewood in rural dwellings in the epidemiology of Chagas' disease in Costa Rica. *Am. J. Trop. Med. Hyg.* **33,** 232–235

# 6

# BEETLES (*Coleoptera*)

WILLIAM L. KRINSKY

*Beetles constitute* the largest order of insects but are of relatively minor public health or veterinary importance. Adults and larvae of a few species occasionally bite, but more species secrete chemicals that can irritate the skin and eyes of humans and other animals. Beetles found in stored products can cause inhalational allergies, and some species found in dung and stored products act as intermediate hosts for helminths that cause pathology in domestic and wild animals. Many dung-inhabiting beetles are beneficial in interrupting the life cycles of mammalian parasitic worms and in acting as predators or parasitoids of pestiferous flies that breed in excrement. A few beetle species are ectoparasites or mutualistic symbionts on mammals, and a few are known to temporarily invade the skin of mammals.

## TAXONOMY

The order Coleoptera is divided into four suborders: Archostemata, considered the most primitive; Adephaga, named for its carnivorous members; Myxophaga, which are algae-eaters; and Polyphaga, the largest suborder, encompassing 90% of beetle families, in which species with diverse feeding habits are grouped. The number of beetle families varies between 135 and 170, depending on whether family designations are based on larval or adult morphology (Crowson 1981, Lawrence and Newton 1995, Downie and Arnett 1996). About 112 families include species that occur in North America.

More than 300,000 species of beetles have been described, representing 30–40% of all known insects. About 25,000 species of beetles occur in the United States and Canada (White 1983, Arnett 1990). Fewer than 100 species worldwide are known to be of public health or veterinary importance. Most of these are in the suborder Polyphaga. The species that have the greatest impact on the health of human and domestic animals are in the following families: Meloidae (blister beetles), Oedemeridae (false blister beetles), Staphylinidae (rove beetles), Tenebrionidae (darkling beetles), Dermestidae (larder beetles), and Scarabaeidae (scarab or dung beetles).

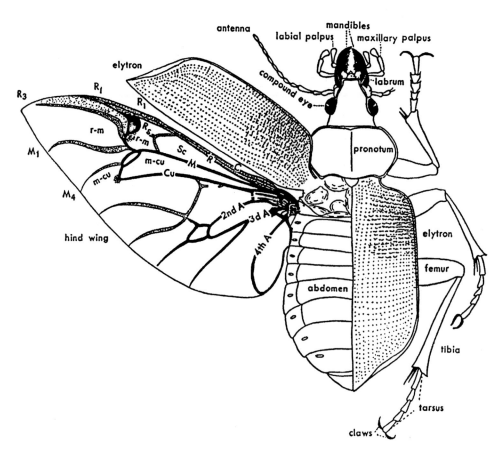

**FIGURE 6.1**  A representative adult beetle (Carabidae), dorsal view, with left elytron and wing spread. A, anal vein ($2^{nd}$, $3^{rd}$, $4^{th}$); C, costa; Cu, cubitus; M, media ($M_1$ = $1^{st}$ branch, etc.); R, radius ($R_1$, $1^{st}$ branch, etc.); $R_s$, radial sector; Sc, subcosta; m-cu, medio-cubital cross-vein; r-m, radio-medial cross-vein. (Modified from Essig, 1942.)

## MORPHOLOGY

Adult beetles are distinguished from all other insects by the presence of hardened fore wings called *elytra* (singular *elytron*) that cover and protect the membranous hind wings (Fig. 6.1). Coleoptera means "sheath-winged" in Greek. The size range of beetles is impressive, varying from 0.25 to 150 mm; however, most species are 2–20 mm long. Black and brown are the most common colors seen in the Coleoptera, but exquisite bright colors, including metallic and iridescent hues, occur, especially in tiger beetles, ground beetles, plant beetles, metallic wood-boring beetles, long-horned beetles, and lady beetles. Beetles vary in shape from elongate, flattened, or cylindrical to oval or round. Their bodies are often hardened, like the elytra, but some families, such as the blister and false blister beetles, have soft elytra and soft body parts that are pliable and sometimes described as leather-like.

The head of a beetle is usually conspicuous, and almost all beetles have some form of biting or chewing mouthparts. Even in specialized species adapted for piercing and sucking plants, the mandibles are retained and are functional. The antennae vary greatly in shape from filiform to pectinate to clavate or clubbed and are usually composed of 11 visible segments. Two compound eyes are present in most species, and ocelli are rarely present.

Part of the thorax is visible dorsally as the pronotum, just posterior to the head (Fig. 6.1). The divisions of the thorax are usually evident only on the ventral side. The legs vary greatly in shape, from thick paddles in swimming species to slender, flexible forms in running species. The paired elytra cover the folded, membranous pair of hind wings. They usually overlay the dorsum of the abdomen and often are all that is visible in the abdominal region when a beetle is viewed from above. In most species, the elytra are raised during flight. Some beetles have no hind wings and are flightless, and some beetles have very short elytra so that the abdominal tergites are visible dorsally (e.g., rove beetles and some blister beetles). Most beetles have eight visible abdominal tergites which can be seen when the elytra and hind wings are raised.

*Defensive glands* that secrete substances to repel predators are best developed in beetles in the suborder Adephaga. They are generally present as *pygidial glands* that open dorsally near the end of the abdomen. Secretions from these glands in the Adephaga are not known to cause notable ill effects in mammals. Within the Polyphaga, pygidial glands occur in a few families, such as the Tenebrionidae. In tenebrionids, the pygidial glands produce secretions that can deter small mammals and cause human skin irritation. The pygidial secretions of most other polyphagan species are not known to affect vertebrates.

Beetles that contain chemicals that are especially irritating to humans and other animals have toxic substances dispersed throughout their bodies rather than sequestered in specialized glands. The blister beetles, paederine rove beetles, and lady beetles fall within this group.

## LIFE HISTORY

All beetles exhibit *holometabolous* development. Eggs are laid singly or in clusters on or in soil, living or dead plant matter, fabrics, water, and carrion and, rarely, on living animals. The *larvae* of most beetles have a distinct head with simple eyes (ocelli) and chewing, mandibulate mouthparts, and the abdomen has 8 to 10 segments. Beetle larvae exhibit diverse morphological types, from elongate-flattened forms (*campodeiform*) to cylindrical-flattened forms (*elateriform*), caterpillar-like forms (*eruciform*), and somewhat C-shaped soft forms (*scarabaeiform*). The larval body type usually is consistent in a particular family of beetles. In a few families, however, the larval form may vary from instar to instar in a given species, a life history progression called *hypermetamorphosis*. Certain blister beetle larvae, including scavengers in bee nests and ectoparasites or endoparasites of other insects, are hypermetamorphic. They emerge from the eggs as active campodeiform larvae, then molt into eruciform and scarabaeiform stages. Most beetle larvae molt at least three times before transforming into pupae.

Although most temperate species undergo only one generation a year, species in warmer climates are often multivoltine. Depending upon the species, any developmental stage may overwinter, but overwintering most often occurs in the pupal or adult stage. Most species exhibit diapause in one or another stage, and those that have developmental cycles exceeding one warm season usually have an obligatory *diapause,* initiated by changes in photoperiod and temperature. Most adult beetles live for weeks to as long as a year. However, adults of some species may live for years, spending much of their lives in diapause during periods when food is scarce.

## BEHAVIOR AND ECOLOGY

Beetles live within all terrestrial and freshwater habitats. The great variation in beetle feeding behavior, whether saprophagous, herbivorous, carnivorous, or omnivorous, reflects the extremely diverse habitats in which these insects live. However, their mouthparts play a minor role in causing discomfort to humans and other animals. Beetle defense mechanisms, which involve the shedding or secretion of physically or chemically irritating materials, and beetle behavior that puts the insects in contact with developmental stages of parasitic helminths and vertebrates can lead to public health and veterinary problems.

Larder or pantry beetles (Dermestidae) are ubiquitous in human and domestic animal environments, where the larval and adult beetles eat stored food, food debris, dead insects, and other organic matter. Setae that cause human skin irritation or act as respiratory allergens are loosely affixed to the larvae of many species. The setae are elaborately barbed so that their firm adherence to many substrates, including human skin, causes them to be dislodged from the crawling, living larvae. In some species, the larvae actively raise the abdomen and make striking movements in response to touch. Other active defensive behaviors are seen in blister beetles, some chrysomelid plant beetles, long-horned beetles, and lady beetles that exude irritant chemicals from the femoro-tibial joints of the legs or from glandular openings around the mouthparts when the beetles are handled or threatened. This *reflex bleeding* repels predators. One of the most dramatic defensive maneuvers is the explosion of boiling hot, acrid quinones from the anal glands of carabid beetles called bombardier beetles. These forceful expulsions, which are aimed with extreme accuracy at potential predators, cause minimal damage to humans and other large animals but can cause physical and chemical burns in insects and small vertebrates (Evans 1975).

Most of the beetles that serve as *intermediate hosts* of helminths parasitic in domestic animals and humans are grain or dung feeders. These species ingest helminth eggs present in animal feces or fecal-contaminated food. Because of the proximity of the beetles to feeding animals, whole adult beetles are often incidentally ingested by potential vertebrate hosts.

The tendency of many beetles to fly to artificial lights puts them in contact with human and domestic animal habitats and increases the chances of vertebrate contact with species that may be the sources of skin irritations, allergies, or helminthic infestations.

# PUBLIC HEALTH IMPORTANCE

Human health problems caused by beetles include skin, eye, ear, and nose irritations, respiratory allergies, and minor gastrointestinal discomfort. Beetle families known to cause public health problems are listed in Table I.

The greatest human discomfort associated with beetles is caused by vesicating species that secrete irritating chemicals when the insects are handled or accidentally contact human skin or sense organs. Blister beetles, false blister beetles, some rove beetles, and some darkling beetles have these irritants in their secretions, hemolymph, or body parts. Larvae of larder beetles are covered with hairs that can act as skin or respiratory allergens.

Invasion of body tissues by beetle larvae is called *canthariasis*, whereas invasion of such tissues by adult beetles is called *scarabiasis*. These forms of infestation occur most often in tropical regions. Most clinical cases involve *enteric canthariasis* that results from the ingestion of foodstuffs infested with beetles, or the accidental ingestion of infested materials by children. Dermestid larvae, such as *Trogoderma glabrum* and *T. ornatum*, have been associated with enteric canthariasis in infants who showed signs of extreme digestive discomfort, which in one case was the result of ulcerative colitis. It is unlikely that larvae were the cause of the latter condition, although larval hairs may have exacerbated the symptoms. Larvae were recovered from the stools of these patients and from the dry cereal they ingested. Other grain-infesting beetles, such as *Tenebrio molitor* and *T. obscurus,* have been accidentally ingested without causing noticeable symptoms.

Rarely, adult and larval beetles have been recovered from human nasal sinuses, and larvae have been recovered from the urethra. Small beetles in various families

**TABLE I**

**Beetle Families of Medical/Veterinary Importance, Listed in Order of Relative Importance**

| Family | Common name | Clinical importance |
|---|---|---|
| Meloidae | Blister beetles | Cause eye irritation and blisters on skin; can poison and kill horses that ingest them. |
| Staphylinidae | Rove beetles | Paederine species cause skin and eye lesions and can poison livestock that ingest them; large species are known to bite humans; species attracted to dung feed on fly eggs, larvae, and pupae and are thereby beneficial in reducing pestiferous fly populations. |
| Scarabaeidae | Dung beetles and chafers | Spines cause irritation when adults enter ears; intermediate hosts of helminths; dung feeders are potential disseminators of pathogens; some dung feeders are beneficial in removing dung that is the source of pestiferous flies and that is infested with intermediate stages of vertebrate worm parasites. |
| Tenebrionidae | Darkling beetles and grain beetles | Cause skin and eye irritation; larvae and adults contain inhalational allergens; grain-feeding species are intermediate hosts of helminths and potential disseminators of pathogens. |
| Dermestidae | Larder beetles, pantry beetles, hide beetles, carpet beetles | Larval setae can cause skin, eye, ear, and nose irritation or gastrointestinal discomfort if ingested; larvae and adults can cause inhalational allergies; grain-feeding species are intermediate hosts of helminths; carrion-feeding species are potential disseminators of pathogens. |
| Histeridae | Hister beetles | Beneficial as predators of fly eggs and larvae developing in avian and mammalian manure. |
| Oedemeridae | False blister beetles | Cause skin and eye irritation. |
| Carabidae | Ground beetles | Intermediate hosts of poultry tapeworms. |
| Silphidae | Burying beetles or carrion beetles | Potential disseminators of pathogens. |
| Corylophidae | Minute fungus beetles | Cause eye lesions. |
| Coccinellidae | Ladybird beetles or ladybugs | Secretions can cause skin discoloration and irritation. |
| Cleridae | Checkered beetles | Can bite humans, causing temporary distress. |
| Cerambycidae | Long-horned beetles | Larger species can bite humans and other animals, causing temporary discomfort. |
| Merycidae | Old World cylindrical bark beetles | Can bite humans, causing temporary distress. |
| Curculionidae | Weevils | Grain-inhabiting species can cause inhalational allergies. |

have been known to fly or crawl into human eyes and ears. Some of these cause minor physical irritation, while others may cause extreme burning sensations, presumably due to chemicals exuded by the insects.

Painful, but temporary, eye lesions caused by tiny *Orthoperus* species (<1 mm long) in the family Corylophidae have been seen in eastern Australia, where the condition has received several names: *Canberra eye, Christmas eye,* and *harvester's keratitis.*

More than 40 species of beetles have been associated with human allergic reactions that result from inhaling beetle parts (e.g., larval setae) or excreta (Bellas 1989). Agricultural and research workers are most often affected by *inhalational allergies,* because most of the beetle species involved occur in large numbers in stored products. Dermestid beetles (*Trogoderma angustum*), tenebrionids (*Tenebrio molitor* and *Tribolium* species), and grain weevils (*Sitophilus granarius*) have been incriminated in many cases of respiratory distress, such as asthma.

Beetles serve as *intermediate hosts* for more than 50 parasitic worms, including tapeworms (Cestoda), flukes (Trematoda), roundworms (Nematoda), and thorny-headed worms (Acanthocephala) (Hall 1929, Cheng 1973). These worms primarily parasitize nonhuman hosts. Only a few species, such as the rodent tapeworms *Hymenolepis nana* and *H. diminuta* and the *Macracanthorhynchus* species of acanthocephalan parasites, occasionally infest children. The intermediate hosts of *Hymenolepis* species are grain beetles (Tenebrionidae), and *Macracanthorhynchus* species undergo development in dung beetles (Scarabaeidae). Children become infested because of their poor hygienic practices or by accidental ingestion of the beetles. Intentional ingestion of living tenebrionids for medicinal purposes in Malaysia is also a potential route for human infestation with rodent tapeworms (Chu *et al.* 1977).

Many beetles, such as scarabs, silphids, and dermestids that feed on dung and carrion have the potential to be mechanical *vectors of pathogens,* such as salmonellae and anthrax bacilli. Although there is experimental evidence for maintenance and excretion of some of these microbes by beetles, given the limited sizes of the inocula and the limited contact between humans and scavenger beetles, there is no indication that these beetles play a role in direct transmission to humans.

Many families of beetles include species known to occasionally cause bites to humans. This may happen when the beetles are accidentally handled or when the beetles occur in such large numbers that many fly or crawl onto the body. Entomologists and others who pick up beetles are the persons most often bitten. Long-horned beetles (Cerambycidae), checkered beetles (Cleridae), rove beetles (Staphylinidae), and cylindrical bark beetles (Merycidae) are among those that have been reported as biting. The somewhat painful bites usually leave few or no skin marks and do not cause any long-lasting discomfort. Long-horned beetles feed on wood in their immature stages and are found as adults on flowers, dead and dying trees, and freshly cut timber. The larger species of rove beetles that can bite are predaceous on fly larvae and are often found on carrion or dung. Cylindrical bark beetles and checkered beetles are found under bark associated with wood-boring insects or fungus.

Population increases and mass migrations of checkered beetles (Cleridae), flat grain beetles (Cucujidae), and ground beetles (Carabidae) have all caused annoyance at times by their sheer numbers and, in some cases, by the strong odors of their defensive secretions.

## MELOIDAE (BLISTER BEETLES)

Blister beetles (Fig. 6.2) occur worldwide. Most, if not all, contain the terpene *cantharidin* ($C_{10}H_{12}O_4$), which can cause skin irritations. People usually develop blisters within 24 hr of contacting the secretions of these beetles or the body fluids from crushed beetles (Fig. 6.3). Often this is accompanied by tingling or burning sensations. The blisters may progress to vesicular dermatitis with itching and oozing lesions. At least 20 species of meloids have been associated with dermatitis (Table II). Cantharidin is present in the hemolymph and in the clear, yellow secretion that is exuded at the joints of the legs of these beetles by reflex bleeding. Reptiles and some predaceous insects are repelled by the fluid. Although cantharidin is irritating to humans, the chemical acts as a meloid courtship stimulant that is secreted by male accessory glands and passed to the female during copulation. The males, being the only source of cantharidin, generally have the highest concentrations of the chemical, with levels in female beetles varying with their mating histories. The meloid spermatophore is rich in cantharidin, and the eggs also contain the substance, presumably to deter predators.

*Blister-beetle dermatitis* has been reported in Europe, Asia, Africa, North America, and Central America. The most famous blister beetle is the "*Spanish fly*" *Lytta vesicatoria* of the Mediterranean region, an insect that has been erroneously touted as a human *aphrodisiac.* Cantharidin is poisonous to humans and other animals when ingested and may cause kidney damage and death. Ingestion of powder made by grinding up dried beetles or any other source of cantharidin produces extremely toxic effects on the urogenital system. The resulting inflammation causes painful urination, hematuria, and persistent penile erection (priapism), a condition mistakenly associated with increased sexual stimulation. Like many other naturally occurring toxins, cantharidin has been prescribed for centuries as a cure for various

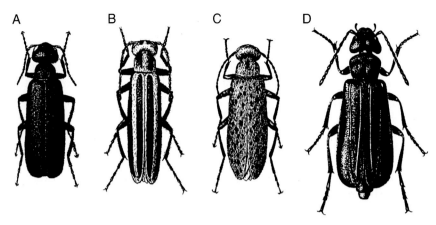

**FIGURE 6.2**   Blister beetles (Meloidae). A, Black blister beetle (*Epicauta pennsylvanica*); B, Striped blister beetle (*E. vittata*); C, Spotted blister beetle (*E. maculata*); D, European "Spanish fly" (*Lytta vesicatoria*). (A–C, modified from White, 1983; D, from Harde, 1984.)

ailments, but it has never been proven to have a therapeutic effect.

Meloid species most often associated with skin lesions in the United States and Mexico are members of the genus *Epicauta*. These include the *striped blister beetle* (*Epicauta vittata*) in the eastern states, the *black blister beetle* (*E. pennsylvanica*) found throughout most of the country, and the *spotted blister beetle* (*E. maculata*) found in the western states (Fig. 6.2). Other species of *Epicauta* cause similar problems in India and Africa (Table II). The genus *Cylindrothorax* occurs over a vast area of the Old World, including Africa, the Near East, India, and parts of Southeast Asia. African species that cause blistering include *Cylindrothorax bisignatus*, *C. dusalti*,

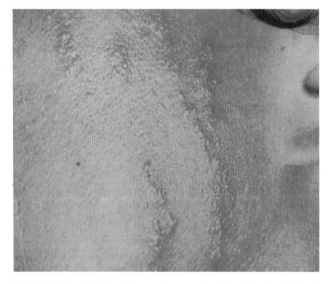

**FIGURE 6.3**   Vesicular skin reaction caused by blister beetle, left cheek of human. (From Weinberg *et al.*, 1975, reproduced with permission of the McGraw-Hill Companies.)

**TABLE II**

**Species of Meloidae Reported to Cause Blistering of Human Skin**

| Species | Geographic occurrence of clinical reports |
|---|---|
| *Lytta vesicatoria* | Europe |
| *L. phalerata* | China |
| *Epicauta cinerea* | United States (southwestern) |
| *E. flavicornis* | Senegal |
| *E. hirticornis* | India |
| *E. maculata* | United States (western) |
| *E. pennsylvanica* | United States, Mexico |
| *E. sapphirina* | Sudan |
| *E. tomentosa* | Sudan |
| *E. vestita* | Senegal |
| *E. vittata* | United States (eastern) |
| *Cylindrothorax bisignatus* | South Africa |
| *C. dusalti* | Senegal, Mali |
| *C. melanocephalus* | Gambia, Senegal |
| *C. picticollis* | Sudan |
| *C. ruficollis* | Sudan, India |
| *Mylabris bifasciata* | Nigeria |
| *M. cichorii* | India |
| *Psalydolytta fusca* | Gambia |
| *P. substrigata* | Gambia |

Modified from Alexander (1984), with information from Selander (1988).

*C. melanocephalus,* and *C. picticollis.* A *Lytta* species in China also has been associated with human dermatitis (Table II).

Blister beetles are found most often on flowers or foliage, where the beetles feed on pollen and other plant tissues. *Epicauta* species are usually abundant where grasshoppers flourish because the larvae of these meloids feed on grasshopper eggs. Most people who develop blister beetle lesions are agricultural workers or soldiers on maneuvers in areas where the beetles are common. Retention of cantharidin in frogs and birds that prey upon meloids may lead to human poisoning when these predators are used as human food. Nineteenth century medical reports of priapism in French legionnaires traced the cause of this clinical problem to the soldiers' ingestion of frogs that had eaten meloids. Humans have also developed signs of cantharidin poisoning following ingestion of cooked wild geese (Eisner *et al.* 1990).

## OEDEMERIDAE (FALSE BLISTER BEETLES)

False blister beetles in the genera *Oxycopis* (Fig. 6.4), *Oxacis* and *Alloxacis* are known to cause vesicular or bullous dermatitis in the United States, Central America, and the Caribbean region. *Sessinia kanak,* a species that is commonly attracted to lights in the Solomon Islands, and *S. lineata,* a New Zealand species, cause similar irritating lesions. Blistering has been observed in people exposed to large numbers of swarming *Eobia apicifusca* in Australia. False blister beetles are attracted to flowers, where they feed on pollen. Immediate burning of the skin following contact with *Sessinia* species swarming around coconut flowers has been reported on the Line Islands, south of Hawaii. As in meloids, *cantharidin* is the toxic substance in all of these oedemerids.

## STAPHYLINIDAE (ROVE BEETLES)

Rove beetles (Fig. 6.5) in the genus *Paederus* contain *pederin* ($C_{25}H_{45}O_9N$), a toxin more potent than that of

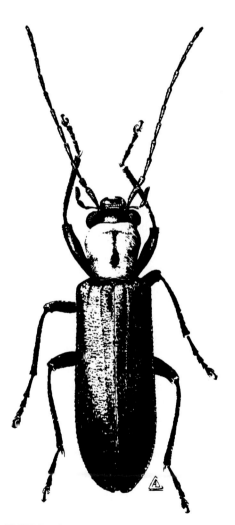

**FIGURE 6.4**  False blister beetle, *Oxycopis mcdonaldi* (Oedemeridae). (From Arnett, 1984.)

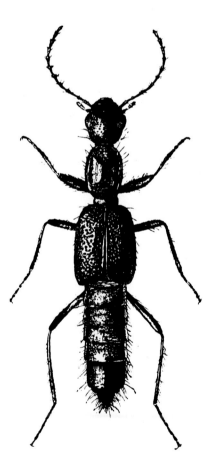

**FIGURE 6.5**  Blister beetle, *Paederus sabaeus* (Meloidae), West Africa. (From Patton and Evans, 1929.)

*Latrodectus* spider venom, and the most complex nonproteinaceous insect defensive secretion known. The beetles, which are mostly 7–13 mm long, are found in North, Central, and South America, Europe, Africa, Asia, and Australasia. Unlike most rove beetles, which are dull-colored, many *Paederus* species have an orange pronotum and orange basal segments of the abdomen, which contrast sharply with the often blue or green metallic elytra and brown or black coloration of the rest of the body. This color pattern may be a form of warning (*aposematic*) coloration, but a defensive function for pederin has not been demonstrated.

At least 20 of the more than 600 described species of *Paederus* have been associated with dermatitis (Table III). Skin reactions to the beetles, named *Ch'ing yao ch'ung*, were described in China as early as AD 739. Most cases of dermatitis have involved tropical species, including *Paederus fuscipes* (widespread from the British

TABLE III

**Species of *Paederus* (Staphylinidae) Reported to Cause Skin Lesions in Humans**

| Species | Geographic occurrence of clinical reports |
|---|---|
| *Paederus alternans* | India, Vietnam, Laos |
| *P. amazonicus* | Brazil |
| *P. australis* | Australia |
| *P. brasiliensis* | Brazil, Argentina |
| *P. columbinus* | Brazil, Venezuela |
| *P. cruenticollis* | Australia |
| *P. eximius* | Kenya |
| *P. ferus* | Argentina |
| *P. fuscipes* | Italy, Russia, Iran, India, China, Taiwan, Japan, Thailand, Vietnam, Laos, Indonesia |
| *P.* nr. *fuscipes* | Papua New Guinea |
| *P. ilsae* | Israel |
| *P.* nr. *intermedius* | Philippines |
| *P. laetus* | Guatemala |
| *P. melampus* | India |
| *P. ornaticornis* | Ecuador |
| *P. puncticollis* | Uganda |
| *P. riparius* | Russia |
| *P. rufocyaneus* | Malawi |
| *P. sabaeus* | Sierra Leone, Nigeria, Zaire, Cameroon, Namibia, Tanzania, Uganda |
| *P. signaticornis* | Guatemala, Panama |
| *P. tamulus* | China |
| *Paederus* spp. | Malaysia, Ceylon |

Modified from Frank and Kanamitsu (1987).

Isles east across Central Asia to Japan and southeast to Australia), *P. sabaeus* (Africa), *P. cruenticollis* and *P. australis* (Australasia), *P. signaticornis* (Central America), and *P. columbinus* and *P. brasiliensis* (South America). Species in South American countries are known by various names, such as *bicho de fuego, pito, potó, podó,* and *trepa-moleque.*

Unlike blister beetles, rove beetles do not exhibit reflex bleeding as a defensive reaction. Pederin contacts human skin only when a beetle is brushed vigorously over the skin or crushed. Because of their general appearance or misunderstandings about their etiology, the resulting skin lesions have been called *dermatitis linearis, spider-lick* (India and Sri Lanka), and *whiplash dermatitis.* The dermatitis may develop on any part of the body; however, exposed areas such as the head, arms, hands, and legs are most often affected. Mirror-image lesions may form where one pederin-contaminated skin surface touches another.

Unlike meloid-induced dermatitis, which develops within 18–24 hr after contact, the paederine-induced reaction of itching and burning usually occurs 24–72 hr after contact with the beetle's body fluid. The affected skin appears reddened and vesicles form about 24 hr after the initial response. The vesicles may coalesce into blisters and become purulent, producing a reaction that is often more severe than that seen following exposure to meloids. The itching may last for a week, after which the blisters crust over, dry, and peel off, leaving red marks that may persist for months. Rubbing the eyes with beetle fluid or contaminated hands, or beetles flying or crawling into eyes, can cause pain, marked swelling of the eyelids and conjunctivae, excessive lacrimation, clouding of the cornea, and inflammation of the iris (iritis). Such ocular lesions seen in East Africa have been called *Nairobi eye.* Although eye involvement often is very irritating, permanent damage is not common.

Rove beetles live in vegetable debris and under stones and other materials, such as leaf litter. They are predaceous on insects and other arthropods, or they may eat plant debris. Paederine staphylinids are most abundant in areas of moist soil, such as irrigated fields and other crop lands, where the adult beetles feed on various herbivorous insects. Consequently, agricultural workers and others working in fields and grassy areas are often affected. Because the beetles are attracted to lights, workers on brightly lit oil rigs and people occupying lighted dwellings in tropical areas are also commonly affected.

## TENEBRIONIDAE (DARKLING BEETLES)

Darkling beetles (Fig. 6.6) produce defensive secretions containing *quinones.* Adults of *Blaps* species found in

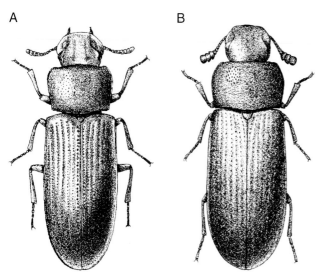

**FIGURE 6.6**   Darkling beetles (Tenebrionidae). (A) Confused flour beetle (*Tribolium confusum*); (B) Red flour beetle (*T. castaneum*). Defensive secretions containing quinones can cause skin irritation. (From Gorham, 1991.)

the Middle East and Europe secrete these chemicals that cause burning, blistering, and darkening of the skin. Adult beetles of some cosmopolitan *Tribolium* species, including *Tribolium confusum* and *T. castaneum*, have been associated with severe itching. North American desert species in the genus *Eleodes*, when threatened, take a characteristic headstand pose and exude various quinones that repel small predators and cause mild irritation to humans who handle these beetles. Darkling beetles are found in diverse habitats, including under logs and stones, in rotting wood and other vegetation, in fungi, in termite and ant nests, and among debris in and outside of homes. Most species live in dry, often desert, environments, while pest species are found in stored products, such as grain and cereals. Most tenebrionids are scavengers on decaying or dry plant material, but a few feed on living plants.

## DERMESTIDAE (LARDER BEETLES)

Larvae of larder beetles, or *pantry beetles* (Fig. 6.7), are covered with barbed and spearlike setae that may cause allergic reactions in the form of pruritic, papulovesicular skin lesions. Dermestid larvae often are found living in household furnishings, such as carpets, rugs, and upholstery, or stored clothing of individuals suffering from these reactions. Larder beetles are named for their common occurrence as pantry pests, but they may also be found in grain storage facilities, in bird and mammal nests and burrows, and on carrion. The larvae and adults are mostly scavengers on decaying or dry plant and animal matter.

Dermestid larvae and adults are known to have crawled into human ears, causing itching and pain. The spearheaded setae of dermestid larvae have been observed on numerous occasions on cervical (Papanicolaou) smear slides and in sputum samples. In all of these cases the setae appear to have been contaminants that were not associated with any pathological changes in the patients (Bryant and Maslan 1994).

## SCARABAEIDAE (SCARAB BEETLES)

In some tropical regions where human and animal excrement are abundant in the vicinity of dwellings, scarab beetles living in the dung are sometimes accidentally ingested by young children. These beetles appear in the newly passed stools of children and may disperse from the excrement in a noisy fashion that has been described in Sri Lanka as *beetle marasmus* ("kurumini mandama"). Although some of these beetles may infest the fecal matter as it is passed or after it reaches the ground, it is quite likely, as local physicians claim, that the scarab beetles (e.g. *Copris* spp., Fig. 6.8) pass through the alimentary tract and remain alive, causing little or no discomfort to the children. Evidence for such durability among the scarabs comes from cases in which frogs, horses, and cattle have ingested scarabs, which then worked their way through the stomach wall and remained alive until the hosts were killed. In Asia and Africa, humans sometimes become infested with dung beetles (*Onthophagus* and *Caccobius* spp.) when these scarabs enter the anus and live within the rectum, causing physical discomfort and damage to the mucosa.

Large numbers of the adult scarab beetles *Cyclocephala borealis* and *Autoserica castanea* invaded the ears of 186 boy scouts sleeping on the ground at a jamboree in Pennsylvania in 1957. The beetles caused pain and some slight bleeding as a result of the tearing action of their tibial spines. After the beetles were removed, there were very few cases of secondary infection (Mattuck and Fehn 1958).

## COCCINELLIDAE (LADY BEETLES)

Lady beetles, also called *ladybird beetles*, have been cited most often as causing prickling or slight stinging sensations, followed by the formation of mild erythematous lesions. These beetles may nip at the skin; however, given their small size, it is more likely that their defensive secretions cause the discomfort. The alkaloid secretions produced by reflex bleeding from the legs and around the

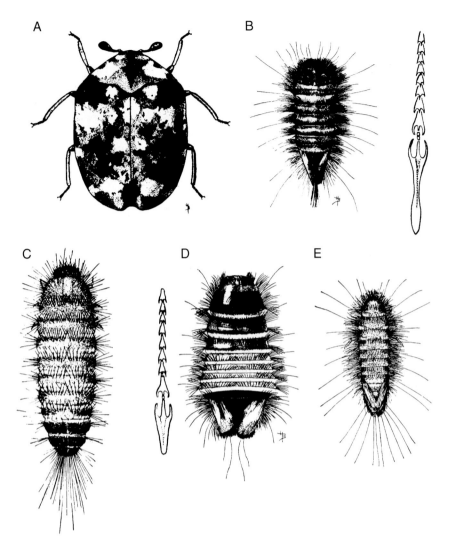

**FIGURE 6.7**    Dermestid beetles (Dermestidae). (A) Furniture carpet beetle (*Anthrenus flavipes*), adult; (B) same, larva, with hastate seta (enlarged); (C) Cabinet beetle (*Trogoderma* sp.), adult; (D) Varied carpet beetle (*A. verbasci*), larva, with hastate seta (enlarged); (E) Common carpet beetle (*A. scrophulariae*), larva. (From Sweetman, 1965.)

mouthparts do stain human skin and can cause mild irritation. Lady beetles are common in gardens in warm weather, where the beetles feed on aphids and scale insects on a variety of plants. Lady beetles also occur in large aggregations as they are crawling and flying from overwintering clusters.

## VETERINARY IMPORTANCE

Although not generally appreciated even by entomologists, beetles are involved in a variety of problems of a veterinary nature. These include toxicity to domestic animals on ingestion, mechanical transmission of disease agents,

intermediate hosts for helminthic parasites, direct injury to animals by ectoparasitic species, and structural damage to poultry facilities. Beetles also can be beneficial by playing an important role in the recycling of animal dung and as natural control agents, especially for dung-breeding flies.

### INGESTION OF TOXIC BEETLES

Several *blister beetles* in the family Meloidae pose a hazard to livestock that feed on forage in which the living beetles are abundant. Horses that ingest quantities of these beetles are especially susceptible to *cantharidin poisoning*. Dead beetles and beetle parts retain their cantharidin

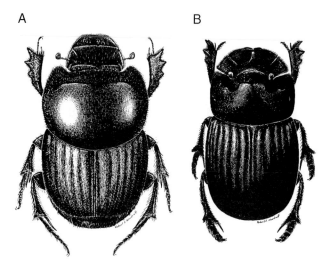

FIGURE 6.8 Scarab beetles (Scarabaeidae). (A) *Onthophagus polyphemi;* (B) *Copris minutus.* (From Woodruff, 1973.)

content so that forage crops that are harvested for livestock feed continue to be a source of the toxin. Blister beetles in alfalfa (*Medicago sativa*) fields contain enough cantharidin to provide lethal doses to horses that feed on this material when it is used as hay. Species that pose problems in the United States include the *striped blister beetle* (*Epicauta vittata*), the *black blister beetle* (*E. pennsylvanica*) (Fig. 6.2), the *margined blister beetle* (*E. pestifera*), and the *three-striped blister beetle* (*E. lemniscata*), as well as *E. fabricii, E. occidentalis,* and *E. temexa.* Individual beetles contain <0.1 to >11 mg of cantharidin, equivalent to <0.1 to >12% of their dry weights, with males of several species averaging >5%. The minimum lethal dose for a horse is about 1 mg/kg, which means that depending on the size of a horse and the cantharidin content of the beetles ingested, anywhere from 25 to 375 beetles are sufficient to cause death (Capinera *et al.* 1985).

Horses have been poisoned by eating forage in the southern, midwestern, and western United States. Poisoning is not limited to a particular geographic area because contaminated forage is transported over long distances. Affected animals exhibit moderate to severe clinical signs, ranging from depression to shock that may be followed by death. Abdominal distress (colic), anorexia, depression, and oral irritation are commonly observed. These signs are accompanied by markedly decreased serum calcium and magnesium. Accelerated heart rate (tachycardia), increased respiratory rate (tachypnea), and increased creatinine kinase activity are indicative of severe toxicosis that is likely to lead to death. In most cases in which death has occurred, a horse has succumbed within 48 hr of the onset of clinical signs. Horses are

the most commonly affected because the kinds of forage they are fed are most often contaminated with beetles. Although the ruminant digestive tract is less susceptible to cantharidin poisoning, goats, sheep, and cattle have also died from cantharidin toxicosis.

Horses and cattle have also been poisoned by *pederin* following ingestion of the tropical staphylinid *P. fuscipes.* Pederin can cause severe damage to the mucosa of the alimentary tract.

Historically, there are reports of chickens, ducklings, goslings, and young turkeys dying as a direct result of ingesting the *rose chafer* (*Macrodactylus subspinosus*), a member of the Scarabaeidae (Lamson 1922). Although this North American species is abundant in the summer months, modern enclosed poultry production facilities may have greatly reduced the incidence of such poisonings.

## TRANSMISSION OF PATHOGENS

Darkling beetles (Tenebrionidae) inhabiting farm buildings can be mechanical vectors of animal pathogens. Tenebrionid beetles infesting feed in chicken houses may become infected with *Salmonella* bacteria passed in feces from infected chickens. Both larval and adult forms of the *lesser mealworm beetle* (*Alphitobius diaperinus*) (Fig. 6.9) have been found to maintain viable pathogens (e.g., *Salmonella typhimurium* and *S. chester*) on their external surfaces and in their digestive tracts. The bacteria survive for days after infection and are disseminated via beetle excreta. In chicken-breeding facilities, grain beetles are potential disseminators of these pathogens

FIGURE 6.9 Lesser mealworm, *Alphitobius diaperinus* (Tenebrionidae), larvae and adults in poultry litter. (Photo by G. R. Mullen.)

that can infect both chicks and adult birds. These organisms can cause gastroenteritis in human consumers. The lesser mealworm also is regarded as a potential disseminator of other bacteria (*Escherichia, Bacillus, Streptococcus*), fungi (*Aspergillus*), and the viruses causing *Marek disease, Newcastle disease, fowlpox, avian leucosis,* and *infectious bursitis* (Gumboro disease). In addition, oocysts formed by protozoans in the genus *Eimeria* are ingested by lesser mealworm beetles and when the infected beetles are ingested by birds, the latter develop *avian coccidiosis,* a serious disease of poultry.

## INTERMEDIATE HOSTS OF PARASITES

Tapeworms (cestodes), flukes (trematodes), roundworms (nematodes), and thorny-headed worms (acanthocephalans) of many species that infest domestic and wild animals use beetles as intermediate hosts. Animals become infested by ingesting parasitized beetles that contaminate feed or bedding (tenebrionids, carabids) or that are attracted to animal dung (scarabaeids), or by ingesting water in which infective beetles have disintegrated.

Two *tapeworms* that infest the small intestines of poultry are the *broad-headed tapeworm* (*Raillietina cesticillus*) and *Choanotaenia infundibulum*. Both parasites cause enteritis and hemorrhaging in chickens, turkeys, pheasants, and guinea fowl. A few tenebrionids and scarabaeids and more than 35 species of carabid beetles, notably in the genera *Amara* and *Pterostichus,* are intermediate hosts for *R. cesticillus* (Cheng 1973). Some tenebrionid and dermestid species, including the lesser mealworm beetle, are intermediate hosts for *C. infundibulum*. Proglottids or tapeworm eggs ingested by beetle larvae or adults develop into cystercerci (encysted larvae) that can then infest birds that eat the beetles. Chicks are most susceptible to serious infestations and often die from worm burdens.

The *beef tapeworm* (*Taenia saginata*) can use dung beetles and carabids as intermediate hosts, although they are not essential for transmission. Beetles associated with infective dung or debris can ingest proglottids or eggs, as in the case of poultry worms. Cattle and humans infested with the tapeworm may show mild symptoms such as weight loss, abdominal pain, and increased appetite.

The *dwarf tapeworms* (*Hymenolepis nana* and *H. diminuta*) that usually infest rodents, especially rats and mice, can infest humans when the intermediate host beetles are accidentally ingested. *Tenebrio molitor* may act as an intermediate host for *H. nana,* although this worm is readily transmitted directly from one vertebrate host to another. Several species of tenebrionids (*Tenebrio* spp. and *Tribolium* spp.) are required intermediate hosts for *H. diminuta*. Larval and adult beetles infesting grain and cereals ingest worm eggs that develop into cysticercoid stages that infest rodents or humans, usually children, who ingest the beetles. Dwarf tapeworms produce minimal symptoms in rodents and people, although heavy infestations in children may cause abdominal pain, diarrhea, convulsions, and dizziness.

Beetles are known to be intermediate hosts for only a few *trematodes*. These are parasites of frogs that become infested by ingesting parasitized dytiscid beetles and pose no problem for other vertebrate animals.

Many *nematodes* infest livestock and wildlife, but only a few use beetles as intermediate hosts. Spirurid nematodes of various species infest livestock and, rarely, humans. *Physocephalus sexalatus* and *Ascarops strongylina* eggs develop in many species of scarabaeid dung beetles (*Geotrupes* spp., *Onthophagus* spp., and *Scarabaeus* spp.) that then may be ingested by pigs. Both wild and domestic swine can be infested with these stomach worms, which can cause digestive problems in heavily infested young animals. *Gongylonema pulchrum* is a parasite of the upper digestive tract of sheep, cattle, goats, and other ruminants, as well as horses, dogs, and humans. The worms burrow in the mucosa and submucosa of the oral cavity and esophagus and may cause bleeding, irritation, numbness, and pain in the mouth and chest. Scarabaeid and tenebrionid beetles serve as intermediate hosts for the larvae. *Physaloptera caucasica,* another spirurid, often parasitizes monkeys in tropical Africa, where humans are also commonly infested. This nematode causes digestive distress by infesting the alimentary tract from the esophagus to the terminal ileum. Scarabaeid dung beetles are its intermediate hosts.

The *acanthocephalans,* aptly named for their thorny heads, include species found worldwide infesting swine, rodents, and carnivores, such as dogs. *Macracanthorhynchus hirudinaceus,* which attaches to the small intestines of swine, causes enteritis and produces intestinal nodules that lower the value of these tissues when they are sold to make sausage casings. Eggs of this parasite are ingested by scarab beetle larvae of species of various genera (*Phyllophaga, Melolontha, Lachnosterna, Cetonia, Scarabaeus,* and *Xyloryctes*), including May and June beetles, leaf chafers, dung beetles, and rhinoceros beetles. Infested beetle larvae, as well as the pupae and adults that develop from them, are infective to both pigs and humans. Humans and pigs often show no symptoms. However, in cases of heavy infestations, both human and porcine hosts may experience digestive problems, such as abdominal pain, loss of appetite, and diarrhea, which can lead to emaciation.

Two other acanthocephalan worms that parasitize the small intestines of their hosts use scarab beetles or tenebrionids as intermediate hosts. They are *M. ingens,* which infests raccoons and occasionally dogs and humans, and *Moniliformis moniliformis,* a parasite of rodents and dogs.

## Nest Associates and Ectoparasites

In addition to those beetles that occasionally invade the alimentary tracts or sense organs of animals, other species have evolved in close association with mammals as nest dwellers or ectoparasites. These species typically have reduced eyes and wings or have lost these structures completely. The family Leptinidae, known as mammal nest beetles, includes *Platypsyllus* spp., which live as larvae and adults on beavers, and *Leptinus* and *Leptinillus* spp., which live as adults on various small rodents. These species feed on skin debris or glandular secretions and have been associated with skin lesions on their hosts. Two other beetle groups, the Staphylinidae (Amblyopinini) and the Languriidae (Loberinae), have species that live on rodents and a few other mammals. These staphylinids, and possibly the languriids, appear to be mutualistically symbiotic with their mammalian hosts. The beetles infest mammalian fur to gain access to their prey, which are mammalian ectoparasites, such as fleas and mites, that live in rodent nests (Ashe and Timm 1987, Durden 1987).

Some scarabaeid beetles are adapted to living in the fur around the anus of certain mammals. These beetles cling to the fur except when they leave to oviposit in the dung. Some of these scarabs (*Trichillium* spp.) are found on sloths and monkeys in South America and others (*Macropocopris* spp.) on marsupials in Australia.

Larvae and adults of the *lesser mealworm beetle* (*Alphitobius diaperinus*) have been found boring into and living in the scrotum of a rat, and feeding on sick domestic chicks and young pigeons. Similarly, the *hide beetle* (*Dermestes maculatus*) can feed on living poultry and has caused deep wounds in adult turkeys. In laboratory experiments, lesser mealworm beetles killed snakes and a salamander, all of which were devoured by the mealworms. The voracious and aggressive behavior of this commonly abundant tenebrionid makes it a significant pest in poultry houses.

In addition to their direct attacks on birds, the lesser mealworm and hide beetle larvae are major causes of structural damage to poultry houses. After reaching their final instar, the larvae migrate into the insulation of poultry houses to seek pupation sites. The larval tunnels and holes produced in insulation and wood framing cause enough damage to alter temperature regulation in the houses, which reduces the efficiency of poultry production (Axtell and Arends 1990).

## Dung Beetles and Biocontrol

Many beetles that are attracted to avian and mammalian excrement should be viewed as beneficial insects. *Scarabaeid beetles* (including coprines) remove large quantities of mammalian dung by scattering or burying the material during feeding or reproduction. The rapid removal of dung helps reduce the development of parasitic worms and pestiferous cyclorrhaphan flies that require dung for their survival and reproduction, and it also opens up grazing land that would be despoiled by the rotting excrement. *Staphylinid beetles* and *histerid beetles* that are attracted to mammalian and avian dung directly reduce muscoid fly populations by feeding on the immature stages of these flies and indirectly reduce these populations by introducing their phoretic mites, which prey upon fly eggs in the excrement.

Dung beetle diversity is greatest in tropical regions, such as Africa, with its abundance of herbivores. More than 2000 scarabaeid species in many genera (e.g., *Onthophagus, Euoniticellus,* and *Heliocopris*) are known to feed and reproduce in dung in Africa. Less diverse dung feeders, such as *Aphodius, Onthophagus, Canthon,* and *Phanaeus* species, provide the same benefits in the United States. In Australia, the development of extensive cattle farming resulted in the production of millions of tons of dung that was not naturally removed, because the native coprophagous beetles were adapted to feeding only on marsupial dung. Within the last few decades, introductions of African beetles by sterile breeding programs have established several coprine species that have helped to open up grazing lands ruined by dung accumulation and to reduce the breeding source of the pestiferous bush fly (*Musca vitustissima*).

*Staphylinid beetles* of several species in the genus *Philonthus* feed as both larvae and adults on fly larvae living in animal excrement. These beetles are maintained as components in biological control programs against the face fly and horn fly. Staphylinid species of *Aleochara* are also helpful in reducing dung-breeding fly populations because the parasitoid larvae of these beetles penetrate fly puparia and destroy the fly pupae. *Histerid beetles,* especially *Carcinops* species, are found in confined animal production facilities, such as poultry houses. The larvae and adults of these beetles feed on eggs and larvae of muscoid flies. Any beetle species observed in animal production facilities should be identified to assess whether its presence is beneficial or detrimental to the maintenance of sanitary conditions and animal health.

## PREVENTION AND CONTROL

Preventing public health and veterinary problems associated with beetles requires education about which species are harmful. Recognition of meloid, paederine, and oedemerid beetles allows one to immediately wash skin surfaces and eyes that come into contact with the beetles, thereby removing the chemicals that cause dermatitis or inflammation of the eyes. With the exception of the smallest species, beetles that are attracted to lights may be prevented from reaching humans or other animals by using screens and bed netting.

Control of vesicating beetles occurring in natural and cultivated vegetation can be achieved with pesticides; however, the wide area over which these chemicals must be broadcast generally makes such control impractical. Human exposure can be prevented by combining education about the problem with personal protective measures and removal of extraneous vegetation and decaying organic matter from around agricultural fields and dwellings.

Prevention of blister beetle toxicosis of farm animals involves care in the handling of forage crops. Harvesting hay at times when meloid beetles are rare, such as in late fall in temperate climates, helps prevent contamination of dried, stored forage with dead beetles. Similarly, harvesting alfalfa before it produces the blooms that attract meloids, or raking hay more frequently after it is cut and allowing it to dry longer before it is conditioned or crimped, will allow beetles to leave the hay before it is baled.

Preventing and controlling dissemination of pathogens and transmission of helminths of veterinary importance can be achieved by a combination of strict sanitary and cultural practices. Removal of dung and organic waste from animal enclosures, as well as sterilization of manure before it is used as fertilizer, helps to interrupt the transmission cycle of parasites by reducing the chances of beetles ingesting worm eggs. Rotation of pastured animals also can limit contact between the definitive hosts and intermediate beetle hosts. Increased abundance of scarabaeid dung beetles that aids in the rapid removal of dung, by both ingestion and burial, has been found beneficial in reducing infestations with intestinal nematodes that do not use beetles as intermediate hosts, but that are transmitted from animal to animal by dung ingestion (Fincher 1975).

Control of destructive poultry-house beetles requires constant monitoring for the insects and strict sanitation.

Pesticides provide only temporary control and are most beneficial when applied to soil that larvae may burrow in to pupate. Careful personal hygiene, use of gowns and masks in beetle-rearing facilities, and regular vacuuming of floors, floor coverings, and furniture in domestic settings help prevent exposure to dermestids and other beetles that can cause allergic responses.

## REFERENCES AND FURTHER READING

Alexander, J. O. (1984). Arthropods and human skin. Springer-Verlag, Berlin.

Archibald, R. G., and King, H. H. (1919). A note on the occurrence of a coleopterous larva in the urinary tract of man in the Anglo-Egyptian Sudan. *Bull. Entomol. Res.* **9**, 255–256.

Arnett, R. H., Jr. (1973). The beetles of the United States (a manual for identification). American Entomological Institute, Ann Arbor.

Arnett R. H., Jr. (1984). The false blister beetles of Florida (Coleoptera: Oedemeridae). Fla. Dept. Agric. & Consumer Serv., Entomol. Circ. 259.

Arnett, R. H., Jr. (1990). Present and future of systematics of the Coleoptera in North America. *In:* Systematics of the North American insects and arachnids: status and needs (M. Kosztarab and C. W. Schaefer, eds). pp. 165–173. Virginia Polytechnic Institute and State University, Blacksburg.

Ashe, J. S., and Timm, R. M. (1987). Predation by and activity patterns of 'parasitic' beetles of the genus *Amblyopinus* (Coleoptera: Staphylinidae*). J. Zool. Lond.* **212**, 429–437.

Avancini, R. M. P., and Ueta, M. T. (1990). Manure breeding insects (Diptera and Coleoptera) responsible for cestoidosis in caged layer hens. *J. Appl. Entomol.* **110**, 307–312.

Axtell, R. C., and Arends, J. J. (1990). Ecology and management of arthropod pests of poultry. *Annu. Rev. Entomol.* **35**, 101–126.

Barrera, A. (1969). Notes on the behaviour of *Loberopsyllus traubi*, a cucujoid beetle associated with the volcano mouse, *Neotomodon alstoni* in Mexico. *Proc. Entomol. Soc. Wash.* **71**, 481–486.

Bellas, T. E. (1989). Insects as a cause of inhalational allergies: a bibliography 1900–1987. Canberra: CSIRO Div. Entomol.

Blodgett, S. L., and Higgins, R. A. (1990). Blister beetles (Coleoptera: Meloidae) in Kansas alfalfa: influence of plant phenology and proximity to field edge. *J. Econ. Entomol.* **83**, 1042–1048.

Blume, R. R. (1985). A checklist, distributional record, and annotated bibliography of the insects associated with bovine droppings on pastures in America north of Mexico. *Southwest. Entomol. Suppl.* **9**, 1–55.

Bryant, J., and Maslan, A. M. (1994). Carpet beetle larval parts in Pap smears: report of two cases. *South. Med. J.* **87**, 763–764.

Capinera, J. L., Gardner, D. R., and Stermitz, F. R. (1985). Cantharidin levels in blister beetles (Coleoptera: Meloidae) associated with alfalfa in Colorado. *J. Econ. Entomol.* **78**, 1052–1055.

Cheng, T. C. (1973). General parasitology. Academic Press, New York.

Chu, G. S. T., Palmieri, J. R., and Sullivan, J. T. (1977). Beetle-eating: a Malaysian folk medical practice and its public health implications. *Trop. Geogr. Med.* **29**, 422–427.

Clausen, C. P. (1940) (reprinted 1972). Entomophagous insects. Hafner, New York.

Crook, P. G., Novak, J. A., and Spilman, T. J. (1980). The lesser mealworm, *Alphitobius diaperinus*, in the scrotum of *Rattus norvegicus*, with notes on other vertebrate associations (Coleoptera, Tenebrionidae; Rodentia, Muridae). *Coleopts. Bull.* **34**, 393–396.

Crowson, R. A. (1981). The biology of the coleoptera. Academic Press, London.

De las Casas, E., Harein, P. K., Deshmukh, D. R., and Pomeroy, B. S. (1976). Relationship between the lesser mealworm, fowl pox, and Newcastle disease virus in poultry. *J. Econ. Entomol.* **69**, 775–779.

Downie, N. M., and Arnett, R. H. (1996). The beetles of Northeastern North America. Vol. 1 and 2. Sandhill Crane Press, Gainesville.

Durden, L. A. (1987). Predator-prey interactions between ectoparasites. *Parasitol. Today* **3**, 306–308.

Eisner, T., Conner, J., Carrel, J. E., McCormick, J. P., Slagle, A. J., Gans, C., and O'Reilly, J. C. (1990). Systematic retention of ingested cantharidin in frogs. *Chemoecology* **1,2**, 57–62.

Essig, E. O. (1942). College entomology. MacMillan Company, New York.

Evans, G. (1975). The life of beetles. Hafner, New York.

Fincher, G. T. (1975). Effects of dung beetle activity on the number of nematode parasites acquired by grazing cattle. *J. Parasitol.* **61**, 759–766.

Fincher, G. T. (1994). Predation on the horn fly by three exotic species of *Philonthus. J. Agric. Entomol.* **11**, 45–48.

Frank, J. H., and Kanamitsu, K. (1987). *Paederus,* sensu lato (Coleoptera: Staphylinidae): natural history and medical importance. *J. Med. Entomol.* **24**, 155–191.

Geden, C. J., Stinner, R. F., and Axtell, R. C. (1988). Predation by predators of the house fly in poultry manure: effects of predator density, feeding history, interspecific interference and field conditions. *Environ. Entomol.* **17**, 320–329.

Guglick, M. A., Macallister, C. G., and Panciera, R. (1996). Equine cantharidiasis. *Compend. Cont. Educ. Pract. Vet.* **18**, 77–83.

Hall, M. C. (1929). Arthropods as intermediate hosts of helminths. Smithsonian, Washington, DC.

Hanski, I., and Cambefort, Y. (eds.) (1991). Dung beetle ecology. Princeton University Press, Princeton, NJ.

Harde, K. W. (1984). A field guide in colour to beetles. Octopus Books Ltd., London.

Helman, R. G., and Edwards, W. C. (1997). Clinical features of blister beetle poisoning in equids: 70 cases (1983–1996). *J. Am. Vet. Med. Assoc.* **211**, 1018–1021.

Lamson, G. H., Jr. (1922). The rose chafer as a cause of death of chickens. *Storrs Agric. Exp. Sta. Bull.* **110**, 118–135.

Lawrence, J. F., and Newton, A. F. (1995). Families and subfamilies of Coleoptera. *In:* Biology, phylogeny, and classification of Coleoptera: papers celebrating the 80th birthday of Roy A. Crowson (J. Pakaluk and S. A. Slipinski, eds.). pp. 779–1006. Muzeum i Instytut Zoologii PAN, Warsaw.

Legner, E. F. (1995). Biological control of Diptera of medical and veterinary importance. *J. Vector Ecol.* **20**: 59–120.

Liggett, H. (1931). Parasitic infestations of the nose. *J. Am. Med. Assoc.* **96**, 1571–1572.

Marshall, A. G. (1981). The ecology of ectoparasitic insects. Academic Press, London.

Mattuck, D. R., and Fehn, C. F. (1958). Human ear invasions by adult scarabaeid beetles. *J. Econ. Entomol.* **51**, 546–547.

McAllister, J. C., Steelman, C. D., and Skeeles, J. K. (1994). Reservoir competence of the lesser mealworm (Coleoptera: Tenebrionidae) for *Salmonella typhimurium* (Eubacteriales: Enterobacteriaceae). *J. Med. Entomol.* **31**, 369–372.

Patton, W. S., and Evans, A. W. (1929). Insects, ticks, mites and Venomous animals. School of Tropical Medicine, Liverpool.

Rajapakse, S. (1981). Letter from Sri Lanka: beetle marasmus. *Br. Med. J.* **283**, 1316–1317.

Samish, M., Argaman, Q., and Perlman, D. (1992). The hide beetle, *Dermestes maculatus* DeGeer (Dermestidae), feeds on live turkeys. *Poultry Sci.* **71**, 388–390.

Schmitz, D. G. (1989). Cantharidin toxicosis in horses. *J. Vet. Intern. Med.* **3**, 208–215.

Schroeckenstein, D. C., Meier-Davis, S., and Bush, R. K. (1990). Occupational sensitivity to *Tenebrio molitor* Linnaeus (yellow mealworm). *J. Allergy Clin. Immunol.* **86**, 182–188.

Selander, R. B. (1988). An annotated catalog and summary of bionomics of blister beetles of the genus *Cylindrothorax* (Coleoptera: Meloidae). *Trans. Am. Entomol. Soc.* **114**, 15–70.

Southcott, R. V. (1989). Injuries from Coleoptera. *Med. J. Aust.* **151**, 654–659.

Sweetman, H. L. (1965). Recognition of structural pests and their damage. Wm. C. Brown Co., Dubuque.

Théodoridès, J. (1950). The parasitological, medical and veterinary importance of Coleoptera. *Acta Trop.* **7**, 48–60.

Waterhouse, D. F. (1974). The biological control of dung. *Sci. Am.* **230**, 100–109.

Weatherston, J., and Percy, J. E. (1978). Venoms of Coleoptera. *In:* Arthropod venoms (S. Bettini, ed.). pp. 511–554. Springer-Verlag, Berlin.

Weinberg, S., Leider, M., and Shapiro, L. (1975). Color atlas of pediatric dermatology. McGraw-Hill Book Company, New York.

White, R. E. (1983). A field guide to the beetles of North America. Houghton Mifflin, Boston.

Whitmore, R. W., and Pruess, K. P. (1982). Response of pheasant chicks to adult lady beetles (Coleoptera: Coccinellidae). *J. Kansas Entomol. Soc.* **55**, 474–476.

Woodruff, R. E. (1973). Scarab beetles of Florida (Coleoptera: Scarabaeidae), Part 1. Arthropods of Florida, vol. 8, Division of Plant Industry, Florida Department of Agriculture, Gainesville.

# 7

# FLEAS (*Siphonaptera*)

LANCE A. DURDEN AND ROBERT TRAUB

*Fleas are* morphologically unique ectoparasites that are unlikely to be confused with any other arthropods. They are a monophyletic group that has evolutionary ties with the mecopteroid insect orders Mecoptera and Diptera. Fleas evolved from winged ancestors during the late Jurassic or early Cretaceous Period 125–150 million years ago in parallel with marsupial and insectivore hosts. As a group, they have principally evolved as parasites of mammals on which 94% of known species feed, representing 15 families and more than 200 genera of fleas. The remaining 6%, representing 5 families and 25 genera, are ectoparasites of birds.

Coevolution has molded many host-flea associations, as reflected by their host specificity and the morphological adaptations of some fleas that conform to the morphology of the host skin, fur, or feathers. Although many flea species do not cause significant harm to their hosts in nature, most species that feed on humans and their companion animals are of medical or veterinary importance.

## TAXONOMY

There are approximately 2500 species and subspecies of fleas that are currently placed in 15 families and 220 genera (Lewis 1993a, 1998). Many of these species have been catalogued by Hopkins and Rothschild (1953–1971), Mardon (1981), Traub *et al.* (1983), and Smit (1987). Except for the work by Traub *et al.* (1983), these works are part of an eight-volume series published by the British Museum (Natural History), now The Natural History Museum, London. Another series of publications that addresses the geographical distribution, host preferences, and classification of the world flea fauna are those of Lewis (1972–1993a). A publication by Ewing and Fox (1943) on North American fleas is largely outdated, and no modern text covering the fleas of this region has been published. However, Holland (1985) has produced an excellent guide to the fleas of Canada, Alaska, and Greenland which can be used for those species that also occur in the continental United States. An earlier work by Fox (1940) and a key by Benton (1983) are useful for identifying fleas from the eastern United States; Benton (1980) also has provided an atlas outlining the distribution of the fleas of this region. An older work by Hubbard (1947) addresses the fleas of western North America, whereas Lewis *et al.* (1988) provide a guide to the fleas of the Pacific Northwest. An updated series of identification guides for North American fleas has been initiated (Lewis and Lewis 1994). Although flea larvae are usually difficult to assign to genus or species, Elbel (1991) provides a useful guide for identifying the larval stages of some flea taxa.

Most fleas of medical or veterinary importance are members of the families Ceratophyllidae, Leptopsyllidae, Pulicidae, or Vermipsyllidae. Occasionally members of other families, notably the Ctenophthalmidae and Rhopalopsyllidae, also feed on humans and domestic animals. Table I shows the family-level classification for the flea species discussed in this chapter.

TABLE I

**Classification of Flea Species Mentioned in the Text**

**Family Ceratophyllidae**
  *Ceratophyllus gallinae* (European chicken flea)
  *C. niger* (western chicken flea)
  *Nosopsyllus fasciatus* (northern rat flea)
  *Orchopeas howardi*
  *Oropsylla montana*

**Family Ctenophthalmidae**
  *Stenoponia tripectinata*

**Family Ischnopsyllidae**
  *Myodopsylla insignis*

**Family Leptopsyllidae**
  *Leptopsylla segnis* (European mouse flea)

**Family Pulicidae**
  *Cediopsylla simplex* (rabbit flea)
  *Ctenocephalides canis* (dog flea)
  *C. felis* (cat flea)
  *Echidnophaga gallinacea* (sticktight flea)
  *E. larina*
  *E. myrmecobii*
  *Euhoplopsyllus glacialis*
  *Hoplopsyllus anomalus*
  *Pulex irritans* (human flea)
  *P. simulans*
  *Spilopsyllus cuniculi* (European rabbit flea)
  *Tunga monositus*
  *T. penetrans* (chigoe)
  *Xenopsylla astia*
  *X. bantorum*
  *X. brasiliensis*
  *X. cheopis* (Oriental rat flea)

**Family Pygiopsyllidae**
  *Uropsylla tasmanica*

**Family Vermipsyllidae**
  *Dorcadia ioffi*
  *Vermipsylla alakurt* (alakurt flea)

Modified from Lewis, 1993a.

Flea classification is based almost exclusively on the chitinous morphology of cleared adult specimens. Male fleas probably have the most complex genitalia in the animal kingdom, and the morphology of the sclerotized parts of these organs is important in most systems of flea classification. Although various classification schemes have been proposed for fleas, one that is widely used today is detailed by Lewis (1998). In this classification, the order Siphonaptera is divided into 15 families, the larger of which are the Ceratophyllidae (540 species), Ctenophthalmidae (744 species), Ischnopsyllidae (135 species), Leptopsyllidae (346 species), Pulicidae (207 species), Pygiopsyllidae (185 species), and Rhopalopsyllidae (145 species).

## MORPHOLOGY

Adult fleas are small (1–8 mm), wingless, almost invariably bilaterally compressed, and heavily chitinized (Fig. 7.1). Many species bear one or more combs, or *ctenidia*, each appearing as a row of enlarged, sclerotized spines (Figs. 7.1 to 7.3). A comb on the ventral margin of the head is called a *genal ctenidium,* whereas a comb on the posterior margin of the prothorax is called a *pronotal ctenidium.* Additional cephalic or abdominal ctenidia occur in some fleas. Smaller rows of specialized setae or bristles adorn various body regions of many fleas. The nature of the ctenidia and specialized setae often reflect the vestiture or habits of the host, especially in host-specific fleas. They aid in preventing dislodgement of fleas from the hair or feathers of the host. It also has been suggested that ctenidia may protect flexible joints.

An important sensory feature of adult fleas is the *sensilium* (the *pygidium* of older works), present on abdominal tergum 9 or 10 (Fig. 7.1). This sensory organ aids fleas in detecting air movement, vibrations, and temperature gradients; in some species it also facilitates copulation. It plays an important role in host detection and in initiating escape responses. Just anterior to the sensilium in most fleas are the stout, paired *antesensilial setae* (*antepygidial bristles*) situated on the posterior margin of tergum 7. Many adult fleas, especially those of diurnal hosts, possess well-developed *eyes* (Fig. 7.3), which are actually clusters of ocelli. Eyes are well developed in most adult fleas of medical or veterinary importance (Figs. 7.2 and 7.3). Short, clubbed, three-segmented *antennae* are held inside protective grooves called *antennal fossae* on the sides of the head which prevent antennal damage as the flea moves through the pelage of its host.

The *mouthparts* of adult fleas are well adapted for piercing and sucking (Figs. 7.3 and 7.4). After a suitable feeding site has been located by the sensory *labial palps,* three

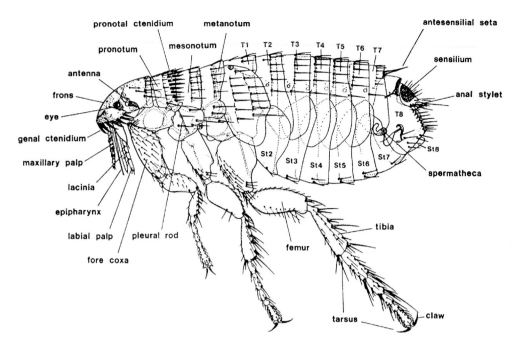

**FIGURE 7.1**  Morphology of generalized adult flea, female. T, abdominal tergites; St, abdominal sternites. (From Lewis, 1993b.)

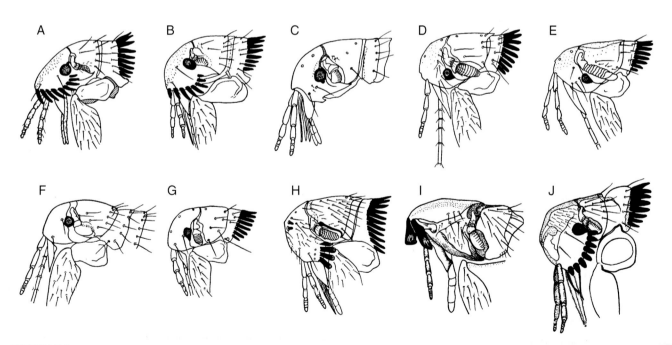

**FIGURE 7.2**  Morphology of the head and prothorax of representative adult fleas of medical and veterinary importance. (A) Cat flea (*Ctenocephalides felis*); (B) Dog flea (*C. canis*); (C) Human flea (*Pulex irritans*); (D) Northern rat flea (*Nosopsyllus fasciatus*); (E) *Oropsylla montana,* a North American rodent flea; (F) Oriental rat flea (*Xenopsylla cheopis*); (G) *Hoplopsyllus anomalus,* a North American rodent flea; (H) European mouse flea (*Leptopsylla segnis*); (I) *Myodopsylla insignis,* a North American bat flea; (J) Rabbit flea (*Cediopsylla simplex*). (From Matheson, 1950.)

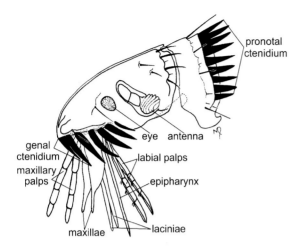

**FIGURE 7.3**   Lateral aspect of the head of a cat flea (*Ctenocephalides felis*) showing mouthparts, combs (ctenidia), antenna, and eye. (Original by Margo Duncan.)

slender, elongate structures collectively called *stylets* or the *fascicle* are used to pierce the host skin. The three stylets consist of two lateral, bladelike *maxillary laciniae* and the central *epipharynx* (Fig. 7.3). The laciniae penetrate the host skin, and the tip of the epipharynx enters a host capillary. A *salivary canal* is formed by the closely appressed medial surfaces of the two laciniae. A *food canal* is formed at the confluence of the laciniae with the epipharynx (Fig. 7.4). *Anticoagulants,* including the antiplatelet enzyme *apyrase,* other salivary components, and sometimes allergens or pathogens, are introduced into the bite wound via the salivary canal while host blood is imbibed through the food canal. In some sedentary fleas, such

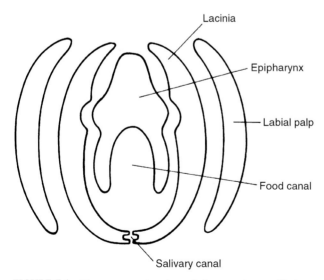

**FIGURE 7.4**   Transverse section through flea mouthparts. (Redrawn from Askew, 1973.)

as sticktights (Pulicidae) and alakurts (Vermipsyllidae), which remain attached to the host for long periods, the mouthparts are elongate and barbed and also function as host-attachment devices.

Internally, the *alimentary tract* of fleas (Fig. 7.5) consists of an anterior *pharynx* which leads to the elongate *esophagus* and then to the *proventriculus* at the junction of the *foregut* and *midgut*. The proventriculus is armed with rows of spines which can be drawn together to prevent regurgitation of a blood meal from the midgut. The midgut expands to accommodate large blood meals but it lacks distensible diverticula or caeca. Many fleas imbibe larger volumes of blood than their midgut can accommodate and must void blood-rich feces during or soon after feeding. Four excretory *Malpighian tubules* radiate from the junction of the midgut with the *hindgut*.

Male flea *genitalia* are morphologically complex (Fig. 7.6). The major structures are the *claspers*, which are used to help secure the female during mating; the often highly specialized *aedeagus;* and the *penis rods,* which are partly inserted into the female opening during mating. The major components of the female genitalia are the *vagina*, the *spermathecal duct*, and the *spermatheca;* sperm are stored in the spermatheca between matings. During copulation, the antennae of the male are held erect and are used to grasp the female.

Flea eggs are small (0.1–0.5 mm), ovoid, and pearly white (Fig. 7.7). Flea *larvae* are elongate, legless, and eyeless, with numerous stout body setae, especially on their abdominal segments. They possess a well-developed head capsule armed with chewing mandibles (Fig. 7.8) and a pair of mandibular *silk glands* that produce silk for constructing the *pupal cocoon*. Most flea larvae are small, wormlike, and highly active, with voracious appetites. Although few flea larvae can be identified to species, most can be assigned to the family level based on the arrangement of head papillae (small, finger-like projections), setae, and sense organs.

Flea pupae are *exarate* (i.e., have externally visible appendages) (Fig. 7.9) and are typically surrounded by a loose silken cocoon which is secreted by the last larval instar. Because of the silk's sticky nature, debris from the substrate often adheres to the cocoon and helps to camouflage it (Fig. 7.10). Many adult fleas possess an anterior *frontal tubercle* on the head which aids them in tearing free from the cocoon during emergence; after use in this manner, the frontal tubercle breaks off in some species.

## LIFE HISTORY

Fleas are *holometabolous* insects, with an egg, larval (typically consisting of three instars), and pupal stage (Figs. 7.7

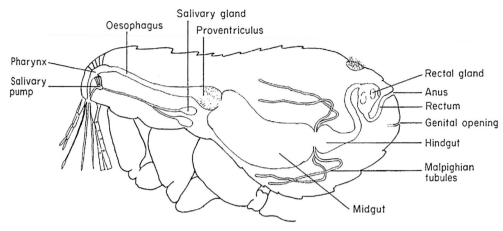

**FIGURE 7.5** Generalized internal anatomy of an adult flea, showing mouthparts and alimentary tract. (From Marshall, 1981.)

to 7.10). Gravid females of most flea species that have been studied can produce hundreds of eggs during their lifetime. Eggs typically hatch in about 5 days. Autogeny, or laying fertile eggs prior to ingestion of a blood meal, is not known to occur in fleas. The eggs are sticky and may adhere briefly to the host pelage; however, they usually drop into the host nest or bedding material, where they hatch a few days later. Some fleas oviposit directly on leaves or debris in host nests. Females of *Stenoponia*

*tripectinata,* a Palearctic rodent flea, glue their eggs to the nesting material.

Most flea *larvae* feed on organic matter in the nest or bedding materials of their hosts. Adult cat fleas, dog fleas, European rabbit fleas, and representatives of several other species void blood-rich fecal pellets during feeding, which in turn provide a nutritious food source for the larvae. Larvae of the northern rat flea aggressively prod adult fleas until they excrete blood-rich feces, which the

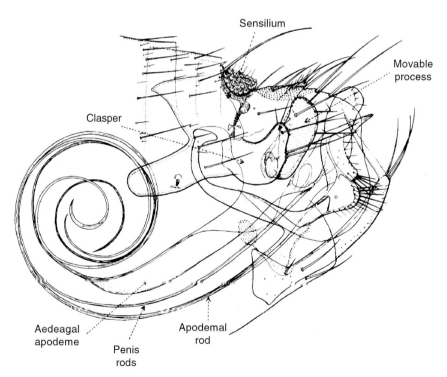

**FIGURE 7.6** Genitalia of a male flea, the western chicken flea *(Ceratophyllus niger)*. (Modified from Smit, 1957.)

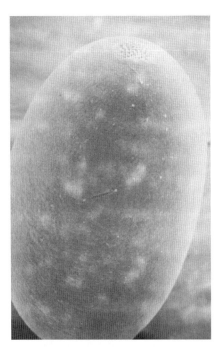

**FIGURE 7.7**    Immature stages of the cat flea (*Ctenocephalides felis*): egg. (Courtesy of Nancy C. Hinkle and National Pest Control Association.)

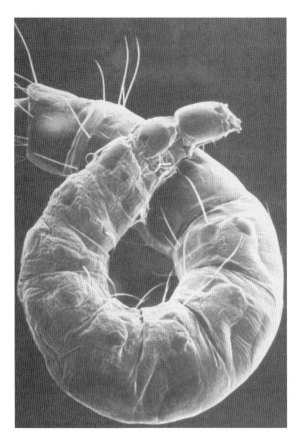

**FIGURE 7.8**    Immature stages of the cat flea (*Ctenocephalides felis*): larva. (Courtesy of Nancy C. Hinkle and National Pest Control Association.)

larvae then ingest. Some flea larvae supplement their diet by feeding on other small arthropods in the host nest, and cannibalism among flea larvae appears to be common.

Duration of the *pupal stage* usually lasts 1–2 weeks but is influenced by ambient temperature and host availability. Eclosed adult fleas of several species can remain within the cocoon as *pre-emergent adults* until suitable host or environmental cues stimulate their emergence. Pre-emergent adult cat fleas can remain quiescent inside cocoons for 4–5 months to avoid desiccation or other environmental extremes that would kill free-living fleas.

Many fleas, including most of those of medical or veterinary importance, undergo continuous generations under favorable conditions. The cat flea is a good example. Indoors, the *generation time* for this flea is usually about 1 month but can be as short as 20 days. Other fleas, such as alakurts associated with migrating ungulates in Asia, and species parasitic on migrating birds are more likely to pass through just one generation per year in synchrony with host availability. Some fleas, especially in temperate regions, may undergo four or five generations each summer but fewer or none during the winter. Host availability clearly affects the number of generations in many fleas. Longevity of fleas in the absence of available hosts is greater at low temperatures and high humidity, such as during winter in temperate regions. Under optimal conditions, adult fleas of certain species may survive away from the host for more than a year.

Specialized or unusual life cycles have evolved in many fleas, including some species of medical or veterinary significance. Females of the genera *Tunga* and *Neotunga*, for example, burrow into host dermal tissue, where they undergo a dramatic size increase (up to 1000-fold) accompanied by extensive morphological degeneration. This type of growth, called *neosomy*, involves major integumental chitin synthesis during the adult stage. The genital opening of the female protrudes through the pore in the host skin to facilitate mating with the free-living males; fertilized eggs are likewise extruded through this opening. Because of the great size increase of neosomic females, they are able to produce many relatively large eggs. In some cases this has led to a reduction in the number of larval instars from three to two; further modifications are exhibited by *Tunga monositus*, a parasite of New World rodents, in which neither larval instar feeds. The chigoe (*Tunga penetrans*) is an important human parasite that belongs to this group of fleas.

Adult females of some fleas oviposit randomly into the environment, where the resulting larvae must search for organic matter suitable to eat. Examples are vermipsyllid fleas in the genera *Vermipsylla* and *Dorcadia*, called

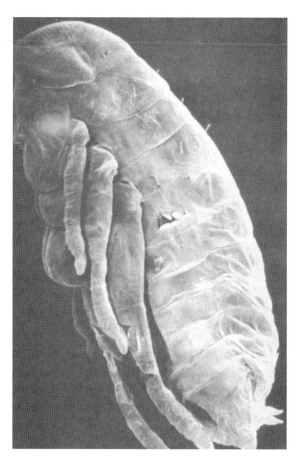

**FIGURE 7.9**   Immature stages of the cat flea (*Ctenocephalides felis*): pupa. (Courtesy of William H. Kern, Jr. and National Pest Control Association.)

**FIGURE 7.10**   Immature stages of the cat flea (*Ctenocephalides felis*): cocoon covered with debris from substrate. (Courtesy of Nancy C. Hinkle and National Pest Control Association.)

*alakurts,* that feed on large ungulates, remaining attached for several days. At the other end of the spectrum, females of *Uropsylla tasmanica* cement their eggs to the fur of their Australian hosts (dasyurid marsupials) and the larvae burrow into host skin, where they subsist as subdermal parasites. Mature *U. tasmanica* larvae drop to the ground, where they spin a cocoon and pupate in a manner typical of most other fleas. Larvae of the hare-infesting *Euhoploplsyllus glacialis* are ectoparasitic, and those of some other fleas feed on host carcasses or even on the superficial tissues of moribund hosts.

## BEHAVIOR AND ECOLOGY

Fleas have evolved a plethora of specialized behaviors and ecologies to locate and exploit their hosts. *Host-finding behavior* is extremely important for adult ectoparasites, such as fleas, in which the immature stages typically occur off the host. Important stimuli used by fleas for host location include host body warmth, air movements, substrate vibrations, sudden changes in light intensity, and odors of potential hosts or their products (e.g., carbon dioxide, urine). The sensilium, antennae, and eyes are important organs used by fleas to detect potential hosts. In cases in which adult fleas emerge from their cocoons in close association with their host, locating a food source is not difficult. However, fleas of other groups of hosts, such as ungulates or migrating birds, typically must employ more elaborate strategies for this purpose. These include jumping towards dark or moving objects and moving towards warmth and $CO_2$ sources.

Some fleas are stimulated to emerge from their pupal case and cocoon by mechanical compression and vibrational stimuli, which often indicate the presence of a potential host. This response is especially noticeable in flea-infested human premises that have been temporarily vacated for weeks or months. When humans or pets return to the premises, these stimuli are largely responsible for synchronized emergences of adult fleas from their cocoons.

While some fleas spend much of their adult lives in the host pelage, most species visit the host principally to feed. In fact some nidicolous (nest-associated) fleas (e.g., *Conorhinopsylla, Megarthroglossus,* and *Wenzella* spp.) spend very little time on the active host, living instead in crevices in the nest and feeding when the host is asleep. Nidicolous habits have evolved in several families of fleas.

Once a flea locates a host, *feeding* is initiated by cues such as body warmth, skin secretions, and host odors. Sensory structures of the maxillary and labial palps aid in selecting a feeding site. The labium and labial palps then guide the stylet-like mouthparts into the host skin. Most fleas are *capillary feeders;* when the tip of the stylet bundle pierces a capillary, feeding is facilitated by contraction of powerful cibarial and pharyngeal muscles. Many fleas possess symbiotic bacteria or fungi in the midgut that aid in digestion of the blood meal.

*Mating behavior* in most fleas that have been studied follows a distinct sequence of events. When the male and female approach one another, the male touches the female with his maxillary palps and his antennae become erect. The male then moves behind the female, lowers his head, and pushes his body beneath her while grasping her with his antennae by sucker-like organs along their inner surfaces. Next, the male raises the apex of his abdomen, partially secures the female with his claspers, and extrudes his penis rods and/or aedeagus to initiate copulation. Sperm deposited into the female are stored in her spermatheca until her eggs are ready for fertilization.

*Locomotory behavior* in adult fleas usually involves walking or running on the substrate or through host pelage. However, jumping is the mode of locomotion for which fleas are best known; this provides both an important means of escape and a way to reach hosts. Fleas jump using a modification of the flight mechanism of their winged ancestors. In addition to using muscles derived from subalar and basalar flight muscles, they have retained the wing-hinge ligaments, which have been displaced midlaterally due to lateral compression of the flea body. The jump is not propelled by direct muscle action, but rather by the sudden expansion of discrete pads of a highly elastic protein in the pleural arch called *resilin.* This remarkable protein can store and release energy more efficiently than any synthetic rubber and more quickly than any muscle tissue. The properties of resilin are unaffected by temperature, enabling fleas to jump even in subfreezing conditions.

Prior to jumping, the flea typically crouches, compresses its resilin pads, and keeps them compressed using one or more catch mechanisms. At takeoff the tergo-trochanteral depressor muscle relaxes to release the catch, allowing the resilin pads to expand rapidly, transferring energy to the hind legs. This results in an acceleration of about $200\,g$, catapulting fleas of some species more than 30 cm in about 0.02 sec. While airborne, the flea somersaults, holding its middle or hind legs aloft to use as grappling hooks for snagging a host or the substrate. After landing, the muscles are rapidly readjusted in preparation for another jump. By repeating this action, the Oriental rat flea can make up to 600 jumps per hour for 72 hr without rest (Rothschild 1973).

Nest-associated fleas typically have reduced jumping abilities because they have less resilin in the pleural arch and have undergone secondary atrophy of jumping muscles. This appears to be adaptive in ensuring that these fleas do not leap out of a nest into an unfavorable environment.

Flea populations may be naturally regulated in several ways. Hosts are often efficient groomers and are able to significantly reduce flea populations on their bodies. Cats, for example, have been shown to remove up to 18% of their fleas within 24 hr. Natural *predators,* such as certain mesostigmatid mites, pseudoscorpions, beetles, ants, and other arthropods, feed on fleas, especially the immature stages in host nests, thereby decreasing their numbers. Various *parasites* also contribute to flea mortality. These include the plague bacillus, *Yersinia pestis,* the protozoan *Nosema pulicis,* the nematode *Steinernema carpocapsae,* and parasitoids such as the pteromalid wasp *Baraimlia fuscipes.*

Environmental factors are often important in determining the abundance of fleas in different habitats or geographical regions. These factors are often related to climate, weather, or soil conditions, such as relative humidity, temperature, and soil moisture content. Favorable environmental conditions, such as host abundance and availability, plentiful food for larvae, high relative humidity, and mild temperatures, promote high populations of many flea species. Because the immature stages typically occupy different niches from adult fleas, the ecological requirements of one or more of the immature stages, rather than of the adult, may be limiting factors that do not permit a species to become established or abundant under certain conditions.

Hormones can play an important role in synchronizing the development of fleas with that of their hosts. The life cycle of the rabbit fleas *Spilopsyllus cuniculi* and *Cediopsylla simplex* (Fig. 7.2J), for example, is mediated by host hormones imbibed with the host blood. These fleas can reproduce only after feeding on a pregnant doe. In this way, the emergence of adult fleas is synchronized with that of a litter of rabbits. Reproductive hormones (e.g., corticosteroids and estrogens) in the blood of the pregnant doe stimulate maturation of the ovaries and oocytes in feeding female fleas and testicular development in the males. The adult fleas are ready to mate when the rabbit litter is born. Flea mating and oviposition occurs after they have transferred onto the newborn young.

The resulting flea larvae feed on organic matter in the nest debris. The next generation of adult fleas appears 15–45 days later, in time to infest the host littermates before they disperse from the burrow. Rothschild and Ford (1973) provide more details on this fascinating flea–host relationship.

# FLEAS OF MEDICAL-VETERINARY IMPORTANCE

### Human flea   *(Pulex irritans)*

This flea will feed on humans and is capable of transmitting pathogens of medical importance. However, it is more commonly an ectoparasite of swine and domestic cats and dogs in most parts of the world. Although *P. irritans* is currently an infrequent parasite of humans in developed countries, this has not always been the case. *P. irritans* has a patchy but cosmopolitan distribution and often occurs in remote and isolated areas. Adults of this species lack both genal and pronotal ctenidia (Fig. 7.2C). *P. simulans* is a closely related species that parasitizes large mammals, including wild canids and domestic dogs, and sometimes people, in the New World. Older records (before 1958) from this region are unreliable and could refer to either *P. irritans* or *P. simulans.*

### Cat flea   *(Ctenocephalides felis)*

The cat flea occurs worldwide and is currently the most important flea pest of humans and many domestic animals. It is primarily a nuisance because it feeds not only on domestic and feral cats, but also on humans, domestic dogs, and several livestock species. It also parasitizes wild mammals such as opossums and raccoons. This ectoparasite is the most common flea on dogs and cats in most parts of the world. Some strains of the cat flea appear to have adapted to hosts such as horses or goats. Cases of severe anemia associated with huge numbers of cat flea bites have been recorded for these and other domestic animals.

Female cat fleas in most populations produce larger numbers of fertile eggs if they take their blood meals from cats rather than other host species. Under optimal conditions a female cat flea can lay about 25 eggs per day for a month, contributing to very high densities of fleas in a relatively short time. Adult cat fleas have well-developed genal and pronotal ctenidia (Figs. 7.2A and 7.3) and can be distinguished from the dog flea (*C. canis*) by the longer head and longer first spine in the genal comb in *C. felis*. For further details on the biology of the cat flea, see Dryden (1993) and Rust and Dryden (1997).

### Dog flea   *(Ctenocephalides canis)*

This flea (Fig. 7.2B) is much less common on domestic dogs in most parts of the world than it was in previous decades. Instead, the cat flea has become the most common flea on domestic dogs. No satisfactory explanation for this change has been documented; perhaps cat fleas can outcompete dog fleas under stress from modern pesticide applications. Nevertheless, dog fleas persist worldwide and remain as the predominant fleas on dogs in Ireland, Israel, and a few other countries. Dog fleas also parasitize wild canids such as foxes, coyotes, and wolves, on which they can be relatively common.

### Oriental rat flea   *(Xenopsylla cheopis)*

This flea is the principal vector of the agents of plague and murine typhus throughout many of the tropical and subtropical parts of the world. Some other species of *Xenopsylla* also are vectors of the plague bacillus. Although it is most common on domestic rats, it also will readily feed on humans, dogs, cats, chickens, and other hosts especially if rats become scarce. Like the human flea, adults of *X. cheopis* lack both a pronotal and a genal ctenidium (Fig. 7.2F).

### European rabbit flea   *(Spilopsyllus cuniculi)*

Originally from Europe, this flea has accompanied its host the European rabbit as it has been introduced throughout the world, either inadvertently or as a laboratory animal. It is an example of a sedentary or *sticktight* flea; adults attach to the host for long periods using their elongate mouthparts to anchor them in host skin and to feed. This flea typically attaches to the ears of rabbits, where a rich peripheral blood supply provides easily accessible blood meals. Adults have a genal ctenidium with a row of five blunt spines oriented almost vertically on the head and a well-developed pronotal ctenidium.

### Sticktight flea   *(Echidnophaga gallinacea)*

As indicated by its name, this is another sticktight flea. It is distributed globally wherever chickens have been introduced as domestic animals. This flea usually attaches semipermanently around the head, especially on the wattle, of chickens (Fig. 7.11). Many additional hosts are also parasitized by *E. gallinacea*, including other domestic birds (e.g., turkeys, quail), domestic rats, dogs, cats, and occasionally humans. Adults of this small flea are easily recognized by their sharply angled squarish head and the absence of both pronotal and genal ctenidia.

### Chigoe   *(Tunga penetrans)*

This flea, also called the *jigger* or *sand flea,* has major medical and veterinary significance because it burrows into the tissues of humans and some domestic

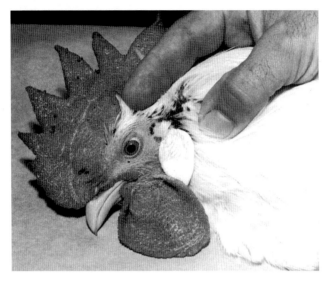

**FIGURE 7.11**    Sticktight fleas (*Echidnophaga gallinacea*), attached to head and neck of rooster. (Courtesy of US Department of Agriculture.)

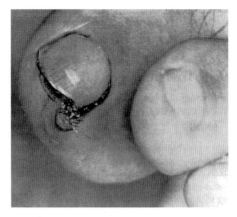

**FIGURE 7.13**    Tungiasis; female *Tunga penetrans* embedded in human toe with flea eggs around the lesion. (From Ibáñez-Bernal and Velasco-Castrejón, 1996.)

animals. In addition to being very small (ca. 1 mm in length), the free-living adult chigoe lacks pronotal and genal ctenidia and has a sharply angled head. It is widely distributed in tropical and subtropical regions. The life cycle of the immature stages and the male of *T. penetrans* does not deviate significantly from that of most fleas. Initially, the female is free-living but soon invades the host skin. Once embedded, she begins to swell by imbibing host fluids, often expanding about 80-fold to reach the size of a pea after 8–10 days (Figs. 7.12 and 7.13). She maintains an opening to the exterior, through which she respires, mates with a free-living male, and expels her eggs. The male possesses the longest intromittent organ relative to body size in the animal kingdom and mates from an inverted position.

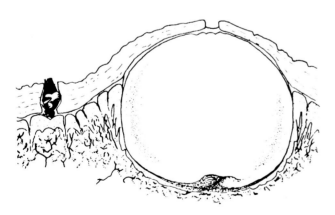

**FIGURE 7.12**    Chigoe (*Tunga penetrans*), adult females embedded in host skin; abdominal segments starting to swell in specimen at left; greatly enlarged, gravid female at right. (From Harwood and James, 1979.)

Eggs usually are expelled onto sandy soils, including coastal beaches (hence the term *sand flea*), frequented by potential hosts where the larvae complete their development. There are only two larval instars. Development from egg to adult usually takes 4–6 weeks, but sometimes it takes only 3 weeks under optimal conditions.

**Northern rat flea**    *(Nosopsyllus fasciatus)*

This is a common flea of domestic rats, especially in temperate and northern regions of the world. Although it will bite humans and can transmit several zoonotic pathogens, such as the agents of plague and murine typhus, it is far less important in this respect than the Oriental rat flea. Occasionally, it parasitizes other rodents or domestic mammals. Adults possess a well-developed pronotal comb but lack a genal comb (Fig. 7.2D).

**European chicken flea**    *(Ceratophyllus gallinae)*

This ectoparasite of feral and domestic birds, especially chickens, originated in Europe but has spread with poultry operations throughout the world. Because *C. gallinae* can feed on so many bird species, it is often difficult to completely eradicate. In contrast to *E. gallinacea*, this flea is highly mobile on the host and can be especially common in host nesting material. Adults have a distinct pronotal ctenidium but lack a genal ctenidium.

**European mouse flea**    *(Leptopsylla segnis)*

This cosmopolitan flea typically parasitizes the house mouse (*Mus musculus*), including laboratory colonies. Rarely, large populations of *L. segnis* cause host anemia or other problems in mouse-rearing facilities. Adults of this flea possess a well-developed pronotal ctenidium and a vertical genal ctenidium consisting of four bluntly rounded spines (Fig. 7.2H).

TABLE II
Pathogens Transmitted by Fleas

| Disease agent | Disease | Vector(s) | Host(s) | Geographic area |
|---|---|---|---|---|
| Virus | | | | |
| Myxoma virus | Myxomatosis | *Spilopsyllus cuniculi* | Rabbits | Europe, Australia |
| Bacteria | | | | |
| *Coxiella burnetii* | Q fever | Several fleas | Mammals | Global |
| *Francisella tularensis* | Tularemia | Several fleas | Mammals | Global |
| *Rickettsia typhi* | Murine typhus | *Xenopsylla, Ctenocephalides* | Mammals | Global |
| *R. prowazekii* | Sylvatic epidemic typhus | *Orchopeas howardi* | Flying squirrels, humans | North America |
| *Salmonella enteriditis* | Salmonellosis | *Pulex, Xenopsylla* | Humans | Eurasia |
| *Staphylococcus aureus* | Staphylococcal infection | Several fleas | Mammals | Global |
| *Yersinia pestis* | Plague | Mainly *Xenopsylla* | Humans, rodents, cats | Global |
| Protozoa | | | | |
| *Trypanosoma lewisi* | Murine trypanosomiasis | *Nosopsyllus, Xenopsylla* | Rats | Global |
| *T. nabiasi* | Rabbit trypanosomiasis | *S. cuniculi* | Rabbits | Global |
| Nematoda | | | | |
| *Acanthocheilonema reconditum* | Canine filariasis | *Ctenocephalides* | Carnivores | Global |
| Cestoda | | | | |
| *Dipylidium caninum*[a] | Double-pored tapeworm | *Ctenocephalides* | Dogs, cats, humans | Global |
| *Hymenolepis diminuta*[a] | Rodent tapeworm | *Nosopsyllus, Xenopsylla* | Rodents, humans | Global |
| *H. nana*[a] | Dwarf tapeworm | *Nosopsyllus, Xenopsylla* | Rodents | Global |

[a] Fleas are not vectors for these pathogens but instead serve as intermediate hosts.

## PUBLIC HEALTH IMPORTANCE

Fleas are of public health significance for several reasons. Many species are annoying biters that can cause considerable discomfort, sometimes leading to secondary infections of bite wounds. The bites of some species can cause *dermatitis* or *allergic reactions*. Allergic responses also can result from contact with, or inhalation of, flea products (e.g., larval exuviae). Females of the *chigoe* actually invade human skin tissue, especially of the feet and toes, and cause painful lesions that are prone to serious secondary infection. Other fleas are intermediate hosts of tapeworms that can parasitize humans. Fleas also serve as vectors of the causative agents of several important zoonotic diseases, such as murine typhus and plague (Table II).

Flea bites (Fig. 7.14) can cause intense irritation for several days. Bites are characterized by a tiny purplish spot, or *purpura pulicosa*, surrounded by slightly swollen skin called *roseola pulicosa*. The vast majority of flea bites experienced by humans are due to the *cat flea*. This flea is an unrelenting biter which generally attacks humans on the ankles (Fig. 7.14), although other parts of the body may be affected. Women tend to be more commonly bitten than men, suggesting a hormonal association between this species and human females. In addition to the annoyance it causes, *Ctenocephalides felis* is a proven vector of the causative agent of endemic typhus. The cat flea is discussed in more detail with respect to its veterinary importance.

The *human flea* also is an annoying biter of people in various parts of the world. The closely related *P. simulans* sometimes infests households and causes dermatitis in humans in western North America. Several other species of fleas may bite humans to the point of annoyance, including the *dog flea*, the *sticktight flea*, the *northern rat flea*, and several species of *Xenopsylla*, including the *Oriental rat flea* (Table I). Also, the *squirrel flea Orchopeas howardi* and some bird fleas belonging to the genus *Ceratophyllus*, including the *European chicken flea*, occasionally bite humans.

Members of households and adjacent premises harboring pets or domestic rodents can be especially prone to

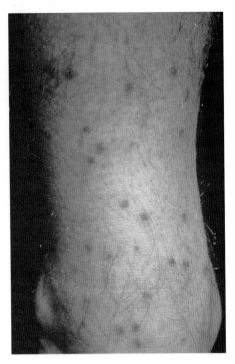

**FIGURE 7.14**   Multiple cat flea bites on human ankle. (Photo by Elton J. Hansens.)

flea bites. *Cat fleas* and *dog fleas* readily bite humans, especially if flea populations are large or if pets are temporarily removed. Rodents inhabiting households also can be a source of fleas that can bite humans. Fleas typically abandon dead hosts; if domestic rats die in wall voids, basements, or other poorly accessible structures, their fleas may seek human hosts. A similar situation can occur involving *squirrel fleas,* whose hosts often nest in attics or eaves of houses. Basements, garbage, and pet food supplies and feeding bowls also can attract scavenging mammals such as opossums, raccoons, and skunks. These animals may leave behind cat fleas and other fleas that will bite humans. Flea bites represent occupational hazards on many farms, in barns, rabbit hutches, and poultry operations. The fleas in such cases may come directly from livestock or indirectly from domestic rodents attracted to food supplies.

## FLEA-ASSOCIATED ALLERGIES

*Flea-bite dermatitis* usually occurs in persons that have become hypersensitive to flea saliva. In sensitized individuals, bite sites typically develop into papules, causing a form of *papular urticaria*, often with associated wheals, especially in children. In more serious cases, skin scaling, hardening, or discoloration can occur. In adults, it is usually the distal extremities that are involved, whereas in children the entire body may be affected. With time and repeated exposure to flea bites, hyposensitization may reduce the severity of the dermatitis without medical intervention. The administration of corticosteroids or desensitizing antigens can be helpful for some hypersensitive individuals.

People can become sensitized to flea feces and particles of exoskeletons upon contacting or inhaling them in house dust. Adult fleas have been identified as the source of some of these allergens (Baldo 1993). Airborne larval exuviae also have been implicated as triggers of asthmatic symptoms. Relief from these allergic responses may be achieved by administering a course of desensitizing antigens to the patient.

## PLAGUE

Plague is caused by infection with *Yersinia pestis,* a gram-negative coccobacillus bacterium. The disease also is referred to as the *black death,* and, in Francophone countries, as *la peste.* The organism was first isolated in 1894 when Swiss bacteriologist Alexandre Yersin cultured it from sick patients at the Pasteur Institute in Hong Kong. Although plague is thought to have originated in Asian gerbils, it is typically maintained in urban areas in commensal rodents, especially the black rat (*Rattus rattus*) and Norway rat (*R. norvegicus*). The pathogen is transmitted to these rodents by fleas, especially the Oriental rat flea and other members of the genus *Xenopsylla.* The role of *Xenopsylla cheopis* as a vector of *Y. pestis* was definitively demonstrated by French physician/bacteriologist Paul-Louis Simond during his plague studies in Pakistan (then India) in 1898. Plague is the most significant flea-borne disease in human and wild mammal populations.

Plague *pandemics* have had a significant impact on human civilization and have claimed more human lives than all wars ever fought. At least three major pandemics have been documented. The first, sometimes called *Justinian's plague,* originated in AD 541 in Africa and later spread throughout Mediterranean Europe, killing an estimated 40 million people during the sixth and seventh centuries.

The second pandemic, usually referred to as the *black death,* originated in central Asia in the 14th century and spread to Europe as a consequence of developing trade routes between these two continents. In 1347, crew members of an Asian trading vessel that docked in Sicily were suffering from a mysterious disease which was later identified as plague. Over the next 5 years, plague spread throughout most of Europe from this point of entry, with devastating consequences; by 1352, at least 25 million people had died in Europe alone. This pandemic continued for more than 200 years, with the disease appearing or reappearing in different areas of Europe. London experienced a major epidemic in 1348, followed by another in 1665, with foci of the disease persisting in the city throughout this time period.

The third major pandemic of plague spread across the globe in the late 1800s after originating from a focus in China's Yunnan Province in 1855. From 1896 to 1948, this pandemic accounted for 12 million deaths in India alone. Several plague foci initiated during the peak of this pandemic still persist today. In the United States, for example, plague bacilli were introduced with infected ship-borne rats in 1899 in San Francisco and from there eventually spread to at least 14 western states and 2 Canadian provinces. Plague continues to persist in native rodents and fleas in most of these areas of western North America. Some researchers, however, contend that plague was already enzootic in North America prior to 1899 and that the apparent spread of the disease merely reflected intensified surveillance efforts. Accounts of plague include works by Stark *et al.* (1966), Bahmanyar and Cavanaugh (1976), Twigg (1978), Barnes (1982), Duplaix (1988), Mee (1990), Craven *et al.* (1993), Poland *et al.* (1994), Madon *et al.* (1997), and Goddard (1999).

Nucleotide sequencing of ribosomal RNA in *Y. pestis* shows a correlation between the *geographical distribution* of genetic strains, or biovars, of *Y. pestis* and their spread during the three pandemics. There are three biovars of *Y. pestis*: biovar Antiqua occurs in Africa, biovar Medievialis mainly in central Asia, and biovar Orientalis in Europe, Asia, Africa, North America, and South America.

Today plague occurs as fairly discrete foci in various parts of Asia, southern and northwestern Africa, South America, and western North America. Recent outbreaks have surfaced in Brazil, Democratic Republic of the Congo, Ecuador, India, Madagascar, Malawi, Mongolia, Peru, South Africa, Tanzania, and Vietnam. Globally, nearly 19,000 human cases of plague were reported to the World Health Organization from a total of 20 countries during 1984–1994. During 1970–1994, a total of 334 cases of indigenous plague were reported for the United States; the peak years were 1983 and 1984, with 40 and 31 cases, respectively (Craven *et al.*, 1993). Eighty percent of cases in the United States have occurred in Arizona, New Mexico, or Colorado.

In addition to bites from infected fleas, plague infections can result from direct contact with moribund or dead mammals infected with *Y. pestis* or, rarely, from inanimate objects harboring the pathogen. In such cases the pathogen typically enters the body through skin lesions. Inhalation infection also can occur from aerosolized *Y. pestis*.

Two *ecological forms of plague* are recognized: *urban plague*, carried by domestic rats and their fleas in cities and towns, and *wild-rodent plague* (*sylvatic plague*, derived from the Latin *silva*, meaning trees; *campestral plague*; *rural plague*), maintained in several species of mammals (mainly rodents) and their fleas in rural areas away from human populations. Over 200 species of rodents and other mammals (e.g., certain carnivores) may serve as reservoir hosts of wild-rodent plague. In North America, ground squirrels, rock squirrels, chipmunks, and prairie dogs are particularly important, whereas in Asia, gerbils and susliks (ground squirrels) typically fill this role. Similarly, various gerbils and the peridomestic rat *Mastomys natalensis* are important reservoirs in parts of Africa. The sigmodontine rat *Zygodontomys brevicauda* is prevalent in most South American foci. Plague-infected tree squirrels have been found in some towns in the western United States. In an attempt to limit future spread, plague is presently one of only three internationally quarantinable infections.

In many plague-endemic regions, wild-rodent plague persists enzootically in discrete rodent populations. Under certain conditions, the disease can become epizootic and spread to commensal rats to trigger urban plague. Some populations of reservoir hosts are refractory to infection with *Y. pestis*, while others are highly susceptible. This is reflected by large-scale die-offs in infected prairie dog (*Cynomys* spp.) towns in North America. Intermediate stages of susceptibility to plague exist between these two extremes in other reservoir populations. In most regions where plague persists there are distinctly different species of enzootic and epizootic rodent reservoir hosts. Most carnivores, especially felids, are susceptible to infection with *Y. pestis*. The disease is often severe in domestic cats, which can serve as a source of infected fleas to households. Bites, scratches, or inhalation of infectious aerosols from infected cats also can disseminate *Y. pestis*. Human plague cases acquired from domestic cats have increased in recent years in the United States.

Although the Oriental rat flea and other *Xenopsylla* species are important *vectors* of *Y. pestis*, there are at least 125 species of fleas that are capable of transmitting the pathogen. In North America, several flea species can transmit the plague bacterium to native rodents; *Oropsylla montana* (Fig. 7.2E) and perhaps *Hoplopsyllus anomalus* (Fig. 7.2G) are the more important among these. In Russia and northern Asia, fleas belonging to the genera *Citellophilus*, *Neopsylla*, and *Ctenophthalmus* are significant enzootic vectors within rodent communities. In addition, the widespread fleas *P. irritans* and *N. fasciatus* are capable of transmitting plague bacilli. Because *X. cheopis* survives poorly in cool climates, historical reevaluations suggest that, contrary to former dogma, *P. irritans*, rather than *X. cheopis*, may have been the principal flea vector of *Y. pestis* in the great plague epidemics of northern Europe.

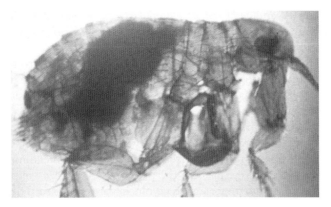

**FIGURE 7.15**    Blockage of midgut of Oriental rat flea (*Xenopsylla cheopis*) by mass of *Yersinia pestis*, the causative agent of plague. (Courtesy of US Public Health Service.)

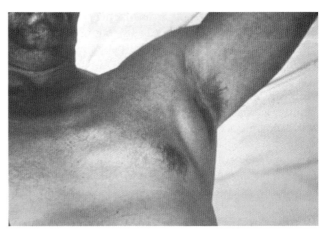

**FIGURE 7.16**    Plague patient with enlarged axillary lymph node, or bubo, characteristic of bubonic plague. (Courtesy of US Public Health Service.)

A susceptible flea typically becomes infected after imbibing plague bacilli in its blood meal from an infected host. The bacterium invades the flea midgut, where under suitable conditions it multiplies rapidly, often culminating in complete blockage of the gut anterior to the proventricular spines. This *proventricular blockage* (Fig. 7.15) results from clumping of the bacteria 8–51 days after ingestion. Although some gut blockages may clear spontaneously, persistent blockage of the flea gut is central to efficient transmission of *Y. pestis.* Fleas with blockages are incapable of ingesting a blood meal from a host. Feeding attempts by these fleas result in the drawing of blood into the flea esophagus, followed by regurgitation of the blood meal into the host. This regurgitation is caused by the elastic recoil action of the esophagus when resistance from the proventricular blockage is reached. Infection results when plague bacilli are regurgitated with the blood meal into the host.

Fleas with gut blockages are hungry and make repeated, aggressive feeding attempts. This can result in the infection of several different hosts and amplification of an epidemic. A single *Y. pestis* gene can determine whether or not proventricular blockage forms in infected *X. cheopis,* whereas another gene determines whether a flea midgut infection will develop. Unless the gut blockage clears, the infected flea ultimately succumbs to starvation, dehydration, or toxicity from bacterial metabolites, itself becoming a victim of plague. As with most other pathogens transmitted by fleas, plague bacilli do not pass through the gut wall of infected fleas to invade the hemocoel, salivary glands, or other organs.

Key environmental parameters influence the establishment of plague foci throughout the world. For example, *X. cheopis* is principally a denizen of drier habitats, and it is in these zones that the major plague foci have persisted. Where environmental factors tend to keep the number of flea species in a given area low, plague is generally absent or rare. If the ambient temperature exceeds 28°C, plague-infected fleas often can clear their guts of blockages, and the disease does not develop to epidemic proportions.

There are three recognized clinical types of plague infection: bubonic, septicemic, and pneumonic plague. The most common of these is *bubonic plague,* which usually results from the bites of infected fleas but can result from handling infectious mammalian carcasses. This type of plague is characterized by grossly enlarged, tender, peripheral lymph nodes called *buboes* (singular *bubo*). They usually occur in the axillary or inguinal region (Fig. 7.16) and are typically teeming with plague bacilli.

In *septicemic plague* the pathogen initially bypasses or overwhelms the peripheral lymph nodes and invades deeper recesses of the body. Although internal buboes may develop, they are not easily detected. Instead, the bloodstream is invaded rapidly by the bacterium, and capillary walls start to leak, often turning the skin black. The absence of external buboes to aid diagnosis, coupled with swift invasion of the blood, make this form of plague especially severe, with many patients succumbing to fatal septicemia.

The most life-threatening form of the disease is *pneumonic plague,* in which patients have a lung infection of *Y. pestis* and can cough or sneeze viable bacteria into the air. Inhalation of *Y. pestis* by susceptible individuals results in the pulmonary form of this disease. Also, bubonic or septicemic plague can progress to pneumonic plague. Without prompt and aggressive medical attention, pneumonic plague is invariably fatal; some untreated patients die within a day of inhaling the pathogen. The severe pathogenicity of *Y. pestis* is largely caused by endotoxins and exotoxins released by the dividing bacilli.

Flea-transmitted infection typically results in *classic bubonic plague,* in which buboes develop after an incubation period of 2–6 days. Accompanying symptoms are severe headache, fever, and shaking chills. Without treatment, most patients deteriorate rapidly, with a typical mortality rate of 50–60%. Septicemic plague is a particularly dangerous form of the disease because the incubation time is only 2–5 days and external buboes are absent. Pneumonic plague has a very short incubation period (1–3 days) and may spread rapidly from one victim to another. Overwhelming pneumonia characterized by coughing, bloody sputum, chills, and fever usually result in death within 3 days unless specific medication is administered within 15–18 hr of the onset of symptoms. Pneumonic plague spreads directly, and usually rapidly, from person to person without the involvement of flea vectors. In some cases, humans have contracted pneumonic plague after inhaling aerosolized bacilli expelled by infected household cats. Accurate diagnosis is important in identifying plague-infected patients. Various biochemical, serological, chromatographic, and staining tests, or lysis with a specific bacteriophage, are typically employed to detect *Y. pestis* or specific antibody directed against it in humans, animals, or fleas. DNA probes and polymerase chain reaction (PCR) techniques used to amplify specific nucleotide sequences of *Y. pestis* are becoming routine screening tools in many public health laboratories. A fiberoptic biosensor also has been developed to detect a specific antigen of *Y. pestis.*

The treatment of plague patients usually involves immediate hospitalization, isolation, and administration of broad-spectrum antibiotics. Formalin-inactivated plague vaccines are available. They are not, however, totally effective. None is currently protective against pneumonic plague in humans, and most must be administered in multiple doses or at regular intervals to ensure protection.

Efforts to control plague typically involve removal of wild-rodent reservoir hosts and/or their fleas. In areas of potential plague activity, samples of rodent blood and tissues often are collected in order to monitor plague in reservoir host populations. Fleas also can be collected and screened for *Y. pestis.* If samples are positive, then rodent and flea control measures should be considered.

## MURINE TYPHUS

Murine typhus, also known as *endemic typhus* or *Mexican typhus,* is caused by infection with the rickettsial organism *Rickettsia typhi* (formerly *R. mooseri*). Although this zoonosis is typically maintained in commensal rats by flea transmission, humans occasionally are infected. Murine typhus is one of the most prevalent rickettsial diseases of humans, even though it is underdiagnosed and its importance is generally unappreciated. *R. typhi* is a small, obligate, intracellular bacterium that can cause mild febrile infection in humans. It usually is transmitted via infected flea feces. When the bite site of an infected flea is scratched, rickettsiae gain access to the host through abraded skin. Under experimental conditions, however, some fleas also can transmit this pathogen via their bite. Reviews on the ecology and epidemiology of murine typhus have been provided by Traub *et al.* (1978), Azad (1990), Rawlings and Clark (1994), Azad *et al.* (1997), and Goddard (1998).

The geographical distribution of murine typhus is almost global. Although it occurs on all continents except Antarctica, its importance as a human pathogen has diminished in recent years. Significant foci persist, however, especially in Indonesia, the People's Republic of China, Thailand, North Africa, and Central America. In the United States, the annual number of human cases has decreased from more than 5000 in 1945 and 1946 to 20–80 per year from 1958 to the present. This zoonosis was formerly widespread throughout the southern and southwestern United States. Currently in the United States it is principally recorded in Texas, where 200 human cases were diagnosed in the 5-year period 1980–1984. Several cases are usually reported annually from California and Hawaii.

Murine typhus is maintained primarily in a cycle that involves commensal rodents of the genus *Rattus* and their ectoparasites, especially fleas of the genus *Xenopsylla.* Humans are typically infected when feeding fleas void infectious feces on their skin. The black rat (*Rattus rattus*) and Norway rat (*R. norvegicus*) are the principal *reservoirs* of *Rickettsia typhi.* Infections also have been recorded in many other mammals, including other commensal rats (*Rattus* spp.) worldwide, bandicoot rats (*Bandicota* spp.) on the Indian subcontinent, house mice (*Mus musculus*) worldwide, the oldfield mouse (*Peromyscus polionotus*) in the southern United States, the giant pouched rat (*Cricetomys gambianus*) in Africa, the house shrew (*Suncus murinus*) in the Old World, domestic cats worldwide, and the Virginia opossum (*Didelphis virginiana*) in North America. Within the last 20 years, peridomestic mammals such as opossums and feral cats and dogs have become more frequently recorded reservoirs of murine typhus in Texas and southern California. Field infection rates in commensal rats of up to 46% have been reported in Burma (Myanmar), Egypt, and Ethiopia, and rates of up to 94% have been reported in some cities in Texas. New World strains of *R. typhi* are much less virulent (ca. 2% mortality rate in humans) than Old World strains (up to ca. 70% mortality rate).

Commensal rats almost invariably are the most important reservoirs and amplifying hosts of *R. typhi.* Infection in these rats is not fatal; instead, they display a persistent transient rickettsemia. This is important in extending the period during which ectoparasites, especially fleas, can feed on infective hosts. Because seropositive Virginia opossums have been associated with human cases in some regions of the United States, it appears that opossums also can be important reservoir hosts.

At least 11 species of fleas belonging to 9 different genera have been found to be infected with *R. typhi* in nature. *X. cheopis* is the most important vector. Other *vectors* are *X. astia*, *X. bantorum*, *X. brasiliensis*, *Ctenocephalides felis*, *Pulex irritans*, *L. segnis*, and *N. fasciatus*. Except for *C. felis* and *P. irritans*, all of these fleas are common ectoparasites of commensal rodents. Human cases of murine typhus usually coincide with population peaks of *X. cheopis* on rats. The number of cases generally declines or disappears after this flea has been controlled by chemical applications or rodent removal. Infection rates of field-collected *X. cheopis* in hyperendemic regions typically are 50–70%.

Infection of a flea occurs when rickettsiae are ingested while the flea is feeding on a host that has *R. typhi* circulating in its blood. The ingested rickettsiae then invade the midgut epithelial cells of the flea and start to replicate by transverse binary fission. Progressive rickettsial infection spreads rapidly until most or all of the midgut cells are infected after 7–10 days. Ultimately, infectious rickettsiae are released from these cells and liberated into the gut lumen, from which they are excreted in the feces. *X. cheopis* fleas are typically infective about 10 days after an infectious blood meal. Infective fleas can transmit the pathogen for at least another 40 days, during which time the rickettsial titer in the fleas remains at a stable, high level. Infected fleas survive with a persistent *R. typhi* infection and demonstrate no obvious pathological effects. This contrasts with the related pathogen, *R. prowazekii*, which causes a fatal infection in its louse vector. Because *X. cheopis* can maintain and transmit *R. typhi* transovarially, this flea may be both a reservoir and a vector of murine typhus rickettsiae.

Although modes of *R. typhi transmission* other than via infected flea feces are known, their significance in nature remains unclear. Because *X. cheopis* has been shown to transmit *R. typhi* by bite in the laboratory, other fleas also may be capable of transmitting *R. typhi* by bite. The possibility of aerosol transmission from infective flea feces has been suggested, involving inhalation of the aerosolized pathogen by susceptible mammals.

*R. typhi* has been detected in ectoparasites other than fleas. Because most of these arthropods do not bite humans, their presumed role is in transmitting *R. typhi* enzootically among commensal rats. Ectoparasites in this category include the sucking lice *Hoplopleura pacifica* and *Polyplax spinulosa*, the mesostigmatid mites *Laelaps echidninus* and *Ornithonyssus bacoti*, and the chigger *Ascoshoengastia indica*. Although the human body louse (*Pediculus*

*humanus humanus*) is an experimental vector of *R. typhi*, it apparently is not involved in natural transmission cycles.

The *diagnosis* of human infection usually involves the demonstration of seroconversion against *R. typhi* or isolation of the bacterium. Recent advances in *R. typhi* detection in fleas include the development of an enzyme-linked immunosorbent assay and of a technique to demonstrate a 434–base pair nucleotide sequence of the *R. typhi* genome using a PCR assay. These techniques are useful in patient diagnosis and in screening potential reservoir hosts.

## OTHER FLEA-BORNE RICKETTSIAL AGENTS

In addition to the causative agent of murine typhus, several other rickettsial agents may be transmitted to humans by fleas. One of these, *Coxiella burnetii*, the agent of *Q fever*, can be transmitted not only by fleas but also by other blood-feeding arthropods, infected mammalian tissues, infective fomites, or aerosol. *Sylvatic epidemic typhus* (*sporadic epidemic typhus*) is a curious but potentially serious disease that occasionally is diagnosed in humans in the United States. The agent of this disease is *R. prowazekii*, which causes classic epidemic typhus transmitted to humans by the body louse, *P. h. humanus* (see Chapter 4). However, flying squirrels (*Glaucomys volans*), rather than humans, are the reservoir hosts of sylvatic epidemic typhus. Flying-squirrel fleas, especially the widespread squirrel flea *Orchopeas howardi*, and lice also harbor the causative rickettsiae. The exact mode of transmission to humans is unknown, but because ectoparasites of flying squirrels rarely feed on humans, it is hypothesized that under certain conditions infective rickettsiae in the feces of fleas and lice become aerosolized and may be inhaled by humans. Infected flying squirrels are not adversely affected. These squirrels sometimes are closely associated with humans through their predilection for constructing nests in attics or eaves of houses. McDade (1987) provides additional discussion of this rickettsial zoonosis.

A recently discovered rickettsial agent is *R. felis* (formerly named the "ELB agent" for EL Laboratories in Soquel, CA. This rickettsia has been found in Virginia opossums and cat fleas (which are common ectoparasites of this opossum) in the United States and has been shown to have caused infection in humans that are serologically positive for infection with *R. typhi*, the etiologic agent of murine typhus. Definitive demonstration of infection by either *R. felis* or *R. typhi* involves PCR amplification of specific nucleotide primers. Thus, it is likely that some human infections serologically attributed to *R. typhi* are actually caused by *R. felis*. *R. felis* is also of interest because it is transmitted transovarially in fleas, and fleas rather than mammals may be the reservoirs for the organism. Azad *et al.* (1997) provide further information on *R. felis* infections. In addition to these rickettsiae, some other fleas are known to harbor symbiotic rickettsiae about which little is currently known.

## OTHER FLEA-BORNE PATHOGENS

Table II lists other pathogens known to be transmitted by fleas. Most of these microorganisms principally occur in the flea gut rather than the salivary glands or other organs. It has been suggested that this is why fleas are ineffective vectors of viruses (Bibikova 1977). However, it is now known that murine typhus rickettsiae can escape the flea gut and multiply in other organs. Transmission of these gut-localized pathogens therefore occurs either by regurgitation (anterior station) or defecation (posterior station) during or soon after flea feeding. The apparent ease with which fleas can acquire and harbor a wide variety of infectious agents indicates why these insects play a major role in the maintenance and epidemiology of enzootic infections among rodents and other mammals. Many of these pathogens can produce disease in humans and domestic animals if these fleas, or bridge vectors, feed on these hosts.

### Bacteria

In addition to flea-borne rickettsial organisms, the following bacterial agents cause diseases that affect humans: *Francisella tularensis* causing *tularemia*, *Salmonella enteriditis* causing *salmonellosis*, and *Staphylococcus aureus* causing *staphylococcal infection*. All of these agents also can be transmitted by other means, such as other ectoparasites and contact or aerosol exposure to infective fomites and mammalian tissues. All three infections are widespread, and the degree of involvement of fleas in transmission varies regionally. Another bacterial agent that may be transmitted by fleas is *Bartonella* (formerly *Rochalimaea*) *henselae*, the agent of *cat scratch disease*. Infections by this zoonotic agent typically cause swollen regional lymph nodes. Long-term bacteremia can occur in inapparently infected cats which appear to be important reservoir hosts. More than 20,000 cases of human infection occur annually in the United States. Infection with *B. henselae* also can cause fever, hepatitis, endocarditis, *bacillary angiomatosis*, and *bacillary peliosis*. The last two conditions manifest as vascular proliferations and are most commonly seen in immunocompromised persons such as HIV-positive individuals. Cat fleas can transmit *B. henselae* by bite to cats under laboratory conditions. Although cat fleas may also be capable of transmitting this pathogen to humans, a scratch from an infected cat appears to

be the usual mode of transmission. A related organism, *B. clarridgeiae*, can also occur in cats and cause infection in humans, but the potential role of fleas as vectors of this agent has not been determined.

The following zoonotic bacterial agents also have been detected in fleas: *Borrelia burgdorferi*, the etiologic agent of *Lyme disease; B. duttoni*, an agent of *relapsing fever; Listeria monocytogenes*, the agent of *listeriosis; Y. pseudotuberculosis*, causing *pseudotuberculosis (yersiniosis); Erysipelothrix rhusiopathiae*, causing *erysipelas;* and *Brucella abortis* causing *brucellosis.* However, fleas are not known to be vectors of these agents.

### Viruses

Although several viral pathogens of humans have been isolated from, or detected in, fleas, the role of fleas in their transmission is either unknown or considered to be incidental. These viruses include those that cause *lymphocytic choriomeningitis, tick-borne encephalitis, Russian spring–summer encephalitis,* and *Omsk hemmorhagic fever.*

It should be emphasized that demonstration of a pathogen within an arthropod does not necessarily imply that it is a vector of the agent. Further information on the possible involvement of fleas in the transmission of some of the above-mentioned pathogens is provided by Jellison (1959), Bibikova (1977), Hopla (1982), Traub (1985), and Hopla and Hopla (1994).

## TUNGIASIS

Tungiasis is the pathological condition resulting from infestation by fleas belonging to the genus *Tunga*. Although there are several species of *Tunga*, only the *chigoe* (*T. penetrans*) attacks humans. *T. penetrans* occurs in many tropical and subtropical zones but is especially common in the New World tropics, the West Indies, tropical Africa, and southern India. The first record of this flea was in 1492 from crewmen of Christopher Columbus stationed in Haiti. It apparently spread from the New World to other areas of the world by shipping commerce, being first recorded on the African continent in 1732 as a consequence of the slave trade.

Females of *T. penetrans* usually invade a site between the toes, beneath the toe nails, or on the soles of the feet. Other sites may include the arms, especially around the elbow, and genital region in heavy infestations. Skin invasion by this flea can cause painful, subcutaneous lesions that often lead to more serious medical complications. The embedded chigoe (Fig. 7.13) invariably causes intense irritation and can result in secondary infections that ooze pus. When several chigoes attack an individual host at the same time, ulcerations often develop as the resultant lesions coalesce. Tetanus, cellulitis, and impaired blood flow to the site often lead to gangrene and may necessitate amputation of toes or, sometimes, an entire foot. Chigoe lesions therefore should receive prompt medical attention. Although the flea can be removed using a sterile needle or scalpel, it is important that lesions be thoroughly cleaned and dressed to avoid infection. This also applies to embedded dead fleas, which may rapidly cause affected tissues to fester and ulcerate if left untreated. The best defense against tungiasis is to avoid walking barefoot on beaches and other sandy soils in endemic regions where this flea develops.

## FLEAS AS INTERMEDIATE HOSTS OF HELMINTHS

Certain fleas are intermediate hosts for the cysticercoid stage of three species of tapeworms that occasionally infest humans. The most important of these is the *double-pored tapeworm* (*Dipylidium caninum*), the adults of which normally parasitize dogs. Gravid proglottids are released by *D. caninum* adults in the gut of the definitive host and are voided in the feces. The subsequently expelled eggs are ingested by flea larvae; the chewing mandibles of the larvae enable them to ingest the eggs, whereas the sucking mouthparts of adult fleas do not. Fleas such as *Ctenocephalides felis*, *C. canis*, and *Pulex irritans* play a significant role as intermediate hosts for this tapeworm. The dog chewing louse (*Trichodectes canis*) occasionally ingests *D. caninum* eggs and also can serve as an intermediate host.

The tapeworm develops slowly in flea larvae but rapidly in flea pupae. Cysticercoids can be seen in the body cavity of larvae and pupae, where they remain through development of the flea to the adult stage. Some flea mortality occurs in the pupal stage due to this helminth. Infestation of the human (definitive) host occurs when a person incidentally ingests an infested flea. The cysticercoid is liberated from the flea by digestive enzymes, after which it everts and attaches to the gut of its new host. Children playing with pets are especially susceptible to infestation by this tapeworm.

Two other tapeworms that utilize fleas as intermediate hosts are the *rodent tapeworm* (*Hymenolepis diminuta*) and the *dwarf tapeworm* (*H. nana*). Both infest rodents and occasionally parasitize humans, especially children. The development and transmission of these two cestodes are similar to that for *D. caninum*. Both *H. diminuta* and *H. nana* form viable cysticercoids in several species of fleas, especially *C. canis*, *P. irritans*, *X. cheopis*, and *N. fasciatus*. They also infest several other arthropods, notably coprophagous beetles.

The zoonotic nematode *Trichinella spiralis*, which causes *trichinosis*, has also been found in fleas, although this is assumed to represent an accidental association.

## VETERINARY IMPORTANCE

Several species of fleas are important ectoparasites of domestic and wild animals. Emphasis here is given to those which infest pets and livestock. Many fleas associated with domestic animals merely cause a nuisance through their biting activity; they also may cause flea-bite dermatitis, allergies, and anemia when present in large numbers. Other fleas, such as sticktights and chigoes, embed their mouthparts or entire bodies in mammalian or avian tissues, causing local inflammation and other problems. Some fleas are intermediate hosts of helminths that parasitize domestic animals, whereas others transmit pathogens such as viruses and trypanosomes to their hosts.

The *cat flea* (*C. felis*) is an extremely important ectoparasite, not only of cats and dogs, but also several other mammals, including opossums, cattle, horses, sheep, goats, rabbits, and monkeys. Some populations of *C. felis* have adapted to certain hosts, such as dogs or cattle, and show a preference for feeding on these species. Occasionally cat fleas infest goats, lambs, calves, or other ungulates in large numbers and can cause anemia or even death. Individual pets, especially cats and dogs, may support hundreds or thousands of cat fleas. Because the larvae thrive on blood-rich fecal pellets voided by adult fleas on the host, it is important to vacuum or treat areas where pets rest or sleep to reduce flea numbers. The *dog flea* (*C. canis*) is a relatively infrequent ectoparasite of dogs, with established populations persisting in only a few regions such as Austria, Ireland, and New Zealand. Almost invariably, fleas associated with dogs are *C. felis*. Further details on the biology of fleas associated with cats and dogs are provided by Dryden (1993) and Rust and Dryden (1997).

Several species of fleas are parasites of domestic and laboratory rats and mice. These include the *Oriental rat flea* (*X. cheopis*), the *northern rat flea* (*N. fasciatus*), and the *European mouse flea* (*Leptopsylla segnis*). Flea infestations of these rodents are usually more important with respect to potential transmission of pathogens rather than their discomforting bites. The *European rabbit flea* (*Spilopsyllus cuniculi*) is a parasite of the European rabbit (*Oryctolagus cuniculus*) throughout much of the world, where it has been introduced as a game or small-livestock animal. Since this is the laboratory rabbit commonly used in scientific studies, the European rabbit flea occasionally is recorded in animal research facilities. This flea is commonly a pest in rabbit hutches and where European rabbits are raised commercially for food in many parts of the world. *S. cuniculi* usually attaches to the ears, where it embeds its mouthparts deeply and for long periods, causing host irritability and ear scabbing.

In Central Asia, the *alakurt fleas Dorcadia ioffi* and *Vermipsylla alakurt* parasitize ungulates, especially horses, sheep, and yaks. These fleas often occur in very large numbers on these hosts and can cause anemia, hair loss, retarded growth, unthriftiness, and occasionally death, especially in newborn lambs.

Other fleas that are annoying biters of domestic mammals include *P. simulans,* the *human flea* (*P. irritans*), and the *sticktight flea* (*Echidnophaga gallinacea*), all of which may be recovered from cats or dogs. *P. simulans* and the human flea can be important ectoparasites of dogs and swine, while the sticktight flea infests domestic rats and several other mammals.

Several species of fleas feed on birds. At least three of these are important pests to the poultry industry. The *sticktight flea* is principally a poultry pest in the subtropical and tropical regions of the New World. These small fleas typically attach to the nonfeathered areas of birds such as the head, comb, wattle (Fig. 7.11), and anus. Large flea populations can cause anemia. Feeding sites can become ulcerated; when this occurs around the eyes, blindness often results and the host is unable to feed. Secondary infections may develop. The *European chicken flea* (*Ceratophyllus gallinae*) is a nonsedentary ectoparasite of domestic fowl in several parts of the world, including eastern North America. In western North America the *western chicken flea* (*C. niger*), another nonsedentary species, is a parasite of domestic fowl and several species of wild birds. All of these poultry fleas can cause host emaciation and reduced egg production when they occur in large numbers.

## FLEA-BITE DERMATITIS

Allergic skin reactions to flea bites are a common problem of domestic animals, especially household pets. Hypersensitivity to saliva from feeding fleas is usually more apparent in pets than in humans because larger numbers of pets are bitten by fleas. A single flea bite can trigger an acute, sometimes chronic, dermatitis in hypersensitive dogs or cats. Incessant scratching and skin irritation, especially during the warmer months, often reflects this condition. In cats, flea-bite dermatitis usually manifests as purplish papules that are often covered with crusts; in dogs, crusts are typically absent. In both cats and dogs, lesions are usually concentrated on the rump and inner thighs, with accompanying fur loss from frequent scratching. Cats sometimes also have a ring of crusts around the neck or a general pruritus. Diligent flea control is important in combating this condition. Administration of corticosteroids or a course of desensitizing antigens is another treatment option. Except in severe cases, the hypersensitivity often resolves after repeated flea bites as the host gradually becomes desensitized to antigens in flea saliva.

## TUNGIASIS

Some domestic animals, especially hogs and occasionally dogs, are parasitized by *Tunga penetrans* causing tungiasis. Infestations in hogs primarily affect the feet but also the snout, teats, legs, and scrotum. Infestations of the teats can result in restricted milk flow in nursing sows and starvation of the piglets. Swine are reservoirs of tungiasis, which can be transferred to humans in some tropical climates. There are at least eight other species of fleas belonging to the genus *Tunga* which burrow into host tissues. Females of each of these species are subdermal parasites which mostly attack New World rodents. Hopkins and Rothschild (1953) discuss other species of *Tunga*.

## MYXOMATOSIS

Myxomatosis is primarily a disease of the European rabbit caused by infection with the *myxoma virus*. The virus causes benign fibromas in its natural rabbit hosts in California, Central America, and South America. However, in the European rabbit, a severe and usually fatal infection with enlarging skin lesions and generalized viremia occurs. The myxoma virus was introduced to Australia in 1950 and to Europe in 1953. The aim of these introductions was to control burgeoning populations of European rabbits.

The virus is mechanically transmitted to rabbits by various blood-feeding arthropods, particularly mosquitoes. However, the European rabbit flea (*S. cuniculi*) is also a proven vector, at least in Britain, where this flea occurs naturally, and in Australia, where it was introduced in 1966. Although it is an inefficient vector, an Australian sticktight flea (*E. myrmecobii*) also can transmit the myxoma virus to rabbits. Infection with this virus apparently does not adversely affect these flea vectors. As with most other flea-transmitted pathogens, myxoma virus remains confined to the gut and mouthparts of *S. cuniculi*. Survival of the virus for 3–4 months in infected fleas has been demonstrated.

Because strains of the virus differ in virulence while rabbit populations differ in their susceptibility to this pathogen, the success of this virus in controlling rabbits has been variable. When the virus was first introduced to Australia and Europe it was very effective in culling wild rabbits; today, however, many rabbit populations in both

Australia and Europe have developed a resistance to several strains of the virus.

An overview of flea-transmitted myxomatosis is provided by Mead-Briggs (1977), and details on myxomatosis in general are provided by Fenner and Ross (1994).

## MURINE TRYPANOSOMIASIS

*Trypanosoma lewisi* is the causative agent of murine trypanosomiasis in domestic rats throughout much of the world. It is principally transmitted by the Northern rat flea (*N. fasciatus*) and the Oriental rat flea (*X. cheopis*). Fleas imbibe trypanosomes while feeding on infected rats; the pathogen remains in the flea midgut where development occurs. Within 6 hr after ingestion, the trypanosomes invade midgut epithelial cells, transform into pear-shaped forms, and begin to divide. The parasitized gut cells rupture after 18 hr to 5 days to release the trypanosomes; these then either invade new epithelial cells to repeat the process or move posteriorly to the rectum and anus. Trypanosomes in this "rectal phase" are voided in the flea feces. The trypanosomes enter their rat hosts when the latter lick and scratch their fur during grooming, representing a classic example of posterior-station transmission. Murine trypanosomiasis is usually a benign infection in rats. However, the *T. lewisi*-flea-rat system has been used as a laboratory model for devising therapies and studying the development of immunity against more virulent trypanosome species that are pathogenic to humans and domestic animals.

At least nine species of trypanosomes other than *T. lewisi* are transmitted to rodents by fleas. Rodent trypanosomes with confirmed flea transmission cycles include *T. musculi* (synonym: *T. duttoni*) of house mice, *T. rabinowitschi* of hamsters, *T. neotomae* of wood rats, and *T. grosi* of the European wood mouse (*Apodemus sylvaticus*). *T. nabiasi* is one of two species of trypanosomes known to be transmitted to rabbits by fleas. Fleas are also suspected as vectors of trypanosomes associated with some birds, shrews, voles, and lagomorphs.

## OTHER FLEA-BORNE PATHOGENS AND PARASITES

Many of the flea-borne pathogens listed in Table II cause diseases in humans, with wild or domestic animals serving as reservoirs. These include plague, tularemia, murine typhus, Q fever, and sylvatic epidemic typhus. Infections of domestic animals with most of these pathogens can be nonapparent, febrile, or fatal, depending on the host species, its health, and the strain of pathogen involved.

Cats, for example, are typically susceptible to most strains of plague, whereas dogs usually are not.

Other pathogens of veterinary importance that have been isolated from, or detected in, fleas include *lymphocytic choriomeningitis* virus, which affects many mammals, especially rodents, *feline leukopenia* virus, and the following bacterial agents: *Borrelia burgdorferi*, the causative agent of *Lyme disease*; *Listeria monocytogenes*, the agent of *listeriosis*, mainly in ungulates; *Brucella abortis*, an agent of *brucellosis*, mainly in bovines; *Pseudomonas mallei*, the agent of *glanders* in equines; and *P. pseudomallei*, the agent of *melioidosis* in several mammals. However, the role of fleas as significant vectors of these pathogens is doubtful or undetermined. Other microorganisms known to occur in fleas and which may be transmitted to vertebrates include *haemogregarine sporozoans*, various rickettsial organisms, and miscellaneous *symbionts*. The protozoan *Hepatozoon erhardovae* is transmitted to European voles (*Clethrionomys* spp.) by at least five species of fleas. The parasite reproduces sexually in the hemocoel of fleas, where it develops to the sporocyst stage; transmission to voles occurs when they eat infected fleas during grooming. The related *Hepatozoon* species *H. pitymysi* and *H. sciuri*, which parasitize North American voles, Eurasian voles, and North American squirrels, respectively, have also been detected in fleas and are thought to be transmitted in a similar way.

## FLEAS AS INTERMEDIATE HOSTS OF HELMINTHS

The *double-pored tapeworm* (*Dipylidium caninum*) normally develops as an adult parasite in the intestines of dogs, cats, and some wild carnivores. The most important intermediate flea hosts are the cat flea and dog flea, although the human flea can also serve in this capacity. In tropical Africa, a warthog flea (*E. larina*) is sometimes responsible for *D. caninum* infestations in domestic dogs. Infestations are usually initiated when animals consume parasitized fleas while grooming.

Two species of tapeworms that typically infest rats and mice as adults are the *rodent tapeworm* (*Hymenolepis diminuta*) and the *dwarf tapeworm* (*H. nana*). Rat fleas, especially the Oriental rat flea and the northern rat flea, serve as intermediate hosts. Infestations are initiated when infested fleas are eaten by the definitive rodent hosts.

The onchocercid nematode, *Acanthocheilonema* (formerly *Dipetalonema*) *reconditum*, which causes a relatively benign form of *canine filariasis* in many parts of the world, has been found in several species of fleas. The cat flea and dog flea are considered to be the principal vectors. Transmission of mature larvae by these fleas occurs by bite. Dogs, jackals, and hyenas are the principal definitive hosts of *A. reconditum*.

Several other species of helminths have been isolated from wild-caught fleas, and fleas have been found to serve as suitable intermediate hosts under laboratory conditions. However, the importance of fleas in maintaining these pathogens in nature is unknown. For example, the *trichina worm* (*Trichinella spiralis*) has been found encysted in the Oriental rat flea in India; this helminth normally encysts in muscle tissue of rats and hogs, causing *trichinosis*.

## PREVENTION AND CONTROL

Various methodologies are used to control fleas or to protect humans and other animals from flea bites. Frequent vacuuming in homes, especially in areas where pets rest or sleep, helps to remove immature fleas; steam cleaning of carpets is even more effective. Household foggers which produce a mist of insecticide in closed, temporarily vacated rooms often are effective in combating household flea infestations. Treatment of flea-infested premises or domestic animals with various insecticides generally provide good flea control. Commercially available products include botanical derivatives, carbamates, organophosphates, pyrethroids, boron compounds, and diatomaceous earth. Some nonchemical techniques are effective in reducing flea populations in homes, especially on or around pets. *Sticky traps* and *pan traps* are included in this category. Pan traps are trays of colored detergent water, often with an attractant light source, in which fleas drown. A flea light-trap fitted with a green-yellow filter with a transmittance spectrum centered at 515 nm is effective against cat fleas.

Several botanical derivatives, such as the pyrethrins, have low mammalian toxicity and are useful as flea powders for dusting flea-infested pets. Products such as flea soaps, which contain fatty acids, and flea shampoos can be used to bathe pets. Bathing removes or drowns many fleas, while those fleas that survive often desiccate because their integumental waxes have been removed by the detergents. Flea combs and flea collars for pets are also effective if used correctly. These collars should not be worn by humans (e.g., on the ankles), because they can cause skin irritation and allergic reactions. Some progress has been made in developing vaccines against fleas, mainly using midgut antigens of the cat flea to induce an immune response in the host. In several trials, dogs, cats, and rabbits that were experimentally challenged with cat-flea antigens had significantly more dead or reproductively compromised fleas than did nonvaccinated animals.

*Insect growth regulators* (IGRs), especially formulations of methoprene, pyriproxyfen, and fenoxycarb, currently are popular flea-control weapons because they have low mammalian toxicity. At low concentrations, these compounds interfere with flea development and eventually (after 1–2 months) provide high levels of flea control. IGRs can be applied as dusts or in pet shampoos. The orally administered systemic parasiticide ivermectin and the growth regulators cryomazine and lufenuron also reduce flea populations on domestic animals, especially cats and dogs. Similarly, the topically applied flea neuroinhibitor imidacloprid, a synthetic nitromethylene compound, is efficacious against cat fleas on pets. Two biological control agents, the *parasitic nematode Steinernema carpocapsae* and the *entomopathogenic fungus Beaveria bassiana*, reduce cat flea numbers under laboratory conditions and show promise as future control agents.

Personal protectants such as those containing DEET (*N*, *N*-diethyl-m-toluamide) or permethrin are often very helpful in reducing the number of flea bites. Permethrin should be applied only to clothing and not directly on the skin. Although banned for use in the United States, DDT (dichlorodiphenyltrichloroethane) is still used to control outbreaks of plague or murine typhus in some parts of the world. As with the use of other insecticides, there is a constant risk that fleas may develop resistance to these chemicals.

Plague outbreaks are usually followed by public education and area-wide programs to remove rodent hosts and flea vectors. Control programs for murine typhus typically involve eliminating the flea vectors or rodent reservoirs by insecticide applications and trapping, respectively. Rodent harborages and access of these reservoir hosts to houses should be eliminated where feasible. Dusting rodent burrows with insecticides or providing rodent bait stations spiked with either rodenticides or flea-control agents can be effective in killing the rodent hosts or fleas, respectively. Frequent surveillance of rodent and flea populations in plague-endemic regions often allows control measures to be implemented before human cases occur. Outbreaks of murine typhus may be handled in a similar manner, although there is greater emphasis on rodent control because the reservoir hosts are more likely to be commensal rats.

An ineffective approach to flea control is the use of ultrasonic repellent devices. No fleas that have been tested have shown responses to ultrasound or to devices incorporating it. Nor has the oral intake of garlic or of B-complex vitamins, including Brewer's yeast, been proven to reduce flea populations on pets, despite claims about their effectiveness. Flea control strategies are addressed by MacDonald (1995), Rust and Dryden (1997), Hinkle *et al.* (1997), and Dryden (1999).

# REFERENCES AND FURTHER READING

Askew, R. R. (1973). Parasitic insects. Heinemann, London.

Azad, A. F. (1990). Epidemiology of murine typhus. *Annu. Rev. Entomol.* **35**, 553–569.

Azad, A. F., Radulovic, S., Higgins, J. A., Noden, B. H., and Troyer, J. M. (1997). Flea-borne rickettsioses: ecologic considerations. *Emerg. Infect. Dis.* **3**, 319–327.

Bahmanyar, M., and Cavanaugh, D. C. (1976). Plague manual. World Health Organization, Geneva.

Baldo, B. A. (1993). Allergenicity of the cat flea. *Clin. Exp. Allergy* **23**, 347–349.

Barnes, A. M. (1982). Surveillance and control of bubonic plague in the United States. *Symp. Zool. Soc. Lond.* **50**, 237–270.

Benton, A. H. (1980). An atlas of the fleas of the eastern United States. Marginal Media, Fredonia, NY.

Benton, A. H. (1983). An illustrated key to the fleas of the eastern United States. Marginal Media, Fredonia, NY.

Bibikova, V. A. (1977). Contemporary views on the interrelationships between fleas and the pathogens of human and animal diseases. *Annu. Rev. Entomol.* **22**, 23–32.

Craven, R. B., Maupin, G. O., Beard, M. L., Quan, T. J., and Barnes, A. M. (1993). Reported cases of human plague infections in the United States, 1970–1991. *J. Med. Entomol.* **30**, 758–761.

Dryden, M. W. (1993). Biology of fleas of cats and dogs. *Comp. Cont. Educ. Pract. Vet.* **15**, 569–579.

Dryden, M. W. (1999). Highlights and horizons in flea control. *Comp. Cont. Educ. Pract. Vet.* **21**, 296–298, 361–365.

Duplaix, N. (1988). Fleas. The lethal leapers. *Natl. Geogr.* **173**, 672–694.

Elbel, R. E. (1991). Order Siphonaptera. *In:* Immature insects, Vol. 2 (F. W. Stehr, ed.). pp. 674–689. Kendall Hunt, Dubuque, IA.

Ewing, H. E., and Fox , I. (1943). The fleas of North America. Classification, identification, and geographic distribution of these injurious and disease-spreading insects. US Department of Agriculture. Miscellaneous publication No. 500. US Govt. Printing Office, Washington, DC.

Fenner, F., and Ross, J. (1994). Myxomatosis. *In:* The European rabbit. The history and biology of a successful coloniser (H. V. Thompson and C. M. King, eds.). pp. 205–239. Oxford University Press, Oxford.

Fox, I. (1940). Fleas of eastern United States. Iowa State College Press, Ames.

Goddard, J. (1998). Fleas and murine typhus. *Infect. Med.* **15**, 438–440.

Goddard, J. (1999). Fleas and plague. *Infect. Med.* **16**, 21–23.

Harwood, R. F., and James, M. T. (1979). Entomology in human and animal health (7th ed.). Macmillan, New York.

Hinkle, N. C., Rust, M. K., and Reierson, D. A. (1997). Biorational approaches to flea (Siphonaptera, Pulicidae) suppression—present and future. *J. Agric. Entomol.* **14**, 309–321.

Hinnebusch, J. B., Perry, R. D., and Schwan, T. G. (1996). Role of the *Yersinia pestis* hemin storage (hms) locus in the transmission of plague by fleas. *Science* **273**, 367–370.

Holland, G. P. (1985). The fleas of Canada, Alaska and Greenland (Siphonaptera). Memoirs of the Entomological Society of Canada, No. 130. Entomological Society of Canada, Ottawa.

Hopkins, G. H. E., and Rothschild, M. (1953–1971). An illustrated catalogue of the Rothschild collection of fleas (Siphonaptera) in the British Museum (Natural History) (Vols. 1–5). British Museum (Natural History), London.

Hopla, C. E. (1982). Arthropodiasis. *In:* CRC handbook series in zoonoses, Section C: parasitic zoonoses. Vol. 3 (J. H. Steele, ed. in chief). pp. 215–247. CRC Press, Boca Raton, FL.

Hopla, C. E., and Hopla, A. K. (1994). Tularemia. *In:* Handbook of zoonoses, Section A: bacterial, rickettsial, chlamydial, and mycotic (2nd ed., G. W. Beran, ed. in chief). pp. 113–123. CRC Press, Boca Raton, FL.

Hubbard, C. A. (1947). Fleas of western North America. Iowa State College Press, Ames.

Ibáñez-Bernal, S., and Velasco-Casrejón, O. (1996). New records of human tungiasis in Mexico (Siphonaptera: Tungidae). *J. Med. Entomol.* **33**, 988–989.

Jellison, W. L. (1959). Fleas and disease. *Annu. Rev. Entomol.* **4**, 389–414.

Lewis, R. E. (1972–1993a). Notes on the geographic distribution and host preferences in the order Siphonaptera. Parts 1–8. *J. Med. Entomol.* **9**, 511–520 (1972); **10**, 255–260 (1973); **11**, 147–167, 403–413, 525–540, 658–676 (1974); **22**, 134–152 (1985—with J. H. Lewis); **30**, 239–256 (1993a).

Lewis, R. E. (1993b). Fleas (Siphonaptera). *In:* Medical insects and arachnids (R. P. Lane and R. W. Crosskey, eds.). Chapman & Hall, London. pp. 529–575.

Lewis, R. E. (1998). Résumé of the Siphonaptera (Insecta) of the world. *J. Med. Entomol.* **35**, 377–389.

Lewis, R. E., and Lewis, J. H. (1994). Siphonaptera of North America north of Mexico. *J. Med. Entomol.* **31**, 82–98 (Vermipsyllidae and Rhopalopsyllidae), 348–368 (Ischnopsyllidae), 795–812 (Hystrichopsyllidae, *s. str.*).

Lewis, R. E., Lewis, J. H., and Maser, C. (1988). The fleas of the Pacific Northwest. Oregon State Univ. Press, Corvallis.

MacDonald, J. M. (1995). Flea control: an overview of treatment concepts for North America. *Vet. Dermatol.* **6**, 121–130.

Madon, M. B., Hitchcock, J. C., Davis, R. M., Myers, C. M., Smith, C. R., Fritz, C. L., Emery, K. W., and O'Rullian, W. (1997). An overview of plague in the United States and a report of investigations of two human cases in kern County, California. *J. Vector Ecol.* **22**, 77–82.

Mardon, D. K. (1981). An illustrated catalogue of the Rothschild collection of fleas (Siphonaptera) in the British Museum (Natural History). Vol. 6. Pygiopsyllidae. British Museum (Natural History), London.

Marshall, A. G. (1981). The ecology of ectoparasitic insects. Academic Press, London.

Matheson, R. (1950). Medical entomology (2nd ed.). Cornell Univ. Press, Ithaca.

McDade, J. E. (1987). Flying squirrels and their ectoparasites: disseminators of epidemic typhus. *Parasitol. Today* **3**, 85–87.

Mead-Briggs, A. R. (1977). The European rabbit, the European rabbit flea and myomatosis. *In:* Applied biology. Vol. 2 (T. H. Coaker, ed.). pp. 183–261. Academic Press, London.

Mee, C. L., Jr. (1990). How a mysterious disease laid low Europe's masses. *Smithsonian* **20**(11), 66–79.

Perry, R. D., and Fetherston, J. D. (1997). *Yersinia pestis*—etiologic agent of plague. *Clin. Microbiol. Rev.* **10**, 35–66.

Poland, J. D., Quan, T. J., and Barnes, A. M. (1994). Plague. *In:* Handbook of zoonoses, Section A: bacterial, rickettsial, chlamydial and mycotic (2nd ed., G. W. Beran, ed.-in-chief). pp. 93–112. CRC Press, Boca Raton, FL.

Rawlings, J. A., and Clark, K. A. (1994). Murine typhus. *In:* Handbook of zoonoses, Section A: bacterial, rickettsial, chlamydial and mycotic (2nd ed., G. W. Beran, ed. in chief). pp. 457–461. CRC Press, Boca Raton, FL.

Rothschild, M. (1973). The flying leap of the flea. *Sci. Am.* **229**, 92–100.

Rothschild, M., and Ford, B. (1973). Factors influencing the breeding of the rabbit flea (*Spilopsyllus cuniculi*): a spring-time accelerator and a kairomone in nestling rabbit urine (with notes on *Cediopsylla simplex,* another "hormone bound" species). *J. Zool.* **170**, 87–137.

Rust, M. K., and Dryden, M. W. (1997). The biology, ecology, and management of the cat flea. *Annu. Rev. Entomol.* **42,** 451–473.

Smit, F. G. A. M. (1957). Handbooks for the identification of British insects. Siphonaptera. Vol. 1, Pt. 16. Royal Entomological Society, London.

Smit, F. G. A. M. (1987). An illustrated catalogue of the Rothschild collection of fleas (Siphonaptera) in the British Museum (Natural History). Vol. 7. Malacopsylloidea. Oxford Univ. Press, Oxford, and British Museum (Natural History), London.

Stark, H. E., Hudson, B. W., and Pittman, B. (1966). Plague epidemiology. US Department of Health, Education, and Welfare, Public Health Service, Centers for Disease Control and Prevention, Training Branch, Atlanta.

Traub, R. (1985). Coevolution of fleas and mammals. *In:* Coevolution of parasitic arthropods and mammals (K. C. Kim, ed.). pp. 295–437. Wiley, New York.

Traub, R., Rothschild, M., and Haddow, J. (1983). The Rothschild collection of fleas. The Ceratophyllidae: key to genera and host relationships with notes on their evolution, zoogeography and medical importance. Cambridge Univ. Press, Cambridge.

Traub, R., and Starcke, H. (eds.) (1980). Fleas. Proceedings of the International Conference on Fleas, Ashton, England, June 1977. Balkema, Rotterdam.

Traub, R., Wisseman, Jr., C. L., and Azad, A. F. (1978). The ecology of murine typhus—a critical review. *Trop. Dis. Bull.* **75,** 237–317.

Twigg, G. I. (1978). The role of rodents in plague dissemination: a worldwide review. *Mamm. Rev.* **8,** 77–110.

World Health Organization (1980). Plague surveillance and control. *WHO Chron.* **34,** 139–143.

# 8

# FLIES (*Diptera*)

ROBERT D. HALL AND REID R. GERHARDT

*The Diptera,* or "true flies," are one of the largest and most diverse orders of insects, both morphologically and biologically. The order name means "two-winged" and refers to the fact that the hind pair of wings is greatly modified and reduced. The number of described species worldwide is estimated to be 120,000 or more. There are perhaps 20,000 species of Diptera in the Nearctic Region, a significant proportion of which is cataloged (Stone *et al.,* 1965). Although flies with medical or veterinary significance constitute only a small fraction of these numbers, their diversity is impressive, ranging from mosquitoes to wingless ectoparasites, larvae that parasitize various animals, and species that help to decompose carrion or feces.

No other group of insects has as much impact on human and animal health as do the Diptera (Tables I and II). Mosquitoes, black flies, and biting midges annoy outdoor enthusiasts as well as livestock, pets, and other domestic or wild animals. Filth flies associated with cattle, hog, and poultry operations can annoy nearby residents and are frequently the focus of litigation. The ubiquitous house fly is an effective mechanical vector of many pathogens associated with enteric diseases. The depredation of blood-sucking and myiasis-producing flies has an adverse effect on the productivity and profitability of animal agriculture worldwide.

No other group of insects exhibits the number or diversity of vector relationships that have evolved among the Diptera (Table I). The two-volume treatise *Flies and Disease,* by Greenberg (1971, 1973), provides an exhaustive list of fly-pathogen associations. Mosquitoes stand as archetypical vectors, being associated with such historically notorious diseases as malaria, encephalitis, yellow fever, and human filariasis. The story of the United States Yellow Fever Commission in Cuba in 1900 and the names Carlos Finlay, Walter Reed (for whom the United States Army Medical Center in Washington, DC, is named), and L. O. Howard are familiar to most students of medicine. Such is the importance of mosquitoes that some institutions offer a separate course in *culicidology,* the study of mosquitoes. Insect-vectored tropical diseases such as malaria, filariasis, leishmaniasis,

TABLE I

Major Fly-Borne Diseases and Related Problems Affecting Human Health

| Family | Diseases and other health-related problems | Geographic occurrence |
|---|---|---|
| Psychodidae | Bartonellosis | Andes Mountains of Columbia, Eduador and Peru |
| | Leishmaniasis | New World tropics; Old World tropics and temperate regions |
| | Sand fly fever | Mediterranean area to southern China and India |
| Culicidae | Dengue fever | Widespread between latitudes 40°N and 40°S |
| | Encephalitis | Widespread |
| | Filariasis | Tropics and Mediterranean area |
| | Malaria | Widespread in humid tropics |
| | Yellow fever | Widespread in humid tropics |
| Simuliidae | Onchocerciasis | Tropical Africa and Americas |
| Tabanidae | Loiasis | Tropical Africa |
| | Tularemia | Widespread in Northern Hemisphere |
| Chloropidae | Conjunctivitis | United States (southern) and Mexico; Orient |
| Muscidae | Enteric diseases | Worldwide |
| Glossinidae | Trypanosomiasis | Tropical Africa |
| Calliphoridae | Enteric disease | Worldwide |
| | Myiasis | Worldwide |
| Sarcophagidae | Myiasis | Worldwide |
| Oestridae | Myiasis | Worldwide |

TABLE II

Major Fly-Borne Diseases and Related Problems Affecting Livestock, Poultry, and Other Domestic or Wild Animals

| Family | Diseases and other health-related problems | Geographic occurrence |
|---|---|---|
| Psychodidae | Leishmaniasis | New World tropics; Old World tropics and temperate regions |
| Ceratopogonidae | Bluetongue | Widespread |
| Culicidae | Malaria | Widespread in tropics |
| | Dirofilariasis | Widespread in tropics and temperate regions |
| | Encephalitis | Widespread |
| | Fowlpox | Widespread |
| | Yellow fever | Widespread in humid tropics |
| Simuliidae | Leucocytozoonosis | Widespread, especially North America |
| | Feeding damage | Worldwide |
| Tabanidae | Anaplasmosis | Widespread |
| | Tularemia | Widespread in Northern Hemisphere |
| | Exsanguination | Worldwide |
| Muscidae | Annoyance | Worldwide |
| | Bovine pinkeye | Northern Hemisphere (widespread) |
| | Exsanguination | Worldwide |
| Glossinidae | Nagana | Tropical Africa |
| Calliphoridae | Myiasis | Worldwide |
| Sarcophagidae | Myiasis | Worldwide |
| Oestridae | Myiasis | Worldwide |

TABLE III

Taxonomic Classification and Families of Diptera of Interest to Medical and Veterinary Entomologists

| Higher taxa | Family | Common names |
|---|---|---|
| Suborder Nematocera | Tipulidae | Crane flies |
| | Bibionidae | March flies |
| | Mycetophilidae | Fungus gnats |
| | Sciaridae | Darkwinged fungus gnats |
| | Psychodidae[a] | Moth flies, sand flies |
| | Chaoboridae | Phantom midges |
| | Culicidae[a] | Mosquitoes |
| | Simuliidae[a] | Black flies |
| | Ceratopogonidae[a] | Biting midges |
| | Chironomidae | Chironomid midges |
| Suborder Brachycera, | | |
| Infraorder Tabanomorpha | Tabanidae[a] | Horse flies, deer flies |
| | Rhagionidae | Snipe flies |
| | Athericidae | Athericid flies |
| | Stratiomyidae | Soldier flies |
| Infraorder Asilomorpha | None | None |
| Infraorder Muscomorpha | | |
| Division Aschiza | Phoridae | Humpbacked flies |
| | Syrphidae | Flower flies, hover flies |
| Division Schizophora | | |
| Section Acalyptratae | Piophilidae | Skipper flies |
| | Drosophilidae | Small fruit flies, vinegar flies |
| | Chloropidae | Chloropid flies, eye gnats |
| Section Calyptratae | Muscidae[a] | House flies, stable flies, and allies |
| | Glossinidae[a] | Tsetse |
| | Calliphoridae[a] | Blow flies |
| | Sarcophagidae[a] | Flesh flies |
| | Oestridae[a] (including Cuterebridae, Gasterophilidae, and Hypodermatidae) | Bot flies, warble flies |
| | Hippoboscidae[a] | Louse flies |
| | Nycteribiidae | Spiderlike bat flies |
| | Streblidae | Bat flies |

[a] Addressed in separate chapters.

and onchocerciasis currently affect almost half a billion humans worldwide, with about 3.5 billion rated at risk.

Flies are occasionally of direct use to humans. Knowledge of the taxonomy and biology of some necrophilous species makes them useful under certain circumstances in determining how long a body has been dead. This subspecialty of medical entomology, called *medicocriminal* or *forensic entomology,* is readily accepted in judicial circles.

Additional information regarding the medical and veterinary importance of Diptera is provided by Horsfall (1962), Smith (1973), Harwood and James (1979), Williams *et al.* (1985), and Lancaster and Meisch (1986).

# TAXONOMY

The order Diptera is divided by most authorities into two suborders: the Nematocera and the Brachycera (Table III). The Nematocera are typified by mosquitoes and other flies with conspicuously long antennae. The Brachycera include horse flies, deer flies, house flies, and other flies with short antennae. The Brachycera are subdivided into three infraorders: Tabanomorpha, including the horse flies and deer flies; Asilomorpha, robber flies and their relatives, which are generally of no medical or veterinary concern; and Muscomorpha, or "circular-seamed" flies, often called Cyclorrhapha. The Muscomorpha infraorder in turn is divided into the Aschiza and Schizophora, and the latter is divided into two sections:

the Acalyptratae and Calyptratae. This taxonomic scheme is essentially that proposed by McAlpine *et al.* (1981b) and followed by Borror *et al.* (1989). A catalog of the Diptera of America north of Mexico is provided by Stone *et al.* (1965).

Various keys are available for identifying adult flies. Keys to the families and genera of most Nearctic Diptera are presented in McAlpine (1981a). The flies of western North America are treated by Cole (1969). The key in Borror *et al.* (1989) is adequate for identification of most North American Diptera to the family level. The larvae of many Diptera can be identified to family with the aid of Teskey (1981a) and Foote (1991); those of synanthropic species are treated by Dusek (1971). Furman and Catts (1982) present a very usable key to both adults and larvae of medically important flies, particularly in the United States, and James (1947) covers flies that cause myiasis in humans. For identification of taxa outside the Nearctic Region, students should refer to Lindner's (1949) series on Palearctic Diptera and to Zumpt (1965) for Old World myiasis-causing flies.

## MORPHOLOGY

Nematoceran larvae range in length from only a few millimeters to many centimeters, depending on the species, and are usually distinguished by having a conspicuous head capsule with opposable mandibles that move in a pincer-like horizontal plane (Fig. 8.1). The general body shape ranges from minute and eel-like in the Ceratopogonidae to large and fleshy in the Tipulidae. Some nematocerans have thoracic *prolegs* (e.g., Chironomidae and Simuliidae) and others have caudal structures (e.g., Simuliidae) which assist in attachment to substrates. Although the early instars of many aquatic species depend on cuticular respiration, the later instars generally respire via gills or have various adaptations that permit them to obtain atmospheric air. Mosquito larvae, for example, are highly adapted, air-breathing nematocerans that hang from the water's surface film by respiratory siphons or specialized abdominal setae.

Tabanomorpha and Asilomorpha larvae have fang-like mandibles that move in a vertical plane; the head capsule is frequently described as "incomplete posteriorly," meaning that only the anterior parts are sclerotized (Fig. 8.2). The latter character is best seen in specimens that have been cleared in potassium hydroxide or lactophenol. Horse fly larvae are good examples of this group. They often have posterior respiratory tubes.

*Muscomorpha larvae* lack a sclerotized head capsule (Fig. 8.3A) and are commonly known as *maggots*. At the narrow, anterior end of the 12-segmented larva is

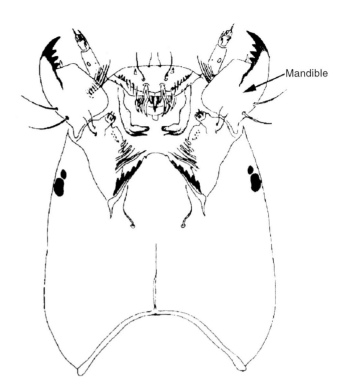

**FIGURE 8.1** Representative head capsule of nematoceran larva, with opposable mandibles; chironomid midge (Chironomidae), ventral view. (Redrawn from Merritt and Cummins, 1996.)

the *cephalopharyngeal skeleton* (Fig. 8.3C), which usually bears one or two mouth hooks used for feeding and in assisting the insect in movement. The caudal end of the maggot is broader and bears the posterior *spiracular plates* (Fig. 8.3E); like the cephalopharyngeal skeleton, they often are valuable for identification. The segments of the maggot typically bear spines in regular patterns (Fig. 8.3D), and the larvae of some species may possess structures that vary from simple setae to large protuberances. Others, such as cattle grubs and bot flies, are rounded and

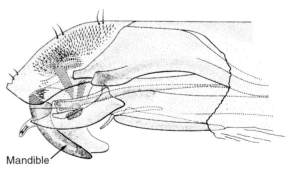

**FIGURE 8.2** Lateral view of anterior part of *Tabanus marginalis* larva (Tabanidae), showing incomplete head capsule and vertical, fanglike mouth hook. (From McAlpine *et al.*, 1981b.)

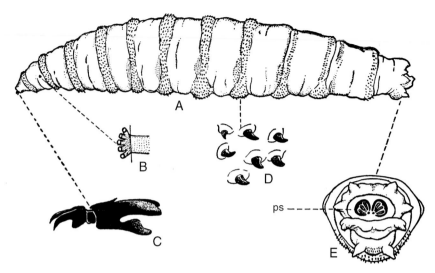

**FIGURE 8.3** Blow-fly larva, *Chrysomya bezziana* (Calliphoridae). (A) Complete larva; (B) anterior spiracle; (C) cephalopharyngeal skeleton; (D) spines; (E) caudal end with pair of spiracular plates. (From James, 1947.)

robust, and their cuticle is frequently armed with stout spines. They range up to several centimeters in length.

Nematocera adults possess elongate, filamentous *antennae* composed of six or more segments (Fig. 8.4A and 8.4B). The antennae usually are longer than the length of the head and thorax combined. A notable exception is the family Simuliidae, in which the antennae are short and compact (Fig. 8.4B). In those groups that feed on blood, only the females display this behavior, doing so by means of piercing/sucking *mouthparts* as in mosquitoes.

Tabanomorpha and Asilomorpha adults are characterized by relatively short *antennae* bearing a terminal *annulus*, or *stylus* (Fig. 8.4C). In general, these are large, robust flies. Like the Nematocera, only the females feed on blood. Members of the Tabanidae are good examples, being typically large, active flies whose females aggressively pursue blood meals. Their *mouthparts* are adapted for lacerating skin to feed on blood that pools at the wound site.

Muscamorpha adults have *antennae* that are *aristate*, bearing a large dorsal bristle (*arista*) on the apical antennal segment (Fig. 8.4D). Division Aschiza, typified by the phorids and syrphids, includes those Muscomorpha lacking a *frontal suture*, or *lunule*. Diptera in the Shizophora have a frontal suture (Fig. 8.5); this group includes a large number of species generally known as the *muscoid flies*. The Schizophora group is perhaps the most taxonomically complex group of Diptera. Members of the Acalyptratae, the *acalyptrate* muscoid flies, lack a dorsolateral seam on the second antennal segment, whereas this seam is present in the Calyptratae, the *calyptrate* muscoid flies (Fig. 8.4D). Calyptrate muscoid flies possess posterobasal wing lobes called *calypters* (Fig. 8.6), which cover the

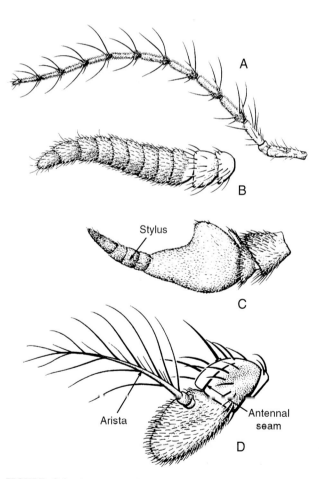

**FIGURE 8.4** Antennae of adult flies. (A) Tipulidae (*Tipula*); (B) Simuliidae (*Cnephia*); (C) Tabanidae (*Tabanus*); (D) Drosophilidae (*Drosophila*). (From McAlpine *et al.*, 1981b.)

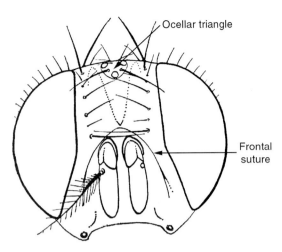

**FIGURE 8.5**   Frontal view of head of female fly, showing frontal suture and ocellar triangle at vertex. (From Greenberg, 1971.)

halteres. Included in the Calyptratae are the hippoboscoid flies, which are sometimes secondarily wingless.

The *mouthparts* of blood-feeding muscomorphan adults are of the piercing/sucking type. In contrast to other Diptera, both male and female Calyptratae suck blood in those species that exhibit this feeding style (e.g., horn flies and stable flies). Other species generally possess mouthparts that permit liquid food materials to be lapped or sponged. The latter type of mouthparts in some species have structures sclerotized enough to scarify tissue during feeding activities (e.g., the face fly).

The functional pair of wings in the Diptera arises from the mesothorax. The metathoracic wings are modified to form a pair of knobbed balancing organs known as *halteres* (Fig. 8.6). The wing venation is highly variable between groups and provides valuable taxonomic characters

for distinguishing the families. Many dipteran adults have characteristic wing patterns, including species of biting midges, deer flies, and horse flies,.

The adults of most Diptera possess distinct compound eyes; ocelli are present in a triangle on the vertex of many species (Fig. 8.5). Adults are identified easily to sex, because most species exhibit some degree of sexual dimorphism. Nematoceran males often possess densely plumose antennae, and the females of blood-sucking species bear stylet-like mouthparts. The eyes of brachyceran males typically meet along the dorsal midline of the head (*holoptic*), whereas the eyes of females are more widely separated (*dioptic*). The female abdomen ends in an ovipositor (larvipositor in some species), whereas the male abdomen typically bears distinct *genitalia* at the terminus. In the males of some Nematocera and Brachycera, the genital segments rotate a half-turn shortly after the adult fly emerges; thus, the genital capsule appears "upside down" in adults of those species. In the Schizophora, this rotation continues through a full circle, so that the genital capsule is in its normal position. A morphological approach to identification that has proven useful in the Diptera, particularly with the Muscomorpha, is the characteristic appearance of *male genitalia*. In many species the aedeagus, claspers, and associated structures are unique. "Pulling the tail" of male flies is a technique used by dipterists that permits detailed examination of the genital structures. Descriptions of species in some families, such as the Sarcophagidae, are based in large part on male specimens.

McAlpine (1981c) and Teskey (1981b) present comprehensive reviews of the morphology and related terminology of diptera adults and larvae, respectively.

## LIFE HISTORY

The Diptera are *holometabolous*. Most dipteran females lay eggs and are thus *oviparous*. Others are *ovoviviparous*, hatching their eggs internally and thus producing motile early-instar larvae. Such flies are called *larviparous*, as represented by flesh flies (Sarcophagidae). In a few dipteran groups, the developing larvae are retained within the female's body until they are ready to pupate. These flies are called *pupiparous* and include the louse flies (Hippoboscidae) and tsetse flies (Glossinidae). The number of offspring produced per female by larviparous and pupiparous species is low compared to oviparous and ovoviviparous species.

Many dipteran species inhabit aquatic or semiaquatic environments during their immature stages. Typical examples are mosquitoes, black flies, and most horse flies and deer flies. The females of many of these

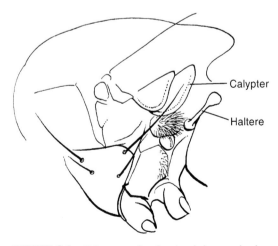

**FIGURE 8.6**   Calypterate fly, showing haltere and calypters. (From Greenberg, 1971.)

hematophagous flies are capable of producing an initial batch of eggs before obtaining a blood meal; this is known as *autogeny*. In contrast, those species that must feed on blood prior to their producing eggs are referred to as *anautogenous*.

While the number of larval instars varies within the Diptera, it remains generally constant for a given species. Mosquitoes and most other nematocerans have four larval instars, whereas most muscoid Diptera pass through three observable larval instars, with a fourth instar, the *prepupa*, occurring cryptically inside the pupal case. The muscoid instars usually can be distinguished morphologically: the first instar lacks anterior spiracles and generally has only one slit in each caudal spiracular plate; second and third instars bear anterior spiracles and have two and three slits, respectively, in the caudal spiracular plate. These slits are lacking in some groups; instead, the spiracular plate has many small openings (e.g., cattle grubs).

The *pupae* of Nematocera are *obtect*, with the appendages and other external body structures of the developing adult being discernible externally. Pupae are typically immobile; significant exceptions are the pupae of mosquitoes and a few other nematoceran families that can move by means of caudal paddles. The Brachycera have *coarctate* pupae, in which the pupa is encased within the hardened exuviae of the penultimate larval instar. The latter structure, called a *puparium*, is most frequently brown in color and is often said to resemble a "pill." It retains many morphological features of the larval integument. Adult flies emerge from the pupal case by employing hydrostatic pressure from the hemolymph to generate splits along predetermined lines. The head of most Schizophora has an eversible sac, the *ptilinum*, which facilitates the fly's escape from the puparium. After emergence, the ptilinum retracts through a fissure proximate to the lunule at the antennal bases.

# BEHAVIOR AND ECOLOGY

An aspect of fly behavior of particular interest to medical and veterinary entomologists is the *host-finding* capabilities of blood-feeding species. While various mechanisms have been described, they generally fall into two categories: olfactory and visual. A common *olfactory* cue used by blood-feeding insects, including many true flies, is the relative titer of carbon dioxide ($CO_2$) in the atmosphere surrounding, or downwind from, the host. If the goal of a mobile parasite is to locate a warm-blooded animal, exhaled $CO_2$ can serve as a cue for recognizing and locating potential hosts. A practical result is the widespread use of dry ice or bottled $CO_2$ to improve trapping success for common blood-feeding flies such as mosquitoes,

black flies, and no-see-ums. Other chemicals (e.g., mercaptans, octenol, and lactic acid) are used as olfactory cues by certain species. The principle means by which most insect repellents work is by inhibiting olfactory perception, thereby disrupting normal host-seeking behavior.

Visual host-finding cues are employed effectively by some flies, notably the Tabanidae. While entomologists have not been able to prove conclusively what horse flies and deer flies actually "see," there is little question that the blood-seeking females are sensitive to black-body radiation outside the spectrum visible to humans. It has been theorized in some cases that such females sense warmth against a cool background, in the manner that thermal-vision cameras are able to scan houses for heat leaks, crops for disease-induced stress, and nocturnal battlefields for invading personnel. The shape and size of hosts also may be important to some flies as they visually recognize or orient to certain host animals. In many instances, olfactory and visual cues presumably complement each other.

Another important aspect of fly behavior is the female's ability to identify an environment suitable for development of her offspring. As with host-finding behavior, olfaction can play an important role. Necrophilous blow flies and flesh flies appear quickly after an animal dies; olfaction is almost certainly their major cue, even though the odor may not be detectable by humans. Similarly, face flies appear at cattle dung pats almost immediately after cattle defecate. Most flies, in common with many other types of insects, can perceive chemical cues at a level many orders of magnitude greater than that of humans. The females of other dipteran species similarly locate appropriate breeding sites. As examples, salt marsh mosquitoes and tree-hole mosquitoes must select aquatic habitats suitable for their eggs. The females of some blow fly and flesh fly species are highly attracted to human feces, and female screwworms are readily drawn to sores or wounds on living hosts.

Flies of medical and veterinary importance afford excellent examples of both *K*- and *r*-strategies in their life history. A few dipterans are known as *K-strategists*, typified best by tsetse, sheep keds, and other members of the Pupipara. The symbol *K* represents the carrying capacity of the environment. These flies have longer life cycles, produce fewer offspring, and are particularly influenced by density-dependent mortality factors. More commonly, pest flies are *r-strategists*, in which large numbers of offspring are produced, with each individual having a relatively small chance of survival. The symbol *r* denotes the instantaneous rate of increase for a population. These flies typically exhibit rapid growth, short life cycles, and high mortality attributable mainly to density-independent factors. House flies and other filth-breeding species, as well as mosquitoes, serve as good examples.

# FAMILIES OF MINOR MEDICAL OR VETERINARY INTEREST

The major families of Diptera of medical/veterinary importance are treated in separate chapters of this book. The following discussion is provided for 13 other families of minor medical/veterinary importance which include species that can cause problems for humans and other animals.

## TIPULIDAE (CRANE FLIES)

Adults are slender-bodied flies, 5–60 mm in length. They have long stiltlike legs, lack ocelli, and have a V-shaped mesonotal suture (Fig. 8.7A). Many species are attracted to light and readily enter houses, where they may be mistaken for large mosquitoes. Some are known to feed on nectar, but none bite or are able to feed on blood. Tipulid larvae have a distinct head capsule which can be retracted into the anterior thoracic segments (Fig. 8.7B). They are found in a wide range of aquatic and semiaquatic habitats and are commonly collected at the margins of streams and ponds and in moist leaf litter. A few species occur in dry soil, where the larvae may be pests of grain and turf crops by feeding on the roots. Most species in temperate areas have one or two generations a year, with four larval instars and a brief pupal stage. The length of the life cycle varies from 6 weeks to 4 years, the latter being typical of some Arctic species.

The Tipulidae is a very large, cosmopolitan family of Diptera with over 60 genera and 1500 species described in North America. Keys to both the adults and larvae of the Nearctic genera are provided by Alexander and Byers (1981) and Byers (1984). Adult and larval ecology are presented in Knizesk and Sullivan (1984) and Freeman (1967), respectively.

## BIBIONIDAE (MARCH FLIES)

March flies are dark-colored flies varying in size from small to moderately large (4–10 mm). The adults (Fig. 8.8) generally can be distinguished from other Nematocera by the lack of a V-shaped suture on the mesonotum, the presence of ocelli, antennae inserted below the eyes, and the presence of tibial spurs and pulvilli. Adults usually emerge in the spring and feed on flower nectar and pollen. The larvae are scavengers and are found mostly in decaying organic materials such as forest litter, manure, and soils rich in humus. Some species cause damage to the roots of cultivated plants, especially cereal and grass crops.

Adults of the *love-bug* (*Plecia nearctica*) often emerge in large swarms along the Gulf and South Atlantic coasts

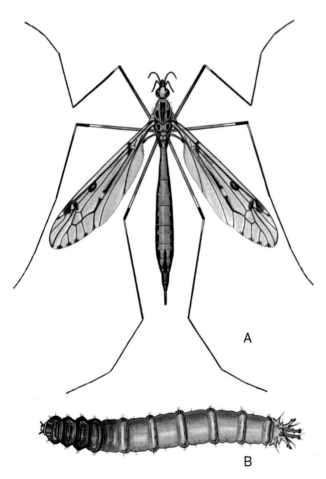

**FIGURE 8.7** Tipulidae (*Tipula*). (A) Adult; (B) larva. (From McCafferty, 1981.)

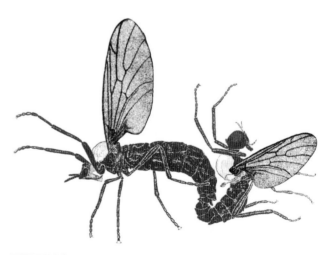

**FIGURE 8.8** Bibionidae, love bug (*Plecia nearctica*); pair of adults *in copula*, female at left. (From Leppla *et al.*, 1975.)

of the United States, mainly during May and September. Larvae are found in aggregations under moist, decaying materials, including leaves, grass clippings, Spanish moss, and manure. They are most often seen as copulating pairs and may remain *in copula* for several days, even while feeding together on flowers. When locally abundant, flying pairs can pose a hazard to automobile travelers by obscuring vision, clogging radiators, and occasionally damaging automobile finishes. Large flights occur in the United States along the Gulf Coast of Florida westward to Louisiana and eastern Texas and southward to Central America (Denmark and Mead, 1992). *P. nearctica* also occurs in large numbers along the Atlantic Coast of Georgia and southern South Carolina Taxonomic keys for the larvae and adults of the six North American bibionid genera are provided in Hardy (1981). Denmark and Mead (1992) provide keys and review the biology and ecology of Nearctic *Plecia*.

## SCIARIDAE (DARKWINGED FUNGUS GNATS)

Darkwinged fungus gnats are 1–11 mm in length and closely resemble the Mycetophilidae except that their eyes meet above the base of the antennae. The adults (Fig. 8.9A) are usually encountered in moist, shady habitats. The larvae (Fig. 8.9B) feed on a wide range of materials, including fungi, decaying plants, manure, and, in some cases, the roots of greenhouse plants, soybeans, and clovers. *Lycoriella mali* is a major pest of commercial mushrooms, feeding on compost and all stages of mushrooms. *Bradysia* species are known to infest greenhouses, where they damage plant roots and consume fungi in potting soil; they also transmit spores of plant-parasitic fungi of the genus *Pythium*. There are four larval instars. The adult-to-adult life cycle lasts 15–49 days in some of the economically important species. Like the mycetophilids, the sciarids may emerge inside houses from ornamental plantings and potted plants.

Sciarids pose no medical problems except for rare reports of household pets becoming ill after eating adult flies. In one Florida case, a 4-month-old dog died after ingesting large numbers of an unidentified sciarid species during an unusually large emergence in early May. The dog exhibited seizures and shock and was comatose by the time it was seen by a veterinarian. The dog died a short time later after experiencing extensive internal hemorrhaging and hepatic toxicosis. Examination of the stomach contents revealed several hundred sciarid adults (G. R. Mullen, personal communication).

There are more than 100 Nearctic species, for which keys to the genera of adults and larvae are found in

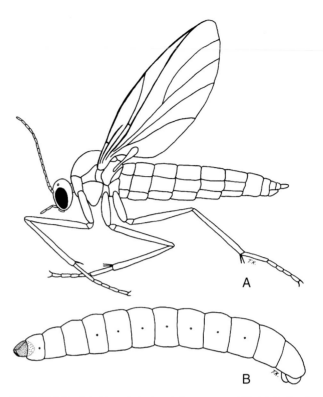

**FIGURE 8.9** Sciaridae, darkwinged fungus gnat. (A) Adult female; (B) larva. (Original by Takumasa Kondo.)

Steffan (1981). The ecology of some species is presented in Madwar (1937).

## CHAOBORIDAE (PHANTOM MIDGES)

Adults are small (1.4–10 mm in length), mosquito-like midges without the elongate proboscis and abundant wing scales characteristic of the Culicidae (Fig. 8.10A). Eggs of members of the common genus *Corethrella* are laid on the surface of water and hatch within 2–4 days. The larval stage averages 15–32 days, and the pupa is active and lasts 3–6 days. The transparent larvae (Fig. 8.10B) are aquatic and are found commonly in lentic habitats (e.g., large lakes, small pools, bogs, small ponds). The larvae of all 19 species in the 5 North American genera are predators which grasp their prey with prehensile antennae. Prey include small crustaceans and aquatic insect larvae, including mosquitoes, which they sometimes eliminate from restricted habitats.

While most adults do not feed on blood, females of the genus *Corethrella* have toothed mandibles and have been found with avian and mammalian blood in their digestive tracts (Williams and Edman, 1968). *Corethrella brakeleyi* and *C. wirthi* have been observed feeding on tree frogs (*Hyla* spp.) (McKeever, 1977). *Corethrella* females are

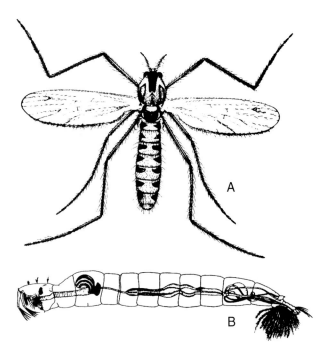

**FIGURE 8.10** Chaoboridae, the Clear Lake gnat (*Chaoborus astictopus*). (A) Adult female; (B) larva. (From Herms, 1937.)

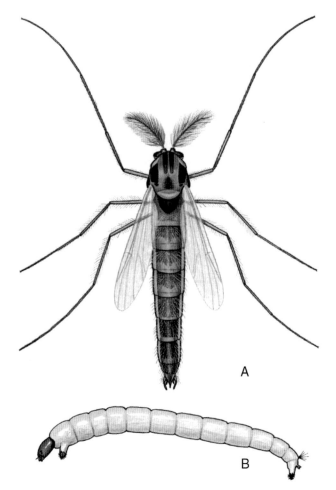

**FIGURE 8.11** Chironomidae. (A) Adult male (*Chironomus* sp.); (B) larva (*Pseudodiamesa* sp.). (From McCafferty, 1981.)

attracted to the calls of male tree frogs (McKeever and French, 1991), to which *C. wirthi* can transmit a *Trypanosoma* species (Johnson *et al.,* 1993). They do not feed on female frogs, which do not call. The *Clear Lake gnat* (*Chaoborus astictopus*) is an inhabitant of large lakes and impoundments in the western United States. Large numbers emerge synchronously in the spring and are attracted to lights in residential and resort areas, where they can cause annoyance (Herms, 1937; Linquist and Deonier, 1942).

Generic keys for chaoborid larvae and adults are found in Cook (1981). Keys to United States species of *Corethrella* are found in Stone (1968).

## CHIRONOMIDAE (CHIRONOMID MIDGES)

Adult chironomid midges (Fig. 8.11A) are 1–10 mm long, with slender legs, narrow, scaleless wings, and plumose antennae in the adult males. They are often mistaken for adult mosquitoes but lack the long proboscis and are unable to feed on blood. Adults are short-lived, living only a few days to several weeks. Some imbibe honeydew and other natural sugars, but some take no food at all as adults. Most chironomid larvae are aquatic or semiaquatic and construct tubes in, or attached to, the substrate. They are often the most abundant benthic organisms and occur in all types of habitats, including rivers,

streams, lakes, ponds, water supplies, and sewage systems. Chironomid larvae are cylindrical and have paired prolegs on the prothoracic and last abdominal segments (Fig. 8.11B). The head is heavily sclerotized and nonretractile. They have no spiracles. Many species, however, have a hemoglobin-like substance in their hemolymph and are called *bloodworms* because of their pink or red color. Most species are detritus feeders that graze on aquatic substrates. Others filter drifting food particles from the water with strands of saliva or are predators on other chironomid larvae or oligochaete worms.

In addition to being mistaken for adult mosquitoes, chironomids can pose other medical and economic problems. Inhabitants of localities where large, synchronous emergences occur can develop allergies to the larval hemoglobin that is carried over from the larva to the adult and becomes airborne as the bodies of the adults decompose (Cranston, 1988). Larval hemoglobin also can induce allergies in workers who process bloodworms into fish food for aquaria. Large chironomid emergences from polluted bodies of water are common and may cause local annoyance to humans, in addition to economic damage to machinery, paint finishes, automobiles, and airplanes (Ali, 1991). Large numbers of adult midges can discourage tourism and cause contamination of materials in food-processing, pharmaceutical, and manufacturing plants. Larvae that occur in water-storage and water-distribution systems can pass through taps into homes (Bay, 1993).

The Chironomidae are a large family distributed worldwide, with more than 130 genera and 700 species in North America (Oliver, 1981). Armitage *et al.* (1995) give an overall account of the biology and ecology of chironomids.

## RHAGIONIDAE (SNIPE FLIES)

Adult snipe flies (Fig. 8.12A) are 4–15 mm in length, with long legs, often spotted wings, and distal antennal flagellomeres forming a slender stylus (Fig. 8.12A). Most prey on other insects, except that females of *Symphoromyia* in western North America and *Spaniopsis* in Australia suck blood. Larvae (Fig. 8.12B) are predatory and are usually found near the surface of moist soil in meadows and steep, well drained slopes, usually associated with mosses, woodland grasses, willows, and/or alders.

In California, *Symphoromyia* adults are active from April through mid-July. They readily attack humans, deer, cattle, and horses, usually inflicting a painful bite around the head. Most of the species studied appear to be anautogenous and univoltine. While they may be annoying to humans, livestock, and wildlife, they have not been implicated in the transmission of any disease organisms (Hoy and Anderson, 1978). In Yellowstone National Park, biting activity starts in early July and continues until early September. Horses, mule deer, and humans are often attacked by swarms of females in localized areas along trails, with relatively fewer attacks outside these areas (Burger, 1995). Human responses to bites range from mildly annoying to very painful, with rare incidences of anaphylactic shock (Turner, 1979). For further information on the taxonomy and biology of the genus *Symphoromyia*, see Turner (1974), James and Turner (1981), and Burger (1995).

## ATHERICIDAE (ATHERICID FLIES)

Adult athericids (Fig. 8.13A) are 7–8 mm long and resemble the rhagionids. They differ from snipe flies by the presence of a strongly developed subscutellum, with the $R_1$ cell being closed at the wing margin, and the absence of spurs on the foretibia. Larvae (Fig. 8.13B) inhabit flowing water, where they prey on other insect larvae. There is apparently one generation per year. Some adults prey on insects, but females of *Suragina* species are known to suck blood from humans, cattle, and some

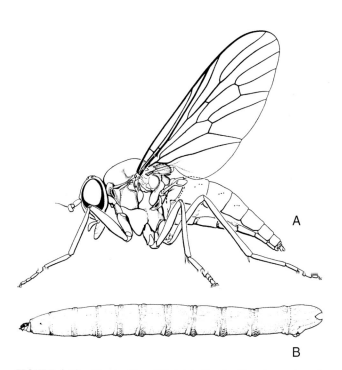

**FIGURE 8.12** Rhagionidae. (A) Adult female (*Symphoromyia* sp.); (B) larva (*Rhagio* sp.). (From McAlpine *et al.*, 1981b.)

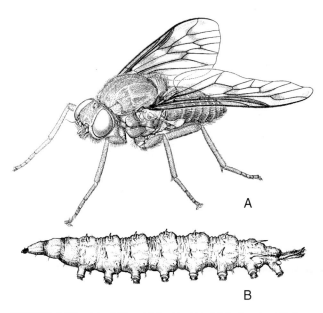

**FIGURE 8.13** Athericidae (*Atherix*). (A) Adult female; (B) larva. (From McAlpine *et al.*, 1981b.)

cold-blooded vertebrates (Hoy and Anderson, 1978). The family has only six Nearctic species, with all known species occurring in Texas and Mexico. Of these, three belong to the blood-feeding genus *Suragina*. Keys to the North American species are provided in Webb (1977, 1981).

## STRATIOMYIDAE (SOLDIER FLIES, LATRINE FLIES)

Adult soldier flies (Fig. 8.14A) vary from 2 to 20 mm in length. Their wings are distinctive by having all branches of the radius thickened and crowded toward the costal margin, ending before the apex of the wing. Body color may be yellow, green, blue, or black, and sometimes metallic. Many adults visit flowers, cattails, or other emergent aquatic vegetation. Larvae (Fig. 8.14B) are elongate and dorso-ventrally flattened and have a toughened or leathery integument with small, closely spaced calcareous tubercles. Many larvae are aquatic, living in a wide range of shallow, lentic habitats where they breathe at the surface through posterior spiracles. Others are terrestrial, breeding in animal wastes, decaying plants and animals, or soil, where they feed on roots of grasses.

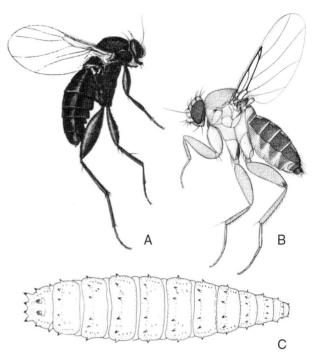

**FIGURE 8.15** Phoridae. (A) Coffin fly (*Conicera tibialis*), adult female; (B) *Megaselia scalaris*, adult female; (C), larva (*Megaselia*). (From Smith, 1986.)

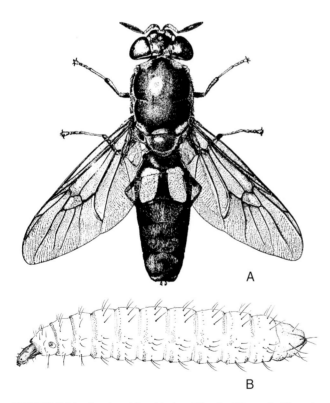

**FIGURE 8.14** Stratiomyidae, black soldier fly (*Hermetia illucens*). (A) Adult female; (B) larva. (From Gagné, 1987.)

The *black soldier fly* (*Hermetia illucens*) (Fig. 8.15A) is the stratiomyid best known to medical and veterinary entomologists and sanitary engineers. The adults are about 20 mm long, bluish-black, with yellowish-white tarsi and two lateral, translucent spots on the second abdominal segment. Mature larvae are about 20 mm long, flattened dorso-ventrally, and dull tan in color, with a narrow head bearing eye spots. They develop in a broad spectrum of decaying materials, including fruits, vegetables, human and animal wastes, and carrion. Eggs are laid in masses on the substrate and hatch in 4 days. The five larval instars last a total of about 14 days, and pupation occurs inside the last larval integument, lasting about 2 weeks. Black solider fly larvae can become abundant in sewer processing plants with trickle filters, where they may be numerous enough to block the system. In caged-layer poultry manure, large populations of larvae can churn the manure and cause it to become liquefied and thus unsuitable for house fly larvae (Sheppard, 1983). In addition to helping to control house flies, this process reduces the total volume of manure. Mature larvae also may be processed into animal food (Sheppard *et al.*, 1994). Larvae occasionally are eaten by humans in overripe fruit or undercooked meat that can result in intestinal myiasis (James, 1947). James (1960, 1981) discusses the biology of and provides keys to larvae and adults.

## Phoridae (Humpbacked Flies, Scuttle Flies)

Adult phorids are 0.5–5.5 mm long with an enlarged thorax, giving them their characteristic humpbacked appearance (Fig. 8.15A and 8.15B). The hind femora are flattened, and the major bristles of the head and legs are feathered. They run in short, quick bursts and are usually found in damp places near larval habitats. Larvae (Fig. 8.15C) are less than 10 mm long, lack an apparent head, and possess abdominal projections that range from being inconspicuous to large and plumose. Larval habitats are extremely varied. They include all kinds of decomposing plant and animal matter, fungi, bird nests, feces, dead insects, sewage treatment beds, and commercial mushrooms. Some larvae are internal parasitoids of other arthropods or live as commensals with social insects.

*Megaselia scalaris* (Fig. 8.15B) is the phorid of most medical importance. The female lays eggs in fruits and vegetables, feces, and decaying plant and animal matter. Sporadic cases of *facultative human myiasis* caused by *M. scalaris* have been documented in many areas of the world; they include cutaneous, pneumonic, nasal, gastrointestinal, urogenital, and ophthalmic myiasis (Carpenter and Chastain, 1992). Phorid larvae also are commonly associated with decomposing animal remains, where they tend to be late invaders after the calliphorid flies have pupated (Smith, 1986). This fly is often a problem around mausoleums and mortuaries, where the larvae develop in burial crypts, producing large numbers of adults (Katz, 1987). A small, black, European species called the *coffin fly* (*Conicera tibialis*) (Fig. 8.15A) is commonly associated with interred human remains that have been underground for a year (Smith, 1986).

There are ca. 350 species and 48 genera of phorid flies in North America. Keys to adults in the Nearctic region are provided in Peterson (1987). The biology, ecology, and keys for identification of Phoridae are compiled in Disney (1994).

## Syrphidae (Flower Flies, Hover Flies)

Adults of this family vary in length from 4 to 25 mm and are distinguished by the presence of a spurious vein between the radius and media. Many are boldly marked with black and yellow transverse bands and are effective wasp mimics. Others, including *Eristalis* and *Eristalinus*, which are called *drone flies*, are covered with fine yellow hairs and resemble honey bees or bumble bees (Fig. 8.16A). Most adults are strong fliers and are often seen hovering near flowers, where they feed on nectar. They neither bite nor are capable of stinging. Syrphid larvae

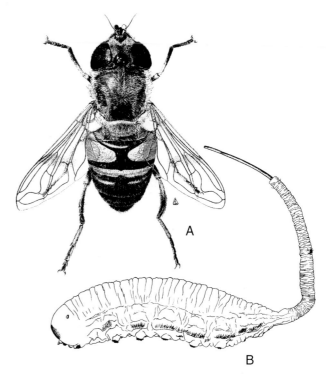

**FIGURE 8.16** Syrphidae, drone fly (*Eristalis tenax*). (A) Adult female; (B) larva, rat-tailed maggot. (From Gagné, 1987.)

are quite varied in form and feeding habits. Some are slug-like and live exclusively in nests of social insects. The most common larval forms are strongly flattened and are predaceous on aphids and other plant-feeding insects. The larvae of *Eristalis* and *Eristalinus* species are aquatic. They are known as *rat-tailed maggots* because of their long, retractable caudal segment bearing the posterior spiracles, which can be extended to 2–3 times the length of the body (Fig. 8.16B). This extensible air tube allows the aquatic larvae to breathe air from the surface while inhabiting highly polluted water. Rat-tailed maggots, especially *Eristalis tenax*, are often found in manure-polluted water in and around confined livestock operations. They are common in wastewater treatment lagoons for livestock and in human wastewater treatment facilities. Occasionally *E. tenax* causes *gastrointestinal* and *urogenital myiasis* in humans. There are over 900 species and more than 90 genera of syrphids in the Nearctic Region (Vockeroth and Thompson, 1987).

## Piophilidae (Skipper Flies)

Adult piophilids (Fig. 8.17A) are small (ca. 5 mm in length), dark, acalypterate flies which are usually shiny black with strong black bristles. The vermiform larvae (Fig. 8.18B) live in a variety of dead plant and animal materials, including carrion, bones, hides, fungi, and

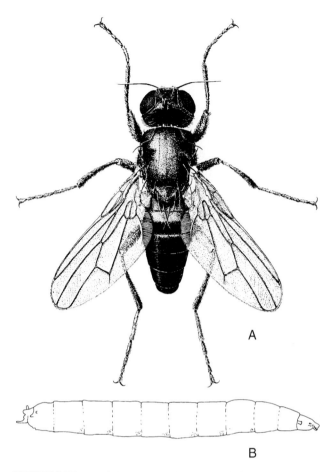

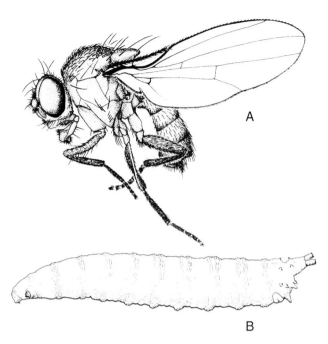

**FIGURE 8.18**   Drosophilidae, *Drosophila melanogaster*. (A) Adult female (from Gagné 1991); (B) larva (from Wheeler, 1987).

**FIGURE 8.17**   Piophilidae, cheese skipper (*Piophila casei*). (A) Adult female (from Gagné, 1991); (B) larva (from Smith, 1986).

stored food products of animal origin. The species most likely to come to the notice of medical or veterinary personnel is the cosmopolitan *cheese skipper, Piophila casei* (Fig. 8.17A). It is a pest of stored food, particularly cheeses and cured hams. The common name derives from the larva's ability to catapult itself into the air by assuming an O-shape by seizing its anal papillae with its mandibles and abruptly releasing its hold. Cheese skippers that are consumed by humans in contaminated food have been responsible for numerous cases of *gastrointestinal myiasis* (James, 1947). Cheese skipper larvae (Fig. 8.17) sometimes colonize corpses in situations where the larger calliphorid and sarcophagid flies are denied access. There are 14 genera containing about 60 species in the Nearctic Region (McAlpine, 1987).

## DROSOPHILIDAE (SMALL FRUIT FLIES)

Also commonly referred to as *vinegar flies,* these are generally small insects (1–6 mm), typically with red eyes. The adults (Fig. 8.18A) are found around the larval habitats of decaying vegetation, plant sap, fungi, and ripe fruit. Larvae are maggot-like with stalked posterior spiracles (Fig. 8.18B). Most feed on yeast and other microorganisms in the decaying substrate. Some are leaf miners, while others are parasitoids or predators of Homoptera.

Drosophilids are familiar in most households, flying around or crawling on overripe fruit. *Drosophila melanogaster* is a common laboratory animal used extensively in genetic research. Although the flies are generally harmless, some species (especially *D. repleta*) are a potential means for mechanical transmission of pathogens when they breed in animal feces (Greenberg, 1973; Harrington and Axtell, 1994). *Drosophila* species occasionally are found in the putrid effluents from corpses. *D. funebris* has been reported to cause *intestinal myiasis* in humans (James, 1947). There are 17 genera and ca. 175 North American species (Wheeler, 1987).

## CHLOROPIDAE (GRASS FLIES, EYE GNATS)

Adults (Fig. 8.19A) are small (1.5 to 5 mm in length) with few large bristles and a prominent break in the costal vein of the wing just mesad of the subcostal junction. Many adults are commonly found in grasses and other low vegetation, or visiting flowers. Larvae (Fig. 8.19B) lack an apparent head and have posterior spiracles and palmate anterior spiracles. Most larvae are phytophagous, feeding

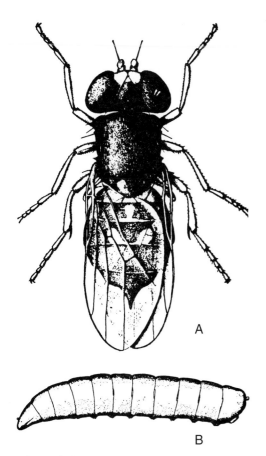

**FIGURE 8.19** Chloropidae, eye gnat (*Liohippelates pusio*). (A) Adult female; (B) larva. (From Herms, 1939.)

on stems, roots, and root hairs of grasses. The frit fly (*Oscinella frit*) and the wheat stem maggot (*Meromyza americana*) are important agronomic pests of grain crops. Other species are saprophytes, feeding mostly on decaying vegetable matter in soils, while a few are predators or gall formers.

Others are known as *eye gnats,* including *Liohippelates* species in North and South America and *Siphunculina* species in Asia. Eye gnats are attracted to humans and other mammals, where they hover about the face, body orifices, and open wounds. The genus *Liohippelates* (formerly included in the genus *Hippelates*) occurs throughout much of North America (Sabrosky, 1980). Several species are particularly abundant in some of the sandy-soil regions of the southeastern United States (e.g., *Liohippelates pusio* and *L. pallipes*) and irrigated areas of southern California (*L. collusor*). The larvae feed on decaying organic matter in soil and can be particularly abundant in humus-enriched, cultivated soil or turf in sandy soils. The life cycle is about 2 weeks, and there are multiple generations each year. The adults hover around the head of humans, causing annoyance, especially when they fly into

eyes, nostrils, or mouths. They also are commonly found on domestic animals, especially on areas soiled with urine or manure (Greenberg, 1971).

*Liohippelates* species have been implicated in the mechanical transmission of several organisms that cause diseases in humans and livestock. *Treponema pertenue,* the spirochete that causes *yaws,* has been shown to be transmitted by *L. flavipes* in Jamaica and other Caribbean and South American locales (Kumm, 1935). *Human acute conjunctivitis* (*pinkeye*), caused by several bacterial species, is noticeably more prevalent during outbreaks of *Liohippelates* in the United States and *Siphunculina* in the Orient (Dow and Hines, 1957; Greenberg, 1973). *Liohippelates* species also have been implicated in the spread of the causative organisms of *vesicular stomatitis* in livestock and *streptococcal infections* of human skin (Francy *et al.,* 1988; Taplin *et al.,* 1967). *Brazilian purpuric fever* is a fulminating, highly fatal, bacterial disease of children caused by *Haemophilus influenzae* biotype *aegyptius,* which produces acute conjunctivitis in children (Brazilian Purpuric Fever Study Group, 1992; Harrison *et al.,* 1989). *L. puruanus* and *Hippelates neoproboscideus* have been implicated as mechanical vectors of *Haemophilus influenzae* biotype *aegyptius* (Tondella *et al.,* 1994).

There are 55 genera and about 270 described species in the Nearctic Region (Sabrosky, 1987). Keys to *Liohippelates* species are provided by Sabrosky (1980). There are no effective area-wide methods for controlling *Liohippelates* species. Temporary relief from their annoyance is provided by protective head nets and insect repellents containing *N, N*-diethyl-m-toluamide (DEET).

## PUBLIC HEALTH IMPORTANCE

On a global scale, the Diptera are the most important order of insects affecting human health (Table I). Mosquitoes are the foremost group because of their role as *vectors* of more pathogenic organisms than any other fly. The adverse impact of malaria, mosquito-borne arboviruses (e.g., yellow fever, dengue, and encephalitis), and metazoan infections such as filariasis on humans worldwide currently exceeds that of recently publicized ailments such as Lyme disease or acquired immune deficiency syndrome. Other human-related vector–pathogen relationships involving the Diptera are exemplified by sand flies and sand-fly fever, bartonellosis, and leishmaniasis; by black flies and onchocerciasis (river blindness); and by tsetse and African trypanosomiasis (sleeping sickness). On a global scale, there are about 270 million humans infected with malaria, 90 million with lymphatic filariasis, 17 million with onchocerciasis, and 12 million with leishmaniasis. In total, almost 3.5 billion humans are currently rated as at risk from fly-borne pathogens.

The Diptera also figure prominently in public health in regard to *filth flies,* which are associated with materials such as dung, carrion and garbage. House flies, stable flies, and blow flies are a few examples. Beginning shortly after Pasteur's formulation of the germ theory in the late 1870s, the link between muscoid flies and human enteric complaints was gradually understood. During the Spanish–American War (1898), this relationship became evident to United States troops in Cuba when "white-legged" muscoid flies roaming on food in mess tents were recognized as the same individuals that previously had been noted on lime-doused feces or corpses. Leland O. Howard, in what was to become one of his major entomological contributions, directed public attention to such filth-breeding flies and almost succeeded in his quest to rename the house fly the "typhoid fly" (Howard, 1911). Many other microorganisms causing enteric disease, such as *Shigella* and *Entamoeba,* can be transmitted in a similar manner. The impact of filth flies on public health was particularly severe in the days before sanitary plumbing, effective pest management, and the ready availability of vaccines. Only the advent of the Salk vaccine for poliomyelitis in the 1950s halted long-term research on possible filth-fly involvement with the etiology of that disease.

Accounts of the enormous numbers of flies associated with corpses on battlefields are virtually as old as warfare itself; those from World War I in France and Belgium and from World War II in the Pacific region are particularly compelling. The beneficial use of certain necrophilous blow flies as *surgical maggots* stems almost directly from battlefield observations. Although *allantoin,* a natural antibiotic secreted by blow fly maggots, has been supplanted by more effective synthetic drugs, the use of surgical maggots remains an effective option for treating wounds, especially when they involve bone infections that are refractory to blood-borne chemotherapy.

## VETERINARY IMPORTANCE

The well-being of wild and domestic animals is directly affected by the Diptera in many ways (Table II). As *vectors* of pathogens, flies are responsible for spreading viruses such as those that cause bluetongue disease of sheep and cattle and hemorrhagic disease of deer; rickettsial infections such as anaplasmosis; protozoans such as those causing avian malaria; and metazoan infections such as canine heartworm. *Trypanosoma*-caused nagana vectored by tsetse has eliminated most animal agriculture throughout large areas of Africa.

In many parts of the world, infestations of living tissue with fly larvae, called *myiasis,* can be a problem for livestock and other domestic or wild animals. At one time, myiasis caused by screwworms constituted a major impediment to cattle, hog and sheep production in the southern United States, Mexico, and Central America and was a major mortality factor among wildlife, especially deer. With the success of sterile-male releases and other antiscrewworm measures, this species is no longer a significant problem in North America. Other myiasis-causing flies in North America include cattle and reindeer grubs, sheep nose bots, deer and reindeer nose bots, rabbit and rodent bots, and stomach bots of horses and other equids.

In most localities, species of blow flies that cause myiasis have a direct summertime impact on most types of livestock and pets. These flies can invade wounds, sores, or body orifices, such as unhealed navels of newborns and vaginal tissues of postpartum females. The condition is perhaps best recognized by the sheep industry as *fly-strike* or *sheep strike* and has had an extensive impact, particularly in Australia and New Zealand.

Blood-feeding flies can affect the productivity and profitability of livestock operations by causing *exsanguination* and, in extreme cases, *anemia.* While the biting rates of diurnal species are obvious to livestock producers (e.g., horse flies), those of crepuscular or nocturnal species are largely unappreciated. Livestock on pasture, or in production systems where large numbers of animals are artificially confined, may be subject to intense biting rates from flies. Bunching, kicking, and other *avoidance behavior* by animals under attack by biting flies can interfere with grazing time, feed consumption, and efficiency of energy conversion.

The enormous numbers of nonbiting flies associated with livestock and livestock facilities can constitute an annoyance factor. The feeding spots and fecal spots made by flies can create sanitary and aesthetic problems. In the poultry industry, *fly specks* on chicken eggs are a major economic problem. Flies produced in one location may affect animals in another. For example, stable flies emanating from cattle feedlots may emigrate to proximate farmsteads, where they feed on pets and companion animals. Flies such as stable flies and eye gnats breeding adjacent to dog kennels may annoy dogs and other animals even at considerable distances from the breeding sites.

## PREVENTION AND CONTROL

In addition to insecticidal applications against immature or adult flies, strategies for suppression of fly-caused problems typically include personal protection. Limiting access of flies to humans or other animals by window screening, mosquito netting, or other physical barriers remains one of the cheapest and most effective forms of control. The invention and ready availability of standard 16-mesh window screening, although considered mundane by most, should actually rank as a genuine marvel of the 20th century. Retreating behind such screening affords humans the opportunity to eat without fly-caused contamination and incessant biting by several species of blood-feeding flies. Similarly, chemical barriers to fly bites afforded by *repellents* are an important part of arthropod-borne pathogen management. Repellents are at the core of preventive medicine with respect to many fly-borne pathogens, especially in protecting military personnel. Typified best by mosquito repellents, these personal protectants include various formulations of DEET. The insecticide permethrin has repellent activity when used as a clothing treatment. Repellents also are often applied to livestock in an effort to reduce biting rates of flies and the annoyance of other, nonbiting, species.

A major emphasis of both civil and military contingency planning is preparation for the management of filth-fly populations in the event of natural or human-caused disasters. Proceeding from the axiom that, all other factors being equal, elimination of fly-breeding sites will eliminate fly populations, a major goal of vector ecologists and sanitary engineers is to reduce the amount of substrate capable of supporting nuisance species. Examples of such *environmental hygiene* range from the virtual replacement of outdoor human privies and pit latrines by sanitary sewage systems in urban and suburban areas to large-scale programs designed to facilitate effective manure management in areas where livestock are aggregated in large numbers.

Feces from cattle, hogs, and poultry are excellent breeding media for many species of noxious flies, most importantly the house fly. Before the advent of automobiles, dung from horses was an important source of house flies and stable flies. Dung mixed with other materials such as hay, straw, mud, and wood shavings is called *manure*. Like dung, manure is often an excellent fly-breeding medium. A biologically inescapable byproduct of livestock production is the generation of significant amounts of dung and manure. House flies in particular tend to disperse from their breeding areas, frequently invading surrounding neighborhoods, where they stain and speck paint, cause annoyance, and may start local fly populations. Modern systems of livestock production therefore depend on effective removal of dung and manure; when these fail, the problem of "rural flies in the urban environment" often winds up in today's courts.

# REFERENCES AND FURTHER READING

Alexander, C. P., and Byers, G. W. (1981). Tipulidae. In "Manual of Nearctic Diptera" (J. F. McAlpine *et al.*, Eds.), Vol. 1, pp. 153–190. Monograph No. 27. Research Branch, Agriculture Canada, Ottawa.

Ali, A. (1991). Perspectives on management of pestiferous Chironomidae: (Diptera), an emerging global problem. *Journal of the American Mosquito Control Association* 7, 260–281.

Armitage, P. D., Cranston, P. S., and Pinder, L. C. V. (1995). "The Chironomidae: Biology and Ecology of Non-Biting Midges." Chapman & Hall, New York.

Bay, E. C. (1993). Chironomid (Diptera: Chironomidae) larval occurrence and transport in a municipal water system. *Journal of the American Mosquito Control Association* 9, 275–284.

Borror, D. J., Triplehorn, C. A., and Johnson, J. F. (1989). "An Introduction to the Study of Insects," 6th ed. Saunders, Philadelphia.

Brazilian Purpuric Fever Study Group. (1992). Brazilian purpuric fever identified in a new region of Brazil. *Journal of Infectious Diseases* 165(Suppl. 1), S16–S19.

Burger, J. (1995). Yellowstone's snipe fly summer. *Yellowstone Science* 3(2), 2–5.

Byers, G. W. (1984). Tipulidae. In "An Introduction to the Aquatic Insects of North America" (R. W. Merritt and K. W. Cummins, Eds.), pp. 491–511. Kendall/Hunt, Dubuque, IA.

Carpenter, T. L., and Chastain, D. O. (1992). Faculative myiasis by *Megaselia* sp. (Diptera: Phoridae) in Texas: a case report. *Journal of Medical Entomology* 29, 561–563.

Cole, F. R. (1969). "The Flies of Western North America." Univ. of California Press, Berkeley.

Cook, E. F. (1981). Chaoboridae. In "Manual of Nearctic Diptera" (J. F. McAlpine *et al.*, Eds.), Vol. 1, pp. 335–339. Monograph No. 27. Research Branch, Agriculture Canada, Ottawa.

Cranston, P. S. (1988). Allergens of non-biting midges (Diptera: Chironomidae): a systematic survey of chironomid haemoglobins. *Medical and Veterinary Entomology* 2, 117–127.

Denmark, H. A., and Mead, F. W. (1992). Lovebug, *Plecia nearctica* Hardy (Diptera: Bibionidae). Entomology Circular No. 350, Florida Department of Agriculture and Consumer Services.

Disney, R. H. L. (1994). "Scuttle Flies: The Phoridae." Chapman & Hall, London.

Dow, R. P., and Hines, J. D. (1957). Conjunctivitis in southwest Georgia. *Public Health Reports* 72, 441–448.

Dusek, J. (1971). Key to larvae. In "Flies and Disease" (B. Greenberg, Ed.), Vol. 1, pp. 163–199. Ecology, Classification, and Biotic Associations. Princeton Univ. Press, Princeton, NJ.

Foote, B. A. (1991). Order Diptera. In "Immature Insects" (F. W. Stehr, Ed.), Vol. 2, pp. 690–699. Kendall/Hunt, Dubuque, IA.

Francy, D. B., Moore, C. G., Smith, G. C., Jakob, W. L., Taylor, S. A., and Calisher, C. H. (1988). Epizootic vesicular stomatitis in Colorado, 1982: isolation of virus from insects collected along the northern Colorado Rocky Mountain front range. *Journal of Medical Entomology* 25, 343–347.

Freeman, B. E. (1967). Studies on the ecology of larval Tipulinae (Diptera, Tipulidae). *Journal of Animal Ecology* 36, 123–146.

Furman, D. P., and Catts, E. P. (1982). "Manual of Medical Entomology." Cambridge Univ. Press, Cambridge.

Gagné, R. J. (1987). Diptera. In "Insect and Mite Pests in Food: An Illustrated Key." U.S. Department of Agriculture. Agriculture Handbook Number 655.

Greenberg, B. (1971). "Flies and Disease. Vol. 1. Ecology, Classification and Biotic Associations." Princeton Univ. Press, Princeton, NJ.

Greenberg, B. (1973). "Flies and Disease. Vol. 2. Biology and Disease Transmission." Princeton Univ. Press, Princeton, NJ.

Hardy, D. E. (1981). Bibionidae. In "Manual of Nearctic Diptera" (J. F. McAlpine *et al.,* Eds.), Vol. 1, pp. 217–222. Monograph No. 27. Research Branch, Agriculture Canada, Ottawa.

Harrington, L. C., and Axtell, R. C. (1994). Comparisons of sampling methods and seasonal abundance of *Drosophila repleta* in caged-layer poultry houses. *Medical and Veterinary Entomology* **8,** 331–339.

Harrison, L. H., Da Silva, G. A., Pitmann, M., Fleming, D. W., Vranjac, A., Broome, C. V., and The Brazilian Purpuric Fever Study Group. (1989). Epidemiology and clinical spectrum of Brazilian purpuric fever. *Journal of Clinical Microbiology* **27,** 599–604.

Harwood, R. F., and James, M. T. (1979). "Entomology in Human and Animal Health." Macmillan, New York.

Herms, W. B. (1937). The Clear Lake gnat. *California Agriculture Experiment Station Bulletin* 607: 22 pp.

Herms, W. B. (1939). "Medical Entomology." Macmillan, New York.

Horsfall, W. R. (1962). "Medical Entomology. Arthropods and Human Disease." Ronald Press, New York.

Howard, L. O. (1911). "The House Fly, Disease Carrier. An Account of its Dangerous Activities and the Means of Destroying It." Stokes, New York.

Hoy, J. B., and Anderson, J. R. (1978). Behavior and reproductive physiology of the blood-sucking snipe flies (Diptera: Rhagionidae: *Symphoromyia*) attacking deer in Northern California. *Hilgardia* **46,** 113–168.

James, M. T. (1947). "The Flies That Cause Myiasis in Man." US Department of Agriculture. Miscellaneous Publication No. 631. 175 pp.

James, M. T. (1960). The soldier flies or Stratiomyidae of California. *Bulletin of the California Insect Survey* **6,** 79–122.

James, M. T. (1981). Stratiomyidae. In "Manual of Nearctic Diptera" (J. F. McAlpine *et al.,* Eds.), Vol. 1, pp. 497–511. Monograph No. 27. Research Branch, Agriculture Canada, Ottawa.

James, M. T., and Turner, W. J. (1981). Rhagionidae. In "Manual of Nearctic Diptera" (J. F. McAlpine *et al.,* Eds.), Vol. 1, pp. 483–488. Monograph No. 27. Research Branch, Agriculture Canada, Ottawa.

Johnson, R. N., Young, D. G., and Butler, J. F. (1993). Trypanosome transmission by *Corethrella wirthi* (Diptera: Chaoboridae) to the green treefrog, *Hyla cinerea* (Anura: Hylidae). *Journal of Medical Entomology* **30,** 918–921.

Katz, H. (1987). Managing mausoleum pests. *Pest Control Techniques* **15,** 72–74.

Knizesk, H. M., and Sullivan, D. J. (1984). Temporal distribution of crane flies (Diptera: Tipulidae) in a southern New York woodland. *Canadian Entomology* **116,** 1137–1144.

Kumm, H. W. (1935). The natural infection of *Hippelates pallipes* Loew with the spirochaetes of yaws. *Transactions of the Royal Society of Tropical Medicine and Hygiene* **29,** 265–272.

Lancaster, J. L., and Meisch, M. V. (1986). "Arthropods in Livestock and Poultry Production." Halsted, New York.

Leppla, N. C., Carlysle, T. C., and Guy, R. H. (1975). Reproductive Systems and the Mechanics of Copulation in *Plecia nearctica* Hardy (Diptera: Bibionidae). International Journal of Insect Morphology & Embryology. 4:299–306.

Lindner, E. (1949). "Die Fliegen der Palaerktischen Region, Handbuch." E. Schweizerbartische Verlagsbuchhandlung, Stuttgart.

Linquist, A. W., and Deonier, C. C. (1942). Flight and oviposition habits of the Clear Lake gnat. *Journal of Economic Entomology* **35,** 411–415.

Madwar, S. (1937). Biology and morphology of the immature stages of Mycetophilidae (Diptera, Nematocera). *Philosophical Transactions of the Royal Society of London B Biological Sciences* **227,** 1–110.

McAlpine, J. F. (1981a). Key to families—adults. In "Manual of Nearctic Diptera" (J. F. McAlpine *et al.,* Eds.), Vol. 1, pp. 89–124. Monograph No. 27. Research Branch, Agriculture Canada, Ottawa.

McAlpine, J. F., Peterson, B. V., Shewell, G. E., Teskey, H. J., Vockeroth, J. R., and Wood, D. M., Eds. (1981b). "Manual of Nearctic Diptera." Vol. 1. Monograph No. 27. Research Branch, Agriculture Canada, Ottawa.

McAlpine, J. F. (1981c). Morphology and terminology—adults. In "Manual of Nearctic Diptera" (J. F. McAlpine *et al.,* Eds.), Vol. 1, pp. 9–63. Monograph No. 27. Research Branch, Agriculture Canada, Ottawa.

McAlpine, J. F. (1987). Piophilidae. In "Manual of Nearctic Diptera" (J. F. McAlpine *et al.,* Eds.), Vol. 2, pp. 845–852. Monograph No. 28. Research Branch, Agriculture Canada, Ottawa.

McCafferty, W. P. (1981). "Aquatic Entomology." Jones and Bartlett Publishers, Boston, MA.

McKeever, S. (1977). Observations of *Corethrella* feeding on tree frogs (*Hyla*). *Mosquito News* **37,** 522–523.

McKeever, S., and French, F. E. (1991). *Corethella* (Diptera: Corethellidae) of eastern North America: laboratory life history and field responses to anuran calls. *Annals of the Entomological Society of America* **84,** 493–497.

Merritt, R. W., and Cummins, K. W. (1996). "An Introduction to the Aquatic Insects of North America." Kendall/Hunt, Dubuque, Iowa.

Oliver, D. R. (1981). Chironomidae. In "Manual of Nearctic Diptera" (J. F. McAlpine *et al.,* Eds.), Vol. 1, pp. 423–458. Monograph No. 27. Research Branch, Agriculture Canada, Ottawa.

Peterson, B. V. (1987). Phoridae. In "Manual of Nearctic Diptera" (J. F. McAlpine *et al.,* Eds.), Vol. 2, pp. 689–712. Monograph No. 28. Research Branch, Agriculture Canada, Ottawa.

Sabrosky, C. W. (1980). New genera new and combinations in Nearctic Chloropidae (Diptera). *Proceedings of the Entomological Society of Washington* **82,** 412–429.

Sabrosky, C. W. (1987). Chloropidae. In "Manual of Nearctic Diptera" (J. F. McAlpine *et al.,* Eds.), Vol. 2, pp. 1049–1067. Monograph No. 28. Research Branch, Agriculture Canada, Ottawa.

Sheppard, D. C. (1983). House fly and lesser house fly control utilizing the black soldier fly in manure management systems for caged laying hens. *Environmental Entomology* **12,** 1439–1442.

Sheppard, D. C., Newton, G. L., Thompson, S. A., and Savage, S. (1994). A value added manure management system using the black soldier fly. *Bioresource Technology* **50,** 275–279.

Smith, K. G. V., Ed. (1973). "Insects and Other Arthropods of Medical Significance." Publication 720. British Museum (Natural History), London.

Smith, K. G. V. (1986). "A Manual of Forensic Entomology." Cornell Univ. Press, Ithaca, NY.

Steffan, W. A. (1981). Sciaridae. In "Manual of Nearctic Diptera" (J. F. McAlpine *et al.,* Eds.), Vol. 1, pp. 247–255. Monograph No. 27. Research Branch, Agriculture Canada, Ottawa.

Stone, A. (1968). The genus *Corethrella* in the United States (Diptera: Chaoboridae). *Florida Entomologist* **51,** 183–186.

Stone, A., Sabrosky, C. W., Wirth, W. W., Foote, R. H., and Coulson, J. R. (1965). "A Catalog of the Diptera of America North of Mexico." Agriculture Handbook, US Department of Agriculture, No. 276. Agricultural Research Service, US Department of Agriculture, Washington, DC.

Taplin, D., Zaias, N., and Rebell, G. (1967). Infection by *Hippelates* flies. *Lancet* **2,** 472.

Teskey, H. J. (1981a). Key to families—larvae. In "Manual of Nearctic Diptera" (J. F. McAlpine *et al.,* Eds.), Vol. 1, pp. 125–147. Monograph No. 27. Research Branch, Agriculture Canada, Ottawa.

Teskey, H. J. (1981b). Morphology and terminology—larvae. In "Manual of Nearctic Diptera" (J. F. McAlpine *et al.,* Eds.), Vol. 1, pp. 65–88. Monograph No. 27. Research Branch, Agriculture Canada, Ottawa.

Tondella, M. L. C., Paganelli, C. H., Bortolotto, I. M., Takano, O. A., Trino, K., Brandileone, M. C. C., Mezzacapa, N. B., Vieira, V. S. D., and Perkins, B. A. (1994). Isolamento de *Haemophilus aegyptius* associado a febre purpurica Brasileira, de cloropideos (Diptera) dos generos *Hippelates* e *Liohippelates*. *Revista Instit. Med. Trop. Sao Paulo* **36,** 105–109.

Turner, W. J. (1974). A revision of the genus *Symphoromyia* Frauenfeld (Diptera: Rhagionidae). I. Introduction. Subgenera and species-groups. Review of biology. *Canadian Entomologist* **106,** 851–868.

Turner, W. J. (1979). A case of severe human allergic reaction to bites of *Symphoromyia* (Diptera: Rhagionidae). *Journal of Medical Entomology* **15,** 138–139.

Vockeroth, J. R., and Thompson, F. C. (1987). Syrphidae. In "Manual of Nearctic Diptera" (J. F. McAlpine *et al.,* Eds.), Vol. 2.

pp. 713–743. Monograph No. 28. Research Branch, Agriculture, Ottawa, Canada.

Webb, D. W. (1977). The Nearctic Athericidae (Insecta: Diptera). *Kansas Entomological Society* **50,** 473–495.

Webb, D. W. (1981). Athericidae. In "Manual of Nearctic Diptera" (J. F. McAlpine *et al.,* Eds.), Vol. 1, pp. 479–482. Monograph No. 27. Research Branch, Agriculture Canada, Ottawa.

Wheeler, M. R. (1987). Drosophilidae. In "Manual of Nearctic Diptera" (J. F. McAlpine *et al.,* Eds.), Vol. 2, pp. 1011–1018. Monograph No. 28. Research Branch, Agriculture Canada, Ottawa.

Williams, J. A., and Edman, J. D. (1968). Occurrence of blood meals in two species of *Corethrella* in Florida. *Annals of the Entomological Society of America* **61,** 1336.

Williams, R. E., Hall, R. D., Broce, A. B., and Scholl, P. J., Eds. (1985). "Livestock Entomology." Wiley, New York.

Zumpt, F. (1965). "Myiasis in Man and Animals in the Old World." Butterworth's, London.

# 9

# MOTH FLIES AND SAND FLIES (*Psychodidae*)

LOUIS C. RUTLEDGE AND RAJ K. GUPTA

*Members of* the Psychodidae are primitive Diptera of the suborder Nematocera. The family is widely distributed in natural, agricultural, and urban environments of tropical, subtropical, and temperate climates. Representatives occur in all zoogeographic regions and many terrestrial biomes, including desert, grassland, chaparral, and forest. Some species occur in mountainous regions at high altitudes.

The subfamily Psychodinae includes nonbiting species known as *moth flies*. Certain species of *Psychoda* and *Telmatoscopus,* known as *drain flies* or *filter flies,* are common pests in buildings and in and around sewage treatment plants. Larvae of two species of *Psychoda* have been implicated in myiasis.

The subfamily Phlebotominae includes biting species known as *sand flies.* Species of *Lutzomyia* and *Phlebotomus* are important biting pests and vectors of agents causing sand fly fever, Changuinola fever, vesicular stomatitis, Chandipura virus disease, bartonellosis, and leishmaniasis in humans, domestic animals, and wildlife. A source of confusion is that the name "sand fly" is sometimes applied to black flies (family Simuliidae) and biting midges (family Ceratopogonidae). In the vernacular the name "sand flea" is applied to all three groups, in addition to chigoes (true fleas).

## TAXONOMY

The family Psychodidae includes six subfamilies (Duckhouse, 1973), only two of which include species of known public health and veterinary importance: the Psychodinae and Phlebotominae. *Sycorax silacea* of the subfamily Sycoracinae is a vector of a filarial parasite (*Icosiella neglecta*) of the frog *Rana esculenta,* a species commercially raised for human consumption in France.

Adults of Psychodinae (Fig. 9.1) have relatively short mouthparts, short antennal segments, and short legs and hold their wings level or sloping downward (rooflike)

**FIGURE 9.1**  Moth fly (*Psychoda* sp.), adult female. (From McCafferty, 1981.)

**FIGURE 9.2**  Sand fly (*Phlebotomus* sp.). (A) Adult female; (B) larva. (From Patton and Evans, 1929.)

when at rest. The mandibles are rudimentary or absent. Two genera, *Psychoda* and *Telmatoscopus,* include species of public health and veterinary importance. Both genera are widely distributed throughout the world. About 80 species occur in North America. Quate and Vockeroth (1981) have reviewed the Nearctic Psychodinae.

Adults of Phlebotominae (Fig. 9.2A) have relatively long mouthparts, long antennal segments, and long legs and hold the wings sloping upward (troughlike) when at rest. The mandibles are well developed. The subfamily includes about 600 species distributed globally between about 50° N and 40° S. About 380 species occur in the New World, including 14 in the United States and Canada. There are no records from Alaska or Hawaii. Canadian records include only British Columbia, Alberta, and Ontario. The New World genus *Lutzomyia* and the Old World genus *Phlebotomus* include species of public health and veterinary importance. Members of the New World genera *Brumptomyia* and *Warileya* and the Old World genus *Sergentomyia* are zoophilic and rarely bite humans.

The genus *Lutzomyia* was reviewed by Young and Duncan (1994), and the genus *Phlebotomus* was reviewed by Lewis (1982). The Phlebotominae of North America, north of Mexico, were reviewed by Young and Perkins (1984). Some authors treat the Phlebotominae as a separate family, the Phlebotomidae, and some treat the subgenus *Psychodopygus* of *Lutzomyia* as a separate genus. There is evidence of cryptic species in some taxa. Many new approaches are being applied to problems of classification and identification; these include electron

micrography, karyotyping, nucleic acid probes, isoenzyme electrophoresis, polymerase chain reaction assay, and analysis of cuticular hydrocarbons.

## MORPHOLOGY

### PSYCHODINAE

Mature *larvae* of the Psychodinae are elongate, legless, and up to 6 mm long (Fig. 9.3). The larvae of many genera (e.g., *Pericoma*) have long spines or feathery processes along the body, but these are poorly developed in *Psychoda* and *Telmatoscopus*. In *Psychoda* and *Telmatoscopus* the body is fusiform or subcylindrical, with three thoracic and nine abdominal segments. The segments are secondarily divided into annuli, with two annuli comprising the thoracic and first abdominal segments and three comprising abdominal segments 2–7. The dorsal cuticle has

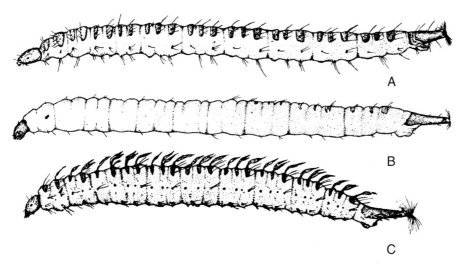

**FIGURE 9.3** Moth fly larvae. (A) *Psychoda* sp.; (B) *Telmatoscopus* sp.; (C), *Pericoma* sp. (From Johansen, 1934.)

minute spines and narrow, transverse, sclerotized plates on the annuli of the segments. The head is well developed, with short antennae, lateral eyespots, and strong mandibles. There are two pairs of spiracles: an anterior pair on the prothorax and a posterior pair at the tip of a rigid siphon terminating the abdomen. The posterior spiracles are surrounded by four lobes bearing water-repellent hairs. Larvae are grayish, with dark head, dorsal plates, and siphon.

The *pupae* of moth flies may be free-floating or attached to the substrate. Attached forms are erect, with exuviae of the last larval instar adhering to the caudal end. The pupa is obtect, with visible appendages of the head and thorax held closely to the body. The thorax bears a pair of tubelike respiratory organs, and the abdomen bears numerous setae and spines.

Adults of moth flies are usually less than 5 mm long, densely hairy, and grayish, brownish, or yellowish (Fig. 9.1). The long, 12- to 16-segmented antennae are similar in males and females. The segments are closely covered with short setae and each bears a whorl of long setae. The palpi are long, recurved, and four-segmented, with scattered setae. Mandibles are rudimentary or absent. Ocelli are absent. The wings are large, broadly ovate to elliptic or pointed, and densely hairy, with no cross veins beyond the basal area. The abdomen has six to eight apparent segments.

## PHLEBOTOMINAE

Eggs of the Phlebotominae are about 400 $\mu$m long, elongate, dark brown, and shiny, with fine surface markings. Markings useful in classification and identification of species include irregular patterns, polygons, ellipses, ridges, pits, and mountain- or volcano-like features (Feliciangeli *et al.*, 1993).

Mature *larvae* of the Phlebotominae are elongate, legless, and up to 5 mm in length (Fig 9.2B). The head, thorax, and abdomen bear numerous, prominent, clavate setae that are used in classification and identification. The well-developed head bears eyespots, short antennae, and heavy, toothed mandibles, which oppose a heavy, platelike, serrate labium. The thoracic and abdominal segments are secondarily divided into annuli, with two annuli comprising the thoracic and first abdominal segments and three comprising abdominal segments 2–7. Abdominal segments 1–8 each bear a medioventral proleg, or *pseudopodium*. There are two pairs of spiracles: an anterior pair on the prothorax and a posterior pair on the greatly reduced abdominal segment 9, which bears two or four long, conspicuous caudal setae adjacent to the spiracles. Larvae are whitish, with a dark head and dark caudal setae.

Phlebotomine *pupae* attach to substrates in an erect position with the exuviae of the last larval instar attached at the caudal end. They can be distinguished from pupae of the Psychodinae by the clavate body setae and long caudal setae of the larval exuviae. Appendages of the head and thorax are visible and closely appressed to the body.

Phlebotomine adults (Fig. 9.2A) are usually less than 5 mm long, densely hairy, and grayish, brownish, or yellowish in color. The head is small and hypognathous, eyes conspicuously dark, and ocelli absent. The long, slender, 12- to 16-segmented antennae are similar in males and females. The segments are closely covered with short setae, and each segment has a whorl of long setae. The thorax is strongly humped. The wings are large, broadly ovate to elliptical or pointed, densely hairy, and without cross veins beyond the basal area. The abdomen is six- to eight-segmented. The male genitalia are large and conspicuous.

The *mouthparts* form a short probocis, bearing long, recurved, five-segmented palps with scattered setae. The mouthparts of the female consist of six broad, knifelike stylets (labrum, paired mandibles and maxillae, and hypopharynx), which are held within the fleshy labium when not in use. The mandibles and maxillae are toothed distally. The mandibles cut the skin with scissors-like and sawing movements while the maxillary teeth engage the sides of the wound and hold the mouthparts in place. Blood is taken from a subcutaneous pool produced by injury to the vessels. The food canal is formed by apposition of the labrum above and the hypopharynx, which contains the salivary duct, below. Males lack mandibles and do not bite, but some have been observed to take blood from wounds made by the females.

# LIFE HISTORY

## PSYCHODINAE

Moth flies breed in aquatic and semiaquatic habitats. Breeding sites of *Psychoda* and *Telmatoscopus* include seashores, margins of streams and ponds, rice fields, ditches, tree holes, sumps, drains, sewers, cesspools, septic tanks, urinals, waste lagoons, sewage treatment plants, and outfalls. The eggs are deposited in gelatinous masses of 20–100 and hatch in about 2 days. The larvae (Fig. 9.3) develop in aquatic surface films, floating algae, mud, manure, and similar wet or moist organic media, where they feed on fungi, microorganisms, and decaying organic matter. The larval period is 9–15 days, and the pupal period is 1–2 days. Larvae of *Psychoda alternata* are highly tolerant of pollution, low pH, low dissolved oxygen, and high temperatures.

## PHLEBOTOMINAE

Sand flies breed in humid, terrestrial habitats. *Breeding sites* include cracks and crevices of soil, manure, rocks, masonry, rubble, forest litter, tree hollows, tree crotches, termite mounds, animal burrows, nests, poultry houses, barns, stables, homes, privies, cesspools, cellars, wells, and other dark, moist locations where organic material is present. Several important neotropical species breed in the litter of the forest floor (e.g., *Lutzomyia gomezi, L. panamensis, L. pessoana,* and *L. trapidoi*). Two important Eurasian species, *Phlebotomus papatasi* and *P. argentipes,* breed in organic soil in and around stables, barns, and houses. *P. perfiliewi* breeds in farm manure in Italy, and *P. caucasicus* breeds in rodent burrows in central Asia.

Sand flies may be autogenous or anautogenous. Females of *autogenous* species complete the first gonotrophic cycle without taking blood but require one or more blood meals to complete each subsequent cycle. Females of *anautogenous* species require one or more blood meals to complete each cycle, including the first. Multiple blood meals in a single cycle have been demonstrated in several species. Females of most species complete multiple cycles. Females of *P. argentipes* may complete as many as four cycles during their normal life span.

Eggs are deposited in small, scattered groups and hatch in 4–20 days. *P. papatasi* illustrates the typical pattern of egg maturation following a blood meal. Ingested blood cells begin to break down 6–18 hr after feeding. The peritrophic envelope matures at 24 hr. Digestion, absorption, and assimilation of the blood, and maturation of the eggs within the ovarian follicles, are usually completed in 5–8 days. Approximately 30–60 eggs are produced in each gonotrophic cycle.

Phlebotomine *larvae* feed on decaying organic matter and fungi. The period of larval development varies from 30 to 60 days, and there are four larval instars. In climates with a cold winter or a long hot or dry season, there may be a diapause or quiescence lasting up to nearly a year in the egg stage or in the fourth larval instar. In *P. papatasi* the proportion of diapausing larvae increases from a low level in the summer to a high level in the fall and is independent of temperature. The pupal period is 7–8 days. The life span of the adult varies from 2 to 6 weeks.

# BEHAVIOR AND ECOLOGY

## PSYCHODINAE

Adult moth flies are common at lights and in and around breeding sites where suitable resting places are available. Resting sites include buildings, drains, sewers, cesspools, septic tanks, sewage treatment plants, and other humid, protected sites. Adult moth flies feed on nectar and septic fluids. They walk with a characteristic, hesitating motion. Flight is noiseless, consisting of short, discrete hops, usually only a few centimeters long.

## PHLEBOTOMINAE

Phlebotomine adults are found in and around the breeding sites where suitable resting places are available. Resting sites include forest litter, tree trunks and tree hollows, leaves of plants, caves, excavations, burrows and nests, livestock pens, buildings, cracks and crevices of rocks, masonry and ruins, and other dark, humid, protected sites. Neotropical forest species may be found in forest litter (e.g., *L. trapidoi*), on understory plants (e.g., *L. pessoana*), on the trunks of trees (e.g., *L. trinidadensis*), or in the forest canopy (e.g., *L. rorotaensis*). The Asian *P. caucasicus* is commonly found in animal burrows. Certain species are more or less peridomestic, including *L. longipalpis, L. verrucarum, P. argentipes,* and *P. papatasi*. In southeastern Asia, *P. argentipes* is found in dark corners of houses and behind hanging clothing and pictures.

Sand flies have a hesitating, running motion and fly readily. Their flight is noiseless, consisting of short, discrete hops, often only a few centimeters long, and slow, steady flights of longer duration. Flight is inhibited by wind and rain; *P. orientalis* does not fly at wind speeds above 15 kph. The flight ranges of neotropical forest species are usually less than 200 m, but some species migrate daily between the forest floor and canopy. Flight ranges of *P. argentipes* and *P. orientalis* may be 500 m or more, and those of *L. longipalpis* and *P. caucasicus* may be 1000 m or more. Longer records include a female *P. ariasi* recaptured 2.2 km from a release point in France and a male *P. perniciosus* captured on the island of Jersey (UK) off the coast of France, 25 km from the nearest source on the mainland.

Adult sand flies feed on nectar, fruit juices, sap, honey dew, and other sugar sources. Females of *P. papatasi* are known to pierce the stems and leaves of plants to obtain sap. Female sand flies feed on blood in addition to sugars. Blood feeding is limited to areas of exposed skin such as the ears, eyelids, nose, feet, and tail. Males of *L. longipalpis* fly to the hosts of blood-feeding females, establish territories at blood-feeding sites on the host, and defend territories against other males. Males emit a terpenoid pheromone from glands on the abdomen to attract females, which then feed and mate within the males' territories. Mating is preceded by elaborate courtship behavior.

The biting activity of most species of sand flies occurs during twilight or darkness. Some, however, bite freely in daylight, including several medically important vector species. *L. panamensis, L. pessoana, L. sanguinaria,* and *L. trapidoi* of the tropical lowlands of Panama bite only at temperatures above 20°C, while *L. verrucarum* of the cool mountain valleys of Peru bites at temperatures as low as 10°C. Some species are endophilic (e.g., *L. verrucarum* and *P. papatasi*), and others are exophilic (e.g., *L. trapidoi* and *P. perniciosus*).

Most sand flies have broad *host ranges*, but some are narrowly restrictive. *L. gomezi* is known to feed on birds and on five different orders of mammals, while *L. vexator* feeds exclusively on lizards and *L. vespertilionis* feeds

exclusively on bats. *P. papatasi* is anthropophilic throughout its range. *P. argentipes* is anthropophilic in some areas but zoophilic, feeding preferentially on cattle, in others. Light-trap collections in Panama indicate the presence of a primarily anthropophilic association represented by *L. gomezi, L. panamensis,* and *L. dysponeta* and a primarily zoophilic association represented by *L. carpenteri, L. triramula* and *L. camposi*. In some locations, the anthropophilic and zoophilic associations alternate by season.

Phlebotomine species often exhibit characteristic *seasonal* and *biotopic patterns*. In tropical areas, populations of most species increase during or shortly after the rainy season. In India, populations of *P. argentipes* and *P. papatasi* decrease in the dry season and increase during the monsoon season. In Africa, population densities of *P. duboscqi* and *P. martini* vary seasonally with those of the rodents on which they feed. Sand fly populations in Panama are strongly correlated with the degree of development of the forest cover, increasing from grassy to secondary forest biotopes and from secondary to mature forest biotopes. Similar situations exist in Kenya, where populations of *S. bedfordi* and *S. antennata* increase from thickets to open-canopy forests and from open-canopy forests to closed-canopy forest. In Nigeria, populations of *S. bedfordi* and *S. antennata* increase from the open plains to more heavily vegetated habitats.

# PUBLIC HEALTH IMPORTANCE

## PSYCHODINAE

Larvae of *Psychoda* and *Telmatoscopus* occur in filter beds and settling tanks of sewage-treatment and water-treatment plants, where they feed on algae, fungi, bacteria, and protozoa. The adults are often annoying pests in the neighborhood of the treatment facilities. Similar nuisance problems have been reported in connection with turf production in Florida and greenhouse operations in California. *Psychoda* and *Telmatoscopus* also may be pests in and around homes and buildings, where they emerge from sumps, sink and floor drains, sewers, cesspools, septic tanks, aquariums, and other breeding sites. Outbreaks of moth flies have been associated with *bronchial asthma* in susceptible individuals. A case of *urinary myiasis* due to *Psychoda albipennis* has been reported from Scotland, and a case of *enteric myiasis* due to *P. alternata* has been reported from Japan.

*P. alternata* is widely distributed in North and South America, Europe, Asia, Africa, and Australia and is the most common species associated with sewage treatment plants worldwide. *P. albipennis* and *P. severini* also are common in sewage treatment plants in Europe. *Telmatoscopus albipunctatus* is widely distributed in the United States and Canada. *P. cinerea* and *P. pacifica* occur in the eastern and western United States, respectively.

## PHLEBOTOMINAE

Sand flies can be annoying *biting pests* in places where they are abundant. They may bite or probe several times before and after feeding, each time causing a sharp, pricking sensation. Residents of Peru have difficulty sleeping in areas that are highly infested with *L. verrucarum*. In one study, the mean biting rate was estimated to be 20–50 bites per person per night. One individual received an estimated 300 bites in a single night. Other highly anthropophilic species are *L. diabolica* in the United States; *L. gomezi, L. olmeca, L. panamensis, L. pessoana, L. sanguinaria, L. trapidoi,* and *L. ylephiletor* in Central America; *L. wellcomei* in Brazil; and *Phlebotomus sergenti* and *P. papatasi* in the Old World.

The initial bites received by an individual typically induce *sensitization,* resulting in immediate or delayed skin reactions to subsequent bites. The reaction to the bite of *P. papatasi* is a pink or red papule about 2–3 mm in diameter and 0.5 mm high, which remains prominent for 4–5 days before gradually disappearing. Moderate to severe itching usually occurs. Individuals that become *hypersensitive* often develop hives, with pronounced swelling of the eyelids and lips if those sites are bitten. Prolonged exposure to sand fly bites results in eventual *desensitization.* Chronically exposed individuals living in areas with high sand fly populations may exhibit no reaction to their bites.

Many species of *Lutzomyia* and *Phlebotomus* are *vectors* of viral, bacterial, and protozoan pathogens of humans (Table I). Zoophilic species, including species of *Brumptomyia, Warileya,* and *Sergentomyia,* as well as *Lutzomyia* and *Phlebotomus,* may be involved in the maintenance of zoonotic diseases.

## SAND FLY FEVER

The *sand fly fever group* of viruses (Bunyaviridae: *Phlebovirus*) includes at least seven viral serotypes that have been isolated from humans: the *Alenquer, Candiru, Chagres,* and *Punta Toro* viruses of Central and South America and the *Naples, Sicilian,* and *Toscana* viruses of southern Europe and North Africa, eastward to China. Sand fly fever, also known as *phlebotomus fever* and *pappataci fever* or *papatasi fever,* is a self-limited, influenza-like, nonrespiratory illness of 2–5 days' duration. Acute onset, fever, malaise, nausea, headache, and retro-ocular,

lower back, and muscular pain are characteristic. Weakness and mental depression may persist after recovery. Encephalitis may occur following infection with Toscana virus. The intrinsic incubation period of the virus is 3–4 days. Virus is present in the blood from 1 day before to 2 days after onset of fever. The suspected *reservoirs* are rodents and primates.

Known *vectors* of sand fly fever group viruses are *L. trapidoi* and *L. ylephiletor* in the New World, and *P. papatasi, P. perfiliewi,* and *P. perniciosus* in the Old World. The extrinsic incubation period of the virus is about 7 days, after which time the sand fly is infective for life. Transovarial transmission has been demonstrated in several species. In Europe, epidemics of sand fly fever commonly occur in the summer and fall, corresponding with two separate generations of *P. papatasi.*

## CHANGUINOLA FEVER

*Changuinola virus* (Reoviridae: *Orbivirus*) occurs in Central and South America and has been associated with clinical illness in Panama. Symptoms are similar to those of sand fly fever. The virus has been isolated from *L. umbratilis.* Sloths are suspected reservoirs.

## VESICULAR STOMATITIS

Three serotypes of *vesicular stomatitis virus* (Rhabdoviridae: *Vesiculovirus*) that cause febrile disease in humans are believed to be transmitted by sand flies. The *Alagoas, Indiana,* and *New Jersey serotypes* are widely distributed in tropical and temperate areas of North and South America and have been repeatedly isolated from *L. shannoni, L. trapidoi, L. ylephiletor,* and unidentified sand flies. *L. shannoni* is a proven vector of the New Jersey serotype among feral pigs on Ossabaw Island, GA. Transovarial transmission has been demonstrated in several species of sand flies, including *L. shannoni, L. trapidoi,* and *L. ylephiletor.*

Vesicular stomatitis virus has been isolated from a number of biting and nonbiting insects, in addition to sand flies. Repeated isolations have been made from black flies (Simuliidae), and outbreaks of the disease in cattle have been associated with high populations of black flies. Laboratory-infected *Simulium vittatum* transmit the virus to mice experimentally.

Symptoms of vesicular stomatitis in humans are similar to those of sand fly fever. Opossums, monkeys, porcupines, raccoons, bobcats, horses, pronghorns, cattle, sheep, and swine are suspected reservoirs. Antibodies occur in domestic and wild dogs.

**TABLE I**
**Sand Fly-Borne Diseases of Humans**

| Disease | Causative agent | Geographic distribution | Reservoirs | Sand fly vectors |
|---|---|---|---|---|
| Sand fly fever (New World) | Sand fly fever virus (Alenquer, Candiru, Chagres, Punta Toro serotypes) | Panama, Colombia | Rodents, Primates | *Lutzomyia trapidoi, L. ylephiletor* |
| Sand fly fever (Old World) | Sand fly fever virus (Naples, Sicilian, Toscana serotypes) | Tropical and subtropical Europe, Asia, northern Africa | Rodents (Muridae) | *Phlebotomus papatasi, P. perfiliewi, P. perniciosus* |
| Changuinola fever | Changuinola fever virus | Central and South America | Sloths | *Lutzomyia umbratilis* |
| Vesicular stomatitis | Vesicular stomatitis virus (Alagoas, Indiana, New Jersey serotypes) | Tropical, subtropical, and temperate North and South America | Opossums, monkeys, porcupines, raccoons, bobcats, horses, swine, pronghorns, cattle, sheep | *L. shannoni, L. trapidoi, L. ylephiletor* |
| Chandipura virus disease | Chandipura virus | India, West Africa | Hedgehogs | *P. papatasi* |
| Bartonellosis | *Bartonella bacilliformis* (bacterium) | Colombia, Ecuador, Peru | None | *L. verrucarum, L. columbiana* |
| Cutaneous leishmaniasis (New World) | *Leishmania amazonensis,[a,b] L. braziliensis,[b] L. colombiensis, L. guyanensis,[a,b] L. lainsoni, L. mexicana, L. naiffi, L. panamensis,[b] L. peruviana, L. shawi, L. venezuelensis* | Tropical and subtropical Central and South America, Mexico, United States (Texas) | Opossums, monkeys, sloths, armadillos, anteaters; various rodents (Sciuridae, Heteromyidae, Muridae, Dasyproctidae, Capromyidae, Echimyidae); mongooses, canines, cats, raccoons, horses | *L. anduzei, L. anthophora, L. ayacuchensis, L. ayrozai, L. carrerai, L. christophei, L. diabolica, L. flaviscutellata, L. gomezi, L. hartmanni, L. intermedia, L. llichyi, L. llanosmartinsi, L. migonei, L. nuneztovari, L. olmeca, L. ovallesi, L. panamensis, L. paraensis, L. peruensis, L. pessoai, L. reducta, L. spinicrassa, L. squamiventris, L. townsendi, L. trapidoi, L. trinidadensis, L. ubiquitalis, L. umbratilis, L. verrucarum, L. wellcomei, L. whitmani, L. ylephiletor, L. youngi, L. yucumensis* |
| Cutaneous leishmaniasis (Old World) | *L. aethiopica, L. killicki, L. major,[b] L. tropica[a]* | Tropical and subtropical Europe, Asia, and Africa | Monkeys, rodents (Sciuridae, Muridae, Cricetidae), dogs, hyraxes | *P. aculeatus, P. alexandri, P. ansarii, P. duboscqi, P. guggisbergi, P. longipes, P. papatasi, P. pedifer, P. rossi, P. salehi, P. sergenti* |
| Visceral leishmaniasis (New World) | *L. chagasi[c]* | Tropical and subtropical Central and South America | Opossums, canines | *L. antunesi, L. evansi, L. longipalpis* |
| Visceral leishmaniasis (Old World) | *L. archibaldi, L. donovani,[b,c] L. infantum[b,c]* | Tropical and subtropical Europe, Asia, and Africa | Canines, rats (Muridae) | *P. ariasi, P. alexandri, P. argentipes, P. caucasicus, P. celiae, P. chinensis, P. kandelakii, P. langeroni, P. longicuspis, P. longiductus, P. martini, P. neglectus, P. orientalis, P. perfiliewi, P. perniciosus, P. smirnovi, P. tobbi, P. transcaucasicus, P. vansomerenae* |

The reservoirs and sand fly vectors listed include both known and suspected species.

[a] Also can cause visceral infections.
[b] Also can cause muco-cutaneous infections.
[c] Also can cause cutaneous infections.

## CHANDIPURA VIRUS DISEASE

This is a phlebotomine-borne arboviral disease of humans in India and west Africa caused by *Chandipura virus* (Rhabdoviridae: *Vesiculovirus*). Symptoms are similar to those of sand fly fever. Encephalitis may occur following infection. Hedgehogs are suspected *reservoirs* in Nigeria. Chandipura virus has been isolated from sand flies (*Phlebotomus* spp.) in India and Senegal. *P. papatasi* is a suspected vector in India. Transovarial transmission of the virus has been demonstrated in *P. papatasi*.

## BARTONELLOSIS

Bartonellosis is a disease of humans caused by a hemotrophic bacterium, *Bartonella bacilliformis* (Fig. 9.4), transmitted by sand flies. It occurs sporadically and in epidemics at 500–3000 m in the Andes mountain valleys of Peru, Ecuador, and Colombia and at low elevations in the coastal provinces of Ecuador. *B. bacilliformis* is a small, motile, aerobic, gram-negative coccobacillus which attaches to red blood cells and invades the endothelial cells of the human host.

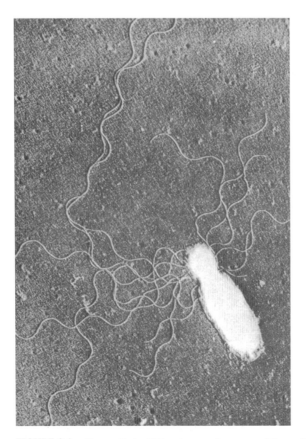

**FIGURE 9.4**   *Bartonella bacilliformis,* causative agent of Oroya fever and verruga peruana. (Courtesy of US Armed Forces Institute of Pathology, (AFIP) 75-8592.)

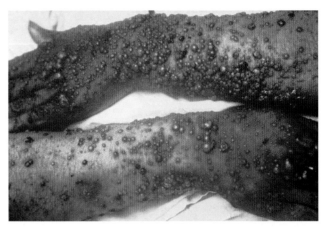

**FIGURE 9.5**   Verruga peruana, or Peruvian wart; nodular lesions on hands and forearms of Peruvian patient. (From Young and Duncan, 1994.)

Bartonellosis occurs in two distinct clinical forms: an acute, febrile anemia (*Oroya fever*) and a chronic cutaneous eruption (*verruga peruana* or *Peruvian wart*). Bartonellosis is also known as *Carrion's disease,* named in honor of Daniel Alcides Carrion, a Peruvian medical student who lost his life in 1885 proving that Oroya fever and verruga peruana are different manifestations of the same disease. Oroya fever often precedes verruga peruana by several weeks or months. It is characterized by fever, headache, muscle and joint pain, enlargement of the lymph nodes, weakness, anemia, pallor, and sometimes delirium and coma. The case-fatality rate of untreated Oroya fever is 10–40%. Verruga peruana has a pre-eruptive stage characterized by muscle, bone, and joint pain, which is followed by a disseminated or nodular eruption (Fig. 9.5). The lesions occur in clusters, usually on the face and limbs; they are typically wartlike, 0.2–4 cm in diameter, and sometimes can enlarge and ulcerate. Verruga peruana may persist for months or years but is seldom fatal.

The intrinsic incubation period of bartonellosis is usually 2–3 weeks. The agent appears in the blood before the onset of illness and may persist for years afterward. The vectors are believed to be *L. verrucarum* and *L. columbiana*. Reservoirs are unknown.

## LEISHMANIASIS

Leishmaniasis is a complex of sand fly-borne diseases widely distributed in tropical and subtropical areas of North and South America, Europe, Asia, and Africa (Fig. 9.6). The ecology of leishmaniasis varies widely in different areas. In central Asia it occurs in semiarid and arid situations. In the American tropics it is primarily a

**FIGURE 9.6** Geographic distribution of human leishmaniases.

forest disease. In Peru it occurs in villages and farms of high mountain valleys. In the Mediterranean region and the Middle East it is primarily urban, and in Africa it is primarily rural. Worldwide, about 12 million people have leishmaniasis, and about 350 million people are at risk of acquiring it.

Leishmaniasis is caused by numerous species of protozoan parasites in the genus *Leishmania* (Table I). Most authors now recognize two subgenera, *Leishmania* and *Viannia,* and accept the elevation of former subspecies to species rank within these subgenera. Although the taxonomy of *Leishmania* is unsettled, the use of modern techniques is contributing significantly to identification of *Leishmania* species and our understanding of their systematics. Species determinations often require culturing the pathogen, followed by immunological, molecular, and biochemical assays.

In the *vertebrate host,* the parasites develop intracellularly in macrophages of the reticuloendothelial system and in circulating monocytes. This stage of development is known as the *amastigote* (Fig. 9.7A, B). Amastigotes are round or oval, 3–7 µm in diameter, with a round nucleus, rod-like kinetoplast, and rudimentary internal flagellum. The parasites multiply by binary fission, producing 50–200 new parasites that rupture the host cell and invade, or are taken up by, other cells.

In the *sand fly host* the parasites develop extracellularly in the alimentary canal from ingested amastigotes. Two morphological forms have been described: the *promastigote* and the *paramastigote*. Promastigotes and paramastigotes are pleomorphic, or variable in shape, with a free flagellum arising anteriorly. Promastigotes are elongate or pear-shaped and 5-24 µm long, with the kinetoplast situated anterior to the nucleus (Fig. 9.7C, E, F, G). Paramastigotes are round or oval and 3–7 µm in diameter, with the kinetoplast situated lateral to the nucleus (Fig. 9.7D, H). Promastigotes and paramastigotes may attach to the lining of the alimentary tract (*haptomonad* phase) or remain free-swimming (*nectomonad* phase). Haptomonads attach to cuticular surfaces of the foregut and hindgut by means of hemidesmosomes formed within the flagellar tip (Fig. 9.7G, H). Nectomonads may temporarily attach in the midgut by interdigitation of the flagellum with the epithelial microvilli.

In the blood meal, amastigotes transform into short promastigotes and mature into long promastigotes within 3 days of ingestion. Initially, the peritrophic envelope surrounding the blood meal prevents establishment of the parasites in the gut. After natural or parasite-facilitated degradation of the envelope, the parasites develop as promastigotes and paramastigotes in the midgut (suprapylarian species, subgenus *Leishmania*) or in both hindgut and midgut (peripylarian species, subgenus *Viannia*) and subsequently migrate to the foregut. Multiplication by binary fission occurs at many points in the life cycle. Evidence of nuclear fusion also has been reported.

Certain relatively short, slender, highly motile nectomonad promastigotes with very long flagella are regarded as infective, or *metacyclic,* forms (Fig. 9.7F).

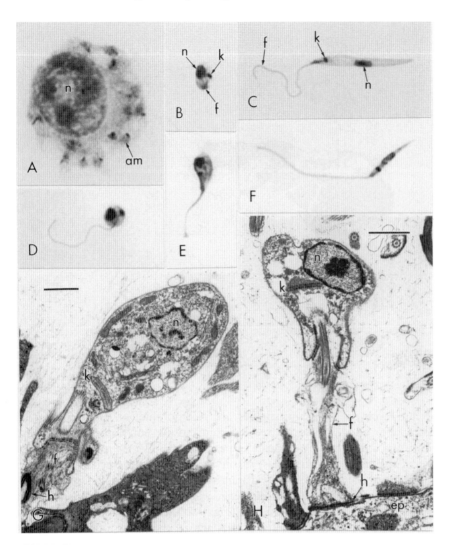

**FIGURE 9.7**   Developmental forms of *Leishmania* in sand flies. (A–F) Light micrographs. (G–H) electron micrographs. (A) Ingested macrophage containing amastigotes; (B) amastigote; (C) elongate nectomonad promastigote; (D) nectomonad paramastigote; (E) pear-shaped haptomonad promastigote dissected from foregut intima; (F) metacyclic promastigote; (G) pear-shaped haptomonad promastigote attached in hindgut; (H) haptomonad paramastigote attached in foregut. am, amastigote; ep, epithelium; f, flagellum; h, hemidesmosome; k, kinetoplast; n, nucleus. Bars 1 μm. (Photos by L. L. Walters; (B) Walters *et al.*, 1989; (C) and (G) Walters *et al.*, 1992; (D) and (E), Walters *et al.*, 1993b; (H) Walters, 1993a.)

These develop in the midgut, foregut, and mouthparts and are transmitted to new hosts during blood feeding. Transmission may be promoted by parasite-induced injury to the mouthparts and/or blocking of the foregut by congregated parasites. Infection of the new host is facilitated by vasodilatory peptides in the sand fly saliva. The extrinsic incubation period is both species- and temperature-dependent and ranges from 4 to 17 days.

Leishmaniasis occurs in two principal clinical forms, known as *cutaneous leishmaniasis* and *visceral leishmaniasis*. While a given species of *Leishmania* typically produces one or the other clinical form, some, including

*Leishmania amazonensis, L. chagasi,* and *L. guyanensis* in the New World and *L. donovani, L. infantum,* and *L. tropica* in the Old World, can produce both (Table I).

## Cutaneous Leishmaniasis

Cutaneous leishmaniasis begins with a macule that develops into a nodular lesion at the site of inoculation by the bite of an infective sand fly. The intrinsic incubation period may be a week to many months. Multiple lesions may develop at sites of multiple bites. Lesions may remain nodular or develop into sharply demarcated,

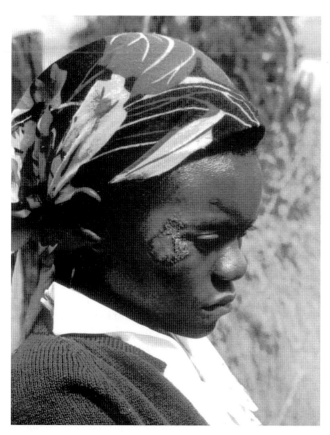

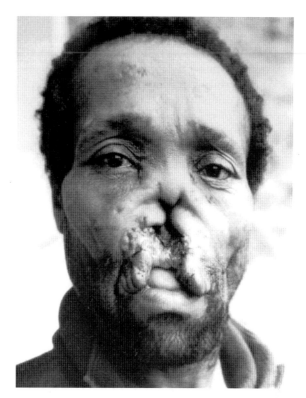

**FIGURE 9.9**  Mucocutaneous form of leishmaniasis, destruction of oral and nasal tissues of man in Bolivia. (From Walton and Valverde, 1979)

**FIGURE 9.8**  Cutaneous leishmaniasis, granulomatous and necrotic lesion on the face of a 16-year-old girl in Kenya. (From Mehrbrahtu *et al.*, 1992)

granulomatous ulcers that discharge necrotic material (Fig. 9.8). New areas of the body become involved by extension of the primary lesions or by metastasis via the blood or lymph (*disseminated cutaneous leishmaniasis*). Lesions of the mucous membranes of the nose, mouth, and pharynx (*mucocutaneous* or *nasopharyngeal leishmaniasis*) may develop after the primary lesion has healed or in the absence of a recognized primary lesion (Fig. 9.9). Species with known potential to produce mucocutaneous infections are indicated in Table I. The term *diffuse cutaneous leishmaniasis* is applied to a progressive, anergic, nonulcerative condition due to defective cell-mediated immunity.

Several clinical manifestations of cutaneous leishmaniasis have acquired specific common names: In the Old World, the condition characterized by single or multiple cutaneous ulcers due to *L. tropica* or *L. major* has been called *oriental sore, tropical sore, Aleppo boil, Baghdad boil, Jericho boil,* or *Delhi boil.* The form of lesion caused by *L. tropica* in the Mediterranean region and eastward to India has been called the *classical form* or *dry form,* and the form of lesion caused by *L. major* in Central Asia has been called the *wet form.* In Central America, the condition characterized by single or multiple ulcers on the face or ears due to *L. mexicana* is known as *chiclero ulcer.* In French Guiana, the condition characterized by moderate ulcers due to *L. amazonensis* or *L. guyanensis* is known as *pian bois.* In Peru and Ecuador, the condition characterized by numerous, small, benign lesions due to *L. peruviana* is known as *uta.* In South America, mucocutaneous leishmaniasis due to *L. amazonensis* or *L. braziliensis* is known as *espundia.*

Cutaneous leishmaniasis may be self-limiting or chronic. Mucocutaneous leishmaniasis persists for many years and ultimately may be fatal. Chiclero ulcer and diffuse cutaneous leishmaniasis are also chronic, and diffuse cutaneous leishmaniasis is resistant to treatment. Infections of *L. tropica* may recur at or near the site of the healed ulcer after apparent cure, a condition clinically known as *leishmaniasis recidivans* or *chronic relapsing leishmaniasis.* Skin lesions of leishmaniasis are prone to secondary infections by bacteria and fungi, and infestation by fly larvae (myiasis) may occur. Disfiguring scars remain after healing.

Known and suspected *vectors* and *reservoirs* of cutaneous leishmaniasis are shown in Table I. Transmission of *L. major* by *P. caucasicus* in semiarid regions of central Asia is a classic example of an endemic zoonosis.

In this area, *P. caucasicus* breeds in burrows of gerbils (*Rhombomys opimus*) and ground squirrels (*Spermophilopsis leptodactylus*) and transmits the agents of cutaneous leishmaniasis from animal to animal and from animals to humans. When gerbils were eradicated from the vicinity of a construction camp in Turkestan, leishmaniasis disappeared from the construction workers.

Cutaneous leishmaniasis due to *L. mexicana* occurs rarely, but widely, in south-central Texas (Bexar, Cameron, Gonzales, Uvalde, and Wells counties) and the adjoining states of Coahuila, Nuevo Leon, and Tamaulipas in Mexico. Typical and diffuse forms have been reported. *Lutzomyia anthophora* is believed to transmit the disease among woodrats (*Neotoma micropus*), and *L. diabolica* is suspected of transmission from woodrats to humans. Seropositive coyotes and an infected cat have been found in southern Texas.

### Visceral Leishmaniasis

Visceral leishmaniasis is also known as *kala-azar* (from the Hindi for "black fever") and *dumdum fever* (from Dum Dum, the location of a former British arsenal near Calcutta). It is a chronic systemic disease that begins with an inconspicuous cutaneous lesion at the site of inoculation by the bite of an infective sand fly. From this site the parasites are distributed through the body in the bloodstream, producing chronic fever, wasting, marked hepatosplenomegaly (especially splenomegaly), pancytopenia, and hypergammaglobulinemia. Incomplete syndromes are common, and early clinical manifestations are variable. Untreated visceral leishmaniasis is usually fatal. *Leishmania tropica*, however, may produce inapparent or subclinical visceral infections, a condition called *viscerotropic leishmaniasis*. The intrinsic incubation period is usually 2 to 4 months. Cutaneous lesions may appear after apparent recovery or cure and may persist for up to 20 years in the absence of treatment. Such lesions are known as *post-kala-azar cutaneous leishmaniasis* or *dermal leishmanoid*.

Known and suspected *vectors* and *reservoirs* of visceral leishmaniasis are shown in Table I. In India, Nepal, and Bangladesh, humans and sand flies are the only known hosts. Transmission of *L. donovani* by *P. argentipes* in India is a classic example of an epidemic disease. An epidemic in Assam State in the 1890s depopulated whole villages and decimated populations over large areas. Subsequently, large epidemics occurred in Assam State, Bihar State, and the Bengal region. A resurgence of visceral leishmaniasis occurred in the 1970s when the World Health Organization program of spraying homes with dichlorodiphenyltrichloroethane (DDT) for eradication of malaria was discontinued. In Bihar State, 100,000 cases were reported in 1977.

## VETERINARY IMPORTANCE

The Psychodinae have no known veterinary importance. The Phlebotominae are undoubtedly pests of livestock, pets, and wildlife in places where they are abundant, although their contribution to the overall economic loss caused by biting arthropods has not been determined. In addition, they are known to transmit leishmanial agents that infect dogs and cats and may play a role in the transmission of vesicular stomatitis virus among livestock.

### LEISHMANIASIS

The veterinary forms of leishmaniasis are *canine leishmaniasis* and *feline leishmaniasis*. Both domestic dogs and cats are susceptible to cutaneous leishmaniasis. The lesions usually occur on the nose and ears. Dogs are also susceptible to visceral leishmaniasis and may be important reservoirs, but cats are rarely infected and do not show signs of disease. The incubation period may be months or years. Infection in dogs is prevalent in Brazil, China, and the Mediterranean region.

In the United States, autochthonous cases in dogs have been reported from Oklahoma, Kansas, and Ohio. The parasites have been variously identified as *L. chagasi*, *L. infantum*, and *L. mexicana*. The sand fly *vectors* are unknown.

Beginning in 1999 an outbreak of visceral leishmaniasis occurred among foxhounds in a foxhunting club in New York, eventually resulting in 20 fatalities. Inquiry uncovered a prior outbreak among foxhounds in a foxhunting club in Michigan in 1989. Subsequent investigation found seropositive dogs in clubs located in 21 US states and the province of Ontario, Canada. Since known cases have been limited to foxhounds, it has been suggested that direct dog-to-dog transmission occurs during annual foxhound shows.

### VESICULAR STOMATITIS

Vesicular stomatitis is an acute, febrile, weakening, viral disease of horses, cattle, swine, and occasionally sheep and goats. It is characterized by small, superficial, erosive blisters that form in and about the mouth and on the feet, teats, and occasionally other parts of the body. Because the symptoms closely resemble those of foot-and-mouth disease, vesicular stomatitis also is known as *pseudo foot-and-mouth disease*. Susceptibility depends on the host animal's immune status. Nonimmune animals are 100% susceptible, and up to 90% develop clinical disease. The intrinsic incubation period is 2–8 days. The disease is usually self-limiting, with recovery in about 2 weeks, but recrudescence or reinfection may occur. Economic losses are due to the reduced condition of infected animals, reduced meat and milk production, and secondary bacterial infections. In the 1950s vesicular stomatitis virus was regarded as a potential biological warfare agent to incapacitate draft animals.

The *Indiana, New Jersey,* and *Alagoas serotypes* of the vesicular stomatitis virus are known to cause disease in domestic animals in North and South America. Vesicular stomatitis is endemic in tropical regions, but it tends to be epidemic in temperate regions. In tropical regions it occurs year-round, but in the United States it occurs primarily in late summer and early fall. Opossums, monkeys, porcupines, raccoons, bobcats, and pronghorns are suspected reservoirs. Antibodies occur in domestic and wild dogs.

*Lutzomyia shannoni* is a proved vector of the New Jersey serotype among feral pigs on Ossabaw Island, GA, and *L. shannoni, L. trapidoi,* and *L. ylephiletor* are suspected vectors in other areas. Vesicular stomatitis virus has been isolated from a number of biting and nonbiting insects, in addition to sand flies. Repeated isolations have been made from black flies (Simuliidae), and outbreaks of the disease in cattle have been associated with high populations of black flies. Laboratory-infected *Simulium vittatum* transmit the virus to mice experimentally.

# PREVENTION AND CONTROL

## PSYCHODINAE

Moth fly larvae can be eliminated from drains in buildings and homes by cleaning the drain, mechanically or with a drain cleaner or strong disinfectant, and flushing with hot water. Larvae can be eliminated from filters of sewage treatment plants by flooding for 24 hr. Flooding does not affect the eggs and must be repeated periodically for continuous control. Larvae also can be eliminated from the filters by addition of insecticides to the flow if an appropriate insecticide, formulation, and dose are chosen to avoid harm to the filters and the downstream environment. Destruction of large numbers of larvae in sewage treatment plants may create an odor problem in the neighborhood of the plant due to decomposition. *Clostridium bifermentans* serovar *malaysia* has been reported to be highly toxic to larvae of *Psychoda alternata.*

Adult moth flies can be controlled by application of insecticides to resting sites on structures and surrounding areas. Organochlorine, organophosphate, thiocyanate, pyrethroid, and growth-regulator compounds are effective against moth fly larvae and adults, but few products have been registered for this use due to the relatively small demand.

## PHLEBOTOMINAE

Methods for investigating suspected breeding sites of phlebotomine sand flies include the direct examination of soil and litter, extraction by Berlese funnel, wet sieving or flotation, and emergence trapping. Surveillance and collection methods for adult sand flies include trapping and aspiration from resting sites, humans, and bait animals. Effective trap designs include light traps, bait traps, sticky traps, and flight traps. Smoke, insect repellent spray, or a twig or stick can be used to flush sand flies from inaccessible resting sites for collection.

Insect repellents and protective clothing are effective personal protectants. Sand flies do not bite through clothing; however, *L. verrucarum* is reported to crawl beneath clothing and bed sheets to reach the skin. Long sleeves, trousers, and socks should be worn in areas where sand flies are active. Head nets, gloves, repellent-treated net jackets and hoods, and repellent lotions and sprays provide effective protection for the face, hands, arms, and other exposed areas of the body.

Outdoor sleeping areas should be high, open, breezy, dry, and as far from potential breeding sites as possible. Bed nets should be used when sleeping in areas where sand flies are present. Standard 16- and 18-mesh bed nets, head nets, window screens, and screen doors do not exclude sand flies, but both standard and large-mesh screens and nets are effective when treated with contact insecticides or repellents. Fine-mesh nets and screens are little used because they impede circulation of the air. Sand flies will bite through untreated nets in contact with the skin.

Ultraviolet *electrocutor traps* have been recommended for control of *Phlebotomus papatasi.* Indoor flight activity can be reduced by use of electric fans. Sand flies do not usually fly into the upper stories of buildings.

The use of *insecticides* can be helpful in reducing sand fly numbers. Natural and synthetic pyrethroid and thiocyanate aerosols provide effective control indoors when applied twice daily, at night and in the early morning. Smoke from "mosquito coils" containing pyrethroids is also effective. Organochlorine, organophosphate, carbamate, and pyrethroid insecticides are effective for residual control of adult sand flies; however, resistance to organochlorine insecticides has been reported in *P. papatasi.* Insecticides should be applied to inside walls and ceilings, to the inside and outside of screens and doors, and to 1–2 feet of the outer wall around window and door casements and along foundations. Wall hangings impregnated with pyrethroids have been used for sand fly control in homes in Kenya. Area treatments may be needed out-of-doors. Area treatments should be directed toward potential breeding and resting sites, such as outcroppings, fences, walls, buildings, caves, burrows, and tree trunks.

Another approach for effective control is the elimination of breeding sites and resting sites of sand flies. Breeding and resting sites of *P. papatasi* in and around homes can be destroyed by filling cracks and crevices of walls, ceilings, and floors and by clearing and rolling, tamping, or paving outdoor areas. In Italy, transmission of leishmanial agents by *P. perfiliewi* has been reduced by locating piles of farm manure where the sand flies breed at safe distances from homes. In Kenya, the transmission of leishmanial agents by *P. martini* has been reduced by destroying all termite mounds, in which this species breeds, within 20 m of homes. In Panama and French Guiana, leishmaniasis has been reduced by eliminating breeding and resting sites of sand flies by deforestation.

Control of animals that serve as *reservoirs* for leishmanial organisms may be feasible in some areas. Elimination of dogs is an effective preventive measure where dogs have been implicated as reservoirs. Reduction of feral dog and jackal populations by disposal of offal from slaughter houses and poultry farms has significantly reduced the incidence of visceral leishmaniasis in Iraq. *P. papatasi* has been controlled in Jordan and Russia by digging, plowing, or flooding gerbil burrows.

The use of parasites and predators in sand fly control has not been demonstrated. The bacteria *Bacillus sphaericus* and *B. thuringiensis* var. *israelensis* (*Bti*, or serotype H12), and certain gregarine protozoa and tylenchid nematodes, are potential agents.

The immunization of humans by inoculation of living *Leishmania tropica*, called *leishmanization*, has been employed for control of cutaneous leishmaniasis in Iran, Israel, and Russia.

# REFERENCES AND FURTHER READING

Alexander, B. (1995). A review of bartonellosis in Ecuador and Colombia. *American Journal of Tropical Medicine and Hygiene* **52**, 354–359.

Barker, D. C. (1987). DNA diagnosis of human leishmanias. *Parasitology Today* **3**, 177–184.

Berman, J. D. (1997). Human leishmaniasis: clinical, diagnostic, and chemotherapeutic developments in the last 10 years. *Clinical Infectious Diseases* **24**, 684–703.

Beverly, S. M., Ismach, R. B., and McMahon-Pratt, D. (1987). Evolution of the genus *Leishmania* as revealed by comparisons of nuclear DNA restriction fragment patterns. *Proc. Natl. Acad. Sci. U.S.A.* **84**, 484–488.

Chang, K.-P., and Bray, R. S., Eds. (1985). "Leishmaniasis." Elsevier, London.

Chaniotis, B. (1978). Phlebotomine sand flies (family Psychodidae). In "Surveillance and Collection of Arthropods of Veterinary Importance" (R. A. Bram, Ed.), pp. 19–30. Agriculture Handbook No. 518. US Department of Agriculture, Animal and Plant Health Inspection Service, Washington, DC.

Comer, J. A., and Tesh, R. B. (1991). Phlebotomine sand flies as vectors of vesiculoviruses: a review. *Parassitologia* **33**(Suppl. 1): 143–150.

Cupolillo, E., Grimaldi, G., and Momen, H. (1994). A general classification of New World *Leishmania* using numerical zymotaxonomy. *American Journal of Tropical Medicine and Hygiene* **50**, 296–311.

Desjeux, P. (1991). "Information on the Epidemiology and Control of the Leishmaniases by Country and Territory. World Health Organization, Geneva. Report WHO/LEISH/91.30.

Duckhouse, D. A. (1973). Family Psychodidae. In "Catalog of the Diptera of the Oriental Region" (M. D. Delfinado and D. E. Hardy, Eds.), Vol. 1, pp. 226–244. Univ. Press of Hawaii, Honolulu.

Endris, R. G., Perkins, P. V., Young, D. G., and Johnson, R. N. (1982). Techniques for laboratory rearing of sand flies (Diptera: Psychodidae). *Mosquito News* **42**, 400–407.

Feliciangeli, M. D., Castejon, O. C., and Limongi, J. (1993). Egg surface ultrastructure of eight New World phlebotomine sand fly species (Diptera: Psychodidae). *Journal of Medical Entomology* **30**, 651–656.

Grimaldi, G., and Tesh, R. B. (1993). Leishmaniases of the New World: current concepts and implications for future research. *Clinical Microbiology Reviews* **6**, 230–250.

Grimaldi, G., Tesh, R. B., and McMahon-Pratt, D. (1989). A review of the geographic distribution and epidemiology of leishmaniasis in the New World. *American Journal of Tropical Medicine and Hygiene* **41**, 687–725.

Johansen, O. A. (1934). "Aquatic Diptera, Part I, Nemocera, Exclusive of Chironomidae and Ceratopogonidae." Agricultural Experiment Station Memoir 164. Cornell University, Ithaca, NY.

Killick-Kendrick, R. (1987). Studies and criteria for the incrimination of vectors and reservoir hosts of the leishmaniases. In "Proceedings of the International Workshop on Control Strategies for the Leishmaniases, Ottawa, 1–4 June 1987." pp. 272–280. International Development Research Centre, Ottawa, Canada.

Killick-Kendrick, R. (1990a). Phlebotomine vectors of the leishmaniases: a review. *Medical and Veterinary Entomology* **4**, 1–24.

Killick-Kendrick, R. (1990b). The life-cycle of *Leishmania* in the sandfly with special reference to the form infective to the vertebrate host. *Annales de Parasitologie Humaine et Comparee* **65**(Suppl. 1), 37–42.

Killick-Kendrick, M., and Killick-Kendrick, R. (1991). The initial establishment of sandfly colonies. *Parassitologia* **33**(Suppl. 1), 315–320.

Kreutzer, R. D., Morales, A., Cura, E., Ferro, C., and Young, D. G. (1988). Brain cell karyotypes of six New World sand flies (Diptera: Psychodidae). *Journal of the American Mosquito Control Association* **4**, 453–455.

Kreutzer, R. D., Souraty, N., and Semko, M. E. (1987). Biochemical identities and differences among *Leishmania* species and subspecies. *American Journal of Tropical Medicine and Hygiene* **36**, 22–32.

Lewis, D. J. (1975). Functional morphology of the mouth parts in New World phlebotomine sandflies (Diptera: Psychodidae). *Transactions of the Royal Entomological Society of London* **126**, 497–532.

Lewis, D. J. (1982). A taxonomic review of the genus *Phlebotomus* (Diptera: Psychodidae). *Bulletin of the British Museum of Natural History* **45**, 121–209.

Lewis, D. J., Young, D. G., Fairchild, G. B., and Mintter, D. M. (1977). Proposals for a stable classification of the phlebotomine sandflies (Diptera: Psychodidae). *Systematic Entomology* **2**, 319–332.

Magill, A. J. (1995). Epidemiology of leishmanias. *Dermatology Clinics* **13**, 505–523.

McCafferty, W. P. (1983). "Aquatic Entomology." Jones and Bartlett Publishers, Sudbury, MA.

Mehbratu, Y. B., Lawyer, P. G., Ngumbi, P. M., Kirigi, G., Mbugua, J., Gachihi, G., Wasunna, K., Pamba, H., Sherwood, J. A., Koech, D. K., and Roberts, C. R. (1992). A new rural focus of cutaneous leishmaniasis caused by *Leishmania tropica* in Kenya. *Transactions of the Royal Society for Tropical Medicine and Hygiene* **86**, 381–387.

Nieves, E., and Pimenta, P. F. P. (2000). Development of *Leishmania* (*Viannia*) *braziliensis* and *Leishmania* (*Leishmania*) *amazonensis* in the sand fly *Lutzomyia migonei* (Diptera: Psychodidae). *Journal of Medical Entomology* **37**, 134–140.

Patton, W. S. (1929). "Insects, Ticks, Mites, and Venomous Animals of Medical and Veterinary Importance." H. R. Grubb, London.

Peters, W., and Killick-Kendrick, R., Eds. (1987). "The Leishmaniases in Biology and Medicine." 2 Vols. Academic Press, London.

Quate, L. W., and Vockeroth, J. R. (1981). Psychodidae. In "Manual of Nearctic Diptera" (J. F. McAlpine, B. V. Peterson, G. E. Shewell, H. J. Tesky, J. H. Vockeroth, and D. M. Wood, Eds.), Vol. 1, pp. 293–300. Agriculture Canada Monograph 27. Canadian Government Publishing Center, Quebec.

Ready, P. D., Fraiha, H., Lainson, R., and Shaw, J. J. (1980). *Psychodopygus* as a genus: reasons for a flexible classification of the phlebotomine sandflies (Diptera: Psychodidae). *Journal of Medical Entomology* **17**, 75–88.

Reif, J. S. (1994). Vesicular stomatitis. In "Handbook of Zoonoses" (G. S. Beran, Ed.), 2nd ed., pp. 171–181. Section B: Viral. CRC Press, Boca Raton, FL.

Rioux, J. A., Lanotte, G., Serres, E., Pratlong, F., Bastien, P., and Perieres, J. (1990). Taxonomy of *Leishmania*. Use of isoenzymes. Suggestions for a new classification. *Annales de Parasitologie Humaine et Comparee* **65**, 111–125.

Rutledge, L. C., Ellenwood, D. A., and Johnston, L. (1975). An analysis of sand fly light trap collections in the Panama Canal Zone (Diptera: Psychodidae). *Journal of Medical Entomology* **12**, 179–183.

Rutledge, L. C., Walton, B. C., and Ellenwood, D. A. (1976). A transect study of sand fly populations in Panama (Diptera: Psychodidae). *Environmental Entomology* **5**, 1149–1154.

Sacks, D. L. (1989). Metacyclogenesis in *Leishmania* promastigotes. *Experimental Parasitology* **69**, 100–103.

Schlein, Y., and Jacobson, R. L. (1994). Some sandfly food is *Leishmania* poison. *Bulletin of the Society for Vector Ecology* **19**, 82–86.

Schultz, M. G. (1968). A history of bartonellosis (Carrion's disease). *American Journal of Tropical Medicine and Hygiene* **17**, 503–515.

Silva, O. S., and Grunewald, J. (2000). Comparative study of the mouthparts of males and females of Lutzomyia migonei (Diptera:

Psychodidae) by scanning electron microscopy. *Journal of Medical Entomology* **37**, 748–753.

Tesh, R. B. (1988). The genus *Phlebovirus* and its vectors. *Annu. Rev. Entomol.* **33**, 169–181.

Walters, L. L. (1993a). *Leishmania* differentiation in natural and unnatural sand fly hosts. *Journal of Eukaryotic Microbiology* **40**, 196–206.

Walters, L. L. (1993b). Life cycle of *Leishmania major* (Kinetoplastida: Trypanosomatidae) in the neotropical sand fly *Lutzomyia longipalpis* (Diptera: Psychodidae). *Journal of Medical Entomology* **30**, 699–718.

Walters, L. L., Chaplin, G. L., Modi, G. B., and Tesh, R. B. (1989). Ultrastructural biology of *Leishmania* (*Viannia*) *panamensis* (= *Leishmania braziliensis panamensis*) in *Lutzomyia gomezi* (Diptera: Psychodidae): a natural host–parasite association. *American Journal of Tropical Medicine and Hygiene* **40**, 19–39.

Walters, L. L., Irons, K. P., Modi, G. B., and Tesh, R. B. (1992). Refractory barriers in the sand fly *Phlebotomus papatasi* (Diptera: Psychodidae) to Infection with *Leishmania panamensis*. *American Journal of Tropical Medicine and Hygiene* **46**, 211–228.

Walton, B. C., and Valverde, L. C. (1979). Racial differences in espundia. *Annals of Tropical Medicine and Parasitology* **73**, 23–29.

Ward, R. D. (1976). A revised numerical chaetotaxy for neotropical phlebotomine sandfly larvae (Diptera: Psychodidae). *Systematic Entomology* **1**, 89–94.

White, G. B., and Killick-Kendrick, R. (1975). Polytene chromosomes of the sandfly *Lutzomyia longipalpis* and the cytogenetics of Psychodidae in relation to other Diptera. *Journal of Entomology Series A* **50**, 187–196.

World Health Organization. (1990). "Control of Leishmaniasis: Report of a World Health Organization Expert Committee." World Health Organization Technical Report Series 793. World Health Organization, Geneva.

Young, D. G., and Duncan, M. A. (1994). Guide to the identification and geographic distribution of *Lutzomyia* sand flies in Mexico, the West Indies, Central and South America (Diptera: Psychodidae). *Memoirs of the American Entomological Institute* **54**, 1–881.

Young, D. G., and Perkins, P. V. (1984). Phlebotomine sand flies of North America (Diptera: Psychodidae). *Mosquito News* **44**, 263–304.

# 10

# BITING MIDGES (*Ceratopogonidae*)

GARY R. MULLEN

*Biting midges* are minute blood-sucking flies represented by only a few of the many genera in the family Ceratopogonidae. They are commonly known as *no-see-ums* owing to their small size and the fact that they often go unnoticed despite the discomforting bites which they cause. Another name for this group, especially in the northeastern United States, is *punkies*. It is derived from a Dutch corruption of the Algonquin Indian root "punkwa," which means "ashlike," referring to the appearance of the fly as it is biting. The associated burning sensation is likened to that of a hot ash from a fire on contact with the skin. The early French Canadians called them *brulôt*, from "bruler," meaning "to burn." They also are called *sand flies*, particularly in the coastal areas of the southeastern United States, the West Indies, and adjacent parts of the Caribbean and Latin America. This name should not be confused with the same term applied to phlebotomine flies of the family Psychodidae. Along the Gulf Coast of Alabama and Florida, local residents refer to biting midges as *five-O's* because of their biting activity which commences late in the afternoon about five o'clock. Other names for biting midges in various parts of the world include *moose flies* in Alaska, *jejenes* in Latin America, *maruins* in Brazil, *kuiki* in India, *makunagi* and *nukaka* in Japan, *nyung noi* in Laos, *agas* and *merutu* in Indonesia, *merotoe* in Sumatra, and *no-no's* in Polynesia.

Biting midges can be annoying pests of humans and both domestic and wild animals. In addition to the discomfort which they cause, biting midges serve as vectors of a number of viruses, protozoans, and nematodes. Among the more important viral diseases are Oropouche fever in humans, bluetongue disease and epizootic hemorrhagic disease (EHD) in ruminants, and African horsesickness (AHS) in equines. Blood protozoans transmitted by biting midges cause diseases in poultry, whereas certain nematodes are the cause of mansonellosis in humans and of onchocerciasis in various domestic and wild animals.

## TAXONOMY

The Ceratopogonidae are represented worldwide by approximately 78 genera and more than 4000 described species. Thirty-six genera and more than 600 species occur in North America. Ceratopogonids are divided into four subfamilies: the Leptoconopinae, Forcipomyiinae, Dasyheleinae, and Ceratopogoninae. With the exception of the Dasyhelinae, each subfamily includes species that feed on vertebrate blood. Only four genera are known to attack humans and other animals. The most important genus in this respect is *Culicoides,* with over 1000 described species. It includes most of the troublesome species throughout the world and those which serve as the principal vectors of animal disease agents. *Leptoconops,* represented by ca. 80 species, occurs primarily in the subtropics and tropics; a few species are annoying biters in the Caribbean area and along the coast of the southeastern United States. *Forcipomyia* species in the subgenus *Lasiohelea* (ca. 50 species) attack vertebrates, particularly in subtropical and tropical rain forests. *Austroconops macmillani,* a blood feeder and the only known species in its genus, has been reported in Western Australia. No member of the Dasyheleinae is of medical or veterinary importance.

Because of its importance in the transmission of animal viruses in North America, the *Culicoides variipennis* complex warrants special comment. For many years, *C. variipennis* was thought to consist of five subspecies: *C. v. albertensis, C. v. australis, C. v. occidentalis, C. v. sonorensis,* and *C. v. variipennis.* However, based on morphological and electrophoretic analyses, this complex is now regarded as three species (*C. occidentalis, C. sonorensis,* and *C. variipennis*), with *C. v. albertensis* and *C. v. australis* being synonyms of *C. sonorensis* (Holbrook *et al.,* 2000). *C. sonorensis,* rather than *C. variipennis,* as widely reported in the literature, is now recognized as the principal vector of the viruses causing bluetongue disease and epizootic hemorrhagic disease in North American ruminants.

Keys to the genera of ceratopogonid adults are provided by Wirth *et al.* (1974) and Downes and Wirth (1981). For keys to adults of North American *Culicoides* species, see Jamnback (1965), Battle and Turner (1971), and Blanton and Wirth (1979). Generic keys for larvae are provided by Glukhova (1977, 1979). For larval keys to North American species of *Culicoides,* see Jamnback (1965), Blanton and Wirth (1979), and Murphree and Mullen (1991). Major taxonomic works for identification of ceratopogonid fauna in other parts of the world include Glukhova (1989; former Union of Soviet Socialist Republics), Wirth and Hubert (1989; Southeast Asia), Glick (1990; Kenya and Ethiopian Region), and Boorman (1993). For a world list of the species, subspecies, and varieties within the genus *Culicoides,* see Boorman and Hagan (1996).

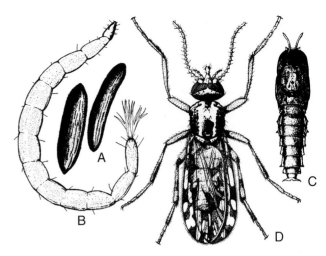

**FIGURE 10.1**    Developmental stages of the salt-marsh biting midge *Culicoides furens.* (A) eggs; (B) larva; (C) pupa; (D) Adult female. (Modified from Hall, 1932)

## MORPHOLOGY

Ceratopogonid larvae (Fig. 10.1B), as represented by *Culicoides* species, are typically long and slender, ranging from 2 to 5 mm in length when mature. The body is translucently whitish, in contrast to the yellow to brownish head capsule (Fig. 10.2). The thorax often is marked by a characteristic pattern of subcutaneous pigmentation. Thoracic and abdominal segments are similar in size, contributing to their elongate, cylindrical body shape. Although larvae of other genera may possess distinctive setae and abdominal projections, the larval chaetotaxy of *Culicoides* and related genera is generally inconspicuous, except for four pairs of setae which may be apparent at the caudal end. These setae are especially long in tree-hole species and are believed to help increase larval mobility. A pair of narrow, bifid anal papillae which function in osmoregulation can be everted through the anus; in most preserved specimens, however, they are retracted into the rectum. Larvae generally lack spiracles and are dependent on cutaneous respiration. Whereas *Culicoides* and *Leptoconops* larvae lack thoracic and abdominal appendages, *Forcipomyia* larvae possess a well-developed ventral prothoracic proleg and associated apical hooklets or setae.

The mouthparts are characterized by a pair of mandibles which are not opposable; they move vertically or partially rotate while the larva feeds and are used to scrape, tear, or seize items, depending on the species involved. Located within the buccal cavity is a complex, sclerotized internal structure called the epipharynx that is best observed in cleared, slide-mounted specimens. It

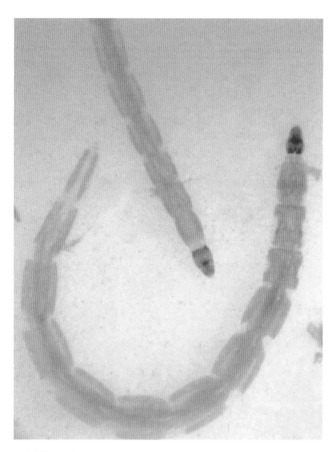

**FIGURE 10.2** *Culicoides variipennis* larvae, fourth instar. (Photo by Bradley A. Mullens.)

consists of a pair of lateral arms and a median region supporting two to four combs which overly one another. The epipharynx is rocked back and forth by muscles attached to the lateral arms and functions by helping to shred solid food and move food items posteriorly into the alimentary tract. In species which feed primarily on detritus and microorganisms, the combs apparently serve to strain material entering the mouth cavity. The number of pharyngeal combs and the degree of sclerotization of the epipharynx are highly variable, reflecting the diversity of ingested food items and feeding behaviors exhibited by ceratopogonid larvae.

*Pupae* are typically brownish in color, with a pair of relatively short but conspicuous prothoracic respiratory horns arising at the anterior end (Fig. 10.1C). Close inspection reveals numerous, tiny spiracular openings at the tip. The respiratory tubes repel water, enabling aquatic forms to hang at the water surface, where they can obtain air during metamorphosis to the adult stage. A pocket of air beneath the developing wings provides additional buoyancy to keep the pupa at the water surface. Cuticular

features in the form of tubercles, spines, and setae provide valuable taxonomic characters for identification of pupae to species.

Adult *Culicoides* midges (Fig. 10.1D) are tiny, usually 1–2.5 mm in body length. Their *mouthparts* are adapted for biting or piercing tissues and are especially well developed in blood-sucking species (Fig. 10.3). In females, the mouthparts are surrounded by a fleshy extension of the labium called a *proboscis*, which is relatively short, about as long as the head. It consists of an upper labrum/epipharynx, a pair of bladelike mandibles, a pair of laciniae (maxillae), and a ventral hypopharynx bearing a median, longitudinal groove along which saliva is passed as the female feeds. The mandibles bear a row of teeth along the inner edge near the tip, which is used to lacerate the skin while biting. The mouthparts of males are generally reduced and are not used in blood feeding.

Associated with the mouthparts are a pair of 5-segmented maxillary palps. The third segment is typically enlarged and bears a specialized group of sensilla located in a depression, or *sensory pit*, that serves as a sensory organ. The adult antennae are 15-segmented and consist of a basal scape, an enlarged pedicel containing Johnston's organ, and 13 flagellomeres. The antennal segments bear differing numbers of small sensory pits (sensilla coeloconica), the number and pattern of which provide important taxonomic characters. The number of segments bearing sensory pits appears to be correlated with host feeding; species which feed primarily on birds generally have more sensory pits than those which feed on mammals. In males, flagellomeres 1–8 possess whorls of long setae, which increase their sensitivity as mechanoreceptors and give them their plumose appearance. The wings possess a characteristic venation that distinguishes the ceratopogonids from other groups of flies. More important, however, are the distinctive wing patterns of the genus *Culicoides*, which are the basis for most species determinations in this large and important group. The darker areas of the wings are not pigmented, but represent the density of tiny setae (micro- and macrotrichia) on the wing surface.

## LIFE HISTORY

Adult females typically require a blood meal in order to develop their eggs (Fig. 10.1A). Some, however, are autogenous and carry over enough nutrients from the larval stage to develop eggs during the first gonotrophic cycle without feeding on blood. Development of the eggs usually requires 7–10 days but may be as short as 2–3 days. The eggs are deposited in batches on moist substrates. The number of eggs per female varies from 30 to 450

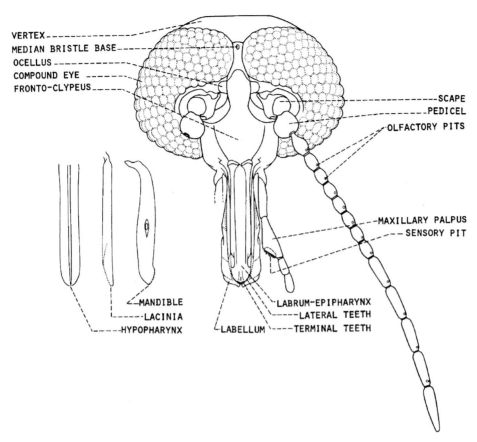

VERTEX
MEDIAN BRISTLE BASE
OCELLUS
COMPOUND EYE
FRONTO-CLYPEUS

SCAPE
PEDICEL
OLFACTORY PITS

MAXILLARY PALPUS
SENSORY PIT

MANDIBLE
LACINIA
HYPOPHARYNX

LABRUM-EPIPHARYNX
LATERAL TEETH
LABELLUM    TERMINAL TEETH

**FIGURE 10.3**    Morphology of head, mouthparts, and other associated structures of a female biting midge, *Culicoides* species. (From Blanton and Wirth, 1979.)

or more depending on the species and size of the blood meal. Autogenous females tend to produce fewer eggs. The eggs are small and elongate (250–500 $\mu$m in length), are often banana-shaped, and are covered with minute projections which apparently function in plastron respiration. They are white when first deposited but gradually turn brown. The eggs hatch in 2–7 days. The *larvae* develop through four instars, with a developmental time that varies from 2 weeks to more than a year, reflecting different species, latitudes, and times of the year. Many species overwinter as larvae and thus commonly pass 7 or 8 months of the year in this stage. In other cases, larvae become dormant during the hot summer months, prolonging their developmental time. Larval development of some Arctic species may take as long as 2 years. Pupation generally occurs near the surface of the substrate where the prothoracic horns of the *pupae* can penetrate the water film. Pupae of *Culicoides* species which develop in water-filled tree holes may remain afloat at the water surface, loosely adhering to the sides of the tree cavity.

Overwintering larvae pupate in the spring or early summer, producing the first generation of adults. Autogenous

females usually oviposit about a week following emergence; thereafter they must obtain a blood meal each time before they can develop another batch of eggs. Newly emerged, anautogenous females oviposit following their initial blood meal. The small percentage of the females which are successful in obtaining a second blood meal can produce a second batch of eggs, but they seldom do so a third time under field conditions. In the laboratory, however, *C. variipennis* is capable of completing up to seven gonotrophic cycles. Longevity of captive adults varies from 2 to 7 weeks, with most individuals probably surviving only a few weeks at most under natural conditions. The *generation time* may be as short as 2 weeks for members of the *C. variipennis* complex, but more typically it is 6 weeks or longer for most species.

Although some species are univoltine, most biting midges are multivoltine, producing two or more generations per year. Because of overlapping generations and multiple oviposition cycles by individual females, populations of a given species may be present throughout the warm months of the year. Usually, however, each species exhibits a general seasonal pattern with characteristic

**FIGURE 10.4** Sampling organically enriched substrate of freshwater marsh for ceratopogonid larvae. (Photo by G. R. Mullen.)

**FIGURE 10.5** Typical breeding site of *Culicoides sonorensis* in wet, manure-contaminated soil surrounding leaking water trough. (Photo by G. R. Mullen.)

peaks of adult abundance. Some species are abundant only in the spring (e.g., *C. biguttatus, C. niger, C. travisi*), whereas others may exhibit high spring populations and be present in lower numbers throughout the summer and fall (e.g., *C. spinosus*). Others tend to be abundant throughout the spring, summer, and fall (e.g., *C. crepuscularis, C. furens, C. haematopotus, C. stellifer, C. venustus*). Still others are bivoltine, with peaks in the spring and fall (e.g., *C. hollensis*).

## BEHAVIOR AND ECOLOGY

Ceratopogonid larvae develop in a wide range of aquatic and semiaquatic habitats ranging from the Tropics to the Arctic tundra. *Leptoconops* species occur primarily in sandy or claylike alkaline soils of arid regions and along tidal margins or coastal marshes and beaches. *Forcipomyia* species are generally found in mosses and algae in shallow water and in more terrestrial habitats, such as rotting wood. The larval habitat of *Austroconops* is unknown. It is difficult to make generalizations about the breeding sites of *Culicoides* species except to say that they occur primarily in organically rich substrates (Fig. 10.4). As a group, they utilize a broad diversity of habitats, including freshwater marshes and swamps; shallow margins of ponds, streams, and rivers; bogs and peat lands; tree holes and other natural cavities in rotting wood; tidal marshes and mangroves; and more specialized habitats, such as rotting cacti, animal manure (Fig. 10.5), and highly alkaline or saline inland pools.

The diversity of ceratopogonid larvae is similarly reflected in their *feeding behavior*. Many are predaceous, feeding on protozoans, rotifers, oligochetes, nematodes, immature stages of insects, and various other small aquatic or semiaquatic invertebrates. Others feed on detritus, bacteria, fungi, green algae, diatoms, and other organic materials. Based on feeding experiments and direct observations, it is apparent that many species are omnivorous and feed opportunistically on a variety of food items. Members of the *C. variipennis* complex, for example, are generally reared on a diet of microorganisms; however, they can also complete their development when fed only nematodes. For most species, the natural diet and nutritional requirements remain unknown, precluding the establishment of laboratory colonies for most of the economically important species. Among the North American species of *Culicoides* which have been successfully colonized are *C. furens, C. guttipennis, C. melleus, C. mississippiensis, C. variipennis* complex, and *C. wisconsinensis*.

Despite their small size, ceratopogonid larvae often can be recognized by their serpentine locomotion consisting of side-to-side lashing movements of the body as they propel themselves through the water. *Culicoides* larvae are generally considered to be good swimmers, especially the later instars. *C. circumscriptus* can lash back and forth an estimated 9 cycles/sec, while members of the *C. variipennis* complex are capable of sustained, directed swimming at speeds up to 1.7 cm/sec. Species such as *C. denningi*, which burrow in the bottom of streams and rivers, are excellent swimmers, enabling them to make their way to shore, where they pupate.

Males typically emerge a short time before females and are ready to mate by the time the females are produced. Mature sperm are already present within 24 hr of eclosion. Unlike mosquitoes, ceratopogonid males do not undergo permanent rotation of the genitalia. Instead, the genital structures are temporarily rotated 180° to facilitate clasping the female in an end-to-end position just before mating takes place.

Mating usually involves swarming in which large numbers of males form aerial aggregations, often near water or in open areas near potential breeding sites. Females fly into the swarm, where males recognize them as being the same species by their wing-beat frequency. Sex pheromones have been shown to be involved in some species. If a female is receptive, she couples with the male and typically drops to the ground or vegetation, where copulation takes place. A few species mate without forming swarms. In such cases, the male and female locate one another by crawling about on the ground or some other substrate where coupling occurs. In other cases, both sexes are attracted to a host, where the male seeks out, and mates with, the female shortly after she has taken a blood meal. Most species are believed to mate only once, although members of the *C. variipennis* complex and others may mate repeatedly. Sufficient sperm from a single mating is stored by females of the *C. variipennis* complex to fertilize up to three batches of eggs.

As adults, both males and females feed on nectar of flowering plants. This serves as an energy source for flight activity and increased longevity, especially in females. Only females feed on vertebrate blood (Fig. 10.6). As in other hematophagous insects, usually blood is required for egg development and subsequent oviposition. Females are *pool feeders*. They lacerate the skin and underlying capillaries with the serrated tips of their mandibles, causing blood to seep into the surrounding tissues. From there it is drawn into the foregut by action of the pharyngeal pump and passed back into the midgut. After feeding, the blood-laden female flies to nearby vegetation or another sheltered site, where she rests for several days while her eggs develop.

Many species of biting midges feed primarily on mammals, whereas others feed preferentially on birds, reptiles, or amphibians. Those which feed on a given class of hosts often show preferences for certain groups within that class, such as small versus large mammals or certain types of birds. While some are quite host-specific, others are considered generalists and may feed, for example, on both birds and mammals, depending on host availability.

In general, *Culicoides* adults are crepuscular or nighttime feeders, whereas *Leptoconops* adults tend to be more active during the daytime. The activity periods of most biting midges occur during twilight, particularly an hour before to an hour after sunrise and sunset. While some species exhibit bimodal activity at dawn and dusk (e.g., *C. barbosai, C. furens, C stellifer*) or during morning and late afternoon (e.g., *Leptoconops bequaerti*), others tend to be active during only one of these periods (e.g., *C. debilipalpis* in the early dawn hours; *C. furens, C. paraensis,* and *L. linleyi* in late afternoon.). Certain species will readily bite during the daytime and can be particularly annoying in late afternoon. Activity periods for a given species may vary at different times of the year, reflecting seasonal changes in temperature and light intensity. The effects of daily and seasonal temperatures on flight activity are evident among salt-marsh species along the Florida coast. Whereas *C. barbosai, C. floridensis,* and *L. bequaerti* are not active below 14°C (57°F), *C. mississippiensis* remains active all year, even at winter temperatures as low as 4°C (37°F).

In addition to temperature, a number of other factors can influence flight activity by biting midges (Kettle, 1972; Lillie *et al.*, 1987). They include light intensity, lunar cycles, relative humidity, changes in barometric pressure, and other weather conditions. Wind velocity is especially important. Because of their small size, most species do not tolerate appreciable air movement and are seldom troublesome at wind speeds above 2.5 m/sec. Such velocities interfere with normal flight activity and the ability to orient to a host. There are exceptions, however, such as *L. bequaerti,* which will continue to bite at wind speeds of 5.0 m/sec or more.

Biting midges are most abundant in proximity to productive breeding sites. From there they disperse into surrounding areas in search of mates and suitable hosts. How far they travel is highly variable, depending on their success in finding a mate, availability of hosts, and prevailing weather conditions. Mark-recapture studies in which adults are released at a given location and are subsequently recovered by trapping at increasing distances from the release point, indicate that the mean distance traveled by many *Culicoides* females is about 2 km. The distance traveled by males is usually much shorter, often less than half that of females of the same species. The flight distance for any individual, however, can be much lower or higher than this mean value implies. Members of the *C. variipennis* complex have been recovered up to 2.8 km from release sites within 12 hr and nearly 5 km

**FIGURE 10.6**    *Culicoides sonorensis,* adult female feeding on human arm. (Photo by P. Kirk. Visscher.)

within 36 hr. Salt-marsh species such as *C. mississippiensis* have been shown to disperse more than 3 km within 24 hr, whereas *L. kerteszi* in semiarid regions of the southwestern United States has been reported to fly 15 km or more in this same time period.

## PUBLIC HEALTH IMPORTANCE

Most complaints about biting midges relate to the annoyance caused by their persistent biting. This is especially a problem in coastal areas, where salt-marsh species are notorious pests, often creating great discomfort for local residents and beach-goers and discouraging tourism during the summer months. In the United States, *C. furens, C. hollensis,* and *C. melleus* are the most troublesome species attacking humans along the Atlantic Coast. *C. mississippiensis* is a problem along the Gulf Coast, whereas *C. barbosai* is commonly a problem near mangroves of southern Florida. Certain *Leptoconops* species are especially pestiferous along coastal beaches, e.g., *L. linleyi* along the Gulf Coast and *L. bequaerti* in the Caribbean, whereas members of the *L. kertszi* group are annoying biters in semiarid regions of the southwestern United States.

A common pest throughout much of the eastern United States is *C. paraensis,* a tree-hole species which readily bites humans, especially during the late afternoon and early evening. It causes considerable discomfort to hunters, campers, and hikers and is typically the species involved in complaints by homeowners while picnicking or trying to work in their yards bordering deciduous woods.

Because of their small size, biting midges frequently are overlooked, their bites often being blamed on mosquitoes. Reactions to their bites generally consist of a localized stinging or burning sensation and a well-defined reddened area around the bite site without the formation of a wheal. The discomfort usually lasts from only a few minutes to a few hours. In individuals who develop hypersensitivity, the bites may continue to itch for 2 or 3 days. In the Tropics, certain *Leptoconops* and *Lasiohelea* species can cause more severe reactions resulting in blisters and serous exudates at the bite sites of sensitized individuals.

Viruses and filarial nematodes are the only disease agents known to be transmitted to humans by the bite of ceratopogonid midges (Tables I and II). They are primarily subtropical or tropical in distribution, with no associated diseases having been reported in North America. Although several viruses have been isolated from *Culicoides* adults, Oropouche virus is the only significant viral agent transmitted to humans by ceratopogonid midges. Biting midges also transmit three filarial nematodes which

infest humans, causing a disease known as *mansonellosis* or *mansonelliasis.* The causative agents are *Mansonella ozzardi* in the Americas, *M. perstans* in Africa and South America, and *M. streptocerca* in Africa. Linley (1983) has provided an excellent overview of the various human pathogens and parasites transmitted by this group of flies.

### OROPOUCHE FEVER

Oropouche fever is caused by a virus in the Simbu group, family Bunyaviridae. Since it was first isolated from a charcoal worker in Trinidad in 1955, this virus has been documented in numerous epidemics in the Amazon region of Brazil. Outbreaks prior to 1980 were largely restricted to Pará State, where over 165,000 human cases of this disease were estimated to have occurred between 1961 and 1980. Since that time, outbreaks have been reported in the Brazilian states of Amazonas, Goias, and Marahnao and in the Amapa Territory. These epidemics have taken place primarily in urban areas, where surveys indicate that up to 44% of local populations have been seropositive for antibodies to the virus.

Oropouche virus causes a nonfatal, acute febrile illness with general muscular and joint pains usually lasting 2–5 days. More than half of the cases involve symptoms such as headaches, dizziness, photophobia, and severe myalgia and arthralgia, which can lead to prostration in some cases. Recurrence of symptoms often prolongs the illness up to 2 weeks. The incubation period for this disease is believed to be 4–8 days.

The epidemiology of Oropouche fever is complicated by multiple strains of the virus and uncertainty about which animals serve as reservoirs. Antibody levels in potential reservoir hosts tend to be highly variable, with some evidence to indicate that urban and sylvatic cycles are involved. During nonepidemic periods, several species of monkeys have been found to have high antibody levels, implicating them as important reservoirs. Other likely reservoirs in the sylvatic cycle are wild birds and sloths. During epidemics, high antibody levels among various carnivores and domestic birds suggest that they play a role as reservoirs in urban outbreaks of this disease.

The principal vector in urban outbreaks is *C. paraensis.* This forest species breeds in tree holes and decaying cacao and calabash pods. It readily feeds on humans both inside and outside houses. Infected *C. paraensis* females can transmit the virus as early as 4–6 days following a blood meal. Although Oropouche virus also has been isolated from naturally infected mosquitoes such as *Culex quinquefasciatus, Aedes serratus,* and *Coquillettidia venezuelensis,* the role of these species as vectors remains uncertain.

TABLE I

**Arboviruses of Medical and Veterinary Interest Transmitted by Biting Midges (*Culicoides* species)**

| Virus | Vertebrate host | Geographic area | Known or suspected vectors |
|---|---|---|---|
| **Bunyaviridae** | | | |
| *Bunyawera group* | | | |
| Lokern | Lagomorphs (*Lepus, Sylvilagus*) | North America | *C. variipennis* complex, *C.* (*Selfia*) spp. |
| Main Drain | Lagomorphs (*Lepus*) | North America | *C. variipennis* complex |
| *Simbu group* | | | |
| Aino | Cattle, sheep, buffalo | Japan | *C. brevitarsis* |
| Akabane | Cattle, sheep, goats, horses, buffalo, camels | Africa, Middle East, Japan, Australia | *C. brevitarsis* |
| Buttonwillow | Lagomorphs (*Lepus, Sylvilagus*) | United States | *C. variipennis* complex |
| Douglas | Cattle, sheep, goats, horses, buffalo deer | Australia, New Guinea | *C. brevitarsis* |
| Oropouche | Humans, forest primates, sloths | South America, Caribbean | *C. paraensis* |
| Peaton | Cattle | Australia | *C. brevitarsis* |
| Sabo | Cattle, goats | Nigeria | *C. imicola* |
| Sango | Cattle | Nigeria, Kenya | *Culicoides* spp. |
| Sathuperi | Cattle | Nigeria, Kenya, India | *Culicoides* spp. |
| Shamonda | Cattle | Nigeria | *C. imicola* |
| Shuni | Humans, cattle | Nigeria, South Africa | *Culicoides* spp. |
| Thimiri | Birds | Egypt, India, Australia | *C. histrio* |
| Tinaroo | Cattle, sheep, goats, buffalo | Australia | *C. brevitarsis* |
| Utinga | Sloth | Panama, Brazil | *C. diabolicus* |
| Utive | Sloth | Panama | *C. diabolicus* |
| *Other Bunyaviridae* | | | |
| Crimean-Congo hemorrhagic fever group | Humans, cattle | Africa, Asia | *Culicoides* spp. (primarily ticks) |
| Dugbe (Nairobi sheep disease group) | Humans, cattle | Africa | *Culicoides* spp. (primarily ticks) |
| Rift Valley fever | Humans, cattle, buffalo, sheep, goats, antelope, camels | Africa | *Culicoides* spp. (primarily mosquitoes) |
| **Reoviridae** | | | |
| African horsesickness | Horses, mules | Africa, Middle East, India, Europe, Asia | *C. imicola* |
| Bluetongue | Cattle, sheep, other domestic and wild ruminants | Africa, Middle East, Europe, Japan, Australia, North America, South America | *C. fulvus, C. gulbenkiani, C. imicola, C. insignis, C. milnei, C. obsoletus, C. variipennis* complex |
| Epizootic hemorrhagic disease | Deer | North America, Africa, Asia, Australia, | *C. variipennis* |
| Equine encephalosis | Cattle | Africa, Australia | Unknown |
| *Palyam group* | | | |
| Abadina | Unknown | Nigeria | *Culicoides* spp. |
| Bunyip Creek | Cattle, buffalo, sheep, deer | Australia | *C. brevitarsis* *C. oxystoma* |
| Chuzan (Kagoshima, Kasba) | Cattle | Japan | *C. oxystoma* |
| CSIRO village | Cattle, buffalo | Australia | *C. brevitarsis* |
| D'Aguilar | sheep Cattle, | Australia | *C. brevitarsis* |
| Marrakal | Buffalo? | Australia | *C. oxystoma* *C. peregrinus* |
| Nyabira | Cattle | Zimbabwe | *Culicoides* sp.? |
| *Wallal group* | | | |
| Mudjinbarry | Marsupials | Australia | *C. marksi* |
| Wallal | Marsupials | Australia | *C. marksi* |

(*Continues*)

TABLE I (*Continued*)

| Virus | Vertebrate host | Geographic area | Known or suspected vectors |
|---|---|---|---|
| *Warrengo group* | | | |
| Mitchell River | Cattle, marsupials | Australia | *Culicoides* spp. |
| Warrengo | Cattle, marsupials | Australia | *C. dycei, C. marksi* (also mosquitoes) |
| **Rhabdoviridae** | | | |
| Bovine ephemeral fever | Cattle | Africa, Asia, Australia | *Culicoides* spp. |
| Kotonkan | Cattle, sheep, rats, hedgehogs | Africa | *Culicoides* spp. |
| Tibrogargan | Cattle, water buffalo | Australia | *C. brevitarsis* |

## OTHER VIRAL AGENTS

Several other arboviruses have been isolated from *Culicoides* adults. Four of them are members of the Bunyaviridae: Crimean-Congo, Rift Valley fever, and Dugbe and Shuni viruses. In addition, three mosquito-borne viruses which cause eastern equine encephalitis, Japanese B encephalitis, and Venezuelan equine encephalitis have been isolated from *Culicoides* and *Lasiohelea* species. There is no evidence, however, to indicate that biting midges play a significant role in transmission of any of these viruses.

## MANSONELLOSIS

Three filarial nematodes in the genus *Mansonella* cause infestations in humans, called mansonellosis (Table II). Cases occur widely throughout the tropical and subtropical regions of both the Old World and New World where *Culicoides, Forcipomyia,* and *Leptoconops* species serve as arthropod vectors. Although infestations involving these parasites are generally mild or asymptomatic, they sometimes cause serious medical problems.

TABLE II

**Filarial Nematodes Transmitted by Biting Midges to Humans and Domestic Animals: Vectors Include** *Culicoides, Forcipomyia,* **and** *Leptoconops* **Species**

| Nematode | Vertebrate host | Geographic area | Known or suspected vectors |
|---|---|---|---|
| *Mansonella ozzardi* | Humans | South America, Caribbean Basin | *C. barbosai, C. furens C. paraensis, C. phlebotomus, Leptoconops bequaerti* |
| *M. perstans* | Humans | Sub-Saharan West Africa; Central Africa to Kenya and Mozambique Northern coast of South America; Caribbean Islands | *C. austeni C. grahamii C. inornatipennis Culicoides* spp. |
| *M. streptocera* | Humans | West and Central Africa (rain forests) | *C. austeni, C. grahamii* |
| *Onchocerca cervicalis* | Horses | North America Australia | *C. variipennis C. victoriae Forcipomyia townsvillensis* |
| *O. gibsoni* | Cattle | India, Sri Lanka, Malaysia, northern Australia, South Africa | *C. pungens Culicoides* spp. |
| *O. gutturosa* | Cattle | Australia | *Culicoides* spp. |
| *O. reticulata* | Horses, ponies | Australia | *C. nubeculosus C. obsoletus* |
| *O. sweetae* | Water buffalo | Unknown | Unknown |

**FIGURE 10.7**   Filarial nematode (*Mansonella ozzardi*), microfilarial stage. (Courtesy of Lea & Febiger.)

### Mansonella ozzardi

*M. ozzardi* is the only native New World cerato-pogonid-borne nematode of humans (Fig. 10.7). It is indigenous to the Americas, occurring in the Amazon Basin (Brazil); along the northern coast of South America (Colombia, Venezuela, Guyana, Suriname, and French Guiana); on Trinidad, Haiti, and other islands of the West Indies; Panama; and parts of Peru, Bolivia, and Argentina. It particularly affects coastal fishing communities near breeding sites of associated vectors. The infection rate among local inhabitants is highly variable, ranging from as low as 5% or less in northern Brazil and some of the Caribbean islands to over 95% among Amerindians in Colombia and Venezuela. Infection rates are generally highest among men and women in older age groups, reflecting chronic exposure to infection in endemic areas.

Infections with *M. ozzardi* usually do not result in significant pathological effects. The microfilariae typically remain in the capillaries of the skin and surrounding dermal tissues, where they cause relatively little harm. Surveys are usually conducted by taking skin biopsies or blood samples and examining them for the presence of microfilariae. The adult worms are found primarily in fat tissue associated with the peritoneum and various body cavities, occasionally causing conjunctivitis and swelling of the eyes. In some cases this nematode can cause more serious problems, such as severe joint pains, eosinophilia,

enlargement of the liver, and blockage or inflammation of the lymphatic vessels, resulting in conditions similar to bancroftian filariasis and elephantiasis. Ivermectin has been successfully used in treatment of *M. ozzardi* cases, whereas the widely used filarial nematocide diethylcarbamazine is ineffective in killing this parasite.

Vectors of *M. ozzardi* include both biting midges and black flies, with different taxa playing important roles in different areas. *Culicoides furens* and *C. phlebotomus* are the principal vectors in Haiti and Trinidad, respectively. Other species which support development of microfilariae to infective larvae and are generally considered to play a secondary role in transmission are *C. barbosai*, *C. paraensis*, and *L. bequaerti*. After ingestion by a biting midge as it feeds on an infected host, microfilariae are carried into the midgut, where they penetrate the midgut wall and make their way to the thoracic muscles within 24 hr. There they develop to third-stage larvae during the next 6 to 9 days before moving to the head and mouthparts. Infective third-stage larvae enter the bite wound when the midge subsequently feeds on another host. Typically only one to three larvae successfully complete development to the infective stage in a host insect, regardless of the number of microfilariae initially ingested.

The role of black flies as vectors of *M. ozzardi* remains unclear. Species in the *Simulium amazonicum* group and the *S. sanguineum* group have been found to be naturally infected with this nematode and probably play a role in transmission, particularly in the Amazon Basin. Other species incriminated as potential vectors based on field collections and experimental infection studies include *S. sanchezi* and *S. pintoi*. Despite earlier suggestions that there may be two different forms or species of nematode involved, one transmitted by biting midges and the other by black flies, recent evidence indicates that they are morphologically identical and represent a single species.

### Mansonella perstans

This nematode (formerly placed in the genera *Acanthocheilonema*, *Dipetalonema*, and *Tetrapetalonema*) is the most widely distributed of the three human filarial nematodes transmitted by biting midges. It is indigenous to the Old World, where it occurs in sub-Saharan Africa, extending primarily from West African countries bordering the Gulf of Guinea (Ivory Coast, Nigeria, and Equatorial Guinea) and from Gabon and Angola east through Central Africa to Kenya and Mozambique. Infection rates are commonly 50% or higher in some communities. *M. perstans* was introduced to Central and South America with the slave trade and now occurs along the northern coast of South America (Colombia, Venezuela, Guyana, Suriname, and French Guiana), in the Yucatán area of Mexico, and on Trinidad and other Caribbean Islands. Prevalence of infection exceeding 50% has been reported among the Curripaco Indians of Venezuela. Although there is evidence to suggest that *M. perstans* represents a complex of species, this issue remains unresolved.

*M. perstans* is typically regarded as nonpathogenic. The microfilariae remain primarily in the circulating blood, whereas the adult worms occur freely in the body cavities. Some infested individuals develop problems such as joint pains, fever, fatigue, transient edema, elephantoid scrota, mild urticarial skin reactions, and eosinophilia. Various ocular problems, including swelling of the eyelids, excessive lacrimation, pruritus, and conjunctival granulomas or nodules, have been reported. The latter is the result of adult worms coiled within the connective tissue of the conjunctiva, causing a condition known as *bulge-eye* or *bung-eye*. Adult worms also have been removed from connective tissue of the pancreas, kidneys, rectum, and mesenteric lymph nodes of infested patients, with little evidence of serious harm. Mebendazole has been used successfully in treating *M. perstans* cases, whereas diethylcarbamazine and ivermectin are ineffective in killing either the microfilariae or adult worms.

The principal vectors of *M. perstans* have not been identified, particularly in the New World, where they remain virtually unknown. In Africa several *Culicoides* species have been implicated as vectors based on natural infections and support of development to the infective stage in experimental studies. They include *C. austeni*, *C. grahamii*, and *C. inornatipennis* as probable vectors and *C. hortensis*, *C. krameri*, *C. kumbaensis*, *C. milnei*, *C. pycnostictus*, *C. ravus*, *C. rutshuruensis*, and *C. vitshumbiensis* as possible vectors. As in the case of *M. ozzardi*, the microfilariae of *M. perstans* move from the midgut of the biting midge to the thoracic musculature, where they complete their development to infective, third-stage larvae 8–10 days after the infective blood meal.

### Mansonella streptocerca

This filarial nematode (formerly placed in the genera *Dipetalonema* and *Tetrapetalonema*) occurs only in the rain forests of West and Central Africa, extending from the Ivory Coast and Burkina Fasso to the Congo and Zaire. Little information on prevalence is available for this species, although a figure of 13–14% has been reported in certain villages of the Central African Republic based on peripheral blood smears. Although it is regarded as nonpathogenic to humans, *M. streptocerca* occasionally causes mild skin reactions due to activity of microfilariae in dermal tissues, usually involving the trunk and upper arms. Adult worms typically occur subcutaneously in upper parts of the body. Diethylcarbamazine is effective as a treatment. *C. grahamii* is regarded as the principal vector.

## VETERINARY IMPORTANCE

Biting midges serve as vectors of more than 35 arboviruses which infect domestic animals (Table I). Only a few of these viruses cause significant clinical disease. Cattle, sheep, and horses usually are the most seriously affected. The majority of these viral agents are members of the Reoviridae and Bunyaviridae, including the pathogens which cause bluetongue disease, epizootic hemorrhagic disease, and African horsesickness. Two other families of viruses with which *Culicoides* species have been implicated as vectors are the Rhabdoviridae and Poxviridae. For most of these viral agents, the principal ceratopogonid species involved as vectors remain largely unknown. Other disease agents transmitted by biting midges include blood protozoans of birds, such as *Haemoproteus meleagridis* in turkeys, *Leucocytozoon caulleryi* in chickens, and the nematode *Onchocerca cervicalis* in horses. Biting midges also can cause discomforting skin reactions in horses known as equine allergic dermatitis.

### BLUETONGUE DISEASE

Bluetongue disease is caused by an orbivirus in the family Reoviridae which infects ruminants, notably sheep and cattle. This disease occurs in many parts of the world, including temperate and southern regions of North America; parts of Central America and South America bordering the Caribbean; and Europe, the Middle East, Asia, Australia, and southern Africa. It was first recognized in sheep and cattle in South Africa in the early 1930s. In the United States bluetongue (BLU) virus was first isolated in 1952 in Texas from sheep exhibiting a condition known as *soremuzzle*. It is now known to occur throughout most of the southern and western states, where prevalence of antibody to BLU virus in cattle is commonly 20–50%. No cases have been reported in Alaska or Hawaii. Canada remains largely bluetongue-free; detection of infected animals has been reported there only in the Okanagan Valley of British Columbia during localized outbreaks in 1975 and 1987. Both of these instances apparently originated from cattle imported from the United States.

Occurrence of clinical bluetongue disease in cattle tends to be sporadic, often involving only one or a few animals in a given herd. Occasionally, however, epizootics do occur. Following outbreaks in Cyprus in 1943, Turkey in 1944, and Israel in 1951, major epizootics occurred in Europe in the late 1950s, where an estimated 179,000 sheep died in Spain and Portugal. The mortality rate among infected animals was 75%. BLU virus has since been introduced to the Caribbean islands, bordering

**FIGURE 10.8** Sheep with bluetongue disease; infected host with characteristic arched back, tender hooves, and hanging head. (Courtesy of US Department of Agriculture, Animal and Plant Health Inspection Service.)

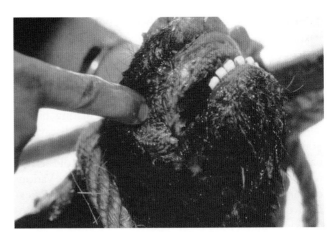

**FIGURE 10.10** Oral and muzzle lesions in a Black Angus calf suffering from bluetongue disease. (Photo by Lloyd L. Lauerman.)

countries of South and Central America, and Australia, where it was first detected in 1974.

Serosurveys for detection of antibodies to BLU virus indicate that most sheep and cattle which are exposed to the virus do not develop clinical signs. As a result, animals typically remain asymptomatic and often serve as unrecognized sources of infection for other animals. However, under circumstances which are poorly understood, some animals develop varying degrees of illness, ranging from mild infections to acute, fatal disease. In more severe cases, animals develop lesions about the mouth and muzzle (Fig. 10.10), with ulceration and sloughing of skin tissues, inflammation of the coronary band at the base of the hooves, lesions between the toes

(Fig. 10.11), respiratory difficulties due to accumulation of fluids in the lungs, and internal hemorrhaging. The term "bluetongue" gets its name from the dusky blue appearance of the tongue and mucosal membranes lining the mouth, resulting from cyanosis. Acutely infected animals often exhibit lameness and a characteristic arched back resulting from efforts to keep weight off of their painful hooves (Figs. 10.8 and 10.9). Death results primarily from congestion of the lungs and massive internal hemorrhaging.

Reproduction is also affected and can result in underweight calves at birth, congenital deformities, stillbirths, and abortions. The severity of reproductive impact is due in large part to the time during gestation when the infection occurs. Infections in the early stages of fetal

**FIGURE 10.9** Black Angus calf in a late stage of bluetongue disease, with general depression, hanging head, labored breathing, and difficulty standing. (Photo by Lloyd L. Lauerman.)

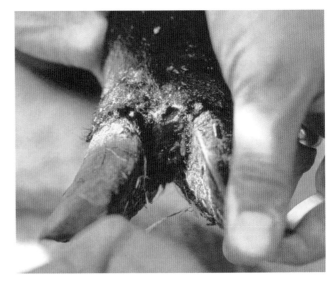

**FIGURE 10.11** Hoof lesions in a Black Angus calf in a late stage of bluetongue disease. (Photo by Lloyd L. Lauerman.)

development can result in aborted or stillborn calves. Infections which occur later in gestation are more likely to result in congenital deformities and underweight calves at birth. The virus is also found in semen of infected bulls. Consequently, restrictions are placed on the exportation of semen and live animals from endemic bluetongue areas for artificial insemination and other breeding purposes. The resulting economic impact on the livestock industry in the United States for mandatory testing of animals and losses in the foreign market is substantial, totaling millions of dollars annually.

BLU virus represents an *antigenic complex* with at least 25 currently recognized serotypes, which vary significantly in their pathogenicity. Five serotypes are known to occur in the United States, with serotypes 10, 11, 13, and 17 being the most widely distributed. The most recently introduced serotype is BLU-2, which was first isolated from cattle in Florida in 1982. It is reported to have caused clinical disease and death in cattle only in Alabama. Based on surveys of sentinel cattle and sheep in the Caribbean Basin, eight serotypes of BLU virus are known to occur in that region (serotypes 1, 3, 4, 6, 8, 12, 14, and 17). However, despite widespread infections of animals, notably cattle, clinical disease is largely absent in the Caribbean area.

The common occurrence of multiple serotypes of BLU virus in the same geographic area has hampered the use of immunization for protecting livestock from infection. Polyvalent vaccines have been developed but generally have not been effective. Natural immunity is acquired by sheep following their recovery from this disease but is limited to the particular serotype with which the animal was infected. Cattle apparently do not develop significant immunity following infection.

The primary mode of transmission of BLU virus is by the bite of infected *Culicoides* midges. Based on experimental studies with members of the *C. variipennis* complex, the virus acquired while the midge feeds on a viremic animal invades the salivary glands of the midge and there multiplies. The intrinsic incubation period is 10–20 days, after which the midge can transmit the virus at subsequent feedings. Following infection, the *Culicoides* host remains infective throughout its life. There is no evidence that either transovarial or transstadial transmission occurs. Infection rates are highly variable, depending on the *Culicoides* species and the geographic populations involved. Selective breeding in laboratory colonies has produced both susceptible and highly resistant lines, indicating the complexity of factors influencing the vector competence of *Culicoides* species involved in the epizootiology of this disease.

Transmission of BLU virus also occurs venereally via semen of infected rams and bulls. When introduced into the female genital tract, the virus potentially can infect the adult animal and, if she is pregnant, the developing embryo or fetus. No methods have been developed to destroy the virus in semen of infected animals. The virus can survive indefinitely in frozen semen samples.

Biting midges of the genus *Culicoides* are the only known *vectors* of BLU virus. Only a few species appear to be responsible for the major epizootics of bluetongue which have occurred worldwide. The two most important vectors are *C. imicola* in Africa and the Middle East and *C. sonorensis* in the southern and western United States. *C. imicola* also has become established in Europe, where it now occurs in Portugal, Spain, and some of the Mediterranean islands. Although other *Culicoides* species have been found to be naturally infected with BLU virus, their importance as vectors remains uncertain. Some of them are likely to serve as secondary vectors which can augment transmission of the virus during local outbreaks.

Known or suspected vectors in South Africa are *C. imicola, C. bolitinos, C. gulbenkiani, C. magnus, C. pycnostictus, C. zuluensis,* and members of the *C. shultzei* group. In Australia *C. actoni, C. brevitarsis, C. fulvus,* and *C. wadai* are considered to be likely vectors, with *C. brevipalpus, C. oxystoma,* and *C. peregrinus* possibly playing minor roles in transmission. In addition to *C. imicola,* the following species are regarded as potential vectors in Europe and the Middle East: *C. nubeculosus, C. obsoletus,* and *C. pulicaris.* In North America, suspected species other than the principal vector *C. sonorensis* include *C. insignis, C. debilipalpis, C. obsoletus,* and *C. stellifer.* The most likely vectors in the Caribbean Basin appear to be *C. filarifer, C. insignis,* and *C. pusillus.*

## EPIZOOTIC HEMORRHAGIC DISEASE

EHD is very similar to bluetongue in many respects, the major difference being that it occurs primarily in wild ruminants, notably deer. It is caused by an orbivirus very closely related to BLU virus. Two strains or serotypes of the virus are recognized in North America, designated EHD-1 and EHD-2. EHD-1, known as the New Jersey strain, was first isolated from white-tailed deer in New Jersey during an outbreak in 1955. EHD-2, commonly referred to as the Alberta strain, was first isolated during an epizootic in that Canadian province in 1962. At least eight other EHD serotypes have been isolated in South Africa, Nigeria, Australia, and Japan.

The clinical signs in EHD cases are virtually indistinguishable from bluetongue. Isolation and identification of the etiologic agent is usually required to determine with certainty which virus is involved. Because of the similarities of these two diseases in wild ruminants, cases are often referred to simply as *hemorrhagic disease*. It also is

FIGURE 10.12    White-tailed deer fawn infected with epizootic hemorrhagic disease virus. Note hanging head and protruding tongue. (Photo by G. R. Mullen.)

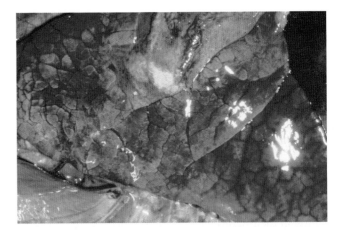

FIGURE 10.14    Extensive hemorrhaging and edema of lung tissue in white-tailed deer that died of epizootic hemorrhagic disease. (Courtesy of Alabama Veterinary Diagnostic Laboratory, Auburn, AL.)

referred to as *black tongue disease* by deer hunters in the southeastern United States.

Clinical disease in white-tailed deer and other ruminants varies from sudden death without apparent signs of illness to mild infections from which animals fully recover. Typically the disease is characterized by rapid onset of fever, loss of appetite, disorientation and weakness, a hanging head, labored breathing with the tongue often protruding (Fig. 10.12), swelling of the head and neck, arched back, and painful hooves (Fig. 10.13). As the virus multiplies in endothelial cells lining the blood vessels, it spreads to various organ systems, causing extensive internal hemorrhaging (Fig. 10.14), intravascular coagulation, and thrombosis. In acute cases, death usually occurs in 4–10 days following the initial infection. In those animals which survive, recovery can be prolonged and debilitating, resulting in permanent lameness due to deformed hooves (Fig. 10.15) and difficulty eating due to damage to the oral tissues.

EHD is the most important infectious disease in wild deer populations in the United States. It primarily affects white-tailed deer, causing sporadic die-offs. Mule deer, pronghorns, and domestic cattle also can develop fatal infections but do so less commonly. Other wild ruminants which have been found to be infected during EHD epizootics include elk, bison, bighorn sheep, Rocky Mountain goats, and several species of exotic animals, such as

FIGURE 10.13    White-tailed deer buck in late stage of epizootic hemorrhagic disease, with characteristic tender hooves, difficulty walking, arched back, laid-back ears, and general depression. (Photo by G. R. Mullen.)

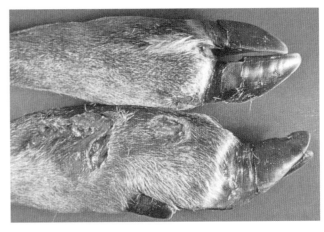

FIGURE 10.15    Foot lesions, swelling, and deformed hooves in white-tailed deer with epizootic hemorrhagic disease. (Courtesy of Southeastern Cooperative Wildlife Disease Study, Athens, GA.)

yak and ibex. Wapiti and moose do not appear to be adversely affected by this virus.

Although this disease is endemic throughout the United States where white-tailed deer populations are established, it is more prevalent in the Southeast, Midwest, and Northwest and along the Pacific Coast. *Epizootics,* with sudden die-offs in local deer herds, tend to occur in more temperate areas, whereas asymptomatic and subclinical infections are more common in the coastal endemic areas of the southeastern states, where infection rates may be as high as 70% or more. Outbreaks of EHD have also been reported in the western provinces of Canada, notably in southeastern Alberta (1962), the Okanagan Valley of British Columbia (1975), and southern Saskatchewan (1986–1987). In each case, the source of infection has been attributed to *Culicoides* from adjacent endemic areas in the United States. The only serotype isolated in Canada has been EHD-2.

Cattle are commonly exposed to EHD virus. Based on serologic surveys, infections in cattle are widespread throughout the United States. In most cases these are silent infections or involve only mild clinical disease. Occasionally, however, epizootics do occur in cattle, as in central Oregon in 1969 and eastern Tennessee in 1972. Infections of cattle with EHD virus also have been reported in the South American countries of Guyana, Suriname, and Colombia and in Taiwan, Malaysia, and Indonesia.

Biting midges of the genus *Culicoides* are the only known *vectors* of EHD virus. The most important species and only proven vector in North America is *C. sonorensis.* Other species have been implicated as potential vectors based on isolations of the virus from field-collected midges and limited experimental studies. The high prevalence of seropositive deer for EHD virus in areas where members of the *C. variipennis* complex are uncommon or absent supports the belief that other *Culicoides* species are involved in transmission of this virus in the United States. The vectors in other parts of the world remain unknown.

## AFRICAN HORSESICKNESS

AHS is a viral disease of horses, donkeys, and mules which can be highly fatal in susceptible animals (Fig. 10.16). It is known by various names, including *la pesta equine* (Spain), *pesta ecvina* (Romania), *equine plague,* and *horse sickness fever.* The disease was first recognized in South Africa in the early 1700s, with the etiologic agent being first isolated from infected horses nearly two centuries later, in 1899. It occurs throughout sub-Saharan Africa and the Arabian Peninsula, extending intermittently into southwestern Asia and southern Europe, where epizootics have occurred in recent years.

**FIGURE 10.16** Horse that died with African horsesickness; death was attributed to pulmonary edema. (Courtesy of USDA-APHIS, Foreign Animal Disease Diagnostic Laboratory, Plum Island, NY.)

The etiologic agent of AHS is an orbivirus in the family Reoviridae, closely related to the viruses that cause bluetongue and EHD. Nine AHS serotypes are recognized, all nine of which occur in South Africa. Eight of the nine serotypes were recovered from just seven horses during an outbreak in Nigeria in 1974–1975. The occurrence of multiple serotypes in a given geographic region and simultaneous infections of animals with more than one serotype underscore the epizootic complexity of this disease.

Four clinical forms of AHS are recognized: pulmonary (peracute), cardiac (subacute), mixed pulmonary-cardiac (acute), and horse sickness fever. The *pulmonary form* is the most fatal, with mortality rates as high as 95%. Clinical signs develop within 3–5 days of the initial infection. The onset of symptoms is sudden, usually beginning with fever followed by congestion of the mucous membranes of the eyes, nose, and mouth. Animals sweat profusely, experience increased respiratory rates, and cough spasmodically due to the accumulation of fluids in the lungs. Froth is commonly emitted from the nostrils in the terminal stage. Death usually occurs within a few days of the onset of clinical signs.

The *cardiac form* is similarly characterized by initial fever and congestion of the mucous membranes following an incubation period of 7–14 days. Animals subsequently develop extensive subcutaneous edema which is often apparent in the neck and jugular area, in the muscles along the back and hips, around the eyes and eyelids, and in the jaws. Other signs include depression and petechial hemorrhages on the underside of the tongue. Infected animals continue to feed and drink throughout the course of the disease. Death usually occurs in 4–8 days following the onset of fever, with mortality rates approaching 50%. The *mixed pulmonary-cardiac form* of AHS is characterized by clinical signs associated with each of the previous two syndromes. The onset of symptoms typically occurs

5–7 days following infection, with death ensuing 3–6 days later. The mortality is approximately 80%, intermediate between the pulmonary and cardiac forms. *Horsesickness fever* is the mildest form of AHS. Infected animals usually recover following a low-grade fever, congested mucous membranes, loss of appetite, and mild depression over a 1-week period.

The principal vertebrate hosts of AHS virus are wild and domestic equids such as zebras, horses, and mules. Donkeys are largely resistant but occasionally develop clinical disease. In endemic areas, however, native breeds seldom exhibit overt signs of infection, apparently having developed a natural or acquired immunity. Most AHS outbreaks have occurred in European breeds of equids introduced to endemic areas or as a result of exposure of susceptible equids to infected animals imported from endemic areas in Africa and parts of the Middle East. Other animals in which the presence of antibodies indicates exposure to AHS virus are goats, sheep, domestic cattle, buffaloes, dromedaries, and elephants. None of these hosts develop more than mild clinical signs but may serve as potential reservoirs for the virus. Infected dogs, however, can develop clinical disease and are believed to be important reservoirs in urban areas. Six strains of AHS virus have been isolated from street dogs in Egypt, where a number of dogs have died after consuming uncooked meat of infected horse carcasses. The progression of the disease in dogs is similar to the pulmonary form in horses.

Until the mid 1900s, *epizootics* of AHS were largely confined to South Africa. Beginning in 1944, with an outbreak among horses in several Middle East countries, major epizootics have occurred in other parts of Africa, the Middle East, India, and Europe. The most devastating outbreak occurred in 1959–1960, in which over 300,000 horses died or had to be destroyed in six Middle East countries and Cyprus, Afghanistan, Pakistan, and India. Although cases of AHS were reported in Spain as early as 1965, the most severe epizootic to occur in Europe took place in Spain and Portugal in 1987–1990, in which more than 160 horses, mules, and donkeys died or had to be destroyed. Some of those animals were valuable thoroughbred horses participating in international equestrian competitions being held in Spain at the time. Ten zebras imported to a zoological park near Madrid, Spain, from Namibia in southern Africa are believed to have been the source of the infection. The virus was subsequently transmitted by indigenous *Culicoides* populations to Portugal and Morocco. Wind-borne midges, particularly *C. imicola*, are believed to have played a role in spreading the virus. Wind dispersal of *Culicoides* vectors may help to explain the spread of AHS virus from endemic areas of Africa, causing outbreaks in various parts of the Middle East, Cyprus and Turkey in the Mediterranean region, the point is that all 3 areas are in the Mediterranean region (i.e., the parts of the Middle East bordering the Mediterranean and Cyprus and Turkey) and the Cape Verde Islands off the northwestern coast of Africa.

There presently is no cure for this disease, leaving supportive therapy as the only means of treatment. Commercially available vaccines, however, have been helpful in protecting equids from infection in areas of Africa where AHS is endemic. Annual vaccinations are effective in maintaining immunity, reflecting the natural and complete immunity acquired by animals chronically exposed to this virus over extended periods of time. Regular vaccination of susceptible equids and strict control of the movement of unvaccinated animals is currently the only practical means of containing this disease.

The major *vectors* of AHS virus are *Culicoides* species. *C. imicola* is believed to be the most important species involved in transmission of this virus in Africa and the Middle East. Since the first isolation from field-collected *C. imicola* during an outbreak in South Africa, other species have been implicated as vectors. A few species have been shown to support replication of AHS virus following experimental inoculation, with members of the *C. variipennis* complex, for example. The virus has been successfully transmitted 12–13 days after an infected blood meal. Some mosquitoes also are believed to be potential vectors even though the virus has not been isolated from them under field conditions. *Cx. pipiens*, *Ae. aegypti*, and *Anopheles stephensi* have been experimentally infected with the virus, but there is no strong evidence to indicate that these particular mosquitoes are natural vectors.

AHS virus has been isolated from naturally infected camel ticks (*Hyalomma dromedarii*) in Egypt, raising a question about possible involvement of this tick as a secondary vector. The brown dog tick *Rhipicephalus sanguineus* has been shown to be capable of biologically transmitting AHS virus between horses and dogs. However, the virus has not been isolated from naturally infected ticks of this species.

## BLOOD PROTOZOANS

Biting midges are biological vectors of a number of protozoans called *haemosporidians*, which are blood parasites of reptiles, birds, and mammals. Three genera which are transmitted by biting midges are *Haemoproteus* (Fig. 10.17), *Hepatocystis*, and *Leucocytozoon*. Most of the species are avian parasites which cause little or no apparent harm to their hosts. A few, however, such as *Haemoproteus meleagridis* of turkeys and *L. caulleryi* of chickens, can cause significant problems for poultry producers.

Haemosporidians transmitted by biting midges are related to malarial parasites (*Plasmodium* species) with which they share a similar life cycle and developmental stages. While feeding on an infected vertebrate host, female midges ingest red blood cells containing

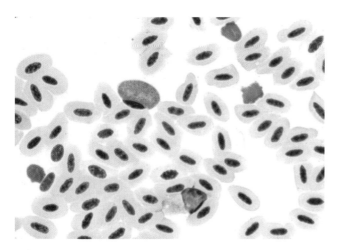

**FIGURE 10.17** *Haemoproteus* sp., avian protozoan developing in blood cells of mourning dove, transmitted by biting midges. (Photo by Mary E. Hayes/Rogers.)

*gametocytes*, the sexual stage of the parasite. In the midgut of the midge, the gametocytes are released, where they unite to form a motile zygote, the *ookinete*. The ookinete typically penetrates the peritrophic membrane and midgut tissue to form a cystlike structure, or *oocyst*, on the outer midgut wall. Within the oocyst, *sporozoites* are produced asexually, eventually rupturing from the mature oocysts into the hemocoel, where they make their way to the salivary glands and accumulate there. The sporozoite is the infective stage which

is transmitted via the saliva to suitable hosts when the biting midge subsequently blood feeds. Development of the parasite in *Culicoides* species usually takes about 6–10 days.

Upon entering the vertebrate host, sporozoites invade cells of fixed tissues, notably the endothelium of various organs, and myofibroblasts, precursor cells that form muscle fibers. There they undergo one or more cycles of asexual reproduction, called *schizogony*, to produce *merozoites*. The merozoites then invade the blood and penetrate circulating erythrocytes. There they develop into gametocytes, thereby completing the life cycle.

The species of blood protozoans which are known to be transmitted by biting midges are summarized in Table III. With the exceptions of *Haemoproteus kochi*, which parasitizes Old World monkeys, and *H. brayi*, which parasitizes Malaysian squirrels, these haemosporidians are parasites of birds. The species are primarily members of the genus *Haemoproteus*, all of which are transmitted by *Culicoides* spp. Relatively few details are known about most of these arthropod-borne haemosporidians and their associated *Culicoides* vectors. What is known is based primarily on studies of the following two species, which are parasites in poultry.

### Haemoproteus meleagridis

*H. meleagridis* is primarily a parasite of wild and domestic turkeys. It also can cause at least transient infections

**TABLE III**
**Protozoans Transmitted by Biting Midges**

| Protozoan | Vertebrate hosts | Geographic area | Known or suspected *Culicoides* vectors |
|---|---|---|---|
| *Haemoproteus* | | | |
|   *H. danilewskyi* | Crows, jays (Corvidae) | North America | *C. crepuscularis, C. sphagnumensis C. stilobezzioides* |
|   *H. desseri* (*H. handai*) | Parakeets (Psittacidae) | Thailand | *C. nubeculosus* (experimental) |
|   *H. fringillae* | Finches, sparrows (Fringillidae) | North America | *C. crepuscularis, C. stilobezzioides* |
|   *H. mansoni* (*H. canachites*) | Grouse (Tetraonidae) | North America | *C. sphagnumensis* |
|   *H. meleagridis* | Turkey (Meleagrididae) | North America | *C. edeni, C. arboricola, C. haematopotus, C. hinmani, C. knowltoni* |
|   *H. nettionis* | Ducks, geese (Anatidae), other waterfowl | Canada | *C. downesi* |
|   *H. velans* | Woodpeckers (Picidae) | North America | *C. sphagnumensis* |
| *Hepatocystis* | | | |
|   *H. brayi* | Squirrels (Sciuridae) | Malaysia | *Culicoides* spp. |
|   *H. kochi* | Monkeys (*Cercopithecus*) | Kenya | *C. adersi* |
| *Leucocytozoon caulleryi* | Chickens | Southeast Asia, Japan | *C. arakawae, C. circumscriptus, C. guttifer, C. schultzei* |

in pheasants and chukars but apparently does not infect chickens, guinea fowl, bobwhite quail, and other gallinaceous birds.

This parasite is generally regarded as nonpathogenic. Even in cases in which large numbers of circulating red blood cells are infected with gametocytes, birds usually exhibit few signs of stress or other pathologic effects. This suggests that compensatory mechanisms are operative in which the replacement rate of erythrocytes is sufficient to maintain a stable hematocrit despite high parasitemia. In other cases, however, there is evidence to indicate that *H. meleagridis* does harm its avian hosts, especially domestic turkeys. Heavy infections can result in anemia, reduced weight gain and growth rates, inflammation of skeletal and cardiac muscles, lameness, damage to the spleen and liver, and a wasting condition associated with chronic infections. Young birds are particularly vulnerable.

Five *Culicoides* species have been identified as vectors of *H. meleagridis* based primarily on studies in Florida. *C. edeni* is regarded as the most important vector, with *C. hinmani, C. arboricola, C. haematopotus,* and *C. knowltoni* playing secondary roles in transmission. Other species, such as *C. baueri, C. nanus,* and *C. paraensis,* have been shown to support development of the parasite only to the oocyst stage. Transmission of *H. meleagridis* occurs throughout the year in southern Florida, whereas it is limited to the warmer months of the year throughout the rest of the United States where turkeys occur.

### Leucocytozoon caulleryi

This is the only *Leucocytozoon* species known to be transmitted by biting midges. It has been recognized for many years as causing a serious poultry disease of chickens in Japan and Southeast Asia, where it is known as *poultry leucocytozoonosis* and by the earlier name *Bangkok hemorrhagic disease* where it occurred in Thailand. The principal vector of *L. caulleryi* is *C. arakawae,* which commonly breeds in rice paddies.

## EQUINE ONCHOCERCIASIS

Equine onchocerciasis is caused by the filarial nematode *Onchocerca cervicalis* (Fig. 10.18), the most widely distributed nematode transmitted to domestic animals by biting midges. Horses are the only known host. Although it occurs worldwide, most of the problems associated with this nematode have been reported in the United States and Australia, where it commonly causes dermatitis. Various names which refer to infestations by *O. cervicalis* include *cutaneous equine onchocerciasis, equine ventral midline dermatitis, equine nuchal disease,* and *fistulous withers.*

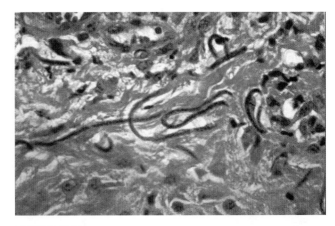

**FIGURE 10.18**   *Onchocerca cervicalis,* histological preparation showing microfilariae in skin of infested horse. (From Montes and Vaughan, 1983.)

Prevalence of *O. cervicalis* is high in many regions of the United States, with up to 85% or more of older horses having been reported infected with this parasite in New York, Kentucky, and the Gulf Coast states.

Adult worms occur primarily in the nuchal ligament of the neck and between the shoulder blades, or withers. Microfilariae produced by the females move to the skin, where they are active in the dermal tissues, often eliciting a host response in the form of localized inflammation and pruritus. The highest concentrations of microfilariae tend to be along the ventral midline of the horse. High numbers of microfilariae also may occur in skin of the inner thighs, chest region, withers, and eyelids. The density of microfilariae in skin tissue varies seasonally, being highest during the spring and summer months and lowest in the winter, when they move to the deeper dermal layers. This is correlated with the seasonal activity of most biting midges.

Horses which become sensitized to *O. cervicalis* develop various types of skin lesions, including depigmentation, pruritus, scaling, and hair loss. This usually occurs on the face, chest, withers, and ventral midline where microfilariae are most abundant. Ocular lesions have also been reported. Diagnosis is based on clinical signs and the detection of microfilariae in skin biopsies. Treatment with ivermectin has been found to be effective in killing microfilariae, but not the adults. In most cases the skin lesions show significant improvement, or are completely resolved, within a few weeks following treatment.

Members of the *Culicoides variipennis* complex are the only known vector of *O. cervicalis* in North America. Based primarily on laboratory studies, *C. victoriae, Forcipomyia townsvillensis,* and the black fly *Austrosimulium pestilens* also have been identified as potential vectors of *O. cervicalis* in Australia. Since these biting midges tend to ingest very few microfilariae while feeding on an infected animal, only one or two infective third-stage larvae are typically found in field-collected flies.

## OTHER FILARIAL NEMATODES

At least three other *Onchocerca* species which infest bovine and equine hosts are believed to be transmitted by biting midges (Table II): *O. gibsoni* of cattle in Southeast Asia, Malaya, Australia, and South Africa; and *O. gutturosa* of cattle and *O. reticulata* of horses and ponies in Australia. They are considered to be nonpathogenic.

## EQUINE ALLERGIC DERMATITIS

Horses exposed to bites of certain *Culicoides* species commonly exhibit an allergic skin reaction. This typically occurs as a seasonal dermatitis affecting the withers, mane, tail, and ears. The back, ventral midline, and other body regions also can be affected, presumably reflecting the feeding sites of different biting midges involved. Equine allergic dermatitis was first attributed to *Culicoides* bites in Australia in the early 1950s, where it was known as *Queensland itch*. It is now known to occur widely throughout the world and is known by various names such as *sweet itch, summer recurrent dermatitis, summer eczema, equine Culicoides sensitivity, Dhobie itch* (Philippines), and *Kasen disease* (Japan). A similar, seasonal dermatitis in response to *Culicoides* bites also has been reported in sheep.

The dermal response apparently is a sensitivity reaction to components of salivary fluids introduced to the bite wound while the flies are feeding (Fig. 10.19).

**FIGURE 10.19** Allergic dermatitis in neck region of horse in response to injections of *Culicoides* extracts. (Courtesy of Yehuda Braverman, Kimron Veterinary Institute, Israel.)

Normal, unsensitized horses usually react to these bites by developing small welts with relatively little associated discomfort. Sensitized horses, however, react more severely by developing intense local inflammation and pruritus; this can result in irritability, rubbing and scratching of involved areas, open wounds, and secondary infections. Ponies are especially sensitive. Affected animals often are unsuitable for riding and, in the case of show horses, may decrease substantially in commercial value because of their irritable behavior, hair loss, and skin blemishes.

Once sensitized, horses experience either an immediate hypersensitivity response, peaking within 4 hr, or a delayed hypersensitivity response in which large welts develop after 24 hr, with inflammation persisting up to 3 weeks or more. There is good evidence to show that *Culicoides*-induced hypersensitivity is a polygenic hereditary trait which predisposes certain animals to this response. This sensitivity occurs primarily in older horses, usually after 4–5 years of age.

A number of *Culicoides* species have been implicated as the cause of equine allergic dermatitis. Most are based on correlations between seasonal occurrences of the midges and clinical signs, biting sites on horses, and positive reactions to intradermal injections of horses with extracts of the respective biting midges. The following species are suspected of being involved: *C. insignis, C. obsoletus, C. spinosus, C. stellifer,* and *C. venustus* in the United States; *C. pulicaris* in England; *C. nubeculosus* and *C. punctatus* in Ireland; *C. chiopterus, C. impunctatus,* and *C. obsoletus* in Norway; *C. imicola* in Israel; and *C. brevitarsis* in Australia.

Treatments for equine allergic dermatitis in the form of antihistamines and corticosteroids usually provide only temporary relief of symptoms. Desensitization of animals with injections of *Culicoides* extracts has not proved to be effective. Horse owners in areas where this condition is recognized as a problem should avoid breeding their animals with lineages of known sensitivity. Insecticides applied directly to horses to repel or kill biting midges afford some protection and can substantially reduce the severity if administered on a regular basis throughout the fly season. Ivermectin, however, is ineffective. Stabling horses at night or pasturing them away from the attack of biting midges can also help to alleviate the problem.

## PREVENTION AND CONTROL

Larviciding generally has not been effective in reducing populations of biting midges. Often the breeding sites are not easily located and may be so dispersed that the application of insecticides to kill the immature stages is not practical. In some situations modifications of the habitat can help to reduce breeding sites by filling low-lying areas, diking, and regulating water levels to disrupt breeding and larval development. Eliminating seepage areas and leaking water troughs in or around livestock facilities can discourage the breeding of important species like *C. sonorensis*. Proper maintenance of farm ponds and fluctuation of the water level in dairy ponds and waste lagoons can help to reduce the number of adult biting midges which emerge. Disking low-lying crop lands and using appropriate irrigation schedules have been effective in reducing adult populations of some pest species.

Adulticides have been used with limited success in suppressing adults. To be effective they are usually applied as mists or fogs in the evening hours when the insects are most active. In coastal areas where problems can be especially severe, aerial applications of ultra-low-volume formulations of insecticides to salt marshes bordering populated areas can help to provide some relief. Ground applications using truck-mounted mist sprayers for control of biting midges are often conducted in conjunction with municipal mosquito control programs in problem areas.

Individual protection of humans and other animals is often the only practical means of discouraging ceratopogonid midges from biting. Scheduling outdoor activities to avoid the peak biting periods of troublesome species is advisable. Animals such as horses can be stabled at night to protect them from species which do not readily enter buildings and shelters to feed. The mesh sizes of most window and door screens are not effective in excluding biting midges, especially in the case of species such as *C. furens* and *C. paraensis*, which will enter buildings in search of hosts. Insect repellents applied to exposed skin and the use of jackets and other clothing impregnated with repellents such as *N*,*N*-diethyl-m-toluamide (DEET) can afford effective protection from the bites of many of the more troublesome species.

## REFERENCES AND FURTHER READING

Akiba, K. (1960). Studies on the *Leucocytozoon* found in the chicken in Japan. II. On the transmission of *Leucocytozoon caulleryi* by *Culicoides arakawae*. *Japanese Journal of Veterinary Science* **22**, 309–317.

Akiba, K. (1970). Leucocytozoonosis of chickens. *National Institute of Animal Health Quarterly* **10**(Suppl.), 131–147.

Anderson, G. S., Belton, P., and Kleider, N. (1993). Hypersensitivity of horses in British Columbia to extracts of native and exotic species of *Culicoides* (Diptera: Ceratopogonidae). *Journal of Medical Entomology* **30**, 657–663.

Atchley, W. R., Wirth, W. W., Gaskins, C. T., and Strauss, S. L. (1981). "A Bibliography and Keyword Index of the Biting Midges (Diptera: Ceratopogonidae)." Bibliographies and literature of agriculture, No. 13. US Department of Agriculture, Science and Education Administration, Washington, DC.

Atkinson, C. T. (1988). Epizootiology of *Haemoproteus meleagridis* (Protozoa: Haemosporina) in Florida: potential vectors and prevalence in naturally infected *Culicoides* (Diptera: Ceratopogonidae). *Journal of Medical Entomology* **25**, 39–44.

Atkinson, C. T. (1991). Vectors, epizootiology, and pathogenicity of avian species of *Haemoproteus* (Haemosporina: Haemoproteidae). *Bulletin of the Society of Vector Ecology* **16**, 109–126.

Barber, T., and Jochim, M. J. (1985). Bluetongue and related orbiviruses. Proceedings of an international symposium. Monterey, California, January 16–20, 1984. *Progress in Clinical Biological Research* **178**, 1–746.

Battle, F. V., and Turner, E. C., Jr. (1971). "A Systematic Review of the Genus *Culicoides* (Diptera: Ceratopogonidae) of Virginia." The Insects of Virginia No. 3. Virginia Polytechnic Institute and State University Bulletin No 44, Blacksburg, VA.

Blanton, F. S., and Wirth, W. W. (1979). "The Sand Flies (*Culicoides*) of Florida (Diptera: Ceratopogonidae). Arthropods of Florida and Neighboring Land Areas." Vol. 10. Florida Department of Agriculture and Consumer Affairs, Gainesville.

Boorman, J. (1993). Biting midges (Ceratopogonidae). In "Medical Insects and Arachnids" (R. P. Lane and R. W. Crosskey, Eds.), pp. 288–301. Chapman & Hall, London.

Boorman, J., and Hagan, D. V. (1996). A name list of world Culicoides (Diptera: Ceratopogonidae). *International Journal of Dipterology* **7**, 161–192.

Desser, S. S., and Bennett, G. F. (1993). The genera *Leucocytozoon*, *Haemoproteus*, and *Hepatocystis*. In "Parasitic Protozoa," 2nd ed. (J. P. Kreler, Ed.), Vol. 4, pp. 273–307.

Downes, J. A., and Wirth, W. W. (1981). Ceratopogonidae. In "Manual of Nearctic Diptera" (J. F. McAlpine *et al.*, Eds.), Vol. 1, Monograph 17, pp. 393–421. Research Branch, Agriculture Canada, Ottawa.

Dulac, G. C., Sterritt, W. G., Dubuc, C., Afshar, A., Myers, D. J., Taylor, E. A., Jamieson, B. R., and Martin, M. W. (1991). Incursions of orbiviruses in Canada and their serologic monitoring in the native animal populations between 1962 and 1991. In (T. E. Walton and B. I. Osburn, Eds.), pp. 120–128. "Bluetongue, African Horsesickness, and Related Orbiviruses." CRC Press, Boca Raton, FL.

Eberhard, M. L., and Orihel, T. C. (1984). The genus *Mansonella* (syn. *Tetrapetalonema*). A new classification. *Annales de Parasitologie Humaine et Comparee* **59**, 483–496.

Foil, L., Stage, D., and Klei, T. R. (1984). Assessment of wild-caught *Culicoides* (Ceratopogonidae) species as natural vectors of *Onchocerca cervicalis* in Louisiana. *Mosquito News* **44**, 204–206.

Gibbs, E. P. J., and Greiner, E. C. (1988). Bluetongue and epizootic hemorrhagic disease. In "The Arboviruses: Epidemiology and Ecology" (T. P. Monath, Ed.), Vol. 2, pp. 39–70. CRC Press, Boca Raton, FL.

Gibbs, E. P. J., and Greiner, E. C. (1994). The epidemiology of bluetongue. *Comparative Immunology, Microbiology, and Infectious Diseases* **17**, 207–220.

Glick, J. I. (1990). *Culicoides* biting midges (Diptera: Ceratopogonidae) of Kenya. *Journal of Medical Entomology* **27**, 87–195.

Glukhova, V. M. (1977). The subgeneric classification of the genus *Culicoides* Latreille, 1809 (Diptera: Ceratopogonidae), including morphological characters of the larvae (in Russian). *Parazitologicheskii Sbornik* **27**, 112–128.

Glukhova, V. M. (1979). "Larval Midges of the Subfamilies Palpomyiinae and Ceratopogoninae of the Fauna of the U. S. S. R." (in Russian). Nauka, Leningrad.

Glukhova, V. M. (1989). "Blood-Sucking Midges of the Genera *Culicoides* and *Forcipomyia* (Ceratopogonidae) of the Fauna of the U. S. S. R." (in Russian). Nauka, Leningrad.

Greiner, E. C. (1995). Entomological evaluation of insect hypersensitivity in horses. *Veterinary Clinics of North America, Equine Practice* **11**, 29–41.

Greiner, E. C., Fadok, V. A., and Rabin, E. B. (1990). Equine *Culicoides* hypersensitivity in Florida: biting midges aspirated from horses. *Medical and Veterinary Entomology* **4**, 375–381.

Greiner, E. C., Mo, C. L., Tanya, V., Thompson, L. H., and Oviedo, M. T. (1991). Vector ecology of bluetongue viruses in Central America and the Caribbean. In (T. E. Walton and B. I. Osburn, Eds.), pp. 320–324. "Bluetongue, African Horsesickness, and Related Orbiviruses." CRC Press, Boca Raton, FL.

Hall, D. G. (1932). A new biting *Culicoides* from saltmarshes in the southeastern states. *Proceedings of the Entomological Society of Washington* **34**, 88–89.

Halldorsdottir, S., and Larsen, H. J. (1991). An epidemiological study of summer eczema in Icelandic horses in Norway. *Equine Veterinary Journal* **23**, 296–299.

Hess, W. R. (1988). African horse sickness. In "The Arboviruses: Epidemiology and Ecology" (T. P. Monath, Ed.), Vol. 2, pp. 1–18. CRC Press, Boca Raton, FL.

Holbrook, F. R., Tabachnick, W. J., Schmidtmann, E. T., McKinnon, C. N., Bobian, R. J., and Grogran, W. L. (2000). Sympatry in the *Culicoides variipennis* complex (Diptera: Ceratopogonidae): a taxonomic reassessment. *Journal of Medical Entomology* **37**, 65–76.

Hunt, G. J. (1994). "A Procedural Manual for the Large-Scale Rearing of the Biting Midge, *Culicoides variipennis* (Diptera: Ceratopogonidae)." US Department of Agriculture, Agricultural Research Service, ARS 121, National Technical Information Service, Springfield, VA.

Jamnback, H. A. (1965). The *Culicoides* of New York State (Diptera: Ceratopogonidae). *New York State Museum of Science Service Bulletin* **399**, 1–154 pp.

Jones, R. H., Luedke, A. J., Walton, T. E., and Metcalf, H. E. (1981). Bluetongue in the United States: an entomological perspective toward control. *World Animal Review* **38**, 2–8.

Kettle, D. S. (1965). Biting ceratopogonids as vectors of human and animal diseases. *Acta Tropica* **22**, 356–362.

Kettle, D. S. (1972). The biting habits of *Culicoides furens* (Poey) and *C. barbosai* Wirth and Blanton. III. Seasonal cycle, with a note on the relative importance of ten factors that might influence the biting rate. *Bulletin of Entomological Research* **61**, 565–576.

Kettle, D. S. (1984). "Medical and Veterinary Entomology." Wiley, New York.

LeDuc, J. W., Hoch, A. L., Pinheiro, F. P., and Travassos da Rosa, A. P. A. (1981). Epidemic Oropouche virus disease in northern Brazil. *Bulletin of the Pan American Health Organization* **15**, 97–193.

Lillie, T. H., Kline, D. L., and Hall, D. W. (1987). Diel and seasonal activity of *Culicoides* spp. (Diptera: Ceratopogonidae) near Yankeetown, Florida, monitored with a vehicle-mounted insect trap. *Journal of Medical Entomology* **24**, 503–511.

Linley, J. R. (1983). Biting midges (Diptera: Ceratopogonidae) and human health. *Journal of Medical Entomology* **20**, 347–364.

Linley, J. R. (1985). Biting midges (Diptera: Ceratopogonidae) as vectors of nonviral animal pathogens. *Journal of Medical Entomology* **22**, 589–599.

Linley, J. R., and Adams, M. (1972). A study of the mating behavior of *Culicoides melleus* (Diptera: Ceratopogonidae). *Transactions of the Royal Entomological Society of London* **124**, 81–121.

Lubroth, J. (1988). African horsesickness and the epizootic in Spain 1987. *Equine Practice* **10**, 26–33.

Lubroth, J. (1991). The complete epidemiologic cycle of African horse sickness: our incomplete knowledge. In "Bluetongue, African Horse Sickness, and Related Orbiviruses" (T. E. Walton and B. I. Osburn, Eds.), pp. 197–204. CRC Press, Boca Raton, FL.

Mellor, P. S. (1990). The replication of bluetongue virus in *Culicoides* vectors. *Current Topics in Microbiology and Immunology* **162**, 143–161.

Miura, Y., Goto, Y., Kubo, M., and Kono, Y. (1988). Isolation of Chuzan virus, a new member of the Palyam subgroup of the genus *Orbivirus*, from cattle and *Culicoides oxystoma* in Japan. *American Journal of Veterinary Research* **49**, 2022–2025.

Mo, C. L., Thompson, L. H., Homan, E. J., Oviedo, M. T., Greiner, E. C., González, J., and Sáenz, M. R. (1994). Bluetongue virus isolations from vectors and ruminants in Central America and the Caribbean. *American Journal of Veterinary Research* **55**, 211–215.

Montes, L. F., and Vaughan, J. T. (1983). "Atlas of Skin Diseases of the Horse." W. B. Saunders, Philadelphia.

Mullen, G. R., and Hribar, L. J. (1988). Biology and feeding behavior of ceratopogonid larvae (Diptera: Ceratopogonidae) in North America. *Bulletin of the Society of Vector Ecology* **13**, 60–81.

Mullens, B. A. (1991). Integrated management of *Culicoides variipennis*: a problem of applied ecology. In (T. E. Walton and I. Osburn, Eds.), pp. 896–905. "Bluetongue, African Horsesickness, and Related Orbiviruses." CRC Press, Boca Raton, FL.

Murphree, C. S., and Mullen, G. R. (1991). Comparative larval morphology of the genus *Culicoides* Latreille (Diptera: Ceratopogonidae) in North America with a key to species. *Bulletin of the Society of Vector Ecology* **16**, 269–399.

Ottley, M. L., Dallemagne, C., and Moorhouse, D. E. (1983). Equine onchocerciasis in Queensland and the Northern Territory of Australia. *Australian Veterinary Journal* **60**, 200–203.

Pearson, J. E., Gustafason, G. A., Shafer, A. L., and Alstad, A. D. (1991). Distribution of bluetongue in the United States. In (T. E. Walton and B. I. Osburn, Eds.), pp. 128–138. "Bluetongue, African Horsesickness, and Related Orbiviruses." CRC Press, Boca Raton, FL.

Pinheiro, K. F. P., Travassos da Rosa, A. P. A., Travassos da Rosa, J. F. S., *et al.* (1981). Oropouche virus I. A review of clinical, epidemiological, and ecological findings. *American Journal of Tropical Medicine and Hygiene* **30**, 149–160.

Tabachnick, W. J. (1992). Genetic differentiation among populations of *Culicoides variipennis* (Diptera: Ceratopogonidae), the North American vector of bluetongue virus. *Annals of the Entomological Society of America* **85**, 140–147.

Tabachnick, W. J. (1996). *Culicoides variipennis* and bluetongue-virus epidemiology in the United States. *Annu. Rev. Entomol.* **41**, 23–43.

Thomas, F. C. (1981). Hemorrhagic disease. In "Diseases and Parasites of White-Tailed Deer" (W. R. Davidson, F. A. Hayes, V. F. Nettles, and F. E. Kellogg, Eds.). Miscellaneous Publication No. 7. 87–96. Tall Timbers Research Station, Tallahassee, FL.

Walton, T. E., and Osburn, B. I. (1991). "Bluetongue, African Horsesickness, and Related Orbiviruses." CRC Press, Boca Raton, FL.

Ward, M. P. (1994). The epidemiology of bluetongue virus in Australia—a review. *Australian Veterinary Journal* **71**, 3–7.

Wirth, W. W., Dyce, A. L., and Peterson, B. V. (1985). An atlas of wing photographs, with a summary of the numerical characters of the Nearctic species of *Culicoides* (Diptera: Ceratopogonidae). *Contributions of the American Entomological Institution* **22**(4), 46 pp.

Wirth, W. W., Dyce, A. L., and Spinelli, G. R. (1988). An atlas of wing photographs, with a summary of the numerical characters of the Neotropical species of *Culicoides* (Diptera: Ceratopogonidae). *Contributions of the American Entomological Institution* **25**, 1–72.

Wirth, W. W., and Hubert, A. A. (1989) The *Culicoides* of Southeast Asia (Diptera: Ceratopogonidae). *Memoirs of the American Entomological Institution* **44**, 1–508.

Wirth, W. W., Ratanaworabhan, N. C., and Blanton, F. S. (1974). Synopsis of the genera of Ceratopogonidae (Diptera). *Annales de Parasitologie Humaine et Comparee* **49**, 595–613.

# 11

# BLACK FLIES (*Simuliidae*)

PETER H. ADLER AND JOHN W. McCREADIE

*As small,* powerful fliers adapted for blood feeding, black flies can be formidable pests of humans, domestic animals, and wildlife, impacting virtually all facets of outdoor life. They are distributed worldwide, with the exception of Antarctica and some oceanic islands. Their distribution is largely influenced by the availability of flowing water, which is required for development by the immature stages. Many of the worst pest species breed in large rivers, some of which can produce nearly a billion flies per kilometer of riverbed per day. Other pest species inhabit the small, but myriad, streams of heavily wooded terrain, making management efforts difficult.

Often ranked third worldwide among arthropods in importance as vectors of disease agents, black flies also are among the few arthropods that have killed animals by exsanguination during massive attacks. Even when not biting, their persistent swarming behavior can create an intolerable nuisance as the blood-seeking females dart into facial orifices and crawl on the skin. As often is the case, the behavior of a minority defines the reputation of the group. So it is with black flies, for only about 10–20% of the world's species are actually pests of humans and their animals. But among these species are the vectors of the agents of human onchocerciasis and mansonellosis, bovine onchocerciasis, and avian leucocytozoonosis. The majority of species, however, go unnoticed, either because they do not feed as adults or because their hosts are of little economic concern.

## TAXONOMY

More than 1700 species of black flies have been described worldwide. Many other species are known but unnamed, and additional species undoubtedly remain to be discovered. The Palearctic Region contains the most described species, about 600, followed by the Neotropical Region, with approximately 400 species. The Nearctic Region has about 255 known species, although not all have been formally described.

Two subfamilies comprise the Simuliidae. The most primitive subfamily Parasimuliinae includes four described and one undescribed species endemic to the Pacific Northwest. The females of these species do not have biting mouthparts. The subfamily Simuliinae contains all remaining species and is divided into two tribes, the *Prosimuliini* and the *Simuliini,* the latter including the majority of pest species. The most universally accepted classification system below the tribal level is summarized by Crosskey and Howard (1997), who recognize 23 extant genera in the subfamily Simuliinae. Eleven genera in the subfamily are found in North America. The largest genus of black flies is *Simulium,* which contains 41 subgenera and many of the species of economic importance.

The morphological uniformity of black flies creates difficulty for species identification. For this reason, a holistic approach to identification is typically used, relying on characteristics from eggs, larvae, pupae, males, females, and the polytene chromosomes, as well as distributional and ecological information. The need for accurate identification, particularly in programs for pest and vector management, has driven the taxonomy of black flies. As a result, black flies are taxonomically one of the best known groups of arthropods at the species level; for example, about 98% of North American species are known as larvae and pupae.

More than 150 identification keys exist for black flies in various parts of the globe. Crosskey and Howard (1997), in their inventory of the black flies of the world, provide a comprehensive list of identification keys by zoogeographic region. Keys to the genera or subgenera of adults, pupae, and larvae of the Nearctic Region are provided by Peterson (1996). Regional keys for Nearctic species are available for the southeastern United States (Stone and Snoddy, 1969), northeastern North America (Adler and Kim, 1986; Davies *et al.,* 1962), the western United States (Peterson and Kondratieff, 1995), and western Canada (Currie 1986; Fredeen, 1985a). The most comprehensive treatment of the Palearctic fauna is by Rubtsov (1956). Keys to the supraspecific taxa of the Australasian and Afrotropical Regions are given by Crosskey (1967, 1969, respectively), and of the Neotropical Region by Vargas and Díaz Nájera (1957) and Coscarón (1987). Comprehensive keys for the Oriental Region are lacking, but the key by Takaoka and Davies (1995) provides a useful starting point.

The giant polytene chromosomes (usually $n = 3$), which are best developed in the larval silk glands, provide a highly useful tool for discovering and identifying species (Fig. 11.1). Giant chromosomes, particularly their banding patterns, often reveal that many black flies regarded as single species are actually complexes of two or more species. These *cryptic species,* or *sibling species,* as they often are called, are reproductively isolated and each is biologically unique. Their existence has far-reaching implications for biological studies and population management of pests and vectors. For example, *Simulium damnosum,* the black fly known for much of the 20th century as a

**FIGURE 11.1**    Polytene chromosomes from larval silk glands of the black fly *Twinnia magadensis.*

vector of the agent of human onchocerciasis or river blindness, consists of as many as 40 distinct species, not all of which are vectors of the pathogen. The existence of *homosequential sibling species* that are identical in both morphology and chromosomal banding patterns increases the taxonomic complexity of the family. *Cytotaxonomy* of black flies has been reviewed for the world fauna (Rothfels, 1979) and for some vector species (Procunier, 1989).

## MORPHOLOGY

The immature stages of black flies are adapted for aquatic life, although the nonmobile pupa also has terrestrial adaptations that are useful if the water recedes. The egg is roughly oval or triangular, with rounded angles. It has a glutinous outer layer and a smooth, pigmented inner shell. A *micropyle,* consisting of a simple hole in the egg for the entry of sperm, is present in some species but not others.

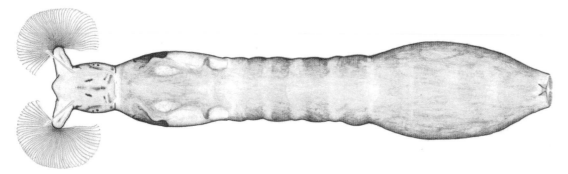

**FIGURE 11.2** Larva of the North American black fly *Simulium luggeri* (original by Lawrence W. Zettler).

The larva (Fig. 11.2) hatches with the aid of an *egg burster,* a small tubercle on the dorsum of the head capsule. The basic larval design consists of a well-sclerotized head capsule bearing an anterior pair of *labral fans,* and an elongate body with one *thoracic proleg* and a terminal *abdominal proleg.* Rows of tiny hooks on the prolegs enmesh with silk pads spun from the pair of larval *silk glands* and applied to a substrate. These silk glands extend from the anterior of the head into the posterior portion of the abdomen where they enlarge and double back on themselves. The adhesiveness of the silk is correlated with the velocity of the flowing water to which each species is adapted.

While clinging to a pad by its posterior proleg, the larva extends its body to filter feed. The prominent labral fans, each with about 20–80 individual rays bearing microtrichia (minute hairs) on their inner surface, are used to filter particulate matter from the water current. Larvae of some species (e.g., *Gymnopais* spp.) that live in habitats, such as glacial meltwaters, with little suspended food have lost the labral fans over evolutionary time. These species rely on their mandibles, specialized labrum, and hypostoma to scrape food from the substrate.

Additional features of the head and body are conspicuous and taxonomically important. The *antennae,* which consist of three articles and a terminal cone sensillum, are elongate, slender, and variously pigmented. A pair of dark *eyespots* is prominent on each side of the head capsule. Pigmentation patterns of the head capsule and body and the shape of the *postgenal cleft,* an area of weakly sclerotized cuticle on the ventral side of the head capsule, are important for interpreting the taxonomy of the family. The anteroventral portion of the head capsule bears the *hypostoma,* an anteriorly toothed plate used in conjunction with the mandibles to cut strands of silk and to scrape food from the substrate. Mature larvae are recognized by the presence of a prominent, dark gill histoblast on each side of the thorax.

The pupa (Fig. 11.3), which resembles an adult with its appendages held close to the body, is housed in a silk

*cocoon.* Cocoons are shapeless sacs in the evolutionarily basal species but are well-formed, slipper- or boot-shaped coverings, sometimes bearing anterior processes and lateral windows, in the more derived species. The pupa is held firmly in its cocoon by numerous anteriorly directed sets of hooklets. A pair of conspicuous *gills* arises from the thorax. The gills are among the most taxonomically useful and fascinating structures in any life stage. They vary in arrangement from thick, clublike structures to clusters of 2 to more than 100 slender filaments.

**FIGURE 11.3** Pupae of the North American black fly *Simulium vittatum* (copyright Dwight R. Kuhn).

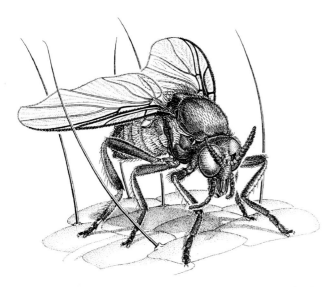

**FIGURE 11.4** Holarctic black fly *Simulium tuberosum*, female. (Copyright Roelof Idema.)

The adult black flies (Fig. 11.4) are characterized by a small but robust body, conical or beadlike *antennae* with seven to nine flagellomeres, and an arched thorax bearing a pair of wings that typically span 6–10 mm and have thickened veins near the leading margin. Most species are blackish, but orange, yellow, and variously patterned species also exist. Males of nearly all species have *holoptic eyes* that occupy most of the head and meet at the midline. Male eyes consist of enlarged dorsal facets, in addition to the typical-sized ventral facets, an arrangement that enhances the ability of males to locate females entering a mating swarm from overhead. Females have smaller, *dichoptic eyes* separated by the frons (Fig. 11.4).

The *mouthparts* arise ventrally from the head. A conspicuous pair of long maxillary palps attaches near the base of the proboscis. The third segment accommodates the *sensory vesicle* (Lutz's organ), which has many chemosensilla that detect odors such as carbon dioxide. The labium forms the back of the proboscis and envelops the other mouthparts, including the minutely serrated *mandibles* and the toothed *laciniae*, with a pair of large, fleshy lobes called the labella. The mouthparts of the male are similar to those of the female, except the mandibles and laciniae are not adapted for blood feeding and, therefore, do not bear teeth.

The stout thorax bears a pair of wings, either smoky or hyaline but never patterned. The venation, including the setation, is taxonomically important at the generic level. The color patterns of the legs and thoracic scutum are useful for species identification. The *tarsal claws* exist in one of three conditions. Species that feed on mammals have either a simple, unarmed claw or a minute tooth at the base of each claw. Bird feeders are endowed with a large thumblike lobe at the base of each claw. The abdomen is weakly sclerotized except the genitalia, which are of the utmost importance in the identification of species. To interpret the taxonomically important characteristics of the genitalia of both males and females, the abdomens must be treated with a clearing agent such as potassium hydroxide or hot lactic acid and slide mounted in glycerin.

## LIFE HISTORY

Immature black flies are found in virtually any water that flows, even if only imperceptibly and temporarily, from the smallest trickles to the largest rivers. Most species occupy specific habitats, and some higher taxa are characteristic of particular environments. For example, members of the genus *Gymnopais* occupy small, icy streams of the Far North, species of *Simulium* (subgenus *Hemicnetha*) live on the lips of waterfalls and in swift rocky flows, members of the *S. noelleri* species group are found below impounded waters, and species of *Simulium* (subgenus *Psilozia*) usually are found in warm, highly productive streams and rivers with open canopies.

Each species of black fly has a specific pattern of seasonal occurrence. Nearly all species in the tribe Prosimuliini are univoltine, completing a single generation annually. The tribe Simuliini contains both univoltine and multivoltine species. Some of these multivoltine species can complete seven or more generations per year in areas of North America with mild climates. In certain tropical areas of the world, some species (e.g., members of the *S. damnosum* complex) might cycle through more than 20 generations each year.

Eggs typically cannot resist desiccation, although some species (e.g., *Austrosimulium pestilens*) can survive in moist soil of dry streambeds for several years, hatching when streams are inundated. During the summer, eggs of multivoltine species (e.g., *S. vittatum*, *S. damnosum*) can hatch in fewer than 4 days. In northern temperate regions, univoltine species (e.g., *Prosimulium* spp.) often spend the warm months as eggs, whereas multivoltine species spend the cold months as eggs. Accordingly, the potential for long-term survival of eggs must be considered in management programs. Eggs of some species (e.g., *S. rostratum*) remain viable in the laboratory just above freezing for up to 2 years.

The larval stage lasts from about a week, or even less, to nearly half a year, depending on species, stream temperature, and food availability. At one extreme are the larvae of some species in the West African *S. damnosum* complex that complete development in 4 days. At the other extreme are the larvae of many univoltine, temperate species that hatch in the fall, develop during the winter, and pupate in the spring. The number of larval instars varies from 6 to 11, depending on species and environmental conditions, such as food supply.

Final-instar larvae typically move to slower water before pupating in a silk cocoon that is spun on a substrate. Some species (e.g., *Prosimulium magnum*) pupate in masses, but most pupate individually. The duration of the pupal stage depends largely on temperature and species, lasting from several days to a few weeks. When the adult is ready to emerge, it expels air from its respiratory system, thus splitting the pupal cuticle along the dorsal eclosion line.

The newly emerged adult, partially covered in air, rises to the surface of the water with enough force to break the water–air interface. It then seeks a resting site, often streamside, to tan and harden. Adults generally live less than a month, during which time mating, sugar feeding, host location, blood feeding, and oviposition must be accomplished. Crosskey (1990) provides a full and detailed treatment of simuliid life history and bionomics.

# BEHAVIOR AND ECOLOGY

After hatching, early instars often disperse to more suitable sites for development. Larvae lead a largely sessile life attached to silk pads on substrates such as stones, trailing vegetation, sticks, aquatic plants, and leaf packs. The larvae of about 30 species, mostly in tropical Africa, are obligatorily phoretic, anchoring themselves to the bodies of larval mayflies and freshwater crabs. When disturbed, the larva repositions itself by looping over the substrate in inchworm-like fashion or by releasing itself from the silk pad and drifting downstream, often on a lifeline of silk. Downstream drift is usually greatest around dusk and during the night; its extent and timing should be considered in management programs.

The majority of larval life is spent feeding, usually by passively filtering suspended matter from the current or actively grazing adherent material from the substrate. The larvae of some species are also predaceous, actively consuming small invertebrates such as chironomid midges. Larvae that filter their food lean with the current and twist their bodies longitudinally 90–180°. In this position, one labral fan receives particulate matter that is resuspended by vortices arising from the substrate, while the other fan receives material from the main flow. Larvae filter particles that are about 0.09 to 350 μm in diameter, with the majority of ingested particles less than 100 μm in diameter. Larval diet consists of detritus, bacteria, small invertebrates, larval fecal pellets, and algae, with gut contents largely reflecting particle size and composition of

available material suspended in the water. Feeding efficiency, i.e., the ability to remove particles from the water column, is very low, typically less than 2%. Retention of material in the gut varies from 20 min to more than 2 hr, depending mainly on larval age, species, and water temperature.

The distribution patterns of larvae and pupae are associated with a variety of environmental factors. Distributions in a small section of stream or on a specific substrate are customarily referred to as *microdistributions*. Factors influencing microdistributions are those that vary over a few centimeters or meters, including substrate texture, water depth, hydrodynamics, and interactions with other organisms. Microdistributions are species specific. For example, last instars of *S. truncatum* and *S. rostratum*, two morphologically similar, boreal species often found in the same section of stream, select different microhabitats on the basis of water velocity. The patterns of larval dispersion on a substrate are either spaced (e.g., *S. vittatum*), with a well defined area surrounding each larva, or clumped (e.g., *S. noelleri*), with each larva occupying only enough space to attach its silk pad. Larvae with spaced patterns vigorously defend their space from other larval black flies.

*Macrodistributions* encompass a scale of many meters to hundreds of kilometers. The most important factors influencing macrodistributions are stream size, substrate, water velocity, temperature, water chemistry, food quality and quantity, and the presence of lake outlets. Within a stretch of stream, species distributions can be predicted by gradients of physical and chemical factors such as temperature and oxygen. Larval densities are usually greatest within a short distance downstream of impoundment outflows. For example, densities as high as 1.2 million larvae per square meter have been recorded for *S. noelleri* at lake outlets in Europe. Some species rarely are found far from lake outlets and, therefore, species assemblages at these outflows are often distinct from those farther downstream. Distributions of species among streams can be predicted by factors such as stream width. For example, species in genera such as *Twinnia, Gymnopais, Greniera,* and *Stegopterna,* as well as many species of *Prosimulium* and *Simulium,* occur in trickles and small streams. Species such as *Metacnephia lyra, Simulium jenningsi, S. arcticum, S. reptans,* and members of the genus *Ectemnia* occupy large streams and rivers. The influence of stream size and impoundment outflows on the distribution of black flies is important throughout most of the world. On larger scales, biogeographic factors are often useful in predicting species distributions. Streams in one ecoregion (e.g., mountains) tend to be more similar to one another than to streams in a different ecoregion (e.g., coastal plain), with respect to physical, chemical, and riparian characteristics. Simuliid faunas also show significant differences among ecoregions. The major factors associated with distributions of black flies are summarized by Adler and McCreadie (1997).

After emergence, adults of most species of black flies undertake short *dispersal flights,* usually less than 5 km. Males disperse to find mates and a source of sugar, whereas females of most species have the additional need to find hosts for blood and sites for oviposition. Although exceptions have been reported, black flies are diurnal fliers, generally taking to the wing when temperatures exceed 10°C. Local meteorological conditions can modify or even halt flight, but the primary factors that control daily flight patterns are wind, light, and temperature. Most species show a propensity to be on the wing at particular times of the day, these times varying with species, sex, physiological state, season, and the nature of the activity.

Some species of black flies (e.g., pestiferous members of the *S. arcticum* complex) also undertake flights in search of hosts or breeding sites. These long-range movements are wind assisted, with the direction of movement being controlled by prevailing winds. Movements of hundreds of kilometers by some species (e.g., members of the *S. damnosum* complex) have implications for control strategies, requiring that breeding grounds remote from problem areas be treated. Long-distance flights typically occur in species that feed on mammals, especially those that inhabit open areas such as savannas or prairies.

The universal energy source used by males and females for flight is sugar. Adults are opportunistic in their choice of carbohydrate sources, using floral or extrafloral nectar, plant sap, and honeydew. Water markedly increases longevity, and a 10% sugar solution further increases longevity. The sugar meal is stored in the crop and passed to the gut as needed for digestion.

Mating is necessary for all but about 10 parthenogenetic species (e.g., *P. ursinum*). These *parthenogenetic species* lack males and are triploid and northern in distribution. In the sexual species, mating occurs shortly after emergence. Males use a variety of strategies to encounter females. The most commonly reported method is the formation of *precopulatory swarms.* These aerial swarms usually form 2–3 m above ground, either beside or above a marker. Swarm markers tend to be visually apparent aspects of the environment, such as a tree branch, rock, waterfall, or host. Females enter the swarm, sometimes immediately after emergence, and are seized by males. Coupled pairs fly out of the swarm or fall to the ground or lower vegetation. Some species (e.g., *S. decorum*) do not form swarms, but instead couple on the ground during large, synchronous emergences. Males also might perch on vegetation and seize passing females. Visual cues mediate mating, but contact pheromones might also play

an important role. Black flies generally are refractive to mating under laboratory conditions, which has impeded attempts to colonize most species and to elucidate details of mating behavior. The long-term successful colonization of *S. vittatum* is a notable exception (Cupp and Ramberg, 1997).

Copulation lasts from a few seconds (e.g., *S. vittatum*) to 2 hr (e.g., *Gymnopais* spp.). During copulation, the male passes a package of spermatozoa, the spermatophore, to the female. The tip of the spermatophore is opened enzymatically by the female, and sperm move into the female's storage structure, the spermatheca. Stored sperm are released to fertilize eggs as they are being deposited.

A blood meal is required for the females of more than 90% of the world's simuliid species to mature the eggs. Males do not feed on blood. The females of some species, however, are *autogenous;* that is, they are able to produce eggs without taking blood. Females of those species that never take blood have feeble, untoothed mouthparts unable to cut host skin. These females are obligatorily autogenous, relying on energy acquired during the larval stage for all of their egg production. Females of species with biting mouthparts are *anautogenous,* that is, they mature their eggs with the aid of a blood meal. Nonetheless, the females of some species with biting mouthparts can mature the first batch of eggs without a blood meal (facultative autogeny) if conditions for larval growth have been optimal. In these facultatively autogenous species, however, each subsequent batch of eggs requires a blood meal. Each ovarian or gonotrophic cycle (i.e., maturation of an egg batch), varies from about 2 days to 2 weeks, depending on the species and ambient temperature. Most females probably do not survive long enough to complete more than two or three ovarian cycles. Because the transmission of pathogens is usually horizontal, passing from host to host via the simuliid vector, anautogenous females have a greater potential than facultatively autogenous females to acquire and transmit disease agents.

More than 90% of the world's simuliid species are hematophagous, taking blood from warm-blooded vertebrates (e.g., Anderson and DeFoliart, 1961; Bennett, 1960). The majority of simuliid species in the world probably feed on mammals (*mammalophily*), although those that feed on birds (*ornithophily*) also are common. About two-thirds of the blood-feeding species in North America are principally mammalophilic and the other third are mainly ornithophilic. A number of these species (e.g., *S. johannseni, S. venustum*), however, feed on both mammals and birds. *Host specificity* varies from highly specific in species such as *S. euryadminiculum,* which feed only on loons, to those such as *S. rugglesi,* which have been recorded feeding on nearly 30 different host species. Most simuliid species attack thinly haired or sparsely feathered regions of the host body and areas that are difficult for the host to groom. Thus, mammals often are attacked along the ventral region of the body and inside the ears. Birds are attacked especially on the neck, bases of the legs, and around the eyes. Humans are bitten wherever flesh is exposed, although specific areas are often attacked, such as along the hairline (e.g., *S. venustum*), the arms and hands (e.g., *S. parnassum*), the upper torso (e.g., members of the *S. ochraceum* and *S. oyapockense* complexes), and the ankles and feet (e.g., those in the *S. damnosum* and *S. metallicum* complexes).

A number of *host attractants* have been identified. Carbon dioxide released from the host, as well as color, shape, and size of the host, provide some of the major cues and attractants that females use to locate an appropriate blood source (Sutcliffe, 1986). Traps used to monitor female populations often exploit these cues. For example, sticky silhouettes and carbon dioxide in gaseous or dry-ice form often are used to monitor females.

Biting and engorging require a series of appropriate cues, especially temperature and various phagostimulants, such as adenosine phosphates, in the host's blood. When the fly begins to bite, the labella are withdrawn, and small teeth and spines at the apex of the labrum and hypopharynx pull the host skin taut (Sutcliffe and McIver, 1984). The serrated mandibles cut the host flesh, allowing the labrum and hypopharynx to enter the wound, along with the laciniae, which are armed with backwardly directed teeth that anchor the mouthparts. Blood from the wound forms a small pool that is drawn up the food channel formed when the mandibles overlap the labral food canal. Because of their method of feeding from pooled blood, black flies are termed pool feeders or *telmophages*. Saliva is applied to the host flesh via a salivary groove along the anterior surface of the hypopharynx. Various salivary components promote local anesthesia, enhance vasodilation, inhibit platelet aggregation, and prevent clotting (Cupp and Cupp, 1997). Chemosensilla on the mouthparts help determine that blood will be directed to the midgut.

Female black flies are determined feeders. Once the host skin is penetrated, females typically do not leave until they are satiated. Because most black flies are not nervous, easily interrupted feeders, they make poor mechanical vectors of pathogens. Most species feed for about 3 to 6 min, taking approximately their own weight in host blood.

Most blood-feeding activity is restricted to outdoor settings (*exophily*), with females infrequently entering shelters to feed. This behavioral trend has implications for vector control. For example, residual house treatments effective for the control of mosquitoes are of no use for controlling black flies. Biting activity occurs within certain optimal ranges of temperature, light intensity, wind speed, and humidity, with optima differing for each species. Given the appropriate range of meteorological conditions, many species bite throughout the day. Other species show a particular pattern of biting activity, such as a single peak in the morning (e.g., members of the *S. exiguum* complex) or a bimodal pattern with peaks in the morning and early evening (e.g., those in the *S. damnosum* complex). For all black flies, feeding typically is restricted to the hours of daylight and dusk. A rapid decrease in air pressure, combined with increased cloud cover, produces a sudden flush of biting activity, to which anyone caught outdoors during these conditions can attest.

Most female black flies can produce a batch of about 100–600 eggs, although the number varies from 25 (some *Gymnopais* spp.) to about 800 (some *Simulium* spp.). Females of some species can produce several of these egg batches in a lifetime, depending on the number of blood meals and how long the female lives. *Oviposition* usually occurs in the late afternoon and early evening. Eggs are deposited freely into the water during flight (e.g., *S. venustum* complex) or are attached in strings or masses to substrates such as rocks and trailing vegetation at the water line (e.g., *S. vittatum*). Some species, however, oviposit in moist fissures in riverbanks (e.g., *S. posticatum*) or in streamside mosses (e.g., some *Prosimulium* spp.).

## PUBLIC HEALTH IMPORTANCE

The importance of black flies to humans centers largely around the pestiferous habits of the blood-seeking females and the disease agents they transmit. The human disease agents transmitted by black flies are those that cause onchocerciasis in the tropics of Africa and Central and South America, and mansonellosis in southern Panama and the western Amazon Region. No other human pathogen or parasite is known definitively to be transmitted by black flies, and no endemic simuliid-borne disease of humans has been reported from North America.

The biting and nuisance problems inflicted by black flies have had severe consequences for most outdoor activities, including agriculture, forestry, industrial development, military exercises, mining, and tourism. Industrial and recreational development in some regions of Canada and Russia has been impeded or halted by overwhelming attacks from black flies. Actual monetary losses due to biting and nuisance problems in different sectors of the economy, although significant and sometimes crippling, are poorly documented.

## BITING AND NUISANCE PROBLEMS

The black flies that bite humans (i.e., anthropophilic species) constitute 10% or less of the total simuliid fauna in any zoogeographic region (Table I), with some areas of the world being nearly free of biting problems. No black

TABLE I

### Species of Black Flies Regarded as Significant Biting and Nuisance Pests of Humans, Livestock, and Poultry

| Species | Geographic Region |
| --- | --- |
| **Humans** | |
| *Austrosimulium ungulatum* | New Zealand |
| *Prosimulium mixtum* group | Eastern North America |
| *Simulium amazonicum* complex | South America (Amazonian Region) |
| *S. arakawae* | Japan |
| *S. bonaerense* | Argentina |
| *S. buissoni* | Marquesas Islands |
| *S. cholodkovskii* | Russia |
| *S. decimatum* | Russia, northern North America |
| *S. jenningsi* | Eastern North America |
| *S. parnassum* | Eastern North America |
| *S. penobscotense* | Northeastern North America |
| *S. posticatum* | England |
| *S. quadrivittatum* | Central America |
| *S. sanguineum* | Northwestern South America |
| *S. tescorum* | Southwestern United States |
| *S. venustum* complex | North America |
| **Livestock** | |
| *A. pestilens* | Australia (Queensland) |
| *Cnephia pecuarum* | United States (Mississippi River Valley) |
| *S. arcticum* complex | Western Canada |
| *S. cholodkovskii* | Russia |
| *S. chutteri* | South Africa |
| *S. decimatum* | Russia |
| *S. equinum* | Europe, Russia |
| *S. erythrocephalum* | Europe |
| *S. incrustatum* | Paraguay |
| *S. jenningsi* | Eastern North America |
| *S. lineatum* | Europe and Russia |
| *S. luggeri* | Western Canada |
| *S. maculatum* | Russia |
| *S. ornatum* complex | Europe, Russia |
| *S. reptans* | Europe, Russia |
| *S. vittatum* complex | North America |
| **Poultry** | |
| *C. ornithophilia* | Eastern North America |
| *S. meridionale* | North America |
| *S. rugglesi* | North America |
| *S. slossonae* | Southeastern United States |

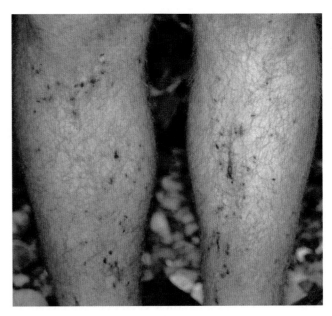

**FIGURE 11.5**   Bite wounds on human legs, caused by a North American black fly of the *Simulium venustum* complex.

fly is known that feeds exclusively on humans. In North America, where the name "black fly" originated, fewer than 60 species have been recorded to bite humans. Less than one-third of these hold any real status as biting pests, but those that do bite regularly can be unrelenting in their attacks. Individual reactions to bites vary from a small red spot at the puncture site, often with initial streaks of oozing blood (Fig. 11.5), to an enlarged swelling the size of a golf ball (Stokes, 1914). Swelling from bites around the eyes can impede vision, and bites on the limbs can impair walking.

A general syndrome, sometimes called *black fly fever*, is common in areas such as northeastern North America, where biting problems can be intense. It is characterized by headache, nausea, fever, and swollen lymph nodes in the neck. Many people experience some itching, intensified by scratching the wound. Severe allergic reactions, including asthmatic responses, are infrequent; however, medical treatment, including hospitalization, is sometimes necessary (Gudgel and Grauer, 1954). No human deaths from simuliid bites have been recorded in the 20th century, although anecdotal accounts suggest that an unclothed human can be exsanguinated in about 2 hr in some areas of Russia. Exposure to fierce attacks of biting and swarming black flies can severely affect a person's emotional state and produce short-term psychological effects that reduce individual efficiency.

Many species of black flies are attracted to humans but do not bite, or they bite infrequently in proportion to the number of flies actually attracted. These species can create enormous *nuisance problems*. One such species is

*S. jenningsi,* a major pest in North America. Females of this species sometimes bite humans and occasionally cause allergic reactions, but they are more of a nuisance because of their habit of swarming about the head and entering the eyes, ears, nose, and mouth. Outdoor activities in afflicted areas, such as Pennsylvania, can become unbearable as the females ceaselessly swarm around the head. More than US$5 million is spent annually in the management of *S. jenningsi.*

Occasional nuisance problems have been caused by large numbers of flies attracted to incandescent lights and by mating swarms that form over bicycle and foot paths at about the same height as a person walking or riding a bicycle. These kinds of problems usually are caused by members of the North American *S. vittatum* species complex, which breed abundantly in human-altered habitats, such as lake outlets and polluted waters.

## HUMAN ONCHOCERCIASIS

The greatest public health problem associated with black flies is onchocerciasis or *river blindness,* a tropical disease caused by the filarial nematode *Onchocerca volvulus,* which is transmitted solely by black flies during blood feeding. In the Old World, river blindness is found in 27 countries in the central belt of Africa, with small foci in southern Yemen. In the New World, where the disease possibly was introduced during the slave trade, its distribution is patchy, with foci in northern Brazil, Colombia, Ecuador, Guatemala, Mexico, and Venezuela. The World Health Organization (1995) conservatively estimated that about 17.7 million people are infected (17.5 million in Africa and Yemen; 140,500 in tropical America), with approximately 270,000 cases of microfilarial-induced blindness and another half million individuals with severe visual impairment. Research on the disease and its vectors has generated a massive literature (Muller and Horsburgh, 1987). Excellent reviews, from which the treatment below largely is drawn, are provided by Shelley (1988b), Crosskey (1990), and the World Health Organization (1995).

*O. volvulus* typically is found only in humans (definitive host) and adult flies of the genus *Simulium* (intermediate host). Various strains of *O. volvulus* are recognized, such as forest and savanna strains in West Africa, and these form highly compatible parasite–vector complexes with distinct clinical facies. Once the female black fly ingests a blood meal from an infected human, the *microfilariae* (220–360μm long) penetrate the gut of the fly and make their way to the thoracic flight muscles. Once in the thoracic muscles, the microfilariae lose their motility and transform to first-stage larvae, which then molt to become second-stage larvae. The final molt in the fly produces the infective third-stage larvae, which migrate to the fly's head and mouthparts. Vector incrimination is based on the presence of third-stage larvae in the head capsules of female black flies. In West Africa, DNA tests allow animal parasites and the human parasites of savanna and forest to be distinguished. Development in the black fly, which is influenced by ambient temperature, typically requires 6–12 days, but the time between successive blood meals taken by the fly is usually 3–5 days. Consequently, the infective larvae will be passed to a human host no earlier than the third blood meal when the fly is about 8–10 days old.

In humans, the infective larvae molt to the fourth larval stage within about a week. One more molt yields juvenile adults, which grow to mature adult worms over the next 12–18 months and begin reproducing. Adults typically become encapsulated in fibrous nodules that vary in size from about 0.5–10.0 cm and can be subcutaneous or deep in muscular and connective tissues; they cause no inflammatory response and no great discomfort. Mating between the small male worms (3–5 cm long) and the large females (30–80 cm) occurs in the nodules. Adult female worms can produce microfilariae for up to 14 years. These microfilariae migrate from the nodules to the skin, where they can be acquired by a vector, as well as to the eyes and various other organs (e.g., liver) of the human host. A diagnostic clinical feature of onchocerciasis is the presence of hundreds of microfilariae in skin snips.

River blindness is essentially a rural disease, afflicting those people most vulnerable to both the medical consequences and social stigmas of infection. Symptoms of the disease depend on factors such as geographical location, microfilarial transmission rates, and frequency of reinfection. Where transmission rates are low, the disease can be asymptomatic. With heavy infections, however, the classical manifestations of the disease appear, i.e., dermal changes, lymphatic reactions, nodules, and ocular disturbances. Other than the nodules in which the adults are enveloped, all symptoms are caused by the microfilariae.

Large numbers of microfilariae migrating throughout the dermis cause horrific itching that can lead to bleeding, secondary bacterial infections, inability to sleep, fever, headache, and even suicide. In addition to itching, chronic infections in Africa and Yemen can cause dermal lesions, patches of depigmentation ("leopard skin"), fibrosis, and loss of elasticity. In Yemen, the itching symptoms of the disease are known as *sowda.* In Central America, two unique, chronic skin conditions occur—a painful, reddish rash on the face (*erisipela de la costa*) and lesions associated with reddish skin on the trunk and arms (*mal morado*). The lymphatic nodes also can be affected, especially in the groin and thighs; combined with loss of skin elasticity, the result is a condition known as *hanging groin.*

Migrating microfilariae also enter the eye, resulting in a severe ocular pathology that can involve all tissues of the eye. Ocular problems are associated with the presence of both live and dead microfilariae, and they manifest themselves in many forms, including cataracts, retinal hemorrhages, corneal opacities, secondary glaucoma, sclerosing keratitis, and optic neuritis. Various forms of visual impairment occur, such as night blindness and reduction in peripheral vision, but the most severe consequence is irreversible blindness with complete loss of light perception (Fig. 11.6). Blindness usually takes years to occur; at age 20, for example, it is rare in infected people, but at 50 years of age, half of the infected victims can be blind. The incidence of blindness is highest in the savannas of West Africa, with some villages experiencing 15% blindness. At these high levels of disease, the village is often abandoned. Outside West Africa, ocular pathology is rare.

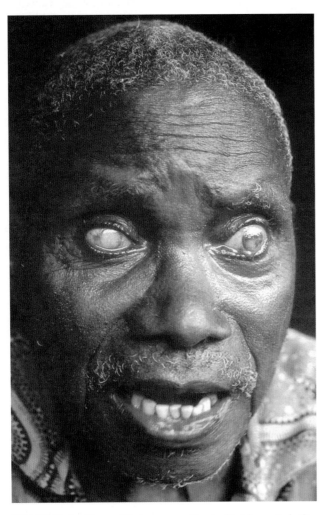

**FIGURE 11.6**    Human blindness caused by the filarial nematode *Onchocerca volvulus*, transmitted by black flies of the *Simulium damnosum* complex in West Africa; note opacity of cornea due to damage by microfilariae. (Copyright Eric Poggenpohl.)

At least 25 species of *Simulium* are known vectors of *O. volvulus* (Table II). Most of these vectors are members of species complexes, and considerable taxonomic work is still needed to resolve all of the vector species in areas such as East Africa and the Americas. In West Africa and Yemen, all vectors are members of the *S. damnosum* species complex and include *S. damnosum sensu stricto* and *S. sirbanum*, the principal vectors associated with the savanna form of the disease and ocular pathology. The vectors in East Africa are members of the *S. damnosum* complex, the *S. neavei* group, and *S. albivirgulatum*. In the Americas, at least 9 species or species complexes are vectors, the most important of which are members of the *S. exiguum*, *S. guianense*, *S. metallicum*, *S. ochraceum*, and *S. oyapockense* complexes; their importance, however, varies with location. Because the vectors in South America are more widespread than is onchocerciasis, the disease is predicted to spread as humans continue to push into undeveloped areas (Basáñez *et al.*, 2000).

An understanding of the unique life history and behavior of each vector species is key to the control of onchocerciasis. Breeding sites of the vectors, for example, represent the targets for control. The immature stages of the *S. damnosum* complex primarily inhabit swift sections of medium to large rivers, from dry savannas to forest highlands, depending on the species. Larvae and pupae of species in the *S. neavei* group live primarily in perennial, shaded forest streams, where they have an obligatory phoretic relationship with river crabs. In the New World, members of the *S. metallicum* and *S. ochraceum* complexes breed in large and small streams, respectively, that drain forested mountain slopes, whereas members of the *S. exiguum* and *S. oyapockense* complexes breed in large rivers of the rain forest. Although each species breeds in a specific habitat, the adults of some species can travel great distances beyond their natal waterways. The adults of *S. sirbanum* and *S. damnosum sensu stricto*, for example, can travel more than 500 km, assisted by seasonally changing winds. In the wet season, moist monsoon winds from the southwest move flies in a northeastwardly direction. In the dry season, winds from the northeast assist flies in their reverse, southwestwardly flights. Continual reinvasions by vectors, therefore, occur with each season in both the northern and southern parts of West Africa and must be considered in control efforts.

## MANSONELLOSIS

The filarial nematode *Mansonella ozzardi* is the causal agent of mansonellosis, a questionably pathogenic disease of humans. It is transmitted by at least four species of black flies in the Neotropical rain forests of Brazil, Colombia, Guyana, Venezuela, and southern Panama

**TABLE II**
**Disease Agents Transmitted by Black Flies**

| Disease agent | Vectors | Hosts | Geographic areas | Select references |
|---|---|---|---|---|
| **Protozoa** | | | | |
| *Leucocytozoon ziemanni* | *S. aureum* complex, *S. vernum* group | Owls | North America | Fallis *et al.*, 1974 |
| *L. dubreuili* | *P. decemarticulatum, C. ornithophilia, S. aureum* complex, *S. vernum* group | Thrushes | North America | Fallis *et al.*, 1974 |
| *L. cambournaci* and *L. icteris* | *P. decemarticulatum, C. ornithophilia, S. anatinum, S. aureum* complex, *S. euryadminiculum, S. venustum* complex, *S. vernum* group | Sparrows, blackbirds | North America | Fallis *et al.*, 1974 |
| *L. lovati* | *S. aureum* complex, *S. vernum* group | Grouse | North America | Greiner, 1991 |
| *L. neavei* | *Simulium* spp., especially *S. adersi* | Guinea fowl | Eastern Africa | Fallis *et al.*, 1974 |
| *L. sakharoffi* | *P. decemarticulatum, S. aureum* complex *S. angustitarse* | Corvids | North America | Fallis *et al.*, 1974 |
| *L. schoutedeni* | *Simulium* spp., especially *S. adersi* | Chickens | Eastern Africa | Fallis *et al.*, 1974 |
| *L. simondi* | *Cnephia ornithophilia, S. anatinum, S. fallisi, S. rendalense, S. rugglesi, S. venustum* complex | Ducks, geese | North America, Norway | Fallis *et al.*, 1974, Greiner, 1991 |
| *L. smithi* | *S. aureum* complex, *S. congareenarum, S. jenningsi* group, *S. meridionale, S. slossonae,* possibly *S. ruficorne* group | Turkeys | North America, introduced to Africa | Skidmore, 1932, Noblet *et al.*, 1975 Greiner, 1991 |
| *L. tawaki* | *Austrosimulium ungulatum* | Penguins | New Zealand | Allison *et al.*, 1978 |
| *Trypanosoma confusum* | *Prosimulium decemarticulatum, Simulium* spp. | Birds | North America | Bennett, 1961 |
| *T. corvi* | *Simulium latipes* | Kestrels | England | Dirie *et al.*, 1990 |
| *T. numidae* | *Simulium* spp., especially *S. adersi* | Chickens, guinea fowl | Eastern Africa | Fallis *et al.*, 1973 |
| **Filarial nematodes** | | | | |
| *Dirofilaria ursi* | *Simulium venustum* complex | Bears | North America | Addison, 1980 |
| *Mansonella ozzardi* | *S. amazonicum* complex *S. argentiscutum, S. oyapockense* complex, *S. sanguineum* | Humans | Northern South America, Panama | Shelley, 1988a |
| *Onchocerca cervipedis* | *Prosimulium impostor, S. decorum, S. venustum* complex | Deer, Moose | North America | Pledger *et al.*, 1980 |
| *O. dukei* | *S. bovis* | Cattle | Africa | Wahl and Renz, 1991 |
| *O. gutturosa* | *S. erythrocephalum, S. bidentatum* | Cattle | Japan, Ukraine | Crosskey, 1990, Takaoka, 1999 |
| *O. lienalis* | *S. erythrocephalum, S. jenningsi, S. ornatum* complex, *S. reptans, S. arakawae, S. daisense, S. kyushuense* | Cattle | North America, Russia, western Europe, Japan | Lok *et al.*, 1983 Crosskey, 1990 Takaoka, 1999 |
| *O. ochengi* | *S. damnosum* complex | Cattle | West Africa | Wahl *et al.*, 1998 |
| *O. ramachandrini* | *S. damnosum* complex | Warthogs | West Africa | Wahl, 1996 |
| *O.* possibly *skrjabini* | *S. aokii, S. arakawae, S. bidentatum, S. daisense* | Japanese deer | Japan | Takaoka, 1999 |
| *O. tarsicola* | *Prosimulium tomosvaryi S. ornatum* complex | Deer | Western Europe | Crosskey, 1990 |
| *O. volvulus* | Africa: *S. albivirgulatum, S. damnosum, S. dieguerense, S. ethiopiense, S. kilibanum, S. konkourense, S. leonense, S. mengense, S. neavei, S. rasyani, S. sanctipauli, S. sirbanum, S. soubrense, S. squamosum, S. woodi, S. yahense;* Americas: *S. callidum, S. exiguum* complex, *S. guianense* complex, *S. incrustatum, S. limbatum, S. metallicum* complex, *S. ochraceum* complex, *S. quadrivittatum, S. oyapockense* | Humans | Africa, Central America, South America | Shelley, 1988b, Crosskey, 1990, World Health Organization, 1995 |
| *Splendidofilaria fallisensis* | *S. anatinum, S. rugglesi* | Ducks | North America | Anderson, 1968 |

(Table II). Black flies first were incriminated as vectors of *M. ozzardi* in 1959 and were subsequently confirmed experimentally as vectors in 1980. Mansonellosis also is found in the Caribbean Islands, where only ceratopogonid midges (*Culicoides* species) are known to transmit the causal agent (Shelley, 1988a).

Adult nematodes of *M. ozzardi* occur in the subcutaneous tissues of humans, and the microfilariae are found principally in the peripheral blood, where they are acquired by blood-feeding females. The life cycle of the nematode in black flies is similar to that of *O. volvulus*. A number of mammals and some birds and amphibians can be infected, but humans are the only significant reservoirs. In some highly endemic areas (e.g., Colombia), up to 70% of the human population can be infected. Mansonellosis generally is viewed as causing little or no pathology, but some reports have indicated that joint pains, headaches, hives, and pulmonary symptoms are associated with infections (Klion and Nutman, 1999). Effective treatment of infections is generally lacking.

## OTHER DISEASES
## RELATED TO BLACK FLIES

Because black flies that feed on humans also feed on other hosts, the potential exists for certain disease-causing agents of domestic and wild animals to be transferred to humans. Black flies, for example, have been implicated as mechanical vectors of the bacterial agent of *tularemia* in the United States and Russia, suggesting that occasional cases of transmission of this pathogen to humans might occur. Similarly, *eastern equine encephalitis virus* in the United States and *Venezuelan equine encephalitis virus* in Colombia have been isolated from several species of black flies, suggesting at least the potential for transmission of these pathogens to humans. Black flies in the Marquesas Islands have been implicated in the indirect transmission of *hepatitis B virus* by causing numerous, itching lesions on the skin (Chanteau *et al.*, 1993). Direct transmission of the virus by black flies also is theoretically possible.

Several additional diseases might be related to biting black flies. One such disease is *endemic pemphigus foliaceus* or *fogo selvagem*, a potentially lethal, autoimmune, blistering skin affliction. The disease is centered among poor, outdoor laborers in certain regions of Brazil (Eaton *et al.*, 1998). Further work is needed to determine if black flies are the causal agents of the disease. Another affliction possibly associated with black flies in the New World is *thrombocytopenic purpura*, a disorder in which the platelet count is reduced. Again, more data are needed before black flies can be linked to the cause.

## VETERINARY IMPORTANCE

The veterinary impact of black flies is manifested through pathogen transmission, biting, and nuisance swarming. Filarial nematodes, protozoans, and possibly several viruses are transmitted to animals. The most insidious parasites are those that cause leucocytozoonosis in domestic ducks, geese, and turkeys.

Deaths of birds and livestock have resulted from attacks by large numbers of black flies. Livestock under persistent attack sometimes stampede, trampling young animals, crashing into structures, and tumbling from precipices. Suffocation has been blamed for some deaths, with so many flies clogging the respiratory passages that breathing can become severely impaired. Deaths also have been attributed to respiratory tract infections caused by inhalation of flies. If enough blood is withdrawn, it may become too thick to transport oxygen efficiently, thereby killing the animal by *exsanguination*. Perhaps the most common cause of mortality can be attributed to the actual bites of the flies or, more specifically, to toxemia and acute shock caused by the various salivary components that are injected during blood feeding.

More difficult to assess in economic terms, but equally harmful to the livestock and poultry industries, are the effects of harassment through biting and swarming (Fig. 11.7, Table I). Biting is often aimed at weakly protected areas of the body, such as the ears, neck, and ventral midline. Persistent attacks by black flies can cause unruly host behavior, weight loss, reduced egg and milk production, malnutrition in young animals, dermatitis and epidermal necrosis, impotence in bulls, delayed pregnancies, and possibly stress-related diseases such as pneumonia. Actual monetary losses are not well documented but can be great. The beef and dairy industries

**FIGURE 11.7** Cattle under attack by black flies of the *Simulium arcticum* complex on prairie, Alberta, Canada. (Photo by J. A. Schemanchuk, Department of Agriculture and Agri-Food, Government of Canada)

of Saskatchewan, for example, lost more than 3 million dollars in 1978 from attacks by *S. luggeri* (Fredeen, 1985b).

An odd, but indirect, nuisance problem mediated occasionally by black flies involves the Neotropical human bot fly *Dermatobia hominis*. Female bot flies capture hematophagous arthropods, including black flies, to which they glue their eggs. Once the carrier has landed on a host, the larvae of the bot fly hatch and bore into the host skin, causing myiasis. At least one species of black fly (*S. nigrimanum*) that feeds on domestic animals has been used as a carrier.

## BOVINE ONCHOCERCIASIS

Black flies transmit at least four species of filarial nematodes (genus *Onchocerca*) to cattle in the Afrotropical, Nearctic, and Palearctic Regions (Table II). *O. lienalis* is the most widespread of these filarial parasites. *S. jenningsi* is its primary vector in the United States, whereas the *S. ornatum* complex is a principal vector in the Old World. The microfilariae of *O. lienalis* are concentrated in the umbilical region of the host. They are ingested during blood feeding and transmitted to a new host after they have developed to the infective third stage in the simuliid vector. The percentage of infected cattle is often quite high, but symptoms and general impact on the host are usually not overt. Infected animals sometimes show dermatitis and inflammation of the skin and connective ligament. *O. gutturosa*, a Palearctic species, often has been confused with *O. lienalis*. Its microfilariae occur in the skin of the neck and back of the host. It has been confirmed from Japan and perhaps the Ukraine, where *S. bidentatum* and *S. erythrocephalum*, respectively, have been implicated in its transmission. Elsewhere, ceratopogonid midges (*Culicoides*) are vectors. In West Africa, *O. ochengi* is transmitted to cattle by members of the *S. damnosum* complex, and *O. dukei* is transmitted by *S. bovis*. Both of these *Onchocerca* species can create nodules, either dermal (*O. ochengi*) or subcutaneous (*O. dukei*), in the inguinal region of the host. Economic losses resulting from bovine onchocerciasis rarely have been assessed, although a few reports have indicated that the quality of hides can be reduced.

At least four additional species of *Onchocerca* are transmitted by black flies to nonbovine hosts (Table II). *O. cervipedis* in North America and *O. tarsicola* in Europe infect the subcutaneous connective tissues, mainly in the legs, of deer and moose; consequently, they are sometimes called legworms. More than 60% of a host population can be infected. *O. ramachandrini* is a parasite of warthogs and is transmitted by members of the *S. damnosum* complex in West Africa. *Onchocerca* possibly *skrjabini* parasitizes Japanese deer and is transmitted by at least four species in Japan. The vectors of 10 or more *Onchocerca* species that infect domestic animals remain unknown but might include black flies.

## LEUCOCYTOZOONOSIS

At least 10 species of protozoans in the genus *Leucocytozoon* are transmitted to birds by black flies, causing a malaria-like disease, leucocytozoonosis (Table II). The disease is known colloquially as *turkey malaria, duck malaria,* or *gnat fever*. Each species of *Leucocytozoon* is specific to one or a few families of birds. Vectors are known for only about 11 of the world's described species of *Leucocytozoon*, and all but one vector are black flies. Only 2 species of the parasite are of major economic concern, and both occur in North America. *Leucocytozoon simondi* is specific to ducks and geese, and its primary vectors are *S. anatinum* and *S. rugglesi*. *L. smithi* is specific to turkeys and is transmitted primarily by *S. meridionale* and *S. slossonae*.

*Leucocytozoon* species undergo a complex malaria-like life cycle. Gametocytes in the blood of an avian host are acquired by a female black fly. The parasite then undergoes both asexual and sexual development over a period of 3–4 days in the fly. During a subsequent blood meal, the fly transmits the parasites, as sporozoites, to another bird, which serves as a host for asexual development and gametocyte production (Fig. 11.8).

Leucocytozoonosis can be fatal in poultry, but its effects on wild hosts, with the exception of some populations of Canada geese (Herman *et al.*, 1975), generally are less apparent. Birds with chronic infections have weakened immune systems and reduced reproduction. Severe infections produce emaciation, dehydration, and convulsions that lead to death. Internally, the liver and spleen of moribund hosts are enlarged, the heart muscle is pale, and the lungs are congested (O'Roke, 1934).

The disease has had devastating effects on the poultry industry throughout much of North America (Noblet *et al.*, 1975). Entire flocks have been killed and production facilities shut down in areas such as Nebraska, South Carolina, and Manitoba. The United States Agricultural Research Service estimated an annual average loss of nearly three-quarters of a million dollars in the United States from 1942 to 1951 as a result of leucocytozoonosis in domestic turkeys. The last major outbreaks of the disease in domestic turkeys were in the 1970s. Turkeys now are raised primarily in poultry houses, reducing the incidence of disease because black flies generally do not venture inside shelters.

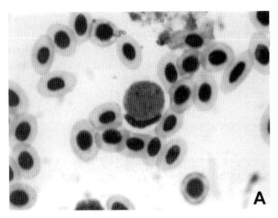

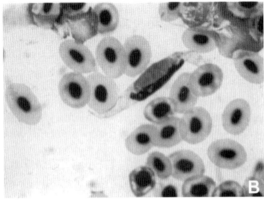

**FIGURE 11.8**    Gametocytes of the protozoan *Leucocytozoon simondi* in blood cells of a mallard duck. (A) Round form; (B) elongate form. (Photo by E. C. Greiner.)

## OTHER PARASITES AND PATHOGENS OF VETERINARY IMPORTANCE

Black flies transmit additional parasites to wild animals (Table II). The protozoan *Trypanosoma confusum* is specific to birds in North America and is transmitted when infected fecal droplets from the black fly contaminate the bite. Birds of numerous families serve as hosts. Other species of bird trypanosomes (e.g., *T. corvi*) are believed to cause infections when the birds consume infected black flies or eat other birds that have been infected. The filarial nematodes *Splendidofilaria fallisensis* and *Dirofilaria ursi* are transmitted to ducks and black bears, respectively. The effects of these protozoan and filarial parasites on their wild hosts are poorly known.

Evidence is mounting that several North American species of simuliids, such as *Simulium notatum* and *S. vittatum*, naturally transmit vesicular stomatitis virus to livestock, primarily cattle, horses, and pigs (Schmidtmann *et al.*, 1999). The virus causes lesions in various epithelial tissues, especially in the mouth. Millions of dollars can be lost during epizootics. Laboratory experiments have

shown that a viremic host is not necessary for a female black fly to become infected; rather, flies can become infected by feeding on the same host with an infected black fly (Mead *et al.*, 2000).

Additional parasites of wildlife have been associated with black flies. The recent discovery of minute nematodes of the family Robertdollfusidae in the guts of African black flies suggests that these parasites might be transmitted to wildlife (Bain and Renz, 1993). *Bunyaviruses, eastern equine encephalitis virus,* and *snowshoe hare virus* have been isolated from several North American black flies. Minimal mechanical transmission has been demonstrated for *Whataroa virus* in laboratory mice in New Zealand (Austin, 1967) and for *myxomatosis* in rabbits in Australia (Mykytowycz, 1957). These examples suggest that much is yet to be learned about the vector potential of black flies among wildlife.

## SIMULIOTOXICOSIS

Attacks by black flies have, at times, been so massive and virulent that livestock have been killed. Many of the deaths probably result from acute toxemia and anaphylactic shock caused by the toxins introduced with the saliva as black flies are feeding. The diseased condition, either temporary or terminal, that results from the bites of black flies is known as *simuliotoxicosis,* a term first used to describe the toxic effects of simuliid bites on reindeer (Wilhelm *et al.*, 1982). Cattle, especially calves, are vulnerable to simuliotoxicosis, but horses, mules, sheep, goats, and pigs also have been affected. Susceptible animals succumb in less than 2 hr. Some immunity is apparent in animals living in afflicted areas. The biochemical nature of simuliotoxicosis requires more investigation.

Most of the species responsible for simuliotoxicosis breed in large rivers, from which the adults emerge in astronomical numbers. They include *A. pestilens* in Queensland (Australia); *Cnephia pecuarum* in the Mississippi River Valley (United States); *S. colombaschense* along the Danube River in central Europe; a member of the *S. arcticum* complex on the Canadian prairies; and *S. erythrocephalum,* the *S. ornatum* complex, and *S. reptans* in central Europe.

One of the worst attacks in recorded history killed about 22,000 animals in 1923 along Europe's Danube River in the southern Carpathian Mountains (Ciurea and Dinulescu, 1924). Prodigious attacks in this region during the 1700s prompted Empress Maria Theresa of the Old Austro-Hungarian Empire to order one of the first biological studies of black flies, which eventually was published in 1795. On the Canadian prairies, thousands of livestock were killed from about 1886 into the 1970s by a member of the *S. arcticum* complex (Fredeen, 1977).

Massive mortality due to attacks by *C. pecuarum* occurred in the United States during and immediately after the Civil War when the levees of the Mississippi River deteriorated, allowing the river to overflow and create extensive breeding areas for this species (Riley 1887).

Simuliotoxicosis on a large scale is now rare, mainly because the former breeding sites of most of the responsible species have been altered by pollution, impoundment, and land development. Some of these species, however, still create nuisance problems for livestock, with occasional deaths in localized areas of their ranges.

## PREVENTION AND CONTROL

Management of black flies typically is aimed at the larval stage, in large part because in this life stage the pest species are concentrated in easily identifiable, often very specific habitats. Although adulticiding has sometimes offered temporary relief, it is typically more costly and has been used less frequently than larviciding. It usually has involved both aerial and ground fogging with dichlorodiphenyltrichloroethane (DDT) or permethrin products. Current efforts to manage black flies in the adult stage are restricted primarily to the application of repellents and pour-on insecticides.

The use of *chemical insecticides* in managing black flies dates to the dawn of the 20th century, reaching a peak from the mid-1940s into the 1970s, when DDT was the principal means of control against both larvae and adults. The development of resistance and the undesirable effects on nontarget organisms led to the abandonment of DDT and the search for surrogate compounds, the most prominent of which were methoxychlor (chlorinated hydrocarbon) and temephos (organophosphate). These compounds, as well as insect growth regulators, were not selective and, therefore, had negative impacts on nontarget organisms. The use of chemical insecticides to manage black flies became infrequent toward the end of the 20th century, although compounds such as methoxychlor and temephos continued to be used in a few areas of the world.

Black flies worldwide are managed primarily through the use of the entomopathogenic bacterium *Bacillus thuringiensis* var. *israelensis* (*Bti*, serotype H14), which is aimed at the larval stage. The actual killing agent is an endotoxin in the parasporal inclusions that disrupts the cells of the highly alkaline larval midgut. The efficacy and environmental safety of *Bti* are so superb that most other means of population suppression and management have decreased greatly since the commercial *Bti* product entered the scene in the early 1980s (Gray *et al.*, 1999; Molloy, 1990). *Bti* can be applied by hand or

aircraft. North America's largest suppression program for black flies is operated by the state of Pennsylvania (United States), which treats waterways for *S. jenningsi* in more than half of its counties.

Natural enemies exert some control in most populations of black flies, but attempts to mass produce them have not been made since the 1970s (Laird, 1981). Commonly encountered parasites include mermithid nematodes, microsporidian protozoa, the chytrid fungus *Coelomycidium simulii*, and several viruses. The prevalence of infection with these parasites and pathogens is usually less than 10% of a population. Infections typically slow development, however, so that parasitized larvae become relatively more frequent in a population over time as healthy individuals pupate first.

*Mermithid nematodes* probably hold the greatest promise for biological control of black flies. However, until more can be learned about their taxonomy and host specificity and how to cultivate them economically for mass release, they are unlikely to be useful in integrated pest management programs. Preparasitic mermithid nematodes crawl on stream substrates and use a protrusible stylet to penetrate the host body. As the mermithids mature, they can be seen through the host integument, coiled within the abdomen. Mermithids either exit and kill the host larva or pass into the adult, exiting shortly thereafter. Postparasitic worms molt to adults, mate, and deposit eggs in the streambed.

Patent infections with microsporidia are recognized by the large, irregular cysts that distort the larval host abdomen. Life cycles of *microsporidia* that attack black flies are poorly known, although transovarial transmission has been documented. Larvae with patent infections of the *fungus C. simulii* are packed with minute, spherical thalli throughout their bodies. Thalli produce spores that are released into the water column after death of the host. Two common viruses that infect larvae are iridescent virus, which imparts an overall blue or violet cast, and cytoplasmic polyhedrosis virus, which creates white bands around the midgut. Many predators consume black flies; most are typically opportunistic.

Physical control of the breeding habitat is occasionally effective in reducing pest populations, usually when the pest species is concentrated in a restricted area, such as directly downstream of an impoundment. In these situations, attachment sites (e.g., trailing vegetation) can be removed, or water levels can be altered to strand larvae above the water line.

Personal protection for humans involves primarily the use of repellents, both natural and synthetic, that are applied directly to the skin or impregnated in clothing. Among the more effective repellents are those with *N,-N*-diethyl-m-toluamide (DEET) as the active ingredient. Wearing light-colored clothing and minimizing openings in the clothing, such as buttonholes, through which black flies can gain access to skin, is standard practice when entering areas where black flies are a problem. Fine-mesh head nets are effective in areas where pest populations are intolerable. Many additional means of protection can be found in the annals of folklore, but the utility of most remains suspect.

Various techniques have been devised to protect livestock, ranging from the use of *smudges* (i.e., smoldering fires that produce dense smoke) to the application of repellent substances and the use of shelters. Repellent products for livestock historically involved oils and greases, often laced with turpentine or other plant-derived products. Among the more commonly used repellents in recent times are permethrin solutions and *eartags* containing ivermectin. Various pour-on and spray formulations of insecticides and repellents are available commercially. White petroleum jelly can be applied inside the ears of horses to reduce biting problems. Providing shelters is an effective means of protecting livestock and poultry because most species of black flies rarely enter enclosures. Housing turkeys in shelters, for example, significantly reduced the prevalence of leucocytozoonosis in these birds. Providing the entries of shelters with self-application devices for repellents provides an added dose of protection.

## WHO ONCHOCERCIASIS CONTROL PROGRAMME

The largest management program in the world for black flies has been the World Health Organization's Onchocerciasis Control Programme (OCP) in West Africa. Its history, as briefly summarized below, has been written by numerous authors (e.g., Anonyme, 1985; Davies, 1994; World Health Organization, 1995). The initial foundations for the program were laid in 1968, and in 1975 the program launched its first aerial treatments for the control of onchocerciasis. The goal of the OCP was to eliminate onchocerciasis as a major public health threat in 7 West African countries: Benin, Burkina Faso, Ghana, Ivory Coast, Mali, Niger, and Togo. The program later was expanded to include the countries of Guinea, Guinea-Bissau, Senegal, and Sierra Leone, thus covering a total of 11 countries and 50,000 km of rivers. It was directed at the vectors of onchocerciasis, namely members of the *S. damnosum* species complex. The primary strategy of the OCP has been a massive aerial larviciding program aimed at reducing adult vector populations, thus interrupting transmission. Maintaining vectors at a sufficiently low number for a sufficiently long time prevents new cases of transmission while worms in the human reservoir die out, thus breaking the disease cycle. Given the longevity

of adult worms, control programs in endemic areas must be maintained for approximately 15 years to eliminate the worm from the human reservoir (Plaisier *et al.*, 1991; Remme *et al.*, 1990). Prior to the OCP, aerial application of DDT was the main means of control, but by 1970 resistance had begun to develop. From 1975 into the 1980s, the OCP applied primarily temephos to the rivers. The first appearance (1980) of resistance to this compound by the vectors in the OCP area (Guillet *et al.*, 1980) eventually led to the rotation of six insecticides, including *Bti.*

Vector control was integrated with an ivermectin chemotherapy program for the human reservoir in 1988. *Ivermectin,* originally developed for veterinary purposes, reduces the number of microfilariae in the skin, so that ingestion of sufficient microfilariae by the vectors becomes difficult. This microfilarialcidal drug, however, does not kill the adult worms. A single oral dose of Mectizan (the formulation of ivermectin for humans) every 6–12 months is not only nontoxic at levels higher than prescribed dosages, but it also is sufficient to kill microfilariae in the skin and eyes and reverse progression of the disease. Dying microfilariae, however, can cause temporary adverse reactions in patients. Mass distribution of ivermectin has been possible through the humanitarian efforts of numerous organizations, including Merck and Co., which decided in 1987 to donate all ivermectin tablets for the worldwide treatment of onchocerciasis for as long as necessary. The possible development of resistance to ivermectin by *O. volvulus* must be closely monitored.

By 1995, vector control had interrupted transmission in about 90% of the original OCP area, protecting more than 30 million people from infection and sparing 100,000 from blindness at a cost of about US$360 million. The combined use of ivermectin and weekly insecticide treatments of larval breeding sites was predicted to free the current OCP area of onchocerciasis by 2002, at which time OCP was scheduled to terminate, with residual activities being transferred to the participating countries.

## REFERENCES AND FURTHER READING

Addison, E. M. (1980). Transmission of *Dirofilaria ursi* Yamaguti, 1941 (Nematoda: Onchocercidae) of black bears (*Ursus americanus*) by blackflies (Simuliidae). *Canadian Journal of Zoology* **58**, 1913–1922.

Adler, P. H., and Kim, K. C. (1986). The black flies of Pennsylvania (Simuliidae, Diptera): bionomics, taxonomy, and distribution. *Pennsylvania State University Agricultural Experiment Station Bulletin* **856**, 1–88.

Adler, P. H., and McCreadie, J. W. (1997). The hidden ecology of black flies: sibling species and ecological scale. *American Entomologist* **43**, 153–161.

Allison, F. R., Desser, S. S., and Whitten, L. K. (1978). Further observations on the life cycle and vectors of the haemosporidian *Leucocytozoon tawaki* and its transmission to the Fjordland crested penguin. *New Zealand Journal of Zoology* **5**, 663–665.

Anderson, J. R., and DeFoliart, G. R. (1961). Feeding behavior and host preferences of some black flies (Diptera: Simuliidae) in Wisconsin. *Annals of the Entomological Society of America* **54**, 716–729.

Anderson, R. C. (1968). The simuliid vectors of *Splendidofilaria fallisensis* of ducks. *Canadian Journal of Zoology* **46**, 610–611.

Anonyme. (1985). 10 années de lutte contre l'onchocercose en Afrique de l'Ouest. Bilan des activités du Programme de Lutte contre l'Onchocercose dans la Région du Bassin de la Volta de 1974 à 1984. OMS/OCP, Genève 1985.1A. World Health Organization.

Austin, F. J. (1967). The arbovirus vector potential of a simuliid. *Annals of Tropical Medicine and Parasitology* **61**, 189–199.

Bain, O., and Renz, A. (1993). Infective larvae of a new species of Robertdollfusidae (Adenophorea, Nematoda) in the gut of *Simulium damnosum* in Cameroon. *Annales de Parasitologie Humaine et Comparée* **68**, 182–184.

Basáñez, M. G., Yarzábal, L., Frontado, H. L., and Villamizar, N. J. (2000). *Onchocerca-Simulium* complexes in Venezuela: can human onchocerciasis spread outside its present endemic area? *Parasitology* **120**, 143–160.

Bennett, G. F. (1960). On some ornithophilic blood-sucking Diptera in Algonquin Park, Ontario, Canada. *Canadian Journal of Zoology* **38**, 377–389.

Bennett, G. F. (1961). On the specificity and transmission of some avian trypanosomes. *Canadian Journal of Zoology* **39**, 17–33.

Chanteau, S., Sechan, Y., Moulia-Pelat, J.- P., Luquiaud, P., Spiegel, A., Boutin, J.- P., and Roux, J.- F. (1993). The blackfly *Simulium buissoni* and infection by hepatitis B virus on a holoendemic island of the Marquesas Archipelago in French Polynesia. *American Journal of Tropical Medicine and Hygiene* **48**, 763–770.

Ciurea, I., and Dinulescu, G. (1924). Ravages causés par la mouche de Goloubatz en Roumanie: des attaques contre les animaux et contre l'homme. *Annals of Tropical Medicine and Parasitology* **18**, 323–342.

Coscarón, S. (1987). "El Género *Simulium* Latreille en la Región Neotropical: Análisis de los Grupos Supraespecíficos, Especies Que los Integran y Distribución Geográfica (Simuliidae, Diptera)." Museum Paraense Emílio Goeldi, Belém.

Crosskey, R. W. (1967). The classification of *Simulium* Latreille (Diptera: Simuliidae) from Australia, New Guinea and the Western Pacific. *Journal of Natural History* **1**, 23–51.

Crosskey, R. W. (1969). A re-classification of the Simuliidae (Diptera) of Africa and its islands. *Bulletin of the British Museum (Natural History) Entomology Supplement* **14**, 1–195.

Crosskey, R. W. (1990). "The Natural History of Blackflies." Wiley, Chichester.

Crosskey, R. W., and Howard, T. M. (1997). "A New Taxonomic and Geographical Inventory of World Blackflies (Diptera: Simuliidae)." The Natural History Museum, London.

Cupp, E. W., and Cupp, M. S. (1997). Black fly (Diptera: Simuliidae) salivary secretions: importance in vector competence and disease. *Journal of Medical Entomology* **34**, 87–94.

Cupp, E. W., and Ramberg, F. B. (1997). Care and maintenance of blackfly colonies. In "The Molecular Biology of Insect Disease Vectors: A Methods Manual" (J. M. Crampton, C. B. Beard, and C. Louis, Eds.), pp. 31–40. Chapman & Hall, London.

Currie, D. C. (1986). An annotated list of and keys to the immature black flies of Alberta (Diptera: Simuliidae). *Memoirs of the Entomological Society of Canada* **134**, 1–90.

Davies, D. M., Peterson, B. V., and Wood, D. M. (1962). The black flies (Diptera, Simuliidae) of Ontario. Part I. Adult identification and distribution with description of six new species. *Proceedings of the Entomological Society of Ontario* **92**, 70–154.

Davies, J. B. (1994). Sixty years of onchocerciasis vector control: a chronological summary with comments on eradication, reinvasion, and insecticide resistance. *Ann. Rev. Entomol.* **39**, 23–45.

Dirie, M. F., Ashford, R. W., Mungomba, L. M., Molyneux, D. H., and Green, E. E. (1990). Avian trypanosomes in *Simulium* and sparrowhawks (*Accipiter nisus*). *Parasitology* **101**, 243–247.

Eaton, D. P., Diaz, L. A., Hans-Filho, G., dos Santos, V., Aoki, V., Friedman, H., Rivitti, E. A., Sampaio, S. A. P., Gottlieb, M. S., Giudice, G. J., Lopez, A., Cupp, E. W., and the Cooperative Group on Fogo Selvagem Research. (1998). Comparison of black fly species (Diptera: Simuliidae) on an Amerindian reservation with a high prevalence of fogo selvagem to neighboring disease-free sites in the state of Mato Grosso do Sul, Brazil. *Journal of Medical Entomology* **35**, 120–131.

Fallis, A. M., Desser, S. S., and Khan, R. A. (1974). On species of *Leucocytozoon*. *Advances in Parasitology* **12**, 1–67.

Fallis, A. M., Jacobson, R. L., and Raybould, J. N. (1973). Experimental transmission of *Trypanosoma numidae* Wenyon to guinea fowl and chickens in Tanzania. *Journal of Protozoology* **20**, 436–437.

Fredeen, F. J. H. (1977). A review of the economic importance of black flies (Simuliidae) in Canada. *Quaestiones Entomologicae* **13**, 219–229.

Fredeen, F. J. H. (1985a). The black flies (Diptera: Simuliidae) of Saskatchewan. *Saskatchewan Cultural Record, Museum of Natural History Contributions* **8**, 1–41.

Fredeen, F. J. H. (1985b). Some economic effects of outbreaks of black flies (*Simulium luggeri* Nicholson & Mickel) in Saskatchewan. *Quaestiones Entomologicae* **21**, 175–208.

Gray, E. W., Adler, P. H., Coscarón-Arias, C., Coscarón, S., and Noblet, R. (1999). Development of the first black fly (Diptera: Simuliidae) management program in Argentina and comparison with other programs. *Journal of the American Mosquito Control Association* **15**, 400–406.

Greiner, E. C. (1991). Leucocytozoonosis in waterfowl and wild galliform birds. *Bulletin of the Society of Vector Ecologists* **16**, 84–93.

Gudgel, E. F., and Grauer, F. H. (1954). Acute and chronic reactions to black fly bites (*Simulium* fly). *Archives of Dermatology and Syphilology* **70**, 609–615.

Guillet, P., Escaffre, H., Ouédraogo, M., and Quillévéré, D. (1980). Mise en évidence d'une résistance au téméphos dans le complexe *S. damnosum* (*S. sanctipauli* et *S. soubrense*) en Côte d'Ivoire (zone du programme de lutte contre l'onchocercose dans la région du bassin de la Volta). *Cahiers O.R.S.T.O.M. Série Entomologie Médicale et Parasitologie* **23**, 291–299.

Herman, C. M., Barrow, J. H., and Tarshis, I. B. (1975). Leucocytozoonosis in Canada geese at the Seney National Wildlife Refuge. *Journal of Wildlife Diseases* **11**, 404–411.

Klion, A. D., and Nutman, T. B. (1999). Loiasis and *Mansonella* infections. In "Tropical Infectious Diseases: Principles, Pathogens, and Practice" (R. L. Guerrant, D. H. Walker, and P. F. Weller, Eds.), pp. 861–872. Churchill Livingstone, Philadelphia.

Laird, M., Ed. (1981). "Blackflies: The Future for Biological Methods in Integrated Control." Academic Press, New York.

Lok, J. B., Cupp, E. W., and Bernardo, M. J. (1983). *Simulium jenningsi* Malloch (Diptera: Simuliidae): a vector of *Onchocerca lienalis* Stiles (Nematoda: Filarioidea) in New York. *American Journal of Veterinary Research* **44**, 2355–2358.

Mead, D. G., Ramberg, F. B., Besslesen, D. G., and Máre, C. J. (2000). Transmission of vesicular stomatitis virus from infected to noninfected black flies co-feeding on nonviremic deer mice. *Science* **287**, 485–487.

Molloy, D. (1990). Progress in the biological control of black flies *Bacillus thuringiensis israelensis,* with emphasis on temperate climates. In "Bacterial Control of Mosquitoes and Black Flies: Biochemistry, Genetics, and Applications of *Bacillus thuringiensis israelensis* and *Bacillus sphaericus*" (H. de Barjac and D. J. Sutherland, Eds.), pp. 161–186. Rutgers Univ. Press, New Brunswick, NJ.

Muller, R., and Horsburgh, R. C. R. (1987). "Bibliography of Onchocerciasis (1841–1985)." C. A. B. International Institute of Parasitology, United Kingdom.

Mykytowycz, R. (1957). The transmission of myxomatosis by *Simulium melatum* Wharton (Diptera: Simuliidae). *CSIRO Wildlife Research* **2**, 1–4.

Noblet, R., Kissam, J. B., and Adkins, T. R., Jr. (1975). *Leucocytozoon smithi:* incidence of transmission by black flies in South Carolina (Diptera: Simuliidae). *Journal of Medical Entomology* **12**, 111–114.

O'Roke, E. C. (1934). A malaria-like disease of ducks caused by *Leucocytozoon anatis* Wickware. *University of Michigan School of Forestry and Conservation Bulletin* **4**, 1–44.

Peterson, B. V. (1996). Simuliidae. In "An Introduction to the Aquatic Insects of North America" (R. W. Merritt and K. W. Cummins, Eds.), 3rd ed., pp. 591–634. Kendall/Hunt, Dubuque, IA.

Peterson, B. V., and Kondratieff, B. C. (1995). The black flies (Diptera: Simuliidae) of Colorado: an annotated list with keys, illustrations and descriptions of three new species. *Memoirs of the American Entomological Society* **42**, 1–121.

Plaisier, A. P., Van Oortmarssen, G. J., Remme, J. H. F., and Habbema, J. D. F. (1991). The reproduction lifespan of *Onchocerca volvulus* in West Africa savanna. *Acta Tropica* **48**, 271–284.

Pledger, D. J., Samuel, W. M., and Craig, D. A. (1980). Black flies (Diptera: Simuliidae) as possible vectors of legworm (*Onchocerca cervipedis*) in moose of central Alberta. *Proceedings of the North American Moose Conference Workshop* **16**, 171–202.

Procunier, W. S. (1989). Cytological approaches to simuliid biosystematics in relation to the epidemiology and control of human onchocerciasis. *Genome* **32**, 559–569.

Remme, J. H. F., De Sole, G., and Van Oortmarssen, G. J. (1990). The predicted and observed decline in the prevalence and intensity of onchocerciasis infection during 14 years of successful vector control. *Bulletin of the World Health Organization* **68**, 331–339.

Riley, C. V. (1887). Report of the entomologist. *United States Department of Agriculture Report* **1886**, 459–592.

Rothfels, K. H. (1979). Cytotaxonomy of black flies (Simuliidae). *Annu. Rev. Entomol.* **24**, 507–539.

Rubtsov, I. A. (1956). "Blackflies (fam. Simuliidae). Fauna of the USSR." New Series No. 64, Insects, Diptera 6 (6). Akad. Nauk SSSR, Moscow. [In Russian; English translation: 1990. Blackflies (Simuliidae) (2nd ed.). Fauna of the USSR. Diptera, 6 (6). Brill, Leiden.].

Schmidtmann, E. T., Tabachnick, W. J., Hunt, G. J., Thompson, L. H., and Hurd, H. S. (1999). 1995 epizootic of vesicular stomatitis (New Jersey serotype) in the western United States: an entomologic perspective. *Journal of Medical Entomology* **36**, 1–7.

Shelley, A. J. (1988a). Biosystematics and medical importance of the *Simulium amazonicum* group and the *S. exiguum* complex in Latin America. In "Biosystematics of Haematophagus Insects" (M. W. Service, Ed.), pp. 203–220. Clarendon, Oxford.

Shelley, A. J. (1988b). Vector aspects of the epidemiology of onchocerciasis in Latin America. *Annu. Rev. Entomol.* **30**, 337–366.

Skidmore, L. V. (1932). *Leucocytozoon smithi* infection in turkeys and its transmission by *Simulium occidentale* Townsend. *Zentralblatt fuer Bakteriologie Parasitenkunde Infektionskrankheiten I (Originale)* **125**, 328–335.

Stokes, J. H. (1914). A clinical, pathological and experimental study of the lesions produced by the bite of the "black fly" (*Simulium venustum*). *Journal of Cutaneous Diseases* **32**, 751–769, 830–856.

Stone, A., and Snoddy, E. L. (1969). The black flies of Alabama (Diptera: Simuliidae). *Auburn University Agricultural Experiment Station Bulletin* **390**, 1–93.

Sutcliffe, J. F. (1986). Black fly host location: a review. *Canadian Journal of Zoology* **64**, 1041–1053.

Sutcliffe, J. F., and McIver, S. B. (1984). Mechanics of blood-feeding in black flies (Diptera, Simuliidae). *Journal of Morphology* **180**, 125–144.

Takaoka, H. (1999). Review on zoonotic *Onchocerca* species and their insect vectors in Japan (in Japanese, with English summary). *Medical Entomology and Zoology* **50**, 1–8.

Takaoka, H., and Davies, D. M. (1995). "The Black Flies (Diptera: Simuliidae) of West Malaysia." Kyushu Univ. Press, Japan.

Vargas, L., and Díaz Nájera, A. (1957). Simúlidos mexicanos. *Revista del Instituto de Salubridad Enfermedades Tropicales* **17**, 143–399.

Wahl, G. (1996). Identification of a common filarial larva in *Simulium damnosum* s. l. (Type D, Duke, 1967) as *Onchocerca ramachandrini* from the wart hog. *Journal of Parasitology* **82**, 520–524.

Wahl, G., and Renz, A. (1991). Transmission of *Onchocerca dukei* by *Simulium bovis* in North-Cameroon. *Tropical Medicine and Parasitology* **42**, 368–370.

Wahl, G., Ekale, D., and Schmitz, A. (1998). *Onchocerca ochengi:* assessment of the *Simulium* vectors in North Cameroon. *Parasitology* **116**, 327–336.

Wilhelm, A., Betke, P., and Jacob, K. (1982). Simuliotoxikose beim Ren (*Rangifer tarandus*). In "Erkrankungen der Zootiere. Verhandlungbericht des XXIV Internationalen Symposium über die Erkrankungen der Zootiere" (R. Ippen and H. D. Schräder, Eds.), pp. 357–360. Akademie-Verlag, Berlin.

World Health Organization. (1995). Onchocerciasis and its control: report of a WHO-Expert Committee on Onchocerciasis Control. *WHO Technical Report Series* **852**, 1–103.

# 12

# MOSQUITOES (*Culicidae*)

WOODBRIDGE A. FOSTER AND EDWARD D. WALKER

*Since ancient* times, mosquito bites or habitats have been associated with human disease, and, in 1878, mosquitoes were the first arthropods formally incriminated as intermediate hosts of vertebrate parasites. During the past century of research, it has become established that mosquitoes are the most important arthropods affecting human health. They attain their greatest impact as vectors for the organisms causing such well-known human diseases as malaria, filariasis, encephalitis, yellow fever, and dengue. These afflictions are especially severe in developing regions of the tropics. They cause early death and chronic debilitation, which strain the resources of health services and reduce human productivity, thereby perpetuating economic hardship.

Mosquito-borne diseases also persist in industrialized temperate countries. Yet, human discomfort from bites is often the chief concern. In the United States,

hundreds of millions of dollars are spent annually to control them for this reason alone. Additionally, large populations of mosquitoes can cause intense irritation and extensive blood loss to livestock and wildlife, resulting in reduced productivity and even death.

Mosquitoes occur in practically every region of every continent in the world except Antarctica. They develop in an extremely broad range of biotic communities: arctic tundra, boreal forests, high mountains, plains, deserts, tropical forests, salt marshes, and ocean tidal zones. Greatest species diversity occurs in tropical forests, but extremely high densities of mosquitoes are common even in the species-poor biomes, such as the tundra. Many species have benefitted from human alteration of the environment, and a few have become domesticated. Because of their immense importance, mosquitoes have been the subject of many major books. Among the more important ones that deal exclusively with mosquito biology are Christophers (1960), Clements (1992, 1999), Forattini (1962, 1965), Gillett (1971), Bock and Cardew (1996), Horsfall (1955), Lounibos et al. (1985), Mattingly (1969), and Service (1990, 1993a). *Journal of the American Mosquito Control Association* is devoted mainly to studies of mosquitoes. A substantial proportion of the scientific articles in the *Journal of Medical Entomology* and *Medical and Veterinary Entomology* also report mosquito research. *Wing Beats* is a trade magazine dedicated to mosquitoes.

# TAXONOMY

The family Culicidae, derived from *culex*, the Latin name for "gnat," is a member of one of the main stocks of Nematocera, the infraorder Culicomorpha. It consists of two superfamilies that include all of the piercing/sucking nematocerans, both predators and blood-feeding biters. The superfamily Chironomoidea comprises the families Chironomidae and Thaumaleidae, which have nonpiercing mouthparts, and Simuliidae and Ceratopogonidae, which pierce either vertebrates or invertebrates. The superfamily Culicoidea comprises the Dixidae, Corethrellidae, Chaoboridae, and Culicidae, the second and fourth of which feed on vertebrate blood. Several of these families are superficially similar. However, among all the culicomorphs, the long proboscis of mosquitoes is distinctive. It is considered the most specialized of biting mouthparts among Nematocera and indicates a long and close association of mosquitoes with vertebrate animals. Wood and Borkent (1989) provide an overview of nematoceran phylogeny and classification.

Culicidae consists of about 3200 recognized species. The largest number remaining to be discovered probably inhabits tropical rain forests, where faunas are more diverse but less well surveyed than temperate regions. Species that have been studied intensively often reveal that they consist of complexes of closely related species, indicating that many reproductively isolated and niche-specific forms remain to be identified or are undergoing speciation. Current culicid classification (Table I) recognizes three subfamilies: Anophelinae, Culicinae, and Toxorhynchitinae. Anophelinae and Toxorhynchitinae sometimes are considered to be primitive groups; other authorities view them as specialized derivations of some Culicinae-type ancestor. Recent cladistic analysis of morphological and nucleotide-sequence data supports the idea that the Anophelinae are only distantly related to the other two subfamilies and that the Toxorhynchitinae do not merit subfamily status (Harbach and Kitching, 1998). Anopheline eggs bear characteristic floats, their larvae lack air tubes, and adults have elongate palps in both sexes. Typical culicine and toxorhynchitine larvae have air tubes, and adult females have short palps. Toxorhynchitines are all predaceous as larvae, are unusually large, and have a curved proboscis suited for feeding on only nectar.

There are 38 genera of mosquitoes, 34 of which are in the subfamily Culicinae. Culicines are organized into 10 tribes, the most diverse of which are Aedini and Sabethini in terms of numbers of genera and species worldwide. The 14 genera in North America north of Mexico, and the number of species in each, are *Anopheles* (16), *Aedes* (7), *Ochlerotatus* (69), *Psorophora* (15), *Haemagogus* (1), *Culex* (29), *Deinocerites* (3), *Culiseta* (8), *Coquillettidia* (1), *Mansonia* (2), *Orthopodomyia* (3), *Wyeomyia* (4), *Uranotaenia* (4), and *Toxorhynchites* (1) (Darsie and Ward, 1981).

Three important species groups of mosquitoes worldwide are the *Anopheles gambiae* and *Culex pipiens* complexes and the *Aedes* subgenus *Stegomyia*. The *An. gambiae* complex of Africa consists of six species. Two of these, *An. gambiae* and *An. arabiensis,* are important vectors of malaria and lymphatic filariasis. *An. arabiensis*

## TABLE I
**Classification of Culicidae**

| Subfamily | Tribe | Genera |
|---|---|---|
| Anophelinae | | *Anopheles* (*An.*), *Bironella* (*Bi.*), *Chagasia* (*Ch.*) |
| Culicinae | Aedeomyiini | *Aedeomyia* (*Ad.*) |
| | Aedini | *Aedes* (*Ae.*), *Ochlerotatus* (*Oc.*), *Verrallina* (*Ve.*), *Ayurakitia* (*Ay.*), *Armigeres* (*Ar.*), *Eretmapodites* (*Er.*), *Haemagogus* (*Hg.*), *Heizmannia* (*Hz.*), *Opifex* (*Op.*), *Psorophora* (*Ps.*), *Udaya* (*Ud.*), *Zeugnomyia* (*Ze.*) |
| | Culicini | *Culex* (*Cx.*), *Deinocerites* (*De.*), *Galindomyia* (*Ga.*) |
| | Culisetini | *Culiseta* (*Cs.*) |
| | Ficalbiini | *Ficalbia* (*Fi.*), *Mimomyia* (*Mi.*) |
| | Hodgesiini | *Hodgesia* (*Ho.*) |
| | Mansoniini | *Coquillettidia* (*Cq.*), *Mansonia* (*Ma.*) |
| | Orthopodomyiini | *Orthopodomyia* (*Or.*) |
| | Sabethini | *Sabethes* (*Sa.*), *Wyeomyia* (*Wy.*), *Phoniomyia* (*Ph.*), *Limatus* (*Li.*), *Trichoprosopon* (*Tr.*), *Shannoniana* (*Sh.*), *Runchomyia* (*Ru.*), *Johnbelkinia* (*Jb.*), *Isostomyia* (*Is.*), *Tripteroides* (*Tp.*), *Malaya* (*Ml.*), *Topomyia* (*To.*), *Maorigoeldia* (*Mg.*) |
| | Uranotaeniini | *Uranotaenia* (*Ur.*) |
| Toxorhynchitinae | | *Toxorhynchites* (*Tx.*) |

The classification of all mosquitoes into 3 subfamilies, 10 tribes of Culicinae, and 38 genera is based on Knight and Stone (1977). In parentheses are the two-letter generic abbreviations recognized by the American Mosquito Control Association and used in several journals and books.

tends to occur in somewhat drier regions than does *An. gambiae.* Both prefer to bite humans, but *An. gambiae* is more anthropophilic, endophilic, and endophagic, and therefore it is the more important vector. The *Cx. pipiens* complex is a ubiquitous group of closely related domestic and peridomestic species. The medically most important taxa worldwide are the temperate species *Cx. pipiens,* the *northern house mosquito,* and the tropical and subtropical *Cx. quinquefasciatus* (= *fatigans*), the *southern house mosquito.* Their ranges are overlapping in the central latitudes of the United States, where they commonly hybridize. They are vectors of several human pathogens, such as St. Louis encephalitis (SLE) virus and lymphatic filariasis. *Cx. molestus* is a name sometimes applied to a variant of *Cx. pipiens,* which is facultatively autogenous and often breeds in subterranean water. *Cx. pallens,* apparently a stable hybrid of *Cx. pipiens* and *Cx. quinquefasciatus,* occurs in temperate China and Japan, whereas *Cx. globocoxitus* and *Cx. australicus* inhabit Australia.

Several brightly marked *Aedes* species in the large subgenus *Stegomyia* are medically important, including *Aedes aegypti* and *Ae. albopictus. Ae. aegypti,* the *yellow fever mosquito,* has a worldwide distribution in the tropics and subtropics. It is the primary vector of both DEN and urban YF viruses. It exists in at least two forms, *aegypti* and *formosus,* considered to be either subspecies or separate species. *Ae. aegypti formosus* is the original feral form and is found in large parts of interior Africa. It has a black body, develops in tree holes, feeds on a wide variety of animals, and rarely enters houses. It has adapted to some domestic situations in Africa, where it develops in rain-filled containers. *Ae. a. aegypti* is a paler, brownish-black domestic form. It occurs mainly in coastal regions of Africa and is distributed throughout much of southern Asia and most warmer parts of the New World, including the southern United States. In Africa it has become independent of rain, developing in hand-filled water jars without regard to season. On other continents, where it does not compete with *Ae. a. formosus,* it utilizes both rain-filled and hand-filled containers. Some authorities recognize a still paler and more domestic type of *Ae. a. aegypti* as the subspecies *Ae. a. queenslandensis,* but this is probably only a localized variant.

*Ae. albopictus,* the *Asian tiger mosquito,* is similar to *Ae. aegypti,* occupies the same kinds of containers, and also transmits DEN virus. It was largely confined to Asia, where it occurs in tropical and subtropical rural settings. It readily oviposits in tree holes. A cold-hardy, egg-diapausing strain of this mosquito has been carried from northern Japan to other parts of the world by the trade in used automobile and truck tires. The first established population was detected in Texas in 1985. It has since spread through much of the southern, central, and eastern United States, including foci in the upper Midwest,

much farther north than the nondiapausing *Ae. aegypti.* It also has gained a foothold in several other parts of the world. In most of its range in the southern United States, *Ae. albopictus* has replaced *Ae. aegypti* as the predominant mosquito in artificial containers in suburban and rural environments.

Other important members of the subgenus *Stegomyia* include *Ae. africanus, Ae. bromeliae,* and *Ae. luteocephalus,* which transmit YF virus in parts of Africa, and *Ae. polynesiensis* and *Ae. pseudoscutellaris,* which transmit lymphatic filariasis in South Pacific islands.

Keys to the mosquito genera worldwide were provided by Mattingly (1971). Keys for the identification of species of restricted geographical regions are available for most states, provinces, and many countries throughout the world. These include many fine handbooks that also present biological and medical information on individual species. Good examples of statewide handbooks are written for New Jersey (Headlee, 1945), California (Bohart and Washino, 1978), Indiana (Siverly, 1972), Minnesota (Barr, 1958), New York (Means, 1979), Florida (Darsie and Morris, 1998), and Alaska (Gjullin *et al.,* 1961). United States regional handbooks include the southeastern United States (King *et al.,* 1960) and the northwestern states (Stage *et al.,* 1952). Some handbooks include keys to pupae, and the handbook for Illinois (Ross and Horsfall, 1965) is noteworthy in particular for its egg keys.

The most recent comprehensive treatments of North American species are Wood *et al.* (1979), which contains keys to larvae and adults of Canada, plates of taxonomic structures for each species, distribution maps, and biological information; and Darsie and Ward (1981), which covers all of North America north of Mexico and has illustrated keys and distribution maps. These works were preceded by Carpenter and LaCasse (1955), which contains formal descriptions, biology, and meticulously crafted full-page plates of adults of each species. A thorough treatment of North American genera was presented by Stone (1981). Other parts of the world covered by notable works include the South Pacific (Belkin, 1962), United Kingdom (Marshall, 1938), the Neotropical Region (Lane, 1953), and Japan and Korea (LaCasse and Yamaguti, 1950). For details on morphological terminology and anatomical features of mosquitoes, Harbach and Knight (1980) is recommended. Members of species complexes are often indistinguishable morphologically. Specialists have overcome some of these problems by using chromosome banding patterns, isozyme profiles, or DNA probes and DNA restriction fragment patterns to distinguish these species from one another. These methods for identification are not yet simple enough, or widely enough available, to be used routinely in field work.

All known mosquito species in the world are listed in *A Catalog of the Mosquitoes of the World* (Knight and Stone, 1977), plus its four supplements ( Gaffigan and Ward, 1985; Knight, 1978; Ward, 1984, 1992). This work provides the taxonomic history and current standing of all recognized species, their distributions by country, and references to the general literature and to taxonomic works for all regions of the world. The most recent large treatment of a major geographic area is the 12-volume catalog on the Australasian Region, edited by Debenham, Hicks, Lee, and others (1980–1989). Original systematic studies of mosquitoes are published in several scientific journals, but the one devoted solely to this subject, *Mosquito Systematics,* was subsumed under the *Journal of the American Mosquito Control Association* in 1995. Two long series of valuable papers of international scope on mosquito taxonomy and distribution have been published in the journals *Contributions of the American Entomological Institute: Contributions to the Mosquito Fauna of Southeast Asia* and *Mosquito Studies.* Preferred common names of mosquito species are listed by the Entomological Society of America (Bosik, 1997) and by Pittaway (1992). An internationally accepted set of two-letter abbreviations for all mosquito genera is shown in Table I. These abbreviations appear in most mosquito publications and are used in this chapter.

## MORPHOLOGY

The eggs of most mosquitoes are elongate, ovoid, or spindle-shaped; others are spherical or rhomboid. The outermost layer of the egg shell, or *chorion* (Fig. 12.1), often has intricate surface structures and patterns diagnostic of the particular species. The chorions of *Anopheles* species have unique, transparent, air-filled compartments flanking the egg that serve as *floats* (Fig. 12.1A). Eggs of *Anopheles, Toxorhynchites, Wyeomyia, Aedes, Ochlerotatus, Psorophora,* and *Haemagogus* species are laid individually, whereas in *Culex, Culiseta, Coquillettidia,* and *Mansonia* species, they are attached together in a single clump, forming a floating *egg raft* (Fig. 12.2A) or a submerged cluster (Fig. 12.2B). *Culex* eggs have a cup-shaped *corolla* at one end (Fig. 12.1B), allowing them to sit vertically on the water surface in a raft (Fig. 12.2A); the upper ends have apical droplets with a chemical thought to maintain the raft upright.

Mosquito *larvae*, commonly known as *wigglers* or *wrigglers,* pass through four instars, which closely resemble one another except for their size. Larvae are rich in taxonomic characteristics that are easy to see on slide-mounted specimens (Fig. 12.3). The head is defined by

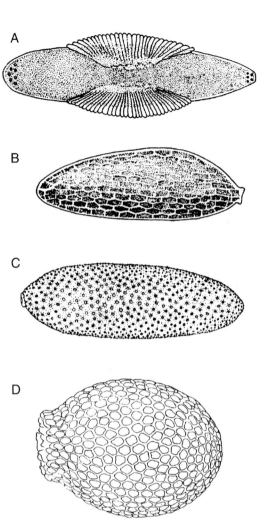

**FIGURE 12.1**    Eggs of mosquitoes, showing variations in shape and chorionic sculpturing. (A) *Anopheles;* (B) *Culex;* (C) *Aedes aegypti;* (D) *Toxorhynchites brevipalpis.* (A and B from Ross, 1947; C and D from Harbach and Knight, 1980.)

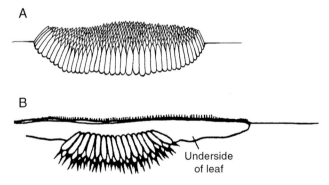

**FIGURE 12.2**    Mosquito egg rafts and clusters. (A) Floating egg raft of *Culex restuans;* (B) submerged egg cluster of *Mansonia,* attached to underside of floating leaf. (A, from Ross, 1947; B, from Gordon and Lavoipierre, 1962.)

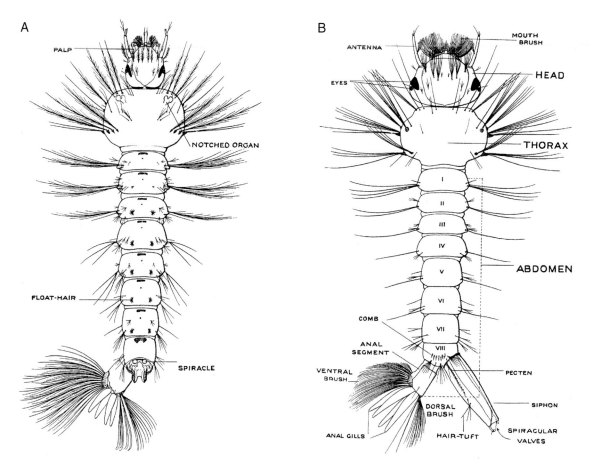

**FIGURE 12.3**   External anatomy of mosquito larvae, dorsal view, with anal segment and siphon at posterior end rotated to provide better view. (A) Anopheline form (*Anopheles maculipennis*); (B) culicine form (*Aedes cinereus*). (From Marshall, 1938.)

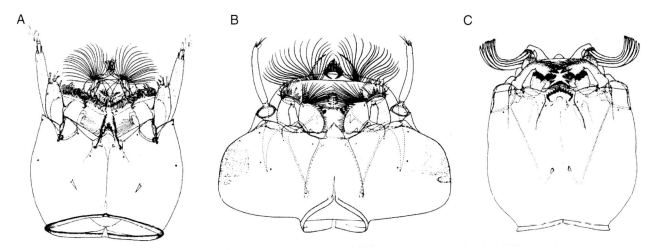

**FIGURE 12.4**   Heads of mosquito larvae, postero-ventral view. (A) Anopheline form (*Anopheles quadrimaculatus*); (B) culicine form (*Ochlerotatus fulvus pallens*); (C) toxorhynchitine form (*Toxorhynchites brevipalpis*). The lateral palatal brushes of most larvae are used to generate water currents for filter feeding; in *Toxorhynchites* they are modified for seizing prey. (From Harbach and Knight, 1980.)

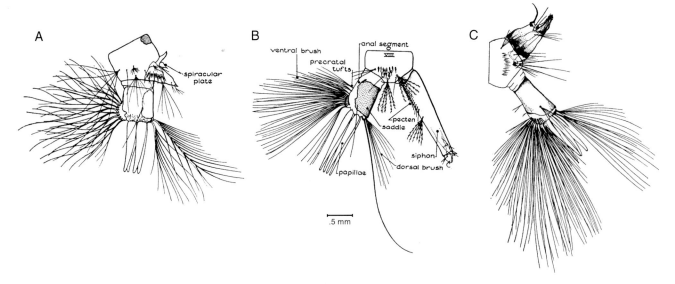

**FIGURE 12.5** Terminal segments of mosquito larvae. (A) Anopheline form (*Anopheles earlei*), lacking a siphon; (B) culicine form (*Ochlerotatus fitchii*), showing elongate siphon; (C) culicine form (*Coquillettidia perturbans*), showing short, stout siphon suited for piercing and clinging to plants. (From Barr, 1958.)

a distinct capsule bearing a pair of "eyes" composed of clusters of lateral ocelli, a pair of antennae of variable shape and length, and chewing mouthparts bearing a variety of brushes, combs, and sweepers used in feeding (Fig. 12.4). The lateral *palatal brushes* on the labrum create water currents that draw floating or suspended particles toward the mouth. Sweepers and brushes on the mandibles, and brushes on the maxillae, are thought to collect and pack particles to create a bolus of food in the pharynx. In predatory larvae, the mandibles and/or maxillae are heavy and sharply toothed for seizing or holding prey. The thorax is wide, with three indistinct, legless segments.

The larval abdomen is narrower than the thorax, is cylindrical, and is composed of eight apparent segments, the second to last being a composite of segments 8 and 9. A pair of spiracles opens on the dorsal side of this segment. In culicines and toxorhynchitines the spiracles open at the end of the *respiratory siphon,* an elongate air tube extending dorsally. The siphon of *Coquillettidia* and *Mansonia* is short, ending in a heavily sclerotized point with a dorsal sawlike edge that is used to pierce and remains lodged in plant tissue. In anophelines the siphon is lacking, and the spiracles are borne on a short spiracular plate. Segment 10, the *anal segment,* extends ventrally at an angle from the rest of the abdomen. It typically bears four *anal papillae* used primarily in osmoregulation. The terminal region of the larva bears several structures useful in identification (Fig. 12.5). They include *comb scales* on segment 8, *pecten spines* on the siphon, a *saddle* sclerite encircling the anal segment, and various tufts and brushes of setae. Some of

these terminal structures apparently are used to groom the mouthparts when the larva bends its body around to form a loop.

Larval *internal anatomy* conforms to the general insect plan. The alimentary tract is almost straight, the only notable features being eight large gastric caeca at the

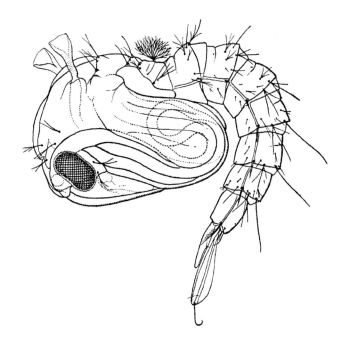

**FIGURE 12.6** Mosquito pupa. Lateral view of *Anopheles gambiae* in resting position at water surface; presence of adult structures visible within pupal cuticle. (From Smart, 1948.)

junction of the foregut and midgut in the thorax and five Malpighian tubules at the midgut–hindgut junction. Because most of the cuticle is semitransparent, two large tracheal trunks are obvious, extending forward from the spiracles to the thorax.

Mosquito *pupae,* commonly known as *tumblers,* are comma-shaped, with the head and thorax fused to form a cephalothorax and the abdomen curled beneath it (Fig. 12.6). Projecting from the dorsal mesothorax is a pair of respiratory tubes, or *air trumpets,* through which the pupa obtains oxygen at the water surface. Within the cephalothorax the developing appendages of the adult

head and thorax usually can be seen coiled ventrally; they envelop an air pocket, the *ventral airspace,* that provides buoyancy to help maintain the pupa at the water surface when resting. At the end of the abdomen two broad *paddles* are attached to the eighth segment. The pupa can flex its abdominal segments, causing the paddles to flap downward, propelling it through the water when it is disturbed.

Adult mosquitoes are slender, with thin legs and narrow, elongate wings (Fig. 12.7). The body surface is covered with scales, setae, and fine pile, creating the characteristic markings and colors of each species.

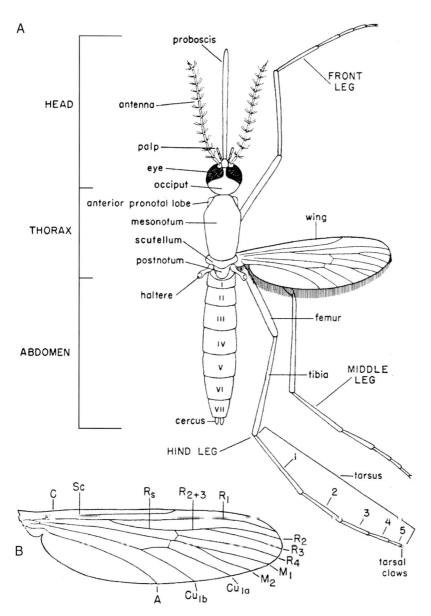

**FIGURE 12.7**  External anatomy of mosquito adult. (A) Generalized adult, dorsal view; (B) wing, showing typical venation and vein nomenclature. (From Ross and Horsfall, 1965.)

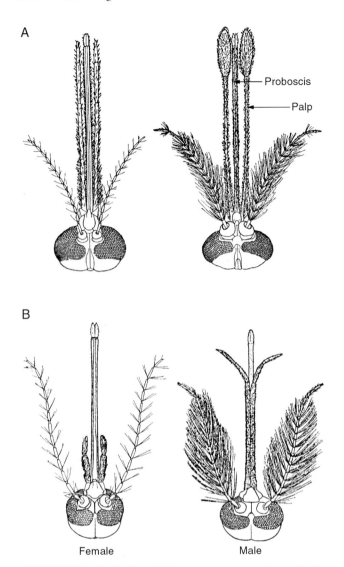

A

Proboscis

Palp

B

Female

Male

**FIGURE 12.8** Heads of anopheline and culicine mosquitoes, females (left) and males (right); males, typically, with plumose antennae. (A) *Anopheles* (anopheline), both males and females with palps about as long as the proboscis; male with plumose antennae and tips of palps broadened; (B) *Culex* (culicine), females typically with short palps and males with long, curved, or brushlike palps. (From Gordon and Lavoipierre, 1962.)

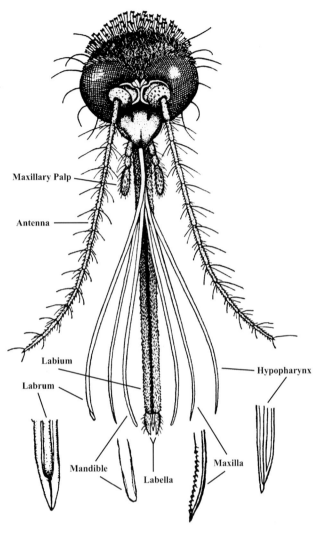

Maxillary Palp

Antenna

Labium

Labrum

Hypopharynx

Mandible

Labella

Maxilla

**FIGURE 12.9** Mouthparts of adult female mosquito, showing labium, splayed stylets, and variations in structure of their tips. (From Matheson, 1944.)

The two compound eyes, each represented by 350–900 ommatidial lenses, wrap around the front and sides of the head. The antennae arise between the eyes, are long and filamentous, and are usually sexually dimorphic. In species in which sound is used to locate females in flight, the flagellum of the male antenna has whorls of much longer fibrillae, giving it a plumose appearance (Fig. 12.8). The pedicel at the base of the antenna is a large globular structure that contains *Johnston's organ*, a mass of radially arranged mechanoreceptors that respond to vibrations of the flagellum induced by sound. In addition to the long fibrillae, the antenna has a variety

of sensory structures, including those for detecting host odors.

The mosquito *proboscis* is prominent, projecting anteriorly at least two-thirds the length of the abdomen. It consists of the basic complement of insect mouthparts: the labrum, paired mandibles, hypopharynx, paired maxillae, and labium. The first four structures have evolved into fine *stylets*, forming a tightly fitting *fascicle* that in females is used to penetrate host skin (Fig. 12.9). The fascicle is cradled within the groove of the large and conspicuous *labium* (Fig. 12.10), which comprises the bulk of the proboscis. The tip of the labium bears two small taste-sensitive labellar lobes and a short, pointed *ligula* (function unknown) between them. Of the fascicle of stylets, the *hypopharynx* and *mandibles* are narrowly pointed at their tips, whereas the *maxillae* end in serrated blades.

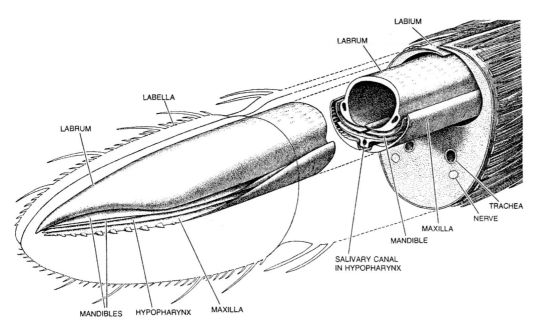

**FIGURE 12.10** Fascicle of stylets of adult female mosquito. Mouthparts near tip of proboscis, showing natural arrangement of stylets in a single bundle, or fascicle, within a groove in the labium, which forms a sheath. (From Jones, 1978; illustration by Tom Prentiss.)

Both mandibles and maxillae puncture the skin and advance the fascicle into the host's tissue. A salivary channel runs the length of the hypopharynx, delivering saliva to the tissue during probing. The *labrum* is curled laterally to form a *food canal* for drawing the host's blood or a sugar solution up the proboscis. In males, and in females of non–blood-feeding species, the mandibles and maxillae have atrophied, so they cannot pierce skin. In both sexes of *Toxorhynchites*, the nonpiercing proboscis is curved downward (Fig. 12.11). Maxillary palps arise at

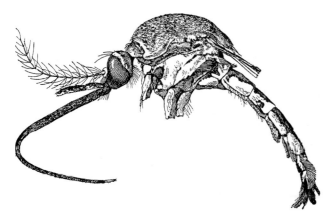

**FIGURE 12.11** *Toxorhynchites* sp., adult female; legs not shown. The form of the palps and antennae in females and males is similar to that of culicines; however, the proboscis of both sexes is bent downward at an angle of 90° or more. (From Smart, 1948.)

the base of the proboscis and bear several kinds of sensilla. Although there are many exceptions, palps usually are short in female culicines and toxorhynchitines but longer than the proboscis in most male culicines and toxorhynchitines and also in both sexes of anophelines (Fig. 12.8).

The mosquito *thorax* forms a single, relatively rigid muscle-filled locomotor unit with obscured segmentation. The mesothorax and metathorax each have a pair of lateral *spiracles*. The slender legs are attached close together on the underside of the thorax by elongate, downward-projecting coxae; the tarsi are tipped with two claws and a central pad, the *empodium*. The *wings* are narrow, have a distinctive pattern of veins, and bear scales along the veins and the hind margin, the latter forming a fringe. The *halteres,* tiny modified hind wings used in flight control, are located right behind the insertions of the wings.

The *abdomen* is clearly segmented and capable of extensive expansion and some movement, owing to the membranous areas between each set of tergites and sternites. This allows for expansion of the abdominal wall to accommodate large blood and sugar meals and developing clutches of eggs. Abdominal segments 5–8 are progressively smaller, so that the abdomen tapers toward the posterior end. Segment 9 is quite small and bears the *cerci,* the postgenital lobe of the female, and the *claspers* and other genitalic structures, or *terminalia,* of the male (Fig. 12.12). At emergence, the male genitalia are inverted. During the first hours of adulthood, segments 8

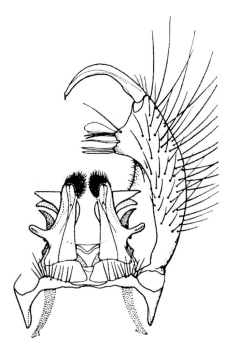

**FIGURE 12.12**  Male genitalia of *Culex quinquefasciatus,* showing principal copulatory structures used in male taxonomic identification. The gonocoxite and gonostylus on the left side have been omitted. (From Ross and Roberts, 1943.)

and ♂ of males together rotate 180° to reach the mature position. The complex and varied male genitalia are a useful source of characteristics for species identification.

Located within the thorax are a pair of three-lobed *salivary glands,* whose ducts join anteriorly to form a common salivary duct that enters the hypopharynx (Fig. 12.13). In males, these glands produce saliva used only in sugar feeding; in females, some portions are devoted to sugar feeding and others to blood feeding. The *foregut,* which begins in the head with the muscular *cibarium* and *pharynx,* pumps food up the labral food canal. The tubular esophagus extends through the cervix, or neck, into the thorax. There it is modified to form three *diverticula,* including two small dorsal outpocketings and a large ventral *crop;* the crop extends through the thorax and expands to form a large sac within the abdomen. Imbibed sugar solutions are stored in these diverticula and pass, a little at a time, through the *proventricular valve* into the midgut. A blood meal, on the other hand, passes directly into the widened posterior *midgut,* or stomach (Fig. 12.13). There it becomes surrounded by a semipermeable, saclike *peritrophic membrane* secreted by the midgut epithelium. The resulting blood bolus then is digested and absorbed. The pyloric valve separates the midgut from the *hindgut;* five *Malpighian tubules* empty into the hindgut just beyond the valve in the pyloric chamber. The anterior portion of the hindgut is tubular and loosely coiled; the posterior part is enlarged to form a bulbous *rectum* with large papillae projecting into it. The papillae probably are involved in salt ion reabsorption.

Paired *gonads* are located in the posterior one-third of the abdomen. The *testes* of males contain packets of sperm in various stages of maturation. A duct extending posteriorly from each testis widens to form a *seminal vesicle,* which stores mature sperm. The two seminal vesicles lie together and unite posteriorly to form the *ejaculatory duct.* Two large *accessory glands* open into this duct, which leads to the *aedeagus.* In females, the reproductive system consists of a pair of ovaries and accessory

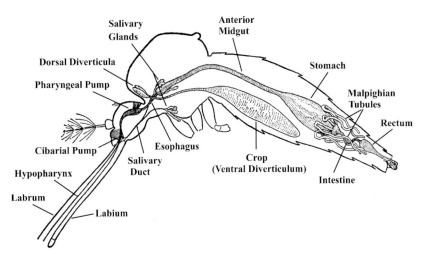

**FIGURE 12.13**  Digestive system of the adult mosquito. Semidiagrammatic view of major structures, including the salivary glands, foregut–midgut junction, and rectum. (From Snodgrass, 1959.)

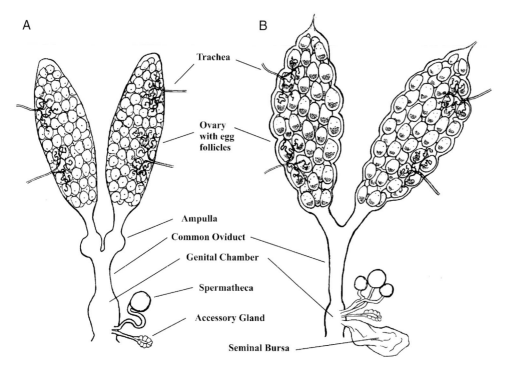

**FIGURE 12.14** Female reproductive system of the mosquito. (A) Anopheline form, based on *Anopheles gambiae*. (B) Culicine form, based on *Aedes aegypti*. (Original by W. A. Foster.)

structures (Fig. 12.14). Each ovary includes a few dozen to over 200 *polytrophic ovarioles,* the egg-forming units. A duct from each ovary extends posteriorly, and the two unite to form a common *oviduct,* which connects to the gonopore via a *genital chamber.* Opening into the genital chamber by tiny ducts are one to three sperm-storing *spermathecae,* small accessory glands, and the *seminal bursa* (lacking in most *Anopheles*), which receives semen from the male during mating.

## LIFE HISTORY

The *holometabolous* life cycle of mosquitoes is completed in two different environments: one aquatic, the other terrestrial. The larvae and pupae develop in a wide range of aquatic habitats. These include temporary surface water (e.g., tidal pools in salt marshes, rain pools, and flood water), permanent surface water (e.g., pools, streams, swamps, and lakes), and diverse natural and artificial water-holding containers (e.g., tree holes, leaf axils, fruit husks, mollusk shells, drinking water pots, and discarded tires). An extensive analysis and classification of larval habitats has been presented by Bates (1949), Mattingly (1969), Laird (1988), and Service (1993b). The only absolute requisite of all development sites is that they maintain at least a film of water for the duration of the larval

and pupal periods. However, individual species tend to oviposit, and therefore develop, in sites with characteristic physical and chemical properties. Adults are much more mobile than the immatures, but they also tend to occupy characteristic resting, foraging, and overwintering habitats.

Mosquito eggs are laid either on or in water, or on solid substrates that are likely to become inundated. Females in the subfamilies Anophelinae, Toxorhynchitinae, and most of the Culicinae in the tribe Sabethini scatter their eggs individually on the water surface, whereas those in the tribe Aedini (e.g., *Aedes, Ochlerotatus, Psorophora,* and *Haemagogus*) attach their eggs individually onto a substrate that later will become inundated with water. Clumped eggs are laid in boatlike rafts on the water surface by several genera of Culicinae (e.g., *Culex, Culiseta, Coquillettidia, Uranotaenia, Armigeres,* and *Trichoprosopon*) or in a radial cluster attached underwater to vegetation in the case of *Mansonia.*

At first the eggs are white, but most turn dark within hours as the chorion tans. In lowland tropics, subtropics, and summers of some temperate regions, eggs usually complete embryonic development within 2–3 days after being laid but may take up to a week or more in cool climates. Larvae hatch soon after embryogenesis in species that lay their eggs directly in water, including all Anophelinae, Toxorhynchitinae, and most tribes of Culicinae. The best known exceptions are members of the genera *Aedes, Ochlerotatus,* and other Aedini that typically develop in temporary water. Their eggs are laid on solid substrates out of water, and the larvae within them remain *quiescent* until inundated. The eggs can tolerate periods of cold and desiccation and may remain viable for years. Hatching usually occurs at warm temperatures after the eggs have been submerged and microbial activity has caused the oxygen level in the water to drop.

Depending on the species and particular conditions of the water, most mosquito *larvae* spend most of their time either at the water surface or at the bottom of the water column, coming to the surface for air only occasionally or not at all. At ideal conditions of food and temperature (26–28°C), the entire larval phase of *Ae. aegypti,* a tropical and subtropical mosquito, may last as few as 5–6 days. The first three instars are completed in about 1 day each and the fourth lasts about 3 days. In males these periods are slightly shorter, so the males pupate about 1 day earlier than females. Larvae of many species grow even faster, as when the water is heated by direct sunlight, whereas others develop slowly. *Toxorhynchites* and *Wyeomyia* species usually take 2–3 weeks even under ideal conditions. At cooler temperatures, or when food is scarce, growth becomes slower and can practically cease, with larvae remaining alive for months. Larvae of some species that inhabit high latitudes or high altitudes, or that develop in the early spring in temperate regions, have growth thresholds close to freezing and can tolerate even temporary entrapment in solid ice. This is typical of the snowpool *Ochlerotatus* species and of mosquitoes that overwinter as larvae, such as *Wyeomyia smithii* in pitcher plants and *Orthopodomyia alba* in tree holes.

The *pupa* spends nearly all of its time at the water surface. By the time it has molted to form a pharate adult within the pupal cuticle, it is very dark. In warm water the entire pupal stage typically lasts about 2 days in both sexes. In some mosquitoes, such as *Toxorhynchites* and *Wyeomyia* species, the shortest pupal periods may be 5–6 days. In all species the pupal period lasts longer at lower temperatures.

Adult males tend to emerge earlier than females, because of their shorter larval growth periods. As *adult emergence* approaches, the pupa remains stationary at the water surface, and the abdomen gradually straightens over 10–15 min. The adult emerges from the pupal cuticle by ingesting air, causing the cephalothorax to split and the adult to rise up out of the cuticle and stand on the water surface. The entire process takes only a few minutes. The newly emerged adult is capable of short flights a few minutes later but cannot sustain long flights for many hours until after the cuticle becomes fully sclerotized. Lipids and glycogen, carried over from larval reserves, provide sufficient energy for a few days of flight and survival.

It is typically during the first 3–5 days of adult life that both sexes obtain sugar from plant nectar or honeydew, become sexually mature, and then mate. In some species (e.g., *Culiseta inornata, Wy. smithii,* and *Deinocerites cancer*) *sexual maturation* is complete at the time of emergence or only a few hours later, and mating occurs almost immediately. Mosquitoes typically first feed on sugar to obtain enough energy for sexual maturation and for the flight necessary for mating, dispersal, and finding vertebrate blood. Natural sugar is taken repeatedly throughout adult life by both sexes of most species. Females typically mate only once. Males can inseminate several females before their supplies of mature sperm and accessory gland secretion become depleted. The semen supply is replenished in a few days.

Amorphous masses of *fat body* line the inner walls of the abdomen. The fat body synthesizes and stores both glycogen for flight and lipids for maintenance, using the digestive products of sugar and blood meals. Glycogen also is stored in the fibrillar flight muscles of the thorax, serving as a source of energy for immediate flight if the sugars in the crop and hemolymph have been exhausted.

Only females feed on vertebrate blood. In most mosquitoes, ingestion and digestion of a *blood meal* initiates egg development by stimulating a cascade of hormones from the brain and ovaries. The large amount of protein contained in hemoglobin and the blood serum provides the amino acids for synthesizing *vitellogenin,* the proteinaceous precursor of egg yolk. The protein also serves as the substrate for building lipid and glycogen, which contribute both to the egg yolk and to the maternal energy reserves used for survival and flight. A blood meal will stimulate egg development only if it is sufficiently large and if the female's ovarian follicles have reached the resting stage, at which point they are considered to be *gonoactive.* If a female has had poor larval nutrition, the follicles may not have reached the resting stage, and she will be unable to develop any eggs until having ingested sugar or a preliminary blood meal. Such a *gonoinactive* female, needing food to bring the ovarian follicles to the resting stage, is sometimes said to be "pre-gravid." Details of the hormonal control of these processes are discussed by Brown and Lea (1990), Hagedorn (1994, 1996), and Klowden (1996).

In most species, females are *anautogenous;* the egg follicles remain in the resting stage until a blood meal is taken. Following each blood meal, the female develops one mature clutch of eggs, exhibiting what is known as *gonotrophic concordance.* However, females of *autogenous* species or populations can develop eggs without a blood meal; among these there are obligate and facultative types. A *facultatively autogenous* female typically develops only the first clutch of eggs without blood; she does so only if she emerges with sufficient reserves and cannot readily find blood. Thereafter, a blood meal is

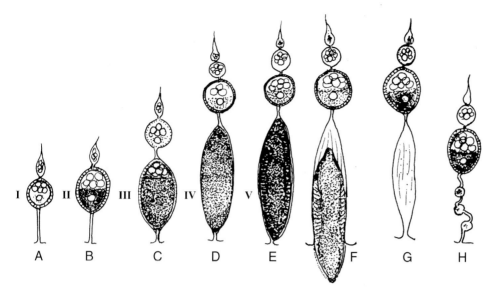

**FIGURE 12.15** Egg follicle development in mosquitoes, showing stages in the development of an egg follicle within a single ovariole of the ovary. A–E, Stages I–V of Christophers in an *Anopheles* female. After oviposition (F), the ovariole stalk remains swollen for a while, called the "sac stage" (G). Note: formerly, the sac was thought to shrink to a dilatation; now it is believed that the dilatations (three of them shown in H) generally form from follicles that fail to develop into eggs after a blood meal and are then resorbed. (From World Health Organization, 1975.)

required for each gonotrophic cycle. Species that are *obligately autogenous* have atrophied feeding stylets, never take blood, and subsist entirely on their larval reserves and plant sugar. Autogeny has been reviewed by O'Meara (1985). At the other extreme, there are some anautogenous species that take blood not only at the beginning of each gonotrophic cycle but also once or twice during egg development. These *supplementary blood meals* can provide extra energy, acting as a substitute for sugar (e.g., domestic *Ae. aegypti, An. gambiae*).

Ordinarily, all eggs develop synchronously and become mature in 2–5 days after blood feeding at favorable temperatures. During this time, the most advanced *follicle* within each ovariole passes through a series of five easily observed physiological stages (Fig. 12.15), originally described by Christophers (1911) and concisely summarized by Clements (1992). These stages comprise four physiological phases. The follicles develop synchronously, beginning with the *previtellogenic phase* (stages G through II). The follicle typically stops growing at stage IIa or IIb, the *resting stage*, until the female takes a blood meal. Within hours of blood feeding, the follicle enters the *initiation phase* (stage IIIa), when yolk (vitellogenin) synthesis by the fat body begins, followed by the *trophic phase* (stages IIIb and IV), the main period of yolk incorporation and follicle growth. In the *posttrophic phase*, the egg has reached its mature form, and the *chorion*, or egg shell, is formed. The eggs are then ready to be oviposited. By this time, the next follicle in each ovariole either has progressed to stage I or already has reached the resting stage (stage II). In the first case, it awaits oviposition before developing to the resting stage. Once oviposition has occurred and these new follicles are in the resting stage, the next blood meal is necessary for subsequent follicle development. The entire cycle of egg production, from blood meal through oviposition, is the *gonotrophic cycle*.

The number of ovarioles that produce mature eggs depends on the sizes of the female body, energy reserves, and blood meal. The follicles that do not pass beyond stage II degenerate. When most or all the ovarioles contain mature eggs, the ovaries may occupy nearly the entire volume of the distended abdomen. The gravid female fertilizes one egg at a time, as each passes down the oviduct to be oviposited on the water or a damp substrate.

When all eggs have been fertilized and expelled during oviposition, the ovaries return to their pre-trophic size, but the tracheae on the ovaries, which had been tightly coiled as *tracheal skins* before the eggs developed, become stretched and straightened. In *Anopheles* species, a swelling at the base of each lateral oviduct, the *ampulla*, becomes permanently stretched during the first oviposition. These signs serve to distinguish *parous* females, those that have completed at least one gonotrophic cycle, from *nulliparous* females, those that have not.

The number of completed gonotrophic cycles also can be determined. According to current interpretations, each ovariole ovulating a mature egg is left with an *egg sac*, which becomes reduced to a zone of granules in the calyx, the ovariole's connection to the lateral oviduct. Furthermore, a *dilatation* is formed in the stalk of each ovariole where a follicle has degenerated after a blood meal, instead of developing into an egg (Fig. 12.15). Thus, a count of the maximum numbers, per ovariole, of dilatations in the stalk and zones of granules in the calyx yields an estimate of the number of gonotrophic cycles completed. This *physiological age grading* can provide the medical entomologist with valuable information on the age of individuals and the age structure of a mosquito population. Details of these processes and their interpretation and application are given by Detinova (1962), Sokolova (1994), Fox and Brust (1994), and Hoc (1996).

*Univoltine* mosquito species complete only one generation per year. This occurs either if the developmental time is slow in relation to the season favorable for development or if the life cycle includes an obligate form of *diapause,* a compulsory phase of arrested development. *Bivoltine* and *multivoltine* species can complete two or more generations, respectively, during each breeding season, but the number actually completed may depend on temperature, available larval habitats, or available hosts. Mosquitoes pass through the winter or dry season as eggs, larvae, or adults, depending on the species and the climate. In cold climates, *overwintering* takes place in a state of diapause.

## BEHAVIOR AND ECOLOGY

Eggs that are laid on or in water generally are not resistant to desiccation and hatch shortly after embryogenesis, provided that they are wet and not too cold. This is typical of *Anopheles, Culex, Culiseta,* and *Toxorhynchites* species. *Aedes, Ochlerotatus, Psorophora,* and *Haemagogus* eggs, on the other hand, typically are laid on damp substrates, display great resistance to desiccation, and remain quiescent for months or years after embryogenesis until they receive a hatching stimulus. Sometimes moisture by itself is sufficient to induce hatching. Usually, however, the requisite stimulus is a reduction of dissolved oxygen in the water caused by microbial activity and decomposition of organic matter. Among quiescent eggs that are eventually submerged, only a portion of a single egg clutch may hatch during any one inundation, resulting in *installment hatching.* This apparently is the combined result of intrinsic variations among eggs in their hatching-stimulus thresholds and of local variations in microbial activity, causing differences in oxygen tension around the eggs. Even during a single inundation, hatching may not occur all at once but over a period of many days.

When an egg is ready to hatch, the first-instar larva uses a dorsal hatching spine on its head, the *egg breaker* or *egg burster,* to apply pressure to a preformed weakness in the chorion. This causes the chorion to pop open at one end, and the larva wriggles free. Because the eggs of *Culex, Culiseta,* and *Coquillettidia* usually stand vertically on the water surface in rafts, the larvae develop inside them with their anterior end oriented downward and hatch directly into the water.

Mosquito *larvae* are not buoyant and must, at rest, be suspended at the surface by special hairs and spiracular structures that cling to the surface tension while obtaining oxygen directly from the air. Culicinae typically migrate up and down in the water column, so they occur both at the surface and at the bottom of a body of water, depending on the availability of food. At the surface the tip of the siphon opens above the surface film, and the larvae hang diagonally downward most of the time (Fig. 12.16B). *Mansonia, Coquillettidia,* and some *Mimomyia* species are unusual in remaining submerged throughout larval and pupal development, with their siphons embedded in the tissues of aquatic plants from which they derive some oxygen (Fig. 12.16C). Mosquitoes that live in water-filled leaf axils (e.g., *Wyeomyia* spp.) are adept at flattening themselves against vertical surfaces and maneuvering in narrow spaces. Anopheline larvae spend most of their time at the water surface, often close to vegetation or floating material. They are able to remain suspended horizontally at the surface (Fig. 12.16A) due to pairs of dorsal palmate setae (float hairs) on several abdominal segments (Figs. 12.3A).

Larvae propel themselves by a back-and-forth lashing movement of the abdomen. Anopheline larvae usually swim horizontally at the surface film. When larvae of typical culicine mosquitoes are feeding below the surface, they periodically swim actively back to the surface to obtain oxygen. However, in many microenvironments dissolved oxygen also is absorbed from the water through the cuticle, requiring infrequent trips to the surface by some species.

Mosquito larvae feed on a variety of organic detritus, suspended material, and small organisms in their aquatic habitats. The organisms include bacteria, protists, fungi, algae, microinvertebrates, and small macroinvertebrates; the organic detritus usually consists of dead plant material and dead macroinvertebrates. They collect these food items in five basic ways: filtering, gathering, scraping, shredding, and preying. *Filterers* generate water currents with their lateral palatal brushes on the labrum, drawing suspended particles though fine combs, where they are collected and directed to the mouth. *Gatherers* use their mouthparts in a similar manner, but only after stirring up the particles from solid surfaces. *Scrapers* obtain food by scraping it off solid surfaces, whereas *shredders* gnaw, chew, and bite off pieces of organic matter. *Predators* grasp insects and other small, mobile prey in their large and sharp mandibles or maxillae (e.g., some *Psorophora* spp.) or with long, curved palatal brushes (e.g., *Toxorhynchites*) (Figs. 12.4C and 12.17). Most species use more

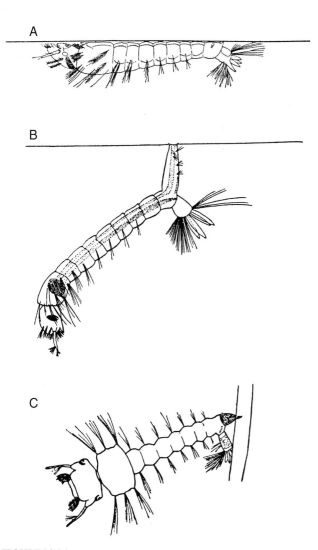

**FIGURE 12.17** *Toxorhynchites amboinensis* larva feeding on larva of *Culex pipiens*. This predaceous species has been used in biological control trials and naturally exerts a damping effect on populations of pest and vector mosquitoes. (Photo by W. A. Foster.)

**FIGURE 12.16** Resting and feeding positions of mosquito larvae. (A) *Anopheles,* showing horizontal position at the water surface, dorsal side up; larva has rotated its head 180° so that the ventral side of the head is uppermost and the mouthparts are applied to the water surface for filter feeding on floating detritus; (B) *Culex,* showing typical diagonal position while suspended from water surface; (C) *Mansonia,* attached to submerged part of aquatic plant (stem or root) by its siphon. (A and B, from Ross, 1947; C, from Gordon and Lavoipierre, 1962.)

than one of the above techniques. *Anopheles* primarily filter-feed at the water surface by rotating their heads 180° so that the oral opening becomes dorsal. Many *Aedes, Ochlerotatus,* and *Culex,* on the other hand, filter-feed near the surface but also gather, scrape, or shred organic matter at the bottom, depending on food availability. *Coquillettidia* and *Mansonia,* which are anchored to submerged vegetation, employ a combination of filter-feeding, gathering, and scraping techniques within their immediate surroundings. Larval feeding has been reviewed by Merritt *et al.* (1992).

Mosquito *pupae* normally remain motionless at the water surface with the tips of their thoracic air trumpets in contact with the air. Like larvae, they dive when disturbed, propelling themselves with their caudal paddles by extending the abdomen, then snapping it back inward toward the cephalothorax. Pupae of most species are buoyant, due to the ventral airspace beneath the cephalothorax, and rise to the surface without swimming. They remain submerged by repeatedly swimming downward or by wedging or lodging themselves under debris. After sufficient submergence time, they lose their buoyancy as their air supply dwindles, and they must swim actively to the surface. Pupae of a few mosquitoes (e.g., *Limatus* spp.) are never buoyant and can keep from sinking only by clinging to the surface film, much as most mosquito larvae do. The plant-piercing pupae of *Mansonia* and *Coquillettidia* species do not rise to the water surface until they release their attachment to plants when ready for adult emergence.

Upon emergence from the pupal stage, adults typically seek shelter in vegetation, cavities, and other resting sites, where they remain except during periods of activity. When resting, they typically are positioned head-up on vertical surfaces, with forelegs and midlegs on the substrate and hindlegs raised (Fig. 12.18). Culicines and toxorhynchitines hold the abdomen in various positions, but the proboscis is always at an angle to it (Fig. 12.18A).

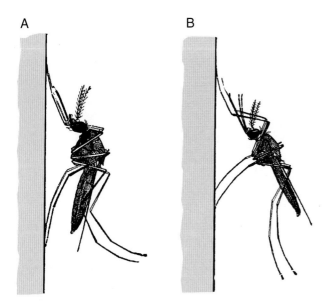

**FIGURE 12.18** Comparison of typical positions of mosquito adults resting on a vertical surface. (A) Culicine; (B) anopheline. Culicines typically hold the abdomen parallel to the substrate or pointed toward it, and the proboscis and abdomen form an angle. Anophelines characteristically tilt the body at a sharp angle to the substrate, with the abdomen pointing away from it, and with the proboscis and abdomen in line. (From Marshall, 1938.)

Anophelines, on the other hand, hold the proboscis and abdomen in line, oblique to the substrate (Fig. 12.18B). This distinctive position also is apparent during feeding. While at rest, adults perform various stereotyped grooming movements and frequently wave their hindlegs.

Each mosquito species has a characteristic pattern of *diel activity*, under the control of an endogenous circadian rhythm that is entrained by the daily light–dark cycle. There are generally one or two *flight periods* each day, characterized as being diurnal, nocturnal, or crepuscular (dawn and dusk). During these periods, both sexes will take flight without external cues. It is likely that mosquitoes have a generalized search pattern during *foraging flights*, responding to stimuli associated with mating sites, sugar sources, hosts, or oviposition sites as they encounter them, depending on their needs. Mosquito species vary in the habitats where they forage for mates and food. Some fly over varied terrain; others tend to be active in either wooded or open areas; still others perform all activities close to larval and resting sites. There is some evidence that adult females become familiar with local habitats that provide both food and oviposition sites and tend to remain there.

*Dispersal* of some species is only a few dozen meters from their larval habitats, for most species less than 2 km. Such ranges are typical of domestic species and result from random elements in their repeated foraging flights for hosts, sugar sources, resting sites, and oviposition

sites. Other species enter a specific dispersal mode that is wind-assisted or light-directed and carries them dozens or even hundreds of kilometers from their origins. These one-way movements are most obvious following massive adult emergences of species with quiescent eggs, such as the salt-marsh mosquito *Ochlerotatus taeniorhynchus* after high tides and floodwater mosquitoes such as *Aedes vexans* in bottomlands after heavy rain. The average dispersal distance of such broods is difficult to determine accurately, because efforts to recapture marked specimens must be made over vast areas, and relatively few are caught. Some salt-marsh species make extended, round-trip migrations to complete their gonotrophic cycles when breeding sites and blood feeding sites are many kilometers apart. Average flight speed also is difficult to determine under natural conditions. *Ae. aegypti* can fly upwind at air speeds up to 5.4 km/hr, while other species in a dispersing mode are estimated to fly much faster. However, their ground speed drops nearly to zero as headwind velocity approaches their maximum air speed. Thus mosquitoes tend to avoid flight under windy conditions, except when a tailwind assists their flight. Mosquito dispersal in its various forms has been reviewed by Service (1997).

Mating usually takes place a few days after adult emergence. The males typically form *flight swarms* at particular times at *swarm markers* (prominent objects or other contrasting features of the environment). Each male follows a looping flight path over the marker. In species such as *Ae. aegypti* and *Ae. albopictus,* the female's preferred host serves as the swarm marker. When a female enters a swarm, males detect the characteristic frequency of her wing beat and her position with their plumose antennae and Johnston's organs. Her tone varies from about 150 to 600 Hz, depending on the temperature and her size and species. It is about 100–250 Hz lower than the males' flight sound. The male turns toward the female, pursues her, and couples with her. Swarms are usually species-specific, but mixed swarms occur. If the female is of another species, males either do not respond to her flight tone or release her upon detecting that she lacks the appropriate species-specific contact pheromones. Otherwise, he may attempt copulation. Successful copulation usually occurs only with the first conspecific male to orient to the female, venter-to-venter, clasping her genitalia in his. The couple drifts or flies from the swarm, often shifting to an end-to-end position, with the female flying forward and the male facing backward, clinging to her only by his genitalia. Mating may be completed in the air or on vegetation. Copulation lasts from 12 sec to several minutes.

There are many exceptions to this standard method of mating. Males of *Deinocerites* and *Opifex* guard pupae at the water surface and mate with the females as they emerge. Males of *Cs. inornata* remain at the emergence site for long periods and locate newly emerged females at random while crawling about, recognizing them by a specific contact pheromone on their legs. A male *Eretmapodites chrysogaster* will follow a female in tandem flight to a host, wait beside her while she takes a blood meal, then copulate when she is finished. Males of the sabethine genera *Limatus, Sabethes, Wyeomyia,* and *Topomyia* locate females at rest on vines, sticks, and tree trunks and perform a variety of leg-waving and genitalic courtship rituals before insemination. Several aspects of mosquito mating behavior have been reviewed by Downes (1969).

During copulation the male deposits a mixture of sperm and accessory gland secretion in the female's seminal bursa or genital chamber. The semen often produces a distinct seminal mass in the bursa or genital chamber, sometimes called a *mating plug.* It disappears in 1–2 days and therefore does not, by itself, prevent subsequent insemination. Within an hour the sperm move into the spermathecae, where they are stored. At least in culicine mosquitoes, the swollen bursa itself and, later, a substance in the accessory fluid called *matrone* cause the female to become unreceptive to males. Substances in the accessory fluid also can affect feeding behavior and promote egg development and oviposition. A single insemination usually is sufficient for the life of a female. Most evidence indicates that females only rarely receive sperm from more than one male.

Adult mosquitoes of both sexes of most species regularly feed on plant sugar throughout life, but only females feed on vertebrate blood. Water presumably is taken from the surface of moist substrates as well as in sugar meals and blood meals. Field studies and experiments indicate that some species are guided in their foraging flights by specific visual features along the horizon, or they fly along the edges of tree stands bordering open terrain. Others apparently simply fly crosswind or downwind, depending on wind speed. When they can see likely sources of sugar or blood, they alter their flight paths to move directly toward the object. If they detect odors of flowers or hosts in the absence of visual information about a food source, they turn to fly upwind within the downwind drifting *odor plume,* eventually arriving at the odor source.

*Sugar feeding* starts soon after emergence, usually before females begin responding to host stimuli. Sugar is taken frequently by both sexes throughout adult life, both between and during gonotrophic cycles. Females of some domestic species take sugar infrequently or never (e.g., *Ae. aegypti* and *An. gambiae* in some localities); they typically live in close association with their hosts and utilize blood for both energy and reproduction. *Sugar sources* include floral and extrafloral nectaries, homopteran honeydew, spoiled or damaged fruit, tree sap, and damaged

or even undamaged plant leaves and stems. *Malaya* species solicit regurgitated nectar and honeydew from ants. Nectar and honeydew are the most important sugar sources. They contain not only sugar but also amino acids; these are insufficient to initiate egg development but probably promote longevity. Most mosquitoes obtain nectar from a variety of plant species; others seem to be fairly specific in their choices, and a few are important pollinators of particular plant species whose flowers they visit. Mosquitoes generally feed from fragrant, light-colored, clustered flowers with short corollas that allow easy access to the nectar. Details of sugar feeding have been reviewed by Foster (1995).

Female mosquitoes rarely begin responding to vertebrate hosts and taking *blood meals* until at least 1–3 days after adult emergence and often not until after mating and sugar feeding. Their *hosts* include all classes of vertebrates: mammals, birds, reptiles, amphibians, and even amphibious fish. They have been reported to take hemolymph from other insects, but perhaps this occurs only when the insects have been contaminated with vertebrate odors. *Host specificity* and host preference vary widely. Some species feed almost entirely on members of one genus of animal; others opportunistically attack members of two or three vertebrate classes. Blood-meal identification methods, used to determine mosquito–host specificity, have been reviewed by Washino and Tempelis (1983). Host specificity is a function of both the mosquito's innate host preference and the hosts available to the mosquito when and where it is active. Some species forage over a broad range of habitats. Others are active principally in either wooded or open areas or remain close to sites of larval development or adult resting. Still other species attack their hosts in rather narrowly defined zones within a habitat. For example, within tropical forests, different species feed in different strata: ground level, intermediate levels, and just below the leaf canopy.

*Host-finding behavior* in mosquitoes involves the use of volatile chemicals to locate vertebrate hosts. Carbon dioxide, lactic acid, and octenol are among the best-documented *host attractants.* Other skin emanations also are known to be important, because odors from live hosts are always more attractive than any combination of these chemicals in a warm, humid airstream. Fatty acids produced by the normal bacterial flora of the skin are particularly effective in attracting *An. gambiae* to human feet. Mixtures of these fatty acids probably play a major role in attracting most mosquitoes. Subtle differences in these odors of different host species and even different individuals undoubtedly play a role in host preference. These odors commonly have a combined effective range of 7–30 m, but the range can be up to 60 m for some species. Vision also is important in orienting to hosts, particularly for diurnal species and especially in an open environment and at intermediate or close ranges. Dark, contrasting, and moving objects are particularly attractive. As the female approaches to within 1–2 m of a potential host, chemical and visual cues are still important, but convective heat and humidity surrounding the body also come into play. Odor, carbon dioxide, heat, and humidity all are detected by sensilla on the antennae and palps. Host-finding behavior in mosquitoes was reviewed by Bowen (1991). Specific behavioral and physiological aspects of attraction have been reviewed by a series of authors in the proceedings of a symposium (Anon. 1994).

If the suite of host stimuli is acceptable, the female attempts to land on the host animal, often preferring certain body parts, such as the head or legs. Upon landing, she proceeds through four phases of *feeding behavior:* exploration, penetration and vessel-seeking, imbibing, and withdrawal. She typically remains motionless for a few seconds, then begins exploratory movements, including contacting the skin surface with her proboscis in probing motions. If the host is not suitable, she may wander for a considerable time and leave without feeding. Even on a suitable animal she usually explores at least briefly before selecting a spot that is likely to be well vascularized. Probing activity is stimulated by heat, moisture, and probably also by chemicals on the surface of the skin. As in the case of airborne attraction, these stimuli are detected by antennal and palpal receptors, but receptors on the proboscis, tarsi, and elsewhere on the legs apparently also are important.

Mosquitoes can feed from a variety of skin surfaces, including the moist skin of frogs and the scaly legs of reptiles and birds. They can penetrate mucus, matted hair, light layers of feathers, and heavy cloth such as denim, provided it is not thicker than the length of the proboscis. Once a feeding site is selected, the fascicle of stylets pierces the skin while the labium serves as its guide and is bent backwards without penetrating (Fig. 12.19). The maxillae and mandibles on each side of the fascicle alternately slide by each other in quick stabbing/puncturing movements. While they do this, the tissue is gripped with the backward-directed maxillary teeth as the stylets penetrate epidermal and subepidermal tissue.

*Saliva* flows from the tip of the hypopharynx as the flexible end of the fascicle bends at sharp angles, probing in various directions within the subepidermal tissue in search of a small arteriole or venule. The saliva contains an antihemostatic enzyme, *apyrase,* which inhibits platelet aggregation and causes randomly punctured vessels to bleed freely into the surrounding tissue spaces. This makes it easier for the mosquito to locate a vessel and shortens the total time on the host. The saliva also contains *anticoagulants,* which facilitate vessel location and blood ingestion by preventing the blood from clotting. Sensilla on the labrum and in the cibarium apparently

**FIGURE 12.19**  Female mosquito (*Aedes aegypti*), showing position of mouthparts during blood-feeding. The fascicle is exsheathed from the labium along part of its length while the labium buckles backward, allowing the fascicle to pierce the skin in search of a blood vessel. (Photo by W. A. Foster.)

detect plasma and cellular factors, including adenyl nucleotides, such as adenosine triphosphate, which help the mosquito to locate a blood vessel and stimulate ingestion. Upon finding a vessel, the female slips the tip of her fascicle into the lumen and draws blood up through the food canal by pumps in the cibarium and pharynx. The blood accumulates in the midgut, allowing the mosquito to engorge fully in 1–4 min. During this time, the female begins to extract water from the blood meal and may deposit small droplets of urine on the host's skin. In some *Anopheles* species, copious fluid excretion begins early, including some blood cells from the accumulating meal, and it appears red. This is due to removal of liquid from the meal directly into the hindgut, a process known as *prediuresis,* rather than by way of the hemolymph and malpighian tubules. When abdominal stretch receptors signal the presence of sufficient blood in the midgut, the female pushes with her forelegs to withdraw her stylets and flies away. Usually she is too heavy to fly far until a substantial amount of water and salt in the blood meal has been excreted, after 1–2 hr.

While digesting the meal and developing eggs, females locate species-characteristic *resting sites* and may remain there until the eggs are mature. However, females of many species are known to leave their resting sites during each daily activity period throughout the gonotrophic cycle. These flights allow them to obtain sugar meals or supplementary blood meals, to relocate closer to an oviposition site, or perhaps simply to find a more suitable resting site. In at least some species, a hormone from the ovaries in the trophic phase inhibits the mosquito's responsiveness to host attractants by blocking host-odor receptors on the antennae, provided she has substantial energy reserves (see Klowden, 1996 for a brief review).

*Oviposition* generally occurs during the same part of the day as mating and feeding. Gravid females locate and evaluate suitable sites by using chemical and visual cues, including organic chemicals, salts, high humidity, dark cavities, and reflective surfaces. The organic chemical cues are derived from decaying organic matter, microorganisms, the chemical byproducts of larvae or pupae that have previously developed there, and the presence of mosquito eggs that have been deposited by other females. The apical droplets on the eggs of *Cx. quinquefasciatus* contain an oviposition-attractant pheromone.

Within each genus of mosquito, there is considerable variation among species in their *oviposition-site preferences* and, therefore, their larval habitats. In general, *Anopheles* species occur in permanent or semipermanent water, such as the edges of lakes, ponds, streams, and pools; others develop in temporary rain puddles, leaf axils, and tree holes. *Culex* typically lay eggs in permanent or semipermanent pools, ponds, and water containers. Several medically important species of both *Anopheles* and *Culex* develop in large bodies of surface water and take advantage of irrigated fields and of reservoirs created by dams. *Culiseta* are found in several kinds of permanent surface pools; some species have very narrow requirements. *Coquillettidia* and *Mansonia* oviposit in permanent bodies of water that contain floating or emergent aquatic plants to which the immatures can attach. *Aedes, Ochlerotatus, Psorophora,* and *Haemagogus* species lay their eggs on damp surfaces where they will be inundated by temporary water or a rising water level. Aedine species and ecologically similar mosquitoes form two general categories, according to typical habitat: (1) *floodwater mosquitoes,* which include floodplain species, salt-marsh species, and snowpool and spring species; some floodplain species have become prolific in rice fields and other forms of irrigation; and (2) *container mosquitoes,* including leaf-axil species, tree-hole species, and artificial-container species. Several medically important species utilize both tree holes and artificial containers. Among genera of minor importance, *Toxorhynchites* lay their eggs only in natural and artificial containers in wooded areas; *Wyeomyia* oviposits primarily in leaf axils; and *Deinocerites* oviposits exclusively in crab holes in tidal mudflats and mangrove swamps.

The distribution of mosquito eggs reflects the availability, size, and stability of the larval habitats used by a species. Though mosquito life histories are highly variable, the oviposition behavior of mosquitoes tends to follow along taxonomic lines. Most mosquitoes fall into one of three behavioral categories: (1) *Eggs laid out of water.* Species in this group may distribute eggs of a single clutch broadly among several potential development sites, particularly if those sites are common but small. Container species such as *Ae. aegypti, Ae. albopictus,* and *Haemagogus* species deposit the eggs at varying distances above the water line, and at least *Ae. aegypti* lays only a portion of the clutch in each water container. Floodwater species such as *Ae. vexans, Ochlerotatus dorsalis,* and the salt-marsh–inhabiting *Oc. taeniorhychus* generally scatter their eggs widely over areas where water will accumulate, inserting them into crevices of drying mud or plant debris in low ground. (2) *Eggs placed on or in water.* Mosquitoes in this category lay the entire clutch in a clump at one site while standing on the water surface or on floating vegetation. The egg rafts of *Culex, Culiseta, Coquillettidia,* and *Uranotaenia* species are formed between the female's hindlegs as she deposits each egg on end in the water, one against the next. Some *Armigeres* species suspend the egg raft above the water with their hindlegs while forming it and then carry it with them before placing it on the water. *Trichoprosopon digitatum* females stand guard over the raft until the eggs hatch. *Mansonia* species prepare their egg clusters underwater, attached to a plant, while standing on the floating leaves of aquatic plants. Exceptional species in various genera deposit egg rafts on top of floating vegetation, lay their eggs singly underwater on the sides of rock pools, or enter beetle holes in bamboo and extrude the eggs in ribbons. (3) *Eggs dropped onto water.* Species in this group oviposit aerially while hovering. *Anopheles* drop all of them at one site or distribute them among several smaller sites. *Toxorhynchites* and most culicine species in the tribe Sabethini (e.g., *Wyeomyia* and *Sabethes*) propel a few eggs into each of many container habitats with a flick of the abdomen, often through very small openings in tree limbs or bamboo. If a mosquito cannot find suitable oviposition sites when the eggs are mature, it may lay them in suboptimal situations or retain them until a suitable site is found. Retained eggs gradually lose their viability over several weeks or months. An extensive review of oviposition behavior is given by Bentley and Day (1989).

Mosquito *dormancy* occurs in all but those tropical and subtropical habitats that provide conditions for year-round larval development. The life stage that becomes dormant depends on the severity of a region's winter or dry season and also on the species. Species of *Aedes, Ochlerotatus, Psorophora,* and *Haemagogus,* which all have quiescent eggs, typically overwinter (*hibernate*) or oversummer (*aestivate*) as eggs. Larvae serve as the dormant stage in mosquitoes whose adult activity is precluded seasonally but whose breeding sites are protected from severe cold or complete drying. When adults overwinter as the dormant stage, typically in *Anopheles* and *Culex,* they seclude themselves in well-protected harborages, or *hibernacula.* Prior to dormancy, mosquitoes often enter *diapause,* a physiological state of arrested development that is induced or broken only by specific environmental cues. *Facultative egg diapause,* a feature of multivoltine species, is induced by exposure of the pupae or adult females to lowered temperatures and short photoperiods. They lay diapausing eggs, which will not hatch until the day length is appropriate, even during unseasonably warm periods in autumn, winter, or early spring, when the resulting larvae and adults might not survive. *Obligate egg diapause* occurs in univoltine species, regardless of preceding conditions, and is maintained despite warm, long-day conditions. This is typical of snowpool and spring *Ochlerotatus* species in cold and temperate climates. Diapause is broken after the eggs have been subjected to winter conditions (in the case of obligate diapause) and when favorable temperatures and long days resume. *Larval diapause* is similar to facultative egg diapause in its induction and termination. Diapausing larvae feed and grow little or not at all, and they do not molt.

Temperate species destined to overwinter as adult females emerge in a state of *reproductive diapause* induced by larval and pupal exposure to shortening photoperiod and cool temperatures. Although these females mate, their egg follicles do not reach the resting stage, despite frequent sugar feeding and accumulation of extensive fat reserves. Fattened female *Culex* species that hibernate through hard winters forego all further feeding until the onset of spring, whereas in milder climates they periodically leave their overwintering sites to take sugar meals. Although diapausing *Culex* adults rarely feed on blood, some *Anopheles* species may take blood meals fairly regularly from hosts near these sites. They develop no eggs, however, exhibiting *gonotrophic dissociation.* Other overwintering *Anopheles* continue to feed and develop eggs, but these are not laid. Similarly, some tropical *Anopheles* take blood repeatedly during the dry season and remain continually gravid because there is nowhere to oviposit. These phenomena are sometimes referred to as *gonotrophic discordance,* a term that also applies to the taking of nonvitellogenic or otherwise supplementary blood meals, mentioned previously.

## PUBLIC HEALTH IMPORTANCE

Mosquitoes are of public health significance because they feed on human blood. Blood feeding compromises skin, presenting the possibility of secondary infection with bacteria. It introduces foreign proteins with saliva that stimulate histamine reactions, causing localized irritation, and that may be antigenic, leading to hypersensitivity, and it allows for acquisition and transmission of microorganisms that cause infection and disease in humans, domestic animals, and wild animals. Mosquito-borne diseases are caused by three groups of pathogens: viruses, malaria protozoans, and filarial nematodes. Mosquitoes are not known to transmit pathogenic bacteria to humans, with the exception of mechanical transmission of the causative agents of tularemia (*Francisella tularensis*) and anthrax (*Bacillus anthracis*).

## MOSQUITO BITES

In addition to the tremendous impact of mosquitoes on human health as vectors of disease pathogens, the bites themselves are important. Aside from the annoying flight and buzzing sound, a single bite can be irritating and a distracting nuisance. In Rangoon, Myanmar, *Cx. quinquefasciatus* has been estimated to have densities of 15 million per square kilometer, and residents in poor districts receive 80,000 bites by this species per year. In Burkina Faso, in West Africa, residents of cities are estimated to experience 25,000 bites by this species per year. In northern Canada, the spring melt of snow brings with it hordes of snowpool *Ochlerotatus;* counts on an exposed human forearm can be as high as 280–300 bites per minute. It has been estimated that this rate of biting could reduce the total blood volume in a human body by half in 90 min, unless protective measures are taken.

As with the other blood-feeding arthropods, the wound created at the bite site may allow secondary infection by bacteria, which can be exacerbated by scratching. In the absence of prior exposure to mosquitoes, a bite rarely produces more than a temporary tingling or burning sensation and sometimes a tiny spot of blood on or just beneath the surface of the skin. After one or more previous exposures to mosquito bites, the proteins in mosquito saliva, which are injected both before and during feeding, normally stimulate development of immunity so that subsequent bites give rise to one or both of two general kinds of allergic response: *immediate reactions* and *delayed reactions.* The immediate reaction, called *type I hypersensitivity,* is an inflammation of the skin known as *wheal and flare.* It usually starts within minutes of the bite and lasts a few hours at most. The typical delayed reaction, designated *type IV hypersensitivity,* involves a *cellular immune response* caused by lymphokines that are secreted by antigen-sensitized T cells. Both delayed and immediate reactions result in itching, redness, and swelling. The typical delayed reaction takes about 1 day to develop, may last for up to a

TABLE II

Selected Mosquito-Borne Viruses of Importance to Humans or Domestic Animals

| Family (genus) | Virus | Distribution |
|---|---|---|
| Togaviridae (*Alphavirus*) | Eastern equine encephalomyelitis | Americas |
| | Venezuelan equine encephalomyelitis | Central America, South America (northern), Mexico, United States (Florida) |
| | Western equine encephalomyelitis | North America, Mexico, South America (eastern) |
| | Chikungunya | Asia, Africa, Philippines |
| | O'nyong-nyong | Africa |
| | Ross River | Australia, New Guinea, Fiji, American Samoa |
| | Semliki Forest | Africa, Asia, Philippines |
| | Mayaro | South America (northern), Trinidad |
| Flaviviridae (*Flavivirus*) | Dengue (four serotypes) | Tropics, especially southern Asia and Caribbean |
| | Yellow fever | Africa, Central and South America |
| | St. Louis encephalitis | Americas |
| | Murray Valley encephalitis | Australia, New Guinea |
| | Japanese encephalitis | Asia (eastern), Philippines |
| | West Nile | Africa, Europe, Israel, Asia |
| | Ilheus | Central and South America |
| | Rocio | Brazil |
| | Wesselsbron | Africa, Asia (southern) |
| Bunyaviridae (*Bunyavirus*) | Bunyamwera group | |
| | Bunyamwera | Africa |
| | Germiston | Africa |
| | Ilesha | Africa |
| | Wyeomyia | Central America |
| | Group C | |
| | Apeu | South America |
| | Itaqui | South America |
| | Marituba | South America |
| | Murutucu | South America |
| | Oriboca | South America |
| | Restan | South America |
| | Madrid | Central America |
| | Nepuyo | Central and South America |
| | Ossa | Central America |
| | California group | |
| | California encephalitis | United States (western) |
| | Jamestown Canyon | North America |
| | La Crosse encephalitis | United States (eastern) |
| | Inkoo | Finland |
| | Tahyna | Europe |
| | Guaroa | South America |
| Bunyaviridae (*Phlebovirus*) | Rift Valley fever | Africa (northern, eastern) |

week, and tends to result in a larger wheal with a deeper discoloration.

## MOSQUITO-BORNE VIRUSES

Among the more than 520 viruses associated with arthropods and registered in the *International Catalogue of Arthropod-Borne Viruses* (Karabatsos, 1985), somewhat less than half have biologic relationships with mosquitoes, and about 100 infect humans. The term *arbovirus* is a contraction of "arthropod-borne virus" and has no strict taxonomic meaning. The most significant mosquito-borne viruses causing human illness belong to four genera in three families (Table II): the *Togaviridae,* genus *Alphavirus;* the *Flaviviridae,* genus *Flavivirus;* and the *Bunyaviridae,* genera *Bunyavirus* and *Phlebovirus.* These taxa have replaced categories in the older arbovirus literature, i.e., group A viruses for the alphaviruses, group B for the flaviviruses, and Bunyamwera supergroup for the bunyaviruses. Some of the arboviruses infect both humans and domestic animals and cause disease in both. A

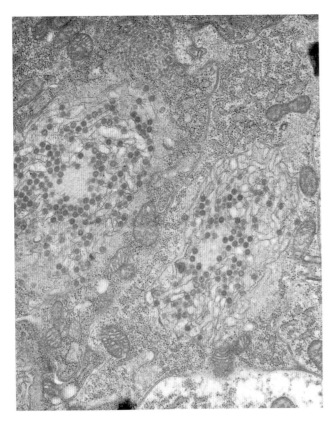

**FIGURE 12.20**    Rift Valley fever virus virions in salivary gland tissues of *Culex pipiens* female, 42,000×. (Courtesy of K. Lerthudsnee and W. S. Romoser.)

representative arbovirus, Rift Valley fever (RVF) virus, is shown in Fig. 12.20.

Aside from their genomic organization and morphology, mosquito-borne viruses may be viewed in terms of the kind of disease symptoms they cause. In humans, generally speaking, the mosquito-borne viruses cause infection with either no apparent symptoms or acute disease of systemic febrile illness (fever), encephalomyelitis (inflammation of the brain and spinal cord), hemorrhagic fever (bleeding and fever), or febrile myalgia and arthralgia (fever with muscle and joint pain, or arthritis). The case mortality (fatality) rate tends to be low for fevers, although morbidity (illness) may be high. For the encephalitis and hemorrhagic fevers, morbidity and mortality may range from low to high, depending upon the virus and factors such as age. After the acute phase, humans either recover fully or show various sequelae, such as neurological problems after acute encephalitis. Long-term, chronic infection with the mosquito-borne viruses does not occur in humans, although the consequences of infection may be long-lasting.

Table II is a list of the more important mosquito-borne arboviruses and their geographic distribution. The viruses are generally classified hierarchically as follows: *serogroup, complex, type, subtype,* and *variety*. The word

*strain* is used to refer to different viruses of the same variety that were isolated from different locations or different biologic sources, or that show minor differences in antigenicity or genotype, but not enough to justify elevating them to the varietal level. For example, Altamont virus is an eastern New York State *strain* of La Crosse virus, which along with Snowshoe hare virus is a *variety* of virus called La Crosse, which is a *subtype* of California encephalitis, the *type* virus of the California encephalitis virus *complex* within the California *serogroup* in the genus *Bunyavirus,* family Bunyaviridae. The relationships among the viruses are determined on the basis of similarities and differences in antigenic reactions to antibodies in immunological tests, and on genetic relationships determined by molecular analyses (i.e., nucleotide sequences or oligonucleotide fingerprint patterns). Arbovirus classification is presented by Calisher and Karabatsos (1988). Karabatsos (1985) lists their histories and basic antigenic properties.

Mosquito-borne viruses multiply in both invertebrate and vertebrate cells. Many arboviruses cause cytopathic effects and cell destruction in vertebrate cells; in invertebrate cells the same viruses typically cause a chronic cellular infection without cytopathology. Competent mosquitoes become infected when they feed on blood of a viremic vertebrate host in which there is sufficient circulating virus in its blood to provide an infectious dose to the mosquito. After blood has entered the midgut, the virions bind to and then pass through the microvillar membrane and into midgut epithelial cells. Within these cells, viruses replicate, and virions bud off from the cells, pass through the basal lamina, and enter the hemolymph. Virions disseminate throughout the body of the mosquito and may infect and replicate in a variety of tissues, including salivary glands, fat body, ovaries, and nerves. A mosquito with a salivary-gland infection (Fig. 12.20) may transmit infectious virions during salivation as it probes the tissues of another vertebrate host. In some mosquitoes and for some viruses, *transovarial transmission* of virions occurs from the female mosquito to her progeny, and females of the next generation can transmit the virus orally without having become infected by a prior blood meal. Also, *venereal transmission* of some arboviruses from male to female mosquito has been documented experimentally. The rate of virus infection and dissemination in mosquitoes is temperature-dependent; higher temperature results in shorter extrinsic incubation. Virus infection may harm mosquitoes in some cases, and they become infected for life.

Because the flaviviruses, alphaviruses, and bunyaviruses have RNA genomes and can replicate in both invertebrate and vertebrate hosts, these viruses have a high capacity for rapid evolution into antigenically variable strains of varying virulence. This capacity for change is important, because it may result in emergence of highly virulent,

TABLE III

**Relationships of Selected Alphaviruses to Mosquito Vectors and Vertebrate Reservoir Hosts**

| Virus | Vector(s) | Vertebrate reservoirs |
|---|---|---|
| Eastern equine encephalomyelitis | *Culiseta melanura, Ochlerotatus sollicitans, Coquillettidia perturbans, Culex nigripalpus, Culex (Melaniconion)* spp. | Birds |
| Western equine encephalomyelitis | *Cx. tarsalis, Culex (Melaniconion)* spp., *Ae. albifasciatus, Ochlerotatus melanimon, Ochlerotatus dorsalis* | Birds, lagomorphs |
| Venezuelan equine encephalomyelitis[a] | *Culex (Melanoconion)* spp.; *Aedes, Ochlerotatus, Psorophora, Anopheles, Mansonia* spp. | Rodents |
| Chikungunya | *Ae. aegypti, Ae. africanus* | Primates, including humans |
| O'nyong-nyong | *Anopheles* spp. | Humans |
| Ross River | *Cx. annulirostris, Ochlerotatus vigilax, Ae. polynesiensis* | Humans, rodents |

[a] See Table IV for elaboration of this virus complex.

epidemic strains; indeed, there is evidence for this process in recent history (see O'Nyong-Nyong virus and Rocio virus, below). Whether rapid evolution of arboviruses leads inevitably to coevolution of viruses and mosquito hosts is debatable, because rapid evolution implies capacity to shift to new hosts, whereas coevolution implies adaptation to specific hosts through reciprocal selection and perhaps cospeciation.

The literature on the mosquito-borne viruses is voluminous. Extensive reviews are presented in the multivolume series edited by Monath (1988a), in Strickland (1991), in the sections on Togaviridae, Flaviviridae, and Bunyaviridae by Fields *et al.* (1996), and in Reeves (1990), as well as in references mentioned under specific groups below and at the end of this chapter.

## Togaviridae (*Alphavirus*)

The Togaviridae, genus *Alphavirus*, contains seven antigenic complexes of viruses involving 37 types, subtypes, and varieties distributed worldwide, 35 of which have been isolated from mosquitoes. Many of these are medically important. Table III summarizes relationships between primary mosquito vectors and important alphaviruses, whereas Fig. 12.21 shows their distribution in the Western Hemisphere.

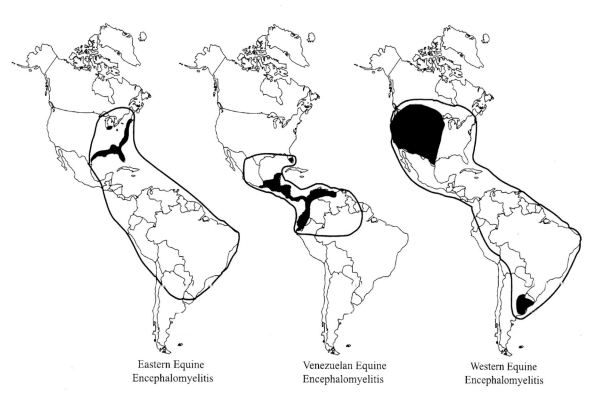

Eastern Equine Encephalomyelitis     Venezuelan Equine Encephalomyelitis     Western Equine Encephalomyelitis

**FIGURE 12.21** Epidemic zones of mosquito-borne alphaviruses that cause encephalitis in the New World. (Reconstructed from Mitchell, 1977, and other sources.)

**Eastern equine encephalomyelitis (EEE) virus**   This virus is distributed in South America, Central America, the Caribbean basin, and eastern North America. Analyses of geographic strains of EEE virus have revealed two varieties: South American and North American/Caribbean. The virus has been isolated from many different states in the United States, but most cases of human or equine disease are in coastal states from Massachusetts to Louisiana, an area of upstate New York near Syracuse, a swamp focus in east central Ohio, southern Michigan, and part of northern Indiana.

EEE virus is one of the most pathogenic among all of the mosquito-borne encephalitis viruses. In humans, disease caused by EEE virus infection results in high morbidity and mortality. The type and severity of illness in humans depends upon the age and health status of the individuals. Children, the elderly, immunocompromised individuals, and sometimes apparently healthy adults develop acute encephalitis with high fever, drowsiness, lethargy, vomiting, convulsions, and coma. Mortality rates among clinical cases exceed 50%. Individuals who survive infection often show neurologic sequelae, although some survivors recover completely, sometimes showing rapid and dramatic improvement from coma.

In the eastern United States, EEE virus occurs in a bird–mosquito enzootic cycle in swamps that support the biology of the enzootic vector, *Culiseta melanura*. The swamps comprising EEE foci are characterized in the northern distribution of its range by northern white cedar, black spruce, tamarack, or red maple trees, typical of swamp or bog ecosystems. Larvae of *Cs. melanura* occur in water-filled cavities underneath raised tree hummocks, water-filled depressions formed by uprooted trees, and in holes in bog mats. Adults feed on birds in the swamp and may leave swamps for open areas to locate hosts, returning later to oviposit. *Cs. melanura* females are highly efficient vectors of EEE virus and transmit it primarily to swamp-dwelling passeriform birds. More than 48 species of wild, native birds have shown evidence of infection. The mechanism of overwintering of EEE virus is unknown. Recent studies in New Jersey indicate that the virus may recrudesce in resident birds in the spring, whereas other studies incriminate reptiles as overwintering hosts. Still other scenarios suggest that this virus is introduced by migrating birds from southern regions where viral transmission may occur year-round.

In some summers, for reasons possibly related to weather patterns and density of bird and mosquito hosts, EEE virus in its enzootic swamp setting becomes amplified to high levels so that epizootics and epidemics develop. Certain mosquito species function as bridge vectors, especially *Ochlerotatus sollicitans* in coastal areas and *Coquillettidia perturbans* in inland areas. They acquire virus infection by feeding on viremic birds, later blood-feeding on mammals and transmitting the virus

to them. In Central and South America, EEE virus apparently circulates among rodents and birds through mosquitoes in the *Culex* (*Melanoconion*) group; however, the relationships among vectors and vertebrate hosts involved in enzootic, epizootic, and epidemic cycles of EEE virus in these regions are poorly known compared to North America.

Epidemics and concurrent epizootics have been documented in the eastern United States since the 1930s, generally involving cases in horses, pheasants, and humans. The virus was first isolated from brains of horses that died during a 1933 epizootic along the eastern coast of the United States, in Virginia, Delaware, and Maryland. Disease in humans was first recognized in 1938 in Massachusetts. Generally, outbreaks involve many horse cases and very few human cases. Nearly all outbreaks involving human cases have occurred along the eastern coast of the United States. From 1964 to 1995, a total of 151 human cases of EEE infection were reported in the United States, with an average of about 5 cases (range, usually 0 to 14) per year. The greatest number of human cases occurred in 1959, when 36 were documented, mainly in New Jersey. Morris (1988) reviewed the ecology, epidemiology, and vector relationships of EEE.

**Western equine encephalomyelitis (WEE) virus**   This virus is a complex of six types and six subtypes. All are mosquito-borne, with the exception of *Fort Morgan virus*, which is associated with cliff swallows and their parasitic cimicid bugs (*Oeciacus vicarius*) in western North America. The other viruses are widely distributed across the high plains of the United States and Canada, in California, and through Central and South America. The WEE virus type that occurs in North America has been responsible for acute encephalitis in horses and humans. An apparently less pathogenic virus, *Highlands J virus*, is closely related to Fort Morgan virus, yet it exists in the same basic North American enzootic cycle as EEE virus.

Disease caused by WEE virus in humans is less severe than that caused by EEE virus, and some infections are inapparent. Symptoms are generally similar, with acute onset of meningitis or encephalitis with headache, fever, and drowsiness and with coma and death in severe cases. Fatality rates are in the range of 5% or less. Most morbidity and mortality occur in infants rather than teens, adults, or the elderly. Neurologic sequelae are often evident in survivors.

In North America, WEE virus apparently exists enzootically in at least two different cycles. The primary cycle involves passerine and columbiform birds (especially house sparrows, house finches, blackbirds, orioles, and mourning doves), with *Cx. tarsalis* functioning as the enzootic, epizootic, and epidemic vector. Nestling birds, which are more exposed to mosquito bites, may be more important as virus amplifier hosts than adult birds. The other cycle involves jackrabbits as vertebrate hosts and *Ochlerotatus melanimon* or *Oc. dorsalis* as vectors in California, Utah, and Colorado. Ground and tree squirrels also may function as vertebrate hosts in some areas, and in the central United States *Oc. trivittatus* may be a secondary vector. In Central and South America, WEE virus apparently circulates in nature among *Culex* (*Melanoconion*) mosquitoes, although a cycle involving *Ochlerotatus albifasciatus* and lagomorphs has been elucidated in Argentina.

Epizootics and epidemics of WEE in North America appear to be related to cumulative summertime precipitation and wintertime snowpack development, both of which can increase populations of *Cx. tarsalis* and favor virus transmission. The larvae of this mosquito are often associated with agricultural irrigation and vernal flooding. Areas that normally are dry can produce large populations of *Cx. tarsalis* if irrigation activities result in the accumulation of pools of still water long enough to support larval development. For example, *Cx. tarsalis* populations burgeon after winter rain and vernal snow melt result in inundation of saltwater marshes along the north shore of Salton Sea in southern California. WEE virus is detected in *Cx. tarsalis* populations about 2 months after these populations increase, during March through June. The virus spreads to upland sites to the northwest of the sea, where mosquitoes are produced primarily from poorly maintained irrigation systems. The movement of WEE virus along this corridor is probably due to dispersal of infected *Cx. tarsalis* rather than to movement of birds.

The mechanism by which WEE virus overwinters at this site, or is introduced there, is not known.

WEE virus was first isolated in 1930 from the brain of a dead horse in the San Joaquin Valley of California. It later was identified as the causative agent of human disease in a child who died of encephalitis in the San Joaquin Valley of California in 1938. Since that time, WEE virus has been implicated in epizootics in horses and concurrent epidemics in humans, with cases numbering from hundreds to thousands in some instances. Large outbreaks occurred in 1941 and 1975 in the Red River Valley in Minnesota and North Dakota and in Manitoba (Canada), in 1952 in the Central Valley of California, and in 1965 in Hale County in western Texas. There are horse cases almost every summer within the range of the virus, but epizootics do not occur every summer. From 1964 to 1995, a total of 639 human cases of WEE infection were reported in the United States, with an average of about 20 cases (range, 0–172) per year. Reisen and Monath (1988) provided a review of WEE.

### Venezuelan equine encephalomyelitis (VEE) virus

This is a complex of 12 viruses that cause disease in humans and equids (horses, burros, and mules) and occurs in northern South America, Central America, and Mexico, occasionally extending into Texas (Walton and Grayson, 1988). These viruses exist as either enzootic or epizootic varieties and strains, with overlapping or disjunct geographic distributions and with variable vector and vertebrate–host relationships. Table IV

TABLE IV

Venezuelan Equine Encephalomyelitis Virus Complex: Classification, Mosquito Associations, and Geographic Distribution of Virus Varieties

| Subtype | Variety | Name | Vector associations | Geographic distribution |
|---------|---------|------|---------------------|--------------------------|
| I | A–B[a] | — | *Aedes, Ochlerotatus, Psorophora, Mansonia, Anopheles, Deinocerites pseudes* | Central America, South America (northern) |
| | C[a] | — | (Same as IA–B) | Central America, South America (northern) |
| | D | — | *Culex* (*Mel.*) *ocossa* *Cx.* (*Mel.*) *panocossa* | Central America, South America |
| | E | — | *Cx.* (*Mel.*) *taeniopus* | Central America |
| | F | — | Unknown | Brazil |
| II | | Everglades | *Cx.* (*Mel.*) *cedecei* | United States (southern Florida) |
| III | A | Mucambo | *Cx.* (*Mel.*) *portesi* | South America (northern) |
| | B[b] | Tonate | Unknown | South America (northern) |
| | C | — | Unknown | Peru |
| IV | | Pixuna | Unknown | Brazil |
| V | | Cabassou | Unknown | French Guiana |
| VI | | — | Unknown | Argentina |

[a] Virulent to equids and humans and involved in epizootics. The other subtypes and varieties are enzootic.

[b] A variety related to IIIB, Bijou Bridge virus, is not listed here. It is associated with cliff swallow bugs (*Oeciacus vicarius;* Hemiptera: Cimicidae) in western North America.

shows the classification and vector associations of these viruses.

Many VEE "enzootic" virus subtypes and varieties exist in cycles involving rodents and mosquitoes of the *Culex* (*Melanoconion*) group, such as *Cx. ocossa*, *Cx. panocossa*, and *Cx. taeniopus* in Central and South America. Rodents in the genera *Sigmodon*, *Oryzomys*, *Zygodontomys*, *Heteromys*, *Peromyscus*, and *Proechimys* are important vertebrate hosts; birds, opossums, and bats also may be reservoir hosts. The ecology of the epizootic viruses is quite different. A large number of species of mosquitoes in several different genera (see Table IV) have been implicated as vectors of the epizootic/epidemic virus strains. Equids attain sufficient viremia to infect these mosquitoes. VEE epidemics can be maintained by mosquito-equid-mosquito transmission, unlike WEE and EEE epidemics, in which equids are for the most part dead-end hosts. Wading birds, particularly green herons, have been incriminated as vertebrate hosts of epizootic strain IA-B in Panama, with the crab hole mosquito *De. pseudes* functioning as vector. Persistence of epizootic strains of VEE in interepidemic periods is not well understood; thus their emergence in epidemics among equids and humans is difficult to predict.

The single representative of the VEE viruses in the United States, other than during epizootics in Texas, occurs in the Everglades region of southern Florida. Called *Everglades virus*, it is associated with *Cx.* (*Melanoconion*) *cedecei* (formerly, *Cx. opisthopus*) as vector and cotton rats (*Sigmodon hispidus*) and cotton mice (*Peromyscus gossypinus*) as vertebrate hosts. The zoonotic setting is the hardwood hammocks of the Everglades, where mosquito and rodent habitats overlap. Serosurveys of Seminole and Miccosukee Indians in these regions have shown that many Indians have antibodies to VEE virus, but there have been very few cases of human disease attributable to this virus.

Humans infected with an epizootic or certain enzootic strains of VEE virus may show no symptoms, only mild flulike symptoms, or severe encephalitis with acute onset of vomiting, headache, seizures, and fever. Symptoms tend to be most severe in children. During epidemics, the mortality rate is typically less than 1%, although in some epidemics the mortality rate has been considerably higher.

The VEE viruses in Central America and northern South America have been intensively studied because of the history of epidemics among equids and humans in these regions. Outbreaks of VEE have occurred periodically in South America, Central America, and Mexico since the 1930s. The first VEE virus was isolated in 1938 from a dead horse in Venezuela. In 1969, a large outbreak of VEE involving both equids and humans in Central America spread northward through Mexico in the next 2 years, moving into Texas in 1971. Cases continued in Mexico through 1972. There were thousands of both horse and human cases throughout this region during that time, but epizootic virus activity did not occur again there until an outbreak in Venezuela in 1992–1993 and in Chiapas, Mexico, in 1993. More recently, an outbreak of VEE occurred in northern Colombia in 1995. The rapidity of spread of these outbreaks over large geographic areas undoubtedly is due to the role of horses as competent reservoir hosts.

**Chikungunya (CHIK) virus**    This virus occurs in eastern Africa and parts of India and southeastern Asia (Jupp and McIntosh, 1988). It generally does not cause the encephalitis-type symptoms characteristic of EEE, WEE, and VEE viral infections, but rather causes a DEN-like arthralgic illness of fever, rash, and severe pain in the joints. In Africa, CHIK virus infects nonhuman primates such as vervet monkeys (*Cercopithecus aethiops*) and baboons (*Papio ursinus*), with *Ae. africanus*, *Ae. luteocephalus*, *Ae. opok*, *Ae. furcifer*, *Ae. taylori*, and *Ae. cordellieri* as enzootic vectors in savannah and forest cycles. In humans it produces viremia sufficient to infect *Ae. aegypti*, the vector in urban areas.

*Ae. aegypti* is the vector of CHIK virus in urban India and Asia, where it causes epidemics of arthralgic disease during rainy seasons. A 1994 epidemic in Vellore, southern India, showed that human cases increased during August and September, reaching a peak in October, as the human-biting frequency and viral infection rate of *Ae. aegypti* increased. The epidemic lasted about 5 months and affected 44% of the city's population. This epidemic and others, along with experimental studies, indicate that interrupted feeding resulting in partial blood meals may facilitate both mechanical and biological transmission of CHIK virus, thus rapidly amplifying the virus in human populations.

### Other Alphaviruses

There are other important alphaviruses that occur endemically and epidemically and cause fever, arthralgia, and other symptoms in humans.

**O'nyong-nyong (ONN) virus**    This is an antigenic subtype of CHIK virus that is transmitted among humans by *Anopheles* species in widespread parts of Africa. The vectors are the same ones that transmit human malaria parasites (i.e., *An. gambiae* and *An. funestus*). A large epidemic occurred from 1959 through the 1960s, infecting about 2 million people in Uganda, Kenya, Tanzania, Malawi, Zambia, and Mozambique. The virus, whose name comes from an Acholi African word meaning "weakening of the joints," was first isolated from humans during this time and was later isolated from *An. funestus* in 1974. Only isolated cases occurred from that time until an epidemic in Uganda in 1997. Neither a vertebrate animal host nor the mechanism of persistence of ONN virus between these epidemics is known. The reservoir host is probably humans. A closely related virus, called *Igbo Ora*, also is transmitted by *Anopheles* mosquitoes and occurs in parts of West Africa, where it was associated with an outbreak of CHIK-like disease in the Ivory Coast in 1984. ONN and Igbo Ora are the only arboviruses causing human disease that have *Anopheles* mosquitoes as the primary vectors.

**Sindbis (SIN) virus**  This virus is distributed widely in Eastern Europe, Scandinavia, the former Soviet Union, Asia, Africa, the Middle East, and Australia. The virus is a member of the WEE virus complex. It was originally isolated from *Cx. univittatus* in Egypt in 1952 and has been associated with a human disease of rash, fever, and muscle and joint pain in Uganda, South Africa, and Australia. Birds are vertebrate hosts. A subtype of SIN virus is *Ockelbo*, which is distributed in Sweden, Finland, and northern Russia. Ockelbo is transmitted by *Cs. ochroptera*, and possibly *Culex* and *Aedes* species, and has been associated with human disease similar to that caused by SIN virus.

**Ross River (RR) virus**  This virus occurs in Australia, Fiji, and the Cook Islands, where it causes an illness known as *epidemic polyarthritis*, consisting of fever, rash, and arthralgia (Kay and Aaskov, 1988). In Australia, RR virus is transmitted by *Ochlerotatus camptorhynchus, Oc. vigilax*, and *Cx. annulirostris*, whereas in the islands the vectors are *Ae. aegypti* and *Ae. polynesiensis*. The virus was first isolated from *Oc. vigilax* in 1963. The vertebrate reservoir hosts are unknown, but in Australia they may be marsupials. *Barmah Forest virus* is another mosquito-borne alphavirus in Australia that causes symptoms similar to RR virus.

**Mayaro virus**  This virus occurs in the Caribbean and parts of South America. In humans the illness is similar to CHIK. The virus was first isolated from febrile patients in Trinidad in 1954. Marmosets are the reservoir hosts of the virus, and *Haemagogus* mosquitoes are vectors.

### Flaviviridae (*Flavivirus*)

The Flaviviridae, genus *Flavivirus*, contains eight antigenic complexes plus many unassigned viruses involving 70 types, subtypes, and varieties distributed worldwide. Some of these have mosquitoes as vectors, while others are associated with ticks or with rodents or bats. The flavivirus diseases include some of the most dangerous and historically significant infections of humans. Table V

**TABLE V**

Relationships of Selected Flaviviruses to Mosquito Vectors and Vertebrate Reservoir Hosts

| Virus | Vector(s) | Vertebrate reservoirs |
| --- | --- | --- |
| Yellow fever | *Aedes aegypti* | Humans in urban environments |
| | *Ae. africanus* | Monkeys |
| | *Ae. bromeliae* | Monkeys |
| | *Ae. furcifer* | Monkeys |
| | *Ae. luteocephalus* | Monkeys |
| | *Ae. metallicus* | Monkeys |
| | *Ae. taylori* | Monkeys |
| | *Ae. vittatus* | Monkeys |
| | *Haemagogus* spp. | Monkeys |
| | *Sabethes* spp. | Monkeys |
| Dengue | *Ae. aegypti* | Humans |
| | *Ae. albopictus* | Humans (monkeys?) |
| | *Ochlerotatus niveus* group | Monkeys |
| | *Ae. scutellaris* | Humans |
| | *Ae. polynesiensis* | Humans |
| | *Ae. pseudoscutellaris* | Humans |
| | *Ae. rotumae* | Humans |
| Japanese encephalitis | *Culex tritaeniorhynchus* | Birds, pigs |
| | *Cx. gelidus* | Birds |
| | *Cx. vishnui* complex | Birds |
| St. Louis encephalitis | *Cx. pipiens* | Birds |
| | *Cx. quinquefasciatus* | Birds |
| | *Cx. tarsalis* | Birds |
| | *Cx. nigripalpus* | Birds |
| Murray Valley encephalitis | *Cx. annulirostris* | Birds |
| West Nile | *Culex* spp. | Birds |
| | *Cx. univittatus* | Birds |
| | *Cx. modestus* | Birds |

shows the mosquito–vector and vertebrate–host associations of some of the more important mosquito-borne flaviviruses. Among them are YF, DEN, Japanese encephalitis (JE), SLE, Murray Valley encephalitis (MVE), West Nile (WN), and Ilheus viruses.

### Yellow Fever

This disease is caused by YF virus and occurs over broad portions of lowland equatorial Africa and South and Central America (Fig. 12.22), either as isolated cases or epidemics (Monath, 1988b). YF virus exists principally in two epidemiological forms: an enzootic form, maintained in monkey populations by forest mosquitoes in a *sylvan cycle* and responsible for most of the isolated human cases (*jungle yellow fever*); and an epidemic form, spreading rapidly through human populations by the domestic form of *Aedes aegypti* (Fig. 12.23) in an *urban cycle*. The urban vector is particularly efficient because it readily enters and typically rests in houses, feeds almost exclusively on

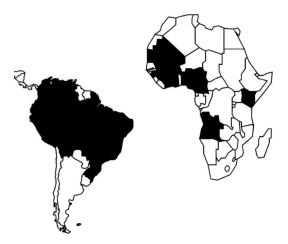

**FIGURE 12.22**   Geographic occurrence of human cases of yellow fever in South America and Africa officially reported to the World Health Organization, 1982–1991. (Reconstructed from Monath and Heinz, 1996, and other sources.)

humans, and oviposits in artificial containers. It frequently takes two or three blood meals per gonotrophic cycle to supplement its energy reserves in lieu of sugar feeding.

In previous centuries, *urban yellow fever* affected subtropical and temperate regions of North America, with devastating effects. It remains a serious cause of mortality, particularly in village settings in tropical Africa, and is a constant threat in South America. There are sporadic, annual cases and occasional epidemics in Africa and 50–300 cases of jungle yellow fever annually in South America. Jungle yellow fever occurred in Brazil (1932), Panama, and northward into Central America (1948–1955), and there were several episodes (including some urban transmission) in Trinidad (1954, 1959, 1978–1979). Epidemics in Africa have involved diverse areas, including Ethiopia (1960–1962), the Gambia (1978), Ghana

(1979–1980, 1983, 1993, 1996), Benin (1996), Burkina Faso (1983), Nigeria (1986, 1994), and Kenya (1992–1993).

Yellow fever is a hemorrhagic disease. Mortality in infected humans ranges from 5 to 75% or more. After infection by mosquito bite, there is an incubation period of 3–6 days, followed by sudden high fever (>104°F or 40°C), headache, nausea, and pain. Humans are viremic during this initial acute phase only for about 3 days. The virus is viscerotropic and causes parenchymal cell necrosis in the liver, resulting in elevated blood bilirubin levels and jaundice, thus the name "yellow fever." Jaundice leads to systemic toxemia. Hemorrhage manifests as bleeding gums, easy bruising, and sloughing of the stomach lining, thus causing a characteristic *black vomit*. Delirium and coma often precede death.

The causative agent of yellow fever is the prototype virus of the Flaviviridae and the first arbovirus to be associated with human disease. Although the disease was first recognized in the New World in the 17th century, the virus was isolated first in 1927 in Ghana, Africa, when the blood of a man with yellow fever was inoculated into rhesus monkeys. The virus apparently originated in Central Africa among monkeys (e.g., *Cercopithecus* spp., family Cercopithecidae) and mosquitoes dwelling in the forest canopy. In this environment, the vector is the monkey-feeding *Ae. africanus,* a relative of *Ae. aegypti.* There also is evidence for infection with chimpanzees (*Pan troglodytes,* family Pongidae), baboons (*Papio,* family Cercopithecidae) and bushbabies (*Galago,* family Lorisidae). Generally, Old World monkeys do not suffer mortality from infection. In western Africa, where YF virus also is enzootic, the likely sylvatic vectors are *Ae. vittatus, Ae. metallicus, Ae. furcifer, Ae. taylori,* and *Ae. luteocephalus,* which develop in tree holes and sometimes other natural containers and rock pools.

Human disease occurs when humans enter the habitat where the mosquito–monkey cycle is ongoing and are bitten by infected mosquitoes. Alternatively, infected monkeys or baboons sometimes enter the human habitat on raids in gardens or banana plantations, bringing YF virus to mosquitoes at the interface between forest and human habitation. The African mosquito *Ae. bromeliae* (formerly *Ae. simpsoni*), the larvae of which develop in water-filled leaf axils of plants such as banana, occupies this interface and frequently has become involved in forest-edge and rural transmission of YF virus to humans. *Ae. bromeliae* sometimes has served as the primary vector in massive rural epidemics. Thus, it functions both as a bridge vector between sylvatic and rural cycles and as an interhuman vector in that peridomestic environment. It was the principal vector during the 1960–1962 epidemic in Ethiopia, in which about 100,000 people became infected and 30,000 died.

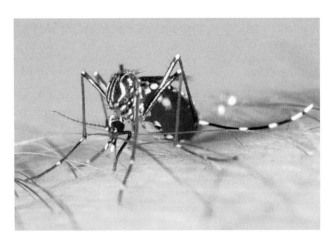

**FIGURE 12.23**   *Aedes aegypti* female feeding on blood. This is the primary vector of dengue and urban yellow fever. (Photo by R. G. Hancock.)

The spread of both the domestic form of *Ae. aegypti* and YF virus from Africa to other parts of the world apparently occurred within the last 400 years. Trading and slaving ships, with their potentially virus-infected cargoes of slaves, and with water barrels as a mosquito development site, greatly facilitated spread and establishment of both vector and virus. The need for slave labor in the newly established sugar cane/molasses/rum economy of the Caribbean region undoubtedly promoted movement of YF virus about the New World. Oftentimes, crews became ill with YF while their ships were in transit. Epidemics regularly occurred in port cities in both West Africa and in coastal South America, North America, and the Caribbean. *Ae. aegypti* also moved into the Arabian peninsula and Indian subcontinent, and then to Asia and the Pacific region, probably via dhow traffic along sea trade routes. YF has not become established in Asia, even though vector-competent mosquitoes and humans occur there.

Yellow fever epidemics have occurred in the New World over a period of three centuries, beginning in the mid 1600s. Epidemics occurred in such places as Barbados and Trinidad (Caribbean), Havana (Cuba), Yucatan (Mexico), Guadeloupe (Caribbean), and Guayaquil (Ecuador); and in Charleston (South Carolina), Mobile (Alabama), Pensacola (Florida), and other areas of the United States as far north as Boston and New York. In the 1700s and 1800s, epidemics continued in ports of tropical and temperate America, including an outbreak in Philadelphia (Pennsylvania) in 1793, where some 4,000 deaths occurred in a population of 55,000. A large outbreak in Haiti in 1802 decimated the French military force there, causing Napoleon to abandon his New World ambitions and contributing to the Louisiana Purchase by the United States government. New Orleans (Louisiana) had regular epidemics of yellow fever from 1796 through 1905. The shipping blockade enforced by the navy on the federalist side of the United States' civil war (1861–1865) prevented yellow fever in New Orleans during that period. Probably, the blockade stopped importation of infected ship crews. An epidemic in the Mississippi River valley in 1878, extending north as far as Gallipolis, Ohio, caused over 13,000 deaths. The last epidemic in the United States was in New Orleans in 1905, involving 3,402 cases and 452 deaths.

As YF virus invaded the New World, it became established in a mosquito–monkey cycle in forested parts of Central and South America. In the sylvatic cycle there, the vectors are forest canopy–dwelling mosquitoes, particularly *Haemagogus* species and *Sabethes chloropterus,* whose larvae develop in tree holes. New World monkeys in the family Cebidae are highly susceptible to infection. During epizootics, howler monkeys (*Alouatta* spp.), squirrel monkeys (*Saimiri* spp.), spider monkeys (*Ateles* spp.),

and owl monkeys (*Aotus* spp.) may show considerable mortality in their populations. However, capuchin monkeys (*Cebus* spp.) and woolly monkeys (*Lagothrix* spp.) circulate mosquito-infective viremias, but they do not die from infection. Therefore, in this cycle noticeable die-offs of monkeys may or may not precede sylvatic transmission of YF virus to humans. Sylvan yellow fever in the Americas can lead to urban outbreaks when humans enter the tropical forests where transmission is ongoing, become infected by mosquito bite, and then return to villages or cities. If *Ae. aegypti* is present in these settlements, it can initiate an epidemic of urban yellow fever.

The history of the discovery that YF virus is transmitted by mosquitoes is intriguing. The discovery is particularly important because YF was the first arbovirus to be recognized as a mosquito-borne agent, and mosquito control measures imposed quickly afterward caused a dramatic reduction in this devastating disease. Although suspicion that the agent causing yellow fever might be transmitted by mosquitoes can be traced to several independent sources in the 1800s, it was the intuition of the Cuban physician Carlos Finlay, followed by the research activities of the United States Yellow Fever Commission in Havana, Cuba, in 1900, that resulted in experimental evidence that *Ae. aegypti* was the vector. This commission was composed of the US Army officers Walter Reed, James Carroll, Jesse Lazear, Aristides Agramonte, and others. After consulting with Finlay, this team carried out a series of experiments that demonstrated the transmissibility of the agent from infected to uninfected humans by mosquito bite, after a suitable incubation period in mosquitoes. During this work, James Carroll allowed himself to be bitten by an infected mosquito and later developed yellow fever, but he recovered. Jesse Lazear was accidentally bitten, and he died of the disease.

The team's findings stimulated William Gorgas, a physician in the US Army, to impose a control program against *Ae. aegypti*, resulting in the elimination of urban yellow fever in Havana and, soon after, in the Panama Canal Zone. Later, the Rockefeller Foundation sponsored teams of biomedical scientists and public health practitioners to begin intensive studies of yellow fever by establishing a research center in Guayaquil, Ecuador, in 1918. This foundation eventually established research institutes in Brazil, Nigeria, Uganda, the United States, and elsewhere, and supported research and disease control programs in many sites. For a time, the causative agent of yellow fever was thought to be a spirochete. Within 10 years of isolation of the virus in 1927, an attenuated, live vaccine was produced which was shown to provide excellent protection against infection. Max Theiler was awarded the Nobel Prize in Medicine for his efforts in this regard. In many areas of South America, antimosquito programs resulted in virtual elimination of urban outbreaks, although sylvan transmission continued. Despite these early advances, cases of jungle and rural yellow fever, and epidemics in Africa, continue to occur. The reintroduction and resurgence of *Ae. aegypti* populations throughout Central and South America and the Caribbean islands in recent decades, and the immense growth of cities that provide a habitat for it, have created an increased potential for urban epidemics.

## Dengue

This disease is caused by DEN virus, represented by four closely related serotypes called Dengue 1, 2, 3, and 4 (Gubler, 1988; Gubler and Kuno, 1997). The disease in humans is either classic dengue fever or the more severe dengue hemorrhagic fever or dengue shock syndrome. The DEN viruses are transmitted by mosquitoes, principally *Ae. aegypti*. The current distribution of DEN viruses includes Southeast Asia, the south Pacific, the Caribbean basin, Mexico, Central America, and South America. However, epidemics of dengue have occurred elsewhere in the past, including the United States, Japan, Australia, Greece, and both eastern and western Africa. Figure 12.24 shows its current distribution and the regions with projected risk for occurrence, given the appropriate climate and availability of vectors. Dengue is commonly reported as an introduced disease in the United States, but indigenous transmission apparently has occurred in Texas in 1980, 1986, and 1995. All four

serotypes now occur in the Western Hemisphere. In addition to large-scale epidemics in the Americas, there have been recent, large outbreaks involving dengue hemorrhagic fever in Africa, China, Taiwan, India, Maldives, and Sri Lanka. In the hyperendemic areas of southeastern Asia, such as Thailand and the Philippines, the severe forms of the disease have become more common and appear in epidemics at 3- to 5-year intervals.

Dengue is characterized by fever, rash, severe headache, and excruciating pain in muscles and joints, earning it the name *breakbone fever*. Clinical disease develops 5–8 days after the bite of an infected mosquito. Often the disease runs a mild course of about a week, leading to complete recovery. However, dengue can become severe in cases of *dengue hemorrhagic fever* (DHF) and *dengue shock syndrome* (DSS), both of which generally occur in children and may be fatal. These manifestations were first observed as complications of dengue fever in 1954 and are characterized by blotchy rash, bleeding from the nose and gums, and shock.

The increased frequency of DHF and DSS in Southeast Asia and parts of the Caribbean and Latin America has stimulated a debate in the biomedical community regarding the mechanisms by which DHF emerges during epidemics of classic dengue fever. One hypothesis is that there are variable forms of the different serotypes of DEN viruses, some of which are more pathogenic than others. Another idea is that people of different races differ in their propensity to develop severe symptoms and

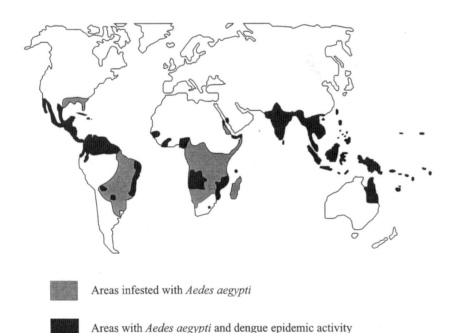

◼ Areas infested with *Aedes aegypti*

◼ Areas with *Aedes aegypti* and dengue epidemic activity

**FIGURE 12.24**   Geographic distribution of dengue and dengue hemorrhagic fever. (Reconstructed from World Health Organization, Vector Biology and Control Division, 1989, and other sources.)

that the virus has been spreading into more susceptible populations. Still another hypothesis is that, as human populations in tropical cities in Asia and the Americas increase and as dengue epidemics in general increase in frequency, there are simply more cases with the noticeable manifestations of DHF and DSS. The fourth hypothesis, and the one currently viewed as most likely, is that prior exposure to one serotype, followed by exposure to another serotype within a critical period of 5 years, leads to the development of hemorrhagic fever and, in some cases, shock syndrome. The more recent epidemics taking place in various parts of the world, especially in the Americas, indicate that DEN has replaced YF as the major urban, epidemic flavivirus of importance worldwide.

The main epidemic vector of DEN viruses, *Ae. aegypti* (Fig. 12.23), is ideal, because it commonly rests indoors, feeds preferentially on humans, has a tendency to take supplementary blood meals, and often moves from one residence to another as it oviposits. The larvae develop in such vessels as water barrels and jars, potted-plant containers, cemetery urns, and discarded tires. The close proximity of these larval habitats to human dwellings further facilitates *Ae. aegypti*–human contact and allows large mosquito populations to develop. To the extent that larval sites are hand-filled, DEN transmission is independent of rainfall patterns. Indeed, in some places where water stores are replenished independently of rainfall, epidemics may occur in the hot, dry season when temperatures are higher and the extrinsic incubation period of the viruses in the mosquitoes is shortened. However, in areas where breeding containers depend primarily on rainwater, dengue epidemics occur during rainy seasons.

Other vectors of DEN viruses are *Ae. albopictus* (Fig. 12.25) in rural areas of Southeast Asia and *Ae. polynesiensis, Ae. scutellaris, Ae. pseudoscutellaris*, and

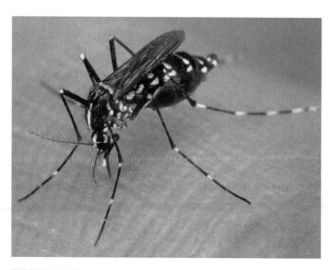

**FIGURE 12.25** Asian tiger mosquito (*Aedes albopictus*), female feeding on human. (Photo by W. A. Foster.)

*Ae. rotumae* in the Pacific region. The role of the newly introduced *Ae. albopictus* in the New World as a vector of DEN viruses remains to be determined. In peninsular Malaysia, a series of studies showed that DEN viruses circulated between monkeys and mosquitoes of the *Ochlerotatus niveus* group, suggesting the possibility of an enzootic sylvatic cycle. However, humans generally are considered to be the only vertebrate host in situations where monkeys do not occur, such as congested urban slums in huge tropical cities, including Bangkok, Manila, Jakarta, Caracas, and Guayaquil. Thus, a mosquito–human–mosquito cycle of DEN transmission is the usual means of virus maintenance and epidemic spread. Transovarial transmission of some of the DEN viruses has been demonstrated in *Ae. aegypti* and *Ae. albopictus*, and therefore mosquitoes may be reservoirs, particularly in periods of low-level transmission among humans.

### Japanese Encephalitis Virus Complex

Another important group of mosquito-borne flaviviruses is the Japanese encephalitis virus antigenic complex, including West Nile, Japanese encephalitis, St. Louis encephalitis, and Murray Valley encephalitis viruses. Some authorities refer to this complex as the *West Nile antigenic complex*. These viruses occur in widely separated geographic regions but show similarities in the nature of their enzootic cycles. Each has *Culex* vectors (Table V) and birds as vertebrate reservoirs. In the case of JE virus, pigs often serve as amplifying hosts. Figure 12.26 shows the worldwide distribution of this complex of viruses. The most important of these in terms of human morbidity and mortality is JE virus. However, all cause human illness of varying severity, depending upon the virus and age and health of the person.

**West Nile (WN) virus** This virus is widely distributed in Africa, the Middle East, Europe, parts of the former Soviet Union, India, and Indonesia (Fig. 12.26) (Hayes, 1988). It has been the cause of endemic and epidemic fever, myalgia, and rash, especially in children in the Middle East, and has caused encephalitis in some instances. Seroprevalence of antibody to WN virus in Egypt can exceed 60% in adults. Recent epidemics have occurred in Israel (1950s), southern France (1962), South Africa (1974 and 1983–1984), Romania (1996), and other eastern European countries. WN virus was first isolated from the blood of a febrile man in Uganda in 1937. Like the other members of this complex, the primary mosquito vectors of WN virus are in the genus *Culex*, particularly *Cx. pipiens* and *Cx. univittatus*. Other species also may be important. As with the other viruses in this complex, birds are vertebrate reservoirs and amplifying hosts during epidemics. In 1999, WN virus was introduced through unknown means into New York City and was linked to human encephalitis in 61 confirmed cases (mostly in the

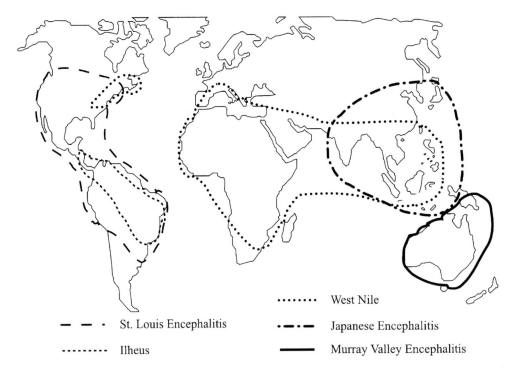

........ West Nile

— — · St. Louis Encephalitis      ·—·— Japanese Encephalitis

········· Ilheus      ———— Murray Valley Encephalitis

**FIGURE 12.26**   Geographic distribution of mosquito-borne flaviviruses that cause encephalitis. (Reconstructed from Mitchell, 1977, and other sources.)

city borough of Queens); there were 7 fatalities. Initially, this outbreak of disease was thought to be due to St. Louis encephalitis (see below). Other unusual features of the outbreak were the large numbers of birds, especially crows (*Corvus brachyrhynchos*), that died. The vectors were likely *Cx. pipiens* and other *Culex* species, although WN virus later was isolated from other mosquito species as well. A serosurvey conducted by the US Centers for Disease Control and Prevention in the Queens area indicated that as many as 1256 people had actually been infected during the course of the outbreak and that 239 (19%) of these people may have experienced clinical illness due to infection. Of interest was that the virus successfully overwintered. A dead, virus-positive hawk and virus-positive overwintering *Culex* females were found in early 2000. Subsequently, virus transmission resumed in the summer of 2000. There were at least 18 human cases in this region in 2000, including 1 death. Also, many wild bird deaths were linked to virus infection, and associations between the virus and several more species of mosquitoes were discovered. Virus activity spread northward to New England, southward to North Carolina, and westward to Pennsylvania and western New York state. Whether WN virus will become firmly established and more widespread in the western hemisphere is not known, but the fact that overwintering occurred and virus distribution increased geographically are signs that it is establishing in this new geographic setting.

**Japanese encephalitis (JE) virus**   JE is a severe disease of acute encephalitis, with children and the elderly primarily affected and with mortality rates reaching >25% of those with overt disease. Many infections are asymptomatic or mild. Survivors often show neurologic sequelae. The disease is distributed throughout the rice-growing areas of Asia, from Japan south to Papua New Guinea and west to India and Nepal (Fig. 12.26). In Japan, Japanese encephalitis epidemics have occurred in August and September in many different years since its discovery there. The virus first was isolated from the brain of a human who died of encephalitis in 1935, then from brain tissue of a horse in 1937, and from *Cx. tritaeniorhynchus* in 1938. Numbers of cases have declined in Japan, Korea, and Taiwan, because of vector control, vaccination, and changes in agricultural practices (Burke and Leake, 1988). The enzootic cycle of JE virus involves *Cx. tritaeniorhynchus* and several other *Culex* species, including *Cx. vishnui, Cx. pseudovishnui, Cx. fuscocephala*, and *Cx. gelidus,* all of which are associated with rice culture. Wading birds such as herons are important enzootic reservoirs. Pigs are important amplifier hosts in rural, rice-growing areas where swine are kept. Japanese encephalitis is probably the most important of the mosquito-borne encephalitis diseases, owing to its epidemic nature, widespread distribution, and large number of humans who acquire infection, die, or recover, yet suffer neurologic sequelae.

**St. Louis encephalitis (SLE) virus** This virus was identified as the causative agent of disease during an outbreak of encephalitis-like illness in the United States, in Paris, Illinois, in 1932 and St. Louis, Missouri, in 1933. It was isolated from a patient with encephalitis in the Yakima Valley of Washington in the early 1940s and found to be a frequent cause of human illness in the Central Valley of California in the 1930s and 1940s. This virus is distributed widely in North America and also occurs in parts of Mexico, Central America, the Caribbean, and South America to Argentina (Fig. 12.26). The encephalitic illness caused by SLE virus shows a bimodal age distribution, with children and elderly people most frequently affected. Attack rates during epidemics range from 5 to 800 per 100,000 population, depending upon location, year, strain virulence, and population immunity due to earlier epidemics. In the eastern United States, mortality rates have ranged from about 3 to 20% of laboratory-diagnosed cases, but in the western United States mortality rates are lower.

In North America, three enzootic cycles of SLE virus have been described. In the eastern United States, north of Florida but including Texas, the primary vectors are *Culex pipiens* and *Cx. quinquefasciatus* (Fig. 12.27). The former mosquito occurs in a more northerly distribution, whereas the latter is more southerly, with a hybrid zone at about the latitude of Memphis, Tennessee. Females of both species feed on birds. In addition, *Cx. quinquefasciatus* females frequently feed on mammals as the summer progresses. Whether these mosquitoes alone or other vectors function in transmission to humans depends upon the abundance of these two species and other competent vectors. House sparrows are important vertebrate amplifier hosts in peridomestic settings. In the

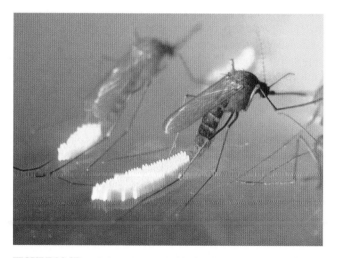

**FIGURE 12.27** *Culex quinquefasciatus* females laying eggs in form of floating rafts; these are vectors of filariasis and St. Louis encephalitis, among other diseases. (Photo by W. A. Foster.)

western United States, a mosquito–bird–mosquito cycle similar to that of WEE virus, involving *Cx. tarsalis*, has been elucidated. In addition, in California both *Cx. pipiens* and *Cx. quinquefasciatus*, and possibly *Cx. stigmatosoma*, function secondarily as either enzootic or epidemic vectors. In Florida, *Cx. nigripalpus* apparently is the enzootic, epizootic, and epidemic vector of SLE virus. In Latin America and in the Caribbean basin, SLE virus has been isolated from many different species of *Culex, Sabethes, Mansonia, Wyeomyia*, and other genera and from a wide variety of birds and mammals. In these areas, human St. Louis encephalitis generally is rare. The mechanism of virus overwintering in North America is not well known. There is some evidence of virus persistence in overwintering, diapausing female mosquitoes.

The history of St. Louis encephalitis in North America has been that of epidemics, either local or widespread, with intervening years when there was apparently no virus activity or were no epidemics, and either no case or a few isolated human cases. The first epidemic, in the early 1930s in St. Louis, Missouri, USA, was accompanied by hot and dry weather, which favored the development of populations of mosquitoes of the *Cx. pipiens* complex, the larvae of which develop in water rich in sewage. The epidemic involved about 1100 human cases and 200 deaths. Since that time, there have been some 50 outbreaks of St. Louis encephalitis in the United States. Cases during three recent decades are shown by state in Fig. 12.28. Human cases also have occurred in Manitoba and Ontario, Canada. These epidemics have been both rural and urban. A very large outbreak occurred in 1975, involving 30 states and the District of Columbia, with over 1800 cases reported. More recently, epidemics have occurred in such disparate locations of the United States as Pine Bluff (Arkansas), Florida, Los Angeles (California), Houston (Texas), New Orleans (Louisiana), and Grand Junction (Colorado). A total of 4437 human cases of SLE infection were reported to the United States Centers for Disease Control and Prevention from 1964 to 1995, with an average of 139 cases per year (range, 4–1967). Monath (1980) and Tsai and Mitchell (1988) have reviewed the ecology and public health significance of St. Louis encephalitis.

**Murray Valley encephalitis (MVE) virus** This virus has been associated with encephalitis-type illness in humans in eastern and western parts of Australia and in New Guinea (Marshall, 1988) (Fig. 12.26). Mortality rates vary greatly during outbreaks, ranging from 18 to 80%. Epidemics of encephalitis in Australia in 1917, 1918, 1922, and 1925 were probably caused by MVE virus. It was isolated from the brain of a human in 1951 and from a pool of *Cx. annulirostris* in 1960. Later epidemics occurred in Australia in 1956, 1971, 1974, 1978, 1981,

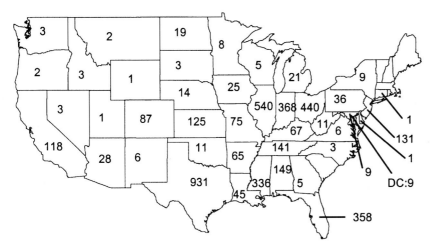

**FIGURE 12.28** Distribution of human cases of St. Louis encephalitis in the United States, reported between 1964 and 1993. (Reconstructed from Monath and Heinz, 1996, and other sources.)

and 1984. The enzootic cycle involves *Cx. annulirostris* as vector and birds as vertebrate reservoir and amplifier hosts. Epidemics appear to be associated with excessive rainfall, which allows an increase of mosquito populations to high densities and the immigration of wading birds. *Kunjin virus* is closely related to MVE virus. It is also mosquito-borne, but it is associated rarely with human disease in Australia.

**Other Flaviviruses**

There are many other mosquito-borne flaviviruses of importance to human health.

**Rocio virus**   This virus occurs in Brazil (Iversson, 1988). It appeared suddenly in São Paulo State in 1975 and caused an encephalitis-type illness in about 1000 cases, with some fatalities. Most cases occurred among young, male agricultural workers. Only a single case has been reported since. The vector and vertebrate host relationships for this virus are poorly known, although *Psorophora ferox* and *Ochlerotatus scapularis* have been incriminated as vectors, based on virus isolation in the field and laboratory transmission experiments. A closely related pathogen is *Ilheus virus*, which occurs in Trinidad, Panama, and parts of South America (Fig. 12.26). It has been associated with about 10 documented human cases of illness and has been isolated from mosquitoes in 8 different genera, but mostly from *Psorophora* species.

**Bunyaviridae (*Bunyavirus and Phlebovirus*)**

The Bunyaviridae include the genera *Bunyavirus* and *Phlebovirus*. The viruses in the genus *Bunyavirus* comprise a complex and diverse group of more than 153 viruses in 16 serogroups distributed worldwide

(Table VI). Mosquitoes or biting midges serve as vectors, and small mammals, ungulates, or birds are the vertebrate hosts. *Phlebovirus* contains 37 viruses, most of which have phlebotomine sand fly vectors. It also includes the important mosquito-borne RVF virus.

Among the 30 viruses associated with human or animal diseases in the genus *Bunyavirus* are those in the *California serogroup* (Eldridge, 1990; Grimstad, 1988). It consists of four virus complexes: California encephalitis, Melao, Trivittatus, and Guaroa. In the *California encephalitis virus complex,* there is one virus type, California encephalitis virus, and seven subtypes and varieties, including the medically important La Crosse and Tahyna viruses. *Melao virus complex* consists of seven subtypes and varieties, whereas *Trivittatus virus* and *Guaroa virus* form monotypic complexes within the serogroup. Viruses in the California serogroup characteristically are transmitted by *Ochlerotatus* species, but other genera such as *Culiseta* and *Anopheles* may be involved. Of the 14 virus types, subtypes and varieties in the California serogroup, 9 occur only in North America: CE, *La Crosse encephalitis,* Snowshoe hare, *San Angelo, Jamestown Canyon, Jerry Slough, South River, Keystone,* and *Trivittatus.* The others, including *Guaroa, Serra do Navio, Melao, Lumbo,* Tahyna, *Lumbo,* and *Inkoo,* are distributed in various places worldwide. Table VII lists the California serogroup viruses and indicates their geographic distribution, mosquito vectors, and vertebrate host relationships. The viruses in this serogroup were isolated and described during the period from 1943 through the 1970s, culminating in a monograph by Calisher and Thompson (1983).

**California encephalitis (CE) virus**   This is the prototype virus for the complex of the same name and was the first of the California serogroup viruses to be associated with human disease, involving three cases in California in

TABLE VI
Family Bunyaviridae

| Serogroup | Virus complex | Distribution of types, subtypes, and varieties |
|---|---|---|
| Bunyamwera | Bunyamwera | Africa, Europe, Asia, Canada, United States, Madagascar, Mexico |
| | Kairi | South America |
| | Main Drain | United States (western) |
| | Wyeomyia | South America, Panama, Brazil |
| Anopheles A | Anopheles A | Colombia, Brazil, Argentina, Trinidad |
| | Tacaiuma | Colombia, Brazil, |
| Anopheles B | Anopheles B | Colombia, Brazil |
| Bwamba | Bwamba | Africa |
| Group C | Caraparu | South America, Panama |
| | Madrid | Panama |
| | Marituba | South America, Trinidad |
| | Oriboca | South America |
| California | California encephalitis | United States, Canada, Europe, Africa, Tajikstan |
| | Melao | South America, Trinidad, Panama, United States |
| | Trivittatus | United States |
| | Guaroa | Colombia, Brazil, Trinidad |
| Capim | Capim | Brazil |
| | Guajara | Brazil, Panama, Guatemala |
| | Bush Bush | Trinidad, Brazil, Guatemala, Panama |
| | Acara | Brazil, Panama, Trinidad |
| | Benevides | Brazil |
| Gamboa | Gamboa | Panama, Surinam, Ecuador |
| | Alajuela | Panama, Ecuador, Argentina |
| Guama | Guama | South America, United States (Florida) |
| | Bertioga | Brazil |
| | Bimiti | South America (northern) |
| | Catu | South America (northern) |
| | Timboteua | Brazil |
| Koongol | Koongol | Australia, New Guinea |
| Minatitlan | Minatitlan | Mexico, Guatemala, Ecuador |
| Olifantsvlei | Olvantsvlei | Africa |
| | Botambi | Central African Republic, Ivory Coast |
| Patois | Patois | United States (Florida), Mexico, Central America, Ecuador |
| | Zegla | United States (Florida), Mexico, Central America |
| Simbu | Simbu | Africa |
| | Akabane | Japan, Australia, Kenya |
| | Manzanilla | Trinidad, Panama, French Guiana, Africa, India, Thailand, Cyprus, United States, Central African Republic, Brazil, Australia |
| | Sathuperi | India, Nigeria, Australia |
| | Shamonda | Nigeria, Kenya |
| | Shuni | Nigeria, South Africa, Japan, Australia, India |
| | Thimiri | India, Egypt |
| Tete | Tete | South Africa, Egypt, Nigeria, Central African Republic, Cyprus, Italy, Japan |
| Turlock | Turlock | North America, South America, India, Malaysia, Australia, Central Europe |
| | M'poko | Central African Republic, Nigeria |
| Unassigned | Kaeng Khoi | Thailand |

Serogroups, complexes, and geographic distribution of representative virus types in the genus *Bunyavirus*. Abstracted from Calisher and Karabatsos (1988).

TABLE VII

**California Serogroup (*Bunyavirus*, Bunyaviridae): Classification, Mosquito Associations, Vertebrate Hosts, and Geographic Distribution of Virus Types and Subtypes**

| Virus type and subtype | Variety | Vector associations | Vertebrate associations | Geographic distribution |
|---|---|---|---|---|
| California encephalitis | | | | |
| California encephalitis | | *Ochlerotatus melanimon* *Ochlerotatus dorsalis* | Lagomorphs | United States (western, southwestern) |
| Inkoo | | *Ochlerotatus* and *Aedes* spp. | Lagomorphs? | Finland |
| La Crosse | La Crosse | *Ochlerotatus triseriatus* | Sciurid rodents, foxes | United States (eastern) |
| | Snowshoe hare | *Ochlerotatus stimulans* group *Ochlerotatus canadensis* *Culiseta inornata* | Lagomorphs | United States (northern), Canada |
| San Angelo | | *Aedes, Anopheles, Psorophora?* | Unknown | United States (southwestern) |
| Tahyna | Tahyna | *Ae. vexans, Cs. annulata* | Lagomorphs | Europe, Tajikistan, Azerbaijan |
| | Lumbo | *Ae. pembaensis?* | Unknown | East Africa |
| Melao | Melao | *Ochlerotatus scapularis?* | Unknown | Trinidad, Brazil, Panama |
| | Jamestown Canyon[a] | *Cs. inornata* *Ochlerotatus communis* group *Ochlerotatus provocans* *Ochlerotatus abserratus* *Ochlerotatus intrudens* *Ochlerotatus stimulans* group *Anopheles* spp. | Deer | United States, Canada |
| | South River | (same as Jamestown Canyon virus?) | Deer? | United States (northeastern) |
| | Keystone | *Ochlerotatus atlanticus,* *Ochlerotatus tormentor* | Lagomorphs, Cotton rats | United States (coastal, eastern) |
| | Serra do Navio | *Ochlerotatus fulvus* | Unknown | Brazil (Amapa state) |
| Trivittatus | | *Ochlerotatus trivittatus* | Lagomorphs | United States |
| Guaroa | | *Anopheles* spp. | Unknown | Panama, Colombia, Brazil |

[a] Jerry Slough virus is a strain of Jamestown Canyon virus in the western United States.

1943. It was isolated from *Ochlerotatus melanimon* and *Cx. tarsalis* at that place and time. Extensive studies of the mosquito and vertebrate-host relationships of this virus in California have shown that *Oc. melanimon* and *Oc. dorsalis* are the principal vectors and that the virus is transmitted transovarially by these species. Serologic surveys have implicated jackrabbits, cottontail rabbits, California ground squirrels, and kangaroo rats as vertebrate hosts. The virus also has been isolated in New Mexico, Utah, and Texas. However, since the time of original discovery, CE virus only rarely has been associated with human disease.

**La Crosse (LAC) virus**    This is the most important human pathogen in the California serogroup, causing an acute, febrile illness in children. Most cases are subclinical or mild, but some progress to severe encephalitis and, rarely, death. In 1964, LAC virus was isolated from preserved brain tissue of a child who had died of encephalitis in 1960 in the vicinity of La Crosse, Wisconsin (USA). The virus currently is distributed in the eastern United States, including the midwestern states bordering the Great Lakes, east to New York and Pennsylvania, south to West Virginia and North Carolina, and west

to Texas. However, most human cases occur in West Virginia, Wisconsin, Illinois, Indiana, and Ohio.

The disease tends to be highly focal within its known range, such that particular regions or towns are known to be endemic. Prevalence varies regionally. In the United States, there were 2245 cases of La Crosse encephalitis reported to the Centers for Disease Control and Prevention from 1964 to 1995, with an average of 70 per year (range, 29–160). In Ohio, where cases are particularly well documented, there was an average of 26 cases per year between 1963 and 1995. La Crosse encephalitis probably is underreported to public health agencies.

The principal vector of LAC virus is the eastern tree hole mosquito, *Ochlerotatus triseriatus*. The virus is transmitted both horizontally, to sciurid rodents, particularly chipmunks and squirrels, and vertically, from female mosquitoes to their progeny. The discovery of transovarial transmission of LAC virus was one of the first documentations of this phenomenon in mosquitoes and revealed an overwintering mechanism for LAC virus. It also demonstrated that vertebrate reservoirs were not always essential to the persistence of mosquito-borne viruses in nature and that the mosquito itself could be a reservoir host. Thus an infected female is able to transmit the virus at its first blood feeding without previously having taken an infectious blood meal. Another important new finding was that *Ae. triseriatus* males, infected transovarially, transferred LAC virus to females via mating (i.e., venereal transmission).

Epidemiologic investigations of cases of encephalitis or aseptic meningitis of unknown origin often reveal La Crosse encephalitis in areas where previously it was unknown. Such investigations almost always reveal populations of *Ochlerotatus triseriatus* in the immediate vicinity where infection was thought to occur, such as backyards or wooded areas where children play. Water-filled artificial containers, particularly discarded tires, have become important habitats for *Oc. triseriatus* larvae and provide a link between the sylvan La Crosse cycle and humans. In Ohio and New York State, LAC virus also has been isolated repeatedly from *Ochlerotatus canadensis;* however, the role of this mosquito as a vector to humans and its role in an enzootic cycle are not well understood.

**Snowshoe hare (SSH) virus** This pathogen is closely related to La Crosse virus, but its ecology is very different. It originally was isolated from the blood of a snowshoe hare in Montana in 1958. Lagomorphs (hares and rabbits) are the enzootic vertebrate hosts. SSH virus is distributed in the northern parts of the United States and in Canada, where it has been isolated from a variety of *Ochlerotatus* species and from *Culiseta inornata.* Even though SSH virus is very similar antigenically to La Crosse virus, human disease rarely has been documented except in Ontario, Quebec, and Nova Scotia (Canada), where 10 cases of an encephalitis-like illness have been attributed to SSH virus.

**Keystone (KEY) virus** This virus is a subtype in the Melao virus complex, first isolated in 1964 from a collection of blood-fed *Ochlerotatus atlanticus* and *Oc. tormentor* in Florida. It is not considered to be a human pathogen. It occurs along the eastern seaboard of the United States, where it has been isolated from *Oc. atlanticus*, *Oc. tormentor*, *Oc. infirmatus*, and other mosquitoes. Transovarial transmission of the virus has been demonstrated for *Oc. atlanticus* in the field. Gray squirrels and cottontail rabbits in northern coastal areas, and cottontail rabbits and cotton rats in Florida and Texas, have been identified as vertebrate hosts.

**Trivittatus (TVT) virus** This virus was first isolated from *Ochlerotatus trivittatus* in North Dakota in 1948. It also has been isolated from other mosquitoes, including *Oc. infirmatus* in the southeastern United States, where *Oc. trivittatus* is absent. Transovarial transmission has been demonstrated in the latter species. Trivittatus virus shows a widespread distribution in the eastern half of the United States. Cottontail rabbits are vertebrate hosts.

**Jamestown Canyon (JC) virus** This is another subtype in the Melao virus complex. It was originally isolated from *Culiseta inornata* in Colorado in 1961. Since that time it has been isolated in both Canada and the United States from *Ochlerotatus, Culiseta,* and *Anopheles* species in regions from Alaska east to Ontario and New England, south to Maryland, and in western and southwestern states, including California. The principal vectors are *Ochlerotatus* species with univoltine life cycles (i.e., snowpool and spring species). An antigenic variant known only from California is *Jerry Slough virus,* which is transmitted by *Cs. inornata.* In the eastern United States, a variety or strain of JC subtype is *South River virus.* Transovarial transmission of JC virus has been demonstrated in some mosquito species. Its vertebrate hosts are large wild ungulates, especially deer. JC virus has been associated with encephalitis-type illness in humans in Ontario, New York, and Michigan.

**Tahyna (TAH) virus** This is a subtype of the CE virus complex, distributed widely in Europe and parts of western Asia. It has been associated with human febrile and central nervous system illnesses in France, the former Czechoslovakia, and Tajikistan. Foci are now known from Finland south to Tajikistan. Although the prevalence of infection in humans is poorly known, serosurveys in the Rhine River valley of Germany documented antibody to TAH virus in up to 23% of humans living in the area. In the former Czechoslovakia, TAH virus was implicated in 1% of febrile illnesses of children in an endemic area and in 20% of central nervous system illnesses. This virus was first isolated from *Ae. vexans* and *Ochlerotatus caspius* in Slovakia in 1958. The mosquito vectors are *Ae. vexans, Oc. caspius,* and *Cs. annulata.* Hares and pigs are vertebrate reservoir hosts. Lumbo virus is a variety of TAH and occurs in parts of Africa.

**Rift Valley fever (RVF) virus** This pathogen (Fig. 12.20) is classified with viruses in the genus *Phlebovirus* in the Bunyaviridae (Meegan and Bailey, 1988). It is distributed in eastern Africa north to Egypt, and in parts of West Africa, where it has been associated with large outbreaks of acute illness in livestock (see Veterinary Importance, below). Humans may become infected by mosquito bite or, more commonly, by contact with virus-contaminated blood or through inhalation of virus in aerosols during slaughter of livestock. Humans rarely die of infection but develop an illness including fever, headache, myalgia, retinitis, and, in rare cases, liver involvement. Large epizootics and epidemics have occurred in South Africa (1950–1951, 1953), Zimbabwe (1968–1969), Egypt (1977–1978, 1993), Mauritania (1987), and Kenya and Somalia (1997–1998). These outbreaks generally involved thousands to hundreds of thousands of cases in livestock. The epidemic in Egypt in 1977–1978 probably involved some 200,000 human cases and 600 deaths. The epidemic in Mauritania involved about 1,000 human cases and 50 deaths.

A variety of mosquito vectors have been associated with RVF virus, including *Cx. pipiens* in the Egyptian outbreak. In parts of the eastern African savannah, an enzootic cycle of RVF virus has been identified in and around *dambos*, low-lying temporary wetlands. Prolonged rainfall floods these areas and allows development of large populations of *Ae. mcintoshi* and other *Aedes* species. These mosquitoes maintain RVF virus through transovarial transmission and transmit it to domestic and wild ungulates that come to the dambos for water, functioning as enzootic vectors. As amplification ensues, epizootic vectors such as *Cx. theileri* become important in transmission to domestic livestock.

## MALARIA

Malaria is one of the most widespread and prevalent of infectious human diseases. It is caused by sporozoan protists that infect blood tissues and other organs of the body, primarily the liver. The organisms are transmitted by *Anopheles* mosquitoes. The word "malaria" derives from the Italian *mala aria* or "bad air." Another term for malaria is *paludism* from the French *paludisme* and the Spanish *paludismo*. Both of these words are derived from the Latin *palus*, meaning "swamp." The connotation is that malaria was contracted through association with swamps and inhaling the bad air emanating from them. In English-speaking countries, malaria was called *ague* from the Old French *agu* ("sharp"), from the context of the Latin *febris acuta*, or acute fever. Ague referred to the cyclic fevers and chills, or *paroxysms*, which are characteristic of malaria.

The organisms that cause human malaria are protozoans of the genus *Plasmodium*, family Plasmodiidae (order Haemosporidida, class Haemosporidea, phylum Sporozoa). They are obligate intracellular parasites. There are four species of human malaria: *Plasmodium falciparum*, causing *malignant tertian malaria; P. vivax*, causing *benign tertian malaria; P. malariae*, causing *quartan malaria;* and *P. ovale*. The first two species are widespread in the tropics, whereas *P. vivax* also occurs in some temperate areas. *P. malariae* also is distributed widely but less commonly, and *P. ovale* is rare, occurring mainly in Africa.

Currently, 1.6 billion people are at direct risk of malaria infection via mosquito bite. Globally, an estimated 300–500 million cases occur annually, including perhaps 100 million cases and >1 million deaths per year in sub-Saharan Africa. Travelers may become infected during visits to endemic areas. Malaria currently occurs in sub-Saharan Africa and parts of northern Africa; the Middle East to Iran, Afghanistan, and Pakistan; India and Sri Lanka; parts of China and Southeast Asia; Indonesia, the Philippines, Irian Jaya, and New Guinea; and Latin America from Mexico through Central America to most of the northern half of South America (Fig 12.29).

The degree of endemicity, or disease prevalence, depends upon a variety of factors, including the species of malaria present, environmental and social factors, and species of vectors present. Human malaria is thought not to have occurred in the Western Hemisphere prior to the period of European exploration, colonization, importation of African slaves, and establishment of intercontinental trade.

**FIGURE 12.29**    Geographic distribution of human malaria. (Reconstructed from World Health Organization, Vector Biology and Control Division, 1989, and other sources.)

The literature on human malaria and its vectors is voluminous. A comprehensive source is Gilles and Warrell (1994). Other sources include Boyd (1949), Macdonald (1957), Molineaux and Gramiccia (1980), Wernsdorfer and McGregor (1988), and Strickland (1991).

## Plasmodium Life Cycle

The complex life cycle of *Plasmodium* species involves both sexual and asexual reproduction. The sexual phase (*gametogony*) begins in the blood of the human host and is completed within the lumen of the midgut of the mosquito. The first phase of asexual reproduction (*sporogony*) occurs on the outer wall of the midgut. The second phase of asexual reproduction occurs first in the liver and later in the blood of the human host. The process in both sites is termed *schizogony* or *merogony*. Sporogony is often referred to as the *exogenous phase* of malaria parasite development, because it occurs outside of the human host. Conversely, merogony is often referred to as the *endogenous phase* of development within the human host. These two developmental phases are depicted in Fig. 12.30 and discussed in more detail below.

An *Anopheles* female becomes infected with malarial parasites when she ingests blood containing red blood cells that are infected with *gametocytes*, specifically the sexual *microgametocyte* and *macrogametocyte* stages of the parasite. A microgametocyte bursts from its host red blood cell within the blood meal in the midgut lumen of the mosquito, where it extends four to eight flagella-like forms called *microgametes* in a process termed *exflagellation*. A macrogametocyte sheds the erythrocytic membrane and transforms into a single mature *macrogamete*. One microgamete locates and fertilizes a macrogamete, forming a diploid *zygote*. The zygote transforms into a motile *ookinete*. The ookinete passes through the peritrophic membrane, then through the midgut epithelial cell membrane, and forms an *oocyst* between the midgut epithelial cells and the basement membrane of the epithelium. A single, malaria-infected *Anopheles* may have few to hundreds of oocysts, depending on the original number of gametocyte-infected red blood cells in the blood meal and on the number of macrogametocytes that become fertilized. During sporogony, the encysted parasite becomes haploid again and undergoes multiple mitoses until the oocyst contains thousands of motile *sporozoites*. The oocyst bursts, releasing sporozoites, which make their way to the salivary glands and penetrate the secretory cells. The sporozoites accumulate within these cells, with some also passing into the salivary ducts. The mosquito then is infective. The sporozoites enter a human host when the mosquito probes and the salivary gland

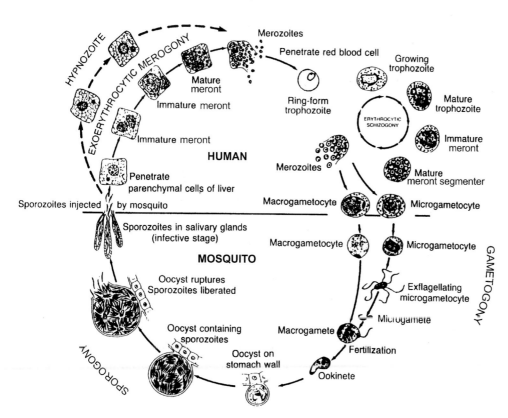

**FIGURE 12.30** Life cycle of *Plasmodium vivax* in human and *Anopheles* hosts. (Modified from Strickland, 1991.)

cells release saliva into the skin prior to ingesting blood. The period of time between ingestion of gametocytes and infection of the salivary glands with sporozoites is the *extrinsic incubation period.*

In the human host, sporozoites migrate in the blood to the liver within minutes of entering subdermal capillaries. They invade liver parenchymal cells, where they typically form a *primary tissue meront.* In the case of *P. vivax* and *P. ovale* they may form a *hypnozoite,* a quiescent or resting stage of the parasite. Inside each meront, *merozoites* develop through the process of *exoerythrocytic merogony.* The meront then bursts, releasing merozoites into the bloodstream, where they circulate and invade red blood cells. The merozoites form meronts in the red blood cells, where they produce more merozoites through a process called *erythrocytic merogony.* An invasive merozoite, once inside a red blood cell, transforms into a *trophozoite,* which utilizes hemoglobin for nutrients, then into a *segmenter* form distinguished by dark dots of heme in the red blood cell, and finally into a mature meront, which produces still more merozoites. Merozoites released from infected red blood cells invade new red blood cells, where the process of erythrocytic merogony begins anew. Other invasive merozoites form microgametocytes or macrogametocytes within the blood cells, which will mature into microgametes and macrogametes if they are ingested by *Anopheles* mosquitoes during blood feeding.

## Clinical Disease

Malaria is characterized by sudden paroxysms of fever and chills, which recur at highly predictable intervals, often in the afternoon. Other acute symptoms include headache, lethargy, fatigue, and profuse sweating after each bout of fever. After infection of erythrocytes by merozoites, erythrocytic merogony leads to the synchronous rupture of the erythrocytes and release of new merozoites, toxins, and heme digestion products. This event in the circulatory system prompts each episode of chills and fever. The next episode occurs in 48 or 72 hr, depending upon the species of *Plasmodium.*

Malarial infections in humans can result in severe illness and sometimes death. However, the particular symptoms, including timing and severity, vary with the species of *Plasmodium.* The most severe form of malaria is caused by *P. falciparum,* whose merozoites invade both young and old red blood cells. Over time, repeated reinvasions and mass destruction of red blood cells may lead to high parasitemia, severe anemia, and anoxia of tissues. In some cases, hemolysis results in a condition of hemoglobinuria, *blackwater fever,* when the urine contains hemoglobin and turns reddish brown. Toxins from dead red blood cells stimulate macrophages to produce chemicals such as *tumor necrosis factor* and other cytokines, which cause characteristic malaria symptoms such as fever. In

*falciparum malaria,* infected red blood cells stick to the vascular epithelium of capillaries in organs including the brain, impeding blood flow and causing a serious and sometimes lethal condition called *cerebral malaria.* Because of this affinity for internal organs, only very young trophozoites and gametocytes are common in peripheral blood. Malaria caused by *P. falciparum* is called *malignant tertian malaria* because of the severity of symptoms and because of the typical 48-hr interval between paroxysms. The term "tertian" for a 2-day cycle originated from counting the day when the paroxysm occurs as the first day, so that the next paroxysm occurs on the third. Left untreated, nonfatal infections with *P. falciparum* last 5 months or more, depending on the immune status of the individual.

*P. vivax* malaria is called *benign tertian malaria* because symptoms are less severe than *P. falciparum* malaria, and death rarely occurs. Paroxysms occur on a 48-hr cycle. In this type of malaria, the merozoites invade only immature red blood cells, called *reticulocytes,* which typically comprise less than 6% of the total red blood cell count in circulation. Thus *vivax malaria,* compared to falciparum malaria, has less severe symptoms of anemia and toxemia, making death unlikely. The infected red blood cells do not stick to the epithelial lining of capillaries as they do in falciparum malaria. Vivax malaria can evolve into chronic infection with development of an enlarged spleen, or *splenomegaly.* However, persons infected with other malarias also may have enlarged spleens, as these organs work to replace red blood cells lost to infection. The hypnozoite stage of *P. vivax* provides a mechanism for the parasite to overwinter in humans in temperate areas with short transmission seasons. The period between infection and onset of symptoms can last up to 9 months, and untreated infection persists in the body for many months to many years, with relapses recurring at irregular intervals after initial infection and acute onset of disease.

*P. ovale* is an uncommon tertian malaria with milder symptoms. Its course of infection is similar to that of *P. vivax.*

*P. malariae,* which causes *quartan malaria,* differs from *P. vivax* and *P. falciparum* in that the parasites invade only mature erythrocytes. Therefore, symptoms can be more severe than in vivax malaria in the acute phase. However, infections tend to develop more slowly and become chronic. Malaria caused by *P. malariae* has a 72-hr erythrocytic cycle. The term *quartan* refers to the 4 days included within one cycle, with a 3-day interval from the beginning of the first paroxysm to the beginning of the next. Recrudescences of *P. malariae* may occur in individuals up to 50 years after initial infection, owing to low levels of parasitemia that increase under periods of immunosuppression.

After a person is bitten and inoculated with sporozoites and exoerythrocytic merogony commences in the

TABLE VIII

*Anopheles* Vectors of Human Malaria Parasites in 12 Epidemiologic Zones: Subgenera, Species, and Geographic Distributions are Given

| Malaria epidemiologic Zone | *Anopheles vectors* |
|---|---|
| North American | Subgenus *Anopheles*: *freeborni, punctipennis, quadrimaculatus*<br>Subgenus *Nyssorhynchus*: *albimanus* |
| Central American | Subgenus *Anopheles*: *aztecus, pseudopunctipennis, punctimacula,*<br>Subgenus *Nyssorhynchus*: *albimanus, albitarsis, allopha, aquasalis, argyritarsis, darlingi* |
| South American | Subgenus *Anopheles*: *pseudopunctipennis, punctimacula*<br>Subgenus *Nyssorhynchus*: *albimanus, albitarsis, aquasalis, argyritarsis, braziliensis, darlingi, nuneztovari*<br>Subgenus *Kerteszia*: *bellator, cruzii* |
| North Eurasian | Subgenus *Anopheles*: *atroparvus, messeae, sacharovi, sinensis*<br>Subgenus *Cellia*: *pattoni* |
| Mediterranean | Subgenus *Anopheles*: *atroparvus, claviger, labranchiae, messeae, sacharovi*<br>Subgenus *Cellia*: *hispaniola, pattoni* |
| Africo-Arabian | Subgenus *Cellia*: *hispaniola, multicolor, pharoensis, sergentii* |
| Africo-Tropical | Subgenus *Cellia*: *arabiensis, christyi, funestus, gambiae, melas, merus, moucheti, nili, pharoensis,* |
| Indo-Iranian | Subgenus *Anopheles*: *sacharovi*<br>Subgenus *Cellia*: *annularis, culicifacies, fluviatilis, pulcherrimus, stephensi, superpictus, tesselatus* |
| Indo-Chinese hills | Subgenus *Anopheles*: *nigerrimus*<br>Subgenus *Cellia*: *annularis, culicifacies, dirus, fluviatilis, maculatus, minimus* |
| Malaysian | Subgenus *Anopheles*: *campestris, donaldi, letifer, nigerrimus, whartoni*<br>Subgenus *Cellia*: *aconitus, balabacensis, dirus, flavirostris, leucosphyrus,*<br> *ludlowae, maculatus, minimus, philippinensis, subpictus, sundaicus* |
| Chinese | Subgenus *Anopheles*: *anthropophagus, sinensis*<br>Subgenus *Cellia*: *pattoni* |
| Australasian | Subgenus *Anopheles*: *bancrofti*<br>Subgenus *Cellia*: *annulipes, farauti, karwari, koliensis, punctulatus, subpictus* |

Modified from Macdonald, 1957, and others.

liver, symptoms do not appear until days to weeks later (up to a month in *P. malariae*), when erythrocytic merogony begins in the blood. In *P. vivax* and *P. ovale*, if the sporozoites develop into hypnozoites in the liver cells, relapses are possible long after inoculation and initial onset of symptoms, with an intervening period of no apparent symptoms of infection. For *P. falciparum* and *P. malariae*, there are no persistent exoerythrocytic stages of the parasites, and relapses do not occur. However, infection with *P. malariae* may recrudesce years after initial infection owing to persistent erythrocytic infections. Therefore, in human malaria there is a clear distinction between relapse and recrudescence of infection. The course of infection of malaria in humans varies with many factors, including history of past exposure; presence of antibodies; age, health, and nutritional status; and genetic resistance factors such as the sickle-cell anemia trait, Duffy-negative blood type, certain hemoglobin types such as hemoglobin S and fetal hemoglobin, and deficiency of the erythrocytic enzyme glucose-6-phospate dehydrogenase.

## Mosquito Vectors and Epidemiology

Many different species of *Anopheles* mosquitoes are competent vectors of malaria organisms (Table VIII). However, most *Anopheles* species are not, because of variation in host-selection patterns, longevity, abundance, and vector competence. In North America, *An. quadrimaculatus*, which forms a complex with four more localized but nearly identical species (Reinert *et al.*, 1997), is the principal vector of malaria in the eastern two-thirds of the continent. It develops along the edges of permanent pools, lakes, and swamps that provide relatively clean, still, sunlit water, with lush emergent vegetation, marginal brush, or floating debris to provide partial shade and protection from wave action. In western North America, *An. freeborni* is the main vector, an inhabitant of clear water in open, shallow, sunlit pools, ponds, ditches, and seepage areas that are partially shaded by vegetation. *An. hermsi* also is a vector in California.

Other important vectors include *An. albimanus* in Central America, *An. darlingi* in South America,

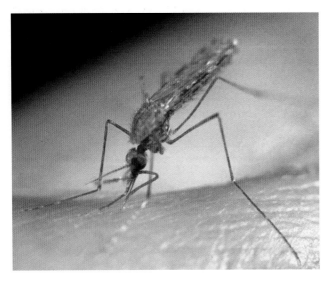

**FIGURE 12.31**  *Anopheles gambiae* female feeding on blood. This is the major vector of malaria in Africa. (Photo by W. A. Foster.)

*An. gambiae* (Fig. 12.31) and *An. funestus* in Africa, *An. culicifacies* in Asia, and *An. dirus* in Southeast Asia. *An. gambiae* is considered the most important of all, because of its involvement in such large numbers of malaria cases and deaths, mainly in Africa. This species lives in close association with humans, on which it primarily feeds, and can complete a gonotrophic cycle in only 2 days. During the rainy seasons, larvae develop in a wide variety of sunlit surface pools, many of which are associated with human activity. These include borrow pits, roadside ditches, wheel ruts, and the hoof prints of domestic animals. Larval development normally takes only about 1 week.

Malaria has been viewed in the context of stable or unstable transmission, reflecting in part attributes of *Anopheles* species that affect their *vectorial capacity*. These include density, longevity, tendency to feed on humans, and duration of the extrinsic incubation period of the parasite in the vector. *Stable malaria* is most often associated with *P. falciparum* infection in highly endemic settings. It is characterized by low fluctuations in parasite incidence in human and vector populations, high prevalence, and high seroprevalence for antibodies. Epidemics are unlikely under these conditions, even though transmission continues at high rates. In such settings, vectors tend to be highly anthropophagic, exhibit greater longevity, and have relatively low, stable densities but still exhibit considerable seasonal variation. *Unstable malaria* tends to be associated with *P. vivax* infections in endemic settings of high fluctuation in disease incidence. Vectors tend to be zoophagic, have seasonally profound variation in population densities, have low or nondetectable field infection rates, and may have shorter longevity than do those in stable malaria settings. Epidemics can occur in conditions of unstable malaria if environmental changes favor increased vector–human contact, e.g., during civil strife, following water projects such as dams or irrigation schemes, or when a new vector is introduced into an area.

## Historical Perspective

After the development of the germ theory of disease by Louis Pasteur, the French-Algerian physician Charles Louis Alphonse Laveran examined and described malarial organisms in the red blood cells of his patients in 1870. This finding, along with the work of Patrick Manson on filarial nematodes and mosquitoes in China, inspired Ronald Ross, then a physician in British colonial India, to examine the hypothesis of mosquito transmission of malaria parasites in the 1890s. His persistent and careful experimentation and observation with both human and bird malarias, using *Anopheles* and *Culex* mosquitoes, respectively, provided conclusive proof that mosquitoes transmit *Plasmodium* species by bite. In concurrent research, Giovanni Batista Grassi and colleagues demonstrated transmission of *P. falciparum* by *An. maculipennis*–complex mosquitoes in the environs of Rome, Italy. Ross was awarded the Nobel Prize for Medicine in 1902.

Malaria was formerly endemic in many temperate areas of the United States, particularly in the South and Southeast. Malaria became epidemic after the Civil War, as malaria-infected soldiers returned to their homes and brought the infection with them to their local communities. Malaria was an important rural disease in the eastern and southern states, California, and other areas of the United States through the 1930s but gradually disappeared by the 1940s. This was due to a combination of antimosquito measures, improved medical care, a higher standard of living, and transformation of marshes and swamps to agricultural land, largely through organized ditching efforts. Changes in lifestyle because of technological advances such as window screens and the invention of the radio, television, and air conditioning also contributed to the decline in malaria. Boyd (1941) reviewed the history of malaria in the United States and, to a brief degree, elsewhere in the New World.

Roughly 1000 cases of malaria are introduced into the United States each year. In addition, cases involving local or indigenous transmission occur sporadically, including recent outbreaks in California (1988, 1989, 1990), Florida (1990, 1996), Michigan (1995), New Jersey (1993), New York (1993), and Texas (1995). These incidents were due to introductions of infected humans into areas with competent *Anopheles* vectors. However, *airport malaria* has occurred near major international airports (e.g., London–Heathrow and Paris–DeGaulle) where infected mosquitoes have been imported on aircraft from endemic regions.

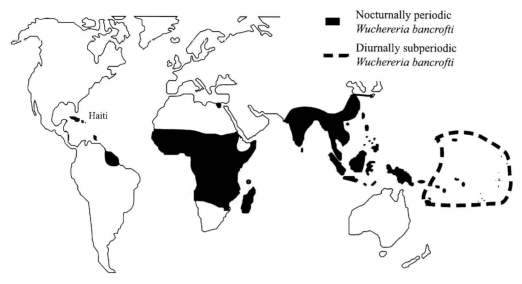

**FIGURE 12.32** Geographic distribution of human lymphatic filariasis caused by *Wuchereria bancrofti,* showing the difference between nocturnal and diurnal periodicity. (Reconstructed from Strickland, 1991, and other sources.)

## FILARIASIS

Filariasis is the infection of vertebrate tissues by filarial nematodes or roundworms (phylum Nematoda, order Spirurida, superfamily Filarioidea, family Onchocercidae). Mosquito-borne filarial nematodes are associated with acute and chronic human disease, termed *lymphatic*

*filariasis,* which is widespread in tropical and subtropical regions (Grove, 1990). The three causative agents of lymphatic filariasis are *Wuchereria bancrofti, Brugia malayi,* and *B. timori.*

The areas of the world endemic for lymphatic filariasis (Figs. 12.32 and 12.33) include parts of western, central, and southern Africa; parts of northeastern South America

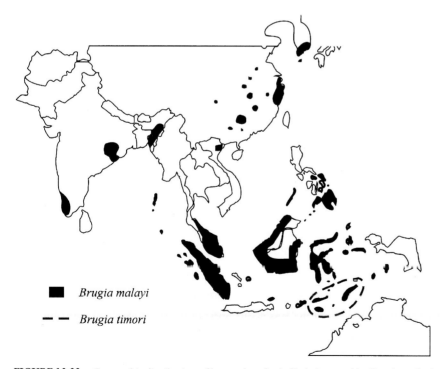

**FIGURE 12.33** Geographic distribution of human lymphatic filariasis caused by *Brugia malayi* and *B. timori.* (Reconstructed from Strickland, 1991, and other sources.)

(principally Brazil, Surinam, and French Guyana), the Dominican Republic, and Haiti; southern and eastern India, southeastern Asia, eastern China, and southern Japan; the Malay archipelago, Indonesia, the Philippines, Irian Jaya, and Papua New Guinea; and many island groups of the south Pacific Ocean, including Melanesia, Micronesia, and Polynesia. Within the United States, filariasis was locally endemic in Charleston, SC, but the disease disappeared there in the late 1930s. It disappeared at about the same time from northern Australia. It no longer occurs in regions of the Mediterranean basin and on the Arabian Peninsula. Recently, however, incidence of lymphatic filariasis has increased in the Nile River Delta of Egypt.

Within the area of current distribution, there are an estimated 905 million people at risk of contracting lymphatic filariasis, and there are some 128 million active infections. Of these, about 115 million are caused by *W. bancrofti*, the causative agent of *Bancroftian filariasis,* which is widespread in both the Old World and New World tropics (Fig. 12.32). Another 13 million cases are caused by *B. malayi*, the causative agent of *Brugian filariasis* or *Malayan filariasis,* which is restricted to southeastern Asia (Fig. 12.33). About 43 million people have chronic symptoms of elephantiasis, hydrocele, or lymphedema (see below). Another Brugian filariasis, *Timorian filariasis,* is caused by infection with *B. timori* and occurs in localized foci among southern islands of Indonesia.

### Filarial Life Cycle

Infection with filarial nematodes in humans begins when infective *third-stage larvae* enter the skin at the site of the mosquito bite, molt twice, and migrate to the lymphatic vessels and lymph nodes, particularly of the lower abdomen. There, the nematodes develop to the adult stage. Female worms (80–100 mm long at maturity for *W. bancrofti* and about half of that length for *B. malayi*) release active, immature worms called *microfilariae* into the peripheral circulatory system. A female may release 50,000 or more microfilariae each day. Microfilariae of *W. bancrofti* are 250–300 $\mu$m long and about 7–9 $\mu$m wide; those of *B. malayi* are somewhat shorter and thinner. Presence of microfilariae in the blood is called *microfilaremia* and first appears about 6 months to 1 year after adult worms become established in the lymphatic system. An infected human may be microfilaremic for more than 10 years. The density of microfilariae in peripheral blood is highly variable, but it can range from 1 to over 500 microfilariae per 20 mm$^3$.

The appearance of microfilariae in the peripheral blood has a 24-hr cycle; i.e., they exhibit *diel periodicity.* If the microfilariae completely disappear from the peripheral circulation at some time during the day, they are said to be *periodic.* If the microfilariae fluctuate in density during a 24-hr period but are detectable at all times, they are said to be *subperiodic.* In most areas, the microfilariae appear only at night and are transmitted by mosquitoes that have night-biting habits. These are the *nocturnally periodic* forms of *W. bancrofti* and *B. malayi.* Both *W. bancrofti* and *B. malayi* also have *nocturnally subperiodic* and *diurnally subperiodic* forms, although nocturnally subperiodic bancroftian and diurnally subperiodic Brugian forms have very restricted distributions. In all the subperiodic forms, the microfilariae appear in the peripheral blood of the human host mainly in the evening and night (nocturnally subperiodic) or mainly during the daytime (diurnally subperiodic). These nematodes are associated with two or more species of mosquito vectors, which differ in their typical biting times and whose combined diel patterns of man-biting density are matched by the periodicity of microfilariae. Both nocturnally periodic *W. bancrofti* and nocturnally periodic *B. malayi* are now considered to be strictly human pathogens. However, *B. malayi* in its subperiodic form is a zoonosis, with both leaf monkeys and humans as reservoirs, and with domestic cats and other carnivores also implicated as hosts. *B. timori* is nocturnally periodic and has no animal reservoir.

Development of *W. bancrofti* and *B. malayi* in mosquitoes is similar. Microfilariae ingested with the mosquito blood meal usually shed their outer, sheath-like membrane as they penetrate through the midgut epithelium. Some microfilariae retain the sheath during penetration. The microfilariae move to the indirect flight muscles of the thorax, penetrate individual cells, and transform to a short *sausage stage,* the $L_1$ or first-stage larva. They molt to more slender $L_2$, or second-stage, larvae and then again to the elongate, filariform $L_3$, or third-stage, infective larvae (about 1.5 mm long). These larvae leave the thoracic flight muscles, traverse the hemocoel of the mosquito, enter the lumen of the mouthparts, and eventually arrive at the apex of the proboscis. When the mosquito blood-feeds, the $L_3$ larvae exit through the cuticle of the labium, crawl onto the skin, and enter through the hole made by the mosquito during feeding. Therefore, technically the transmission method may be said to be contaminative, rather than inoculative. Heavy infections of larval nematodes can be fatal to the mosquito.

Many species of mosquitoes are refractory to filarial development, owing to genetic factors and to their ability to mount adequate immune responses, whereas other species are susceptible to parasite infection and support the development described above. The relationships between *W. bancrofti* and certain mosquito species exhibit evidence of local adaptation. For example, in West Africa, *Cx. quinquefasciatus* is not a competent vector of

*W. bancrofti,* whereas *Anopheles gambiae* is competent there. By contrast, in India, *Cx. quinquefasciatus* is a competent vector and most *Anopheles* species are not. Thus, there is geographic variation in susceptibility of mosquitoes to filarial nematodes.

## Clinical Disease

Generally, a case of human infection does not occur after the bite of a single, infective mosquito. Rather, it results after accumulation of hundreds to thousands of such infective bites, under conditions in which there is a high probability of parasite maturation and mating. Humans may show disease symptoms without having microfilaremia, or they may have microfilaremia without showing signs of disease. Lymphatic filariasis has both acute and chronic manifestations in the human host. The disease may cause chronic debilitation in untreated cases. *Acute lymphatic filariasis* is characterized by episodes of fever, swelling, pain, and inflammation in the affected lymph nodes and lymph vessels, a condition called *adenolymphangitis.* The episodes may last for several days and incapacitate the affected individual because of the local and systemic effects. Over time, deep abscesses may develop at the sites of inflammation. Dermal ulcers may form through the skin over these sites, and secondary bacterial infection may ensue.

In the *chronic phase,* which often occurs years after onset of acute symptoms, the pathology may involve accumulation of lymphatic fluid (*lymphedema*) in the limbs, breasts, vulva, and scrotum, resulting in swelling and enlargement. In the scrotum, this condition is termed *hydrocele.* The grotesque distentions and thickening, folding, and nodulation of the skin, notably the lower limbs, is a condition called *elephantiasis.* Bacterial and fungal infections at affected sites exacerbate these conditions. Appearance of lymph fluid in urine may occur as a consequence of the disruption of the abdominal lymphatic vessels and leakage of lymph fluid into the urinary tract, causing urine to appear whitish, a condition called *chyluria.* In contrast with *W. bancrofti,* infection with *B. malayi* is not associated with scrotal distension, but rather involves only the limbs. Hypersensitivity to parasite-associated antigens also may be part of the syndrome of lymphatic filariasis. It is mediated through elevated immunoglobulin antibodies IgE and IgG4 and is characterized by increased production of eosinophils, coughing, and shortness of breath. This type of filarial disease is called *tropical eosinophilia.*

Lymphatic filariasis has important social implications in communities where it is endemic. Acutely affected individuals are often feverish and in pain, and they may have difficulty working and thus suffer economic loss. Hard work may bring on attacks of filarial fever, which

require rest for recovery. In a study conducted in Ghana, West Africa, the episodes of acute adenolymphangitis lasted about 5 days, with 3 days of incapacitation, and occurred during those months of the year when peak agricultural work was required and when mosquito transmission of infective stage larvae was highest (Gyapong *et al.,* 1996). It is likely that chronically infected individuals are immunologically sensitized to their worm infections, and exposure to new L$_3$ larvae results in hypersensitive reactions such as adenitis. Chronically affected individuals with symptoms of scrotal hydrocele and elephantiasis may have difficulty in their social and personal lives and may suffer incontinence and impotence. Although hydrocele can be treated through fluid aspiration from the scrotum, the gross distention of elephantiasis is more difficult to remedy even with surgery.

## Mosquito Vectors and Epidemiology

The filarial nematodes that cause lymphatic filariasis have evolved associations with mosquitoes in the genera *Culex, Mansonia, Aedes, Ochlerotatus,* and *Anopheles* (Table IX). This probably has occurred through a process of adaptive radiation from the original *Anopheles* vectors in Southeast Asia. A very likely scenario for *W. bancrofti* is that it arose as a human pathogen in forested regions of Indonesia and perhaps other parts of Southeast Asia, where the *An. umbrosus* group of mosquitoes serves as vectors of *W. kalimantani.* This filarioid nematode is a parasite of leaf monkeys (also called brow-ridged langurs), *Presbytis cristata.* Similarly, *B. malayi* infection in humans probably evolved from subperiodic *B. malayi* infections in leaf monkeys (*Presbytis* spp.), with *An. hyrcanus* as the vector. Human infection and new disease foci probably arose in forests and along forest ecotones with these same *Anopheles* vectors. Through time, as people developed agricultural systems and migrated to other regions, both *W. bancrofti* and *B. malayi* adapted to these new settings and the competent mosquito vectors there. In some parts of Southeast Asia, members of the *Ochlerotatus niveus* group are important vectors of the subperiodic form of *W. bancrofti,* including *Oc. niveus* in Thailand; this system may have been the origin of the subperiodic strains that radiated to the Pacific regions.

Transmission of *W. bancrofti* occurs in both urban and rural areas. The primary mosquito vector in urban areas is *Cx. quinquefasciatus.* It is the most important vector of nocturnally periodic *W. bancrofti* in the Americas and parts of Africa and Asia, particularly India. It feeds opportunistically at night on both mammals and birds. This mosquito occurs abundantly in areas with poor sanitation, open sewers, untreated waste water, and pit latrines, which provide the high organic content and low oxygen characteristic of the larval habitat. For this reason,

**TABLE IX**

**Mosquito Vectors of Filarioid Nematodes of Humans: Geographic Distribution and Associations with Periodicity of Microfilaremia**

| Geographic region | Filarioid species | Periodicity | Mosquito vectors |
|---|---|---|---|
| Neotropical | *Wuchereria bancrofti* | Nocturnally periodic | *Anopheles aquasalis, An. bellator, An. darlingi, Ochlerotatus scapularis, Culex quinquefasciatus, Mansonia titillans* |
| Afrotropical | *W. bancrofti* | Nocturnally periodic | *An. funestus, An. gambiae, An. arabiensis, An. bwambae, An. melas, An. merus, An. nili, An. pauliani, Cx. quinquefasciatus* |
| Middle Eastern | *W. bancrofti* | Nocturnally periodic | *Cx. molestus* |
| Oriental | *W. bancrofti* | Nocturnally periodic | *An. anthropophagus, An. kweiyangensis, An. nigerrimus, An. letifer, An. whartoni, An. aconitus, An. flavirostris, An. minimus, An. candidiensis, An. balabacensis, An. leucosphyrus, An. maculatus, An. philippinensis, An. subpictus, An. vagus, Ochlerotatus niveus, Oc. togoi, Oc. poicilius, Cx. bitaeniorhynchus, Cx. sitiens complex, Cx. pallens, Cx. quinquefasciatus, Ma. uniformis* |
| | *W. bancrofti* | Nocturnally subperiodic | *An. sinensis complex, Ochlerotatus harinasutai* |
| | *Brugia malayi* | Nocturnally periodic | *An. barbirostris, An. campestris, An. donaldi, An. anthropophagus, An. kweiyangensis, An. nigerrimus, Ochlerotatus togoi, Ma. uniformis, Ma. bonnae, Ma. dives* |
| | *B. malayi* | Nocturnally subperiodic | *An. sinensis complex, Ma. uniformis, Ma. bonnae, Ma. annulata, Ma. indiana, Ma. dives* |
| | *B. timori* | Nocturnally periodic | *An. barbirostris* |
| West Pacific | *W. bancrofti* | Nocturnally periodic | *Cx. pallens, Oc. poicilius, Ochlerotatus togoi* |
| | *B. malayi* | Nocturnally subperiodic | *Ochlerotatus togoi* |
| Papuan | *W. bancrofti* | Nocturnally periodic | *An. bancrofti, An. punctulatus, An. farauti, An. koliensis, Cx. annulirostris, Cx. bitaeniorhynchus, Ma. uniformis* |
| South Pacific | *W. bancrofti* | Nocturnally subperiodic | *Ochlerotatus samoanus* |
| | | Diurnally subperiodic | *Ochlerotatus fijiensis, Oc. oceanicus, Oc. vigilax group, Aedes futunae, Ae. polynesiensis, Ae. pseudoscutellaris, Ae. tabu, Ae. tongae, Ae. upolensis* |
| | *B. malayi* | Nocturnally periodic | *Ochlerotatus oceanicus* |
| | | Nocturnally subperiodic | *Ochlerotatus oceanicus* |

*Cx. quinquefasciatus* is often more abundant during parts of the year when water stagnates from lack of rain. In the Nile River Delta of Egypt, *Cx. molestus,* a name applied to the autogenous variant of *Cx. pipiens,* is the primary vector. In rural settings, nocturnally periodic bancroftian and Brugian filariases are transmitted by *Anopheles* mosquitoes, which are also nocturnally active. Often, the same *Anopheles* species that transmit *W. bancrofti* or *B. malayi* in an area also are responsible for local malaria transmission (e.g., *An. darlingi* in South America and *An. gambiae* and *An. funestus* in parts of West and East Africa, respectively, for Bancroftian filariasis; and *An. sinensis* in rice-growing areas of China, for Brugian filariasis).

Nocturnally periodic Brugian filariasis occurs in rural parts of southern India, Malaysia, the Philippines, and Indonesia; there the nocturnally active *Mansonia annulifera* and *Ma. uniformis,* and also *Anopheles* species, are vectors near rice fields and open swamps. The nocturnally subperiodic form occurs in swamp forest areas of Southeast Asia and Indonesia, involving *Ma. bonneae* and *Ma. dives,* which are nocturnally active but also feed during the day within the swamp forests. All of these *Mansonia* species are associated with particular kinds of plants, where the larvae and pupae attach to their submerged roots and stems. In the Pacific region, where many island groups are endemic for *W. bancrofti,* the primary vectors are day-biting *Aedes* and *Ochlerotatus* species, but nocturnal biters also are involved, and the form of the parasite is diurnally subperiodic. *An. barbirostris* is the vector of *B. timori.*

The endemicity of mosquito-borne filariasis depends on a high and steady rate of transmission of infective-stage larvae in the human population. In endemic areas, the inoculation rate (parasite transfer rate) can range as high as hundreds of infective bites and thousands of larval inoculations per person per year. The estimated number of $L_3$ larvae transmitted per person per year is called the *annual transmission potential.* With each infective bite, only a few $L_3$ larvae actually enter the skin.

These larvae must then develop further and migrate to a person's lymphatic system, where mature male and female nematodes mate and initiate microfilarial production. Accumulation of thousands of infective bites over months or years eventually results in an infection of mature worms in a human, who normally then will have a microfilaremia and possibly chronic disease. Most microfilariae entering the circulatory system are never ingested by a mosquito, while those that are ingested become infective-stage larvae only in a competent vector and only if the individual mosquitoes survive the extrinsic incubation period. Furthermore, many infective-stage larvae fail to reach a new human host or fail to mature if they do. The inefficiency of transmission and parasite perpetuation is compensated by the prodigious production of microfilariae and the long life of adult worms.

### Historical Perspective

Association of infection with filarial nematodes and lymphatic filariasis was first established in the late 1800s. Our understanding of the natural history of lymphatic filariasis is related intimately to the initial discovery of a link between human pathogens and insect vectors. During 1877–1878, Patrick Manson, working in China as a medical officer for the Chinese Imperial Customs Service, conducted experiments on the development of filarial nematodes. Manson had already discovered that the microfilariae occurred in the peripheral blood only at night. He speculated that this was timed to coincide with the night-time biting activity of mosquitoes. After feeding mosquitoes (*Cx. quinquefasciatus*) on his gardener, who had a microfilaremia, and then dissecting the mosquitoes on successive days, he found that the worms developed within the mosquitoes into longer, different forms. He speculated that the mosquito functioned as a kind of "nurse" for the filarial worms, so that when a mosquito died on the water after laying an egg raft, the worms entered the water and were later infective to a person drinking it. At that time, it was not known that mosquitoes could bite more than once during their lives, so the principle of transmission by bite was not established. Yet, the idea that mosquitoes could function as intermediate hosts for a human pathogen was founded through Manson's experiments.

## VETERINARY IMPORTANCE

Aside from their importance as vectors of disease agents of animals, mosquitoes are a cause of irritation, blood loss, and allergic reactions. They not only are annoying, but also disrupt normal behavior of livestock and companion animals. Large swarms may cause livestock to discontinue feeding and to seek relief. Increased scratching behavior

may result in skin abrasions, hair loss, and secondary infection with bacteria at the bite and scratch sites. For cattle, mosquito bites can result in decreased weight gains and milk production and prompt producers to alter pasturing practices. Deaths of cattle due to anemia and stress have been reported.

### MOSQUITO-BORNE VIRUSES OF ANIMALS

Mosquito-borne viruses affecting domesticated animals include the groups of alphaviruses that are associated with the equine encephalitides (EEE, WEE, and VEE), all of which cause an acute encephalitis with high fever in equids (horses, donkeys, mules). The history, distribution, vector relationships, and vertebrate reservoir hosts of these viruses are discussed above under Public Health Importance. Other mosquito-borne viruses of veterinary significance include Japanese encephalitis virus, Rift Valley fever virus, Wesselsbron virus, fowlpox virus, and myxomatosis virus. *Equine infectious anemia virus* (EIA), a lentivirus in the family Retroviridae, may be mechanically transmitted by mosquitoes, but its more important mechanical vectors are larger biting flies (deer flies, horse flies, and stable flies).

### Eastern equine encephalomyelitis (EEE) virus

This virus is an important cause of mortality of horses and other equids, caged pheasants, whooping cranes, and emus. It occurs in endemic areas of the United States in Texas, along the Gulf coast and Atlantic seaboard to Massachusetts, and at inland sites in upstate New York, Ohio, Michigan, Indiana, Georgia, and Alabama. Horses rapidly succumb to infection after a short incubation period of 2–5 days. They exhibit abnormal behavior and high fever, then drop to the ground and lapse into coma before death (Fig. 12.34). Few horses survive infection involving these acute symptoms. Viral infection in the brain shows characteristic lesions in nerve tissue, accompanied by perivascular cuffing with macrophages; viral antigen is detectable in neurons (Fig. 12.35). In pheasant flocks, often a single infected, sick bird will be pecked by other birds, thus transmitting the virus directly to healthy birds without mosquito bite. During such occurrences, called *epiornitics,* thousands of pheasants in a single outdoor pen may die, yet none of the pheasants in adjacent pens become infected. Aside from equids and exotic birds, EEE viral infection has been reported in young dogs and pigs.

Cases of EEE in horses in cool temperate climates tend to occur in mid to late summer and early fall, whereas in milder climates horse cases begin to occur earlier in the spring and summer. In the tropics and subtropics, cases may occur year-round. Horse deaths due to EEE viral infection are an important indicator of virus activity promoted by bridge vectors in endemic areas. Rapid,

**FIGURE 12.34**   Horse dying from infection with eastern equine encephalomyelitis virus in Michigan outbreak in 1980. (Photo by H. D. Newson.)

differential diagnosis of horse cases is crucial if these animals are to be used as sentinels for potential transmission of EEE virus to humans.

### Western equine encephalomyelitis (WEE) virus

As with EEE virus, the primary epidemic host of significance for WEE virus is the horse. Human cases are rare. Since the first isolation of WEE virus from the brain of a dead horse in 1930 in the San Joaquin Valley of California, this virus has been implicated in epizootics in horses, with cases numbering from hundreds to thousands in some instances. Large outbreaks occurred in 1941 and 1975 in the Red River Valley in Minnesota, North Dakota, and Manitoba (United States and Canada), in 1952 in the Central Valley of California, and in 1965 in Hale County in western Texas. There are horse cases almost every summer within the range of the virus, but epizootics do not always occur. For both EEE and

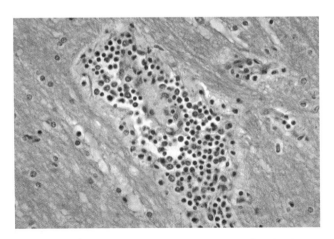

**FIGURE 12.35**   Section of horse brain (cerebrum) infected with eastern equine encephalomyelitis virus, showing neutrophil invasion around capillary. (Photo by J. D. Patterson.)

WEE viruses, immunization has reduced the frequency of horse cases.

### Venezuelan equine encephalomyelitis (VEE) Complex

Encephalomyelitis in equids caused by viruses of the VEE complex occurs in northern South America, Central America, and Mexico. The epizootic viruses are transmitted by many species of mosquitoes (see Table IV) among horses, burros, and mules. These animals develop a viremia sufficient to infect the mosquitoes. Consequently, VEE epidemics can be maintained by transmission between mosquitoes and horses, differing in this regard from WEE and EEE, for which horses are largely dead-end hosts. An epidemic strain of VEE virus was first isolated in 1938 from a horse in Venezuela. In 1969, a large outbreak of VEE involving both equids and humans in Central America spread northward in the succeeding 2 years through Mexico, and in 1971 it spread across the border into Texas (Pan American Health Organization, 1972). Cases continued in Mexico through 1972. There were thousands of horse cases throughout this region during that time. Epizootic virus activity did not occur again in the region until an outbreak in Venezuela in 1992–1993 and in Chiapas, Mexico, in 1993. Another outbreak of VEE occurred in northern Colombia and Venezuela in 1995. The rapidity of spread of these outbreaks over large geographic areas is undoubtedly due to the role of both horses and birds as competent reservoir hosts. It is expedited by the evacuation of horses, already infected but not yet ill, away from an epizootic area.

### Japanese encephalitis (JE) virus

Encephalitis caused by JE virus occurs in widespread parts of Asia, including Malaysia and Indonesia. In Japan, epizootics and epidemics have occurred in August and September in many years since the discovery of this disease in 1935 in that country. The virus was isolated from brain tissue of a horse in 1937. JE virus causes acute infection in horses and swine. It is particularly an economic problem because of the importance of swine as a food source and market commodity in rural Asia. Pigs develop viremia sufficient for mosquito transmission, therefore serving as important amplifying hosts, and may develop encephalitic symptoms. Transplacental infection causes stillbirth and abortion. Infected boars may become sterile.

### Rift Valley fever (RVF) virus

This pathogen has caused epizootics of acute illness, elevated rates of abortion, and death in cattle, goats, and sheep in Egypt and parts of sub-Saharan Africa.

Outbreaks in Egypt and Mauritania were particularly noteworthy. The virus is both viscerotropic and neurotropic in these animals. Their viremias are of sufficient titer to infect mosquitoes. Disease outbreaks generally have involved thousands to hundreds of thousands of livestock cases, causing substantial economic losses.

### Wesselsbron (WSL) virus

This is a flavivirus with distribution in parts of sub-Saharan Africa, Madagascar, and Thailand. It causes a disease similar to that of RVF in sheep and goats and also causes a mild illness in cattle. Infected ewes may abort their fetuses, and lambs suffer high mortality. Humans infected with Wesselsbron virus may develop a febrile illness with rash, fever, and myalgia. The virus is transmitted by *Aedes* species, including *Ae. mcintoshi* and *Ae. circumluteolus* in South Africa.

### Fowlpox virus

This virus belongs to a group of poxviruses that infect vertebrates and invertebrates and are classified within the family Poxviridae. Among the poxviruses are those in the bird-infecting genus *Avipoxvirus,* such as fowlpox, canarypox, and pigeonpox viruses. Mosquitoes may mechanically transmit the avipox viruses by contamination of mouthparts and subsequent transfer of infectious virions to noninfected birds. Fowlpox is an important disease of domestic fowl, particularly chickens. It causes development of papules along the comb and beak. While probing these papules, mosquitoes may contaminate their mouthparts with virions. If disturbed during feeding, they may move to another animal to feed, thus transferring the virus to a new host. Another form of fowlpox virus is transmitted directly by droplets of pus containing the virus.

### Myxoma virus

This is a leporivirus and the causative agent of *myxomatosis,* an enzootic disease of lagomorphs in parts of South America and the western United States. It is transmitted mechanically by the bite of arthropods, principally fleas. These viruses produce dermal vascularized tumors. When mosquitoes, fleas, or black flies probe these tumors, the mouthparts become contaminated with virus particles. Later, if the mosquitoes probe another uninfected lagomorph, that animal may become infected. Natural infections of myxoma virus occur without acute disease in rabbits of the genus *Sylvilagus* in South America and California. However, Old World rabbits (*Oryctolagus cuniculus*) are highly susceptible to infection and generally die. Outbreaks of acute disease among domesticated Old World rabbits have been documented in South America and California.

Myxoma virus was introduced into Australia in the 1950s as a means of controlling introduced European rabbits, a pest in that country. The virus spread rapidly through the rabbit populations via mechanical transmission by fleas, mosquitoes, and other means and greatly reduced rabbit populations there. The mosquito vectors in Australia are *Cx. annulirostris, An. annulipes,* and *Aedes* species.

## Nonhuman Malarias

Many *Plasmodium* species infect animals other than humans, including reptiles, birds, rodents, and nonhuman primates.

### Reptilian Malarias

The malarias of reptiles, also called *saurian malarias,* are caused by a group of 29 *Plasmodium* species. They infect a wide range of lizards and some snakes in 15 families (Telford, 1994). Vectors are biting midges, phlebotomine sand flies, and *Culex* mosquitoes. Haemoproteid and leucocytozooid malarias also occur in reptiles, but their vectors have not been established.

### Avian Malarias

Malarial infection of birds is widespread geographically (Van Riper *et al.,* 1994). Parasites in three common genera of hemosporine blood parasites of birds (*Hepatocystis, Haemoproteus,* and *Leucocytozoon*) are transmitted by biting midge, louse fly, and black fly vectors, respectively. The avian malarias in the genus *Plasmodium* are all mosquito-borne. *Plasmodium* species that infect birds have been important research models for studying malaria. Indeed, the original observations by Ronald Ross on the role of mosquitoes as malaria vectors were made with bird malaria.

Currently, about 30 species of avian *Plasmodium* are recognized. However, the taxonomic status of some species is uncertain and others remain to be described. Among the important species that cause disease in domestic fowl or wild birds are *Plasmodium gallinaceum* (sometimes called *chicken malaria*), *P. hermansi* (a parasite of wild and domestic turkeys in the United States), *P. relictum, P. lophurae, P. cathemerium, P. circumflexum,* and *P. elongatum.* As with human malarias, there is variation in life cycles and pathogenesis of the avian malarias. This variation is related to intrinsic qualities of the species and to variation in susceptibility among host species, age, and general health status.

A bird becomes infected after inoculation of sporozoites from an infective mosquito. Merogony occurs in bone marrow, in endothelial cells, and in the erythrocytes.

In acute infections, these parasites may cause severe anemia, damage to bone marrow tissues, and other pathology that may result in death. Younger birds tend to be more susceptible to overt illness than older birds.

Although *Anopheles* mosquitoes can be competent laboratory vectors for some bird malarias, field and laboratory data show that culicines in the genera *Culex*, *Culiseta*, *Aedes*, and *Ochlerotatus* are the natural vectors. In Africa, *Ae. aegypti* is an important local vector of *P. gallinaceum* to chickens. The impact of bird malaria on natural bird populations is poorly known. It was introduced into Hawaii along with exotic birds and *Culex* mosquitoes and is thought to be responsible for the reduction and extinction of native bird populations there. Bird malaria occasionally has been documented as the cause of morbidity and mortality among penguins in zoos.

### Rodent Malarias

The 12 *Plasmodium* species infecting rodents, called rodent or *murine malarias*, all occur in Africa and Asia. The vectors are assumed to be *Anopheles* mosquitoes, but in most cases the vector species is unknown. *P. berghei*, *P. vinckei*, *P. yoelli*, *P. chabaudi*, and *P. aegyptensis* parasitize African murine rodents. The first two are transmitted by *An. dureni* in Zaire, and *P. vinckei* is transmitted by *An. cinctus* in Nigeria. *P. atheruri* infects the African brush-tailed porcupine (*Atherurus africanus*) and is transmitted by *An. smithii*. *P. anomaluri*, *P. landauae*, and *P. pulmophilum* occur in African flying squirrels (*Anomalurus* spp.); *An. machardyi* is the probable vector of the *P. atheruri*. The three species of *Plasmodium* found in Asian flying squirrels are *P. booliati*, *P. watteni*, and *P. incertae*. The significance of rodent malarias to the health and population dynamics of their natural hosts is largely unknown, although the prevalences of infection can be high. They have become important laboratory models for human malaria, particularly in host immunological responses, drug screening studies, and vaccine development. Cox (1993) provides a succinct review of the rodent malarias.

### Primate Malarias

The nonhuman primate malarias are caused by a group of 25 *Plasmodium* species, many of which are closely related to the human malarias (Collins and Aikawa, 1993). Seven of them infect lemurs in Madagascar and are poorly known. All 18 others have life cycles similar to those of the human malarias. Most have a tertian periodicity, but two species (*P. brasilianum* and *P. inui*) are quartan and one (*P. knowlesi*) is quotidian; i.e., it has a periodicity of 1 day. Probably all are transmitted by *Anopheles* mosquitoes, but for 10 of them the vector species are unknown.

Of the 18 well-known primate malaria species, 13 occur in southern or southeastern Asia, where macaques, langurs (leaf monkeys), gibbons, and orangutans are the vertebrate hosts. These plasmodia include *P. pitheci* in the orangutan, the first nonhuman primate malaria to be described; *P. knowlesi*, a macaque parasite that has become an important laboratory model for development of human vaccines; and *P. cynomolgi*, a parasite of macaques and langurs that serves as an important model for human *P. vivax* malaria. The vectors of these Asian primate malarias include *An. hackeri*, *An. dirus*, *An. balabacensis*, *An. elegans*, and *An. introlatus*.

Three primate malarias occur in Africa, where *P. gonderi* infects mangabeys and mandrills and *P. reichenowi* and *P. schwetzi* infect chimpanzees and gorillas. Their natural vectors are unknown. Two *Plasmodium* species infect nonhuman primates in South America. *P. simium* infects howler monkeys and woolly spider monkeys in Brazil. It is similar to the human parasite *P. malariae*. *P. brasilianum* infects a wide range of New World monkeys in the family Cebidae, including howler monkeys, spider monkeys, woolly spider monkeys, titis, capuchins, woolly monkeys, bearded sakis, and squirrel monkeys. It is similar to the human parasite *P. vivax*. Both South American species are transmitted by *An. cruzii*, which also is an important vector of human malaria in parts of South America. The larvae inhabit water-filled leaf axils of bromeliad plants at heights of 5 m or more, where the adults are likely to encounter arboreal primates.

Many species of *Anopheles* are competent laboratory vectors of primate malarias, including *An. stephensi*, *An. maculatus*, *An. gambiae*, and *An. dirus*, which serve as vectors of human malaria. *P. knowlesi* can infect humans experimentally and can be transmitted by the bite of *An. dirus* to other humans. *P. cynomolgi* has infected laboratory workers, and experimental studies showed that mosquito transmission from monkeys to humans, and from humans to humans, can occur. *P. brasilianum* also infects humans. Human malaria due to infection with *P. brasilianum* and *P. simium* possibly occurred as a zoonosis in the New World prior to the arrival of Europeans, with *An. cruzii* acting as the vector. Alternatively, these two simian parasites might be derived from human *Plasmodium* species to which they are closely related. Coatney *et al.* (1971) reviewed the infectivity of nonhuman primate malarias to humans, and Collins and Aikawa (1993) reviewed the primate malarias.

## DOG HEARTWORM

Dog heartworm is caused by the mosquito-borne filarial nematode *Dirofilaria immitis*, a member of the family

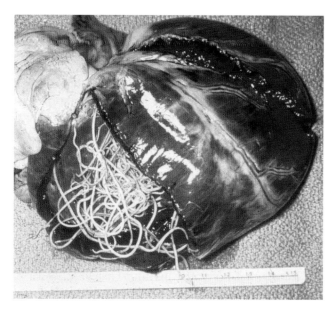

**FIGURE 12.36** *Dirofilaria immitis* adults in right ventricle of dog heart. (Photo from H. D. Newson.)

Onchocercidae (Boreham and Atwell, 1988). Adult *D. immitis* occupy the right ventricle of the canine heart and the pulmonary arteries (Fig. 12.36). The worms are 12–31 cm long and form aggregations of up to 50 or more individuals. In large aggregations, infection may extend to the right atrium. Contrary to popular belief, heartworm disease in dogs is not simply a consequence of a heavy worm burden in the ventricle resulting in impedance of blood flow. Rather, it is the result of deleterious changes in the endothelium and integrity of the walls of the pulmonary arteries, leading to pulmonary hypertension and right ventricular hypertrophy. These pathologic changes cause decreased cardiac output to the lungs, weakness, lethargy, chronic coughing, and ultimately congestive heart failure. Dogs may die if left untreated.

The life cycle of *D. immitis* involves canids and mosquitoes. Dogs become infected by the bite of a mosquito whose labium carries third-stage larvae. These larvae break out of the labium while it is bent during feeding and are deposited onto the dog's skin, along with a small droplet of mosquito hemolymph from the ruptured labium. Only about 10% of the larvae successfully enter the skin, generally through the hole made by the mosquito's fascicle. They remain *in situ* subcutaneously, where they molt to fourth-stage larvae. The larvae then migrate to other subcutaneous, adipose, or muscle tissues and molt again to a fifth-stage larva. These worms, now approximately 18 mm long, enter the venous circulation and become established in the heart and pulmonary arteries. Generally, the fifth-stage larvae reach the heart at about 70–90 days after infection.

In the heart and pulmonary arteries, the fifth-stage larvae develop into sexually mature adults. After mating, at 6–7 months, the females begin to release into circulation the microfilariae, active embryonic life stages about 300 $\mu$m long and 7 $\mu$m wide. The microfilaremia varies considerably, from 1,000 to 100,000 microfilariae per milliliter. It is nocturnally subperiodic, with peak concentrations occurring in the peripheral blood in the evening. Some dogs never develop microfilaremia, even though they support *D. immitis* adults and may have patent disease. These dogs are said to have *occult infections.*

Mosquitoes become infected with *D. immitis* when they imbibe blood from a microfilaremic dog. In an average blood meal of 5 $\mu$l, a mosquito may ingest between 5 and 500 microfilariae. Within 48 hr of ingestion, microfilariae migrate posteriorly in the midgut lumen to the Malpighian tubules and then into the distal cells of these tubules, where they develop intracellularly to "sausage forms" or first-stage larvae, taking about 4 days at 26°C. Some remain trapped in the midgut. If more than a few begin to develop in the tubules, the mosquito is likely to be killed. The first-stage larvae molt to the second stage at about 8–10 days after ingestion. As they continue to grow they cause swelling and distention of the Malpighian tubules. At 12–14 days after ingestion, they molt to the third stage. These forms break out of the Malpighian tubules and migrate through the hemolymph to the head and base of the mouthparts, then into the interior of the labium. The mosquito is then infective. The rate of these developmental processes is temperature dependent and varies with factors affecting competence for parasite development.

Vectors of *D. immitis* differ with geographic region; many mosquito species in several genera are competent to transmit it. Grieve *et al.* (1983) listed 20 species field-caught in the United States, in the genera *Aedes, Ochlerotatus, Psorophora, Anopheles,* and *Culex,* in which infective-stage larvae of *D. immitis* have been detected.

## OTHER FILARIAL
## NEMATODES OF ANIMALS

Other species of *Dirofilaria* infect mammals. These include *D. ursi,* a bear parasite transmitted by the black fly *Simulium venustum; D. roemeri,* a wallaroo (a type of small kangaroo) parasite transmitted by the horse fly *Dasybasis hebes;* and the following mosquito-transmitted *Dirofilaria* species: *Dirofilaria repens* in canids; *D. carynodes* and *D. magnilarvatum* in monkeys; *D. scapiceps* in rabbits; *D. tenuis* in raccoons; and *D. subdermata* in porcupines.

In addition to *Dirofilaria* species, a large number of filarial nematodes in other genera of the Onchocercidae infect wild and domestic animals. Vectors include mosquitoes and a wide range of other blood-feeding Diptera, lice, fleas, mites, and ticks. The mosquito-borne onchocercid nematodes include species in the following genera: *Aproctella, Breinlia, Brugia, Cardiofilaria, Conispiculum, Dirofilaria, Deraiophoronema, Folyella, Loiana, Molinema, Pelecitus, Oswaldofilaria, Saurositus, Skrjabinofilaria, Waltonella,* and *Wuchereria* (Anderson, 1992; Bain and Chabaud, 1986; Hawking and Worms, 1961; Lavoipierre, 1958). *Brugia pahangi* of jirds (*Meriones*) is an important laboratory organism for studies on filariasis. *B. malayi* develops in the peritoneal cavity of gerbils, providing a laboratory infection model.

## PREVENTION AND CONTROL

The four overlapping aims of mosquito control are to prevent mosquito bites, keep mosquito populations at acceptable densities, minimize mosquito-vertebrate contact, and reduce the longevity of female mosquitoes. All of these actions minimize the annoying and harmful effects of bites and blood loss and interrupt pathogen transmission. The eradication of either mosquito species or their associated diseases is no longer viewed as a viable objective, except in small, isolated regions or in the case of recent invasions. Two exemplary failures of the eradication approach, on a grand scale, were the World Health Organization's global malaria eradication program and the Pan American Health Organization's attempt to eradicate *Ae. aegypti* from the Western Hemisphere. A notable exception was the successful elimination of the African immigrant *An. gambiae* from Brazil. The more realistic objective of modern mosquito control programs is *integrated pest management* to reduce mosquito abundance and disease prevalence, using prudent combinations of methods.

*Personal protection* is the most direct and simple approach to prevention. Outdoor exposure can be avoided during peak mosquito activity, and window screens can prevent mosquito entry into houses and animal shelters. Head nets reduce annoyance and prevent bites about the face and neck. Bed nets, impregnated with synthetic pyrethroid and strung over beds at night, repel mosquitoes and kill those that land on the nets. Impregnated mesh suits with hoods work similarly and can be worn over clothing. Other insecticidal devices create a repellent smoke or vapor that reduces mosquito attack in the immediate vicinity. Chemical repellents applied to skin or clothing prevent mosquitoes from landing or cause them to leave before probing. The most common one is *N, N*-diethyl-m-toluamide, or DEET.

*Organized control* provides efficient, area-wide mosquito management at local, regional, or national levels. In the United States, mosquito programs typically are county-level abatement districts. These focus on the control of nuisance and vector species, but they often also participate in surveillance for mosquito-borne disease pathogens. National organizations are usually parts of ministries of health and coordinate their disease and vector control efforts at that level. Especially in developing countries, there is now increasing emphasis on community cooperation, low technology, sustainability, and the integrated use of a variety of control tools that are adapted to local customs, conditions, and resources.

*Habitat modification* is a traditional and reliable tool in mosquito management. Adult resting places can be rendered unsuitable by *harborage alteration*. Changes in larval habitat that prevent oviposition, hatching, or larval development are called *source reduction*. Water is altered or eliminated in a variety of ways. This includes plastic foam beads that provide a floating barrier over latrine water, underground sewage lines, land drainage through ditches or underground tile pipes, waste tire shredding, trash-container disposal and natural container elimination, lids for water-storage barrels, vegetational changes in ponds, altered flow of tidal water through salt marshes, and water-level manipulation in reservoirs and rice fields. Each method is designed to interfere with specific features of a mosquito's natural history. Through appropriate application of ecological principles and an intimate knowledge of mosquito behavior and life cycles, desirable natural wetlands and newly created ones can be modified to minimize mosquito production while benefitting other wildlife.

*Biological control* of mosquitoes by predators or parasites has been studied extensively and has been reviewed by Chapman (1985), Beaty and Marquardt (1996), and others. Aerial predators, such as dragonflies, birds, and bats, receive much attention but do not specialize in adult mosquitoes and have little if any effect on their densities. Most efforts have been directed at the larval stage. Aquatic predators, both naturally occurring and introduced, include the mosquito fish (*Gambusia affinis*) and killifish (*Fundulus* spp.). Other fish, such as grass carp, e.g., *Tilapia* and *Cyprinus*, remove aquatic vegetation that provides harborage for larvae. Invertebrate predators include the predatory mosquito *Toxorhynchites*, several families of aquatic bugs and beetles, predatory copepods, hydras, and turbellarian flatworms; however, none has been implemented with great success. There have been attempts to develop the use of parasites and pathogens of mosquito larvae as control agents, including the nematode *Romanomermis culicivorax*; protozoans such as the ciliates *Lambornella* and *Tetrahymena*; the gregarine sporozoan *Ascogregarina*; and the microsporidian

*Nosema.* Fungal pathogens include *Coelomomyces, Lagenidium, Culicinomyces,* and *Metarhizium.* Viruses pathogenic to larvae include the iridescent viruses, densonucleosis viruses, polyhedrosis viruses such as the baculoviruses, and entomopox viruses. Generally, the above-mentioned parasites or pathogens of mosquito larvae are still in experimental stages of development, or they have limited effectiveness and have not been used in operational programs.

An exception is the bacterium *Bacillus thuringiensis israelensis,* or *Bti,* which has been developed into commercial formulations since its original discovery in 1975. It is used extensively in mosquito control programs. Larvae die when they ingest crystalline, proteinaceous toxins produced by the bacterial cells during sporulation. The bacterium *B. sphaericus* has a similar mode of action but is more specific. It is particularly effective against *Culex* larvae, and it is more persistent in water and more tolerant of water with a high organic content than is *Bti.*

*Genetic control,* a biological control category using a variety of genetic methods, has been successful against some pests; however, its use against mosquito vectors of disease remains experimental. There are two hypothetical approaches: release of sterilized males or incompatible strains, resulting in attrition of the natural population, and replacement of natural vector populations with species or strains that are poor vectors or are not susceptible to infectious agents. These methods have been reviewed by Rai (1996).

*Chemical control* is achieved with insecticides against either larvae or adults. *Larvicides* are placed in water where larvae develop or where water will accumulate and provide habitat for larvae. Formerly used larvicides included inorganic compounds such copper arsenate, fuel oil, and organochlorine chemicals such as dichlorodiphenyltrichloroethane (DDT) and dieldrin. Currently, categories of registered larvicides are light mineral oils, organophosphates, and insect-growth regulators. Rapidly degradable oils spread over the water surface, penetrating the tracheal systems of larvae and pupae and suffocating them. Organophosphates, such as temephos, malathion, and chlorpyrifos, function as nerve poisons. The insect-growth regulator methoprene is a mimic of juvenile hormone and interferes with metamorphosis and emergence. The specific kind and formulation (dust, powder, water-soluble liquid, emulsion, oil-soluble liquid, granule, pellet, briquet) of the larvicide recommended depends on the biology of the target mosquito, the kind and size of habitat, the method of application, the chemical composition of the water, and the presence of nontarget organisms that might be adversely affected. Some can be formulated for slow release from a carrier. These may be applied to dry ground, releasing the active ingredient when inundated.

*Adulticides* are applied to surfaces where adults will rest or in the air where they fly. Residual insecticides applied to resting surfaces may retain their toxicity for days to months. They were central to the global malaria eradication program, in which DDT spraying of the inner walls of human dwellings at 6-month intervals killed all mosquitoes landing on these walls before or after taking blood. In areas where the vectors bit humans primarily indoors, this effectively interrupted most malaria transmission until mosquito populations developed resistance to the insecticide or when programs were abandoned. This approach is still used widely in some areas. Residual adulticides also can be used outdoors on vegetation or structures that serve as harborages. They tend to have short-term effects, because sunlight, wind, and rain cause the insecticide to degrade.

Adulticides intended for direct contact between airborne droplets and the mosquito are of two types: *thermal fogs* and *low-* or *ultra-low-volume* (ULV) *sprays.* Both can be applied from hand-carried equipment, motor vehicles, or aircraft. Thermal fogging involves mixing an insecticide with a combustible liquid such as kerosene. The mixture is heated, creating a fog of insecticide that drifts through the area to be treated. The ULV approach involves special nozzles and pumps that dispense fine droplets of insecticide, forming a mist that passes through the target area. Currently, insecticides registered for use in fogs and low-volume sprays are organophosphates, carbamates, pyrethrins, and synthetic pyrethroids. *Resistance* to insecticides is an important consequence of their use and has developed in many mosquito populations. The mechanisms of physiological resistance have been well characterized biochemically and genetically. Behavioral resistance also can develop. This is typically a change in adult feeding or resting behavior, so that mosquitoes no longer contact insecticide residues. Resistance has been reviewed by Ferrari (1996).

*Surveillance,* which is at the core of effective mosquito control programs, determines mosquito distribution and abundance and degree of pathogen activity. The goal is to provide data so that control agencies can take action to prevent mosquito-related problems from occurring. Unfortunately, there have been few control programs establishing *action thresholds* for mosquito density or infection rate, the levels of threat at which controls should be initiated. More often, action is based on human perception of a pest problem, conditions similar to past experience with disease outbreaks, or first detection of pathogen activity. Surveillance strategies and techniques for mosquito-borne encephalitis viruses have been presented by Moore *et al.* (1993) and Moore and Gage (1996). Bruce-Chwatt (1980) and Sasa (1976) reviewed traditional techniques for detecting malaria and filarial parasites, respectively, and several new ones are in use.

Although all methods of reducing vector populations can lower the incidence of mosquito-borne disease, quantitative models of the dynamics of disease transmission have become important tools for setting realistic control objectives. They allow programs to focus efforts on parts of the pathogen-transmission system most vulnerable to attack. Useful references on this subject are Ross (1911), Macdonald (1957), Molineaux and Gramiccia (1980), Fine (1981), Koella (1991), and Dye (1992). The *Ross-Macdonald equation* describes the case reproduction rate, or the total number of new cases of a disease arising from a single infective case in a totally susceptible population. The *vectorial capacity equation* expresses that function on a daily basis using entomological parameters. Although the vectorial capacity measure is not epidemiologically comprehensive, it allows a comparison of the relative importance of different vectors and provides estimates of *critical vector density*, the adult mosquito density below which the case reproduction is less than 1 and the disease should die out. It also illustrates, mathematically, that even small changes in the interval between bites on susceptible hosts (which is squared) or in the longevity of vectors (which changes exponentially) cause large changes in transmission rates. The latter relationship has been critical in mounting effective disease-control operations, which target older females, rather than just female density in general. More complex models sometimes show good agreement between predicted and observed results in extensive field studies of malaria ( Koella, 1991; Molineaux and Gramiccia, 1980) and may become useful in establishing action thresholds. A simple and direct measure of transmission is the *entomological inoculation rate* (Onori and Grab, 1980), which is the product of the vector's human-biting rate and proportion of vectors that are infective.

Vaccines and drugs are important tools in protecting or treating humans and other animals susceptible to mosquito-borne disease. They serve not only to protect the individual but also to reduce transmission to others. Vaccines are available for several arboviral diseases, including YF and JE for humans, and eastern, western, and Venezuelan encephalitis for equids. These vary in the duration of protection they provide. An experimental human vaccine against eastern equine encephalitis has been produced. None currently exists for the DEN viruses. Human malaria vaccines are under development, and some field trials have achieved limited success, but their wide-scale efficacy remains uncertain. The three kinds of malaria vaccines being considered use antigens from sporozoites, blood stages, or gametes; the last kind is called a *transmission-blocking vaccine* because the human antibodies take effect against stages that form within the midgut of the mosquito. Among drugs, there exists a wide spectrum of *antimalarials* used for prophylaxis,

therapy, or both. The most commonly used chemoprophylactic is chloroquine, against which there is now widespread resistance in *P. falciparum* and some *P. vivax* populations. Mefloquine is prescribed for areas with resistant populations. Strickland (1991) presented a detailed review of malaria chemotherapy and chemoprophylaxis. For lymphatic filariasis, diethylcarbamazine (DEC) is the standard chemotherapy, which reduces microfilaremia but does not kill adult worms. Advanced disease manifestations (e.g., elephantiasis) cannot be reversed, except by surgery, but sustained mass treatment of human populations can drive transmission to zero. This was achieved in parts of China in 1 year by the use of DEC-fortified cooking salt. Owing to the longevity of adult worms, mass treatment for 5–10 years is necessary to completely break the infection cycle in a community. Ivermectin and albendazole are two other drugs showing efficacy in lymphatic filariasis cases. Both DEC and ivermectin are used as chemoprophylaxis against dog heartworm infections.

# REFERENCES AND FURTHER READING

Allan, S. A., Day, J. F. and Edman, J. D. (1987). Visual ecology of biting flies. *Annu. Rev. Entomol.* **32**, 297–316.

American Mosquito Control Association. (1979). ''Mosquitoes and Their Control in the United States.'' American Mosquito Control Association, Fresno, CA.

Anderson, R. C. (1992). ''Nematode Parasites of Vertebrates: Their Development and Transmission.'' C. A. B. International, Wallingford.

Anonymous. (1994). Attractants for mosquito surveillance and control: a symposium. *Journal of the American Mosquito Control Association* **10**, 253–338.

Asman, S. M., McDonald, P. T., and Prout, T. (1981). Field studies of genetic control systems for mosquitoes. *Annu. Rev. Entomol.* **26**, 289–318.

Bain, O., and Chabaud, A. G. (1986). Atlas des larves infestantes de filaires. *Tropical Medicine and Parasitology* **37**, 301–340.

Barr, A. R. (1958). ''The Mosquitoes of Minnesota (Diptera: Culicidae: Culicinae).'' Agriculture Experiment Station Bulletin No. 228. Univ. Minnesota, Agriculture Experiment Station, St. Paul.

Bates, M. (1949). ''The Natural History of Mosquitoes.'' Macmillan, New York. (1965 edition: Harper & Row, New York).

Beaty, B. J., and Marquardt, W. C., Eds. (1996). ''The Biology of Disease Vectors.'' Univ. Press of Colorado, Niwot.

Beaty, B., Miller, B. R., Shope, R. E., *et al.* (1982). Molecular basis of bunyavirus per os infection of mosquitoes: role of the middle-sized RNA segment. *Proc. Natl. Acad. Sci. U. S. A.* **79**, 1295–1297.

Belkin, J. N. (1962). ''The Mosquitoes of the South Pacific (Diptera: Culicidae).'' Univ. California Press, Los Angeles.

Belkin, J. N., Schick, R. X., Galindo, P., and Aitken, T. H. (1965). Mosquito studies (Diptera: Culicidae) I. A project for a systematic study of the mosquitoes of Middle America. *Contributions of the American Entomological Institute* **2**, 1–17.

Bentley, M. D., and Day, J. F. (1989). Chemical ecology and behavioral aspects of mosquito oviposition. *Annu. Rev. Entomol.* **34**, 401–421.

Besansky, N. J., Finnerty, V., and Collins, F. H. (1992). A molecular genetic perspective on mosquitoes. *Advances in Genetics* **30**, 123–184.

Bock, G. R., and Cardew, G., Eds. (1996). "Olfaction in Mosquito-Host Interactions." Wiley, Chichester.

Boddy, D. W. (1948). An annotated list of the Culicidae of Washington. *Pan-Pacific Entomologist* **24**, 85–94.

Bohart, R. M., and Washino, R. K. (1978). "Mosquitoes of California." Division of Agricultural Science, Univ. of California Press, Berkeley.

Boreham, P. F. L., and Atwell, R. B., Eds. (1988). "Dirofilariasis." CRC Press, Boca Raton, FL.

Bosik, J. J. (1997). "Common Names of Insects and Related Organisms." Entomological Society of America, Lanham, MD.

Bowen, G. S., and Francy, D. B. (1980). Surveillance. In "St. Louis Encephalitis" (T. P. Monath, Ed.), pp. 473–499. American Public Health Association, Washington, DC.

Bowen, M. F. (1991). The sensory physiology of host-seeking behavior in mosquitoes. *Annu. Rev. Entomol.* **36**, 139–158.

Boyd, M. F. (1941). An historical sketch of the prevalence of malaria in North America. *American Journal of Tropical Medicine* **21**, 223–244.

Boyd, M. F., Ed. (1949). "Malariology: A Comprehensive Survey of All Aspects of This Group of Diseases from a Global Standpoint." Saunders, Philadelphia.

Bradley, T. J. (1987). Physiology of osmoregulation in mosquitoes. *Annu. Rev. Entomol.* **32**, 439–462.

Brown, M. R., and Lea, A. O. (1990). Neuroendocrine and midgut endocrine systems in the adult mosquito. *Advances in Disease Vector Research* **6**, 29–58.

Bruce-Chwatt, L. J. (1980). "Essential Malariology." Heinemann, London.

Burke, D. S., and Leake, C. J. (1988). Japanese encephalitis. In "The Arboviruses: Epidemiology and Ecology (T. P. Monath, Ed), Vol. 3, pp. 63–92. CRC Press, Boca Raton, FL.

Burkot, T. R., and Graves, P. M. (1994). Human malaria transmission: reconciling field and laboratory data. *Advances in Disease Vector Research* **10**, 149–182.

Calisher, C. H., and Karabatsos, N. (1988). Arbovirus serogroups: definition and geographic distribution. In "The Arboviruses: Epidemiology and Ecology" (T. P. Monath, Ed.), Vol. 1, pp. 19–57. CRC Press, Boca Raton, FL.

Calisher, C. H., and Thompson, W., Eds. (1983). "California Serogroup Viruses." A. R. Liss, New York.

Carlson, J., Olson, K., Higgs, S., and Beaty, B. (1995). Molecular genetic manipulation of mosquito vectors. *Annu. Rev. Entomol.* **40**, 359–388.

Carpenter, S. J. (1941). "The Mosquitoes of Arkansas." Arkansas State Board of Health, Little Rock.

Carpenter, S. J., and LaCasse, W. J. (1955). "Mosquitoes of North America (North of Mexico)." Univ. California Press, Berkeley.

Centers for Disease Control. (1977). "Mosquitoes of Public Health Importance and Their Control." Health and Human Services Publication No. (Centers for Disease Control) 82-8140. US Department of Health and Human Services, Atlanta.

Chapman, H. C. (1966). "The Mosquitoes of Nevada." US Department of Agriculture and Univ. Nevada, Carson City.

Chapman, H. C., Ed. (1985). "Biological Control of Mosquitoes." Bulletin 6. American Mosquito Control Association, Fresno, CA.

Christophers, S. R. (1911). Development of the egg follicle in Anophelines. *Paludism* **2**, 73–88.

Christophers, S. R. (1960). "*Aedes aegypti* (L.). The Yellow Fever Mosquito. Its Life History, Bionomics and Structure." Cambridge Univ. Press, London.

Clements, A. N. (1963). "The Physiology of Mosquitoes." Pergamon, New York.

Clements, A. N. (1992). "The Biology of Mosquitoes. Vol. 1. Development, Nutrition and Reproduction." Chapman & Hall, New York.

Clements, A. N. (1999). "The Biology of Mosquitoes. Vol. 2. Sensory Reception and Behaviour." CABI Publ., Wallingford.

Coatney, G. R., Collins, W. E., Warren, Mc. W., and Contacos, P. G. (1971). "The Primate Malarias." US Govt. Printing Office, Washington, DC.

Collins, W. E., and Aikawa, M. (1993). Plasmodia of nonhuman primates. In "Parasitic Protozoa" (J. P. Kreier, Ed.), 2nd ed., Vol. 5, pp. 105–134. Academic Press, New York.

Corbet, P. S., Williams, M. C., and Gillett, J. D. (1961). O'nyong nyong fever: an epidemic virus disease in East Africa: IV. Vector studies at epidemic sites. *Transactions of the Royal Society of Tropical Medicine and Hygiene* **55**, 463.

Cox, F. E. G. (1993). Plasmodia of rodents. In "Parasitic Protozoa" (J. P. Kreier, Ed.), 2nd ed., Vol. 5, pp. 49–104. Academic Press, New York.

Cox, G. W. (1944). "The Mosquitoes of Texas." Texas State Health Department, Austin.

Curtis, C. F., Ed. (1990). "Appropriate Technology in Vector Control." CRC Press, Boca Raton, FL.

Dale, P. E. R., and Hulsman, K. (1990). A critical review of salt marsh management methods for mosquito control. *Aquatic Sciences* **3**, 281–311.

Darsie, R. F., Jr. (1951). Pupae of the culicine mosquitoes of the northeastern United States (Diptera, Culicidae, Culicini). *Memoirs of the Cornell Agriculture Experiment Station Bulletin* **304**, 1–67.

Darsie, R. F., Jr. (1989). Keys to the genera, and to the species of five minor genera, of mosquito pupae occurring in the Nearctic region (Diptera: Culicidae). *Mosquito Systematics* **21**, 1–10.

Darsie, R. F., Jr., and Morris, C. D. (1998). Keys to the adult females and fourth instar larvae of the mosquitoes of Florida (Diptera, Culicidae). Bulletin of the Florida Mosquito Control Association No. 1. Florida Medical Entomology Laboratory, Institute of Food and Agricultural Sciences, Univ. Florida, Vero Beach.

Darsie, R. F., Jr., and Ward, R. A. (1981). Identification and geographical distribution of the mosquitoes of North America, north of Mexico. *Mosquito Systematics Supplement* **1**, 1–313.

Davidson, E. W., and Becker, N. (1996). Microbial control of vectors. In "The Biology of Disease Vectors" (B. J. Beaty and W. C. Marquardt, Eds.), pp. 549–663. Univ. Press of Colorado, Niwot.

de Barjac, H., and Sutherland, D. J., Eds. (1989). "Bacterial Control of Mosquitoes and Black Flies." Rutgers Univ. Press, New Brunswick, NJ.

Debenham, M. L., Ed. (1987a). "Culicidae of the Australasian Region. Vol. 4. Nomenclature, Synonymy, Literature, Distribution, Biology and Relation to Disease: Genus *Aedes*, Subgenera *Scutomyia*, *Stegomyia*, *Verrallina*." Entomology Monograph No. 2 (in part). Australian Government Publication Service, Canberra.

Debenham, M. L., Ed. (1987b). "Culicidae of the Australasian Region. Vol. 5. Nomenclature, Synonymy, Literature, Distribution, Biology and Relation to Disease: Genus *Anopheles*, Subgenera *Anopheles*, *Cellia*." Entomology Monograph No. 2 (in part). Australian Government Publication Service, Canberra.

Debenham, M. L., Ed. (1988a). "Culicidae of the Australasian Region. Vol. 6. Nomenclature, Synonymy, Literature, Distribution, Biology and Relation to Disease: Genera *Armigeres*, *Bironella* And *Coquillettidia*." Entomology Monograph No. 2 (in part). Australian Government Publication Service, Canberra.

Debenham, M. L., Ed. (1988b). "Culicidae of the Australasian Region. Vol. 9. Nomenclature, Synonymy, Literature, Distribution, Biology and Relation to Disease: Genus *Culex* (Subgenera *Lutzia*, *Neoculex*, Subgenus Undecided), Genera *Culiseta*, *Ficalbia*, *Heizmannia*, *Hodgesia*, *Malaya*, *Mansonia*." Entomology Monograph No. 2 (in part). Australian Government Publication Service, Canberra.

Debenham, M. L., Ed. (1988c). "Culicidae of the Australasian Region. Vol. 10. Nomenclature, Synonymy, Literature, Distribution, Biology and Relation to Disease: Genera *Maorigoeldia, Mimomyia, Opifex, Orthopodomyia, Topomyia, Toxorhynchites.*" Entomology Monograph No. 2 (in part). Australian Government Publication Service, Canberra.

Debenham, M. L., Ed. (1989a). "Culicidae of the Australasian Region. Vol. 7. Nomenclature, Synonymy, Literature, Distribution, Biology and Relation to Disease: Genus *Culex*, Subgenera *Acallyntrum, Culex.*" Entomology Monograph No. 2 (in part). Australian Government Publication Service, Canberra.

Debenham, M. L., Ed. (1989b). "Culicidae of the Australasian Region. Vol. 8. Nomenclature, Synonymy, Literature, Distribution, Biology and Relation to Disease: Genus *Culex*, Subgenera *Culiciomyia, Eumelanomyia, Lophoceraomyia.*" Entomology Monograph No. 2 (in part). Australian Government Publication Service, Canberra.

Debenham, M. L., Hicks, M. M., and Griffiths, M., Eds. (1989). "Culicidae of the Australasian Region. Vol. 12. Summary of Taxonomic Changes, Revised Alphabetical List Of Species, Supplementary Bibliography, Errata and Addenda, Geographic Guide to Species, Synopsis of Disease Relationships, Indexes." Entomology Monograph No. 2 (in part). Australian Government Publication Service, Canberra.

DeFoliart, G. R., Grimstad, P. R., and Watts, D. M. (1987). Advances in mosquito-borne arbovirus/vector research. *Annu. Rev. Entomol.* **32**, 479–505.

Detinova, T. S. (1962). Age-grading methods in Diptera of medical importance. *World Health Organization Monograph Series* **47**, 1–216.

de Zulueta, J. (1994). Malaria and ecosystems: from prehistory to posteradication. *Parassitologia* **36**, 7–15.

Dickinson, W. E. (1944). The mosquitoes of Wisconsin. Bulletin of the Public Museum, City of Milwaukee 8(3), 269–365.

Dixon, R. D., and Brust, R. A. (1972). Mosquitoes of Manitoba. III. Ecology of larvae in the Winnipeg area. *Canadian Entomologist* **104**, 961–968.

Downes, J. A. (1969). The swarming and mating flight of Diptera. *Annu. Rev. Entomol.* **14**, 271–298.

Dye, C. (1992). The analysis of parasite transmission by bloodsucking insects. *Annu. Rev. Entomol.* **37**, 1–19.

Eldridge, B. F. (1990). Evolutionary relationships among California serogroup viruses (Bunyaviridae) and *Aedes* mosquitoes. *Journal of Medical Entomology* **27**, 738–749.

Evenhuis, N. L., and Gon, S. M., III. (1989). Family Culicidae. In "Catalog of the Diptera of the Australasian and Oceanian Regions" (N. L. Evenhuis, Ed.), pp. 191–218. Bishop Museum Press, Honolulu.

Ferrari, J. A. (1996). Insecticide resistance. In "The Biology of Disease Vectors" (B. J. Beaty and W. C. Marquardt, Eds.), pp. 512–529. Univ. Press of Colorado, Niwot.

Fields, B. N., Knipe, D. N., and Howley, P. M., Eds. (1996). "Fields Virology," 3rd ed., Vols. 1–2. Lippincott-Raven, Philadelphia.

Fine, P. E. M. (1981). Epidemiological principles of vector-mediated transmission. In "Vectors of Disease Agents" (J. J. McKelvey, B. F. Eldridge, and K. Maramorosch, Eds.), pp. 77–91. Praeger, New York.

Foote, R. H., and Cook, D. R. (1959). "Mosquitoes of Medical Importance." Department of Agriculture, Agriculture Handbook No. 152. Agricultural Research Service, Washington, DC. pp. 1–158.

Forattini, O. P. (1962, 1965). "Entomologia Medica. Vol. 1: Parte Geral., Diptera, Anophelini; Vol. 2: Culicini: *Culex, Aedes* e *Psorophora;* Vol. 3: Culicini: *Haemagogus, Mansonia, Culiseta,* Sabethini. Toxorhychitini. Arboviruses. Filariose bancroftiana. Genetica." Universidade de Sao Paulo, Faculdade de Higiene e Saude Publica.

Foster, W. A. (1995). Mosquito sugar feeding and reproductive energetics. *Annu. Rev. Entomol.* **40**, 443–474.

Fox, A. S., and Brust, R. A. (1994). How do dilatations form in mosquito ovarioles? *Parasitol. Today* **10**, 19–23.

Gaffigan, T. V., and Ward, R. A. (1985). Index to the second supplement to "A catalog of the mosquitoes of the world," with corrections and additions. *Mosquito Systematics* **17**, 52–63.

Gartrell, F. E., Cooney, J. C., Chambers, G. P., and Brooks, R. H. (1981). TVA mosquito control 1934–1980—experience and current program trends and developments. *Mosquito News* **41**, 302–322.

Gerhardt, R. W. (1966). "South Dakota Mosquitoes and their Control." Agriculture Experiment Station Bulletin 531, South Dakota State Univ., Brookings.

Gilles, H. M., and Warrell, D. A. (1994). "Bruce-Chwatt's Essential Malariology," 3rd ed. Little, Brown, & Co., Boston.

Gillett, J. D. (1971). "The Mosquito: Its Life, Activities, and Impact on Human Affairs." Doubleday, New York.

Gillies, M. T. (1988). Anopheline mosquitos: vector behaviour and bionomics. In "Malaria Principles and Practice of Malariology" (W. H. Wernsdorfer and I. McGregor, Eds.), Vol. 1, pp. 453–485. Churchill Livingstone, Edinburgh.

Gillies, M. T., and Coetzee, M. (1987). A supplement to the Anophelinae of Africa south of the Sahara (Afrotropical region). *Publications of the South African Institute for Medical Research* **55**, 1–143.

Gillies, M. T., and de Meillon, B. (1968). The Anophelinae of Africa south of the Sahara (Ethiopian geographical region). *Publications of the South African Institute for Medical Research* **54**, 1–343.

Gjullin, C. M., Sailer, R. I., Stone, A., and Travis, B. V. (1961). "The Mosquitoes of Alaska." Agriculture handbook, US Department of Agriculture, No. 182. Agricultural Research Service, Washington, DC.

Gordon, R. M., and Lavoipierre, M. M. J. (1962). "Entomology for Students of Medicine." Blackwell Sci., Oxford.

Grieve, R. B., Lok, J. B., and Glickman, L. T. (1983). Epidemiology of canine heartworm infection. *Epidemiological Review* **5**, 220–246.

Grimstad, P. R. (1988). California group virus disease. In "The Arboviruses: Epidemiology and Ecology" (T. P. Monath, Ed.), Vol. 2, pp. 99–136. CRC Press, Boca Raton, FL.

Grove, D. (1990). "A History of Human Helminthology." C. A. B. International, Wallingford.

Gubler, D. J. (1988). Dengue. In "The Arboviruses: Epidemiology and Ecology" (T. P. Monath, Ed.), Vol. 2, pp. 223–260. CRC Press, Boca Raton, FL.

Gubler, D. J., and Bhattacharya, N. C. (1974). A quantitative approach to the study of bancroftian filariasis. *American Journal of Tropical Medicine and Hygiene* **23**, 1027–1036.

Gubler, D. J., and Clark, G. G. (1994). Community-based integrated control of *Aedes aegypti:* a brief overview of current programs. *American Journal of Tropical Medicine and Hygiene* **50**, 50–60.

Gubler, D. J., and Kuno, G., Eds. (1997). "Dengue and Dengue Hemorrhagic Fever." CAB International, Wallingford.

Gutsevich, A. V., Monchadskii, A. S., and Shtakel'berg, A. A. (1971). "Mosquitoes, Family Culicidae. Fauna of the U. S. S. R.: Diptera 3(4) Academy of Science, Zoological Institute, Leningrad. (English translation 1974, Israel Program for Scientific Translations, Jerusalem.)

Gwadz, R., and Collins, F. H. (1996). Anopheline mosquitoes and the agents they transmit. In "The Biology of Disease Vectors" (B. J. Beaty and W. C. Marquardt, Eds.), pp. 73–84. Univ. Press of Colorado, Niwot.

Gyapong, J. O., Gyapong, M., and Adjei, S. (1996). The epidemiology of acute adenolymphangitis due to lymphatic filariasis in northern Ghana. *American Journal of Tropical Medicine and Hygiene* **54**, 591–595.

Hagedorn, H. H. (1994). The endocrinology of the adult female mosquito. *Advances in Disease Vector Research* **10**, 109–148.

Hagedorn, H. H. (1996). Physiology of mosquitoes. In "The Biology of Disease Vectors" (B. J. Beaty and W. C. Marquardt, Eds.), pp. 273–297. Univ. Press of Colorado, Niwot.

Harbach, R. E., and Kitching, I. J. (1998). Phylogeny and classification of the Culicidae (Diptera). *Systematic Entomology* **23**, 327–370.

Harbach, R. E., and Knight, K. L. (1980). "Taxonomist's Glossary of Mosquito Anatomy." Plexus, Marlton, NJ.

Harbach, R. E., and Knight, K. L. (1981). Corrections and additions to Taxonomists' Glossary of Mosquito Anatomy. *Mosquito Systematics* **13**, 201–217.

Hardy, J. L. (1988). Susceptibility and resistance of vector mosquitoes. In "The Arboviruses: Epidemiology and Ecology" (T. P. Monath, Ed.), Vol. 1, pp. 87–126. CRC Press, Boca Raton, FL.

Harris, K. F., Ed. (1985–). "Advances in Disease Vector Research (formerly Current Topics in Disease Vector Research)." Vol. 1–. Springer-Verlag, New York.

Harrison, G. (1978). "Mosquitoes, Malaria and Man: A History of the Hostilities since 1880." Murray, London.

Hawking, F., and Worms, M. (1961). Transmission of filarioid nematodes. *Annu. Rev. Entomol.* **6**, 413–432.

Hawley, W. A. (1988). The biology of *Aedes albopictus*. Journal of the American Mosquito Control Association 4(Suppl. 1), 1–40.

Hawley, W. A., Reiter, P., Copeland, R. S., Pumpuni, C. B., and Craig, G. B., Jr. (1987). *Aedes albopictus* in North America: probable introduction in tires from northern Asia. *Science* **236**, 1114–1116.

Hayes, C. G. (1988). West Nile fever. In "The Arboviruses: Epidemiology and Ecology" (T. P. Monath, Ed.), Vol. 5, pp. 59–88. CRC Press, Boca Raton, FL.

Headlee, T. J. (1945). "The Mosquitoes of New Jersey and Their Control." Rutgers Univ. Press, New Brunswick, NJ.

Hearle, E. (1926). "The Mosquitoes of the Lower Fraser Valley, British Columbia, and Their Control." Report No. 17. National Research Council, Ottawa.

Hicks, M. M., Ed. (1989). "The Culicidae of the Australasian Region. Vol. 11. Nomenclature, Synonymy, Literature, Distribution, Biology and Relation to Disease: Genera *Tripteroides, Uranotaenia, Wyeomyia, Zeugnomyia*." Entomology Monograph No. 2. Australian Government Publication Service, Canberra.

Higgs, S., and Beaty, B. J. (1996). Rearing and containment of mosquito vectors. In "The Biology of Disease Vectors" (B. J. Beaty and W. C. Marquardt, Eds.), pp. 595–605. Univ. Press of Colorado, Niwot.

Hoc, T. Q. (1996). Application of the ovarian oil injection and ovariolar separation techniques for age grading hematophagous Diptera. *Journal of Medical Entomology* **33**, 290–296.

Hopkins, C. C., Hollinger, F. B., Johnson, R. F., Dewlett, H. J., Newhouse, V. F., and Chamberlain, R. W. (1975). The epidemiology of St. Louis encephalitis in Dallas, Texas, 1966. *American Journal of Epidemiology* **102**, 1–15.

Hopkins, G. H. E. (1952). "Mosquitoes of the Ethiopian Region I. Larval Bionomics of Mosquitoes and Taxonomy of Culine Larvae," 2nd ed. British Museum (Natural History), London.

Horsfall, W. R. (1955). "Mosquitoes. Their Bionomics and Relation to Disease." Ronald Press, New York.

Horsfall, W. R., Fowler, H. W., Jr., Moretti, L. J., and Larsen, J. R. (1973). "Bionomics and Embryology of the Inland Floodwater Mosquito *Aedes vexans*." Univ. Illinois Press, Urbana.

Iversson, L. B. (1988). Rocio encephalitis. In "The Arboviruses: Epidemiology and Ecology" (T. P. Monath, Ed.), Vol. 4, pp. 77–92. CRC Press, Boca Raton, FL.

Johnston, R. E., and Peters, C. J. (1996). Alphaviruses. In "Fields Virology" (B. N. Fields, D. M. Knipe, P. M. Howley, *et al.*, Eds.), 3rd ed., pp. 843–898. Lippincott, Philadelphia.

Jones, J. C. (1978). The feeding behavior of mosquitoes. *Scientific American* **238**, 138–148.

Jupp, P. G., and McIntosh, B. M. (1988). Chikungunya virus disease. In "The Arboviruses: Epidemiology and Ecology" (T. P. Monath, Ed.), Vol. 2, pp. 137–158. CRC Press, Boca Raton, FL.

Karabatsos, N., Ed. (1985). "International Catalogue of Arthropod-Borne Viruses," 3rd ed. American Society for Tropical Medicine and Hygiene, San Antonio.

Kay, B. H., and Aaskov, J. G. (1988). Ross River virus (epidemic polyarthritis). In "The Arboviruses: Epidemiology and Ecology" (T. P. Monath, Ed.), Vol. 4, pp. 93–112. CRC Press, Boca Raton, FL.

Kettle, D. S. (1995). "Medical and Veterinary Entomology," 2nd ed. CAB International, Wallingford.

King, W. V., Bradley, G. H., Smith, C. N., and McDuffie, W. C. (1960). "A Handbook of the Mosquitoes of the Southeastern United States." Agriculture Handbook, US Department of Agriculture, No. 173. Agricultural Research Service, Washington, DC.

Klowden, M. J. (1996). Vector behavior. In "The Biology of Disease Vectors" (B. J. Beaty and W. C. Marquardt, Eds.), pp. 34–50. Univ. Press of Colorado, Niwot.

Knight, K. L. (1978). "Supplement to a Catalog of the Mosquitoes of the World (Diptera: Culicidae)." Thomas Say Foundation, Supplement to Vol. 6. Entomological Society of America, College Park, MD.

Knight, K. L., and Stone, A. (1977). "A Catalog of the Mosquitoes of the World (Diptera: Culicidae)," 2nd ed. Thomas Say Foundation. Vol. 6. Entomological Society of America, College Park, MD.

Knight, K. L., and Wonio, M. (1969). "Mosquitoes of Iowa (Diptera: Culicidae)." Agricultural and Home Economic Experiment Station. Special Report No. 61. Iowa State University of Science and Technology, Ames.

Koella, J. C. (1991). On the use of mathematical models of malaria transmission. *Acta Tropica* **49**, 1–25.

LaCasse, W. J., and Yamaguti, S. (1950). "Mosquito Fauna of Japan and Korea." Office of the Surgeon, Hq. 8th US Army, Kyoto.

Lacey, L. A., and Undeen, A. H. (1986). Microbial control of black flies and mosquitoes. *Annu. Rev. Entomol.* **31**, 265–296.

Laird, M. (1988). "The Natural History of Larval Mosquito Habitats." Academic Press, New York.

Laird, M., and Miles, J. W., Eds. (1983). "Integrated Mosquito Control Methodologies. Vol. 1. Experience and Components from Conventional Chemical Control." Academic Press, New York.

Laird, M., and Miles, J. W., Eds. (1985). "Integrated Mosquito Control Methodologies. Vol. 2. Biocontrol and Other Innovative Components, and Future Directions." Academic Press, New York.

Lane, J. (1953). "Neotropical Culicidae." (3 vols.). Univ. of Sao Paulo, Sao Paulo.

Laven, H. (1967). Eradication of *Culex pipiens fatigans* through cytoplasmic incompatibility. *Nature* **216**, 383–384.

Lavoipierre, M. M. (1958). Studies on the host-parasite relations of filarial nematodes and their arthropod hosts. II. The arthropod as a host to the nematode; a brief appraisal of our present knowledge, based on a study of the more important literature from 1878 to 1957. *Annals of Tropical Medicine and Parasitology* **52**, 326–345.

Lee, D. J., Hicks, M. M., Griffiths, M., Russell, R. C., and Marks, E. N. (1980). "The Culicidae of the Australasian Region. Vol. 1." Entomology Monograph No. 2 (in part). Australian Government Publication Service, Canberra.

Lee, D. J., Hicks, M. M., Griffiths, M., Russell, R. C., and Marks, E. N. (1982). "The Culicidae of the Australasian Region. Vol. 2. Nomenclature, Synonymy, Literature, Distribution, Biology and Relation to Disease: Genus *Aedeomyia*, Genus *Aedes* (Subgenera *Aedes, Aedimorphus, Chaetocruiomyia, Christophersiomyia, Edwardsaedes* and

*Finlaya*)." Entomology Monograph No. 2 (in part). Australian Government Publication Service, Canberra.

Lee, D. J., Hicks, M. M., Griffiths, M., Russell, R. C., and Marks, E. N. (1984). "The Culicidae of the Australasian Region. Vol. 3. Nomenclature, Synonymy, Literature, Distribution, Biology and Relation to Disease: Genus *Aedes*, Subgenera *Geokusea, Halaedes, Huaedes, Leptosomatomyia, Levua,Lorrainea, Macleaya, Mucidus, Neomelanoconion, Nothoskusea, Ochlerotatus, Paraedes, Pseudoskusea, Rhinoskusea*." Entomology Monograph No. 2 (in part). Australian Government Publication Service, Canberra.

Lindsay, S. W., and Gibson, M. E. (1988). Bednets revisited: old idea, new angle. *Parasitology Today* **4**, 270–272.

Lounibos, L. P., Rey, J. R., and Frank, J. H., Eds. (1985). "Ecology of Mosquitoes: Proceedings of a Workshop." Florida Medical Entomology Laboratory, Vero Beach.

Lu, B. L., and Su, L. (1987). "A Handbook for the Identification of Chinese Aedine Mosquitoes." (In Chinese.) Science Press, Beijing.

Lu, B. L., Chen, B. H., Xu, R., and Ji, S. (1988). "A Checklist of Chinese Mosquitoes (Diptera: Culicidae)." (In Chinese, English introduction.) Guizhu People's Publ. House, Beijing.

Macdonald, G. (1957). "The Epidemiology and Control of Malaria." Oxford Univ. Press, London.

Mail, G. A. (1934). "The Mosquitoes of Montana." Agriculture Experiment Station Bulletin No. 288. Montana State College, Bozeman.

Marshall, I. D. (1988). Murray Valley and Kunjin encephalitis. In "The Arboviruses: Epidemiology and Ecology" (T. P. Monath, Ed.), Vol. 3, pp. 151–190. CRC Press, Boca Raton, FL.

Marshall, J. F. (1938). "The British Mosquitoes." British Museum (Natural History), London.

Matheson, R. (1944). "Handbook of the Mosquitoes of North America," 2nd ed. Comstock, Ithaca, NY.

Mattingly, P. F. (1969). "The Biology of Mosquito-Borne Disease." Allen & Unwin, London.

Mattingly, P. F. (1971). Contributions to the mosquito fauna of Southeast Asia. XII. Illustrated keys to the genera of mosquitoes (Diptera: Culicidae). *Contributions of the American Entomological Institute* **7**, 1–84.

Mattingly, P. F. (1973). Culicidae (Mosquitoes). In "Insects and Other Arthropods of Medical Importance" (K. G. V. Smith, Ed.), pp. 37–107. British Museum (Natural History), London.

McDonald, J. L., Sluss, T. P., Lang, J. D., and Roan, C. C. (1973). "Mosquitoes of Arizona." Agriculture Experiment Station Technical Bulletin No. 205. Univ. Arizona, Tucson.

McIver, S. B. (1982). Sensilla of mosquitoes (Diptera: Culicidae). *Journal of Medical Entomology* **19**, 489–535.

McKelvey, J. J., Eldridge, B. F., and Maramorosch, K., Eds. (1981). "Vectors of Disease Agents. Interactions with Plants, Animals, and Man." Praeger, New York.

McKiel, J. A., Hall, R. R., and Newhouse, V. F. (1966). Viruses of the California encephalitis complex in indicator rabbits. *American Journal of Tropical Medicine and Hygiene* **15**, 98–102.

Means, R. G. (1979). "Mosquitoes of New York. Part I. The Genus *Aedes* Meigen with Identification Keys to Genera of Culicidae; Part II. Genera of Culicidae Other Than *Aedes* Occurring in New York." University of the State of New York, State Education Department, State Science Service, New York State Museum, Albany.

Meegan, J. M., and Bailey, C. L. (1988). Rift Valley fever. In "The Arboviruses: Epidemiology and Ecology" (T. P. Monath, Ed.), Vol. 4, pp. 51–76. CRC Press, Boca Raton, FL.

Meola, R., and Readio, J. (1988). Juvenile hormone regulation of biting behavior and egg development in mosquitoes. *Advances in Disease Vector Research* **5**, 1–24.

Merritt, R. W., Dadd, R. H., and Walker, E. D. (1992). Feeding behavior, natural food, and nutritional relationships of larval mosquitoes. *Annu. Rev. Entomol.* **37**, 349–376.

Minar, J. (1991). Family Culicidae. In "Catalogue of Palearctic Diptera. Vol. 2. Psychodidae—Chironomidae" (A. Soos and L. Papp, Eds.), pp. 73–113. Elsevier, Amsterdam.

Mitchell, C. J. (1977). Arthropod-borne encephalitis viruses and water resource developments. *Cahiers O.R.S.T.O.M. Serie Entomologie Medicale et Parasitologie* **15**, 241–250.

Mitchell, C. J. (1983). Mosquito vector competence and arboviruses. *Current Topics in Vector Research* **1**, 63–92.

Molineaux, L., and Gramiccia, G. (1980). "The Garki Project." World Health Organization, Geneva.

Monath, T. P., Ed. (1980). "St. Louis Encephalitis." American Public Health Association, Washington, DC.

Monath, T. P., Ed. (1988a). "The Arboviruses: Epidemiology and Ecology." Vols. 1–5. CRC Press, Boca Raton, FL.

Monath, T. P. (1988b). Yellow fever. In "The Arboviruses: Epidemiology and Ecology" (T. P. Monath, Ed.), Vol. 3, pp. 139–231. CRC Press, Boca Raton, FL.

Monath, T. P., and Heinz, F. X. (1996). Flaviviruses. In "Fields Virology" (B. N. Fields, D. M. Knipe, P. M. Howley, *et al.*, Eds.), 3rd ed., pp. 961–1034. Lippincott, Philadelphia.

Moore, C. G., and Gage, K. L. (1996). Collection methods for vector surveillance. In "Biology of Disease Vectors" (B. J. Beaty and W. C. Marquardt, Eds.), pp. 471–491. Univ. Press of Colorado, Niwot.

Moore, C. G., McLean, R. G., Mitchell, C. J., Nasci, R. S., Tsai, T. F., Calisher, C. H., Marfin, A. A., Moore, P. S., and Gubler, D. J. (1993). "Guidelines for Arbovirus Surveillance in the United States." Centers for Disease Control and Prevention, US Department of Health and Human Services, Fort Collins, CO.

Morris, C. D. (1988). Eastern equine encephalomyelitis. In "The Arboviruses: Epidemiology and Ecology" (T. P. Monath, Ed.), Vol. 3, pp. 1–20. CRC Press, Boca Raton, FL.

Morris, C. D., Baker, R. H., and Opp, W. R., Eds. (1992). "H. T. Evans' Florida Mosquito Control Handbook." Florida Mosquito Control Association, Fort Myers.

Muirhead-Thomson, R. C. (1951). "Mosquito Behaviour in Relation to Mosquito Transmission and Control in the Tropics." Arnold, London.

Muirhead-Thomson, R. C. (1968). "Ecology of Insect Vector Populations." Academic Press, New York.

Muirhead-Thomson, R. C. (1982). "Behaviour Patterns of Blood-Sucking Flies." Pergamon, Oxford.

Nasci, R. S., and Miller, B. R. (1996). Culicine mosquitoes and the agents they transmit. In "The Biology of Disease Vectors" (B. J. Beaty and W. C. Marquardt, Eds.), pp. 85–97. Univ. Press of Colorado, Niwot.

Nayar, J. K. (1982). "Bionomics and Physiology of *Culex nigripalpus* (Diptera: Culicidae) of Florida: An Important Vector of Diseases." Florida Agriculture Experiment Stations, Bulletin No. 827. Institute of Food and Agricultural Sciences, University of Florida, Gainesville.

Nayar, J. K., Ed. (1985). "Bionomics and Physiology of *Aedes taeniorhynchus* and *Aedes sollicitans,* the Salt Marsh Mosquitoes of Florida. Florida Agriculture Experiment Stations, Bulletin No. 852. Institute of Food and Agricultural Sciences, University of Florida, Gainesville.

Nedelman, J. (1990). Gametocytemia and infectiousness in falciparum malaria: observations and models. *Advances in Disease Vector Research* **6**, 59–89.

O'Meara, G. F. (1985). Ecology of autogeny in mosquitoes. In "Ecology of Mosquitoes: Proceedings of a Workshop" (E. P. Lounibos, J. R. Rey, and J. H. Frank, Eds.), pp. 459–471. Florida Medical Entomology Laboratory, Vero Beach.

Onori, E., and Grab, B. (1980). Indicators for the forecasting of malaria epidemics. *Bulletin of the World Health Organization* **58**, 91–98.

Ottesen, E. A., and Ramachandran, C. P. (1995). Lymphatic filariasis infection and disease: control strategies. *Parasitology Today* **11**, 129–131.

Owen, W. B., and Gerhardt, R. W. (1957). The mosquitoes of Wyoming. Publication 21(3), pp. 71–141. Univ. Wyoming, Laramie, WY.

Pampana, E. (1963). "A Textbook of Malaria Eradication." Oxford Univ. Press, Oxford.

Pan American Health Organization. (1972). "Venezuelan Encephalitis." Pan American Sanitary Bureau, Scientific Publication No. 243. Pan American Health Organization, Washington, DC.

Patarroyo, M. E., Amador, R., Clavijo, P., *et al.* (1988). A synthetic vaccine protects humans against challenge with asexual blood stages of *Plasmodium falciparum* malaria. *Nature* **332**, 158–161.

Peters, W. (1985). The problem of drug resistance in malaria. *Parasitology* **90**, 705–715.

Pittaway, A. R. (1992). "Arthropods of Medical and Veterinary Importance: A Checklist of Preferred Names and Allied Terms." CAB International, Wallingford.

Pratt, H. D., Barnes, R. C., and Littig, K. S. (1963). "Mosquitoes of Public Health Importance and Their Control." Publication No. 772. Public Health Service, Washington, DC.

Rai, K. S. (1991). *Aedes albopictus* in the Americas. *Annu. Rev. Entomol.* **36**, 459–484.

Rai, K. S. (1996). Genetic control of vectors. In "Biology of Disease Vectors" (B. J. Beaty and W. C. Marquardt, Eds.), pp. 564–574. Univ. Press of Colorado, Niwot.

Raikhel, A. S. (1992). Vitellogenesis in mosquitoes. *Advances in Disease Vector Research* **9**, 1–39.

Reeves, W. C., Ed. (1990). "Epidemiology and Control of Mosquito-Borne Arboviruses in California, 1943–1987. California Mosquito and Vector Control Association, Sacramento.

Reinert, J. F., Kaiser, P. E., and Seawright, J. A. (1997). Analysis of the *Anopheles* (*Anopheles*) *quadrimaculatus* complex of sibling species (Diptera: Culicidae) using morphological, cytological, molecular, genetic, biochemical, and ecological techniques in an integrated approach. Journal of the American Mosquito Control Association 13(Suppl.), 1–102.

Reisen, W. K., and Monath, T. P. (1988). Western equine encephalomyelitis. In "The Arboviruses: Epidemiology and Ecology" (T. P. Monath, Ed.), Vol. 4, pp. 89–137. CRC Press, Boca Raton, FL.

Rempel, J. G. (1950). A guide to the mosquito larvae of western Canada. *Canadian Journal of Research Section D Zoological Sciences* **28**, 207–248.

Rempel, J. G. (1953). The mosquitoes of Saskatchewan. *Canadian Journal of Zoology* **31**, 433–509.

Restifo, R. A. (1982). "Illustrated Key to the Mosquitoes of Ohio [adapted from Stojanovich (1960, 1961)]." Ohio Biology Survey, Biological notes, No. 17. College of Biological Sciences, Ohio State Univ., Columbus.

Ribeiro, J. M. C. (1987). Role of saliva in blood feeding by arthropods. *Annu. Rev. Entomol.* **32**, 463–478.

Ross, E. S., and Roberts, H. R. (1943). "Mosquito Atlas. Part I: The Nearctic *Anopheles,* Important Malaria Vectors of the Americas and *Aedes aegypti, Culex quinquefasciatus;* Part II: Eighteen Old World Anophelines Important to Malaria." American Entomological Society, the Academy of Natural Sciences, Philadelphia.

Ross, H. H. (1947). "The Mosquitoes of Illinois (Diptera, Culicidae)." Illinois Natural History Survey, Bulletin Vol. 24(1), Urbana.

Ross, H. H., and Horsfall, W. R. (1965). "A Synopsis of the Mosquitoes of Illinois (Diptera. Culicidae)." Illinois Natural History Survey, Biological Notes No. 52, Urbana.

Ross, R. (1911). "The Prevention of Malaria," 2nd ed. Murray, London.

Rozeboom, L. E. (1942). "The Mosquitoes of Oklahoma." Technical bulletin No. T-16. Oklahoma Agriculture Experiment Station, Oklahoma State University, Stillwater.

Russell, P. F. (1955). "Man's Mastery of Malaria." Oxford Univ. Press, Oxford.

Russell, P. F., Rozeboom, L. E., and Stone, A. (1943). "Keys to the Anopheline Mosquitoes of the World, with Notes on Their Identification, Distribution, Biology, and Relation to Malaria." American Entomological Society, the Academy of Natural Sciences, Philadelphia.

Rutschky, C. W., Mooney, T. C., Jr., and Vanderberg, J. P. (1958). "Mosquitoes of Pennsylvania. An Illustrated Key to Species with Accompanying Notes on Biology and Control." Agriculture Experiment Station Bulletin No. 630. Pennsylvania State Univ., University Park.

Sasa, M. (1976). "Human Filariasis: A Global Survey of Epidemiology and Control." Tokyo Univ. Press, Tokyo, Japan.

Service, M. W., Ed. (1988). "Biosystematics of Haematophagous Insects." Clarendon, Oxford.

Service, M. W. (1989). "Demography and Vector-Borne Diseases." CRC Press, Boca Raton, FL.

Service, M. W. (1990). "Handbook of the Afrotropical Toxorhynchitine and Culicine Mosquitoes, Excepting *Aedes* and *Culex.*" British Museum (Natural History), London.

Service, M. W. (1993a). "Mosquito Ecology: Field Sampling Methods," 2nd ed. Elsevier, New York.

Service, M. W. (1993b). Mosquitoes (Culicidae). In "Medical Insects and Arachnids" (R. P. Lane and R. W. Crosskey, Eds.), pp. 120–240. Chapman & Hall, New York.

Service, M. W. (1997). Mosquito (Diptera: Culicidae) dispersal—the long and short of it. *Journal of Medical Entomology* **34**, 579–588.

Siverly, R. E. (1972). "Mosquitoes of Indiana." Indiana State Board of Health, Indianapolis.

Smart, J. (1948). "Insects of Medical Importance." British Museum (Natural History), London.

Snodgrass, R. E. (1959). "The Anatomical Life of the Mosquito." Smithsonian Miscellaneous Collections. Vol. 139, No. 8. Baltimore Press, Baltimore.

Snow, K. R. (1990). "Mosquitoes." Naturalists' Handbooks 14. Richmond Publ., Slough, UK.

Sokolova, M. I. (1994). A redescription of the morphology of mosquito (Diptera: Culicidae) ovarioles during vitellogenesis. *Bulletin of the Society for Vector Ecology* **19**, 53–68.

Soper, F. L. (1963). The elimination of urban yellow fever in the Americas through the eradication of *Aedes aegypti. American Journal of Public Health* **53**, 7–16.

Soper, F. L., and Wilson, D. B. (1943). "*Anopheles gambiae* in Brazil, 1930–1940." Rockefeller Foundation, New York.

Stage, H. H., Gjullin, C. M., and Yates, W. W. (1952). "Mosquitoes of the Northwestern States." Agriculture Handbook, US Department of Agriculture, No. 46. US Department of Agriculture, Washington, DC.

Steffan, W. A., and Evenhuis, N. L. (1981). Biology of *Toxorhynchites. Annu. Rev. Entomol.* **26**, 159–181.

Stojanovich, C. J. (1960). "Illustrated Key to Common Mosquitoes of Southeastern United States." Cullom & Ghertner, Atlanta.

Stojanovich, C. J. (1961). "Illustrated Key to Common Mosquitoes of Northeastern North America." Emory Univ. Branch, Cullom & Ghertner, Atlanta.

Stone, A. (1981). Culicidae. In "Manual of Nearctic Diptera" (J. F. McAlpine *et al.*, Eds.), Vol. 1, Chapt. 25, pp. 341–350. Research Branch, Agriculture Canada, Ottawa.

Stone, A., and Delfinado, M. D. (1973). Family Culicidae. In "A Catalog of the Diptera of the Oriental Region. Vol. 1. Suborder Nematocera" (M. D. Delfinado and D. E. Hardy, Eds.), pp. 266–343. Univ. Press of Hawaii, Honolulu.

Strickland, G. T., Ed. (1991). "Hunter's Tropical Medicine," 7th ed. Saunders, Philadelphia.

Tabachnick, W. J. (1994). Genetics of insect vector competence for arboviruses. *Advances in Disease Vector Research* **10**, 93–108.

Tate, H. D., and Gates, D. B. (1944). "The Mosquitoes of Nebraska." Agriculture Experiment Station Bulletin No. 133. Univ. Nebraska, Lincoln.

Telford, S. R., Jr. (1994). Plasmodia of reptiles. In "Parasitic Protozoa" (J. P. Kreier, Ed.), 2nd ed., Vol. 7, pp. 1–72. Academic Press, New York.

Tempelis, C. H. (1975). Host-feeding patterns of mosquitoes, with a review of advances in analysis of blood meals by serology. *Journal of Medical Entomology* **11**, 635–653.

Tsai, T. F., and Mitchell, C. J. (1988). St. Louis encephalitis. In "The Arboviruses: Epidemiology and Ecology" (T. P. Monath, Ed.), Vol. 4, pp. 113–143. CRC Press, Boca Raton, FL.

Tulloch, G. S. (1939). A key to the mosquitoes of Massachusetts. *Psyche* **46**, 113–136.

Turell, M. J. (1988). Horizontal and vertical transmission of viruses by insect and tick vectors. In "The Arboviruses: Epidemiology and Ecology" (T. P. Monath, Ed.), Vol. 1, pp. 127–152. CRC Press, Boca Raton, FL.

United States Department of Agriculture, Animal and Plant Health Inspection Service. (1973). "The Origin and Spread of Venezuelan Equine Encephalomyelitis." APHIS 91-10. Animal and Plant Health Inspection Service, Hyattsville, MD.

United States Public Health Service and Tennessee Valley Authority. (1947). "Malaria Control on Impounded Water." US Govt. Printing Office, Washington, DC.

Van Dine, D. L. (1922). "Impounding Water in a Bayou to Control Breeding of Malaria Mosquitoes." Bull. No. 1098, US Department of Agriculture. US Govt. Printing Office, Washington, DC.

Van Riper, C., III, *et al.* (1994). Plasmodia of birds. In "Parasitic Protozoa" (J. P. Kreier, Ed.), 2nd ed., Vol. 7, pp. 73–140. Academic Press, New York.

Wallis, R. C. (1960). "Mosquitoes in Connecticut." Connecticut Agriculture Experiment Station Bulletin No. 632. New Haven.

Walton, T. E., and Grayson, M. A. (1988). Venezuelan equine encephalomyelitis. In "The Arboviruses: Epidemiology and Ecology" (T. P. Monath, Ed.), Vol. 4, pp. 203–231. CRC Press, Boca Raton, FL.

Ward, R. A. (1984). Second supplement to "A catalog of the mosquitoes of the world (Diptera: Culicidae)." *Mosquito Systematics* **16**, 227–270.

Ward, R. A. (1992). Third supplement to "A catalog of the mosquitoes of the world (Diptera: Culicidae)." *Mosquito Systematics* **24**, 177–230.

Ward, R. A., and Darsie, R. F. (1982). Corrections and additions to the publication, *Identification and Geographical Distribution of the Mosquitoes of North America, North of Mexico. Mosquito Systematics* **14**, 209–219.

Washino, R. K., and Tempelis, C. H. (1983). Mosquito host bloodmeal identification: methodology and data analysis. *Annu. Rev. Entomol.* **28**, 179–201.

Watts, D. M., Pantuwatana, S., DeFoliart, G. R., Yuill, T. M., and Thompson, W. H. (1973). Transovarial transmission of LaCrosse virus (California encephalitis group) in the mosquito, *Aedes triseriatus. Science* **182**, 1140–1141.

Wernsdorfer, W. H., and McGregor, I., Eds. (1988). "Malaria: Principles and Practice of Malariology." 2 Vols. Churchill Livingstone, New York.

White, G. B. (1980). Family Culicidae. In "Catalogue of the Diptera of the Afrotropical Region" (R. W. Crosskey, Ed.), pp. 114–148. British Museum (Natural History), London.

Wilkerson, R. C., and Strickman, D. (1990). Illustrated key of the female anopheline mosquitoes of Central America and Mexico. *Journal of the American Mosquito Control Association* **6**, 7–34.

Wood, D. M., and Borkent, A. (1989). Phylogeny and classification of the Nematocera. In "Manual of Nearctic Diptera" (J. F. McAlpine, Ed.), Vol. 3, Monograph No. 32, pp. 1333–1370. Research Branch, Agriculture Canada, Ottawa.

Wood, D. M., Dang, P. T., and Ellis, R. A. (1979). "The Insects and Arachnids of Canada. Part 6. The Mosquitoes of Canada. Diptera: Culicidae." Canadian Government Publishing Centre, Hull, Quebec.

Woodring, J. L., and Davidson, E. W. (1996). Biological control of mosquitoes. In "The Biology of Disease Vectors" (B. J. Beaty and W. C. Marquardt, Eds.), pp. 530–548. Univ. Press of Colorado, Niwot.

World Health Organization. (1973). "Manual on Larval Control Operations in Malaria Programmes." Offset Publication No. 1. World Health Organization, Geneva.

World Health Organization. (1975). "Manual on Practical Entomology in Malaria." Parts I and II. Offset Publication No. 13. World Health Organization, Geneva.

World Health Organization. (1982). "Manual on Environmental Management for Mosquito Control." Offset Publication No. 66. World Health Organization, Geneva.

World Health Organization. (1991). "Prospects for Malaria Control by Genetic Manipulation of Its Vectors." Unpublished document TDR/BCV/MAL-ENT/91.3. World Health Organization, Geneva.

World Health Organization, Vector Biology and Control Division. (1989). "Geographical Distribution of Arthropod-Borne Diseases and Their Principal Vectors." World Health Organization/VBC Publication 89.967, Geneva.

Wright, J. W., and Pal, R., Eds. (1967). "Genetics of Insect Vectors of Disease." Elsevier, New York.

Yamaguti, S., and LaCasse, W. J. (1951). "Mosquito Fauna of North America." Parts I–V. Office of the Surgeon, Hq., Japan Logistical Command, APO 343

# 13

# HORSE FLIES AND DEER FLIES (*Tabanidae*)

BRADLEY A. MULLENS

*Because of* their fairly large size, striking appearance, and diurnal biting habits, horse flies and deer flies (Fig. 13.1) are familiar to most people who have livestock or engage in outdoor activities. Diversity within the family is greatest in the tropics, but moist temperate regions typically have a rich fauna as well. Tabanids are present on every continent except Antarctica and have managed to colonize remote islands such as the Galápagos and the Melanesian Archipelago. Large seasonal populations of some species occur as far as 60° N latitude, but they disappear above the tree line.

The eyes of many species, when alive, are brilliantly patterned with shades of green, yellow, orange, and violet. Some species with strikingly green eyes are commonly called *greenheads,* and others are called *yellow flies* due to their yellow bodies. The *blue-tail fly* of American folk music probably was *Tabanus atratus,* a large black species which has a blue cast to the abdomen. Common names also may reflect times or places where these biting flies are found, but the source of some terms is obscure. *Cleg, breezefly, whamefly, marchfly, May fly,* and *mango fly* are but a few of the many colloquial names used. Other common names, like *bulldog* and *gadfly,* reflect the persistent annoyance of tabanids in seeking a blood meal.

The term *horse fly* is applied to relatively large species of tabanids, typically 10–30 mm in length. They are a serious nuisance to livestock and can mechanically transmit several significant animal pathogens, including those which cause surra, anaplasmosis, and equine infectious anemia. Even moderate numbers of flies feeding on livestock can result in significant production losses. A few horse flies readily bite people, examples being the infamous *greenheads* ( *T. nigrovittatus* and *T. simulans*) in the coastal regions of the eastern United States.

The smaller tabanid species called *deer flies* typically are 6–10 mm long. In contrast to horse flies, they frequently attack humans. Fortunately, there are just a few human diseases known to be associated with deer flies. The most important tabanid-transmitted human diseases are loiasis and tularemia. Outdoor activity and tourism suffer in areas where tabanid populations are high, although such losses are hard to quantify.

A

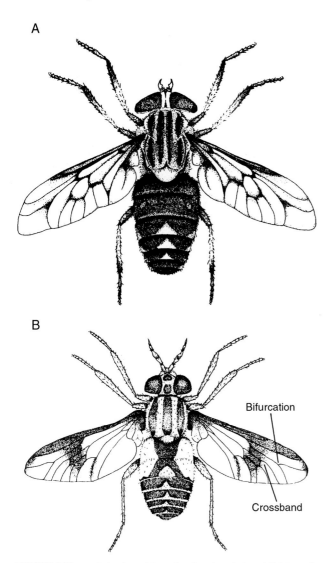

B

Bifurcation

Crossband

**FIGURE 13.1**    Adult tabanid flies, females, dorsal view. (A) Horse fly, *Tabanus trimaculatus;* (B) deer fly, *Chrysops callidus.* Note distinctive wing venation, including bifurcation of vein near wing tip, and the darkened crossband in deer flies. (Original by S. J. M. Hope.)

## TAXONOMY

The family Tabanidae includes approximately 4300 species and subspecies in 133 genera worldwide. Of these, 335 species in 25 genera are found in the Nearctic Region (Burger, 1995). While the temperate fauna is well known, the tropical fauna has been less studied; this is particularly true for the immature stages. Most pest species in North America are members of the genera *Chrysops, Hybomitra,* and *Tabanus.*

The family Tabanidae is divided into three subfamilies (Fairchild, 1969; Mackerras, 1954) (Table I). The subfamily Pangoniinae is regarded as ancestral, containing fascinating but poorly known genera such as *Stonemyia*

**TABLE I**

**Subfamilies, Tribes, and Selected Genera of Tabanidae in North America: The Largest Genera in Terms of Numbers of Species are *Chrysops, Tabanus,* and *Hybomitra***

| Taxon | No. species |
| --- | --- |
| Subfamily Pangoniinae | |
| Tribe Pangoniini | |
| Genus *Apatolestes* | 13 |
| Genus *Stonemyia* | 6 |
| Tribe Scionini | |
| Genus *Goniops* | 1 |
| Subfamily Chrysopsinae | |
| Tribe Bouvieromyiini | |
| Genus *Mercomyia* | 2 |
| Tribe Chrysopsini | |
| Genus *Silvius* | 12 |
| Genus *Chrysops* | 83 |
| Subfamily Tabaninae | |
| Tribe Diachlorini | |
| Genus *Diachlorus* | 1 |
| Genus *Chlorotabanus* | 1 |
| Genus *Leucotabanus* | 2 |
| Tribe Haematopotini | |
| Genus *Haematopota* | 5 |
| Genus *Tabanus* | 107 |
| Genus *Atylotus* | 14 |
| Genus *Hybomitra* | 55 |

and *Goniops,* many of which are not known to feed on blood. Most of the economically important tabanids are members of the two other subfamilies. Tabanids in the subfamily Chrysopsinae are called *deer flies;* nearly all are members of the genus *Chrysops,* which includes more than 80 Nearctic species. The term "deer fly" also is applied to members of the genus *Silvius,* a few species of which can be quite pestiferous on humans and animals in the western United States.

Members of the Tabaninae are the most evolutionarily derived. This subfamily includes the *horse flies,* represented by *Tabanus,* which has 107 Nearctic species, and *Hybomitra,* with 55 Nearctic species. Species of *Haematopota,* together with *Tabanus* and *Hybomitra,* are important pests in the Old World. Only five species of *Haematopota* occur in the Nearctic region, where *Haematopota americana* is the only species known to be a pest of mammals.

Burger (1995) compiled a complete catalog of species of Nearctic Tabanidae. Pechuman and Teskey (1981) provide generic keys to larvae, pupae, and adults of Nearctic tabanids. There are several regional keys for identification of adults of North American tabanids at the species level: California (Middlekauf and Lane, 1980), New York (Pechuman, 1981), Illinois (Pechuman *et al.,* 1983), Tennessee (Goodwin *et al.,* 1985), Texas

(Goodwin and Drees, 1996), and Canada and Alaska (Teskey, 1990). Most of these works include valuable information on biology and ecology. Immatures of Nearctic tabanids are more difficult to identify than adults. Many North American species remain undescribed, and the immature stages are less likely to be encountered by the casual collector. Immatures of some species can be identified using keys or references found in Pechuman *et al.* (1983), Goodwin *et al.* (1985), and Teskey (1990).

Taxonomic references for other regions include the following: Europe (Chvala *et al.*, 1972), Neotropics (Fairchild, 1986; Fairchild and Burger, 1994), Australia (Mackerras, 1954), Ethiopian region (Oldroyd, 1954–1957), the former Soviet Union (Olsufiev, 1977), and Japan (Takahasi, 1962). Immature stages of Palearctic Tabanidae are treated by Andreeva (1990).

## MORPHOLOGY

Tabanid *larvae* (Fig. 13.2) are spindle-shaped (fusiform) and generally whitish in color, although some are shades of brown or green. Mature larvae of common species typically measure 15–30 mm in length, but some larger tabanid larvae may be as long as 60 mm. The head capsule is incomplete and partially sclerotized. The mandibles are strong, parallel, and ventrally curved and are used to capture and subdue prey. The larval cuticle has distinctive longitudinal striations and often exhibits species-specific pubescence patterns that give some tabanid larvae a mottled appearance. Abdominal segments have lateral and

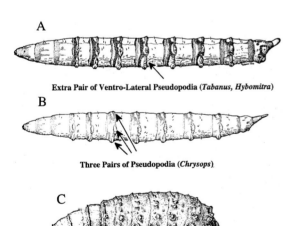

**A**

**Extra Pair of Ventro-Lateral Pseudopodia (*Tabanus, Hybomitra*)**

**B**

**Three Pairs of Pseudopodia (*Chrysops*)**

**C**

**FIGURE 13.2** Larvae of Tabanidae. Typical semiaquatic larval forms are shown in A and B. Note the extra pair of pseudopodia useful in differentiating the common horse fly genera Tabanus and Hybomitra from the common deer fly genus Chrysops. Larva of the unusual terrestrial genus Goniops is shown in C. From the Manual of Nearctic Diptera. Reproduced with the permission of the Minister of Public Works and Government Services Canada, 2002.

ventral *pseudopods* for locomotion (three pairs in *Chrysops* spp. and four pairs in Tabaninae). Larvae of the more terrestrial species tend to be relatively stocky, with short pseudopods. Species adapted to a fully aquatic existence in streams (e.g., *T. fairchildi*) have elongated pseudopods armed with cuticular, recurved, distal hooks. Semiaquatic larvae, represented by the majority of tabanid species, have intermediate characteristics. Located in the anterodorsal portion of the anal segment is a pear-shaped structure called *Graber's organ*. It is unique to tabanid larvae and its function is unknown. Two main tracheal trunks run the length of the body, terminating in a dorsally directed respiratory siphon. A terminal spine is present on the siphon of some species.

Tabanid *pupae* are usually tan or brown, with the eyes, legs, and wing pads visible externally. A fringe of spines on the posterior margin of many abdominal segments and a starlike series of three or four pairs of caudal projections called *pupal asters* are useful in identification.

Tabanid adults are stout-bodied flies. They generally can be distinguished as horse flies or deer flies based on several morphological characteristics (Table II). The antennae are prominent and extend anteriorly. The flagellum, with four to eight flagellomeres, usually is enlarged at the base in Tabaninae but is only slightly enlarged in *Chrysops* species (Fig. 13.3). The eyes often consist of large ommatidial facets dorsally and smaller facets ventrally. This arrangement is believed to enhance visual acuity in locating potential mates. The male has *holoptic eyes*, which occupy most of the head, touching each other medially. The female has *dichoptic eyes* that are smaller than those of the male and that are separated by the frons. The frons of most species is covered by very fine pubescence. Slightly raised, sometimes bare areas of cuticle, called the *median callus* and *basal callus*, aid identification. Color patterns of the eyes sometimes provide useful taxonomic characteristics. Unfortunately, eye patterns and colors disappear when a specimen dries; however, the basic pattern

**TABLE II**

**Morphological Characteristic Used to Differentiate Adult Horse Flies and Deer Flies**

| Characteristic | Horse flies (e.g., *Tabanus*) | Deer flies (e.g., *Chrysops*) |
| --- | --- | --- |
| Body length | 10–30 mm | 6–11 mm |
| Antennae | Short, base of flagellum greatly enlarged | Long, base of flagellum not greatly enlarged |
| Ocelli | Vestigial or lacking | Present |
| Wings | Clear, uniformly cloudy, or spotted | Distinctly banded |
| Apical spurs on hind tibiae | Lacking | Present |

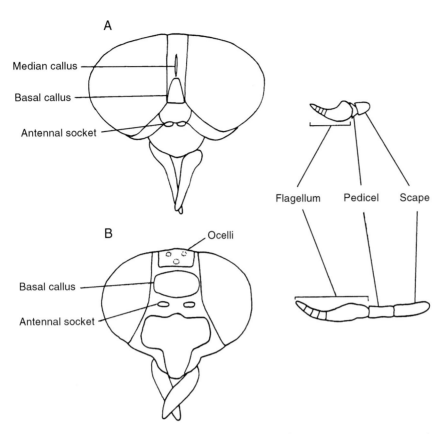

**FIGURE 13.3**    Morphology of head and antennae of tabanid flies, noting important taxonomic characters. (A) Horse fly, *Tabanus* sp.; (B) deer fly, *Chrysops* sp.

often can be restored by rehydration of the specimen. The Pangoniinae and Chrysopsinae possess well developed *ocelli* at the vertex of the frons, whereas the ocelli of Tabaninae are vestigial or absent. *Hybomitra* species usually exhibit a raised, denuded *ocellar tubercle* which is lacking in *Tabanus* species.

The maxillary palps are two-segmented and enlarged at the base of the apical palpomere. The proboscis is stout and includes toothed, bladelike mandibles and maxillary laciniae which lacerate the skin and capillary beds during blood-feeding. The female hypopharynx is rigid, with the salivary duct opening at the tip to introduce saliva into the feeding wound. Blood is drawn up between the labellar lobes into the food canal between the labrum and hypopharynx. This feeding method is known as pool feeding, or *telmophagy*. Males, which do not feed on blood, lack mandibles, recurved teeth on the laciniae, and a rigid hypopharynx.

The thorax is stout, with a prominent *notopleural lobe* and strong flight muscles. Legs are also stout, with fairly prominent tibial spurs. *Apical spurs* are present on the hind tibia of the Pangoniinae and Chrysopsinae but are lacking in the Tabaninae. The wing venation is quite consistent within the Tabanidae; a key feature is the widely divergent R4 and R5 veins, which fork and enclose the apex of the wing. Wing membranes are clear in some species and variously darkened in others, providing useful taxonomic characters, particularly for *Chrysops* species. Wings of *Haematopota* species (clegs) are mottled, while those of most other Tabaninae are not. The abdomen of tabanids is as wide as the thorax, slightly compressed dorso-ventrally, and often distinctly colored or patterned.

Internally, tabanids have a large crop for water and sugar storage. Blood is directed to the midgut. As with many other blood-feeders, tabanids can rapidly eliminate excess fluids from the blood meal via the anus. Genitalia are fairly simple structurally in both sexes and are of little use in routine species identifications.

## LIFE HISTORY

Tabanid eggs are 1–3 mm long and are deposited in masses (Fig. 13.4). The female usually lays 100–800 eggs in a single mass; the numbers vary substantially with the species and size of the blood meal. Some species lay several smaller batches of eggs, particularly in captivity;

**FIGURE 13.4** Female horse fly, *Tabanus imitans,* ovipositing on plant stem. (Photo by S. McKeever.)

**FIGURE 13.5** Two egg masses of *Chrysops callidus,* deposited on vegetation. Note the single layer of eggs typical of most deer flies. (Photo by L. L. Pechuman.)

*Atylotus thoracicus* has been observed to deposit eggs singly on sphagnum moss in the laboratory. Eggs are white when laid but darken to grey, brown, or black within several hours. Egg masses most often are found on leaves or stems of emergent vegetation at the edges of ponds (lentic habitats) or streams (lotic habitats) or on leaves or bark of trees overhanging the water. The larvae of Nearctic species of *Chrysops* have been categorized as 65% lentic and 18% lotic, with 13% of the species being found in both lentic and lotic habitats. Stream-dwelling species often deposit their eggs above the waterline on stones in the stream where water flow is moderate. Some species can be terrestrial and may be found in fairly dry soil (e.g., *T. abactor, T. sulcifrons, T. subsimilis*). Terrestrial species usually lay their eggs on vegetation or in leaf litter. *Apatolestes actites* is known to oviposit in crustacean burrows in beach habitats.

Many *Chrysops* species deposit eggs in a single layer, e.g., *Chrysops callidus* (Fig. 13.5), whereas others deposit eggs in tiers, e.g., *C. cincticornis* (Fig. 13.6). Species of *Tabanus* and *Hybomitra* lay their eggs in tiers, commonly three or four tiers high. Such masses taper in pyramid

fashion from base to apex. The exact shape of the egg mass often reflects the oviposition substrate; e.g., the same species may deposit an elongate egg mass on a grass stem or a broader one on a deciduous leaf. Females of an unusual and primitive tabanid, *Goniops chrysocoma,* lay eggs on the underside of a leaf and secure themselves above the mass using the tarsal claws. They remain with the mass until it hatches, buzzing noisily if they are disturbed. The female dies soon after the eggs hatch.

Embryogenesis typically requires 5–12 days at temperatures of 21–24°C and is both temperature- and species-dependent. Egg hatch can occur within 2–3 days at

**FIGURE 13.6** Egg mass of *Chrysops cincticornis.* Note the multiple layers of eggs typical of certain deer flies and most horse flies. (Photo by E. J. Hansens.)

temperatures of 30–35°C. First-instar larvae are equipped with an *egg burster,* a projection on the head capsule with which they split the chorion, and then they drop to the water or moist substrate below. They molt once, apparently without feeding, before beginning to move about in the substrate.

Tabanid *larvae* are found in a wide variety of aquatic and semiaquatic habitats. These include mud or saturated vegetation in marshes or near pond or creek margins, under stones in and along streams, and in terrestrial habitats such as under forest litter. Some species are common in a variety of semiaquatic habitats (e.g., *T. punctifer*). Others are quite specific; larvae of *Leucotabanus annulatus,* for example, are found only in rotten tree stumps. Tabanid larvae are general predators that feed on a variety of invertebrates such as larvae of chironomid midges and crane flies, or annelids. Horse fly larvae also are cannibalistic, a factor which influences their densities and distribution and complicates efforts to rear them in the laboratory. It is unusual to find high densities of tabanine larvae in nature. Nonetheless, densities of larvae in flooded hardwood forests in Louisiana have been estimated at about $10/m^2$, leading to high adult populations. In contrast to tabanine larvae, *Chrysops* larvae apparently are not as cannibalistic and may be found at high field densities. *Chrysops* larvae probably are predaceous and feed primarily on invertebrates, although some authors have speculated that they feed on detritus.

Tabanids undergo 6–13 larval molts and overwinter as larvae. Most temperate species have one generation per year (univoltine), whereas others may produce two or more generations (multivoltine). Particularly large species, such as *T. atratus* and *T. calens,* may spend 2 or 3 years as larvae. Development may be prolonged for 3 years or more in very cold, seasonally dry, or otherwise unfavorable conditions. In the spring, the larvae leave the water-saturated soil to pupate above the waterline. Pupal periods vary with species and temperature but typically last 4–21 days.

Most temperate species have very distinctive *seasonal flight periods* which vary little from year to year. In the eastern United States and Canada, for example, *Hybomitra lasiophthalma* is one of the first pestiferous horse flies to emerge. It begins activity as early as March in Louisiana or as late as June in southern Canada and is abundant for 3–6 weeks, and then adults disappear until the following spring. In contrast, *T. subsimilis* has been collected in Tennessee from mid May through early October. Some tabanids can develop from egg to adult in as little as 6 weeks, and a few species have multiple broods within a season. A prolonged period of adult emergence, or the presence of unrecognized sibling species, can contribute to what appears to be a long adult flight period.

Many tabanids are *anautogenous* and require a single large blood meal in order to develop a batch of eggs. Blood-meal size varies from 20 to 25 mg for many *Chrysops* species to almost 700 mg for *T. atratus.* Other species, such as *T. nigrovittatus,* are obligately *autogenous* in the first gonotrophic cycle and seek a blood meal for each subsequent cycle. Still others seem to be facultatively autogenous, probably reflecting genetic variability and the carry-over of available nutrients from the larval stage. Following a blood meal, egg development is believed to be typically 3–4 days. Time to oviposition may be several days longer for some species, especially under laboratory conditions, where the full range of oviposition cues may be lacking. One California species, *A. actites,* apparently undergoes two autogenous gonotrophic cycles, a rarity among families of hematophagous Diptera.

## BEHAVIOR AND ECOLOGY

The biology and behavior of tabanid *larvae* are generally poorly known. They are laborious to rear in the laboratory due to their long developmental times and predaceous and cannibalistic habits, and no tabanid species has been successfully colonized. As soil dwellers they are difficult to observe and sample. Tabanid larvae are rarely free-swimming in nature, but some, such as *T. punctifer,* are buoyant and can swim effectively by repeatedly flipping the rear half of the body and propelling themselves in short, gliding spurts.

On contacting a prey item, a tabanid larva will strike, often with an audible "click," seizing the prey with its mandibles. Tabanid larvae are capable of capturing prey larger than themselves, and prey struggles usually cease very quickly. A toxin likely is involved in prey capture, but this has not been conclusively demonstrated.

Prior to *pupation,* larvae generally seek out drier soil, such as above the waterline at the edge of ponds or streams. Pupation occurs near the soil surface, with the head end oriented upward. Larvae of a few species, including *T. atratus* in the New World and several other *Tabanus* spp. in the Old World, construct a *mud cylinder* above the waterline. They spiral downward 5–13 cm to delineate the perimeter of the 3- to 9-cm diameter cylinder and then burrow into the center to form a *pupation tunnel.* This unusual behavior may preserve the structural integrity of the pupation tunnel and facilitate future emergence or yield a drier pupation site in periodically flooded habitats. Mortality during tabanid egg and larval stages is high. A production ratio of only three pupae per egg mass has been calculated for *H. bimaculata* in Swiss bog-meadow habitats.

The biology and behavior of tabanid adults are better understood than of the immatures. The sex ratio at emergence is approximately 1:1, and emergence of males precedes that of females by 1 day to a few days. An important activity for both sexes is *carbohydrate feeding,* which provides energy for general body maintenance, flight, and mating. Sugars are obtained at floral or extrafloral nectaries or other natural plant sugar deposits. *T. nigrovittatus* adults feed on aphid honeydew. Males, in particular, engage in "dipping" behavior, touching the surface of pools of water with the mouthparts while in flight. This may serve to fill the crop with enough water to allow flies to regurgitate on honeydew deposits. The ingested sugars replenish energy reserves expended during daily flight and mating activities.

Tabanid mating occurs in flight, especially in the morning, but has never been induced in the laboratory. This is the key barrier to colonization. Most observations are of individuals or small groups of males (Tabaninae) hovering within 3 m of the ground along forest roads or ecotone areas or above natural features (e.g., plant clumps) which serve as markers. Some species apparently hover above treetops or forest canopies as high as 90–100 m above ground level. Males of other species, such as *C. fuliginosus* and *H. illota,* perch on vegetation and other objects. In both case, males detect and pursue passing females. They also chase conspecific males and other passing insects. Males of *H. hinei* have been observed to pursue 8-mm beads shot past them at speeds of 27–30 m/sec.

Males of some species are thought to exhibit *territoriality.* However, what may appear to be agonistic interactions between males may actually be normal pursuit behaviors directed toward any appropriate-sized object moving through their response zone. Individual males do not necessarily use a particular site continuously over time. The occurrence of *male aggregations* at the tops of hills, known as "hilltopping" behavior, is common for some species. The larger eyes of males reflect a mating strategy dominated by visual cues. The larger dorsal ommatidia may be more sensitive to ultraviolet light, allowing the male to detect a fast-moving female against the sky, and the smaller ventral ommatidia may be used to resolve visual details. While pheromones are suspected for a few tabanid species, this has yet to be proved.

Adult feeding activity is typically diurnal but occasionally crepuscular or nocturnal. It is affected by changes in environmental conditions, particularly temperature and barometric pressure. Species that feed diurnally generally attack hosts throughout the daytime, with discernible periods of higher activity. Females of *T. wilsoni* and *T. pallidescens* tend to feed near midday, whereas *T. abactor* feeds more frequently in late afternoon or early evening. Species which feed during crepuscular periods, particularly dusk, include *Chlorotabanus crepuscularis,*

**FIGURE 13.7** *Tabanus sulcifrons* feeding on a cow. Note the blood droplet from a prior bite wound. (Photo by B. A. Mullens.)

*L. annulatus, T. moderator,* and *T. equalis.* Some crepuscular species feed into the early night.

Tabanid females usually mate before they seek a vertebrate host. Males do not feed on blood. Most species, particularly the Tabaninae, feed on large mammals such as cattle, horses, and deer (Fig. 13.7). Deer flies often attack large mammals, including humans, but there also are records of *Chrysops* species feeding on ravens, crows, ducks, and robins. Reptiles such as turtles are occasionally attacked.

Tabanids are strong fliers and readily disperse several kilometers in short-term flights. Adult *dispersal* probably is influenced by host availability. Mark-release-recapture studies with tabanids may yield 3–6% recovery, which is very high for insects. This suggests that local populations tend to remain in a given area, with dispersal occurring in a series of short flights. Marked *T. abactor* females have been shown to return to a host at the same site where they had obtained a blood meal 3–4 days earlier.

The *attack rates* by *Chrysops* species, and to a lesser extent by *Tabanus* and *Hybomitra,* vary substantially in different habitats. Many *Chrysops* species, for example, tend to frequent forest edges or ecotones, attacking in large numbers a host entering these specific areas from adjacent open fields. Dark-colored hosts, or even dark areas on a black and white animal such as a Holstein cow, are often favored for attack.

Many tabanids are selective in attacking specific body regions of their hosts (Fig. 13.8), regardless of color.

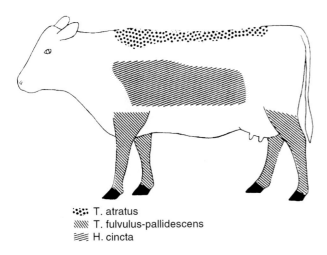

:•:• T. atratus
\\\\\ T. fulvulus-pallidescens
▨▨▨ H. cincta

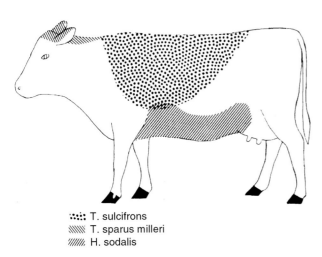

:•:• T. sulcifrons
\\\\\ T. sparus milleri
▨▨▨ H. sodalis

**FIGURE 13.8**    Feeding sites on cattle characteristic of individual tabanid species. H, *Hybomitra*; T, *Tabanus*. (From Mullens and Gerhardt, 1979.)

Species-specific attack behaviors probably contribute to the disproportionate collections of certain tabanid species in traps. Deer flies usually feed high on the body, especially on the head or shoulders, and are poorly represented in canopy or box traps, which require entry from below. Horse flies feed in various regions, depending on the species. Legs are favored feeding sites for many horse fly species that attack livestock. Feeding-site selection appears to reflect resource partitioning in both time and space, thereby reducing competition for hosts among tabanid species. Competition both within and between species is mediated by intensified host defensive behaviors, such as kicking. In fact, daily animal movement patterns and herd structure may be substantially influenced by fly attack. Animals in larger groups, especially individuals distant from the herd perimeter, tend to suffer fewer bites.

Field studies have documented the importance of *persistent feeding behavior* in mechanical transmission of disease agents. It is difficult to imagine a better group of potential mechanical vectors than horse flies. They are large, painful biters which are frequently interrupted in the act of feeding. If disturbed or dislodged, they will return to the same host or one nearby within seconds. Their success in initial feeding attempts often is poor. It has been estimated that only 10% of horse flies successfully feed to repletion during the initial attempt on cattle, although this varies among species.

The propensity of horse flies to return to the same host animal or transfer to another depends on the fly species and the distance between animals. Larger tabanids are more likely to transfer. It has been estimated that almost 90% of certain horse flies that attack horses will return to the original horse if other hosts are more than 35 m away. The moderate-sized tabanid *T. fuscicostatus* has an average of 10 nl of blood residue on its mouthparts following feeding. Biting rates and interrupted feeding are important factors that influence the transmission of disease agents.

Tabanids respond to volatile compounds that serve as *chemical host cues*. The best known and most effective attractant is $CO_2$. Species vary in their response to $CO_2$, but even a low release rate of 100 ml/min can result in a two- to fourfold increase in collections of tabanid females in traps. Compounds such as 1-octen-3-ol, ammonia, and phenols have been shown to be attractive to certain tabanids and may act synergistically with $CO_2$.

In addition, visual cues such as shape, color, and movement of potential hosts are very important. Shades of blue, black, or red are particularly attractive to tabanids, while contrast with the background and reflectance also play a role in orientation of tabanids to their hosts. Many imaginative traps have been designed to collect tabanids based on their visual response and orientation behavior. Among the more widely used devices are box traps and canopy traps, both of which collect primarily host-seeking females. The effectiveness of canopy traps is enhanced by incorporating movement in the form of a suspended, reflective black sphere which responds to air movements. The distinctive eye patterns of many species are due to corneal structures acting as interference filters. Older host-seeking females of some species have been shown to be relatively more sensitive to green wavelengths. This presumably aids in discriminating hosts against a background of green vegetation.

Once tabanid females have located a suitable host, they alight and begin probing. Initial *blood-feeding* attempts are painful and often result in vigorous host response and attempts to dislodge the fly. Flies frequently persist and attack repeatedly. Once blood flow begins, tabanids resist being dislodged, even to the point of sustaining direct

TABLE III
Disease Agents Transmitted by Tabanids

| Disease agent | Vectors | Geographic occurrence | Transmission |
|---|---|---|---|
| **Viruses** | | | |
| Equine infectious anemia | *Tabanus, Hybomitra, Chrysops* spp. | Worldwide | Mechanical |
| Bovine leukemia | *Tabanus* spp. | Worldwide | Mechanical |
| Hog cholera | *Tabanus* spp. | Worldwide; eradicated from North America, Australia, New Zealand, South Africa | Mechanical |
| **Bacteria/rickettsia** | | | |
| *Anaplasma marginale* | *Tabanus* spp. | Worldwide (Tropics, Subtropics) | Mechanical |
| *Francisella tularensis* | *Chrysops* spp. | North America, Russia, Japan | Mechanical |
| *Bacillus anthracis* | *Tabanus, Haematopota, Chrysops* spp. | Worldwide | Mechanical |
| **Protozoa** | | | |
| *Besnoitia besnoiti* | *Tabanus, Atylotus* spp. | South America, Southern Europe, Africa, Asia, | Mechanical |
| *Trypanosoma evansi* | *Tabanus, Haematopota, Chrysops* spp. | South America, North Africa, Asia, India | Mechanical |
| *T. vivax* | *Tabanus* spp. | South America, Africa | Mechanical |
| **Filarial nematodes** | | | |
| *Loa loa* | *Chrysops* spp., especially *C. dimidiatus, C. silaceus* | Central Africa | Biological |
| *Elaeophora schneideri* | *Hybomitra, Tabanus* spp. | North America, southern Europe | Biological |

strikes by an animal's tail, feet, or head. Salivary anticoagulants maintain blood flow, sometimes for several minutes after feeding ceases. One such anticoagulant is *chrysoptin*, a salivary protein which blocks platelet aggregation. It is not unusual for other flies, such as the house fly and face fly, to gather around tabanid feeding wounds to imbibe blood flowing from the wound site.

# PUBLIC HEALTH IMPORTANCE

In most temperate areas, adult tabanids are primarily nuisance pests of humans. In this regard they can pose economically significant problems for local tourism. The painful bites, sometimes exceeding 10 per minute, can entirely prevent recreational outdoor activity. Horse fly larvae can be local pests by inflicting painful bites to the feet of people working in rice paddies. If handled carelessly the larvae will bite defensively but rarely can penetrate the skin of human fingers.

Tabanids transmit some pathogens and parasites biologically, in which cases the disease agent replicates and/or develops within the fly for a period of time prior to transmission (e.g., the filarial nematode *Loa loa*). More commonly, however, tabanids transmit pathogens mechanically via contaminated blood on their mouthparts. Although many pathogenic viruses, bacteria, protozoa, and filarial nematodes have been recovered from tabanids, documentation of transmission is relatively uncommon, in part because of the difficulties in working with tabanids in the laboratory. Many of the disease associations, particularly those involving viruses and bacteria recovered from tabanids, need to be viewed cautiously, as they may be relatively insignificant epidemiologically.

Fortunately, there are relatively few human pathogens transmitted regularly by Tabanidae, as reviewed by Krinsky (1976) and Foil (1989). The more significant tabanid-associated diseases are shown in Table III.

## LOIASIS

The most important tabanid-transmitted disease agent of humans is the *African eyeworm, L. loa*, which causes *human loiasis*. This filarial nematode is biologically transmitted by *Chrysops* species in equatorial rain forests of western and central Africa. *Simian loiasis* is caused by a closely related form (*L. loa papionis*) and also involves *Chrysops* species as vectors. As with other filarial nematodes, transmission of *L. loa* is cyclodevelopmental, requiring the fly as an intermediate host. Interestingly, *Chrysops atlanticus* and several other common deer flies in the southeastern United States have been shown to support development of *L. loa*.

Adult nematodes live in subcutaneous tissues of the vertebrate host, particularly the thorax, scalp, axillary regions, or eyes. They are 2–7 cm long and produce inflammatory responses as they move through these tissues. If they remain in one area for a time, localized enlargements known as *Calabar swellings* occur; these swellings disappear when the nematode leaves. The common name "eyeworm" is due to adult nematodes sometimes moving across the conjunctiva of the eye (Fig. 13.9). There, or just beneath the skin surface, they often are clearly visible and sometimes can be surgically removed. Migrating nematodes can cause considerable pain, in addition to discoloration and bruising of the affected tissues, particularly evident in the eye.

Mature *L. loa* adults mate and produce *microfilariae*. Females of *Chrysops* species ingest the microfilariae with blood when feeding on an infected person. The microfilariae penetrate the midgut and develop in the abdominal fat bodies or, sometimes, the thorax. There they molt to second-stage larvae and eventually move to the head and mouthparts as infective third-stage larvae ($L_3$). In the laboratory, infected deer flies may produce 100 or more *L. loa* infective-stage larvae per fly. This process is temperature-dependent and requires at least 7–10 days. On a subsequent feeding, the infective larvae escape from the fly mouthparts and enter a new host through the bite wound during blood-feeding.

The primary vectors of *L. loa* were incriminated in a series of studies in the 1950s in the Congo region of equatorial Africa, which includes parts of Zaire, Congo, Gabon, Cameroon, and southern Nigeria (see Krinsky, 1976). *C. silaceus* and *C. dimidiatus* are particularly attracted to people near fire. From 80 to 90% of their blood meals are obtained from humans, and in some hyperendemic areas 90% of the people harbor microfilariae or exhibit loiasis symptoms. Infection rates of 0.5–1.0% of these two *Chrysops* species have been reported in central Africa, although infection rates in parous flies alone may be much higher. Depending on the geographic location, people can be subjected to as many as 2000–3000 bites per year, i.e., several infective bites per month during the rainy season.

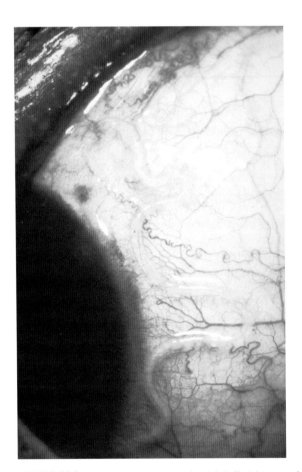

**FIGURE 13.9**    African eyeworm, *Loa loa*. Adult filarial nematode just beneath conjunctiva of human eye, Cameroon. (Courtesy of US Armed Forces Institute of Pathology, AFIP 73-6654.)

## TULAREMIA

Tularemia, sometimes called "rabbit fever" or "deer fly fever," is a zoonosis caused by the bacterium *Francisella tularensis*. The name *rabbit fever* reflects the fact that a common method of transmission, particularly in past decades, is through cuts on the hands of hunters and other people handling infected wild rabbits. In the central United States where most American tularemia cases occur, transmission is usually via ticks or direct animal contact. Transmission occurs less commonly by ingestion or aerosol exposure to the bacteria. The epidemiological role of tabanids as vectors in the central United States is unknown. However, transmission by tabanids has been well documented in parts of western North America and is suspected in parts of Russia. Periodic outbreaks of tularemia in Utah since the early 1900s have been linked convincingly to deer fly bites, particularly those of *C. discalis*, hence the name *deer fly fever*.

At the site of bacterial introduction, a distinctive lesion develops with an ulcerated, pinkish pit in the center and a raised, ridgelike wheal around the perimeter (Fig. 13.10). Bacterial septicemia and resultant fever cause severe illness and occasionally death if the person is not adequately treated with antibiotics. Often the initial lesions are found on the head or upper torso, where deer flies commonly

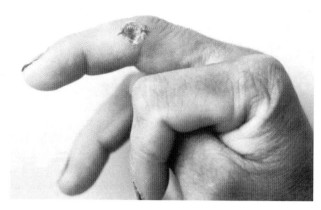

**FIGURE 13.10**  Tularemia lesion on middle finger of human hand, caused by the bacterium *Franciscella tularensis.* (Courtesy of Y. Ohara.)

bite people. The ability of *C. discalis* to acquire and later transmit the bacterium to humans is dependent on the propensity of this deer fly to feed on other animals, particularly rabbits, which serve as a pathogen reservoir. Occasionally other biting flies, including the horse fly *T. punctifer*, also can transmit *F. tularensis.* Such transmission is likely to be mechanical, with the bacterium being introduced into a bite wound via contaminated mouthparts. Feces of deer flies which are fed the bacterium experimentally have been shown to be infective when the feces is rubbed into abraded skin.

## OTHER TABANID-TRANSMITTED HUMAN PATHOGENS

*Anthrax,* a potentially dangerous disease of humans, still occurs worldwide, including localized areas in the United States. Although tabanids are capable of mechanically transmitting the causal bacterium, *Bacillus anthracis,* this mode of transmission is minor in the epidemiology of the disease. Tabanids also have been implicated as possible vectors of *Borrelia burgdorferi,* the spirochete that causes *Lyme disease.* While most transmission is accomplished by ticks, tabanids are suspected of contributing to transmission of *B. burgdorferi* in some parts of Europe and North America. This has not been well documented, however, and requires further study.

# VETERINARY IMPORTANCE

Owing to their painful, persistent biting behavior, tabanids are significant pests of livestock, particularly cattle and horses, and are extremely bothersome to many wildlife species. Heavy attack by tabanids can cause direct reductions in weight gains of beef cattle, reduced milk yield, reduced feed-utilization efficiencies, and hide damage from the feeding punctures. Cattle protected from tabanid attack in screened enclosures have been shown to gain up to 0.1 kg/day more than control animals exposed to tabanids. Such direct losses can be exacerbated by a concomitant reduction in feed-utilization efficiency of up to 17%. A daily loss of 200 ml of blood per animal may be common during periods of intense tabanid attack.

Tabanids serve as vectors of a number of disease agents of animals, including viruses, bacteria, protozoans, and nematodes (Table III). For an overview of these pathogens, see Krinsky (1976).

## SURRA AND RELATED TRYPANOSOMIASES

One of the more serious disease agents of livestock transmitted by tabanids is *Trypanosoma evansi,* the causal agent of *surra.* It is morphologically indistinguishable from *T. brucei.* Unlike tsetse fly (*Glossina* species) transmission of several other *Trypanosoma* species, including *T. brucei,* tabanid transmission of *T. evansi* is mechanical. Surra also is transmitted mechanically by vampire bats in parts of South America, where the disease is known as *murrina.* Surra affects a variety of wild and domestic mammals in northern Africa, southern Asia, the Philippines and Indonesia, and parts of Central and South America. It was apparently introduced into the Western Hemisphere by Spaniards in the 16th century via infected horses. Untreated infections are usually 100% fatal in horses, elephants, and dogs. The disease is serious but chronic in camels, which are thought to be the original host. Cattle and buffalo, in contrast, are not severely affected and may remain asymptomatic for months. Relatively resistant animals may serve as reservoirs of infection. Symptoms of infection are similar to those of other trypanosomiases. An old Arab name for surra, dating from the early 1900s or before, is *mard el debab,* which means "sickness of the gadflies."

Particularly in northern South America, mechanical transmission by tabanids of a related pathogen, *T. vivax,* is a serious problem for sheep producers and, to a lesser extent, for cattlemen. *T. equinum* infects horses in South America, where it causes a disease called *mal de caderas,* with symptoms similar to surra. In parts of Africa, mechanical transmission by tabanids of species such as *T. brucei,* normally transmitted by tsetse flies, may be significant, but the epidemiological importance of this is not well documented.

*T. theileri* causes widespread, generally nonpathogenic infections of cattle and wild hosts such as deer. This trypanosome is found commonly in the hindguts of tabanids but is absent from the salivary glands. Transmission therefore occurs primarily through feces entering the bite wound or perhaps by crushing or ingestion of infected tabanids by the animal. In the latter case, infection must occur through abrasions or breaks in the skin or through oral mucosa. Research on the epizootiology of this and other trypanosome diseases is complicated by the presence of *Blastocrithidia* species, nonpathogenic trypanosomes which occur naturally in tabanids and are easily confused with pathogenic trypanosomes.

## EQUINE INFECTIOUS ANEMIA

Equine infectious anemia (EIA), commonly known as *swamp fever*, is a serious viral disease of horses and other equids (Fig. 13.11). It is a febrile illness which causes lethargy, weight loss, and sometimes death in affected animals. Different strains of EIA virus differ in pathogenicity, and infected animals differ significantly in how they are affected. Acutely infected animals nearly always are killed fairly quickly. Chronically infected animals eventually die of complications, and inapparent carriers may live a number of years with few obvious health problems. This disease is found in many areas of the world and is common in the southeastern United States. A series of studies conducted mostly in Louisiana provides a good model for understanding mechanical transmission of pathogens by tabanids (Foil, 1989; Foil and Issel, 1991), especially EIA and other retroviruses such as bovine or feline leukemia viruses.

Because virus infectivity declines rapidly on the insect mouthparts, rapid and frequent transfer of vectors between hosts is essential for significant transmission. Even though the amount of blood transferred by an individual fly is low, the potential for transmission is high when multiplied by a large number of persistent flies feeding per unit of time.

## ANAPLASMOSIS

Ticks and tabanids are the primary vectors of *Anaplasma marginale*, a rickettsia which causes anaplasmosis in cattle. This disease is most prevalent in the tropics and subtropics, including Africa, Australia, and the Americas. In the United States, the incidence of disease is highest in the southeastern states. Whereas calves seldom are affected severely, adult cattle show marked anemia, fever, and weight loss. Mortality may be as high as 50%.

While ticks may transmit the rickettsia biologically, mechanical transmission by tabanids is probably more important overall. Experimental transmission has been accomplished with interrupted feedings by *Tabanus* species, with infected flies being able to transmit *A. marginale* up to 2 hr after feeding.

## ELAEOPHOROSIS

A filarial nematode of domestic animals and wild ruminants, *Elaeophora schneideri* (Fig. 13.12) is widely distributed in North America (Pence, 1991). Most research on prevalence and epidemiology has been done in the Rocky Mountain states and Texas. Infection rates in mule deer (*Odocoileus hemionus*) often exceed 50%. Mule deer apparently are the normal reservoir host in western North America and rarely show any pathological effects. Compared with mule deer, white-tailed deer (*O. virginianus*) are infected at a lower level (2–10%) in many regions of the southern United States and sometimes show

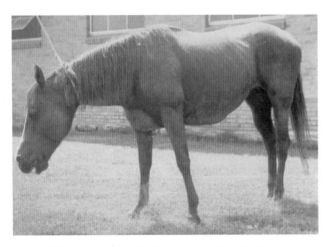

**FIGURE 13.11**   Horse suffering from equine infectious anemia. (Photo by W. V. Adams, Jr.)

**FIGURE 13.12**   Arterial worm, *Elaeophora schneideri*. Adult filarial nematode in carotid artery of white-tailed deer. (Courtesy of Southeastern Cooperative Wildlife Disease Study, University of Georgia.)

clinical signs of infection. Domestic sheep and goats may harbor the nematodes but generally do not show notable pathology. Some wild ruminants, particularly elk and moose, can be quite severely affected; up to 90% or more of the elk in parts of Arizona and New Mexico have been shown to harbor the nematode. In susceptible hosts, *E. schneideri* causes obstruction of arterial flow, hence the common name *arterial worm*. Reduced blood flow results in dermatosis, sloughing of distal tissues such as ear or antler tips, blindness, neurological disease, and sometimes death.

At least 16 species of tabanids in the genera *Tabanus, Hybomitra,* and *Silvius* have been implicated in transmission of *E. schneideri*. In the United States, prevalence of nematodes in horse flies captured from hyperendemic areas of New Mexico has been as high as 10–20%. Prevalence in other states, such as parts of southern Montana and South Carolina, is less than 1% in field-collected flies. In a given region, one or two species, such as *T. laticornis* in New Mexico, generally seem to be responsible for most of the transmission. Following ingestion of microfilariae by a tabanid, nematode development proceeds in the fat body and hemocoel. Infective larvae ($L_3$) are found in the mouthparts after about 14 days.

## OTHER PATHOGENS OF VETERINARY IMPORTANCE

Several other pathogens are transmitted mechanically by tabanids to livestock and wild animals. Most of them, however, also may be transmitted in other ways. Notable among these are the viruses that cause *bovine leukemia* and *hog cholera* and the sporozoan genus *Besnoitia* that causes *besnoitosis*. The role of tabanids in transmitting these pathogenic agents is usually very secondary.

There are several other interesting, but economically minor, disease agents transmitted to wild animals by tabanids. A protozoan parasite of turtles, *Haemoproteus metchnikovi*, is transmitted by deer flies (*Chrysops* spp.), and tabanids biologically transmit the filarial nematode *Dirofilaria roemeri*, a pathogen of kangaroos and wallabies.

## PREVENTION AND CONTROL

Tabanid control is difficult to achieve. A given area typically has multiple species with different seasonal occurrences and biological characteristics. Typical host contact is only about 4 min per fly during blood feeding, which may occur only once every 3–4 days. Short-term control

on livestock for several days may be achieved through use of *insecticides*, but insecticide sprays often are not particularly effective. Aerial applications of pyrethroid insecticides can suppress local populations of certain tabanids over a period of several hours to perhaps a few days, but they are considered far less effective than similar applications made for mosquito control. Use of insecticides for control of larvae or pupae, which are typically inaccessible in soil, is generally ineffective and can result in environmental damage. Insect *repellents* applied to human skin or impregnated into clothing may offer temporary relief for humans, but some individuals report poor results when using them.

Providing animals with structures for shelter, or pasturing them away from pasture-forest ecotones, can help to reduce tabanid biting intensity. Because tabanids prefer to fly around rather than over vegetation or screen *barriers* over 2 m high, such barriers may help to reduce tabanid access to livestock.

Another approach to tabanid control is *water management* in areas where drainage or manipulation of water levels is feasible. This must be done carefully, however, since such practices can actually enhance tabanid populations. Larvae of the salt marsh greenhead *T. nigrovittatus*, for example, are more abundant in better drained sites, and ditching salt marshes can increase its populations. Even in restricted habitats such as livestock ponds in pastures, water management is not always feasible. Heavy rains and flooding during periods when tabanids are pupating can kill the pupae, resulting in fewer adults a few weeks later. In situations where oviposition sites are limited, properly timed removal of emergent vegetation on which tabanid eggs are deposited can result in a significant reduction in the number of eggs and resultant larvae.

The use of traps for tabanid adults provides another potential control technique. In some coastal areas of the eastern United States, box traps are widely used for suppression of the salt marsh greenheads *T. nigrovittatus* and *T. conterminus* (Fig. 13.13). These flies are obligately autogenous, and responding females already have laid a complement of eggs. Under certain conditions, the traps may distract enough host-seeking adults from humans to result in temporary relief. Similar efforts have been made on a smaller scale using canopy traps, but with limited success. Where large, fairly mobile populations of horse flies occur, even large collections of adults may result in questionable suppression. In one Louisiana study, 95,000 tabanids were captured with 20 $CO_2$-baited sticky traps from a pasture area over a period of several days, but this resulted only in very temporary reduction in numbers of tabanids attacking cattle. The development of better attractants for adult tabanids could enhance the control potential of traps.

**FIGURE 13.13**    Box trap for collecting tabanid adults along salt marsh of Atlantic Coast, New Jersey. (Photo by E. J. Hansens)

The use of *biological control agents* also offers some potential for reducing tabanid populations. All stages of tabanids are subject to mortality by predators, including ladybird beetle larvae preying on eggs, wading birds feeding on larvae, and dragonflies and certain solitary wasps attacking adults. A few species of bembicine wasps (Sphecidae, Bembicinae) called *horse guards* specialize in capturing adult tabanids on which they rear their young. They tend to fly around pastured animals where they seize tabanids as they attempt to feed. Two species of horse guards in the southeastern United States are *Strictia carolina* and *Bembix texana*. Cannibalism among horse fly larvae also may be important in biological control.

Tabanid eggs are parasitized by wasps in the families Trichogrammatidae and Scelionidae, which sometimes causes high egg mortality (>50%). Tabanid larvae are parasitized by flies in the Tachinidae and Bombyliidae, and pupae are parasitized by wasps in the Diapriidae and Pteromalidae. Tabanid larvae also are subject to mortality from nematode parasites in the family Mermithidae. At least one mermithid species, *Pheromermis myopis,* parasitizes other invertebrates which are likely to be eaten by a tabanid larva, thereafter killing the larva while completing its development. A number of fungal, bacterial, and protozoan pathogens also are known from tabanids. The bacterium *Bacillus thuringiensis* and the fungus *Metarhizium anisopliae* have been tested against tabanid larvae in Russia.

# REFERENCES AND FURTHER READING

Allan, S. A., Day, J. F., and Edman, J. D. (1987). Visual ecology of biting flies. *Annu. Rev. Entomol.* **32**, 297–316.

Anderson, J. F. (1985). The control of horse flies and deer flies (Diptera: Tabanidae). *Myia* **3**, 547–598.

Andreeva, R. V. (1984). "The Ecology of Horse Fly Larvae and Their Parasitoses." Naukova Dumka, Kiev.

Andreeva, R. V. (1989). The morphological adaptations of horse fly larvae (Diptera: Tabanidae) to developmental sites in the Palearctic Region and their relationship to the evolution and distribution of the family. *Canadian Journal of Zoology* **67**, 2286–2293.

Andreeva, R. V. (1990). "Identification of the Larvae of Horse Flies of the European Part of the USSR, Caucasus and Central Asia." Naukova Dumka, Kiev.

Auroi, C. (1983). [The life cycle of *Hybomitra bimaculata* (Marqu.) (Diptera: Tabanidae). III. Pupation, emergence, blood meal and oogenesis]. *Mittellungen der Schweizerischen Entomologischen Gesselschaft* **56**, 343–359.

Burger, J. F. (1995). Catalog of Tabanidae (Diptera) of North America north of Mexico. *Contributions on Entomology International* **1**, 1–100.

Burger, J. F., Lake, D. J., and McKay, M. L. (1981). The larval habitats and rearing of some common *Chrysops* species (Diptera: Tabanidae) in New Hampshire. *Proceedings of the Entomological Society of Washington* **83**, 373–389.

Chippaux, J.-P., Bouchité, B., Demanov, M., Morlais, I., and LeGoff, G. (2000). Density and dispersal of the loiasis vector *Chrysops dimidiata* in southern Cameroon. *Medical and Veterinary Entomology* **14**, 339–344.

Churchill, D. R., Morris, C., Fakoya, A., Wright, S. G., and Davidson, R. N. (1996). Clinical and laboratory features of patients with loiasis (*Loa loa* filariasis) in the U. K. *Journal of Infection* **33**, 103–109.

Chvala, M., Lyneborg, L., and Moucha, J. (1972). "The Horse Flies of Europe (Diptera, Tabanidae)." Entomological Society of Copenhagen, Denmark.

Cooksey, L. M., and Wright, R. E. (1987). Flight range and dispersal activity of the host-seeking horse fly, *Tabanus abactor* (Diptera: Tabanidae) in north central Oklahoma. *Environmental Entomology* **16**, 211–217.

Cooksey, L. M., and Wright, R. E. (1989). Population estimation of the horse fly, *Tabanus abactor* (Diptera: Tabanidae) in north central Oklahoma. *Environmental Entomology* **16**, 211–217.

Dukes, J. C., Edwards, T. D., and Axtell, R. C. (1974). Distribution of larval Tabanidae (Diptera) in a *Spartina alterniflora* salt marsh. *Journal of Medical Entomology* **11**, 79–83.

Fairchild, G. B. (1969). Climate and the phylogeny and distribution of Tabanidae. *Bulletin of the Entomological Society of America* **15**, 7–11.

Fairchild, G. B. (1986). Tabanidae of Panama. *Contributions of the American Entomological Institute* **22**, 1–139.

Fairchild, G. B., and Burger, J. F. (1994). "A Catalog of the Tabanidae (Diptera) of the Americas South of the United States." Memoirs of the American Entomological Institute No. 55. Associated Publishers, Gainesville, FL.

Foil, L. D. (1989). Tabanids as vectors of disease agents. *Parasitology Today* **5**, 88–96.

Foil, L. D., and Issel, C. J. (1991). Transmission of retroviruses by arthropods. *Ann. Rev. Entomol.* **36**, 355–381.

Goodwin, J. T., and Drees, B. M. (1996). The horse flies and deer flies (Diptera: Tabanidae) of Texas. *Southwestern Entomology* **20**(Suppl.) 1–140.

Goodwin, J. T., Mullens, B. A., and Gerhardt, R. R. (1985). "The Tabanidae of Tennessee." Tennessee Agriculture Experiment Station

Bulletin 642. University of Tennessee, Institute of Agriculture, Knoxville, Tennessee.

Grevelink, S. A., Youssef, D. E., Loscalzo, J., and Lerner, E. A. (1993). Salivary gland extracts from the deer fly contain a potent inhibitor of platelet aggregation. *Proc. Natl. Acad. Sci. U. S. A.* **90**, 9155–9158.

Hayes, R. O., Doane, O. W., Jr., Sakolsky, K., and Berrick, S. (1993). Evaluation of attractants in traps for greenhead fly (Diptera: Tabanidae) collections on a Cape Cod, Massachusetts, salt marsh. *Journal of the American Mosquito Control Association* **9**, 436–440.

Hollander, A. L., and Wright, R. E. (1980). Impact of tabanids on cattle: blood meal size and preferred feeding sites. *Journal of Economic Entomology* **73**, 431–433.

Krinsky, W. L. (1976). Animal disease agents transmitted by horse flies and deer flies (Diptera: Tabanidae). *Journal of Medical Entomology* **13**, 225–275.

Lane, R. S., Anderson, J. R., and Philip, C. B. (1983). Biology of autogenous horse flies native to coastal California: *Apatolestes actites* (Diptera: Tabanidae). *Annals of the Entomological Society of America* **76**, 559–571.

LePrince, D. J., and Lewis, D. J. (1986). Sperm presence and sugar feeding patterns in nulliparous and parous *Tabanus quinquevittatus* Wiedemann (Diptera: Tabanidae) in southwestern Quebec. *Annals of the Entomological Society of America* **79**, 912–917.

Mackerras, I. M. (1954). The classification and distribution of Tabanidae (Diptera). I. General review. *Australian Journal of Zoology* **2**, 431–454.

Magnarelli, L. A., Anderson, J. F., and Barbour, A. C. (1986). The etiologic agent of Lyme disease in deer flies, horse flies, and mosquitoes. *Journal of Infectious Diseases* **154**, 355–358.

McElligott, P. E. K., and Lewis, D. J. (1996). Distribution and abundance of immature Tabanidae (Diptera) in a subarctic Labrador peatland. *Canadian Journal of Zoology* **74**, 1364–1369.

McKeever, S., and French, F. E. (1997). Fascinating, beautiful blood feeders—deer flies and horse flies, the Tabanidae. *American Entomologist* **43**, 217–226.

McMahon, M. J., and Gaugler, R. (1993). Effect of salt marsh drainage on the distribution of *Tabanus nigrovittatus* (Diptera: Tabanidae). *Journal of Medical Entomology* **30**, 474–476.

Middlekauf, W. W., and Lane, R. S. (1980). "Adult and Immature Tabanidae (Diptera) of California." Bulletin of the California Insect Survey No 22. Univ. of California Press, Berkeley.

Mullens, B. A., and Gerhardt, R. R. (1979). Feeding behavior of some Tennessee Tabanidae. *Environmental Entomology* **8**, 1047–1051.

Noireau, F., Nzoulani, A., Sinda, D., and Itoua, A. (1990). Transmission indices of *Loa loa* in the Chaillu Mountains, Congo. *American Journal of Tropical Medicine and Hygiene* **43**, 282–288.

Oldroyd, H. (1954–1957). "Horseflies of the Ethiopian Region," Vols. 1–3. British Museum of Natural History, London.

Olsufiev, N. G. (1977). "Horse Flies. Family Tabanidae. Fauna of the USSR." New Series No. 113, Insects, Diptera 7(2). Nauka, Leningrad.

Pechuman, L. L. (1981). "The Horse Flies and Deer Flies of New York (Diptera: Tabanidae)." Search Agriculture, Cornell University Agriculture Experiment Station No. 18. Cornell University, Ithaca, New York.

Pechuman, L. L., and Teskey, H. J. (1981). Tabanidae. In "Manual of Nearctic Diptera" (J. F. McAlpine, Ed.), Vol. 1, pp. 463–478. Monograph 27. Agriculture Canada Research Branch, Ottawa.

Pechuman, L. L., Webb, D. W., and Teskey, H. J. (1983). "The Diptera, or True Flies, of Illinois. I. Tabanidae." Illinois Natural History Survey Bulletin No. 33. Illinois Department of Energy and Natural Resources, Champaign, Illinois.

Pence, D. B. (1991). Elaeophorosis in wild ruminants. *Bulletin of the Society of Vector Ecology* **16**, 149–160.

Perich, M. J., Wright, R. E., and Lusby, K. S. (1986). Impact of horse flies (Diptera: Tabanidae) on beef cattle. *Journal of Economic Entomology* **79**, 128–131.

Poinar, G. O., Jr. (1985). Nematode parasites and infectious diseases of Tabanidae (Diptera). *Myia* **3**, 599–616.

Ralley, W. E., Galloway, T. D., and Crow, G. H. (1992). Individual and group behavior of pastured cattle in response to attack by biting flies. *Canadian Journal of Zoology* **71**, 725–734.

Roberts, L. W., and Wilson, B. H. (1967). Predation of horse flies by two bembicine wasp species in certain areas of southern Louisiana. *Journal of Economic Entomology* **60**, 412–415.

Schutz, S. J., and Gaugler, R. (1989). Honeydew-feeding behavior of salt marsh horse flies (Diptera: Tabanidae). *Journal of Medical Entomology* **26**, 471–473.

Smith, S. M., Turnbull, D. A., and Taylor, P. D. (1994). Assembly, mating, and energetics of *Hybomitra arpadi* (Diptera: Tabanidae) at Churchill, Manitoba. *Journal of Insect Behavior* **7**, 355–383.

Stoffolano, J. G., Jr., and Lin, L. R. S. (1983). Comparative study of the mouthparts and associated sensillae of adult male and female *Tabanus nigrovittatus* (Diptera: Tabanidae). *Journal of Medical Entomology* **20**, 11–32.

Sutton, B. D., and Carlson, D. A. (1997). Cuticular hydrocarbon variation in the Tabanidae (Diptera): *Tabanus nigrovittatus* complex of the North American Atlantic coast. *Annals of the Entomological Society of America* **90**, 542–549.

Takahasi, H. (1962). "Fauna Japonica. Tabanidae (Insecta: Diptera)." Biogeographical Society of Japan, National Science Museum, Tokyo.

Teskey, H. J. (1990). "The Insects and Arachnids of Canada, Part 16. The Horse Flies and Deer Flies of Canada and Alaska (Diptera: Tabanidae)." Agriculture Canada Publication 1838. Minister of Supply and Services Canada, Ottawa.

Waage, J. K., and Davies, C. R. (1986). Host-mediated competition in a bloodsucking insect community. *Journal of Animal Ecology* **55**, 171–180.

Wilkerson, R. C., Butler, J. F., and Pechuman, L. L. (1985). Swarming, hovering, and mating behavior of male horse flies and deer flies (Diptera: Tabanidae). *Myia* **3**, 515–546.

Wilson, B. H. (1968). Reduction of tabanid populations on cattle with sticky traps baited with dry ice. *Journal of Economic Entomology* **61**, 827–829.

# 14

# MUSCID FLIES (*Muscidae*)

ROGER D. MOON

*The family* Muscidae includes significant blood-feeding parasites, vectors of disease agents, and species that simply annoy humans and domestic animals. These flies and others in related families are often called *synanthropic flies,* species that exploit foods and habitats created by agriculture and other human activities. Muscid flies and their relatives can be grouped according to their habitat affinities. There are *filth flies,* such as the house fly, whose adults and immatures occur in a variety of filthy organic substrates, including latrines, household garbage, manure, and manure-soiled animal bedding. A subset of the filth flies are *dung flies,* such as the horn fly, whose immatures occur exclusively in cattle droppings. Another group is the *sweat flies,* whose adults feed persistently on perspiration.

Muscid flies also can be grouped by the nature of their mouthparts. The *nonbiting muscid flies* have sponging mouthparts used to ingest liquids from inanimate substrates and animal tissues. These mouthparts are soft, fleshy, and incapable of penetrating skin. In contrast, *biting muscid flies* have piercing/sucking mouthparts that pierce the skin to obtain blood.

Useful reviews of the literature on muscid flies include a two-volume treatise on the biology and disease associations of synanthropic flies (Greenberg, 1971, 1973) and a monograph on the identification and biology of immature muscid flies (Skidmore, 1985). A comprehensive review of the veterinary effects and control of muscid flies and other arthropods on livestock is provided by Drummond *et al.* (1988). Additional reviews and bibliographies concentrate on selected species and their close relatives: the house fly (Thomas and Skoda, 1993; West, 1951; West and Peters, 1973), the stable fly (Morgan *et al.,* 1983a; Petersen and Greene, 1989; Thomas and Skoda, 1993; Zumpt 1973), the horn fly (Bruce, 1964; Morgan and Thomas, 1974, 1977), and the face fly (Krafsur and Moon, 1997; Morgan *et al.,* 1983b; Pickens and Miller, 1980).

## TAXONOMY

The Muscidae include approximately 9000 species in 190 genera. The North American fauna contains about

TABLE I

**Important Muscid Pests of Humans and Domestic Animals in North America**

| Ecological group | Common name | Scientific name | Hosts for adults |
| --- | --- | --- | --- |
| Filth flies | House fly[a] | *Musca domestica* | None required |
| | Stable fly[a] | *Stomoxys calcitrans* | Cow, horse, dog, humans, and others |
| | Garbage flies | *Hydrotaea* (=*Ophyra*[b]) *aenescens, H. ignava* \* (=*Ophyra leucostoma*) | None required |
| | False stable fly and relatives | *Muscina stabulans M. levida* (= *assimilis*[c]) | None required |
| | Little house fly | *Fannia canicularis* | None required |
| | Latrine fly and relatives | *F. scalaris, F. femoralis* | |
| Dung flies | Horn fly[a] | *Haematobia irritans irritans* | Cow, bison, horse |
| | Face fly | *Musca autumnalis* | Cow, bison, horse |
| Sweat flies | Sweat flies | *Hydrotaea meteorica, H. scambus,* and others | Large mammals |

[a] Species introduced to North America from the Old World; others are cosmopolitan or native to the New World.

[b] According to Huckett and Vockeroth (1987).

[c] According to Skidmore (1985).

700 species in 46 genera. Fortunately, only a few of these genera contain important medical or veterinary pests. The important North American taxa are listed in Table I. Five of them have been introduced from the Old World through human commerce. Outside North America, the same species, or close relatives with similar life cycles, habits, and ecology, may be encountered.

Important muscid flies (Table I) occur in two subfamilies, the Muscinae and the Fanniinae. Important nonbiting Muscinae are the house fly, the garbage flies, the false stable fly and relatives, the face fly, and the sweat flies. The important biting Muscinae are the stable fly and horn fly. A third biting species in North America, the moose fly (*Haematobosca stimulans*), formerly *Haematobia alcis*, occurs exclusively on the moose (*Alces alces*). The second subfamily, the Fanniinae, are represented by the nonbiting little house fly and its relatives (*Fannia* spp.). Although some authors consider the Fanniinae to be a separate sister family (Fanniidae), it is treated here a subfamily of Muscidae.

Adults and larvae of the North American Diptera can be identified to family using keys in McAlpine *et al.* (1981), and adult Muscidae can be keyed to genus using McAlpine *et al.* (1987) and references therein. The Muscidae of North America are cataloged in Stone *et al.* (1965). Other aids for identification are Skidmore's (1985) keys and descriptions of larvae and pupal cases (*puparia*) and James' (1947) classic treatment of adults and larvae of flies in Muscidae and other families that cause myiasis.

# MORPHOLOGY

The life stages of a typical muscid fly consist of egg, larva, pupa, and adult (Fig. 14.1). Eggs are similar to those of closely related families. They may occur singly or in groups. They are generally creamy in color, 0.8–2.0 mm long, elongate/ovate in shape, and concave dorsally where two ribs form hatching pleats (Fig. 14.2).

Larvae of muscid flies and related families are known as *maggots*, and there are three instars in all species. The body is tapered, with the head and mouth hooks at the pointed end and the anus and spiracles at the blunt end (Fig. 14.1). The head is greatly reduced; it lacks eyes and has minute antennae that resemble papillae. The thorax is legless and has a pair of lateral *prothoracic spiracles*. There are eight segments of the abdomen, and each is marked ventrally with transverse rows of spines forming *creeping welts*.

Although the head lacks a sclerotized capsule, it is supported internally by a sclerotized *cephalopharyngeal skeleton* (Fig. 14.3). This complex structure is partially visible through the integument of live larvae and is best visualized in cleared, slide-mounted specimens. The size, shape, and arrangement of elements of the cephalopharyngeal skeleton are useful in the identification of larvae. Paired *mouth hooks*, reduced from ancestral mandibles, can be extended and retracted from the oral cavity. They help in crawling, burrowing, and tearing into food and other substrates. Internal *dental sclerites, accessory sclerites,* and *pharyngeal sclerites* make up the rest

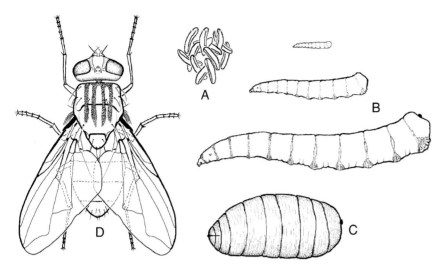

**FIGURE 14.1** *Life cycle of house fly (Musca domestica). (A) Eggs; (B) larvae, three instars, redrawn from Skidmore (1985); (C) pupa; (D) adult, traced from drawing by F. Gregor in Greenberg (1971).*

of the cephalopharyngeal skeleton and are variously modified for muscle attachment and feeding.

Most muscid larvae occur in wet substrates, where they filter particles of food from the substrate. These filter feeders possess a porous, ventral *sieve mechanism* between their pharyngeal sclerites (Fig. 14.3A). Exceptions are *Muscina* species and *Hydrotaea* species, whose third instars are facultative predators. The sieving mechanism is absent in these predatory forms (Fig. 14.3B).

Characteristics of the *spiracles* (Figs. 14.4 and 14.5) are useful for determining the species and instar of muscid larvae. Paired spiracles occur on the prothorax and at the end of the abdomen. *Prothoracic spiracles* are absent on first instars, whereas they are present on second and third instars. The shape and number of *spiracular tubercles* (spiraculare lobes) vary considerably in different species.

Structures associated with the paired *caudal spiracles* on the abdomen are of greatest taxonomic value. Each *spiracular plate* (Fig. 14.4) consists of a *peritreme* forming the plate's perimeter, two or three *slits* that are the

openings for gas exchange, and a *scar* that is a remnant from a previous molt. First and second instars have two slits, whereas third instars have three. An exception is the horn fly, in which both the second and third instars have three slits. Peritreme shape, position of the scar, and shape and orientation of the slits are all useful in identifying muscid larvae (Fig. 14.5).

The muscid pupa (Fig. 14.1C) occurs in a case, the *puparium*, that forms during a process called *pupariation*, which involves contraction and hardening of the third instar's integument (Frankel and Bhaskaran, 1973). During subsequent apolysis, or separation of the new pupal integument inside the puparium, a space forms around the pupa. The larval cephalopharyngeal skeleton remains attached inside the puparium's cephalic cap, and

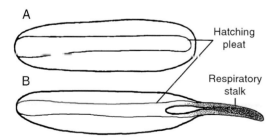

**FIGURE 14.2** Eggs of muscid flies, showing hatching pleat. (A) House fly (*Musca domestica*); (B) face fly (*Musca autumnalis*), with distinctive respiratory stalk.

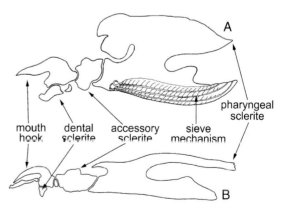

**FIGURE 14.3** Cephalopharyngeal skeletons of muscid fly larvae. (A) Filter-feeding larva, characterized by sieve–like structures; (B) Predatory larva, with pair of strong mouth hooks. (Redrawn from Skidmore, 1985.)

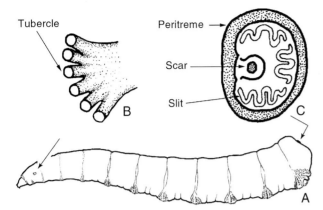

FIGURE 14.4 Respiratory spiracles of house fly (*Musca domestica*), third-instar larva. A, Larva, showing location of anterior pair of spiracles on first thoracic segment and posterior pair of spiracular plates on caudal segment (arrow); B, Anterior spiracle, enlargement showing multiple openings to trachea on individual tubercles; C, Posterior spiracular plate, showing sclerotized ring (peritreme) surrounding spiracular area, three sinuous spiracular slits, and scar left by previous molt. (Redrawn from Skidmore, 1985)

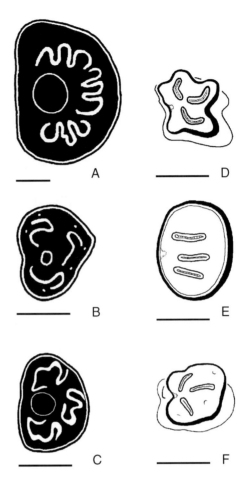

FIGURE 14.5 Caudal spiracular plates of third-instar larvae of six important muscid flies; only the right plate of each pair is shown. (A) Face fly (*Musca autumnalis*); (B) stable fly (*Stomoxys calcitrans*); (C) horn fly (*Haematobia irritans*); (D) false stable fly (*Muscina stabulans*); (E) black garbage fly (*Hydrotaea ignava*); (F) *Hydrotaea* sp. Bar = 0.2 mm. (Redrawn from Skidmore, 1985)

prothoracic and posterior spiracles remain embedded in the wall of the puparium. Thus, when an adult specimen has been associated with its puparium, distinguishing traits of the adult, the puparium, and the mature larva can be defined and used to identify the species from a specimen in any of the three stages.

Adult muscid flies (Fig. 14.1) are 4–12 mm long, with wings longer than the abdomen. Integument colors vary from brownish gray to black, often with dark longitudinal stripes on the thorax, called *vittae*, and dark spots or blotches on the abdomen. The head (Fig. 14.6) has three ocelli and a prominent pair of compound eyes, which in males are *holoptic*, nearly meeting at the dorsal midline; in females they are *dichoptic*, or more widely separated. Each antenna consists of a *scape*, a *pedicel*, and an *arista*. The arista arises from the pedicel as an undivided flagellum homologous to the flagellum of nematocerous flies. In most muscid flies, the arista is a single hair with fine setae along its shaft. A *ptilinal suture* encircles the bases of the antennae. The ptilinal suture is a remnant of the *ptilinum*, an eversible sac used by the emerging adult to break open and exit through the cephalic cap of the hardened puparium. In most muscid flies, a pair of strong bristles called *oral vibrissae* project ventrally from the lower edge of the face toward the mouthparts.

Mouthparts vary considerably among species. Generally they consist of a *proboscis* with a basal *rostrum*, a slender *haustellum*, and a terminal *labellum* (Fig. 14.6). The *maxillary palps* arise from the rostrum and appear to be one-segmented. The haustellum is formed by three structures held in union (Fig. 14.6C): an anterior *labrum*, a slender *hypopharynx*, and a posterior *labium*. The labium encloses both the labrum and hypopharynx and terminates in a two-lobed *labella*. Saliva flows from the *salivary glands* through the salivary canal in the hypopharynx that terminates in the *prestomum*, or preoral cavity, at the center of the labella. A *cibarial pump* inside the head creates suction that draws liquids through the prestomum and up through the *food canal* between the labrum and labium.

Important structural differences distinguish the labellae of nonbiting and biting muscid flies (Elzinga and Broce, 1986). Among nonbiting species, the labella is an enlarged, fleshy, two-lobed structure (Figs. 14.6 and 14.7A). On the mesal surface of each lobe are *pseudotracheae*, rows of fine setae used to scrape food and direct fluid toward the prestomum. The labella also houses mechanoreceptors, chemoreceptors, and *prestomal teeth*. These teeth may be short or elongate and dentate in some *Musca* species and *Hydrotaea* species. In contrast, mouthparts of the biting species (Figs. 14.7B, and 14.8) are adapted for piercing or tearing host skin to obtain

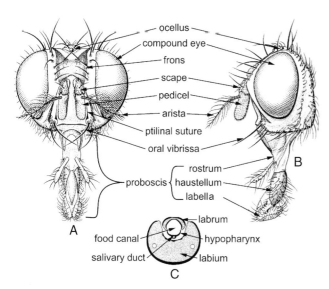

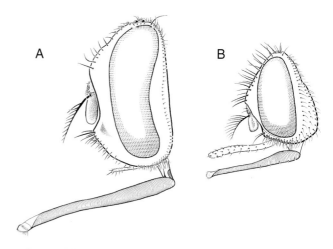

**FIGURE 14.6** Head and mouthparts of a nonbiting muscid fly, adult. (A) Anterior view; (B) lateral view; (C), cross-section of haustellum, showing relationship of the individual mouthparts, food canal, and salivary duct. (Adapted from original drawings by R. Idema, McAlpine *et al.*, 1981. Reproduced with permission of Minister of Public Works and Government Services, Canada.)

**FIGURE 14.8** Heads of two muscid flies with biting mouthparts; in both cases the mouthparts are held in a horizontal position beneath the head when not feeding. (A) Stable fly (*Stomoxys calcitrans*); (B) horn fly (*Haematobia irritans*). (Redrawn from Edwards *et al.*, 1939. © The Natural History Museum, London.)

blood. Their labellar lobes are comparatively small, but their prestomal teeth are sharp, sclerotized, and greatly enlarged.

Other morphological characters of the thorax and wings are typical of muscid flies. On the thorax (Fig. 14.9), the *hypopleuron* is bare, entirely lacking bristles and setae. The first *anal wing vein* (vein a1, Fig. 14.10), including a faint trailing fold if present, vanishes before it reaches the wing margin. The combination of these characteristics, along with aristate antennae, a ptilinal

suture, and usually robust oral vibrissae distinguish members of the Muscidae from all other flies.

The abdomen is reduced to five visible segments in both sexes, with succeeding segments 6–12 modified into eversible *reproductive terminalia*. The male possesses an *aedeagus*, or intromittent organ, which when at rest is rotated 180° and is partially enclosed in a genital pouch. In the female, the terminalia are modified into a tubular, telescoping *ovipositor* that is extended to lay eggs. Chemoreceptors and mechanoreceptors occur on the terminalia and also on the antennae, labellum, and tarsi.

The digestive system of adult muscid flies (Fig. 14.11) is much like that of other Diptera. Saliva is produced in

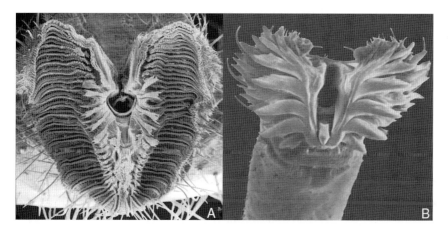

**FIGURE 14.7** Scanning electron micrographs of everted labella at tip of proboscis of two muscid flies, adults. (A) Nonbiting fly (face fly, *Musca autumnalis*), showing parallel arrangement of feeding channels (pseudotracheae) and relatively small prestomal teeth surrounding opening to food canal (prestomum); (B), Biting fly (stable fly, *Stomoxys calcitrans*), lacking pseudotracheae and with enlarged prestomal teeth used to cut through host skin to feed. (Micrographs by A. B. Broce.)

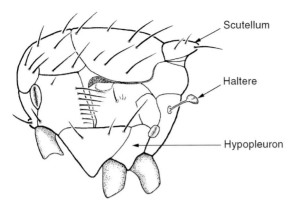

**FIGURE 14.9** Lateral view of thorax of adult muscid fly, with wings removed; anterior to left; showing typical sclerites, scutellum, hind wing reduced to form a haltere, and a taxonomically important sclerite, the hypopleuron. (Redrawn from Borror, *et al.*, 1989, with permission of Wadsworth, an imprint of the Wadsworth Group, a division of Thomson Learning.)

the *salivary glands,* which extend posteriorly from the head into the abdomen. Ingested foods that are nutrient-rich, such as blood and dung fluids, are routed via the *proventriculus* of the *foregut* to the *midgut,* where digestion occurs. Fluids that are dilute, such as plant nectar and milk, are shunted for temporary storage to the

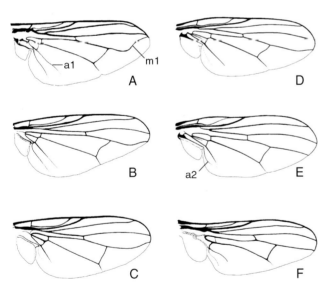

**FIGURE 14.10** Wings and wing venation of important muscid flies; right wing shown in each case. (A) House fly (*Musca domestica*); (B) stable fly (*Stomoxys calcitrans*); (C) horn fly (*Haematobia irritans*); (D) false stable fly relative (*Muscina levida*); (E) little house fly (*Fannia canicularis*); (F) black garbage fly (*Hydrotaea ignava*). (A), (B) and (F) redrawn from Axtell, 1986, all rights reserved, copyright © Novartis; (C) redrawn from Lane and Crosskey, 1993 © The Natural History Museum, London; (D) and (E) redrawn from McAlpine *et al.*, 1987, with permission of Minister of Public Works and Government Services, Canada.

*diverticulum,* typically a ventral, saclike extension of the esophagus. Contents of the diverticulum are regurgitated through the proboscis onto the fly's substrate and then reingested to aid in evaporation. Waste products and undigested food ultimately pass through the *hindgut* and out the *anus* as drops of liquid fly feces. The brown *fly specks* that collect on feeding and resting substrates consist of droplets of two substances—fly vomit and feces.

## LIFE HISTORY

All of the important North American muscid species are *oviparous,* meaning that females deposit fertilized eggs in the environment before they hatch. Some members of the Muscidae in the Old World are larviparous. *Oviposition substrates* vary among the different species. Filth flies oviposit in organic debris that is wet enough to support aerobic microbial fermentation. Common substrates are human feces and garbage and decomposing organic matter such as rotting algal mats, piles of lawn clippings, and food-processing wastes. Where livestock and poultry are confined, attractive substrates include *manure* (a mixture of aging feces and urine), soiled bedding (bedding, manure, and feed), and wet, rotting feeds such as hay, silage, and grain. Filth flies commonly exploit the kinds of wastes that accumulate around human habitations and animal-confinement facilities.

In contrast to the filth flies, the muscid dung flies and sweat flies lay their eggs in a much narrower range of substrates. Dung flies oviposit on or into cattle dung pats within minutes to a day or so after the animal defecates. The sweat flies oviposit in plant litter and decomposing dung in grasslands and forests.

Larvae of all of the important muscid flies burrow, feed, and develop in their respective ovipositional substrates. However, *larval feeding* differs among species and even among instars of the same species (Skidmore, 1985). Most muscid flies are *saprophages* that feed by filtering bacteria, yeasts, and other small organic particles suspended in their semiliquid habitat. All three instars of the house fly, the stable fly, *Fannia* species, the horn fly, and the face fly are saprophages. In contrast, the third instars of the garbage flies, *Muscina* species, and the sweat flies are *facultative predators.* These larvae can mature as saprophages, but they will switch to predation and consume other soft-bodied insects if these are available. These facultative predators contribute to the natural biological control of other flies that occur in the same habitats.

Once mature, third-instar muscid larvae cease feeding, empty their alimentary tract by defecating, and enter a wandering phase before they pupate. Depending on the

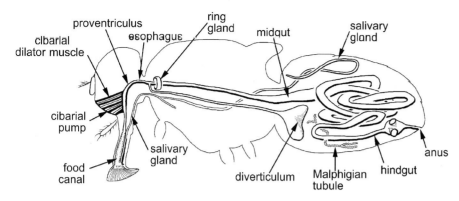

**FIGURE 14.11** Digestive system of adult muscid fly; left salivary gland and distal ends of Malphigian tubules omitted for clarity. (Redrawn from Patton, 1929)

species and substrate moisture, they may disperse to adjacent drier locations or pupate directly in their larval medium.

*Developmental times* from egg to adult of the different muscid flies range from 1 to 6 weeks during the summer (Table II). Most of the species develop fastest at temperatures of 27–32°C and all virtually cease activity at temperatures below 10°C. Sustained temperatures beyond these limits are usually lethal, with the exceptions of cold-tolerant species that overwinter as larvae or pupae. Developmental times also can be affected by the supply of food. For example, when crowded or otherwise starved, filth flies will delay pupation to achieve a minimum size. The dung flies have a different strategy; their larvae sacrifice size and will pupate earlier, resulting in adults that are smaller and less fecund than their better-nourished counterparts. Regardless of larval conditions, emerging adults typically occur in an equal sex ratio (1:1).

The *reproductive capacity* of a muscid fly population is determined in part by fecundity, which is the number of eggs that the average female can produce per batch. The house fly and other muscid filth flies are the most fecund, whereas the dung flies produce fewer eggs per batch (Table II). Reproductive capacity is also determined by the length of the gonotrophic cycle, which is how many days a female requires to develop and deposit a batch of eggs, and by longevity, which governs how long she will live and how many cycles she can complete. Under summer conditions, muscid flies can develop a new batch of eggs every 2–5 days. The average female probably lives long enough to produce one or two batches, although longevity is difficult to measure under field conditions.

## BEHAVIOR AND ECOLOGY

Because adult muscid flies emerge with little stored energy and nutrients, they must find water, salts, carbohydrates, and protein if they are to survive and reproduce. The nonbiting species obtain sugars from off-host

**TABLE II**

**Life History Attributes of Important North American Muscid Flies, Compiled from Skidmore (1985) and Other Sources**

| Name | Days from egg to adult | | Eggs per cycle | Generations per year | Overwintering stages |
| | Minimum | Normal | | | |
|---|---|---|---|---|---|
| House fly | 7 | 10–21 | 120–150 | Multiple | All? |
| Stable fly | 12 | 15–30 | 80–100 | Multiple | All? |
| Garbage flies | 9 | 14–25 | 70–190 | Multiple | Larva |
| False stable fly and relatives | 10 | 14–40 | 98–150 | Multiple | Adult in diapause |
| Little house fly and relatives | 22 | 25–50 | 58–72 | Multiple | Larva? |
| Horn fly | 9 | 10–20 | 20–32 | Multiple | Pupa in diapause |
| Face fly | 7 | 12–28 | 16–42 | Multiple | Adult in diapause |
| Sweat flies | 25 | 35–300 | 17–150 | 1–2 | Larva |

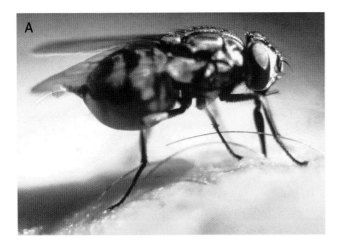

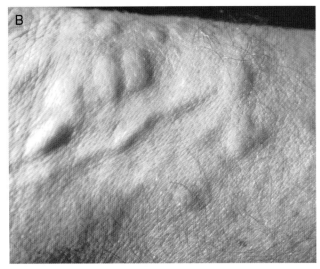

**FIGURE 14.12** Stable fly (*Stomoxys calcitrans*) feeding on human. (A) Fully fed female, abdomen distended with blood; (B) resultant bite reaction in form of welts. (Photos by Elton Hansens)

sources such as nectar from plants and honeydew from sap-sucking aphids and scale insects. The species that feed on vertebrates obtain the bulk of their protein from blood, serum, saliva, mucus, and lachrymal secretions. Both sexes of the biting horn fly and stable fly obtain nearly all their nutrients from blood.

The *feeding behavior* of a nonbiting fly differs from that of biting species. A nonbiting fly opens its labella and presses it against the substrate. If the food is not liquid, the fly releases enzyme-laden saliva and repetitively opens and closes the labella to scarify solid food into the saliva. The suspension is then drawn along the pseudotracheae into the food canal. Feeding by nonbiting flies can be characterized as a process of salivating, scrubbing, and sucking; they cannot physically penetrate skin. When a biting species feeds (Fig. 14.12), it presses its proboscis against skin and rapidly opens and closes the labella, directing the prestomal teeth in a downward and outward rasping motion. Once the skin is penetrated, the teeth

anchor the proboscis while blood flows into the subsurface lesion and up the food canal.

All of the important muscid flies are *anautogenous,* meaning that females require protein to complete their first gonotrophic cycle. Protein stimulates yolk synthesis and allows maturation of eggs. In the nonbiting species, eggs in the ovarioles mature in synchrony 2–5 days after protein is first obtained. Development of a subsequent batch is arrested hormonally until the preceding batch has been laid. There are corresponding cycles of attraction to different substrates, first to sources of protein and then to oviposition sites as eggs mature. In the horn fly and stable fly, eggs develop asynchronously, and feeding and oviposition are distributed more evenly in time.

Behavior associated with mating differs among species. Males of *Fannia* species and some of the sweat flies and garbage flies hover in swarms, usually in locations shaded by trees or the roofs and eves of buildings. These males are attracted to females that fly into the swarm. Once coupled, a mating pair will fall to the ground and complete copulation. Males of the other muscid flies do not form swarms. Instead, they generally perch or rest in sunny locations on substrates such as tree trunks, fence posts, and rocks, and the males intercept passing females. Females of all the important muscid flies typically store enough sperm from a single mating to fertilize all the eggs they can produce during the remainder of their lives.

Activities and locations of adult muscid flies vary markedly with time of day. All the important species are active during daylight hours, and almost all are inactive at night. Activities include flying, host location, feeding, mating, and ovipositing. Sight and olfaction are used to locate hosts and oviposition substrates. Most muscid flies are *exophilic,* being reluctant to enter buildings. A few species are more *endophilic* and will enter buildings. Species that feed on animals may be on a host as briefly as a few minutes, just long enough to obtain available foods. The flies leave their hosts when replete and rest in the surrounding environment while digestion proceeds. Because feeding times are much shorter than digestion times, the adults on a host at any instant are likely to be only a small fraction of all the adults present in the host's environment.

Muscid flies apparently choose daytime resting sites, in part, according to their needs for thermoregulation. They rest in sunny sites when the air temperature is below about 20°C and in shady sites when temperatures exceed about 30°C.

An exception to the generalized pattern of daytime activity and host-visiting behavior occurs in the horn fly. Once a host is located, the adult remains on its host almost continuously, except when disturbed or laying eggs. Horn flies feed and oviposit at all hours of the day and night.

The *flight range* of muscid flies is extensive. Detectable numbers of all the important North American species have been collected more than 5 km from known or presumed points of origin. Large numbers of stable flies can appear on beaches 10 or more miles downwind from the nearest likely breeding sites.

The *seasonal patterns* in abundance and age structure of adult subpopulations vary among species, years, and locations. In localities with cold winters, population growth outside buildings is restricted to a distinct breeding season—the warmer, wetter months of spring, summer, and autumn. In these cases, populations grow to a single peak of abundance, normally in early autumn. A notable exception is the house fly, which can breed continuously in heated buildings. In warmer climates, the breeding season for most species is longer and may be continuous year-round. For example, adults of the house fly, *Fannia* species, and horn fly occur throughout the year in the southeastern United States and southern California. Densities of adults have two seasonal peaks, with growth phases in spring and autumn separated by periods of decline during summer and winter.

Most muscid flies of medical/veterinary importance are *multivoltine*, developing through two or more generations per breeding season (Table II). These generations usually overlap, so recruitment of new adults is continuous; eggs, larvae, pupae, and adults of all ages are present simultaneously throughout most of the breeding season. Population growth within the breeding season is influenced by availability of breeding media, by weather and its effects on survival of immature stages, and by the fly reproductive rate. Survival of larvae is enhanced if their breeding habitat remains wet enough to support filter feeding, yet dry enough to allow aerobic respiration. Suitable moisture levels are about 30–75%. Substrate moisture is critical because the saprophagous larvae feed by filtering particles suspended in their medium.

Muscid flies *overwinter* in different ways (Table II). The house fly and the stable fly breed continuously in frost-free southern regions of North America. Breeding by these flies is restricted to the warmer months in more northern latitudes, because they lack a stage that can endure temperatures below freezing for much more than a day. It was once thought that these two filth flies, lacking a freeze-tolerant life stage, died out each winter in temperate latitudes and were repopulated each spring from milder regions. However, it is now known that local populations persist through winter in protected, semiheated substrates associated with humans and livestock. Regional repopulation does occur, however, with the bush fly in Australia, where immigrants disperse southward from more northerly latitudes that remain warm during winter (Hughes, 1977).

Some other muscid flies of medical/veterinary importance overwinter in *diapause,* a state of developmental arrest typically associated with a tolerance for freezing. The face fly and *Muscina* species overwinter as adults. In autumn, these flies enter hibernacula, such as occur under bark of dead trees and siding on buildings, and emerge the next spring to begin reproduction. The horn fly, in contrast, overwinters in temperate regions as a diapausing pupa. Garbage flies, *Fannia* species, and sweat flies are thought to overwinter as larvae, but further study is needed to determine if they exhibit a true diapause or are in a simpler state of cold-tolerant quiescence.

# SPECIES OF MEDICAL/ VETERINARY IMPORTANCE

Adults and larvae of some important muscid flies can be identified tentatively from external characteristics and from features of their behavior, habitat, and geographic location.

### House fly (*Musca domestica*)

This nonbiting filth fly occurs on all continents except Antarctica. It is native to the Afrotropical and Oriental regions and was probably introduced into the Americas by Europeans during colonial times. Adults are gray and black flies, 6–9 mm long, with four black vittae on an otherwise gray thorax (Figs. 14.1 and 14.13A). The wing has a sharp forward bend in vein M1 (Fig. 14.10A). The abdomens of typical females are checkered gray and black at the dorsal midline and creamy yellow on the sides, which in North America is sufficient to distinguish this species from the face fly. Larvae have large caudal spiracles that resemble back-to-back D's, and the slits are sinuous (Fig. 14.4B).

Immatures can be found in a wide variety of decaying organic substrates. Major breeding sites include human garbage dumps, open privies, livestock manure, soiled bedding, poultry litter, and wastes around fruit and vegetable processing plants. Breeding continues year-round in tropical and subtropical regions but is interrupted by winter in temperate regions. From a public health standpoint, the house fly is probably most significant as a nuisance and potential vector of enteric pathogens. Although the house fly can become quite abundant where livestock, poultry, and companion animals are housed, its direct effects on animal health are comparatively unimportant.

### Bazaar fly (*Musca sorbens*)

This nonbiting filth fly is the most abundant synanthropic muscid fly in many parts of the Afrotropical, Oriental, and Pacific regions. It was introduced through commerce into Hawaii and probably would flourish elsewhere in tropical latitudes of the Americas. Greenberg's (1971) key provides characteristics to

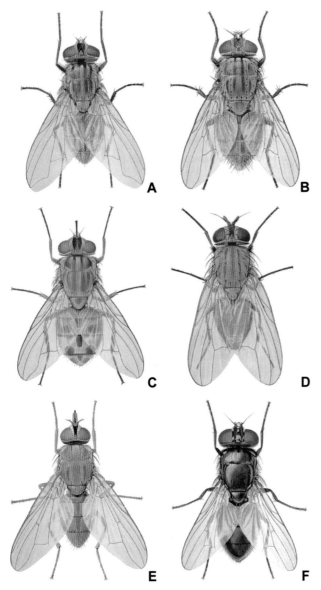

**FIGURE 14.13**   Muscid flies, adult females. (A) house fly (*Musca domestica*); (B) false stable fly (*Muscina stabulans*); (C) stable fly (*Stomoxys calcitrans*); (D) little house fly (*Fannia canicularis*); (E) horn fly (*Haematobia irritans*); (F) black garbage fly (*Hydrotaea ignava*). Not to scale. (original drawings by F. Gregor, and published in Greenburg, 1971, reprinted by permission of Princeton University Press)

distinguish the bazaar fly from other *Musca* species in the Afrotropical and Oriental regions.

The species recognized as *M. sorbens* before 1970 apparently consists of a complex of at least three species that are partially distinguishable by the ratio of the width of the male's frons (area between compound eyes) to the width of the head (including eyes). The "broad-frons" form is known correctly as the bazaar fly (*M. sorbens*), and this species occurs from Africa east through the Orient and on many Pacific islands. A "narrow-frons" form occurs in Australia and Papua

New Guinea and is considered a distinct species, the bush fly (*M. vetustissima*). A second "narrow-frons" form (*M. biseta*) coexists with the bazaar fly in Africa and eastward, but further study is needed to resolve distinctions between *M. biseta* and the bush fly in southern Asia and the Orient (Pont, 1991).

Adult bazaar flies are strongly exophilic, being far less inclined than the house fly to enter buildings. Larvae have been recorded in unburied human stools and dog feces and, less commonly, in feces of other animals, in carrion, and in garbage. The bazaar fly is important to public health, but it is probably unimportant to the health of domestic animals.

**Bush fly** (*Musca vetustissima*)

This nonbiting dung fly occurs in Australia, where it is a major nuisance to humans and livestock. It is related closely to the bazaar fly and keys out as *M. sorbens* in Greenberg (1971). Adults are attracted to large mammals as sources of fluids for nourishment and feces for oviposition. Several authors have speculated that the bush fly was originally associated with aboriginal encampments, but that its abundance increased when domestic cattle were imported. Larvae have been recorded from the feces of a wide variety of large mammals, but in nature cattle dung pats are overwhelmingly the most productive. Breeding is continuous in subtropical Australia, and southward migrations serve to repopulate temperate Australia and Tasmania each spring (Hughes, 1977).

**Face fly** (*Musca autumnalis*)

This nonbiting dung fly is native to Europe and central Asia and was introduced into North America before 1952 (Krafsur and Moon, 1997). It occurs in all southern Canadian provinces and in the United States north of Arizona−Georgia (35°N). The adult resembles the house fly (Figs. 14.1 and 14.13A), is 6−10 mm long, has four black vittae on an otherwise gray thorax, and has a sharp forward bend in wing vein m1 (Fig. 14.10A). The male's abdomen has a distinct, black longitudinal band along the midline and bright yellow sides. The female has a characteristic yellow patch on the ventrolateral aspect of the first visible abdominal segment; the remaining segments are gray−black to the ventral midline. The egg has a distinct brown−black respiratory stalk (Fig. 14.2B). Mature larvae are bright yellow with black, D-shaped spiracular plates (Fig. 14.5A), and puparia are white due to calcification.

During the fly breeding season, adult face flies occur around grazing cattle and horses. Their larvae develop exclusively in fresh cattle dung pats. In autumn, newly emerged adults enter diapause, aggregate on buildings, and eventually accumulate behind siding, in wall voids, in attics, and occasionally in interior rooms. Thousands of flies can occur in such aggregations. The face fly was first recognized as a nuisance in European households due to its overwintering habits. It became recognized as a pest of cattle and horses after it was introduced into North America.

### Cluster fly (Pollenia rudis)

The cluster fly is discussed here because it often occurs along with the face fly in household infestations. This calliphorid fly is native to Europe and North Africa and was introduced into North America, where it now occurs in all southern provinces of Canada and throughout the United States. The adult is 7–9 mm long, the abdomen is completely black with silvered checking, and crinkly golden hairs occur on the head and thorax. Cluster fly larvae are internal parasites of earthworms (*Allolobophora* spp., Lumbricidae) and produce two to four generations per year. Adults can be a nuisance in households, but they do not affect domestic animals.

### Stable fly (Stomoxys calcitrans)

This biting filth fly (Figs. 14.12 and 14.13C) is native to Africa, Europe, Asia, and the Orient and was probably introduced into the Americas and Australia during colonial times. At least three common names are used regionally for the stable fly. It is known as the *beach fly* because of outbreaks on recreational beaches, the *dog fly* because it pesters dogs, and the *lawn-mower fly* because larvae have been found in damp, matted grass on the undersides of lawn mowers. The stable fly is also misleadingly called the *biting house fly* because of its superficial resemblance to the house fly.

The adult is 5–7 mm long, has seven circular black spots on an otherwise gray abdomen (Fig. 14.13C), and has a piercing/sucking proboscis with short maxillary palps (Fig. 14.8A). Larvae and pupae have uniquely shaped, subtriangular posterior spiracles (Fig. 14.5B) that are far apart; the horizontal space between them is greater than twice a plate's width. Larvae occur in decaying fibrous substrates such as straw bedding, wet hay, algal mats, and wet grass clippings. Other larval habitats include accumulations of manure from dairy and beef cattle, mixtures of soil and partially composted bedding and animal manure, and byproducts of crop processing, such as peanut hulls, beet pulp, and sugarcane bagasse. Breeding is continuous in tropical and subtropical climates, and the species is thought to overwinter as immatures in colder climates. Stable flies are important to public health because they will attack and annoy people, but they are much more important from a veterinary perspective.

### Horn fly (Haematobia irritans irritans) and buffalo fly (Haematobia irritans exigua)

These biting dung flies were once recognized as two separate species: the horn fly (*H. irritans*) and the buffalo fly (*H. exigua*). However, Zumpt (1973) concluded that they are subspecies of *H. irritans*, based mainly on subtle morphological differences and allopatric distributions. The horn fly is native to northern Africa, Europe, and central Asia and was introduced into North America from Europe in the middle 1880s. It has since spread to all cattle-producing regions in the Americas, including Hawaii. The buffalo fly is native to southern Asia, the Orient, Indonesia, and several Pacific islands. It spread through commerce into New Guinea and Australia before 1840. It is possible that the two subspecies intergrade in parts of Asia.

The adult horn fly (Fig. 14.13E) is 3–5 mm long and has a piercing/sucking proboscis (Fig. 14.8B). The maxillary palps are held appressed to the haustellum and are almost as long as the haustellum. Wing vein m1 is gently curved (Fig. 14.10C). Adults of both subspecies are specific to cattle, bison, and water buffalo; aberrant hosts include horses and other large mammals. The flies occur mainly on the withers, back, and sides but will move to the belly when weather is hot. Females of both flies lay their eggs exclusively under edges of dung pats, usually within minutes of host defecation. For unknown reasons, dung from horses, sheep, and other large mammals is unsuitable. Horn flies occur in far greater numbers on grazing cattle than on animals confined in drylots or indoors. Reproduction is continuous and populations are multivoltine. The horn fly overwinters as a diapausing pupa in temperate latitudes. Neither subspecies poses any threat to human health, but both species are serious economic pests of grazing beef and dairy cattle.

### False stable fly (Muscina stabulans) and its relatives

These nonbiting filth flies include the false stable fly, which has been spread worldwide through commerce, and another 10 species that occur mainly in the Holarctic region. The name *M. assimilis*, used widely in older literature for a relative of the false stable fly, has been relegated to a synonym of *M. levida* (Skidmore, 1985). The false stable fly (Fig. 14.13B) and its relatives are stout flies, 8–12 mm long, with brown–black bodies and a rounded bend in wing vein m1 (Fig. 14.10D). The tip of the scutellum of the false stable fly is red–orange. The posterior spiracular plates are roughly circular; they are separated by one plate's width, and the slits are bowed and arranged radially (Fig. 14.5D). Third-instar larvae are facultatively predatory, and adults overwinter in a prereproductive diapause. These species can affect public health, but they are not thought to affect the health of domestic animals.

### Little house fly (Fannia canicularis) and its relatives

There are about 100 species of these nonbiting filth flies in North America (Chillcott, 1961) and

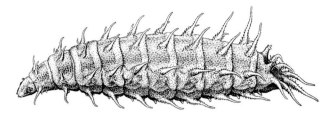

**FIGURE 14.14** Little house fly (*Fannia canicularis*), third-instar larva. (From McAlpine *et al.*, 1981 reproduced with permission of Minister of Public Works and Government Services, Canada)

additional ones in Latin America, Africa, Europe, and Asia. The little house fly (*F. canicularis*) and the *latrine fly* (*F. scalaris*) have spread by commerce throughout the world. *Fannia* species are 5–8 mm long, with dark thoraces and abdomens variously marked with yellow (Fig. 14.13D). The arista lacks setae and the second anal vein (a2) curves toward the first anal vein (a1) (Fig. 14.10E). Larvae and puparia have characteristic lateral and dorsal processes (Fig. 14.14) whose function is unknown. Males form mating swarms in shady locations, and it is this swarming behavior that most often brings them into contact with people. The little house fly is probably the most endophilic and commonly encountered species of this genus in North America. The latrine fly is more exophilic. Although these flies are most noticeable where domestic and zoo animals are confined, they are not important to public and veterinary health.

**Garbage flies** (*Hydrotaea* spp.)

There are seven known species of garbage flies, and at least one species occurs in every biogeographic region. This group of nonbiting filth flies, once placed in the genus *Ophyra*, has been merged into *Hydrotaea* (Huckett and Vockeroth, 1987). Accordingly, the scientific names of the common species have changed. The *black garbage fly*, known in older literature as *Ophyra leucostoma*, is now *Hydrotaea ignava*. It is native to the Old World and has been introduced into North America (Skidmore, 1985). The *black dump fly*, formerly *O. aenescens*, is now *H. aenescens*. It is native to the New World, occurs in the eastern Pacific Islands, including Hawaii, and has been introduced into Europe.

The garbage flies are 4–7 mm long, with shiny black thoraces and abdomens (Fig. 14.13F). Wing vein m1 is virtually straight (Fig. 14.10F). Posterior spiracles of mature larvae and puparia are roughly circular; they are separated by less than one plate's width and have slightly curved slits that barely diverge from a faint scar (Fig. 14.5E). Adults are strongly exophilic. Larvae have been recorded in a great variety of filthy substrates, including carrion. Third instars are facultative predators and will consume larvae of other flies that cohabit their breeding medium. These filth flies pose a modest threat to public health, but they are not known to harm domestic animals.

**Sweat flies** (*Hydrotaea* spp.)

About 50 species of sweat flies occur in the Palearctic region, and fewer species occur in the remaining biogeographic regions. Sweat flies are gray to black and 3–8 mm long, with an arista that lacks setae. Although the female has no simple distinguishing characters, the male has a ventral notch or depression at the distal end of the fore femur. Third-instar larvae are facultative predators. Their spiracular plates are stalked, with radially arranged slits (Fig. 14.5F).

Females of 6 of the 24 North American species, including *H. meteorica* and *H. scambus*, are persistent in their attempts to imbibe perspiration and secretions from the eyes, nostrils, lips, and other parts of mammalian hosts. The remaining North American species are apparently not attracted to animals (Huckett, 1954). In Europe, the *sheep head fly* (*H. irritans*) is a primary pest of sheep, cattle, and deer.

## PUBLIC HEALTH IMPORTANCE

Muscid flies affect people most frequently as nuisances, occasionally as vectors of pathogenic organisms, and rarely as agents of human myiasis. The cosmopolitan house fly and stable fly are of greatest medical significance. Other notable examples are the bazaar fly in Africa, Asia, and Pacific islands, including Hawaii; the bush fly in Australia; and *Stomoxys nigra* and *S. sitiens* in Africa and Asia.

Filth flies pose particular risks as mechanical vectors of pathogens that cause *enteric disease* in humans. Among the 1.3 million cases of notifiable infectious diseases reported in 1998 to the United States Centers for Disease Control and Prevention, approximately 74,000 (6%) were enteric infections causing diarrhea or dysentery. These diseases arise from direct or indirect fecal contamination of food and water. Globally, the World Health Organization reports that diarrhea and dysentery account for more childhood deaths and morbidity than any other infectious diseases.

Enteric diseases are caused by certain bacteria, viruses, and protozoa. The bacteria include *Escherichia coli*, *Salmonella* species, and *Shigella* species; the viruses include Cocksackie, enterovirus 72 (hepatitis A), and enteric cytopathogenic human orphan viruses; and the protozoa include *Chilomastix*, *Cryptosporidium*, *Entamoebae*, and *Giardia* species. Infections range in severity from benign to fatal, being most severe among children, the elderly, and others who are infirm. Common sources of enteric pathogens are food and water contaminated with feces from infected people or animals, or indirectly via hands, utensils, and flies.

Greenberg (1971, 1973) summarized the extensive literature on pathogens associated with muscid flies. Evidence is strong that filth flies in particular are *mechanical vectors*. Mouthparts, tarsi, and gastrointestinal tracts become contaminated when the flies feed on contaminated substrates. Upon dispersal, the flies can inoculate new substrates with contaminated tarsi, mouthparts, fly vomit, and feces.

The medical significance of *filth flies* at a given time and place depends on which flies and people are involved and on circumstances in which flies and people come into contact. A substantial majority of people in the United States and Canada now live and work in urban and suburban settings where indoor and outdoor environments are essentially free of filth flies. Exceptions are rural settings lacking adequate sanitation systems or neighboring mismanaged livestock and poultry operations. Intolerance for flies is, in part, the basis for municipal health codes used to enforce proper management of organic wastes on the affected premises. Sanitary standards established by the mid-1900s have dramatically reduced the epidemiological importance of filth flies in many parts of the developed world. Too often, however, basic sanitation and filth-fly management are unsatisfactory due to poverty, famine, or war. Under these circumstances, filth flies can reach tremendous densities, breeding in and around accumulated human waste and carrion.

The following muscid flies warrant attention with regard to public health.

### House fly  *(Musca domestica)*

The house fly is the most common cause of fly annoyance in North America. Adults aggregate around garbage, compost piles, and other food sources, and they readily enter buildings. House flies are conspicuous when alighting directly on people, crawling on human food, or resting on walls, windows, and ceilings. These substrates become soiled with fly specks, dried droplets of fly vomit, and feces.

In a classic pair of experiments, Watt and Lindsay (1948) and Lindsay *et al.* (1953) provided strong evidence that the house fly is a significant vector of enteric pathogens. They controlled filth flies with residual insecticides in selected towns in southern Texas and southern Georgia and left neighboring towns untreated as controls. Fly surveillance in the treated and untreated towns showed that treatments greatly reduced the densities of house flies and other species.

Surveillance of the residents in the treated towns showed concurrent declines in the incidence of diarrhea in people of all ages and in isolates of *Shigella* from children under 10 years of age.

Two more recent studies have confirmed the importance of the house fly in the epidemiology of enteric diseases. Intensive trapping to remove flies at two military field bases in Israel caused declines in fly populations at mess tents, concurrent declines in frequencies of diarrhea and shigellosis among base recruits, and declines in rates of seroconversion for antibodies to *Shigella* and enterotoxigenic *E. coli* (Cohen *et al.*, 1991). Elsewhere, village-wide spraying of six Pakistani villages during two consecutive fly seasons reduced house fly populations by 95% and lowered the incidence of childhood diarrhea by 23% (Chavasse *et al.*, 1999).

These studies provide strong evidence that house flies can be important routes for spread of fecal-borne pathogens. Prudence dictates that the house fly and other filth flies should be controlled through sanitation in the synanthropic environment and that they should be prevented from contacting human food at all points of production, distribution, preparation, and consumption.

### Bazaar fly  *(Musca sorbens)*

This nonbiting, synanthropic fly is common in Africa, Asia, and many Pacific islands. Adults feed persistently at the eyes, noses, and mouths of people (Fig. 14.15) and other large mammals. The flies are also conspicuous wherever human food is exposed outdoors. Fortunately, the species is strongly exophilic. Greenberg

**FIGURE 14.15**  Aggregating bazaar flies (*Musca sorbens*) on human hosts. (Photo by R. Lewis and D. Dawnway, with permission, © The Natural History Museum, London.)

(1971, 1973) summarized the extensive literature that associates the bazaar fly and its close relatives with human pathogens. Most notably, these flies are strongly suspected of mechanically transmitting enteric pathogens and the causal agents of acute bacterial conjunctivitis and trachoma. A recent study involving paired villages in The Gambia (Emerson *et al.*, 1999) showed that community spraying, which reduced bazaar fly populations by around 75%, lowered incidence of trachoma eye disease (caused by *Chlamydia trachomatis*) by 75% and the incidence of childhood diarrhea by 22%.

### Bush fly (*Musca vetustissima*)

The earliest European travelers in Australia recorded the annoying presence of the bush fly. This nonbiting dung fly, like the closely related bazaar fly, is strongly exophilic and is a probable irritant to humans almost anywhere in Australia. Flies that are attracted to people swarm around the head, feed at eyes and nostrils, and settle on the head, back, and shoulders. Once on hosts, the flies are peculiarly sedentary. More than a casual brush of the face with the hand is required to dislodge them, leading to a hand gesture that is humorously called an "Aussie salute." Larvae are known to occur in human and animal feces, so adults are a potential mechanical vector of enteric pathogens. Furthermore, the propensity of the adults to feed at a host's eyes makes the bush fly a prime suspect in transmission of eye pathogens (Greenberg, 1971).

### Face fly (*Musca autumnalis*) and cluster fly (*Pollenia rudis*)

The face fly and the cluster fly are two of several species of flies that can be a nuisance in households during winter and early spring. Other species include various blow flies (Calliphoridae), *Muscina* species, and *Ceroxys latiusculus* (Otitidae). Overwintering flies in buildings can be activated by heaters or warm weather and become attracted by light to inhabited rooms. Often people first take notice when live and dead specimens occur at sunny windows. Dead flies that have accumulated over years in an infested building can attract dermestid beetles. Although the flies and beetles can be a source of allergens, these insects do not pose any other known medical threat.

### Stable fly (*Stomoxys calcitrans*)

The stable fly is an important nuisance in outdoor environments throughout North America. This fly will readily attack people, usually on the lower part of the legs, causing a searing pain with each probe of its bayonet-like proboscis (Fig. 14.12A). It does not take many stable flies to disrupt activities of sunbathers, anglers, and others seeking outdoor leisure. Outbreaks have been recorded in the United States at tourist spots in the Great Lakes area, the Atlantic seaboard, and the Gulf Coast. Annoyance by stable flies is not confined to resorts and beaches; the flies can occur wherever people, fly breeding sites, and favorable weather coincide.

Adoption of the United States Declaration of Independence on July 4, 1776, by delegates to the Continental Congress may have been hastened by stable flies. According to Fuller (1913), debate on the Declaration drafted by Thomas Jefferson and his committee might have lasted much longer were it not for torment from stable flies. Jefferson noted the weather was oppressively warm that day in Philadelphia, and the meeting room was next to "...a stable, whence the hungry flies swarmed thick and fierce, alighting on their legs and biting hard through their thin silk stockings. Treason was preferable to discomfort." Clearly Jefferson had a wit, but he also knew enough entomology to infer that the nearby stable was the source of the flies.

As with any blood-feeding arthropod, stable flies provide an opportunity for transmission of blood-borne human pathogens. Experimental evidence suggests that the fly can acquire animal pathogens as mouthpart contaminants. Ingested particles can remain viable in the lumen of a fly's gut and diverticulum for hours to several days. However, none of these pathogens infects the fly, so only mechanical transmission would be possible. Experimental evidence using animal disease models in realistic settings suggests that the stable fly is not a vector of any consequence to human health.

### False stable fly (*Muscina stabulans*) and its relatives

The false stable fly and its relatives are common around filthy habitats, including latrines, household wastes, and accumulations of animal manure. The adults have feeding habits similar to those of the house fly and present similar risks for mechanical transport of food-borne pathogens. These flies remain outdoors and rarely feed on human food. However, they do feed and defecate on fruit and serve as potential vectors wherever breeding sites are near open-air markets and roadside fruit stands. Larvae of the false stable fly and of *M. levida* have been involved in rare cases of intestinal and urinary myiasis.

### Little house fly (*Fannia canicularis*) and its relatives

These nonbiting filth flies can become nuisances when swarms occur inside inhabited buildings. Hovering *Fannia* species often occur at head height indoors, where they can be particularly distracting and bothersome.

Adults of both sexes can be contaminated with pathogenic microbes from filthy larval breeding sites such as latrines, rotting garbage, and poultry litter. It is important, therefore, to exclude *Fannia* species from areas where human food is prepared or consumed. Nonetheless, *Fannia* species generally pose less of a health hazard than house flies because *Fannia* species rarely land and feed on human food.

In the western United States, *Fannia* species in the *benjamini* group are commonly attracted to human sweat and mucus. One species in this group, *F. thelaziae,* is a developmental vector for the mammalian *eyeworm, Thelazia californiensis.* Definitive hosts of this nematode include deer, canids, horse, rabbit, sheep, and black bear; people are rare, accidental hosts. Females of *T. californiensis* live in their host's lachrymal ducts, where they cause mild irritation and ophthalmia (Soulsby, 1965). Eggs are shed and hatch in eye fluids. First-stage larvae are ingested by eye-feeding flies, penetrate the midgut, and develop further in the fly's haemocoel. After 2–4 weeks' extrinsic incubation, infectious third-stage nematodes exit the fly's mouthparts when the vector feeds on another host.

The little house fly and the latrine fly have been involved in cases of intestinal, aural, and urinary myiasis of people. Most of the cases are thought to have arisen from eggs laid on clothing or bedding soiled with human feces.

### Garbage flies *(Hydrotaea* spp.*)*

Garbage flies and their larvae are common around municipal garbage dumps, compost sites, poultry houses, hog barns, and dairies. As occurs with other flies from these kinds of environments, garbage flies can be contaminated with microbial pathogens. However, garbage flies are more sedentary than house flies and are far less inclined to enter buildings and contaminate human food.

### Sweat flies *(Hydrotaea* spp.*)*

Very little is known about the medical importance of sweat flies in North America. Females of six North American species, including *H. meteorica* and *H. scambus,* are persistent in their attempts to obtain perspiration and secretions from the eyes, nostrils, lips, and other parts of their hosts. Because sweat flies are exophilic and occur most frequently in wooded areas, they are encountered by people in wooded parks, golf courses, and similar outdoor habitats. Except for their annoyance, sweat flies are not regarded as medically significant.

## VETERINARY IMPORTANCE

Muscid flies affect the health and comfort of domestic and wild animals. Domestic hosts include cattle, sheep, goats,

**FIGURE 14.16**  Holstein cattle bunching in response to attack by flies. (Photo by R. Moon)

horses, dogs, pigs, and poultry. Wildlife hosts include bison, elk, deer, moose, and rabbits in North America, and other mammals elsewhere in the world. Muscid flies cause discomfort, injure skin, affect growth and thriftiness, and transmit pathogenic viruses, bacteria, helminths, and cestodes. Repeated feedings in localized areas of skin can lead to secondary infections and scabs on the ears, legs, back, and other body regions of affected animals. Feeding by nonbiting muscid flies can retard the healing of wounds caused by biting arthropods and other agents. Muscid fly larvae also can be involved in cases of secondary myiasis.

Animals display a variety of *aversive responses* to attack by biting and nonbiting flies. As examples, horses will stamp and switch their tails, dogs will cower under cover, and cattle in a herd will mill together, bunched in a rosette formation with tails outward (Fig. 14.16). Frequencies of these and other aversive behaviors increase with fly density. Aversive behaviors can be disruptive and interfere with the handling, working, and showing of animals.

Biting muscid flies cause host vital signs to elevate (Schwinghammer *et al.*, 1986a,b). These changes, if prolonged, may be accompanied by changes in water and nitrogen balance. The net effect can be to reduce the amount of metabolic energy available for growth and lactation and to reduce the efficiency with which animals convert their feed into animal products. Livestock owners recognize a condition called *fly worry,* where animals appear irritated and generally unthrifty.

Economic effects of muscid flies on livestock and poultry industries of the United States are substantial. Estimates indicate that the stable fly and the horn fly alone cause annual losses of US$1.3 billion in reduced yields and increased production costs for beef and dairy industries (Drummond, 1987). Losses attributable to the face fly are US$123 million annually, resulting from the role of this fly in the epizootiology of bovine *pinkeye.* Costs incurred to manage the house fly and other filth flies around livestock and poultry operations have not been estimated. They are no doubt substantial and will probably increase as suburban development continues to expand into traditionally agricultural lands (Thomas and Skoda, 1993).

### House fly (*Musca domestica*)

This cosmopolitan fly is often the most abundant insect where livestock, poultry, or companion animals are housed. Adults occur on virtually all substrates surrounding the animals, including feed, feces, vegetation, and the walls and ceilings of buildings. Adults also occur directly on animals, where they feed on available blood, sweat, tears, saliva, and other body fluids. In response to the fly annoyance, animals flap their ears, shake their heads, and avoid pen locations where flies are particularly abundant. Beyond these behavioral symptoms, however, house flies appear to cause no measurable harm. Even when present in large numbers, house flies cause little or no adverse effects on animal growth or feed conversion in cattle, pigs, and other animals. Thus, house flies have much less impact on these animals than on the health and comfort of people living in the vicinity.

House flies can be significant mechanical vectors of enteric pathogens. The adults feed on feces and manure and foul their environment with fly specks. These habits degrade the appearance of facilities and contribute to microbial contamination of eggs (Fig. 14.17) and milk at points of production.

House flies are also developmental hosts for *Habronema muscae* and *Draschia megastoma*, two spirurid nematodes that cause gastric and cutaneous forms of *habronemiasis* in horses. In gastric infections, female worms invade the mucosa of the horse stomach and lay eggs that eventually pass out in the feces. Fly larvae become infested by ingesting these eggs. First-stage

**FIGURE 14.17** Chicken egg speckled with vomit and feces from house fly (*Musca domestica*). (Photo by R. Williams)

nematode larvae pass through the maggot's midgut into the haemocoel and subsequently metamorphose into infectious third instars while the maggot metamorphoses into an adult. After the fly emerges, the infectious third-stage larvae migrate through the thorax and eventually reach the mouthparts. A new gastric infestation in a horse can arise as nematodes exit the mouthparts of flies that feed around the host's mouth or if the horse ingests an infected fly in its feed. The nematodes eventually mature and become established in the mucosa. A new cutaneous infestation occurs if an infested fly feeds on the host's skin. Cutaneous infestations are a dead end for these nematodes because larvae in skin do not develop to maturity. Bush flies and bazaar flies are also hosts for *H. muscae* and *D. megastoma*. Further details of the worms' life cycles and the development of habronemiasis are provided in Soulsby (1965).

The house fly is also a developmental host for a *chicken tapeworm, Choanotaenia infundibulum*. The prevalence of this tapeworm is greatly reduced where chickens are housed in elevated cages, which prevents the birds from eating infected fly larvae and pupae.

House fly larvae have been recorded in cases of secondary wound myiasis. Females attracted to purulent wounds can feed and oviposit, and subsequent larvae feed on wound discharges and retard healing. Cases have been reported from nearly all species of domestic animals.

### Bush fly (*Musca vetustissima*)

As adults, the bush fly aggregates around large mammals, including cattle and horses. These nonbiting dung flies feed on facial and urogenital secretions and on serum and blood from wounds. Irritation by feeding flies can lead to skin lesions around the eyes and vulva, particularly of horses, and can retard wound healing. Annoyance by bush flies can induce animals to bunch and mill about, but the economic consequences are not documented. Studies in Australia in the first half of the 1900s suggested that the bush fly is a developmental host for the equine parasites *H. muscae* and *D. megastoma*. Their extrinsic life cycle in the bush fly is the same as in the house fly.

### Face fly (*Musca autumnalis*)

Adult face flies are conspicuous on cattle (Fig. 14.18), bison, and horses, swarming around their heads and feeding at eyes, faces, and wounds. Hosts respond by blinking their eyes, flapping their ears, shaking their heads, and switching their tails. However, modest numbers of face flies do not appear to affect the thriftiness of grazing dairy and beef cattle. Experimental dairy herds protected with repellents grazed as much, grew as fast, and produced as much milk as unprotected

**FIGURE 14.18**  Face flies (*Musca autumnalis*) on head of a Hereford cow. (Photo by H. J. Meyer)

herds. In other experiments with beef cattle, steers in screen cages with populations of face flies consumed as much feed and grew as fast as steers in cages without flies.

The face fly is much more important as a vector of bovine and equine pathogens. Of great concern to North American cattle producers is *pinkeye* (Fig. 14.19). This eye disease, also known as *infectious bovine keratoconjunctivitis*, is caused by the bacterium *Moraxella bovis*. Symptoms include reddened conjunctiva, excessive tearing, photophobia, opacity, and ulceration of the cornea. Pinkeye may involve one or both eyes and is most frequent among calves of white-faced breeds. Expenses associated with pinkeye involve surveillance and treatment of affected animals and retarded growth and blindness in cases that go undetected.

Face flies are mechanical vectors of *M. bovis*, as evidenced by isolation of the pathogen from face flies collected near infected cattle. In laboratory studies, viable bacteria have been recovered from the tarsi, mouthparts, diverticula, and regurgitant of flies exposed to bacterial cultures several hours earlier. Thus, *M. bovis* can acquire the bacterium from cattle, and the bacterium can remain viable for several hours on and in contaminated flies. Face flies also may create avenues of infection when they scarify the host conjunctiva while feeding. Furthermore, by stimulating host bunching, face flies may contribute to direct eye-to-eye spread of the bacterium among herdmates. Although the face fly is potentially important in the spread of *Moraxella*, it is

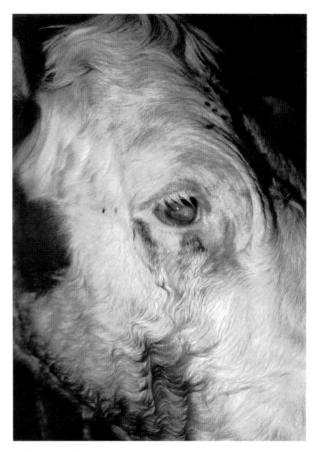

**FIGURE 14.19**  Pinkeye in a Hereford cow. Note opacity of cornea, reddened conjunctiva, and tearing below eye. (Photo by R. Moon)

not necessary that the face fly be present to have outbreaks of the disease. Pinkeye was known to occur in the United States at least 50 years before face fly was introduced.

The face fly is also a developmental host for several spirurid nematodes. *Eyeworms* in the genus *Thelazia* live in the lachrymal ducts of horses, cattle, and other mammals. In North America, the face fly transmits *T. lacrymalis* among horses and transmits *T. gulosa* and *T. skrjabini* among cattle. Fortunately, infections are benign in both hosts. The life cycles of these nematodes are similar to that of *T. californiensis*, which is transmitted to humans and other animals by *F. thelaziae*. The filarial nematode *Parafilaria multipapillosa* infests domestic and wild equids in southern Europe, northern Africa, the Middle East, central and southern Asia, and South America (Soulsby, 1965). It is transmitted in Russia by *Haematobia atripalpis*, but vectors elsewhere have not been identified.

In Sweden and South Africa, the face fly and closely related species are intermediate hosts of another filarial nematode, *P. bovicola*. This worm causes *bovine parafilariasis,* also known as *green muscle disease,* named after the appearance of subcutaneous green carcass lesions trimmed at slaughter. Following mating, female nematodes become established under the skin of the back or sides of infected hosts. There they create bleeding points, holes in the skin that exude blood and serum that attract nonbiting flies. Appearance of bleeding points coincides with the presence of vectors in the spring. Microfilariae in the exudates are ingested by the flies, penetrate the midgut, and undergo development for 2–3 weeks in the fly haemocoel. During this extrinsic incubation period, the nematode metamorphoses to the infectious third stage and then migrates through the thorax to the fly's proboscis. New hosts become infected when the fly feeds on another animal. In a new host, *P. bovicola* takes about 10 months to reach the host's back, where it matures to the adult. For further details on the biology of this parasite, see Bech-Nielsen *et al.* (1982).

In South Africa where *P. bovicola* is endemic, the worms are transmitted by three endemic *Musca* species, all in the subgenus *Eumusca.* Four other *Musca* species in other subgenera are not competent vectors, for reasons that are unclear. *P. bovicola* was recently introduced into Sweden and France, where the South African vectors are absent. In the worm's new range, it is transmitted exclusively by the face fly, itself a member of *Eumusca.* If *Parafilaria* were to be imported into North America, its establishment and spread would be almost certain because the face fly is already present.

### Stable fly  (*Stomoxys calcitrans*)

This cosmopolitan species attacks most large mammals, including domestic cattle, horses, donkeys, dogs, swine, sheep, goats, and camels. Wild hosts include Bovidae, Cervidae, Equidae, Canidae, and Felidae, both in their native ranges and in zoos. The stable fly often is abundant where mammals are confined indoors and outdoors. Confined animals attract and sustain immigrating adults, and organic debris associated with the animals promotes local breeding.

Stable flies attack large animals on the legs (Fig. 14.20), sides, back, and belly, whereas small ruminants and dogs are attacked most frequently on the legs, head, and ears. Individual stable flies typically feed once per day and remain on their host for 2–5 min, just long enough to obtain a blood meal. Engorged flies can be found on nearby vegetation, fences, and walls. Stable fly bites are painful, and localized wounds can coalesce to form scabs that are slow to heal, especially when aggravated by host scratching and rubbing. Such lesions commonly occur on the tips of dogs' ears and elsewhere on other hosts where hair is short or naturally parted.

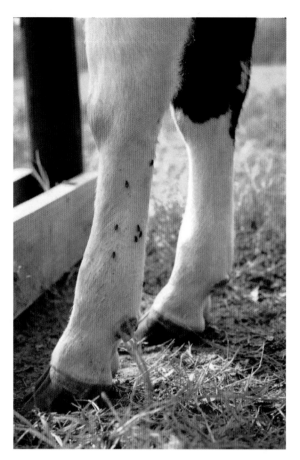

**FIGURE 14.20**    Stable flies (*Stomoxys calcitrans*) on the foreleg of a Holstein calf. (Photo by G. R. Mullen)

Behavioral responses to fly attack include a variety of aversive behaviors and physiological responses. Attacked animals are likely to stamp and kick their legs, which makes dairy cows difficult to milk and horses difficult to groom and show. Unrestrained cattle, horses, and small ruminants commonly will bunch when attacked. This behavior, combined with leg stamping and tail switching, is a clear indicator of stable fly activity.

Experiments with penned beef cattle have shown that irritation by stable flies causes cattle to consume less feed, to grow more slowly, and to convert feed into body mass less efficiently. These effects are greater when weather is hot and humid, presumably because bunching in response to fly attack interferes with the ability of the animals to dissipate excess heat. From an economic perspective, stable flies increase beef production costs because affected cattle require more time and more feed to reach slaughter weight. It is likely that stable flies similarly affect growing beef calves and lactating dairy cows. As a general guideline, economic losses in feedlots are likely to occur whenever the average number of stable flies per foreleg is three or more (Catangui *et al.*, 1997). Stable fly control is usually warranted when bunching, stamping, and tail switching are excessive.

The stable fly is not an important vector of animal pathogens. Experimental evidence has shown that it is possible for the stable fly to mechanically transmit retroviruses that cause *equine infectious anemia* in horses and *bovine leukosis* in cattle. In nature, however, the role of stable flies as vectors of these agents is negligible. Tabanid flies are far more important than stable flies in the spread of equine infectious anemia virus, and with bovine leukosis virus, transplacental transmission and transmission during vaccination, tattooing, and rectal palpation are much more important in establishing infection than biting flies.

The stable fly also is a developmental vector for *Habronema microstoma*, a spirurid nematode that causes gastric and cutaneous forms of *habronemiasis* in horses throughout the world. The gastric form of infestation is benign, whereas the cutaneous form presents conspicuous granular lesions known as *summer sores*. Onset of summer sores coincides with the stable fly breeding season. These diseases, the parasite that causes them, and the roles played by the stable fly in their transmission are very similar to the situations with *D. megastoma*, *H. muscae*, and the house fly (Soulsby, 1965).

**Horn fly** *(Haematobia irritans irritans) and buffalo fly (Haematobia irritans exigua)*

Of all parasitic arthropods, these two biting dung flies (Fig. 14.21) have the greatest effects on the health and productivity of cattle. Both sexes of the flies feed frequently each day, consuming an average of 10 $\mu$l of blood per fly per day (Kuramochi, 1985). At this rate, a cow with an exceptionally large population of 3000 flies would lose about 30 ml of blood each day, a small amount given that the blood volume of an adult cow is about 25 liters. Nonetheless, the bites are painful and irritating, and feeding lesions become cosmetic defects in tanned and dyed leather. Feeding by the buffalo fly on zebu cattle in Australia can lead to scabs on their hosts' withers and faces. Infested hosts react to horn flies and buffalo flies by licking their backs, twitching their flanks, switching their tails, and kicking at their bellies with their hind legs. These defensive behaviors usually suggest that an animal is being attacked by the horn fly or another biting fly.

Studies in the United States and Canada demonstrate that control of horn flies on mother cows can lead to a 12% increase in the average daily growth rate of nursing calves (Drummond, 1987). Similarly, growth rates of yearling stocker cattle and lactation rates of dairy cows may increase by about 16%. However, the size of the benefit of horn fly control has varied among studies, perhaps due to differences in densities of flies, degree of control, and presence of other biting flies and internal parasites. Benefits also may vary with weather, availability of forage, and growth potential determined by cattle genotype, age, and condition when flies are present.

Increases in animal performance following horn fly control make sense in light of animal metabolism, behavior, and energy budgets. When attacked by horn flies, stanchioned steers have elevated heart beats, respiratory rates, rectal temperatures, urine production, and urine nitrogen concentration (Schwinghammer *et al.*, 1986b). Pastured steers switch tails more frequently, spend less time grazing, and spend more time walking and resting during the day. These metabolic and behavioral responses suggest that horn flies increase the amount of energy spent by cattle in defending themselves against flies, leaving less energy available for growth. With nursing calves, the response to horn fly control is likely to be reflected in increased milk production by their dams; horn flies occur mainly on the cows and only incidentally on their calves.

Progressive management programs for beef cattle usually rely on static thresholds to judge if control of the horn fly will be economically justified. Measured as an average number of horn flies per animal side, recommended thresholds range from 25 flies per side in Alberta (Canada) to 100 in Nebraska and 250 in Texas. When densities exceed these thresholds, it is likely that increases in calf or steer growth rates in response to fly control during the fly season will more than pay for the cost of treating the cattle, whatever method is used.

The horn fly is a developmental vector for *Stephanofilaria stilesi,* a spirurid nematode that causes

**FIGURE 14.21**  Clusters of horn flies (*Hematobia irritans*) on the back of an Angus bull. (Photo by H. J. Meyer)

*stephanofilariasis* in cattle. This is a form of granular dermatitis that occurs mainly on the belly, scrotum, prepuce, and udder. This nematode is most prevalent in the western United States and Canada, but it is also recorded from cattle in the Old World. Mature *S. stilesi* occur in the skin. First-stage larvae are acquired by feeding horn flies, and the nematodes metamorphose to the third stage in the fly's haemocoel before being introduced into the definitive host when the fly feeds at a later time. Extrinsic incubation is about 3 weeks. Other species of *Stephanofilaria* occur in the Old World, where they are thought to be transmitted by other species of muscid flies (Soulsby, 1965).

### Sweat flies    *(Hydrotaea spp.)*

A few species of these flies feed on large mammals. In Europe, the *sheep head fly* (*H. irritans*) swarms and feeds at the faces of sheep, cattle, and deer. Affected sheep develop a condition known as *head fly disease*. Irritated sheep scratch, rub, and open wounds that are further aggravated by the flies. The disease is worst among horned animals of open-faced breeds lacking wool on their faces and among flocks grazing in wooded pastures. Larvae of sheep head fly are sparsely dispersed in soil and decomposing plant litter in grasslands and forests. Larvae of other species have been recorded from undisturbed cattle dung pats (Robinson and Luff, 1979).

The sheep head fly and a complex of other European sweat flies also feed at cow teats. Circumstantial evidence suggests that teat-feeding sweat flies may be mechanical vectors of *Actinomyces* (formerly *Corynebacterium*)*pyogenes*, the putative cause of *summer mastitis* in pastured dairy cattle. It is likely that *Hydrotaea* species are secondary vectors of this pathogen. The importance of sweat flies as vectors of these pathogens in North America is not well documented.

## PREVENTION AND CONTROL

Three general approaches are used to avoid or reduce problems caused by muscid flies: (1) preventing breeding (this can be either indirect, by making candidate media unavailable or unsuitable for survival of preadult stages, or direct, by killing immatures before they can develop to adults), (2) killing adults before they cause harm or produce offspring, and (3) excluding adults with screens and other barriers. A variety of methods can be used to accomplish these objectives (Drummond *et al.*, 1988).

The best approach is to use several methods simultaneously in an *integrated pest management program* to achieve desired levels of control in poultry houses, stables, and dairies (Axtell, 1986). For example, *sanitation* and *surveillance* of adult abundance are commonly used in combination. When densities exceed the tolerance threshold, sanitation can be increased and adulticides can be used to keep flies below intolerable densities. Choices among alternative practices are determined by effectiveness against the target insect, practicality in a given situation, costs of the practices in materials and labor, and environmental acceptability.

Emphasis should be placed on *source reduction* wherever possible. Housing for people or animals should be designed to limit accumulation of fly breeding media. Particular attention should be given to locations where human and other animal feces, domestic garbage, and rotting animal feed accumulate. The crucial first step to preventing enteric diseases is to prevent filth flies from breeding near human communities. The best defense is a closed sewage system or privy that will exclude ovipositing flies from reaching human excrement. Curtis (1989) presents designs of privies that do not require running water.

Facilities should be designed to minimize the labor required to maintain adequate sanitation. In livestock and poultry housing, lanes, alleys, and pens where manure can collect should be made easy to scrape. Feed and water should be provided in separate areas, if possible. Straw bedding for animals is particularly difficult to handle and is a notorious source of filth flies, so alternatives such as sawdust, sand, or washable mats should be considered.

In practice, even well-designed facilities have residual places in corners, around feeders, or along fence lines where organic debris can accumulate and fly breeding can occur. These places should be inspected regularly. Waste disposal should involve proper burial, spreading in a thin layer (<3 cm) on open fields, submersion in water, or aerobic composting. Compost piles must be turned frequently to keep the material hot and in a state of active fermentation. Special attention should be given to seepage areas that can form at the margins of compost piles if the material is not contained in bunkers with vertical sides.

Many *beneficial organisms* such as predators, parasites, and natural competitors occur in the breeding media of muscid flies. These natural biological control organisms kill developing fly eggs, larvae, and pupae. The faunas in poultry litter and feedlot manure are best known (Axtell, 1986; Rueda and Axtell, 1985). Important groups include nymphs and adults of predatory mites, larvae and adults of predatory beetles, predatory third-instar larvae of *Hydrotaea* spp. and *Muscina* spp., and adults and larvae of parasitic wasps. The latter group, called *parasitoids*, can be particularly effective. Once a female wasp finds a host, she drills into the host and deposits one or more eggs

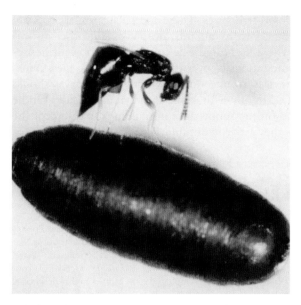

**FIGURE 14.22**  Parasitoid wasp (*Muscidifurax zaraptor*), female, stinging pupa of house fly. (Photo by V. Cervenka)

(Fig. 14.22). The offspring eventually consume the host and emerge as adults. Pupal parasitoids are species that attack and emerge from host pupae, whereas larval–pupal parasitoids attack larvae and emerge from pupae. The parasitoids most frequently encountered in poultry and cattle manure are pupal parasitoids in the genera *Muscidifurax* and *Spalangia* (Pteromalidae). Other genera and families of parasitic wasps and beetles are prominent in dung pats, and these are mainly larval–pupal parasitoids.

Populations of beneficial insects and mites can be favored by keeping potential fly breeding media as dry as possible. Soil should be sloped to drain water away from possible breeding areas, and waterers should be kept in good repair. Certain species of parasitoids are available commercially, and these can be released to augment natural populations. However, the cost-effectiveness of releasing parasitoids in commercial poultry and livestock facilities has been questioned (Thomas and Skoda, 1993).

In emergencies, *larvicides* can be sprayed directly into infested breeding media to kill fly larvae before adults emerge. Alternatively, larvicides can be administered to animals as feed additives or boluses. The active ingredients in these formulations pass through the animal's digestive system to create an insecticidal residue in feces or soiled bedding. Limitations of feed-through larvicides are that they are effective only against flies breeding in feces and bedding, but not in other substrates, and that some larvicides can disrupt natural biological control. Whatever application method is used, larvicidal residues need to be considered when disposing of treated media.

Management of adult flies is accomplished mainly with *traps* and *adulticides* (Drummond *et al.,* 1988).

Inside closed buildings, house flies and *Fannia* species can be killed with sticky traps, light traps, sugar- and pheromone-based insecticidal baits, and adulticides formulated as knockdown or residual sprays. With the exception of baits, the same methods can be effective against adult stable flies. The use of *space sprays* of ingredients with short half-lives, such as synergized pyrethrins, can be effective when applied as mists or fogs in closed spaces. These materials have a rapid knockdown effect on flies contacted directly with the spray droplets. In contrast, *residual sprays* of more persistent insecticides, such as the pyrethroids and some organophosphates, can be applied as coarse sprays to structural surfaces. These formulations provide a more prolonged effect because the residues remain toxic to flies that later walk or rest on the treated surfaces. In outdoor situations, residual sprays should be directed at fly resting sites such as building walls, fence lines, and vegetation where flies seek shelter during hot weather. To limit costs and to retard development of insecticide resistance, residual sprays should be used sparingly and only when necessary.

Traps are generally effective in closed environments, but they can be overwhelmed by immigration from outside sources. Options in outdoor environments are more limited. Walk-through traps (Fig. 14.23) can be used to collect and kill muscid flies from pastured cattle. These traps are most effective against the horn fly in situations where host animals are forced to pass through the traps on a daily basis. This is accomplished by placing traps in an entryway of a fenced enclosure surrounding water or feed supplement or in the doorway of a milking parlor.

Materials formulated as *topical insecticides* can be applied directly to animals. A variety of compounds can be applied as sprays, pour-ons, dusts, or wipe-ons. Some

**FIGURE 14.23**  Cow emerging from walk-through trap used to control flies on cattle. Note baffles in side and fabric draped over exiting cow. (Photo by H. J. Meyer)

**FIGURE 14.24**   Dust bag positioned in fence line, used for self-application of topical insecticides for control of flies on cattle. (Photo by H. J. Meyer.)

of the compounds are formulated into plastic, slow-release *ear tags*, whereas others can be dispensed in self-applicators such as *oilers, back rubbers*, and *dust bags* (Fig. 14.24). To be most effective, self applicators should be maintained in areas where animals are forced to use them on a daily basis.

Topical insecticides usually are ineffective against house flies, stable flies, and face flies because these species spend so little time directly on animals. In contrast, topical insecticides are more effective against adult horn flies because they remain continuously on their hosts. Insecticidal ear tags were adopted widely in the cattle industry in the 1980s because they were inexpensive, required little labor to apply at turnout for spring grazing, and provided several months of effective horn fly control. However, their success was a mixed blessing. Pyrethroid insecticides were the active ingredients in the first widely used tags. Unfortunately, horn flies developed resistance to those compounds in the first 3–4 years of tag use. In response, manufacturers substituted organophosphate insecticides into new tags. If the new tags are used as widely, it is likely that horn flies will evolve resistance to the organophosphate compounds as well.

Chemical *repellents* can be applied by hand directly to individuals to provide temporary relief from muscid flies and other pests. Repellents function mainly by interfering with host-finding behaviors, and less so as toxicants. A variety of formulations are marketed mainly for use on horses and companion animals. Effectiveness varies with weather and level of animal activity. Few repellents are effective against muscid flies for more than a few hours.

The most effective way to prevent the house fly and the other filth flies from entering buildings is *adult exclusion* with door and window screens. Double doorways or positive-pressure air doors can further reduce fly entry into closed structures. These approaches are appropriate at entrances to restaurants, hospitals, and other institutions where flies cannot be tolerated. To prevent household infestations of overwintering adult face flies, cracks and crevices around doors, windows, and eaves should be sealed tightly with caulk. Residual insecticides can be sprayed on the sunny sides of buildings to intercept the flies as they arrive in autumn.

## REFERENCES AND FURTHER READING

Axtell, R. C. (1986). "Fly Control in Confined Livestock and Poultry Production." Technical Monograph. Ciba-Geigy, Greensboro, NC.

Bech-Nielsen, S., Bornstein, S., Christensson, D., Wallgren, T.- B., Zakrisson, G., and Chirico, J. (1982). *Parafilaria bovicola* (Tubangui 1934) in cattle: epizootiology-vector studies and experimental transmission of *Parafilaria bovicola* to cattle. *American Journal of Veterinary Research* **43**, 948–954.

Borror, D. J., Triplehorn, C. A., and Johnson, N. F. (1989). "An Introduction to the Study of Insects." Saunders, Philadelphia.

Bruce, W. G. (1964). "The History and Biology of the Horn Fly, *Haematobia irritans* (Linnaeus): With Comments on Control." North Carolina Agriculture Experiment Station Technical Bulletin No. 157. North Carolina Agricultural Experiment Station and Entomology Research Division, Agricultural Research Service, US Department of Agriculture, Raleigh.

Burger, J. F., and Anderson, J. R. (1974). Taxonomy and life history of the moosefly, *Haematobosca alcis*, and its association with the moose, *Alces alces shirasi*, in Yellowstone National Park. *Annals of the Entomological Society of America* **67**, 204–214.

Catangui, M. A., Campbell, J. B., Thomas, G. D., and Boxler, D. J. (1997). Calculating economic injury levels for stable flies (Diptera: Muscidae) on feeder heifers. *Journal of Economic Entomology* **90**, 6–10.

Chavasse, D. C., Shier, R. P., Murphy, O. A., Huttly, S. R. A., Cousens, S. N., and Akhtar, T. (1999). Impact of fly control on childhood diarrhoea in Pakistan: community-randomised trial. *Lancet* **353**, 22–25.

Chillcott, J. G. (1961). A revision of the Nearctic species of Fanniinae (Diptera: Muscidae) from North America. *Canadian Entomologist Supplement* **14**, 1–295.

Cohen, D., Green, M., Block, C., Slepon, R., Ambar, R., Wasserman, S. S., and Levine, M. M. (1991). Reduction of transmission of shigellosis by control of houseflies (*Musca domestica*). *Lancet* **337**, 993–997.

Curtis, C. F., Ed. (1989). "Appropriate Technology in Vector Control." CRC Press, Boca Raton, FL.

Drummond, R. O. (1987). Economic aspects of ectoparasites of cattle in North America. In "The Economic Impact of Parasitism in Cattle. Proceedings of a Symposium, XXIII World Veterinary Congress, Montreal" (W. H. D. Leaning and J. Guerrero, Eds.) pp. 9–24. Veterinary Learning Systems, Lawrenceville, NJ.

Drummond, R. O., George, J. E., and Kunz, S. E. (1988). "Control of Arthropod Pests of Livestock: A Review of Technology." CRC Press, Boca Raton, FL.

Edwards, F. W., Oldroyd, H., and Smart, J. (1939). "British Blood-Sucking Flies." British Museum (Natural History), London.

Elzinga, R. J., and Broce, A. B. (1986). Labellar modifications of Muscomorpha flies (Diptera). *Annals of the Entomological Society of America* **79**, 150–209.

Emerson, P. M., Lindsay, S. W., Walraven, G. E. L., Faal, H., Bogh, C., Lowe, K., and Bailey, R. L. (1999). Effect of fly control on trachoma and diarrhoea. *Lancet* **353**, 1401–1403.

Frankel, G., and Bhaskaran, G. (1973). Pupariation and pupation in cyclorrhaphous flies (Diptera): terminology and interpretation. *Annals of the Entomological Society of America* **66**, 418–422.

Fuller, H. B. (1913). Myths of American history. *Munsey's Magazine* May, 278–284.

Greenberg, B. (1971). "Flies and Disease. Vol. 1. Ecology, Classification and Biotic Associations." Princeton Univ. Press, Princeton, NJ.

Greenberg, B. (1973). "Flies and Disease. Vol. 2. Biology and Disease Transmission." Princeton Univ. Press, Princeton, NJ.

Hall, R. D. (1984). Relationship of the face fly (Diptera: Muscidae) to pinkeye in cattle: a review and synthesis of the relevant literature. *Journal of Medical Entomology* **21**, 361–365.

Huckett, H. C. (1954). A review of the North American species belonging to the genus *Hydrotaea* Robineay-Desvoidy. *Annals of the Entomological Society of America* **47**, 316–342.

Huckett, H. C., and J. R. Vockeroth. (1987). Muscidae. In "Manual of Nearctic Diptera" (J. F. McAlpine, Ed.), Vol. 2, Monograph 28, Chapt. 105, pp. 1115–1131. Agriculture Canada, Ottawa.

Hughes, R. D. (1977). The population dynamics of the bushfly: the elucidation of population events in the field. *Australian Journal of Ecology* **2**, 43–54.

James, M. T. (1947). "The Flies That Cause Myiasis in Man." Miscellaneous Publication 631. US Department of Agriculture, Washington, D.C.

Krafsur, E. S., and Moon, R. D. (1997). Bionomics of the face fly, *Musca autumnalis. Annu. Rev. Entomol.* **42**, 503–523.

Kuramochi, K. (1985). Studies on the reproductive biology of the horn fly, *Haematobia irritans* (L.) (Diptera: Muscidae). II. Effect of temperature on follicle development and blood meal volume of laboratory-reared flies. *Applied Entomology and Zoology* **20**, 264–270.

Lane, R. P., and Crosskey, R. W., Eds. (1993). "*Medical Insects and Arachnids.*" Chapman and Hall, London.

Lindsay, D. R., Stewart, W. H., and Watt, J. (1953). Effect of fly control on diarrheal disease in an area of moderate morbidity. *Public Health Reports* **68**, 361–367.

McAlpine, J. F., *et al.*, Eds. (1981). "Manual of Nearctic Diptera." Vol 1, Monograph 27. Agriculture Canada, Ottawa.

McAlpine, J. F., *et al.*, Eds. (1987). "Manual of Nearctic Diptera." Vol. 2, Monograph 28. Agriculture Canada, Ottawa.

Morgan, C. E., and Thomas, G. D. (1974). Annotated bibliography of the horn fly, *Haematobia irritans* (L.), including references on the buffalo fly, *H. exigua* (de Meijere), and other species belonging to the genus *Haematobia*. Agricultural Research Service, Miscellaneous Publication No. 1278. pp. 1–134. United States Department of Agriculture, Washington, D.C.

Morgan, C. E., and Thomas, G. D. (1977). Supplement I: annotated bibliography of the horn fly, *Haematobia irritans* (L.), including references on the buffalo fly, *H. exigua* (de Meijere), and other species belonging to the genus *Haematobia*. Agricultural Research Service, Miscellaneous Publication No. 1278. pp. 1–38. United States Department of Agriculture, Washington, D.C.

Morgan, C. E., Thomas, G. D., and Hall, R. D. (1983a). "Annotated Bibliography of the Stable Fly, *Stomoxys calcitrans* (L.), Including References on Other Species Belonging to the Genus *Stomoxys*" Missouri Agriculture Experiment Station Bulletin No. 1049. University of Missouri, Columbia.

Morgan, C. E., Thomas, G. D., and Hall, R. D. (1983b). Annotated bibliography of the face fly, *Musca autumnalis* (Diptera: Muscidae). *Journal of Medical Entomology* (Suppl. 4), 1–25.

Paterson, H. E., and Norris, K. R. (1970). The *Musca sorbens* complex: the relative status of the Australian and two African populations. *Australian Journal of Zoology* **18**, 231–245.

Patton, W. S. (1929). "Insects, Ticks, Miles and Venomous Animals of Medican and Veterinary Importance. Part I—Medical." H. B. Grubb, Ltd., Croydon.

Petersen, J. J., and Greene, G. L., Eds. (1989). "Current Status of Stable Fly (Diptera: Muscidae) Research." Miscellaneous Publication No. 74. Entomological Society of America, Lanham, MD.

Pickens, L. G., and Miller, R. W. (1980). Biology and control of the face fly, *Musca autumnalis* (Diptera: Muscidae). *Journal of Medical Entomology* **17**, 195–210.

Pont, A. C. (1991). A review of the Fanniidae and Muscidae (Diptera) of the Arabian Peninsula. In "Fauna of Saudi Arabia" (W. Buttiker and F. Krupp, Eds.), Vol. 12, pp. 312–365. Pro Entomologia c/o Natural History Museum, Basle, Switzerland.

Robinson, J., and Luff, M. L. (1979). Population estimates and dispersal of *Hydrotaea irritans* Fallen. *Ecological Entomology* **4**, 289–296.

Rueda, L. M., and Axtell, R. C. (1985). Guide to common species of pupal parasites (Hymenoptera: Pteromalidae) of the house fly and other muscoid flies associated with poultry and livestock manure. Technical Bulletin No. 278. North Carolina Agriculture Research Service, Raleigh, NC.

Schwinghammer, K. A., Knapp, F. W., Boling, J. A., and Schillo, K. K. (1986a). Physiological and nutritional response of beef steers to infestations of the stable fly (Diptera: Muscidae). *Journal of Economic Entomology* **79**, 1294–1298.

Schwinghammer, K. A., Knapp, F. W., Boling, J. A., and Schillo, K. K. (1986b). Physiological and nutritional response of beef steers to infestations of the horn fly (Diptera: Muscidae). *Journal of Economic Entomology* **79**, 1010–1015.

Skidmore, P. (1985). The biology of the Muscidae of the world. *Dr. W. Junk Series Entomologica* **29**, 1–550.

Soulsby, E. J. L. (1965). "Textbook of Veterinary Clinical Parasitology. Vol. 1. Helminths." Davis, Philadelphia.

Stone, A., Sabrosky, C. W., Wirth, W. W., Foote, R. H., and Coulson, J. R. (1965). "A Catalog of the Diptera of America North of Mexico." Agriculture Handbook No. 276. US Department of Agriculture, Washington, DC.

Thomas, G. D., and Skoda, S. R., Eds. (1993). "Rural Flies in the Urban Environment?" North Central Regional Research Publication No. 335, Institute of Agriculture and Natural Resources. Univ. of Nebraska, Lincoln.

Watt, J., and Lindsay, D. R. (1948). Diarrheal disease control studies. I. Effect of fly control in a high morbidity area. *Public Health Reports* **63**, 1319–1334.

West, L. S. (1951). "The House Fly. Its Natural History, Medical Importance, and Control." Comstock, Ithaca, NY.

West, L. S., and Peters, O. B. (1973). "An Annotated Bibliography of *Musca domestica* Linnaeus." Dawsons, Folkstone, UK.

Zumpt, F. (1973). "The Stomoxyine Biting Flies of the World." Fischer, Stuttgart.

# 15

# TSETSE FLIES (*Glossinidae*)

WILLIAM L. KRINSKY

*Tsetse flies* (Fig. 15.1) are obligate blood-sucking flies of medical and veterinary importance because they transmit trypanosomes that cause African sleeping sickness in humans and cause nagana in livestock. Fossil tsetse flies in the Florissant shale of Colorado in the western United States indicate that this family was present in the Western Hemisphere as recently as 26 million years ago. Tsetse flies now occur in the tropical and subtropical regions of sub-Saharan Africa (ca. 15° N to 26° S). Recently, isolated populations of two species of tsetse flies were observed in southwestern Saudi Arabia (Elsen *et al.,* 1990).

Tsetse (pronounced *tsé-tsee*) commonly is used as both a singular and plural term to denote one or more individuals or species of these flies. Although the origin of the name is obscure, it was used as early as the 19th century by the Tswana people living along the edge of the Kalahari Desert. "Tsénse," the Mozambique word for "fly," as well as other similar sounding African names meaning "fly," are apparently onomatopoetic terms derived from imitations of the unique buzzing sound made by the adult flies (Austen, 1903).

Tsetse generally are considered one of the greatest factors affecting the course of economic and social development in Africa. The morbidity and mortality caused by African sleeping sickness continues to be significant. Nagana, which has stifled agricultural productivity for decades, still stands as a major deterrent to the development of animal agriculture on that continent.

There is an extensive literature on tsetse, but a few monographic works provide particularly useful introductions to the field. The classic work by Buxton (1955) reviews the natural history of tsetse and provides a detailed historical account of the diseases associated with it. Mulligan (1970) includes an historical perspective in addition to an overview of the biology of tsetse and its parasites, pathology of these parasites in humans and domestic animals, treatment, and control. The historical, social, and economic effects of tsetse in five different African regions are extensively reviewed by Ford (1971), and the impact of tsetse on African rural development is discussed by Jordan (1986). A comprehensive monograph was written by Leak (1999).

**FIGURE 15.1**   Adult tsetse fly (*Glossina* sp.) on rabbit. (Courtesy of The Rockefeller Foundation.)

## TAXONOMY

Tsetse were formerly included in their own subfamily, Glossininae, or the Stomoxyini of the Muscidae because of the resemblance of tsetse to the stable fly and other biting muscids. However, because of their unique antennal structure, tsetse are now placed in their own family, Glossinidae. The reproductive and morphological similarities of tsetse to the keds and other hippoboscid flies has led to placement of Glossinidae within the Hippoboscoidea (McAlpine, 1989). Glossinidae includes the single genus *Glossina* with 23 species, 6 of which are further divided into 14 subspecies (Gouteux, 1987; Potts, 1973). *Glossina* means "tongue fly," in reference to its prominent proboscis. Keys to species and subspecies are included in Jordan (1993). *Glossina* species are arranged in three subgenera (*Austenina, Nemorhina,* and *Glossina*) that correspond roughly with groups of species found in different ecological settings. The subgenera often are cited by their group names, each designated by one of the better-known species in each subgenus, i.e., the *fusca* group (*Austenina*), the *palpalis* group (*Nemorhina*), and the *morsitans* group (*Glossina*). Species in the *fusca* group are most often found in forested habitats, such as rain, swamp, and mangrove forests. Species in the *palpalis* group occur among vegetation around lakes and along rivers and streams. The *morsitans* group, with the exception of the forest-dwelling *Glossina austeni*, occurs in open country and is most often found in dry thickets, scrub vegetation, and areas of savanna woodland (commonly composed of *Berlinia, Isoberlinia,* and *Brachystegia* species).

The *geographical distributions* of the three taxonomic groups are shown in Fig. 15.2. The *palpalis* group, which includes *G. palpalis, G. tachinoides, G. fuscipes,* and two less well known species, occurs primarily along watercourses in western and central Africa. The *morsitans* group of savanna species, which includes *G. morsitans, G. pallidipes, G. longipalpis, G. swynnertoni,* and *G. austeni,* is primarily central and southeastern in distribution. The *fusca* group, which includes *G. fusca, G. tabaniformis, G. medicorum, G. longipennis, G. brevipalpis,* and eight other species, is found in forested areas that overlay most of the western and central African distribution of the *palpalis* group.

## MORPHOLOGY

*Glossina* species are tan or brown flies which range in length from 6 to 14 mm, excluding the proboscis. Members of the *fusca* group, which is considered phylogenetically primitive, are the largest, being 9.5–14 mm long. The *palpalis* and *morsitans* group species are small to medium in size, about 6.5–11 mm long. Species in the *palpalis* group generally have a uniformly dark brown abdominal tergum, and the dorsal aspect of each hind tarsal segment is dark brown or black. Species in the *morsitans* group usually have dark segmental bands on the abdomen, and only the distal segments of the hind tarsi are darkened dorsally.

Tsetse adults are characterized by several distinctive morphological features. These include the shape of the proboscis, the position and branching of the fringe on the arista of their antenna, and the wing venation and folding pattern. The swollen, bulbous base of the *proboscis* that lies under the head is very different from the angled and thinner bases of the proboscises of the Stomoxyini. When the fly is not feeding, the proboscis extends directly forward between the palps in front of the head (Fig. 15.1). The proboscis (Fig. 15.3) is composed of two elongate, stylet-like mouthparts: the labrum and hypopharynx. The stylets are protected ventrally by the labium. The labellum at the tip of the labium is armed with teeth for cutting into host skin. The labrum, bounded by the hypopharynx and the labium, forms the food canal through which blood is drawn as the fly feeds (Fig. 15.3). The hypopharynx has a hollow central portion that forms the salivary canal through which saliva is secreted into the feeding site.

The three-segmented *antennae* arise on the frons, just below the ptilinal suture, as in muscoid flies. The first segment is very small; the second is at least 2–4 times larger than the first and generally about as long as wide; the third is very elongately oval to pea pod–shaped and bears the distinctive arista. The arista has a conspicuous fringe of hairs along its dorsal surface, and these hairs have small branch hairs, which are not found on any other aristate fly (Fig. 15.4). The large brown or reddish eyes are separated

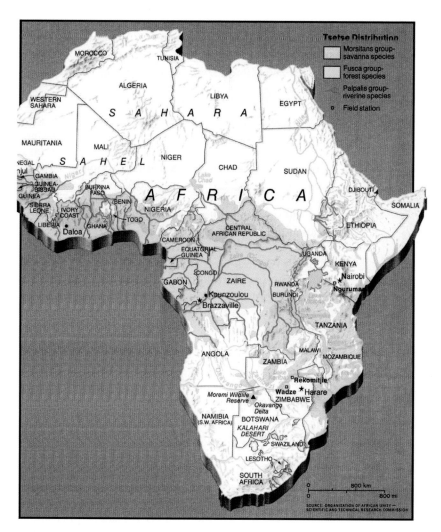

**FIGURE 15.2** Distribution of the following *Tsetse* species groups in Africa: *morsitans* group (savanna); *fusca* group (forest); *palpalis* group (riverways). (Used with permission of the National Geographic Society.)

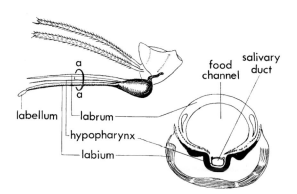

**FIGURE 15.3** Details of proboscis and palps of *Glossina* species, with palps separated from the proboscis (left) and cross section about midway along length (at *a*) of proboscis (right). (From Potts, 1973; after Newstead *et al.*, 1924.)

in both sexes and comprise most of the posterior portion of the head.

The base of the thorax is only slightly wider than the width of the head across the eyes. The thorax tapers to a waistlike constriction at the level of the scutellum. The wings vary from hyaline to dusky depending on the species. They are folded scissors-like over the back, with the tips extending slightly beyond the end of the abdomen. The tsetse wing has a distinctive, hatchet-shaped discal cell. This is formed by the fourth (medial) vein that curves anteriorly to produce a wing cell (discal cell) resembling the elongate handle of a hatchet attached to a thickened blade (Fig. 15.5).

The base of the abdomen is about equal in width to that of the head and thorax. Male tsetse can be readily distinguished from females by the presence of a prominent button-like hypopygium on the ventral surface of

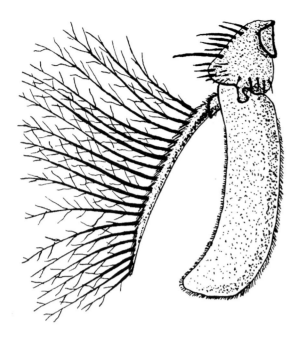

**FIGURE 15.4** Antenna of *Glossina fuscipleuris,* showing plumose setae on arista, characteristic of tsetse fly adults. (After Zumpt, 1936.)

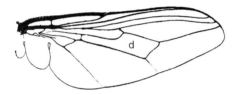

**FIGURE 15.5** Wing structure and venation of adult tsetse flies (*Glossina*), showing characteristic hatchet-shaped discal cell (*d*). (Modified from Potts, 1973.)

logical details of both male and female genitalia provide taxonomic characteristics that are used for distinguishing tsetse species.

The *alimentary tract* is adapted for hematophagy. The strong musculature of the pharynx forms a cibarial pump used for imbibing blood. The proventriculus secretes a peritrophic membrane that lines and protects the midgut. The midgut contains symbionts (Enterobacteriaceae) that provide compounds associated with vitamin B metabolism. Females devoid of these symbionts are unable to reproduce. The reproductive tract of the female fly is unusual compared to that of most oviparous dipterans; it is very similar, however, to the reproductive system seen in the other hippoboscoid families (Hippoboscidae, Streblidae, and Nycteribiidae). Each of the two ovaries has only two ovarioles. The ovarian ducts form a common duct that expands to form a uterus in which one embryo at a time is retained during development. Associated with the uterus are a pair of specialized branched glands that produce nutrients for the developing tsetse larva. Because of this function, they are commonly called *milk glands.*

## LIFE HISTORY

Tsetse adults of both sexes bite vertebrates and imbibe blood, the fly's only food. Unfed females are sexually receptive about 1 day after emergence from the puparium, whereas male tsetse require several blood meals before they are fully fertile. At close range, the male visually locates a female and, once contact is made, a pheromone in the cuticle of the female stimulates mating. The female endocrine system will induce ovulation only if

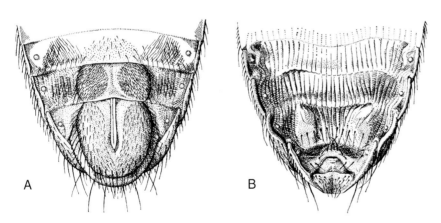

**FIGURE 15.6** Posterior ends of abdomens of *Glossina* adults, ventral view, showing sexual differences. (A) Male, with knoblike appearance of hypopygium drawn up into the abdomen; (B) female, lacking knoblike hypopygium. (From Potts, 1973.)

mating lasts longer than an hour. Sperm are transferred in a spermatophore and are stored in the female's spermathecae. Once inseminated, the female remains fertile for life and rarely mates more than once in nature.

About 9 days after copulation, the first ovulation of a single egg occurs, and sperm are released through the spermathecal duct by dilation of a sphincter. The egg is positioned with the micropyle against the spermathecal duct opening, allowing for fertilization. The fertilized egg moves posteriorly into the uterus, where hatching occurs about 4 days later. The first-instar *larva* uses an "egg tooth" on its anterior end to rupture the chorion of the egg. The larva is retained in the uterus, where it is held against the uterine wall by a supporting structure called the *choriothete*. Secretions from the milk glands pool around the larval mouth and are easily ingested. The developing larva molts twice within the uterus, becoming a second-instar larva 1 day after hatching and a third instar about 1.5 days later. The third-instar larva is fully developed about 2.5 days after the second molt, at which time it occupies most of the female's abdomen and is about equal in weight to the rest of the female's body. The female continues to ingest blood, albeit in progressively smaller amounts, as the larva grows.

About 9 days after ovulation, the fully developed third-instar larva is deposited on the ground by the female. Shortly thereafter, the female ovulates again (within as little as 1 hr after *larviposition*). A well-nourished female, after this first larviposition, will deposit a third-instar larva about every 7–11 days, depending on the ambient temperature. The average interval for all tsetse species is 9–12 days. The ovaries, and the ovarioles in each, alternate in releasing a single egg at each ovulation, starting with the right ovary. Follicular relicts seen in dissected flies reflect the ovulation history of individual females and can help in estimating the longevities of wild-caught female flies.

Tsetse females generally live for about 20 to 40 days, but may have a maximum life span of 3–4 months. The males typically mate only once or twice during their lives and apparently survive in the wild for 2–3 weeks (Glasgow, 1963; Potts, 1973). More accurate estimates of longevity will probably become possible with newly developed fluorescence techniques that measure accumulated pteridines in tsetse head capsules (Lehane, 1991).

The full-size third-instar larva is cream colored and oval-shaped. It measures 3–8.5 mm in length, depending upon the species, and has two prominent black knobs at the posterior end (Fig. 15.7). These conspicuous knobs are respiratory lobes that function only during, and for a short time following, intrauterine life. The active larva is deposited on the ground, usually in loose soil shaded by trees or other vegetation. The larva, which is negatively

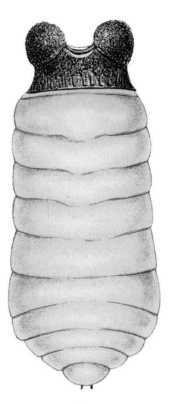

**FIGURE 15.7** Larva of tsetse fly (*Glossina morsitans*). (From Newstead *et al.*, 1924.)

phototactic and positively thigmotactic, quickly burrows to 1.5–2.5 cm below the soil surface. Within a few hours of deposition, the larval integument hardens and darkens, and the third-instar larva becomes an immobile brown to black *puparium*. About 2–4 days later, molting occurs within the puparial case and a true pupa is formed. A key for the identification of puparia to species is given by Jordan (1993).

Adult flies emerge about 30 days after formation of the puparium. As in all other cyclorrhaphan flies, eclosion involves the breaking of the circular puparial cap by a ptilinum. The teneral adult pushes its way to the surface of the substrate, where it rests for a short time, usually less than an hour, before it can fly. The teneral fly does not fully harden and the thoracic flight muscles do not completely develop until about 9 days later, after the fly has had at least a few blood meals (Glasgow, 1963; Lehane, 1991; Potts, 1973).

The low reproductive rate in tsetse is compensated by the extreme protection given to each larva by the female, by virtue of the viviparous mode of development. However, the low reproductive rate makes the impact of any loss of female flies greater than in species that mass produce eggs.

# BEHAVIOR AND ECOLOGY

Although tsetse are found over an area estimated to be at least 10 million square kilometers, the distribution of the flies is discontinuous. The areas they inhabit may extend to several hundred kilometers and form what have been traditionally called *fly belts*. Within these belts are patches of forest and bush where environmental conditions, such as shade and high humidity, are suitable for tsetse survival and reproduction. Local residents living in their vicinity are often aware of these areas of high tsetse concentrations. One or more species of tsetse usually are found where woody vegetation is at least 4.5 m high. In many cases, Africans can predict the presence of particular species of tsetse by observing the types of shrubs and trees that occur in a given habitat. Rather than representing direct associations of tsetse species with specific plants, the plant communities observed probably reflect differences in a variety of microhabitat factors that directly affect the survival of tsetse, such as the water content of the soil, the availability of mammalian hosts, and the occurrence of natural predators. Remotely sensed satellite data that provide identification of different types of vegetation over large geographic areas have been used to estimate distributions of different species of tsetse (Rogers *et al.*, 1994).

Tsetse flies are restricted in northern Africa by desert conditions, and in southern Africa, by the deserts of Namibia and Botswana and their lower ambient temperatures. Tsetse live in areas where the annual rainfall is at least 0.5 m per year. They require temperatures between 16 and 40°C, with optimal development occurring at 22–24°C; for this reason, the flies are not found at elevations above ca. 1500 m.

The potential difficulty of males and females finding each other in low-density populations is apparently overcome in some species by the attraction of both males and females to large moving animals. Mating usually occurs on or in the vicinity of a host. Once they have mated, however, females and males tend to be more attracted to stationary animals. Tsetse feed on an array of hosts including reptiles and mammals, but rarely birds. Individual species and species groups have definite host preferences. These preferences are of considerable epidemiological significance in relation to the reservoir hosts of the pathogenic trypanosomes transmitted by the flies to humans and domestic animals.

*Host preferences* vary among tsetse species. Members of the *palpalis* group feed mostly on reptiles (*e.g.*, crocodiles and monitor lizards) in their riverine and lacustrine habitats and on bushbuck, oxen, and occasionally smaller mammals and humans that visit these watering spots. Species of the *morsitans* group, living in scattered patches of vegetation in open country, feed mostly on the mammals of the savanna. In addition to showing a strong preference for warthogs, the savanna-dwelling tsetse feed on a diversity of mammalian species, including bushbuck, buffalo, giraffe, kudu, rhinoceros, duiker, bushpig, and oxen. The one forest-dwelling species in the savanna group, *G. austeni*, feeds almost exclusively on suids such as bushpigs and forest hogs. The *fusca* group feeds on a variety of host species, including bushbuck, buffalo and other cattle, giraffe, rhinoceros, elephant, hippopotamus, bushpig, river hog, porcupine, aardvark, and even the ostrich (Lehane, 1991). Humans are not the preferred hosts of any of these fly species. In some cases, people in the vicinity of other mammals will actually repel tsetse, whereas hungry flies will suck blood from humans who enter tsetse habitat.

*Host attraction* and *host recognition* are mediated by visual and olfactory cues. Their vision enables tsetse to react to a herd of moving cattle as far away as 180 m. The attraction of both male and female flies to large moving objects accounts for the common occurrence of tsetse attacking occupants of trucks and tourists in jeeps on safaris. When landing on the sides of vehicles, male flies that land with their heads directed upward are more likely to be hungry than those that land with their heads directed downward (Newberry, 1982). Tsetse species in the *morsitans* group, living in open spaces, have shown the greatest attraction to host odors. Certain tsetse species are attracted to components of ox breath, such as carbon dioxide, acetone and octenol, and phenols found in mammalian urine (Willemse and Takken, 1994). Host odors have been shown to be attractive to tsetse from distances up to 100 m away.

Although tsetse feed mostly in the daylight, feeding does occur at night, as in the case of *G. medicorum*, which feeds on the nocturnal aardvark. In general, tsetse adults are most active in the morning and late afternoon. They rarely fly for more than 30 min a day and are known to disperse up to about 1 km/day. They spend most of their time resting on vegetation. Some species, such as *G. morsitans*, rest as high as 12 m above the ground, while others, such as *G. pallidipes*, are seldom found above 3 m. When seeking a host, *Glossina* species can fly very rapidly, reaching speeds up to 6.5 m/sec (ca. 25 km/hr).

Host behavioral differences may account in part for the feeding preferences shown by tsetse species. Mammals that are heavily fed upon and irritated by other kinds of biting flies sometimes react with strong defensive behaviors, such as muscle twitching and rapid tail movements, that repel tsetse. Tsetse are more prone to start feeding on calm animals and often seem to prefer to feed on a host that is in the shade. The latter may be an adaptation to avoid reaching lethal body temperatures and may serve as a means of avoiding predation during feeding, or just after, when the fly takes off and alights a short distance from its host (Glasgow, 1963).

Upon landing on a host, a tsetse fly grips the skin with its claws and applies pressure to the skin surface with its proboscis. The teeth and rasps on the labellum aid the labium in penetrating the skin. Strong back-and-forth movements of the fly's head cause the labium to rupture one or more capillaries in the skin, resulting in a hemorrhage within the bite site. The blood is rapidly sucked into the food canal of the labrum by the negative pressure produced by the cibarial pump in the fly's head. Saliva is pumped intermittently through the salivary canal of the hypopharynx into the wound. The saliva contains anticoagulant substances, including an antithrombin and an apyrase that inhibit platelet aggregation. As in other hematophagous insects that have anticoagulins in their saliva, tsetse flies presumably benefit from these substances by their role in increasing blood flow at the feeding site, thereby reducing feeding time and the vulnerability of the fly to host defenses. If a tsetse fly is disturbed while penetrating the skin, it will rapidly withdraw the proboscis and fly away; however, once feeding begins, a tsetse is less likely to react to movement and physical stimuli that would normally cause it to escape (Glasgow, 1963; Lehane, 1991).

Tsetse engorge fully within about 1–10 min, the length of time depending in large part on how quickly the labium is able to rupture a capillary. The actual penetration of host skin occurs quite rapidly, whether it involves the thick hide of a rhino or a thin artificial feeding membrane. During feeding, a clear fluid is excreted from the anus. A tsetse fly imbibes about 0.03 ml of blood and when fully engorged weighs about 2–3 times its unfed body weight (Fig. 15.8). The ungainly fully fed insect slowly flies from the host (ca. 1.6 m/sec) and lands on a nearby tree or other substrate. There the fly continues to excrete anal fluid as a means of ridding itself of excess water while concentrating its blood meal. About 40% of the blood meal weight is lost in the first 30 min after feeding. The rapid loss of excess fluid that begins during blood feeding helps the fed fly regain flight agility as quickly as possible. This helps it evade the defensive movements of the host and destruction by predatory flies, other arthropods, and vertebrate predators. A larva being carried by a female is especially vulnerable to loss just after the female has taken on the extra burden of a blood meal and has lost much of her maneuverability. Complete digestion of the blood meal occurs by about 48 hr. The interval between blood meals varies, with a mean of 3–5 days.

Because tsetse feed exclusively on blood, their main source of energy is derived from protein. They depend on the amino acid proline as the major energy source for flight. The energy is produced by the partial oxidation of proline to alanine in flight muscle (Glasgow, 1963; Lehane, 1991). The unique metabolism of tsetse flies enables them to live in dry habitats in which blood is their sole source of nutrition and water and to develop massive thoracic flight muscles that allow them to fly with heavy loads of blood and/or an internally developing larva.

## PUBLIC HEALTH IMPORTANCE

For centuries, tsetse flies have had a great impact on human health in Africa, both as efficient vectors of trypanosomes that cause extreme human suffering in the form of African sleeping sickness and as vectors of trypanosomes that kill nonnative animals, preventing the development of animal domestication. The exclusion of cattle from most of the African continent has prevented their use as draught animals, as sources of human dietary protein from meat and milk, and as sources of manure and transport. Although human disease has diminished, domestic animal disease continues to inhibit agricultural productivity and economic development.

### African Sleeping Sickness

Trypanosomes were first associated with African sleeping sickness in 1903, after the British Tsetse Fly Commission, composed of David Bruce, David Nabarro, Aldo Castellani, and others, was sent to Africa to investigate outbreaks among British colonists. The clinical presentation of the disease was different in west African and east African countries, with the East African form being much more severe. The trypanosomes isolated from the two forms of African sleeping sickness were named

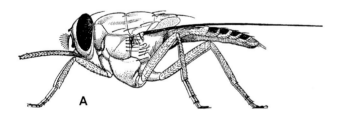

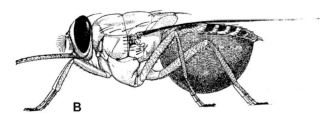

**FIGURE 15.8**  Tsetse fly (*Glossina morsitans*), female. (A) Before feeding; (B) after feeding. (From Austen, 1903.)

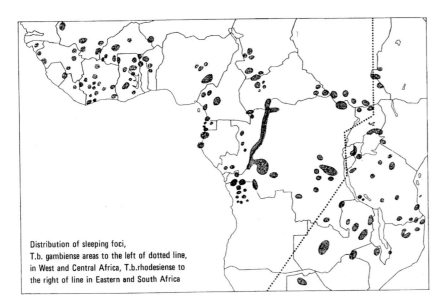

Distribution of sleeping foci,
T.b. gambiense areas to the left of dotted line,
in West and Central Africa, T.b.rhodesiense to
the right of line in Eastern and South Africa

**FIGURE 15.9**   Distribution of foci of African sleeping sickness in humans in central and western Africa. (Courtesy of World Health Organization, 1989.)

to reflect their geographic distributions (Fig. 15.9). The West African form was named *Trypanosoma gambiense* for Gambia. It still occurs in that country, east through central Africa to northern Uganda, and south to Zaire and northern Angola. The East African form was named *T. rhodesiense* for Rhodesia, now Zimbabwe. This trypanosome now occurs in Sudan, Uganda, Kenya, Tanzania, southern Zaire, Zambia, Malawi, and Mozambique, as well as Zimbabwe and Botswana.

The two species of trypanosomes pathogenic to humans are morphologically identical and microscopically indistinguishable from *T. brucei*, the species that causes some of the trypanosomal disease seen in domestic animals in Africa. For taxonomic purposes, *T. gambiense* and *T. rhodesiense* are considered subspecies of *T. brucei*, namely, *T. brucei gambiense* and *T. brucei rhodesiense* (Hoare, 1972). However, for convenience in the medical and parasitological literature, the subspecies names are often retained as species designations, as in the original *T. gambiense* and *T. rhodesiense*.

The term *African sleeping sickness* refers to the drowsy to comatose condition of acutely ill patients. This disease occurs in two clinical forms that result from differences in the pathogenicity of the trypanosomes transmitted by tsetse flies in West and East Africa. *West African trypanosomiasis* is a chronic illness involving mental deterioration and progressive weakening, and *East African trypanosomiasis* is an acute, rapidly fatal, febrile illness characterized by myocarditis and meningoencephalitis. The West African disease has been known since the 14th century, when it was described by the Arab writer al-Qalqashandi. The first medical description of the disease was written in 1734 by John Atkins, who had served as a British Navy

surgeon on slave ships traveling from West Africa to the West Indies. His description of the clinical signs of the disease were accurate, but his ideas concerning its cause created an extremely prejudiced view of those afflicted with the "Negro lethargy." Atkins attributed the disease to the "natural weakness of the brain... brought about by lack of use" (McKelvey, 1973).

The first recorded large-scale epidemics of sleeping sickness occurred in the middle to late 19th century during the period of active European exploration and colonization of the African continent. Controversy exists over the role of British and other imperialistic activities in either stimulating, or just recognizing and alleviating, these outbreaks. If nothing else, expanded navigation of waterways such as the Congo River, which led to increased trade and development by late 19th century colonists, facilitated the spread of tsetse and sleeping sickness from western and central Africa to eastern regions of the continent. Although there is strong historical evidence for this pattern of dispersal, the extreme differences in the clinical presentations and pathogeneses of West and East African trypanosomiases lead to the more probable explanation that the geographical forms of the parasite evolved independently as humans were exposed under different ecological conditions to parasites harbored by wild ungulate mammals.

More than three quarters of a million people died from sleeping sickness in Africa between 1896 and 1906. Currently, the number of Africans succumbing annually to trypanosomiasis is in the thousands. An estimated 50 million Africans in 38 countries are at risk of infection by exposure to tsetse flies, with more than 25,000 individuals being infected annually. Imported cases in the

United States, mostly involving tourists to East Africa, have numbered only about 15 in the last 25 years, but the risk of serious illness or death is great because of the American medical community's unfamiliarity with the disease.

In both West and East African trypanosomiasis, trypanosomes injected into the skin by a tsetse fly reproduce locally in connective tissue to form a nodule or chancre, called a *trypanoma*.

## West African Trypanosomiasis

In the Western form, the skin lesion and associated erythematous swelling occur within a week or so after the bite. The trypanosomes then spread to the lymphatics and the resulting lymphadenopathy in the back of the patient's neck is called *Winterbottom's sign*. This distinctive sign is diagnostic for the disease in a person who has been exposed to tsetse. Urticaria and rashes are common in this chronic form of the disease. The illness progresses as the trypanosomes continue to multiply in the lymphatic and circulatory systems. After months or years, the parasites enter the central nervous system and produce symptoms such as behavioral and personality changes, hallucinations and delusions, drowsiness by day (i.e., sleeping sickness), tremors, and stupor. In untreated patients, stupor is often followed by convulsions and, inevitably, by death.

## East African Trypanosomiasis

In the East African form, there is acute onset of fever, headache, and dizziness within a few days after the fly bite. There is usually little or no lymph node involvement. Instead, early circulatory system disease that includes myocardial and pericardial inflammation becomes clinically apparent as tachycardia and arrhythmias. Immune complexes composed of trypanosomal variant antigens and complement-fixing host antibodies stimulate release of proteolytic enzymes. Either the latter or toxin production, in combination with host autoantibodies, causes damage to red blood cells, brain and heart tissue, and other organs. Anemia, thrombocytopenia, and disseminated intravascular coagulation, followed by renal disease, may precede localization of the trypanosomes in the blood vessels of the central nervous system. Hemorrhages, edema, and thrombosis leading to neuronal degeneration are common following inflammation of the cerebral arteries. Damage to brain tissue results in the convulsions and other signs seen in the West African form, but much sooner, with death often occurring in weeks to months. Definite diagnosis of either form of the disease requires observation or isolation of trypanosomes from blood, cerebrospinal fluid, scrapings from a trypanoma, lymph node aspirates, or bone marrow.

## Life Cycle of Trypanosomes

The *developmental cycles* of *T. gambiense* and *T. rhodesiense* in the flies appear to be identical, even though the tsetse vector species often are different. Trypanosomes ingested by a tsetse fly in a blood meal pass into the midgut, where their life cycle continues. The relatively short trypanosomes that enter the fly transform into thin procyclic forms, which then multiply and become *trypomastigotes* in about 3–4 days. These forms then multiply for about 10 days before they move to the posterior part of the midgut. There, they pass either through the peritrophic membrane or around its posterior open end, into the ectoperitrophic space. The trypomastigotes move anteriorly in the ectoperitrophic space of the midgut and reach the junction with the proventriculus about 5 days later. The parasites elongate and penetrate the soft basal ring of the peritrophic membrane to return to the endoperitrophic space. The trypomastigotes, having returned to the endoperitrophic space of the proventriculus, migrate anteriorly through the esophagus and pharynx and enter the proboscis. They then enter the open end of the hypopharynx and migrate posteriorly through the salivary ducts into the salivary glands. In the lumen of the glands and attached to the epithelium, the trypomastigotes transform into *epimastigotes,* which multiply and form short *metatrypanosomes* that are infective to vertebrates. The metatrypanosomes are injected by *salivarian transmission* when the infected tsetse feeds on another vertebrate host.

An alternative trypanosome life cycle has been observed in some flies. In these flies, migration of trypomastigotes from the midgut to the salivary glands involves a different anatomical route. The trypomastigotes that have moved anteriorly in the ectoperitrophic space of the midgut enter the midgut cells, penetrate the wall of the midgut, and enter the hemocoel. From there, they migrate anteriorly and penetrate the hemocoel side of the salivary glands to reach the lumen. Whether the parasites move anteriorly via the gut lumen or the hemocoel, the complete trypanosome cycle in the fly usually takes 11–38 days, but it may be as long as 80 days (Hoare, 1972; Lehane, 1991).

In nature, very small numbers of tsetse may be infective (e.g., with *T. brucei,* < 0.4%) and still maintain high rates of trypanosome transmission. The physiological and ecological factors that attract tsetse flies to hosts enable continued cycling of the trypanosomes between insect and vertebrate hosts. In epidemic situations, tsetse and other blood-sucking flies, such as tabanids and stomoxyines, may transmit trypanosomes from person to person by mechanical transmission.

Epidemiologically, the West African and East African forms of the disease are different. The species of tsetse flies that act as vectors are different. The vertebrate host species and the degree to which humans are part of the natural life cycles of the trypanosomes are different.

In West Africa, the major *vectors* of sleeping sickness are tsetse species in the *palpalis* group. The medically most important species in this group include *Glossina palpalis*, *G. tachinoides*, and *G. fuscipes*. These species are found in shaded forested areas close to rivers, streams, and lakes. In West African trypanosomiasis, transmission to humans usually involves a solely human–tsetse cycle without any nonhuman reservoir hosts, although some domestic and wild animals are known to be infected with *T. gambiense*. Therefore, West African trypanosomiasis is considered an *anthroponosis*. Humans are most often infected by exposure within a peridomestic environment that encompasses forested waterways near their dwellings. Activities such as washing, bathing, gathering water for cooking, and wood-gathering for fires and construction purposes place humans in the riverine habitats of the *palpalis* group of tsetse flies.

In East Africa, the major vectors of sleeping sickness are tsetse species in the *morsitans* group. The medically most important species in this group include *G. morsitans*, *G. pallidipes*, and *G. swynnertoni*. These species are found in vegetation such as tall grasses, thickets, and small groups of trees which occur in open savannas. An exception to the vector distinction between East and West Africa is the occurrence of *G. fuscipes* in Uganda and Kenya along the northern shore of Lake Victoria, where outbreaks of East African trypanosomiasis are associated with this *palpalis*-group species. In East African trypanosomiasis, transmission involves a wild animal/tsetse/human cycle. Antelopes, the bushbuck (*Tragelaphus scriptus*), and the hartebeest (*Alcelaphus buselaphus*), are natural reservoirs for the trypanosome. Humans are incidental hosts. Therefore, East African trypanosomiasis is a zoonotic disease that can be designated an *anthropozoonosis* (Baker, 1974). In East Africa, African men and tourists generally have the highest risk of infection because it is the men who are the hunters and honey-gatherers, and it is the tourists on safaris who most often enter the savanna areas where game animals and the *morsitans*-group species thrive.

The distribution of human trypanosomiasis in Africa is not as widespread as the distribution of tsetse species. The reason is that many tsetse species do not readily feed on humans and that many potential vector species inhabit areas where there is little or no contact between the flies and humans.

Besides transmitting trypanosomes, tsetse flies have other direct, but minor, effects on public health. Some people, who are particularly sensitive to arthropod *antigens,* develop large skin rashes when bitten, and *anaphylactic reactions* to tsetse bites are known. However, as with most hematophagous species that are efficient vectors of human pathogens, people who are bitten usually feel little pain or only slight irritation.

# VETERINARY IMPORTANCE

Tsetse-borne trypanosomiasis in both domestic and wild, nonhuman animals is caused by a number of trypanosome species. Most wild hosts in Africa are immune to these parasites or have inapparent or mild infections. Infections that cause disease in wildlife are rare because wild animals that do develop pathological changes following infection are species that are rarely fed upon by tsetse (Ford, 1971). However, more than 30 species of wild animals native to Africa harbor trypanosomes that are pathogenic when transmitted to domestic animals. The disease associated with any of these infections is called *nagana*.

## NAGANA

Nagana, which killed camels and horses used by 19th-century missionaries, is now considered to have been a major factor in halting the spread of Islam through sub-Saharan Africa. The disease was known to 19th-century European explorers in Africa, who similarly lost large numbers of pack animals, such as horses, mules, and oxen. In 1895, David Bruce recognized an association between the disease and tsetse flies. He named the new disease *nagana*, which is a Zulu word for "low or depressed in spirits" (McKelvey, 1973). Bruce also identified the etiologic agent that was later named *Trypanosoma brucei* in his honor. In 1909, Kleine demonstrated the biological transmission of trypanosomes by tsetse, which led to the elucidation of the life cycle of the trypanosome in the fly by Robertson in 1913. Bruce's earlier observations had involved only short-term mechanical transmission of the parasite.

Nagana continues to have a major impact in preventing the development of commercial domestic animal production over about one-third of the African continent. The scarcity of domestic animals results in a severe lack of animal protein for use as human food, a lack of draught animals for use in crop production, and the absence of manure suitable for use as fertilizer. At present, about 40 million cattle and millions of sheep, goats, horses, mules, pigs, and camels are at risk of infection in Africa. Unlike African sleeping sickness, in which human disease does not occur over the entire distribution of tsetse vectors, nagana occurs wherever tsetse are found, in addition to other areas where infection can be maintained by mechanical transmission by biting flies other than tsetse.

**FIGURE 15.10** A naturally infected yearling cow sick with nagana (bovine trypanosomiasis), showing stunted growth and characteristic "nagana" pose. (From Murray *et al.*, 1979, with permission of the Food and Agriculture Organization of the United Nations.)

Chronic disease involving anemia and weakness is common (Fig. 15.10). Affected animals have reduced muscle mass, pendulous fluid-filled abdomens, scurfy skin, rough coats, and enlarged lymph nodes. Chronic fever and watery diarrhea are common. Most organ systems are infected, and enlargement of the heart, lungs, liver, and spleen is often seen at necropsy. Chronically infected animals are unsuitable for use as food or as suppliers of manure for agricultural use. Early death may result from secondary infections. For further information on the clinical pathology of nagana, see Losos and Ikede (1972) and Jubb *et al.* (1985).

The six species of trypanosomes that cause nagana are listed with some of their wild and domestic animal hosts in Table I. The trypanosomes are identifiable by differences in morphology and antigenicity, and the trypanosome species differ to some extent in the anatomical sites where they develop in the tsetse vector. Hoare (1972) reviewed the morphology and life cycles of the tsetse-borne trypanosomes of Africa. *T. brucei*, a parasite of diverse domestic and wild mammals, has a developmental cycle in its tsetse vector identical to that of its human forms, *T. gambiense* and *T. rhodesiense*. It undergoes development during migration from the proboscis to the midgut and subsequently to the salivary glands. The vector cycle of *T. suis*, a parasite of pigs, is almost identical to the latter forms and occupies the same anatomical sites. The vector cycles of *T. congolense*, a parasite of all domestic mammals, antelopes, and other wildlife, and *T. simiae*, found in pigs, cattle, horses, camels, and warthogs, are almost identical to each other. These trypanosomes are never found in the salivary glands of the fly. Development is restricted

TABLE I

Tsetse Vectors and Vertebrate Hosts of Human and Animal Trypanosomiases (From Hoare, 1972)

| Disease | Disease agent | Vectors | Hosts |
|---|---|---|---|
| West African sleeping sickness | *Trypanosoma gambiense* | *Glossina fuscipes, G. palpalis, G. tachinoides* | Humans |
| East African sleeping sickness | *T. rhodesiense* | *G. morsitans, G. pallidipes, G. swynnertoni* | Humans, antelopes (bushbuck, hartebeest) |
| Nagana | *T. brucei* | *G. fuscipes, G. longipalpis, G. morsitans, G. palpalis, G. pallidipes, G. tachinoides* | All domestic mammals; antelopes (e.g., impala, hartebeest, wildebeest); warthog, hyena, lion |
| | *T. suis* | *G. brevipalpis, G. vanhoofi* | Suids (domestic pigs, warthogs) |
| | *T. congolense* | *G. morsitans* group; *G. brevipalpis, G. fuscipes, G. palpalis, G. tachinoides, G. vanhoofi* | All domestic mammals, elephant, zebra, antelopes (e.g., impala, hartebeest, duiker, gnu); giraffe, bushpig, hyena, lion |
| | *T. simiae* | *G. austeni, G. brevipalpis, G. fusca, G. fuscipleuris, G. longipalpis, G. morsitans, G. pallidipes, G. palpalis, G. tabaniformis, T. tachinoides, G. vanhoofi* | Domestic pig, warthog, camel, horse, cattle |
| | *T. uniforme* | *G. fuscipes, G. palpalis* | Cattle, goats, sheep, antelopes (e.g., bushbuck, situtunga, waterbuck); buffalo, giraffe |
| Nagana or souma | *T. vivax* | *G. morsitans* group; *G. fuscipes, G. palpalis, G. tachinoides, G. vanhoofi* | Domestic mammals (esp. cattle, horses, mule); wild bovids, zebra, antelopes (e.g., impala, hartebeest, gnu); giraffe, warthog, lion |

to the midgut, proventriculus, esophagus, and the food canal of the labrum, from which infective forms pass into a vertebrate host. In tsetse infections with *T. vivax* and *T. uniforme,* parasites of diverse vertebrate hosts (Table I), development of the parasite is even more restricted, being limited to the proboscis, where the trypanosomes are found in the labrum and hypopharynx. The only tsetse-borne trypanosome found outside of Africa is *T. vivax,* which also occurs in Central and South America. As in Africa, this trypanosome can be transmitted mechanically among cattle by tabanids and other hematophagous flies in Latin America.

The only nonmammalian trypanosome transmitted by tsetse is *T. grayi,* a parasite of crocodiles. Its development in the fly is restricted to the posterior midgut, hindgut, and rectum. Infective metatrypanosomes are transmitted to crocodiles when they ingest infected flies.

The *tsetse vectors* of the trypanosomes causing nagana include species of the *palpalis, morsitans,* and *fusca* groups (Table I). *Glossina* species of all three groups transmit *T. congolense, T. simiae,* and *T. vivax.* Species in both *palpalis* and *morsitans* groups transmit *T. brucei.* Only tsetse species of the *palpalis* group transmit *T. uniforme,* and only *Glossina* species of the *fusca* group transmit *T. suis.* Transmission of the latter two trypanosomes by single vector group species restricts the distribution of *T. uniforme* to areas near waterways and of *T. suis* to dense forested habitats, where their respective vectors live.

The wild animal hosts listed in Table I are reservoir hosts for the trypanosomes that cause disease in the domestic animals listed. Recognition of the geographic and ecological distribution of reservoir hosts and specific tsetse vector species can help determine where introduction of domestic animals is most likely to succeed; however, after the wide areas inhabited by the diverse reservoir and vector species are excluded, little habitat remains for the maintenance of healthy domestic animals.

## PREVENTION AND CONTROL

Several approaches have been taken to prevent African sleeping sickness and nagana and to control tsetse vectors. These include intensive treatment and isolation of infected human and domestic animal hosts to try to break transmission cycles, use of trypanotolerant animals for agricultural purposes, laboratory research on development of vaccines for human and nonhuman hosts, and chemical and ecological attacks on the tsetse flies themselves.

The drugs used to treat human patients have not changed very much since the early 1900s. They include compounds, such as arsenicals, that kill the trypanosomes but cause severe side effects by interfering with the normal metabolism of the patients. Suramin, a sulfated naphthylamine, is now the drug of choice for the hemolymphatic stage of human trypanosomiasis. The arsenical melarsoprol, known as Mel B, is the drug of choice for destroying trypanosomes in the central nervous system stage of both forms of the disease. Samorin is given to cattle as a preventative, and diaminazene aceturate is effective in treating infected domestic animals. Although marketed exclusively for veterinary purposes, the latter compound has been administered to large numbers of patients in Africa and has been effective in the treatment of early stages of both forms of human trypanosomiasis.

Maintenance of noninfected human or domestic animal populations requires regular surveillance for trypanosomes. Examining blood for trypanosomes is still very common. However, use of the recently developed Card Agglutination Trypanosomiasis Test for antibodies is proving more efficient for surveillance under field conditions.

The use of *trypanotolerant breeds* of domestic animals is being studied, and new agricultural practices are being developed which enable Africans to breed native, normally wild, trypanotolerant animals for food and other domestic purposes. Indigenous cattle breeds, such as N'dama, which tolerate nagana trypanosomes without developing disease, are the focus of investigations to determine mechanisms of tolerance, with the hope of selective breeding or genetically engineering new tolerant breeds. Trypanotolerant game animals such as antelopes are being assessed for use in meat production. Trypanotolerant animals require careful handling because stress can trigger disease or death from their infections.

The development of an effective vaccine for domestic animals or humans has been thwarted by the presence of variant surface glycoproteins on the trypanosomes that enable the parasites to change their antigenicity in response to each wave of antibodies during the course of infection.

Tsetse populations have been directly attacked with *insecticides* and, historically, indirectly attacked by destroying tsetse resting and oviposition sites and wild hosts. The direct approach, used in attempts to eradicate the fly, involves aerial spraying and ground spraying from backpacks and trucks. The indirect approach, which generally would not be tolerated today, involved cutting or burning vast areas of vegetation, to destroy adult resting sites and puparia, and game hunting, to rid large areas of sources of blood for tsetse. Recent development involving building and paving, with its concomitant destruction of vegetation, has had similar effects to those from purposeful clearing of tsetse habitats. In order to be effective, most direct attacks on tsetse flies require frequent repetition of treatments, which are either costly or harmful to the environment. One direct approach that has been considered for general use is to treat cattle with an insecticide when they are run through an acaricide dip to kill ticks. In that way, cattle that are attacked by tsetse can also help to remove flies from the environment.

Another approach to tsetse control is the mass rearing and release of *sterile male flies.* In experimental tests, eradication has been achieved in fenced grazing areas, as large as several hundred square miles, surrounded by tsetse-free buffer zones. The buffer zones are continually monitored for tsetse, and tsetse eradication is maintained with aerial or ground insecticides or insecticide-impregnated attractant baits to keep fertile flies from migrating into the eradication area. The extreme effort and expense required to maintain such tsetse-free zones is impractical for large-scale animal production and the risks associated with releasing laboratory-bred flies that are capable vectors are too high for general use in Africa.

Any effective means of controlling tsetse flies requires constant monitoring of fly populations. Surveillance for tsetse originally involved African youths (''fly boys'') walking around a designated route of about 1.6 km called a *fly round* and catching flies with hand nets. This method evolved into walking with an ox as bait (ox round), bicycling over longer fly rounds, or riding on the back of trucks and catching tsetse attracted to the vehicles. Because of the low density of tsetse, fly-round surveillance over small areas, sometimes combining hand-netting with sprayer backpacks, became a useful control technique.

Surveillance by *adult trapping* has a long history involving different trap designs and identification of tsetse attractants. As in many other hematophagous flies, the attraction of tsetse to dark objects and to carbon dioxide, as well as other products of mammalian metabolism, has been put to practical use in designing tsetse traps. Tsetse behavior in relation to trapping methods and host odors has recently been reviewed by Colvin and Gibson (1992), Willemse and Takken (1994), and Green (1994).

Studies of traps and baits for surveillance have led to improvements that now make control with these devices the method of choice. Much of the work on attractants and trapping over the last 25 years has been done by Glyn Vale and his colleagues in Zimbabwe. His studies and those of others have led to the development of traps of various kinds, such as those resembling hosts (e.g., Morris trap), biconical designs (e.g., Challier and Laveissiére trap), and those with square or rectangular cloth targets. One of the most effective targets is black and blue and is baited with attractant components of ox breath or urine. *Attractants* include acetone, 1-octen-3-ol, and phenols (4-methyl- and 3-n-propyl). The target is designed so it can be used either with an electrocution device or an insecticide. An unattended trap charged with a residual insecticide can be used to remove flies from

the environment for 12–18 months, long enough to eradicate local populations of tsetse. Other methods being studied involve targets impregnated with insect growth regulators or chemosterilants and combinations of these substances with tsetse sex pheromone. Former large-scale use of insecticides and animal baits has given way to less costly, nonpolluting, and more efficient selective control with traps or targets. The possibility that traps can be maintained at little cost by local landowners, such as pastoral farmers, removes the need for extensive interference by foreign agencies and places the responsibility for maintenance on those whose economic well-being is directly related to the success of the trapping program.

Natural enemies of tsetse include puparial parasites, such as ants and beetles, over 20 species of wasps, and at least 10 species of bombyliids (*Thyridanthrax* spp.). Predators of adults include spiders, odonates, asilids, and sphecid and vespid wasps. Field studies of predation of *G. pallidipes* puparia in Kenya suggest that more than 20% of all puparia may be killed by predators during their buried, 30-day developmental period. The parasites and predators of tsetse are reviewed by Mulligan (1970), Laird (1977), and Leak (1999).

The presence of tsetse has helped to preserve the wildlife of Africa by preventing domestic animal production and consequent overgrazing that has occurred in some tsetse-free areas. A debate continues between those who view tsetse as an environmental benefit and those who see tsetse flies as a barrier to agricultural and other forms of development considered essential for the future economic, social, and political success of the African continent.

# REFERENCES AND FURTHER READING

Aksoy, S. (1995). *Wigglesworthia* gen. nov., and *Wigglesworthia glossinidia* sp. nov. Taxa consisting of the mycetome-associated, primary-endosymbionts of tsetse flies. *International Journal of Systematic Bacteriology* **45**, 848–851.

Austen, E. E. (1903). ''A Monograph of the Tsetse-Flies [Genus *Glossina*, Westwood].'' Longmans & Co., London (reprinted 1966 by Johnson Reprint Corp.).

Baker, J. R. (1974). Epidemiology of African sleeping sickness. In ''Trypanosomiasis and Leishmaniasis with Special Reference to Chagas' Disease.'' Ciba Foundation Symposium, New Series, No. 20. Elsevier, Amsterdam.

Beard, C. B., O'Neill, S. L., Mason, P., Mandelco, L., Woese, C. R., Tesh, R. B., Richards, F. F., and Aksoy, S. (1993). Genetic transformation and phylogeny of bacterial symbionts from tsetse. *Insect Mol. Biol.* **1**, 123–131.

Buxton, P. A. (1955). ''The Natural History of Tsetse Flies.'' H. K. Lewis & Co., Ltd., London.

Colvin, J., and Gibson, G. (1992). Host-seeking behavior and management of tsetse. *Annu. Rev. Entomol.* **37**, 21–40.

Elsen, P., Amoudi, M. A., and Leclercq, M. (1990). First record of *Glossina fuscipes fuscipes* Newstead, 1910 and *Glossina morsitans submorsitans* Newstead, 1910 in southwestern Saudi Arabia. *Annales de la Société Belge de Médecine Tropicale* **70,** 281–287.

Ford, J. (1971). "The Role of the Trypanosomiases in African Ecology– A Study of the Tsetse Fly Problem." Clarendon, Oxford.

Glasgow, J. P. (1963). "The Distribution and Abundance of Tsetse." Macmillan, New York.

Gouteux, J. P. (1987). Une nouvelle glossine du Congo: *Glossina* (*Austenina*) *frezili* sp. nov. (Diptera: Glossinidae). *Tropical Medicine and Parasitology* **38,** 97–100.

Green, C. H. (1994). Bait methods for tsetse fly control. *Advances in Parasitology* **34,** 229–291.

Hoare, C. A. (1972). "The Trypanosomes of Mammals." Blackwell Sci., Oxford.

Jordan, A. M. (1986). "Trypanosomiasis Control and African Rural Development." Longman, London.

Jordan, A. M. (1993). Tsetse-flies (Glossinidae). In "Medical Insects and Arachnids" (R. P. Lane and R. W. Crosskey, Eds.), pp. 333–338. Chapman & Hall, London.

Jordan, A. M. (1995). Control of tsetse flies (Diptera: Glossinidae) with the aid of attractants. *Journal of the American Mosquito Control Association* **11,** 249–255.

Jubb, K. V. F., Kennedy, P. C., and Palmer, N. (1985). "Pathology of Domestic Animals." Vol. 3. Academic Press, San Diego.

Laird, M., Ed. (1977). "Tsetse: The Future for Biological Methods in Integrated Control." International Development Research Centre, Ottawa.

Leak, S. G. A. (1999). "Tsetse Biology and Ecology: Their Role in the Epidemiology and Control of Trypanosomosis." CABI, New York.

Lehane, M. J. (1991). "Biology of Blood-sucking Insects." Harper Collins, London.

Losos, G. J., and Ikede, B. O. (1972). Review of pathology of diseases in domestic and laboratory animals caused by *Trypanosoma congolense, T. vivax, T. brucei, T. rhodesiense* and *T. gambiense*. Veterinary Pathology (Suppl.)9, 71 pp.

McAlpine, J. F. (1989). Phylogeny and classification of the Muscomorpha. In "Manual of Nearctic Diptera" (J. F. McAlpine and D. M. Wood, Eds.), Vol. 3, pp. 1397–1518. Monograph No. 32, Agriculture Canada, Ottawa.

McKelvey, J. J., Jr. (1973). "Man against Tsetse: Struggle for Africa." Cornell Univ. Press, Ithaca.

Mulligan, H. W., Ed. (1970). "The African Trypanosomiases." Wiley, New York.

Murray, M., Morrison, W. I., Murray, P. K., Clifford, D. J., and Trail, J. C. M. (1979). Trypanotolerance -a review. *World Animal Review* (FAO) **31,** 2–12.

Newberry, K. (1982). The behaviour of *Glossina morsitans morsitans* near roads. *Transactions of the Royal Society of Tropical Medicine and Hygiene* **76,** 281–282.

Newstead, R. [with Evans, A. M., and Potts, W. H.] (1924). Guide to the study of tsetse-flies. *Liverpool School of Tropical Medicine Memoir* (*New Series*) No. 1.

Pépin, J., and Milord, F. (1994). The treatment of human African trypanosomiasis. *Advances in Parasitology* **33,** 1–47.

Potts, W. H. (1973). Glossinidae (tsetse-flies). In "Insects and Other Arthropods of Medical Importance" (K. G. V. Smith, Ed.), pp. 209–249. British Museum (Natural History), London.

Quinn, T. C. (1996). African trypanosomiasis (sleeping sickness). In "Cecil Textbook of Medicine" (J. C. Bennett and F. Plum, Eds.), 20th ed., pp. 1896–1899. Saunders, Philadelphia.

Rogers, D. J., and Randolph, S. E. (1985). Population ecology of tsetse. *Annu. Rev. Entomol.* **30,** 197–216.

Rogers, D. J., Hendricks, G., and Slingenbergh, J. H. W. (1994). Tsetse flies and their control. *Revue Scientifique et Technique Office International des Epizooties* **13,** 1075–1124.

Vale, G. A. (1993). Development of baits for tsetse flies (Diptera: Glossinidae) in Zimbabwe. *Journal of Medical Entomology* **30,** 831–842.

Willemse, L. P. M., and Takken, W. (1994). Odor-induced host location in tsetse flies (Diptera: Glossinidae). *Journal of Medical Entomology* **31,** 775–794.

World Health Organization (1989). Geographical distribution of arthopod-borne diseases and their principal vectors. Manual WHO 89.967. Vector Biology and Control Division, Geneva.

Zumpt, F. (1936). "*Die Tsetsefliegen*." Fischer, Jena.

Zumpt, F. (1973). "The Stomoxyine Biting Flies of the World." Fischer, Stuttgart.

# 16

# MYIASIS (*Muscoidea, Oestroidea*)

E. PAUL CATTS AND GARY R. MULLEN

*Myiasis is* the invasion of a living vertebrate animal by fly larvae. This invasion may or may not be associated with feeding on the tissues of the host. Myiasis-causing flies are represented by a diversity of species. Some are rarely involved in myiasis, whereas for others it is the only way of life. Many of these same fly species also feed on carrion. Among flies, dietary proteins are required for growth, egg production, and development. Proteins may be obtained by adult flies, by their larvae, or by both. In the case of larval diets, proteins are assimilated, stored, and carried through the pupal stage for subsequent use by the repro ducing adult fly. A larval diet rich in proteins dictates that there is less need for adults to seek proteins. Thus myiasis is a means of exploiting a rich protein source by the larva for its own growth and, in some cases, for reproduction by the adult.

Myiasis can be classified based on the degree to which fly species are tied to a host. Three types of myiasis generally are recognized: *accidental, facultative,* and *obligatory myiasis.* In accidental myiasis, also called *pseudomyiasis,* the fly larvae involved normally are not parasitic but under certain conditions may become so.

*Accidental myiasis* generally results when fly eggs or larvae contaminate foods ingested by a host animal. An example of this can result from the ingestion of food contaminated with eggs or larvae of pomace flies and fruit flies (*Drosophila* spp.). The 50 or so fly species involved include those which typically are free-living in all stages and rarely are parasitic. In most cases these flies pass unharmed through the host's alimentary tract, but they can cause discomfort, nausea, diarrhea, and a plethora of related problems on their way through. In some cases, symptoms can be severe. Invasion of the alimentary tract can occur in two ways: either through ingestion of con taminated food or by retroinvasion through the host's anus. There is some doubt as to whether or not these cases are true myiasis because there is scant evidence that any fly development takes place after the ingested eggs hatch.

*Facultative myiasis* involves larvae which can be either free-living saprophages or parasites. These flies are op portunistic, having the ability to exploit living tissue. An example of facultative myiasis is the invasion of open sores on livestock by maggots of blow flies which normally fre quent carrion.

Kettle (1995) recognizes three types of facultative myiasis: *primary myiasis,* involving those species which can initiate myiasis; *secondary myiasis,* involving species which continue myiasis, but only after it is started by primary species; and *tertiary myiasis,* involving species which join the primary and secondary species just prior to host death. Facultative myiasis species are the evolutionary bridge linking saprophagous feeders to those restricted to feeding on living tissues.

Many of these facultative myiasis flies are able to shift from dead to living tissue and back again with alacrity (e.g., *Cochliomyia macellaria, Wohlfhartia nuba*). In a sense, these are borderline parasites that are capable of invading a sick or injured host and continuing their larval development after the death of that host. The adult flies are attracted to open wounds or chronic surface sores with purulent exudates.

In *obligatory myiasis* the maggots of the fly species involved are always parasitic; they require a living host for their development. Examples are primary screwworms and bot flies. Included here are those species which cause temporary obligatory myiasis, such as nestling maggots and floor maggots. In this type of myiasis, the maggots do not keep continual contact with their host. Occasional parasitism of atypical hosts by obligate myiasis-producing flies is called *incidental myiasis.*

Myiasis also can be categorized in relation to the site of maggot invasion or subsequent development in the host. Thus, the descriptives *gastrointestinal, urogenital, ocular, nasopharyngeal, auricular,* and *cutaneous* are antecedent to the word myiasis, indicating the general site of maggot infestation.

*Gastrointestinal myiasis* refers to fly larvae in the alimentary tract of a host. This can be accidental myiasis, such as the ingestion of false stable fly eggs or larvae in uncooked fruits, or obligatory myiasis, such as the development of stomach bot flies in a horse. *Enteric myisasis* refers specifically to the intestinal tract. *Urogenital myiasis* is the invasion of the urethra and/or genitalia by fly larvae. This can occur when a host is debilitated and the urogenital openings exposed or when the host has a urogenital infection producing exudates that attract flies. Cases of urogenital myiasis usually involve blow flies and flesh flies. *Ocular myiasis* is the invasion of eye tissues by fly larvae; most cases are caused by the sheep nose bot fly (*Oestrus ovis*) and infrequently by rodent bot flies (*Cuterebra* spp.). *Nasopharyngeal myiasis* is the invasion of nasal and deep oral cavities and recesses by fly larvae. As with urogenital myiasis, this often is associated with a microbial infection but also can be caused by nose bot flies in healthy hosts. *Auricular myiasis* is the invasion of ears by fly larvae, usually caused by blow flies or flesh flies. *Cutaneous myiasis* involves invasion of the skin, usually by blow flies, flesh flies, screwworms, or certain bot flies. When cutaneous myiasis is associated with a break, laceration, or open sore in the host's skin, it is called *traumatic myiasis.* Cases of *temporary myiasis* involve intermittent contact between a fly larva and its host.

Myiasis apparently has evolved along different lines in different groups of flies. Gastrointestinal myiasis and urogenital myiasis, for example, appear to represent a transition from species contaminating host foods to those associated with host excretions. They usually involve free-living species. Obligatory cutaneous myiasis appears to have evolved from carrion-breeding flies and from predaceous flies which prey on them. The origin of obligatory nasopharyngeal myiasis is less clear but probably is linked with host secretions associated with upper respiratory infections. Ocular and auricular myiasis are characteristically either accidental or incidental in nature, resulting in damage to tissues at those respective sites. Temporary myiasis appears to have evolved from nest associates or lair-frequenting scavenger species which fed on organic morsels.

## TAXONOMY

The vast majority of species involved in myiasis are members of two superfamilies and six families of calypterate flies: Muscoidea (Anthomyiidae, Fanniidae, and Muscidae) and Oestroidea (Calliphoridae, Sarcophagidae, and Oestridae). A dozen other families in eight superfamilies include species reported to cause myiasis; however, with the exception of the nest skipper fly (Neottiphilidae) and 10 species of Australian frog flies (Chloropidae), these cause accidental myiasis only. In contrast, all species of bot flies (Oestridae) cause obligatory myiasis. Myiasis-causing species among the muscids, calliphorids, and sarcophagids are typically facultative or obligatory myiasis producers. Table I shows the taxonomic relationships and associated types of myiasis for those flies known to invade living hosts.

The Anthomyiidae are a large family with more than 100 genera worldwide. Although members are called root maggot flies, the larvae occur in a wide range of habitats other than roots. Cladistically, the fanniids are a sister group to the muscids and often are included as a subfamily of the Muscidae. Most muscids are house fly–like in appearance, having a rather drab coloration, with dorsal longitudinal stripes on the thorax. Muscidae is a very large family with worldwide distribution that includes species typically associated with excrement and decaying plant matter.

TABLE I

**Taxonomic Relationships of Flies Known to Cause Myiasis: The Types of Myiasis in Each Case Are Indicated in Parentheses as Accidental (A), Facultative (F), or Obligatory (O)**

| Superfamilies and families | Common names | Genera and species | Type of myiasis |
|---|---|---|---|
| Tipuloidea | | | |
|   Tipulidae | Crane flies | Unspecified | Gastrointestinal (A) |
| Psychodoidea | | | |
|   Psychodidae | Moth flies | *Psychoda* (3 spp.) | Gastrointestinal (A) |
| Stratiomyoidea | | | |
|   Stratiomyidae | Black soldier fly | *Hermetia illucens* | Gastrointestinal (A) |
| Asiloidea | | | |
|   Therevidae | Stiletto flies | *Thereva* sp. | Gastrointestinal (A) |
| Platypezoidea | | | |
|   Phoridae | Humpback flies, scuttle flies | *Megaselia* (3 spp.) | Gastrointestinal (A), traumatic (F) |
| Syrphoidea | | | |
|   Syrphidae | Flower flies | *Eristalis* (2 spp.) | Gastrointestinal (A) |
| Tephritoidea | | | |
|   Piophilidae | Cheese skipper fly | *Piophila casei* | Gastrointestinal (A) |
|   Neottiophilidae | Nest skipper fly | *Neottiophilum praeustrum* | Cutaneous (O) |
| Ephydroidea | | | |
|   Drosophilidae | Pomace flies, fruit flies | *Drosophila melanogaster* | Gastrointestinal (A) |
|   Chloropidae | Australian frog flies | *Batrachomyia* (10 spp.) | Cutaneous (O) |
| Muscoidea | | | |
|   Anthomyiidae | Anthomyiid flies | *Hylemya* (2 spp.) | Gastrointestinal (A) |
|   Fanniidae | Lesser house fly | *Fannia cuicularis* | Gastrointestinal, urogenital, traumatic (A) |
| | Latrine fly | *F. scalaris* | |
|   Muscidae | House fly | *Musca domestica* | Gastrointestinal, urogenital, traumatic (A) Gastrointestinal, traumatic (A) |
| | False stable fly | *Muscina stabulans* | Gastrointestinal, traumatic (A) |
| | — | *Muscina* (2 spp.) | Gastrointestinal, traumatic (A) |
| | — | *Hydrotaea rostrata* | |
| | Tropical nest flies | *Passeromyia* (3 spp.) | Cutaneous (O) |
| | Tropical nest flies | *Mydaea* (25 spp.) | Cutaneous (O) |
| | Neotropical nest flies | *Neomusca* (*Philornis* (35 SPP.) | Cutaneous (O) |
| Oestroidea | | | |
|   Calliphoridae | Green bottle flies | *Bufolucilia* spp. | Traumatic, cutaneous, gastrointestinal, nasopharyngeal Traumatic, cutaneous, gastrointestinal, nasopharyngeal auricular (F) |
| | | *Lucilia* spp. | |
| | | *Phaenicia* spp. | |
| | Blue bottle flies | *Calliphora* spp. | Traumatic, cutaneous, gastrointestinal, nasopharyngel auricular (F) |
| | | *Eucalliphora* spp. | |
| | | *Paralucilia* sp. | |
| | | *Protophormia* sp. | |
| | | *Cynomya* sp. | |
| | Black blow fly | *Phormia regina* | |
| | Primary screwworm | *Cochliomyia hominivorax* | Traumatic (O) |
| | Secondary screwworm | *C. macellaria* | Traumatic (F) |
| | Old World screwworm, | *Chrysomya bezziana* | Traumatic (O) |
| | — | *Chrysomya* (9 spp.) | Traumatic (F) |
| | Bird blow flies | *Protocalliphora* (≈90 spp.) | Traumatic (O) |
| | Tumbu fly | *Cordylobia anthropophaga* | Traumatic (O) |

*(Continues)*

**TABLE I** (*Continued*)

| Superfamilies and families | Common names | Genera and species | Type of myiasis |
|---|---|---|---|
| | African mouse/Lund's fly | *Cordylobia* (2 spp.) | Traumatic (O) |
| | Congo floor maggot | *Auchmeromyia senegalensis* | Traumatic (O) |
| | Africa suid maggots | *Auchmeromyia* (4 spp.) | Traumatic (O) |
| | — | *Pachychoeromyia praegrandis* | Traumatic (O) |
| | Asian deer/water buffalo skin maggots | *Trypocalliphora linderi* *Boopona* (4 spp.) | Traumatic (O) |
| | Indian elephant skin maggot | *Neocordylobia* (2 spp.) | Traumatic (O) |
| | | *Elephantoloemus indicus* | Traumatic (O) |
| Sarcophagidae | | | |
| | Flesh flies | *Sarcophaga* spp. | Traumatic (F), gastrointestinal (A) |
| | | *Wohlfahrtia* (4 spp.) | Traumatic (O) |
| | Lizard flesh fly | *Anolisomyia* | Cutaneous (O) |
| | Turtle flesh fly | *Cistudinomyia cistudinis* | Cutaneous (O) |
| | Toad flesh fly | *Notochaeta bufonovoria* | Cutaneous (O) |
| | Lizard egg fly | *Eumacronychia nigricornis* | Cutaneous (O) |
| | Sea turtle egg fly | *E. sternalis* | Cutaneous (O) |
| | Terrapin egg fly | *Metoposarcophaga importuna* | Cutaneous (O) |
| Oestridae | Bot flies | 29 genera, 166 spp. | Cutaneous, gastrointestinal, nasopharyngeal (O) |
| | | | Cutaneous (O) |
| Cuterebrinae | New World skin bot flies | 6 genera, 83 spp. | Cutaneous (O) |
| | Rodent and rabbit bot flies | *Cuterebra* (72 spp.) | Cutaneous (O) |
| | Howler monkey bot | *Allouattamyia baeri* | Cutaneous (O) |
| | Tórsalo, human bot fly | *Dermatobia hominis* | Cutaneous (O) |
| | — | *Pseudogametes* (2 spp.) | Cutaneous (O) |
| | — | *Montemyia* sp. | Cutaneous (O) |
| | — | *Rogenhofera* (6 spp.) | |
| Hypodermatinae | Old World skin bot flies | 7 genera, 27 spp. | Cutaneous (O) |
| | Common cattle grub | *Hypoderma lineatum* | Cutaneous (O) |
| | Northern cattle grub | *H. bovis* | Cutaneous (O) |
| | Reindeer grub | *H. tarandi* | Cutaneous (O) |
| | — | *Hypoderma* (3 spp.) | Cutaneous (O) |
| | Old World rodent grubs | *Oestromyia* (5 spp.) | Cutaneous (O) |
| | — | *Oestroderma* sp. | Cutaneous (O) |
| | — | *Pallisiomyia* sp. | Cutaneous (O) |
| | — | *Pavlovskiata* sp. | Cutaneous (O) |
| | — | *Porchinskia* (7 spp.) | Cutaneous (O) |
| | — | *Przhevalskiana* (6 spp.) | Cutaneous (O) |
| Oestrinae | Nose bot flies | 9 genera, 34 spp. | Nasopharyngeal (O) |
| | Deer nose bots | *Cephenemyia* (8 spp.) | Nasopharyngeal (O) |
| | Camel nose bot | *Cephalopina titillator* | Nasopharyngeal (O) |
| | — | *Gedoelstia* (2 spp.) | Nasopharyngeal (O) |
| | — | *Kirkioestrus* (2 spp.) | Nasopharyngeal (O) |
| | Sheep nose bot | *Oestrus ovis* | Nasopharyngeal, ocular (O) |
| | — | *Oestrus* (5 spp.) | Nasopharyngeal (O) |
| | African elephant throat bot | *Pharnygobolus africanus* | Nasopharyngeal (O) |
| | — | *Pharygomya* (2 spp.) | Nasopharyngeal (O) |
| | Horse nose bot fly | *Rhinoestrus purpureus* | Nasopharyngeal (O) |
| | | *Rhinoestrus* (10 spp.) | Nasopharyngeal (O) |
| | Kangaroo throat bot fly | *Tracheomyia macropi* | Tracheal (O) |

(*Continues*)

TABLE I (*Continued*)

| Superfamilies and families | Common names | Genera and species | Type of myiasis |
|---|---|---|---|
| Gasterophilinae | Stomach bot flies | 5 genera, 15 spp. | Gastrointestinal (O) |
| | Horse stomach bots | *Gasterophilus* (9 spp.) | Gastrointestinal (O) |
| | Common horse bot | *G. intestinalis* | Gastrointestinal (O) |
| | Throat horse bot | *G. haemorrhoidalis* | Gastrointestinal (O) |
| | Nose horse bot | *G. nasalis* | Gastrointestinal (O) |
| | Broad-bellied horse bot | *G. nigricornus* | Gastrointestinal (O) |
| | Dark-winged horse bot | *G. pecorum* | Gastrointestinal (O) |
| | Black elephant stomach bot | *Cobboldia elephantis* | Gastrointestinal (O) |
| | Blue elephant stomach bot | *Platycobboldia loxodontis* | Gastrointestinal (O) |
| | Green elephant stomach bot | *Rhodhainomyia roverei* | Gastrointestinal (O) |
| | Rhinoceros stomach bots | *Gyrostigma* (3 spp.) | Gastrointestinal (O) |
| Unplaced genera | African elephant skin bot | *Neocuterebra sqamosa* | Cutaneous (O) |
| | African elephant foot bot | *Ruttenia loxodontis* | Cutaneous (O) |

The superfamily Oestroidea is composed of six families: Calliphoridae, Sarcophagidae, Oestridae, Tachinidae, Rhinophoridae, and Mystacinobiidae. The first five of these include parasitic species. Tachinids and rhinophorids parasitize invertebrate hosts, mostly other insects, and are not involved in myiasis. Adult mystacinobiids are phoretic, not parasitic, on bats.

Members of the Calliphoridae are called *blow flies*. This large family includes over 1000 species worldwide. Most have a polished or metallic blue, green, or bronze appearance and also are called *bottle flies*. There is confusion and continuing speculation concerning the phylogenetic relationships within this family. The major subfamilies include Calliphorinae, Chrysomyinae, Mesembrinellinae, Polleniinae, and Rhiniinae. Other subfamilies contain specialized genera or genera with limited distributions. The largest subfamily, Rhiniinae, includes about 40 genera of Old World blow flies. The Mesembrinellinae are large, showy tropical blow flies, whereas the Calliphorinae are the more temperate-climate bottle flies. The subfamily Chrysomyinae includes the economically important *screwworms* and the bird-nest blow flies. Members of the subfamily Rhiniinae are of no significance regarding myiasis.

The Sarcophagidae are the *flesh flies,* with some 2000 species distributed worldwide. Many of these are parasitic on hymenopterous hosts or are predaceous on other insects, but a few cause myiasis. Members of two of the four subfamilies include myiasis-causing species, the Sarcophaginae and Miltogramminae.

The Oestridae are the *bot flies,* with fewer than 200 species worldwide. Adult flies are robust and hairy, with a general appearance of honey bees or bumble bees. Larvae of all of these cause obligatory myiasis. They are placed in four subfamilies: Cuterebrinae, Hypodermatinae,

Oestrinae, and Gasterophilinae. Phylogenetic relationships among these subfamilies are unclear. They are grouped together mostly for convenience and because of the similar form and manner of larval development.

For general taxonomic information and keys for identifying adults and larvae of muscoid and oestroid flies, see Curran (1965), Furman and Catts (1982), Greenberg (1971), James (1947), Lane and Crosskey (1993), McAlpine *et al.* (1981, 1987, 1989), Smith (1986), and Zumpt (1965). Additional sources for identifying the immature stages are Liu and Greenberg (1989) and Stehr (1991). The Calliphoridae and Oestridae are treated by Hall (1948) and Papavero (1977), respectively, whereas calliphorid eggs can be identified with Greenberg and Singh (1995).

## MORPHOLOGY

Members of the superfamilies Muscoidea and Oestroidea are distinguished from other calypterate flies by possessing both a *ptilinal suture* and *facial lunule* (Fig. 16.1B, C); the *antennal pedicel,* with a complete dorsal seam and flagellum composed of a single compound segment bearing an *arista* or *stylus;* and a bulbous, greater *ampulla* below the wing base (Fig. 16.1A). All oestroid flies have a vertical row of bristles on the thoracic *meron* (Fig. 16.1A). The Muscoidea are distinguished from the Oestroidea by lacking bristles on the thoracic meron even though the rest of the body is bristled.

Larvae of the muscoid and oestroid families are called *maggots.* The larvae of muscoid flies usually lack armature, or body spines, whereas oestroid larvae possess armature ranging from sparse segmental belts of spines to

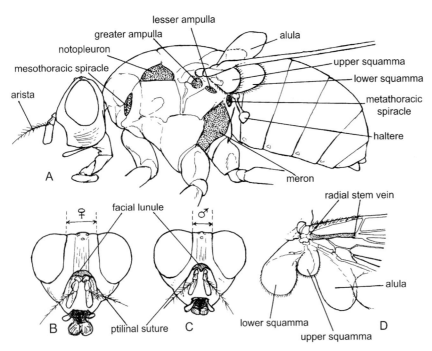

**FIGURE 16.1** External morphology of representative adult muscoid fly. (A) Lateral view; (B) frontal view, female; (C) frontal view, male; (D) dorsal view of wing base. (Original by E. P. Catts.)

a rather complete spiny vesture. The three larval instars of both superfamilies can be distinguished by the form and number of slits in the *posterior spiracular plates* and by the size and shape of the internal *cephalopharyngeal skeleton* (Fig. 16.2). The body form of muscoid maggots is peglike and gradually tapers from a blunt posterior end to a pointed anterior end. Larvae in the genus *Fannia* are atypical, bearing segmental protuberances. Larvae of blow flies and flesh flies are more cylindrical than those of muscoid flies; they are tapered more abruptly at each end. Larvae of these two families can be distinguished by the slant of the inner posterior spiracular slits. The slits of blow flies slant toward the midline, whereas those of flesh flies slant away from the midline.

Mature bot fly maggots, some of which are called *grubs*, are much larger and thicker in cylindrical form than those of other oestroids. Except for the larvae of Gasterophilinae and Cuterebrinae, which possess elbowed or highly sinuous posterior spiracular slits, respectively, oestrid larvae have porous posterior spiracular plates that mask the more typical arrangement of slits.

The *puparium*, which contains the developing pupa, is formed from the integument of the last larval instar. All puparia except those of oestroids can be separated at the family level by examining remnant larval features (mouthparts and posterior spiracles). In contrast, oestrid puparia exhibit overall morphological forms which characterize each subfamily.

Among the adult muscoids, the wing venation of anthomyiids differs from that of muscids by the length of the anal vein. Some muscid genera (e.g. *Morellia, Neomyia,* and *Hydrotaea*) have a metallic or polished abdomen, superficially resembling blow flies. Most adult muscoid flies have sucking, rasping, or sponging mouthparts. A few (e.g., *Stomoxys, Haematobia*) possess a hardened labium as a piercing structure.

Among the oestroid flies, calliphorid adults have well-developed sponging-sucking mouthparts, the thoracic meron has a vertical row of bristles, and the postscutellum is small. Typically these flies are metallic blue, green, or bronze in body coloration. Members of the major subfamilies Chrysomyinae and Calliphorinae differ in the presence or absence of hairs on the *radial stem vein* of the wing (Fig. 16.1D).

Adult sarcophagids also have sponging mouthparts, the thoracic meron is bristled, the postscutellum is small or absent, and three or four *notopleural setae* are usually present. Body coloration is typically gray and black, giving them the general appearance of large houseflies. Their color patterns, however, are more distinct than those of muscoid flies. The abdomen typically displays a tessellated, or checkerboard, pattern of gray and black.

Unlike other oestroid flies, adult oestrids lack head and body bristles, except for the bristles on the meron. They are medium to large, hairy flies that resemble bees. They possess small antennae and relatively small

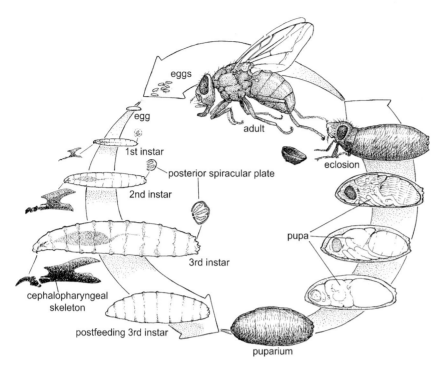

**FIGURE 16.2**  Typical life cycle of muscoid and oestroid flies. (Original by E. P. Catts.)

sponging-sucking mouthparts which in some instances are atrophied and nonfunctional. All except the members of the Gasterophilinae have large wing *squammae* (Fig. 16.1D).

## LIFE HISTORY

The life history of muscoid and oestroid flies follows the typical holometabolous and cyclorrhaphous pattern of four stages: egg, larva (three instars), pupa (in a puparium), and adult (Fig. 16.2). Male flies are usually smaller and emerge earlier than females. All of these flies form the puparium free from the larval substrate. Generally, given an adequate larval diet, more time in the developmental life cycle of an individual is spent in the pupal stage than in that of either the egg or larva. Among the oestrid flies, the cuterebrines have a long pupal stage that allows overwintering free from the host. Other bots have a prolonged larval period as an adaptation for overwintering within the protection of their host. Pupation is typically preceded by a wandering, postfeeding period by the last larval instar.

Adult longevity among these flies varies depending on environmental conditions. Longevity ranges from as little as 1 day to several days among stomach bot flies, to 1 or 2 weeks for other bot flies, to 1 month or more among muscoids and other oestroids. In general, bot fly adults have a shorter life because they do not feed. All myiasis-causing flies either oviposit or larviposit, but the degree of embryo development within the egg prior to deposition differs. Usually eggs are fertilized at the time of oviposition. Consequently, some, such as muscids and most oestroids, lay eggs which require an additional period of embryonic development before hatching (i.e., *oviposition*). In contrast, the nose bot flies and flesh flies retain fertilized eggs in an expanded pouch, or uterus. They deposit active larvae within membranous "shells," which hatch immediately (i.e., *larviposition*).

In most cases, eggs or larvae are laid in contact with the larval nutrient substrate. However, in a few cases (e.g., cuterebrines and nest maggots) the first-instar larvae must locate and invade their nutrient resources.

After hatching, the larvae begin to feed by scratching the nutrient substrate with their hooklike mandibles. This loosens fragments of substrate which are bathed in the maggot's salivary enzymes and are sucked up as a nutrient soup. Only the cephalopharyngeal skeleton is fixed in size at the time of molting relative to the total length and girth of the feeding larva. The ratio between skeleton and body sizes indicates the larval age within each instar. The body integument remains pliable, facilitating considerable growth and expansion during each stadium. While inside a living host, larvae are relatively protected from exploitation by predators or parasites, but they must cope with hosts' immune defenses. This is an advantage over free-living related species or siblings developing in carrion, where competition and exploitation are major risks.

The next period in development is that of the wandering, mature, postfeeding larva. Larvae may continue to feed on available nutrients even after reaching and passing a critical size necessary for wandering and subsequent *pupation*. This critical size for some species is reached early in the third stadium so that, if the nutrient source becomes exhausted, the result can be very small adult flies. The duration of wandering is an innate characteristic for each species and is influenced by environmental factors such as heat, moisture, and light. It may be shortened considerably if a suitable pupation site is obtained early on. Once a site is found the larva becomes immobile and contracts in size. The resulting puparium is formed by the synchronized, concerted action of muscular and cuticular structures, resulting in a rigid, hardened protective shell formed from the cuticle of the last larval instar.

Adult flies typically emerge from the puparium in the morning hours. The teneral adult exits the puparium through a predetermined portal formed by the opening of an anterior operculum or by detachment of the anterior end of the puparium. After a few hours, adults are ready to take to flight in search of a mate and, in the case of females, a suitable adult and larval nutrient source. *Mating* usually occurs after adult flies have matured for a period of a few days to about 1 week. Most adult female myiasis-causing flies require additional nutrients and a period of ovarian development of several more days. However, nonfeeding bot flies do not require more nutrients.

Flies which lay eggs that require time to complete their embryonic development outside of the female are *oviparous*. In bot flies, egg development is initiated during the larval stage and continues through pupation, so that eggs can be deposited with less delay. The more primitive cuterebrine bot flies require about 5 days between eclosion and oviposition. In contrast, cattle grubs and stomach bots have fully developed eggs at eclosion and their females are ready to oviposit as soon as mating is completed. As already noted, nose bot flies and flesh flies retain eggs *in utero* so that their eggs hatch immediately following oviposition. This delayed oviposition is correctly termed *ovoviviparity* but usually is referred to as *larviposition*. Unlike other oestroids, the bot flies have only one egg-producing (gonotrophic) cycle. Although subsequent cycles do not occur, among the cattle grubs each ovarian follicle may develop two eggs simultaneously, thus doubling the egg production of the one gonotrophic cycle. This limitation to a single cycle among bot flies reflects their shorter, nonfeeding adult life span. In contrast, muscoid and other oestroid flies (calliphorids and sarcophagids) can hold their unfertilized eggs for several weeks until a suitable larval breeding medium is found. Following

oviposition or larviposition they are able to initiate additional gonotrophic cycles.

The primary activities of male muscoid and oestroid flies are to locate and copulate with receptive females. Males are capable of multiple matings during their life but play no further role in reproduction or dispersal of progeny. Typically females mate only once. Dispersal of progeny is accomplished by the female fly and by the mobile host. This is another peculiar aspect of obligatory myiasis because the female fly usually attacks more than one host, thus distributing her progeny. The larvae, when mature, drop from their ambulatory hosts and thus can be scattered over an extensive area.

A number of vernacular terms are associated with myiasis-causing flies. Some refer to the adult flies or, more often, their larvae, and others refer to the pathology which they cause. The term *maggot* is of Scandinavian origin and is used for the larval stage of muscoid, calliphorid, and sarcophagid flies. The term *bot* is applied only to oestrid larvae and is derived either from a Gaelic word, *botus,* meaning belly worm, or from the Italian *botta,* which refers to the cutaneous ulcer caused by these flies. This ulcer, or open cyst, is termed a *warble,* which is derived from the Scandinavian *varbulde,* meaning boil. In the southeastern United States warbles have been called *wolves* since colonial times in reference to the boil-like cuterebrine infections in rodents and rabbits. *Grub* is a term applied to cattle bots (as well as to scarab beetle larvae). This term probably is derived from the Indo-European word *ghrebl,* meaning to scratch, scrape, or bore into. The common name of *screwworm* probably describes the threaded appearance of the maggot stage of these blow flies and the belief that they twist and bore their way into host flesh.

When flies oviposit on a host or on carrion the action is referred to as *striking* or *fly strike*. This is an English term originating from the Latin *stringere,* meaning to touch lightly or brush against. Its use probably was more in reference to the ovipositing of hovering nose bot flies or stomach bot flies.

A host or other substrate on which fly eggs have been laid is said to be *fly blown*. The use of "blow" in this context comes from the Old English term *blawan* and probably refers to the production of gas from bloated carrion containing maggots. This is also the source of the name *blow fly* for those flies which are most conspicuous at carrion, the Calliphoridae.

## ECOLOGY AND BEHAVIOR

The environmental constraints on myiasis-causing flies during the free-living periods of their life cycle are the same as for other Diptera: primarily moisture and heat. These constraints are of little importance during the parasitic period of their life cycle because both moisture and heat are provided by the host at levels well within the tolerance limits of eggs and larvae. Some myiasis-causing species lay their eggs free from the host and thus are adapted to a wider range of environmental conditions (e.g., cuterebrines and nest maggots). Muscoid and oestroid larvae are adapted in form and behavior to life in a moist, organic substrate ranging from wet feces to living tissue. Constraints on larval development, unlike on adults, stem more from host resistance and inter- or intraspecific competition.

Facultative myiasis-causing flies generally develop in carrion and feces as massed aggregations of maggots. In such situations the increased temperature that allows for rapid larval development mostly comes from metabolic heat produced by the clustered maggots themselves. For these species, a living, protein-rich, moist, warm host substrate merely substitutes for the nonliving substrate. Facultative myiasis often results in the death of the host either by direct effects of maggots or by the indirect effect of stress which predisposes the host to predation or disease. Low populations of maggots, however, can have therapeutic benefits and have been used in treating deep wounds and sores.

For flies that cause facultative myiasis, finding a receptive mate generally is not difficult. The synchronous emergence of sibling adults developed from the same mass of maggots puts newly emerged males and females in close proximity. These species also find their oviposition sites by responding to the same stimuli that attract them to any nonliving resource (e.g., fetid or putrid odors, purulent discharges, and/or accumulations of animal excretions). They show very little discrimination regarding host species. Host size and habitat are the principal limiting factors. Even though their resources have a patchy distribution, these flies are capable of responding quickly to chemical stimuli at very low concentrations in order to locate widely scattered resources.

The greater longevity of muscoid flies, blow flies, and flesh flies also requires that they imbibe fluids to maintain an internal water balance and that they obtain an energy source beyond that acquired during larval development. The energy source is often honeydew and plant juices, especially nectar. This is why adults of these flies also occur on flowering plants and are important pollinators for many plants, including certain crops. Most female flies also require a protein supplement to complete oogenesis. This is supplied by a wide array of sources (e.g., blood secretions, excretions, wounds, and carrion).

The obligatory myiasis-causing screwworms (e.g., primary screwworm, Old World screwworm, and wound

flesh fly) are closely related to the facultative myiasis-causing species. The obligatory species are simply precocious carrion flies which attack their resource while it is still living. However, because their progeny usually are scattered widely, the adults aggregate at certain flowering shrubs, where they mate and feed. Their egg development, mating, and indiscriminate oviposition on any suitable host are similar to that of facultative myiasis-causing flies normally occurring in carrion. For the obligatory myiasis-causing oestrid flies, mating, host acquisition, and host–parasite interactions are more complicated. Because adults usually emerge from scattered sites, contact between males and females requires *aggregation behavior*. Species in each of the oestrid subfamilies aggregate at certain topographic sites, where mating takes place. This probably is the case for all oestrids. Some bot flies also mate near potential hosts. Where hosts are plentiful, aggregation behavior can cause crowding of adult bot flies, thus exposing the flies to predation and to less-than-ideal environmental conditions for survival. At these sites bot flies exhibit a male spacing behavior which counteracts crowding, interferes with intraspecific pairing, minimizes predation, and ensures that all available aggregation sites are occupied. Because of their short life span, adult oestrid flies do not feed, but live on fat reserves accumulated as larvae.

Bot flies display a high degree of *host specificity*, with each species parasitizing only one or, rarely, a few host species. Host species other than the native hosts are either intolerant or refractory to parasitism by a given species of bot fly. Although other host species are susceptible to bot fly invasion, this does not mean that they are suitable hosts or that mature larvae and adults are produced. Susceptible, but unsuitable, hosts tend to be species that are unrelated phylogenetically and ecologically to the native host species. Thus, dogs and cats are susceptible, unsuitable hosts for cuterebrine bot infections, whereas a given mouse species can be completely refractory to infection by the bot fly of a related mouse species. This high degree of host specificity reflects a long coevolutionary relationship between bot flies and their respective hosts.

Each bot fly species typically develops only at a specific site in its native hosts (e.g., nose bots, stomach bots, and foot bots). Even among the cutaneous bot flies (excluding *Dermatobia)* the location of warbles on the host occurs at specific anatomical sites. In nonnative and atypical hosts, however, this site occurrence can be erratic, often with grave consequences for the host and its parasites.

*Host–parasite interaction* involving bot flies is a delicate balance of tolerance by the host and limited pathology by the parasite. It also involves immunosuppression induced by some larvae. In bot fly infestations, excessive parasite burden usually is avoided by limiting the number of larvae to which a host is exposed. Bot fly females rarely dump their total reproductive complement on a single host; if this occurs, the host is likely to succumb to predators or other parasites due to the stress imposed by the developing bot larvae. Species that cause facultative myiasis show little adaptation to neutralizing host defenses. When present in large numbers, however, rapidly developing larvae overcome the nonspecific host response to tissue invasion characterized by self-grooming and inflammation. In cases of floor maggots and nest maggots, host resistance is avoided by the nature of their interrupted contact with the host.

The number of bot fly larvae infesting an individual host varies from one to perhaps several hundred, depending on the fly species involved. This is far less than the thousands of maggots that can make up the maggot mass in other myiasis-causing flies. Although much larger than other myiasis-causing flies, oestrid larvae appear to cause much less stress to their host. Although the reasons are unclear, this has the advantage of reducing the likelihood that the host will die during their development. Unlike oestrid larvae, most other myiasis-causing larvae can complete their development in the dead host.

In the case of facultative myiasis caused by nest maggots, floor maggots, and screwworms, most hosts show little evidence of *host resistance*. However, in the case of oestroid parasitism, hosts do show resistance but appear to be very tolerant of parasites. Immune resistance by hosts to bot fly–caused myiasis apparently lasts only a short time following an infestation.

First-instar bot larvae move from their point of host invasion to the site of development in their host. Studies have shown that this movement within host tissues is typically along predetermined pathways within connective tissue. This keeps the larvae from prolonged contact with the host's hemopoietic defenses and helps to explain the seeming lack of immune resistance. The wall of the warble is formed by the host's response to foreign body invasion. The warble wall more or less confines the developing larvae in a multilayered pocket of fibrous connective tissue. The wall is characterized by the presence of giant cells typical of a chronic inflammatory response. In addition, the developing larva secretes a bacteriostatic substance which prevents secondary microbial infection of the warble. In a native host, once the mature larva has dropped free, the empty warble collapses and heals rapidly.

Some hosts show exaggerated, usually futile, *behavioral responses* to certain bot flies in the act of ovipositing. For example, cattle panic when under attack by adults of the northern cattle grub (*Hypoderma bovis*). They respond by headlong flight with tail erect, called *flagging*, apparently attempting to outdistance the hovering fly. This response also is called *gadding*, a term derived from the Old English *gad*, a sharply pointed stick used to goad or prod livestock. Reindeer and caribou show a similar response to attacks by *H. tarandi*. In contrast, cattle display far less concern for the attacks by adults of the common cattle grub (*H. lineatum*). Gadding behavior also is shown by horses when under attack by the stomach bot fly (*Gasterophilus nasalis*), whereas horses usually pay little heed to oviposition by the common horse bot fly (*G. intestinalis*).

Another behavioral avoidance response is shown by deer and sheep when under attack by nose bot flies. A threatened host will lower its head and press its muzzle into a clump of grass or against the ground to deter strikes by the hovering female flies (e.g., *Oestrus ovis, Cephenemyia* spp.).

## MYTHICAL BELIEFS

A number of misconceptions have developed concerning bot flies and myiasis. One is the notion that adults of deer nose bots (*Cephenemyia* spp.) are capable of flying at speeds in excess of the speed of sound. The belief had its origin from a single questionable field observation of the deer nose bot fly coursing by a man standing on a hilltop. This tale gained some ill-founded credibility from entomologists and was published as fact for some time. The reasoning was that the bot fly needed to fly fast in order to overtake a swift running host. This reasoning ignored the fact that deer are incapable of running speeds in excess of 80 km/hr (50 miles/hr) and that the fly usually oviposits while the host is standing still. Although there have been no definitive studies of this subject, field observers conclude that *Cephenemyia* species probably cannot exceed a flight speed of 48 km/hr (30 miles/hr).

Another myth is that the larva of the rodent bot fly *Cuterebra emasculator* emasculates its squirrel and chipmunk hosts by consuming the host's testes as it develops. This tale arose because the specific site for larval development by this species is the inguinal area. Studies have shown, however, that the enlarging bot larva and warble in the scrotal sac of male hosts merely prevents the seasonal descent of gonads into the scrotum from the host's body cavity and does not cause sterility.

Because of the conspicuous, and sometimes self-destructive, gadding response of cattle and horses under certain bot fly attacks, the belief arose that the bot flies involved were stinging their hosts. The extensible terminal segments of the abdomen of *Gasterophilus* females suggest a formidable sting. However, these flies contact only individual host hairs during oviposition. They do not injure or even touch the skin surface. Hosts are frightened by the buzzing sound made by these flies as they hover, and when one individual in a herd bolts, the herd instinct prompts the same reaction from others.

There is a common belief in the southeastern United States that the flesh of squirrels and rabbits infected with wolves (*Cuterebra* spp.) is unfit to eat. Because of this, the entire squirrel carcass is often discarded. This too is a myth. In fact, skin-bot larvae are eaten as a delicacy by subarctic human cultures which traditionally herd reindeer in which bots commonly occur. Although the host flesh surrounding the warble may be discolored, there is no health hazard in eating either it or the bot larva. Squirrels and rabbits have long been a staple in the fall and winter diet of rural people in the United States. In some states the legal hunting season even has been postponed to begin after the bot season has ended to avoid the wasteful discarding of bot-infested game.

## FLIES INVOLVED IN MYIASIS

Flies which typically develop in dung or decaying plant matter generally are involved in gastrointestinal myiasis. A few muscids also are adapted for blood-feeding on nestling birds. The carrion-breeding flies and their more fastidious relatives, screwworms and flesh flies, are adapted for both facultative and obligatory myiasis. The bot flies (Oestridae) are all obligatory myiasis producers.

The families and groups of flies that follow include species which cause myiasis in humans and other animals. A more extensive listing of families, species, and the types of myiasis that they cause is presented in Table I. For reviews of this subject, see James (1947), Zumpt (1965), Leclercq (1969), and Hall (1995).

### STRATIOMYIDAE (SOLDIER FLIES)

The only species in this family that reportedly causes myiasis is the *black soldier fly* (*Hermetia illucens*). Although originally a New World species, it is widely distributed in warmer temperate and tropical areas of the world. The

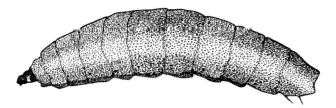

FIGURE 16.3   Black soldier fly, *Hermetia illucens*, larva (Stratiomyidae). (Original by E. P. Catts.)

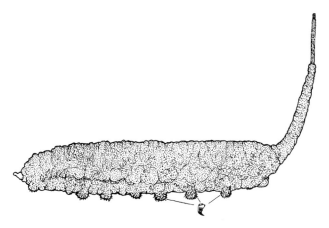

FIGURE 16.4   Rat-tailed maggot, *Eristalis tenax* (Syrphidae). (Original by E. P. Catts.)

larvae (Fig. 16.3) develop in decaying fruits, vegetables, and other plant material, decomposing animal carcasses, and excrement. Only a few human cases of intestinal myiasis involving *H. illucens* have been documented. At least one case resulted in rather severe enteric disturbances, apparently due to the large size and vigorous activity of the larvae (James, 1947).

### SYRPHIDAE (FLOWER FLIES, HOVER FLIES, RAT-TAILED MAGGOTS)

This is a large family (180 genera, 6000 species) that includes only a few taxa that cause gastrointestinal myiasis. Adults also are called *drone flies* because of their beelike appearance and resemblance to honeybee drones. The terms *flower flies* and *hover flies* refer to their common habit of visiting flowers for nectar and pollen and their ability to hover motionless in flight. The larvae of *Eristalis* (Fig. 16.4) and other aquatic genera are called *rat-tailed maggots*, referring to the long, telescopic, three-segmented respiratory tube at their posterior end by which they breathe at the water surface.

The syrphid species most frequently involved in myiasis is *Eristalis tenax*. Its larvae develop in sewage, liquefied excrement, decaying animal carcasses, and other decomposing plant and animal material of a liquid consistency. A number of human cases of gastrointestinal myiasis have been reported, with live larvae being passed in stools. Two other *Eristalis* species have been identified in human myiasis: *E. arbustorum* in Europe and *E. dimidiata* in the United States (James, 1947). Leclercq (1969) described several human cases involving rat-tailed maggots, including a man in Germany who passed more than 40 *Eristalis* larvae in one day and another 10–30 larvae

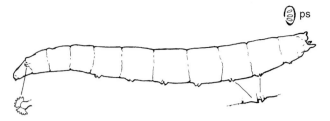

**FIGURE 16.5**    Cheese skipper, *Piophila* sp., larva (Piophilidae). (Original by E. P. Catts.)

**FIGURE 16.6**    Pomace fly or fruit fly, *Drosophila* sp., larva (Drosophilidae). (Original by E. P. Catts.)

on each of the following 6 days. *Eristalis* larvae also have been the cause of vaginal myiasis in cattle.

## PIOPHILIDAE (SKIPPER FLIES)

This is a small family of about 70 species in 35 genera worldwide. Females of the *common cheese skipper* (*Piophila casei*) oviposit on putrid, dried, cured, or smoked meats and cheeses, typically depositing 400–500 eggs per female. Adult cheese skippers are small (3–5 mm), slender, glossy black flies with yellow on the lower face and part of the legs. Larvae are slender, cylindrical, white, and truncated caudally with three pairs of short caudal protuberances, the ventral pair being the largest (Fig. 16.5). Larvae require about 5 days to develop under warm conditions. In temperate regions the mature larva overwinters. The name *skipper* originates from the ability of the larva to flex head to tail in a circle and, following total-body muscular contraction and release, the larva propels itself off the substrate for a considerable distance (up to 24 cm). This behavior is used as a means of escape when the larva is disturbed or is dispersing to suitable pupation sites. The pupal stage lasts about 5 days. The life cycle, egg to egg, can be completed in as little as 2 weeks.

*P. casei* is a widely distributed species which commonly infests cured meats and cheeses and dried fish. It probably is the species most commonly involved in gastrointestinal myiasis of humans. The tendency of people to leave cured meats and cheeses unrefrigerated makes these foods available to gravid females for oviposition. These flies can survive the rigors of alimentary tract passage and can even pupate and emerge as adults prior to leaving the host. The related *P. vulgaris* and *Stearibia* species are common inhabitants of dried carrion.

## NEOTTIOPHILIDAE (NEST SKIPPER FLIES)

This small Palaearctic family includes two genera: *Neottiophilum* and *Actinoptera*. The larvae cause cutaneous

myiasis by sucking the blood of nesting birds. Larvae attack with their mandibles and can penetrate deeply enough into their host to cause septicemia and death. Larvae pupate in the host's nest in the fall and emerge as adults in the spring. The adult is a yellow-brown fly, about 7–8 mm in length, with wings "pictured" with a few brown spots. Hosts typically are passeriform birds. Local strains of *Neottiophilum praeustum* show narrow host specificity to different avian species.

## DROSOPHILIDAE (POMACE FLIES, VINEGAR FLIES, FRUIT FLIES, AND WINE FLIES)

This is a large family (3000 species in 60 genera) of small red-eyed flies (1.6 mm) whose adults favor the odors of overripe or fermenting plant products, usually fruits. Larvae feed on microorganisms found in such substrates. The genus *Drosophila* is the largest and includes more than half of the species in this family. Larvae have posterior spiracles on paired caudal protuberances (Fig. 16.6) which also are evident in the puparia. The best-known species is the highly domesticated "kitchen gnat" *Drosophila melanogaster*. An additional seven species also are locally common domestic pests (e.g., *D. busckii*, *D. funebris*, *D. hydei*, *D. immigrans*, *D. repleta*, *D. simulans*, and *D. virilis*). The life cycle for these species is typically 12–14 days, making these small flies useful as biological models in studies of genetics, physiology, cytology, and population dynamics. Because of their attraction to fruits and vegetables, these species can cause accidental gastrointestinal myiasis.

## CHLOROPIDAE (GRASS FLIES AND AUSTRALIAN FROG FLIES)

The genus *Batrachomyia* includes 10 species whose larvae occur individually in swollen, subcutaneous pockets on the body (not the legs) of Australian frogs. The adult flies are yellow-brown in color and possess hairy eyes. Adults feed on plant juices. Their eggs require high humidity and are laid near, but not on, the host. After moving to a frog host, the larvae attack and appear to feed on blood, reaching a length of 10 mm when fully mature.

**FIGURE 16.7** Australian frog fly, *Bratrachomyia* sp., larva (Chloropidae). (Original by E. P. Catts.)

The mature larvae are peculiar in appearance, having paired anterior and posterior "tentacles," each bearing a spiracle (Fig. 16.7). Seasonal prevalence in frog populations can be as high as 25%, with a parasite load of one to four maggots per host. Death results in about 10% of the frogs at the time of larval drop.

## ANTHOMYIIDAE (ROOT MAGGOTS)

The Anthomyiidae and the following two closely related families, Fanniidae and Muscidae, are quite similar in their development, behavior, and occasional association with cases of gastrointestinal myiasis. Both the immatures and adults are similar in appearance to muscids. As the name "root maggot" implies, most anthomyiid larvae feed on plants. The anthomyiid genus *Hylemya*, however, is very large, with 180 species occurring in a diversity of habitats in the Nearctic region. The gastrointestinal myiasis which they can cause generally results from the ingestion of larvae infesting vegetables or fruits.

## FANNIIDAE (FANIID FLIES)

Four species of *Fannia* have been reported to cause myiasis: *little housefly* (*F. canicularis*), *latrine fly* (*F. scalaris*), *F. incisurata*, and *F. manicata*. Adults of these flies look like small, slender house flies. They are drab gray in color and lack black stripes on the thorax. The larvae commonly occur in feces, on rotting fruits or bulbs, and in bird nests. Larvae occasionally occur in older, somewhat dried carrion. The larvae have a characteristic, fringed appearance (Fig. 16.8) that easily distinguishes them from other muscoid maggots. The fringes apparently allow these maggots

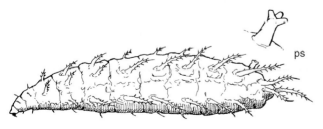

**FIGURE 16.8** *Fannia* sp., larva (Fanniidae); ps, posterior spiracle. (Original by E. P. Catts.)

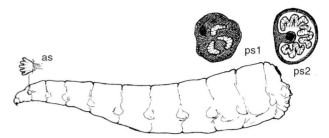

**FIGURE 16.9** Typical muscid larva (Muscidae), third instar. as, anterior spiracle; ps1, posterior spiracular plate, house fly (*Musca domestica*); ps2, posterior spiracular plate, false stable fly (*Muscina stabulans*). (Original by E. P. Catts.)

to float in a near-liquid medium. The life cycle requires about 1 month. Although cases of enteric and urethral myiasis in humans have been documented (James, 1947; Leclercq, 1969), the involvement of *Fannia* species in myiasis is rare.

## MUSCIDAE (DUNG FLIES)

The large family Muscidae includes at least seven genera in which species cause myiasis (Table I). Muscid larvae (Fig. 16.9) develop in a wide diversity of decaying organic matter, usually of plant or fecal origin. Occasionally they develop in old or buried carrion that is unsuitable for blow-fly exploitation. All stages of myiasis-causing muscids are typically house fly-like in appearance. Exceptions are *Neomyia* and *Hydrotaea* species, in which some adults have metallic coloration similar to blow flies, and the nest flies, which are larger and yellow to yellow-brown in color. Gastrointestinal myiasis caused by muscids usually results from oviposition on wet foods. It also may result from retroinfection through the host's anus following fly attraction to foul odors or soiling by feces. Urogenital myiasis may occur in association with purulent discharges, urine-soaked clothing, and secondary microbial infections.

The genus *Musca* includes about 60 species, which are confined mostly to the Old World. The two most important species that have invaded the New World are the *house fly* or *typhoid fly* (*Musca domestica*) (Fig. 16.9) and the *face fly* (*M. autumnalis*) in temperate regions. In the Old World tropics, the prevalent species is the *bazaar fly* (*M. sorbens*). Both the house fly and bazaar fly oviposit in a wide range of wet, decaying organic matter. Greenberg (1971) suggests that the house fly and bazaar fly were originally adapted, as larvae, to develop in wet ungulate feces. They do show preference for accumulations of animal excrement, especially that from the horse, cow, human, pig, and poultry. Their egg complement ranges from 1000 to 3000 per female and is laid over the adult life span in clusters of 120–150 eggs. The quality of larval and adult diets largely determines the number of eggs produced by any one fly. Maggots develop rapidly in 3–5 days under wet, warm conditions but are intolerant of desiccation. In the tropics, their life cycle can be as brief as 10–12 days, whereas 3 weeks is more typical in other regions. All stages can overwinter, but in colder areas there is a dramatic winter die-off. Larvae of *Musca* species that invade wounds feed primarily on necrotic tissues. *Musca* species with rasping-sucking mouthparts that feed on blood are not known to cause myiasis.

The genus *Muscina* includes eight species, three of which are implicated in accidental myiasis (*Muscina stabulans, M. assimilis,* and *M. pabulorum*). The *false stable fly* (*M. stabulans*) (Fig. 16.9) is the most important and is involved primarily in gastrointestinal myiasis. Occasionally their maggots occur in fetid sores or wounds. Adults of this species look much like house flies, except that they are usually larger and more robust. They are attracted to, and feed on, plant juices, rotting fruits, and insect-excreted honeydew. Females oviposit by scattering their 140–200 eggs on the surface of overripe, decaying fruit. They also oviposit on accumulations of dead insects or feces, usually from human sources, and on buried carrion. Early-instar larvae are saprophagous but become predaceous as they mature. Third-instar larvae prey on smaller maggots. This transition from a saprophagous to a predaceous habit has two advantages over species whose larvae remain saprophagous: first, the maturing larvae can store protein resources obtained from their prey to be used by the adult in reproduction; and second, this habit enables this species to exploit a wider range of protein-poor resources as a larval substrate. Larval development varies from 2 to 3 weeks. They usually overwinter as pupae.

*Hydrotaea* species are metallic-colored muscids usually found in association with feces and older carrion. Like *M. stabulans,* their dark, cream-colored larvae (up to 15 mm long) become predaceous on other maggots. In Australia, *Hydrotaea rostrata* has been implicated as a tertiary invader in myiasis of sheep but appears to feed only on necrotic tissues.

### Tropical Nest Flies

These muscid flies include the genera *Passeromyia, Mydaea,* and *Philornis.* Species in these three genera usually parasitize the young of cavity-nesting birds. Although few details are known about these flies, their biologies appear to be similar and are discussed collectively here. Species in the first two genera are widely distributed in the Ethiopian, Oriental, and Australian regions. The latter genus is found in the New World Tropics. Currently *Philornis* is the valid generic name for those flies formerly listed in the genus *Neomusca.*

Adults of these tropical nest flies feed on plant juices and wet feces. Gravid females oviposit in bird-nest debris near nesting birds. After the eggs hatch, the maggots crawl to the nestlings, scratch the skin with their mouth-hooks, and imbibe blood. As they feed, they continue to scratch and penetrate host tissues. In heavy infestations they can penetrate the host body cavity, with fatal results for the bird. The larvae develop rapidly, in less than 1 week, before leaving the host to pupate in the nest debris. Postfeeding larvae of *Philornis* species exude a frothy, sticky, salivary spittle which coats the puparium and to which camouflaging debris adheres. In spite of this defensive measure, pupae are subject to attack by parasitic wasps.

## CALLIPHORIDAE (BLOW FLIES, CARRION FLIES, FLOOR MAGGOTS, NEST MAGGOTS, AND SCREWWORMS)

The most generalized of the six families of Oestroidea is the Calliphoridae, with over 1000 species. Among the members of this large family there is a transition from the facultative myiasis habit by a large number of normally saprophagous species to obligatory myiasis by a relatively small number of species (ca. 100). The larvae typically feed on wet living or dead flesh. Desiccation is detrimental to both egg and larval survival. The following discussion treats the more important, widely distributed, and common genera of myiasis-causing calliphorids.

### Carrion-Associated Blow Flies

These are the showy metallic *blue-bottle flies, green-bottle flies,* and *black blow flies* (Fig. 16.10). They include members of the genera *Calliphora, Chrysomya, Cynomya, Eucalliphora, Lucilia, Paralucilia, Phaenicia, Phormia,* and *Protophormia,* which are commonly associated with dead animal tissues, or carrion. The Old World genus *Chrysomya* also includes one species that causes obligatory myiasis. The duration of the life cycle of these flies in carrion differs among species but, in general, roughly one-third of the time is spent as the eggs and larvae, one-third as pupae, and one-third as adults from emergence to mating and oviposition. Life cycles typically take 3 to 4 weeks but are prolonged by cold temperatures. These flies are attracted to fetid, purulent open sores and chronic nasopharyngeal or urogenital infections. Heavy larval infestations often result in death of the host, after which the maggots continue to feed on the resulting carrion.

The body form of most of these calliphorid larvae is typical of members of this family (Fig. 16.11) and is

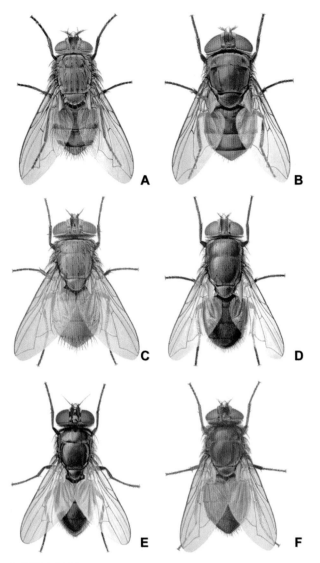

**FIGURE 16.10** Blow flies that cause myiasis, adults (Calliphoridae). (A) *Calliphora vicina;* (B) *Chrysoma megacephala;* (C) *Phaenicia sericata;* (D), *Phormia regina;* (E), *Ophyra leucostoma* (Muscidae); (F) *Protophormia terranovae.* (From Greenberg, 1971.)

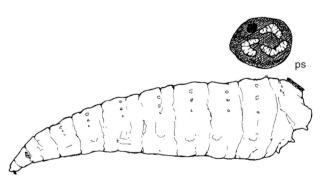

**FIGURE 16.11** Typical blow fly larva (Calliphoridae), third instar. ps, posterior spiracular plate. (Original by E. P. Catts.)

distinguished by the arrangement and shapes of their spiracles and cephalopharyngeal skeleton. The larvae of a few genera, however, are atypical in possessing numerous girdling bands of fleshy processes (e.g., *Chrysomya* spp.) similar to those of the genus *Fannia*. Like *M. stabulans*, the maggots of the *hairy maggot blow fly* (*Chrysomya rufifacies*) can switch from a myiasis to a predatory role. Postfeeding maggots drop free from the host or leave the carcass of a dead host to wander in search of pupation sites.

**Toad blow flies** *(Bufolucilia spp.).*
These flies include several species that cause primary obligatory myiasis in amphibians. Although they are fairly common in the Old World, *Bufolucilia silvarum* is the only North American species. The female lays her eggs on the back and flanks of a living toad, a risky activity for any fly. The resultant larvae invade the host's nasal passages or eye sockets, where they develop rather quickly, in less than 1 week. The host dies at about the time of larval maturation, after which some larvae may remain to feed on the dead host. Maggots thus also occur on dead amphibians. The metallic green adults are attracted to other carrion as well.

**Screwworms** *(Cochliomyia, Chrysoma spp.)*
The *New World screwworm* (*American screwworm*) (*Cochliomyia hominivorax*) and the *Old World screwworm* (*Chrysomya bezziana*) cause obligatory myiasis and can be of major economic importance. Both are *primary screwworms* with similar biology. Females, when gravid, are attracted to fresh open wounds on any warm-blooded animal. The female quickly oviposits 100–300 eggs on the dry perimeter of the chosen site, with each female producing up to 3000 eggs in her lifetime. Female flies commonly feed at the wound, thus obtaining protein for producing their next egg mass. After a brief incubation period of 10–20 hr, the eggs hatch and the larvae begin feeding on the open wound. The maggots develop rapidly to maturity in 4–12 days and then drop to the ground to pupate. After about a week as pupae, adults emerge, mate at aggregation sites, and after several days seek a protein meal and new living host. The entire cycle (egg to egg) takes 2–3 weeks.

Infested wounds range from mere thorn scratches or insect and tick bite marks to gaping lacerations. Infestations of the umbilicus of newborn calves by *Cochliomyia hominivorax* and *Chrysomya bezziana* are common. Livestock husbandry operations such as castrating, dehorning, branding, and shearing also cause wounds subject to invasion. Untreated screwworm cases can be fatal, due to the invasion of host's vital organs, septicemia caused by feeding maggots, or secondary infections. The maggots literally eat the host

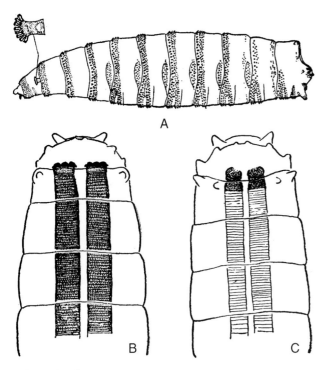

**FIGURE 16.12**  New World screwworm larvae (Calliphoridae), third instars. (A,B) Primary screwworm (*Cochliomyia hominivorax*), with enlargement of anterior spiracle and darkened portion of large tracheal trunks from posterior spiracles extending into three or four abdominal segments; (C) secondary screwworm (*C. macellaria*), with darkened portion of the tracheal trunks restricted to part of one abdominal segment. (Modified from James, 1947.)

alive. In areas of mild winters, adults can be active during warm spells. In temperate regions screwworm attacks are restricted to the warm seasons, but in the Tropics they are more or less continuous.

*Cochliomyia hominivorax* (Fig. 16.12A, B) is a major livestock pest, especially to cattle in the Neotropics. Although formerly it ranged throughout tropical and temperate regions of the New World, innovative control measures using male sterilization and baiting of females have eliminated it from the Nearctic. *Chrysomya bezziana* is distributed widely in Southeast Asia, New Guinea, and Africa. It is the Afro-Asian counterpart of *Cochliomyia hominivorax* and also is primarily important as a parasite of livestock. Like the American screwworm, this species is attracted to dry, open wounds as well as to most body openings. Generally eggs are deposited in the late afternoon such that their development is completed by the next morning, thereby avoiding lethal exposure to direct sunlight and drying. The rapidly developing maggots consume flesh in localized sites, often penetrating *en masse* deep into the host tissues, with fatal consequences.

Certain advantages are gained by these flies attacking only living hosts. There is less interspecies competition than among carrion-exploiting species, and the constant body heat of the host enhances maggot development. Additionally, the host tissues are less acidic and more digestible to the maggots than those of fresh carrion.

Species of *secondary screwworms* are close relatives of primary screwworms but cause facultative, rather than obligatory, myiasis. They are termed ''secondary'' because they infest wounds only after invasion by primary myiasis-causing flies. The most important species are *C. macellaria* (Fig. 16.12C) and *Chrysomya megacephala*. The latter species has become established in the Neotropics and appears to outcompete the former species. Another related blow fly is *C. rufifacies*. It is a widely distributed carrion-breeding fly that commonly shifts from being a scavenger in carrion to a predator of other maggots. However, on the island of Maui in Hawaii, maggots of *C. rufifacies* have been known to cause primary myiasis in the umbilicus of newborn calves, with prevalence of infestations as high as 30%.

**Nest blow flies**  *(Protocalliphora* spp.*)*
These flies are members of the large genus *Protocalliphora* (90 species), Holarctic flies whose maggots (Fig. 16.13) are obligatory parasites of nestling birds. The intermittent, temporary blood-sucking habits of these maggots are similar to those of the tropical muscid nest maggots. The maggots attach to the host by means of their mouthhooks to feed and then drop free to hide in nest debris while they digest their blood meals. Direct adverse effects on the nestlings are apparent only when maggot numbers are high. There is evidence, however, that their frequent bleeding of the host prolongs nestling development. The longer the time that birds stay in the nest, the more they are at risk of predation. Maggots thus can indirectly influence nestling success. Many species of *Protocalliphora* show high host specificity, indicating a long relationship with their respective hosts. These flies seem to favor cavity-nesting birds. Pupation and oviposition take place in the host nest. How the adult fly locates a specific host's nest is unknown.

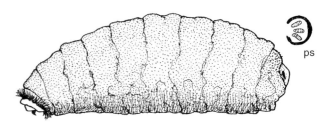

**FIGURE 16.13**  Nest blow fly, *Protocalliphora* sp., third-instar larva (Calliphoridae). ps, posterior spiracular plate. (Original by E. P. Catts.)

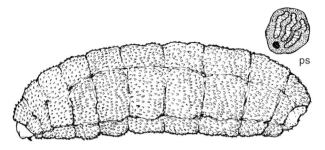

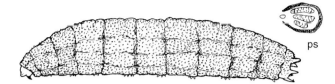

**FIGURE 16.15** Congo floor maggot, *Auchmeromyia senegalensis* (Calliphoridae), third instar. ps, posterior spiracular plate. (Original by E. P. Catts.)

**FIGURE 16.14** Tumbu fly, *Cordylobia anthropophaga*, third-instar larva (Calliphoridae). ps, posterior spiracular plate. (Original by E. P. Catts.)

**Tumbu fly** *(Cordylobia anthropophaga)*

The tumbu fly belongs to a small group of African blow flies whose maggots develop in warble-like cysts in hosts ranging from rodents to dogs and people. Rodents are assumed to be their primitive hosts, the flies having become secondarily adapted to other species. Except for their smaller size, these flies have biologies similar to the New World bot flies (Cuterebrinae), in many ways suggesting convergent evolution or a distant relationship.

The adult tumbu fly is yellow-brown in color with two rather broad, but variable, dorsal thoracic stripes. Mature maggots are up to 15 mm in length and densely, but incompletely, covered with small, backwardly directed, single-toothed spines. The posterior spiracles have a weakly sclerotized peritreme and three sinuous slits (Fig. 16.14). Adults feed on decaying fruits, carrion, and feces. Females are shade-loving and deposit eggs singly in dry sand or dirt contaminated with host urine or feces. Females also are attracted to dry, urine-soiled diapers or clothing. They lay up to 500 eggs over their lifespan of 2–3 weeks. The eggs hatch after several days, following which the first-instar larvae wait in the dry substrate for a host. Contact with a host stimulates the maggots to attach and penetrate the skin. Maggots develop in shallow warbles within or just beneath the skin in about 7–10 days and drop free to the ground to pupate in surface debris. Adults emerge after another several weeks. Although the tumbu fly invades a wide range of hosts, its successful development varies significantly among different host species. The domestic dog is an important reservoir, but maggots develop best in native rodents.

**Congo floor maggot** *(Auchmeromyia senegalensis)*

The Congo floor maggot (formerly *Auchmeromyia luteola*), is one of four or five species in this genus causing obligatory, temporary myiasis similar to that caused by *Protocalliphora* species. They all occur in Africa south of the Sahara, and most are associated with the burrows of larger mammals such as warthogs. *A. senegalensis*, however, appears to prefer humans as hosts. Adults are yellow-brown in color with markings similar to the tumbu fly. They feed on rotting fruits and feces (e.g., human, monkey, pig). Females can lay up to 300 eggs. The maggots (Fig. 16.15) spend most of their time hidden in loose dirt and floor debris of native huts. At night they crawl to sleeping hosts, scrape or otherwise break the skin, and suck the oozing blood. After feeding for about 20 min the maggots return to hide in debris until the next night. They require two blood meals for each of the three instars and pupate in debris. The life cycle takes about 10 weeks.

**Deer and water buffalo skin maggots** *(Boopona* spp.*)*

These myiasis-causing flies include four species of yellow–brown blow flies whose maggots parasitize the skin of the back and feet of cervids and bovids in eastern Europe, Asia, and the Orient. In the case of cervids, they also attack the soft, developing antler buds. Their eggs are attached singly to hairs of the host and require 3–5 days to develop prior to hatching. These maggots invade the host skin, where they develop individually in warble-like boils in about a 1-week period. Mature maggots drop from the host to the ground to pupate, and adults emerge in 2–3 weeks.

**Elephant skin maggot** *(Elephantolomeus indicus)*

This species is a small, orange–brown blow fly whose maggots develop only in warble-like boils in the skin of the Asian elephant. Little is known of its biology.

## SARCOPHAGIDAE (FLESH FLIES)

Species of this large, widely distributed family are classified into two subfamilies: the Miltogrammatinae, which, with few exceptions, are obligatory parasitoids of insects and other arthropods; and the Sarcophaginae, with necrophagous species that include facultative and obligatory parasites causing myiasis. Adults (Fig. 16.16) are typically medium to large, black and gray flies with longitudinal thoracic stripes and a checkered, or tessellated, abdominal pattern. All sarcophagid species are

**FIGURE 16.16**   Red-tailed flesh fly, *Sarcophaga hemorrhoidalis* (Sarcophagidae), adult. (From Greenberg, 1971.)

*larviparous;* the gravid female retains the eggs in an expanded, bilobed, uterine pouch until they are ready to hatch. Females produce 30–200 larvae, depending on the species involved. Flesh fly larvae (Fig. 16.17) are more robust than blow fly larvae and possess paired mandibles in all instars. Their posterior spiracles are recessed in a deep cavity, and the inner slit of each spiracle is parallel to, or slants away from, the ventral midbody line.

*Sarcophaga* species usually are associated with carrion or feces but can cause facultative wounds and accidental gastrointestinal myiasis. About 20 *Sarcophaga* species have been incriminated in cases of gastrointestinal

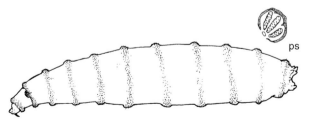

**FIGURE 16.17**   Flesh fly, *Sarcophaga* sp. (Sarcophagidae), third-instar larva. ps, posterior spiracular plate. (Original by E. P. Catts.)

myiasis. One widely distributed species is the *red-tailed flesh fly* (*Sarcophaga haemorrhoidalis*), which frequents feces and is attracted indoors by fecal odors. Only a few species in other sarcophagine genera have been recorded as causing myiasis of the gastrointestinal or wound type.

Few flesh flies cause obligatory myiasis. The most widespread and important species are in the miltogrammine genus *Wohlfahrtia*. They include the Old World species *W. magnifica* and the New World species *W. opaca* and *W. vigil*. These three species have evolved as primary invaders. Their females larviposit at moist body openings and at fresh wounds or scratches. Larvae can even penetrate thin, unbroken skin. The maggots burrow into the subcutaneous tissue to feed, inducing the formation of a boil-like cyst around groups of larvae with a small, common breathing pore opening to the outside.

*Wohlfahrtia* adults resemble large house flies with very distinct, longitudinal, thoracic stripes. Unlike other flesh flies, these have the abdomen clearly marked with black spots. A gravid *Wohlfahrtia* female produces 120–170 larvae. In a host, the maggots grow rapidly and can cause considerable tissue destruction. After about 1 week, maggots drop to the ground to pupate and can overwinter in this stage.

The closely related *W. nuba* is not parasitic, but feeds only on necrotic flesh. It has been used successfully in treating ragged, infected wounds, similar to the use of *Phaenicia sericata* (Calliphoridae). Most other *Wohlfahrtia* species are scavengers, but all are probably capable of at least facultative myiasis.

A group of four unrelated genera of flesh flies cause facultative, and apparently obligatory, myiasis of certain amphibians and reptiles. In some cases they also attack amphibian and reptilian eggs, killing the developing embryos. These genera are *Anolisimyia*, *Cistudinomyia*, *Eumacronychia*, and *Metoposarcophaga*. Little is known of their biology.

## OESTRIDAE (BOT FLIES)

Bot flies are the most highly evolved group of obligate myiasis-causing parasites of mammals. They are treated as four distinct subfamilies in the Oestridae. The most primitive are the Cuterebrinae, the *New World skin bot flies*. Their counterparts are the Hypodermatinae, the *Old World skin bot flies*. The *nose bot flies* are in the Oestrinae, with their probable center of origin being in Africa. The remaining subfamily is Gasterophilinae, the *stomach bot flies*, which also appears to have evolved in Africa. All four of these subfamilies were recognized previously as families and are treated as such in earlier literature. The subfamilies of Oestridae can be separated as third-instar larvae by their general appearance and by the form of their caudal spiracular openings. Bot-fly maggots are thick, robust, grublike larvae with moderate to heavy spiny armature. As with most other flies discussed in this chapter, they pass through three larval instars, and they drop to the ground to pupate. Except for members of the Cuterebrinae, all bot flies overwinter as larvae in the host. The cuterebrines typically overwinter as diapausing pupae free from the host, although there are exceptions in subtropical and tropical regions.

Bot flies differ from other obligatory myiasis producers in several ways. First, the adults do not feed or take in nutrients. Most of them have only rudimentary, nonfunctional mouthparts and are unable to feed. Those with functional mouthparts and an associated alimentary tract probably only imbibe water to maintain an internal fluid balance. Second, bot flies either show a high degree of host specificity or they parasitize only a small group of related hosts. Although some bot fly maggots occasionally occur on atypical hosts, the susceptibility of a host does not necessarily imply the suitability of that host for normal, or successful, bot fly development. Third, bot fly maggots show a marked level of site specificity in a normal host. In abnormal hosts, site specificity can be erratic and can lead to dire results for both host and parasite. Fourth, the site of invasion by the first-instar bot maggot generally is not the site of maggot development. With the exception of *Dermatobia*, first-instar maggots of oestrid flies move from the point of invasion to a different site for further development. Interaction between the host and its developing bot fly maggots is generally benign, with the associated pathology and parasite burden being tolerated well by native, coevolved hosts. Bot fly maggots generally cause little injury to their hosts at low to moderate population levels.

Humans are not among the normal hosts for any bot fly species, including the so-called human bot fly (*Dermatobia hominis*). However, people may become incidentally infested by bot flies under certain circumstances. In such cases, the associated pathology tends to be more severe than that of their normal hosts.

Burrowing first-instar bot fly larvae occasionally cause paralysis or death of the host. Developing larvae located in warbles at critical sites such as around the eyes and on the feet can increase the risk of predation by interfering with the host's ability to see or escape. Small mammalian hosts encumbered by an ever-enlarging cluster of warbles also may have difficulty in foraging.

Another characteristic of bot flies is that the bee-like adults usually aggregate at specific topographic sites for pairing and copulating. Favored sites are hilltops, cliff faces, steep slopes, prominent rocks or trees, and streambeds. Male flies remain at these sites throughout their brief life, but females leave the sites soon after mating to search for suitable hosts or oviposition sites.

The major importance of bot flies is the economic losses that they cause in livestock operations (e.g., cattle, sheep, goats, reindeer, and horses). Secondary microbial infection of the bot warble is rare because bot fly maggots produce a bacteriostatic secretion as they develop. However, after the larva exits the warble, other myiasis-causing flies may exploit the empty wound. Bot fly maggots cannot complete their development in a dead host. If the host dies, so do its bots.

The evolutionary history of bot flies is not known but warrants comment. Zumpt (1965) proposed two possible routes for bot fly evolution. One route is through blood-sucking larvae such as nest maggots or floor maggots. The other is through carrion-breeding species and screwworms. Both alternatives seem plausible with regard to skin bots, but they do not explain how the more internally adapted groups such as nose bots and stomach bots originated. The nose bots may have evolved from myiasis-causing flies that were attracted to mucopurulent nasal secretions in hosts suffering from respiratory infections. Stomach bots, on the other hand, may have originated from fly species infesting decaying, fermenting forage, a diet favored by many large herbivores. In any case this obligate form of parasitism appears to have arisen independently among different groups of flies associated with certain hosts.

## NEW WORLD SKIN BOT FLIES (CUTEREBRINAE)

There are 6 genera and 83 species in this subfamily of bot flies, all restricted to the Western Hemisphere. The largest genus is *Cuterebra* (ca. 70 species), which includes the largest bot flies. Some of these robust, thick-bodied bot flies (Fig. 16.18) are up to 30 mm in length. Their

**FIGURE 16.18** Rodent bot, *Cuterebra fontinella*, adult female (Oestridae, Cuterebrinae); reared from white-footed mouse, *Peromyscus leucopus*. (Photo by Sturgis McKeever.)

**FIGURE 16.19**    Cotton mouse, *Peromyscus gossypinus*, with mature rodent bot, *Cuterebra* sp., or "wolf"; posterior end of bot, with posterior spiracles exposed, projecting from location at base of host tail. (Photo by G. R. Mullen.)

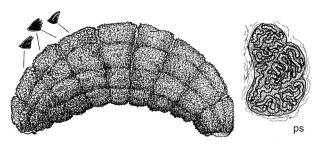

**FIGURE 16.20**    Rodent bot, *Cuterebra* sp. (Oestridae, Cuterebrinae), third-instar larva. ps, posterior spiracular plate. (Original by E. P. Catts.)

Generally there is little economic importance associated with *Cuterebra* species, although they can be a seasonal problem in commercial rabbit operations. Sport hunters often discard bot-infested squirrels (Fig. 16.21) and rabbits in the erroneous belief that the carcasses are spoiled by the presence of these maggots. A few

normal hosts are rodents (e.g., *Microtus, Neotoma,* and *Peromyscus* spp.) and lagomorphs (e.g., *Lepus* and *Sylvilagus* spp.). At temperate latitudes cuterebrine maggots show seasonal peaks in prevalence. For example, 40% prevalence in *Peromyscus* populations is not unusual during late summer. Nonnative rodents and rabbits (e.g., *Mus, Rattus, Cricetus,* and *Oryctolagus* spp.) also are parasitized, but in these hosts the pathology is more severe and can lead to death of both the host and parasite.

*Cuterebra* species oviposit in areas close to the center of host activity (e.g., near nests and lairs or along runs). After about 1 week, the eggs hatch in response to a sudden increase in temperature, normally indicating a nearby warm host. First-instar maggots adhere to the host pelage, crawl to natural body orifices of the head, and penetrate the mucosal tissue at such sites as the mouth and nose. After about 1 week in the pharyngeal areas of the host, the maggots actively burrow through sheets of host connective tissue to a species-specific, cutaneous site for maggot development (Fig. 16.19). Once there, the maggot cuts an opening through the skin and molts within the newly formed warble. Depending on the bot species involved, maggot development at this site requires 3–8 weeks, during which time the much enlarged maggot can increase some 100,000-fold in weight. When mature, the third-instar larva (Fig. 16.20) backs out through the warble pore and drops to the ground to pupate. After the bot exits, the collapsed warble heals quickly, usually without secondary infection. In cool, temperate regions it is the pupa that diapauses and overwinters, and there is but one adult flight season per year. Warmer areas probably have two flight seasons, where adults are on the wing during late spring and summer. The adult life span is about 2 weeks.

**FIGURE 16.21**    Multiple "wolves" of squirrel bot (*Cuterebra* sp.) in shoulder area of gray squirrel, *Sciurus carolinensis*. (Courtesy of Department of Pathobiology, Auburn University College of Veterinary Medicine.)

*Cuterebra* species have been colonized in the laboratory and can serve as natural bot-host models for the study of bots affecting livestock.

### Tórsalo *(Dermatobia hominis)*

The *tórsalo* is a Neotropical species that occurs widely from southern Mexico to Argentina. Although primarily a pest of cattle, it also infests humans, monkeys, sheep, dogs, other domestic and wild mammals, and occasionally birds (e.g., macaws). This is the only bot fly that frequently parasitizes humans, hence its alternative common name, the *human bot fly.*

It is a woodland species encountered along forest margins of river valleys and lowlands. It is unusual among cuterebrine flies because of its unique oviposition behavior and means of egg dispersal. Rather than deposit eggs directly on a host, the adult female (Fig. 16.22) captures various zoophilic or anthropophilic arthropods, usually dipterans, and glues her eggs in clusters (15–45 eggs) to their abdomen (Fig. 16.23). Embryonization requires 5–15 days. These egg carriers, or "porters," subsequently transport the eggs to a vertebrate host, where they hatch while the arthropod feeds. Among the more common porters are day-flying mosquitoes (particularly *Psorophora* spp.) and muscid flies (e.g., *Sarcopromusca, Stomoxys,* and *Synthesiomyia* spp.). The newly emerged larvae enter the skin either through the bite puncture or via hair follicles, soft folds of skin, or areas of moist skin in contact with clothing or bedding.

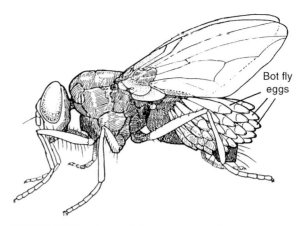

**FIGURE 16.23** A muscid fly "porter" (*Sarcopromusca arcuata*) to which a tórsalo bot fly (*Dermatobia hominis*) has attached her eggs. (Original by E. P. Catts.)

Development occurs at the point of entry, forming a boil-like pocket, or *furuncular lesion,* where the larva (Fig. 16.24) undergoes three or four instars. This development usually takes 5–10 weeks but sometimes takes as long as 3 months or more. During this time the narrower posterior end of the larva is extended into the opening at the skin surface where it exchanges air in

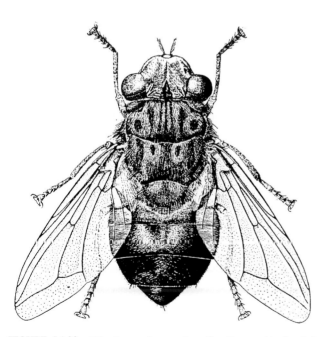

**FIGURE 16.22** Tórsalo, or human bot fly, *Dermatobia hominis* (Oestridae, Cuterebrinae), adult female. (From James, 1947.)

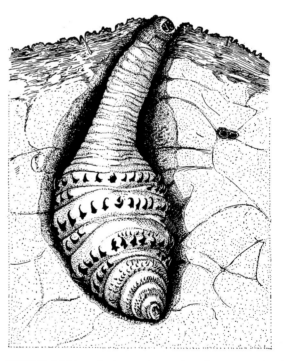

**FIGURE 16.24** Tórsalo, or human bot fly, *Dermatobia hominis* (Oestridae, Cuterebrinae), mature larva in human skin, with posterior spiracles exposed through hole in skin surface. (From Craig and Faust, 1940.)

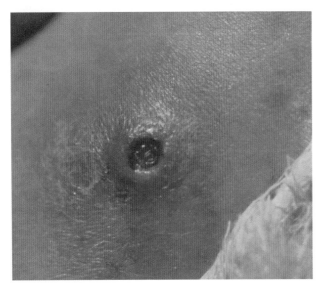

**FIGURE 16.25**    Tórsalo, or human bot fly, *Dermatobia hominis;* posterior end of maturing larva visible in hole made by larva in skin on human forearm. Note inflammation and swelling at site of wound. (Photo by Ronald D. Cave.)

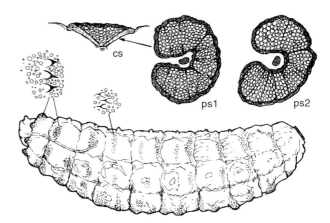

**FIGURE 16.26**    Cattle grubs, *Hypoderma* spp. (Oestridae, Hypodermatinae), third instars. cs, cross section of posterior spiracle of *Hypoderma bovis,* showing depressed center which contrasts with flat spiracle of *H. lineatum;* ps1, northern cattle grub (*H. bovis*), third instar; ps2, common cattle grub (*H. lineatum*), third instar. (Original by E. P. Catts.)

respiration (Fig. 16.25). After the larva matures, it enlarges the opening and drops to the ground to pupate. The pupal stage lasts 14–24 days.

### Other cuterebrine flies

The remaining four genera of cuterebrine flies are tropical and include only a few species. Most of them parasitize rodents, and little is known about their biology. However, the genus *Alouattamyia*, with the single species *Alouattamyia baeri*, parasitizes howler monkeys (*Alouatta* spp.) in Central and South America. This is the only bot fly specific to a primate host. Warbles of this species usually are located in the cervical and axillary regions of their arboreal hosts.

## OLD WORLD SKIN BOT FLIES (HYPODERMATINAE)

These flies are the Eurasian counterpart of the New World skin bots. There are 8 genera and 30 species occurring in rodents, deer, goats, and cattle. All but 2 species are found only in Asia, Europe, and Africa. The most widespread and important species are in the genus *Hypoderma*, which includes 7 species, 5 causing myiasis in cervids and 2 in bovids.

### Cattle grubs (*Hypoderma* spp.)

The *northern cattle grub* (*Hypoderma bovis*) and the *common cattle grub* (*H. lineatum*) (Fig. 16.26) are Holarctic in distribution, having been introduced

wherever cattle are raised. They are major economic pests of domestic cattle. Losses include damage to hides and self-injury by hosts during headlong flights of panic or gadding (Fig. 16.27), in futile attempts to escape ovipositing flies. Thus the name *gad fly* is often used to refer to the adults. Adults of *Hypoderma* species also are called *heel flies*, referring to the defensive behavior of cattle in kicking up their hooves.

The biology of *H. bovis* and *H. lineatum* is very similar. In late spring and early summer, adult females (Fig. 16.28) of both species glue their eggs directly to host hairs (Fig. 16.29). *H. lineatum* usually oviposits on the lower body regions of standing or resting hosts,

**FIGURE 16.27**    Gadding behavior of calf in response to attack by heel fly (*Hypoderma bovis*), with tail raised and calf kicking up heels of hind legs while running frantically about. (Photo by J. Weintraub, Agriculture and Agri-Food Canada, Lethbridge.)

**FIGURE 16.28** Heel fly, *Hypoderma bovis* (Oestridae, Hypodermatinae), adult female. (Courtesy of USDA-ARS, Livestock Insect Research Laboratory, Kerrville, TX.)

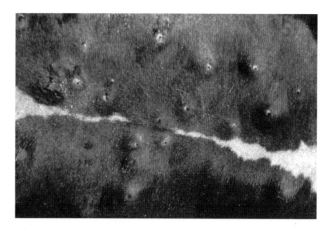

**FIGURE 16.30** Multiple warbles along back of cow, caused by *Hypoderma bovis* (Oestridae, Hypodermatinae). (Photo by J. Weintraub, Agriculture and Agri-Food Canada, Lethbridge.)

whereas *H. bovis* oviposits in the same regions on active hosts. Presence of the latter species is what causes cattle to gad. After an incubation period of 3–7 days, eggs hatch and the first-instar larvae crawl to the base of the hairs on which the eggs were glued. They then penetrate the host skin using their hooklike mouthparts and proteolytic enzymes. A 4- to 6-month period of burrowing follows as the larvae make their way between sheets of connective tissue within the host. During the winter, first-instar larvae of *H. lineatum* eventually cluster along the esophagus, whereas the larvae of *H. bovis* cluster along the spinal column. With the onset of spring the larvae leave these sites and move to the host's back, where they cut a hole, the warble pore, and develop through two subsequent larval instars.

A boil-like warble develops around the enlarging maggot (Fig. 16.30). Mature maggots back out of the warble pore and drop to the ground to pupate. As with all bots, it is at this time of dropping from the host that maggots are most vulnerable to predation by birds, rodents, and insectivores. The pupal stage lasts 1–3 months. Adult flies are on the wing throughout late spring and early summer.

Another *Hypoderma* species worthy of note is *H. tarandi* (formerly *Oedemagena tarandi*). In the arctic and subarctic regions, *H. tarandi* causes cutaneous myiasis in the backs of reindeer and caribou, similar to that caused by cattle grubs. Heavier infestations generally occur in yearling fawns rather than in other age groups. Over time infested animals slowly develop partial immunity to these parasites. *Oestromyia* species (Fig. 16.31) commonly cause cutaneous myiasis in wild rodents and lagomorphs in Europe, Asia, and the Far East.

**FIGURE 16.29** Eggs of heel fly, *Hypoderma bovis* (Oestridae, Hypodermatinae), attached to body hair of cow. (Courtesy of Agriculture and Agri-Food Canada, Lethbridge.)

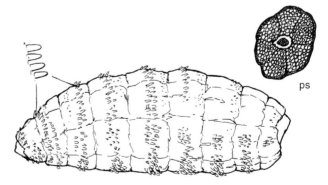

**FIGURE 16.31** *Oestromyia* sp. (Oestridae, Hypodermatidae), third-instar larva; causes myiasis in wild rodents and lagomorphs. ps, posterior spiracular plate. (Original by E. P. Catts.)

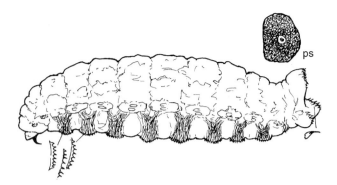

**FIGURE 16.32**   Kangaroo throat bot, *Tracheomyia macropi* (Oestridae, Oestrinae), third-instar larva; develops in trachea of the red kangaroo, Australia. ps, posterior spiracular plate. (Original by E. P. Catts.)

## NOSE BOT FLIES (OESTRINAE)

This subfamily of oestrid flies includes 9 genera and 34 species which parasitize members of the mammalian orders Artiodactyla, Perissodactyla, and Proboscidea (elephants). Most nose bot flies are African or Eurasian in distribution; an exception is the Holarctic genus *Cephenemyia*. The genera *Oestrus* (6 species) and *Rhinoestrus* (11 species) comprise the majority of species in this group. Another species, *Tracheomyia macropi* (Fig. 16.32), develops in the trachea of the red Kangaroo in Australia and is the only native bot of that continent.

Nose bot flies differ from other bots in that their eggs develop *in utero*. The first-instar larvae are squirted by the hovering female directly into the muzzle or eye of the host. The larvae crawl down the throat to enter tracheal branches of the lungs, but soon return to the nasal sinuses or pharyngeal region of the host to complete their development. As with other bots in native hosts, there is little pathology at moderate parasite levels. However, purulent mucous exudates associated with an abundance of maggots may lead to respiratory complications or to secondary fly attack. While hosts are under attack by adult nose bots, they stop grazing and attempt to thwart the attack by pushing their muzzles into bushes or clumps of grass. Following development, mature larvae are sneezed from the nostrils of the host, causing some temporary suffering during this time. Occasionally a few larvae may become lodged in the nasal sinuses and can cause the death of their host. After a pupal period of 4–6 weeks, adults emerge and seek a mate at aggregation sites. Adults generally are univoltine in cold regions and at least bivoltine in tropical and warm temperate areas. This indicates that larval development can be delayed during winter and accelerated during summer. Overwintering takes place within the host, as is the case of most other bot flies.

The most widely distributed and economically important species is the *sheep nose bot* (*Oestrus ovis*),

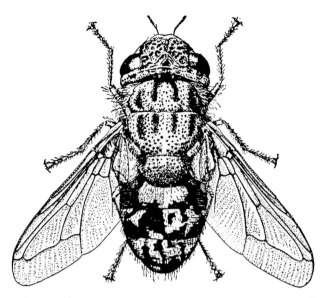

**FIGURE 16.33**   Sheep bot fly, *Oestrus ovis* (Oestridae, Oestrinae), female. (From James, 1947.)

which parasitizes domestic and wild sheep and goats (Figs. 16.33 and 16-34). Females produce about 500 progeny, which they larviposit in batches of 10–20 at a time. The usual host burden is 12–24 maggots, with gradual attrition reducing this number to fewer than 10 survivors by the time the maggots mature. Other *Oestrus* species parasitize antelopes in Africa.

The *horse nose bot* (*Rhinoestrus pupureus*) is distributed widely throughout Eurasia, Africa, and the Orient. It is most prevalent in Asia, where high population levels of these bots in domestic horses have been recorded (>700 maggots in a single host). High parasite loads can cause death of the host. Ocular myiasis in people who live near, or handle, horses is not uncommon. The general life history of the horse nose bot is like that of *O. ovis*. *Rhinoestrus* includes 11 species, 4 in equids and 7 others

**FIGURE 16.34**   Sheep bot, *Oestrus ovis* (Oestridae, Oestrinae), third-instar larva. (Courtesy of P. Scholl.)

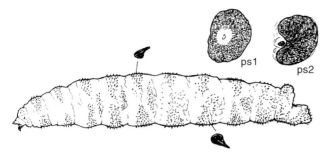

**FIGURE 16.35** Deer nose bot, *Cephenemyia jellisoni* (Oestridae, Oestrinae), third-instar larva. ps, posterior spiracular plate. (Original by E. P. Catts.)

in specific nonequine hosts (e.g., giraffe, antelope, bush-pig, and warthog).

Seven species in the genus *Cephenemyia* (Fig. 16.35) parasitize cervids (deer). Only one, *Cephenemyia trompe*, is distributed throughout the northern Holarctic region. The others are confined to either the Nearctic or Palearctic areas. All are *deer nose bots*. The four common species in North America are *C. phobifer* to the east of the Continental Divide and *C. apicata, C. jellisoni,* and *C. pratti* to the west. Although *C. trompe* is named the *reindeer throat bot,* it occurs in deer, moose, and caribou, as well as reindeer. The life history of these bots is like that described for other nose bots. Mature maggots usually crowd the retropharyngeal pouches in the throat of their host. On completing their development they crawl to the anterior nasal passages, where they are expelled by sneezing and pupate on the ground under surface debris.

The incidental occurrence of several oestrine species in the eyes of atypical hosts suggests that the orbit also may serve as a target for certain larvipositing bot flies. In Africa, first-instar larvae of *Gedoelstia cristata* regularly occur in the eyes of native wildebeest and hartebeest hosts. Later instars are found in the nasal cavities. Ocular invasion by this species in domestic sheep, goats, cattle, and horses produces gross pathological lesions and high levels of mortality. *Cephalopina titillator* (Fig. 16.36) is

the nose bot of camels and dromedaries from Central Asia to Africa. Its life history is like that of other nose bots.

## STOMACH BOT FLIES (GASTEROPHILINAE)

Adult flies of this group, represented by 15 species, resemble honey bees in their general size and color. The largest genus is *Gasterophilus,* the *horse stomach bot flies,* with 9 species, 3 of which have worldwide distribution (*Gasterophilus intestinalis, G. nasalis,* and *G. haemorrhoidalis*). These parasites are common companions of horses and donkeys wherever these hosts occur. Few horses live to old age without having carried their load of stomach bots along the way.

Following oral entry, developing maggots attach to the gastrointestinal mucosa (Fig. 16.37), causing inflammation, sloughing of tissue, and ulcerations. They do not cause warble formations. Burrowing of first-instar larvae in the mouth lining, tongue, and gums can produce pus pockets, loosen teeth, and cause loss of appetite of the host. Producing a gadding response to hovering, ovipositing females can cause self-inflicted injuries as well. However, *G. nigricornis* does not hover but alights on the host's cheek to oviposit, and *G. pecorum* oviposits on leaves and stems of potential host forage.

Most of these flies glue their eggs to the hair shafts of the host's body. Because of their brief life span of a few days at most, females can deposit all of their eggs within a couple of hours if hosts are available and the weather is mild. Their maggots develop in the stomach and intestines of equids, elephants, and rhinoceroses, suggesting a very old coevolutionary relationship to these

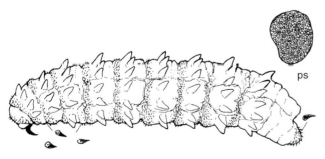

**FIGURE 16.36** Camel nose bot, *Cephalopina titillator* (Oestridae, Oestrinae), third-instar larva. ps, posterior spiracular plate. (Original by E. P. Catts.)

**FIGURE 16.37** Common horse stomach bot, *Gasterophilus intestinalis* (Oestridae, Gasterophilinae); larvae attached to mucosa and inner surface of stomach of heavily infested horse. (Photo by Martin Hall, © The Natural History Museum.)

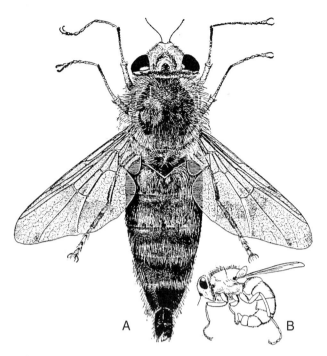

**FIGURE 16.38**   Common horse bot fly, *Gasterophilus intestinalis* (Oestridae, Gasterophilinae), adult female. (A) Dorsal view; (B) lateral view. (From James, 1947.)

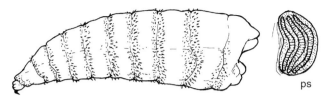

**FIGURE 16.39**   Common horse stomach bot, *Gasterophilus intestinalis* (Oestridae, Gasterophilinae), third-instar larva. ps, posterior spiracular plate. (Original by E. P. Catts.)

distantly related mammalian groups. Eggs which are ingested with forage or wetted by self-grooming of the front legs hatch after a brief incubation. Eggs located on the host's head hatch spontaneously. Upon burrowing into the oral tissues for a brief period, first-instar larvae eventually are swallowed and attach to the stomach or intestinal wall. The site of attachment is specific for each fly species. Maggots overwinter in the gastrointestinal tract. After larval development is completed, the mature larvae are expelled with the host feces during the warmer seasons. Pupation occurs in the soil soon after larvae drop from the host, with the pupal stage lasting about 3 weeks. Adult flies emerge, mate, and quickly resume activity near potential equine hosts. If hosts are not available, the bot flies move to high points to aggregate and mate, following which the females initiate a longer-distance search for hosts.

The *common horse stomach bot fly* (*G. intestinalis*) (Figs. 16.38 and 16.39) is worldwide in distribution and is the predominant species in North America. It prefers to oviposit on the lower forelegs of horses. The two other species in North America are the *throat horse bot* (*G. nasalis*) and the rarer *nose horse bot* (*G. haemorrhoidalis*). The former oviposits on the hairs of the chin and lower jaw and the latter on the hairs of the nose and lips. The *dark-winged horse bot* (*G. pecorum*) is the most commonly encountered species in Eurasia and Africa. It is the most pathogenic fly in this genus and can cause host fatalities

resulting from constricted swelling of the esophagus due to attached maggots.

Other genera of stomach bots include *Gyrostigma* (three species) in rhinoceroses and *Cobboldia* (three species) parasitizing African and Indian elephants. Their life histories are similar to that of *Gasterophilus*. Two other obligate-myiasis fly species are associated with the African elephant. Both are cutaneous parasites. *Ruttenia loxodontis* develops in warble-like skin boils on the buttocks and flanks of elephants, whereas *Neocuterebra squamosa* develops in shallow ulcers in the skin crevices of the elephant's feet. Both produce a *pseudowarble* during their development. The phylogenetic relationship of these unplaced genera to other bot flies is uncertain. Zumpt (1965) suggests that both represent early gasterophilines.

## OTHER OESTROID FLIES

There are three other families of flies included in the Oestroidea in which the species do not parasitize vertebrate animals: Tachinidae, Rhinophoridae, and Mystacinobiidae. All members of the Tachinidae are obligate parasitoids of other insects. None is involved in myiasis. The Rhinophoridae are a small sister group of the tachinids which parasitize isopods. Mystacinobiidae, a sister group of the calliphorids, includes but a single species in New Zealand that is coprophagous on bat guano and phoretic on the bats themselves. None of these families includes myiasis-causing species.

## PUBLIC HEALTH IMPORTANCE

Myiasis is a relatively uncommon ailment among people worldwide, occurring only seasonally and sporadically in temperate regions and being associated with wet seasons in the tropics. Most human cases of facultative or obligatory myiasis are only temporary or are aborted because humans are unsuitable hosts (e.g., nose or stomach bots). Maggots usually do not complete their development in people because it is interrupted by self-grooming or medical treatment. Infants and infirm or debilitated older age groups are generally at greater risk because of associated difficulties in maintaining a minimal level of personal hygiene. Soiling of bedding or clothing with excrement can result in invasion through urogenital or anal sites. *Gastrointestinal myiasis* usually results from the ingestion of eggs or maggots with infested foods, commonly causing general malaise, nausea, vomiting, cramps, and diarrhea. Although such cases are seldom serious, adverse effects can be prolonged and may require purging of the alimentary tract to resolve the problem.

Facultative or obligatory *wound myiasis* is more serious because of the potential for rapid and indiscriminate destruction of tissues. Infirm persons are at particular risk, as are people who work and live in close proximity to livestock operations where carrion and suitable hosts are more available. Myiasis involving flesh flies and bot flies occasionally occurs in people. The former usually involves infants and the latter, livestock handlers and field workers. *Ocular myiasis* is the most common form of human myiasis, for which numerous cases are recorded. It is usually caused by the *sheep nose bot* when a gravid female larviposits directly into an eye while hovering. As many as 50 first-instar *O. ovis* maggots have been removed from the eye following the strike of a single fly. Although the larvae do not actually penetrate the eye tissues, they can cause extreme irritation as they crawl about on the eyeball. They are best removed by flushing the eye with fluids or by gentle suction.

An aberrant type of *cutaneous myiasis* in humans is caused by the first-instar larvae of *Gasterophilus* and *Hypoderma* species. In such cases, larvae hatching from eggs laid on a person penetrate the skin and wander in the epidermis, producing visible, sinuous, inflamed tracks accompanied by irritation and pruritus. This is called *cutaneous larval migrans*. If cases involving *horse bot flies* (*Gasterophilus* species) are not treated, the larvae eventually die. However *cattle grubs* (*Hypoderma* species) can cause much more severe pathology in humans, sometimes with fatal results. The larvae are able to penetrate and burrow into deeper tissues with the potential for serious consequences because they are not well adapted to human hosts. More cases involve *H. bovis* than *H. lineatum* and tend to be more severe in children. Both have been reported as causing ocular myiasis and skin warbles, the latter usually in the neck and shoulders. Several human cases involving prolonged, but temporary, paralysis have been reported.

*New World bot flies* also parasitize humans. More than 60 cases of human cutaneous myiasis caused by *Cuterebra* species have been documented in North America. Clinical signs appear about 10–14 days after the bot fly eggs have been deposited on the skin in outdoor situations. The individual's body heat stimulates the eggs to hatch, following which the first-instar larvae, after contacting the host, enter the skin through cuts, abrasions, or natural body openings. Larval development usually progresses to the second and, occasionally, the third instar. The danger is that in the meantime the wandering first-instar larvae may invade deeper tissues, leading to dire consequences if they should penetrate the cranial cavity, eyes, or vital organs.

The tórsalo, or *human bot fly,* causes cutaneous myiasis in people throughout its range in Central and South America. Lesions generally occur on the exposed extremities, but they also may involve the scalp, forehead, external ears, eyelids, lower back, and thighs. Skin lesions at the site of larval development are relatively minor during the early stages. As the larva grows, however, the site becomes intensely itchy and exudes serosanguineous fluids. About a week before emergence of the mature larva, the lesion becomes tender, moderately swollen, and generally more painful due to the spines of the larva irritating the skin as it moves or rotates within. Upon emergence of the larva the lesion usually heals spontaneously in about a week. If the larva dies before emergence or the wound becomes infected, serious complications can result. These include foul discharges and fetid odors that may attract other skin-invading parasites (e.g., screwworms), leading to cases of secondary myiasis. Occasionally, *D. hominis* infestations of the head have resulted in penetration of brain tissue, particularly in infants and young children, causing convulsions and death (Rossi and Zucoloto, 1972).

Larvae of *D. hominis* can be extracted by physically squeezing them out or killing them by injection with an anesthetic and surgically removing them. They also can be induced to exit the skin by blocking off their air supply with strips of pork fat (''bacon therapy''), lard, petroleum jelly, soft beeswax, and other materials applied to the lesion. In attempts to reach air, the larvae penetrate the material and are extracted when it is removed.

## CLINICAL USE OF MAGGOTS

A few myiasis-causing flies feed only on necrotic tissue as they develop. They are species such as *Phaenicia sericata* that are usually associated with carrion. In hospitals the larvae may be deliberately used in the treatment of deep wounds and sores in situations in which surgery is impractical. This approach has been used by people since ancient times for treating deep, chronic, or extensive surface wounds. Anthropological evidence shows that such widely located cultures as the Mayans of Central America, the aboriginals of Australia, and the hill people of Myanmar (formerly North Burma) all used maggots for such *wound therapy.* Later, with the advent of explosive artillery shells, bombs, and grenades in warfare, battlefield injuries became much more difficult to treat using conventional surgery. During the Napoleonic War, the American Civil War, and World War I, military physicians observed that wounds inadvertently infested with maggots healed quickly with minimal scarring. This led to the deliberate use of maggots, hatched from surface-sterilized eggs, to remove necrotic tissue in treating chronic wounds, bed sores, severe burns, and bone infections. The procedure was called *maggot therapy* and gained wide use. Maggot therapy was a common procedure in medicine until the general use of antibiotics replaced the practice in the 1940s. Now, following a 50-year hiatus, the clinical use of *surgical maggots* in treating such maladies is being practiced again in cases where antibiotics are ineffective and where surgery is either impractical or is refused by the patient.

Three species of blow flies have been commonly used in treating wounds: *Lucilia illustris, Phaenicia sericata*, and *Phormia regina*. Usually *P. sericata* is the species of choice. Care must be taken to ensure that maggots from the population selected for colonization feed only on dead tissues. Once the laboratory colony is established, eggs are sterilized in an antiseptic bath (e.g., sodium hypochlorite, formalin, or hydrogen peroxide) and held for hatching. First-instar maggots are transferred to the patient's wound, where they are confined with a sterile screen-dressing fastened to the surrounding healthy tissue with adhesive.

The maggots act as microsurgeons, removing necrotic tissue along with debris and associated microorganisms. In moving about the wound while feeding, maggots stimulate the formation of desirable granulation tissue and secrete calcium carbonate, ammonia, and allantoin (a substance that promotes wound healing). Together they produce a more alkaline milieu less conducive to bacterial growth. The maggots develop at a predictable rate and can be removed and counted as they mature. Healing of maggot-treated wounds is very rapid with less scarring than conventional surgery. There is no need to use anesthesia during this process because the active maggots cause little discomfort except for occasional tingling sensations. Another advantage is the much lower cost of the entire procedure, even when including the cost of maintaining a fly colony.

## VETERINARY IMPORTANCE

Livestock and wildlife are at greater risk of attack by myiasis-causing flies than are people. This is a result of greater exposure and the tendency to have more untreated open wounds as sites for fly exploitation. The seasonal incidence of myiasis in nonhuman hosts increases during calving, foaling, and lambing, when young animals are less resistant in their behavioral and humoral response to maggot invasion. Also the availability of placental carrion and the exposed umbilicus of newborn animals enhances related fly activity. Open wounds associated with shearing, branding, dehorning, and castrating operations increase the vulnerability of animals to attack. Blow flies also oviposit on fecal- or urine-soiled pelage. The exploitation of scrapes, scratches, and other skin wounds by myiasis-causing carrion flies, flesh flies, and screwworms can have a major economic impact on livestock operations throughout the world, especially in tropical and warm temperate latitudes.

*Phaenicia pallescens* is an important cause of myiasis in sheep in Australia and South Africa. It especially attacks Merino and other breeds which have numerous skin folds and crusted wool that retain excrement, causing *fleece rot*. Common sites of attack are the anal and genital regions of sheep. Prophylactic measures include keeping these body regions shorn or surgically removing skin in these areas to limit fouling, termed *Mule's operation*. *P. pallescens* generally invades only fresh carrion at warm latitudes, where it is not associated with myiasis. However, in Australia, a race of this species is known to cause secondary facultative myiasis in sheep. This race of fly also exhibits a greater developmental success when infesting live sheep than when infesting dead sheep as carrion. Annual economic losses to blow-fly myiasis in Australia and other sheep-raising countries are substantial. *Pissle strike* is another form of myiasis in Australia in which blow flies infest the urethral orifice of a ram's penis, attracted by the surrounding dense, urine-soaked fleece.

In terms of gross pathology, mortality, and economic losses, the most important myiasis-causing flies worldwide are the *screwworms*. Untreated screwworm cases nearly always lead to death of the host. The adult flies produced in one infested animal serve as a source for subsequent cases. Fatalities are due to the invasion of vital organs, septicemia resulting from large maggot-infested wounds, or predation of the debilitated host. Prior to the development and implementation of screwworm control measures, annual losses attributed to screwworms amounted to several hundred million dollars annually in North America alone. Because wild vertebrates also are subject to screwworm attack, they constitute an important source of these parasites for domestic species. Any open wound, no matter how small, is a potential site for oviposition by primary screwworms. The treatment of wounds, whether to prevent or remove screwworms, is a costly procedure in time and labor, especially when it involves livestock on open range.

The *torsalo* is an important pest of domestic cattle throughout tropical areas of Central and South America. Horses and other equids, however, are seldom bothered.

Clusters of warbles can occur anywhere on the host, but only rarely are they secondarily invaded by other myiasis-causing flies. Heavily infested cattle become weak and emaciated and may have difficulty walking because of *Dermatobia hominis* larvae infesting their legs and feet. Animals can become seriously crippled when wounds become infected, especially when retaining dead larvae. Deaths of heavily infested cattle and dogs have occurred in such cases. Certain cattle breeds (e.g., Brahma, Zebu) appear to be differentially resistant to tórsalo infestations and show lower infestation levels than do Herefords and dairy breeds.

Economic losses to *cattle grubs* due to poor weight gains, reduced milk production, and damage to hides are estimated as high as US$100 million annually in North America (Drummond *et al.*, 1988). This figure does not include the cost of controlling cattle grubs. A grub burden of less than 50 maggots per host appears to have little effect on the vitality or weight increase of infested cattle. Economic losses also are caused in cattle when open warbles and scars of healed warbles along the host's backline decrease the value of hides used in making leather. The green, or raw, hide constitutes about 10% of the total carcass value, and *shot-holed hides* (warble-scarred) can result in considerable reduction in grower profits. Warbles caused by cattle grubs and tórsalo larvae in the backs of slaughtered animals require excessive trimming of discolored flesh and jellied fat surrounding the warbles, further contributing to decreased profits due to labor costs and loss of meat. Horses also are commonly infested by cattle grubs. Because of the discomfort they cause to the backs of infested horses, the use of these animals as saddle mounts is precluded during certain times of the year.

*Gadding* by cattle in response to *Hypoderma* females, either in enclosures or on open range, can result in serious injuries and even death of animals. Such injuries in turn are open to invasion by other myiasis-causing flies. Headlong galloping by panic-stricken cattle can bruise udders and, in the case of pregnant cows, cause spontaneous abortion. The accidental death of an animal killed in a fall incurred while fleeing from a bot-fly attack has been blamed at times on large predators subsequently discovered feeding on the remains. In the Neotropics, because of the egg-porter phenomenon, hosts take no evasive action to avoid adult tórsalo flies.

*Nose bots* in sheep cause relatively low economic losses compared to cattle grubs. Unless the nose bots are very numerous, the value gained by the costly handling of sheep to administer controls is questionable. At times individual sheep may die as a result of unusually large numbers of maggots (e.g., 100–300 per host), on the order of 10 times the usual host burden. Significant weight gains and increased fleece length have been reported in Russia and South Africa as a result of controlling nose bots in lambs. Organophosphate and avermectin insecticides are effective against nose bots.

*Horse stomach bot flies* are ubiquitous companions of horses throughout the world. Chronic and repeated infestations can result in loosened teeth due to larval tunneling in the gums and in ulcerations of the gastric lining. Eventually this will interfere with forage ingestion and digestion. However, most horses tolerate a maggot burden of 100 or so with no apparent ill effects. Depending on the value of the individual animal, whether a race horse or cow pony, a family pet or a "candidate for the glue factory," the economic loss is widely variable. Different breeds of the domestic horse also react differently to egg laying by the hovering adults of the common stomach bot fly. Thoroughbreds, American saddle horses, and Arabians, for example, are much disturbed by the hovering fly and will take evasive action that can cause self-injury. At the other extreme, Shetland ponies, Morgans, and draft breeds appear to ignore the ovipositing flies. All breeds, however, react in wild panic to ovipositing *G. nasalis* females.

Among the horse stomach bot flies, *G. pecorum* is the most pathogenic. Its maggots often attach *en masse* in the oral cavity and upper alimentary tract. This can result in obstruction caused by the enlarging maggots attached to the soft palate and esophagus and the associated inflammation. Heavy populations hinder swallowing, prevent food passage, and can cause severe digestive problems in the host, sometimes with fatal results.

The possible use of cuterebrine bots as biocontrol agents of rodents has been suggested. However, the associated pathology in native hosts is minimal, and bot populations appear to have little effect on rodent numbers.

## PREVENTION AND CONTROL

There are three major approaches for controlling myiasis: avoiding contact between potential hosts and myiasis-causing flies; early treatment of wounds to prevent myiasis; and reduction or elimination of myiasis fly populations. In the area of human health, there is general reliance on hygiene and medical or surgical intervention. In veterinary cases the most common approach is the use of insecticides, especially systemic compounds, administered to the host.

Preventing unnecessary or avoidable outdoor exposure of humans during fly seasons is one obvious way to control facultative or obligatory myiasis. Attractive odors associated with urine, feces, vomitus, nasal secretions, and purulent sores invite fly strike. Healthy adults are able to react to flies attempting to feed or oviposit, but infants and elderly individuals often are either unaware or unable to take evasive measures. Myiasis-causing flies sometimes enter buildings when attracted by similar odors. Management of geriatric-care facilities should routinely include measures such as screening, diligent resident hygiene, and treatment of sores to avoid attraction and contact with flies. Securing foods in fly-proof containers, under screens, or under refrigeration is the best means of preventing gastrointestinal myiasis. Foods which have been exposed, even when fly eggs are present, can be made safe by freezing or thorough cooking.

The risk of human myiasis is greater in livestock areas. People working or residing close to livestock operations (e.g., stables, dairies, feedlots, and poultry houses) are most at risk. Flies or bee-like insects hovering around a person or alighting at open sores should be suspect. Immediate evasive responses usually will drive the fly away. Where the tórsalo fly occurs in the Neotropics, insect repellents can be used to reduce the attacks of blood-seeking insects carrying tórsalo eggs.

Human myiasis can be prevented by early treatment of open wounds with an antiseptic salve and protective dressing. Wound dressings should be changed regularly and not be allowed to become excessively soiled by wound exudates. Maggots in warbles can be induced to back at least partway out of the warble by covering the pore with a thick smear of petroleum jelly. The protruding maggot then can be grasped with forceps or fingers and slowly extracted. Sometimes surgical removal is necessary; this entails snipping the rim of the warble pore to allow greater ease of extraction. Maggots in eyes, nasal sinuses, or the auditory meatus should be removed by flushing with saline or a diluted antiseptic such as carbolic acid. Gastrointestinal myiasis can be treated with purges or emetics, although several treatments may be necessary to remove all maggots.

Preventive measures for the control of myiasis in livestock, zoo animals, and wildlife mainly involve the removal of carrion that may attract flies, early treatment of open wounds to prevent fly strike, and minimizing injuries during the peak fly season. Husbandry practices that can help to reduce the incidence of myiasis include regular inspection of animals to detect and treat all wounds; deep burial (ca. 1.5 m) or cremation of carrion; limiting of dehorning, castrating, and branding to nonfly seasons; removal of sharp objects and other materials than can cause wounds (e.g., horns, barbwire, thorny shrubs); and effective control of biting flies and ticks. The application of salvelike ointments, or "smears," can help to prevent fly strike or to kill maggots that are already present. Smears contain larvicides such as benzol or lindane and repellents such as pine-tar oil.

For many years *dipping vats* have been used throughout the world to treat livestock for arthropod pests, including cattle grubs, screwworms, and other myiasis-causing flies.

This entails forcing animals to swim or otherwise move through a water-filled pit containing insecticides or acaricides. This approach, however, has been largely replaced by other methods, such as insecticide-impregnated *ear tags*. The tags slowly release low levels of chemicals that afford protection for extended periods. Another slow-release device developed for cattle grub control is the insecticide-impregnated *ankle band* attached around the back legs of cattle. Systemic insecticides such as chlorinated hydrocarbons, organophosphates, carbamates, and sulfur compounds kill cattle grubs before they reach the back of the host. Systemic materials can be administered by *injection* (e.g., avermectins), applied to the back as *pour-ons* to be absorbed by the bloodstream, as *boluses* placed in the animal's stomach to provide slow release of systemic compounds, or in *mineral blocks* ("salt licks") for *ad libitum* consumption by livestock and wildlife species. Whereas systemic organophosphates are effective only against prewarble larvae, avermectins are effective against all larval instars.

Other methods used to control myiasis-causing flies that attack livestock are *low-volume sprays, oral drenches, power dusters,* and self-treatment devices such as suspended *dust bags* and *back rubbers.* In the case of "fleece worm" blow flies, the topical application of insecticides to body areas subject to soiling with maggot secretions is an effective, albeit labor-intensive, practice. The use of *biocontrol agents,* such as parasitic pteromalid wasps that attack fly pupae, and the physical removal of adult flies by *trapping* can also play a role in integrated pest management programs for some muscoid species that cause myiasis.

The use of *genetically altered flies* has met with some success in reducing local populations of certain myiasis-causing calliphorid species. This approach shows promise in Australia, where *Lucilia cuprina* (formerly *Phaenicia cuprina*) causes annual losses estimated at $120 million (Australian) due to secondary myiasis in sheep. Radiation treatment has been used to induce genetic translocations in chromosomes of a laboratory strain of *L. cuprina* that makes mutant male flies impotent. The fewer daughters that are produced from mutant male and wild female matings are able to reproduce normally. However, the second-generation daughters carry about 85% of the induced mutations. The result has been a reduction of *L. cuprina* populations by about half in the first generation and by increased proportions in subsequent generations.

## SCREWWORM ERADICATION PROGRAM

The idea of using *sterilized males* to eradicate populations of wild flies was first conceived by E. F. Knipling in the

1930s. It was not until 1954–1955, however, that the first successful field test was achieved with the eradication of the primary screwworm *Cochliomyia hominivorax* on the island of Curacao (Netherlands Antilles). This entailed the mass production of sterile males from gamma-irradiated pupae which were then released to mate with wild females. Eggs resulting from these matings do not hatch, thereby reducing fly populations. Subsequent successes in eradicating *C. hominivorax* were achieved in the southeastern United States in 1959, in the rest of the United States by 1966, and in Puerto Rico in 1974. Following the establishment of the joint Mexico–United States Commission on Screwworm Eradication in 1972, a cooperative program was begun in 1975 to eliminate *C. hominivorax* in Mexico as far south as the Isthmus of Tehuantepec. Hundreds of millions of flies were mass-reared weekly in facilities at Mission, TX, and Tuxtla Gutierrez, Chiapas (Mexico), for aerial releases in the targeted areas. By 1985 the screwworm had been successfully eliminated from most of southern Mexico and Central America, with a sterile-male barrier being established at the Darien Gap, the isthmus at the border between Panama and Colombia. Efforts are underway to construct a new mass-rearing fly facility in Panama and to extend the eradication effort into Colombia and other South American countries. In the meantime, sterile males have been successfully used in areas where *C. hominivorax* has reinvaded former eradication zones and where it has been introduced for the first time. *C. hominivorax* was detected for the first time in Libya in 1988, for example, where it was eradicated within a few months following the release of sterile males in 1991. For further details on the history of the screwworm eradication program, see Meyer (1996).

Another approach to screwworm control has been the development of an attractant called *swormlure* that simulates the odor of animal wounds (Cunningham *et al.*, 1992; Snow *et al.*, 1982). This has been combined as a bait with pesticides in a pelletized form that can be applied by aircraft to attract and kill gravid females. Called the Screwworm Adult Suppression System, this approach has been successfully used to complement sterile-male release efforts by killing wild female flies that have not mated with irradiated males.

## CATTLE GRUB CONTROL

The best approach to area-wide control of cattle grubs is integrated management programs enlisting the cooperation of all producers in the targeted region. The most effective method to date is use of the systemic, antiparasitic compounds known as avermectins. Unlike traditional systemic insecticides that kill only the migrating larvae of cattle grubs, avermectins are also effective in killing second- and third-instar larvae after they have formed warbles. Sterile-male release technology also has shown promise but is limited by inherent logistic problems in the mass rearing of *Hypoderma* species for this purpose. By combining the use of systemic compounds and sterile-male releases, cattle grub populations and the associated economic losses have been dramatically reduced in North America through the joint United States–Canada Cattle Grub Project initiated in 1982 (Klein *et al.*, 1990; Kunz *et al.*, 1984). Likewise, significant reductions in cattle grub problems have been achieved in Great Britain and several European countries (Boulard and Thornberry, 1984; Wilson, 1986). Although experimental vaccines against *Hypoderma* species have been developed, they have not been widely field tested. They may, however, play a greater role in the future as one more component of integrated management programs. For further information on the biology and control of cattle grubs, see Scholl (1993).

## REFERENCES AND FURTHER READING

Baird, C. R., Podgore, J. K., and Sabrosky, C. W. (1982). *Cuterebra* myiasis in humans: six new case reports from the United States with a summary of known cases (Diptera: Cuterebridae). *Journal of Medical Entomology* 19, 263–267.

Baird, J. K., Baird, C. R., and Sabrosky, C. W. (1989). North American cuterebrid myiasis. *Journal of the American Academy of Dermatology* 21, 763–772.

Baumgartner, D. L. (1988). Review of myiasis (Insecta: Diptera: Calliphoridae, Sarcophagidae) of Nearctic wildlife. *Wildlife Rehabilitation* 7, 3–46.

Boulard, C., and Thornberry, H., Eds. (1984). "Warble Fly Control in Europe." Balkema, Rotterdam/Boston.

Brewer, T. F., Wilson, M. E., Gonzalez, E., and Felsenstein, D. (1993). Bacon therapy and furuncular myiasis. *Journal of the American Medical Association* 270, 2087–2088.

Catts, E. P. (1982). Biology of New World bot flies: Cuterebridae. *Annu. Rev. Entomol.* 27, 313–338.

Catts, E. P. (1994). Sex and the bachelor bot fly. *American Entomologist* 40, 153–160.

Chernin, E. (1986). Surgical maggots. *Southern Medical Journal* 79, 1143.

Craig, C. F., and Faust, E. C. (1940). "Clinical Parasitology," 2nd ed. Lea & Febiger, Philadelphia.

Cunningham, E. P., Abusowa, M., Lindquist, D. A., Sidahmed, A. E., and Vargas-Teran, M. (1992). Screwworm eradication programme in North Africa. *Revue d'Elevage et de Medicine Veterinaire des Pays Tropicaux* 45, 115–118.

Curran, C. H. (1965). "The Families and Genera of North American Diptera," 2nd ed. Tripp, Woodhaven, NY.

Drummond, R. O., George, J. E., and Kunz, S. E. (1988). "Control of Arthropod Pests Of Livestock: A Review of Technology." CRC Press, Boca Raton, FL.

Ferrar, P. (1987). A guide to the breeding habits and immature stages of Diptera Cyclorrhapha. *Entomography* 8, pp. 83–98. Brill, Leiden.

Furman, D. P., and Catts, E. P. (1982). "Manual of Medical Entomology," 4th ed. Cambridge Univ. Press, Cambridge, UK.

Graham, O. H. (1985). Symposium on Eradication of the Screwworm from the United States and Mexico." Miscellaneous Publication No. 62. Entomological Society of America, College Park, MD.

Greenberg, B. (1971). "Flies and Disease. Vol. 1. Ecology, Classification and Biotic Associations." Princeton Univ. Press, Princeton, NJ.

Greenberg, B., and Singh, D. (1995). Species identification of calliphorid (Diptera) eggs. *Journal of Medical Entomology* **32**, 21–26.

Hall, D. G. (1948). "The Blow Flies of North America." Thomas Say Foundation, Entomological Society of America, College Park, MD.

Hall, M. (1995). Myiasis of humans and domestic animals. *Advances in Parasitology* **35**, 257–334.

Hendrix, C. M., King-Jackson, D. A., Wilson, M., Blagburn, B. L., and Lindsay, D. S. (1995). Furunculoid myiasis in a dog caused by *Cordylobia anthropophaga. Journal of the American Veterinary Medicine Association* **207**, 1187–1189.

James, M. T. (1947). "The Flies that Cause Myiasis in Man." US Department of Agriculture, Miscellaneous Publication 631. U. S. Government Printing Office, Washington, DC.

Kettle, D. S. (1995). "Medical & Veterinary Entomology." Oxford Univ. Press, Oxford, UK.

Klein, K. K., Fleming, C. S., Colwell, D. D., and Scholl, P. J. (1990). Economic analysis of an integrated approach to cattle grub (*Hypoderma* spp.) control. *Canadian Journal of Agricultural Economics* **38**, 159–173.

Kunz, S. E., Drummond, R. O., and Weintraub, J. (1984). A pilot test to study the use of the sterile insect technique for eradication of cattle grubs. *Preventive Veterinary Medicine* **2**, 523–527.

Lane, R. P., and Crosskey, R. W. (1993). "Medical Insects and Arachnids." Chapman & Hall, London.

Leclercq, M. (1969). "Entomological Parasitology—The Relation between Entomology and the Medical Sciences." Chapt. 5, Myiases. Pergamon, Oxford.

Liu, D., and Greenberg, B. (1989). Immature stages of some flies of forensic importance. *Annals of the Entomological Society of America* **82**, 80–93.

McAlpine, J. F., *et al.*, Eds. (1981, 1987, 1989). "Manual of Nearctic Diptera," Vols. 1–3. Research Branch Agriculture Canada, Canadian Government Printing Center, Quebec, Monographs 27, 28, and 32, respectively.

Meyer, N. V. (1996). "History of the Mexico-United States Screwworm Eradication Program." Vantage Press, New York.

Norris, K. R. (1965). The bionomics of blow flies. *Annu. Rev. Entomol.* **10**, 47–68.

OISTROS. An irregularly published newsletter dealing exclusively with current research on the Oestroidea; Swedish Museum of Natural History, Stockholm, Sweden.

Papavero, N. (1977). "The World Oestridae (Diptera), Mammals and Continental Drift." Junk, The Hague.

Reames, M. K., Christensen, C., and Luce, E. A. (1988). The use of maggots in wound debridement. *Annals of Plastic Surgery* **21**, 388.

Rosen, I. J., and Neuberger, N. (1977). Myiasis *Dermatobia hominis,* Linn: report of a case and review of literature. *Cutis* **19**, 63–66.

Rossi, M. A., and Zucoloto, S. (1972). Fatal cerebral myiasis caused by the tropical warble fly, *Dermatobia hominis. American Journal of Tropical Medicine and Hygiene* **22**, 267–269.

Sabrosky, C. W. (1986). North American species of *Cuterebra,* the rabbit and rodent bot flies (Diptera: Cuterebridae). Thomas Say Publication, Entomological Society of America, College Park, MD.

Sabrosky, C. W., Bennett, G. F., and Whitworth, T. L. (1990). "Bird Blowflies (*Protocalliphora*) in North America (Diptera: Calliphoridae)." Random House (Smithsonian Inst. Press), New York.

Sancho, E. (1988). *Dermatobia,* the Neotropical warble fly. *Parasitology Today* **4**, 242–246.

Scholl, P. J. (1993). Biology and control of cattle grubs. *Annu. Rev. Entomol.* **39**, 53–70.

Scholl, P. J., Colwell, D. D., Weintrub, J., and Kunz, S. E. (1986). Area-wide systemic insecticide treatment for control of cattle grubs, *Hypoderma* spp. (Diptera: Oestridae): two approaches. *Journal of Economic Entomology* **79**, 1558–1563.

Scott, H. G. (1963). "Myiasis: Epidemiologic Data on Human Cases (North America North of Mexico: 1952–1962 Inclusive)." US Department of Health, Education, and Welfare, Centers for Disease Control and Prevention, Atlanta, GA.

Sherman, R. A., and Pechter, E. A. (1988). Maggot therapy: a review of the therapeutic applications of fly larvae in human medicine, especially for treating osteomyelitis. *Medical and Veterinary Entomology* **2**, 225–230.

Smith, K. G. V. (1986). "A Manual of Forensic Entomology." British Museum (Natural History), London.

Snow, J. W., Siebenaler, A. J., and Newell, F. G. (1981). "Annotated Bibliography of the Screwworm, *Cochliomyia hominovorax* (Coquerel)." US Department of Agriculture, ARM-S-14. Agricultural Reviews and Manuals, Southern Series No. 14.

Snow, J. W., Coppedge, J. R., Broce, A. B., Goodenough, J. L., and Brown, H. E. (1982). Swormlure: development and use in detection and suppression systems for adult screwworm (Diptera: Calliphoridae). *Bulletin of the Entomological Society of America* **28**, 277–284.

Stehr, F. W. (1991). "Immature Insects," Vol. 2. Kendall/Hunt, Dubuque, IA.

Stone, A., Sabrosky, C. W., Wirth, W. W., Foote, R. H., and Coulson, J. R. (1965). "A Catalogue of Diptera of America North of Mexico." Agriculture Handbook 276. US Department of Agriculture, Washington, DC.

Wall, R., French, N. P., and Morgan, K. L. (1995). Population suppression for control of the blowfly *Lucilia sericata* and sheep blowfly strike. *Ecological Entomology* **20**, 91–97.

Wilson, G. W. C. (1986). Control of warble fly in Great Britain and the European community. *Veterinary Record* **118**, 653–656.

Zumpt, F. (1965). Myiasis in man and animals in the Old World. Butterworth's, London.

# 17

# LOUSE FLIES, KEDS, AND RELATED FLIES (*Hippoboscoidea*)

JOHN E. LLOYD

*The Families* Hippoboscidae, Streblidae, and Nycteribiidae are obligate, blood-feeding ectoparasites. The Hippoboscidae are variously called *louse flies, bird flies, feather flies, spider flies, flat flies, tick flies, ked flies,* and *keds.* Most species in this family are restricted to a narrow range of hosts. Approximately three-fourths of the known species are ectoparasites of birds, whereas the remainder occur on a variety of mammals other than bats. The Streblidae, called the *streblid bat flies* or *bat flies,* and the Nycteribiidae, called the *nycteribiid bat flies* or *spider-like bat flies,* are ectoparasites of bats and are rarely encountered except by individuals working with bats. For further information on the Hippoboscoidea, the reader is referred to the monographs and other works by Bequaert (1942,

1953–1957), Maa (1963, 1966, 1969, 1971), Maa and Peterson (1987), Theodor and Oldroyd (1964), Peterson and Wenzel (1987), and Wenzel and Peterson (1987).

Although worldwide in distribution, most species of Hippoboscidae are tropical and subtropical in both the Old and New Worlds. The Paleotropics are richer in hippoboscids than any other region. Some hippoboscids may be temporary summer residents of temperate regions due to the migratory habits of their hosts. A few species (e.g., the ''grouse fly'' *Ornithomya fringillina*) are restricted to temperate regions.

The Streblidae are largely New World and tropical and subtropical in distribution. Relatively few species occur in the warm temperate zones. Most members of the Nycteribiidae are found in the Old World and occur primarily in the tropical and subtropical regions.

## TAXONOMY

The Hippoboscidae, Streblidae, and Nycteribiidae represent three families of the superfamily Hippoboscoidea (order Diptera, suborder Cyclorrhapha), although some authors consider these families to be subfamilies of the Hippoboscidae. Because of similarities in mechanisms of feeding and reproduction as well as specialized morphological characteristics, the Hippoboscidae, Streblidae, and Nycteribiidae have been considered by some to be monophyletic and are included in the group Pupipara (Theodor, 1964). An opposing view is that Pupipara is neither an appropriate name nor a natural group. Because the third-instar larva and not the pupa is deposited by the female, these insects are actually larviparous and not pupiparous. Furthermore, the similarities between the Hippoboscidae and the other two families more likely reflect convergent evolution, in which case the group should be considered diphyletic in origin. Based on their morphology, the Hippoboscidae appear to be most nearly related to the tsetse flies, genus *Glossina* (family Glossinidae). The Streblidae and Nycteribiidae have common structural characteristics and a common origin. They do not, however, share these similarities with the Glossinidae.

There are approximately 19 genera and 150 described species in the family Hippoboscidae. Thirteen genera containing 31 species and 2 subspecies have been reported from the Nearctic region (Maa and Peterson, 1987). Pfadt and Roberts (1978) have presented a list of the louse flies recorded from the United States with their hosts and distribution, both in the United States and worldwide. Maa (1966, 1969) has divided the Hippoboscidae into three subfamilies. The Ornithomyinae include most of the hippoboscid species. All but seven are parasites of birds. The Lipopteninae contains approximately 34 species, all parasitic on mammals. The Hippoboscinae contains 8 species, of which 7 infest mammals and 1 infests birds.

Wenzel and Peterson (1987) recognized 25 genera and 149 species of Streblidae in the New World and 6 genera and 72 species in the Old World. Theodor (1967) recognized 2 genera and 22 species of Nycteribiidae in the New World and 9 genera and 170 species in the Old World, whereas Peterson and Wentzel (1987) recognized 12 genera and 256 species of Nycteribiidae worldwide.

# MORPHOLOGY

All members of the Hippoboscoidea are morphologically adapted for an ectoparasitic existence among the hairs or feathers of their hosts. Certain parts of the exoskeleton have become modified, mainly by fusion and reduction or atrophy, in response to permanent ectoparasitism.

## HIPPOBOSCIDAE

Adults of this family (Fig. 17.1) vary in size from 1.5 to 12.0 mm. The body is dorsoventrally flattened, with the depressed head, thorax, and abdomen giving these

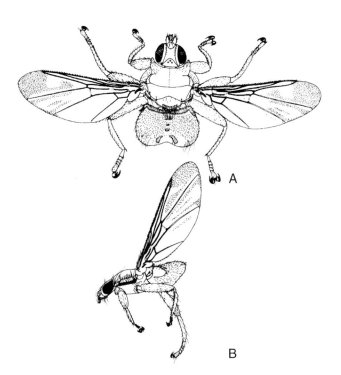

**FIGURE 17.1**    *Ornithomya avicularia*, adult female (Hippoboscidae). (A) Dorsal view, with wings spread; (B) lateral view, showing dorsoventral flattening. (From Hutson, 1984.)

insects their louselike appearance. The mouthparts are directed forward rather than downward. The abdominal integument is soft and flexible, allowing for stretching and distension of the abdomen during larval development within the female and while feeding.

The legs of hippoboscids are generally robust, with enlarged femora, flattened tibiae, and short, compact tarsi with one or more basal teeth. The legs tend to be shorter and stouter with heavier tarsal claws in species that infest mammals than in those that infest birds. According to Bequaert (1953), hippoboscids that infest birds have legs that are adapted for scurrying swiftly forward, backward, and sideways amidst the soft feathers. In species parasitizing mammals the legs are built more for grasping and clinging to the skin and the coarse hairs of the pelt.

The compound eyes are generally well developed in those genera which have functional wings. The eyes are greatly reduced in genera which have small, nonfunctional wings and in those which lose their wings after reaching the host. The sheep ked, *Melophagus ovinus*, which spends its entire life in the wool of sheep, has small compound eyes with relatively few ommatidia. Ocelli are present in several genera of Hippoboscidae but absent in others. The *antennae* are small and immovable and are located in deep antennal sockets.

The Hippoboscidae are vessel feeders (*solenophages*), with both sexes being obligate blood feeders. The *proboscis* is strongly sclerotized. Its base is partially retracted

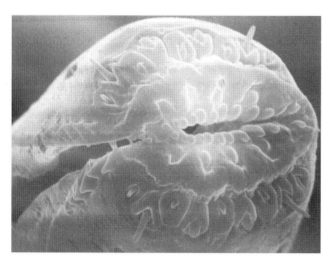

**FIGURE 17.2** Tip of proboscis of the sheep ked, *Melophagus ovinus* (Hippoboscidae), showing labella armed with teeth for cutting into host tissue. (Courtesy of J. E. Lloyd.)

**FIGURE 17.3** Representative adult female of bat flies (Streblidae), ectoparasitic on bats. (From Furman and Catts, 1982.)

into a pouch on the ventral side of the head when not in use. At rest, the distal portion is concealed in the palpal sheath. The structure of the proboscis resembles that of blood-sucking muscid flies. The labium is the principal piercing structure. The labrum and hypopharynx lie in a dorsal groove of the labium and together these form the food channel. The labella at the tip of the labium are armed with teeth (Fig. 17.2).

Adults of most species of Hippoboscidae have relatively long, broad fore wings. The hind pair of *wings* is represented by the *halteres,* characteristic of dipterans. At rest, the fore wings lie flat over each other on the abdomen like closed scissors blades, similar to tsetse flies. In both *Lipoptena* and *Neolipoptena* the newly emerged adult has fully functional wings; this winged adult is referred to as a *volant.* After reaching the host, the wings of these insects break off at the base, leaving a stump. The first blood meal from the host stimulates physiological changes in the fly, including histolysis of flight muscles and growth of leg muscles, to accommodate the subsequent parasitic life of the adult.

Both birds and mammals harbor a few species of Hippoboscidae with reduced wings that are not used for flight. Bequaert (1953) noted that at least 4 genera and 15 species have reduced wings (*subapterous*). The halteres, however, are not appreciably reduced in these species. The reduced wings are immovable and cover the halteres, which they may help protect. Both the fore wings and halteres are reduced to small stumps and are nonfunctional in the genus *Melophagus.* This genus is possibly the most specialized in the family. The adults of *Melophagus* species emerge from the puparium with rudimentary, nonfunctioning flight muscles which later atrophy.

## STREBLIDAE

This family includes a wide variety of forms exhibiting highly specialized adaptations for parasitic life. The adults of most species are 1.5–2.5 mm in length; a few Neotropical species may be 0.75–5.0 mm (Wenzel and Peterson, 1987). Several genera are flattened dorsoventrally and superficially resemble hippoboscids or nycteribiids (Fig. 17.3). A few are flattened laterally and resemble fleas. Although most streblids are winged and very mobile, there may be different degrees of wing reduction even within a single genus. The head is usually small in the Streblidae, and the antennae are inconspicuous. The Streblidae lack ocelli, and the compound eyes are absent or reduced to one or only a few facets. The marked reduction of the eyes of both Streblidae and Nycteribiidae appears to be associated with their occurrence on bats as nocturnal hosts.

## NYCTERIBIIDAE

Species in this family vary in body length from 1.5 to 5.0 mm. They are an older group than the Streblidae and have become structurally much more modified. The adults are wingless but still possess halteres. Antennae are moderately large in relation to the size of the head and, as in the Hippoboscidae, the antennae of nycteribiids are usually located in antennal pits. The eyes may be absent or are reduced to one, two, or, rarely, four facets.

The dorsal plates of the thorax are reduced, and the head and legs of nycteribiids are displaced dorsally. This articular displacement of the legs, together with the complete loss of wings, give the nycteribiids their spider-like appearance (Fig. 17.4). The head is narrower than those of the hippoboscids and streblids and, in the resting position, is folded back so that its dorsal surface is in contact with the dorsum of the thorax. The head is rotated forward and downward, through about 180°, in order to feed.

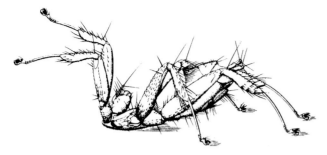

**FIGURE 17.4**  *Nycteribia pedicularia*, spider-like bat fly (Nycteribiidae), ectoparasitic on bats, female, lateral view. (From Hutson, 1984.)

## LIFE HISTORY

The Hippoboscoidea are *larviparous*. They exhibit a form of viviparity called *adenotrophic viviparity*. A single egg is passed to the uterus, where it embryonates and hatches. The egg contains sufficient yolk to nourish the embryo until hatching. The two subsequent *larval instars* remain in the uterus, where they are nourished by a pair of accessory glands, or *milk glands*, that empty into the uterus. The glands are very similar in structure to those in female tsetse flies, although their secretions are slightly different.

Parturition occurs when the larva is fully developed but prior to formation of the *puparium*. The term *prepupa* has been applied to this stage because its structure is similar to that of the third-instar larva. It has ceased to feed, but histolysis of larval organs and formation of the true pupa have not yet started. Shortly after the larva emerges, its integument hardens to form a puparium. In some species, the larva may remain in the uterus until after internal pupal transformations have been initiated. Most puparia are deposited or dropped in the roost, nest, bedding, or elsewhere in proximity to the host. The puparium of the sheep ked is unusual in that it is attached by the female to the fleece of the host. The adult fly emerges after a period of several weeks to several months, depending upon the species and temperature.

## BEHAVIOR AND ECOLOGY

Both sexes of hippoboscoid flies feed as ectoparasites on the blood of birds or mammals. *Host specificity* varies considerably among different groups. Some are restricted to a single host species. Others are restricted to a genus or to several related genera of hosts, whereas still others are generalists that feed on a relatively wide range of host taxa.

Hippoboscidae occur on 18 orders of birds and 5 orders of mammals. No species occurs on both birds and mammals. Nor, with the exception of *Hippobosca*, does any genus occur on both birds and mammals. Host specificity is more marked in species parasitic on mammals than in those parasitic on birds. Apterous species and those with reduced wings, or those which have lost their wings altogether, tend to be most host-specific. In addition, the more advanced or specialized species tend to be more host-specific.

Members of the Streblidae and Nycteribiidae are exclusively parasites of bats (order Chiroptera). No species of either family is known to occur naturally on both of the chiropteran suborders, Megachiroptera and Microchiroptera. Host specificity varies widely within the streblid bat flies and nycteribiid bat flies from one to many host species. According to Wenzel *et al.* (1966), New World streblids, which tend to be host-specific, have become adapted to living and feeding on particular body regions. Individual species are restricted to the wing membranes, head, or trunk. Some bats (e.g., *Phyllostomus hastatus*) commonly harbor three or four species at the same time, with most hosts having at least two species.

Nycteribiidae, and most species of Streblidae and Hippoboscidae, deposit their offspring away from their hosts. The fully developed third-stage larva is dropped to the ground, litter, or nesting material; deposited in a preferred site; or attached to the host or other suitable substrate. Female nycteribiid and streblid flies leave their hosts to deposit larvae in the vicinity of bat roosts. This includes bat-roost surfaces, walls of caves, and branches or leaves of trees.

In the Hippoboscinae, the freshly deposited larva of *Melophagus* is covered with a secretion which hardens upon drying and glues the puparium to the hairs of the host. *Neolipoptena* and *Lipoptena*, which shed their wings after reaching the host, also larviposit on the host. These larvae are not fastened to the host and eventually drop to the ground. Most *Hippobosca* species larviposit away from the host in some favored location, as does the pigeon fly *Pseudolynchia canariensis*. Many species of Hippoboscidae that feed on nesting birds larviposit in nesting materials, from which their puparia may be collected.

Winged streblids move readily from one bat to another within a roost. Similarly, winged hippoboscids are very mobile. Newly emerged *Lipoptena* and *Neolipoptena* often swarm in large numbers in search of a host at certain seasons. These volants have functional wings that break off near the base after the host is reached. Once on the host, adults move swiftly among feathers or hair and are difficult to collect. The relatively slow-moving sheep ked is an exception.

## COMMON SPECIES OF HIPPOBOSCIDS

A number of louse flies in the genus *Hippobosca* are of particular interest to veterinary entomologists. Most

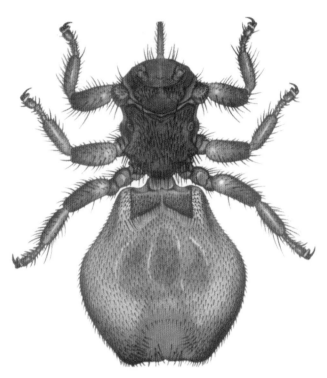

**FIGURE 17.5** Sheep ked, *Melophagus ovinus* (Hippoboscidae), female, dorsal view. (Courtesy of Cornell University Agricultural Experiment Station.)

occur in Europe, Africa, and Asia. Occasional introductions have been made into the United States with the importation of zoo animals. With the exception of the *ostrich louse fly* (*Hippobosca struthionis*), they are primarily parasites of mammals. The sheep ked (*M. ovinus*) is a parasite of sheep and is considered one of the most important insect pests of sheep in many areas of the world.

**Sheep ked** (*Melophagus ovinus*)

The sheep ked (Fig. 17.5) is a wingless ectoparasite that spends its entire life on domestic sheep. It is worldwide in distribution except in tropical regions, where it occurs only in the cooler highlands. It probably was introduced into the United States in the 15th century, shortly after the European discovery of the New World. Often called the *sheep tick* by sheep producers, it is found on both range and farm flocks of sheep. The sheep ked is of considerable economic importance and is generally regarded as the most damaging ectoparasite of sheep in North America. A relative of the sheep ked, *M. montanus*, occurs on Dall's sheep (*Ovis dalli*).

Much of what is known about the life history of hippoboscid flies is based on studies of *M. ovinus*. After a period of 7–8 days of feeding and growing in the uterus of the female, the fully developed larva is deposited and cemented to the sheep's wool. Members of the genus *Melophagus* are the only hippoboscids to attach their larvae to the host. The reddish, barrel-shaped puparium

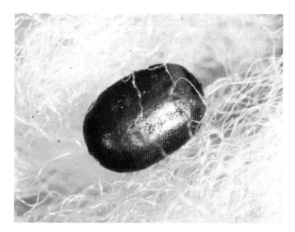

**FIGURE 17.6** Puparium of sheep ked, *Melophagus ovinus* (Hippoboscidae), adhering to sheep wool. (Courtesy of J. E. Lloyd.)

(Fig. 17.6) is fully formed within 12 hr of parturition. In Wyoming, Swingle (1913) determined that the duration of the pupal stage varied from 19–23 days in summer to 20–36 days in winter. The variation of this period was attributed to differences in temperature and the distance of the pupa from the skin of the host. Slightly shorter periods have been reported in areas that are warmer than Wyoming. Swingle further indicated that the duration of the pupal stage is unlikely to be less than 19 days even in the warmest climate and may increase to 40–45 days in the winter.

Teneral females of the sheep ked mate within a day of eclosion, although the first larva is not deposited for at least 12 days. This period includes 6–7 days for larval maturation. Although one mating provides sufficient sperm for a lifetime, repeated matings usually occur when multiple males are present. A female sheep ked normally lives about 4 months and produces 10–12 larvae during its lifetime. Some, however, can live for 6 months or more and can produce 15–20 larvae. The male life span is slightly shorter, approximately 2–3 months.

Larviposition by the sheep ked tends to occur on lower body parts, especially under the neck and in the breech area. In unshorn sheep, adults are consistently most numerous in the rib area. Contrary to popular belief, there is no daily or seasonal movement from one location to another on the host. In the spring and summer, sheep keds are more likely to be found on the underside of sheep that have been recently shorn and on young lambs with a very short fleece. Often they can be found in tufts of longer fleece missed by the shears. Numbers of keds tend to be greater on younger animals.

Sheep keds generally live for only a few days if removed from the host. However, they may live up to 5 days in wool in the laboratory and, when kept under cool and moist conditions away from the host and fleece, they may live even longer. Their vigor and ability to relocate a host diminish the longer they remain separated from a host animal.

Although sheep keds that become dislodged from their host have the ability to locate a host from the ground, transfer of sheep keds is primarily by animal-to-animal contact. Newborn lambs become infested with keds directly from their mothers soon after birth. Within a flock, transfer occurs when sheep keds move to the tips of the fleece in response to increasing air temperature and possibly in response to the brisk movement of sheep that accompanies flocking behavior. Air temperature usually must be 21°C (70°F) or above before many sheep keds are observed on the surface of the fleece. At 27–58°C (80–90°F) sheep keds are common on the outer wool surface. Thus, transfer between animals is more likely, and occurs more rapidly, in summer than in winter.

Like many ectoparasites, populations of *M. ovinus* exhibit *annual fluctuations* in their numbers. Although minor variations have been reported from different parts of the world, populations of the sheep ked tend to be highest from late winter to early spring and lowest in summer. In Wyoming, ked numbers on ewes tend to increase from September to February. Periods of high populations are extended on rams, pregnant ewes, and undernourished animals, with increases in numbers being prolonged for several weeks in pregnant ewes (Nelson and Qually, 1958). On newborn lambs, which receive only teneral keds from their mothers, numbers of sheep keds increase from their birth in early Spring to a couple months later, when populations begin their normal decline. Seasonal decline of sheep ked populations is attributed to *acquired resistance* (Baron and Nelson, 1985; Nelson and Bainborough, 1963). This resistance is apparently caused by a long-lasting, cutaneous, arteriolar vasoconstriction that cuts off much of the capillary blood flow to the upper dermis. Keds are unable to obtain sufficient blood and die of starvation.

Sheep keds feed approximately every 24–36 hr, with the feeding time increasing to 2-day intervals as the keds become older. The feeding period of an individual ked is typically 5–10 min. Feeding is from larger vessels (30–100 $\mu$m) near the bases of the wool follicles and often near the apocrine glands or sweat glands associated with primary follicles (Nelson and Petrunia, 1969). Penetration of the dermis by the mouthparts is accomplished by rapid and continuous movement of prestomal teeth on the labellum, followed by movement of the entire haustellum. After piercing a blood vessel, the mouthparts are secured in place by the prestomal teeth, which are everted and serve to anchor the labella to the vessel wall.

### Dog fly (*Hippobosca longipennis*)

The dog fly originally was a parasite of wild carnivores in East Africa. It has since become widely distributed in

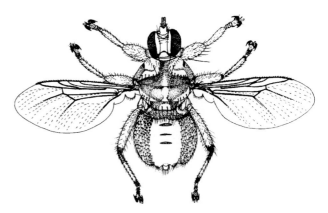

**FIGURE 17.7**   Horse ked, *Hippobosca equina* (Hippoboscidae), female, dorsal view. (From Hutson, 1984.)

association with domestic dogs from southern Europe and the Mediterranean region to China. It appears best adapted to warm and arid climates. Up to one-third of dogs in parts of Egypt are infested with this louse fly. It is found mainly on dogs in the Palaearctic region and on wild carnivores in Africa. It has been recorded from the families Canidae (dogs, foxes), Viverridae (mongoose, civet), Hyaenidae (hyena), and Felidae (cats).

In 1972 *H. longipennis* was introduced into North America on captive cheetahs from Africa. Subsequently the species has been detected in the United States on cheetahs at wild animal or safari parks in California, Texas, Georgia, and Oregon. Efforts were made by officials in each of the affected states to eradicate this ectoparasite before it escaped from its introduced host to domestic pets, livestock, or wildlife. There is no evidence that this species has become established in the United States or elsewhere in the New World.

### Hippobosca equina

This species (Fig. 17.7) is normally a parasite of Equidae (horse, donkey, ass) and is a facultative parasite of cattle. Although widely spread in the Old World (Europe, northern Africa, western Asia), it does not occur on wild hosts. The original hosts are unknown. *H. equina* is a serious and common pest of a wide variety of domestic animals in Egypt (Hafez *et al.*, 1977). It can torment its hosts with painful bites and possibly act as a vector of disease agents, including those that cause piroplasmosis of horses, Q fever, and other types of rickettsioses.

### Hippobosca variegata

Tropical Africa is probably the center of distribution of this species, from which it has spread northward to the Mediterranean and eastward into Asia. It is normally a parasite of the domestic horse and its relatives

(*Equus* spp.) and cattle (*Bos* spp.). *H. variegata* is also reported from camels, dromedaries, and water buffalo in Africa and Asia, but these are considered facultative hosts.

### Deer keds (*Lipoptena* and *Neolipoptena* spp.)

Three species of *Lipoptena* and one of *Neolipoptena* are parasites of deer in North America, where they are called *deer keds*. In western North America *Lipoptena depressa* and *Neolipoptena ferrisi* are frequently found on the same host. The wings of deer keds are deciduous. They are fully developed and functional in the newly emerged adult, or volant, although the wing venation is greatly reduced. The wings are shed, probably within 48 hr after the ked reaches a suitable host. Prepupae are deposited while the ked is on the host but eventually fall to the ground.

### Lipoptena depressa

According to Bequaert (1957), *L. depressa* is the usual and common parasite of several races of *Odocoileus hemionus*. These include the Rocky Mountain mule deer (*O. h. hemionus*), the Columbian black-tailed deer (*O. h. columbianus*), the California black-tailed deer (*O. h. californicus*), and probably the southern black-tailed deer (*O. h. fulginatus*). It also probably parasitizes the western white-tailed deer (*O. virginianus leucurus*). Both deer species are efficient breeding hosts. *L. depressa* is present along the Pacific Coast from southern British Columbia to southern California and as far inland as Alberta (Canada) and western South Dakota and Nebraska. Its range includes most of the Rocky Mountain states but apparently not Arizona and New Mexico.

*L. depressa* has been divided into two subspecies: *L. d. depressa* and *L. d. pacifica* (Maa, 1969). *L. d. depressa* is limited in its distribution to the eastern slope of the Rocky Mountain highlands in western Montana, northern Wyoming, southwestern South Dakota, and northwestern Nebraska. The normal host is the Rocky Mountain mule deer. *L. d. pacifica* is found on the western slopes of the Rocky Mountain lowlands, including British Columbia (Canada) and the states of Washington, Oregon, Idaho, and California. *L. d. pacifica* normally breeds on Columbian black-tailed deer and the western subspecies of white-tailed deer.

*L. depressa* volants will alight on any moving object. Most leave quickly, however, without dropping the wings if they land on an accidental host such as humans or horses. Westrom and Anderson (1992) found that *L. depressa* was a bivoltine species in California, with volants appearing in peak numbers in October and April. On the host, peak populations of apterous adults occurred in midsummer and early winter following adult

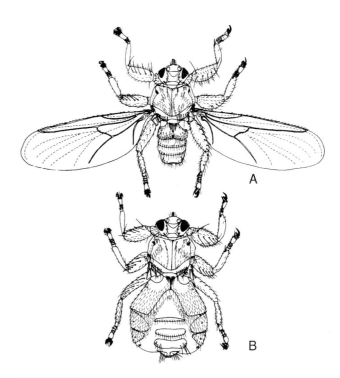

**FIGURE 17.8** Deer ked, *Lipoptena cervi* (Hippoboscidae), females, dorsal view. (A) Winged (alate); (B) wingless (dealate). (From Hutson, 1984.)

flights. Thousands of *L. depressa* may be found on an individual deer, with populations especially heavy in the fall.

### Lipoptena cervi

This deer ked (Fig. 17.8), is a common parasite of the true elk (*Alces alces*), red deer (*Cervus elaphus*), and other species of deer in Europe. In the United States it was first reported from New Hampshire and Pennsylvania in 1907. It now occurs in New York, New Hampshire, Massachusetts, and Pennsylvania. Bequaert (1942) believed that *L. cervi* was introduced into North America by humans with deer from Europe, and the species subsequently spread to white-tailed deer (*O. virginianus*). It also has been reported from another host native to the United States, the wapiti (*C. canadensis*).

Swarms of newly emerged, winged *L. cervi* appear in the fall, but apterous keds may be found on the host year round. In Russia mass emergence is observed late in August and the first part of September (Chistyakov, 1968). They are most active on warm, clear afternoons. They are concentrated in low places protected from the wind, in young deciduous forests. The puparia remain on the ground in areas where their hosts are normally found until they emerge in September. Puparia may be particularly numerous at wallows and places where the hosts rub and shed their winter coats.

### Lipoptena mazamae

This species is a parasite of white-tailed deer and brocket (*Mazama* spp.). The range of this tropical insect extends northward from Argentina through South and Central America into the United States, where it occurs in the states bordering the Gulf of Mexico and along the Atlantic coastal states to South Carolina. In surveys of white-tailed deer in southern Texas and Florida, it was the most prevalent ectoparasite (Forrester *et al.,* 1996; Samuel and Trainer, 1972). Volants may be active in every month from April through November. Numbers on the host appear to be significantly lower in the fall than in the spring or summer. Low populations of this species have been attributed to mortality of the puparia in areas of flooding or high rainfall.

### Neolipoptena ferrisi

This species occurs in the western United States from Canada to Mexico. Its range includes California, Oregon, Washington, the Rocky Mountain States, and as far east as South Dakota. This deer ked occurs on three races of *O. hemionus:* the Rocky Mountain mule deer (*O. h. hemionus*), the coastal black-tailed deer (*O. h. columbianus*), and the California black-tailed deer (*O. h. californicus*). Bequaert (1957) considered *O. hemionus* to be the only true breeding host of this insect and regarded records from western white-tailed deer (*O. virginianus leucurus*) and pronghorn antelope (*Antilocapra americana*) as accidental occurrences.

In dual infestations, *N. ferrisi* is normally outnumbered by *L. depressa. N. ferrisi* tends to be collected most frequently from the anterior regions of the body, with the highest population density on the head, whereas *L. depressa* is collected most frequently from the posterior regions of the body, including the tail (Westrom and Anderson, 1992).

### Pigeon fly (*Pseudolynchia canariensis*)

The *pigeon fly* (Fig. 17.9) is a winged hippoboscid that was introduced into North America at least a century ago. The earliest record in North America is 1896, when it was taken on pigeons at Savannah, GA (Knab, 1916). Its distribution now is nearly cosmopolitan. This is the only species of hippoboscid that parasitizes domesticated birds. In North America, it is found only on domestic pigeons (*Columba livia*). In the Old World it occurs both on domestic pigeons and on birds of several other avian orders. Juvenile birds are more frequently attacked. Heavily infested birds become emaciated and susceptible to

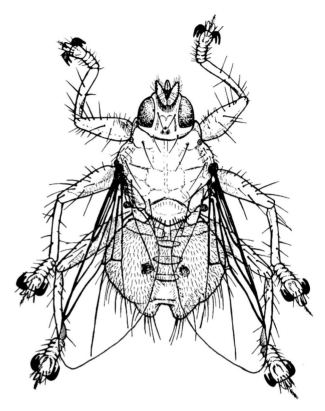

**FIGURE 17.9** *Pseudolynchia canariensis* (Hippoboscidae), female, dorsal view. (From Furman and Catts, 1982.)

secondary infections. The pigeon fly can bite humans and can be irritating to individuals handling domestic pigeons.

## PUBLIC HEALTH IMPORTANCE

Humans are not normal hosts of any hippoboscoid species. Occasionally, however, species such as the sheep ked and the pigeon fly bite humans and can be annoying to those routinely handling sheep or domestic pigeons, respectively. Pigeon flies also can be a problem to inhabitants of buildings that have become infested with feral pigeons. Hippoboscids that parasitize nondomesticated animals will bite, and even occasionally feed on, a human host. For example, *Olfersia coriacea*, normally a parasite of gallinaceous birds, has been reported attaching to and feeding on the back of the neck of a human in Panama (Harlan and Chaniotis, 1983). Deer keds in some parts of the world may constitute an annoyance because they swarm in large numbers and land on humans, getting into the hair and under clothing.

Human reactions to *hippoboscid bites* are variable. A redness and swelling at the bite site is reported by some individuals. Others report that the bite is painful or that there is subsequent irritation at the site of the bite. Individuals have even reported severe pain and swelling that required emergency medical attention. In India *M. ovinus* reportedly causes painful bites to shepherds engaged in shearing (Joseph *et al.,* 1991). On the other hand, the author has been bitten on numerous occasions by the sheep ked and has experienced no associated pain or swelling. In Russia, in the area of St. Petersburg, a dermatitis was reported among more than 300 individuals bitten by *L. cervi* during mass flights of volants in August and early September (Chistyakov, 1968).

The restriction of streblid bat flies and nycteribiid bat flies to bat hosts minimizes the potential for transmission of pathogens to other animals, including humans. Nycteribiidae have not been reported as biting animals other than bats (Marshall, 1970). Streblids, however, apparently will bite humans. Fritz (1983) has suggested that streblids might be important in maintaining and spreading disease-causing organisms among bats. As frequent blood feeders that move readily between bats, streblids have the potential to quickly spread pathogens from one host animal to another.

## VETERINARY IMPORTANCE

Louse flies directly affect their hosts by feeding on blood. Sometimes heavily infested animals become emaciated and susceptible to secondary infections. Juvenile birds and mammals are often more heavily infested with hippoboscids than are older animals of the same species. The body conditions of wintering hosts—birds or mammals—may be worsened by infestation of these parasites. In addition to the discomfort their biting causes, louse flies can be annoying to their hosts simply by crawling about on the body. Louse flies also serve as vectors of pathogens and parasites (Table I) and as disseminators of certain ectoparasitic arthropods. These include mammalian trypanosomes and filarial worms, avian trypanosomes, haemosporinan blood protozoans, lice, and mites.

Baker (1967) published a review of the role played by the Hippoboscidae as *vectors of endoparasites.* All known hippoboscid vectors of parasitic protozoans are members of two subfamilies: the Ornithomyinae on birds and the Lipopteninae on mammals. *Dipetalonema dracunculoides,* a parasitic filarial nematode of dogs and hyenas in the Old World, undergoes cyclical development in the dog fly (*H. longipennis*), which is thought to be its vector. Hippoboscids may transmit filariae of other mammals, particularly those of camels and lemurs, as well as ostriches and other birds. According to Pfadt and Roberts

(1978), the role of hippoboscids as vectors of pathogens is probably much greater than presently known. This may be true of streblids and nycteribiids as well.

The sheep ked transmits *Trypanosoma melophagium,* a nonpathogenic flagellate protozoan of sheep present wherever ked-infested sheep are found. Although it is distributed worldwide, this flagellate protozoan is rarely observed because it is present in relatively small numbers. The trypanosomes are ingested by the keds while they feed on sheep blood. The immature forms of the parasite develop in the posterior midgut of the sheep ked, while infective forms develop in the hindgut and are voided with the feces. They normally do not cross the gut wall into the hemolymph. Flagellates gain entry into the sheep when the keds or their feces are ingested. *L. capreoli,* an ectoparasite of domestic goats and the chamois goat in the Old World, transmits *T. theodori* in a similar manner, as does *Ornithomya avicularia,* which transmits *T. avium,* found in corvid birds. *P. canariensis* and *Stilbometopa impressa* are possible vectors of other avian trypanosomes of pigeons (Columbidae) and quail (Phasianidae).

Several hippoboscid flies have been identified as vectors of *Haemoproteus* species, haemosporidian blood parasites that cause bird malarias. The importance of hippoboscids in the natural transmission of most species of *Haemoproteus* is unknown. It is generally assumed that individual species of *Haemoproteus* are transmitted by either hippoboscid flies or *Culicoides* species (Ceratopogonidae), but not both.

Development of *Haemoproteus* in a hippoboscid vector is similar to that of the mosquito-borne malarial parasites in the genus *Plasmodium.* After microgamete production and fertilization of the macrogamete in the hippoboscid midgut, the zygote develops into a motile ookinete which penetrates and encysts on the outside of the wall of the stomach. The oocyst enlarges and its contents differentiate into sporozoites. The enlarged oocyst bursts to release the sporozoites, some of which enter the salivary glands to be introduced into the next host on which the hippoboscid feeds (Baker, 1967).

The best known *Haemoproteus* species is *Haemoproteus columbae,* which is parasitic in erythrocytes and visceral endothelial cells of the domestic pigeon (*C. livia*). It is transmitted by the pigeon fly, *P. canariensis.* Infections can result in anemia and unthriftiness in pigeons and cause economic losses to pigeon breeders in the form of nestling mortality. Several other species of *Haemoproteus* are transmitted by hippoboscid flies to a variety of avian hosts (Table I). Proven and presumed vectors of avian haemoproteids include species of *Pseudolynchia, Stilbometopa, Icosta, Ornithomya,* and other hippoboscid genera (Baker, 1967; Pfadt and Roberts, 1978).

Although they are neither mechanical nor biological vectors of any important diseases of sheep, *sheep keds* have

TABLE I

Species of Louse Flies (Hippoboscidae) in the United States and Selected Species of Veterinary Importance from Other Regions of the World: Included are Parasites and Disease Agents That They Reportedly Can Transmit

| Species | Hosts | Geographic Distribution | Parasites or disease agents |
|---|---|---|---|
| **Subfamily: Hippoboscinae** | | | |
| *Hippobosca longipennis* | Domestic dogs, hyenas | Southern Europe, northern Africa to China | *Dipetalonema dracunculoides* (filarial nematode) |
| **Subfamily: Lipopteninae** | | | |
| *Lipoptena cervi* | Deer, elk | Northeastern United States | None |
| *L. depressa* | White-tailed deer, mule deer | Western United States | *Corynebacterium lipoptenae* (bacterium; symbiote?) |
| *L. mazamae* | White-tailed deer, brocket | Southeastern United States to South America | None |
| *Melophagus ovinus* | Domestic sheep | Worldwide | *Trypanosoma malophagium, Rickettsia melophagi* (rickettsia; symbiote?), Bluetongue virus? |
| *Neolipoptena ferrisi* | Mule deer | Western United States | None |
| **Subfamily: Ornithomyinae** | | | |
| *Icosta albipennis* | Egrets, ibises, and other wading birds (Ciconiiformes) | Widespread | None |
| *I. americana* | Owls, hawks, grouse, turkeys | Widespread | None |
| *I. angustifrons* | Owls, hawks, falcons | Eastern United States | None |
| *I. ardeae boutaurinorum* | American bittern | Widespread | None |
| *I. hirsuta* | Quails, grouse, sage hens | Western USA | *Haemoproteus lophortyx* |
| *I. holoptera holoptera* | Rails | Eastern and central USA | None |
| *I. nigra* | Hawks, falcons | Widespread | None |
| *I. rufiventris* | Hawks, falcons, owls | Eastern and central USA | *Haemoproteus lopohortyx* |
| *Microlynchia pusilla* | Pigeons, doves, quails, roadrunners | Widespread | *H. columbae, H. maccallumi?, H. sacharovi?* |
| *Olfersia bisulcata* | Vultures | Texas | None |
| *O. fumipennis* | Osprey | Widespread | None |
| *O. sordida* | Pelicans, cormorants | Widespread | None |
| *O. spinifera* | Frigate birds | Florida and Louisiana | None |
| *Ornithoctona erythrocephala* | Hawks, pigeons, others | Widespread | None |
| *O. fusciventris* | Warblers, flycatchers, others | Widespread | None |
| *Ornithoica confluenta* | Egrets, ibises, other wading birds | Florida | None |
| *O. vicina* | Owls, sparrows, others | Widespread | None |
| *Ornithomya anchineuria* | Hawks, crows, sparrows, others | Widespread | None |
| *O. bequaerti* | Small passerine birds | Widespread | None |
| *Pseudolynchia brunnea* | Whip-poor will, nighthawks | Widespread | None |
| *P. canariensis* | Domestic pigeons | Widespread | *H. columbae, H. maccallumi, T. hannai?* |
| *Stilbometopa impressa* | Quails and related game birds | Western USA | *H. lophortyx, Trypanosoma* sp. |
| *S. podopostyla* | Wild pigeons, doves | California | *T. avium* |

Unless otherwise indicated, the parasites and disease agents are all blood protozoans (adapted from Baker, 1967, and Pfadt and Roberts, 1978).

been shown to be capable of transmitting bluetongue virus in experimental studies (Luedke *et al.,* 1965). Such transmission, if it even occurs naturally, is probably only mechanical. In another study sheep keds were unable to transmit *Anaplasma ovis,* the etiologic agent of ovine anaplasmosis, from infected to uninfected sheep (Zaugg and Coan, 1986).

Feeding by the sheep ked can cause a defect in sheepskins called *cockle* or *rib cockle* (Fig 17.10). Blemishes appear at the individual bite sites and are presumed to be the result of an allergic reaction to the salivary secretions of the feeding keds. The result is scattered, dense, brownish nodules in the grain layer of sheepskin, which seriously downgrades both grain and suede types of leather. The nodules of dense fibrous material cannot be flattened out and are impenetrable to dyes (Fig. 17.11). This defect, especially damaging in garment suede, causes economic losses of several million dollars to the leather industry

**FIGURE 17.10**  Feeding damage caused by the sheep ked, *Melophagus ovinus* (Hippoboscidae). Grain side of pickled sheepskin showing pitted surface, or cockle. (Courtesy of USDA.)

in the United States each year. When sheep keds are eliminated, the skin recovers from the effect of the bites, resulting in usable pelts. The length of time required for recovery by the living animal has not been determined but may be several weeks. In New Zealand a similar defect is caused by the sheep biting louse (*Bovicola ovis*).

Results of studies of weight gains and wool growth of sheep parasitized by sheep keds are equivocal. Several reports in the literature indicate no adverse effects due to sheep ked infestations. In a study of the effect of keds on weight gains of feeder lambs in Wyoming, for example, there was no significant difference in gain between ked-infested and uninfested lambs. In that study the number of keds infesting untreated lambs markedly decreased during the period of lamb feeding (Pfadt *et al.*,

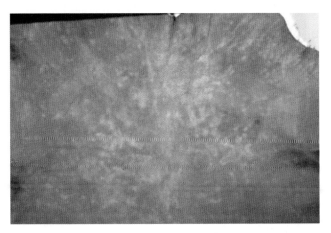

**FIGURE 17.11**  Discoloration and mottling of sheep leather due to feeding by sheep ked, *Melophagus ovinus* (Hippoboscidae). (Courtesy of J. E. Lloyd.)

1953). In a study in Canada, Nelson and Slen (1968) found that ked-free lambs on various diets gained approximately 1.4–3.6 kg (3–8 lb) more than infested lambs and that uninfested yearling ewes produced about 11% more wool than infested ones. In a study in New Mexico, a 2% higher dressing percentage was observed in carcasses of uninfested lambs, and carcass weights were significantly heavier (0.9 kg per animal) (Everett *et al.*, 1969). Fleece length was about 8% longer and the percentage of clean fiber was approximately 7% greater in uninfested lambs, while the difference in clean, dry weight was about 20% in favor of uninfested animals.

Many ranchers in the western United States believe that heavy infestations of sheep ked contribute to the incidence of *back loss*. Back loss is a term referring to the death of ewes that roll onto their backs in an apparent attempt to relieve irritation caused by keds. An animal that becomes stuck and remains on its back will eventually suffocate due to the pressure of its internal organs against the diaphragm.

## PREVENTION AND CONTROL

Control technology has not been developed for the vast majority of the Hippoboscoidea, and most of these parasites are to be endured. The few species that affect domestic animals and birds may be controlled through treatment of the host with insecticide formulations. The pigeon fly, for example, is controlled by periodic cleaning of the pigeon loft and, as necessary, dusting squabs with an insecticidal dust.

The sheep ked is the only species for which an extensive control technology has been developed. *Shearing* prior to lambing can reduce sheep ked populations by approximately 75%. Shearing not only removes many pupal and adult keds with the wool, but also kills many that are cut with the shears. If ewes are not shorn prior to lambing, substantial numbers of keds can transfer from full-fleeced ewes to their lambs where they are not subjected to the hazards of shearing.

*Insecticidal treatment* of sheep in the spring following shearing is a common and effective practice. Best results are achieved when ewes are treated following shearing but before lambing. Often, the ewes are treated as they leave the shearing shed. Several traditional treatment methods include whole-body *sprays*, *dusts*, and *dips*. Fall treatments tend to be less effective, possibly due to the greater length of the fleece at that time.

Other methods of treatment are the *pour-on* and *low-volume spray* applications. Particularly effective are pyrethroid insecticides, which have become popular because of their ease of application. The pour-on method

**FIGURE 17.12**   Pour-on application of insecticide for control of sheep keds. (Courtesy of J. E. Lloyd.)

entails applying a few ounces of insecticide along the backline of each animal (Fig. 17.12). The chemical may be poured from a calibrated dipper or forcefully expelled into the fleece with an application gun. These methods are usually more effective if one waits a few weeks after shearing to permit growth of sufficient wool to retain the liquid formulation.

In the low-volume application, less than an ounce of insecticide is applied to each animal at a pressure of about 50 psi. Large numbers of sheep may be treated in a short period of time by driving the sheep through a spray race equipped with one or more stationary spray nozzles. Animals can be driven through the spray race at approximately one animal per second, providing a treatment method well suited for large range flocks.

Several western states in the United States (Colorado, Idaho, Montana, North Dakota, South Dakota, Utah, and Wyoming) have adopted statewide voluntary ked-free programs. These programs involve regular surveillance of sheep, usually at shearing, and prompt treatment for sheep keds when necessary. These programs are normally coordinated by sheep-producer associations within each state. The first statewide ked-free program in the United States was implemented in 1986 by the Wyoming Woolgrowers Association and currently includes 80% of the sheep in that state.

## REFERENCES AND FURTHER READING

Atkinson, C. T. (1991). Vectors, epizootiology, and pathogenicity of avian species of *Haemoproteus* (Haemosporina: Haemoproteidae). *Bulletin of the Society of Vector Ecology* **16**, 109–126.

Baker, J. R. (1967). A review of the role played by the Hippoboscidae (Diptera) as vectors of endoparasites. *Journal of Parasitology* **53**, 412–418.

Baker, J. R. (1978). Biology of trypanosomes of birds. In "Biology of the Kinetoplastids" (W. H. R. Lulmsden and D. A. Evans, Eds.), Vol. 1, pp. 131–174. Academic Press, London.

Baron, R. W., and Nelson, W. A. (1985). Aspects of the humoral and cell-mediated immune responses of sheep to the ked *Melophagus ovinus* (Diptera: Hippoboscidae). *Journal of Medical Entomology* **22**, 544–549.

Bequaert, J. C. (1942). A monograph of the Melophaginae, or ked-flies, of sheep, goats, deer and antelopes (Diptera, Hippoboscidae). *Entomologica Americana* (new series) **22**, 1–220.

Bequaert, J. C. (1952–1953). The Hippoboscidae or louse-flies (Diptera) of mammals and birds. Part 1. Structure, physiology and natural history. *Entomologica Americana* (new series) **32**, 1–209 (1952), **33**, 211–442 (1953).

Bequaert, J. C. (1954–1957). The Hippoboscidae or louse-flies (Diptera) of mammals and birds. Part II. Taxonomy, evolution and revision of American genera and species. *Entomologica Americana* (new series) **34**, 1–232 (1954), **35**, 233–416 (1955), **36**, 417–611 (1957).

Borror, D. J., Triplehorn, C. A., and Johnson, N. F. (1989). "An Introduction to the Study of Insects." Saunders, Philadelphia.

Chistyakov, A. F. (1968). Skin lesions in people due to bite of *Lipoptena cervi*. *Vestnik Dermatologii i Venerologii* **42**, 59–62.

Cline, L. D., and Throne, J. E. (1993). Seasonal flight activity of *Lipoptena mazmae* (Diptera: Hippoboscidae) in South Carolina. *Entomological News* **104**, 129–132.

Constantine, D. G. (1970). Bats in relation to health, welfare and economy of man. In "Biology of Bats" (W. A. Wimsatt, Ed.), Vol. 2, pp. 319–449. Academic Press, San Diego.

Drummond, R. O., Lambert, G., Smalley, H. E., Jr., and Terrill, C. E. (1981). Estimated losses of livestock to pests. In "CRC Handbook of Pest Management in Agriculture" (D. Pimentel, Ed.), Vol 1, pp. 111–127. CRC Press, Boca Raton, FL.

Evans, G. O. (1950). Studies on the bionomics of the sheep ked, *Melophagus ovinus* L., in West Wales. *Bulletin of Entomological Research* **40**, 459–478.

Everett, A. L., Roberts, I. H., Willard, H. J., Apodaca, S. A., Bitcover, E. H., and Naghski, J. (1969). The cause of cockle, a seasonal sheep-skin defect, identified by infesting a test flock with keds (*Melophagus ovinus*). *Journal of the American Leather Chemists' Association* **64**, 460–476.

Everett, A. L., Roberts, I. H., and Naghski, J. (1971). Reduction in leather value and yields of meat and wool from sheep infested with keds. *Journal of the American Leather Chemists' Association* **66**, 118–130.

Forrester, D. J., McLaughlin, G. S., Telford, S. R., Jr., Foster, G. W., and McCown, J. W. (1996). Ectoparasites (Acari, Mallophaga,

Anoplura, Diptera) of white-tailed deer, *Odocoileus virginianus,* from southern Florida. *Journal of Medical Entomology* **33,** 96–101.

Fritz, G. N. (1983). Biology and ecology of bat flies (Diptera: Streblidae) on bats in the genus *Carollia. Journal of Medical Entomology* **20,** 1–10.

Furman, D. P., and Catts, E. P. (1970). Manual of medical entomology. Fourth edition. Cambridge University Press.

Graham, P. P. H., and Taylor, K. L. (1941). "Studies on Some Ectoparasites of Sheep and Their Control. 1. Observations on the Bionomics of the Sheep Ked (*Melophagus ovinus*)." Council of Scientific and Industrial Research, Pamphlet No. 108, pp. 8, 9–26. Melbourne, Australia.

Hafez, M, Hilali, M., and Fouda, M. (1977). Biological studies on *Hippobosca equina* (L.) (Diptera: Hippoboscidae) infesting domestic animals in Egypt. *Zeitschrift fuer Angewandte Entomologie* **83,** 426–441.

Hagan, H. R. (1951). "Embryology of the Viviparous Insects." Ronald Press, New York.

Hare, J. E. (1945). Flying stages of the deer lousefly, *Lipoptena depressa* (Say), in California (Diptera, Hippoboscidae). *Pan-Pacific Entomologist* **21,** 48–57.

Harlan, H. J., and Chaniotis, B. N. (1983). Report of *Olfersia coriacea* (Diptera: Hippoboscidae) feeding on a human in Panama. *Journal of Parasitology* **69,** 1026.

Heath, A. C. G., Cooper, S. M., Cole, D. J. W., and Bishop, D. M. (1994). Evidence for the role of the sheep biting louse *Bovicola ovis* in producing cockle, a sheep pelt defect. *Veterinary Parasitology* **59,** 53–58.

Hutson, A. M. (1984). "Keds, Flat-Flies and Bat-Flies, Diptera, Hippoboscidae and Nycteribiidae. Handbooks for the Identification of British Insects." Vol. 10, Part 7. Royal Entomological Society of London.

Jobling, B. (1929). A comparative study of the structure of the head and mouthparts in the Streblidae (Diptera: Pupipara). *Parasitology* **18,** 319–349.

Joseph, S. A., Karunamoorthy, G., Ramachandran, P. K., Sukumaran, D., and Rao, S. S. (1991). Studies on the haematophagous arthropods of zoonotic importance in Tamil Nadu. Entomology for Defense Services. In "Proceedings of the Symposium held on 12–14 September 1990," pp. 185–192. Defense Research & Development Establishment, Gwalior, India.

Keirans, J. E. (1975). A review of the phoretic relationship between Mallophaga (Phthiraptera: Insecta) and Hippoboscidae (Diptera: Insecta). *Journal of Medical Entomology* **12,** 71–76.

Knab, F. (1916). Four European Diptera established in North America. *Insecutor Inscitiae Menstruus* **4,** 1–4.

Laidet, M. (1969). L'orinine de la noisillure le Melophage. *Technicuir* **4,** 39–50.

Legg, D. E., Kumar, R., Watson, D. W., and Lloyd, J. E. (1991). Seasonal movement and spatial distribution of the sheep ked (Diptera: Hippoboscidae) on Wyoming lambs. *Journal of Economic Entomology* **84,** 1532–1539.

Lenoble, B. J., and Denlinger, D. L. (1982). The milk gland of the sheep ked, *Melophagus ovinus:* a comparison with *Glossina. J. Insect. Physiol.* **20,** 165–172.

Lloyd, J. E. (1985). Arthropod pests of sheep. In "Livestock Entomology" (R. E. Williams, R. D. Hall, A. B. Broce, and P. J. Scholl, Eds.), pp. 253–267. Wiley, New York.

Lloyd, J. E., Olson, E. J., and Pfadt, R. E. (1978). Low-volume spraying of sheep to control the sheep ked. *Journal of Economic Entomology* **71,** 548–550.

Lloyd, J. E., Pfadt, R. E., and Olson, E. J. (1982). Sheep ked control with pour-on applications of organophosphorus insecticides. *Journal of Economic Entomology* **75,** 5–6.

Luedke, A. J., Jochim, M. M., and Bowne, J. G. (1965). Preliminary bluetongue transmission with the sheep ked *Melophagus ovinus* (L.). *Canadian Journal of Comparative Medicine and Veterinary Sciences* **29,** 229–231.

Maa, T. C. (1963). Genera and species of Hippoboscidae (Diptera): types, synonymy, habitats and natural groupings. *Pacific Insects Monograph* **6,** 1–186.

Maa, T. C. (1966). Studies in Hippoboscidae (Diptera). *Pacific Insects Monograph* **10,** 1–148.

Maa, T. C. (1969). Studies in Hippoboscidae (Diptera). Part 2. *Pacific Insects Monograph* **20,** 1–312.

Maa, T. C. (1971). Studies in batflies (Diptera: Streblidae, Nycteribiidae). Part I. *Pacific Insects Monograph* **28,** 1–248.

Maa, T. C., and Peterson, B. V. (1987). Hippoboscidae. In "Manual of Nearctic Diptera" (J. F. McAlpine, B. V. Peterson, G. E. Shewell, H. J. Teskey, J. R. Vockeroth, and D. M. Wood, Eds.), Vol. 2, pp. 1271–1281. Monograph 28. Research Branch, Agriculture Canada, Ottawa.

Marshall, A. G. (1970). The life cycle of *Basilia hispida* Theodor 1967 (Diptera: Nycteribiidae) in Malaysia. *Parasitology* **61,** 1–18.

Marshall, A. G. (1981). "The Ecology of Parasitic Insects." Academic Press, San Diego.

Nelson, W. A. (1958). Transfer of sheep keds, *Melophagus ovinus* (L.), from ewes to their lambs. *Nature* **181,** 56.

Nelson, W. A., and Bainborough, A. R. (1963). Development in sheep of resistance to the ked *Melophagus vinus* (L.). III. Histopathology of sheep skin as a clue to the nature of resistance. *Experimental Parasitology* **13,** 118–127.

Nelson, W. A., and Petrunia, D. M. (1969). *Melophagus ovinus:* feeding mechanism on transilluminated mouse ear. *Experimental Parasitology* **26,** 308–313.

Nelson, W. A., and Qually, M. C. (1958). Annual cycles in numbers of the sheep ked, *Melophagus ovinus* (L.). *Canadian Journal of Animal Science* **38,** 194–199.

Nelson, W. A., and Slen, S. B. (1968). Weight gains and wool growth in sheep infested with the sheep ked *Melophagus ovinus. Experimental Parasitology* **22,** 223–226.

Peterson, B. V., and Wenzel, R. L. (1987). Nycteribiidae. In "Manual of Nearctic Diptera" (J. F. McAlpine, B. V. Peterson, G. E. Shewell, H. J. Teskey, J. R. Vockeroth and D. M. Wood, Eds.), Vol. 2, pp. 1283–1291. Monograph 28. Research Branch, Agriculture Canada, Ottawa.

Pfadt, R. E., and Roberts, I. H. (1978). Louse flies (family Hippoboscidae). In "Surveillance and Collection of Arthropods of Veterinary Importance" (R. A. Bram, Ed.), pp. 60–71. US Department of Agriculture, Agriculture Handbook No. 518. Animal and Plant Health Inspection Service, Washington, DC.

Pfadt, R. E., Paules, L. H., and DeFoliart, G. R. (1953). Effect of the sheep ked on weight gains of feeder lambs. *Journal of Economic Entomology* **46,** 95–99.

Pfadt, R. E., Lloyd, J. E., and Spackman, E. W. (1973). "Control of Insects and Related Pests of Sheep." Agriculture Experiment Station Bulletin 514R. Univ. of Wyoming, Laramie, Wyoming.

Philips, J. R., and Fain, A. (1991). Acarine symbionts of louseflies (Diptera: Hippoboscidae) *Acarologia* **32,** 377–384.

Samuel, W. M., and Trainer, D. O. (1972). *Lipoptena mazamae* Rodani, 1878 (Diptera: Hippoboscidae) on white-tailed deer in southern Texas. *Journal of Medical Entomology* **9,** 104–106.

Schlein, Y. (1970). A comparative study of the thoracic skeleton and musculature of the Pupipara and the Glossinidae (Diptera). *Parasitology* **60,** 327–373.

Strickman, D., Lloyd, J. E., and Kumar, R. (1984). Relocation of hosts by the sheep ked (Diptera: Hippoboscidae). *Journal of Economic Entomology* **77,** 437–439.

Swingle, L. D. (1913). The life-history of the sheep-tick *Melophagus ovinus*. Agricultural Experiment Station Bulletin 99. Univ. of Wyoming, Laramie, Wyoming.

Theodor, O. (1964). On the relationships between the families of the Pupipara. In "Proceedings of the First Congress on Parasitology, Rome, Italy." pp. 999–1000. Pergamon Press.

Theodor, O. (1967). "An Illustrated Catalogue of the Rothschild Collection of Nycteribiidae (Diptera) in the British Museum (Natural History) with Keys and Short Descriptions for the Identification of Subfamilies, Genera, Species and Subspecies." British Museum (Natural History) Publication 655. British Museum (Natural History), London.

Theodor, O. (1968). New species and new records of Nycteribiidae from the Ethiopian, Oriental, and Pacific regions. *Parasitology* **58**, 247–276.

Theodor, O. (1975). "Fauna Palaestina, Insecta I: Diptera Pupipara." Publication of the Israel Academy of Sciences and Humanities, Section of Sciences. Jerusalem Post Press, Jerusalem.

Theodor, O., and Oldroyd, H. (1964). Hippoboscidae. In "Die Fliegen der Palearktischen Region" (E. Lindner, Ed.), Vol. 65, pp. 1–70.

Wenzel, R. L., and Peterson, B. V. (1987). Streblidae. In "Manual of Nearctic Diptera" (J. F. McAlpine, B. V. Peterson, G. E. Shewell, H. J. Teskey, J. R. Vockeroth, and D. M. Wood, Eds.), Vol. 2, pp. 1293–1301. Monograph 28. Research Branch, Agriculture Canada, Ottawa.

Wenzel, R. L., Tipton, V. J., and Kiewlicz, A. (1966). The streblid batflies of Panama. In "Ectoparasites of Panama" (R. L. Wenzel and V. J. Tipton, Eds.), pp. 405–675. Field Museum of Natural History, Chicago.

Westrom, D. R., and Anderson, J. R. (1992). The distribution and seasonal abundance of deer keds (Diptera: Hippoboscidae) on Columbian black-tailed deer (*Odocoileus hemionus columbianus*) in northern California. *Bulletin of the Society of Vector Ecology* **17**, 57–69.

Zaugg, J. L., and Coan, M. E. (1986). Test of the sheep ked *Melophagus ovinus* (L.) as a vector of *Anaplasma ovis* Lestoquard. *American Journal of Veterinary Research* **47**, 1060–1062.

# 18

# MOTHS AND BUTTERFLIES (*Lepidoptera*)

GARY R. MULLEN

*Many moths* and butterflies are recognized as economic pests of row crops, fruit and shade trees, ornamental shrubs, and other plantings on which their larvae feed. Others are household pests which infest cereals, grains, and other stored products or attack woolen fabrics and other materials of animal origin. Adult moths can also be a nuisance because of their attraction to lights, often entering homes at night. Butterflies, however, are seldom pests. The adults are generally viewed as colorfully attractive insects which are a pleasure to see in flight or visiting flowers for nectar. However, a number of lepidopteran species, notably moths, can cause significant health problems for humans and other animals.

In most cases of a medical-veterinary nature, it is the larval stage that is involved. The larvae of many species are armed with poisonous setae or spines which, on contact with skin, can cause a stinging or burning sensation. Under certain circumstances, stinging caterpillars may be ingested by domestic animals, resulting in gastrointestinal problems. This most commonly occurs when animals such as cattle and horses graze on infested forage. Adults can also contribute to health problems. The inhalation of wing scales and body hairs of adult moths can induce allergic reactions, whereas contact with the silk of certain species can cause allergic responses in sensitized individuals. The most unusual lepidopterans from a medical-veterinary perspective are species which, as adults, feed on animal fluids. Some moths feed about the eyes, whereas others are attracted to wounds and in some cases can actually penetrate the skin of humans and other animals to feed directly on blood.

The general term *lepidopterism* is applied to adverse reactions of humans and other animals to moths and butterflies. The term also is used in a more restricted sense for cases involving only the adults. Reactions to larvae are called *erucism* (L. *eruca,* caterpillar). Most cases involve *urticaria* (L. *urtica,* nettle), a vascular reaction of the skin in the form of papules or wheals, caused by contact with the specialized defensive spines or setae of certain species. On rare occasions, lepidopterous larvae invade animal tissues, causing *scoleciasis* (G. *scolec,* worm).

An excellent overview of the Lepidoptera, including morphology, developmental stages, behavior, and higher taxonomic classification, is provided by Scoble (1992). For information specifically on Lepidoptera-related problems of medical-veterinary significance, the following reviews are recommended: Delgado (1978), Southcott (1978, 1983), Kawamoto and Kumada (1984), and Wirtz (1984).

## TAXONOMY

Classification of the Lepidoptera above the superfamily level is subject to debate among taxonomists. The simple separation of the order into two groups, moths and butterflies, is no longer phylogenetically appropriate. Nonetheless, for the purpose of this chapter, the terms "moths" and "butterflies" will be used in the generally accepted context for the species discussed. Virtually all species of medical-veterinary importance are members of the following four superfamilies: Bombycoidea, Noctuoidea, Papilionoidea, and Zygaenoidea. For details on the higher classification of the Lepidoptera, see Kristensen (1984), Nielsen and Common (1991), and Scoble (1992).

Among the more than 100 lepidopteran families, 14 include species which as larvae cause health-related problems. Twelve of these families are moths and two are butterflies. They represent more than 60 genera and 100 species worldwide. Members of at least 9 families and 42 genera are known to cause medically related problems as larvae in North America. The families most commonly encountered as problems are Limacodidae, Megalopygidae, and Saturniidae. Other important families in various parts of the world are the Arctiidae, Lasiocampidae, and Lymantriidae.

Six families of Lepidoptera that include species in which adults feed on animal wounds and various body secretions are Geometridae, Noctuidae, Notodontidae, Pyralidae, Sphingidae, and Thyatiridae. The species most commonly observed feeding on animals are *Lobocraspis griseifusa* (Noctuidae), *Hypochrosis* species (Geometridae), and *Filodes* and *Microstega* species (Pyralidae) in Southeast Asia. In Africa, *Arcyophora* species (Noctuidae) are more important. The only species which are known to be capable of piercing vertebrate skin are members of the noctuid genus *Calyptra,* also in Southeast Asia.

## MORPHOLOGY

Adult moths and butterflies are easily recognized by their scale-covered wings, wing venation, and long, coiled

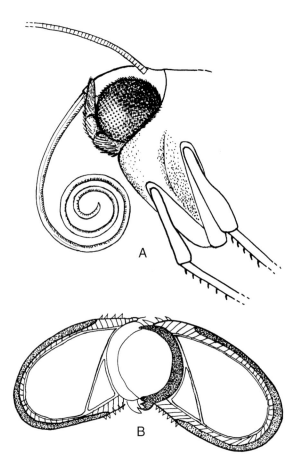

**FIGURE 18.1**    Anterior portion of adult moth, showing location and structure of proboscis. (A) Proboscis coiled beneath head when not feeding; (B) cross-section midway along length of proboscis, showing central food channel formed by the interlocked pair of maxillary galeae. (A, Original by Margo Duncan; B, Redrawn from Scoble, 1992.)

proboscis or feeding tube (Fig. 18.1). The proboscis serves primarily as a means of imbibing fluids such as nectar, fruit juices, honeydew, and water. While feeding, fluid is drawn through a channel formed by the tightly interlocked pair of elongate maxillae (galeae), each of which bears a longitudinal groove along its inner surface (Fig. 18.1B). From there it passes into the alimentary tract by the contraction of pharyngeal muscles. In species which as adults feed on wounds and body fluids of animals, the only significant differences in their feeding mechanism are external modifications, particularly near the tip of the proboscis, to facilitate the rasping or piercing of tissues (Fig. 18.2).

The larvae of moths and butterflies are called *caterpillars.* The larva is typically cylindrical, with a well developed head capsule, three pairs of thoracic legs, and five pairs of fleshy, unsegmented *prolegs,* one pair each on abdominal segments 3–6 and 10 (Fig. 18.3). Such larvae are called *eruciform.* Prolegs usually bear tiny hooks called *crochets* which aid them in clinging to various surfaces while moving about. The mouthparts are adapted

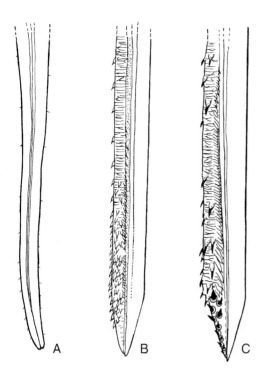

**FIGURE 18.2** Modifications of the distal tip of the proboscis in fruit-piercing and skin-piercing moths. (A) Unmodified proboscis of nectar-sucking moth, *Autographa gamma* (Noctuidae); (B), Fruit-piercing moth, *Scolyopteryx libatrix* (Noctuidae); (C) skin-piercing, blood-sucking moth, *Calyptra eustrigata* (Noctuidae). (Redrawn from Bänziger, 1971.)

for chewing plant material on which the larvae feed. The general body form of larvae is highly modified in some families to the extent that they may not be recognized as lepidopteran larvae by the nonspecialist. Examples include some of the more medically important taxa such as puss caterpillars and hag moths.

Caterpillars which cause dermatitis on contact with vertebrate skin are protected by specialized hairs and spines. In some cases these structures cause simple mechanical injury or irritation when they penetrate the skin. In other cases, they have associated poison glands which

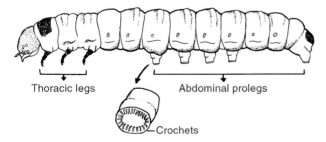

**FIGURE 18.3** External morphology of typical lepidopteran larva, with enlargement of abdominal proleg showing tiny hooks, or crochets. (Redrawn from Romoser, 1981.)

secrete toxic substances that elicit varying degrees of inflammation and swelling at the contact site. The location, numbers, and types of urticating hairs and spines vary significantly among different families and genera (Fig. 18.4).

Setae and spines are produced by specialized *trichogen cells*, literally "hair-forming" cells. These cells secrete multiple layers of cuticle to form the wall structure, differentiated from the surrounding epidermis. Each spine typically articulates with a socket formed by a *tormogen cell*, or "socket-forming" cell. Setae, or hairs, are generally formed by one or a few trichogen cells, whereas the larger, more robust spines are multicellular in origin and are produced by many trichogen cells. These setae and spines usually are innervated and associated with supporting cells and, depending on the taxon, may or may not have poison-secreting cells and tracheoles.

The specialized setae and spines of urticating lepidopterans are highly variable in structure. The following categories are based on Kawamoto and Kamada's (1984) modification of a classification proposed by Kano (1967, 1979), which includes types found in adults as well as larvae. According to their classification, there are two major groups of urticating structures: spicule hairs and spine hairs. *Spicule hairs* are detachable setae which are readily rubbed off or can become airborne to cause dermatitis on contact with animal skin. These hairs are easily shed and often are incorporated into the silk of the pupal cocoon. Although some of them cause only mechanical injury, others contain toxins which can affect vertebrates on contact. *Spine hairs*, on the other hand, cause urticaria only when the caterpillar makes direct contact with the skin. They often have associated poison glands and must be innervated in order for the toxins to be released. The following seven types of spicule hairs and four types of spine hairs are recognized.

## SPICULE HAIRS

**Type 1 (*Euproctis* type).** These are very small hairs (length 50–200 μm; diameter up to 5 μm), each of which has a pointed end that articulates with its own socket. The distal end has multiple small barbs. Upon detaching, the pointed tip of the hair penetrates the skin. They occur as small clusters of 3–15 hairs each in cup-shaped papillae (Fig. 18.5A) on various parts of the caterpillar. Type 1 hairs are characteristic of the brown-tail moth (*Euproctis chrysorrhoea*) and other *Euproctis* species in the family Lymantriidae. The number of these papillae can be extremely large, as in *E. similis* larvae, with an estimated 600,000 spicule hairs, and *E. subflava* larvae, with more than 6 million.

**Type 2 (*Thaumetopoea* type).** These spicule hairs are similar to type 1 in size and shape but are pointed at

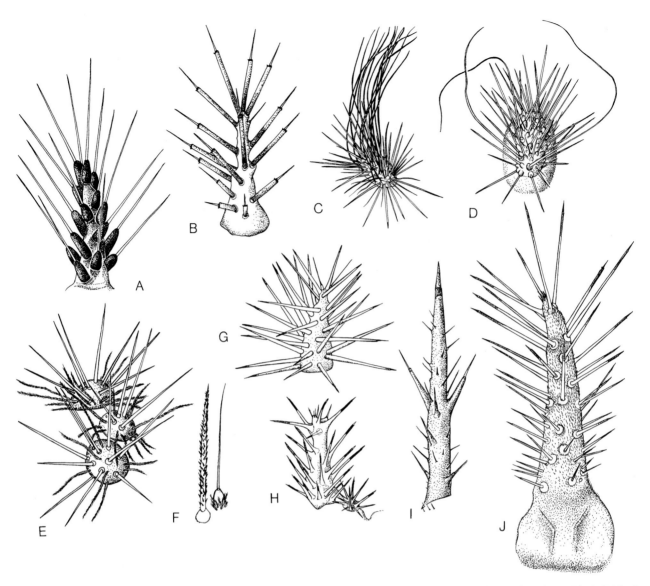

**FIGURE 18.4**    Urticating hairs and spines of North American caterpillars. (A) Pandora caterpillar, *Coloradia pandora* (Saturniidae); (B) buck moth, *Hemileuca maia* (Saturniidae); (C) puss caterpillar, *Megalopyge opercularis* (Megalopygidae); (D) white flannel moth, *Norape ovina* (Megalopygidae); (E) smeared dagger moth, *Acronicta oblinita* (Noctuidae); (F) hag moth, *Phobetron pithecium* (Limacodidae); (G) io moth, *Automeris io* (Saturniidae); (H) spiny oak-slug caterpillar, *Euclea delphinii* (Limacodidae); (I) mourningcloak butterfly, *Nymphalis antiopa* (Nymphalidae); (J) saddleback caterpillar, *Sibine stimulea* (Limacodidae). (Original by Margo Duncan.)

both ends. They are inserted point-downward into individual cuplike sockets and occur only in third-instar and older larvae. They are typical of the family Thaumetopoeidae. The processionary caterpillar (*Thaumetopoea processionea*) is estimated to have over 630,000 of these specialized defensive hairs.

**Type 3 (*Dendrolimus* type).**    These are relatively long (0.5–1.0 mm), slender spicule hairs which have blunt proximal ends and sharply pointed distal tips (Fig. 18.5B). They are loosely articulated with individual sockets and are easily broken off. In addition to mechanical

injury to the skin, they can cause localized reactions upon discharge of toxin when the spicule wall is broken. These urticating hairs occur in adults of the Lasiocampidae, notably on the mesothoracic and metathorax of the genus *Dendrolimus*.

**Type 4 (*Latoia* type).**    These spicule hairs are relatively short (0.5–1.0 mm) and stout with a pointed distal tip and 3–4 basal barbs. They are multicellular in origin, typically having a large poison-secreting cell surrounded by supporting cells. They are characteristic of slug caterpillars of the genus *Latoia*, family Limacodidae,

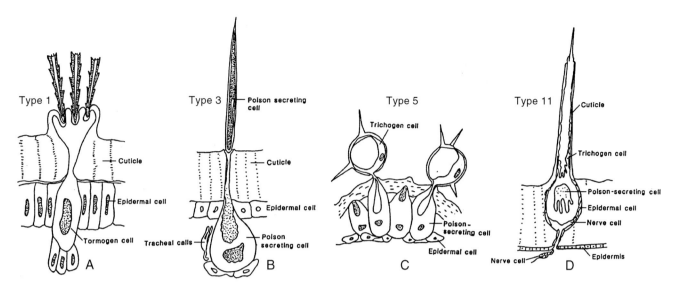

**FIGURE 18.5**   Representative types of spicule hairs and spine hairs in urticating caterpillars. (A) Type 1 spicule hair (*Euproctis*, Lymantriidae); (B) Type 3 spicule hair (*Dendrolimus*, Lasiocampidae); (C) Type 5 spicule hair (starlike, Limacodidae); (D) Type 11 spine hair (*Latoia*, Megalopygidae). (Redrawn and modified from Kawamoto and Kumada, 1984.)

where they occur on the ninth segment of the last larval instar.

**Type 5 (starlike hair).**   These highly specialized urticating hairs are very small, compact, rhomboid-shaped structures (Fig. 18.5C), each produced by a single trichogen cell. Projecting from the outer wall are tiny, pointed spikes or prickles which cause a nettling sensation on contact with skin. These spicule hairs are characteristic of many limacodid larvae, where they are found primarily in clusters on the lateral tubercles.

**Type 6 (brush hair).**   These long (ca. 4.0 mm), slender spicule hairs occur in dense, brushlike clusters, appearing as tufts on the abdominal segments. The basal end of each hair is sharply pointed with many barbed hooks. They are easily detached and commonly incorporated with the silk as larvae spin their cocoons. These urticating hairs are common in the Lymantriidae, including *Calliteara pundibunda* and the Douglas-fir tussock moth *Orygia pseudotsugata*. They also occur in the arctiid genus *Premolis* and in the Australian anthelid known as the hairymarry caterpillar, *Anthela nicothoe.*

**Type 7 (moth spicules).**   These spicule hairs occur only in adult moths. They are similar to type 2 hairs, from which they differ primarily in their sockets being located in papillae. The spicule hairs themselves are very small (170 $\mu$m long, 3–5 $\mu$m in diameter), with sharp tips and tiny barbs on the upper one-third of the shaft. They are characteristic of the genus *Hylesia*, family Saturniidae, and adults of several genera of the Thaumetopoeidae, such as *Anaphe, Epanaphe, Epicoma,* and *Gazalina.*

## Spine Hairs

**Type 8 (primitive type).**   These are structurally similar to normal body hairs, with sharply pointed tips and varying numbers of tiny barbs projecting from their surface. Dermatitis is usually limited to mechanical injury by these hairs because most species lack associated poison-producing cells. Exceptions occur in the genera *Chalcosia* and *Erasmia* (Zygaenidae), in which poison glands are located at the base of the hairs. Type 8 spine hairs are characteristic of larvae of most species of Arctiidae and Noctuidae and of such butterflies as Nymphalidae that are known to cause urticaria.

**Type 9 (simple poisonous spines).**   These are simple hairs similar to type 8 spine hairs except that each possesses a poison-secreting cell. Relatively little is known about the morphology of these spines despite their wide occurrence among lepidopterous larvae. They are probably the most common type of urticating spine in caterpillars and cause more severe skin reactions than simple mechanical injury.

**Type 10 (*Balataea* type).**   These are venomous spines characterized by a bulbous base containing the poison reservoir. The spines are multicellular in origin, involving several trichogen cells, the poison cell, supporting cells, and a nerve cell. Toxin secreted by microvilli of the associated poison cell passes into the reservoir via a small duct. Apparently there is no opening at the tip of the spine; the toxin is released only when the tip of the spine breaks off or when the spines are broken upon penetrating the skin. Type 10 spines typically are found in caterpillars of the zygaenid genera *Balataea* and *Illiberis,* the arctiid genera *Eilema* and *Lithosia,* and first-instar larvae of the gypsy moth *Lymantria dispar* (Lymantriidae).

**Type 11 (*Latoia* type).** These spines are relatively short, stout, and conical, with a bulblike base containing a large poison gland (Fig. 18.5D). There is no opening at the pointed tip. The structure of type 11 spine hairs is more complex than the others, involving not only multiple trichogen cells and poison-secreting cells, but also tracheal and nerve cells. The distinguishing feature is the presence of a diaphragm on which the poison cell rests. When mechanically stimulated on contact with skin, the pressure of the hemolymph acting on the diaphragm causes toxin to be ejected from the tip of each broken spine. The result can be some of the most severe dermal reactions associated with urticating caterpillars. Type 11 spine hairs occur most commonly in the Limacodidae, best represented by *Latoia* larvae. Other limacodid genera with this type of spine are *Microleon, Monema, Parasa,* and *Scopelodes.* Similar urticarial spines are found in larvae of *Automeris io* and *Dirphia* (Saturniidae), *Catocala* (Noctuidae), and the flannel-moth genera *Doratifera* and *Megalopyge* (Megalopygidae).

## LIFE HISTORY

Moths and butterflies undergo holometabolous development. Adults typically deposit their eggs, either singly or in clusters, on host plants, which serve as food for the *larvae.* Soon after egg hatch, the larvae begin feeding, usually on the leaves. They grow rapidly, molting 3 to 10 times, depending on the species, as they consume increasing quantities of foliage or other plant material. When time for *pupation* nears, most species leave the host plant, crawling or descending on a strand of silk to the ground, where they seek protected recesses in which to pupate. Others remain on the host plant. The larvae of most moths spin a protective silken *cocoon* in which pupation takes place. Cocoons may be attached to twigs and leaves or constructed under tree bark, under rocks or ground debris, in the soil or litter, and in other suitable sites. Larvae of moths which do not spin cocoons generally pupate in protected recesses or chambers which they excavate in the soil. The larvae of butterflies do not spin cocoons. Instead they transform into a naked pupa, or *chrysalis,* which usually hangs exposed on the host plant. Although most chrysalides are green, brown, or otherwise cryptically camouflaged, some are attractively colored or ornamented. Most lepidopterans produce one generation per year, with a few taking two or more years to complete their development.

Overwintering usually occurs in the egg or pupal stage.

## BEHAVIOR AND ECOLOGY

Only a few groups of moths have become adapted for feeding directly on secretions and body fluids of live animals. This behavior apparently is derived from the commonly observed habit of adult lepidopterans feeding on animal excretory products such as fecal material, urine-contaminated substrates, and animal secretions such as saliva and nasal mucus smeared on vegetation or other objects. Imbibing fluids from the wounds and eyes of host animals is a behavioral modification requiring relatively minor morphological changes. In those species that feed on eye secretions, the proboscis is frequently moved over the sensitive eye region to irritate the conjunctiva and increase the flow of tears, on which the moths feed. At least one genus (*Calyptra*) includes species that have the ability to actually pierce the skin to feed directly on blood. It has developed unique modifications of the proboscis, notably hooks and erectile barbs (Fig. 18.2C). They are movable by internal hemolymph pressure which, aided by special proboscis motions, enables the proboscis to penetrate vertebrate tissues.

Moths which are attracted to animals are called *zoophilous.* They have been observed feeding on the following animal fluids: lachrymal secretions from eyes, blood and skin exudates associated with wounds, nasal secretions, saliva, perspiration, urine, and droplets of blood extruded from the anus of mosquitoes feeding on host animals. Because lepidopteran adults, with few exceptions, lack proteinases and therefore cannot digest proteins from these fluids, this specialized feeding behavior is believed to serve primarily as a means of obtaining salts. The feeding time for most zoophilous species is usually a few minutes.

Species which feed about the eyes on tears from the lachrymal glands are called *lachryphagous.* They also are called *tear drinkers* and *eye-frequenting moths.* Lachryphagy is the most striking zoophilous behavior, as documented by Bänziger in Southeast Asia. Since the first observation in 1904 of a moth feeding on the eyes of a horse in Paraguay (Shannon, 1928), lachryphagous moths have been reported attacking a wide range of wild and domestic animals, particularly ungulates and elephants. Zebu and water buffalo tend to be particularly favored hosts; other common hosts include horses, mules, tapirs, rhinoceroses, kangaroos, deer, and humans. It appears that many species exhibit a fairly high degree of host specificity at the order level.

Most lachryphagous species settle on the host to feed; others tend to hover or continually flutter their wings, remaining partially airborne while feeding. The proboscis is used to feed on tears flowing from the eyes. In some cases, it is used to irritate the conjunctiva or cornea and to feed directly on the eye tissue itself. Many lachryphagous moths are capable of slipping the proboscis between the closed eyelids of sleeping or dozing animals. Some even continue to feed when the host animal tightly closes the eyelids or blinks as a defensive response to the associated irritation. Other moths may irritate eye tissues, particularly the sensitive inner surface of the lids, with their tarsal claws while they attempt to feed.

Some adult moths are attracted to open wounds, where they imbibe blood and other host fluids. In most cases the feeding behavior is similar to drinking water and other fluids from the surface of mud, fresh animal manure, or honeydew. In other cases the moths actually probe the wound, penetrating damaged tissues to feed on fresh blood. They are variously called *wound-feeding,* *hematophagous,* and *blood-sucking moths.* Only a few taxa of *skin-piercing moths* are capable of piercing intact skin in order to feed.

## URTICATING CATERPILLARS

The three most important lepidopteran families with stinging caterpillars are Limacodidae, Megalopygidae, and Saturniidae. Other families include Lymantriidae, Arctiidae, Lasiocampidae, Noctuidae, Thaumetopeidae, Nymphalidae, and Morphoidae. With the exception of the last two families, all are moths as adults.

### Megalopygidae

Members of this family are called *flannel moths,* referring to the densely hairy adults and larvae. They occur in the Palearctic and Nearctic regions, but especially in South America and the West Indies, where the urticating larvae are known as *tataranas,* meaning "like fire," *cuy machucuy,* and *fire caterpillars.* All megalopygid larvae are protected by poisonous spines concealed beneath the more conspicuous long, fine hairs. These are generally type 11 spine hairs which cause a nettling sensation on contact with skin. Some species can cause particularly severe reactions and present occupational hazards for tree-plantation workers in parts of South America.

Among the 11 species in North America, the *southern flannel moth (Megalopyge opercularis)* is the most commonly encountered by humans. The larva is called a *puss caterpillar,* referring to its very hairy appearance (Fig. 18.6). The dense, fine, silky hairs vary in color from

**FIGURE 18.6** Puss caterpillar, *Megalopyge opercularis* (Megalopygidae). (Photo by Sturgis McKeever.)

tan to dark brown or charcoal gray, with one or more pairs of small, dorsolateral patches of white setae. The hairs at the posterior end form a tail-like tuft, whereas the head is concealed beneath the mouselike pelage. It occurs primarily in the southeastern and south-central United States, where the larvae feed on oaks, hackberry, persimmon, apple, orange, almond, pecan, roses, and other trees and shrubs. The short, toxic spines are arranged in radiating clusters (Fig. 18.4C) on three pairs of elevated, longitudinal ridges along the middorsum and sides of the body. The tips of these spines break off on contact with the skin, releasing toxin from the bulbous cavity at their base. All instars are capable of stinging.

The puss caterpillar causes the most painful and severe reactions among urticating species in the United States. Reactions typically include an initial burning sensation, commonly followed by numbness and occasional localized swelling, nausea, and vomiting. Reddened blotches or mottling develop at the contact site, often associated with a glistening appearance as cell fluids are released at the skin surface. Edema may occur, especially if the wrist or lower arm is involved; in such cases, the entire limb from hand to shoulder may become swollen. This is often accompanied by inflammation of lymphatic vessels and dull, throbbing aches involving the axillary nodes which may persist for 12 hr or more. Stings on the neck can be particularly severe. Occasionally, when populations of *M. opercularis* are unusually high, large numbers of people may be affected. Two such instances have occurred in Texas. One involved hundreds of children and resulted in the closing of public schools (Bishopp, 1923); the other was a widespread outbreak in which over 2100 cases were reported (Keegan, 1963).

Other megalopygid species which as larvae cause urticaria in North America are the *crinkled* or *black-waved flannel moth (Lagoa crispata), yellow flannel moth*

FIGURE 18.7   White-flannel moth, *Norape ovina* (Limacodidae). (Photo by G. R. Mullen.)

(*L. pyxidifera*), and *white flannel moth* (*Norape ovina*) (Fig. 18.7).

## Limacodidae (Cochlidiidae, Eucleidae)

This is a large, mostly tropical and subtropical family of moths which occurs widely throughout the Neotropical, Ethiopian, Indo-Australian, and Palearctic regions. Approximately 50 species occur in North America. The larvae are called *slug caterpillars* or *nettle grubs,* referring to their unusual shape and the fact that most species have stinging spicule hairs. The larvae are usually somewhat flattened or sluglike, with a small retractable head, short thoracic legs, and reduced abdominal prolegs which are modified as suckers. They move in a gliding motion, suggestive of slugs. The poisonous setae are usually type 4 or type 5 spicule hairs, often in the form of starlike clusters of prickles borne on cone-shaped protuberances; the spines break off easily on contact to cause a nettling sensation.

Six species of urticating slug caterpillars occur in North America. The most commonly encountered is the *saddleback caterpillar* (*Sibine stimulea*). It is easily recognized by a dorsal, brown oval spot with a white border, in turn surrounded by a green area suggesting a saddle and saddle blanket (Fig. 18.8). Urticating hairs are borne on two pairs of large, dark brown, fleshy protuberances (Fig. 18.4J), one pair at each end, and on smaller prominences along the sides. In addition, two pairs of rounded lobes bearing specialized, deciduous setae called *calytropes* that cause irritation to the skin are located at the caudal end. The stinging reaction consists of a burning sensation and an erythematous lesion which is usually much less severe than that of the puss caterpillar. The saddleback caterpillar is found on oaks, elms, dogwoods, linden, corn, ixora, asters, blueberries, grapes, and a number of fruit trees such as apple, citrus, pear, plum, and banana.

FIGURE 18.8   Saddleback caterpillar, *Sibine stimulea* (Limacodidae). (Photo by Sturgis McKeever.)

Larvae of the *hag moth* (*Phobetron pithecium*) are sometimes called *monkey slugs*. Their unkempt, haglike appearance is attributed to their lateral fleshy processes of variable lengths which are covered with fine, brown or grayish plumose hairs (Fig. 18.9). The relatively few, tuberculate stinging hairs (Fig. 18.4F) are located at the tips of the processes and laterally on each segment. Contrary to some reports in the literature, the urticarial reaction to hag moth larvae is mild, at most. The urticaria is similar to that caused by the saddleback caterpillar. The larvae feed on ashes, birches, hickories, oaks, chestnut, willows, apple, and persimmon.

Another urticating species is the *spiny oak caterpillar* (*Euclea delphinii*) in the eastern United States

FIGURE 18.9   Hag moth, *Phobetron pithecium* (Limacodidae). (Photo by Sturgis McKeever.)

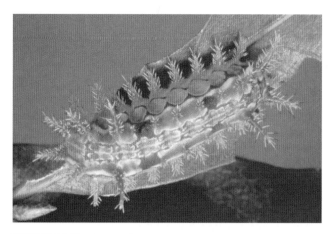

**FIGURE 18.10** Spiny oak-slug caterpillar, *Euclea delphinii* (Limacodidae). (Photo by Sturgis McKeever.)

**FIGURE 18.12** Smaller parasa, *Parasa chloris* (Limacodidae). (Photo by Carl Hanson, courtesy of Smithsonian Institution.)

(Fig. 18.10). The yellow-green larvae feed on a variety of woody plants, including beech, cherry, maple, oak, redbud, sycamore, and willow. In addition to urticating spines on the dorsal and lateral processes (Fig. 18.4H), they possess a pair of caudal patches of densely clustered, brown, deciduous setae (*calytropes*), which can be shed defensively, causing skin irritation. These specialized setae also are incorporated into the silk used in spinning the cocoon, providing protection for the developing pupa.

Two closely related limacodid species with urticating larvae are the *stinging rose caterpillar* (*Parasa indetermina*) (Fig. 18.11) and the *smaller parasa* (*P. chloris*) (Fig. 18.12). *P. indetermina* is found on oaks, hickories, maples, poplars, apple, dogwood, and roses, whereas *P. chloris* occurs most commonly on apple, dogwood, elms, and oaks. Less commonly encountered are the stinging caterpillars of *Nason's slug moth* (*Natada nasoni*) (Fig. 18.13) and *Adoneta spinuloides*. *N. nasoni* feeds

on beeches, hickories, hornbeam, chestnut, and oaks, whereas *A. spinuloides* feeds on beeches, birches, linden, willows, plums, and other trees and shrubs. *Isa textula* can cause a slight urticaria in some individuals; it is found on cherry, maples, and oaks.

Slug caterpillars are common pests among agricultural workers, particularly in Latin America and Southeast Asia. They represent significant occupational hazards for workers in banana and rubber plantations, groves of coconut and oil palms, and other tree crops in the Tropics.

**Saturniidae**

The members of this family are called *giant silk moths* and include some of the largest and most colorful of all adult moths. Two North American species in particular have urticating caterpillars: the io moth and the buck moth. The *io moth* (*Automeris io*) occurs throughout the eastern

**FIGURE 18.11** Stinging rose caterpillar, *Parasa indetermina* (Limacodidae). (Photo by Sturgis McKeever.)

**FIGURE 18.13** Nason's slug moth, *Natada nasoni* (Limacodidae). (Photo by Jerry F. Butler.)

**FIGURE 18.14**  Io moth, *Automeris io* (Saturniidae). (Photo by Sturgis McKeever.)

United States and Canada. The larvae feed on a wide range of host plants; these include trees such as oaks, willows, maples, birches, and elms, in addition to other plants such as corn, clover, and ixora. The caterpillar is pale green and fairly stout, with lateral stripes varying in color from yellow to reddish or maroon (Fig. 18.14). The stinging spines are usually yellow with black tips and are borne on fleshy tubercles along the back and sides (Fig. 18.4G). The sharply pointed tips break off easily on contact, allowing the toxin to penetrate the skin. Other *Automeris* species cause urticaria in South America, especially in Brazil and Peru.

The *buck moth* (*Hemileuca maia*) caterpillar is dark, sometimes almost black, with conspicuous black spines borne on dorsal and lateral tubercles (Figs. 18.4B and 18.15). Contact with the skin causes an immediate net-tling sensation, often followed by local puffiness. The histamine-induced edema resulting from punctures of the skin by individual spines commonly coalesces to form pronounced wheals. This species occurs in the central and eastern United States, where it feeds most commonly on oaks. The adults are unusual in being active fliers during the daytime. Other North American *Hemileuca* species with urticating caterpillars are the *New England buck moth* (*H. lucina*), *Nevada buck moth* (*H. nevadensis*), and *H. oliviae*.

In South America, two species of saturniids are involved in urticarial cases: *Lonomia achelous* in Venezuela and other northern countries, and *L. obliqua*, particularly in southern Brazil. The larvae are gregarious, feeding at night in trees and moving to the trunk or lower branches during the day. Their toxin contains a potent proteolytic enzyme that breaks down fibrinogen in human blood, interfering with its ability to clot. Dermal contact with *Lonomia* larvae causes an immediate burning sensation, often followed within hours by generalized discomfort, weakness, and headache. This is accompanied by hemorrhaging of capillaries near the skin surface (ecchymosis). In severe or untreated cases, there may be profuse bleeding from the nose, ears, intestinal tract, and vagina within 2–10 days after onset of symptoms. Fatalities have been documented involving cerebral hemorrhages and kidney failure.

Other saturniid species which as larvae cause urticaria are *Dirphia multicolor* and *D. sabina* in Brazil and Peru.

## Lymantriidae

Members of this family are called *tussock moths*. The term refers to the larvae, which are typically hairy, with prominent tufts of hairs, or tussocks, on the back. The family occurs throughout the world, but mainly in the Nearctic, Palearctic and Indo-Australian regions. A relatively few species possess urticating hairs. When present they occur as modified, simple hairs arising from cuticular cups clustered in groups at the base of the tussocks. The two most common species that cause urticaria in North America are the brown-tail moth and the gypsy moth. Both are introduced species from Europe which have become serious forest pests in the northeastern United States and the maritime provinces of Canada.

The *brown-tail moth* (*Euproctis chrysorrhoea*) is so called because of the brownish tip of the abdomen which contrasts sharply with the otherwise white body of the adult. It was introduced from Europe into Massachusetts in 1897, where it became a major defoliator of New England woodlands. The larvae feed on many species of trees and shrubs, especially members of the rose family. Host plants include apple, pear, plum, cherry, hawthorns, oaks, willows, bayberry, and roses. The larva has a light brown head and dark brown, almost black, body with broken lines on either side; two prominent dorsal red spots are located near the posterior end. Numerous tubercles on the back and sides bear long, barbed setae, with shorter, brown setae in between. Barbed, stinging setae are borne on tubercles amidst the longer body hairs. These specialized hairs break off easily and cause a

**FIGURE 18.15**  Buck moth, *Hemileuca maia* (Saturniidae). (Photo by Lacy L. Hyche.)

nettling sensation when they contact skin. They also are incorporated into the silken cocoon which the larva spins, thereby protecting the pupa from potential enemies. The adult female possesses similarly specialized hairs on her abdomen which she uses to cover her egg masses while ovipositing. Thus, all developmental stages of the brown-tail moth are provided with stinging hairs which can cause urticaria in humans and other animals.

The *gypsy moth* (*Lymantria dispar*) is similar in many respects to the brown-tail moth. Following its introduction to the United States from Europe about 1868, in an unsuccessful effort to use this species for developing a silk industry in Massachusetts, the gypsy moth escaped and became established in New England. It since has spread widely throughout much of northeastern and Great Lakes area of the United States. The larvae feed on a wide range of deciduous trees, causing extensive damage when their populations are high. They prefer oaks but also attack apple, basswood, alder, birches, boxelder, poplars, willows, hazelnut, mountain-ash, sumac, witch-hazel, and roses. The larvae are quite hairy, with a pair of blue tubercles on each of the thoracic and first two abdominal segments; a pair of red tubercles is present on the next six abdominal segments (Fig. 18.16).

Gypsy moth larvae possess two types of defensive setae. One causes irritation to the skin primarily due to mechanical damage by tiny projections on the long, slender shaft of each seta. The other type is represented by shorter, smoothly tapered setae which arise from a ball-in-socket joint; they are connected with poison glands which apparently produce histamine. Reactions to these stinging hairs vary from mild to moderately severe pruritus with accompanying erythema and papule formation. The onset of discomfort is usually noticed within 8–12 hr after contact, often becoming more pronounced 1–2 days later. Most cases resolve in a few days or up to 2 weeks. Delayed hypersensitivity reactions sometimes result in irritation to the eyes, inflammation of the nasal passages, and shortness of breath. This is especially common in the case of airborne hairs of adult gypsy moths, or

**FIGURE 18.16** Gypsy moth, *Lymantria dispar* (Lymantriidae). (Courtesy of US Department of Agriculture, Forest Service.)

contact with clothes hanging on outdoor lines when this moth is locally abundant. A major infestation of the gypsy moth in the northeastern United States in 1981 resulted in thousands of cases of pruritic dermatitis being reported that year. Like the brown-tail moth, female gypsy moths cover their egg masses with specialized body hairs that can cause urticaria upon contact with skin.

Other lymantriid species which cause urticaria in North America are the *whitemarked tussock moth* (*Orgyia leucostigma*) and the *yellow-tailed moth* or *mulberry tussock moth* (*Euproctis. similis*). Wind-dispersed hairs of *E. similis* resulted in an estimated 500,000 human cases of pruritic dermatitis in Shanghai, China, in 1981. A similar outbreak involving the *Oriental tussock moth* (*E. flava*) affected more than 200,000 people in Japan in 1955. The airborne setae are believed to have originated from larval hairs woven into the cocoon that adhered to the adult moths as they emerged. The *pale tussock moth* (*Dasychira pudibunda*) has been reported as the cause of *hop dermatitis* in Europe.

## Arctiidae

The adults of this family are known as *tiger moths*. The larvae usually are covered with fairly dense hairs of varying colors arising from raised warts, in contrast to the normally bare, shiny head. Included in this group is the familiar "wooly bear" caterpillar of the Isabella tiger moth (*Pyrrharctia isabella*), which, like most arctiids, does not possess urticating hairs or spines. In larvae that cause skin irritation, type 8 spicule hairs are borne on dorsal tufts, partially concealed by the longer body hairs. Members of the following six genera in North America include urticating caterpillars: *Adolia, Callimorpha, Euchaetes, Halysidota, Lophocampa*, and *Parasemia*. Most of these species are relatively uncommon and are only occasionally involved in urticarial cases. The *milkweed moth* (*Euchaetes egle*) is perhaps the best known. Its larvae are common on milkweed and are distinguished by dense tufts of black, yellow, and white hairs. Larvae of the *hickory halysidota* (*Halysidota caryae*), as the name implies, are found on hickories.

Although the family is cosmopolitan, the largest diversity of arctiid moths occurs in the Neotropical and Oriental regions. Where abundant, they can cause occupational erucism among field workers, as in the case of *Premolis semirrufa* in South America.

## Lasiocampidae

Members of this family are commonly called *tent caterpillars* or *lappet moths*. The larvae are usually very hairy and often colorful, with longitudinal stripes. In the case of lappet moths, the larvae are somewhat flattened with hair-covered, fleshy lobes (lappets) on the sides of each segment. They are typically gregarious, forming communal silken webs or "tents" in trees for protection from natural enemies. Their specialized defensive hairs cause only mild, transient discomfort on contact with skin. Urticating hairs are used for strengthening and protecting cocoons and represent another source of contact for humans.

Lasiocampid larvae of only a few North American species reportedly possess urticating hairs. These include the *eastern tent caterpillar* (*Malacosoma americanum*), commonly found on apple and cherry trees, and two lappet moths of the genus *Tolype*: the *large tolype* (*Tolype velleda*) on apple, oak, ash, elm, birch, plum, and other trees, and the *small tolype* (*T. notialis*) on conifers.

Several reports in the older literature refer to the larvae of *Bombyx* species (family Bombycidae) causing urticaria in humans. These include species in Great Britain (Jenkyns, 1886; Long, 1886; Sharp, 1885) and in India, Sri Lanka, and Africa (Castellani and Chalmers, 1913). The species reported in Great Britain as *Bombyx rubi* and *B. quercus* are now recognized as the lasiocampids *Macrothylacia rubi* and *Lasiocampus quercus*. Both are said to cause a nettling sensation and small white blisters in some individuals who handle them. The likelihood is that other old reports of bombycid larvae causing urticaria also refer to lasiocampid species.

## Noctuidae

This is the largest family of Lepidoptera, with over 20,000 species worldwide. The adults are known as *owlet moths* or simply *noctuids*. Only a relatively few species are known to possess stinging hairs or spines. Spine hairs are type 8, with sharp tips that break off and penetrate the skin; the spines may be branched and sometimes form brushes. Urticating species are primarily members of the subfamilies Acronictinae (dagger moths) and Catocalinae (underwings). The larvae of dagger moths are commonly covered with tufts of long hairs, superficially resembling the Arctiidae.

In North America, only larvae of the genera *Acronicta* and *Catocala* cause urticaria. The *smeared dagger moth* (*Acronicta oblinita*), the larva of which is known as the *smartweed caterpillar,* is a pest of apple and other fruit trees; in addition, it is found on elms, oaks, pines, willows, cotton, corn, clover, smartweed, strawberry, and grasses. The *cottonwood dagger moth* (*A. lepusculina*) feeds on cottonwoods, poplars, aspens, birches, and willows. Larvae of several species of *underwing moths* (*Catocala* spp.) are protected by stinging hairs.

## Thaumetopoeidae

The larvae of this family are best known as *processionary caterpillars*. They typically live in communal webs which they leave at night to feed on foliage. When moving, they crawl one behind the other, forming rows or columns and marching in an orderly fashion. They occur primarily in the Palearctic, Oriental, and Ethiopian regions, where they feed on pines and oaks and, less commonly, on cedars and walnut. No processionary caterpillar occurs in North America. The urticating hairs of their caterpillars are type 2 spicule hairs similar to those found in the caudal tufts of adult females. Venom is drawn by capillary action into the tip of the seta from tiny glands in the epidermis. In adults, contractions of the abdomen are sufficient to release these hairs from their sockets. Pruritic dermatitis and urticaria can result from contact with the larvae, airborne setae from the adults, or contact with egg masses in which the barbed hairs from the adult female have been incorporated as a protective covering.

The *oak processionary* (*Thaumetopoea processionea*) is widely distributed in Europe, where it commonly causes urticaria. The larva is covered with long whitish hairs arising from reddish warts, contrasted against a blue–gray coloration above the line of the spiracles and a greenish gray below; velvety black, dorsal patches occur on most of the abdominal segments. The larvae feed primarily on oaks and sometimes on walnut. Other processionary caterpillars which cause discomforting rashes include *Anaphe infracta* in Europe, *T. wilkinsoni,* and several *Thaumetopoea* species in Africa and Madagascar.

## Nymphalidae

The only nymphalid butterfly in North America in which the larva possesses urticating hairs is the *mourningcloak* (*Nymphalis antiopa*). It is an introduced species from Europe which occurs throughout the eastern United States and Canada. The larvae are velvety black, speckled with tiny white dots, with a row of middorsal red spots and several rows of long, branched spines (Fig. 18.17) which are capable of piercing skin. The urticating

**FIGURE 18.17**   Mourningcloak, *Nymphalis antiopa* (Nymphalidae). (Photo by Sturgis McKeever.)

structures are typical type 8 spine hairs. Caterpillars are found on elm, hackberry, poplar, willow, rose, and other common host plants.

## Morphoidae

This family includes the showy, brightly iridescent-blue *morpho butterflies* which occur only in the Neotropics. The larvae of at least seven species are known to cause urticaria: *Morpho achillaena, M. anaxibia, M. cypri, M. hercules, M. laertes, M. menelaus,* and *M. rhetenor* (Rotberg, 1971) Most encounters involve accidentally brushing against the larvae feeding on plants in the families Leguminosae and Menispermaceae. Cases are relatively few. They can occur any time of the year but are most commonly seen during the summer when larval populations are highest. Little is known about the nature of the urticating structures.

## LACHRYPHAGOUS MOTHS

More than 100 species of zoophilous moths have been observed feeding on lachrymal secretions (Figs. 18.18 and 18.19), primarily in Thailand, Malaysia, and other parts of Southeast Asia (see Bänziger references). Most of these moths are members of the Geometridae, Pyralidae, and Notodontidae, with a few species of Noctuidae, Sphingidae, and Thyatiridae.

## Geometridae

Members of only a few of the 2700 genera of geometrid moths are reportedly zoophilous. Nonetheless, this family includes the largest number of lachryphagous taxa,

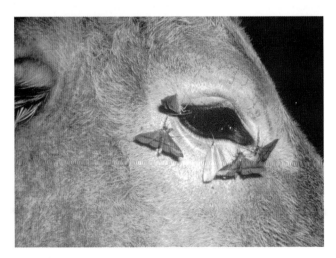

**FIGURE 18.18**  Three species of moths feeding on eye secretions of zebu: *Hypochrosis irrorata* (Geometridae), *Filodes mirificalis* (Pyralidae), and *Lobocraspis griseifusa* (Noctuidae). (Photo by Hans Bänziger.)

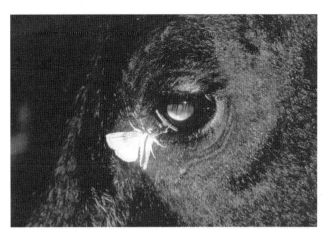

**FIGURE 18.19**  Lachryphagous moth, *Chaeopsestis ludovicae* (Thyatiridae), feeding at eye of zebu. (Photo by Hans Bänziger.)

with more than 50 species in Southeast Asia. As in other zoophilous moths, except some Noctuidae, only the males are attracted to animals. Most of them feed on mammalian body fluids which either drop to the ground or are smeared on vegetation. They have been observed primarily in association with water buffalo, but other ungulates and elephants also are frequently visited. A few species have been reported imbibing droplets of blood extruded by mosquitoes as they feed on host animals. As a group they do not commonly frequent the eyes; however, *Hypochrosis hyadaria, H. flavifusata, Godonela eleonora,* and, to a lesser extent, other *Hypochrosis, Godonela, Scopula, Problepsis,* and *Zythos* species are locally among the more frequent tear drinkers. The only lachryphagous species that has been reported in the United States is the *pectinate euchlaena* or *forked euchlaena* (*Euchlaena pectinaria*), observed feeding on eye secretions of a horse in Arkansas (Selman, 1972).

## Pyralidae

Pyralid moths are second only to the Geometridae in the number of species known to feed on lachrymal secretions. Members of the following genera are zoophilous and to various extents lachryphagous: *Botyodes, Epipagis, Hemiscopis, Lamprophaia, Pagyda, Pyrausta,* and *Thliptoceras. Microstega homoculorum, Filodes mirificalis,* and *Paliga damastesalis* are among the more common visitors of human eyes, while *Thliptoceras* and *Hemiscopis* tend to suck human perspiration. Typically, however, they have been observed feeding on lachrymal and skin secretions of ungulates and elephants.

## Notodontidae

Adult males of at least eight species of the genera *Tarsolepis, Togarishachia,* and *Pydnella* are lachryphagous.

Elephants appear to be their preferred hosts; however, these moths feed on a wide range of other large mammals in Southeast Asia, including water buffalo, zebu, tapir, rhinoceros, deer, and humans. Although they feed primarily on tears, they also have been seen imbibing saliva from around the mouth. They are persistent feeders. Some cause only mild discomfort to their hosts, while others are very irritating.

### Noctuidae

Although only a few species among the more than 3800 genera of noctuid moths are zoophilous, they are behaviorally the most advanced in terms of lachryphagy and locally can be the most frequent tear drinkers. The highly flexible proboscis is swept back and forth across the eye to induce tearing as the moth feeds. The extra length of the proboscis allows these moths to feed between the eyelids of dozing animals and reduces the risk of being dislodged by eyelid movements of wakeful hosts. Both males and females of *Arcyophora* and *Lobocraspis* species are lachryphagous and are the only known tear drinkers capable of digesting proteins contained in lachrymal fluids.

### Sphingidae

*Rhagastis olivacea* in Thailand is the only sphingid moth confirmed as being lachryphagous. It feeds while hovering about the eyes of horses, mules, and humans. It also has been observed inserting its proboscis between the lips and into the nostrils of humans to feed on saliva and nasal secretions; the latter has been described as causing a tickling sensation. Only mild discomfort is experienced when they feed on eyes.

### Thyatiridae

This is a relatively small family with 70 described genera worldwide. Only a few species in the genera *Chaeopsestis* and *Neotogaria* in Thailand and China are known to be zoophilous. They tend to be avid tear drinkers on zebu, horses, and mules, although they also feed on wounds. *Chaeopsestis ludovicae* is the only thyatirid known to feed on humans (Fig. 18.20). It has been observed licking perspiration on human skin and clothes, in addition to imbibing nasal mucus and saliva of human hosts. This moth is an aggressive feeder and can cause considerable pain due to irritation of the conjunctiva and inner surface of the eyelid with its tarsal claws. The discomfort has been likened to a grain of sand being rubbed between the eye and eyelid. Adding to the annoyance is its persistence in fluttering about the eyes and repeated attempts to feed even when the eyelids are tightly closed.

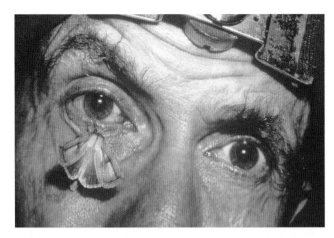

FIGURE 18.20    Tear-drinking moth, *Chaeopsestis ludovicae* (Thyatiridae), feeding from human eye with tip of proboscis just inside lower eyelid. (Photo by Hans Bänziger.)

## WOUND-FEEDING AND SKIN-PIERCING MOTHS

The only lepidopterans known to be capable of piercing animal skin are members of the noctuid genus *Calyptra* in Southeast Asia. Like many geometrids and other zoophilous moths, they tend to be attracted to wounds, open sores, cuts, scratches, scabs, and other skin lesions (Figs. 18.20, 18.21). However, while these other moths imbibe only exposed wound fluids, *Calyptra* spp. are capable of piercing the underlying tissue to draw fresh blood. Only the males are hematophagous. Females are believed to feed almost exclusively on fruits and are able to pierce the outer layers of ripening fruit to reach the

FIGURE    18.21    Two    wound-feeding    moths,    *Hypochrosis pyrrhophaeata* and *Zythos* sp. (Geometridae), feeding at site of host injury. (Photo by Hans Bänziger.)

**FIGURE 18.22** *Hypochrosis hyadaria* (Geometridae) feeding at open wound of zebu. (Photo by Hans Bänziger.)

sugar-rich juices within. Other moths closely related to *Calyptra* are fruit piercers, suggesting that blood-feeding is a relatively recent development in this group, derived from such fruit-piercing behavior.

A number of *Calyptra* species have been observed piercing mammalian skin under natural conditions. Five *Calyptra* species are known to feed on humans: *Calyptra bicolor, C. fasciata, C. ophideroides, C. parva* (Fig. 18.23), and *C. pseudobicolor. Calyptra* species also have been observed piercing the skin of elephants, water buffalo, zebu, Malayan tapir, rhinoceros, deer, antelope, mules, and pigs. The feeding times typically range from 3 to 15 min. The reaction to the proboscis penetrating the skin varies from being barely felt to causing locally intense pain accompanied by a burning sensation. The latter has been attributed to saliva which is introduced as the moth feeds,

**FIGURE 18.23** Skin-piercing moth, *Calyptra parva* (Noctuidae), feeding on human. (Photo by Hans Bänziger.)

whereas the amount of pain is believed to depend on the number of pain receptors which are hit by the piercing stylets. Other associated reactions include localized swelling which may persist for several hours, slight numbness or itching, pressure sensitivity at the bite site, and mild induration the following day.

For further details on the biology and behavior of lachryphagous, wound-feeding and skin-piercing moths, see the references by Bänziger.

## PUBLIC HEALTH IMPORTANCE

The severity of reactions to *urticating caterpillars* is highly variable, depending on the species involved, the degree of contact, and the nature of the toxin. Toxic components commonly include histamine-like or histamine-releasing substances which cause edema and wheal formation at the site of contact; proteolytic enzymes and esterases; peptides and other substances that can increase vascular permeability, destroy blood cells, or cause local necrosis of tissues; and globulins with immunologic properties.

Many cases involve mild, localized dermatitis in the form of a nettling sensation or minor irritation with transient puffiness or redness at the contact site. In more severe encounters, individuals often experience an intense burning sensation with associated wheal formation and persistent erythema. Other cases may include localized numbness, formation of vesicles, nausea, vomiting, or fever. In cases such as those involving the saddleback caterpillar and puss caterpillar, the affected skin tends to glisten after several minutes as fluids from the damaged dermal cells seep onto the skin surface. Without treatment, the burning sensation usually subsides within 30 min to an hour but may persist much longer. Radiating pains and lymphadenopathy may occur in more severe cases involving the limbs. The pain often extends proximally to the axillary or inguinal lymph nodes, sometimes with associated inflammation of the lymphatic vessels which may be visible as reddened traces on the skin surface. Some cases result in dull, throbbing aches in the lymph nodes which can persist as long as 12–24 hr. The contact site may remain discomforting and sensitive to touch for several days thereafter as the skin heals. The pattern of actual contact with the urticating spines may be evident for days, sometimes weeks.

*Occupational erucism* is a common occurrence in tropical countries, notably in South America and Southeast Asia, where field workers are exposed to stinging caterpillars. In addition to causing temporary discomfort and annoyance affecting primarily the arms and hands, chronic contact with some species (e.g., the arctiid *Premolis semirrufa* in Brazil) can result in persistent swelling and fibrous lesions of the joints of the hands and fingers. Workers in saw mills may experience dermatitis as a result of contact with urticating hairs in egg masses on the trunks of trees being cut for timber or stored at mill sites.

Reactions to adult moths usually occur as dermal irritation induced by airborne setae on contact with the skin, a condition called *moth dermatitis*. Most of the reported cases involve *Hylesia* species. The response is similar to that caused by urticating larvae. The most commonly affected parts of the body are the face, neck, and upper limbs. As the setae are rubbed into the skin, they may release histamine-like substances that cause erythema and pruritus. The problem is further aggravated and spread by scratching and sweating or by clothing and bedding contaminated with the irritating hairs. In severe cases, symptoms may persist for several days or longer, sometimes resulting in fever, insomnia, malaise, nausea, vomiting, or muscle spasms.

When airborne body setae or wing scales of moths are drawn into the respiratory tract, they can cause mechanical irritation and inflammation of the nasal passages, pharynx, and trachea. Individuals which become sensitized to these insect parts may develop *inhalation allergies* upon subsequent exposure. Most reported cases involve *Hylesia* species in South America and the Caribbean region. Thousands of microscopic setae from the lateral and ventral areas at the end of the female abdomen become airborne during outbreaks of these moths. Clouds of setae may be released into the air when the moths are attracted to light at night and bump against windows and outdoor light fixtures. Several such instances involving thousands of human cases of moth dermatitis and inhalation allergies have been reported in Peru (Allard and Allard, 1958) and Brazil (Gusmão *et al.*, 1961).

A less common problem is *silk-induced allergies* resulting from contact with the silk of certain moths. This most commonly occurs among workers in the silkworm industry. Individuals involved in processing natural silk can become sensitized to silk proteins, leading to skin irritation, conjunctivitis, allergic rhinitis, and asthma. Others may experience similar allergic responses from contact with silk clothing and silk-filled bed quilts, as reported in China (Wen *et al.*, 1990). For further information on inhalant silk allergies, see Kagan (1990).

The major health concern from *lachryphagous moths* feeding on humans is the irritation of eye tissues caused by microscopic lesions as the tip of the proboscis abrades the eyeball or inner surface of the eyelids. The most discomforting cases are caused by the tarsal claws of species which secure themselves to the eyelids while feeding. Actual damage to the eyes is usually minor, however, and generally heals without becoming infected. The risk of mechanically transmitted agents is greater in the case of

ocular abrasions by the tarsal claws than by the proboscis, simply because the tarsi are more apt to become contaminated with infectious agents. They come in contact with a wider array of animal substances and potential pathogens. Nonetheless, there is no evidence that any human pathogen is transmitted by eye-frequenting moths.

The potential for mechanical transmission of pathogens is greatest in the case of *skin-piercing moths*. The relatively large size of these moths provides a greater surface area for picking up contaminants, while the deeper skin punctures and significant blood flow around the bite wound can be contributing factors. The fact that their bites are more painful than lachryphagous species increases the chances of interrupted feedings and attacks on more than one animal, enhancing prospects for transfer of pathogens. However, there is no documented evidence that any disease agent is transmitted to humans by these moths.

Treatment of urticarial cases includes the mechanical removal of urticating spines or hairs and the application of materials to alleviate the symptoms. The broken spines or hairs can be removed from the skin with fine-tipped forceps or by lightly applying sticky tape to the affected skin surface and lifting the spines or hairs away as the tape is removed. It is important not to rub the spines into the skin in the process of trying to remove them. They also may be removed by submerging the affected areas in warm water to float off the hairs, gently washing the skin surface with running water, or showering. Since components of the venom are generally water-soluble, water helps to dilute and destroy the toxins, thereby reducing their potency.

Other measures taken to relieve discomfort include the local application of ice packs, calamine lotion, and antihistamines. Soaking with warm bicarbonate of soda or solutions of household ammonia is generally ineffective, although relief has been reported in some cases. Mild analgesics such as salicylic acid do not usually provide much relief, either for the localized pain or accompanying headaches.

## VETERINARY IMPORTANCE

Under certain conditions, lepidopterous species can cause veterinary problems. These usually involve the ingestion of urticating caterpillars or the caterpillars of nonurticating species, which contain toxins in their body fluids. The result is irritation and inflammation of the lining of the digestive tract, variously called *erucic stomatitis, erucic gastroenteritis,* and *erucic gastroenterocolitis*. Reactions following ingestion may be immediate or delayed and range from being mild to fatal. Although this type of erucism is most common in grazing cattle, horses, and other herbivores, it also has been reported to cause severe stomatitis in dogs and cats. Canine and feline cases have followed ingestion of caterpillars and leaves contaminated with urticating hairs (Delgado, 1978). In the United States, *Hemileuca maia* and other *Hemileuca* species have been abundant enough to pose serious threats to cattle (Caffrey, 1918). In other cases, cattle have died following ingestion of the larvae or pupae of cabbage butterflies (Pieridae) containing poisonous body fluids that can cause severe enteritis (Delgado, 1978).

Cases of urticaria in domestic animals are uncommon, although they are occasionally observed in horses. Other minor concerns are the suspected involvement of adult moths in transmitting the bacterial agent of *bovine infectious keratitis,* or *pinkeye,* in Uganda (Guilbride *et al.,* 1959) and some moths serving as intermediate hosts for the *rat tapeworm Hymenolepis diminuta* (Belding, 1964). There are also reports of adult moths feeding on blood of chickens in coastal Ecuador, where they are known by the local inhabitants as *chupa-gallina* ("chicken suckers").

Wound-feeding and skin-piercing moths are not known to transmit any animal pathogens. However, their feeding can cause considerable irritation and discomfort, especially when associated moth populations are high and their attacks persistent. Most of the lachryphagous moths cause relatively mild discomfort and are considered more of an annoyance than a serious health problem.

## PREVENTION AND CONTROL

The greatest problems are presented for individuals working under conditions in which urticating caterpillars commonly pose an occupational hazard. The best line of defense is to recognize, and thereby avoid contact with, those species which cause urticaria and other health-related problems. Protective clothing in the form of long-sleeved shirts, pants, gloves, and suitable headwear can greatly reduce the risks involved. In the case of workers exposed to airborne setae, the use of protective eyeglasses, masks, or respirators is recommended.

To reduce the risk of exposure to *Hylesia* and other moths, lights, burning candles, and fires should be extinguished in the evening to discourage the attraction of moths. Light fixtures, windowpanes, window and door frames, and other surfaces with which the moths may come in contact should be wiped clean with a damp cloth. Clothing and bed linen should be washed daily during periods of flight activity by the adult moths. This not only helps to remove urticating setae, but also destroys the water-soluble toxins. Other approaches to reducing the problem include the application of insecticides to kill *Hylesia* adults. It is important to immediately remove the

dead moths in order to eliminate them as a source of more setae. Desensitization of individuals with a series of injections of moth extracts is also a consideration under certain circumstances. The resultant immunity, however, is not permanent.

No practical preventive measures are recommended for protecting animals from lachryphagous, wound-feeding, or hematophagous moths.

## REFERENCES AND FURTHER READING

Allard, R. F., and Allard, H. A. (1958). Venomous moths and butterflies. *Journal of the Washington Academy of Science* **48**, 18–21.

Amarant, T., Burkhart, W., LeVine, H., III, Arocha-Pinango, C. L., and Parikh, I. (1991). Isolation and complete amino acid sequence of two fibrinolytic proteinases from the toxic Saturnid caterpillar *Lonomia achelous. Biochimica et Biophysica Acta* **1079**, 214–221.

Baerg, W. J. (1924). On the life history and the poison apparatus of the white flannel moth, *Lagoa crispata* Packard. *Annals of the Entomological Society of America* **17**, 403–415.

Bänziger, H. (1968). Preliminary observations on a skin-piercing blood-sucking moth (*Calyptra eustrigata*) (Hmps.) (Lep., Noctuidae) in Malaya. *Bulletin of Entomological Research* **58**, 159–163.

Bänziger, H. (1969). The extraordinary case of the blood-sucking moth. *Animals (London)* **12**, 135–137.

Bänziger, H. (1971). Bloodsucking moths of Malaya. *Fauna* **1**, 3–16.

Bänziger, H. (1976). In search of the blood-sucker. *Wildlife (London)* **18**, 366–369.

Bänziger, H. (1980). Skin-piercing blood-sucking moths. III: Feeding act and piercing mechanism of *Calyptra eustrigata* (Hmps.) (Lep., Noctuidae). *Mittellungen der Schweizerischen Entomologischen Gesellschaft* **53**, 127–142.

Bänziger, H. (1986). Skin-piercing blood-sucking moths. IV: Biological studies on adults of 4 *Calyptra* species and 2 subspecies (Lep., Noctuidae). *Mittellungen der Schweizerischen Entomologischen Gesellschaft* **59**, 111–138.

Bänziger, H. (1987). Description of new moths which settle on man and animals in S. E. Asia (genera *Thliptoceras, Hemiscopis, Toxobotys,* Pyarlaidae, Lepid.). *Revue Suisse de Zoologie* **94**, 671–681.

Bänziger, H. (1988a). Lachryphagous Lepidoptera recorded for the first time in Indonesia (Sumatra) and Papua New Guinea. *Heterocera Sumatrana* **2**, 133–144.

Bänziger, H. (1988b). The heaviest tear drinkers: ecology and systematics of new and unusual notodontid moths. *Natural History Bulletin of the Siam Society* **36**, 17–53.

Bänziger, H. (1988c). Unsuspected tear drinking and anthropophily in Thyatirid moths, with similar notes on sphingids. *Natural History Bulletin of the Siam Society* **36**, 117–133.

Bänziger, H. (1989a). A persistent tear drinker: notodontid moth *Poncetia lacrimisaddicta* sp. n., with notes on its significance to conservation. *Natural History Bulletin of the Siam Society* **37**, 31–46.

Bänziger, H. (1989b). Skin-piercing blood-sucking moths. V. Attacks on man by 5 *Calyptra* spp. (Lepidoptera, Noctuidae) in S and SE Asia. *Mittellungen der Schweizerischen Entomologischen Gesellschaft* **62**, 215–233.

Bänziger, H. (1992). Remarkable new cases of moths drinking human tears in Thailand (Lepidoptera: Thyatiridae, Sphingidae, Notodontidae). *Natural History Bulletin of the Siam Society* **40**, 91–102.

Bänziger, H. (1995). *Microstega homoculorum* sp. n.—the most frequently observed lachryphagous moth of man (Lepidoptera,

Pyralidae: Pyraustinae). *Revue Suisse de Zoologie* **102**, 265–276.

Bänziger, H., and Büttiker, W. (1969). Records of eye-frequenting Lepidoptera from man. *Journal of Medical Entomology* **6**, 53–58.

Bänziger, H., and Fletcher, D. S. (1988). Description of five new lachryphagous and zoophilous *Semiothisa* moths from SE Asia, with five new synonymies (Lepid., Geometridae). *Revue Suisse de Zoologie* **95**, 933–952.

Belding, D. L. (1964). Order Lepidoptera. In "Textbook of Parasitology," pp. 825–826. Meredith, New York.

Bettini, S., Ed. (1978). "Arthropod Venoms." Springer-Verlag, Berlin.

Bishopp, F. C. (1923). The puss caterpillar and the effects of its sting on man. US Department of Agriculture, Department Circular No. 288. U.S. Government Printing Office, Washington, DC.

Bucherl, W., Buckley, E. E., and Deulofeu, V., Eds. (1971). "Venomous Animals and Their Venoms. Vol. 3. Venomous Insects." Academic Press, New York.

Büttiker, W. (1967). First records of eye-frequenting Lepidoptera from India. *Revue Suisse de Zoologie* **74**, 389–398.

Büttiker, W., and Bezuidenhout, J. D. (1974). First records of eye-frequenting Lepidoptera from South West Africa. *Journal of the Entomological Society of South Africa* **37**, 73–78.

Caffrey, D. J. (1918). Notes on the poisonous urticating spines of *Hemileuca oliviae* larvae. *Journal of Economic Entomology* **11**, 363–367.

Castellani, A., and Chalmers, A. J. (1913). "Manual of Tropical Medicine." Ballière, Tindall & Cox, London.

Cheverton, R. L. (1936). Irritation caused by contact with the processionary caterpillar (larva of *Thaumetopoea wilkinsoni* Tams and its nest). *Transactions of the Royal Society of Tropical Medicine and Hygiene* **29**, 555–557.

Cock, M. J. W., Godfray, H. C. J., and Holloway, J. D. (1987). "Slug and Nettle Caterpillars: The Biology, Taxonomy and Control of the Limacodidae of Economic Importance on Palms in South-East Asia." CAB International, Wallingford, UK.

Davidson, F. F. (1967). Biology of laboratory-reared *Megalopyge opercularis* Sm. & Abb. Morphology and histology of the stinging mechanism of the larvae. *Texas Journal of Science* **19**, 258–274.

Delgado, A. (1978). Venoms of Lepidoptera. In "Arthropod Venoms" (S. Bettini, Ed.), pp. 555–611. Springer-Verlag, Berlin.

Duarte, A. C., Crusius, P. S., Pires, C. A. L., Schilling, M. A., and Fan, H. W. (1996). Intracerebral haemorrhage after contact with *Lonomia* caterpillars. *Lancet* **348**, 1033.

Epstein, M. E. (1996). Revision and phylogeny of the limacodid-group families, and evolutionary studies on slug caterpillars (Lepidoptera: Zygaenoidea). *Smithsonian Contributions to Zoology* **582**, 1–102.

Foot, N. C. (1922). Pathology of the dermatitis caused by *Megalopyge opercularis*, a Texan caterpillar. *Journal of Experimental Medicine* **35**, 737–753.

Gilmer, P. M. (1925). A comparative study of the poison apparatus of certain lepidopterous larvae. *Annals of the Entomological Society of America* **18**, 203–239.

Gilmer, P. M. (1928). The poison and poison apparatus of the whitemarked tussock moth *Hemerocampa leucostigma* Smith and Abbot. *Journal of Parasitology* **10**, 80–86.

Guilbride, P. D. L., Barber, L., and Kalikwani, A. M. (1959). Bovine infectious keratitis suspected moth-borne outbreak in Uganda. *Bulletin of Epizootic Diseases of Africa* **7**, 149–154.

Gusmão, H. H., Forattini, O. P., and Rotberg, A. (1961). Dermatite provocada por lepidopteros do gênero *Hylesia*. Revista do Instituto de Medicina Tropical de São Paulo **3**, 114–120.

Ishizaki, T., and Nagai, R. (1956). Clinical studies on dermatitis due to *Euproctis flava* (report 1). *Japanese Journal of Sanitary Zoology* **7**, 113.

Jenkyns, M. S. (1886). Urtication by *Bombyx rubi*. *Entomologist* **19**, 42.

Jones, D. L., and Miller, J. H. (1959). Pathology of the dermatitis produced by the urticating caterpillar, *Automeris io*. American Medical Association *Archives of Dermatology* **79**, 81–85.

Kagan, S. L. (1990). Inhalant allergy to arthropods: insects, arachnids, and crustaceans. *Clinical Reviews in Allergy* **8**, 99–125.

Kano, R. (1967). Venomous Lepidoptera. *Japanese Journal of Sanitary Zoology* **18**, 170–171.

Kano, R. (1979). Lepidoptera (butterflies and moths). In "Animals of Medical Importance in the Nansei Islands in Japan" (M. Sasa, H. Takahasi, R. Kano, and H. Tanaka, Eds.), pp. 117–119. Shinjuku Shobo, Japan.

Katzenellengogen, I. (1955). Caterpillar dermatitis as an occupational disease. *Dermatologica* **111**, 99–106.

Kawamoto, F., and Kumada, N. (1984). Biology and venoms of Lepidoptera. In "Handbook of Natural Toxins" (A. J. Tu, Ed.), Vol. 2, pp. 291–330. Dekker, New York.

Keegan, H. L. (1963). Caterpillars and moths as public health problems. In "Venomous and Poisonous Animals and Plants of the Pacific Region (H. L. Keegan and W. V. Macfarlane, Eds.)," pp. 165–170. Pergamon Press, Elmsford, NY.

Kristensen, N. P. (1984). Studies on the morphology and systematics of primitive Lepidoptera (Insecta). *Steenstrupia* **10**, 141–191.

Long, F. R. J. (1886). Urtication by larvae of *Bombyx rubi*. *Entomologist* **19**, 45.

Lucas, T. A. (1942). Poisoning by *Megalopyge opercularis* ("Puss caterpillar"). *Journal of the American Medical Association* **119**, 877–880.

Marshall, G. A. K., Jack, R. W., and Neave, S. A. (1915). A noctuid feeding on the moisture from the eyes of mules. *Proceedings of the Entomological Society of London*, cxvii–cxix.

McGovern, J. P., Barkin, G. B., McElhenney, T. R., and Wende, R. (1961). *Megalopyge opercularis*. Observations of its life history, of its sting in man, and report of an epidemic. *Journal of the American Medical Association* **175**, 1155.

McMillan, C. W., and Durcell, W. R. (1964). Health hazard from caterpillars. *New England Journal of Medicine* **271**, 147–149.

Mills, R. G. (1923). Observations on a series of cases of dermatitis caused by a liparid moth *Euproctis flava* Bremer. *China Medical Journal* **37**, 351–371.

Neustater, B. R., Stollman, N. H., and Manten, H. D. (1996). Sting of the puss caterpillar: an unusual cause of abdominal pain. *Southern Medical Journal* **89**, 826–827.

Nielsen, E. S., and Common, I. F. (1991). Lepidoptera (moths and butterflies). In "The Insects of Australia" (I. D. Naumann, Ed.), 2nd ed., Vol. 2, Chapt. 41, pp. 817–915. Melbourne Univ. Press, Carlton, Victoria, and Univ. College of London Press, London.

Perlman, F., Press, E., Googins, J. A., Malley, A., and Poareo, H. (1976). Tussockosis: reactions to Douglas fir tussock moth. *Annals of Allergy* **36**, 302–307.

Pesce, H., and Delgado, A. (1971). Poisoning from adult moths and caterpillars. In "Venomous Animals and Their Venoms" (W. Brucherl *et al.*, Eds.), Vol. 3, pp. 119–156. Academic Press, New York.

Picarelli, Z., and Valle, J. R. (1971). Pharmacological studies of caterpillar venoms. In "Venomous Arthropods and Their Venoms" (W. Bucherl *et al.*, Eds.), Vol. 3, pp. 103–118. Academic Press, New York.

Romoser, W. S. (1981). "The Science of Entomology," 2nd ed. Macmillan, New York.

Rotberg, A. (1971). Lepidopterism in Brazil. In "Venomous Animals and Their Venoms" (W. Bucherl *et al.*, Eds.), Vol. 3, pp. 157–168. Academic Press, New York.

Rothschild, M., Reichstein, T., von Euw, J., Aplin, R., and Harman, R. R. M. (1970). Toxic Lepidoptera. *Toxicon* **8**, 293–299

Scoble, M. J. (1992). "The Lepidoptera: Form, Function and Diversity." Oxford Univ. Press, London.

Selman, C. L. (1972). Observation of an eye-frequenting geometrid in the United States. *Journal of Medical Entomology* **9**, 276.

Shama, S. K., Etkind, P. H., Odell, T. M., Canada, A. T., Finn, A. M., and Soter, N. A. (1982). Gypsy-moth-caterpillar dermatitis. *New England Journal of Medicine* **306**, 1300–1301.

Shannon, R. C. (1928). Zoophilous moths. *Science* **68**, 461–462.

Sharp, H. (1885). Urtication by larvae of *Bombyx rubi. Entomologist* **18**, 324.

Southcott, R. V. (1978). Lepidopterism in the Australian Region. Records of the Adelaide Children's Hospital **2**, 87–173.

Southcott, R. V. (1983). Lepidoptera and skin infestations. In "Cutaneous Infestations of Man and Animal" (Parish *et al.*, Eds.), pp. 304–343. Praeger, New York.

Sterling, P. H. (1993). Brown-tail: the invisible itch. *Antenna* **7**, 110–113.

Tu, A. T., Ed. (1984). "Insect Poisons, Allergens and Other Invertebrate Venoms." Handbook of Natural Toxins, Vol 2. Dekker, New York.

Wen, C. M., Ye, S. T., Zhou, L. X., and Yu, Y. (1990). Silk-induced asthma in children: a report of 64 cases. *Annals of Allergy* **65**, 375–378.

Wirtz, R. A. (1984). Allergic and toxic reactions to non-stinging arthropods. *Annual Review of Entomology* **29**, 47–69.

Zaias, N., Ioannides, G., and Taplin, D. (1969). Dermatitis from contact with moth (Genus *Hylesia*). *Journal of the American Medical Association* **207**, 525.

# 19

# ANTS, WASPS, AND BEES (*Hymenoptera*)

ROGER D. AKRE AND HAL C. REED

*The social* Hymenoptera—the ants, bees and wasps—are the primary group that constitutes a stinging hazard to humans. Together with the termites they probably constitute over 75% of all the insect biomass (Hölldobler and Wilson, 1990). Most Hymenoptera are solitary species which do not present much of a stinging concern. While hymenopteran stings can cause problems to humans, this order includes very beneficial insects, such as pollinating bees and the parasitoids, that serve as natural biological control agents for many insect pests. Also, ants and wasps are among the foremost predators in regulating insect

populations in some forest communities and agroecosystems.

The Hymenoptera of most concern to human health are the *aculeates,* or stinging hymenopterans, which can use their sting apparatus as either an offensive or defensive weapon. Most aculeates are solitary, nonaggressive species that use their sting and venom primarily to subdue prey. Those that can sting usually cause only moderate discomfort to humans and the pain is of short duration. Many solitary wasps and bees are so small that the sting cannot penetrate human skin except in tender areas. Most serious stings are inflicted by those species that live in colonies, which are defending stores of nectar and pollen or their developing brood. Often their venoms contain chemicals that cause intense pain and that serve as effective deterrents against vertebrate predators.

## TAXONOMY

Worldwide there are 115,000 described species in the order Hymenoptera, but systematists estimate that there may be an additional 200,000 undescribed species. The Hymenoptera are divided into two major suborders: the Symphyta, or sawflies, which are plant feeders, and the Apocrita, which normally feed on other arthropods. In the Apocrita the abdomen is narrowly joined to the thorax ("wasp waist"), whereas in the Symphyta the abdomen and thorax are broadly joined. The Apocrita are further divided into the Terebrantia (Parasitica), which use their ovipositor for egg laying, and the Aculeata, which have the ovipositor modified as a sting. The 50,000–60,000 species of aculeates in the world are classified into eight superfamilies. The work by Goulet and Huber (1993) provides identification of all families of Hymenoptera. The most important groups that cause health-related problems are the Formicoidea, Vespoidea, and Apoidea.

The Formicoidea, or ants, are divided into 11 subfamilies (Hölldobler and Wilson, 1990) (Table I). There are 8,800 described species of ants, although it is estimated that there may be as many as 16,000–20,000 species worldwide. Most ants possess a sting, the exceptions being members of the subfamilies Formicinae and Dolichoderinae. Many of those which lack stings squirt various caustic chemicals onto antagonists for defensive purposes. Excellent sources of information on the taxonomy and biology of this diverse group are Hölldobler and Wilson (1990) and Bolton (1994).

Social wasps of the family Vespidae, represented by 860 species worldwide, are the most important stinging wasps. A few species, including some sphecid wasps (Sphecidae), velvet ants (Mutillidae), and spider wasps (Pompilidae), also occasionally cause stinging problems. A general reference of Hymenoptera (e.g., Gauld and Bolton, 1988; Goulet and Huber, 1993), supplemented with general textbooks on insects, should suffice for most readers wishing to identify solitary wasps to family or lower taxa. Most social vespids in North America are members of two subfamilies: Vespinae, the hornets and yellowjackets; and Polistinae, the paper wasps (Table II). The Vespinae include the hornets *Provespa* in Southeast Asia and *Vespa* in Europe and Asia, and the temperate yellowjackets *Dolichovespula* and *Vespula* of temperate regions. The Polistinae are further divided into four tribes: the Ropalidiini, mostly in the tropics of Africa and Asia; the Epiponini, mostly in tropical Asia, Africa, and South America; mischocyttarini, mostly in tropical America and the Polistini, including the cosmopolitan paper wasps, *Polistes*. Other social vespids in the subfamily Stenogastrinae occur only in Southeast Asia and usually are not a significant stinging hazard (Akre and Reed, 1984).

Some taxonomists place all the bees, totaling more than 20,000 species, in a single family, Apidae. However, we have followed bee taxonomy as presented by Michener (1974), in which bees are regarded as members of the superfamily Apoidea, with nine families. Most of these families consist primarily of solitary or communal species that are rarely a stinging hazard to humans. Most of the stinging bees are social species in the family Apidae, such as the ubiquitous honey bees (*Apis* spp.) and bumble bees (*Bombus* spp.). Some species of sweat bees (*Halictidae*) and carpenter bees (*Anthophoridae*) are minor stinging threats. For further information on the phylogeny and classification of bees, see Roig-Alsina and Michener (1993), Alexander and Michener (1995), and Michener *et al.* (1994).

## MORPHOLOGY

Members of the Hymenoptera range in body length from 0.1 mm for parasitic wasps to over 50 mm for some of the

TABLE I

Subfamilies, Common Names, and Distribution of Ants (Formicidae) Based on Hölldobler and Wilson (1990): Omitted are the rare subfamilies Nothomyrmeciinae, Leptanillinae, and Aneuretinae

| Subfamily | Common name(s) | Distribution | Sting |
|---|---|---|---|
| Ponerinae | Ponerine ants | Mostly tropical, worldwide | Present |
| Myrmeciinae | Bull-dog ants | Mostly Australia | Present |
| Pseudomyrmecinae | Acacia ants | Tropical Asia, Africa, America | Present |
| Myrmicinae | Harvester ants, leaf cutting ants, fire ants, pavement ants, pharaoh ants, others | Worldwide | Present in about half of species |
| Dorylinae (Old World) | Driver ants, safari ants | Asia, Africa | Present |
| Ecitoninae (New World) | Army ants, legionary ants | Mostly tropical, America | Present |
| Dolichoderinae | Argentine ant and others | Worldwide | Absent |
| Formicinae | Thatching ants, carpenter ants, weaver ants, others | Worldwide | Absent |

TABLE II

Subfamilies, Species, Common Names, and Principle Nest Sites of Selected Vespid Wasps in North America

| Species | Common name | Nest site |
|---------|-------------|-----------|
| **Subfamily Polistinae** | | |
| *Polistes* | | |
| *P. apachus* | None | Aerial |
| *P. aurifer* | Golden paper wasp | Aerial |
| *P. annularis* | Spanish Jack | Aerial, river and lake shores |
| *P. carolina* | Red wasp | Aerial, concealed |
| *P. dominulus* | European paper wasp | Aerial, especially around man-made structures |
| *P. exclamans* | Guinea wasp | Aerial, especially around man-made structures |
| *P. fuscatus* | None | Aerial, concealed |
| *P. metricus* | None | Aerial, concealed |
| *P. perplexus* | Red wasp | Aerial, concealed |
| **Subfamily Vespinae** | | |
| *Vespa* | | |
| *V. crabro* | European hornet | Hollow trees, attics |
| *Dolichovespula* | | |
| *D. arenaria* | Aerial yellowjacket | Aerial, trees, structures |
| *D. maculata* | Baldfaced hornet | Aerial, trees |
| *Vespula* | | |
| *V. flavopilosa* | Hybrid yellowjacket | Subterranean |
| *V. germanica* | German yellowjacket | Subterranean, voids in structures |
| *V. maculifrons* | Eastern yellowjacket | Subterranean |
| *V. pensylvanica* | Western yellowjacket | Subterranean |
| *V. squamosa* | Southern yellowjacket | Subterranean |
| *V. sulphurea* | California yellowjacket | Subterranean |
| *V. vulgaris* | Common yellowjacket | Subterranean |

predaceous wasps. The integument of hymenopterans is usually heavily sclerotized, and the pleural sclerites of the thorax are highly modified and fused for strength. Wings are usually well developed in most bees and wasps and in the reproductives, or sexual forms, of ants. However, wings are absent in the nonreproductive worker caste in ants.

In wasps, the first abdominal segment is fused with the thorax as a *propodeum*. The second abdominal segment forms the *petiole* or narrow constriction between the thorax and the more enlarged remaining part of the abdomen called the *gaster*. The prominent petiole of many wasps has given rise to the term "wasp waist" and "wasp-waisted."

Although bees resemble wasps in many features, they generally have more hairs on the body. Most bees have at least a few branched hairs on the legs or gaster and broad hind legs that enable them to gather and transport pollen to their nest. The honey bees and other bees in the family Apidae possess dense rows of branched hairs on the hind legs called *corbicula*, or *pollen baskets*, that can store large amounts of pollen.

All ants are social and evolved from a primitive wasp-like ancestor. They are all readily identified by having a narrow petiole consisting of one or two segments and bearing a *dorsal lobe*. Worker ants have *geniculate* (elbowed) antennae, with the first segment very long.

Many ant species, like many other social insects, have distinct *castes* of *queens*, sterile female *workers*, and *males*. The size variation among ant species is extremely great, ranging from the little black ant *Monomorium minimum* with workers 1.5–2.0 mm long to the comparatively giant ponerine ant *Dinoponera grandis*, which may be up to 34 mm. Some ants such as the primitive, tropical ponerine *Paraponera clavata* exhibit few, if any, morphological or size differences between the queen and the workers; it is *monomorphic*, or represented by only one form. However, most ants are *polymorphic* and are easily separated into castes. Some species of ants also exhibit marked worker polymorphism, with extremely large individuals, usually with large heads, called *majors*. They may also have intermediate-sized individuals and *minors*, or small workers. The minors usually constitute most of the *nurse workers* that take care of the brood, while the majors, or *soldiers*, respond quickly to disturbances to defend the colony. Soldiers of army ants have disproportionally large heads to house the large adductor muscles that operate the equally large, ice-tong–shaped mandibles. The great

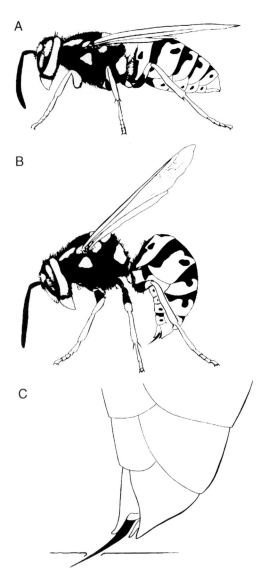

**FIGURE 19.1** Sting action of a yellowjacket (Vespidae) worker. (A) Resting position; (B) sting position with abdomen flexed and lancets of sting extruded; (C) sting at tip of abdomen piercing skin. (From Akre *et al.*, 1981.)

differences in size and morphology among the various castes sometimes cause problems in identification unless the castes are recognized as all belonging to the same species.

The *sting apparatus* is the most important morphological trait of the aculeate Hymenoptera (Fig. 19.1). The basic hymenopteran sting in wasps, bees, and ants is simply a modification of the female egg-laying structure, or *ovipositor*. Males lack such a structure and cannot sting. In ants without a sting, the components of the ovipositor have been reduced or lost; these ants spray defensive secretions from the tip of the gaster into wounds caused by the mandibles.

The visible portion of a typical ovipositor consists of three pairs of elongate structures, or *valves*, which are used to insert the eggs into a substrate such as plant tissue or soil. One pair of valves often serves as a *sheath* and is not a piercing structure; the other two pairs form a hollow shaft, which pierces the substrate by a back and forth sawing motion, with one pair held in position by the other (Fig. 19.2). The eggs pass down through the shaft in most Hymenoptera, except the Aculeata. During oviposition by aculeate hymenopterans, the sting apparatus is flexed up and out of the way and the eggs are passed from the genital opening at the base of the ovipositor. There are usually two *accessory glands* in the female which secrete substances to form egg coverings or to glue the eggs to the substrate.

Although cartoonists delight in portraying the sting of wasps and bees as a constantly protruding spike, the real sting is part of a complex apparatus hidden in a cavity at the end of the abdomen (Fig. 19.3). The morphological differences among the various aculeate groups are minor. The principal components of the sting apparatus are a pair of long, slender *lancets* that are encompassed by a single *stylet*. The ventral edges of the stylet function as a guide rail along which each lancet can slide freely. These structures converge at the base and form a channel

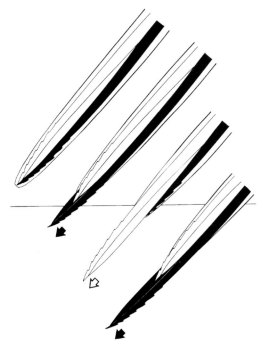

**FIGURE 19.2** Mechanism by which the pair of lancets of hymenopterans penetrates the skin when a victim is stung. The lancets slide alternately back and forth as their serrated tips work their way into the tissue. The opposing surfaces of the lancets are concave, forming a tubular channel through which venom is injected into the wound. (From Akre *et al.*, 1981.)

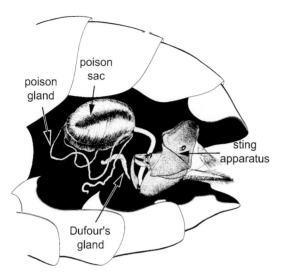

**FIGURE 19.3** Poison gland and associated structures of the sting apparatus of a yellowjacket (Vespidae) worker. (From Akre *et al.*, 1981.)

through which the venom flows. The distal tips of the lancets are armed with barbs in some species to aid their penetration. When in the retracted position the sting apparatus is covered by two membranous *sting sheaths*. Penetration by the lancets and stylet into a victim is accomplished by contraction of muscles connected to large sclerotized plates articulating with the base of the stylets and lancets.

The accessory glands of the female reproductive tract have become modified for specialized purposes in stinging hymenopterans (Fig. 19.3). One has become a *poison gland*. Another, called *Dufour's gland*, produces lubricants and coatings for the eggs, linings for brood chambers, and pheromones in some species. The venom is actually produced in the poison gland consisting of two slender, elongate tubules that empty their products into a prominent, sometimes muscular reservoir, or *poison sac*. This reservoir stores the venom until the insect stings, at which time the venom is ejected through a narrow tubule into the base of the sting apparatus. The morphology of the glands associated with the venom apparatus in stinging Hymenoptera is reviewed by van Marle and Piek (1986).

## LIFE HISTORY

Hymenopterans undergo holometabolous development with egg, larval, pupal, and adult stages. They exhibit *facultative arrhenotoky*, in which males are produced parthenogenetically, enabling them to control the sex of their offspring. Males are haploid and develop from unfertilized eggs, while the diploid female queens and workers develop from fertilized eggs. Colonies of many hymenopterans consist of sterile female workers with a single queen (i.e., *monogyny*) or multiple queens (i.e., *polygyny*). In the temperate zones reproductive males and females are normally produced only once a year, typically in late summer and early fall. These reproductives usually leave the colony for mating purposes. Males typically die soon after the mating season and do not contribute to colony function or maintenance. An inseminated queen usually will overwinter and start a new colony in the spring, as in the case of yellowjackets, paper wasps, many ants, and bumble bees. However, great variation in colony founding modes are seen among the social species. Honey bee colonies, for example, reproduce when a mated queen leaves the parent colony with a group of workers to begin a new hive, a process known as *swarming*. Colonies of honey bees and many ant species exist for more than one year as *perennial* colonies, whereas most other temperate social wasps and bees typically have *annual* colonies. Development of the *queen* among social insects is a result of nutrition and the increased space given to a larva during development. These activities are mediated by pheromones that influence the behavior of the workers toward the developing brood. For example, honey bee larvae destined to become queens are reared in large queen cells and fed a special diet of *royal jelly*, secretions from head glands of *nurse bees*, throughout their development. Rearing of a new queen usually occurs when the queen is aging, her pheromone secretions are waning, or she has died. New queens are also produced in large, healthy colonies of some species (e.g., *Apis* spp.).

## BEHAVIOR AND ECOLOGY

Behavior among hymenopteran species is extremely diverse. It varies from individuals that are *solitary*, with little or no interaction among nesting females, to species that are highly *social*, forming large colonies. Many species have intermediate levels of social behavior in which the parent cares for its offspring and may even share a common nesting area. The highest level of sociality is represented by *eusocial* species, which exhibit a reproductive division of labor, a worker caste caring for the young, and overlapping generations in which the offspring assist the parents in rearing the brood. Ants, honey bees, bumble bees, and vespid wasps are eusocial species.

*Stinging behavior* is one of the intriguing aspects of social hymenopterans. Visual cues, vibrations, and chemicals all play important roles in eliciting a stinging response. Situations in which stinging occurs ordinarily include the disturbance of a nest; threatening a worker by swatting, touching, or squeezing it; and the sensing by a worker of alarm chemicals released by another defending worker. Dark, moving objects are often attacked by honey bees and vespid wasps when defending their colonies. Vibrations of the ground or substrate near the nest stimulate alarm and attack in most species. Pheromones alerting nestmates to defend the colony are a common strategy among many ants, bees, and wasps with large colonies. For example, the honey bee produces several volatile compounds from the sting apparatus and one from the mandibular glands that together act as *alarm pheromones.* These compounds, when released by a stinging bee, alert the colony and help to focus the attack of hive mates so that numerous bees can quickly attack an intruder. Disturbed workers may also attack when they sense certain chemicals on the attacked individual. Honey bee workers, for example, may sting people when they are applying enamel paints containing *isoamyl acetate,* a natural component of the honey bee alarm pheromone.

Foraging wasps and bees are less prone to sting than individuals near or on the nest. However, foragers will inspect humans, especially those wearing sweet-smelling perfumes and hair sprays and bright orange, yellow, and blue clothing. If irritated, they will sting to defend themselves. Yellow construction machinery and the blaze-orange vests worn by many traffic-control flaggers attract foraging yellowjackets, which may result in stings when such individuals swat at yellowjackets to chase them away.

Typical *stinging* behavior of social hymenopterans is illustrated by an attacking yellowjacket (Fig. 19.1). The insect grips the skin firmly with its legs and sometimes with its mandibles, exposes the sting apparatus, then plunges the tip of the interlocked lancets and stylet into the skin with a downward thrust of the abdomen. Simultaneously, the contraction of the muscles surrounding the poison sac forces the venom into the sting bulb and through the channel formed by lancets and shaft, much like a hypodermic needle. In order to penetrate, the lancets are moved forward in alternate strokes, each sliding on its track against the stylet shaft (Fig. 19.2). The tips of the lancets are equipped with tiny barbs in bees, wasps and a few ants to facilitate penetration; they literally saw through the victim's flesh as each lancet in turn is thrust forward and anchored in place by the barbs.

*Sting autotomy* is behavior in which the anchored sting is left embedded in the skin when the insect pulls away (Fig. 19.4). It is best known in honey bees but also occurs in some tropical epiponine social wasps and harvester ants (*Pogonomyrmex* spp.). Although sting autotomy results in evisceration and death of the insect, the sting with the attached poison gland reservoir continues to pump venom into the wound. Sting loss is due to the presence of large, recurved barbs on the lancets and to lines of weakness in the structure of the sting apparatus, causing it to readily detach as the insect pulls away

**FIGURE 19.4**   Honey bee worker (Apidae) tearing sting apparatus from abdomen as it pulls away, leaving the sting embedded in the skin. (Photo by J. Schmidt.)

after stinging. The barbs on a yellowjacket or paper wasp sting are much smaller and less recurved than those of the honey bee; the sting does not normally become anchored in a person's flesh. The wasp usually can quickly withdraw with an upward pull of its abdomen and sting again. However, a few species will sometimes lose their stings when they attack human skin (e.g., *Vespula maculifrons*), and workers of most species of yellowjackets will leave their stingers behind when they sting into thick leather gloves.

Aculeate hymenopterans feed on a variety of prey and plant sources. The *larval food* consists of nectar and pollen in the bees and masarine wasps, seeds in the harvester ants, and various arthropod prey in most wasps and some ants. The *adult food* usually consists of similar materials, but in a liquefied state. Ingesting solid foods is very difficult because adult hymenopterans have a very narrow esophagus that is designed to convey only liquids to the crop in the gaster. Thus, many adults feed on the body fluids of prey items or juices from plant materials which they collect to feed the larvae. In eusocial species, liquids are mutually exchanged among workers and between larvae and workers in a process called *trophallaxis.*

The use of chemical communication is very important in maintaining colony cohesion and activities in the societies of most social Hymenoptera. Many species not only release alarm pheromones, but also lay down trail pheromones to the food site to recruit additional foragers, thereby increasing the efficiency of food retrieval. Additionally, they produce queen pheromones that limit reproduction by workers, and they convey information about the status of the colony by trophallactic exchange of chemical substances among colony members.

# HYMENOPTERA VENOMS

Hymenopteran venoms are rich biochemical mixtures of compounds that paralyze prey, induce pain in large predators, or act as toxicants. Probably fewer than a hundred of these venom compounds have been identified, while many more remain to be discovered and characterized. There are three general categories of venom compounds: (1) small, nonproteinaceous molecules with molecular weights less than 300; (2) peptides with molecular weights 1,500–4,000; and (3) larger proteins and enzymes with molecular weights over 10,000. Compounds of the first category include *histamines, serotonin,* and various *catecholamines* that induce itching, immediate pain, redness, and changes in capillary permeability. The second category includes peptides such as *hemolysins,* which destroy red blood cells and cause pain; neurotoxins; and other pain-inducing compounds, such as *kinins.* The third category (larger proteins and enzymes) generally does not cause pain, but aids in the spread and activity of other venom components. A common example is *hyaluronidase,* which facilitates the spread of toxic components through the tissues. Exceptions are *phospholipases* that are toxic, disrupt the cell membranes, and cause release of pain-inducing agents (Schmidt, 1986a, 1986c, 1992).

Venoms of *solitary wasps* are designed to cause paralysis in insects, spiders, and other arthropods on which they prey. These venoms directly affect the nervous system and cause a general decline in the rate of metabolic processes. The purpose of these venoms is not to cause the death of the prey, but to incapacitate it as food for the larvae. Common components of the venoms of various solitary wasps are *histamines, polyamines,* and substances such as *bradykinins* that cause smooth muscles to contract. Some of their venoms also contain large amounts of *acetylcholine,* as in the case of the sphecid wasp *Philanthus triangulum.* The venoms of solitary wasps generally produce only momentary, slight pain in humans.

## ANT VENOMS

Ant venoms serve a variety of functions, including defense, prey capture, aggregation, trail marking, alarm, and repelling intruders. Only the components of ant defensive venoms which are toxic to vertebrate animals are discussed here. The toxins normally are injected via the sting; however, some ants lack a sting and spray *formic acid* at their attackers (e.g., many formicine ants). Formic acid is a very effective deterrent, especially if sprayed into the eyes or applied directly into wounds made with the mandibles of the ant.

Venoms of the majority of stinging ants and other aculeate Hymenoptera are predominately composed of proteinaceous mixtures of simple organic compounds. *Fire ant venoms,* however, largely consist of *alkaloids* (95%), with only a small proteinaceous component (0.1–1%). The alkaloids cause most of the local sting reactions, whereas the *proteins* contain active allergenic antigens. The alkaloids are methyl-n-alkylpiperidines, called *solenopsins,* and a *piperideine.* The alkaloids are cytotoxic, hemolytic, fungicidal, insecticidal, and bactericidal. The characteristic dermal necrosis that becomes evident at the sting site is due to these alkaloids.

Protein-rich ant venoms are found in most subfamilies of ants, including the Ponerinae, Myrmeciinae, Pseudomyrmecinae, Ecitoninae, and some of the Myrmicinae (Table I). These venoms have not been well investigated because of difficulty in obtaining sufficient quantities of pure venom for analysis. The only studies available are those of primitive ants in the genus *Myrmecia* (Myrmeciinae) and the highly evolved *Myrmica* and *Pogonomyrmex* (Myrmicinae). *Harvester ants (Pogonomyrmex)* have a proteinaceous venom with high amounts of *phospholipase A* and *B, hyaluronidase,* a potent hemolysin called *barbatolysin,* and *histamines* (Schmidt, 1986b). Most ant venoms contain only small amounts of these materials. Other enzymes that have been identified in ant venoms include *acid phosphatase, alkaline phosphatase, phosphodiesterase, lipase, esterase,* and a nonspecific *protease.* The primary function of these compounds is to cause pain, either directly or through tissue destruction.

## VESPID VENOMS

Like those of most stinging social hymenopterans, vespid venoms are designed to cause pain. Their venom typically produces immediate pain, local swelling, and erythema caused by an increase in the permeability of blood vessels at the sting site. The pain often continues for several hours, whereas itching at the sting site may persist for days. Vespid venoms also cause the contraction of smooth muscles, reduced blood pressure, and the release of histamine and other biogenic amines. Hemolysis induced by lytic *peptides* and *phospholipases* may cause kidney damage. There is usually additional damage to surrounding tissues from the products of histolysis. Vespid venoms contain biologically active amines such as *serotonin, histamines, tyramine,* and *catecholamines,* all of which tend to produce pain. *Acetylcholine* has been reported to occur in the venoms of some *Vespa* species. However, the primary pain-causing substances are *kinins.* In addition, the venoms contain mast-cell degranulating peptides called *mastoparans* that cause the release of histamines.

Venoms also contain enzymes that can act as specific allergens and, in some species, neurotoxic compounds. The immediate pain caused by a vespid sting is principally due to serotonin and kinins. Venoms of some vespids contain alarm pheromones that function to alert nestmates to an intruder and focus the defense.

## HONEY BEE VENOM

The venom of honey bees is a complex mixture of proteins, peptides, and small organic molecules (Schmidt, 1992). The most dangerous components for humans are *phospholipases* and *hyaluronidase*. Individuals can become sensitized to these materials and subsequently even die from a serious allergic reaction. Bee venom contains large quantities of a potent membrane-disrupting material called *melittin,* which makes membranes extremely susceptible to attack by phospholipases. Melittin also causes pain, increases capillary blood flow and cell permeability, triggers lysis of red blood cells, and enhances the spread of toxins. The effects of melittin, phospholipase, and a *mast-cell degranulating peptide* cause the release of histamine and serotonin from red blood cells and mast cells. While the components of honey bee venom that cause pain are very different from those in vespid venom, the end results are very similar. Some components of honey bee venom regulate and/or decrease inflammatory responses in some individuals. This perhaps explains why bee venom therapy has been useful in the treatment of certain forms of arthritis. Another component of honey bee venom, a neurotoxin called *apamin,* seems to cause more effects in insects than in humans.

## ANTS

Ants are ubiquitous, occurring throughout most of the world, including most oceanic islands. The only places that they do not exist is above the tree line in mountainous areas and in the Antarctic. Most groups of ants have species that occur worldwide. However, some groups and subgroups are restricted to specific areas; the acacia ants (pseudomyrmecines) and ponerines, for example, occur primarily in the tropics, and bull-dog ants are found only in Australia. The primarily tropical army ants have a limited distribution in the United States, with two species occurring as far north as Iowa.

The only ants of significant medical-veterinary concern in the United States are the *fire ants* (*Solenopsis* and *Wasmannia* spp.) and *harvester ants* (*Pogonomyrmex* spp.). Most other North American ants rarely sting

people, or they are so small that they are incapable of piercing human skin. The *carpenter ants* (*Camponotus* spp.), which are commonly destructive wood pests, lack a sting, as do members of the subfamilies Formicinae and Dolichoderinae. Other ants such as *ponerines* (Ponerinae) and *army ants* (Ecitoninae) are a concern in tropical regions. However, there is one ponerine species occurring in the southeastern United States, *Odontomachus haematoda,* that can cause painful stings to humans. This peculiar ant possesses elongated mandibles that are held open at 180°, and these snap shut quickly to impale prey or enemies on their sharp teeth. It also uses its mandibles to snip and to jump away when threatened.

The life cycle of ants is highly varied. In some species, such as the army ants, whose queens are wingless throughout their life, colony initiation occurs by *budding,* a process whereby a colony divides into two colonies. In contrast, many ant species have winged reproductives, and colony initiation is typically by a single winged queen. At certain times of the year mature colonies produce an abundance of winged males and queens that leave the nest *en masse* on a *nuptial flight.* After mating, the males die and the inseminated queens lose their wings before searching for a suitable nest site. The queen lays eggs in the new nest site and feeds the developing larvae from her food reserves stored as fat and flight muscle. The emerging brood become the workers and take over nest maintenance, foraging, and nursing activities. The small colony grows slowly at first and may take a few years to become mature and produce its own reproductives. A single queen is the rule in some ant species, whereas multiple queens occur in others. Some species, such as the imported fire ants, have both monogynous and polygynous colonies.

Ants nest in a variety of situations. In the case of army ants, there is no physical nest, but only a *bivouac* formed from the ants themselves holding on to one another by their legs to form a large mass. Carpenter ants excavate wood to form cavities for their nest, while many ants make aerial nests of *carton,* a material formed from soil or plant fibers and saliva. Some *Formica* species build large nests up to 1 m high consisting of a mound of small twigs, hence the name "thatching ants." Most ants establish their nests in soil, where they excavate extensive galleries and tunnels. Soil is an ideal nesting material as it moderates temperature extremes, holds moisture, and can be easily shaped by the ants into brood- or food-holding chambers. The various groups of ants have definite preferences for specific types of soil.

Ants can have tremendously populous colonies. Primitive ants tend to have only a few hundred workers (e.g., some ponerines), while other ants may have up to 10,000 workers in their colonies (e.g., harvester ants). The enormous populations of 1−2 million for army ants

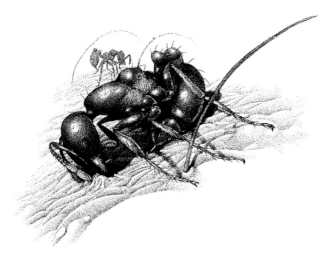

**FIGURE 19.5**    Red imported fire ant (*Solenopsis invicta*) worker, sting-ing human skin. The ant seizes the skin between its mandibles to provide leverage as it flexes the tip of its abdomen forward to penetrate the skin with its sting. (Original by Blair Sampson.)

and 22 million for driver ants (Dorylinae) are impressive. However, colonies of these species are small compared to *megacolonies* of some *Formica* species, which may contain 300 million workers and nearly 1 million queens occupy-ing an area of a few square kilometers. Such polygynous colonies are usually very tolerant of nonnestmate ants, and it is very difficult to determine whether such colonies are separate units or indeed one giant ant colony.

**Fire ants** *(Solenopsis* species*)*

Most ant stinging problems in North America are due to the two species of imported fire ants in the south-ern United States: *Solenopsis invicta* (Fig. 19.5) and *S. richteri* (Fig. 19.6). Less frequent stinging problems are caused by the two native fire ants in the southern United States, *S. geminata* and *S. xyloni,* and the little fire ant *Wasmannia auropunctata.*

The *black imported fire ant* (*S. richteri*) was intro-duced from South America into the United States at

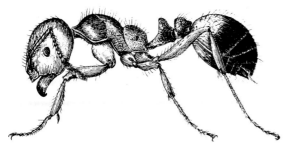

**FIGURE 19.6**    Black imported fire ant (*Solenopsis richteri*), worker. (Courtesy of US Department of Agriculture.)

Mobile, AL, in the early 1900s, followed by the in-troduction of the *red imported fire ant* (*S. invicta*) at the same port in the late 1930s. Since that time few other stinging insects have created more controversy, generated more research, or received more publicity than these two species of ants. They now inhabit a major portion of 12 southern states (Fig. 19.7), with *S. invicta* occupying 95% of the infested area; *S. rich-teri* occurs only in parts of Mississippi, Alabama, and Tennessee. The distributions of these ants probably have reached their northernmost limits where they can survive in the central and eastern United States. There is, however, a possibility that these ants will become established in the western coastal states as far north as Washington. Climatic conditions on the West Coast are suitable, and the westward spread of these ants has been impeded only by the arid regions of western Texas. The red imported fire ant was reported in Nevada in 1993 and in Arizona in 1994. More recently, it has invaded California and Oklahoma.

Fire ants are omnivores and opportunistic feeders. They feed mostly on insects and other arthropods. Fire ants readily feed on seeds of some plants and can af-fect local plant assemblages by transporting viable seeds of other plant species. They also feed on germinating plants, as well as on fruits and roots, and can cause fur-ther damage by girdling tree seedlings.

Fire ants are soil nesters, with most colonies be-ing initiated by a single inseminated queen after a nuptial flight during April–August. A queen makes a burrow 3–12 cm deep and within 24 hr lays her first eggs in a chamber at the end of the burrow. As many as 2,500 colonies can be initiated per hectare, but few of these incipient colonies survive the next winter. Colony growth is rapid and often produces more than 10,000 workers within a year. Some colonies contain as many as tens of thousands to hundreds of thousands of workers within a few years. Polygynous colonies are fairly common. Broods may be produced year-round in the southernmost distributions of the ants, but brood production ceases during the winter months north of 30° N latitude. The size and texture of the above-ground mound is variable depending on soil type, moisture, and vegetation (Fig. 19.8). Mounds in sandy areas are generally flat, whereas those occurring in clay soil may be up to 1 m tall by 1 m in diameter. Large colonies often construct several interconnected mounds.

Fire ants quickly respond to disturbances of their nests and attack intruders in force. When a colony is disturbed, the ants swarm over the intruder until the first worker stings and alarm pheromones are re-leased. This triggers stinging behavior in other work-ers. The workers grasp the skin with their mandibles

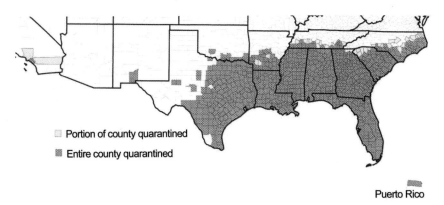

Portion of county quarantined

Entire county quarantined

Puerto Rico

**FIGURE 19.7**    Geographic distribution of the red imported fire ant (*Solenopsis invicta*) in the continental United States, 2000. It also occurs in Puerto Rico. (Courtesy of the US Department of Agriculture/Animal and Plant Health Inspection Service.)

(Fig. 19.5) and sometimes spin in a circle from this attachment, stinging the entire time. The common name "fire ants" reflects the burning sensation caused by their stings.

Fire ants occur in rural, urban, and suburban areas, where they constitute a serious stinging problem. In addition to damaging crops, they often sting farm workers as they harvest crops by hand. Large sun-hardened mounds can damage farm equipment or make it nearly impossible to harvest the crops by machine. High densities of colonies in agricultural areas can result in land devaluation. The impact of the red imported fire ant on recreation is decidedly negative, as tourists tend to avoid areas with heavy infestations (Ervin and Tennant, 1990). Colonies are common in urban yards and parks and can be a serious stinging problem to small children and pets. Colonies are even known to nest in traffic lights, air conditioners, and

other electrical equipment. Foraging ants will enter houses such that stinging incidents can even occur inside homes.

These introduced pests are outcompeting and displacing many native ant species and may even be adversely affecting plant and invertebrate communities. At the same time these ants have had some beneficial effects. They are general predators of a wide variety of crop-damaging insects, as well as of ticks and flies detrimental to livestock and game animals.

The *southern fire ant* (*S. xyloni*) ranges from California to South Carolina and Florida. This ant usually nests in soil, although it also will nest under stones, in the woodwork of houses, and in masonry. Outdoor nests are marked by excavated soil deposited in irregular piles around the entrances. These ants can girdle nursery stock and other plants and will burrow into plant buds, potato tubers, and strawberries.

The *fire ant* (*S. geminata*) ranges from Costa Rica to Texas and South Carolina to Florida. Until the imported fire ants were introduced, this ant was the most common and serious ant pest in Florida. Its nests are built in the soil, producing mounds up to 20 cm high. The biology of this species is similar to *S. invicta*. Like *S. xyloni*, it is only a minor problem and has been largely displaced by *S. invicta* in the United States as *S. invicta* has extended its range.

The *little fire ant* (*W. auropunctata*) is only 1.5 mm long. It nests in soil, in decayed wood, under stones, in cavities in plants, at the bases of trees, and in houses (Smith, 1965). The limits of a nest are hard to determine, suggesting that nests have satellites and that colonies are polygynous. This ant occurs only in Florida and cannot survive in cooler areas of the United States. Unlike the other fire ants, workers of this ant are not aggressive and sting only when trapped in clothing or similar situations. Unfortunately, these ants will nest in

**FIGURE 19.8**    Typical fire ant mound in the piedmont region of the southeastern United States. (Photo by G. R. Mullen.)

houses and infest clothing, beds, and food. Laborers sometimes refuse to work in cropland where these ants are abundant.

For a comprehensive review of fire ants in the United States, see Taber (2000).

### Harvester ants (*Pogonomyrmex* species)

As their common name implies, these ants all regularly use seeds as part of their diet. In addition, *Pogonomyrmex* workers scavenge for dead arthropods. Although there are a number of genera in the subfamilies Ponerinae, Myrmicinae, and Formicinae that comprise the harvester ants, only *Pogonomyrmex* species are of particular stinging concern in North America. Several species occur in North America, but probably only seven species constitute a stinging hazard (Cole, 1968). The more common ones are the *western harvester ant* (*Pogonomyrmex occidentalis*), the *red harvester ant* (*P. barbatus*), the *California harvester ant* (*P. californicus*), and the *Florida harvester ant* (*P. badius*).

*Pogonomyrmex* workers (Fig. 19.9) are large, up to 10 mm in length. Most are light red or brown, although the gaster of some species may be dark brown to black (Cole, 1968). These ants are identified by the presence of a *psammophore*, a fringe of hairs on the underside of the head. These "beards" are used in nest excavation to push material from the nest much like the blade of a bulldozer. Harvester ants are usually slow-moving and cannot walk up slippery vertical surfaces such as glass. Dense concentrations of colonies are common in the western United States, where most species occur.

Harvester ants construct their nests in dry, sandy to hard soils. The entrance to the nests is often marked by a crater or a cone in the center of a slight mound. A pile of small stones usually surrounds the entrance. Some species in hot deserts lack a mound. The nest can be 1–10 m in diameter, with tunnels extending down to 5 m or more. The area around the nest is usually completely devoid of vegetation. Colonies of some species have up to 10,000 workers. Individual colonies often survive for 14–50 years, reaching maximum densities of 80 or more nests per hectare. Foraging trails from individual nests may extend out 60 m. Where nest densities are high, large expanses of ground may have little vegetation. Because of this habit of harvesting seeds and reducing vegetation, they can damage rangeland used for cattle grazing and sometimes become significant pests locally (MacKay, 1990). Nests invariably occur in sunny locations; if the nest becomes shaded by vegetation or human activity, the ants generally move.

Harvester ants sting readily and can inflict intense pain. The incidence of stings is low, however, because their relatively large size and conspicuous nests cause most people to avoid them. Also, their colonies are relatively small compared to some other ant species. An additional source of stings has been ant farms sold in stores throughout the United States. Ant-farm kits usually contain coupons to be exchanged for live ants. In some cases, *Pogonomyrmex* species have been shipped for this purpose. Children have been stung severely when the plastic ant farms containing these ants have been dropped or otherwise broken open.

### Pavement ant (*Tetramorium caespitum*)

This ant (Fig. 19.10) was introduced to North America from Europe. At one time it was especially common only along the Atlantic seacoast, but now it is very common throughout North America. Nests are usually located in soil, but also commonly occur in houses. The pavement ant is omnivorous, being particularly fond of meats and fatty substances. Although it can cause extensive damage to many cultivated plants, it is more of a nuisance than a stinging problem for homeowners.

**FIGURE 19.9** Harvester ant (*Pogonomyrmex* sp.), workers gathering food. (Photo by R. D. Akre.)

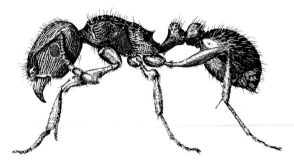

**FIGURE 19.10** Pavement ant (*Tetramorium caespitum*), worker. (Courtesy of the US Department of Agriculture.)

FIGURE 19.11    Pharaoh ant (*Monomorium pharaonis*), worker. (Courtesy of the US Department of Agriculture.)

This small ant (3–3.5 mm long) usually is not capable of penetrating human skin with its sting. Nevertheless, stings have been reported to cause a skin rash in children, but only rarely have they stimulated serious allergic reactions.

### Pharaoh ant *(Monomorium pharaonis)*

This tiny ant (Fig. 19.11) is only about 2 mm long and nests in every conceivable habitat: in soil, in houses, between sheets of paper or linen, and in trunks of clothing, to mention only a few. The pharaoh ant probably is native to Africa and has been widely disseminated by commerce so that it is now found throughout most of the world. It occurs in nearly all cities in the United States. It forms huge, polygynous colonies of more than a million workers and produces brood year-round. This ant is omnivorous, feeding on sugary materials, dead insects, breads, and many other foodstuffs, and, like the pavement ant, it prefers meats and fats. It is among the more difficult of all ants to control.

In addition to being a concern to human health because it stings, the pharaoh ant has been known to infest surgical dressings and intravenous units in hospitals and will attack the delicate tissues of newborn babies, especially the eyelids and navel. Because this ant has been found in bedpans, drains, and washbasins, it can come into contact with disease organisms and has been implicated in the transmission of pathogenic organisms in some hospitals (Edwards, 1986). In tropical regions other ant species have also been implicated as vectors of pathogens in hospitals (Fowler *et al.*, 1993).

# WASPS AND HORNETS

## SOLITARY WASPS

The term "wasp" encompasses a diverse assemblage of hymenopterous groups, including solitary-nesting potter wasps and social wasps such as yellowjackets and paper wasps. Solitary wasps hunt various arthropods as prey to

FIGURE 19.12    A brightly colored velvet ant, *Dasymutilla occidentalis* (Mutillidae), commonly known as "cow killer" because of the intense pain its sting causes. (Photo by E. J. Hansens.)

feed their larvae. Their venom is designed primarily to paralyze their prey and not to cause pain in vertebrates. The female wasp lays an egg on the immobilized prey. Since the prey is still alive, it is essentially preserved until it can be consumed by the developing larva. Only a few solitary wasps are a stinging threat to people.

### Mutillidae

Among the solitary wasps perhaps the best known for their stings are the brightly-colored *velvet ants,* or *mutillids* (Figs. 19.12). Unlike the males, females are wingless and can cause a very painful sting. The stings of large species produce an intense burning sensation and may cause substantial swelling and redness at the sting site. Females often are seen walking over the ground, especially in open sandy areas, searching soil nests of bees, wasps, and other insects to invade and parasitize. One large red and black mutillid, *Dasymutilla occidentalis,* is called the *cow killer* because of its particularly painful sting (Fig. 19.12). For further information on this poorly known group of solitary wasps, including keys to species in the southeastern United States, see Manley (1991).

### Pompilidae

The metallic blue or black *spider wasps* in this family may be seen flying low along the ground in search of spiders, which they attack, sting, and carry to a burrow in the soil where they oviposit on their prey. Some large spider wasps (e.g., *Pepsis* spp.) (Fig. 19.13) can deliver painful stings.

FIGURE 19.13 Spider wasp, *Pepsis* sp. (Pompilidae). (Photo by R. D. Akre.)

## Sphecidae

This family is a diverse and large group of solitary wasps, most of which are not likely to be a stinging threat (Bohart and Menke, 1976). Sphecid wasps such as mud daubers (*Sceliphron* and *Chalybion* spp.) and cicada killers (*Sphecius* spp.) commonly nest near or on human dwellings. Although they appear threatening, stings to humans are relatively infrequent. *Mud daubers*, such as the black and yellow mud daubers (*Sceliphron* spp.) (Fig. 19.14), commonly build their mud nests in attics, carports, and porches and provision them with spiders. These wasps are often evident around nesting sites as they make frequent trips carrying mud to build their nests. They will rarely sting during these flights; in fact, it is difficult to induce mud daubers and other solitary wasps to sting humans or animals even by disturbing them. When stinging does occur, it usually involves the

FIGURE 19.15 Cicada killer, *Sphecius speciosus* (Sphecidae). (Courtesy of the Oklahoma State University Entomology Department.)

wasp being trapped inside clothing, being stepped on, or otherwise being pressed against the victim's skin or clothing. Occasionally these solitary wasps sting careless collectors.

The *cicada killers* (*Sphecius speciosus* and other species) are large black wasps marked with yellow bands (Fig. 19.15). They capture cicadas and carry their stung prey to burrows they have excavated in soil. The female digs a fairly large burrow with a distinct soil mound surrounding the entrance. They often nest in lawns, gardens, or flower beds around human dwellings. The female will not sting unless handled but can inflict a mildly painful sting to people walking barefoot near their nest sites. The males frequently fly around the nesting sites patrolling their territories. During their flights they will approach other insects, birds, and even people that enter their territories. Although they can appear intimidating, these individuals, like all hymenopteran males, cannot sting.

## SOCIAL WASPS (VESPIDAE)

In contrast to the relatively innocuous solitary wasps, social vespids are very defensive and will sting readily. This usually occurs in proximity of their nests but also occurs at foraging sites. Vespid wasps are unique in that they usually fold their wings longitudinally at rest. The yellowjackets and hornets (vespines) and the paper wasps (*Polistes* spp.) are responsible for most stinging cases in North America. Vespine nests consist of horizontal, rounded combs

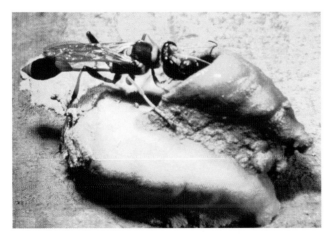

FIGURE 19.14 Mud dauber, *Sceliphron* sp. (Sphecidae), constructing nest. (Photo by R. D. Akre.)

**FIGURE 19.16**    Aerial paper nest of baldfaced hornet, *Dolichovespula maculata* (Vespidae), in tree. (Photo by R. D. Akre.)

attached one below the other, usually covered with a multilayered paper envelope (Akre *et al.*, 1981; Ross and Matthews, 1991). Some yellowjacket species build nests in exposed aerial locations (Fig. 19.16), whereas others build them in subterranean sites (Fig. 19.17). A few construct their nests in both situations. Colony size ranges from fewer than a hundred to several thousand individuals. Colonies are typically annual and are initiated by a single inseminated queen; she is the only member of the colony to overwinter. During early spring (April–June), the queen begins to construct a nest, lays eggs, and then forages for arthropods to feed the developing larvae. After the first group of workers emerge, they assume most colony functions of foraging for food, fiber, and water

and taking care of the larvae. The queen no longer leaves the nest after this time but assumes her primary function of laying eggs. Some species scavenge for dead organic material as a protein source for their larvae. This behavior frequently brings these wasps in contact with humans near garbage or other refuse and at picnic and recreational sites.

Perennial colonies sometimes occur when the new queens that emerge in the fall mate and then rejoin a colony. This is most likely to occur when the foundress queen has died or is losing her influence over the colony. These polygynous colonies can become perennial and contain tens of thousands of workers. Large perennial colonies of *V. vulgaris*, *V. pensylvanica*, *V. germanica*, *V. squamosa*, and *V. maculifrons* have been reported in the United States, usually in subtropical areas such as Florida or in moderate climates like that along the California coast. These colonies often have multiple entrances and more than 80,000 workers. Such nests are dangerous to destroy even by experienced people.

**Yellowjackets** *(Dolichovespula* and *Vespula)*
"Yellowjacket" is an American term that is used for all species of wasps in the genera *Dolichovespula* and *Vespula*. The name refers to the yellow and black patterns of most species (Fig. 19.18); however, some species are black and white, such as the *baldfaced hornet*, *Dolichovespula maculata*. There are 19 species of yellowjackets in North America (6 *Dolichovespula* spp. and 13 *Vespula* spp.). Species are distributed transcontinentally in the United States, but most species occur primarily in the northern areas of the country (Akre *et al.*, 1981). True hornets (*Vespa*) are larger than yellowjackets, with body lengths of 20–25 mm in the workers. There is only one true hornet that occurs

**FIGURE 19.17**    Exposed underground nest of eastern yellowjacket, *Vespula maculifrons* (Vespidae). (Photo by H. Reed.)

**FIGURE 19.18**    Workers of the western yellowjacket, *Vespula pensylvanica* (Vespidae) guarding the entrance of their underground colony. (Photo by Roger D. Akre.)

TABLE III

Comparison of Colony Parameters and Foraging Behavior of Yellowjackets in North America: Colony Decline is Defined as the Period When Reproductives Emerge

| Yellowjacket | Distribution | Foraging behavior | Colony size | Colony decline |
|---|---|---|---|---|
| *Vespula* | | | | |
| germanica | Transcontinental | | 500–5,000 workers | Late September |
| maculifrons | Eastern | Predators and | 500–15,000 cells | to early December |
| pensylvanica | Western | scavengers | perennial colonies | |
| vulgaris | Transcontinental | | 100,000+workers | |
| flavopilosa | Eastern | | 1,000,000 cells | |
| *Vespula* | | | | |
| atropilosa | Continental | Strictly predators | 75–400 workers | Late August |
| acadica | Continental | | 500–2,500 cells | to September |
| consobrina | Continental | | No perennial colonies | |
| *Dolichovespula* | | | | |
| arenaria | Transcontinental | Strictly predators | 100–700 workers | Early July |
| maculata | Transcontinental | | 500–4,500 cells | to September |
| | | | No perennial colonies | |
| *Vespula* | | | | |
| squamosa | Southern United States | Predators and some | 500–4,000 workers | Late August |
| sulphurea | California | scavengers | 500–10,000 cells | to November |
| | | | Some perennial colonies | |

in North America, the *European hornet, Vespa crabro.* This species was accidentally introduced from Europe into the New York area during the mid 1800s. Today it is most common in the Washington, D.C., area, although colonies have been found throughout the eastern United States and west to the Mississippi River.

Only two species of *Dolichovespula* are considered hazardous in North America: the *aerial yellowjacket* (*D. arenaria*) and the *baldfaced hornet* (*D. maculata*). Both species occur primarily in forested areas and are distributed throughout North America; however, the aerial yellowjacket does not occur as far south (Akre *et al.,* 1981). Their large aerial spherical or egg-shaped nests are familiar sights in trees (Fig. 19.16). The aerial yellowjacket occasionally builds nests at or just below ground level. In urban areas nests are often built under the eaves of houses or on nearby bushes and trees, where disturbances of the nests are common. Although the nests are usually obvious, some people fail to notice them. When they are constructed above a doorway, the vibrations from opening and closing the door can excite the wasps to sting. Both species can squirt venom from the sting into the eyes of nest intruders during an attack. The immediate watering of the eyes and the pain can temporarily blind humans. Other defending wasps attack simultaneously and deliver stings. Knowledgeable collectors wear goggles or glasses under a bee veil when collecting or controlling nests of these wasps. Fortunately, the average colony size is relatively

small by vespine standards, consisting of fewer than 400 workers.

*Vespula* species are typically subterranean soil nesters (Fig. 19.17). Ordinarily the queen initiates the nest in an old rodent burrow or in other cavities in soil, logs, or trees. Some species, however, commonly establish colonies in wall voids of buildings and other structural crevices. The paper nests of most yellowjackets are made of gray carton, but those of *Vespula flavopilosa, V. maculifrons,* and *V. vulgaris* consist of tan, fragile carton (Fig. 19.17). Colony cycles of scavenging species extend later into the fall (September–December) than those of the nonscavenging species, by which time their nests are large and contain as many as 5000 workers (Table III). During the production of new males and queens (September and later), the workers become more defensive of the colony. Additionally, as the colony goes into decline, the undernourished workers become more persistent and aggressive at food sites and are more likely to sting with little provocation.

Workers of seven species of *Vespula,* such as the *eastern, western,* and *southern yellowjacket,* scavenge for protein at garbage or picnic sites, where they frequently come into contact with humans (Table III). Scavenging *Vespula* species forage for fresh and decaying tree sap, protein (e.g., insects, vertebrate flesh, processed meats), and carbohydrates from almost any human source (e.g., soda pop, beer, sugary confection). Most workers of the other North American species, such as *V. atropilosa*, lack this foraging habit and therefore are seldom involved in sting cases. When their nests are disturbed, they will sting, but because their colonies are small the result is not as serious as attacks from larger colonies.

The *eastern yellowjacket* (*V. maculifrons*) and the *hybrid yellowjacket* (*V. flavopilosa*) are major picnic and campsite pests, especially in the fall in the eastern United States. Workers of both species will feed on fresh and decaying fruits and vegetables. These two species are responsible for many stings to humans and their pets during outdoor recreational activities such as ball games and harvest celebrations, especially during years of extremely high yellowjacket populations.

The *western yellowjacket* (*V. pensylvanica*) (Fig. 19.18) is the dominant yellowjacket in dry forests of the western United States. It attains high population densities in some years, often correlated with warm spring weather. This commonly results in increased encounters with humans around homes, picnic areas, and campgrounds. During peak populations of *V. pensylvanica*, recreational areas such as resorts, hunting camps, and parks may have to be closed. These "wasp years" often coincide with a high incidence of forest fires, posing a serious stinging problem for firefighters and smoke jumpers. Yellowjacket stings in some years also cause a significant loss of worker time by U.S. Forest Service personnel in the western states. Many people are stung while getting a drink of water from fountains frequented by yellowjackets seeking water. Others accidentally "drink" wasps that have crawled into their beer or soda pop cans. Stings delivered inside the mouth or in the esophagus and stomach can be serious due to resultant swelling of the adjacent trachea and shutting off the air passage. Stinging incidents also occur to fishermen cleaning their catch or to hunters retrieving a deer; a fresh carcass can attract large numbers of scavenging yellowjacket workers. Huge perennial colonies have been reported in Hawaii, where they may disrupt both agriculture and tourism.

The *German yellowjacket* (*V. germanica*) is a species native to Europe which has shown a remarkable propensity for becoming established in temperate areas of the world. It now occurs in many countries, including New Zealand, Australia, South Africa, Chile, and the United States. Colonies tend to be large, especially in warmer areas, where perennial polygynous colonies may contain thousands of workers. The German yellowjacket has been introduced into the United States many times, but it did not become established until the late 1960s. The biotype in North America usually nests inside structures, unlike the European biotype, which typically nests underground. Colonies in buildings benefit from the associated heat and protection and tend to persist very late into the fall and winter months, resulting in larger colonies. At that time of year, workers occasionally chew through walls and emerge inside buildings. The German yellowjacket is becoming more established across eastern and midwestern North America and even the West Coast.

Early in the colony cycle, German yellowjacket workers start to scavenge for protein, liquid carbohydrates, and fruits. They even have been observed feeding their larvae pieces cut from small green apples. This early-season scavenging seems to give these wasps an advantage over other yellowjacket competitors that exclusively spend time capturing insect prey. Consequently, this species has become the dominant species in some areas it has invaded. German yellowjackets have been reported to be more aggressive and more prone to sting than most other yellowjacket species. This perception is attributed in part to their larger colony sizes in comparison to colonies of other sympatric species.

The *common yellowjacket* (*V. vulgaris*) occurs in Europe and across North America. *V. vulgaris* workers are a stinging problem in the same situations as *V. pensylvanica* but more so in moist forests of the United States. Perennial colonies of *V. vulgaris* consisting of 50,000–100,000 workers have posed particularly serious stinging threats in some areas of California.

The *southern yellowjacket* (*V. squamosa*) produces perennial colonies that are particularly common in Texas and Florida (Akre *et al.*, 1981). Some of these colonies consist of more than 100,000 workers and are a serious hazard to humans. One such nest reported from Parrish, Florida, was 3.6 m high by 1.8 m in diameter, built on a broken tree stump. These wasps produce an alarm pheromone that causes stinging attacks on intruders to be particularly vigorous. Workers of some *V. squamosa* colonies are scavengers and are common pests at picnics. A sister species of *V. squamosa*, the *California yellowjacket* (*V. sulphurea*), also scavenges for protein and has been reported as a nuisance for picnickers.

### Hornets (*Vespa* species)

The *European hornet* (*Vespa crabro*) is the largest social wasp in North America (body length > 20 mm), with contrasting brown and deep yellow bands on the gaster. Although *Vespa* hornets are important stinging hazards in Asia, *V. crabro* is seldom responsible for stinging incidents in North America. This is primarily because *V. crabro* is relatively uncommon and usually nests in hollow trees in areas away from human activities. Brown envelope and carton distinguish their nests from the more common gray nests of *D. maculata*. Despite the large size of the workers and the nest, this species is less aggressive than most other North American vespines. A typical colony consists of 200–400 workers at peak populations, but large colonies can reach 1000. Colonies have a long seasonal cycle lasting from early spring into late August–November.

**Paper wasps** (*Polistes* species*)*

Paper wasps are the most common stinging wasps encountered by humans in the southern United States, especially during the summer months. Some paper wasps, such as *Polistes exclamans,* are improperly called "yellowjackets" due to their alternating bands of yellow and dark-brown markings on their gaster. Paper wasps are usually longer and have more slender bodies than yellowjackets. *Polistes* wasps construct paper nests consisting of a single comb with no envelope (Fig. 19.19). Several species (such as *P. exclamans* and the introduced European species, *P. dominulus*) tend to nest near or on buildings and as such come into frequent contact with people. Their nests are often found under roof eaves and window sills; around door frames; and inside garages, storage buildings, clothesline poles, bird houses, wall voids, and attics. More natural nesting sites include trees, shrubs, and cliff overhangs. Some species, such as *P. exclamans,* prefer nesting in exposed sites on structures, while others, such as the *red wasp*

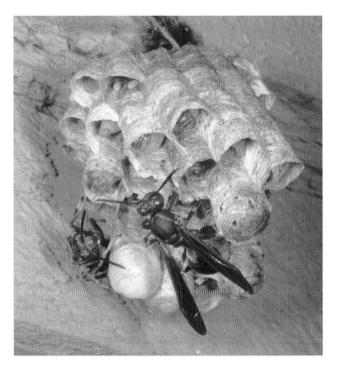

**FIGURE 19.19** Paper wasp, *Polistes metricus* (Vespidae), with two female foundresses on the nest. Note eggs and developing larvae in unsealed cells. (Photo by G. R. Mullen.)

(*P. perplexus*), usually nest in concealed sites such as wall voids and attics (Table II).

Most colonies are small, with fewer than 100 adults and 100–200 cells; larger nests consist of about 400 cells. The annual colonies are initiated by a single foundress or a group of foundresses that compete for reproductive dominance. After the workers emerge, one foundress usually becomes the primary reproductive or queen. Unlike vespine wasps, no queen–worker size dimorphism occurs in paper wasps. Paper wasps forage for caterpillars and thus are excellent natural control agents of many crop pests. Colonies even have been propagated and transplanted into fields to decrease pest populations.

In the fall, paper wasp reproductives sometimes swarm in large numbers at the top of tall towers and buildings, where they may create concerns. However, many of these wasps are males, which cannot sting, whereas the females are primarily involved in mating and aggregation activities and only rarely sting at these sites. Later in the fall and winter the females aggregate in large numbers at hibernation sites such as the attics of houses, apartments buildings, and barns. Contact with humans is common at this time, resulting in stings. Stinging episodes also occur in the spring when the aggregating wasps come out of hibernation. Unlike yellowjackets, paper wasps do not scavenge and rarely sting away from the colony.

*P. annularis* is known as *Spanish jack* to fishermen, boaters, and river dwellers in the Gulf Coast states (United States) because of the occurrence of its nests in shrubs, trees, and other structures along streams and lakes. People are commonly stung when they inadvertently disturb a nest. This large wasp produces the largest colonies of paper wasps in North America and can often inflict multiple stings. Several other species, such as *P. metricus, P. fuscatus, P. aurifer, P. exclamans, P. apachus,* and *P. dominulus,* are also involved in stinging episodes with humans (Table II).

# BEES

## SOLITARY BEES

Most bees are solitary or, at most, communal or semisocial. Some species of the family Halictidae, which include the sweat bees, are in fact true social bees. A few of the solitary communal and social species form dense nesting aggregations, usually in soil, where they excavate cells in which to rear their larvae. Larvae of all bee species feed on pollen and nectar provisioned in these cells. Nesting activity is quite variable, ranging from a single nest to

**FIGURE 19.20**    Sweat bee, *Lasioglossum zephyrus* (Halictidae), on human arm. (Photo by G. R. Mullen.)

**FIGURE 19.21**    Carpenter bee, *Xylocopa virginica* (Anthophoridae), male. Note that abdomen is mostly black and lacking yellow hairs on most segments. (Photo by G. R. Mullen.)

thousands of individual nests concentrated in a given area. Some bees make nests in wood cavities such as the hollow stems of trees and shrubs, burrows of wood-boring beetles, and artificial cavities such as the hollow pipes of wind chimes and keyholes in doors. Few solitary bees are stinging hazards to humans.

## Halictidae

The *sweat bees* (Fig. 19.20) frequently land on people to imbibe perspiration. In response, a person may incite the bee by swatting at it. The stings usually are considered minor irritations and are simply a nuisance.

## Anthophoridae

The *carpenter bees* are similar in size and general appearance to bumble bees. However, they lack the hairy appearance and yellow coloration typical of bumble bees, and the dorsum of the gaster is mostly black (Fig. 19.21). These bees often nest in wood around human dwellings, where they bore round holes in window sills, eaves, railings, fence posts, and other wooden structures (Fig. 19.22). Female carpenter bees rarely sting, and when they do the pain is relatively mild. Males often are seen flying around nesting sites and may appear threatening; however, like all male hymenopterans, they cannot sting. The common carpenter bee in eastern North America is *Xylocopa virginica;* the common species in western North America are *X. californica* and *X. varipuncta.*

## SOCIAL BEES

Most bee-sting cases are due to social bees in the family Apidae, which include bumble bees (Bombinae), honey

bees (Apinae), stingless bees (Meliponinae), and orchid bees (Euglossinae). The *stingless bees* are exclusively tropical and are found in the Old and New World. It is often assumed that these bees pose no hazard to humans since they are stingless. However, they do swarm out of the nest to attack intruders by biting, especially around the eyes, nose, and ears. One species is known as the *fire bee* (*Oxytrigona tataira*) because of the caustic defensive secretions produced by its mandibular glands that cause intense burning pain when applied to the skin. The *orchid bees* are restricted to the New World tropics and are rarely a stinging concern.

**FIGURE 19.22**    Circular entrance to nest of carpenter bee, *Xylocopa virginica* (Anthophoridae), in cedar wood of house eave. (Photo by G. R. Mullen.)

**FIGURE 19.23** Bumble bee, *Bombus* sp. (Apidae), foraging worker. Note conspicuous yellow hairs on abdominal segments. (Photo by T. Kondo.)

**FIGURE 19.24** Exposed subterranean nest of bumble bee, *Bombus* sp. (Apidae), showing both open and sealed cells and surrounding nesting material. (Photo by R. D. Akre.)

**Bumble bees** *(Bombus species)*

These large, hairy, yellow and black bees (Fig. 19.23) pose little stinging hazard for most people, common as they are around flowers. Most of the 400 species are nonaggressive, even when the nest is disturbed. However, a few species can be very aggressive and persistent in their stinging attacks when their nests are threatened. Bumble bee workers can sting an intruder repeatedly.

Bumble bees generally occur in the more temperate areas of North America, where their colony cycle is very similar to that of yellowjackets. Colonies are initiated by an inseminated queen early in the spring, followed by the production of a few hundred workers during the summer. Most nests are established in abandoned rodent burrows, although they also may be built in attics and wall voids of houses. Nests consist of wax pouches or empty cocoons for storing pollen and honey, and wax chambers for rearing brood (Fig. 19.24). Workers forage for pollen and nectar to feed the larvae. These bees are important natural pollinators and also pollinate many agricultural crops, especially cranberries and raspberries.

**Honey bees** *(Apis species)*

Honey bees (Figs. 19.25 and 19.26) are native to Europe, western Asia, and Africa. Four of them are common species that occur from Southeast Asia to Africa and in Europe: the *common honey bee (Apis mellifera)*, *giant honey bee (A. dorsata)*, *Asian honey bee (A. cerana)*, and *dwarf honey bee (A. florea)*. Only the common honey bee occurs in North America. Several subspecies and races of *A. mellifera* were domesticated by people in Europe and have been introduced into nearly every country in the world. This species is an invaluable pollinator of native and commercial plants, especially fruits and vegetables. In the United States, the social gains due to *A. mellifera* from their honey production and their pollination of more than 120 cultivated crops have been estimated to be U.S. $2–18 billion annually.

Honey bees build large nests, called *hives*, consisting of several wax combs of hexagonal cells (Fig. 19.26), in which they rear the brood and store pollen and nectar. Feral or "wild" honey bee colonies construct nests in cavities of hollow trees and in attics and wall voids of human dwellings. These colonies typically have 15,000–30,000 bees, whereas commercial hives are usually larger, with 30,000–50,000 bees. Despite the fact that commercial breeding of honey bees has

**FIGURE 19.25** Honey bee, *Apis mellifera* (Apidae), worker foraging on flower. (Courtesy of *California Agriculture*, University of California.)

**FIGURE 19.26**    Honey bees (*Apis mellifera*) on surface of comb in hive. European honey bee (center), surrounded by Africanized honey bee workers. (Courtesy of Entomological Society of America.)

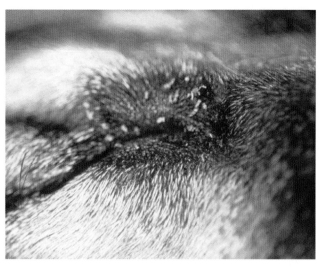

**FIGURE 19.27**    Head of German shepherd dog in fatal case of attack by Africanized honey bees (*Apis mellifera*). Note the concentration of bee stings about the eye. (Photo by J. Schmidt.)

dampened the defensive tendencies of many races, honey bee colonies can still attack and sting intruders in large numbers.

The perennial colonies of honey bees usually survive the winter, and in late spring through early summer they reproduce by *swarming*. Swarms often are seen resting in exposed sites such as trees, shrubs, and under eaves of buildings while they are seeking a suitable cavity in which to establish a new hive. Although these swarms are less defensive than an established colony because they lack brood, stored pollen, and honey, it is best not to approach or disturb them.

Of particular concern in the United States is the northward movement of the *Africanized honey bee* (*A. mellifera scutellata*), an aggressive subspecies of the common honey bee that has spread in recent years through tropical regions of South and Central America. It is most abundant in tropical, humid areas of Africa but extends into arid regions of South Africa. This honey bee was introduced into Brazil in 1956 in an effort to improve the beekeeping industry in Latin America. Captive bees escaped in 1957 when queen excluders were removed from some of the hives. They spread south to Argentina and north through South America, reaching Panama in 1982. The first Africanized honey bee colony trapped in the United States was in 1990 in Texas. As of 1996, colonies have been discovered in south Texas, New Mexico, Arizona, California, and Puerto Rico.

This bee quickly establishes itself in new areas. When foraging or nesting conditions become restrictive, the bees leave their hives in a process called *absconding* and relocate to new nesting sites. Other subspecies of the common honey bee become established more slowly

in new areas and are less likely to abscond when conditions change.

The Africanized honey bee has a propensity for mass-stinging attacks of both people and animals (Fig. 19.27). Detailed, quantitative studies have shown that these bees are much more alert than other subspecies to movement, vibration, and their own pheromones that mediate alarm and attack responses. The alarm compounds are not released in greater quantity by Africanized bees, but rather the threshold of response of the bees to a perceived threat is much lower. Nor is their venom any more potent than that of other *A. mellifera* subspecies; in fact, the composition is nearly identical.

Aside from their behavioral differences, individual Africanized honey bee workers are indistinguishable from workers of the European subspecies of *A. mellifera*. It requires a specialist to examine and measure specific morphological (e.g., wings) and biochemical features of several bees to determine their identity. While the Africanized honey bee stings much more readily than the European honey bee, so far deaths of animals and humans due to *A. mellifera scutellata* are not common. About 350 human deaths in Venezuela between 1975 and 1988 were attributed to Africanized honey bees (Winston, 1992). Since they reached Mexico in 1986, the bees have reportedly killed several dozen people. Annual fatalities due to bee stings in Texas have not increased since this bee arrived there in 1990.

Stinging problems due to the Africanized honey bee are actually less of a concern than the potential disruption to beekeeping and agriculture. The bees are more difficult to manage and store less honey than their

European counterparts. These traits, coupled with their extremely aggressive nature, have raised great concern for beekeeping in North America, which relies upon organized transport of many managed hives for pollination services. Also, the impact on tourism and recreation in Africanized honey bee areas could be considerable.

The future status and distribution of the Africanized honey bees in North America is controversial. Winston (1992) presented several predictions on the northernmost limit of the spread of this bee and stated that the most likely distribution is an Africanized zone extending through the southern one-third of the United States. The U.S. Department of Agriculture in 1984 predicted that the impact on beekeeping and agriculture in the United States will be U.S. $26–58 million in annual losses to beekeepers and an additional $93 million in annual losses due to diminished pollination of crops. The impact may be lessened by modifying hive management practices (e.g., requeening with docile European queens), thereby altering their genetic makeup by interbreeding with European strains. As the bee dispersed throughout tropical and subtropical areas of South and Central America, there were relatively few beekeepers in the areas traversed. However, now that the bees are in the United States, the management practices of thousands of beekeepers are expected to modify the behavior of the bees by interbreeding these bees with the more docile commercial colonies.

## PUBLIC HEALTH IMPORTANCE

Most encounters with venomous arthropods involve stings from ants, wasps, and bees. Probably several million people are stung annually by these insects; most of them do not require professional medical treatment. Some data on sting frequency are available on fire ants. Studies in the southeastern United States have indicated that 30–60% of people are stung by fire ants each year, but only 1–5% of these cases require medical treatment (Lofgren, 1986; Schmidt, 1986b). This rate is higher than that for all other hymenopterans. Stinging hymenopterans also pose a hazard during natural catastrophes such as forest fires and hurricanes. Insect stings, primarily due to yellowjackets, were the single most common cause of nonfatal injuries following Hurricane Hugo, which hit the southeastern coast of the United States in 1993 (Brewer *et al.*, 1994).

A total of 40–50 deaths due to Hymenoptera stings are usually reported in the United States each year (Schmidt, 1986c). However, it is suggested that these figures are very conservative and that a more accurate figure is probably closer to 200 (Akre and MacDonald, 1986). More

people than this die each year as a result of allergic reactions to penicillin or from lightning strikes (Camazine, 1988). Honey bees are believed to cause about half of the annually reported human deaths due to hymenopteran stings in the United States (Schmidt, 1992). This figure may be somewhat misleading because the general public and medical profession do not reliably differentiate between honey bees and other types of bees and wasps. Some deaths attributed to bees probably are due to stings from yellowjackets and paper wasps.

The stings of most social hymenopterans cause intense pain to humans, with reactions to various species differing primarily in intensity or duration. Comparative scales for ranking the severity of pain caused by aculeate hymenopterans have been proposed by Starr (1985) and Schmidt (1986c, 1990b). The intensity of pain caused by most social bees and wasps is similar, and the responses to stings of only a few species, such as the fire and harvester ants, are diagnostic. Also, there exists great variation in responses among individuals to stings by the same species. The location of the sting on the body (e.g., finger vs neck) also may influence individual reactions.

Although most ant stings are painful, those of *Paraponera* and *Pogonomyrmex* species are especially noteworthy. The large ponerine *Paraponera clavata* of Central and South America injects venom that produces intense, debilitating pain lasting several hours. This sting is considered to be the most painful of all Hymenoptera (Schmidt, 1986b, 1990b). The affected area can expand 20–30 cm within an hour of the sting. The pain induced by stings of *P. clavata* serves as a comparative standard for stings delivered by other stinging species (Starr, 1985). Although less severe than that from *Paraponera*, the pain caused by stings from harvester ants (*Pogonomyrmex* spp.) is extremely intense and can last up to several hours. It has been likened to "turning a screw" into the flesh around the sting site and also causing a sensation that has been described as "chilling." The sting is unique in that it induces piloerection (elevation of the hairs) and sweating around the sting site for 4–8 hr (Schmidt, 1986c). The venoms of some harvester ants are more toxic to mice than any other tested insect venom and 8–10 times more toxic than honey bee venom; they are almost as toxic per unit volume as the most venomous snakes (Schmidt and Blum, 1978).

Victims of *fire ant* stings usually experience a temporary burning sensation and discomfort, with some swelling around the sting site. In most cases a characteristic vesicle, 3–5 mm in diameter, containing a clear fluid develops within 6–24 hr at each sting site. The fluid becomes cloudy, forming a sterile white pustule (Fig. 19.28). These pustules usually disappear within a few days. They are replaced by a discolored lesion that results from tissue necrosis caused by the alkaloids in the venom

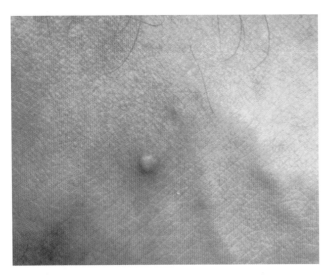

**FIGURE 19.28** Fire ant sting by *Solenopsis invicta* on human ankle, showing characteristic sterile pustule formed at sting site, 24 hr after sting. (Courtesy of M. Horton.)

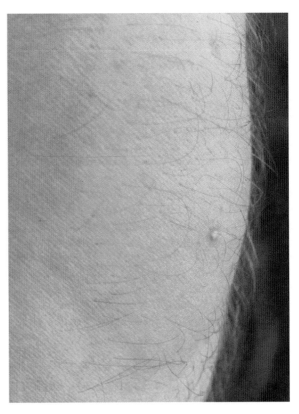

**FIGURE 19.29** Fire ant sting by *Solenopsis invicta* on human forearm, with localized swelling and inflammation, 15 hr following sting. (Photo by G. R. Mullen.)

and that persists for weeks or months. These lesions can be intensely pruritic and are subject to secondary bacterial infections when they are scratched or otherwise broken.

Reactions to insect stings have been variously classified (Reisman, 1994a; Schmidt, 1992) but can be divided into four categories: local, large local, systemic, and toxic reactions (Table IV). *Local reactions* are normal for most people and include immediate pain and/or burning at the sting site, development of a flare and wheal, and swelling that is limited to the sting site (Figs. 19.29 and 19.30). Later, itching at the sting site usually occurs. A *large local reaction* involves painful swelling at the site and even the associated extremity (Fig. 19.31), but this reaction is not considered serious. These are normal responses that are not life-threatening but may be accompanied by systemic, cutaneous reactions (e.g., hives) on parts of the body other than where the sting occurs.

*Systemic reactions* are more generalized responses that induce reactions away from the sting site. The most serious of these are *allergic reactions*, which can be fatal. Allergic reactions typically occur after a second or subsequent sting by the same or closely related species. The first sting causes the body to produce antibodies to specific proteins in the venoms. Later when this *hypersensitive* individual is stung again, the immune system overreacts to the presence of the same foreign venom proteins. These stinging episodes sometimes result in *anaphylaxis*, a sudden drop of blood pressure and respiratory distress

**TABLE IV**
**Descriptions of the Four General Types of Reactions to Hymenopteran Stings**

| Type | Duration | Response |
|------|----------|----------|
| Local | Ca 2 hr | Redness, itching, swelling, pain, wheal forms at sting site |
| Large local | 2–5 days | Painful swelling ($\geq 5$ cm) of sting site or entire extremity. General malaise or ill feeling, palpitation of heart, elevated blood pressure |
| Systemic | 10 min–3 weeks | Reactions to areas beyond the sting site. Normal symptoms plus hives, respiratory distress, laryngeal swelling, gastrointestinal distress, hypotension, cutaneous uticaria, widespread edema, and anaphylactic shock, sometimes leading to death |
| Toxic | Hours to months | Destruction of muscle and red blood cells, blood pressure drop, kidney failure |

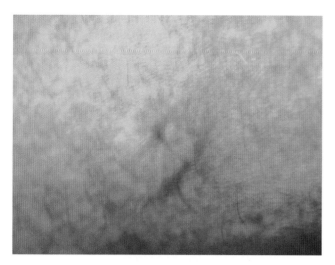

**FIGURE 19.30** Yellowjacket (*Vespula* sp.) sting on human thigh, showing central puncture mark and puffy swelling at sting site, 10 min after sting. (Photo by E. J. Hansens.)

triggered by this immunological response. In some cases this can result in *anaphylactic shock* and death. Such lethal reactions have been reported for stings of vespines, paper wasps, honey bees, and fire ants.

Deaths from hymenopterous stings are usually due to respiratory failure (70%). Other reported causes of mortality are anaphylaxis (15%), cardiovascular collapse (9%), and neurological complications (6%) (Schmidt, 1986b). Death from stings is usually very rapid, often within 20 min of the sting. Nearly 60% of deaths occur in less than 1 hr following envenomization. Therefore, it is imperative in cases of severe allergic responses that treatment be administered as soon as possible after the sting occurs. A few deaths attributed to toxic reactions have been reported in humans and animals receiving hundreds

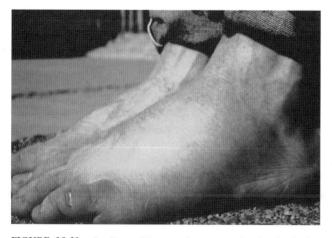

**FIGURE 19.31** Swelling of human foot as result of yellowjacket (*Vespula* sp.) sting. (Photo by R. D. Akre.)

to thousands of stings from the Africanized honey bee or yellowjackets. The victims in these cases usually are either young children or people restrained in some way in close proximity to a disturbed colony.

Although relatively few people are killed by hymenopteran stings, many people suffer some degree of *sting hypersensitivity*. Estimates of the incidence of hypersensitivity reactions, from mild to severe systemic responses, vary from 0.15 to 4% in the United States (Schmidt, 1992). An incidence of 2% in this case represents about 5 million people. Fewer than 1% of fire ant sting victims experience anaphylactic shock when stung, but this may be increasing as more people are stung by these aggressive ants. Vespid wasps cause twice as many allergic reactions as honey bees. However, the importance of the species involved differs with the region. In the United States, for example, yellowjackets cause more serious reactions than do honey bees in the Pacific Northwest and Washington, D.C. area. Paper wasps are more important in Texas and the southwestern states, whereas fire ants cause more allergic reactions and deaths than do honey bees in the southeastern states (Camazine, 1988). Hypersensitive people often suffer great physical stress and illness, while others may experience apprehension and other psychological effects by not knowing when or how serious a subsequent sting might be (Reisman, 1994b).

A unique type of allergy has been associated with honey bees. Family members of beekeepers can develop an intense allergy to constituents of honey bee venom while laundering venom-impregnated garments. Most of these people do not realize they are allergic because they have had little or no direct contact with the bees themselves.

Several approaches can be taken in the *treatment of sting victims*. Embedded stings should be removed as soon as possible to lessen the sting reaction. Conventional advice on sting removal has emphasized that the sting apparatus should be scraped off, not pinched off, as the latter was thought to squeeze more venom into the sting site. Experimental evidence on honey bee stings contradicts this advice and emphasizes the importance of immediate removal of the embedded sting apparatus in order to minimize the sting reaction (Visscher *et al.*, 1996).

The sting site should be washed with soap and water to minimize the possibility of secondary infection. Apply ice packs or a paste made by mixing baking soda or meat tenderizer with water to reduce the amount of venom absorbed, pain, and local swelling. Baking soda apparently helps to neutralize the acidic components of the venom. Meat tenderizers generally contain a proteolytic enzyme (e.g., papain) that breaks down venom proteins. Other home remedies are wet salt, moistened tobacco, 10% ammonia solution, wet aspirin, and commercial sting relievers that help to reduce swelling and pain when applied to the sting sites (Weathersby, 1984). Plastering mud directly on the sting site also works well. Most of these remedies are thought to relieve the effects of the sting by osmotic removal of the venom. The more immediately they are applied, the more effective they are in reducing the intensity of the reaction. Antihistamines are helpful in lessening the swelling in mildly allergic reactions.

People with suspected allergic responses can be tested by means of a skin test involving subcutaneous injection of diluted venom, by a radioallergosorbent test that requires the taking of a blood sample to combine with the venom to detect reactions, or by a blood test using a histamine leukocyte release assay (Goddard, 1996; Piek, 1986a; Reisman, 1994b; Schmidt *et al.*, 1984). All tests should be done in an allergy clinic where immediate medical treatment is available should adverse reactions to the tests occur.

If a serious systemic reaction occurs, a physician's help should be sought immediately. The prompt injection of *epinephrine* (adrenaline) is the initial step to combat a life-threatening anaphylactic reaction. Most physicians recommend that persons with demonstrated hypersensitivity wear an identification tag and carry a small emergency *sting kit* containing *antihistamines* and a syringe of epinephrine. Such kits are available with a doctor's prescription for a modest cost. Hypersensitive people at high risk of a fatal reaction should consider *desensitization therapy*. These individuals can be desensitized by a series of injections using attenuated doses of the appropriate venom. Such treatment gradually builds up the individual's tolerance to the venom and helps to prevent subsequent systemic reactions. Desensitization programs can be expensive and sometimes require many years.

Additional information about hymenopteran venoms, clinical aspects of stings, allergic reactions, and treatment of sting victims is provided by Piek (1986b), Schmidt (1992), Charpin *et al.* (1994), Levine and Lockey (1995), Meier (1995), and Mosbech (1995).

## VETERINARY IMPORTANCE

Comparatively little is known about the effects of hymenopteran stings on farm and game animals except for stings by fire ants. Imported fire ants *Solenopsis invicta* and *S. richteri* commonly attack and kill newborn game animals such as rabbits, deer, quail, and other ground-nesting birds (Lofgren, 1986). Even the native fire ant in the United States, *S. xyloni,* is known to attack and kill newly hatched poultry. Deaths have been reported only rarely, however, among other newborn farm animals. A few accounts of animal deaths have been attributed to stings of the Africanized honey bee, usually involving

dogs (Fig. 19.27) or livestock either restrained or enclosed near a hive. The German yellowjacket can injure the teats of milking cows by biting (Braverman *et al.*, 1991), and the resulting lesions may be associated with outbreaks of mastitis in Israel (Shwimmer *et al.*, 1995). Yellowjackets also have been observed cutting flesh from wounds of horses.

Few published data are available on allergic sting reactions in nonhuman animals. Severe systemic, allergic reactions to stings of bees and wasps have been reported in dogs. Fire ant stings in dogs do not develop the pustules so characteristic of humans and there have been no reports of anaphylaxis in dogs due to fire ant stings (Rakich *et al.*, 1993). Some dogs, when stung on the face by bees or wasps, develop an enormous swelling of the head that may persist for a few days to a week.

Some ant species are intermediate hosts for parasitic helminths of vertebrates. *Formica* species are intermediate hosts for the *lancet fluke* (*Dicrocoelium dendriticum*), which infests the bile ducts of cattle, sheep, pigs, goats, horses, dogs, and, occasionally, humans. Pavement ant workers and *Pheidole* ants also serve as intermediate hosts of the *poultry tapeworms Raillietina tetragona* and *R. echinobothrida* (Harwood and James, 1979; Olsen, 1974).

## PREVENTION AND CONTROL

Avoidance of stinging insects is the best approach to preventing envenomization. People at risk should avoid areas with high concentrations of ants, bees, or yellowjackets and be very observant and alert when working in areas where colonies are likely to occur. Colonies of harvester ants can be easily avoided because the ants are slow moving and their nest mounds in the center of an area devoid of vegetation are very noticeable.

Certain colors of clothing should be avoided. Foraging yellowjackets and bees are attracted to yellows, light blues, and many bright or fluorescent colors. Dark-colored objects are more attractive to attacking wasps. It is advisable to wear long pants, preferably with the legs tucked into boots, and to wear a long-sleeved shirt to protect the arms. Individuals at risk should avoid using perfumes, hair sprays, sweet-smelling lotions, aftershave lotions, hand and body lotions, and even certain suntan lotions. Some of the odors emanating from these materials are highly attractive to flying hymenopterans. Paints containing *isoamyl acetate,* a component of the honey bee alarm pheromone, should be used with care around honey bees. These materials should not be applied outdoors where stinging bees and wasps are likely to be attracted to them.

*Repellents* commonly used to deter mosquitoes and other biting insects are not effective against stinging hymenopterans. However, studies have suggested that lotions containing butyl, isopropyl and octyl palmitates, and methyl myristate may repel foraging yellowjackets from human food sources (Henderson *et al.*, 1993).

Care should be taken not to provoke colonies by approaching too closely or creating *substrate vibrations* which might disturb the nest. Similarly, one should avoid using machines such as power tools and lawnmowers near nests or opening and shutting doors and windows in the vicinity of a nest. Once a colony is aroused, it is best to slowly back out of the area rather than to run. Wasps and bees defending their nests tend to attack any nearby moving object. However, once a person is stung, it is best to leave the area as quickly as possible before the alarmed wasp or bee can alert other nestmates to join the attack.

In cases of *mass stinging attacks,* such as by the Africanized honey bee or perennial colonies of yellowjackets, it is best to run away from the nest. The nose and eyes should be covered without blocking vision to avoid stings directed to one's head. Running, perhaps a quarter to half a mile, will gradually leave the attacking wasps or bees behind. One should not crawl or climb into an enclosed area, such as rock crevices; most serious injuries from multiple stings have occurred when victims were trapped, caged, or otherwise incapacitated.

*Insecticides,* especially aerosol sprays containing a quick knockdown and a long-lasting toxicant, can be used to kill colonies of both yellowjackets and paper wasps. While these aerosols can propel toxicants up to 6–7 m and are designed primarily for treating aerial nests, they work equally well on colonies nesting underground. Nest destruction should be conducted at night, when wasps are least active and all the individuals are in the nest. If insecticides are applied during the daytime, the nest entrance should not be blocked after treatment so that returning wasps can readily enter the nest to contact the toxicant. Long-lasting residual insecticides will kill workers emerging from capped cells after the initial treatment. Some aerosols do not contain a long-lasting toxicant and are not effective on later-emerging adults. Controlling any nest, especially large colonies of paper wasps and yellowjackets, can be difficult and dangerous, so it is often best left to experienced pest control operators. Protective clothing such as a bee suit and veil should be worn when destroying colonies.

Only under certain circumstances is there reason to control *honey bees*. When necessary, local beekeepers may be willing to collect swarms near houses or other sites of human activity. Colonies of honey bees or yellowjackets nesting in the wall voids of houses probably should be destroyed. They can be killed quickly with the application of a quick-knockdown insecticide from a spray can that propels the toxicant up to 6 m. The insecticide stream should be directed into the nest entrance, followed by an application of insecticidal dust into the voids through openings such as electrical plate outlets. Destruction of well-established honey bee colonies in walls of buildings can result in the stored honey dripping through the walls or ceiling. In such cases, the colony and honey combs should be physically removed after the bees are killed or otherwise removed.

Control of most social insects is best effected by the use of *baits* incorporating a slow-acting insecticide. Foragers are attracted to the bait, return to the colony with the toxic bait, and spread the material throughout the colony by trophallaxis. Several toxicants that show promise for control are classified as insect growth regulators (IGRs), but several other slow-acting toxicants with direct poisoning action are also effective. Although ants in heavily used areas such as picnic grounds can be killed by direct application of insecticides to the nest, it is best to use some type of carbohydrate or fatty bait that contains a slow-acting poison. Baits are the control of choice for *pharaoh ants* and *pavement ants* since treatments with insecticidal sprays in buildings tend to cause the ants to disperse, making them more difficult to control. Traps baited with pheromones are being used to attract honey bee swarms to monitor the northward movement of the Africanized honey bees.

Control measures for *fire ants* include the direct application of insecticides to the nest, the use of various formulations of toxic baits and IGRs, and the use of biological control agents such as pathogens and parasitoids (Collins, 1992; Drees and Vinson, 1993). The key to effective control is persistence in treating the new colonies as they appear. Unfortunately, despite extensive efforts, control of fire ants has not been very successful except in limited areas.

The control of scavenging yellowjackets perhaps is best accomplished by a trapping system with attractant baits containing a slow-acting insecticide. Research is ongoing to develop such a commercial bait and trap system. All scavenging *Vespula* species are attracted to meat baits. A simple meat trap with fish or ham suspended over a pan of detergent water can be used to drown thousands of attracted wasps. Workers cut off pieces of meat that are too heavy for them to carry and drop down into the water as they try to fly away. Each day, the drowned wasps should be removed and the detergent water and the meat should be replaced.

Because meat baits spoil and lose their attraction quickly, potential synthetic attractants have been investigated. The western yellowjacket, as well as the California and southern yellowjackets, can be lured into traps containing the synthetic attractant *heptyl butyrate*. Two coattractants, *isobutanol* and *acetic acid,* isolated from fermenting molasses, have been shown to be effective baits for several scavenging yellowjacket species in North America, e.g., eastern, western, southern, and German yellowjackets (Landolt, 1998; Landolt *et al.*, 1999).

More information on avoidance and control measures for the various stinging hymenopterans can be found in Akre and MacDonald (1986) for yellowjackets, Drees and Vinson (1993) and Lofgren (1986) for fire ants, and Vinson (1986) for social insects in general.

## REFERENCES AND FURTHER READING

Akre, R. D. (1982). Social wasps. In "Social Insects" (H. R. Hermann, Ed.), Vol. 4, Chapt. 1, pp. 1–105. Academic Press, New York.

Akre, R. D., and MacDonald, J. F. (1986). Biology, economic importance, and control of yellowjackets. In "Economic Impact and Control of Social Insects" (S. B. Vinson, Ed.), Chapt. 14, pp. 353–412. Praeger, New York.

Akre, R. D., and Reed, H. C. (1984). Biology and distribution of social Hymenoptera. In "Insect Poisons, Allergens, and Other Invertebrate Venoms" (A. T. Tu, Ed.), Handbook of Natural Toxins, Vol. 2, Chapt. 1, pp. 3–47. Dekker, New York.

Akre, R. D., Greene, A., MacDonald, J. F., Landolt, P. J., and Davis, H. G. (1981). "The Yellowjackets of America North of Mexico." US Department of Agriculture Handbook No. 552. US Department of Agriculture, Science and Education Administration, Washington, DC.

Alexander, B., and Michener, C. D. (1995). Phylogenetic studies of the families of short-tongued bees (Hymenoptera: Apoidea). *University of Kansas Science Bulletin* **55**, 377–424.

Banks, B. E. C., and Shipolini, R. A. (1986). Chemistry and pharmacology of honey-bee venom. In "Venoms of the Hymenoptera. Biochemical, Pharmacological and Behavioural Aspects" (T. Piek, Ed.), Chapt. 7, pp. 329–416. Academic, San Diego.

Blum, M. S. (1984). Poisonous ants and their venoms. In "Insect Poisons, Allergens, and Other Invertebrate Venoms. Handbook of Natural Toxins" (A. T. Tu, Ed.), Vol. 2, Chapt. 7, pp. 225–242. Dekker, New York.

Bohart, R. M., and Menke, S. (1976). "Sphecid Wasps of the World: A Generic Revision." Univ. California Press, Berkeley.

Bolton, B. (1994). "Identification Guide to the Ant Genera of the World." Harvard Univ. Press, Cambridge, MA.

Braverman, Y., Marcusfeld, O., Adler, H., and Yakobson, B. (1991). Yellowjacket wasps can damage cow's teats by biting. *Medical and Veterinary Entomology* **5**, 129–130.

Brewer, R. D., Morris, P. D., and Cole, T. B. (1994). Hurricane-related emergency department visits in an inland area: an analysis of the public health impact of Hurricane Hugo in North Carolina. *Annals of Emergency Medicine* **23**, 731–736.

Camazine, S. (1988). Hymenoptera stings: reactions, mechanisms, and medical treatment. *Bulletin of the Entomological Society of America* **34**, 17–21.

Charpin, D., Birnbaum, J., and Vervloet, D. (1994). Epidemiology of hymenoptera allergy. *Clinical and Experimental Allergy* **24**, 1010–1015.

Cole, A. C. (1968). "*Pogonomyrmex* Harvester Ants: A Study of the Genus in North America." Univ. Tennessee Press, Knoxville.

Collins, H. (1992). Control of imported fire ants: a review of current knowledge. US Department of Agriculture Technical Bulletin 1807.

US Department of Agriculture, Animal and Plant Health Inspection Service, Washington, DC.

Drees, B. M., and Vinson, S. B. (1993). Fire ants and their management. Texas Agricultural Extension Service B-1536. Texas A & M University, College Station, Texas. 20 pp.

Edwards, J. P. (1986). The biology, economic importance and control of the Pharaoh's ant, *Monomorium pharaonis* (L.). In "Economic Impact and Control of Social Insects" (S. B. Vinson, Ed.), Praeger, New York.

Ervin, R. T., and Tennant, W. T., Jr. (1990). Red imported fire ants' (*Solenopsis invicta*) impact on Texas outdoor recreation. In "Applied Myrmecology: A World Perspective" (R. K. Vander Meer, K. Jaffe, and A. Cedeno, Eds.), Chapt. 46, pp. 504–510. Westview, Boulder, CO.

Fowler, H. G., Bueno, O. C., Sadatsune, T., and Montelli, A. C. (1993). Ants as potential vectors of pathogens in hospitals in the state of São Paulo, Brazil. *Insect Science and Its Application* **14**, 367–370.

Gauld, I. D., and Bolton, B. (1988). "The Hymenoptera." Oxford Univ. Press, London.

Gillaspy, J. E. (1986). *Polistes* wasps: biology and impact on man. In "Economic Impact and Control of Social Insects" (S. B. Vinson, Ed.), Chapt. 13, p. 332–352. Praeger, New York.

Goddard, J. (1996). "Physician's Guide to Arthropods of Medical Importance," 2nd ed. CRC Press, Boca Raton, FL.

Goulet, H., and Huber, J. T. (1993). "Hymenoptera of the World: An Identification Guide to Families." Research Branch, Agriculture Canada, Ottawa.

Graham, J. M., Ed. (1992). "The Hive and the Honey Bee." Dadant, Hamilton, IL.

Harwood, R. F., and James, M. T. (1979). "Entomology in Human and Animal Health," 7th ed. MacMillan, New York.

Henderson, G., Blouin, D. C., and Jeanne, R. L. (1993) Yellowjacket (Hymenoptera: Vespidae) repellency by natural products of paper wasps and Avon's Skin-So-Soft. *Journal of Entomological Science* **28**, 387–392.

Hoffman, D. R. (1984). Insect venom allergy, immunology, and immunotherapy. In "Insect Poisons, Allergens, and Other Invertebrate Venoms. Handbook of Natural Toxins" (A. T. Tu, Ed.), Vol. 2, Chapt. 6, pp. 187–223. Dekker, New York.

Hölldobler, B., and Wilson, E. O. (1990). "The Ants." Belknap/Harvard Univ. Press, Cambridge, MA.

Landolt, P. J. (1998). Chemical attractants for trapping yellowjackets *Vespula germanica* and *Vespula pensylvanica* (Hymenoptera: Vespidae). *Environmental Entomology* **27**, 1229–1234.

Landolt, P. J., Reed, H. C., Aldrich, J. R., Antonelli, A. L., and Dickey, C. (1999). Social wasps (Hymenoptera: Vespidae) trapped with acetic acid and isobutanol. *Florida Entomologist* **82**, 609–614.

Levine, M. I., and Lockey, R. F., Eds. (1995). "Monograph on Insect Allergy." Dave Lambert, Pittsburgh, PA.

Lofgren, C. S. (1986). The economic importance and control of imported fire ants in the United States. In "Economic Impact and Control of Social Insects" (S. B. Vinson, Ed.), Chapt. 8, pp. 227–256. Praeger, New York.

MacKay, W. P. (1990). The biology and economic impact of *Pogonomyrmex* harvester ants. In "Applied Myrmecology: A World Perspective" (R. K. Vander Meer, K. Jaffe, and A. Cedeno, Eds.), Chapt. 50, pp. 533–543. Westview, Boulder, CO.

Manley, D. G. (1991). "The Velvet Ants (Hymenoptera: Mutillidae) of South Carolina." South Carolina Agriculture Experiment Station Technical Bulletin 1100. Clemson University, Clemson, South Carolina. 55 pp.

Meier, J. (1995). Biology and distribution of hymenopterans of medical importance, their venom apparatus and venom composition. In "Handbook of Clinical Toxicology of Animal Venoms and Poisons"

(J. Meier and J. White, Eds.), Chapt. 21, pp. 331–348. CRC Press, Boca Raton, FL.

Michener, C. D. (1974). "The Social Behavior of the Bees." Belknap/Harvard Univ. Press, Cambridge, MA.

Michener, C. D., McGinley, R. J., and Danforth, B. N. (1994). "The Bee Genera of North and Central America (Hymenoptera Apoidea)." Smithsonian Inst. Press, Washington, DC.

Mosbech, H. (1995). Clinical toxicology of hymenopteran stings. In "Handbook of Clinical Toxicology of Animal Venoms and Poisons" (J. Meier and J. White, Eds.), Chapt. 22, pp. 349–359. CRC Press, Boca Raton, FL.

Nakajima, T. (1984). Biochemistry of vespid venoms. In "Insect Poisons, Allergens, and Other Invertebrate Venoms. Handbook of Natural Toxins" (A. T. Tu, Ed.), Vol. 2, Chapt. 4, pp. 109–133. Dekker, New York.

Nakajima, T. (1986). Pharmacological biochemistry of vespid venoms. In "Venoms of the Hymenoptera. Biochemical, Pharmacological and Behavioural Aspects" (T. Piek, Ed.), Chapt. 6., pp. 309–327. Academic, San Diego.

Olsen, O. W. (1974). "Animal Parasites, Their Life Cycles and Ecology." Dover, New York.

Piek, T. (1984). Pharmacology of Hymenoptera venoms. In "Insect Poisons, Allergens, and Other Invertebrate Venoms. Handbook of Natural Toxins" (A. T. Tu, Ed.), Vol. 2, Chapt. 5, pp. 135–185. Dekker, New York.

Piek, T. (1986a). Venoms of bumble-bees and carpenter-bees. In "Venoms of the Hymenoptera. Biochemical, Pharmacological and Behavioural Aspects" (T. Piek, Ed.), Chapt. 8, pp. 417–424. Academic, San Diego.

Piek, T., Ed. (1986b). "Venoms of the Hymenoptera. Biochemical, Pharmacological and Behavioural Aspects." Academic Press, San Diego.

Rakich, P. M., Latimer, K. S., Mispagel, M. E., and Steffens, W. L. (1993). Clinical and histological characterization of cutaneous reactions to stings of the imported fire ant (*Solenopsis invicta*) in dogs. *Veterinary Pathology* **30**, 555–559.

Reisman, R. E. (1994a). Insect stings. *New England Journal of Medicine* **331**, 523–527.

Reisman, R. E. (1994b). Venom hypersensitivity. *Journal of Allergy and Clinical Immunology* **94**, 651–658.

Rhoades, R. B., Stafford, C. T., and James, F. K., Jr. (1989). Survey of fatal anaphylactic reactions to imported fire ant stings. *Journal of Allergy and Clinical Immunology* **84**, 159–162.

Roig-Alsina, A., and Michener, C. D. (1993). Studies of the phylogeny and classification of long-tongued bees. *University of Kansas Science Bulletin* **55**, 123–162.

Ross, K. G., and Matthews, R. W., Eds. (1991). "The Social Biology of Wasps." Comstock, Ithaca, NY.

Schmidt, J. O. (1986a). Allergy to Hymenoptera venoms. In "Venoms of the Hymenoptera. Biochemical, Pharmacological and Behavioural Aspects" (T. Piek, Ed.), Chapt. 10, pp. 509–546. Academic Press, San Diego.

Schmidt, J. O. (1986b). Chemistry, pharmacology, and chemical ecology of ant venoms. In "Venoms of the Hymenoptera. Biochemical, Pharmacological and Behavioural Aspects" (T. Piek, Ed.), Chapt. 9, pp. 425–508. Academic Press, San Diego.

Schmidt, J. O. (1986c). Hymenoptera envenomation. In "Urban Entomology: Interdisciplinary Perspectives" (G. W. Frankie and C. S. Koehler, Eds.), Chapt. 9, pp. 187–220. Praeger, New York.

Schmidt, J. O. (1990a). Africanized and European honey bee venoms: implications for beekeepers and the public. *American Bee Journal* **130**, 810–811.

Schmidt, J. O. (1990b). Hymenoptera venoms: striving toward the ultimate defense against vertebrates. In "Insect Defenses: Adaptive Mechanisms and Strategies of Prey and Predator" (D. L. Evans and J. O. Schmidt, Eds.), pp. 390–395. State Univ. of New York Press, Albany.

Schmidt, J. O. (1992). Allergy to venomous insects. In "The Hive and the Honey Bee" (J. Graham, Ed.), Dadant and Sons, Hamilton, IL.

Schmidt, J. O., and Blum, M. S. (1978). A harvester ant venom: chemistry and pharmacology. *Science* **200**, 164–166.

Schmidt, J. O., Menke, G. C., Chen, T. M., and Pinnas, J. L. (1984). Demonstration of cross-allergenicity among harvester ant venoms using RAST and RAST inhibition. *Journal of Allergy and Clinical Immunology* **73**(1, part 2), 158.

Schumacher, M. J., and Egen, N. B. (1995). Significance of Africanized bees for public health. *Archives of Internal Medicine* **155**, 2038–2043.

Shipolini, R. A. (1984). Biochemistry of bee venom. In "Insect Poisons, Allergens, and Other Invertebrate Venoms. Handbook of Natural Toxins" (A. T. Tu, Ed.), Vol. 2, Chapt. 2, pp. 49–85. Dekker, New York.

Shwimmer, A., Shpigel, N. Y., Yeruham, I., and Saren, A. (1995). Epidemiological and bacteriological aspects of mastitis associated with yellowjacket wasp teat lesions in Israeli dairy cows. In "Proceedings of the Third IDF International Mastitis Seminar." pp. 100–102. M. Lachmann Printers Ltd., Haifa, Israel.

Smith, M. R. (1965). "House-Infesting Ants of the Eastern United States. Their Recognition, Biology, and Economic Importance." US Department of Agriculture Technical Bulletin 1326. U.S. Department of Agriculture, Agricultural Research Service, Washington, DC.

Spivak, M., Fletcher, D. J. C., and Breed, M. D., Eds. (1991). "The 'African' Honey Bee." Westview, Boulder, CO.

Starr, C. K. (1985). A simple pain scale for field comparison of Hymenoptera stings. *Journal of Entomological Science* **20**, 225–232.

Taber, S. W. (2000). "Fire Ants." Texas A&M Univ. Press, College Station, TX.

Tu, A. T., Ed. (1984). "Insect Poisons, Allergens, and Other Invertebrate Venoms. Handbook of Natural Toxins." Vol. 2. Chapt 2, pp. 49–85. Dekker, New York.

Ulloa-Chacon, P., and Cherix, D. (1990). The little fire ant *Wasmannia auropunctata* (R.) (Hymenoptera: Formicidae). In "Applied Myrmecology: A World Perspective" (R. K. Vander Meer, K. Jaffe, and A. Cedeno, Eds.), Chapt. 27, pp. 281–289. Westview, Boulder, CO.

Vander Meer, R. K., Jaffe, K., and Cedeno, A., Eds. (1990). "Applied Myrmecology: A World Perspective." Westview, Boulder, CO.

van Marle, J., and Piek, T. (1986). Morphology of the venom apparatus. In "Venoms of the Hymenoptera. Biochemical, Pharmacological and Behavioural Aspects" (T. Piek, Ed.), Chapt. 2, pp. 17–44. Academic Press, San Diego.

Vinson, S. B., Ed. (1986). "Economic Impact and Control of Social Insects." Praeger, New York.

Vinson, S. B. (1994). Impact of the invasion of *Solenopsis invicta* (Buren) on native food webs. In "Exotic Ants: Biology, Impact, and Control of Introduced Species" (D. F. Williams, Ed.), Westview, San Francisco.

Vinson, S. B., and Greenberg, L. (1986). The biology, physiology, and ecology of imported fire ants. In "Economic Impact and Control of Social Insects" (S. B. Vinson, Ed.), Chapt. 7, pp. 193–226. Praeger, New York.

Visscher, P. K., Vetter, R. S., and Camazine, S. (1996). Removing bee stings. *Lancet* **348**, 301–302.

Weathersby, A. B. (1984). Wet salt for envenomization. *Journal of the Georgia Entomological Society* **19**, 1–6.

Winston, M. L. (1992). "Killer Bees: The Africanized Honey Bee in the Americas." Harvard Univ. Press, Cambridge, MA.

# 20

# SCORPIONS (*Scorpiones*)

GARY R. MULLEN AND SCOTT A. STOCKWELL

*Scorpions represent* an ancient group of arachnids which is believed to be descended from now-extinct eurypterids living in estuaries and coastal lagoons. Except for their smaller size, scorpions are notably similar in appearance to their marine ancestors, which first crawled onto land as air-breathing arachnids during the late Devonian and early Carboniferous Period 325–350 million years ago. Throughout recorded history scorpions have intrigued human cultures, being revered and attributed special powers by some and feared as sinister and ominous by others. Scorpion images appear as religious symbols, on seals, magical tablets, amulets, and boundary stones, and in the origin stories of ancient civilizations such as the Chaldeans and Egyptians. They also figured prominently in Greek mythology and as one of the 12 constellations or signs of the Zodiac.

Despite the many superstitions and misconceptions about scorpions that persist to this day, their reputation as venomous arthropods is generally overstated. Most scorpions are not aggressive and inflict only minor, transient pain and discomfort when they do sting, typically to defend themselves when threatened. There are, however, 40–50 species worldwide which pose significant health problems. About 25 species are considered to be capable of causing human deaths. Most of them occur in the Tropics and Subtropics or in arid regions of temperate zones.

## TAXONOMY

Scorpion taxonomy has improved significantly in the last decade. In his review of scorpion classification, Sissom (1990) listed 1077 species, 117 genera, and 9 families. Two of those families (Chactidae and Vaejovidae) could not be distinguished in the identification key provided. Since that time, many new families have been proposed, in most cases helping to create a clearer picture of phylogenetic relationships among the genera of scorpions. In their recent catalog of the scorpions of the world, Fet *et al.* (2000) list 16 families, 154 genera, and 1252 species of extant scorpions. We recognize 17 families here but acknowledge the fact that this number will likely increase over the next few years as research at the suprageneric level continues.

At least 98 species belonging to 12 genera and 5 families are known to occur in the continental United States. A single species (*Paruroctonus boreus*) occurs in the extreme southern portions of British Columbia, Alberta, and Saskatchewan, Canada. Mexico is home to at least 178 species spread among 22 genera and 7 families. The catalog by Fet *et al.* (2000) provides the most up-to-date picture of scorpion taxonomy and is an invaluable resource. Keys to most of the genera of scorpions of the world are provided by Sissom (1990). A key to the North American families and genera is provided by Stockwell (1992). For species identification of the North American fauna, see the following regional works: Arizona (Stahnke, 1940), Nevada (Gertsch and Allred, 1965), Utah (Johnson and Allred, 1972), Idaho (Anderson, 1975), California (Hjelle, 1972; Williams, 1976), Baja California and Mexico (Williams, 1980), and Florida (Muma, 1967).

## BUTHIDAE

This is the largest and most widespread scorpion family, with 72 currently recognized genera and at least 528 valid species. Buthids are found throughout the world, with their greatest diversity in the Old World, especially the Afrotropical Region and the southern Palaearctic Region. Most of the scorpions that are dangerously venomous to humans and other animals belong to this family (Table I). The important genera in this respect are *Androctonus* and *Leiurus* in northern Africa and western Asia, *Mesobuthus* in Asia and India, *Parabuthus* in southern Africa, *Centruroides* in North America, and *Tityus* in South America. Numerous other genera may contain members of minor medical importance. Members of this scorpion family are commonly encountered by people and pets, making their status as venomous pests all the more important. The only buthid genus that naturally occurs in North America is *Centruroides*. Members of this genus are crevice dwellers that commonly enter homes. Five species occur in the United States. *Centruroides hentzi* is found throughout Florida and adjoining portions of Alabama and Georgia. *C. guanensis* is common on the islands of the Bahamas and Cuba, but it is also found in the southernmost part of the Florida Peninsula and the Florida Keys. *C. gracilis* is native to Central America and Mexico but has been introduced to many other tropical areas, including Florida. *C. vittatus* (Fig. 20.1) is the most widespread species of scorpion in the United States, with a range that extends from the Rio Grande River in the west to the Mississippi River in the east. This species has been collected as far north as southernmost Nebraska. It is found throughout Texas

## TABLE I
### Dangerous Species of Scorpions Based on the Toxicity of Their Venoms

| Species | LD$_{50}$ | Geographic occurrence |
|---|---|---|
| *Leiurus quinquestriatus* | 0.25 | Turkey, Israel, Egypt, Algeria, Libya, Sudan |
| *Androctonus mauretanicus* | 0.31 | Morocco |
| *A. australis* | 0.32 | Morocco, Algeria, Libya, Tunisia, Egypt |
| *A. crassicauda* | 0.40 | Turkey, Israel, Iraq, Arabian Peninsula |
| *Tityus serrulatus* | 0.43 | Brazil |
| *Centruroides limpidus* | 0.69 | Mexico |
| *A. amoreuxi* | 0.75 | Middle East |
| *Buthus occitanus* | 0.90 | Morocco, Algeria, Jordan, southern Europe |
| *C. exilicauda* | 1.12 | United States, northern Mexico |
| *Parabuthus transvaalicus* | 4.25 | Southern Africa |

The lethal dose is expressed as mg/kg of venom required to kill 50% of the mice (LD$_{50}$) following subcutaneous injection. The lower the LD$_{50}$, the more potent the venom. All of the scorpions listed are members of the family Buthidae (compiled from multiple sources).

and Oklahoma but is also found in adjoining parts of New Mexico, Colorado, Kansas, Illinois, Missouri, Arkansas, and Louisiana. *C. exilicauda* (formerly and occasionally still known as *C. sculpturatus*) is found in most of Arizona, as well as adjoining parts of New Mexico, Utah, Nevada, and California. All species of *Centruroides* can deliver extremely painful stings that may be accompanied by systemic symptoms. However, only *C. exilicauda* is

**FIGURE 20.1**    *Centruroides vittatus* (Buthidae), the most common scorpion in the United States. (Photo by S. A. Stockwell.)

considered dangerous, and then only to very small children. The only other buthid occurring in the United States is the cosmotropical species *Isometrus maculatus*, an Asian species that has been introduced to tropical port cities around the world. This species is very common on the island of Oahu in Hawaii.

## MICROCHARMIDAE

This family contains two genera (each with its own subfamily) of small, tropical, forest-dwelling scorpions from Africa and Madagascar. Traditionally placed in the family Buthidae, their inclusion together in a separate family bears re-evaluation.

## PSEUDOCHACTIDAE

This recently described family contains only one species from central Asia (Uzbekistan and Tajikistan). It possesses an unusual trichobothrial pattern similar to that of the Buthidae.

## CHAERLIDAE

This family is represented by the single genus *Chaerilus*, with 21 described species, none of which are dangerously venomous. These curious scorpions are unique in many ways, but they share some characteristics with the Buthidae. They occur in the Old World in India, Sri Lanka, Nepal, Bangladesh, Myanmar, Malaysia, Singapore, and many islands of the Philippines and Indonesia. None is known to be dangerous.

## CHACTIDAE

This family contains 11 genera and approximately 131 species. None is known to be dangerous. Most of the chactid species are found in South America (Colombia, Venezuela, Trinidad and Tobago, Guyana, French Guiana, Suriname, Ecuador, Peru, and Brazil). A few species are found as far north as Panama and Costa Rica. One peculiar species, *Nullibrotheas alleni*, is endemic to Baja California Sur, Mexico and is the only North American representative of the family. None of these is considered dangerous to humans.

## EUSCORPIIDAE

This small family shares many characteristics with the Chactidae, of which it was once considered a subfamily.

There are 4 genera and 14 species. One genus, *Euscorpius*, with 6 species, is found throughout southern (Mediterranean) Europe and has been variously divided into 3 subgenera and numerous subspecies. The other genera in this family are found in eastern Mexico and Guatemala. *Megacormus* and *Plesiochactas* are closely related epigean forms. The two species of *Troglocormus* are troglobitic, having lost the median eyes, and are found only in caves.

## SCORPIOPIDAE

These scorpions share many similarities with the Euscorpiidae and the Chactidae. This group was once considered to be a subfamily of the Vaejovidae. There are 6 genera and 27 described species. Scorpiopidae may be found at relatively high altitudes throughout their range. They are native to parts of Afghanistan, Pakistan, India, Sikkim, Nepal, China, Bangladesh, Myanmar, Thailand, Laos, Vietnam, Malaysia, Indonesia, and Bhutan. None of them is regarded as dangerous.

## SUPERSTITIONIIDAE

This small family was previously regarded as a subfamily in the Chactidae. They share no characteristics with the Chactidae, but they appear to be related to the Vaejovidae and Iuridae. Superstitioniidae contains four genera and nine species. They are generally small in size. At 9 mm total length, *Typhlochactas mitchelli* is the smallest species of scorpion in the world. The one exception in the family is *Alacran tartarus*, which measures up to 70 mm in length. Most of the species tend to be troglobitic in form, lacking eyes and pigmentation. *Superstitionia donensis* is found in the southwestern United States and is the only species with a full complement of eyes. The other species are all found in Mexico in caves or leaf litter. These scorpions are too small or inaccessible to be dangerous to humans. *Alacran* is found only in caves at depths of up to 800 m.

## TROGLOTAYOSICIDAE

The two monotypic genera in this family were formerly placed in the Superstitioniidae. Both are poorly known and may not be closely related. Both species are troglobites, possessing lateral eyes, but no median eyes. *Troglotayosicus vachoni*, known from a single individual from a cave in Ecuador, shares many characteristics with the Superstitioniidae. *Belisarius xambeui*, from caves in the eastern Pyrenees of Spain and France, was traditionally grouped with *Euscorpius*, with which it bears a superficial resemblance. But the trichobothrial pattern of this species is very primitive and much closer to that of the Vaejovidae, Superstitioniidae, and Iuridae.

## IURIDAE

This interesting, but small, group of scorpions spans three continents. None of them is considered dangerous. There are 6 genera and approximately 21 species. The genera *Iurus* and *Calchas,* each with 1 species, are closely related, are relatively large, and are found in Turkey and Greece (including Samos, Crete, and other islands). The genera *Caraboctonus* (1 species) and *Hadruroides* (9 species) are found in western South America. *Caraboctonus* is found in Chile and Peru; *Hadruroides* is distributed throughout Ecuador (including the Galapagos Islands), Bolivia, Peru, and Chile. There are 8 species in the North American genus *Hadrurus*. Members of this genus are desert-dwelling burrowers found from south-central Mexico to the western United States. They are the largest scorpion species found in North America. Three species occur in the United States in desert areas of Oregon, Idaho, California, Nevada, Utah, Colorado, and Arizona. A close relative of *Hadrurus, Anuroctonus* contains a single species distributed from Baja California, Mexico through California, Nevada, and Utah. *Anuroctonus phaiodactylus* constructs permanent burrows in a variety of habitats throughout its range, but it is associated with canyons, ravines, and hillsides.

## VAEJOVIDAE

This family is composed of 10 genera and approximately 146 species that are restricted mostly to North America. Species of Vaejovidae are found in every conceivable habitat in nearly every state of Mexico and much of the United States, especially the west (Washington, Oregon, Idaho, Montana, Wyoming, California, Nevada, Utah, Colorado, North Dakota, South Dakota, Nebraska, Arizona, New Mexico, and Texas). One species ranges south into Guatemala, while another species can be found in Canada. *Vaejovis carolinianus* occurs in wooded, mountainous areas of the eastern United States (parts of Kentucky, Tennessee, Virginia, North Carolina, South Carolina, Georgia, Alabama, Mississippi, and Louisiana). It is a small (about 2.4 cm), dark scorpion that readily enters homes throughout its range but is particularly prevalent in the Atlanta, GA, area. Other species may enter homes in other areas. In California, species of the genera *Pseudouroctonus* and *Uroctonus* commonly enter

homes. No member of this family poses any appreciable health threat, but stings from these species are more likely to cause minor localized discoloration, swelling, and necrosis than are the more painful stings of buthid scorpions.

## BOTHRIURIDAE

This family of 12 genera and 90 species exhibits a Gondwanan distribution. One genus, *Cercophonius*, with 7 described species, is distributed throughout Australia but is also found on New Caledonia and in northern India. The primitive genus *Lisposoma*, with 2 species, is endemic to Namibia. The remaining genera are distributed throughout western and southern South America (Ecuador, Peru, Brazil, Bolivia, Paraguay, Chile, Argentina, and Uruguay). None of the bothriurids is considered medically important.

## ISCHNURIDAE

Formerly considered a subfamily of the Scorpionidae, the ischnurids are now regarded as a separate family. They range in size from small to very large and typically have a flattened body shape. The claws are massive in comparison to the body, but the metasoma is unusually thin and feeble-looking. Sometimes the metasoma is so short that it cannot reach to the front of the animal. Although capable of burrowing, most of these species are associated with crevice habitats in rocky areas, on trees, under debris, on man-made structures such as stone walls and wooden bridges, etc. The most impressive of the family is *Hadogenes troglodytes* from South Africa, the males of which can attain a body length up to 21 cm. The family contains 8 genera and 115 species. Representatives are widely distributed throughout the Tropics: in the Caribbean, Haiti and Dominican Republic; in Central America, Panama and Cocos Island (Costa Rica); in South America, Peru, Colombia, Venezuela, French Guiana, and Brazil; in central and southern Africa, Cameroon, Democratic Republic of Congo, Gabon, Congo, Malawi, Uganda, Ethiopia, Kenya, Tanzania, Angola, Namibia, Botswana, Zimbabwe, Mozambique, South Africa, Lesotho, Swaziland, Mauritius, Round Island, Seychelles, Zanzibar, and Madagascar; in Asia, China, Korea, Japan, India, Aru Islands, Bangladesh, Myanmar, Thailand, Cambodia, Laos, Vietnam, Malaysia, Indonesia, Philippines, and Papua New Guinea; in Oceania, Federated States of Micronesia, Fiji, French Polynesia, Key Islands, Kiribati, Mariana Islands, Marshal Islands, New Caledonia, Palau, Ponape, Tuvalu, Samoa,

Solomon Islands, Tonga, and Vanuatu; and, of course, Australia. As a group, they are considered relatively harmless. In the case of stings by the giant *Hadogenes* scorpions, the effect may be so slight as to be barely felt. Other species, such as *Opisthacanthus lepturus* from Panama, can deliver a sting that causes soreness in joints as well as mild, localized discoloration, swelling, and necrosis.

## HETEROSCORPIONIDAE

This is a tiny family with one genus (*Heteroscorpion*) and two species that are endemic to Madagascar. This genus was formerly placed among the Ischnuridae, with which they share many characteristics. Their medical importance is not known.

## URODACIDAE

This family, often considered a subfamily of Scorpionidae, contains a single genus (*Urodacus*) endemic to Australia. There are 19 described species. None is known to be dangerous.

## DIPLOCENTRIDAE

This is another comparatively small family, with 8 genera and about 76 described species, which occur primarily in the New World. Exceptions are the genera *Nebo*, found in Syria, Jordan, Lebanon, Israel, Egypt (Sinai), Saudi Arabia, Yemen, and Oman, and two species of *Heteronebo*, both known from the island of Abd-el-Kuri (Yemen). Oddly, the other 12 species of *Heteronebo* are found on various islands in the Caribbean, along with the genera *Oiclus* and *Cazierius* and most of the species in the genus *Didymocentrus*. *Didymocentrus* is also represented in Honduras, El Salvador, Nicaragua, and Costa Rica. Another genus, *Tarsoporosus,* is found in Venezuela and Colombia. The genus *Bioculus* is endemic to Baja California Sur, Mexico, and its associated islands. Widespread through Honduras, Guatemala, Belize, and Mexico, *Diplocentrus,* with approximately 34 described species, is the largest genus in the family. Five species occur in the southern parts of Arizona, New Mexico, and Texas. Diplocentrids are not generally considered dangerous. However, stings from the Middle Eastern species, *Nebo hierichonticus,* may cause mild, local hemorrhages and slight necrosis.

## SCORPIONIDAE

As the oldest recognized scorpion family, this group once contained all known scorpions. Over the years, its scope has been reduced. Nevertheless, the family remains a catch-all, with membership dictated by tradition and not hypothesized phylogenetic relationships. We include two subfamilies. The Scorpioninae, with 4 genera and 105 species, includes some of the world's largest and most formidable-looking scorpions. All are heavy bodied, with large, powerful claws. Some members of the Asian genus *Heterometrus* reach lengths of 16 cm or more. *Pandinus imperator* from West Africa is often cited as one of the largest scorpions, occasionally attaining a body length of 18 cm and weighing up to 32 g as nongravid females. This large, black scorpion is commonly sold in pet stores. The 19 subspecies of *Scorpio maurus* are distributed across north Africa (Senegal, Mauritania, Morocco, Algeria, Tunisia, Libya, and Egypt) and the Middle East (Turkey, Lebanon, Syria, Iraq, Kuwait, Iran, Israel, Jordan, Saudi Arabia, Qatar, and Yemen). The genus *Pandinus* (24 species) is distributed across central Africa (Senegal, Gambia, Guinea-Bissau, Guinea, Ivory Coast, Burkina Faso, Ghana, Togo, Nigeria, Cameroon, Equatorial Guinea, Gabon, Congo, Democratic Republic of Congo, Sudan, Eritrea, Ethiopia, Somalia, Kenya, Tanzania, Malawi, Zimbabwe, and Mozambique). The genus is also represented on the nearby coasts of Saudi Arabia and Yemen. *Opistophthalmus,* with its 50 species, is found throughout southern Africa (Tanzania, Angola, Zambia, Zimbabwe, Mozambique, Namibia, Botswana, Lesotho, and South Africa). The genus *Heterometrus* is distributed through India, Sri Lanka, Bangladesh, Myanmar, Thailand, Cambodia, Laos, Vietnam, Malaysia, Singapore, Indonesia, Brunei, and the Philippines. Although normally not considered dangerous, most species can deliver stings that may cause mild to severe localized discoloration, swelling, and tissue damage. The stings of many species of *Heterometrus* can cause serious localized hemorrhaging and blistering.

Members of the subfamily Hemiscorpiinae superficially resemble members of the Ischnuridae, being flattened and possessing a thin, delicate postabdomen. They are also crevice dwellers like many of the ischnurids. These scorpions probably should be placed in the Ischnuridae or recognized as a family-level taxon. The two genera and seven species of this group are found in Somalia, Eritrea, Saudi Arabia, Yemen, Oman, United Arab Emirates, Iraq, Iran, and Pakistan. *Hemiscorpius lepturus* from Iran is the only nonbuthid species of scorpion reported to cause significant mortality in humans. The venom contains a potent cytotoxin that causes severe tissue damage and necrosis near the sting site, as well as severe systemic symptoms.

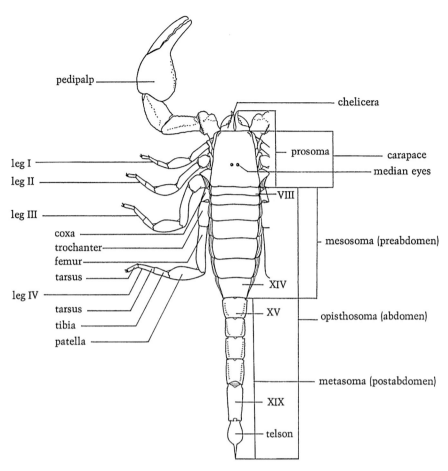

**FIGURE 20.2**   Scorpion morphology, adult, dorsal view. (From Keegan, 1980, with permission of University Press of Mississippi.)

## MORPHOLOGY

The scorpion body (Fig. 20.2) is divided into two major parts: the *prosoma* (cephalothorax) and the *opisthosoma* (abdomen). The opisthosoma is segmented and further divided into the *mesosoma* (preabdomen) and the more slender, tail-like *metasoma* (postabdomen). The metasoma bears at its posterior end a stinging structure called the *telson*.

The dorsal aspect of the prosoma consists of a single sclerotized plate called the *carapace*. It is marked by various furrows, grooves, depressions, and keels that indicate internal apodemes and other surfaces for the attachment of muscles associated with the legs and other appendages. A pair of *median eyes* is located on an *ocular tubercle* along the midline of the carapace. Two additional groups of smaller *lateral eyes* are situated at the anterolateral margins of the carapace. There may be as many as five pairs of lateral eyes, whereas eyes may be lacking altogether in some cave-dwelling species. The ventral aspect of the prosoma consists of a posteromedian *sternum* and the broad

coxae of the legs. The sternum is basically pentagonal in shape but may appear to be more triangular in some taxa.

The mesosoma is divided dorsally into seven apparent segments, each bearing a *tergite*. Ventrally there are five *sternites*, the first four of which bear a pair of *spiracles* (Fig. 20.3). The *genital aperture* is located anteriorly between the coxae of the fourth pair of legs and is covered by a pair of small plates called the *genital opercula*. The genital opercula are commonly fused in females, but not in males. Just behind the sternum is a pair of appendages unique to scorpions, called *pectines* (singular *pecten*) or *comb organs* (Fig. 20.4). These structures function primarily as mechanoreceptors which can sense the nature of the substrate and apparently aid in detecting substrate vibrations. Although quite variable when all scorpion genera are considered, the *pecten* typically consists of three anterior marginal lamellae, the middle lamellae, a row of triangular fulcra, and a posterior series of fleshy lamellae called *pectinal teeth*. The ventral surface of each pectinal tooth is covered with mechanoreceptors in the form of tiny sensory pegs visible only at high magnification. Up to

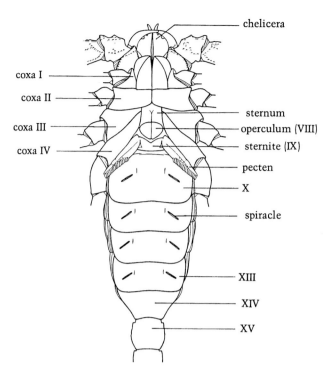

FIGURE 20.3 Scorpion morphology, adult, ventral view, with legs beyond coxae and most of metasoma (postabdomen) not shown. (From Keegan, 1980, with permission of University Press of Mississippi.)

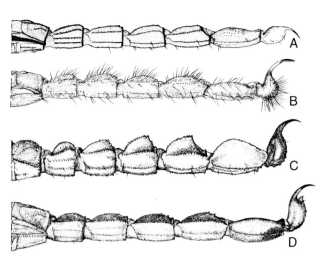

FIGURE 20.5 Metasoma (postabdomen, including terminal telson) of scorpions, showing various morphological features. (A) *Centruroides vittatus* (Buthidae), with relatively slender segments and inconspicuous keels; (B) *Leiurus quinquestriatus* (Buthidae), with slender segments and numerous, long sensory hairs; (C) *Androctonus australis* (Buthidae), with enlarged, robust segments and prominent keels; (D) *Tityus serrulatus* (Buthidae), with well-developed keels and distinct subaculear tubercle on telson. (Adapted from Keegan, 1980, with permission of University Press of Mississippi.)

1200 sensory pegs per pectinal tooth have been reported in *Leiurus quinquestriatus.*

The metasoma, or "tail," is divided into five segments, plus the telson (Fig. 20.5). The segments are well sclerotized and may bear longitudinal ridges, or *keels,* along their dorsal, lateral, and ventral surfaces (Fig. 20.5). The nature and location of these keels can serve as

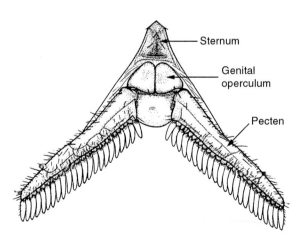

FIGURE 20.4 Sternum, genital opercula, and pectines of *Centruroides vittatus* (Buthidae), located on venter of scorpion at level of third and fourth pairs of legs. (From Keegan, 1980, with permission of University Press of Mississippi.)

important taxonomic characters. The *telson* consists of a bulbous base, called the *vesicle* or *ampulla,* and a curved, sharply pointed terminal spine, the *aculeus.* Just below the aculeus the telson also may bear a small, median *subaculear tubercle* or *accessory spine* (Fig. 20.5D). The vesicle contains a pair of *venom glands* and associated musculature. The venom glands may be simple and saclike or more complexly folded, with pouchlike extensions that greatly increase the surface area of the secretory epithelium. The venom is discharged by contraction of the muscles surrounding the glands which compress the glands against the vesicle wall. The venom is forced out through the pair of *venom ducts* that open near the tip of the aculeus.

The prosomal appendages of scorpions are a pair of chelicerae (Fig. 20.6B), a pair of pincer-like pedipalps (Fig. 20.6A), and four pairs of walking legs. Each *chelicera* consists of three segments. The terminal, third segment serves as a *movable finger* which opposes the second segment, or *hand* (*manus*), bearing an anterior apophysis called the *fixed finger.* Both fingers are armed with teeth which facilitate the grasping and tearing of food. The *pedipalps* are six-segmented, each consisting of a coxa, trochanter, femur, patella (genu), tibia, and tarsus. The distal-most segment (tarsus) forms a *movable finger* which opposes the fixed finger of the tibia, or *hand.* The various numbers and arrangements of keels, tubercles, denticles, granules, and *trichobothria* (sensory hairs) on the

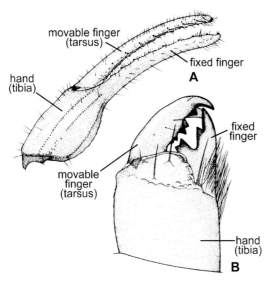

**FIGURE 20.6** *Centruroides vittatus* (Buthidae). (A) Chelate (pincer-like) tibia and tarsus of pedipalp, used for seizing prey; (B) Chelicera, showing the movable tarsus and fixed finger of the tibia for crushing and tearing prey. (Adapted from Keegan, 1980, with permission of University Press of Mississippi.)

hand, patella, and femur of the pedipalp are important taxonomically. The walking legs consist of the same six segments as the pedipalps, plus a seventh terminal segment, the *pretarsus*, which bears a pair of *lateral claws* and a single, small *median claw*. The tarsus is divided into two tarsomeres. The presence of *tibial spurs* and *pedal spurs* on the tarsomeres can be helpful in identifying certain groups of scorpions.

Although scorpions are sexually dimorphic, it generally takes a specialist to reliably distinguish males and females. There are no uniform gender-specific, external morphological characteristics for determining the sex that apply to all scorpions. Instead, determinations are based on comparative morphological differences that often require familiarity with gender-related traits within conspecifics. In general, however, compared to females of the same species, males are smaller and less robust. Males tend to be relatively more slender than females, but there are many exceptions. In some genera, the pedipalp chelae of males are longer and more slender, whereas in other genera they are shorter and thicker. The males of some species have special depressions for accepting the female's pedipalp fingers during mating. Usually, the metasoma of males is relatively longer and more slender than that of females; however, it is sometimes shorter and thicker, being different with each species. The pectines are often strikingly different, with the males having larger pectines with more pectinal teeth. In some cases the males can be recognized by the presence of a pair of *genital papillae* protruding posteriorly from beneath the genital operculum.

## LIFE HISTORY

Although scorpions as a group differ in details relating to their biology and behavior, they follow a similar life-history pattern. Females are viviparous, giving birth to young following a developmental period that varies from 2 to 18 months, depending on the species. Scorpions are unique among arachnids in this respect, the only other known case being one family of mites, the Pyemotidae. The average brood size for all scorpions is about 26, but it can range from as few as 1 up to 105. The sex ratio at birth is about equal, even though this ratio later shifts significantly toward females after reaching maturity. Two species, *Tityus serrulatus* and *Liocheles australasiae*, are reported to be parthenogenic and produce only female offspring.

Following their birth, the newborns immediately crawl onto their mother's back, where they remain through the first instar without feeding. If removed during this period, they die without successfully molting. A major contributing factor seems to be their dependence on obtaining water from their mother's cuticle; they are prone to desiccation due to the permeability of their cuticle at this critical time. In most cases, dispersal of the young occurs shortly after they molt to the second instar, usually within 3–14 days after birth. In other cases, the young may remain in the burrow with their mother, where she cares for them and may even feed them captured prey. This type of social behavior appears to be common in the scorpionoids and has been described in other taxa (e.g., *Euscorpius*).

Most species molt 5 or 6 times before becoming adults, although the number of molts varies from 4 to 9 depending on the species. Males generally undergo fewer molts than females of the same species. No molting occurs in the adults. Maturity is reached in as few as 6 months in some of the smaller species (e.g., *Centruroides* spp.) but may take 3–7 years in some of the largest species (e.g., *Pandinus* spp.). Buthid scorpions develop the fastest, averaging about 18 months, whereas the mean developmental time for all other scorpions is about 3 years. The longevity of most scorpions is probably about 2–5 years. It seems likely that scorpions in temperate environments may take longer to mature (1 molt per year) and thus live longer overall due to short growing seasons. Scorpions may live longer in captivity. The longest reliable report is 96 months for *Pandinus gambiensis*.

## BEHAVIOR AND ECOLOGY

Scorpions are well adapted for surviving in a wide range of habitats, including deserts, grasslands, savannas, and both temperate and tropical forests. In addition, they are found from intertidal zones at sea level to snow-covered mountains at elevations of over 5500 m and in cave systems at depths of more than 800 m. They can tolerate highly varied environmental conditions, including extremes of temperature, both heat and cold, complete emersion in water for hours, and prolonged periods of drought and starvation. In large part, these adaptations are due to behavioral thermoregulation, low metabolic rates, and high efficiency in conserving water. To moderate their body temperatures, scorpions are typically nocturnal, retreating to the protection of burrows and other sheltered sites during the daytime. They experience minimal water loss via their cuticle, spiracles, and book lungs while excreting nitrogenous wastes in almost insoluble forms such as guanine, xanthine, and uric acid. Similarly, their feces are extremely dry.

Most scorpions live on or very near the ground, where they typically are found under objects, in forest litter, or in excavated burrows. The major exception is the large and important family Buthidae, in which the species are often excellent climbers. They commonly are found under the bark of trees, in the tops of palms and other plants, and in crevices of rocky cliffs. Upon entering houses, these species are likely to be seen on the walls and even ceilings, not infrequently gaining access to the upper floors of multistory buildings. Such climbers include some of the most venomous scorpions, notably members of the genus *Centruroides*. Even some of the common vaejovid scorpions, such as *V. carolinianus* in the southeastern United States, are excellent climbers. They frequently enter homes, where they may be seen on walls or clinging upside down on ceilings.

Scorpions feed on a variety of prey, notably soft-bodied insects and arachnids. Heavily sclerotized insects and other invertebrates such as certain isopods are often rejected. Common prey items include spiders, solpugids, other scorpions, millipedes, centipedes, gastropods, and other invertebrates. The larger scorpions will also attack and feed on small vertebrates such as lizards, snakes, and rodents. Owing to poor vision, scorpions depend primarily on their sensory hairs and their ability to sense ground vibrations as a means of detecting, locating, and recognizing suitable prey. Using mechanoreceptors on their tarsi, they can detect potential prey up to 15 cm away. Some arboreal scorpions even can capture flying insects that approach close enough for them to detect via air movements with the trichobothria on their pedipalps.

Scorpions with large, robust pedipalps can often subdue their prey with little or no use of their venom. Smaller species with weaker, more slender pedipalps are far more dependent on stinging their prey upon seizing it with their chelate pincers. The thrust with their stinger is usually carefully delivered to penetrate the softer areas of the prey's integument between sclerites or other harder body parts. After locating a suitable site, sufficient venom is injected to immobilize the prey, following which the sting is withdrawn. This is in strong contrast to the defensive strikes directed toward threatening enemies in which the telson lashes forward to sting its target, inject the venom, and be quickly withdrawn.

Once a scorpion has captured a prey item, it crushes it with the coxal bases of its pedipalps and the first two pairs of legs while tearing at it with its chelicerae. Digestive juices from specialized glands in the gut flow through a channel formed by the coxae of the pedipalps and anterior legs. Following extraoral digestion, the semidigested food is drawn into the mouth, assisted by the chelicerae; undigested parts are trapped by setae in the preoral cavity and expelled. The feeding process is slow, taking as long as 2.5 hr to consume an item such as a blow fly. Owing to the efficiency in storing digested food in the hepatopancreas, a well-fed scorpion can survive for months without further feeding.

When it comes time to seek a mate, the female recognizes conspecific males by a behavior called *juddering*, in which the male displays a series of shaking movements, rocking back and forth with his pectines spread out and quivering. The resulting vibrations are communicated via the substrate to the female. There is also evidence to suggest that pheromones may be involved in sex recognition in at least some species. Having located a female, the male initiates courtship by grasping the female's pedipalps with his own and guiding her through a complex courtship behavior in the form of a mating dance or *promenade*. During the promenade they may engage in cheliceral massages or *kissing*, in which the male grasps the female's chelicerae with his chelicerae and gently kneads them, apparently serving to suppress her aggressiveness. In many species the male actually stings the female, usually in the tibial joint of the pedipalp, where the inserted sting may be held for 3 to 20 min or longer. This also appears to reduce her aggression and render her more docile during the courtship.

Throughout the promenade the male uses his pectines to sweep back and forth across the ground, sensing the substrate to determine if it is suitable for depositing a spermatophore. Having found an acceptable site, he extrudes a complexly structured spermatophore from his genital aperture and attaches it to the substrate in an upright position with the sticky basal plate firmly anchoring it in place. He then guides the female over the spermatophore to make contact with her genital valves. As the spermatophore bends under pressure from her mesosoma, the sperm is released directly into her genital tract. Contact with the spermatophore may last anywhere from a few seconds to several minutes. Once insemination is completed, the male abruptly disengages from the female and effects an immediate escape, lest he be attacked and eaten by his no longer receptive mate.

When the gravid female is ready to give birth, she assumes a position known as *stilting,* in which she raises the anterior part of her body and forms a *birth basket* with her pedipalps and first two pairs of legs. While she maintains this posture, the young emerge one at a time from the genital aperture and drop into the birth basket. From there they clamber onto their mother's back. The birth time for an individual varies from 1 min up to about an hour, with the total birth process lasting from less than 12 hr to as long as 3 days.

More than 150 species of predators have been reported to feed on scorpions. Among the most common are birds and lizards, followed by various mammals, frogs, toads, and snakes. Several invertebrates also are natural enemies, including spiders, solpugids, ants, centipedes, and other scorpions. In many cases scorpions are resistant to their own venoms; however, they readily fall prey to attacks by other species. Scorpions comprise a significant part of the diet of burrowing owls, elf owls, and grasshopper mice. They also are hosts for mermithid nematodes and the ectoparasitic larvae of certain mites.

## PUBLIC HEALTH IMPORTANCE

Scorpion *sting cases* can be categorized as two general types: those involving only localized, transitory symptoms usually lasting from a few minutes to several hours, and those involving systemic reactions. *Localized responses* are characterized by immediate pain followed by moderate swelling at the sting site, often likened to the sting of a wasp or bee. In some cases, the sting may result in a raised, reddened, indurated lesion, even in the case of relatively harmless scorpions (e.g., *Vaejovis carolinianus*). In cases involving cytolytic toxins (e.g., scorpionids and ischnurids), swelling may persist up to 72 hr, followed by development of hemorrhages and blood-filled blisters near the sting site. Sloughing of skin may occur, but this varies greatly in severity. Other localized effects include gooseflesh, sweating, and muscle spasms near the sting site. In cases of buthid stings, pain usually radiates from the site of the sting up the affected limb. The pain tends to concentrate in the joints, especially the armpits and groin, and often crosses from one armpit to the other.

In cases of *systemic reactions,* the clinical signs and symptoms are highly variable, ranging from mild to life-threatening. Systemic reactions commonly are mild and are not necessarily indicators of a serious problem. Often there is no appreciable swelling or discoloration of the skin at the sting site. An intense aching and burning sensation may spread to adjacent tissues, which in turn often

throb, sometimes becoming numb. The acute pain at the sting site turns into a chronic, dull pain accompanied by a feeling of numbness around the edge of the sting site, which may persist for one to several days. Numbness in the face, mouth, and throat is fairly common. Muscles may become spasmatic, resulting in muscular twitching, slurred speech, difficulty swallowing, tightness or cramps in the chest and back, rapid heartbeat, and nausea. Often these systemic responses persist less than an hour after the sting and are not considered serious.

In more severe systemic reactions, neurologic effects can lead to profuse sweating and salivation, restlessness, extreme nervousness, respiratory and cardiovascular problems, mental confusion, and convulsions. As the clinical symptoms indicate, the principal components of the venom of dangerous scorpions are *neurotoxins*. These toxins act on the autonomic, sympathetic, and neuromuscular systems, causing the wide range of systemic reactions reported in sting victims. They act by disrupting the voltage-sensitive sodium and potassium channels of nerves, which in turn causes neural depolarization, prolonged action potentials, repetitive firing, and uncontrolled release of vasodilators and neurotransmitters, which affect virtually every major organ system. The effect on neurotransmitters results in a depletive release of catecholamines (e.g., adrenaline, noradrenaline) that severely damages the heart and other organs. The most commonly reported cause of death in scorpion sting cases is cardiac failure. In other cases, respiratory failure may be the cause, especially in patients with upper respiratory infections or related problems. Death usually occurs several days after envenomation. If symptoms subside during the first 2–12 hr following a sting, the prognosis for recovery is generally good. Mortality rates are quite variable, depending on the species and amount of venom injected. The rates are much higher among children than adults. For further details on the clinical toxicology and symptoms of scorpion stings, see Dehesa-Davila *et al.* (1995) and Ismail (1995).

Scorpion *venom* is a very complex mixture of substances which differs significantly among the various taxa, within families, and among genera. Differences also occur in different geographic populations of the same species and even within the same population. The toxins are low-molecular-weight proteins which are among the most powerful toxins known. They are comparable in some species to the neurotoxins of certain deadly snakes. Two recognized types of neurotoxins are *α-scorpion toxin,* characteristic of the genera *Androctonus* (Fig. 20.7), *Leiurus* (Fig. 20.8), and *Buthus,* and *ß-scorpion toxin,* characteristic of *Centruroides. Tityus* species appear to have both types. The effects of envenomation by any given scorpion species can differ significantly among individual cases, owing to a wide range of contributing factors.

**FIGURE 20.7** *Androctonus australis* (Buthidae), adult. A highly venomous scorpion in North Africa. (Photo by S. A. Stockwell.)

These include the quantity of venom injected and the age, size, and general health of the victim.

The sting of most scorpions usually requires no special treatment, although the application of ice to the sting site helps to relieve local pain. Incisions, as used in cases of poisonous snake bites, should never be made. Nor is the use of most drugs recommended in noncomplicated cases because antihistamines, steroids, analgesics, and sedatives usually have little or no effect. In more severe cases involving systemic reactions, medical attention should be sought immediately.

Substances that have been found to be effective in treating scorpion stings are atropine to counter effects on the parasympathetic system, calcium gluconate given intravenously to relieve muscle spasms, and sodium phenobarbital administered intravenously to prevent convulsions. Insulin has also been reported to be beneficial in treating cases in India. Morphine and Demerol generally are not recommended as pain relievers because of their tendency

**FIGURE 20.8** *Leiurus* sp. (Buthidae). A member of this genus, *L. quinquestriatus*, in North Africa and the eastern Mediterranean region, is regarded as one of the most dangerous scorpions in the world. (Photo by S. A. Stockwell.)

to act as synergists, increasing the toxicity of certain venoms (e.g., *C. exilicauda*).

*Antivenins* are recommended where available. However, caution in the use of antivenins must be noted since many antivenins are of poor quality and often are administered at doses far below the level required to produce effective results. They may adequately neutralize the larger venom peptides but not necessarily the more important small-molecular-weight toxins. To be effective, they must be administered within the 1st hour following the sting. Even then, the cessation of systemic symptoms within about an hour may not necessarily reflect the use of the antivenom, since this is the same period of time that symptoms often subside without treatment of any kind. Another limitation is the fact that most scorpion antivenins are very specific, often for a single species, and are produced only on a limited regional basis (Lucas and Meier, 1995; Theakston and Warrell, 1991).

## SCORPIONS OF MEDICAL IMPORTANCE

Some of the more dangerous species of scorpions for which toxicity data have been reported are shown in Table I. Based on mammalian toxicity, *Leiurus quinquestriatus*, *A. australis*, and *A. mauretanicus* are generally recognized as the most venomous. *L. quinquestriatus* is known in the Middle East and Sahel region of northern Africa as the yellow scorpion and in the Sudan as the Omdurman scorpion. The *Androctonus* species are commonly called fat-tailed scorpions, referring to the marked thickness and width of their postabdominal segments. *A. australis*, which occurs primarily in arid mountainous regions, causes more deaths than any other species in North Africa. Based on the numbers of cases and fatalities, it is probably the most deadly scorpion worldwide. *Buthus occitanus*, a widely distributed species in the Middle East and North Africa, is the only medically important species in southern Europe. Its toxicity varies markedly in different parts of its range, apparently reflecting different subspecies. The Indian red scorpion (*Mesobuthus tamulus*) is the most medically important scorpion on the Indian subcontinent.

The most important venomous genera in the New World are *Centruroides* and *Tityus*. *Centruroides* species occur primarily in Mexico, Central America, and the West Indies. They are often called *bark scorpions* due to their habit of hiding under loose tree bark or in crevices of dead logs and trees. They are often found around domestic settings in piles of wood, stones, bricks, and discarded debris. Stings are likely to occur when they are disturbed in their hiding places or when they enter homes at night in search of prey. *C. exilicauda* (formerly *C. sculpturatus*) is often cited as the most dangerous scorpion in the United States. Deaths from its sting, however, are rare. In fact, the toxicity and effects of its venom are very similar to that of the striped scorpion *C. vittatus*. The latter is the most widely distributed of all American scorpions, occurring in the southern United States from the Mississippi River west to New Mexico, as far north as Kansas and Missouri, and well south into northeastern Mexico.

Among the more than two dozen *Centruroides* species and subspecies in Mexico, the following taxa are of particular concern because of the seriousness of sting cases: *C. elegans, C. exilicauda, C. infamatus, C. noxius,* and *C. suffusus*. All are closely related to one another (Exilicauda group). The most notorious are *C. suffusus* and *C. limpidus,* both of which are capable of causing human deaths. Despite its small size, usually less than 5 cm, *C. noxius* is considered very venomous.

*Tityus* species are similar to *Centruroides* species in size, general appearance, and behavior. As a group of over 100 species, they occur throughout South America and the Caribbean Basin. In most places where they occur, all *Tityus* species are considered dangerous. The most venomous is the Brazilian species *T. serrulatus,* which is common in urban areas and readily enters homes. Second only to *T. serrulatus* in its medical importance is another house-infesting scorpion in Brazil, *T. bahiensis*. Related *Tityus* species which also are highly venomous include *T. cambridgei,* a forest-dwelling species in the Amazon Basin and northern South America, *T. trinitatis,* which can be a serious problem in coconut groves and cane fields in Venezuela and Trinidad, and *T. trivittatus,* a house-infesting scorpion in Argentina.

## VETERINARY IMPORTANCE

There is little evidence to indicate that scorpions pose any significant threat to domestic animals, including cats and dogs. One can only assume that stinging encounters do occur, perhaps even commonly, but that the effects are seldom serious enough to draw attention of the owners.

## PREVENTION AND CONTROL

Pesticides are not generally recommended for controlling scorpions indoors or for preventing their entering homes. Instead, appropriate measures can be taken to "scorpion-proof" buildings or otherwise significantly reduce the prospects of their entering homes. Entry can be discouraged by raising the floor level at least 20 cm above the ground. A single step to reach the threshold is better

than multiple steps and should be separated from the wall of the structure by a gap of 6 cm or more. The installation of a horizontal row of glazed ceramic tiles on the vertical surfaces of steps and around the entire perimeter of a building also will provide a barrier which scorpions cannot readily climb. Smooth exterior wall surfaces, such as planed cement, further impede their climbing ability. Worn weather stripping around doors and windows should be replaced, and potential entry sites around water pipes and electrical conduits in foundations should be sealed. To prevent scorpions from gaining access to the roofs of structures, a row of ceramic tiles also can be applied to the outer walls just below the roof line.

Scorpions can be discouraged from frequenting the immediate vicinity of homes by trimming plantings which touch buildings and removing piles of firewood, lumber, bricks, and other materials which serve as harborage. The use of coarse bark mulches around plants near the foundation of buildings should be avoided for the same reason.

In areas where climbing scorpions commonly infest homes, measures can be taken to reduce the risk of envenomation. A sheet of muslin or other suitable cloth can be suspended from the ceiling over the sleeping quarters to catch any scorpions that might drop from the roof structure. Mosquito netting over beds affords similar protection. Regularly shaking out clothing and footwear before putting them on is highly recommended.

In temperate regions, the greatest number of complaints of scorpions entering homes is often seasonal, most commonly in the early spring and late fall. Heavy or frequent rains in the spring can saturate the soil and ground litter around building foundations, driving scorpions indoors in their efforts to find drier sites. With the onset of colder weather in the late fall, scorpions are similarly apt to find their way indoors while seeking warmer temperatures. Another circumstance which contributes to scorpion problems is the construction of new homes or subdivisions in previously undisturbed woodlands or other habitats where scorpions are abundant. The clearing of such areas and the associated disturbance of the ground litter often causes displaced scorpions to wander extensively. In the process they frequently find their way inside nearby homes. Sealing or blocking possible access sites is the only practical means of preventing their entry.

## REFERENCES AND FURTHER READING

Anderson, R. C. (1975). Scorpions of Idaho. *Tebiwa* **18**, 1–17.
Balozet, L. (1971). Scorpionism in the Old World. In "Venomous Animals and Their Venoms" W. Bücherl and E. E. Buckely, Eds.), Vol. 3, pp. 349–371. Academic Press, New York.
Bettini, S., Ed. (1978). "Arthropod Venoms." Springer-Verlag, Berlin.

Briggs, D. E. G. (1987). Palaeontology: scorpions take to the water. *Nature* **326**, 645–646.

Bücherl, W. (1978). Systematics, distribution, biology, venomous apparatus, etc. of Tityinae: venom collection, toxicity, human accidents and treatment of stings. In "Arthropod Venoms" (S. Bettini, Ed.), pp. 371–378. Springer-Verlag, Berlin.

Cloudsley-Thompson, J. L. (1990). Scorpions in mythology, folklore, and history. In "The Biology of Scorpions" (G. A. Polis, Ed.), pp. 462–485. Stanford Univ. Press, Stanford, CA.

Cloudsley-Thompson, J. L. (1993). Spiders and scorpions (Araneae and Scorpiones). In "Medical Insects and Arachnids" (R. P. Lane and R. W. Crosskey, Eds.), pp. 659–682. Chapman & Hall, London.

Coddington, J. A., Larcher, S. F., and Cokendolpher, J. C. (1990). The systematic status of Arachnida, exclusive of Acari, in North America north of Mexico. In "Systematics of the North American Insects and Arachnids: Status and Needs" (M. Kosztarab and C. W. Schaefer, Eds.), pp. 5–20. Virginia Agricultural Experiment Station Information Series 90-1. Virginia Polytechnic Institute and State Univ., Blacksburg.

Couraud, F., and Jover, E. (1984). Mechanisms of action of scorpion toxins. In "Handbook of Natural Toxins" (A. T. Tu, Ed.), Vol. 2, pp. 659–678. Dekker, New York.

Dehesa-Davila, M., Alagon, A. C., and Possani, L. D. (1995). Clinical toxicology of scorpion stings. In "Handbook of Clinical Toxicology of Animal Venoms and Poisons" (J. Meier and J. White, Eds.), pp. 221–238. CRC Press, Boca Raton, FL.

Diniz, C. R. (1978). Chemical and pharmacological aspects of Tityinae venoms. In "Arthropod Venoms" (S. Bettini, Ed.), pp. 379–394. Springer-Verlag, Berlin.

Efrati, P. (1978). Symptomatology and treatment of Buthinae stings. In "Arthropod Venoms" (S. Bettini, Ed.), pp. 312–316. Springer-Verlag, Berlin.

El-Asmar, M. F. (1984). Metabolic effect of scorpion venom. In "Handbook of Natural Toxins" (A. T. Tu, Ed.), Vol. 2, pp. 551–576. Dekker, New York.

El-Ayeb, M., and Delori, P. (1984). Immunology and immunochemistry of scorpion neurotoxins. In "Handbook of Natural Toxins" (A. T. Tu, Ed.), Vol. 2, pp. 607–638. Dekker, New York.

Ennik, F. (1972). A short review of scorpion biology, management of stings, and control. *California Vector Views* **19**, 69–80.

Fet, V., Sissom, W. D., Lowe, G., and Braunwalder, M. E. (2000). "Catalog of the Scorpions of the World (1758–1998)." The New York Entomological Society, New York.

Gertsch, W. J., and Allred, D. M. (1965). Scorpions of the Nevada test site. *Brigham Young University Science Bulletin, Biology Series* **6**(4), 1–15.

Gertsch, W. J., and Soleglad, M. E. (1972). Studies of North American scorpions of the genera *Uroctonus* and *Vejovis* (Scorpionida, Vejovidae). *Bulletin of the American Museum of Natural History* **148**, 551–608.

Goyffon, M., and Kovoor, J. (1978). Chactoid venoms. In "Arthropod Venoms" (S. Bettini, Ed.), pp. 395–418. Springer-Verlag, Berlin.

Gueron, M., and Ovsychcher, I. (1984). Cardiovascular effects of scorpion venoms. In "Handbook of Natural Toxins" (A. T. Tu, Ed.), Vol. 2, pp. 639–658. Dekker, New York.

Hassan, F. (1984). Production of scorpion antivenin. In "Handbook of Natural Toxins" (A. T. Tu, Ed.), Vol. 2, pp. 577–606. Dekker, New York.

Hjelle, J. T. (1972). Scorpions of the northern California coast ranges (Arachnida: Scorpionida). *Occasional Papers California Academy of Science* **92**, 1–59.

Ismail, M. (1995). The scorpion envenoming syndrome. *Toxicon* **33**, 825–858.

Johnson, J. D., and Allred, D. M. (1972). Scorpions of Utah. *Great Basin Naturalist* **32**, 154–170.

Kaestner, A. (1968). Order Scorpiones, Scorpions. In "Invertebrate Zoology. Vol. 2. Arthropod relatives, Chelicerata and Myriapoda," pp. 101–114. Interscience, New York.

Keegan, H. L. (1980). "Scorpions of Medical Importance." Univ. Press of Mississippi, Jackson.

Lucas, S. M., and Meier, J. (1995). Biology and distribution of scorpions of medical importance. In "Handbook of Clinical Toxicology of Animal Venoms and Poisons" (J. Meier and J. White, Eds.), pp. 205–219. CRC Press, Boca Raton, FL.

Muma, M. H. (1967). "Scorpions, Whip Scorpions and Wind Scorpions of Florida. Arthropods of Florida and Neighboring Land Areas." Vol. 4. Florida Department of Agriculture, Gainesville, FL.

Polis, G. A., Ed. (1990). "The Biology of Scorpions." Stanford Univ. Press, Stanford, CA.

Possani, L. D. (1984). Structure of scorpion toxins. In "Handbook of Natural Toxins" (A. T. Tu, Ed.), Vol. 2, pp. 513–550. Dekker, New York.

Rankin W., and Walls, J. G. (1994). "Tarantulas and Scorpions: Their Care in Captivity." T. F. H. Publ, Neptune City, NJ.

Shulov, A., and Levy, G. (1978). Systematics and biology of Buthinae. In "Arthropod Venoms" (S. Bettini, Ed.), pp. 309–312. Springer-Verlag, Berlin.

Simard, J. M., and Watt, D. D. (1990). Venoms and toxins. In "The Biology of Scorpions" (G. A. Polis, Ed.), pp. 414–444. Stanford Univ. Press, Stanford, CA.

Sissom, W. D. (1990). Systematics, biogeography, and paleontology. In "The Biology of Scorpions" (G. A. Polis, Ed.), pp. 64–160. Stanford Univ. Press, Stanford, CA.

Stahnke, H. L. (1940). The scorpions of Arizona. *Iowa State College Journal of Science* **15**, 101–103.

Stahnke, H. L. (1978). The genus *Centruroides* (Buthidae) and its venom. In "Arthropod Venoms" (S. Bettini, Ed.), pp. 279–308. Springer-Verlag, Berlin.

Stockwell, S. A. (1992). Systematic observations on North American Scorpionida with a key and checklist of the families and genera. *Journal of Medical Entomology* **29**, 407–422.

Theakston, R. D. G., and Warrell, D. A. (1991). Antivenoms: a list of hyperimmune sera currently available for the treatment of envenoming by bites and stings. *Toxicon* **29**, 1419–1470.

Tu, A. T., Ed. (1984). "Insect Poisons, Allergens, and Other Invertebrate Venoms." Handbook of Natural Toxins, Vol. 2. Dekker, New York.

Wainschel, J., Russell, F. E., and Gertsch, W. S. (1974). Bites of spiders and other arthropods. In "Current Therapy" (H. F. Conn, Ed.), pp. 865–867. Saunders, Philadelphia.

Whittemore, F. W., and Keegan, H. L. (1963). Medically important scorpions in the Pacific area. In "Venomous and Poisonous Animals and Noxious Plants of the Pacific Region" (H. L. Keegan and W. V. Macfarlane, Eds.), pp. 107–110. Macmillan, New York.

Williams, S. C. (1969). Birth activities of some North American scorpions. *Proceedings of the California Academy of Science, Series 4* **37**, 1–24.

Williams, S. C. (1976). The scorpion fauna of California. *Bulletin of the Society of Vector Ecology* **3**, 1–4.

Williams, S. C. (1980). Scorpions of Baja California, Mexico and adjacent islands. *Occasional Papers California Academy of Science* **135**, 1–127.

Zlotkin, E., Miranda, F., and Rochat, H. (1978). Chemistry and pharmacology of Buthinae scorpion venoms. In "Arthropod Venoms" (S. Bettini, Ed.), pp. 317–370. Springer-Verlag, Berlin.

# SOLPUGIDS (*Solifugae*)

GARY R. MULLEN

REFERENCES AND FURTHER READING   426

---

*Solpugids are* usually yellow or brownish in color and rather hairy. The body length varies from 1 to 7 cm, with the largest species having a leg span up to 12 cm. The prosoma and opisthosoma are broadly joined, with the latter being visibly segmented (Fig. 21.1). The most prominent structures are the greatly enlarged, powerful pair of chelicerae which are used to seize, crush, and tear apart food. With the exception of *Rhagodes nigrocinctus* in India, solpugids are generally believed to lack distinct venom glands and rely primarily on their size and strength to overpower prey. The pedipalps are long and leglike, each ending in an eversible adhesive organ rather than claws. The first pair of legs is modified as slender tactile organs which are held outstretched as the solpugid moves about. Unique, mallet-shaped structures called *racquet organs* (malleoli) are borne on the underside of the fourth pair of legs in both sexes. They are innervated and function in chemoreception while probing various substrates, presumably to detect chemical cues associated with food and potential mates.

The order Solifugae includes 12 families, approximately 150 genera, and over 900 species worldwide (Punzo, 1998). They occur most commonly in tropical and subtropical deserts in Africa, the Middle East, western Asia, and the Americas. In Africa they also are found in grasslands and forests. They occur in the United States and southern Europe but not in Australia or New Zealand. The 2 major families in North America are the Ammotrechidae and Eremobatidae, together represented by 11 genera and about 120 species. Most of them occur in the western half of the United States. The exception is *Ammotrechella stimpsoni,* which is found under the bark of termite-infested tree stumps in Florida. For a comprehensive treatment of solpugids, including keys to the families and genera worldwide, see Punzo (1998). For further information on solpugids in the United States, see Muma (1951).

Members of this group are variously known as *solpugids, sun spiders, wind spiders, wind scorpions, camel spiders, barrel spiders, false spiders,* and *romans.* Local names in the United States include *bulldozer spiders* in the Big Bend area of Texas and *sand puppies* in Wyoming. They also are known by the British terms *jerrymander* and *jerrymunglum.* In Mexico they are called *mata venado* ("deer killer") in the mistaken belief that they are venomous enough to kill large animals. In southern Africa solpugids are called *hair cutters* and *beard cutters* in the undocumented belief that females are attracted to the hair of sleeping humans and other animals, which they clip with their chelicerae and carry to their burrows or other retreats to line their nests in preparation for egg laying.

Despite their common names, they do not bear a close resemblance to either spiders or scorpions, although they occur primarily in arid habitats where these other

**FIGURE 21.1** Solpugid in desert of southwestern United States. Despite their greatly enlarged pair of chelicerae and formidable appearance, solpugids lack venom glands and are generally harmless to humans. (Photo by Debbie R. Folkerts.)

arachnids are found. They are typically nocturnal, hiding during the day under stones and in crevices or burrowing into loose soil. The name "sun spider" refers to some species that are active during the daytime. The name "wind scorpion" reflects their peculiar, rapid movement as they run about the surface of desert sands hunting prey; they give the appearance of being blown across the sand and have been likened by some to tumbleweeds. The name "camel spider" refers to the arch-shaped plate on the dorsum of the prosoma of many species. Solpugids feed primarily on insects, spiders, and scorpions. The larger species, however, also are known to attack and kill small lizards, mice, and birds.

Solpugids will readily attack humans and other animals when provoked. Despite their formidable appearance and aggressive posturing, their bites usually are not serious. However, the larger species can inflict severe wounds with their powerful chelicerae. In one case involving United States military personnel in the Persian Gulf, an individual was bitten on the lip and required 10 stitches to close the wound (Conlon, 1991). The greatest concern is usually preventing secondary infections, which can lead to painful swellings, necrosis of tissues surrounding the bite site, and gangrene.

## REFERENCES AND FURTHER READING

Aruchami, M., and Sundara-Rajulu, G. (1978). An investigation on the poison glands and the nature of the venom of *Rhagodes nigrocinctus* (Solifugae: Arachnida). *National Academy of Science Letters (India)* **1**, 191–192.

Cloudsley-Thompson, J. L. (1958). "Spiders, Scorpions, Centipedes and Mites." Pergamon, New York.

Cloudsley-Thompson, J. L. (1992). Solifugae and keeping them in captivity. In "Arachnida. Symposium on Spiders and Their Allies, London (1987)" (J. E. Cooper, P. Pearce-Kelly, and D. L. Williams, Eds.), pp. 52–56. Chiron Publ., Keighley.

Conlon J. M. (1991). Vectors & war. Part 2. Desert Storm. *Wing Beats. Florida Mosquito Control Association*, **22**, 16–20, 22.

Hickin, N. E. (1984). Solifugae. In "Pest Animals in Buildings" (N. E. Hicken, Ed.), pp. 85–86. Godwin, London.

Muma, M. H. (1951). The arachnid order Solpugida in the United States. *Bulletin of the American Museum of Natural History* **97**, 35–141.

Punzo, F. (1998). "The Biology of Camel-Spiders (Arachnida, Solifugae)." Kluwer Academic, Dordrecht/Norwell, MA.

# 22

# SPIDERS (*Araneae*)

GARY R. MULLEN

*All spiders,* except the Symphytognathidae and Uloboridae, possess venom glands which are used to subdue captured prey. When threatened, however, spiders will often defend themselves by biting, thereby injecting those same toxins into vertebrate skin. In most cases, the venom produces only mild, localized reactions that do not warrant medical attention. Other spiders have much more potent venoms that can cause severe reactions in bite victims, occasionally resulting in deaths. Only about 60 species of spiders worldwide are considered to have significant medical importance. Among them are only a few genera that are dangerously venomous to humans. Most occur in the Subtropics and Tropics. A few tropical species, however, have extended their ranges into temperate regions, particularly those with Mediterranean-like climates.

Envenomation by spiders is called *araneism,* after Araneae, the arachnid order to which spiders belong.

Separate names, however, are commonly given to bites or syndromes associated with the more dangerous spider genera, each of which is generally characterized by typical clinical signs and symptoms. Examples include *atraxism* (*Atrax* spp.), *cheiracanthism* (*Cheiracanthium* spp.), *latrodectism* (*Latrodectus* spp.), *loxoscelism* (*Loxosceles* spp.), *phoneutriism* (*Phoneutria* spp.), and *tegenariism* (*Tegenaria* spp.). There also are cases in which individuals develop an abnormal fear of spiders such that the mere sight of one can cause panic or hysteria, a condition called *arachnophobia* or, more specifically, *araneophobia.* This should not be confused with the unfortunate disdain that many people have for spiders, often reflecting their upbringing and misconceptions about spiders in general.

For general information on spiders, see Comstock (1948), Gertsch (1979), Foelix (1992, 1996), and Preston-Mafham and Preston-Mafham (1984).

## TAXONOMY

Approximately 3,000 genera and 36,000 species of spiders have been described worldwide. In North America alone there are 64 families, some 500 genera, and 3,400 species (Coddington *et al.,* 1990). Among the more than 100 families of spiders, about 20 families include species that reportedly cause medical concerns when they bite humans and other animals. The 5 most important families are the Dipluridae, Hexathelidae, Theraphosidae, Sicariidae (Loxoscelidae), and Theridiidae. See Bettini and Brignoli (1978) and Ori (1984) for additional venomous spiders representing more than 60 genera worldwide.

The order Araneae is divided into two suborders: the Mesothelae and Opisthothelae. The Mesothelae include the single family Liphistiidae, a small group of primitive spiders in Southeast Asia and the Indo-Malaysian region. The Opisthothelae are composed of two groups: the Mygalomorphae (tarantula-like spiders) and Araneomorphae (all other spiders). The mygalomorphs are the more primitive spiders, represented by trap-door spiders, funnel-web spiders, and tarantulas. They include the largest spiders. The araneomorphs are a very diverse group that include wandering spiders and those which are familiar to most people by the diversity of silken webs which they produce.

Taxonomic catalogs of the world spider fauna are provided by Roewer (1942–1954), Bonnet (1945–1961), Brignoli (1983), and Platnick (1989, 1993). For identification keys to the families and genera of mygalomorphs, see Raven (1985), and for North American families and genera of spiders, see Kaston (1978) and Roth (1993).

The following is a synopsis of the major families of medical-veterinary importance.

## MYGALOMORPH SPIDERS

### Actinopodidae

This small group is closely related to the typical trap-door spiders of the family Ctenizidae, which they resemble both morphologically and behaviorally. They construct vertical, silk-lined burrows in the soil, the opening at the surface of which is covered by a hinged "trap door." Their venom is weakly neurotoxic to vertebrates and causes no necrosis. Of the three recognized genera, only *Actinopus* in Central and South America has been reported as biting humans, producing only local pain and transient muscle contractions.

### Barychelidae

Members of this family are closely related to the Theraphosidae, or tarantulas, and are largely restricted to southern Africa and Australia. *Idiommata blackwelli* occurs widely throughout Australia in dry areas, where it constructs silk-lined burrows provided with a saucer-shaped door that fits tightly into the opening at the ground surface. Its bite is painful, causing local redness and edema in humans. Most encounters occur when wandering males enter homes during the late summer and early fall or when individuals are dislodged from their burrows in new suburban areas when people rake their yards.

### Dipluridae

Known as *sheet-web* or *funnel-web–building tarantulas,* diplurids construct burrows in the ground in a wide range of habitats. The family includes 19 genera. They are particularly abundant in the Southern Hemisphere and Australian region. The venom of *Trechona* species is especially toxic to humans and has been reported to cause human deaths in South America. Two species of particular importance are *Trechona venosa* and *T. zebra,* which occur in tropical forests and coastal areas, where they are encountered on vegetation and along trails. They are very aggressive and, if disturbed, will readily bite. The hexathelid genera *Atrax, Hadronyche,* and *Macrothele* were previously included in this family.

### Hexathelidae

Formerly included in the Dipluridae, some of the members of this family are regarded as among the most venomous of spiders for humans. Among the 11 recognized genera, *Atrax, Hadronyche,* and *Macrothele* are the most dangerous. They are known as *funnel-web spiders* due to their habit of building expansive, funnel-like webs near the entrance to their shallow, silk-lined burrows in the ground, among rocks, or in stumps and rot holes of trees. The most serious bites are caused by *Atrax robustus* and *Hadronyche formidabilis* in Australia.

### Theraphosidae

This is the largest mygalomorph family, with 84 recognized genera. They are best known for their large size and hairy appearance and are familiar to most people as *tarantulas.* As a group they are primarily tropical and subtropical, occurring widely throughout both the Old World and New World, where they are variously known as *bird spiders, bird-eating spiders,* and *monkey* or *baboon spiders.* In North America they extend into the southwestern United States but do not naturally occur east of the Mississippi River. Although the bite of most species is relatively harmless, several genera can cause severe envenomation, particularly in South America, where the genera dangerous to humans are *Acanthoscurria, Pamphobeteus, Phormictopus,* and *Sericopelma.* On the Indian subcontinent, the genus *Poecilotheria* has venom which causes a reaction similar to widow venom, perhaps making it the most dangerous tarantula.

## ARANEOMORPH SPIDERS

### Agelenidae

Agelenid spiders are called *funnel weavers,* not to be confused with the mygalomorph funnel-web spiders (Dipluridae and Hexathelidae). They typically build horizontal sheet webs with a tubular retreat or "funnel" leading into a protected recess. When their webs are constructed in vegetation, they are often called *grass spiders*. A few species occur in homes, especially in basements and cellars, where their chances of encounters with humans are greatest. The only recognized species of medical concern in North America is the *hobo spider, Tegenaria agrestis.* In Europe, *Agelena labyrinthica* and *Coelotes obesus* have been reported to bite humans.

## Araneidae

This is one of the largest families of spiders; it is familiar to most people because of the symmetrical spiral-like webs which they construct for snaring flying insects. Known as *orb weavers,* they are commonly found around homes and other dwellings, where they take advantage of artificial lights to attract prey at night. They seldom bite humans or other vertebrates and usually cause only minor, temporary discomfort when they do. Even the North American *black and yellow garden spider, Argiope aurantia,* produces only localized pain, redness, and edema on the rare occasions in which it has been known to bite. This is a large and colorful species which often attracts the attention of homeowners in the late summer and fall. *A. lobata* in Europe causes a similar, mild reaction. Members of the genus *Mastophora,* known as *bolas spiders,* possess the most potent venoms among araneid spiders and can cause serious medical problems. Their bites can result in generalized pain and swelling, fever, sweating, hemolysis, and necrosis at the bite site; convulsions and deaths have been reported in some cases. Included in this genus are the *podadoras* of Bolivia, Chile, and Peru.

## Clubionidae

Members of this large, diverse family are known as *sac spiders,* referring to the silken tubular retreats which they typically make in rolled-up leaves, other ground litter, and under bark and stones. They are nocturnal, vagrant spiders which commonly are found hunting prey on plants. They incidentally enter houses and other buildings. Only members of the genus *Cheiracanthium* are generally regarded as being venomous enough to warrant medical attention.

## Corrinnidae

This family is closely related to the Clubionidae. It includes *Trachelas* species (e.g., *Trachelas volutus*), which have been reported to bite humans in the United States, causing a stinging sensation and localized erythema and swelling.

## Ctenidae

Members of this family are wandering spiders which do not build webs. Most are of moderate size (1.5–2.5 cm in body length) and occur primarily in ground litter and low vegetation, where they hunt prey. They resemble wolf spiders (Lycosidae) in both their general appearance and behavior. The most venomous taxa of medical concern are members of the South American genus *Phoneutria,* relatively large species (> 3 cm) which can cause severe envenomation. Examples are the *banana spider* (*Phoneutria nigriventer*) of South America and *Elassoctenus harpax* of Western Australia, which reportedly inflict painful bites. A human bite case implicating a *Florida false wolf spider* (*Ctenus captiosus*) has been documented in Florida. The bite was described as a needle-like puncture with subsequent swelling about the site, nausea, dizziness, and flulike symptoms that persisted for several days (C. Moore and G. B. Edwards, personal communication).

## Gnaphosidae

These wandering spiders are commonly found under stones, in rolled leaves, and in ground debris and are known as *ground spiders.* The bites of most gnaphosids are relatively harmless. *Herpyllus blackwalli* and *H. ecclesiasticus* have been reported to cause moderately severe bites in the United States, usually upon entering homes at night.

## Lamponidae

Members of this family are similar to gnaphosids in their habitats and behavior. The Australian *white-tailed spider* (*Lampona cylindrata*) can cause a painful bite with localized inflammation, intensely cyanotic lesions, blistering, persistent ulcerations, and necrosis at the bite site. It is unclear if the damage is due to the venom itself or to associated bacteria that are introduced with the venom (Sutherland, 1987).

## Lycosidae

Commonly known as *wolf spiders,* lycosids represent a highly successful family of hunting spiders which are noted for their relatively large size (up to 4 cm) and hairy appearance. Their posterior median and posterior lateral eyes are greatly enlarged and aid them in capturing prey. Members of the genus *Lycosa* possess cytotoxic venoms which can cause painful bites, and a few cause necrotic skin lesions. Included in this genus are the so-called tarantulas of Europe, such as *Lycosa tarentula* of tarantism fame. Although the bites of many wolf spiders are painful, they generally cause only temporary, local discomfort.

## Oxyopidae

Members of this largely tropical and subtropical family are called *lynx spiders.* They are active hunters which rely on their keen eyesight, speed, and agility to capture prey while climbing in foliage. Although the family is not generally regarded as being medically important, females of the *green lynx spider* (*Peucetia viridans*) are known to forcibly expel venom from their fangs as a defensive response, especially when guarding their egg sacs. Droplets can be squirted up to 20 cm and on contact with human eyes can cause impaired vision and moderately severe conjunctivitis (Fink, 1984; Tinkham, 1946)

## Pisauridae

This family is closely related to the wolf spiders, which they strongly resemble. They occur most frequently near water, where they are adept at moving about on the water surface to capture prey, hence their common name *fishing spiders.* They are also known as *nursery-web spiders* because of the habit of females suspending their egg sacs in a protective silken "nursery" in vegetation and guarding the resultant spiderlings until they disperse. Because of their large size (body length up to 4 cm or more) and powerful chelicerae, they can bite if handled, causing local pain and transient swelling. The bite of the European species *Dolomedes fimbriatus* is reported to cause a reaction similar to that of the agelenid *Coelotes obesus.*

## Salticidae

This is the largest family of spiders, with over 400 genera and over 4000 species widely distributed throughout the world. They are known as *jumping spiders* because of their habit of stalking and pouncing on prey or jumping to escape when threatened. The anterior median eyes are complex and greatly enlarged, providing them with the keenest vision of all spiders. Some of the larger species can be aggressive and inflict painful bites when handled or pressed against the skin. The venom of at least some species contains cytotoxins that cause necrotic lesions at the puncture site, often being slow to heal. The bite of *Phidippus johnsoni* can cause a dull, throbbing pain that may persist for a few hours, in addition to swelling, tenderness, and itching that may last for 1–4 days following the bite (Russell, 1970).

## Segestriidae

Members of this relatively small family live in silken retreats under stones and bark or in crevices of wood and rocks. They are active nocturnal hunters which may enter homes or construct their retreats in and around human dwellings. Despite their large, well-developed chelicerae, they are not very aggressive, rarely bite, and are not considered to be very venomous. Nonetheless, a few cases of human bites by the European species *Segestria florentina* reportedly have involved local pain, redness, and swelling and occasionally nausea and vertigo.

## Sicariidae (Including the Former Loxoscelidae)

The sicariids are a small group of relatively primitive araneomorphs. Included in this family are the *recluse spiders* in the genus *Loxosceles,* which cause a severe necrosis. Approximately 50 species of *Loxosceles* have been described in the Americas and at least 70 worldwide. They are generally similar in appearance and are difficult to recognize from one another by the nonspecialist. They typically are found in ground litter and under bark or stones; a few species occur in caves, and some are decidedly synanthropic, living in close association with humans. *Loxosceles* species are primarily tropical but have been introduced to temperate regions of Europe, Africa, the Americas, and Australia. The genus *Sicarius* has been shown to be also highly toxic in laboratory studies.

## Theridiidae

Members of this large family are called *cobweb weavers* or *comb-footed spiders.* The latter refers to a row of serrated bristles on the hind tarsus which is used to comb the silk from the spinnerets during construction of their irregular webs or wrapping prey. The only genus considered particularly venomous to humans and domestic animals is *Latrodectus,* which includes the *widows* or *shoebutton spiders.* Another genus which is purportedly venomous to humans is *Steatoda.* Several species in South America, including *Steatoda andinus* (formerly placed in *Lithyphantes*) and the well-known *cirari* of Bolivia, Chile, and Paraguay are said to cause serious envenomation (Southcott, 1984). Venom of the Mediterranean species *S. paykulliana* has been shown to be neurotoxic to guinea pigs but this spider has not been reported to bite humans. The cosmopolitan species *S. grossa,* known as the *false black widow,* causes only a local bite reaction with no apparent neurotoxic effects (Maretic and Lebez, 1979).

## Thomisidae

Members of this family are called *crab spiders* because of their generally flattened appearance, laterigrade legs, and crablike gait. They usually are cryptically colored and ambush their prey from camouflaged sites such as tree bark and flower heads. As a group they are considered harmless to humans and other animals. Some species of *Misumenoides,* however, have been suspected of causing relatively minor bites in humans (Hickin, 1984).

# MORPHOLOGY

The body of a spider is divided into two regions: the anterior *cephalothorax* (prosoma), which represents a fusion of the head and thoracic segments, and the *abdomen* (opisthosoma) (Fig. 22.1). The cephalothorax bears the chelicerae, pedipalps, eyes, and legs. The *chelicerae* (singular, chelicera) are paired structures used to seize prey or to bite defensively when threatened (Fig. 22.2). They also serve other functions in different groups of spiders such as digging by burrowing species, transporting prey, and carrying eggs sacs by some spiders. Each chelicera consists of two parts: a stout basal portion (paturon) and a movable *fang.* The fang rests in a groove and is extended when the spider bites or otherwise attempts to grasp something. Near the tip of each fang is a tiny opening to the *venom duct,* which leads from the *venom gland;* the latter is located in the basal part of the chelicera and usually extends back into the cephalothorax. The *mouth* is located just behind the bases of the chelicerae. The chelicerae of mygalomorphs are oriented parallel to the long axis of the body and move parallel to one another in a vertical plane. These spiders strike downward when seizing prey. The chelicerae of araneomorphs are oriented perpendicular to the long axis of the body and are opposed to one another, moving together in a pincer-like motion.

Most spiders have eight eyes located anteriorly on the cephalothorax (Fig. 22.2). They are *simple eyes* (ocelli) usually arranged in two rows. Some spiders lack one or more pairs of eyes, as in *Loxosceles* species, which have only six eyes. In other species, such as wolf spiders and

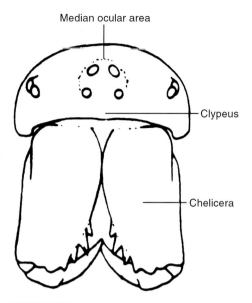

**FIGURE 22.2** Head and chelicerae of representative spider, frontal view. (Modified from Kaston, 1978.)

jumping spiders, some eyes may be enlarged, reflecting their greater visual acuity. The size and arrangement of the eyes often serve as valuable taxonomic characters.

The *palps* (pedipalps) are a pair of six-segmented appendages that arise immediately behind the mouth. They are primarily tactile structures used in sensing the substrate, perceiving contact stimuli from conspecifics, and both detecting and manipulating prey. Whereas the palps of immature spiders and adult females tend to resemble legs, the palps of adult males are modified as copulatory organs. In such cases the terminal segment (palp tarsus) is enlarged with a ventral, bowl-shaped cavity enclosing a complex of specialized structures formed from the pretarsus that serve as an intromittent organ for inseminating females. Adult males usually can be recognized by the swollen terminus of their palps and their often smaller body size relative to females of the same species.

Spiders have four pairs of *legs,* each with seven segments: coxa, trochanter, femur, patella, tibia, metatarsus, and tarsus (Fig. 22.3). Spiders that run about on the

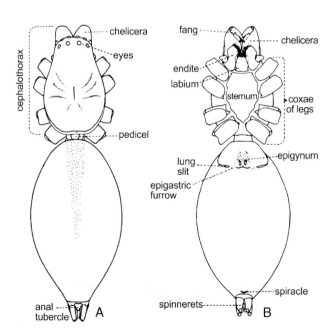

**FIGURE 22.1** Morphology of representative spider. (A) Dorsal view; (B) ventral view. (Modified from Kaston, 1978.)

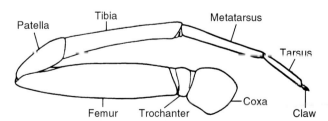

**FIGURE 22.3** Leg of spider showing the seven leg segments. Spiders possess a patella and metatarsus not found in insects. (From Kaston, 1978.)

ground and other substrates without building a trapping web typically have only two tarsal claws on each leg. Many of these hunting spiders possess dense tufts of hairs (*scopulae*) directly beneath the pair of claws or along the ventral side of the tarsus and metatarsus. They provide physical adhesion to facilitate climbing on smooth surfaces and grasping prey. The scopulae are especially prominent in tarantulas. Three tarsal claws are characteristic of web-building spiders. The single median claw on each leg is used to hold onto silken threads by those spiders that hang suspended in their webs.

The *abdomen* is connected with the cephalothorax by a narrow *pedicel*, which provides great flexibility and movement between the two body regions. One or two pairs of slitlike openings to the *book lungs*, the principal respiratory organs in spiders, are located ventrally on the second and third abdominal segments. The more "primitive" spiders tend to retain two pairs of book lungs, whereas most spiders have only one pair. The second pair of book lungs in some spiders is modified to form *tubular tracheae* that open via a *spiracle* or pair of spiracles on the third abdominal segment. Most spiders, however, have only a single spiracle located in front of the spinnerets.

The *genital opening* of both sexes is located ventrally on the second abdominal segment between the book lungs. In females of the more "advanced" spiders, a sclerotized copulatory structure called the *epigynum* is located just in front of the genital opening and leads to the *spermathecae*, where sperm is stored after mating. The presence of the epigynum is helpful in distinguishing adult females from immatures and males and in distinguishing species.

Located at the posterior end of the abdomen are the *spinnerets*, through which silk from several types of internal *silk glands* is extruded via small spigots. Most spiders have three pairs of spinnerets, the size and relative lengths of which provide useful taxonomic characters. In some groups of araneomorph spiders, a sievelike plate of minute spigots called the *cribellum* is present in front of the anterior pair of spinnerets. These taxa, called *cribellate spiders,* also possess a row of specialized setae on the metatarsi of the fourth pair of legs called a *calamistrum*. The calamistrum is used to comb silk from the cribellum by rhythmic movements of the hind legs. The vast majority of araneomorph spiders lack a cribellum and calamistrum and are referred to as *ecribellate spiders.*

## LIFE HISTORY

Much reproductive activity of spiders occurs during the late spring and early summer, when mating takes place and eggs are deposited. Some species, however, are reproductively active in the late summer and fall. Embryonic development commonly takes about 2 weeks, during which time one or two *prelarvae* and a *larva* are formed within individual eggs contained in a silken *egg sac*. The larva subsequently molts and emerges from the egg sac as a *nymph*. The young nymph, or *spiderling*, is a miniature of the adult spider, with functional spinnerets and poison glands. The nymphs undergo 2–12 molts as *juveniles,* depending on spider species, before reaching sexual maturity as adults.

In temperate regions, juvenile spiders are present throughout the summer months, with overwintering typically occurring as late-instar nymphs. However, if egg sacs are constructed by females late in the year (e.g., some orb-weaving spiders), the resultant spiderlings remain inside the egg sac until spring. In other cases, spiders overwinter as mature females. The life cycle for most spiders is 1 year, following which the adults die. Males are shorter lived than females and usually die soon after mating. Some of the more "primitive" spiders live much longer; purse-web spiders (Atypidae) may live 7 years, whereas the larger tarantulas (Theraphosidae) may live 20–30 years, especially in captivity.

## BEHAVIOR AND ECOLOGY

*Mating* behavior in spiders varies greatly from one group to another and in some cases involves complex courtship displays. Prior to mating, the male typically deposits a droplet of sperm from the genital pore onto a small, silken platform called a *sperm web*. The droplet is then drawn into the copulatory organ at the tip of each male palp in a process called *sperm induction*. Sperm is stored in the palp until mating takes place. Following acceptance of a male's advance by a female, *copulation* is accomplished by indirect insemination in which the male transfers the sperm from his palps to the spermathecal ducts on the underside of the female's abdomen. Although the males of many species die soon after mating, others refill their palps with sperm and may mate with other females one or more times thereafter. Contrary to popular belief, most males either walk away or hurriedly retreat without being attacked and eaten by the female.

*Oviposition* typically occurs within a few weeks after mating. The female spins a silken sheet onto which she extrudes fertilized eggs from her genital opening, forming a mass in which the individual eggs are cemented together. She then protects the eggs by spinning multiple layers of silk to form the *egg sac*. More than one egg sac may be produced, as commonly observed in black widow spiders. Although some species of spiders remain with the eggs to protect them until they hatch, most spiders provide no maternal care and leave the eggs to hatch unguarded.

Spiders have been very successful in exploiting a wide range of ecological habitats, including islands. This is accomplished, in part, by their specialized *dispersal* behavior called *ballooning.* Spiderlings, and even adults of many small spiders, are carried aloft on silken threads by which they may be conveyed by wind currents over considerable distances. To become airborne the spider orients into the wind, lifts the tip of its abdomen, and extrudes a strand of silk from the spinnerets. When the strand is long enough to provide enough buoyancy to support its body weight, the spider releases its grip on the substrate and floats away.

Spiders can be categorized into three major groups based on their general behavior and feeding strategies: burrowers, vagrants or wanderers, and web builders. *Burrowing spiders* usually excavate their burrows in the soil of suitable sites, where they remain more or less permanently throughout their lives. They typically capture prey which comes within reach of the burrow entrance or make short excursions to capture food and return to the safety of the burrow. Among this group are many of the ground-dwelling tarantulas, trap-door spiders, and burrowing wolf spiders. *Vagrant spiders* tend to wander extensively and may or may not regularly return to a given location where they have constructed a retreat. They do not produce a silken web for capturing food, but instead hunt or ambush their prey. Examples include many wolf spiders, jumping spiders, sac spiders, gnaphosids, and ctenids. These spiders commonly enter homes during their hunting forays and occasionally bite humans. The brown recluse spider also falls into this group. It differs from the others, however, by actually living indoors rather than wandering inside incidentally, and it makes a flimsy web. *Web-building spiders* construct various silken structures to detect and capture potential prey. The webs may be sheetlike, as in diplurids and agelenids, or more complex trapping webs, like those of comb-footed spiders and orb weavers.

Burrowing and vagrant spiders usually rely on their physical size and strength to capture food, together with sufficiently potent venoms to subdue their prey quickly. The size of acceptable prey items is often positively correlated with the size of the spider itself. Web-building spiders, on the other hand, tend to rely on the use of silk to ensnare or immobilize their prey. They are able to utilize a wider range of prey items, often capturing and feeding on insects and other arthropods much larger than themselves. Those that construct aerial webs also have access to a wider variety of flying insects than do ground-dwelling spiders. The venom of web-building spiders is typically less potent than that of non−web-building species, accounting in part for the fact that bites of even the larger web builders are usually quite harmless.

# PUBLIC HEALTH IMPORTANCE

Recognizing that virtually all spiders possess venom glands, it is not surprising that when they are threatened, species with chelicerae large enough to pierce the skin will bite humans and other animals. In most cases the reaction is minor, usually limited to mild, localized pain and slight to moderate swelling at the bite site. The severity of the reaction is dependent on the species of spider, its size, and the amount of venom injected.

The *venom* varies greatly among different spider taxa in terms of chemical composition and its effects on different animals upon injection. Components include a wide range of proteases, esterases, polyamines, free amino acids, histamine, and specific toxic compounds unique to individual groups or species. Whereas some venoms are primarily cytolytic, causing the destruction of cells and tissues with which they come into contact, others act as neurotoxins or disrupt normal blood functions. For details on the biochemistry and pharmacology of spider venoms, see Bucherl (1971), Bettini (1978), Duchen and Gomez (1984), and Geren and Odell (1984).

Most problems warranting medical attention are not due directly to bites but to secondary infections. In other cases involving the more venomous species, reactions can be much more severe, occasionally causing deaths. See the following sources for information on venomous spiders in different regions of the world: North America (Wong *et al.,* 1987), South America (Lucas, 1988), Europe (Maretic and Lebez, 1979), South Africa (Newlands and Atkinson, 1988), and Australia (Southcott, 1976; Sutherland, 1990).

The following accounts address spider problems of particular medical importance.

## TARANTISM

The term *tarantism* has special significance from a medical viewpoint. It refers to a condition in which individuals allegedly bitten by a ''tarantula'' spider experience a wide range of symptoms, including tremors, hyperactivity, difficulty breathing, muscular rigidity and priapism (painful penile erection in males), sweating, and uncontrolled crying. In its most extreme form it can lead to fainting spells, delirium, and convulsions. Although descriptions of this syndrome can be traced back as early as Aristotle's writings in the fourth century BC, it was most prevalent in Europe during the Middle Ages. It is believed to have been named after Taranto, Italy, where an epidemic of tarantism occurred in 1370. From there the phenomenon spread throughout Italy to present-day Croatia, Spain, and other parts of the Mediterranean. The only cure was thought to be prolonged and vigorous dancing to special, lively music to induce perfuse sweating and eventual collapse from sheer exhaustion. Municipalities sometimes hired musicians to play in shifts for 3–4 days at a time as victims danced themselves into frenzies, seeking relief from their affliction. Not uncommonly this led to mass hysteria among local residents and shameless exhibitionism on the part of some individuals. Some have linked this choreomania to Saint Vitus' dance, a nervous disease with involuntary jerking motions.

The bite of the European wolf spider *L. tarentula,* commonly called the "tarantula," traditionally has been blamed as the cause of tarantism. The reason for this connection is uncertain; in fact, this species seldom comes into contact with people. Its bite causes only mild pain and slight swelling at the bite site and causes none of the neurological effects characteristic of tarantism victims. Convincing evidence suggests that the spider involved was actually a *Latrodectus* species. Even as late as the 1950s, spiders of this genus were called "tarantola" in southern Italy, and cases involving their bites were noted in medical records as a "tarantola bite" or "tarantolism." Today tarantism is regarded as largely a psychosomatic response to real or imagined spider bites, rooted in legend, ignorance, or superstition linked to cases of latrodectism. For more details on the history and nature of tarantism, see Maretic and Lebez (1979).

## TARANTULISM

Tarantulas (Theraphosidae) (Fig. 22.4) are typically ground dwellers living in silk-lined burrows. They often leave their burrows at night to hunt prey; at such times they may enter homes and other shelters or otherwise

**FIGURE 22.4** Mexican red-leg tarantula, *Brachypelma smithi* (Theraphosidae). (From Cooke *et al.,* 1972, courtesy of the American Museum of Natural History.)

come into contact with people. During the mating season, males are more likely to be encountered as they wander in search of females; they are particularly aggressive at this time and are easily provoked. Other circumstances that contribute to human encounters are disturbances of their burrows, land development, and flooding or other natural disasters which tend to displace them. About a dozen genera of tarantulas are considered venomous enough to humans to require medical treatment of bite victims. They are found primarily in the tropics of South America, Africa, and Australia, where most of the serious cases of human envenomation occur.

Despite their large size, powerful fangs, and intimidating appearance, most tarantulas are not very venomous. Bite reactions vary from almost painless to moderately or intensely painful with reddening about the puncture site. The sensation is commonly likened to that of a bee sting, except that the pain is not immediate and develops more slowly. The pain subsides gradually, seldom persisting for more than 30 min. This may be accompanied by a burning sensation, localized swelling, and tightening of the muscles near the bite wound. Secondary infections can be avoided by cleansing the wound and applying a topical antibiotic.

In cases of more dangerous tarantulas, neurotoxic components in the venom can cause severe, sometimes life-threatening reactions. These toxins are designed to act quickly in subduing vertebrate prey such as frogs, lizards, and birds, on which some species feed. When injected into a human bite wound, the venom can cause not only intense pain, but also muscle spasms, edema, inflammation of lymphatic vessels, and systemic reactions that can lead to shock and vascular collapse. Other effects reported in laboratory animals include local necrosis, hemoglobin in the urine, and jaundice, indicating the presence of necrotoxic and haemolytic components in the venom of some species.

Members of the genus *Harpactirella* are among those sometimes known as *baboon spiders.* They occur in South Africa, where they live in silk-lined tunnels under logs, stones, and other debris. They are aggressive hunters, frequently entering homes and animal shelters during their wanderings. The bite of *Harpactirella lightfooti* causes an immediate burning pain followed by paleness, vomiting, and severe systemic reactions that can lead to shock and collapse. The bite is not fatal, with recovery usually occurring within 24 hr. None of the tarantulas in the United States is considered to be dangerously venomous. However, the venom of the *Texas brown tarantula, Aphonopelma hentzi,* contains a necrotoxin that has been shown to damage myocardial tissues in mice.

Many tarantulas possess tiny (0.2–1.2 mm long) specialized *urticating hairs* on their abdomen which are readily detached when stroked with their hind legs

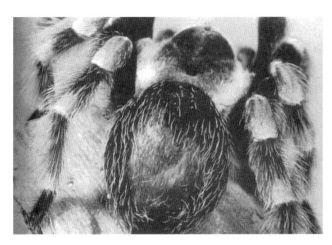

**FIGURE 22.5** Mexican red-leg tarantula, *Brachypelma smithi* (Theraphosidae), with bald patch on dorsal aspect of abdomen where urticating hairs have been defensively flicked off by hind legs. (From Cooke *et al.*, 1972, courtesy of the American Museum of Natural History.)

(Figs. 22.5 and 22.6). Only New World theraphosids in the subfamilies Aviculariinae, Ischnocolinae, and Theraphosinae are known to possess such hairs; the latter includes the genera *Brachypelma* and *Aphonopelma*, representatives of which are commonly sold as pets. These hairs are armed with spines and barbs designed to penetrate vertebrate skin and other tissues with which they come in contact. Some species have up to 10,000 of these hairs per square millimeter, totaling well over 1 million urticating hairs per individual.

Four basic types of hairs are described by Cooke *et al.* (1972). Type I hairs enter the skin at a shallow angle, do not penetrate very deeply, and cause only a mild reaction. This is the only type found in species in the United States.

Type II hairs are not flicked off but are incorporated into the silk lining the tarantula's retreat, eggs sacs, or silk mats used during molting. Type III hairs penetrate the skin up to 2 mm, causing a persistent urticaria and inflammation that may last for 2–3 weeks; this type is characteristic of many Mexican, Caribbean, Central American, and South American species. Type IV hairs cause inflammation of the respiratory tract in small mammals, although little is known regarding their effects on humans. A given species may have more than one type of urticating hair. Bald patches on the back of the abdomen (Fig. 22.5) are usually evidence that a tarantula has defended itself by dislodging these specialized hairs. They are replaced with each molt, even in adult females that continue to molt after reaching maturity.

When threatened, the tarantula flicks a cloud of these hairs at the offender, usually rodents and other small mammals which try to attack them in their burrows. In addition to causing irritation to the skin (Fig. 22.6), they can cause severe inflammation of the eyes, mouth, and respiratory passages, serving as an effective deterrent against predators. The effects are solely mechanical and do not involve chemical substances. In the case of certain tarantulas (e.g., *Megaphobema*, *Pseudotheraphosa*, and *Theraphosa* spp.), the abdominal hairs are incorporated with silk into the egg sacs or the silk mats on which they molt as a defensive barrier to attack by potential predators and parasites.

Humans experience similar reactions to those of other animals when handling or provoking certain tarantulas, including species sold as pets. Common symptoms are urticarial dermatitis, mild edema, and vascular dilation. When the setae come into contact with the eyes, they cause an immediate burning or stinging sensation

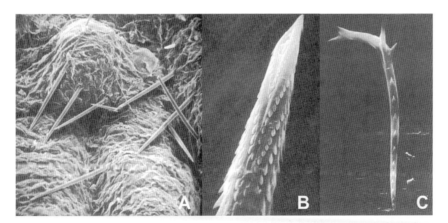

**FIGURE 22.6** Urticating hairs of tarantulas (Theraphosidae). (A) Hairs of *Avicularia surinamensis* embedded in skin of young mouse; (B) basal tip of type II hair of *A. surinamensis*, with backward-directed barbs that help to work the hair into skin; (C) Type IV hair of unidentified tarantula from Chile that causes inflammation of respiratory passages of small mammals. (From Cooke *et al.*, 1972, courtesy of the American Museum of Natural History.)

followed by intense pruritus, lachrymation, swelling of the eyelids, and corneal abrasions. The problem is exacerbated by the natural inclination of the victim to rub the affected areas. Corneal lesions may still be evident 6–9 months after the encounter as the embedded hairs are gradually resorbed. The damage is not permanent, and full visual acuity is gradually restored.

## ATRAXISM

The genera *Atrax* and *Hadronyche* occur only in the eastern half of Australia, including Tasmania, where its members are known as *funnel-web* or *tube-web spiders*. They construct their silken retreats in rock crevices or ground burrows. Two species are considered highly venomous to humans—*Atrax robustus* and *Hadronyche formidabilis*—with the venom of males being more dangerous than that of females.

The *Sydney funnel-web spider, A. robustus* (Fig. 22.7), the more commonly encountered species, is largely restricted to areas within a 160-km radius of Sydney (New South Wales). It is a large species, with males, the larger of the sexes, measuring about 25 cm in length. They construct their tubular webs under logs, amidst rocks, and in various ground debris. They are often found in suburban gardens, where they are attracted to damp, well-watered sites with abundant ground litter. Most bites by this species occur during the summer, when roaming males are most likely to enter homes. Wandering males are extremely aggressive, readily attack when provoked, and account for the majority of cases of human envenomation. The bite produces immediate pain and a wide range of neurological symptoms. These include agitation, anxiety, hypertension, generalized muscular twitching, and irregular heart beat, in addition to pulmonary edema and

**FIGURE 22.7** Sydney funnel-web spider, *Atrax robustus* (Dipluridae), of Australia, female. (From Southcott, 1976.)

intravascular coagulation. Deaths have been reported, notably in children. The principal toxic component of the venom has been named *atraxotoxin*. It acts by stimulating the release of acetylcholine at motor end plates and throughout the autonomic nervous system. The effects are usually transient and reversible, causing no permanent damage.

*H. formidabilis* is called the *North Coast funnel-web spider* or the *tree funnel-web spider*. It occurs in the rain forests of southeastern Queensland and northern New South Wales. The effects of its bite are similar to those of *A. robustus*, although the toxicity of its venom is greater. It is less commonly encountered, however, with relatively few bite cases having been reported. An antivenin directed at the male toxins is available for treating both *A. robustus* and *H. formidabilis* bites.

## PHONEUTRIISM

The genus *Phoneutria* is widely distributed throughout South America, where members are known as *wandering spiders* and *banana spiders*. They are large aggressive spiders which actively hunt at night, feeding on both invertebrate and vertebrate prey. The venom is highly neurotoxic to humans and acts on both the central and peripheral nervous systems, producing a characteristic syndrome. The most dangerous species is *Phoneutria nigriventer*, which occurs throughout Brazil, Uruguay, and Argentina, where cases of human envenomation are quite common. It is a large species, with females attaining body lengths up to 5 cm. Its venom contains a number of pharmacologically active compounds, including histamine and serotonin in addition to neurotoxin. Its bite is extremely painful and causes a number of symptoms such as salivation, sweating, muscular spasms, painful penile erection, and visual problems. Deaths are usually attributed to respiratory failure.

## CHEIRACANTHISM

Only a few members of the Clubionidae, or *sac spiders* (Fig. 22.8), pose health concerns. Most are *Cheiracanthium* species, which as a group occur throughout the Northern Hemisphere. Envenomation usually occurs when they are trapped against the skin after crawling into clothing or footwear or when an individual rolls over onto one while sleeping. The bite typically causes immediate local pain, redness, and formation of a wheal similar to that from the sting of a wasp or bee. In some cases the victim may experience a persistent loss of sensitivity involving the musculature and nerves at the bite site. Tissue surrounding the puncture wound often becomes necrotic,

**FIGURE 22.8** Sac spider, *Clubiona obesa* (Clubionidae), female. (From Gertsch, 1979.)

forming a small skin lesion. In more severe cases, systemic responses may include mild fever, nausea, and loss of appetite. No deaths have been attributed to spiders in this genus.

*Cheiracanthium punctorium* is the species most commonly involved in cases of cheiracanthism in Europe. Its bite causes a painful burning sensation and associated swelling, which may persist for several days. Other reported symptoms include chills, general muscular aches, and tenderness of regional lymph glands. Envenomation by *C. japonicum* in Japan is similar to that by *C. punctorium* but also may involve local petechiae (purplish, hemorrhagic spots on the skin), nausea, vomiting, and, rarely, shock. Reactions to one of the two species found in North America, *C. inclusum,* are painful but relatively mild compared to *C. punctorium* and *C. japonicum;* however, *C. mildei* has been reported to sometimes cause severe necrosis similar to a recluse spider bite. Several of the *Cheiracanthium* species of medical importance are fairly recent introductions from other parts of the world. These include *C. mildei,* introduced to the United States from Europe; *C. mordax,* introduced to Hawaii from Australia; and *C. diversum,* introduced to Hawaii and Fiji from Australia.

## TEGENARIISM

Some agelenid spiders in the genus *Tegenaria* occur in close association with humans. Upon entering homes, they may become established in basements and other relatively dark, damp locations, where they construct sheet webs with a funnel-like retreat characteristic of the Agelenidae. In Europe, these species are called *house spiders.* One species of particular medical importance in North America is the *hobo spider, Tegenaria agrestis,* which was introduced from Europe to the Pacific Coast of the United States some 70 years ago. It now occurs widely in the northwestern United States and adjacent Canada, where it is recognized as a common cause of lesions similar to those induced by the brown recluse spider.

### Hobo spider *(Tegenaria agrestis)*

This species was first reported at Seattle (USA) in 1930 but did not become common in the Pacific Northwest until the 1960s. It was not until the 1980s, however, that it was confirmed as the cause of bites previously attributed to *Loxosceles reclusa.* The hobo spider is now well established in Washington, Oregon, and Idaho, with its distribution extending from Montana, Wyoming, and Utah northward into British Columbia and Alberta to the Alaskan panhandle.

Adults of this spider are 12–18 mm long and brownish, with long, unmarked legs and a herringbone pattern or series of dark chevron-like markings on the dorsal aspect of the abdomen (Fig. 22.10). It occurs in funnel-like webs (Fig. 22.9), particularly in basements and cellars, window wells of homes, and crawl spaces, around house foundations, in wood piles, under rocks and wood used in landscaping, and other suitable sites at ground level. The males tend to wander at night in search of females, at which time they enter homes and are more likely to be encountered than females. This occurs primarily from midsummer through fall. In Europe, *T. agrestis* is found more typically outdoors, apparently being unable to compete well indoors with its close relative *T. gigantea.* The latter is less prone to bite when disturbed and has a milder venom. This probably explains why few cases of *T. agrestis* bites are

**FIGURE 22.9** Female hobo spider, *Tegenaria agrestis* (Agelenidae). in entrance to funnel-web retreat. (From Akre and Catts, 1992, courtesy of Washington State University Cooperative Extension.)

**FIGURE 22.10**   Male hobo spider, *Tegenaria agrestis* (Agelenidae). (From Akre and Catts, 1992, courtesy of Washington State University Cooperative Extension.)

reported among Europeans. Similarly, where *T. gigantea* has become established within the range of *T. agrestis* in North America, the incidence of *T. agrestis* bites tends to be diminished.

The venom of *T. agrestis* is similar to that of *Loxosceles* species in that it is cytolytic, causing necrosis of tissues and slow-healing, ulcerated lesions (Fig. 22.11). The venom of males and juveniles appears to be more potent than that of the adult females. The bite generally is not painful and may result initially in a slight prickling sensation or go unnoticed altogether. Within 30 min, the bite site becomes numb and indurated, surrounded by an expanding reddened area (5–15 cm diameter). This is followed by the formation of pus-filled blisters during the next 1–2 days. The blisters break about a day later, causing the wound to ooze and develop an encrusted, crater-like skin lesion. As

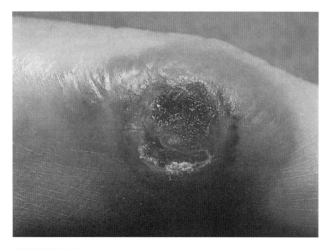

**FIGURE 22.11**   Lesion on human finger caused by bite of hobo spider, *Tegenaria agrestis* (Agelenidae). (From Akre and Catts, 1992, courtesy of Washington State University Cooperative Extension.)

in cases of *Loxosceles* envenomation, a black scablike eschar forms over the wound, and sloughing of dead tissue can result in disfiguring scars. Healing may take 1–2 weeks to several months. In cases in which significant fatty tissue is involved, the wound may become extensive and deep, requiring as long as 2–3 years to heal. Treatment is similar to that recommended for brown recluse spider bites.

Systemic effects occur in approximately 45% of bite cases caused by the hobo spider, about one-third of which may require hospitalization. The earliest indication is usually severe headaches within the first 10 hr which are not relieved by aspirin. These headaches may persist for several days or up to a week. Other associated problems include general muscular weakness, fatigue, sweating, dizziness, nausea, impaired vision, disorientation, and temporary memory loss. Without appropriate medical treatment, severe systemic reactions may result in intractable vomiting, profuse secretory diarrhea, destruction of bone-marrow tissue, anemia, pancytopenia, low platelet counts, and internal hemorrhaging. Although rare, systemic reactions can be fatal (Anonymous, 1996).

For additional information on the hobo spider and its medical importance, see Vest (1987), Akre and Myhre (1991), and Anonymous (1996).

## LOXOSCELISM

The clinical syndrome called *loxoscelism* is caused by the bite of *Loxosceles* species known as *fiddle-back*, *brown recluse*, or *violin spiders*. It is also called *necrotic arachnidism* because of cytolytic components of the venom which cause necrosis of tissues around the bite wound. The common names of these spiders refer to a usually distinct fiddle- or violin-shaped marking on the dorsum of the cephalothorax, the neck of which is directed posteriorly. The base of the "violin" encompasses the eyes and is darkly contrasted against the lighter, general body color (Fig. 22.12). The eyes are distinctive among spiders in that there are only six, rather than the usual eight, and in that they are arranged in three groups of two eyes each (Fig. 22.12). The combination of a violin-shaped marking and this eye pattern distinguishes *Loxosceles* from all other spider genera. The body color and legs are usually a light, tawny brown but may be dark brown or even grayish in some populations. *Loxosceles* species quite closely resemble one another, usually requiring a specialist to make species determinations. The legs are relatively long and slender, making them agile spiders which can move quickly. They are primarily nocturnal hunters, either catching prey that comes into contact with their irregular webs or actively wandering from the security of

**FIGURE 22.12**  Brown recluse spider, *Loxosceles reclusa* (Sicariidae), female, dorsal view. Note the dark, violin-shaped marking on the cephalothorax and arrangement of the six eyes in three groups of two eyes each. (Photo by G. R. Mullen.)

**FIGURE 22.13**  Brown recluse spider, *Loxosceles reclusa* (Sicariidae), female. This spider is typically brown or tawny in color, with relatively long legs. (Photo by Sturgis McKeever.)

their silken retreats to capture food items. They do not wrap their prey but rely on their potent venom to quickly subdue it.

Approximately 70 species of *Loxosceles* have been described, with nearly 50 of them being found in the Americas (Gertsch and Ennik, 1983). The other species occur primarily in Europe and Africa. Fourteen *Loxosceles* species are found in the United States, including *L. arizonica* (Arizona, New Mexico, Texas), *L. blanda* (New Mexico), *L. deserta* (California), *L. devia* (Texas), *L. laeta* (Massachusetts, California), *L. reclusa* (southeastern except Florida, south-central, and midwestern states), *L. rufescens* (scattered localities in eastern states), and *L. unicolor* (southwestern states). *L. laeta* and *L. rufescens* are introduced species from South America and the Mediterranean region, respectively. *L. rufescens* is the most widely distributed member of the genus. It is endemic in southern Europe and northern Africa, from which it has spread to northern Europe and parts of the Middle East and has been introduced to Australia, Japan, Madagascar, and North America. The most important species are *L. reclusa* in North America, *L. laeta* in South America, and *L. rufescens* in the Mediterranean. These species have been introduced into nonnative areas through commerce and vacation travel.

**Brown recluse spider** *(Loxosceles reclusa)*

The brown recluse spider (Figs. 22.12 and 22.13) occurs primarily in the southeastern and central United States. Records outside this area are believed to represent scattered but not well-established populations, unlike *L. rufescens*. It is typically found indoors in warm, dry, undisturbed areas such as closets, attics, basements, storage areas, utility rooms, heated garages, lofts of feed mills, storerooms of broiler houses, and heated warehouses. It also is found hiding in cabinets and furniture, behind baseboards, door facings, and wall hangings, and in crevices and corners of rooms. Particularly common sites to find them are in old boxes and accumulations of materials that have not been disturbed for some time.

When mature, *L. reclusa* females are 7–12 mm in body length but may look much larger because of their long legs. The males are slightly smaller (mean, 8 mm) and are easily recognized by their bulbous pedipalps. Mating occurs from February to October but most commonly in June and July. The inseminated female produces 1–5 egg sacs, each containing 20–50 eggs. The egg sacs are white, about 17 mm in diameter, and flattened on the underside and convex above, and they are constructed in the spider's silken retreat. The spiderlings usually emerge in 3–7 weeks and remain in the web with the female until after the first or second molt. Development is relatively slow, requiring 7–8 months under favorable conditions. During this time they molt another 6 or 7 times, undergoing 8 instars before becoming adults. Adults commonly live up to 2.5 years and have been known to survive 5–10 years under laboratory conditions.

The web of *L. reclusa* is constructed in poorly lighted, undisturbed, out-of-the-way places where the spider spends most of its time. It is a rather irregular, nondescript tangle of silken strands which continues to grow in thickness as new silk is laid down. Freshly deposited silk is sticky but soon becomes covered with dust, contributing to the unkempt appearance of the webbing. In addition to being a retreat, the silk serves to detect the presence of potential prey. When food is scarce, *L. reclusa* will leave the web at night to roam in search of prey. It is under such circumstances that they are most likely to come in contact with humans.

Although being nonaggressive and very retiring, as its common name implies, the brown recluse spider will bite if provoked. Most encounters occur either at night, when a person rolls onto one of them in bed, or when a person puts on clothes or footwear into which the spider has crawled. Often the victim is not aware of the bite until 2–3 hr later, whereas in other cases it may be immediately felt as a stinging sensation. This is usually followed by intense local pain with the formation of a small blister at the bite site. The area around the bite becomes reddened and swollen as the venom seeps into the surrounding tissues, making it very sensitive to touch. The extent of the skin area involved is usually evident within 6–12 hr. The venom is highly cytotoxic, killing any cells it contacts. Within 24 hr the involved skin tissue turns dusky or purplish as the blood supply and oxygen to the affected area are cut off. The result is necrosis of the tissues and formation of a crater-like ulcer (Fig. 22.14) within 3–4 days. Histological

evidence indicates acute injury to the blood vessels and infiltration by white blood cells at the site. *Phospholipase,* a major component of the venom, induces this white blood cell response while also causing platelets in the blood to aggregate and causing the liberation of inflammatory substances that contribute to development of the skin lesion.

The extent of tissue damage is largely dependent on the amount of venom injected at the time of the bite. Small doses can elicit very little response such that many bites go virtually unnoticed or do not result in ulcerations. A high dose, on the other hand, can result in destruction not only of the skin but also the underlying muscles into which the venom seeps due to gravity. The irregular shape of the skin lesion itself also reflects the effect of gravity, which is most evident in bites on the arms and legs. Some of the more severe cases result when the bite occurs in areas associated with fat tissue. The enzyme *sphingomyelinase* in the venom readily destroys lipid cells, causing saponification and extensive damage to the vasculature. This can cause severe tissue damage to the eyelids and face and to "baby fat" of infants.

Healing occurs very slowly, often requiring 6–8 months or more. The edges of the wound become thickened and raised as the central area begins to undergo scar formation. The necrotic tissue gradually sloughs away, often exposing the underlying muscles. As the wound heals from beneath, a black scablike *eschar* develops over the damaged area, protecting it during the healing process (Fig. 22.15). Throughout this period it is important to keep the wound clean to avoid infections, which can significantly prolong the healing

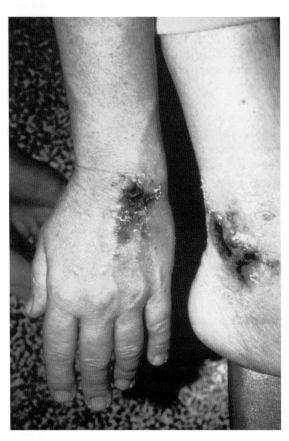

**FIGURE 22.14**    Skin lesions on human hand and ankle caused by bites of brown recluse spider, *Loxosceles reclusa* (Sicariidae). Note the irregular shape of the damaged skin and formation of characteristic eschars, black scablike tissue that forms over the slowly healing bite wound. (From Honig, 1983, with permission of Harcourt Health Communications, Philadelphia.)

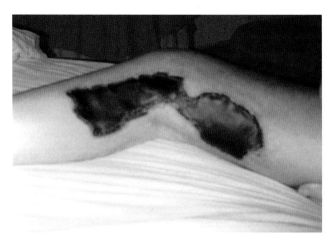

**FIGURE 22.15**    Severe case of envenomation by brown recluse spider, *Loxosceles reclusa,* on inner surface of leg of 19-year-old woman. The bite occurred at night while she was sleeping in bed. The large, black eschar denotes extent of tissue damage, as evident 3 months after the bite, when picture was taken. (Courtesy of Carolyn Grissom, Shelbyville, TN.)

time. The end result is typically a sunken scar varying in size from about 2 cm up to 10 cm or more.

A small percentage of victims of brown recluse spider bites experience systemic reactions, usually within 24–48 hr after envenomation. These may include fever, malaise, nausea, vomiting, joint pains, and a generalized pruritic rash. Occasionally the systemic symptoms can be even more serious in the form of hemolysis, intravascular coagulation, and renal failure.

Treatment of *L. reclusa* bite victims often entails administration of corticosteroids injected directly into the lesion; to be effective, however, this must be done within a few hours after the bite. If a limb is involved it should be elevated to help slow the spread of the venom by gravity. Other recommendations include cleansing the wound with hydrogen peroxide, applying hyperbaric oxygen to the ulcer, and using burn creams to alleviate pain. An alternative approach is early surgical excision of the affected skin in an effort to remove the venom before it can do further damage. Many physicians, however, are reluctant to do this because most bites are not destructive enough to warrant it. Others believe this to be a good approach, feeling that it is best to avoid the risk of disfiguring scars and subsequent skin grafts and other reconstructive surgery to repair more extensive tissue damage that could result. In cases of systemic reactions, the anticoagulant heparin can be administered to reduce the threat of intravascular coagulation. Aggressive therapy to counter hemolysis and the use of dialysis in cases of renal failure also may be required. *Antivenins* are available for treatment of *L. laeta* and other *Loxosceles* species in South America; however, despite the development of antivenin for treatment of *L. reclusa* bites in North America, its production has not proved to be commercially feasible.

Diagnosis of a brown recluse spider bite often can be difficult. Frequently the spider is not seen, or, if it is, it is not recovered for identification. In other cases the reaction is confused with fire ant stings, bee and wasp stings, assassin-bug or bed-bug bites, tick bites, skin abscesses and infections, slow-healing wounds of diabetic patients, pyodermic gangrenosa associated with rheumatoid arthritis patients, and various allergic responses. This is further compounded by the tendency of physicians and the general public to blame "bites" of unknown origin on spiders. Although some diagnostic tests have been developed, none of them have proved to be reliable for clinical use.

### South American violin spider *(Loxosceles laeta)*

*L. laeta* is the largest species in the genus and poses a significant health concern in Central America and South America. It closely resembles *L. reclusa*, from which it is generally distinguished by its more reddish coloration and by the fourth pair of legs of the female being longer than the others. In addition to its common names of *South American violin* or *brown recluse spider*, it is called *arañade los rincones*, or the *corner spider*, because of its occurrence indoors in the corners of rooms. This spider has been introduced to several parts of North America and Europe, where local populations have become established as far north as New England (United States) and Finland. In 1960 an infestation was discovered on the Harvard University campus at Cambridge, Massachusetts (USA), where it was believed to have been present for some 20 years. Established populations of *L. laeta* have been documented at several locations in southern California, where they also have been known to occur since the 1960s. At more northern locales, they tend to occur exclusively indoors, whereas in southern California they occasionally can be found in sheltered places outdoors.

Females produce multiple egg sacs, each containing about 50 eggs, which are deposited in a dense, cottony part of the web, usually at floor level. The number of egg sacs per female varies significantly and may be as high as 15 under laboratory conditions. In natural settings, females produce an average of 3–7 egg sacs following a single mating. Most eggs are produced during the spring and summer (October–January) in South America. The developmental time from egg hatch to adult ranges from as short as 6–8 months to a year or more. The adults are relatively long-lived, with mated and unmated females surviving about 3 and 4 years, respectively. Males live only about half as long as the females. Like other *Loxosceles* species, both sexes of *L. laeta* are able to survive prolonged periods without food and water, reportedly up to 2 years for some females. *L. laeta* often produces extensive webbing that is particularly noticeable in corners of rooms and along floor-level runways that they follow at the base of walls. These are composed of multiple layers of coarse silk, the amount of which reflects the degree of spider activity and duration of the infestation.

For many years before the cause was determined in 1947, skin lesions resulting from the bite of *L. laeta* in South America were known as *gangrenous spot syndrome*. The bite reaction is similar to that of *L. reclusa*, producing a necrotic lesion that heals slowly. However, it is more often accompanied by systemic effects that can be life-threatening. Such cases are referred to as *viscerocutaneous loxoscelism*, in which the lungs, kidneys, liver, and central nervous system may be damaged. The venom causes severe inflammatory, cytotoxic, necrotic, and degenerative changes in tissues, leading to fever, jaundice, blood or hemoglobin in the urine, and sensorial involvement. Recovery from severe cases of viscerocutaneous loxoscelism results in immunity to subsequent envenomation by this spider.

## LACTRODECTISM

The term *latrodectism,* also known as *neuromyopathic araneism,* is a syndrome caused by the bite of any of several *Latrodectus* species. The venom of these spiders contains potent neurotoxins which cause generalized pain, nausea, vomiting, faintness, dizziness, perspiration, and neuromuscular involvement in the form of muscle weakness, stiffness, cramps, tremors, incoordination, numbness or prickling sensations, paralysis, disturbed speech, and difficulty breathing. The main toxic fraction of the venom of *Latrodectus mactans* is a protein called α-latrotoxin, which acts on the motor nerve endings at the neuromuscular junctions. It causes depletion of the synaptic vesicles and the selective release of neurotransmitters that cause contraction of voluntary muscles. The autonomic nervous system is also affected. In severe cases the victims typically experience painful abdominal and leg cramps, profuse sweating, lachrymation, and spasms of the jaw muscles that distort the face and cause grimacing. Although symptoms often appear within 10–60 min, the syndrome may take several hours to develop; symptoms usually persist for 20–48 hr. A diagnostic sign of latrodectism is sweating at the bite site. This may be accompanied by localized swelling and redness, increased blood pressure, and the development of various types of rashes, either generalized or limited to the bite area. The fatality rate is relatively low (<5%) even in untreated cases.

Prompt treatment of *Latrodectus* bite victims significantly reduces the severity of symptoms and promotes recovery. Where practical, a tourniquet can be effective if applied within a few minutes after the bite to reduce spread of the venom. Immersing the body in a hot bath and administering morphine or muscle relaxants help to alleviate the pains and cramps. Particularly effective is the intravenous administration of a 10% solution of calcium gluconate or calcium lactate in saline, which affords welcomed relief in reducing muscle spasms and hypertension. *Antivenin* can be very effective as a treatment, especially in cases involving *L. mactans.* Alcohol should never be taken by the victim at any time during the course of the illness. Children are more likely to have serious reactions to *Latrodectus* bites and should receive medical attention as quickly as possible.

There are 16–30 species of *Latrodectus* worldwide; the number varies according to different taxonomists (Levi, 1959). Although there is evidence to support the recognition of several subspecies of *L. mactans* (e.g., *L. m. hasselti, mactans, menadovi,* and *tredecimguttatus*), these taxa are treated here as separate species. As a group they are medium-sized spiders, seldom more than 1.3 cm in body length, with globose abdomens and relatively long legs. They are generally recognized by their shiny black color and a red or orange hourglass marking on the underside of the abdomen. Some species, however, are more drab in appearance, or they may be colorfully patterned, as in the case of *L. bishopi.* They are known by a variety of common names in different parts of the world. In North America they are usually called *black widow spiders* and, less commonly, *hourglass spiders* or *button spiders.* The term "widow" is derived from the misconception that the females invariably devour the male after mating, whereas the term "button" refers to the resemblance of the shiny black, round abdomen of the female to the buttons on old-fashioned shoes. Other common names include *shoe-button spiders* in South Africa; *jockey* in Arabia; *karakurt* (black wolf) in Russia; *night stinger* or *katipo* in Australia and New Zealand; *la malmignatte* in the Mediterranean region (notably Italy and Corsica); *araña capulina, chintatlahua,* and *viuda negra* in Mexico; *culrouge* and *veinte-cuatro horas* in the West Indies; *lucacha* in Peru; *mico* in Bolivia; *guina* and *pallu* in Chile; and *araña del lino* (flax spider), *araña del trigo* (wheat spider), and *araña rastrojera* (stubblefield spider) in Argentina and other parts of South America.

*Latrodectus* species are shy, retiring spiders which construct their tangled webs of coarse silk in dark, undisturbed places, usually close to the ground. They are especially common under logs and stones and in abandoned animal burrows, crevices in protected earthen banks, and various materials stacked on the ground. Some species, however, build large, irregular aerial webs in shrubs and other vegetation, often up to a meter or more above the ground. The adults are primarily nocturnal, spending most of the day in the security of their silken retreats in protected recesses adjoining the web. Prey consists primarily of medium- to large-sized insects and other arthropods which stumble into the web.

Following mating, the female constructs one or more egg sacs, which she suspends in the web. Each sac typically contains 200–250 eggs. The total number of egg sacs produced by a given female varies considerably among species, with up to 10 for *L. mactans* and 20 for *L. hesperus.* The egg sacs are usually spherical or pyriform, white or grayish, with a tough, tightly woven outer covering. The eggs hatch in 14–30 days. The spiderlings undergo their first molt within the sac 3–4 days after hatching and then emerge via one or more tiny holes which they cut through the silken layers. The spiderlings remain in their mother's web for several weeks before dispersing. They undergo 4–9 molts, depending on the species and sex. Males reach maturity in 2–5 months, whereas females require somewhat longer, usually 3.5–8 months.

Spiderlings of *Latrodectus* species look quite different from the adult females. Young spiderlings are pale colored with light and dark stripes on the abdomen and legs. They often exhibit patterns of white, yellow, and red bands and spots, which are gradually lost in the females as they mature. The males, however, tend to retain the color pattern of the immatures, including the abdominal markings and leg bands. Upon reaching sexual maturity, the males leave their webs to wander in search of a female. Mating takes place in the female's web. Contrary to popular belief, females of most *Latrodectus* species do not kill and devour the males following insemination any more frequently than do most other spiders. In fact, in the case of *L. bishopi* the adult male and female actually live together in the same web.

Five *Latrodectus* species occur in North America: *L. bishopi, L. geometricus, L. hesperus, L. mactans,* and *L. variolus. L. mactans* is the most venomous and widespread of these species, causing most of the cases of lactrodectism requiring medical attention. *L. tredecimguttatus* is the most important species in Europe, whereas *L. curacaviensis, L. hasselti,* and *L. katipo* are important species in South America, Australia, and New Zealand, respectively.

### Red widow (*Latrodectus bishopi*)

This is the most colorful of the North American widow spiders. The cephalothorax and legs are orange or reddish, contrasted with a black abdomen which has red or orange spots with yellow borders. Rarely is the abdomen all black. It lacks the characteristic hourglass found in most *Latrodectus* species. Instead, the marking is pale and usually reduced to a transverse bar or a single triangular spot, representing one-half of an hourglass (Fig. 22.16). The red widow is known to occur only in the sand-pine scrub habitat of peninsular Florida. There it often builds its web in palmettos (Aracaceae), forming its cone-shaped, silk-lined retreat in a folded or rolled leaf frond which it ties together with silk. The snare itself is commonly 1–2 m above the ground, extending as a large, tangled web amidst the fronds to

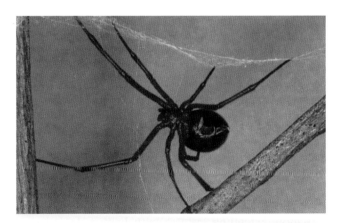

**FIGURE 22.16** Red widow spider, *Latrodectus bishopi* (Theridiidae), female, ventral view. This widow spider is unusual in lacking the red hourglass marking; the marking is typically reduced to a transverse bar or a single, triangular spot. (From Short and Castner, 1992, courtesy of University of Florida-IFAS.)

**FIGURE 22.17** Brown widow spider, *Latrodectus geometricus* (Theridiidae), female, dorsolateral view. (Photo by Sturgis McKeever.)

form, in part, a horizontal silken sheet reminiscent of sheet-web spiders. Very little is known about the potency of *L. bishopi* venom, and no bite cases have been documented.

### Brown widow (*Latrodectus geometricus*)

As its common name implies, *L. geometricus* females are brownish, rather than black, and have a highly variable color pattern that gives it a mottled appearance (Fig. 22.17). Typically there are three pairs of irregular-shaped spots along the dorsal midline of the abdomen. These vary from simple white spots with black borders to multicolored bull's-eye spots marked with white, yellow, orange, reddish brown, tan, gray, or aqua. The hourglass is generally dull orange, complete, and commonly bordered with yellow. Occasionally individuals are nearly black, more closely resembling other widow spiders.

The brown widow occurs most commonly throughout Brazil and along the eastern coast of South America. From there it has spread by cargo ships and other means of commerce to many other parts of the world, and it is now widely distributed in the Tropics. Introductions have led to established populations in peninsular Florida and Hawaii. This spider is found in a wide range of domestic settings in or near buildings. It constructs a relatively small tangled retreat in corners and chinks of brick and cement walls, building foundations, and fences, under overhangs of steps, and along street curbs. In addition to urban areas, *L. geometricus* is common in South America along ocean beaches, where it seems to prefer to construct its web in low, running plants such as the morning-glory *Ipomoea biloba* (Convolvulaceae) above the high-tide mark. It is even more shy than other widow spiders, is not aggressive, and rarely has been recorded biting humans. Some authors regard it as the least dangerous of the five widow spiders in North America.

### Southern black widow (*Latrodectus mactans*)

This is the most notorious of all venomous spiders because of its potent venom, widespread occurrence, and likelihood of coming into contact with humans and domestic animals. It is found throughout eastern North America from southern New England to eastern Mexico. It also occurs in the Bahamas, West Indies, South America, southern Europe, eastern Arabia, India, Australasia, the Pacific region in general, and Hawaii. The adult female typifies the general description of *Latrodectus* in being shiny black with a prominent red hourglass on the underside of the abdomen (Fig. 22.18). It usually has a small red spot just above the spinnerets and occasionally a median series of additional red spots extending anteriorly onto the abdominal dorsum. It is found outdoors in various protected places under rocks, logs, boards, and other ground debris and frequently near buildings. Unlike the other *Latrodectus* species in North America, *L. mactans* also occurs indoors in barns, wood sheds, garages, and various other unheated storage areas. Before the days of indoor plumbing, the widespread use of outdoor privies contributed significantly to the number of human cases of envenomation, especially in males bitten on the genitalia by black widow spiders in webs under the toilet seats.

Although the bite of *L. mactans* may go unnoticed in some cases, it is more commonly felt as a pinprick or an immediate sharp, burning pain with little or no swelling. The pain spreads from the bite site to regional lymph nodes and other parts of the body, usually reaching its maximum intensity in 1–3 hr. Thereafter the pain may be continuous or intermittent, lasting up to 48 hr. The accompanying muscular spasms and cramps, especially in the abdomen and legs, can lead to tightness of the chest, boardlike rigidity of abdominal muscles, and complete prostration. In extreme cases, complications may occur in the form of shock, leukocytosis, and lesions of the liver, spleen, and kidney, evidenced by blood and elevated protein levels in the urine. The death rate can be as high as 4–5% in untreated cases, the highest of all *Latrodectus* species.

### Northern black widow (*Latrodectus variolus*)

The northern black widow closely resembles *L. mactans* in both size and general appearance. It is usually distinguished, however, by the ventral hourglass being divided into two transverse bands and a row of prominent red spots along the dorsal midline of the abdomen. Its geographic range largely overlaps that of *L. mactans* in North America; it occurs widely throughout the eastern United States as far south as western Florida and eastern Texas. It is more common in the northern states, extending into southeastern Canada. Unlike *L. mactans,* it seldom is found in buildings, preferring outdoor situations such as old stumps, piles of dead tree branches, hollow logs, abandoned animal burrows, cavities in rock walls, and under debris. It is common in wooded areas, where it constructs a large tangled web in shrubs and tree branches, sometimes as high as 6 m above the ground.

### Western black widow (*Latrodectus hesperus*)

This species also is very similar in appearance to *L. mactans.* The abdominal dorsum is typically all black, only infrequently having red markings. The ventral red hourglass is well defined and usually complete. It occurs from Oklahoma, Kansas, and central Texas into the southwestern United States and adjacent Mexico. It also is found in Israel as an introduced species. The western black widow utilizes a wider range of habitats than *L. mactans.* It most commonly constructs its web near the ground, in animal burrows, or under various objects. However, it also is found above the ground in shrubs and trees and in such places as grape arbors and bird nests. It is well adapted to semiarid and arid habitats, where it can be found in soil crevices and various desert plants such as agaves and cacti.

### European black widow (*Latrodectus tredecimguttatus*)

*L. tredecimguttatus* is the most common widow spider in Europe. It occurs primarily in the Mediterranean region, extending eastward into Eurasia. In Italy and Corsica it is called *la malmignatte,* and in Russia, *karakurt,* or "black wolf." It occurs exclusively outdoors, where it constructs its web in a variety of herbaceous and shrubby vegetation, including cultivated

**FIGURE 22.18**   Southern black widow spider, *Latrodectus mactans* (Theridiidae), female, ventral view, showing characteristic red hourglass marking. (From Gertsch, 1979.)

crops and arbors. This species poses a significant occupational hazard for farmers and field workers, who are commonly bitten while harvesting or threshing wheat, handling hay, picking fruit, or working in vineyards.

### Araña del trigo (*Latrodectus curacaviensis*)

In South America, *L. curacaviensis* is frequently encountered by humans in buildings, garages, and privies. It also is found in cultivated crops such as wheat, accounting for its common name *araña del trigo*, or *wheat spider*. It closely resembles *L. mactans*, with which it is easily confused. Like *L. mactans*, its bite can cause serious illness and death.

### Australian red-back spider (*Latrodectus hasselti*)

Females of this Australian species are 2–3 cm long. They are black with a prominent red stripe along the dorsal midline of the abdomen (Fig. 22.19), hence the common name *red back*. Some authors regard this spider as a subspecies of *L. mactans*. It occurs in sheltered, preferably dry sites such as hollows of trees and under logs and rocks. It often builds its retreat and web around building foundations and in ventilator gratings, trash cans, and gas-meter boxes. Although this spider is not aggressive, the female causes a painful bite that can lead to systemic envenomation and fatalities (ca. 5%) if untreated. In typical cases, the initial bite is relatively painless, comparable to a pinprick. Thereafter the pain intensifies from a few minutes to a half hour, often accompanied by localized perspiration and edema, nausea, and vomiting. Localized sweating at the bite site is an important diagnostic sign (Wiener, 1961). In untreated systemic cases, recovery is usually protracted and may take 3–4 weeks. The severity of red-back spider bites was dramatically reduced in Australia with the introduction of an antivenom against *L. hasselti* in 1956; since that time, no fatalities have been recorded.

### Katipo spider (*Latrodectus katipo*)

This spider is primarily a coastal species found high on ocean beaches and in river beds under driftwood or at the base of vegetation. It occurs in the Caribbean region (Jamaica, British West Indies) and New Zealand, where it is known as the *New Zealand redback*.

## VETERINARY IMPORTANCE

Under certain circumstances spiders can pose health threats to household pets, livestock, and other domestic animals. Most cases occur either in stables or in pasture situations where animals are grazing. In the latter situation, a high density of a venomous species can cause significant veterinary concerns. Although the evidence of spider bites in nonhuman animals often is circumstantial, enough cases have been documented to show that theridiid and theraphosid spiders are the more common causes of serious spider bites of veterinary importance.

Most cases of envenomation by spiders in grasslands and pastures are caused by *Latrodectus* species. They usually occur in localized areas where populations of certain venomous taxa are high. Outbreaks involving *L. erebus* were reported in the steppes of southern Russia in the 1830s. Grazing animals were severely bitten, causing some to stampede due to the pain and to run until they dropped. Fatalities as high as 12% in sheep, 17% in horses, and 33% in camels were reported (Motchoulsky and Becker, 1855). Notable outbreaks of latrodectism affecting cattle and agricultural workers also occurred in Spain in the 1830s and 1840s and caused deaths in horses, sheep, and other livestock in Chile in the 1870s. For other examples of latrodectism involving goats, sheep, cattle, and horses in Europe, South Africa, and Indonesia, see Maretic and Lebez (1979). Latrodectism does not seem to present a significant problem for livestock in North America or South America.

**FIGURE 22.19** Australian red-back spider, *Latrodectus hasselti* (Theridiidae), female, dorsal view, showing characteristic, wide, reddish stripe along dorsal midline of abdomen. (Photo by G. R. Mullen.)

Envenomation by *Badumna robusta,* a member of the cribellate family Desidae, has been reported as a problem in stables in Queensland, Australia. Known as the *black house spider* or *window spider, B. robusta* is nocturnal and has been known to bite horses while they are bedded down for the night. It can inflict a painful bite with associated inflammation and swelling, usually about the head and neck (Southcott, 1976).

Tarantulas have been implicated in several reported cases of envenomation of pastured animals in Central America in recent years. *Aphonopelma seemani* and *Sphaerobothria hoffmani,* two theraphosid species that construct deep burrows in pasture soils, are believed to be the cause of necrotic lesions in cattle and horses in Costa Rica (Herrero and Bolanos, 1982). These and other theraphosids in the Americas are suspected as the cause of vesicular lesions on the muzzle, hooves, and udder of cattle, horses, and swine. Although the implications are unclear, the venom of *Tarentula cubensis* (Lycosidae) in South Africa has been shown to have pharmacological activity in treating foot problems called *pododermatitis circumspecta* in cattle and gangrenous udder in goats (Stampa, 1986).

Spider-bite cases in cats, dogs, and other household animals are rarely reported. Presumably this reflects their uncommon occurrence and the relatively minor problems that such bites cause, seldom warranting veterinary attention (Taylor and Greve, 1984). Envenomation by adults of *L. mactans* is known to cause acute persistent hypertension in cats. In contrast, extracts of the eggs and spiderlings of the same species have been shown to induce a potentially fatal hypotension in cats, indicating significantly different toxins in the immature and adult spiders (Ebeling, 1978).

## PREVENTION AND CONTROL

The best way to avoid being bitten by spiders is not to handle them or allow direct contact while working in areas where they are likely to occur. Gloves can be worn to protect the hands while gardening, potting plants, stacking or handling firewood, moving rocks and other ground materials, or being involved in various other outdoor activities. Spiders can be discouraged from constructing webs around windows and under eaves of buildings by regular removal of their webs with a broom and turning off unnecessary lights at night that attract insect prey. Lumber, trash, garden materials, or other items piled next to buildings should be removed, and cluttered areas in basements, attics, and outbuildings which can serve as protected sites for spiders should be eliminated. Windows and doors should be tightly closed or adequately screened to prevent wandering spiders from entering homes during their activity periods. Regular removal of cobwebs and vacuuming the corners of rooms, window frames and sills, baseboards, and the underside of furniture help to prevent spiders from establishing themselves indoors. The elimination of household insects on which indoor spiders depend for food also helps to reduce infestations.

If necessary, pesticides registered for indoor control of spiders can be used. They are available in the form of sprays, dusts, aerosols, and fogs which help to reach spiders that may escape sweeping and vacuuming by retreating into cracks and crevices. Products with residual activity can be applied to outdoor surfaces such as under eaves, crawl spaces under houses, and around decks and patios.

To avoid bites by black widow spiders and related *Latrodectus* species, precautions should be taken to minimize contact with them in wood piles, under rocks and logs, in outbuildings, and in other sheltered, outdoor sites. Avoid putting unprotected hands into recesses at ground level, such as water-meter encasements and accumulations of trash and other debris that can serve as retreats. Special precaution should be taken when using outdoor privies and portable toilets to avoid contact with webs of black widow spiders on the underside of toilet seats.

In the case of *Loxosceles laeta, L. reclusa,* and *L. rufescens,* which occur indoors, appropriate measures can be taken to minimize the risk of being bitten in infested premises. Visually inspect clothes closets, water-heater closets, utility rooms, basements, attics, and other storage areas, killing and removing any specimens that are found. Inspect dry, relatively undisturbed hiding places such as old boxes, wooden cabinets, behind wall hangings, and under beds and other furniture. Shake out clothing that has been hanging unused for some time to avoid encounters with spiders in sleeves and trouser legs. Inspect shoes and other footwear before putting them on, and visually check towels or shake them before use. Where infants or small children are involved, cribs or beds should be pulled away from the wall, and the bed clothes should be kept from reaching the floor.

Established *Loxosceles* populations can be difficult to eliminate without the use of pesticides to reach individuals that are not accessible to sweeping and vacuum cleaning. Sprays of appropriate materials can be applied in attics, crawl spaces, along baseboards, in corners of rooms, under furniture, behind cabinets, and in other out-of-the-way places to kill the spiders. Treating any webs that are found with pesticide dusts generally helps to reduce infestations more quickly than applying materials to other surfaces. Sticky traps can be used to catch *Loxosceles* species and to monitor their activity before or after treatments.

# REFERENCES AND FURTHER READING

Akre, R. D., and Catts, E. P. (1992). Spiders. Insect Answers Series, EB1548. Washington State University Cooperative Extension. 8 pp.

Akre, R. D., and Myhre, E. A. (1991). Biology and medical importance of the aggressive house spider, *Tegenaria agrestis*, in the Pacific Northwest (Arachnida: Araneae: Agelenidae). *Melanderia* **47**, 1–30.

Anonymous. (1996). Necrotic arachnidism—Pacific Northwest, 1988–1996. *Morbidity and Mortality Weekly Report* **45**, 433–436.

Bettini, S., Ed. (1978). "Arthropod Venoms. Handbook of Experimental Pharmacology." Vol. 48. Springer-Verlag, New York.

Bettini, S., and Brignoli, P. M. (1978). Review of the spider families, with notes on the lesser-known poisonous families. In "Arthropod Venoms. Handbook of Experimental Pharmacology" (S. Bettini, Ed.), Vol. 48, pp. 103–120. Springer-Verlag, New York.

Bettini, S., and Maroli, M. (1978). Venoms of Theridiidae, genus *Latrodectus*. In "Arthropod Venoms. Handbook of Experimental Pharmacology" (S. Bettini, Ed.), Vol. 48, pp. 149–184. Springer-Verlag, New York.

Bonnet, P. (1945–1961). "Bibliographia Araneorum." Frères Douladoure, Toulouse.

Breene, R. G. (1996). Arachnid common names in North America. *Newsletter of the British Arachnology Society* **77**, 1–4.

Brignoli, P. M. (1983). "A Catalogue of the Araneae Described between 1940 and 1981." Manchester Univ. Press, Dover, NH.

Bucherl, W. (1971). Spiders. In "Venomous Animals and Their Venoms" (W. Bucherl and E. Buckley, Eds.), Vol. 3, pp. 197–277. Academic Press, New York.

Coddington, J. A., Larcher, S. F., and Cokendolpher, J. C. (1990). The systematic status of Arachnida, exclusive of Acari, in North America north of Mexico. In "Systematics of North American Insects and Arachnids: Status and Needs" (M. Kosztarab and C. W. Schaefer, Eds.), pp. 5–20. Virginia Agriculture Experiment Station Information Series 90-1. Virginia Polytechnic Institute and State University, Blacksburg, VA.

Comstock, J. H. (1948). "The Spider Book." Cornell Univ. Press, Ithaca, NY.

Cooke J. A. L., Roth, V., and Miller, F. H. (1972). The urticating hairs of theraphosid spiders. *American Museum Novitiates* **2498**, 1–43.

Duchen, L. W., and Gomez, S. (1984). Pharmacology of spider venoms. In "Handbook of Natural Toxins" (A. T. Tu, Ed.), Vol. 2, pp. 483–512. Dekker, New York.

Ebeling, W. (1978). "Urban Entomology." Univ. California, Berkeley.

Fink, L. S. (1984). Venom spitting by the green lynx spider, *Peucetia viridans* (Araneae, Oxyopidae). *Journal of Arachnology* **12**, 372–373.

Foelix, R. F. (1992). "Biologie der Spinnen." Thieme, Stuttgart.

Foelix, R. F. (1996). "Biology of Spiders, 2nd ed." Oxford Univ. Press, New York.

Geren, C. R., and Odell, G. V. (1984). The biochemistry of spider venoms. In "Handbook of Natural Toxins" (A. T. Tu, Ed.), Vol. 2, pp. 441–481. Dekker, New York.

Gertsch, W. J. (1979). "American Spiders," 2nd ed. Reinhold, New York.

Gertsch, W. J., and Ennik, F. (1983). The spider genus *Loxosceles* in North America, Central America and the West Indies (Araneae, Loxoscelidae). *Bulletin of the American Museum of Natural History* **175**, 263–360.

Gray, M. R. (1988). Aspects of the systematics of the Australian funnel web spiders (Araneae: Hexathelidae: Atracinae) based upon morphological and electrophoretic data. In "Australian Arachnology" (A. D. Austin and N. W. Heather, Eds.), Miscellaneous Publication No. 5, pp. 113–125. Australian Entomological Society, Brisbane.

Gray, M. R., and Sutherland, S. K. (1978). Venoms of Dipluridae. In "Arthropod Venoms. Handbook of Experimental Pharmacology" (S. Bettini, Ed.), Vol. 48, pp. 121–148. Springer-Verlag, New York.

Herrero, M. V., and Bolanos, R. (1982). Life-history and tunnels of 2 'horse-biting' spiders from Costa Rica (Araneae: Theraphosidae). Preliminary observations (in Spanish). *Brenesia* **19-20**, 319–324.

Hickin, N. E. (1984). "Pest Animals in Buildings." Godwin, London.

Honig, P. J. (1983). Bites and parasites. *Pediatric Clinics of North America* **30**, 563–581.

Kaston, B. J. (1970). Comparative biology of American black widow spiders. *Transactions of the San Diego Society of Natural History* **16**, 33–82.

Kaston, B. J. (1978). "How to Know the Spiders," 3rd ed. Wm. C. Brown, Dubuque, IA.

Levi, H. W. (1959). The spider genus *Latrodectus* (Araneae: Theridiidae). *Transactions of the American Microscopical Society* **78**, 7–43.

Levi, H. W. (1970). Probable envenomation by *Chiracanthium mildei*; a spider found in houses. *American Journal of Tropical Medicine and Hygiene* **19**, 729–732.

Lucas, S. (1988). Spiders in Brazil. *Toxicon* **26**, 759–772.

Maretic, A., and Lebez, D. (1979). "Araneism with Special Reference to Europe." Nolit, Belgrade.

McCrone, J. D., and Levi, H. W. (1964). North American widow spiders of the *Latrodectus curacaviensis* group (Araneae: Theridiidae). *Psyche* **71**, 12–27.

Millikan, L. E. (1984). Biology of spiders. In "Mammalian Diseases and Arachnids" (W. B. Nutting, Ed.), Vol. 1, pp. 59–81. CRC Press, Boca Raton, FL.

Motchoulsky, V., and Becker, N. (1855). Bulletin de Moscou, as cited by Maretic, A., and Lebez, D. (1979). "Araneism with Special Reference to Europe." Nolit, Belgrade.

Newlands, G., and Atkinson, P. (1988). Review of southern African spiders of medical importance, with notes on the signs and symptoms of envenomation. *South African Medical Journal* **73**, 235–239.

Ori, M. (1984). Biology of and poisoning by spiders. In "Handbook of Natural Toxins" (A. T. Tu, Ed.), Vol. 2, pp. 397–440. Dekker, New York.

Platnick, N. I. (1989). "Advances in Spider Taxonomy 1981–1987. A supplement to Brignoli's *A Catalogue of the Araneae Described between 1940 and 1981*." Manchester Univ. Press, Manchester.

Platnick, N. I. (1993). "Advances in Spider Taxonomy 1988–1991. With Synonymies and Transfers 1940–1980" (P. Merrett, Ed.). New York Entomological Society, New York.

Preston-Mafham, R., and Preston-Mafham, K. (1984). "Spiders of the World." Facts on File Publications, New York.

Raven, R. J. (1985). The spider infraorder Mygalomorphae (Araneae): cladistics and systematics. *Bulletin of the American Museum of Natural History* **182**, 1–180.

Roewer, C. F. (1942–1954). "Katalog der Araneae." Natura, Bremen.

Roth, V. D. (1993). "Spider Genera of North America with Keys to Families and Genera and a Guide to Literature," 3rd ed. American Arachnological Society, Gainesville, FL.

Russell, F. E. (1970). Bites by the spider *Phidippus formosus* Case history. *Toxicon* **8**, 193–194.

Schenberg, S., and Pereira Lima, F. A. (1978). Venoms of Ctenidae. In "Arthropod Venoms. Handbook of Experimental Pharmacology" (S. Bettini, Ed.), Vol. 48, pp. 217–246. Springer-Verlag, New York.

Schenone, H., and Suarez, G. (1978). Venoms of Scytodidae, genus *Loxosceles*. In "Arthropod Venoms. Handbook of Experimental Pharmacology" (S. Bettini, Ed.), Vol. 48, pp. 247–275. Springer-Verlag, New York.

Short, D. E., and Castner, J. L. (1992). Venomous spiders of Florida. Leaflet No. SP 104. University of Florida, Institute of Food and Agricultural Sciences.

Southcott, R. V. (1976). Arachnidism and allied syndromes in the Australian regions. *Records of the Adelaide Children's Hospital* **1**, 97–187.

Southcott, R. V. (1984). Diseases and arachnids in the tropics. In "Mammalian Diseases and Arachnids" (W. B. Nutting, Ed.), Vol. 2, pp. 15–56. CRC Press, Boca Raton, FL.

Stampa, S. (1986). A field trial comparing the efficacy of sulphamonomethoxine, penicillin, and tarantula poison in the treatment of pododermatitis circumspecta of cattle. *Journal of the South African Veterinary Association* **57**, 91–93.

Sutherland, S. K. (1987). Watch out, Miss Muffet! *Medical Journal of Australia* **147**, 11–12, 531.

Sutherland, S. K. (1990). Treatment of arachnid poisoning in Australia. *Australian Family Physician* **19**, 1, 17, 50–55, 57–61, 62.

Taylor, S. P., and Greve, J. H. (1984). Suspected case of loxoscelism (spider-bite) in a dog. *Iowa State University Veterinarian* **47**, 84–86.

Tinkham, E. R. (1946). A poison-squirting spider. *Bulletin of the US Army Medical Department* **5**, 361–362.

Tu, A. T. (1984). "Handbook of Natural Toxins." Vol. 2. Insect Poisons, Allergens, and Other Invertebrate Venoms. Dekker, New York.

Vest, D. K. (1987). Necrotic arachnidism in the Northwest United States and its probable relationship to *Tegenaria agrestis* (Walckenaer) spiders. *Toxicon* **25**, 175–184.

White, J., Hirst, D., and Hender, E. (1989). 36 cases of bites by spiders, including the white-tailed spider, *Lampona cylindrata*. *Medical Journal of Australia* **150**, 401–403.

Wiener, S. (1961). Red back spider bite in Australia: an analysis of 167 cases. *Medical Journal of Australia* **2**, 44–49.

Wong, R. C., Hughes, S. E., and Voorhees, J. J. (1987). Spider bites. *Archives of Dermatology* **123**, 98–104.

# 23

# MITES (*Acari*)

GARY R. MULLEN AND BARRY M. OCONNOR

*More than* 250 species of mites are recognized as the cause of health-related problems for humans and domestic animals. Types of problems include (1) temporary irritation of the skin due to bites or feeding on host skin, fur, and feathers; (2) persistent dermatitis in response to mites invading the skin or hair follicles; (3) mite-induced allergies; (4) transmission of pathogenic microbial agents and metazoan parasites; (5) intermediate hosting of parasites, notably tapeworms; (6) invasion of respiratory passages, ear canals, and occasionally internal organs; (7) an abnormal fear of mites, or *acarophobia;* and (8) *delusory acariosis,* a psychological condition in which individuals are convinced that they are being attacked by mites when, in fact, no mites are involved. The general term for infestations of animals by mites is called *acarinism,* whereas any disease condition caused by mites is *acariasis* (*acarinosis*).

For an introduction to mites in general, see Krantz (1978), Woolley (1988), or Evans (1992). For major works dealing specifically with taxa of medical-veterinary importance, the following sources are suggested: Hirst (1922), Baker *et al.* (1956), Strandtmann and Wharton (1958), Sweatman (1971), Yunker (1973), Nutting (1984), and Baker (1999).

## TAXONOMY

Based on the classification scheme described by Evans (1992), mites comprise the arachnid subclass Acari, which in turn is divided into two major groups: the Anactinotrichida and Actinotrichida. These are further subdivided into seven orders (Table I). The use of alternative names for these orders and the designation of the orders as suborders by various authors causes understandable confusion for those not familiar with mite classification. For discussions of the higher classification of mites, see van der Hammen (1972), Krantz (1978), Kethley (1982), OConnor (1984), and Evans (1992).

Members of the orders Ixodida, Mesostigmata, Prostigmata, Astigmata, and Oribatida are the cause of animal health problems. Most of them are represented by the 30 families listed in Table II and covered in this chapter. Not included in the list are the many families of ectoparasitic mites found on wild mammals, birds, and

TABLE I

**Higher Classification of Mites as Presented by Evans (1992): Alternative Names for the Respective Orders Are Shown in Parentheses**

Class Arachnida
Subclass Acari

    Superorder Anactinotrichida (Parasitiformes)
      Orders
        Notostigmata (Opilioacarida)
        Holothyrida (Tetrastigmata)
        Ixodida (Metastigmata)
        Mesostigmata (Gamasida)

    Superorder Actinotrichida (Acariformes)
      Orders
        Prostigmata (Actinedida + Tarsonemida)
        Astigmata (Acaridida)
        Oribatida (Cryptostigmata, Oribatei)

TABLE II

**Orders and Families of Mites That Include Species of Medical Veterinary Importance or Interest: Families Are Listed Alphabetically under Each Order**

| | |
|---|---|
| Order Ixodida | Order Astigmata |
|   Argasidae |   Acaridae |
|   Ixodidae |   Analgidae |
| |   Atopomelidae |
| Order Mesostigmata |   Carpoglyphidae |
|   Ascidae |   Cytoditidae |
|   Dermanyssidae |   Dermoglyphidae |
|   Entonyssidae |   Echimyopodidae |
|   Halarachnidae |   Epidermoptidae |
|   Laelapidae |   Gastronyssidae |
|   Macronyssidae |   Glycyphagidae |
|   Raillietidae |   Hypoderatidae |
|   Rhinonyssidae |   Knemidokoptidae |
|   Spinturnicidae |   Lemurnyssidae |
| |   Listrophoridae |
| Order Prostigmata |   Lamninosioptidae |
|   Cheyletidae |   Myocoptidae |
|   Demodicidae |   Pneumocoptidae |
|   Ereynetidae |   Proctophyllodidae |
|   Harpyrhynchidae |   Psoroptidae |
|   Myobiidae |   Pterolichidae |
|   Pyemotidae |   Pyroglyphidae |
|   Psorergatidae |   Rhyncoptidae |
|   Syringophilidae |   Sarcoptidae |
|   Tarsonemidae |   Syringobiidae |
|   Trombiculidae |   Turbinoptidae |
| | |
| | Order Oribatida |
| |   Ceratozetidae |
| |   Galumnidae |
| |   Oribatulidae |
| |   Scheloribatidae |

reptiles, most of which cause little or no significant harm to their hosts. Given the number and diversity of taxa involved, there is no single source to which one can turn for identification of mites of medical-veterinary importance. Reliable species determinations usually require the preparation of slide-mounted specimens for microscopic examination and the assistance of an acarologist who has access to the appropriate taxonomic literature. A reference that the nonspecialist can use for identifying some of the more common mite pests of public health and veterinary importance is the *CDC Pictorial Keys: Arthropods, Reptiles, Birds and Mammals of Public Health Significance* (Pratt and Stojanovich, 1969).

## MORPHOLOGY

The basic body plan of mites is shown in Fig. 23.1. The body is divided into two major regions: the anterior *gnathosoma*, bearing the pedipalps and chelicerae; and the *idiosoma*, the remainder of the body, bearing the legs and eyes (when present). The *pedipalps* are typically five-segmented but may be greatly reduced and highly modified in different groups of mites. The pedipalps are primarily sensory appendages equipped with chemical and tactile sensors that assist mites in finding food and perceiving environmental cues. In some groups they may be modified as raptorial structures for capturing prey or as attachment devices to facilitate clinging to hosts. The

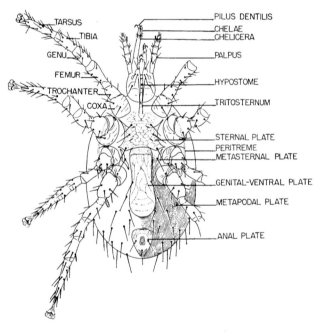

**FIGURE 23.1** Generalized morphology of adult female mite (Mesostigmata, Laelapidae, *Androlaelaps fahrenholzi*), ventral view. (From Baker *et al.*, 1956.)

mouthparts consist primarily of a pair of *chelicerae*, each of which is typically three-segmented and terminates in a *chela*, or pincer. The chela is composed of a fixed digit and a movable digit designed for seizing or grasping. In the case of certain parasitic mites the chelicerae are highly modified as long, slender structures for piercing skin to feed on blood and other host tissues. In some groups, structures associated with the mouthparts may be modified as attachment devices to help secure them to their hosts (e.g., chiggers).

The idiosoma can be divided into several regions. The anterior part bearing the legs is the *podosoma*. The posterior section behind the legs is the *opisthosoma*. Other regions include the *propodosoma*, that portion of the idiosoma bearing the first and second pairs of legs, and the *hysterosoma*, extending from just behind the second pair of legs to the posterior end of the body. The designation of these body regions is helpful to morphologists and taxonomists in locating specific setae and other structures. The size and arrangement of sclerotized plates, the chaetotaxy, and the nature and location of sensory structures on the idiosoma serve as important taxonomic characters.

Mites typically have four pairs of legs as nymphs and adults, but they have only three pairs as larvae. The legs are divided into the following segments: *coxa, trochanter, femur, genu, tibia, tarsus,* and *pretarsus*. The pretarsus commonly bears a pair of claws, a single median *empodium*, and, in certain groups, a membranous *pulvillus*. These structures are highly variable among different groups of mites and aid in movement or in clinging to various surfaces, including hosts.

The *respiratory systems* of mites often include tracheal ducts which supplement the exchange of oxygen, carbon dioxide, and other gases across the body surface. The presence of spiracular openings associated with the tracheal ducts, commonly denoted by a sclerotized plate, or *stigma* (plural stigmata), and their location on the body provide important taxonomic characters for recognizing the acarine suborders. In the Prostigmata, for example, the stigmata are located at the anterior margin of the idiosoma. In the Mesostigmata they are usually located dorsolaterally to the third or fourth pairs of legs, whereas in the Oribatida (Cryptostigmata) they are typically hidden, opening ventrolaterally near the bases of the second and third pairs of legs. Tracheal systems and spiracular openings are lacking in the Astigmata.

*Reproductive structures* are very diverse among mites, providing important characters for distinguishing the sexes and identifying taxa. *Sperm transfer* may be *direct* (e.g., insemination by transfer of sperm via the male *aedeagus* to the sperm storage organ, or *spermatheca*, of the female) or *indirect* (e.g., transfer of sperm via the male chelicerae to the female genital opening).

A few points should be mentioned regarding the internal morphology of mites that are pertinent to species of medical-veterinary importance. The *digestive system* handles primarily liquefied food that has been preorally digested by enzymes secreted in the saliva. The paired *salivary glands* are typically located in the anterior portion of the idiosoma and open via ducts into the mouth region of the gnathosoma. In addition to digestive enzymes, these glands secrete anticoagulants in hematophagous mites. In certain groups (e.g., chiggers) they may produce cementing substances to help anchor the mouthparts in host skin. Also important in certain acarine groups are *coxal glands*. They are derived from excretory structures and serve primarily in osmoregulation. Waste products, in the form of *guanine*, are excreted by one or two pairs of long, slender *Malpighian tubules* that open into the alimentary tract just anterior to the hindgut.

## LIFE HISTORY

The basic developmental stages in the life history of mites are the egg, prelarva, larva, protonymph, deutonymph, tritonymph, and adult. Depending on the taxonomic group, one or more stages may be suppressed, resulting in a wide range of life-history patterns (e.g., chiggers; Fig. 23.2). Eggs may be deposited externally or may be

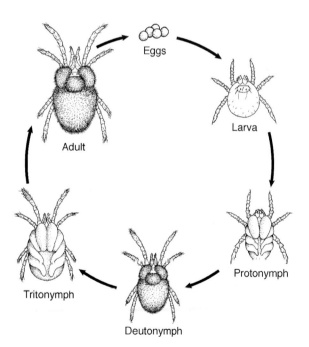

**FIGURE 23.2** Developmental stages in life cycle of chiggers (Trombiculidae). The protonymph and tritonymph are inactive stages passed within the cuticle of the engorged larva and deutonymph, respectively. (Original by Rebecca L. Nims.)

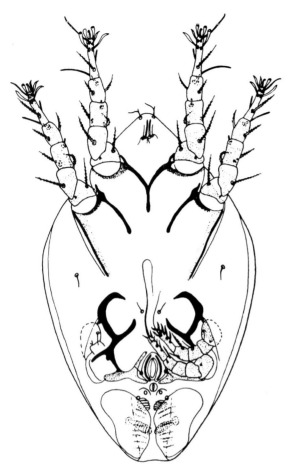

**FIGURE 23.3** Hypopus (deutonymph) of *Glycyphagus hypudaei* (Glycyphagidae), ventral view. This is a specialized, typically phoretic, nonfeeding stage that lacks functional mouthparts. (From Whitaker, 1982; original Fain, 1969.)

retained in the uterus until hatching. The *prelarva* is a nonfeeding, quiescent stage that may or may not have legs, mouthparts, or other distinct external features. The *larva* is typically an active form which molts to produce a nymph. The nymphs usually resemble the adults of a given taxon except for their smaller size, pattern of sclerotization, and chaetotaxy. The deutonymph of certain astigmatid mites is noteworthy in that it is highly modified morphologically as a nonfeeding stage adapted for surviving adverse environmental conditions. Such deutonymphs are called *hypopi* (singular *hypopus*) or *hypopodes*. They often have specialized clasping structures modified from ventral suckers (Fig. 23.3) that enable them to adhere to phoretic hosts, which aid in carrying them to more favorable sites where they can continue their development. Certain of these deutonymphs may become parasites in hair follicles or subcutaneous tissues of mammals and birds.

The developmental times from egg to adult and the number of generations per year are too variable to make meaningful generalizations. It is therefore important to understand the developmental biology and life-history patterns of individual groups and species. For further information on the development and life history of mites, see Krantz (1978), Woolley (1988), Schuster and Murphy (1991), and Houck (1994).

## BEHAVIOR AND ECOLOGY

Because of the diversity of behavioral and ecological aspects of mites, no attempt is made to discuss them here. Instead they are addressed, where relevant, in the accounts of individual groups and species of mites that follow. For an overview of feeding, mating and reproduction, oviposition, and dispersal behavior of mites, see Woolley (1988) and Evans (1992).

## PUBLIC HEALTH IMPORTANCE

Mites can adversely affect human health in many ways. They can infest homes, including carpets, mattresses and bedding, clothing, stored food products, and household pets. Usually they remain unnoticed unless individuals in the household become sensitized and develop various allergies upon subsequent exposure to these mites. Other mites which normally parasitize nonhuman hosts can cause dermatitis in humans when they bite the skin in efforts to feed on blood or other tissues. Most commonly involved in such cases are mite associates of rodents and birds that infest the premises. Such problems typically occur when the natural hosts have died or departed, forcing the mites to seek an alternative food source. A similar situation occurs outdoors when the parasitic larval stage of trombiculid mites, known as chiggers or redbugs, attempts to feed on human skin. Humans are not their normal hosts and often experience intense local skin reactions where these mites attach.

Mites also can pose occupational hazards for farmers, field hands, mill workers, warehouse operators, and others who handle mite-infested materials such as straw, hay, and grains. The mites involved normally feed on fungi, plant materials, or various arthropods; however, on contact with humans, they can pierce the skin, sometimes causing severe dermatitis. Other mites actually invade human skin, either burrowing through cutaneous tissues (e.g., scabies mites) or infesting the hair follicles and associated dermal glands (follicle mites). Infestations of these mites can cause persistent, sometimes severe, forms of dermatitis.

In addition to the temporary discomfort or annoyance that mites can cause, some mites are responsible for more serious or chronic medical problems. A number of species may be inhaled or ingested, causing infestations of the respiratory tract and digestive system. Mites even are reported to have been recovered from bile of patients suffering from chronic cholecystitis (inflammation of the gallbladder) and occasionally from the urinogenital tract.

The most widely recognized mite problems affecting human health are respiratory allergies caused by mites infesting house dust. In sensitized individuals, this can lead to chronic respiratory stress, bronchitis, and asthma. There are very few human diseases, however, that involve pathogens transmitted by mites. The most important is tsutsugamushi disease (scrub typhus or chigger-borne rickettsiosis), which occurs primarily in southeastern Asia, Australia, and Pacific islands. The only significant mite-borne disease in the New World is rickettsialpox, reported in the northeastern United States.

For convenience of discussion, problems of a public health nature caused by mites can be grouped into the following categories: mite-induced dermatitis, respiratory allergies, storage-mite allergies, internal acariasis, mite-borne human diseases, and acarophobia or delusory parasitosis. The mites involved represent at least 18 families in the 3 suborders Mesostigmata, Prostigmata, and Astigmata.

## MITE-INDUCED DERMATITIS

Species in approximately 14 families of mites are known to cause dermatitis in humans. In many cases these represent encounters with species that infest stored products. In other cases, they are ectoparasites of other animals, notably rodents, nesting birds, and poultry. Species in only 2 families (Demodicidae and Sarcoptidae) utilize humans as their normal hosts. Skin reactions to the feeding or burrowing of these mites range from minor, localized irritation at individual bite sites to severe dermal responses in individuals who become sensitized to specific mite antigens. Still others are free-living, predatory mites that may bite on contact with human skin.

### Ascidae

The only ascid species thought to be involved in a human case of dermatitis is *Proctolaelaps pygmaeus,* reported in New Zealand by Andrews and Ramsay (1982). This species is probably cosmopolitan and represents an incidental case. Its bites can cause red, papular lesions where the mite pierces the skin.

### Dermanyssidae

Members of both recognized genera of dermanyssid mites have been reported biting humans: *Dermanyssus*

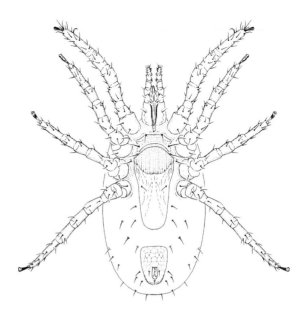

**FIGURE 23.4**  Chicken mite, *Dermanyssus gallinae* (Dermanyssidae), female, ventral view. (Modified from Gorham, 1991 courtesy of the US Department of Agriculture.)

and *Liponyssoides.* They are ectoparasites primarily on wild and domestic rodents and birds. These mites feed on the blood of their hosts by piercing the skin with their long, slender, extrusible chelicerae with highly reduced chelae at their tips. Dermanyssids spend most of their time in the nests of their hosts, crawling onto the animals primarily to feed. When they come into contact with human skin, they are prone to bite, typically causing erythematous papules at each puncture site that are often accompanied by intense itching.

### Chicken mite *(Dermanyssus gallinae)*

Also called the *red poultry mite,* this cosmopolitan species (Fig. 23.4) is the most common dermanyssid mite that bites people. It parasitizes a very broad range of hosts. This mite is especially a problem in the Palaearctic region and in the United States, where most cases occur in poultry houses or around buildings where pigeons, house sparrows, or starlings are nesting. The mites live in nesting materials, where they spend most of their time, moving onto the birds to feed on blood at night. Consequently, workers in poultry operations seldom experience a biting problem while working during the daytime, even when the houses are heavily infested. However, individuals who enter infested buildings at night may be readily bitten. Occasionally, pet canaries and parakeets also serve as sources of human infestations.

The term *pigeon mite* refers to *D. gallinae* when it infests pigeons or rock doves. The mites frequently enter buildings from pigeon roosts or nests. This tends to happen in the late spring and early summer months when the young pigeons fledge and the nests are abandoned, forcing the mites to seek alternative hosts. Although most human bites occur at night, bites may occur during the daytime when buildings are darkened.

A number of cases have been reported in hospitals and other institutional settings where employees and patients have been bitten by *D. gallinae*. The sources of the problem generally can be traced to nesting birds, notably pigeons, on windowsills, on ledges, under eaves, in air-intake ducts, or on air-conditioners mounted on the outside walls. The mites enter rooms around windows and doors, through crevices and cracks, or via ventilation ducts and air-conditioning systems. In other situations, they may drop onto individuals from roosting or nesting birds in ceilings or from overhead sites on porches and walkways near buildings. In such cases, close inspection may reveal mites crawling on clothing, furniture, or bed linens, particularly at night, when the mites are active.

Human infestations with *D. gallinae* have been variously called *chicken tick rash*, *bird mite disease*, *psora dermanyssica*, *pseudogale*, and *gamasidosis*. The term *fowl mite dermatitis* is likewise used but also can be applied to skin reactions caused by other avian mites that attack people.

Most bites tend to occur on the arms and chest protected by clothing, rather than on exposed skin such as the hands and face. Only in exceptional cases do bites occur in the axillary and pubic areas. The bites are usually painful and typically result in red maculopapular skin lesions on the upper portions of the body and extremities. While a human is being bitten, close examination will reveal the mite as a tiny red speck at the center of the papule. Occasionally the bites produce vesicles, urticarial plaques, and diffuse erythema, with dermatographia being frequently seen. In multiple-bite cases, a pruritic rash may develop and persist until the source of the infestation has been eliminated. Itching tends to be most intense at night. The problem is usually resolved by treatment with antihistamines or topically applied steroids, combined with moving individuals from affected areas.

St. Louis encephalitis (SLE), eastern equine encephalitis (EEE), and western equine encephalitis (WEE) viruses have been isolated from *D. gallinae* infesting wild birds. However, conflicting evidence has been reported regarding the ability of *D. gallinae* to transmit any of these viruses among birds or to humans.

### American bird mite *(Dermanyssus americanus)*
This mite is very closely related to *D. gallinae* but only rarely has been reported biting humans. It can cause acute, generalized, eczematous dermatitis which is easily misdiagnosed as other skin disorders unless the presence of mites is confirmed. WEE virus has been isolated from these mites infesting nests of the house sparrow, but its significance in transmission or maintenance of WEE virus is unknown.

### *Dermanyssus hirundinis*
This hematophagous mite is a common ectoparasite of certain birds, especially swallows (*Hirundo* spp.) and the house wren (*Troglodytes aedon*) in North America and Europe. It is not unusual for hundreds or thousands of these mites to infest individual nestling birds. At least one case has been documented in Europe of *D. hirudinis* biting a human and causing urticarial dermatitis (Dietrich and Horstmann, 1983).

### House mouse mite *(Liponyssoides sanguineus)*
This mite (Fig. 23.5), referred to in the earlier literature as *Allodermanyssus sanguineus*, is an ectoparasite of domestic and wild rodents. It commonly parasitizes mice, including the house mouse (*Mus musculus*),

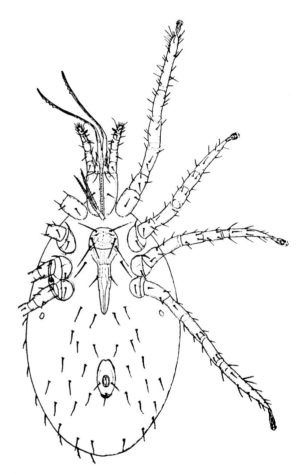

FIGURE 23.5 House mouse mite, *Liponyssoides sanguineus* (Dermanyssidae), female, ventral view. Note the pair of long, attenuated, extruded chelicerae with serrated tips for piercing skin to feed on blood. (Modified from Baker *et al.*, 1956)

in the United States and the spiny mouse (*Acomys* spp.) in North Africa. It occurs less commonly on rats (*Rattus* spp.), voles (*Microtus* spp.), and other rodents in localized areas of eastern North America, Europe, Asia, and Africa. It is primarily of interest to medical entomologists because of its role as the vector of *Rickettsia akari,* the etiologic agent of *rickettsialpox* in humans.

Like most other dermanyssid mites, the house mouse mite lives in nesting materials, where it spends most of its time, crawling onto host animals to feed. Its life cycle and behavior are similar in many respects to those of *D. gallinae.* Females oviposit in rodent nests or along rodent runways 2–5 days after feeding on host blood. The eggs hatch in 4–5 days to produce larvae which do not feed but instead molt to protonymphs about 3 days later. The protonymphal stage lasts 4–5 days, during which time the mite takes a blood meal, usually engorging in less than an hour, and then molts to the deutonymph. The deutonymph lives about 6–10 days and requires a blood meal before transforming to the adult. The developmental time from egg to adult normally takes 2–3 weeks. After feeding, blood-engorged females leave the rodent host and can be found in the nests and runways, along the walls of infested premises, and especially in warmer areas of buildings, such as furnace and incinerator rooms.

## Macronyssidae

Macronyssid mites are blood-feeding ectoparasites on reptiles, birds, and mammals. Five species account for most of the cases of medical interest. Three of these are *Ornithonyssus* species infesting rodents or birds; *Chiroptonyssus* is parasitic on bats, whereas *Ophionyssus* is parasitic on snakes.

### Tropical rat mite (*Ornithonyssus bacoti*)

This cosmopolitan mite (Fig. 23.6) is a parasite of rats, particularly the black rat (*Rattus rattus*), and other rodents in both tropical and temperate regions. In cooler climates this mite occurs only indoors and in nests of wild rodents. Occasionally it also infests carnivores, birds, and humans. When rats are killed or abandon their nests or runways, the mites are left behind and will readily crawl or drop onto humans and other passing animals. Rodents killed by household cats and left near human dwellings also can serve as a source of infestation. The mites are active and can move some distance from their source to enter nearby buildings.

Human bite cases involving the tropical rat mite usually occur in rodent-infested buildings. Using their long, slender chelicerae, they probe the skin in an effort to feed on blood. In some cases they produce a prickling sensation at the bite sites, whereas in other cases

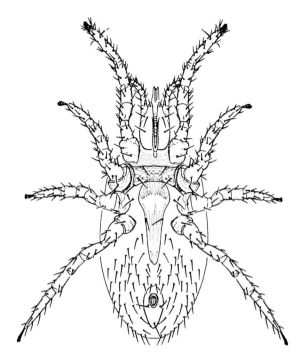

**FIGURE 23.6** Tropical rat mite, *Ornithonyssus bacoti* (Macronyssidae), female, ventral view. (Modified from Gorham, 1991; courtesy of the US Department of Agriculture.)

the bite is painful. Multiple bites are often clustered and subsequently develop into a pruritic, erythematous, papular rash. This may be accompanied by localized swelling and occasional vesicle formation. Although they will bite almost any part of the body, they tend to bite where the clothing is tight; e.g., neck, shoulders, and waist. *O. bacoti* is visible to the unaided eye and may be seen crawling on the clothing or skin, floors, walls, and other structural surfaces.

This mite has been shown experimentally to be capable of being infected with, or transmitting, several human pathogens, including those that cause murine typhus, rickettsialpox, plague, tularemia, and coxsackie virus disease. However, their importance in the epidemiology of these diseases is regarded as negligible. On the other hand, recent evidence supports the possibility that *O. bacoti* may serve as both a vector and reservoir of *Hantaan virus,* the causative agent of *Korean hemorrhagic fever (epidemic hemorrhagic fever)* of humans in Asia.

### Tropical fowl mite (*Ornithonyssus bursa*)

As its common name implies, the tropical fowl mite (Fig. 23.7) is distributed widely throughout subtropical and tropical parts of the world, where it parasitizes various domestic and wild birds. It occurs in the eastern and southern United States, Hawaii, Central America, Colombia, South Africa, India, China, and

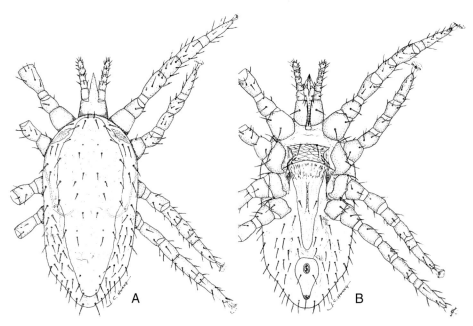

**FIGURE 23.7**  Tropical fowl mite, *Ornithonyssus bursa* (Macronyssidae), female. (A) Dorsal view; (B) ventral view. (From Strandtmann and Wharton, 1958)

Australia. In addition to being a poultry pest, notably attacking chickens, it parasitizes pigeons, house sparrows, grackles, and other wild avian hosts. Human bite cases can usually be traced to nesting birds under eaves, on ledges, and in other building structures. The mites spend most of their time in the nest, moving onto the host to feed.

Human bites typically occur when young birds leave the nest or the nest is otherwise abandoned, compelling the mites to wander in search of alternative hosts. Wild birds carrying the mites commonly infest poultry operations, leading to workers being bitten when they come into contact with infested, commercially produced birds or their nest materials. While less

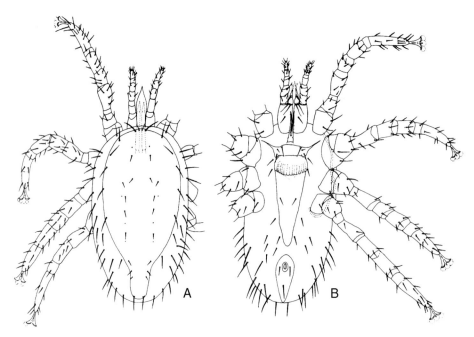

**FIGURE 23.8**  Northern fowl mite, *Ornithonyssus sylviarum* (Macronyssidae), female. (A) Dorsal view; (B) ventral view. (Modified from Strandtmann and Wharton, 1958)

common than cases involving the tropical rat mite, human bites due to the tropical fowl mite can be equally discomforting. The latter tend to be more sharply painful and result in more persistent itching. Although Western Equine Encephalitis virus has been isolated from *O. bursa* in house sparrow nests, there is no evidence that this mite actually transmits the virus. No other human pathogen has been associated with this mite.

## Northern fowl mite (*Ornithonyssus sylviarum*)

The northern fowl mite (Fig. 23.8) is widely distributed throughout temperate regions of the world as a parasite of domestic fowl and wild birds. Although regarded as a major pest of chickens, it occasionally bites people. Most human cases result when poultry workers handle infested birds or when the mite enters buildings from nearby bird nests. While bite cases may occur year-round in commercial poultry operations, most cases in homes and other work places occur about the time young birds fledge and the adults vacate their nests. The bite reaction is similar to that of *O. bursa,* producing red papular skin lesions that are often accompanied by intense itching.

The viruses that cause Western equine encephalitis and St. Louis encephalitis have both been detected in *O. sylviarum* from nests of wild birds in North America. In the case of WEE virus, it has been shown to persist in avian hosts and to be transmitted by bite to other birds. Newcastle disease virus has been detected in *O. sylviarum* following its feeding on infected chickens, but the virus does not establish persistent infection in the mite. The northern fowl mite does not appear to play a significant role in the natural transmission of these or other arboviruses affecting humans.

## Free-tailed bat mite (*Chiroptonyssus robustipes*)

This mite is a common blood-feeding ectoparasite on the Brazilian free-tailed bat (*Tadarida braziliensis*) roosting or nesting in walls and attics of buildings throughout its range in the southern United States, Mexico, and Central America. Occasionally it is found in low numbers on other molossid bats roosting with *T. braziliensis.* Only the protonymphs and adult mites feed on blood and other tissue fluids of their bat hosts. Human bite cases are uncommon and usually involve zoologists handling infested bats or otherwise working with mite-infested bat colonies. The mites do not readily bite and are primarily a nuisance as they actively crawl about on the skin and clothing. On rare occasions *C. robustipes* has been reported to invade the living quarters of homes, where it has bitten the occupants. One such case involved an 18-month-old boy in California who was bitten repeatedly on the face and abdomen, causing a persistent dermatitis. The problem was not resolved until it was discovered that a wall infestation of Brazilian free-tailed bats was the source of mites that were biting the child each time he was bathed in a bathroom sink (Keh, 1974).

## Snake mite (*Ophionyssus natricis*)

This mite (Fig. 23.9) is a common pest of captive snakes and only rarely has been reported biting people.

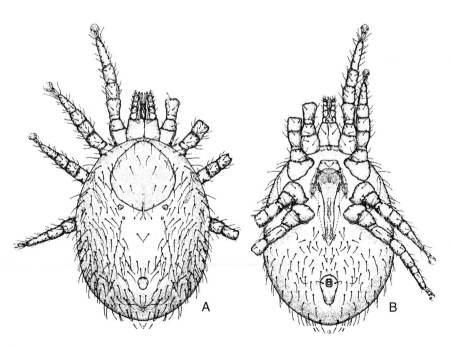

**FIGURE 23.9** Snake mite, *Ophionyssus natricis* (Macronyssidae), female. (A) Dorsal view; (B) ventral view. (Modified from Strandtmann and Wharton, 1958.)

Human bites occur primarily in reptile houses at zoological parks, affecting personnel who handle infested snakes. A well-documented case involved several members of a family in a household where a python was kept as a pet (Schultz, 1975). The family members had experienced skin lesions in the form of a papular rash on the forearms and other parts of the body over a 5-month period before the source of the problem was identified. Mites were observed to be attached to the skin while attempting to feed and also were found in a chair frequented by the snake. Humans do not serve as suitable hosts for this mite. The mites tend to become immobile with their legs curled underneath the body within a few minutes after they begin feeding on human blood; often they do not recover. They do not transmit any known human pathogens.

### Laelapidae

Members of this family include both free-living and parasitic species, often associated with rodents and other nest-building mammals. The only significant laelapid species that may affect human health is the spiny rat mite. Occasionally other species cause temporary discomfort to humans, as reported in possible cases of *Haemogamasus pontiger* causing dermatitis in England (Theiler and Downes, 1973).

   The laelapid *Haemogamasus liponyssoides* is an obligate blood feeder on wild rodents which has the potential for transmitting human disease agents, even though it has not been reported to bite people (Furman, 1959). Other rodent-associated laelapid mites may play a role in the transmission of *Hantaan virus,* the causative agent of *Korean hemorrhagic fever,* based on isolation of this human pathogen from *Laelaps jettmari* in Korea (Traub *et al.,* 1954).

### Spiny rat mite *(Laelaps echidninus)*

   The spiny rat mite is a common hematophagous ectoparasite of domestic rats throughout the tropical and temperate zones. Although it is capable of biologically transmitting disease agents, such as the agent of murine typhus, among wild rodents, its potential role as a vector of human pathogens remains uncertain. *Junin virus,* which causes *Argentinian hemorrhagic fever,* has been isolated from *L. echidninus* and associated rodent hosts in South America (Parodi *et al.,* 1959; Theiler and Downes, 1973).

### Trombiculidae

Larvae of members of the family Trombiculidae are called *chiggers*. This is the only parasitic stage in the life cycle of trombiculid mites. As a group, they feed on a wide variety of vertebrate hosts, including amphibians, reptiles,

**FIGURE 23.10**   Larval stage (chigger) of the harvest mite, *Neotrombicula autumnalis* (Trombiculidae), of Europe. (A) Dorsal view; (B) scutum (dorsal plate), showing characteristic arrangement of setae. (Modified from Baker *et al.,* 1956.)

birds, and mammals, with humans serving only as accidental hosts.

   The life history of trombiculid mites includes the following sequence of stages: egg, prelarva, larva, protonymph, deutonymph, tritonymph, and adult (Fig. 23.2). The *eggs* are typically laid in soil or ground debris. After about 6 days, the egg shell splits to expose an inactive nonfeeding stage, the *prelarva*. After another 6 days, the active six-legged larva (i.e., chigger) is produced (Fig. 23.10). After successfully attaching to a suitable host, the larva generally feeds 3–5 days on the host before dropping to the ground to form an inactive transitional stage, the *protonymph*. This stage in turn develops into the active eight-legged *deutonymph*. The deutonymph subsequently undergoes development as another quiescent stage, the *tritonymph,* to produce the eight-legged *adult*. The deutonymph and adult are free-living predators that feed on small arthropods (e.g., collembolans) and their eggs. The duration of the life cycle requires 2–12 months or longer, depending on the species and environmental conditions. In temperate areas, there may be one to three generations per year, whereas in tropical regions generations may be continuous throughout the year.

   Although trombiculid larvae usually cause little or no apparent harm to their normal hosts, they often cause

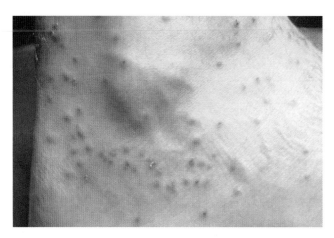

**FIGURE 23.11** Multiple chigger bite lesions on human ankle and foot. (Photo by Elton J. Hansens.)

dermatitis when they attach to and attempt to feed on humans and other atypical hosts. Such an infestation by trombiculid larvae is called *chigger dermatitis,* or *trombiculosis* (*trombidiosis* of the older literature).

Chiggers are just large enough (150–300 $\mu$m) to be visible to the unaided eye. They are yellowish, orange, or red and can be seen on close inspection at the center of the skin lesions they induce. Unfortunately, chiggers are often encountered in large numbers, resulting in multiple bites (Fig. 23.11). Given their preference for attaching where clothing fits snugly against the skin, the bites tend to be concentrated about the ankles, lower legs, and waist and along the elastic borders of undergarments.

Contrary to popular belief, chiggers do not burrow into the skin of their hosts. Instead they attach by piercing the epidermis with their chelicerae and feed externally. Because of their small size and tiny mouthparts, chiggers usually attach where the skin is thin or soft. A preferred site is the opening to hair follicles. There they insert their capitulum to feed on the thin epidermal lining. In humans, this results in inflammation at the point of attachment and localized swelling of the skin around the chigger, giving the mistaken impression that it has burrowed into the skin. Their food consists primarily of partially digested skin cells and lymph broken down by saliva introduced at the attachment site. They do not feed on blood. Feeding is facilitated by formation of a feeding tube, or *stylostome,* produced by the interaction of the saliva with surrounding host tissue.

With the exception of *Leptotrombidium* species, chiggers often do not survive more than 1–2 days on human hosts, due to the adverse host reaction they cause and injury or removal due to scratching. By then, however, the damage is already done, typically producing a discrete, persistent, itching, reddened papule at each attachment site. The lesions persist for several days but may take several weeks to heal if they become secondarily infected. The recovery time can be significantly shortened by prompt treatment to kill the chiggers when they are first detected, generally within 3–6 hr following attachment; application of a topical medication to alleviate itching and prevent infection; and avoidance of scratching or otherwise excoriating the skin.

More than 50 species of trombiculid mites have been recorded attacking humans. Of this number, about 20 species are considered to be medically important, either due to the dermatitis they cause or due to their role in transmission of disease agents. Four species of particular interest are *Eutrombicula alfreddugesi* in North America and South America, *Neotrombicula autumnalis* in Europe, and *Leptotrombidium akamushi* and *L. deliense* in the Orient.

### Eutrombicula alfreddugesi

This is the most common and widespread trombiculid mite in the Western Hemisphere, occurring from Canada to Argentina and in the West Indies. In North America, it and related species that attack humans are known as *red bugs,* especially in the southeastern United States. In Mexico it is called *tlalzahuatl,* and in Mexico and other parts of Latin America it is called *coloradilla* and *bicho colorado. E. alfreddugesi* is parasitic as larvae on a variety of amphibians, reptiles, birds, and mammals. It is particularly common in areas of secondary growth, such as shrub and brush thickets and blackberry and bramble (*Rubus* spp.) patches, along margins of swamps, and ecotones between woodlands and open fields or grasslands. The larvae are present in late summer and early fall in the more temperate parts of its range and throughout the year in the tropics and subtropics, including southern Florida. Although it is the most common cause of chigger dermatitis in the New World, it is not involved in the natural transmission of any human disease agent.

### Neotrombicula autumnalis

Known as the *harvest mite,* this is the most common chigger that attacks humans in Europe and the British Isles (Fig. 23.10). Other names include *aoutat* and *lepte automnal. N. autumnalis* is particularly annoying during harvest time in late summer and fall. The larvae are present from July to the onset of winter, usually reaching peak populations in early September. They tend to be most active on warm, sunny days in grasslands, cultivated grain fields, brush lands, and thickets. The widespread occurrence of this mite reflects its wide variety of natural hosts, especially mammals and certain ground-dwelling birds. Rabbits are a particularly common host. Other hosts include voles, wood mice, hedgehogs, squirrels, cattle, sheep, goats, horses, dogs, cats, pheasants, partridges, chickens, and other domestic fowl.

### *Eutrombicula* spp.

*E. splendens* occurs in the eastern United States from the Gulf Coast north to Massachusetts, Minnesota, and Ontario, Canada. It is especially common in the southeastern United States, where it is second only to *E. alfreddugesi* as the cause of trombiculosis in humans. Although it occurs in drier habitats with *E. alfreddugesi*, it is especially abundant in moist habitats such as swamps, bogs, and low-lying areas with rotting stumps and fallen trees. The larva is parasitic on amphibians, reptiles, birds, and mammals but seems to prefer snakes and turtles as natural hosts. The seasonal occurrence of *E. splendens* is similar to that of *E. alfreddugesi*. Another *Eutrombicula* species that causes chigger dermatitis of humans in the United States is *E. lipovskyi*. It is restricted to moist areas, generally characterized by an abundance of decaying logs and stumps bordering swamps and streams. It occurs from Alabama and Tennessee west to Arkansas, Oklahoma, and Kansas. Reptiles, rodents, and birds serve as hosts.

### *Leptotrombidium* spp.

Several members of the genus *Leptotrombidium* serve as vectors of *Orientia tsutsugamushi*, the causative agent of tsutsugamushi disease. They occur widely throughout Southeast Asia and the southwestern Pacific islands. As a group, the larvae of medically important *Leptotrombidium* species are parasitic primarily on ground-dwelling rodents (e.g., *Rattus*, *Microtus*, and *Apodemus* spp.). Other hosts include insectivores, marsupials, cattle, dogs, and cats. They occur in forests, second-growth areas along the margins of woodlands, in river valleys, and in abandoned agricultural fields where populations of rodents flourish. The principal vectors of the tsutsugamushi disease agent are *L. deliense* in Southeast Asia, the southwestern Pacific islands, and northern Australia; *L. akamushi* in Japan; *L. arenicola* and *L. fletcheri* in the Pacific islands; and *L. pallidum*, *L. pavlovskyi*, and *L. scutellare* in more restricted regions of the Asian mainland, Japan, and Malaysia (see Table III).

## STORED-PRODUCTS MITES

Members of several families of mites that infest unprocessed and processed plant materials can cause human dermatitis and other health-related problems. Most cases involve people handling infested materials such as grains, flour, hay, straw, dried fruits, and vegetables. Others involve processed materials of animal origin such as meats,

**TABLE III**

**Major Chigger Vectors of *Orientia tsutsugamushi*, the Causative Agent of Tsutsugamushi Disease in Humans: All are *Leptotrombium* Species (source, Kawamura *et al.*, 1995)**

| Trombiculid species | Geographic occurrence |
|---|---|
| *Leptotrombidium akamushi* | Japan |
| *L. arenicola* | Malaya, Indonesia, Thailand |
| *L. deliense* | Southeast Asia, China, southwestern Pacific islands, northern Australia, Pakistan |
| *L. fletcheri* | Malaysia, New Guinea, Philippines, Indonesia, Melanesia |
| *L. pallidum* | Japan, Korea, Primorye region of Russia |
| *L. pavlovskyi* | Primorye region of Russia |
| *L. scutellare* | Japan, China, Thailand, Malaysia |

hides, cheeses, dried milk, and other dairy products. Such mite infestations are the cause of *occupational acarine dermatitis* in farmhands, granary operators, warehouse workers, and other personnel.

The stored-products mites responsible for most human cases of acarine dermatitis are members of the families Acaridae, Pyemotidae, and Cheyletidae.

### Acaridae and Other Astigmata

Acarid mites infest a wide range of stored materials such as grains, milk products, dried fruits, straw, and animal hides in both households and commercial storage facilities. They also are common contaminants of culture media in which insects and other invertebrates are reared; in bedding materials for mice, guinea pigs, hamsters, and other vertebrates; and in animal feed and animal-holding cages in pet stores and zoos. Their numbers can build rapidly, especially when the infested materials are damp enough to support the growth of fungi on which they typically feed. Dermatitis occurs when the mites pierce the skin in attempts to feed or obtain moisture. The reaction in some cases also may involve contact allergens.

The most important acarid mite in stored products is *Acarus siro* (Fig. 23.12), a species found throughout most of the world. It is particularly a pest of processed cereal products (e.g., flour), rather than whole grains or hay. The females are 350–650 $\mu$m in length, with a colorless body and yellow-to-brown gnathosoma and appendages. It can develop at temperatures of 24–32°C and a relative humidity greater than 60%. This mite tends to congregate where the relative humidity is 80–85%, at which its reproductive rate is highest. The amount of damage it causes to grains is directly related to the moisture content; the germ is attacked only when the water content is 14% or

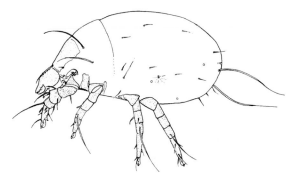

**FIGURE 23.12** *Acarus siro* (Acaridae), female, lateral view. (Modified from Hughes, 1976.)

higher. The dermatitis experienced by food handlers on contact with *A. siro* is commonly known as *grocer's itch*. Other names for dermatitis caused by *A. siro* and related mites are *baker's itch*, *dried-fruit-mite dermatitis* (*Carpoglyphus lactis*, Carpoglyphidae; Fig. 23.13), *wheat pollard itch* (*Suidasia nesbitti*; family Suidasiidae), and *vanillism*, reflecting the product or commodity involved.

A mite closely related to *A. siro* that also causes human dermatitis is *A. farris* (Fig. 23.14). It is a widespread

**FIGURE 23.14** *Acarus farris* (Acaridae), female, ventral view. (Modified from Hughes, 1976.)

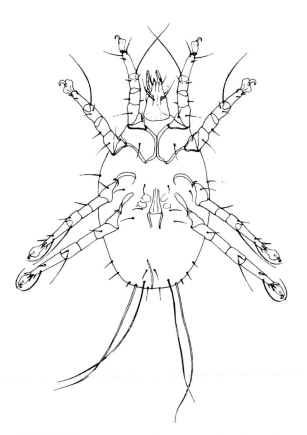

**FIGURE 23.13** *Carpoglyphus lactis* (Carpoglyphidae), male, ventral view. (Modified from Hughes, 1976.)

species that has been reported to cause skin irritation to farm workers handling infested bales of hay in England (Hughes, 1976).

Another common acarid mite in stored products is the cosmopolitan *Tyrophagus putrescentiae*. It is particularly a problem in foods with a high protein and fat content, such as hams, cheeses, nuts, seeds, dried eggs, and fish meal. This mite feeds primarily on fungi (e.g., *Aspergillus*, *Eurotium*, *Penicillium*) that tend to thrive on foods stored at warm temperatures (>30° C) and relatively high humidities (>85%). Under such conditions it can complete its development from one generation to the next in 2–3 weeks. It can be a pest in mycology laboratories, where it often contaminates fungal cultures. The term *mold mite* is commonly used to refer to *T. putrescentiae* and a closely related species, *T. longior* (Fig. 23.15).

In the Tropics, *T. putrescentiae* causes a dermatosis called *copra itch* among workers handling copra, dried coconut kernels from which coconut oil is extracted. In Italy, human cases of cutaneous and respiratory allergies have been attributed to this mite among workers handling raw hams; the mite apparently thrives in the white dust (*ruffino*) that covers hams during the seasoning process (Ottoboni *et al.*, 1989). *T. putrescentiae* occurs throughout much of the world, where it is found in a wide range of situations, including grasslands, soil, old hay,

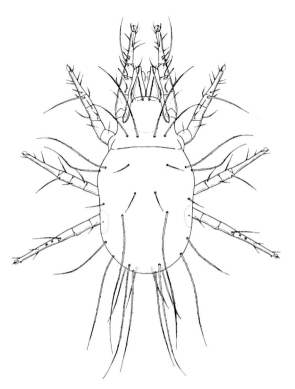

**FIGURE 23.15**  *Tyrophagus longior* (Acaridae), male, dorsal view. (Modified from Hughes, 1976.)

mushrooms, and the nest of bees and ducks. This mite was reported as the cause of human dermatitis in a butcher's shop in Austria, where it was breeding in molds growing on bacon in a poorly ventilated room (Czarnecki and Kraus, 1978).

A few other acarid mites are known to cause human dermatitis. One is *Tyrolichus casei*, reported by Henschel (1929). This is a cosmopolitan species commonly found in stored foods, grains, flour, cheeses, dogmeal, old honey combs, and insect collections (Hughes, 1976). Another is *Suidasia nesbitti*, which occurs in Europe, Africa, North America, and the West Indies. Although it is particularly associated with wheat pollards and bran in England (Hughes, 1976), it also has been reported in rice, in whale meat infested with dermestid beetles, in dried bird skins, and in milking machinery.

### Pyemotidae

Pyemotid mites are ectoparasites of insects that typically attack the larval stage of moths, beetles, and hymenopterans. A few species commonly occur in dried, insect-infested plant products such as hay, straw, and grains. On contact with humans and other animals, these mites cause intense itching when they pierce the skin with their stylet-like chelicerae and inject a toxin produced in their

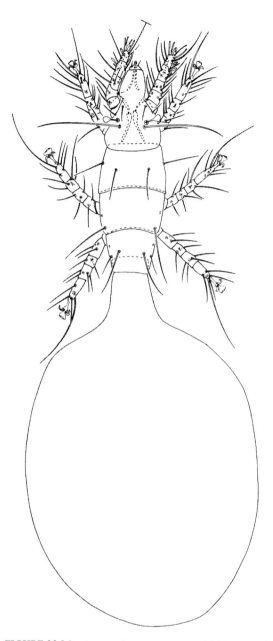

**FIGURE 23.16**  Straw itch mite, *Pyemotes tritici* (Pyemotidae), gravid female, dorsal view. (From Gorham, 1991; courtesy of the US Department of Agriculture.)

salivary glands. It is a very potent neurotoxin which they use to immobilize their insect prey, enabling them to paralyze insects 150,000 times their size (Tomalski and Miller, 1991).

The most important species affecting humans is *Pyemotes tritici* (Fig. 23.16). It is variously known as the *straw itch mite, hay itch mite,* and *grain itch mite,* depending on the plant material with which it is associated. Exposure to *P. tritici* represents an occupational hazard for agricultural workers, sales and stock personnel in farm supply stores, and other individuals in the arts and crafts field who handle wheat, hay, and straw. People handling infested materials usually develop multiple skin lesions in the form of papules or papulovesicles, accompanied by intense itching. Each bite site typically consists of a minute white wheal with a central erythematous area where a tiny vesicle forms. During the early stages, the mite often is visible as a tiny white speck where the vesicle is located. Although lesions can occur on any exposed part of the body, they usually appear on the back, abdomen, and forearms, where contact with infested materials typically takes place. Lesions seldom occur on the face or hands. Heavily infested, or sensitized, individuals may experience other symptoms, including headache, fever, nausea, vomiting, diarrhea, and asthma (Southcott, 1984). Less commonly reported are chills, fever, malaise, and anorexia (Betz *et al.,* 1982).

Two other species of *Pyemotes* reportedly cause human dermatitis. Several people in France developed erythematous lesions and complained of intensely itchy papules after handling dried everlasting flowers (*Helichrysium angustifolium*) infested with *P. zwoelferi* imported from Yugoslavia (Le Fichoux *et al.,* 1980). Other people working in a food-mixing shed of a farrowing house developed a papular rash after contact with grain infested with *P. herfsi* in Czechoslovakia.

### Cheyletidae

Cheyletid mites are mostly free-living predators that commonly feed on other mites and small arthropods in stored products. Occasionally they cause pruritic dermatitis in people handling infested grains and other dried plant materials. The most common cheyletid found in stored products is *Cheyletus eruditus.* This cosmopolitan species has been used commercially as a biological control agent to reduce the numbers of grain mites, notably *Acarus siro* and *Lepidoglyphus destructor,* in granaries and agricultural warehouses. Severe pruritus was reported in a worker at a wholesale florist shop handling fern wreaths imported to the United States from the Philippines (Shelley *et al.,* 1985). The mite involved was apparently *C. malaccensis,* a species previously shown to cause itching papules in humans (Yashikawa *et al.,* 1983). A second cheyletid mite, *Cheyletomorpha lepidopterorum,* also may have been involved.

## Skin-Invading Mites

Representatives of only two families of mites typically invade human skin or associated dermal structures and glands. They are the Demodicidae, or *follicle mites,* and the Sarcoptidae, or *scabies mites.* Whereas only a relatively small number of humans infested with follicle mites develop clinical problems, most individuals who become infested with the human scabies mite experience an annoying, often severe, dermatitis.

### Demodicidae

Members of this family are called *follicle mites.* They are extremely tiny, elongate, annulate mites with very short, stout, three-segmented legs (Fig. 23.17). They lack body setae and possess a pair of tiny, needle-like chelicerae which are used to pierce dermal cells, on which the mites feed. Their minute size and strong reduction of most of the external features represent adaptations for living in the close confines of hair follicles and associated ducts and glands.

Two species of *Demodex* infest humans. *Demodex folliculorum* (Fig. 23.17) occurs primarily in hair follicles, whereas *D. brevis* is generally found in the sebaceous glands that open via ducts into the hair follicles. Both species may infest the same host, appearing together in samples taken from a given individual. Adults of the two

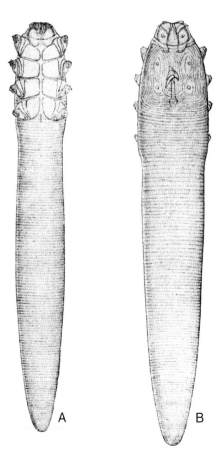

**FIGURE 23.17**    Human follicle mite, *Demodex folliculorum* (Demodicidae) (A) Female, ventral view; (B) male, dorsal view. (From Hirst, 1919.)

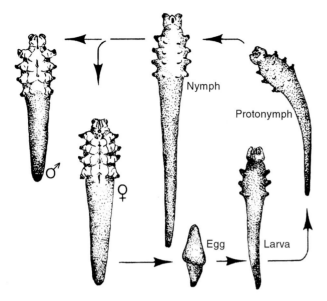

**FIGURE 23.18**   Life cycle of human follicle mite, *Demodex folliculorum* (Demodicidae). (Modified from Nutting, 1984.)

species closely resemble one another but can be distinguished based on the general body shape and relative size of the males and females. *D. folliculorum* females have elongate bodies that are gently tapered from the podosoma to the slender, rounded caudal end. *D. brevis* females have bodies that are usually widened posterior to the podosoma and terminate in a more broadly rounded caudum. The eggs are also distinctive, being spindle-shaped in *D. folliculorum* and oval in *D. brevis.*

The entire life cycle of *D. folliculorum* (Fig. 23.18) and *D. brevis* is spent on their human host. The mites feed by piercing host cells with their styletiform chelicerae and drawing the cell contents into the esophagus with a pumping action of the pharynx. They are highly host-specific and can survive only on humans. Transfer of mites from one individual to another is presumed to occur primarily between mothers and infants during the intimacy associated with facial contact and nursing. Adults of both sexes are readily transferred between hosts at these times. The fact that 90–100% of all humans apparently harbor follicle mites attests to the success with which such transfers are accomplished.

Human follicle mites tend to occur primarily in the regions of the forehead, eyelids, and nose. They also can occur in the eyebrows, the Meibomian glands of the eyelids, perioral mucosa, ear canal, chest, nipples, and other parts of the body (Nutting *et al.,* 1989). In most cases they cause no apparent harm and go virtually unnoticed. Only under unusual circumstances, which remain largely unexplained, do they cause clinical problems that warrant medical attention. Such cases involving dermal reactions to *Demodex* mites are called *demodicosis.* It does

not appear that any specific pathogens are involved, although secondary bacterial infections can aggravate the condition.

In addition to differences in one's body chemistry and immunological responses, certain hormones affect population levels of *Demodex* mites and the development of demodicosis. Populations tend to build as the host matures, leveling off in the middle age groups. Substances such as diethylstilbestrol tend to inhibit mite populations, whereas progesterone and testosterone may promote an increase. The long-term use of topically applied corticosteroids has been correlated with an increased incidence of demodicosis, suggesting a possible link between hormonal levels and the development of inflammatory reactions induced by follicle mites. Cases of human demodicosis can be categorized into five clinical forms: demodex folliculitis, demodex blepharitis, pityriasis folliculorum, demodex granuloma, and human demodectic mange.

*Demodex folliculitis* occurs most commonly on the face, but also on the forearms and chest. It typically causes rosacea-like skin lesions, initially appearing as red follicular papules and tiny pustules. This is the most difficult *Demodex* infestation to diagnose because it is almost indistinguishable clinically from other skin problems such as acne cosmetica, corticosteroid telangiectasia (dilated blood vessels within the skin that have a tortuous appearance), and rosacea. In some cases it may complicate or aggravate preexisting skin conditions. Confirmation of demodicosis is therefore dependent on demonstrating the presence of the mites, all stages of which can be found in the pustule contents.

*Demodex blepharitis,* also known as *ocular demodicosis,* is an inflammation of the hair follicles of the eyelids associated with high populations of demodicid mites. The patient's eyelids typically itch or burn, become reddened, and are often characterized by accumulations of waxy or gelatinous debris at the base of the eyelashes. The presence of mites can usually be detected by plucking affected lashes and examining them under a microscope.

*Pityriasis folliculorum* is an uncommon form of demodicosis which is clinically recognized as dry, scaly skin with brownish or grayish hyperpigmentation and associated pruritus. The condition is often intensified by scratching or shaving, resulting in erythema, excoriation, mottling, and what has been described as a nutmeg-like roughening of the affected skin. It usually occurs on the face and neck in both young and older adults.

*Demodex granuloma* results when follicle mites rupture out of blocked hair follicles into subcutaneous tissues. There they can elicit a response by lymphocytes and histiocytes to form granulomas.

*Human demodectic mange* is the term applied to the remaining form of human demodicosis in which transient infestations of humans by *Demodex* spp. are contracted from other host species. The most common source is dogs and usually involves intimate contact, such as sleeping with a pet. Patches of papules and vesicles develop, accompanied by a burning or itching sensation. Commonly affected areas include the chin, neck, chest, forearms, stomach, and thighs, reflecting the skin surfaces most likely to come into contact when handling household pets.

A positive diagnosis of demodicosis is made by confirming the presence of large numbers of *Demodex* mites directly associated with the affected areas of skin. The mites can be seen by microscopic examination of skin scrapings, pustule contents, cellular debris from hair follicles, and plucking eyelashes in the case of demodex blepharitis. Adhesive cellophane tape, applied to infested areas of the skin, can be used to recover demodicid mites near the follicular orifices or moving on the surface of the skin. A follicular biopsy also can be helpful, in which a quick-setting cyanoacrylate polymer is used to extract the contents of sebaceous follicles (Mills and Kligman, 1983). Various stages of the mites, including eggs, are usually evident.

Cases of human demodicosis can be effectively treated by daily washing of the affected skin with mild alkaline or sulfur soap, followed by application of a mild sulfur lotion sold for this purpose. Other compounds, such as gamma benzene hexachloride (lindane), metronidazole, and physostigmine ophthalmic ointment in blepharitis cases, also are effective. When properly treated, cases often are resolved in 2–3 weeks but may take as long as 2 months. This is not to say that the mites are eliminated; their numbers simply are reduced to lower levels that do not cause pathogenesis. Regular daily washing of the face and eyelids with alkaline soap helps to suppress *Demodex* populations and reduces the risk of developing demodicosis. The use of mascara also seems to retard mite increases. On the other hand, the regular use of medicated creams, skin moisturizers, and topical applications of corticosteroids tend to promote *Demodex* numbers, leading to heavier infestations and increased prospects of related skin problems.

For further information on *Demodex* species and their medical importance, see Desch and Nutting (1972), Nutting (1976a,b,c), English and Nutting (1981), Rufli and Mumculoglu (1981), Franklin and Underwood (1986), and Burns (1992).

## Sarcoptidae

The only mites in this family that commonly infest humans are members of the genus *Sarcoptes,* generally referred to as *scabies mites.* They represent a taxonomic complex of varieties or physiological types of the single species *Sarcoptes scabiei.*

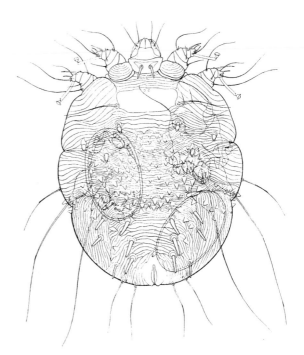

**FIGURE 23.19**  Human scabies mite, *Sarcoptes scabiei* (Sarcoptidae), female, dorsal view. (From Hirst, 1922.)

### Human scabies mite *(Sarcoptes scabiei)*

The form which typically infests people is called the *human scabies mite,* or *human itch mite, S. scabiei* var. *hominis.* This mite is cosmopolitan in distribution and infests human populations of all races as an obligate parasite that lives in the skin. The adults are small (females 350–450 $\mu$m, males 180–240 $\mu$m in length) and rounded in shape, with tiny pointed, triangular spines on their dorsal surface that assist them in burrowing (Fig. 23.19). These spines are more numerous and conspicuous in females than in males. The legs are short, with legs 1 and 2 of the female and legs 1–3 of the male each bearing a terminal sucker. The two hind pairs of legs of the female and the last pair of legs in the male lack a sucker and instead terminate in long setae or bristles.

The adult mites can crawl quite rapidly on the surface of the skin, with females traveling up to 2.5 cm/min. Upon finding a suitable site, the female uses her chelicerae and first two pairs of legs to burrow into the skin, disappearing beneath the surface in about 1 hr. There she waits in this temporary pit, or shallow burrow, for a wandering male to find her, following which mating takes place. The fertilized female then emerges on the skin surface and searches for a site in which to excavate a permanent burrow. She penetrates the skin once again and makes her way down through the stratum corneum, or horny layer of the skin, to its lower boundary with the underlying stratum granulosum. There she excavates a horizontal burrow within the stratum corneum where she will spend the rest of her life, commonly 30 days or more. During this time she continues to extend the length of her burrow by 0.5 mm/day or more, commonly reaching a total length of 1 cm or more. As viewed from the skin surface, fresh burrows appear as tiny, grayish, sinuous lines, with the adult female discernible as a whitish speck at the end of the tunnel.

Within a few hours, the female begins laying eggs in the burrow, producing two to three each day thereafter. The eggs hatch in 3–4 days. The resultant larvae often remain in the burrow for about a day before actively crawling out of the burrow onto the surface of the skin. There they excavate shallow burrows in which they molt to nymphs about 3 days later. The nymphs in turn either remain on the skin surface or dig just beneath the surface, where they molt to adults in 3–4 days. The developmental time from egg to adult typically takes about 10 days for males and 14 days for females.

Although the temporary burrows made by the larvae, nymphs, and virgin females may occur on many parts of the body, the more permanent burrows made by fertilized females tend to be in very characteristic locations. The most frequent sites are folds of the skin about the wrists and in the sides of, or webbing between, the fingers. Other common sites are the elbows, feet, and ankles; axillae; buttocks; penis; scrotum; and, for women, breasts. The location of burrows in infants and young children differs somewhat from that of adults, commonly involving the palms, sides and soles of the feet, and areas about the head and neck. In addition to the rash and discomfort directly associated with the burrows, rashes often occur on other parts of the body and do not correspond with the distribution of the adult female mites. These other rashes are believed to be caused, in part, by the shallow burrowing of the immature stages of *S. scabiei* and temporary burrows made by unfertilized females. Unlike adults, children often develop rashes on the face, chest, and back.

The most common means of transmission is by direct contact between individuals when the mites are crawling on the skin surface. However, transmission also can occur via bed linen, clothing, and other fabrics from infested hosts. The mites are able to survive 2–3 days at room temperatures when the relative humidity is more than 30%. The higher the relative humidity, the higher the survival rate. Larvae of *S. scabiei* can hatch from eggs deposited off the host and infest fomites up to 7 days. However, transmission by fomites generally is not of major importance in temperate regions. *S. scabiei*

from infested horses reportedly has been transmitted to humans via saddle blankets, harnesses, and grooming utensils.

## HUMAN SCABIES

Scabies victims usually experience intense itching, especially at night. The itching typically is out of proportion to the visible signs of the infestation and tends to be aggravated by heat, warm baths, and removal of clothing. The pruritus and rash are attributed to antigens associated with the mite bodies, secretions, and fecal material deposited in the burrows. They stimulate the host's cell-mediated immune response, contributing to the development of acquired immunity to subsequent infestations by *S. scabiei* following initial exposure. This suggests the possibility of vaccines being developed to protect human and other hosts against natural infestations of scabies mites (Arlian *et al.*, 1994a,b). The antigenic nature of the cuticular components of the mites, their secretions, and excretory products helps to explain the persistence of the rash and other clinical signs long after the mites themselves have been killed by acaricidal treatments.

Cases of *human scabies* occur in a number of clinical forms. The most common type is *papular scabies* characterized by erythematous papules that erupt as a generalized pruritic rash on various parts of the body. The accompanying itching usually leads to scratching and excoriation of the affected areas, contributing to an eczema-like condition. Vigorously scratched lesions may become secondarily infected with pyogenic, or pus-forming, bacteria, causing an acute, inflammatory, destructive skin condition called *pyoderma*. In some individuals, tiny vesicles develop in the epidermis in response to burrowing mites. If they become enlarged enough to form macrovesicles, or bullae, they cause what is known as *bullous scabies*. In other cases, the patient may develop *urticarial scabies* in which a histamine-like vascular reaction produces wheals or hives that may be intensely itchy and can obscure the primary cause of the problem.

In a small number of individuals, *S. scabiei* may burrow deeper into the skin, penetrating the dermis and inducing infiltration of lymphocytes. This can lead to the formation of firm, reddish brown, pruritic masses and a condition called *nodular scabies*. The nodules tend to occur most commonly at the elbows, axillary region, groin, and male genitalia, where they may persist for months or even a year or more despite treatment. Mites seldom are recovered from nodules that are more than a month old. Cases ultimately resolve with or without therapy.

One of the more distinctive, yet rare, clinical types of the disease is *crusted scabies,* also called *hyperkeratotic scabies* or *Norwegian scabies.* It is characterized by dry, scaly, or crusted lesions, usually on the hands and feet. Pruritus is typically mild or absent altogether, despite the extremely large numbers of mites, sometimes in the thousands, amidst the overgrowth of keratin tissue in the horny layer of the epidermis. The lack of discomfort and absence of burrows often results in these cases going undiagnosed. This condition is highly contagious and can be spread even on casual contact due to the large numbers of mites involved. Victims thus can serve as silent carriers and are often detected only as a result of clusters of cases of the more common forms of scabies in individuals with whom the source has come in contact, especially in hospitals and other institutional settings. Evidence indicates that the mites even can become airborne along with small scales of skin from the crusted lesions. Crusted scabies is generally associated with immunosuppressed individuals who do not respond normally to infestations of *S. scabiei* or individuals with nervous disorders that render them insensitive to pain, especially skin sensations. They do not experience the usual itching, and their inclination to scratch is suppressed. Consequently, cases of crusted scabies often are associated with the mentally impaired and physically or immunologically compromised patients.

Despite the high host-specificity of the different varieties of *S. scabiei,* many cases have been reported of humans being temporarily infested with scabies mites from other animals. Such cases are referred to as *animal scabies* and *human sarcoptic mange*. Although these cases usually involve dogs, particularly puppies, sources include livestock such as horses, cattle, sheep, goats, camels, and pigs. Such infestations typically result in localized erythematous papules and pruritus at contact sites. The mites do not form burrows and rarely survive to reproduce. Infestations are self-limiting and usually resolve themselves within a few weeks, provided the source is removed to prevent reinfestation. The absence of burrows and the low numbers of mites usually make it difficult to confirm cases by recovering mites from affected individuals. The diagnosis therefore often is based on demonstrating *S. scabiei* infesting the suspected animals involved.

A diagnosis of scabies can be confirmed by demonstrating the presence of *S. scabiei*. The presence of eggs, immature stages, adults, or fecal material from the burrows are all diagnostic. The presence of burrows in characteristic locations such as the wrists, fingers, elbows, and feet are considered nearly pathognomic, i.e., by themselves they virtually confirm the diagnosis. To help in locating burrows, one or two drops of ink can be applied to suspected areas and then wiped off with alcohol after 10 min. The ink is retained in the burrows, making them more discernible. Several techniques have been developed to recover mites from scabies patients for microscopic examination and identification. Adult females can be removed from the blind end of their burrows by using a sharp-pointed scalpel blade to pierce the skin and gently pick out the mite. Alternatively, scrapings can be taken by vigorously scraping the affected skin several times with a sterile scalpel blade. The scraping is then transferred to a glass microscope slide for examination. Even in the absence of adult mites, the oval-shaped eggs (ca. $170 \times 190$ $\mu$m) are often clearly visible, as are the characteristic yellowish brown fecal pellets.

Skin biopsies can be taken and prepared for histological examination. Another method is to place skin scrapings in a small petri dish, or other container, and examine it after 12–24 hr for the presence of mites crawling on the bottom. A centrifuge-flotation method also has been used with some success, especially in cases of crusted scabies or when abundant material from affected areas can be collected. The scrapings are placed in 10% potassium hydroxide or sodium hydroxide and gently heated. The mixture then is added to a saturated sugar solution in a centrifuge tube and spun until any mites or eggs that are present float to the surface. Drops of the surface fluid can be microscopically examined. Eggs and egg shells have been detected by examining suspected skin scrapings in glycerine preparations using fluorescent microscopy.

The most widely used and effective means of treating scabies cases is the topical application of acaricides to the affected areas of skin. Among the more commonly prescribed acaricides are 1% lindane (gamma benzene hexachloride), crotamiton creams and lotions, sulfur applied directly to the skin or used in baths, 5% flower-of-sulfur suspended in lanolin or petrolatum, benzyl benzoate emulsions in the form of a lotion or ointment, and tetrahydronaphthalene with copper oleate. It is recommended that these materials be applied after taking a warm, soapy bath. The number of follow-up applications and the prescribed intervals vary depending on the particular product used. Overtreatment can complicate conditions and should be avoided.

In addition to treating known cases and individuals with whom they recently have had contact, fomites should be treated to disrupt possible transmission. Acaricide sprays containing pyrethrins or 5% lindane are commercially available for this purpose. Laundering clothes, bedding, towels, and other fabrics using the hot cycle of a washing machine is usually adequate to kill *S. scabiei*. Hot ironing and placing items in a freezer for 1 week also is effective in killing them. Clothing and other fomites that cannot be treated (e.g., rugs, couches) should be set aside, if possible, and not touched for 2 weeks. Any scabies mites that may have been present will have died by then.

For further information on human scabies, see Heilesen (1946), Mellanby (1972), and Orkin *et al.* (1977).

## HUMAN NOTOEDRIC MANGE

Humans occasionally become infested with *Notoedres cati*, a sarcoptid mite that causes notoedric mange in cats. Cases in humans are called *human notoedric mange* or *human notoedric scabies*. Following prolonged exposure to infested cats, people may become sensitized to this mite and develop intense pruritus within a few hours of subsequent contact with them. The reaction is induced without the mites actually burrowing. The most common sites of skin lesions are on the hands and legs, reflecting the areas most likely to come into contact with pets. The lesions subside when infested cats are either treated or removed from further contact (Chakrabarti, 1986).

## MITE-INDUCED ALLERGIES

Members of several families of mites can cause allergic responses in humans by either direct contact of mites with the skin or inhalation of mites or mite parts. The most common sources of allergy-inducing mites are stored products and house dust.

## STORAGE MITES

People who handle mite-infested stored products may become sensitized to the mites on subsequent contact, resulting in an immunological response called *storage-mite allergy*. Although the precise nature of the allergens is unknown, these substances include components of both live and dead mites and material produced in the mite alimentary tract. Sensitized persons may experience either contact dermatitis or respiratory allergy, depending on the type of exposure.

*Allergic contact dermatitits* results from exposure to mites in grains, dried fruits, flour, and other stored products, causing itching and redness at the contact sites. The families of mites most commonly involved are the Acaridae, Carpoglyphidae, and Glycyphagidae. In addition, what was probably *Dermatophagoides pteronyssinus*, but reported as *D. scheremetewski* (Pyroglyphidae), has been associated with cases of *feather pillow dermatitis*. Contact with this mite infesting feather pillows is known to cause red papular lesions and pruritus about the scalp, eyes, ears, and nostrils (Aylesworth and Baldridge, 1983; Traver, 1951). A similar allergenic response to *D. farinae* associated with buckwheat-husk pillows has been reported in China (Hong *et al.*, 1987).

*Inhalational allergy* results when airborne mites and associated allergens are drawn into the respiratory tract. The mucosal membranes lining the nasal and bronchial passages become irritated and inflamed, causing allergic rhinitis and asthma. The mucous membranes lining the eyelids also may be affected, causing conjunctivitis. These responses involve a T-cell–type reaction and both immediate and delayed hypersensitivity. Such reactions to mites present an occupational hazard, especially among farmers and other agricultural workers who handle mite-infested grains and other stored materials. Among the more common storage mites that cause inhalational allergy are *Aleuroglyphus ovatus* and *Tyrophagus putrescentiae* (Acaridae), *Lepidoglyphus destructor* (Glycyphagidae), and *Blomia tropicalis* (Echimyopodidae). For further information on storage-mite allergy, see Cuthbert (1990).

## HOUSE-DUST MITES

A major source of human allergens in the home is house dust and its associated mite fauna. Where humidity is sufficiently high, fungi tend to thrive in accumulated dust, providing food for a variety of house-infesting mites that are primarily saprophages or fungivores. Many of these mites are the same species that infest stored products, nests of rodents and birds, and animal litter. When their populations reach high levels in the home, they can cause acute or chronic allergic reactions commonly known as *house-dust allergy*. The principal allergenic components in house dust are mites and mite feces, rather than the dust material itself.

As many as 10 families and 19 species of mites have been recovered from house dust in a single urban community (Tandon *et al.*, 1988), reflecting the diversity of mites that occur in that microhabitat. The most important taxa that cause human allergy are members of the Pyroglyphidae, notably those belonging to the genera *Dermatophagoides* and *Euroglyphus*. These mites typically comprise 90% or more of the mites found in house dust. The other families of mites commonly associated with house dust are the Acaridae, Glycyphagidae, and Cheyletidae, represented by many of the same species that infest stored products. Four of the more common storage mites found in house dust are *Acarus siro, T. putrescentiae, L. destructor,* and *Glycyphagus domesticus* (Fig. 23.20) (Wraith *et al.*, 1979).

The most widespread pyroglyphid species that causes house-dust allergy is the *European house-dust mite* (*D. pteronyssinus*), which thrives in floor dust and the surface dust of mattresses. It is regarded as the most frequently encountered house-dust mite, occurring especially in humid coastal areas of western Europe and North America. This was the first mite to be identified as a cause of house-dust allergy in 1966, shortly after the genus *Dermatophagoides* was first linked to house dust and bronchial asthma. The *American house-dust mite,*

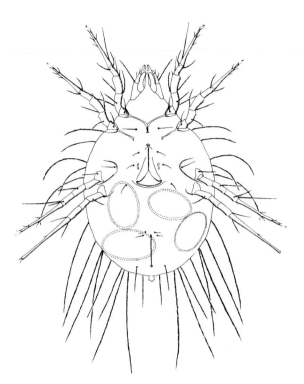

**FIGURE 23.20** *Glycyphagus domesticus* (Glycyphagidae), female with four large, ovoid eggs, ventral view. (From Hughes, 1976.)

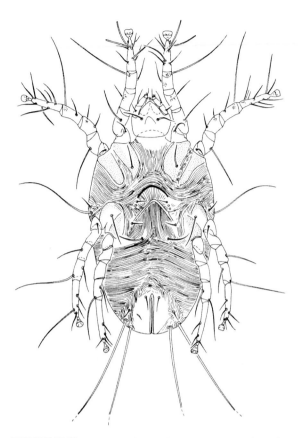

**FIGURE 23.21** American house-dust mite, *Dermatophagoides farinae* (Pyroglyphidae), female, ventral view. (From Gorham, 1991; courtesy of the US Department of Agriculture.)

*D. farinae* (Fig. 23.21), tends to be common in drier regions than *D. pteronyssinus,* such as the more continental-type climates of central Europe and the central United States. It is a frequent inhabitant of dried animal meal (e.g., dog biscuits, poultry feed) and coarsely ground wheat. The common name reflects its more common and widespread occurrence in the United States than in Europe and other parts of the world. The third most common mite known to cause house-dust allergy is *Euroglyphus maynei* (Fig. 23.22). This is a cosmopolitan species frequently implicated in human allergy cases in Europe and Japan. It typically occurs in damper habitats than that of *D. pteronyssinus,* with which it is often associated.

House-dust mites thrive in environments with relative humidities above 65–70%. These mites are dependent on water vapor as their primary source of water, which they extract from the air. They cannot actively survive more than 6–11 days at relative humidities below 50%. They can, however, endure prolonged dry periods by forming desiccation-resistant protonymphs which can survive for months below the critical humidity for the active stages. Their feeding activity, reproductive rate, and amount of fecal material generated are all directly related to humidity levels (Arlian and Hart, 1992). Populations tend to increase beginning in early summer, reach their highest

levels in early fall, and remain relatively constant during the winter months, reflecting indoor humidity.

The developmental times of *Dermatophagoides* and *Euroglyphus* species vary with temperature and humidity. Under favorable conditions at room temperature and a relative humidity of 75%, they typically complete a generation in about 30 days. Females do not lay eggs unless the eggs are fertilized and commonly experience multiple matings. They lay one or two eggs per day during their adult life, which usually lasts 30 or more days. Half or more of the mites in dust samples may be represented by eggs, and, if overlooked, this often leads to underestimates of population sizes when only the nymphs and adults are counted (Colloff and Hart, 1992).

*Dermatophagoides* and *Euroglyphus* species are saprophages which, in the home, feed primarily on fungi growing on the organic components of floor and mattress dust. A number of ingested fungi have been identified in the alimentary tract of house-dust mites, including xerophylic species in the genera *Aspergillus, Eurotium,* and *Wallemia* (Hay *et al.,* 1992). Evidence indicates that the mites feed selectively on different fungi and that species differ in

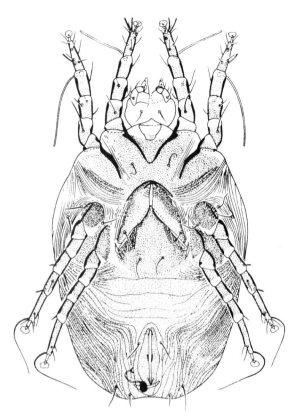

**FIGURE 23.22**   European house-dust mite *Euroglyphus maynei* (Pyroglyphidae), female, ventral view. (From Gorham, 1991; courtesy of the US Department of Agriculture.)

their nutritional value. Some fungi, such as *Aspergillus penicillioides,* actually may be detrimental to mite growth and reproduction (Hay *et al.,* 1993). Sloughed skin scales from humans and household pets serve as nutritional substrates for fungi. Mattresses thus provide particularly favorable sites for mites to develop. Human semen associated with sleeping quarters has been shown to be a dietary supplement for house-dust mites and can significantly increase the number of eggs a female produces (Colloff *et al.,* 1989).

House-dust mites occur in greatest numbers in the more humid living quarters of homes frequented by the occupants, notably wherever dust accumulates in bedrooms and living rooms. Mattresses are especially suitable, apparently due to the accumulation of human squamal cells and other skin debris. Under optimal conditions, as many as 5000 mites have been recovered per gram of mattress dust. The type of floor can influence the species and number of mites. In damp houses, carpeted floors contribute to high populations, whereas in drier homes there may be little difference in the numbers of mites in carpeted and noncarpeted flooring. When humidity levels are high, even floors covered with

linoleum and other plastic materials will support relatively high numbers of house-dust mites. In general, however, drier floors and carpets support higher populations of *D. farinae* than either *D. pteronysinnus* or *E. maynei. E. maynei* has the highest humidity requirements, occurs primarily in mattresses and bedding, and is the least likely to be found in carpets. Although some studies have shown that wool carpets have higher numbers of mites than carpets made from synthetic fibers (e.g., nylon), other studies have shown no significant differences between them.

House-dust mites also may occur in fairly large numbers in other situations in the home. In a survey of household fabrics in Germany, 18% of the mites recovered were found in clothing (e.g., suits) hanging in closets (Elixmann *et al.,* 1991). The same mites also may infest improperly stored food products. In one such case in Alabama (USA), an individual experienced sneezing, intense ocular pruritus, and facial edema within minutes of inhaling a puff of dry pizza-dough mix heavily infested with *D. farinae* (Skoda-Smith *et al.,* 1996).

House-dust mites are now recognized as the primary source of allergens that cause *house-dust allergy,* especially in children and young adults. The most common clinical manifestation is bronchial asthma, characterized by difficulty breathing, inflammation of the nasal passages, and conjunctivitis. This may be accompanied by atopic eczema in some sensitized individuals. Asthmatic attacks tend to occur at night, especially in poorly ventilated bedrooms with old bedclothes and accumulated mattress and floor dust. Occurrence of symptoms is usually seasonal, reflecting the size of the mite populations. The acuteness of the allergic attacks is directly correlated with the number of mites present.

*D. pteronyssinus* generally produces the most potent house-dust allergens. However, a portion of individuals of other species, including *D. farinae, E. maynei,* and certain storage mites, can elicit allergic responses as great as that of *D. pteronyssinus.* Each species of mite appears to have its own species-specific antigens and allergens, with differences between those associated with the mite body and feces. There is significant cross-reactivity among the antigens of different species. This makes it difficult to determine which mite is involved in individual cases, thereby complicating clinical diagnosis and treatment. Diagnostic tests for mite-induced house-dust allergy include skin tests and bronchoprovocation using commercial extracts of individual mite species. Enzyme-linked immunosorbent assays and radioallergosorbent tests have been developed to help in the diagnosis of house-dust mite allergy cases. However, they tend to be less effective than the traditional skin-prick test in identifying people who are only mildly sensitized to mite allergens in house dust.

Several sampling techniques have been developed to determine if house-dust mites are present in the home and, if so, what species they are and their relative numbers. Most of the techniques entail collecting samples of mattress and floor dust with a vacuum device and examining the samples microscopically for the presence of mites. Various flotation and staining methods can be used to facilitate the process. Another approach is the use of a *guanine test* as an indirect means of determining the number of mites present. Guanine is excreted in mite feces and serves as a quantitative index of mite numbers, irrespective of the species. The amount of guanine can be measured using high-performance liquid chromatography, providing a simple, rapid method for determining the amount of mite activity in different parts of the home (Quoix *et al.*, 1993).

## INTERNAL ACARIASIS

Under certain exposure conditions, mites may enter natural body orifices, leading to cases of temporary internal acariasis. These cases commonly involve the ingestion of mites with food and the inhalation of airborne mites or mite-contaminated dust via oral or nasal routes. Mites that are swallowed or inhaled can lead to acariasis involving the alimentary tract, whereas mites that are inhaled also can invade the respiratory tract. Cases of mites infesting the urinary tract are rare. For a general discussion of the mites involved, see Ma and Wang (1992).

*Pulmonary acariasis*, in which mites invade the lungs, occurs most frequently among individuals exposed to mite-infested stored grains and dried herbs. This reportedly can be a serious problem among workers in grain-storage facilities and medicinal-herb warehouses in China (Chen *et al.*, 1990; Li and Li, 1990). Clinical signs and symptoms include cough, expectorated phlegm and blood, difficulty breathing, chest pain, low-grade fever, restlessness, and marked eosinophilia. Pulmonary lesions have been documented on X-ray film as shadows and nodular opacities in lung tissues. The following five families and nine species of mites have been recovered from sputum of affected individuals: Acaridae (*Acarus siro, Tyrophagus putrescentiae, Aleuroglyphus ovatus, Sancassania berlesei* [reported as *Caloglyphus berlesei*], and what was probably *Sancassania mycophaga* [reported as *Carpoglyphus mycophagus*]); Pyroglyphidae (*Dermatophagoides farinae, D. pteronyssinus*); Tarsonemidae (*Tarsonemus granarius*); and Cheyletidae (*Cheyletus eruditus*). It is not clear which of these mites cause the more serious problems. Some species, such as *S. berlesei*, can thrive in exceptionally damp food stores covered with a film of water and may be able to survive for some time in the lungs.

A *Carpoglyphus* species (believed to be *Carpoglyphus lactis*) was associated with a case of pulmonary acariasis in Spain (Taboada 1954), whereas a *Tyrophagus* species and other unidentified mites were recovered from sputum, bronchial washings, and needle-aspirated lung specimens in routine examinations of patients suffering from respiratory ailments (Farley *et al.*, 1989). Mite eggs, larvae, and adults were found in cytology specimens in the latter study, with evidence of their being surrounded by acute inflammatory cells in several cases.

Human cases of *enteric acariasis* occasionally are reported in which mites are found in excreta, suggesting their presence in the digestive tract. In most cases they are acarid mites in the genera *Acarus, Suidasia,* or *Tyrophagus. Suidasia pontifica* (reported as *S. medanensis*) was recovered from feces of a woman and two infants in Mexico (Martinez Maranon and Hoffman, 1976), whereas various stages of an *Acarus* or *Tyrophagus* species, together with eggs, were recovered from bile of a Romanian patient with chronic cholecystitis (Pitariu *et al.*, 1978). It was concluded that the woman probably ingested the mites with her food and that the mites simply aggravated her preexisting cholecystitis by causing inflammation of her digestive tract until the mites were eliminated with the bile. Other cases of enteric acariasis have been reported in children with chronic digestive disorders in Russia (Prisich *et al.*, 1986).

A few cases of *urinary acariasis* have been reported, primarily involving acarid mites in the genus *Tyrophagus.* Two species allegedly recovered from the human urinary tract are *Tyrophagus putrescentiae* and *T. longior* (Harwood and James, 1979). Many, if not most, of these cases appear to be misleading and probably involve contamination of containers in which urine was collected or examined. A possible exception was the recovery of unidentified mites in urinary samples from several patients in Romania during acute attacks involving inflammation of the kidneys and urinary bladder (Pitariu *et al.*, 1979). Numerous acarid mites and their eggs were observed in the urinary sediments; others were dead and encrusted with salts. Whether or not contamination of samples can be ruled out in these cases is unclear. Other cases of urinary acariasis have been reported in Japan (Harada and Sadaji, 1925) and China (Chen *et al.*, 1992; Ma and Wang, 1992).

*T. longior* (Fig. 23.15) occurs primarily in cool, temperate regions of Europe where it infests stored grains, hay, and straw; hay stacks in open fields; cucumber plants, tomatoes, and beets; and poultry litter in broiler houses. Cases of digestive and urinary acariasis in humans involving *T. longior* have been reported (Harwood and James, 1979).

Stored-products mites in the genus *Tarsonemus* (family Tarsonemidae) have been reported to be associated with human dermatitis and other skin disorders (Hewitt *et al.*, 1973; Krantz, 1978; Oehlschlaegel *et al.*, 1983) and to invade various organs and body fluids of humans and other animals (Dahl, 1910). The most commonly implicated species is *Tarsonemus hominis*. It is generally believed that these reports represent contamination of glass slides and other materials used in preparation of tissues for microscopic examinations (Hewitt *et al.*, 1973; Samšiňák *et al.*, 1976).

# MITE-BORNE DISEASES OF HUMANS

Excluding tick-borne diseases, there are only two significant diseases of humans for which mites serve as the principal vectors: rickettsialpox and tsutsugamushi disease.

## RICKETTSIALPOX

Rickettsialpox was first recognized in 1946 during an outbreak in New York City (Huebner *et al.*, 1946). Sporadic cases had been reported as early as 1909 in Washington, DC, and other cities along the northeastern seaboard of the United States. Rickettsialpox is a relatively uncommon illness, with only 800–900 cases having been reported in the United States. Cases occur primarily in urban areas in crowded living quarters infested with the house mouse (*M. musculus*) that serves as the major reservoir. The pathogen is transmitted to humans by the bite of the *house mouse mite, Liponyssoides sanguineus* (Dermanyssidae) (Fig. 23.5). Other countries in which cases of rickettsialpox have been reported are Russia, Korea, and parts of equatorial and central Africa.

The causative agent of rickettsialpox is *Rickettsia akari*. It is a spotted fever group (SFG) rickettsia and is morphologically indistinguishable from *R. rickettsii*, the causative agent of Rocky Mountain spotted fever. The intracellular site in which it replicates in human hosts remains unknown.

Rickettsialpox is usually a mild, nonfatal illness which typically begins with the appearance of a nonpruritic, erythematous papule at each infectious bite site, usually within 24–48 hr of contact with *L. sanguineus*. Soon thereafter a small vesicle forms at the center of the papule, filling initially with a clear, then cloudy, fluid. The vesicle dries, producing first a crusty lesion and then a brown or black scab, or eschar, in the center of a larger, indurated area 0.5–3.0 cm in diameter. These lesions can occur on any part of the body but usually occur on the face, trunk, and extremities. They may occur on the palms and soles and on mucous membranes about the mouth. The latter include the palate and, less commonly, the general

mouth cavity, tongue, and pharynx. Although there usually are only a few, as many as 100 discrete lesions have been reported in some cases.

Systemic symptoms appear about the time that eschars form, 9–14 days after the initial bites. Fever (usually peaking at 38–40° C), headache, and malaise are characteristic and may be accompanied by muscle aches, especially backaches, drenching sweats, and shaking chills. Less common symptoms are cough, runny nose, sore throat, nausea, vomiting, enlarged and tender regional lymph nodes, and abdominal pain. Most cases resolve in 6–10 days without treatment. In some cases, however, headache and lassitude may persist for another 1–2 weeks. Treatment with antibiotics generally alleviates the fever and other symptoms within 48 hr.

Diseases that should be included in the differential diagnosis of rickettsialpox are other members of the SFG rickettsiae, notably Boutonneuse fever, tsutsugamushi disease, Siberian tick typhus, and Queensland tick typhus. They can be distinguished, however, by their geographic occurrence and the clinical nature of the associated skin lesions. The nonrickettsial disease with which rickettsialpox is most commonly confused is chickenpox, caused by a virus. In chickenpox cases, however, the vesicles are not raised on papules, eschars are not formed, and the lesions are much more numerous. The clinical syndrome and a rise in titer of SFG-specific antibodies are generally sufficient to confirm a diagnosis of rickettsialpox. Immunity appears to be complete, perhaps lifelong, following recovery from infection. For additional information on the clinical and diagnostic aspects of this disease, see Brettman *et al.* (1981) and Kass *et al.* (1994).

*L. sanguineus,* the vector of the rickettsialpox agent, is primarily a parasite on the house mouse. The mite is also found on rats (*Rattus* spp.) and voles, although the role of these and other wild rodents in the ecology of this disease is uncertain. Whereas *L. sanguineus* nymphs generally take a single blood meal, adults move onto and off the host to take several blood meals. Most of the time is spent off the host in nests and runways of mouse-infested areas. Where it occurs in human dwellings, the mite seeks the warmth of furnace rooms and incinerators of old buildings, where they may occur in large numbers on the walls and ceilings. Human bites are believed to occur primarily when house mice in apartment buildings become less attractive as hosts, inducing the mites to seek alternate hosts. The occurrence of lymphocytic choriomeningitis in house mice during outbreaks of rickettsialpox in humans has been suggested as a possible factor; such infections cause changes in a mouse's body temperature, perhaps inducing the mites to abandon their natural host (Krinsky, 1983). Starved adults can live 7–8 weeks, whereas blood-fed adults can live 9 weeks or longer.

The only other mite reported as a possible vector of *R. akari* is *Ornithonyssus bacoti,* based on experimental transmission studies using laboratory white mice (Lackman, 1963; Philip and Hughes, 1948).

## TSUTSUGAMUSHI DISEASE

Tsutsugamushi disease is a mite-borne rickettsiosis of humans that is endemic in eastern and southern Asia, the western Pacific region, along the northern coast of Australia (Queensland and Northern Territory), and the Indian subcontinent. Cases may occur as far west as Afghanistan, Pakistan, and neighboring areas of the former Soviet Union. It is also known as *scrub typhus* and *chigger-borne rickettsiosis.* The causative agent is *Orientia tsutsugamushi* (formerly *Rickettsia tsutsugamushi*) transmitted by the bite of trombiculid larvae, or *chiggers* (Fig. 23.10).

Tsutsugamushi disease was recognized as early as the fourth century AD, when it was described in clinical manuals as an illness associated with mites. It was not until 1930, however, that Japanese workers first isolated and identified the pathogen as a rickettsia. This disease first caught the attention of the Western world during World War II when the Allied Forces were severely affected during operations in the Pacific Theater. The number of cases of tsutsugamushi disease exceeded that of direct wartime casualties among the military forces in that region. Fatality rates as high as 27–35% occurred among troops on the islands of Goodenough and Finchhaven in New Guinea (Philip, 1948; Philip and Kohls, 1945). With the advent of effective antibiotics for treatment, the incidence of tsutsugamushi disease decreased dramatically in the region during the late 1940s and 1950s. However, sudden increases have occurred since that time in Japan (ca. 1975), Korea (ca. 1985), and other areas. For comprehensive reviews on this important mite-borne disease, see Traub and Wisseman (1974) and Kawamura *et al.* (1995).

The causative agent of tsutsugamushi disease is considered to be distinct enough from related rickettsial organisms to be placed in its own genus, *Orientia* (Tamura *et al.,* 1995). Like *Rickettsia* species, it is an obligate intracellular parasite that multiplies in the cytoplasm of host cells. The clinical picture is complicated, however, by a multitude of antigenic variants, or strains, that exhibit various degrees of pathogenicity to humans. Among the better characterized strains are *Gillian, Karp, Kato, Kawasaki, Kuroki,* and *Shimokoski.* The relationships among the different strains and their mite vectors remain largely unknown.

The classic form of tsutsugamushi disease varies from a mild to severe illness. It begins with the development of a small papule at the bite site of an infected chigger.

FIGURE 23.23  *Leptotrombidium akamushi* (Trombiculidae), chigger vector of *Orientia tsutsugamushi,* the causative agent of tsutsugamushi disease in Japan. (Modified from Baker *et al.,* 1956.)

The skin reaction varies from hardly noticeable or mildly itchy to painful. The latter discomfort is characteristic of bites of the mite *Leptotrombidium akamushi* (Fig. 23.23) and has been likened to a tiny thorn that has penetrated the skin and induces pain when it is rubbed. This sensation, called *ira* in endemic areas of Japan, usually appears about 10–20 hr after the bite and is believed to be caused by an inflammatory eruption associated with formation of a feeding tube (*stylostome*) by the attached mite. Bites occur most frequently in the folds of soft skin of the axillary region, upper legs, and abdomen. Other common sites include webs between the fingers, skin behind the knees, genitalia, under breasts, and skin constricted by clothing. The bites become ulcerated and form hard, black scabs (*eschars*), typically accompanied by fever and a maculopapular rash.

Following an incubation period of about 10 days (range, 5–20 days), symptoms generally include loss of appetite, fever, headache, muscle aches, and general malaise; regional or generalized lymphadenopathy is also common. The more virulent strains of *Orientia tsutsugamushi* can cause hemorrhaging, intravascular coagulation, and other blood disorders as the rickettsiae multiply in epithelial cells of the vascular system. This can lead to microthrombi in the kidneys, lungs, and heart, contributing to fatalities. Mortality varies widely from 3 to 60%, depending on the strain and geographic region. Cases of tsutsugamushi disease can be treated effectively with antibiotics. However, in severe cases with hemorrhagic complications, heparin therapy and platelet transfusion may be necessary. Immunity following recovery is not lasting, such that reinfections are common in endemic areas.

Although cases of tsutsugamushi disease occur throughout the year, they often are seasonal in certain areas, reflecting the activity of local mite vectors. In some parts of Japan, for example, cases known as *Japanese river fever* tend to occur during the warm, summer months along river terraces where larvae of *L. akamushi* are present. Cases in Japan during the autumn and winter months, which may extend into late spring or early summer, are usually associated with two other chigger species, *L. pallidum* and *L. scutellare*. These seasonal differences are reflected in local Japanese names such as *Umayado disease* for the summer form of tsutsugamushi disease in Kagawa Prefecture and *Shichito fever* for the autumn/winter form in the Hachijo Islands of Izu Shichito. The nonsummer types of tsutsugamushi disease are usually characterized by relatively mild symptoms and low fatality rates. In other subtropical and tropical regions of Southeast Asia and the southwestern Pacific, cases occur independent of seasons, correlated with the presence of chigger vectors that are present year-round (e.g., *L. arenicola, L. deliense, L. fletcheri*).

More than 40 species of trombiculid mites (13 genera) are known or suspected to be vectors of *O. tsutsugamushi*. The major chigger vectors of this pathogen and their geographic occurrence are shown in Table III. The most important genus is *Leptotrombidium*, represented by approximately 25 vector species, most of which belong to the subgenus *Leptotrombidium*. Two or more vector species are known in each of the following 4 genera: *Neotrombicula* (6 spp.), *Ascoschoengastia* (2 spp.), *Euschoengastia* (2 spp), and *Walchia* (2 spp.). Other trombiculid genera that play a role in transmission of *O. tsutsugamushi* to humans are *Acomatacarus, Eutrombicula, Gahrliepia, Leeuwenhoekia, Mackiena, Neoschoengastia, Odontacarus,* and *Shunsennia*.

Chiggers that serve as vectors of tsutsugamushi disease are primarily parasites of wild rodents such as field mice (*Apodemus* spp.), voles (*Microtus* spp.), and rats (*Leopoldamys, Maxomys, Rattus,* and other genera). They also occur on a wide range of birds (including pheasants, pigeons, and chickens) which are susceptible to infection and can develop at least transient rickettsemia. Although it is conceivable that, there is no conclusive evidence to show that animals other than humans are involved in the transmission cycle rodents serve as a source of infection of *O. tsutsugamushi*. Instead, the mites themselves serve as the principal natural reservoirs of the disease agent. Certain strains of trombiculid mites of a given species effectively transmit the rickettsia transovarially and/or transstadially; thus, the pathogen is transmitted from the adult female to her larval offspring and to each developmental stage that follows. Consequently, *O. tsutsugamushi* is passed from generation to generation of mite and is maintained within local mite populations in endemic areas. Were this not the case, this mite-borne rickettsia could not persist in nature. Because trombiculid mites feed on only a single host, they do not have the opportunity to transmit an acquired pathogen at a subsequent feeding. For further information on the ecology and medical aspects of tsutsugamushi disease, see Kawamura *et al.* (1995).

## INTERMEDIATE HOSTS OF HUMAN PARASITES

No metazoan parasites of major health importance to humans involve mites as intermediate hosts. However, mites are hosts for a few *tapeworms* that occasionally infest people. Oribatid mites are intermediate hosts for two *Bertiella* species of anoplocephalan tapeworms. *Bertiella studeri* parasitizes the small intestine of a wide range of Old World primates, including rhesus and cynomolgus monkeys, Japanese macaques, baboons, mandrills, gibbons, orangutans, chimpanzees, and occasionally humans in Asia and Africa. Although several European species of oribatid mites have been shown to support development of *B. studeri* experimentally (Denegri, 1985; Stunkard, 1940), the oribatid species involved in the natural transmission cycle remain unknown. *B. mucronata,* which parasitizes monkeys in South America, develops in the oribatid mites *Dometorina suramerica, Scheloribates atahualpensis,* and other species of the genera *Achiptera, Galumna, Scheloribates,* and *Scutovertex* (Denegri, 1985; Sengbusch, 1977). In the case of both of these tapeworms, wild primates are the primary vertebrate hosts, with human cases occurring where infested primates live in close association with people. They cause no apparent lesions or other harm to their hosts.

Occasionally humans may be parasitized by tapeworms of the genus *Mesocestoides* (family Mesocestoididae) which use mammalian carnivores and charadriiform birds as hosts in North America, Europe, Asia, and Africa. Oribatid mites are believed to play a role as intermediate hosts in the relatively complex life cycles of these cestodes.

## DELUSORY ACARIASIS AND ACAROPHOBIA

Because of their tiny size and the general lack of knowledge about mites by the general public, mites are often mistakenly blamed as the cause of skin problems or bite-like sensations when the underlying cause is unknown. The term for this is *delusory acariasis,* the imagined notion that mites are biting or infesting the skin when in fact they are not. A rational discussion is unlikely to convince individuals involved otherwise. This is a specific type of the more general phenomenon of *delusory parasitosis*. A typical example is attributing various skin conditions among office workers to "paper mites." Although there are no such creatures, it is difficult to dispel the misconception that such mites are involved. Other imaginary mites are "telephone mites" and "cable mites" blamed as the cause of skin irritation among telephone users and computer operators.

The term *acarophobia* refers to an undue fear of mites that can cause psychological stress. This may develop as the result of an actual experience with mites or, more likely, as a consequence of one or more episodes of delusory acariasis.

# VETERINARY IMPORTANCE

Mites have been very successful in exploiting vertebrate hosts. Many are ectoparasites of skin, scales, feathers, or fur, whereas others are endoparasites that have invaded body cavities, respiratory passages, and internal tissues and organs. Some mites are vectors of disease agents of domestic and wild animals, while still others serve as intermediate hosts for animal parasites, notably tapeworms. Occasionally mites are the cause of allergic reactions of pets and other animals. For overviews of mites of veterinary importance, see Hirst (1922), Baker *et al.* (1956), Strandtmann and Wharton (1958), Sweatman (1971, 1984), Yunker (1973), Georgi and Whitlock (1980), Whitaker (1982), Nutting (1984), and Pence (1984). For works of a more regional nature, see Domrow (1988, 1991, 1992) for Australia, Mulla and Medina (1980) for South America, and Cosorabā (1994) for Eurasia.

## MITE-INDUCED DERMATITIS

Four families of ectoparasitic mites commonly cause irritation when they bite host animals to feed on blood, lymph, or skin tissues. Three families are mesostigmatid mites: Dermanyssidae, Macronyssidae, and Laelapidae. The fourth is the prostigmatid family Trombiculidae (chiggers).

### Dermanyssidae

Most dermanyssid mites cause relatively little harm even to heavily infested hosts. *Dermanyssus hirudinis*, for example, has been observed to cause few adverse effects on the survival, growth, or health of house wrens (*Troglodytes aedon*) infested with hundreds or thousands of mites per nestling (Johnson and Albrecht, 1993). Similarly, *D. americanus* and *D. gallinae* seldom cause problems when infesting wild avian hosts. *D. gallinae*, however, commonly causes dermatitis in atypical avian hosts and

domestic mammals and can cause severe infestations and economic losses in domestic chickens.

### Chicken mite (*Dermanyssus gallinae*)

This mite (Fig. 23.4), also known as the *red poultry mite* and *pigeon mite*, is an obligate parasite of wild and domestic birds worldwide, including chickens, pigeons, canaries, parakeets, house sparrows, and starlings. Occasionally it infests dogs, cats, horses, cattle, rodents, rabbits, and other mammals. Cases involving domestic animals usually occur in association with poultry houses or infested bird nests.

The chicken mite is especially a problem in poultry operations. It hides by day in crevices and nesting materials, moving onto the birds to feed at night. Skin lesions in chickens are usually inapparent but may occur as erythematous papules on any part of the body. Chronic or heavy infestations can be debilitating and result in skin irritation, loss of vigor, stunted growth, reduced egg production, anemia, and death due to exsanguination. Newly hatched chicks are particularly vulnerable. Setting hens may be driven from their nests, and susceptibility to disease agents may be significantly increased.

Dogs exposed at night to *D. gallinae* around poultry houses may react adversely to their bites. In addition to developing intense pruritus, they remain awake or howl at night, become less active during the day, and may show signs of depression. Skin lesions include erythematous papules, hyperpigmentation, and scaling, accompanied in some cases by partial loss of hair and slightly enlarged lymph nodes. Removal of an animal from the source of mites usually results in prompt alleviation of symptoms.

*D. gallinae* can develop from egg to adult in as few as 5 days, completing its life cycle in 9–10 days. Mating and oviposition occur off the host. The eggs typically are deposited in groups of 4–7 eggs, with a given female producing a total of 20–24 in her lifetime. The mites usually engorge to repletion at one feeding and lay their eggs at approximately 3-day intervals. The eggs hatch in 1–2 days to produce larvae that do not feed. The protonymphs and deutonymphs, like adults, feed on blood. The adults can endure starvation for several months, enabling them to survive extended periods in abandoned bird nests and unoccupied poultry houses.

### Macronyssidae

Most members of the Macronyssidae are obligate parasites of vertebrates. As a group they appear to have evolved on bats, from which they have secondarily transferred to other mammals, reptiles, and birds. Although most macronyssid mites cause little or no apparent harm

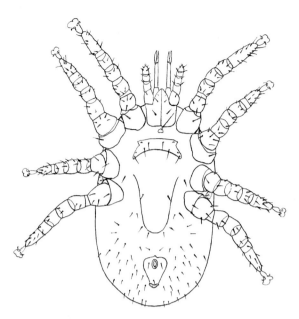

**FIGURE 23.24** *Radfordiella oudemansi* (Macronyssidae), female, ventral view. (From Strandtmann and Wharton, 1958.)

to their bat hosts, *Chiroptonyssus robustipes* has been known to cause the death of heavily infested, captive Brazilian free-tailed bats (*Tadarida brasiliensis*). This mite attaches primarily to the wings, where its feeding can result in increased vascularity and edema at the bite sites, enlargement of lymphatic vessels, hyperkeratosis, and excoriation of the stratum corneum (Sweatman, 1971).

A few macronyssid mites have invaded the oral mucosa of pollen-feeding and fruit-feeding bats (Psyllostomatidae). All are members of the genus *Radfordiella* (Fig. 23.24). In heavily infested bats they can cause bone damage to the hard palate and destruction of gingival tissues, resulting in loss of teeth. The damage is caused by protonymphs. The adult mites presumably are nidicolous and move onto the host only intermittently to feed on parts of the body other than the mouth (Phillips *et al.*, 1969).

### Tropical rat mite (*Ornithonyssus bacoti*)

The tropical rat mite (Fig. 23.6) occurs throughout the world, where it parasitizes primarily rodents. It occasionally infests cats, wild carnivores, chickens and other birds, and humans. Most veterinary problems involving this mite occur in laboratory mice, rats, and hamsters. Heavily infested animals may become debilitated and anemic or experience reduced reproductivity; death can occur in some cases. Infestations usually are recognized by the presence of blood-engorged deutonymphs and adults in the animal bedding, cages, and corners or crevices of cage racks. The tropical rate mite is a vector of *Litomosoides carinii*, a filarial nematode in cotton rats (see under Mite-Borne Diseases).

Blood-fed females lay their eggs in bedding or nest debris of their hosts and in cracks and crevices. The eggs hatch in 1–4 days, producing larvae that molt to protonymphs about 1 day later without feeding. The protonymphs feed on blood, molting to deutonymphs in 1–2 weeks. Deutonymphs molt to adults after 1–2 days without feeding. Development from egg to adult can be completed in 11–16 days under favorable conditions. Adults usually mate within 1–2 days following emergence and can survive several days to a few weeks without a blood meal. The tropical rat mite is found primarily in rodent nests, moving onto the host animals only to feed.

### Tropical fowl mite (*Ornithonyssus bursa*)

The tropical fowl mite (Fig. 23.7) is a common ectoparasite of wild and domestic birds throughout the warmer regions of the world. Domestic or peridomestic birds which may be infested include chickens, ducks, pigeons, starlings, house sparrows, and canaries. Most infestations originate from contact with wild birds or infested nest materials. Heavy infestations in chickens and other domestic fowl can result in anemia, decreased weight gain and egg production, and occasionally death. Newly hatched chicks and young birds are especially vulnerable. Blood-feeding by this mite causes skin irritation that may be intense enough to induce setting hens to leave their nests. Inspection of the plumage will reveal the mites in the down feathers, particularly around or just below the vent. Infested feathers are soiled or dirty in appearance due to the accumulation of mites, exuviae, eggs, and excreta. In the case of young birds, the mite commonly occurs around the eyes and beak.

*O. bursa* may cause more problems for wild birds than previously suspected. In Denmark, for example, nest infestations of the barn swallow (*Hirundo rustica*) have been shown to decrease reproductive success by reducing clutch sizes, nesting periods, and number of fledglings and by lengthening the time between clutches and the incubation period (Möller, 1990). Barn swallows regularly reuse old nests but tend to avoid nests infested with this mite.

The tropical fowl mite lays its eggs on the host or in nesting materials, where they hatch in 2–3 days. The larvae do not feed, whereas the protonymphs and adults feed intermittently on host blood. Although relatively few details about its life history have been reported, this mite is believed to develop from egg to adult in 6–8 days. In the absence of a host, it can survive about 10 days.

### Northern fowl mite (*Ornithonyssus sylviarum*)

The northern fowl mite (Fig. 23.8) is a major pest of chickens and other domestic fowl, particularly in temperate regions of North America and Eurasia. It also is an economic pest of domestic fowl in Australia, New Zealand, and other parts of the world where it has been introduced. In addition to infesting chickens, it is commonly found in the nests of pigeons and various wild birds and as an incidental pest biting rodents, hamsters, and humans.

The greatest economic impact occurs in chicken houses. Initial infestations usually occur via wild birds or newly acquired chickens already infested with the mite. They can then spread throughout even the largest houses within a few weeks. The northern fowl mite causes problems very similar to those caused by *O. bursa*. These include skin irritation without apparent lesions at the bite sites; matted and grayish feathers, especially around the vent where the mites, their exuviae, and feces are concentrated (Fig. 23.25); scaly, scabby, or thickened skin; and general loss of thriftiness. Individual caged birds may have 10,000 or more mites. Heavily infested birds often become anemic, experience decreased weight gains and egg production, and sometimes die. Eggshells may become significantly thickened, and egg production may drop 5–15% compared to uninfested birds. The greatest effect on body weight and efficiency of feed conversion generally occurs when hens are infested with mites before they reach full egg production. Although pathogens of poultry, such as viruses that cause *Newcastle disease* and *fowlpox*, have been recovered from *O. sylviarum* after feeding on infected chickens, there is little or no evidence that the mite transmits these agents when it bites.

The northern fowl mite spends most of its time on the host. It also occurs, however, in nesting materials and nest debris, roosting areas, cracks and crevices in the floors, and walls of chicken houses, where they can be found during the day or night. Oviposition usually occurs on the host about 2 days following complete engorgement with blood. Eggs typically are laid two or three at a time, but there may be as many as five. The eggs hatch in about 24 hr. The protonymph, the only immature stage that feeds on blood, requires at least two feedings before it molts to produce the deutonymph. The complete life cycle is typically 5–7 days, enabling populations to build very rapidly. Survival time without a host is usually 3–4 days but may be as long as 2–3 weeks.

**Snake mite** (*Ophionyssus natricis*)

This mite (Fig. 23.9) is a common parasite on captive snakes, where it is usually found around the eyes and under chin scales. It also may be found on captive lizards and other reptiles. It behaves much like a nest parasite and is believed to have evolved from a mite associated with rodents. The sources of infestation are usually caged snakes in pet stores, zoos, or laboratories. Although *O. natricis* occasionally is found in low numbers (< 100 per host) on wild snakes, its populations on captive snakes may reach hundreds or thousands per host. Heavy infestations cause listless behavior, loss of appetite, anemia, skin irritation, loss of scales, and, in some cases, death. Several hundred mites are sufficient to cause severe anemia and other symptoms.

*O. natricis* is a vector of the bacterium *Aeromonas hydrophila* that causes *hemorrhagic septicemia* in snakes. Infected snakes hemorrhage internally and often die 3–4 days following infection.

The snake mite lays its eggs in crevices, debris, and on rough surfaces of cages, where they hatch in 2–4 days. The larvae do not feed, whereas the protonymphs and adults are obligate parasites that feed exclusively on blood. The life cycle usually is completed in 2–3 weeks at room temperature, with adult females living 5–6 weeks. Females typically feed 2 or 3 times, depositing about 20–25 eggs after each blood meal. See Camin (1953) for further details on the biology and behavior of this mite.

## Laelapidae

Laelapid mites are commonly associated with rodents, other nest-building mammals, and bird nests. Those of medical or veterinary concern belong to the subfamilies Laelapinae and Haemogamasinae, which includes both facultative and obligate parasites that generally cause little or no apparent harm to their hosts. The genera most commonly encountered by veterinarians are *Androlaelaps*, *Haemogamasus*, and *Laelaps*. *Haemogamasus liponyssoides* is an opportunistic blood feeder that inhabits nests of rodents and shrews. Its slender, chelate chelicerae are capable of piercing skin to feed on blood, enabling this mite to feed on laboratory mice if they become infested.

**FIGURE 23.25** Northern fowl mite, *Ornithonyssus sylviarum* (Macronyssidae); heavy infestation of chicken in vent area. (Photo by Jerry F. Butler.)

## Spiny rat mite *(Laelaps echidninus)*

The spiny rate mite occurs throughout most of the world as an ectoparasite associated primarily with the Black and Norway rats. Occasionally it is found on *Sigmodon* and *Rattus* species, the house mouse, and other domestic and wild rodents. It is rarely found on laboratory animals. This mite is generally easy to recognize by its large body size (ca. 1 mm long), heavy sclerotization, and long, stout body setae that give it a spiny appearance. It lives primarily in host bedding or nesting materials, moving onto the host at night to feed. Its chelicerae are not capable of piercing intact skin but instead assist the mite in feeding on lachrymal secretions and blood or serous exudates from abraded skin. Rarely does its feeding cause discernible lesions, although injury to the footpads of suckling mice has been reported. The spiny rat mite is a vector for two blood-protozoan parasites, *Hepatozoon muris* of rats and *H. griseisciuri* of the gray squirrel (*Sciurus carolinensis*).

Regular blood meals are required for *L. echidninus* to survive and reproduce. Blood-fed females give birth to live larvae which do not feed. The protonymphs and deutonymphs both apparently feed similarly to adults, completing their development to adults in 1–3 weeks. The length of their life cycle is variable, requiring at least 16 days. The females can live 2–3 months; without food, however, they survive only about a week.

## Trombiculidae

Although not widely recognized as a problem, larvae of trombiculid mites (*chiggers*) commonly infest domestic animals. Only in cases of heavy infestation or sensitivity reactions are they likely to be brought to the attention of veterinarians. As in humans, the resultant dermatitis is a response to chiggers that are normally parasitic on other host animals. Only incidentally do they attach to atypical hosts such as cats, dogs, sheep, other livestock, or, occasionally, to domestic or pet birds. Most cases involve mild pruritus and are likely to go unnoticed. Cases of heavy infestations, however, can result in severe itching with formation of vesicles and crusty or scabby skin lesions, usually about the head and neck. Large numbers of engorged chiggers may be visible as orange patches associated with the lesions. The chiggers typically remain attached only up to 2–3 days. Treating the lesions with an acaricide and preventing secondary infection usually resolves the problem if the animal is not reinfested.

Some chiggers enter the skin via large hair follicles, crawling down the shaft of the hair, sometimes well beneath the skin surface. An extension of the stylostome, or feeding tube, may extend backward around the mite to form a hyaline capsule. Usually these *capsule-forming chiggers* are completely intradermal and may cause localized inflammation and edema. In some cases they induce formation of cysts at the base of the hair follicles that can lead to secondary infections and slow-healing lesions. *Intradermal chiggers* include members of the trombiculid genera *Cheladonta, Euschoengastia, Gahrliepia, Guntheria, Intercutestrix,* and *Schoutedenichia* (Sweatman, 1971). These mites occur primarily in Southeast Asia, the South Pacific islands, Australia, Africa, and other parts of the Old World. Rodents, shrews, and bandicoots are some of their more common natural hosts.

A few species of chiggers have been identified as the cause of *trombiculosis* in domestic cats. In the United States, *Walchia americana* is known to cause papules on the face, ears, and thoracic areas of cats, in addition to thickening and crusting of the skin on the abdomen and legs. This is accompanied by hyperkeratosis, eosinophilia, and infiltration of mast cells, as evidenced in skin biopsies at the lesion site (Lowenstine *et al.,* 1979). Large numbers of *Eutrombicula alfreddugesi* have been observed as distinct orange patches on the head and ears of a cat in North Carolina (USA), causing inapparent dermatitis (Hardison, 1977). Other chiggers known to infest cats are *Odontacarus adelaidiae, O. australiensis, Schoengastia philippinensis,* and *S. westraliensis* in Australia. Natural hosts for these mites include wallabies, grey kangaroos, and wild pigs (Wilson-Hanson and Prescott, 1985). Other cases in cats involving unidentified chigger species have been reported in Australia. Lesions occurred as pinpoint erythemas and orange crusts on the ears (pinnae), pruritus, papules and orange crusty lesions about the eyes and face, conjunctivitis, and ocular discharges. In one case, swelling and irritation of the perineal region with concomitant inability to pass urine was attributed to trombiculid mites (Wilson-Hanson and Prescott, 1985).

Dogs appear to be less commonly bothered by chiggers. Bite reactions are similar to those of other host animals, with localized redness, pruritus, and development of papules or vesicles at the bite sites. Cases involving heavy infestations may warrant veterinary attention. In Europe, the harvest mite (*Neotrombicula autumnalis*) reportedly has caused nervous symptoms in dogs, including partial paralysis of the limbs and lameness (Prosl *et al.,* 1985).

Virtually all species of livestock are subject to chigger infestations while grazing, walking paths to and from barns, or being held in enclosures or by contact with recently harvested hay or grains infested by these mites. Skin lesions in the form of papules or crusty eruptions can be irritating and lead to self-inflicted skin damage as the host animal rubs and abrades the affected areas. Lesions occur primarily on the lips, muzzle, face, feet, and belly. Pigs have developed a generalized pruritus after feeding on fresh, chigger-infested grains from automatic feeders. Sheep, goats, and cattle are particularly prone to infestations with *N. autumnalis* in Europe during the harvest season; this mite causes pruritus, scabs, and loss of hair, particularly about the head and neck. In Australia, sheep have experienced severe dermatitis on the legs and feet due to infestations by *Eutrombicula sarcina,* a chigger that normally parasitizes kangaroos. An orflike condition in sheep, caused by a *Guntheria* species, has been reported during the summer months in South Africa (Otto and Jordaan, 1992).

Domestic birds such as chickens may become parasitized by chiggers (e.g., *Neoschoengastia americana*), leading to itching and dermatitis. In most cases the mites can be found under the wings or around the vent. Reports of anemia in chickens attributed to heavy infestations of chiggers should be treated with skepticism; chiggers feed on dermal tissues and not on blood. Occasionally other captive birds may be affected. A chronic infestation of canaries by an unidentified trombiculid mite has been reported in Australia. The canaries, in a commercial aviary, developed nonpruritic, subcutaneous swellings of the legs and ventral trunk, with acute inflammation and skin necrosis at the sites of mite attachment (Pass and Sue, 1983).

Wild animals generally do not show adverse reactions to chiggers, even when heavily infested. Occasionally, however, they do react severely to bites of certain species that normally parasitize other hosts. Reactions include formation of vesicular or crusty lesions, slow-healing eschars, localized skin discoloration, and some loss of hair. Examinations often reveal orange or red clusters of mites about the head, ears, neck, axillae, or groin. Sometimes infestations of chiggers about the eyes cause ocular lesions in the form of pruritic eyelids and conjunctivitis. Snakes, lizards, skinks, and other reptiles are parasitized by chiggers, most commonly noticed as orange or red clusters of mites on the head and neck. The host seldom shows apparent harm, even in cases of individual snakes infested with several thousand mites.

## Fur Mites

Certain families of mites are categorized as *fur mites* because they are specially adapted for living in the hair coat of mammalian hosts. They often exhibit striking modifications of the palps, legs, and other body structures for grasping or clinging to hair. The five groups of particular veterinary interest are the cheyletoid families Cheyletidae (including the former family Cheyletiellidae) and Myobiidae and the astigmatid families Listrophoridae, Atopomelidae, and Myocoptidae.

### Cheyletidae

Parasitic cheyletid mites occur on domestic cats, dogs, and rabbits, as well as many wild mammals and birds. They are nonburrowing mites that live in the pelage of their hosts and feed on lymph and other tissue fluids by piercing the epidermis with their stylet-like chelicerae. The enlarged gnathosoma and pair of large, terminal palpal claws give cheyletid mites a characteristic appearance. These structures are used to secure the mites to their hosts and to assist them in inserting their chelicerae. Members of the genus *Cheyletiella* can cause problems that warrant veterinary attention. Although most cases of *cheyletiellosis* go unnoticed, infestations of these mites can cause eczema-like skin conditions, or *cheyletid mange,* with associated pruritus and hair loss. Three *Cheyletiella* species of veterinary importance are *Cheyletiella blakei* of cats (Fig. 23.26), *C. yasguri* of dogs, and *C. parasitivorax* of rabbits (Fig. 23.27). All developmental stages of these mites occur on the host animal. The eggs are glued to hairs but can be dislodged with loose hairs by host grooming. They also can be ingested and passed in the feces. The presence of *Cheyletiella* eggs in cat and dog feces thus serves as evidence of mite infestations even in asymptomatic cases (Fox and Hewes, 1976; McGarry, 1993). Transmission is usually by direct contact with infested animals, including maternal transfer while nursing. Because *Cheyletiella* species can survive up to 10 days or more off a host, animal bedding, household furniture, blankets, and carpets frequented by pets can serve as other sources of these mites. *Cheyletiella* mites are commonly phoretic on cat and dog fleas (*Ctenocephalides* spp.) and also may be transmitted via these ectoparasites.

*C. blakei* usually infests the facial area of cats. Heavy infestations can result in the formation of small, crusty, erythematous papules and loss of hair, accompanied by itching and scratching. Long-haired cats tend to be more commonly infested than short-haired cats and are more likely to be involved in human cases of cheyletiellosis. *C. yasguri* parasitizes domestic dogs, particularly in Europe and North America. It is generally less common than *C. blakei* and only occasionally causes problems warranting veterinary attention. Signs of an infestation are scratching and a mealy or powdery dandruff in the affected areas, commonly the lower back. Heavy infestations can cause scaling, hyperkeratosis and thickening of the skin, erythema, pruritus, and hair loss. Puppies tend to have a higher incidence of *C. yasguri* than do adult dogs and are more likely to exhibit pruritus. This mite can cause dermatitis in humans upon close contact with infested dogs, especially puppies (Fig. 23.28). Acaricidal treatments of dogs and their surroundings are effective in controlling *C. yasguri.* This is especially important in kennels, which serve as a common source of mite infestations.

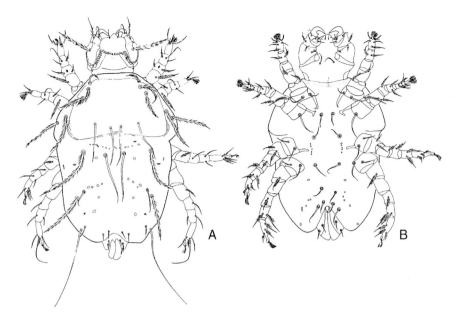

**FIGURE 23.26**    *Cheyletiella blakei* (Cheyletidae), female. (A) Dorsal view; (B) ventral view. (Modified from Domrow, 1991.)

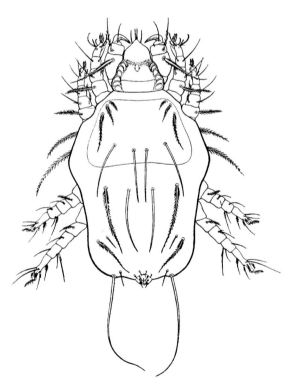

**FIGURE 23.27** Rabbit fur mite, *Cheyletiella parasitivorax* (Cheyletidae), female, dorsal view. (From Hirst, 1922.)

*C. parasitivorax,* the *rabbit fur mite,* is a common parasite of the European rabbit (*Oryctolagus cuniculus*) in North America, Europe, Asia, Australia, and New Zealand. It occurs most frequently in the posterior-back region of infested hosts but also may occur on the face, frontal area, and other parts of the body. High mite populations induce the accumulation of epidermal scales and a scurfy appearance, leading in untreated animals to varying degrees of dermatitis, erythema, thickening of the skin, and hair loss. Severe cases may involve serous

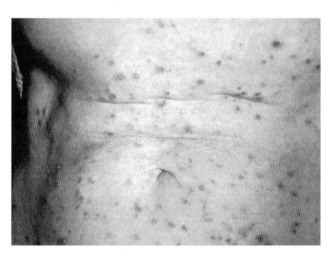

**FIGURE 23.28**    Multiple lesions from bites of *Cheyletiella yasguri* (Cheyletidae) on abdomen of woman following contact with infested puppy. (From Southcott, 1976.)

exudates and hairless patches in which the mites can be found in the disrupted keratin layer amidst epidermal debris. Infestations of *C. parasitivorax* are particularly a problem in commercial rabbit colonies and laboratories where rabbits are closely confined. Wild rabbits seldom exhibit cheyletiellid mange. *C. parasitivorax* is capable of transmitting myxomatosis virus among European rabbits in Australia.

For additional information on *Cheyletiella* species and their veterinary importance, see Smiley (1970) and van Bronswijk *et al.* (1976).

## Myobiidae

Members of the family Myobiidae are obligate parasites of rodents, bats, insectivores, and certain marsupials. They typically grasp the hairs of their host with their forelegs, which are often highly modified for this purpose. The mites move up and down the hair shaft and remain clinging to the hairs as they feed on epidermal fluids. Their chelicerae are long and stylet-like and adapted for puncturing thin epidermal tissues to feed on extracellular fluids. A few species, however, are known to feed on blood (e.g., *Blarinobia simplex* on shrews and *Eadiea brevihamata* on the shrew-mole). Most species cause little apparent discomfort or harm to their hosts, even when mite populations are high. Exceptions of veterinary interest are a few *Myobia* and *Radfordia* species that commonly infest rats and mice, often causing mild dermatitis and scurfiness in laboratory rodents.

The most widely recognized myobiid is the *mouse fur mite, Myobia musculi* (Fig. 23.29). This is a cosmopolitan, ubiquitous species that infests the pelage of both wild and captive house mice (*Mus musculus*). Most of what is known about the development and biology of myobiids is based on this species. Females deposit their eggs singly, gluing them to the bases of hair shafts. The developmental time from egg hatch to adult is about 23 days. All stages, including the larvae, feed on dermal tissue fluids. The host response varies greatly depending on the strain, sex, age, and sensitivity differences of individual mice. Lightly infested hosts are often asymptomatic or exhibit little adverse reaction. In highly sensitive hosts, however, even a few mites can elicit allergic reactions and severe pathologic responses. Heavy infestations can lead to severe dermatitis, with intense pruritus, hair loss, self-inflicted trauma from scratching, and, in some cases, death. This especially can be a problem in laboratory mice colonies where infestations are likely to involve virtually all individuals.

Two *Radfordia* species occur worldwide, infesting the fur of wild and laboratory rodents. The more common is *Radfordia ensifera*, parasitic on rats (Fig. 23.30); the other is *R. affinis,* parasitic on the house mouse. They

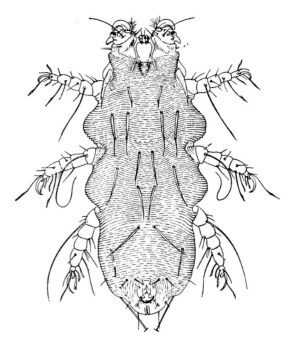

**FIGURE 23.29**   Mouse fur mite, *Myobia musculi* (Myobiidae), female, dorsal view. (From Baker *et al.*, 1956.)

closely resemble *Myobia musculi,* from which they are distinguished by a pair of tarsal claws (rather than one claw) on the second leg. Although they generally cause little pathologic effect, dermatitis and self-inflicted trauma have been associated with heavy infestations of *R. ensifera* on laboratory rats.

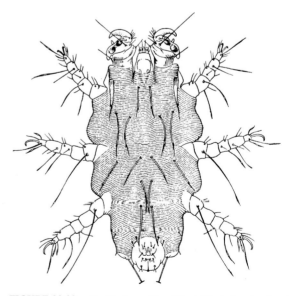

**FIGURE 23.30**   *Radfordia ensifera* (Myobiidae), female, dorsal view. (From Baker *et al.*, 1956.)

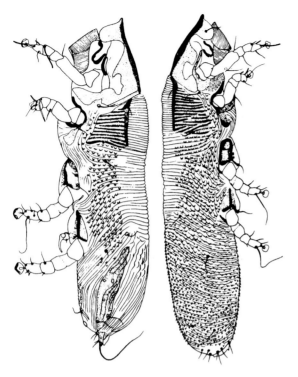

**FIGURE 23.31** *Listrophorus synaptomys* (Listrophoridae), male (left), female (right), lateral views. (From Whitaker, 1982; after Fain *et al.*, 1974.)

## Listrophoridae

Listrophorid mites (Fig. 23.31) are obligate parasites of rodents, lagomorphs, carnivores, and other mammalian hosts in the Old World and New World. They are well adapted for clasping securely to hair shafts, with appendages modified for this purpose. They attach so firmly that they are difficult to remove and can damage the hairs by bending or crimping them. They feed primarily on sebaceous secretions, usually causing little or no apparent harm even when mite numbers are high. Exceptions occur, however, in rodents and rabbits, especially under conditions of confinement or crowding. Heavy infestations in such cases can result in dermatitis, scratching, and hair loss.

Members of the family are cylindrical to laterally flattened and are distinguished by the anterior coxal fields that are expanded and flattened, with grooved surfaces that serve as attachment organs for grasping host hairs. Rabbits infested with *Leporacarus gibbus* may experience pruritus, hair loss, skin abrasion due to scratching, and occasionally damage to the fur as a result of nibbling. Wild populations of *Rattus* and *Mus* harbor species of *Afrolistrophorus* in tropical regions, but these have not been reported from laboratory populations. *Lynxacarus radovskyi*, which infests domestic cats, has been reported

to cause patches of mange and a scurfy appearance of cats in Hawaii (Tenorio, 1974).

## Atopomelidae

Atopomelid fur mites are parasites of marsupials, rodents, insectivores, and primates, primarily in the Southern Hemisphere. They attach to hair shafts with their anterior two pairs of legs. They are distinguished from the Listrophoridae by the lack of a projecting tegmen above the gnathosoma, and the striated forecoxal fields do not project around the host's hair.

Guinea pigs are commonly infested with *Chirodiscoides caviae*, particularly in laboratory colonies. This mite usually attaches to hairs on the back but may occur on any part of the body. Attachment to hairs is facilitated by their striated sternal area and legs 1 and 2, which are flattened and curved as clasping structures. Although most infestations in guinea pigs go unnoticed, hair loss and severe pruritus can occur; however, these are more likely the result of *Trixacarus caviae* (Sarcoptidae).

*Listrophoroides cucullatus* is widely distributed on commensal rats (*Rattus rattus*, *R. norvegicus*), primarily in tropical and subtropical areas. This species is rarely observed in laboratory colonies and is not known to cause damage.

## Myocoptidae

This family is superficially similar to the Listrophoridae. All stages occur in the pelage of their rodent and marsupial hosts, where they attach to hairs with their modified legs 3 and 4. In the female the tibia and tarsus of both pairs of legs fold against the striated inner surfaces of the genu and femur to provide efficient clasping organs. The most common member of this family is *Myocoptes musculinus* (Fig. 23.32), which infests wild and captive house mice throughout the world. It attaches its eggs singly to the lower part of the hair shaft and requires about 14 days to complete its life cycle. These mites feed on superficial epidermal tissues, with infestations usually going unnoticed. In laboratory mice, however, conditions often contribute to a buildup of mite numbers that can cause *myocoptic mange*. This is characterized by pruritus, erythema, development of a dull coat, and thinning of the hair due to physical damage by attached mites and scratching. Signs usually are first noticed on the neck and from there spread to the shoulders, back, and other parts of the body. Myocoptic mange generally is more severe in older mice or those with lower resistance.

A few other myocoptic mites occasionally cause mild dermatitis in rodent hosts. These include *Trichoecius romboutsi*, which is known to infest laboratory mice; *T. tenax*, which infests voles (*Microtus* spp.); and *Sciurocoptes sciurinus*, which infests squirrels.

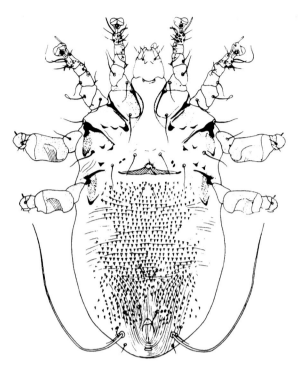

**FIGURE 23.32** *Myocoptes musculinus* (Myocoptidae), female, ventral view. (Modified from Whitaker, 1982; after Fain *et al.*, 1970.)

## FEATHER MITES

Mites representing 33 families and approximately 440 species live on or in the feathers of birds (Gaud and Atyeo, 1996). Their diversity as a group is reflected in their morphological adaptations for exploiting the many microhabitats that feathers provide. These include different types of feathers (e.g., primary and secondary flight feathers, wing coverts, contour feathers) and their location on the feathers. Mites that live on exposed feather surfaces tend to be more sclerotized, with a reduced chaetotaxy and prominent terminal body setae. Those that live in more protected sites (e.g., contour feathers or calmus) are generally less sclerotized, often with distinctly modified body forms and pretarsal structures. The majority of feather mites live on the surface of the feathers, where they feed primarily as saprophages on skin scales, feather debris, and oily secretions. Diatoms, fungal spores, and other organic materials also serve as food. Among the more common examples of such feather mites are members of the families Analgidae (Fig. 23.33), Proctophyllodidae (Fig. 23.34), and Pterolichidae (Fig. 23.35). The analgid mites *Meginia cubitalis* and *M. ginglymura* and the pterolichid mite *Pterolichus obtusus* occasionally cause dermatitis and reduced thriftiness in chickens.

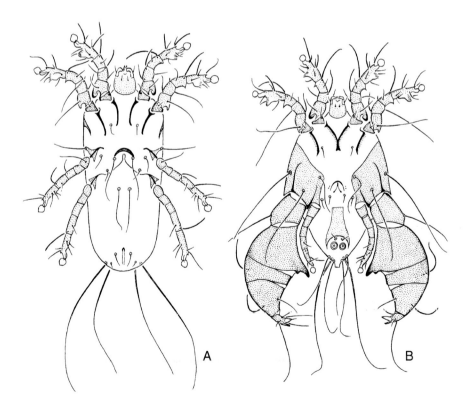

**FIGURE 23.33** *Analges chelopus* (Analgidae), ventral views. (A) Female; (B) male. (From Gaud and Atyeo, 1996.)

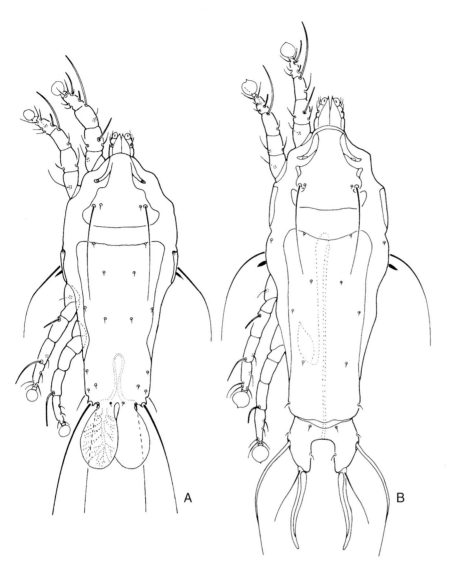

**FIGURE 23.34**  *Proctophyllodes glandarinus* (Proctophyllodidae), dorsal views. (A) Male; (B) female. (From Gaud and Atyeo, 1996.)

Other mites called *quill mites* live inside the base of feathers (calmus), where they feed on feather tissues or by piercing the calmus wall to feed on host fluids. Two such families are the Syringobiidae (Fig. 23.36) and Syringophilidae (Fig. 23.37). Like most feather mites, quill mites rarely cause apparent harm to their hosts and generally are not considered to be economically important. Occasionally, however, heavy infestations of *Syringophilus bipectinatus* cause feather loss in chickens.

## MANGE MITES

The following families include species that cause mange in animals, including livestock, poultry, companion animals, and laboratory colonies: Epidermoptidae, Knemidokoptidae, Laminosioptidae, Demodicidae, Psorergatidae, Sarcoptidae, Psoroptidae, Harpyrhynchidae, and Hypoderatidae.

### Epidermoptidae

Several species of the Epidermoptidae have been reported to cause discomfort and injury to infested birds. Although commonly referred to as feather mites, they are more appropriately called *avian skin mites*, as reflected in the family name. They generally live on the skin surface or in feather follicles, where their feeding can lead to itching, pityriasis (scaly or scabby dermatitis), and various other types of superficial skin lesions. *Epidermoptes bilobatus* (Fig. 23.38) commonly infests galliform birds

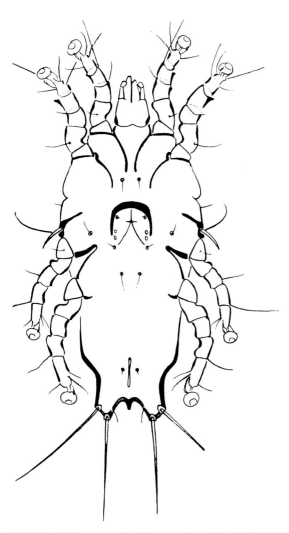

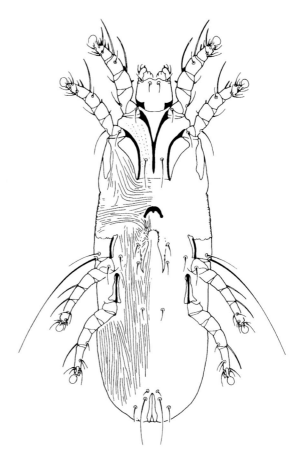

**FIGURE 23.36** *Longipedia tricalcarata* (Syringobiidae), female, ventral view. (From Gaud and Atyeo, 1996.)

**FIGURE 23.35** *Musophagobius cystodorus* (Pterolichidae), female, ventral view. (From Gaud and Atyeo, 1996.)

and occasionally causes pityriasis in chickens, whereas *E. odontophori* has been reported to cause mange in African birds. *Microlichus avus* and *M. americanus* are known to produce crateriform skin lesions and severe mange in several avian species. *Myialges* species (e.g., *Myialges macdonaldi*) may invade the outermost skin layers to produce pityriasis and mange, sometimes severe enough to cause feather loss. Heavy infestations of *Rivoltasia bifurcata* on chickens can result in intense itching and pityriasis, especially involving the head. For further information on these and other epidermoptid genera of veterinary interest, see Fain (1965) and Krantz (1978).

### Knemidokoptidae

Knemidokoptid mites superficially resemble sarcoptids, from which they differ by having short legs without

pretarsi or long setae and by lacking dorsal triangular setae. They invade the feather follicles and skin of wild and domestic birds worldwide, causing *knemidokoptic mange* in some species. Their life cycle is similar to that of *Sarcoptes scabiei*. All stages of these mites occur on the host, and transmission is by direct contact with infested birds. There are several species of veterinary importance: *Knemidokoptes mutans* and *Neocnemidocoptes gallinae* infesting poultry, *K. pilae* infesting parakeets, and *K. jamaicensis* infesting passerine birds, including canaries.

**Scaly-leg mite** (*Knemidokoptes mutans*)
This mite (Fig. 23.39) is a pest of poultry, especially chickens, in North America, Europe, and Africa and probably occurs worldwide. It burrows beneath the epidermal scales of the legs and feet, causing irritation, inflammation, hyperkeratization, formation of vesicles, and encrustations (Fig. 23.40). The crusts may cover entire limbs, hence the term *scaly-leg*, a condition most commonly seen in older birds. In chronic cases, infestations can lead to lameness, deformed legs and feet, and occasionally the loss of digits. The skin of the comb and

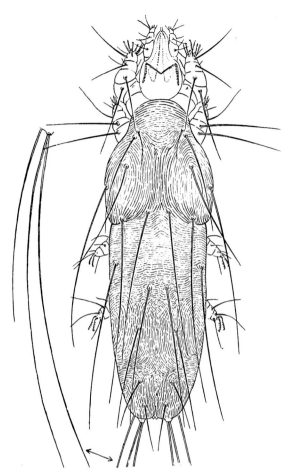

**FIGURE 23.37** *Syringophilus bipectinalis* (Syringophilidae), female, dorsal view. (From Baker *et al.*, 1956.)

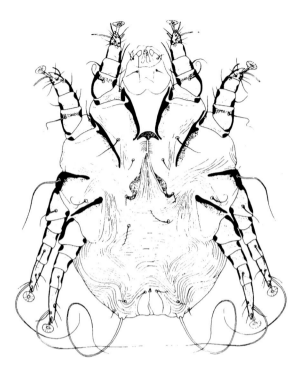

**FIGURE 23.38** *Epidermoptes bilobatus* (Epidermoptidae), female, ventral view. (From Gaud and Atyeo, 1996.)

wattle also may be involved. *K. jamaicensis* causes similar symptoms in wild and cage-reared passerine birds.

### Scaly-face mite (*Knemidokoptes pilae*)

This cosmopolitan mite infests captive parakeets, causing crusty lesions primarily about the face, head, and legs. Lesions usually appear initially in the ceres (at base of the upper beak) and at the corners of the beak where the mites invade the feather follicles and folds of skin. There they form pouchlike cavities, or pits, and a honeycomb pattern that is discernible on close examination. Early signs include whitish excrescences that may spread to the eyes, forehead, and other parts of the body. These form gray–white to yellow encrustations in chronically infested birds. Lesions of the legs and feet, especially in the early stages, can closely resemble those of *K. mutans*. Advanced cases may involve distortions of the beak as the keratinized tissues become overgrown and friable. Even in such cases, there appears to be little or no pruritus, nor is there rubbing or scratching of the beak.

### Depluming itch mite (*Neocnemidocoptes gallinae*)

Unlike the previous two mites, the depluming mite infests the feathered areas of chickens, burrowing into the epidermis at the base of feathers or into the feather shafts. The parts of the body most commonly affected are the head, neck, back, abdomen, and upper legs. The wings and tail are not usually involved. The mites tend to be confined to the stratum corneum, causing hyperkeratosis, thickening and wrinkling of the skin, and sloughing of keratinous layers. Feathers in the affected areas often break off or fall out or may be plucked by the bird, accounting for the mite's common name. Severely infested birds may become emaciated and die. A similar species, *Picicnemidocoptes laevis*, may cause similar symptoms in pigeons and doves.

### Laminosioptidae

The only significant laminosioptid mite of veterinary importance is the fowl cyst mite.

### Fowl cyst mite (*Laminosioptes cysticola*)

This mite (Fig. 23.41) occurs worldwide as a parasite of chickens, pheasants, turkeys, geese, pigeons, and other birds. It invades the skin of its avian hosts to form small, yellowish, subcutaneous nodules, or cysts, up to several millimeters in size. The nodules are formed by calcareous deposits produced around the mites after they have

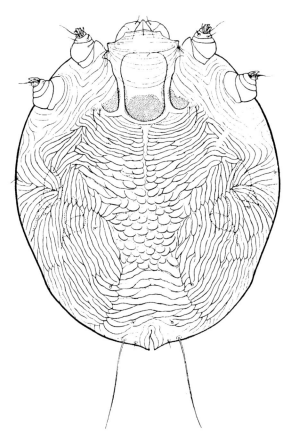

FIGURE 23.39 Scaly-leg mite, *Knemidokoptes mutans* (Knemidokoptidae), female, dorsal view. (From Hirst, 1922.)

FIGURE 23.40 Chicken with crusty, raised scales on feet due to infestation with scaly-leg mite (*Knemidokoptes mutans*). (Photo by Jerry F. Butler.)

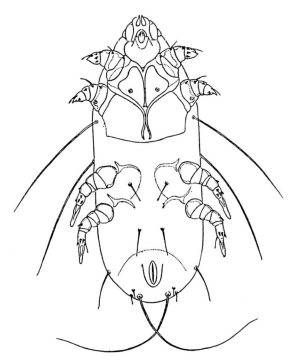

FIGURE 23.41 *Laminosioptes cysticola* (Laminosioptidae), female, ventral view. (From Hirst, 1922.)

died. They are most easily seen in living birds by wetting and parting the breast feathers and sliding the skin back and forth with the fingertips. Occasionally *L. cysticola* causes heavy infestations that can be fatal. A case was reported in West Virginia involving a wild turkey with severe neurologic disease (Smith *et al.*, 1997). The affected turkey held its head bent over its back when at rest and exhibited circling behavior and falling to one side when it attempted to walk. Histopathologic examination revealed *L. cysticola* mites in enlargements of the wing nerves, inflammation of the brain, and numerous mites in the esophagus, small intestine, and other internal tissues. The fowl cyst mite also has been reported to cause granulomatous pneumonia in dogs (Shaddock and Pakes, 1978). Little is known about the life history or transmission of this mite.

### Demodicidae

Mites in this family are highly specialized skin parasites that live in the hair follicles and associated glands of domestic and wild mammals. An infestation of demodicid mites is called *demodicosis*, whereas cases with clinical signs are called *demodectic mange*. The mites are very host-specific and typically occur either in hair follicles or dermal glands. These sites include sebaceous glands, modified sebaceous glands (e.g., meibomian, caudal, preputial, and vulval glands), modified sweat glands (e.g., ceruminous glands and submaxillary skin papillae), and mixed sebaceous/sweat glands (e.g., perianal glands). *Opthalmodex* infests lachrymal ducts. A few species burrow into the skin to form epidermal pits, as in *Demodex criceti* of hamsters. Other demodicid species invade oral tissues of their host, infesting the oral epithelium, tongue, and esophagus of the grasshopper mouse (*Onychomys leucogaster*) in western North America and the oral cavities of bats and lemurs in Europe and Africa. In most cases, demodicid mites cause little or no apparent harm to their hosts. In other cases, infestations can lead to varying degrees of dermatitis and other skin problems.

Demodectic mange is common in dogs; livestock such as cattle, goats, sheep, and swine; wild animals such as foxes, other canids, and rabbits; and occasionally laboratory animals such as hamsters, gerbils, guinea pigs, rats, and mice. It is relatively uncommon in cats and horses.

Injury to the host occurs as the mites puncture with their stylet-like chelicerae the epithelial cells lining the hair follicles and glands to feed on the cell contents. In most cases the host response is only mild to moderate hypertrophy of the affected epithelia. In other cases, marked hypertrophy and cell destruction may occur. The openings of hair follicles or the ducts of glands may become blocked, leading to the formation of dermal papules and nodules. Damage to the follicles can lead to hair loss, whereas secondary bacterial infections may cause inflammation, pruritus, and the formation of pustules. Lesions often occur first on the face and head, spreading from there to other parts of the body.

Two major clinical forms of demodicosis are recognized: squamous and papulonodular. *Squamous demodicosis* is the more common form, characterized by a dry, scaly dermatitis with itching and loss of hair in the affected areas. Secondary infections often result in the rupture of follicular cells, severe inflammation, and purulent exudates. This can be either a localized or generalized skin condition and occurs in all host groups. *Papulonodular demodicosis* occurs when the hair follicles or gland ducts become obstructed and produce palpable, cystlike or nodular swellings in the skin, trapping the mites within. The development of demodectic papules and nodules is most commonly seen in cattle, goats, and pigs. These lesions continue to enlarge as the mites multiply, sometimes reaching several thousand mites per lesion, along with accumulated cellular debris and glandular secretions. These nodules may rupture externally, leading to secondary infections and abscesses. In other cases they may rupture within the skin, introducing the mites to the circulatory and lymphatic systems. There they can cause thromboses and internal infestations.

### Dog follicle mite *(Demodex canis)*

This mite (Fig. 23.42) inhabits the hair follicles and sebaceous glands of dogs throughout the world. It

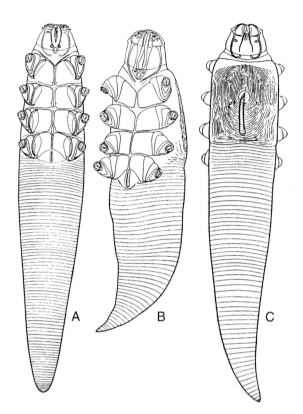

**FIGURE 23.42**   Follicle mites (Demodicidae) of domestic animals. (A) *Demodex canis;* (B) *D. phylloides;* (C) *D. cati.* (From Hirst, 1922.)

completes its life cycle in 3–4 weeks, with the eggs and all developmental stages being found in the follicles or glands. Clinical signs are most common in dogs less than a year old, presumably reflecting an immunodeficient state in young animals. *Canine demodectic mange* (Fig. 23.43) usually appears as mildly erythematous patches about the eyes and corners of the mouth, typically associated with hair loss. From there the infestation may spread to the forelegs and trunk as a typical squamous form of demodicosis. Most cases resolve themselves without treatment. In genetically predisposed or immunodepressed animals and cases of secondary infections, the condition can develop into chronically severe, moist, purulent dermatitis known as *pustular demodicosis*. This is often accompanied by an unpleasant odor variously described as rancid or mousy. If this becomes generalized with intense redness and tenderness of the skin and easily bleeds, it is called *red mange*. Severe cases of red mange also occur in foxes and other wild canids, sometimes leading to deaths in heavily infested animals. Diagnosis of *Demodex* infestations is confirmed by demonstrating the presence of mites in material expressed from hair follicles and in skin scrapings or biopsies of skin lesions. Although most dogs recover from localized infestations even without treatment,

**FIGURE 23.43** Demodectic mange in dog caused by *Demodex canis* (Demodicidae). (Photo by Jerry F. Butler.)

other cases may require extended acaricide applications; some are never resolved despite every treatment effort.

Most dogs apparently become infested with *D. canis* as newborn pups, either while nursing or in other intimate contact with their mother or another infested dog. There is little or no evidence to substantiate reports of prenatal transmission of the mite. Nor is the transfer of *D. canis* between mature animals likely, and this apparently occurs only in unusual circumstances.

### Cat follicle mite *(Demodex cati)*

Occasionally cats develop lesions attributed to *D. cati* and *D. gatoi*. This is usually the squamous form of demodicosis and occurs on the head or as a generalized condition with varying degrees of pruritus. Cases of *feline demodicosis* are believed to be associated with underlying immunosuppressive diseases (e.g., feline leukemia, diabetes mellitus).

### Cattle follicle mite *(Demodex bovis)*

Cattle infested with *D. bovis* commonly develop a papulonodular form of *bovine demodicosis*. Adult female mites deposit their eggs in hair follicles, where their populations may build to hundreds or thousands of mites per follicle as the follicles become dilated to form dermal papules or cysts. As they enlarge, they can be felt beneath the skin even though they may be difficult to see. Some female mites exit the follicular cysts to invade other hair follicles, thereby spreading the infestation. It is presumably at this time that they also are transferred to other animals by intimate contact. It is

postulated that transfer between cattle can occur during copulation.

The lesions tend to be concentrated on the anterior parts of cows, notably the neck, shoulders, and axillary region, but also may occur on the udder. Papular cysts enlarge to form granulomatous nodules when the follicular opening becomes blocked by mite bodies, keratin, and other debris. The occurrence of papulonodular demodicosis in cattle is commonly associated with cows that are stressed by pregnancy or lactation. Individual nodules typically form over the period of a month, then gradually disappear only to be replaced by other developing nodules. Both natural and acquired immunity play a role in reducing mite numbers and associated clinical signs in infested cattle.

The dilated follicles vary in size from that of a pinhead to that of a chicken egg. Not uncommonly, the larger nodules rupture to produce suppurative sores. The puslike exudate containing large numbers of *D. bovis* has the consistency of toothpaste and can serve as a means of transfer of mites to other animals. Skin damage resulting from these ruptured nodules can cause defects in raw leather and significant economic losses to the tanning industry in the form of diminished quality of processed cow hides.

A second species of demodicid mite of bovids, *D. tauri*, has been reported from hair follicles and ducts of sebaceous glands in eyelids of cattle in Czechoslovakia (Bukva, 1986).

### Goat follicle mite *(Demodex caprae)*

Goats develop dermal papules and nodules similar to those in cattle when infested with *D. caprae*. Cases of *caprine demodicosis* occur most commonly in young animals, pregnant does, and dairy goats, the latter presumably reflecting the stress of lactation. Papules usually appear on the face, neck, axillary region, or udder, with a few to several hundred lesions per animal. They are easily palpable in the skin, enlarging to form nodules up to 4 cm in diameter as the mites multiply within. Ruptured nodules tend to suppurate, contributing to transmission of the mites via exudates to other animals. As in dogs, goats have a high incidence of generalized demodicosis that can involve almost any part of the body. If the nodules rupture internally, granulomas develop while phagocytic giant cells of the goat host engulf and destroy the mites. As individual nodules disappear, others are formed. Transmission of *D. caprae* to newborn goats typically occurs within the first day following birth. Other possible means of transfer are parental licking and intimate contact of animals during copulation. Certain breeds of goats (e.g., Saanen) tend to be much more sensitive to demodicosis than are others.

## Psorergatidae

Psorergatid mites are obligate parasites of mammals. These mites live in superficial layers of the skin of their hosts. Their life histories are similar to those of the Demodicidae. Whereas most species cause no apparent harm to their hosts, a few species cause *psorergatid mange* in sheep, rodents, and certain primates. Transmission between animals is by direct contact and transfer of adult females which enter hair follicles, where they deposit their eggs. All developmental stages occur within the follicles where the mites feed by puncturing follicular cells. Mange-inducing *Psorergates* species cause inflammation and enlargement of the hair follicles to form dermal pouches, or pockets, beneath the stratum corneum. The lesions appear as small, white nodules (up to 2 mm in diameter) on the inner skin surface. The nodules contain all stages of the mite and necrotic debris from the destroyed follicles. In rodents, large lesions may form, with mites lining the inner surface of the lesions.

### Sheep itch mite *(Psorobia ovis)*

The psorergatid mite of most veterinary importance is the sheep itch mite, formerly named *Psorergates ovis.* It infests all breeds of domestic sheep, but especially Merinos. Clinical signs include dry scurfy skin, loss of hair, and sometimes erythema accompanied by intense pruritus. Infested animals often are restless and bite or rub the affected areas, damaging the skin and wool. Lesions occur most frequently on the neck and shoulders, and gradually may spread to the face, flanks, thighs, and other parts of the body. The spread of *P. ovis* through a flock is slow and is most evident during the winter months.

### Mouse itch mite *(Psorergates simplex)*

The mouse follicle mite (Fig. 23.44) infests wild and laboratory mice in Europe and North America. Its development and life history are similar to those of *P. ovis.* The dermal pouches and resultant white nodules occur most frequently on the head of infested mice but also may involve the neck, legs, and other parts of the body. Chronic infestations, especially in laboratory mice, can lead to crusty or ulcerous nodules, hypertrophy of skin cells, dermatitis, and hair loss.

   *Psorergates* mites are sometimes associated with cases of *murine ear mange,* evidenced by pale yellow crusts on the inner and outer ear surfaces. Such cases in *Mus musculus* and other rodent hosts probably involve species other than *P. simplex.*

   Other *Psorergates* species that cause lesions in rodents include *P. apodemi, P. dissimilis,* and *P. microti.* Cattle are hosts for *Psorobia bos,* which seldom causes

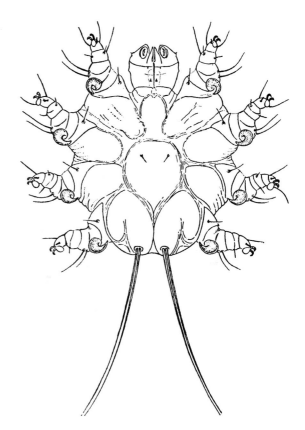

**FIGURE 23.44**   Mouse itch mite, *Psorergates simplex* (Psorergatidae), female, ventral view. (Adapted from Baker *et al.,* 1956.)

apparent lesions or harm to its bovine hosts. Occasionally, however, it is reported to cause psorergatid mange, as in an infested herd of Bonsmara bulls in South Africa (Oberem and Malan, 1984). *P. cercopitheci* reportedly causes mild dermatitis in mangabey monkeys (Sheldon, 1966) and vervet monkeys (Seier, 1985), whereas unspecified psorergatid species are known to cause dermal cysts and crusty skin in platas monkeys (Raulston, 1972) and macaques (Lee *et al.,* 1981).

## Sarcoptidae

The family Sarcoptidae includes three important genera that infest domestic and wild animals: *Sarcoptes, Notoedres,* and *Trixacarus.* The adult females burrow into the epidermis of their hosts, causing varying degrees of dermatitis with accompanying erythema, pruritus, hair loss, scaling, and dermal encrustations characteristic of *sarcoptic mange* (Fig. 23.45). Lesions may occur on any part of the body but often begin in typical locations that depend on the mite species involved.

   For additional information on sarcoptid taxonomy, phylogenetic relationships, and host associations, see Klompen (1992).

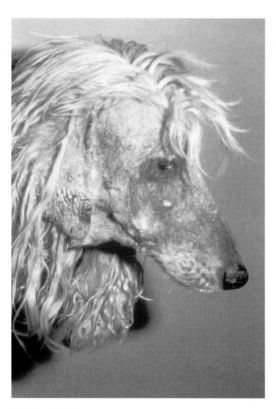

**FIGURE 23.45** Severe case of sarcoptic mange in dog, caused by *Sarcoptes scabiei* var. *canis* (Sarcoptidae). (Photo by John M. MacDonald, Auburn University College of Veterinary Medicine.)

## Scabies mite *(Sarcoptes scabiei)*

Most *Sarcoptes* mites that cause mange in animals are morphologically indistinguishable from the mite that causes human scabies, *S. scabiei*. Their life histories are very similar, with all developmental stages (larva, protonymph, tritonymph, adult) living in burrows formed by the adult females in the stratum corneum, stratum lucidum, and upper Malpighian layer of the skin. The female lays up to three eggs per day in the burrow over a period of 2–3 weeks, following which she dies. Development from egg to adult takes 2–3 weeks and occurs largely within the burrows. Male protonymphs leave the burrow to establish new epidermal tracts, in which they molt to adults. The adult males then mate with females either on the skin surface or in shallow dermal pits.

Host reactions occur primarily in response to the mites and their fecal deposits in the burrows. This usually occurs 3 weeks or more after the initial infestation, with the reaction time becoming much shorter (e.g., a few days) after subsequent exposures. Initial lesions can occur anywhere on the body but are usually localized where the hair tends to be thin, most commonly on the head. From there the infestation can spread quickly to cause a more generalized mange. Infestations generally appear as papular eruptions with erythema, pruritus, and hair loss. As it progresses, skin in the affected areas often becomes thickened, crusted with exudates, and secondarily infected following excoriation of the skin due to scratching and rubbing by the host. Scaly areas around the periphery of infested patches often indicate the spread of mites. In extreme cases, severely sensitized animals may experience weight loss, difficulty eating, impaired hearing, blindness, exhaustion, and death.

Burrows seldom are detectable in nonhuman animals. This makes it difficult to recover mites to confirm their identification. Negative skin scrapings therefore are not conclusive. As a result, the diagnosis of scabies in animals is often presumptive, based on clinical signs and positive responses to acaricide treatments.

The various subspecies, or races, of *S. scabiei* tend to be relatively host-specific, infesting a range of domestic and wild mammals. Transfer of mites occurs among conspecific hosts by direct contact. Transfer between different host species, when it does occur, often results in only temporary infestations, usually limited to a transient, mild dermatitis. Such a reaction in humans (e.g., from *S. scabiei* of dogs or goats) is called *animal scabies*.

Virtually all domestic animals, except cats and guinea pigs, are subject to infestations of *S. scabiei*. Dogs are the most commonly affected (Fig. 23.45). Initial lesions tend to occur on thinly haired parts of the body such as the ear margins, belly, axillary and inguinal regions, elbows, and hocks. If untreated, the infestations generally spread to the head and other parts of the body. In particularly severe cases, dogs may develop thickened, pigmented skin with almost complete hair loss in affected areas of the neck, shoulders, back, trunk, and extremities. The situation often is complicated by secondary infections, self-inflicted trauma in efforts to relieve the itching, emaciation, and sometimes death. Conditions which may be mistaken for canine scabies include seborrhea, eczema, allergic dermatitis, ringworm dermatophytosis, and infestations of other mange mites (notably *Demodex, Notoedres,* and *Otodectes* spp.). Detection of mites requires deep skin scrapings, with best results being obtained from the tips of the ears even in the absence of ear lesions.

Initial lesions in farm animals are usually localized in specific body regions. In sheep, goats, and horses this is typically on the face, head, and neck. In cows, lesions usually first appear on the underside of the neck, inner thigh, brisket, and tail head. In pigs the lesions tend to be more generalized (Fig. 23.46) but are most commonly evident as encrustations and scabs in the ears of chronically infested sows. The highest incidences of *S. scabiei* infestations in farm animals usually occur

**FIGURE 23.46**   Sarcoptic mange in pig, caused by *Sarcoptes scabiei* var. *suis* (Sarcoptidae). Note inflammation of skin and hair loss. (Photo by John M. MacDonald, Auburn University College of Veterinary Medicine.)

in the winter months and are attributed to crowding of animals and loss of condition at that time of the year.

In addition to direct transfer between mature animals, *S. scabiei* also is transmitted from mother to offspring at birth. Pruritus has been reported in 4-day-old piglets born to infested sows. Maternal antibodies to *S. scabiei* are detectable in neonatal pigs within 6 hr, reaching their maximum levels 24–48 hr after birth (Bornstein and Zakrisson, 1993).

Although cases of sarcoptic mange in goats often resolve themselves without developing severe signs, heavily infested goats may exhibit crusty lesions and extensive hair loss around the muzzle, eyes, and ears; lesions on the inner thighs extending to the hocks, brisket, ventral abdomen, and axillary region; dermal thickening and wrinkling on the scrotum and ears; and dry, scaly skin on all parts of the body, especially in areas of hair loss (Kambarage, 1992). Sarcoptic mange also can cause significant problems in camels (Kumar *et al.*, 1992).

*S. scabiei* may also infest laboratory animals. Canine scabies is common in laboratory dogs obtained from commercial suppliers and animal shelters, with the highest incidence in young dogs and short-haired breeds. Laboratory rabbits usually develop initial lesions on the head, ears, and legs, followed by a more generalized dermatitis with associated erythema, pruritus, scaling of skin, and loss of fur. The situation can be complicated by scratching and self-inflicted injuries within the confines of a cage. Other laboratory animals are infrequently or rarely subject to sarcoptic mange. These include mice, hamsters, and guinea pigs.

In addition to human hosts, *S. scabiei* has been reported infesting captive *cynomolgus monkeys* and *siminan primates* (orangutan, gibbons, and chimpanzees). Lesions are characterized by thickening of the skin of the neck, shoulders, back, and sometimes the lower trunk and extremities. This may be associated with other signs such as skin scales, pruritus, hair loss, and emaciation.

For further information on sarcoptic mange in laboratory animals, see Yunker (1973).

Several groups of wild mammals are susceptible to sarcoptic mange by *S. scabiei*. Members of the Canidae and Cervidae are the more commonly infested. Canid hosts are the most severely affected carnivores in North America, notably the *red fox* (*Vulpes fulva*), *gray fox* (*Urocyon cinereoargenteus*), *wolf* (*Canis lupus*), and *coyote* (*C. latrans*). The most severe cases have been reported in red foxes during the winter months. In the northeastern United States, eastern Canada and Russia, epizootic sarcoptic mange has had an economic impact on the pelt industry in some years. Early signs of mite infestations are dry, flaky skin, followed in succeeding weeks and months by crusty lesions, hair loss, eyelid scaling, emaciation, scratching, biting, and even death. Affected areas include the muzzle, neck, shoulders, back, and hind-quarters. Successful transfer of *S. scabiei* from red foxes to domestic dogs, gray foxes, and feral dog/coyote hybrids has been demonstrated experimentally (Stone *et al.*, 1972).

Cervid species that are known to develop sarcoptic mange include *roe deer* (*Capreolus capreolus*) and *red deer* (*Cervus elaphus*) in Europe and *wapiti* (*C. elaphus*) in North America. Lesions in deer occur as encrustations of the outer ear, whereas in wapiti they may appear as moist scabs in the dorsal and lateral thoracic regions. Some cases involve head lesions, impaired vision, blindness, and general debilitation.

Other animals that are known to develop sarcoptic mange are the *fisher* (*Martes pennanti*) and *ferrets* (*Mustela* spp.) in North America, Europe, and Asia; *llamas* (*Lama* spp.) in Central and South America; *Thompson's gazelle* (*Gazella thompsonii*) and *wildebeests* (*Connochaetes taurinus*) in Africa; and *chamois* (*Rupicapra rupicapra*) in Europe and the Caucasus. There is evidence in the case of chamois that mineral imbalances may contribute to the severity of mange cases. Providing mineral blocks or salt licks has been shown to significantly reduce mange in infested animals (Onderscheka *et al.*, 1968). Zoo animals also become infested with *S. scabiei* as reported in the *capybara* (*Hydrochoerus hydrochoeris*), *tapir* (*Tapirus* sp.), and *camel* (*Camelus* sp.) in Poland (Zuchowska, 1991).

Wild primates may develop sarcoptic mange, sometimes in epizootic form. Such epizootics among *chimpanzees* have resulted in noticeable hair loss. Severe cases in white-headed capuchins attributed to *S. scabiei* have been characterized by abscesses, localized hemorrhages, stratified crusts, weight loss, extreme debilitation, epileptic excitations, and death in some cases (Sweatman, 1971). It is probable, however, that another sarcoptid mite was involved.

## Other Sarcoptid Genera

Sarcoptid mites in genera other than *Sarcoptes* also naturally infest primates. These species are very similar morphologically and in the damage they inflict. Typical signs are scabby, encrusted papular lesions, intense pruritus, and hair loss. *Prosarcoptes talapoini* and *P. pitheci* cause mange in *guenons* (*Cercopithecus* spp.) and *baboons* (*Papio* spp.), and *P. scanloni* causes mange in *macaques* (*Macaca* spp.). *Kutzerocoptes gruenbergi* causes a similar condition in *capuchin monkeys* (*Cebus capucinua*). For further information on sarcoptid mites infesting primates, see Fain (1968), Smiley and OConnor (1980), and Klompen (1992).

The following genera of sarcoptid mites are restricted to bats: *Chirophagoides* (New Zealand), *Chirnyssoides* (Neotropics), *Nycteridocoptes* (Europe, Asia, Africa), *Cynopterocoptes* (Asia), *Rousettocoptes* (Asia), *Tychosarcoptes* (Asia), *Teinocoptes* (Africa, Asia, Australia), and *Chirobia* (Africa, Asia). Most species of *Notoedres* are also restricted to bats hosts. Whereas some bat hosts exhibit little adverse reaction, others develop pustules, scabby lesions, or small cornified pouches, nodules, or cysts containing the mites. These lesions can occur on any part of the body, including the ears and wing membranes. In some cases the mites burrow into the skin, particularly the anterior part of the body, or excavate shallow epidermal pits in the stratum corneum where the mites can be found. For further details on sarcoptid mites of bats, see Sweatman (1971).

**Notoedres species** The genus *Notoedres* is similar to *Sarcoptes*. These mites differ from *Sarcoptes*, however, by having the anal opening located dorsally and by lacking dorsal spines. They are skin parasites which typically burrow in the stratum corneum, causing inflammation, crusty lesions, and hair loss known as *notoedric mange*. Common hosts include cats, rats, squirrels, rabbits, and bats. *Notoedres* species have been reported infrequently parasitizing dogs and foxes, lorises (primates), koalas and bandicoots (marsupials), hamsters, and hedgehogs. Most infestations begin on the head, causing a condition called *head mange*, commonly observed in cats and rabbits. From there the mites spread to other parts of the body. Severe cases can lead to dehydration, emaciation, and death of the host. Transmission between animals is by

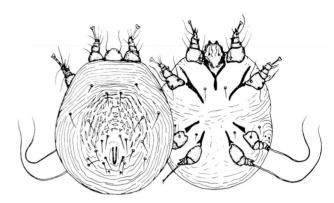

**FIGURE 23.47** *Notoedres cati* (Sarcoptidae), female; dorsal (left) and ventral (right) views. (From Nutting, 1984)

direct contact, and only rarely are humans affected. Diagnoses are based on the clinical pattern of lesions and identification of the mites in skin scrapings. *Notoedres* mites are much more readily recovered in skin scrapings than are *Sarcoptes* mites.

**Notoedric cat mite** (*Notoedres cati*)
This is the common *Notoedres* mite (Fig. 23.47) of domestic cats in North America, Europe, and Africa, although it probably occurs worldwide. It also infests wild cats, laboratory rabbits, and, rarely, dogs, foxes, other canids, and civets. As the adult female burrows in the skin, she deposits eggs, which hatch in 3–4 days. Development from egg to adult requires 6–10 days. Although *Notoedres cati* typically burrows in the stratum corneum and stratum germinativum, it occasionally invades hair follicles and sebaceous glands, causing hyperkeratosis and thickening of the epidermis. Lesions usually appear first on the ears (Fig. 23.48), neck, face, and shoulders but sometimes appear on the ventral abdomen, legs, and genital area, especially in younger animals. The feet and perineum may become involved due to the cat's sleeping position and grooming behavior. Typical signs are intense pruritus, erythema, skin scaling, grayish-yellow crusts, and loss of hair. As infestations progress, the affected skin becomes thickened, folded, and wrinkled. Scratching to alleviate the itching aggravates the condition by excoriating the skin and causing inflammation. Severe chronic cases can lead to systemic debilitation and death. To distinguish cases from other possible skin problems or mite species, the identity of *N. cati* should be confirmed by examining skin scrapings, usually best taken from the ears.

Wild cats known to become infested with *N. cati* include the Siberian tiger (*Felis tigris*) and North American bobcat (*Lynx rufus*). A fatal case of notoedric mange in an adult bobcat was reported in Texas (USA). The cat was emaciated and extremely weak, with hair loss about the head, neck, and shoulders, with

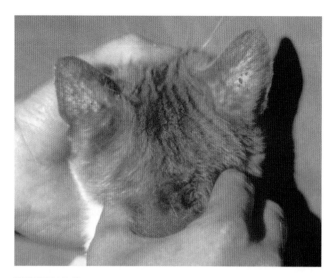

**FIGURE 23.48**    Notoedric mange in cat caused by *Notoedres cati* (Sarcoptidae), showing characteristic skin lesions on outer ears. (Photo by John M. MacDonald, Auburn University College of Veterinary Medicine.)

associated greatly thickened skin and gray encrustations. Bobcat kittens in the same area exhibited similar skin lesions confirmed as notoedric mange (Pence *et al.*, 1982).

**Notoedric squirrel mite** *(Notoedres centrifera)*
This mite, formerly known as *N. douglasi*, causes notoedric mange in North American squirrels (*Sciurus* spp.) and porcupines (*Erethizon dorsatum*). Until the first report of infestations in porcupines (Snyder *et al.*, 1991), it was thought to infest only sciurids. Cases in the United States have been reported in the eastern gray squirrel (*Sciurus carolinensis*) in Massachusetts, the California gray squirrel (*S. griseus griseus*) in California, and the fox squirrel (*S. niger*) in Indiana, Michigan, and West Virginia. Lesions are similar to those of other *Notoedres* species, with thickened and wrinkled skin, scurfy yellowish crusting, and hair loss, usually about the head and neck. Other affected areas include the back, torso, limbs, and base of the tail. In some cases multifocal hyperpigmentation, pinpoint nodules, and microabscesses have been reported, in addition to extensive hair loss, dehydration, emaciation, and death. Significant mortality has occurred in epizootics of notoedric mange among California gray squirrels, attributed in part to impaired vision and disruption of food-seeking ability of severely affected animals. For further information on *N. centrifera*, see Carlson *et al.* (1982) and Kazacos *et al.* (1983).

**Notoedric rat mite** *(Notoederes muris)*
This mite infests rodents and is the cause of *notoedric ear mange*, commonly seen in laboratory rats. It is known to parasitize the Norway rat (*Rattus norvegicus*), black rat (*R. rattus*), multimammate mouse (*Mastomys natalensis*), and certain wild rodents, marsupials, and hedgehogs. Its distribution is apparently cosmopolitan, albeit sporadic. The mite burrows into the stratum corneum where the eggs are deposited and hatch in 4–5 days. The entire life cycle is completed in about 3 weeks. Only occasionally does *N. muris* penetrate the deeper skin tissues. The female lives 2–3 weeks, laying up to three eggs per day. Although the larvae and nymphs may develop in the parent burrow, they also may move onto the skin surface to excavate pits in the stratum corneum which in turn become the entrances to new burrows as they continue to develop (Sweatman, 1971). Mating takes place in the burrow, and transmission between hosts typically involves direct transfer of the active immature stages.

Lesions usually appear several weeks after the initial infestation, in the form of wartlike, horny excrescences and yellowish encrustations on the ears, nose, neck, tail, and sometimes the limbs and genitalia. The involvement of the ears and tail is often diagnostic. The skin becomes greatly thickened, and the hairs appear to become shortened and displaced due to proliferation and cornification of epidermal cells in the affected areas. Severe cases can develop erythema, vesicular or papular lesions, serous exudates, and other complications due to secondary bacterial infections

Related species (*N. musculi*, *N. oudemansi*, *N. pseudomuris*) are known from *Rattus* and *Mus* species, as well as from wild rodents and insectivores. These and other *Notoedres* species of rodents show more restricted geographic ranges. The three mentioned are all reported to cause mange similar to that caused by *N. muris*.

**Trixacarus species**    Two sarcoptid mites in the genus *Trixacarus* can cause severe skin problems called *trixacaric mange* in guinea pigs and laboratory rats. Like those of *Sarcoptes* and *Notoedres* species, *Trixacarus* females burrow in the upper layers of cornified epithelium, where the eggs are deposited and the larvae and nymphs are found. The duration of each developmental stage and the time required to complete the life cycle is unknown. *Trixacarus* species are much smaller than *Sarcoptes scabiei* (females 140–180 vs 400 $\mu$m in *S. scabiei*) and can be distinguished from adults of the latter by the following traits: propodosomal shield reduced to a small, circular plate; elongate, sclerotized striae on dorsal and ventral anterior idiosoma; large, weakly sclerotized denticles on dorsum; long, spinelike dorsal setae; and absence of an ambulacral sucker on leg 4 in the male.

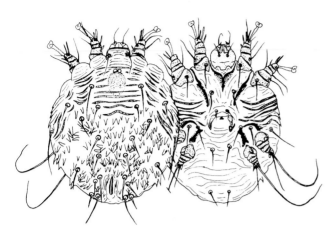

**FIGURE 23.49** *Trixacarus caviae* (Sarcoptidae), female; dorsal (left) and ventral (right) views. (From Nutting, 1984.)

### Trixacarus caviae

This mite (Fig. 23.49) was first described from a guinea pig colony in England in the early 1970s. Subsequently it has been reported causing mange in laboratory and pet guinea pigs in other parts of Europe and in North America. Lesions begin as an erythematous rash that progresses to a pruritic dermatitis, with thickening and wrinkling of the skin and induced scratching. Lesions vary from dry and scaly to moist and crusty, with associated hair loss. Affected areas include the head and neck, shoulders, back, sides, lower abdomen, axillary region, and inner thighs. The intense pruritus can lead to self-inflicted trauma, such as frantic running about in cages and blindly striking objects, loss of condition, lethargy, and grand mal seizures. Death commonly occurs within a few weeks or months in heavily infested, untreated animals.

*T. caviae* is readily transferred on contact with other guinea pigs, including mothers to neonates, causing rapid spread through colonies. Owners of pet guinea pigs can develop papulovesicular lesions and pruritus by direct contact with infested animals held against the skin. This also is occasionally reported as a problem in people working in animal facilities. For further information on *T. caviae*, see Kummel *et al.* (1980) and Zenoble and Greve (1980).

### Trixacarus diversus

This mite is reported to cause severe infestations in the Norway rat, white mice, and hamsters in animal facilities in Europe and in wild field mice (*Calomys musculinus*) in Argentina (Klompen, 1992). Lesions are similar to those described for *T. caviae*, first appearing between the shoulders and from there spreading to the back and sides. Young or weakened animals may die 2–3 weeks following the appearance of skin problems, whereas untreated adult rats are more likely to die after 5–6 weeks (Sweatman, 1971).

## Rhyncoptidae

Rhyncoptid mites are similar to sarcoptid mites but are generally more elongate and are typically restricted to hair follicles. Species of *Audycoptes*, *Saimirioptes*, and most *Rhyncoptes* species occur in monkeys, with other *Rhyncoptes* species in African porcupines, *Caenolestocoptes* species in South American marsupials, and *Ursicoptes* species in bears and raccoons. The primate parasites such as *Rhyncoptes grabberi* from rhesus macaques (Klompen, 1989) are generally innocuous, but mange conditions have been reported associated with *Ursicoptes americanus* in black bears in North America (Yunker *et al.*, 1980) and with *U. procyonis* in North American raccoons.

## Psoroptidae

Mites in the family Psoroptidae are mammalian ectoparasites called *scab mites*. All developmental stages occur on the host. They do not burrow into the epidermis but instead live on the surface of the skin. Some pierce the skin with their chelicerae to feed on lymph, blood, and serous exudates. Others have chelicerae adapted for feeding on sloughed skin scales and other epidermal debris. Feeding injury commonly results in inflammation, pruritus, hair loss, crusting, and scab formation. The host hair often becomes matted and, together with the skin, may be severely damaged due to biting and rubbing by the host against fence posts and other objects. This can result in extensive loss of hair and generalized debilitation, with death occurring in some cases.

Psoroptid mites of veterinary interest parasitize primarily the Artiodactyla, Perissodactyla, and Carnivora. They are also found on certain edentates, marsupials, insectivores, and primates. Members of four genera cause problems for domestic and laboratory animals. *Psoroptes* mites infest cattle, sheep, goats, horses, and rabbits; *Chorioptes* mites are primarily a problem on cattle; *Otodectes* mites cause problems for cats and dogs; and *Caparinia* mites infest wild and captive hedgehogs.

**Psoroptic Scab Mites** For many years it generally has been accepted that there are five *Psoroptes* species (*Psoroptes cervinus*, *P. cuniculi*, *P. equi*, *P. ovis*, and *P. natalensis*) based on host associations, location on the host, and the length of the outer opisthosomal setae ($L_4$) of the adult males (Sweatman, 1958b). There is now compelling evidence to indicate that *P. cuniculi* and *P. ovis* are strains, or ecophenotypic variants, of the same species. Evidence to support this view includes the following:

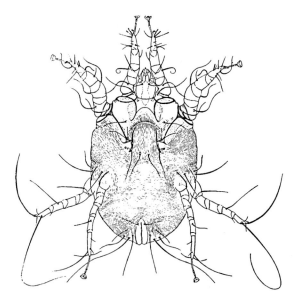

**FIGURE 23.50**    *Psoroptes equi* (Psoroptidae), female, ventral view. (From Baker *et al.*, 1956.)

(1) production of viable offspring by reciprocal genetic crosses between *P. cuniculi* and *P. ovis*, (2) presence of both typical and intermediate forms of these two mites associated with body lesions, and (3) the unreliable nature of the L$_4$ opisthosomal setae as a species-specific character, due to overlapping variations (Bates, 1999; Wright *et al.*, 1983, 1984). The *cuniculi* form typically occurs in the ears, whereas the *ovis* form is characteristically found on other parts of the body, notably the back and flanks. The *ovis* form causes scabs and other epidermal body lesions, whereas the *cuniculi* form does not cause body lesions but instead tends to move to the ears, where it induces aural lesions. For convenience of the following discussion, *P. cuniculi* and *P. ovis* are treated as separate species while a change in their taxonomic status awaits formal recognition.

Mites in the genus *Psoroptes* (Fig. 23.50) cause *psoroptic mange,* a highly contagious form of mange that can spread rapidly by direct transfer of mites between animals or indirectly by rubbing against fence posts and other objects. Infestations tend to be more prevalent during the winter months, usually subsiding or disappearing during the warmer seasons. *Psoroptes* mites cause more severe problems than any of the other psoroptid genera, commonly resulting in economic losses in cattle, sheep, and goat operations.

The mouthparts of *Psoroptes* species are adapted for feeding on the surface of the skin rather than piercing the epidermis. The mites abrade the stratum corneum with their chelicerae, ingesting lipids and other dermal substances. The host responds antigenically to the mites by developing localized perivascular dermatitis and edema.

Hemoglobin and other blood components are ingested by the mites only as a result of small hemorrhages at the skin surface of abraded sites. Their food primarily consists of digested cells of the stratum corneum and skin exudates.

Species of particular veterinary importance are *P. ovis* of sheep and cattle, *P. cervinus* of deer and bighorn sheep, and *P. cuniculi* of rabbits, sheep, and goats. *P. equi* (Fig. 23.50) is a minor pest of horses in England. *P. natalensis* causes mange in cattle, zebu, Indian water buffalo, and horses. It occurs in South Africa, South America (Brazil, Uruguay), New Zealand, and possibly France (Sweatman, 1958b). Other psoroptid mites infest wild animals but seldom cause apparent problems. Exceptions are cases of mange in the African buffalo and in South American primates such as monkeys, marmosets, and tamarins, especially when held in captivity (Sweatman, 1971).

### Sheep scab mite  (*Psoroptes ovis*)

This mite causes psoroptic mange in sheep and cattle known as *sheep scab* and *cattle scab*. It also is called the *psoroptic body mite*. Clinical cases of *P. ovis* are misleadingly called "scabies," a term that is best reserved for mange caused by *S. scabiei*. Although more common in domestic stock, *P. ovis* can be transferred from domestic sheep to wild sheep, causing severe infestations and die-offs in bighorn sheep (*Ovis canadensis*) in the western United States. *P. ovis* tends to occur in the more densely haired or wooly parts of sheep, with initial lesions usually appearing on the back and sides. Heavy crusting and scab formations with associated inflammation, hair damage, and depilation are typical in animals that become antigenically sensitized to this mite (Fig. 23.51). Heavily infested lambs have been found to have thicker than normal pelts, with matted wool and enlarged lymph glands attributed to *P. ovis* (Cochrane, 1994). Pruritus often induces self-inflicted trauma, with extensive wool loss and skin injury due to licking, biting, and rubbing. Mites move to the periphery of the affected areas, thereby spreading over the body surface. In untreated animals, lesions may develop over major portions of the body, causing extensive loss of wool and reduced weight gains as high as 30% (Kirkwood, 1980). In extremely severe cases, sheep may die. In other cases, sheep heavily infested with *P. ovis* may show no gross evidence of skin damage yet exhibit adverse effects on wool quality and general body condition.

Transfer of *P. ovis* among sheep primarily involves mites less than 2 weeks old. Older mites apparently fail to establish infestations even upon successful transfer to a new host. When *P. ovis* is transferred from infested sheep to calves and goats, the mites survive only about

**FIGURE 23.51** Psoroptic mange, or sheep scab, in sheep caused by *Psoroptes ovis* (Psoroptidae). Note extensive hair loss. (Courtesy of the US Department of Agriculture/ARS, Kerrville, TX, M&M 8165.)

a week on the recipient host and do not induce clinical signs, at least not in naive, i.e., previously unexposed, animals (O'Brien *et al.*, 1994).

Infestations of *P. ovis* in cattle cause exudative dermatitis and hair loss similar to that in sheep (Fig. 23.52). The severity varies from mild cases to those that can involve virtually the entire body surface. Systemic effects can include mild anemia with hematologic changes such as marked reductions in lymphocytes, neutrophils, and total white blood cells; increases in plasma proteins and fibrinogen; and bone-marrow

**FIGURE 23.52** Psoroptic mange in calves, caused by *Psoroptes ovis* (Psoroptidae). (Courtesy of the US Department of Agriculture/ARS, K7268-14.)

effects (Stromberg and Guillot, 1987a,b; Stromberg *et al.*, 1986). The degree of these responses is directly correlated with the severity of the associated dermatitis. Infested animals also may experience reduced weight gain, reduced energy conversion rates, and higher maintenance energy requirements (Cole and Guillot, 1987).

Significant differences occur between cattle which have not previously been exposed to *P. ovis* and those that have. In naive animals, skin lesions are slower to appear but progress rapidly. The associated mite populations grow much more quickly, reaching densities 100–1000 times that in previously infested animals. The lower growth rates and lower fecundity in previously exposed cattle are due to both cellular and humoral immune responses involving the development of antibody activity to live *P. ovis* mites infesting the skin. Mites that infest animals with this acquired immunity have much lower ovipositional rates, reflecting decreases in the number of ovigerous females, rather than detrimental effects on egg development (Guillot and Stromberg, 1987). Cattle can become hypersensitive to *P. ovis* antigens, thereby developing severe clinical disease even when infested by relatively modest numbers of mites. This helps to explain why cattle in areas endemic for *P. ovis* generally experience more severe lesions than cattle in nonendemic areas. The severity also is exacerbated by stress caused by stanchioning animals and by extremely cold weather that can contribute to hypothermia and death of infested cattle.

### *Psoroptes cervinus*

Psoroptic mange occurs in bighorn sheep, mule deer, elk, and wapiti in the western United States. The identity of the *Psoroptes* species involved, however, remains unclear. Whereas some reports have called them *P. ovis*, others have recognized them as *P. cervinus*. Often these mites are simply referred to as "*Psoroptes* sp." because of this taxonomic uncertainty. *Psoroptes* mites collected from bighorn sheep apparently do not establish lasting infestations when transferred to domestic sheep and can be established only with difficulty on cattle (Wright *et al.*, 1981). Comparative studies based on antigenic characterization of *Psoroptes* mites from various hosts further suggests that the mite that infests bighorn sheep and mule deer is different from *P. ovis* on cattle (Boyce and Brown, 1991). For the purpose of the discussion here, the mites causing psoroptic mange in bighorn sheep and cervids are called *P. cervinus*, to distinguish them from their closely related, and possibly conspecific, counterparts on domestic sheep and cattle.

Lesions of psoroptic mange in bighorn sheep occur primarily in the ears or on the face and other parts of the head. The affected areas are characterized by yellowish-white scabs of dried serous exudates and crusty, exfoliated epidermal tissue overlying a reddened and raw epidermis. Other clinical signs are hair loss on the head, neck and back; droopy ears; and blockage of the outer ear canal with cerumen and exudates. Lesions on other parts of the body generally are less extensive or severe, with mites being recovered primarily from the head and ears.

*Psoroptes* infestations can be especially severe in desert bighorn sheep (*O. canadensis mexicanus*) in the San Andres Mountains of New Mexico (USA). High mortality attributed to psoroptic mange reduced the population of desert bighorn sheep in the San Andres National Wildlife Refuge from more than 200 to about 25 individuals during the 12-year period 1978–1989 (Hoban, 1990). Serologic surveys using immunologic tests for detection of antibodies to *Psoroptes* mites have shown widespread prevalence of these mites in desert bighorn sheep populations in California (USA), where lesions tend to be mild and confined to the ears (Mazet *et al.*, 1992).

*P. cervinus* also parasitizes elk and wapiti (*Cervus elaphus*) in the western United States, where it is called the *elk scab mite*. Infested elk may develop moist, thick scabs with associated dermatitis and hair loss, especially at the base of the neck and on the dorsal and lateral thorax. Particularly affected are the cows, young males, and calves (Samuel *et al.*, 1991). In wapiti, body lesions occur primarily in the winter months and may involve large areas of the neck, trunk, and upper legs. A wet eczema with overlying scabs and extensive hair loss is similar to that observed in desert bighorn sheep and can be fatal.

**Chorioptic Scab Mites**   Psoroptic mites in the genus *Chorioptes* cause *chorioptic mange* in domestic ungulates, notably cattle, sheep, goats, and horses. The species of greatest importance is *Chorioptes bovis*, which infests each of these hosts. Another species, *C. texanus,* infests the ear canals of reindeer in Canada. As a group, chorioptic mites are primarily parasites of herbivores (Artiodactyla, Perissodactyla, and Lagomorpha), including llama, guanaco, alpaca, and rabbits. They feed on sloughed epidermal tissues, sometimes causing irritation and crusty, pruritic lesions that warrant treatment. For further information on *Chorioptes* species and their host associations, see Sweatman (1957, 1958c).

*Chorioptes bovis*

This mite (Fig. 23.53) occurs primarily on the legs and feet of its hosts, where all of the developmental stages are likely to be found. Eggs are deposited singly at the

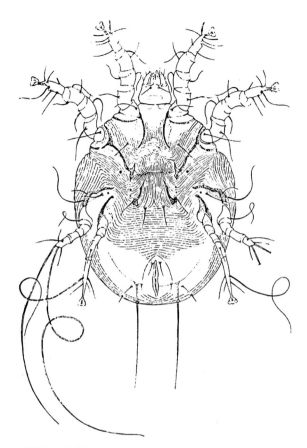

**FIGURE 23.53**   *Chorioptes bovis* (Psoroptidae), female, ventral view. (From Baker *et al.*, 1956.)

rate of one egg per day and are attached with a sticky substance to the host skin. Adult females usually live for 2 weeks or more, producing about 14–20 eggs during this time. The eggs are often clustered as multiple females oviposit in common sites or as females return on successive days to deposit their eggs. The eggs hatch in 4 days. The larval and protonymphal stages last 3–5 days each, whereas the tritonymphal stage takes 7–8 days, with 1 day of quiescence between each developmental stage. The cycle is completed in about 3 weeks. Optimal conditions for development are about 35° C and a relative humidity of 80%.

Most animals infested with *C. bovis* do not exhibit noticeable lesions or unusual discomfort due to this mite, even at relatively high mite densities. As a result, infested sheep and cattle often remain asymptomatic, serving as silent carriers and a source of infestation for other animals. Host reactions are induced only when numbers increase to thousands of mites per host, occasionally causing extensive mange and pruritus. Most of the mites are found on the feet, notably the pasterns, regardless of where lesions appear elsewhere on the body. The irritation in sensitized animals can lead to stamping of feet, rubbing and chewing of legs, and other self-inflicted injury. Body lesions in severe cases are characterized by dermal crusting, erythema, and hair loss.

In cattle, *C. bovis* is found more commonly on the hindfeet than on the forefeet and particularly on the pasterns (between fetlock and hoof). The mites move from the feet to other parts of the body to cause mange of the escutcheon, base of the tail, buttocks, and perineum. This has given rise to names such as *foot mange, leg mange,* and *itchy heel* in referring to chorioptic mange of cattle and sheep. Mite populations are highest during the winter and are especially a problem in housed animals. In the spring, their numbers drop sharply and lesions generally disappear when cattle are turned out to pastures.

Although sheep are commonly parasitized by *C. bovis,* the small crusty lesions are hidden beneath the coat and usually go unnoticed. When clinical cases do occur, they are typically in the form of foot mange, affecting the forefeet. The mites occur about the accessory digits and along the coronary border of the outer claws, often in clusters, causing crusting primarily below the accessory digits and in the interdigital spaces. Infestation rates of 30–60% in sheep have been reported in the United States, Europe, Australia, and New Zealand. Prevalence of chorioptic mange tends to be highest in rams and generally low in ewes and lambs.

*C. bovis* may spread by direct contact of the feet with other parts of the body, notably the upper parts of the hind legs and scrotal area of rams, causing an exudative dermatitis called *scrotal mange.* The scrotal skin develops thick, yellowish, crusty layers as much as 4 cm deep. In severe cases, elevated scrotal temperatures attributed to allergic responses can cause degeneration of the seminiferous tubules, reduced sperm quality, and complete spermatogenic arrest. Testicular weights may become significantly reduced. The effects are reversible, with seminal regeneration and restored sperm production occurring following treatment for mites or spontaneous recovery of infested rams. Prevalence of leg and scrotal mange is usually highest in the fall and winter months and declines in the spring.

Infestation rates of *C. bovis* tends to be higher in goats than in sheep, with up to 80–90% of goats in individual herds being parasitized (Cremers, 1985). As in sheep, the mites occur most commonly on the forefeet of goats, where the largest numbers of mites and lesions are usually associated with the accessory digits and claws. However, they also may occur on the

pastern or higher on the foot. Lesions are generally mild and seldom draw attention.

Chorioptic mange due to *C. bovis* is occasionally observed in horses, with Belgian and Frisian breeds being among the more commonly infested. The mites are largely restricted to the pasterns and are most likely to cause foot mange in the above horse breeds with long-haired feet (Cremers, 1985). Signs of *C. bovis* infestations in horses include stamping of feet and rubbing one foot against the opposite leg or against some object.

**Caparinic Scab Mites** Psoroptid mites of the genus *Caparinia* infest hedgehogs and a few other Old World mammals, causing *caparinic mange.* All active stages of these mites feed on sloughed skin cells and epidermal debris, similar to *Chorioptes* species. Although most infestations tend to go unnoticed, high mite populations can cause severely debilitating conditions and even death of the host. Two species that have drawn particular attention in recent years are *Caparinia tripilis* and *C. erinacei,* both of which parasitize wild and captive hedgehogs.

*C. tripilis,* the *Eurasian hedgehog mange mite,* infests the Eurasian hedgehog (*Erinaceus europaeus*). It was first recognized in Great Britain in the late 1880s, where it is still occasionally reported and has caused the death of at least one captive hedgehog. This mite was introduced to New Zealand on hedgehogs in the 19th century but did not attract attention there until 1955, after hedgehog populations increased dramatically (Brockie, 1974). More recently, *C. tripilis* has been introduced to the United States via breeding colonies of African hedgehogs for sale as pets. An infested colony has been reported in New Mexico (Staley *et al.,* 1994), and the death of a pet hedgehog attributed to *C. tripilis* has been documented in Alabama (Mullen, unpublished data). The mites tend to gather in clusters on their host and invade the skin of the head and ears, flanks, and inner surfaces of the legs. The affected skin becomes dry and scaly, may become thickened and folded, and may crack and bleed, leading to secondary infections. The common association of hedgehog ringworm (*Trichophyton erinacei*) with heavy *C. tripilis* infestations suggests that invasion of the skin by this fungus may contribute to severity of the resultant mange. In severe cases, body spines and hairs may fall out and lesions about the eyes may cause blindness (Brockie, 1974). Heavily infested animals become listless, lose weight, and scratch the affected skin and may abandon their normal nocturnal behavior to become active in the daytime. Male hedgehogs are usually more severely affected than females, with higher mortality occurring in captive than in wild hedgehogs.

*C. erinacei,* the *African hedgehog mange mite,* infests the African hedgehog (*Atelerix albiventris*). Unlike *C. tripilis,* this mite does not form clusters on its hosts and exhibits low pathogenicity, occurring most abundantly on the dorsal parts of the body and rarely on the face. It has been reported parasitizing more than 70% of wild hedgehogs in Kenya (Gregory, 1981).

### Harpyrhynchidae

Harpyrhynchid mites are typically parasites of birds, in which they invade feather follicles and the skin. Evidence of infestations range from small white cysts or lumps associated with individual feathers to large, irregularly lobed, papilloma-like cysts that can occur on any part of the body. The lesions are typically pale yellow, with a dry, granular appearance. Histological preparations of the lesions reveal multiple spaces lined by epidermal cells and packed with large numbers, sometimes thousands, of mites. The most common genus involved is *Harpyrhynchus,* which has been reported to cause disfiguring ruffling of feathers and feather loss in a variety of avian hosts (e.g., lorikeets, warblers, eagles) in Australia, the southwestern Pacific region, and North America. Females of the genus *Ophioptes* excavate small, crater-like pits in the body scales of snakes, where they deposit their eggs. The larvae and nymphs lack legs, feed on host tissues, and develop to adults within the pits. Most reported cases have occurred in colubrid snakes in South America and Australia. Except for localized damage to individual scales, these mites do not cause significant harm to their hosts.

### Hypoderatidae

Hypoderatid mites (previously included in the glycyphagid subfamily Hypodectinae) are parasites of birds, and rarely mammals, in which they develop in subcutaneous fat tissue. These are unusual mites in that they are parasitic as deutonymphs, invading host tissues where they undergo growth and enlargement. The other life stages are nest-inhabiting detritivores or are, rarely, nonfeeding. The most common hypoderatid species is *Hypodectes propus,* which parasitizes pigeons and doves in North America, Europe, and Africa. The mites are visible at necropsy as tiny white nodules embedded in fat tissue just beneath the skin. For further information on hypoderatid mites and their host associations, see Pence *et al.* (1997).

## Mite-Induced Allergies

In addition to sensitization of animals to mites that invade the skin (e.g., demodicid and sarcoptid mites), pets and other domestic animals occasionally develop allergic reactions to nonparasitic mites. Reported cases, for example, have involved mites in the family Acaridae, which infest dry pet foods and livestock or poultry feed. Although the ingestion of even heavily mite-infested feed usually causes no apparent harm to animals, some individuals may become sensitized.

## Internal Acariasis

Approximately 15 families of mites include species that cause internal acariasis in animals of veterinary interest. The most common type is respiratory acariasis, which may involve the nasal passages, nasal sinuses, trachea, bronchi, or pulmonary tissues. Less commonly, mites cause oral, esophageal, gastric, or enteric acariasis, and occasionally they invade other internal organs, the body cavity, lymph, and blood.

Species in five families may be found in oral tissues or the alimentary tract of their hosts. Certain *Radfordiella* species (Macronyssidae) invade the oral mucosa of the gums and hard palate of bats, causing erosion of soft tissues and bone. Heavy infestations can result in significant oral lesions, with loss of teeth and sometimes exposure of the maxillary sinus. A few demodicid species invade the tongue, oral epithelium, and esophagus of bats, lemurs, and mice but rarely cause noticeable problems. Occasionally *Demodex* mites have been recovered from the alimentary tract of dogs, without evidence of penetrating the epithelial lining. *Gastronyssus bakeri* (Gastronyssidae) attaches to the mucosa of the stomach and duodenum of bats, whereas *Paraspinturnix globosus* lives in the anal canal of its bat hosts.

*Cytodites nudus* (Cytoditidae) has been found in the alimentary canal, peritoneum, and body cavity of chickens and is associated with peritonitis, enteritis, and occasionally deaths. At least some records of *Cytodites* outside the respiratory system could reflect gross dissections of infested hosts. This mite normally is found in the air sacs, but these collapse upon dissection, possibly leading to misleading reports of *Cytodites* in surrounding organs.

In unusual cases, mites have been found in other internal organs. For example, all stages of some *Demodex* species have been found in the liver and spleen (Kirk, 1949), whereas *C. nudus* has been recovered from surface tissues of the heart, liver, and kidneys (Baker *et al.,* 1956). *Demodex* species also have been found alive in lymph nodes, lymphatic vessels, and circulating blood. It is presumed that the mites invade the lymphatic and circulatory systems, where extensive destruction of the surrounding dermal tissue occurs in severely infested hosts.

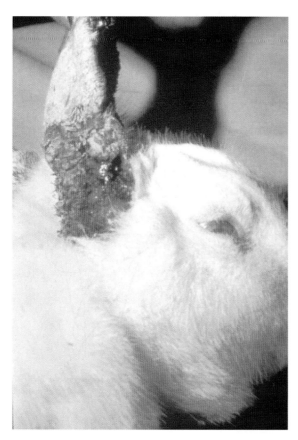

**FIGURE 23.54** Rabbit with ears heavily infested with rabbit ear mite, *Psoroptes cuniculi* (Psoroptidae). Note skin injury, crusty scabs, and resultant bleeding caused by rubbing and scratching to alleviate discomfort. (Photo by John M. MacDonald, Auburn University College of Veterinary Medicine.)

The two most common internal sites of animals that parasitic mites have exploited are the ear canals and respiratory passages.

## EAR MITES

Several species of mites infest the ears of domestic and wild animals, causing problems that may warrant veterinary attention. The families most commonly involved are Psoroptidae, Trombiculidae, and Raillietidae.

### Psoroptic ear mite (*Psoroptes cuniculi*)

Known as the *psoroptic ear mite* or *ear mange mite* of rabbits, *P. cuniculi* causes lesions in laboratory rabbits and commercial rabbit operations (Fig. 23.54). Such infestations are referred to as *psoroptic otoacariasis.* This cosmopolitan mite also infests sheep, goats, and occasionally deer, antelope, and laboratory guinea pigs. Lesions occur primarily in the ears, causing crust formation, malodorous discharges in the external ear canal,

and behavioral responses such as scratching the ears, head shaking, loss of equilibrium, and spasmodic contractions of neck muscles (torticollis). In severe cases *P. cuniculi* infestations may spread to other parts of the body, notably the face, neck, and legs.

*P. cuniculi* lives its entire life under the margins of scabs formed at infested sites. There the eggs are deposited and hatch in 4 days. The complete life cycle takes about 3 weeks. All stages of this nonburrowing mite pierce the stratum corneum to feed on epidermal tissue. Transmission between animals occurs by direct contact.

Infestations of sheep with *P. cuniculi* can cause varying degrees of problems, usually limited to the ears. Lesions in lambs are generally mild and characterized by small, discrete, crusty lesions on the inner surface of the pinnae and around the entrance to the outer auditory canal. Similar lesions may occur at the base of the ears. Severe infestations in older animals can lead to inflammation of the ears (otitis), hematomas, and suppurating abscesses. In other cases sheep may show no clinical signs despite confirmation of mites in the ear canal on otoscopic examination. Surveys of sheep flocks in England have shown that the prevalence of *P. cuniculi* is usually higher in lambs than in adults and that up to 60% of some infested flocks are positive for this mite (Morgan, 1992).

Both domestic and wild goats are subject to *P. cuniculi* infestations. Prevalence rates as high as 80–90% have been reported in dairy goats, including both kids and adults, in the United States (Williams and Williams, 1978). Goats less than 1 year old generally exhibit much higher infestation rates than do older animals. Clinical signs of *P. cuniculi* mites in kids are often observed as early as 3 weeks after birth, reflecting transfer of mites between mother and young. By 6 weeks of age most kids in infested goat herds are likely to harbor these mites. There is no evidence, however, for cross-infections between goats and sheep, even when they are held in common enclosures (Williams and Williams, 1978). Infestations can cause scaling, crusting, inflammation, and hair loss about the ears; accumulations of wax in the external ear canal; and ear scratching, head shaking, and rubbing of the head and ears against objects in an effort to alleviate the discomfort. Chronic infestations also can lead to anemia and weight loss. Extreme cases are sometimes fatal, with death being preceded by circling, violent fits, and other aberrant behavior. Nondomestic goats reportedly parasitized by *P. cuniculi* include the Nubian mountain goat (*Capra ibex nubiana*) and cross-breeds between domestic and mountain goats, such as Yaez.

Cervid hosts of *P. cuniculi* include both free-ranging and captive white-tailed deer (*Odocoileus virginianus*) and mule deer (*O. hemionus*) in North America. Most reports have come from the southeastern United States, where up to 80% of some white-tailed deer populations have been found to be infested. Cases are typically mild, with loss of hair about the ears and base of the antlers and yellow crusty lesions, accumulated cerumen (ear wax), and serous exudates in the external ear. In more severe infestations, the ear canal may become inflamed and infected, leading to pyogenic bacterial otitis and neurologic disorders (Rollor *et al.*, 1978). In such cases, the infected ear canal and tympanic cavity become filled with mucopurulent exudates, with or without damaging the tympanic membrane. This affects the sensory organs in the inner ear, causing neurologic signs such as excessive salivation, circling behavior, difficulty standing, loss of muscular coordination, and torticollis.

*P. cuniculi* is the common ear mange mite of laboratory rabbits that causes *psoroptic ear canker.* Occasionally wild rabbits and hares (e.g., *Lepus europaeus* in Europe) also are affected. Ear infestations are characterized by loose, crusty lesions, excessive cerumen, inflammation, accumulated exudates, and necrotic debris in the external ear. In some heavy infestations, lesions may spread from the ears to other parts of the body, including the face, neck, and genitalia. Often the first signs are behavior such as tilting or shaking of the head, drooping ears, and scratching or self-inflicted trauma to the ears. Inflammation of the middle ear (otitis media) and brown, malodorous discharges in the external auditory canal are common. Cases can become complicated by bacterial infections, causing loss of equilibrium, torticollis and, in some cases, fatal meningitis.

Other animals occasionally infested with *P. cuniculi* are horses, donkeys, mules, antelopes, and guinea pigs. Crusty lesions with accumulated exudates in the ear canal are typical of other hosts. In heavy infestations, mites may spread to other parts of the body (e.g., face, belly, hind legs), causing erythematous and pruritic lesions complicated by secondary bacterial infections. Severe cases of this nature have been reported in the blackbuck antelope (*Antilope cervicapra*) (Wright and Glaze, 1988) and the guinea pig (Yeatts, 1994).

### Otodectic ear mite (*Otodectes cynotis*)

This mite (Fig. 23.55) is known as the *ear mite* or *ear canker mite* of cats and dogs and as the cause of *otodectic mange.* It occurs worldwide and parasitizes other carnivores such as foxes, ferrets, and raccoons. *Otodectes cynotis* is closely related to *Psoroptes* species, which it resembles in size and general appearance. It can be distinguished from *Psoroptes,* however, by its short, unsegmented tarsal stalks supporting the ambulacral suckers in both sexes and by the greatly reduced

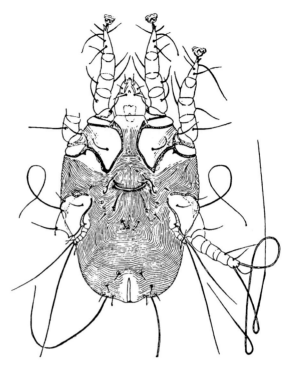

**FIGURE 23.55**   Ear mite of cats and dogs, *Otodectes cynotis* (Psoroptidae), female, ventral view. (From Baker *et al.*, 1956.)

hind pair of legs in the female, which terminate in two long, whiplike setae.

*O. cynotis* typically occurs deep in the external ear canal, where all of the developmental stages are found. Occasionally it secondarily infests other parts of the body, including the head, back, tip of the tail, and feet. It does not burrow into the skin but lives as a surface parasite that may pierce the skin to feed on blood, serum, and lymph. Some workers contend that it feeds more commonly on desquamated epithelial cells and possibly cerumen or other aural exudates. It is believed that development of clinical signs reflects allergic hypersensitivity on the part of the host to antigenic substances introduced while the mites are feeding. This can lead to highly variable responses ranging from asymptomatic or mild cases to severe otitis and convulsive seizures.

The ear canals of animals infested with *O. cynotis* become excessively moistened with accumulations of cerumen and purulent, brown–black exudates resembling coffee grounds. This is accompanied by inflammation and pruritus, usually involving both ears. As a result of intense itching, infested cats and dogs scratch their ears, shake their head or hold it to one side, and may turn in circles. When the ear canal is massaged, the animal typically responds with pleasurable grunting sounds and by thumping its hind leg on the

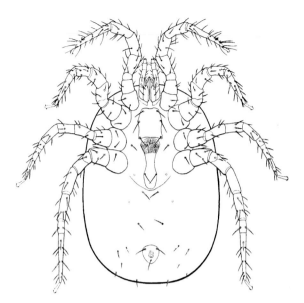

**FIGURE 23.56** *Rallietia auris* (Raillietidae), female, ventral view. (From Hirst, 1922.)

corresponding side. Severe, untreated cases can lead to emaciation, self-induced trauma, spasms, and convulsions, especially in cats. Diagnosis of *O. cynotis* is confirmed by otoscopic examination and by recovering the mite from aural scrapings.

**Cattle and goat ear mites** *(Raillietia* spp.*)*

Mites of the genus *Raillietia* (family Raillietidae) are the only known mesostigmatid species that live in the ear canals of domestic animals. The most widespread species is the *cattle ear mite* (*Raillietia auris*) (Fig. 23.56), which infests dairy and beef cattle in the United States, Europe, western Asia, and Australia. Although it is generally considered to be a relatively harmless mite living in cerumen, *R. auris* can cause blockage of the auditory canal by plugs of paste-like wax. Severe cases can result in inflammation of the ear canal, pus formation, ulcerated lesions, and hemorrhaging, with accompanying hearing loss in some animals. Based on studies of *R. flechtmanni,* which infests cattle and buffalo (*Bubalus bubalis*) in Brazil, adult mites on pasture vegetation enter the ear canals of grazing animals, where they feed, mate, and oviposit. Larvae, upon hatching from the eggs, leave the host to complete their development as nymphs and adults on pasture (Costa *et al.,* 1992). Other *Raillietia* species include *R. caprae* and *R. manfredi* of goats in Brazil and Australia, respectively; *R. acevedoi* of the Alpine ibex (*C. ibex*) in Europe and Asia; *Raillietia* species infest the ear canals of the waterbuck (*Kobus ellipsiprymus*) and Uganda kob (*K. kob*) in Africa; banteng (*Bos javanius*) in Indonesia; and wombat (*Vombatus*

*ursinus*) in Australia. *R. caprae* has been implicated in mycoplasma infections of goats (DaMassa *et al.,* 1992).

## RESPIRATORY MITES

Representatives of several families of mites are specialized, obligate parasites in the respiratory tracts of reptiles, birds, and mammals. They commonly live in the nasal passages and lungs, causing *nasal acariasis* and *pulmonary acariasis,* respectively. Among the taxa of veterinary interest are members of the families Entonyssidae, Rhinonyssidae, Halarachnidae, Ereynetidae, Trombiculidae, Lemurnyssidae, Turbinoptidae, Cytoditidae, Pneumocoptidae, and Gastronyssidae. For keys to species, host lists, and a bibliography for nasal mites of North American birds, see Pence (1975).

### Entonyssidae

These mites are endoparasites that infest the tracheae and lungs of snakes (Fain, 1961). They rarely seem to cause problems for their hosts but occasionally induce congestion of the lungs when mite numbers are high.

### Rhinonyssidae

Rhinonyssid mites are endoparasites in the nasal passages, and occasionally the tracheae, of birds throughout the world. They are common, typically infesting 30–50% of birds examined in local surveys (Domrow, 1969; Pence, 1973; Spicer, 1987). Their chelicerae are reduced, membranous structures that are used to imbibe liquid food, including blood, from their hosts as they crawl about on the mucous membranes lining the nasal airways. Feeding is facilitated by the claws on leg 1 that are used in lieu of the chelicerae to tear or otherwise penetrate respiratory tissues. Such injury can cause rhinitis or sinusitis, especially in heavy mite infestations. In most cases, however, infested birds do not experience apparent respiratory problems.

An exception is *Sternostoma tracheacolum* (Fig. 23.57), which parasitizes the tracheae, bronchi, parenchymal lung tissue, and air sacs of both wild and captive birds. It does not occur in the nasal cavities. Typical hosts are canaries, parakeets, swallows, and finches. This mite is sometimes called the *canary lung mite* because of the respiratory problems it causes in captive canaries. Little is known about the behavior of *S. tracheacolum* except that it crawls freely about in the mucous lining of the trachea and bronchi. The mite also may invade the air sacs and lung tissue, where it dies and disintegrates. Its presence causes inflammation and the development of characteristic nodular lesions containing masses of dead mites and purulent, fibrous exudates. Early

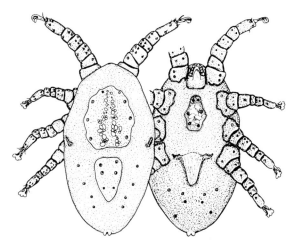

**FIGURE 23.57**    *Sternostoma tracheacolum* (Rhinonyssidae), female; dorsal (left) and ventral (right) views. (From Nutting, 1984.)

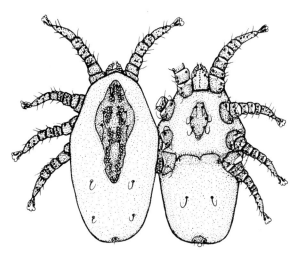

**FIGURE 23.58**    Monkey lung mite, *Pneumonyssus simicola* (Halarachnidae), female; dorsal (left) and ventral (right) views. (From Nutting, 1984.)

signs of respiratory distress include listlessness and difficulty breathing.

Heavy infestations can result in tracheitis, bronchitis, hemorrhaging in parenchymal tissue surrounding the terminal bronchi, small foci of bronchial pneumonia, lung congestion, and pneumonitis. As the damage progresses, infested birds may become emaciated and die. Although wild birds are not as severely affected as captive birds, they can develop bronchopneumonia and inflammation of their air sacs.

### Halarachnidae

Mites in this family are obligate parasites in the respiratory tracts of a variety of mammals. Their hosts include marine mammals, porcupines, squirrels, canids, and nonhuman primates. Halarachnid mites generally are regarded as benign, except for a few species that infest domestic dogs and captive primates. Most species occur in the nasal passages, whereas others may live in sinuses, tracheae, bronchi, or lung tissue. The larvae and adults are the active stages, typically piercing the epithelium with their long chelicerae to feed on lymph and other fluids. Only a few taxa are known to feed on blood. Transmission is presumed to occur by direct transfer of larvae around the host nostrils or by sneezing and coughing of infested animals. Although most infestations are relatively asymptomatic, others can result in inflammation of the respiratory passages, pulmonary nodules, lung congestion, and host death. Diagnosis of halarachnid infestations is usually based on recovery of larvae in tracheobronchial washings or histological examination of pulmonary tissues. The genera of particular veterinary interest are *Pneumonyssus* of Old World monkeys and apes; *Halarachne* and *Orthohalarachne* of seals, sea otters, and walruses; and *Pneumonyssoides* of dogs.

*Pneumonyssus simicola*, the *monkey lung mite* (Fig. 23.58), is the most common halarachnid mite of primates, infesting the lungs of rhesus, cynomolgus, and macaque monkeys in Africa. It is especially a problem in rhesus colonies, in which up to 100% of the individuals may be infested. Although relatively benign in wild hosts, infestations of *P. simicola* in captive primates in laboratories and zoological parks can cause a wide range of respiratory problems, occasionally proving fatal. The damage results directly from mites attached to the bronchiolar walls piercing the surrounding parenchyma to feed on blood, lymph, and pulmonary epithelial cells. The severity depends largely on the number of mites. Low levels of infestation can cause inflammation of the bronchioles and mild coughing or sneezing. As the number of mites increases, lesions are produced in the form of soft, yellowish nodules containing up to 20 mites each. Other lesions appear as pale spots containing golden-brown, needle-like crystals and dark pigments. The latter are believed to be breakdown products of host blood excreted by the mites. Deaths, in cases of massive infestations, are attributed to congestion of the lungs and alveolar collapse. For additional information on *Pneumonyssus* species, see Hull (1970).

*Halarachne* species are nasal mites of earless seals (Phocidae) and the Pacific sea otter (*Enhydra lutris*). Captive sea otters may develop heavy infestations of *Halarachne microungae*, resulting in inflammation of the nasal mucosa, obstruction of nasal passages, destruction of associated bony tissues (turbinates), and pulmonary congestion. More than 3000 mites have been reported infesting one sea otter that had died. For a key to *Halarachne* species and a review of the genus, see Furman

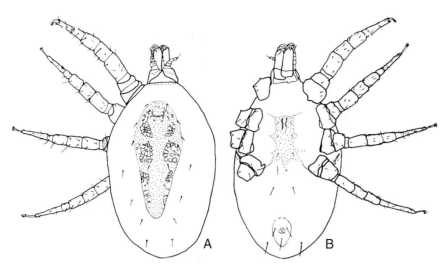

**FIGURE 23.59**  Dog nasal mite, *Pneumonyssoides caninum* (Halarachnidae), male. (A) Dorsal view; (B) ventral view. (From Strandtmann and Wharton, 1958.)

and Murray (1980) Sea lions and fur seals (Otariidae) and walruses (Odobenidae) are hosts for *Orthohalarachne* species. *Orthohalarachne attenuata* infests the nasopharynx, whereas *O. diminuta* infests the lungs. Fur seals often are simultaneously parasitized by both of these mites, with entire populations of seals over 3 months old being infested. Clinical signs are mucus-filled turbinates, nasal discharges, sneezing, coughing, and impaired respiration. Heavy infestations can lead to alveolar emphysema and a predisposition to more serious ailments that can kill host animals (Kim *et al.*, 1980).

*Pneumonyssoides caninum*, the *dog nasal mite* (Fig. 23.59), lives in the nasal sinuses of dogs (Fig. 23.60) in many parts of the world, including the continental United States and Hawaii, Europe, South Africa, and Australia. Details about its life cycle are largely unknown. This mite is regarded as relatively nonpathogenic, with clinical signs of most infestations being limited to excessive nasal secretions and hyperemia of the nasal mucosa. More severe cases may involve listlessness, loss of appetite, tearing of the eyes, chronic sneezing, bronchial cough, and rhinitis or sinusitis (Koutz *et al.*, 1954). There also is evidence that *P. caninum* can penetrate host tissues and move beyond the respiratory system to cause lesions in the liver and kidneys (Garlick, 1977).

### Ereynetidae

Ereynetid mites are primarily free-living detritivores. However, members of two subfamilies are obligate parasites in the respiratory tracts of terrestrial vertebrates. Mites of the subfamily Lawrencarinae infest the nares and nasal passages of amphibians, notably African frogs and toads. Examples are *Lawrencarus eweri* and *Xenopacarus africanus*. Although both feed on tissue fluids, including blood, it is not clear how much harm they cause to their hosts. Mites of the subfamily Speleognathinae live in the mucus-lined nasal passages of a wide range of birds and mammals throughout the world. Occasionally they also invade the lungs. Rarely do they seem to cause harm to their hosts, despite the fact that some species feed on blood (e.g., *Boydaia sturnellus* of meadowlarks in the United States). *Speleognathus australis* parasitizes cattle and bison. For additional information on speleognathine mites, see Lawrence (1952), Clark (1960), Baker (1973), Pence (1973, 1975), Fain and Hyland (1975), and Spicer (1987).

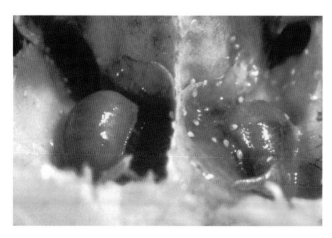

**FIGURE 23.60**  Nasal sinuses of dog infested with *Pneumonyssoides caninum* (Halarachnidae); whitish mites are seen crawling over tissue surface. (Courtesy of Department of Pathobiology, Auburn University College of Veterinary Medicine)

## Trombiculidae

Approximately 20 genera of trombiculid mites are parasitic as larvae (chiggers) in the nasal passages of reptiles, birds, and mammals in both the Old World and New World. Rodents and bats are the most common hosts, parasitized by *Ascoschoengastia, Doloisia, Gahrliepia, Microtrombicula, Schoutedenichia,* and other genera. Other hosts of intranasal chiggers include marsupials (e.g., ring-tailed possum, water opossum, bandicoots), edentates (e.g., armadillos, anteaters), hyraxes (e.g., tree hyrax), lagomorphs (e.g., hares), birds (e.g., sooty tern), felids (e.g., African wild cat), marine iguanas, and sea snakes (Nadchatram, 1970). *Vatacarus* species, which parasitize the last two groups of reptiles, infest not only the nasal fossae, but also the tracheae and lungs of their hosts. The nymphs and adults of intranasal chiggers are free-living and presumably inhabit the nests, dens, and other sheltered locations of their respective hosts. Virtually nothing is known about possible adverse effects that these chiggers may have on their hosts.

## Lemurnyssidae

Mites in this psoroptoid family occur as intranasal parasites of lorisid primates (lorises, bushbabies) in Africa and monkeys (Cebidae) in South America (Fain, 1964b). They live in the nasal fossae, where they apparently cause little or no harm to their hosts.

## Turbinoptidae

Members of this family are exclusively intranasal mites of birds and occur in both the Old World and New World. They infest the nasal fossae without causing apparent harm to their avian hosts. For a review of North American turbinoptid mites, keys to genera and species, and a list of hosts, see Pence (1973, 1975). For African and European species, see Fain (1957, 1970); for species of eastern Australia, see Domrow (1969).

## Cytoditidae

Mites in this family are internal parasites of birds that typically infest the respiratory system but also may invade the peritoneum and visceral organs. Members of the genus *Cytonyssus* are usually found only in the nasal passages, whereas *Cytodites* species often invade the lungs and air sacs. The species of greatest veterinary importance is *Cytodites nudus,* the *air-sac mite* of chickens (Fig. 23.61), which occurs worldwide. Although low-level infestations do not cause apparent harm, heavy infestations can lead to severe clinical signs and occasionally host death. *C. nudus* is found most commonly in the lining of the air sac but also may invade the air passages and lungs, causing accumulation of mucus in the tracheae and bronchi.

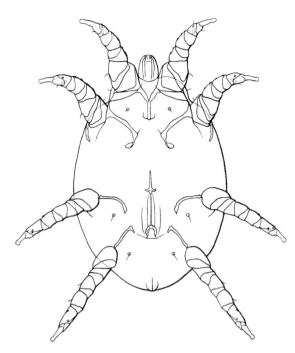

**FIGURE 23.61**    Air-sac mite of chickens, *Cytodites nudus* (Cytoditidae), female, ventral view. (From Hirst, 1922.)

Clinical signs include coughing, obstruction of air flow, pulmonary edema, and pneumonia. Infested birds may exhibit weight loss, general weakness, and loss of balance or coordination. In some cases, *C. nudus* invades the body cavity and visceral organs, including the heart, alimentary tract, liver and kidneys. It shows a particular predilection for the peritoneum. Deaths are usually associated with peritonitis, enteritis, emaciation, and respiratory complications.

*C. nudus* also has been reported infesting canaries and ruffed grouse, and in one case it was reportedly recovered from the peritoneum of a human in Uganda (Castellani, 1907). Symptoms in canaries are similar to those in chickens but may include bulging eyes and sores or swellings at the corners of the beak (Higby, 1946). Little is known about the life history of this mite. Transmission is presumed to be by coughing.

## Pneumocoptidae

Members of this family infest the lungs of rodents. *Pneumocoptes jellisoni* and *P. penrosei* are common pulmonary parasites of prairie dogs (*Cynomys* spp.) and mice (*Peromyscus* and *Onychomys* spp.) in the midwestern United States; *P. banksi* parasitizes the California ground squirrel (*Spermophilus beecheyi*), whereas *P. tiollaisi* parasitizes voles (*Clethrionomys* spp.) in Europe. Rarely do these mites cause apparent harm to their hosts. A possible exception is a case involving a captive black-tailed prairie dog (*Cynomys ludovicianus*) at the Philadelphia

Zoological Gardens that died from acute bronchopneumonia. Large numbers of *P. penrosei* were found infesting the lungs at necropsy, with postmortem evidence of emphysema and dilatation of the bronchi and bronchioles (Wiedman, 1916).

### Gastronyssidae

Gastronyssid mites are intranasal parasites of rodents and bats. Examples include *Opsonyssus* and *Rodhainyssus* species in the nasal fossae of African and Asian bats and *Sciuracarus paraxeri* in the nasal fossae of South African sun squirrels (Fain, 1956, 1964a, 1967). *Opsonyssus* species also have been reported infesting eye orbits, whereas *G. bakeri* lives in the alimentary tract of fruit bats (Pteropodidae). *Yunkeracarus* is common in a diversity of rodents worldwide. Any deleterious effects they may have on their respective hosts remain unknown.

## MITE-BORNE DISEASES

With the exception of ticks and chiggers, only a few species or groups of mites play a significant role in the transmission of disease agents to domestic and wild animals. In many cases the evidence of mite involvement is circumstantial, based primarily on experimental studies. *Dermanyssus gallinae,* for example, has been shown to transmit the viral agents of *fowlpox,* St. Louis encephalitis, and Eastern equine encephalitis among birds, for which mosquitoes are the usual natural vectors. Similarly, *Cheyletiella parasitivorax* is capable of experimentally infecting rabbits with the virus causing *myxomatosis.* Other examples include viruses that cause *Newcastle disease* of birds, involving *Ornithonyssus sylviarum,* and Western equine encephalitis, involving *Dermanyssus americanus.* Perhaps more significant is experimental evidence that *O. bacoti* possibly serves as both a vector and reservoir of *Hantaan virus,* the causative agent of *Korean hemorrhagic fever,* which infects mice and humans (Meng *et al.*, 1991).

The importance of mites in the ecology of these diseases should be viewed with caution. The isolation of viruses and other pathogens from naturally infected mites, in the absence of documented evidence of transmission capabilities and natural host associations, should not be construed as evidence that such mites necessarily play a role in the transmission of these disease agents.

With the exception of rickettsial organisms transmitted by ticks and chiggers, bacterial disease agents of animals seldom involve mites as vectors. Some reports, however, suggest the possible role of hay-infesting mites as reservoirs for the causative agents of *scrapie* in sheep and *bovine spongiform encephalopathy,* also known as *mad cow disease,* in cattle (Wisniewksi *et al.*, 1996). Both diseases are believed to be caused by brain-destroying proteins called prions.

The only significant protozoan disease agents of vertebrates transmitted by mites are blood parasites of the hemogregarine genus *Hepatozoon* that cause *hepatozoonosis.* Their life cycles are similar in many ways to *Plasmodium* species that cause malaria. *Hepatozoon* species parasitize a wide range of hosts, including amphibians (anurans), reptiles, birds, and mammals, in which they undergo development and multiplication in the liver and circulating blood cells. Hematophagous mites ingest infected blood cells while feeding. The *Hepatozoon* parasites are released in the lumen of the mite midgut, where they penetrate the gut wall to enter the hemocoel. There they form oocysts containing sporozoites. Vertebrates become infected by ingesting mites with infective oocysts. Among the more common hosts of *Hepatozoon* species that utilize mites as vectors are rodents (e.g., field mice, voles, rats, squirrels), skinks, and other lizards. Mites involved are typically mesostigmatid species, including the genera *Haemogamasus, Laelaps,* and *Ophionyssus.* Three *Hepatozoon* species are known to infest domestic animals: *Hepatozoon canis* of dogs, *H. felis* of cats, and *H. muris* of laboratory cotton rats. In addition to mites, many hematophagous insects serve as vectors of other *Hepatozoon* species, including fleas, triatomine bugs, sucking lice, mosquitoes, sand flies, and tsetse flies. For further information on *Hepatozoon* species and their hosts, see Smith (1996).

The only notable nematode parasite of domestic animals transmitted by mites is the filarial worm *Litomosoides sigmodontis* (formerly *L. carinii*), the causative agent of *cotton rat filiariasis.* This nematode occurs in the southeastern United States, where it is transmitted among cotton rats (*Sigmodon hispidus*) and squirrels by the tropical rat mite *O. bacoti.* Related *Litomosoides* species occur in South America, where their hosts include wood rats (several spp.), marsh rats (*Holochilus* spp.), and the house mouse (*Mus musculus*). Microfilariae of *L. sigmodontis* in host blood are ingested by *O. bacoti,* where they develop to infective third-stage larvae. Rodents become infested when the mite takes a subsequent blood meal. The nematodes become established in the pleural cavity (also the peritoneal cavity in heavy infestations), where they develop to adults. Laboratory-infected cotton rats subjected to chronic reinfection with *L. sigmodontis* tend to lose weight, become feverish, exhibit shallow respiration, and undergo behavioral changes, e.g., sitting in a haunched position with raised fur.

The tropical rat mite (*O. bacoti*) is a vector of *L. sigmodontis,* a filarial nematode in rats and other wild rodents. The cotton rat (*S. hispidus*) is a common host in the southern United States. The mites ingest microfilariae with blood from infested rats. The microfilariae penetrate the gut wall and move through the hemocoel to

TABLE IV

**Anoplocephalid Tapeworms of Domestic Animals for Which Oribatid Mites Serve as Intermediate Hosts: The Oribatid Genera Listed Include Both Those Which Have Been Found Naturally Infected and Those Which Have Been Shown Experimentally to Support Development of Cysticercoids**

| Tapeworm species | Domestic hosts | Intermediate hosts (oribatid genera) | Geographic occurrence |
|---|---|---|---|
| *Anoplocephala perfoliata* | Horse, donkey | *Achipteria, Carabodes, Ceratozetes, Eremaeus, Galumna, Hermanniella, Liacarus, Liebstadia, Parachipteria, Platynothrus, Scheloribates, Trichoribates, Urubambates, Zygoribatula* | Cosmopolitan |
| *A. magna* | Horse, donkey | *Scheloribates* | Cosmopolitan |
| *Avitellina bangaonensis* | Cattle, goats | Oribatids and/or psocids? | India |
| *A. centripunctata sensu latu* (including *A. goughi, A. lahorea, A. sudanea*) | Sheep (primarily), goat, cattle, buffalo, zebu, camel, other ruminants | *Punctoribates, Scheloribates, Trichoribates,* (also psocids, collembolans) | Europe, Asia, India, Africa |
| *A. chalmersi* | Sheep, goat | Oribatids and/or psocids? | India, Africa |
| *A. tatia* | Goat | Oribatids and/or psocids? | India |
| *A. woodlandi* | Sheep, goat | Oribatids | India |
| *Moniezia autumnalia* | Sheep, cattle | Oribatids? | Bulgaria, Tadzhikistan, Russia |
| *M. benedeni* | Cattle (primarily), water buffalo, bison, sheep, goat, other ruminants | *Achipteria, Ceratoppia Ceratozetes, Galumna, Liebstadia, Oribatula, Pergalumna, Platynothrus, Punctoribates, Scheloribates, Spatiodaemaeus, Trichoribates, Zygoribatula* | Cosmopolitan |
| *M. expansa* | Sheep (primarily), goat, cattle, ibex, gazelle, camel, other ruminants | *Achipteria, Allogalumna ( Galumna?), Cepheus, Ceratoppia, Ceratozetes, Eremaeus, Eupelops, Euzetes, Furcoribula, Galumna, Hermanniella, Liacarus, Oribatella, Oribatula, Parachipteria, Peloptulus, Peloribates, Pergalumna, Platynothrus, Protoribates, Punctoribates, Scheloribates, Scutovertex, Spatiodamaeus, Trichoribates, Unguizetes, Xenillus, Zygoribatula* | Cosmopolitan |
| *M. denticulata* | Cattle, sheep, goat, others | Oribatids | Cosmopolitan |
| *M. neumani* | Sheep | *Punctoribates, Scheloribates, Trichoribates* | ? |
| *Paranoplocephala mamillana* | Horse | *Achipteria, Allogalumna, Ceratozetes, Galumna, Scheloribates* | ? |
| *Stilesia globipunctata* | Sheep, goat, cattle, zebu, gazelle, camel, other ruminants | *Africacarus, Allogalumna (Galumna?), Scheloribates, Zygoribatula; (psocids?)* | Europe (Spain), Asia Minor (Turkey), Asia, Africa |
| *S. hepatica* | Sheep, goat, cattle (rarely), wild ruminants | Oribatids? | Asia, Africa |
| *S. vittata* | camel and dromedary (primarily), sheep, goat | Oribatids? | Uzbekistan, India, Africa (eastern and southern) |
| *Thysaniezia giardi* (syn. *T. ovilla*) | Sheep, goat (primarily), cattle buffalo | *Achipteria, Libstadia, Punctoribates, Scheloribates, Tichoribates, Zygoribatula* | Cosmopolitan |
| *Thysanosoma actinioides* | Sheep, goat, cattle, deer, antelope | Oribatids and/or psocids? | North America, South America (western regions) |

invade the salivary glands, fat-body cells, coxal glands, and glands associated with the female reproductive organs. There they develop to infective third-stage larvae in about 2 weeks. Infective larvae are introduced to rodent hosts when the mite subsequently feeds, developing to adult worms.

## MITES AS INTERMEDIATE HOSTS OF TAPEWORMS

Oribatid mites serve as intermediate hosts for about 27 species (14 genera) of tapeworms in the family Anoplocephalidae. Of this number, approximately 20 species are

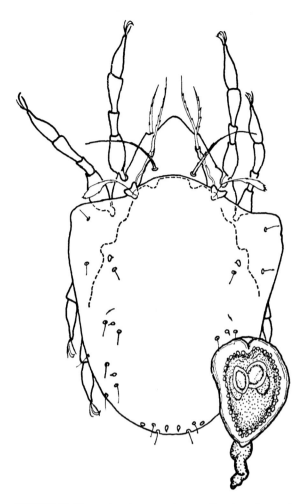

**FIGURE 23.62** Oribatid mite as intermediate host for cysticercoid stage of tapeworms; cysticercoid of tapeworm is depicted at lower right. (From Sengbusch, 1977.)

TABLE V

**Families and Genera of Oribatid Mites Known to Support Development of Anoplocephalid Tapeworms That Parasitize Domestic Animals Based on Allred (1954), Sengbusch (1977), Narsapur (1988), and Denegri (1993): The Approximate Number of Species Is Indicated in Parentheses Following Each Genus**

| | | |
|---|---|---|
| Achipteriidae | Galumnidae | Oribatellidae |
| *Achipteria* (2) | *Allogalumna* | *Oribatella* (1) |
| *Parachipteria* (2) | *Galumna* (15) | Oribatulidae |
| Astegistidae | *Pergalumna* (2) | *Liebstadia* (1) |
| *Furcoribula* (1) | *Pilogalumna* (1) | *Oribatula* (2) |
| Camisiidae | Haplozetidae | *Zygoribatula* (11) |
| *Platynothrus* (1) | *Peloribates* (2) | Pelopidae |
| Carabodidae | Hermanniellidae | *Peloptulus* |
| *Carabodes* (1) | *Hermanniella* (2) | Phenopelopidae |
| *Cepheus* (1) | Liacaridae | *Eupelops* (1) |
| Ceratozetidae | *Adoristes* (2) | Protoribatidae |
| *Ceratozetes* (4) | *Liacarus* (1) | *Protoribates* (1) |
| *Hypozetes* (1) | *Xenillus* (1) | Scheloribatidae |
| *Trichoribates* (3) | Metrioppidae | *Scheloribates* (14) |
| Damaeidae | *Ceratoppia* (1) | *Urubambates* (1) |
| *Spatiodamaeus* (1) | Mochlozetidae | Scutoverticidae |
| Epilohmanniidae | *Unguizetes* (1) | *Scutovertex* (1) |
| *Epilohamminia* (1) | Mycobatidae | Xylobatidae |
| Eremaeidae | *Punctoribates* (3) | *Xylobates* (1) |
| *Eremaeus* (2) | Oppiidae | Family? |
| Euzetidae | *Oppiella* (1) | *Africacarus* (1) |
| *Euzetes* (1) | | |

parasites of domestic animals (Table IV). The most important anoplocephalan genus that infests domestic animals is *Moniezia*, represented by 2 species that parasitize ruminants worldwide. *Moniezia benedeni* is primarily a parasite of cattle, whereas *M. expansa* is primarily a parasite of sheep and goats. These tapeworms are especially prevalent in young host animals less than 6–8 months old. Older animals tend to be less susceptible and after 2 years of age seldom have more than 1 worm or a few worms.

Most anoplocephalid tapeworms cause little apparent harm to their hosts even when the parasite burden is high. In some cases, however, they can cause loss of weight gain, unthriftiness, colic, and intestinal blockage that may require veterinary attention. Sheep are the most adversely affected, whereas cattle and horses seldom experience significant health problems due to cestodes associated with oribatid mites. Heavy infestations can lead

to severe health problems, especially in Asia, involving weight loss, reduced wool yield, anemia, enteritis, diarrhea, intestinal obstruction, toxemia, convulsions, and death. In the case of *Stilesia hepatica*, which infests the bile ducts of sheep, economic losses can result from condemnation of sheep livers at meat inspection. For further details on the biology and pathogenic consequences of anoplocephalid tapeworms infesting domestic animals, see Graber (1960), Soulsby (1965, 1982), Narsapur (1988), and Kauffmann (1996).

Although tapeworms of the genus *Avitellina* are common parasites of domestic ruminants in Europe, Asia, and Africa, they do not appear to significantly harm their hosts. The taxonomy of *Avitellina* remains unstable, with several long-recognized species (e.g., *Avitellina goughi*, *A. lahorea*, *A. sudanea*) being regarded as synonyms of the widespread species *A. centripunctata* by some workers (Raina, 1975) but as valid species by others (Malhotra and Capoor, 1982). At the same time, there is evidence that *A. centripuncta* is represented in Africa as a complex of at least five cryptic species (Ba *et al.*, 1994). This has complicated the interpretation of any meaningful associations of individual oribatid species as intermediate hosts of several *Avitellina* species.

The development and life history of *Moniezia* species are representative of anoplocephalid tapeworms in general. The *proglottids* of adult tapeworms containing eggs are passed in the host feces, contaminating pasture grasses associated with oribatid mites. The mites feed on the eggs by breaking the shell with their chelicerae and ingesting the developing embryo, or *oncosphere*. In the mite, the oncosphere penetrates the midgut wall to enter the hemocoel, where it slowly develops in about 4 weeks to the *cysticercoid* stage (Fig. 23.62). The developmental time, however, varies significantly with different species and environmental temperatures, ranging from 2 to 7 months. Infested oribatid mites are consumed with grasses and other forage by ruminants as they graze. The cysticercoids are released as the mites are digested and attach to the wall of the alimentary tract, or bile ducts in some species, where they grow and mature to *adult tapeworms* in about 5–6 weeks. Mature tapeworms typically live 2–6 months before being spontaneously eliminated. During this time they release egg-filled proglottids that are passed in the host feces.

More than 125 species of oribatid mites, representing 37 genera and 25 families, have been shown to support development of anoplocephalid tapeworms (Denegri, 1993) (Table V). Of this number, only about 45 oribatid species have been found to be naturally infected with cysticercoids. This indicates that many oribatid species are capable of supporting development of cysticercoids when experimentally infected but do not necessarily ingest tapeworm eggs under natural conditions. The most important mite families that serve as intermediate hosts are Ceratozetidae, Galumnidae, Oribatulidae, and Scheloribatidae, apparently reflecting their relatively large size and ability to ingest tapeworm eggs. Other contributing factors are the size and structure of the chelicerae, the natural diet of the mites, and their tendency to move upward from the soil into grasses when ruminants are grazing. This combination of traits helps to explain why *Galumna*, *Scheloribates*, and *Zygoribatula* species are the oribatids most commonly infested with cysticercoids of anoplocephalid tapeworms. For further information on oribatid taxa as intermediate hosts for tapeworms, see Sengbusch (1977), Narsapur (1988), and Denegri (1993).

Heavy tapeworm infestations in ruminants are directly correlated with large oribatid mite populations. Newly seeded and first-year pastures have low mite numbers and result in low infestation rates even in young animals. Older pastures that have been left undisturbed for two or more years support the buildup of oribatid numbers, dramatically increasing the risk of parasitism. To reduce mite populations and avoid this problem, lambs and calves should be started on new pastures, permanent pastures should be cultivated annually, and pastures should be fenced off to prevent livestock from grazing in adjacent rough pasture and woodland.

A few anoplocephalid tapeworms in the genus *Cittotaenia* that utilize oribatid mites as intermediate hosts are parasites of rabbits and hares in North America, Europe, and Asia. Heavy infestations by those tapeworms can cause digestive disturbances, emaciation, and death of hosts. In addition, *Cittotaenia pusilla* occasionally parasitizes laboratory mice and rats in North America, Europe, and Japan. Its natural hosts include the Norway rat, black rat, house mouse, meadow mice (*Microtus* spp.), and other wild rodents. The grain-infesting mite *Glycyphagus domesticus* (family Glycyphagidae) reportedly serves as the intermediate host (Allred, 1954).

# REFERENCES AND FURTHER READING

Allred, D. M. (1954). Mites as intermediate hosts of tapeworms. *Proceedings of the Utah Academy of Science* **31**, 44–51.

Andrews, J. R., and Ramsay, G. W. (1982). A case of papular dermatosis in man attributed to an ascid mite (Acari). *Journal of Medical Entomology* **19**, 111–112.

Arlian, L. G., and Hart, B. J. (1992). Water balance and humidity requirements of house dust mites. *Experimental & Applied Acarology* **16**, 15–35.

Arlian, L. G., Morgan, M. S., Vyszenski-Moher, D. L., and Stemmer, B. L. (1994a). *Sarcoptes scabiei*: the circulating antibody response and induced immunity to scabies. *Experimental Parasitology* **78**, 37–50.

Arlian, L. G., Rapp, C. M., Vyszenski-Moher, D. L., and Morgan, M. S. (1994b). *Sarcoptes scabiei*: histopathological changes associated with acquisition and expression of host immunity to scabies. *Experimental Parasitology* **78**, 51–63.

Aylesworth, R., and Baldridge, D. (1983). *Dermatophagoides scheretewski* and feather pillow dermatitis. *Minnesota Medicine* **66**, 43.

Ba, C. T., Wang, X. Q., Renaud, R., Euzet, L., Marchand, B., and de Meeus, T. (1994). Diversity in the genera *Avitellina* and *Thysaniezia* (Cestoda: Cyclophyllidea): genetic evidence. *Journal of the Helminthological Society of Washington* **61**, 57–60.

Baker, A. S. (1999). "Mites and Ticks of Domestic Animals. An Identification Guide and Information Source." H. M. Stationery Office, London.

Baker, E. W., Evans, T. M., Gould, D. J., Hull, W. B., and Keegan, H. L. (1956). "A Manual of Parasitic Mites." National Pest Control Assoc., New York.

Baker, R. A. (1973). Notes on the internal anatomy, the food requirements and development in the family Ereynetidae (Trombidiformes). *Acarologia* **15**, 43–52.

Bates, P. G. (1999). Inter- and intra-specific variation within the genus *Psoroptes* (Acari: Psoroptidae). *Veterinary Parasitology* **83**, 201–217.

Betz, T. G., Davis, B. L., Fournier, P. V., Rawlings, J. A., Elliot, L. B., and Baggett, D. A. (1982). Occupational dermatitis associated with straw itch mites (*Pyemotes ventricosus*). *Journal of the American Medical Association* **247**, 2821–2823.

Borstein, S., and Zakrisson, G. (1993). Clinical picture and antibody response in pigs infected by *Sarcoptes scabiei* var. *suis*. *Veterinary Dermatology* **4**, 123–131.

Boyce, W. M., and Brown, R. N. (1991). Antigenic characterization of *Psoroptes* spp. (Acari: Psoroptidae) mites from different hosts. *Journal of Parasitology* 77, 675–679.

Brettman, L. R., Lewin, S., Holzman, R. S., Goldman, W. D., Marr, J. S., Kechijian, P., and Schinella, R. (1981). Rickettsialpox: report of an outbreak and a contemporary review. *Medicine (Baltimore)* 60, 363–372.

Brockie, R. E. (1974). The hedgehog mange mite, *Caparinia tripilis*, in New Zealand. *New Zealand Veterinary Journal* 22, 243–247.

Bukva, V. (1986). *Demodex tauri* sp. n. (Acari: Demodicidae), a new parasite of cattle. *Folia Parasitologica* 33, 363–369.

Bukva, V. (1990). Three species of the hair follicle mites (Acari: Demodicidae) parasitizing sheep, *Ovis aries* L. *Folia Parasitologica* 37, 81–91.

Burns, D. A. (1992). Follicle mites and their role in disease. *Clinical and Experimental Dermatology* 17, 152–155.

Camin, J. H. (1953). Observations on the life history and sensory behavior of the snake mite, *Ophionyssus natricis* (Gervais) (Acarina: Macronyssidae). *Chicago Academy of Science Special Publications* 10, 1–75.

Carlson, B. L., Roher, D. P., and Nielsen, S. W. (1982). Notoedric mange in gray squirrels (*Sciurus carolinensis*). *Journal of Wildlife Diseases* 18, 347–348.

Castellani, A. (1907). Note on an acarid-like parasite found in the omentum of a negro (Cytoditidae). *Centralblatt für Bakteriologie, Parasitenkunde und Infektionskrankheiten* I. 43, 372.

Chakrabarti, A. (1986). Human notoedric scabies from contact with cats infested with *Notoedres cati*. *International Journal of Dermatology* 25, 646–648.

Chen, X. B., Sun, X., and Hu, S. F. (1990). Clinical manifestation and treatment of pulmonary acariasis (in Chinese). *Chinese Journal of Parasitology and Parasitic Diseases* 8, 41–44.

Chen, X. B., Fu, C. B., Chen, X. B., and Fu, C. B. (1992). Mites causing pulmonary, intestinal and urinary acariasis. In "Researches of Acarology in China" (X. B. Chen and E. P. Ma, Eds.), pp. 109–113. Chongqing Publ., Chongqing.

Clark, G. M. (1960). Three new nasal mites (Acarina: Speleognathidae) from the grey squirrel, the common grackle, and the meadowlark in the United States. *Proceedings of the Helminthological Society of Washington* 27, 103–110.

Cochrane, G. (1994). Effects of *Psoroptes ovis* on lamb carcasses. *Veterinary Record* 134, 72.

Cole, N. A., and Guillot, F. S. (1987). Influence of *Psoroptes ovis* on the energy metabolism of heifer calves. *Veterinary Parasitology* 23, 285–295.

Colloff, M. J., and Hart, B. J. (1992). Age structure and dynamics of house dust mite populations. *Experimental & Applied Acarology* 16, 49–74.

Colloff, M. J. (G. P. Channabasavana and C. A. Viraktamath, Eds.) Oxford & IBH Publishing Co., New Delhi. (1989). Human semen as a dietary supplement for house dust mites (Astigmata: Pyroglyphidae). In "Progress in Acarology," Vol. 1, pp. 141–146.

Cosoroabă, I. (1994). "Acarologie Veterinaria" (in Romanian). Ceres Publ., Bucharest.

Costa, A. L., Leite, R. C., Faccini, J. L. H., and DaCosta, A. L. (1992). Preliminary investigations on transmission and life cycle of the ear mites of the genus *Raillietia* Trouessart (Acari: Gamasida) parasites of cattle. *Memorias do Instituto Oswaldo Cruz* 87(Suppl. 1), 97–100.

Cremers, H. J. W. M. (1985). The incidence of *Chorioptes bovis* (Acarina: Psoroptidae) on the feet of horses, sheep, and goats in the Netherlands. *Veterinary Quarterly* 7, 283–289.

Cuthbert, O. D. (1990). Storage mite allergy. *Clinical Reviews in Allergy* 8, 69–86.

Czarnecki, N., and Kraus, H. (1978). Milbendermatitis durch *Tyrophagus dimidiatus*. *Zeitschrift fuer Hautkrankheiten* 53, 414–416.

Dahl, F. (1910). Milben als Erzeuger von Zellwucherunge. *Centralblatt für Bakteriologie, Parasitenkunde und Infektionskrankheiten* 53, 524–533.

DaMassa, A. J., Wakenall, P. S., and Brooks, D. L. (1992). Mycoplasmas of goats and sheep. *Journal of Veterinary Diagnostic Investigation* 4, 101–113.

Davis, J. W., and Anderson, R. C., Eds. (1971). "Parasitic Diseases of Wild Mammals." Iowa State Univ. Press, Ames.

Denegri, G. M. (1985). Desarrollo experimental de *Bertiella mucronata* Meyner, 1895. (Cestoda: Anoplocephalidae) de origen humano en su huésped intermediario. *Zentralblatt fuer Veterinaermedizin Reihe B* 32, 498–504.

Denegri, G. M. (1993). Review of oribatid mites as intermediate hosts of tapeworms of the Anoplocephalidae. *Experimental & Applied Acarology* 17, 567–580.

Desch, C., and Nutting, W. (1972). *Demodex folliculorum* (Simon) and *D. brevis* Akbulatova of man: redescription and reevaluation. *Journal of Parasitology* 58, 169–177.

Dietrich, M., and Horstmann, R. D. (1983). Urtikarielle Dermatitis bei Acariasis durch *Dermanyssus hirundinis*. Schwalben-Trugkratze, Pseudoscabies. *Medizinische Welt* 34, 595–597.

Domrow, R. (1969). The nasal mites of Queensland birds (Acari: Dermanyssidae, Ereynetidae and Epidermoptidae). *Linnean Society of New South Wales* 93, 297–426.

Domrow, R. (1988). Acari Mesostigmata parasitic on Australian vertebrates: an annotated checklist, keys and bibliography. *Invertebrate Taxonomy* 1, 817–948.

Domrow, R. (1991). Acari Prostigmata (excluding Trombiculidae) parasitic on Australian vertebrates: an annotated checklist, keys and bibliography. *Invertebrate Taxonomy* 4, 1283–1376.

Domrow, R. (1992). Acari Astigmata (excluding feather mites) parasitic on Australian vertebrates: an annotated checklist, keys and bibliography. *Invertebrate Taxonomy* 6, 1459–1606.

Elixmann, J. H., Jorde, W., and Schata, M. (1991). Incidence of mites in domestic textiles in Germany. *Allergologie* 14, 451–460.

English, F. P., and Nutting, W. B. (1981). Demodicosis of ophthalmic concern. *American Journal of Ophthalmology* 91, 362–372.

Evans, G. O. (1992). "Principles of Acarology." CAB International, Oxon, UK.

Fain, A. (1956). Une nouvelle famille d'acariens endoparasites des chauves-souris: Gastronyssidae fam. nov. *Annales de la Société Belge de Médecine Tropicale* 36, 87–98.

Fain, A. (1957). Les Acariens des familles Epidermoptidae et Rhinonyssidae parasites des fosses nasales d'oiseaux au Ruanda-Urundi et au Congo Belge. *Annales de Musée Royale du Congo Belge, Tervuren Reeks in 80*, pp. 1–176.

Fain, A. (1961). Les acariens parasites endopulmonaires des serpents (Entonyssidae: Mesostigmata). *Institut Royale des Sciences Naturelles de Belgique* 37, 1–135.

Fain, A. (1964a). Chaetotaxie et classification des Gastronyssidae avec description d'un nouveau genre parasite nasicole d'un Ecureuil sudafricain (Acarina: Sarcoptiformes). *Revue de Zoologie et de Botanique Africaines* 70, 40–52.

Fain, A. (1964b). Les Lemurnyssidae parasites nasicoles des Lorisidae africains et des Cebidae sud-américaines. Description d'une espèce nouvelle (Acarina: Sarcoptiformes). *Annales de la Société Belge de Médecine Tropicale* 44, 453–458.

Fain, A. (1965). A review of the family Epidermoptidae Trouessart parasitic on the skin of birds. Parts I–II. Koninklijke Acaemie voor Wetenschappen, Letteren en Schon Kunsten van België. 27(84), 1–176; 5–144.

Fain, A. (1967). Observations sur les Rodhainyssinae acariens parasites des voies respiratoires des chauves-souris (Gastronyssidae: Sarcoptiformes). *Acta Zoologica et Pathologica Antverpiensia* **44**, 3–35.

Fain, A. (1968). Étude de la variabilite de *Sarcoptes scabiei* avec une revision des Sarcoptidae. *Acta Zoologica et Pathologica Antverpiensia* **47**, 3.

Fain, A. (1969). Les deutonymphs hypopiales vivant en association phoretique sur les mammiferes (Acarina: Sarcoptiformes). *Bulletin de Institut Royale des Sciences Naturelles de Belgique* **45**, 1–262.

Fain, A. (1970). Novoeaux acarines nasicoles de le famillie Turbinoptidae (Sarcoptiformes). *Bulletin & Annalen de la Societe Royale Belge d'Entomologie* **106**, 28–36.

Fain, A., and Hyland, K. E. (1975). Speleognathinae collected from birds in North America (Acarina: Ereynetidae). *Journal of the New York Entomological Society* **83**, 203–208.

Fain, A., Munting, A. J., and Lukoschus, F. (1970). Les Myocoptidae parasites de rongeurs en Hollande et en Belgique (Acarina: Sarcoptiformes). *Acta Zoologia et Pathologia Antverpiensia* **50**, 67–172.

Fain, A., Whitaker, J. O., Jr., McDaniel, B., and Lukoschus, F. (1974). *Listrophorus synaptomys*, a new species from *Synaptomys* and *Lemmus*. *Acarologia* **16**, 319–324.

Farley, M. L., Mabry, L. C., and Hieger, L. R. (1989). Mites in pulmonary cytology specimens. *Diagnostic Cytopathology* **5**, 416–426.

Flynn, R. J., Ed. (1973). "Parasites of Laboratory Animals." Iowa State Univ. Press, Ames.

Fox, J. G., and Hewes, K. (1976). *Cheyletiella* infestation in cats. *Journal of the American Veterinary Medical Association* **169**, 332–333.

Franklin, C. D., and Underwood, J. C. (1986). *Demodex* infestation of oral mucosal sebaceous glands. *Oral Surgery, Oral Medicine, Oral Pathology* **61**, 80–82.

Furman, D. P. (1959). Feeding habits of symbiotic mesostigmatid mites of mammals in relation to pathogen-vector potentials. *American Journal of Tropical Medicine and Hygiene* **8**, 5.

Furman, D. P., and Dailey, M. D. (1980). The genus *Halarachne* (Acari: Halarachnidae), with the description of a new species from the Hawaiian monk seal. *Journal of Medical Entomology* **17**, 352–359.

Furman, D. P., and Murray, M. D. (1980). The genus *Halarachne* (Acari: Halarachnidae), with the description of a new species from the Hawaiian monk seal. *Journal of Medical Entomology* **17**, 352–359.

Garlick, N. L. (1977). Canine pulmonary acariasis. *Canine Practice* **4**, 42–47.

Gaud, J., and Atyeo, W. T. (1996). Feather mites of the world (Acarina, Astigmata): the supraspecific taxa. Part I. Text. Part II. Illustrations of feather mite taxa. *Annalen Zoologische Wetenschappen-Koninklijk Museum voor Midden-Africa/Musee Royal de l'Afrique Centrale, Tervuren, Belgium* **277**(1), 1193 pp. **277**(2), 2436 pp.

Georgi, J. R., and Whitlock, J. H. (1980). Arachnids. In "Parasitology for Veterinarians" (J. R. Georgi, Ed.), 3rd ed., Chapt. 3, pp. 41–67. Saunders, Philadelphia.

Gerson, U., Fain, A., and Smiley, R. L. (1999). Further observations on the Cheyletidae (Acari), with a key to the genera of the Cheyletinae and a list of all known species in the family. *Bulletin de l'Institut Royal des Sciences Naturelles de Belgique Entomologie* **69**, 35–86.

Goff, M. L., Loomis, R. B., Welbourn, W. C., and Wrenn, W. J. (1982). A glossary of chigger terminology (Acari: Trombiculidae). *Journal of Medical Entomology* **19**, 221–238.

Gorham, J. R., Ed. (1991). "Insect and Mite Pests in Food: An Illustrated Key." U.S. Department of Agriculture, Agriculture Handbook No. 655. U.S. Government Printing Office, Washington, DC.

Graber, M. (1960). "Symposium on Helminthiasis in Domestic Animals (Exclusive of Poultry)," Nairobi, Kenya, 1959. pp. 81–130. Interagency Committee on Environment and Development, Commission for Technical Cooperation in Africa South of the Sahara, Watergate House, London.

Gregory, M. W. (1981). Mites of the hedgehog *Erinaceus albiventris* Wagner in Kenya: observations on the prevalence and pathogenicity of *Notoedres oudemansi* Fain, *Caparinia erinacei* Fain and *Rodentopus sciuri* Fain. *Parasitology* **82**, 149–157.

Guillot, F. S., and Stromberg, P. C. (1987). Reproductive success of *Psoroptes ovis* (Acari: Psoroptidae) on Hereford calves with a previous infestation of psoroptic mites. *Journal of Medical Entomology* **24**, 416–419.

Harada, S. H., and Sadaji, T. (1925). On a case of mites found in human urine (in Japanese). *Chugai Iji Shimpo* **44**, 859–866.

Hardison, J. L. (1977). A case of *Eutrombicula alfreddugesi* (chiggers) in a cat. *Veterinary Medicine and Small Animal Clinician* **72**, 47.

Harwood, R. F., and James, M. T. (1979). "Entomology in Human and Animal Health," 7th ed. Macmillan Co., New York.

Hay, D. B., Hart, B. J., Pearce, R. B., Kozakiewicz, Z., and Douglas, A. E. (1992). How relevant are house dust mite–fungal interactions in laboratory culture to the natural dust system? *Experimental & Applied Acarology* **16**, 37–47.

Hay, D. B., Hart, B. J., and Douglas, A. E. (1993). Effects of the fungus *Aspergillus penicillioides* on the house dust mite *Dermatophagoides pteronyssinus*: an experimental re-evaluation. *Medical and Veterinary Entomology* **7**, 271–274.

Heilesen, B. (1946). "Studies on *Acarus scabiei* and Scabies." Rosenkilde & Bagger, Copenhagen.

Henschel, J. (1929). Reizphysiologische Untersuchung der Käsemilbe *Tyrolichus casei* (Oudemans). *Zeitschrift für Vergleichende Physiologie* **9**, 802–837.

Hewitt, M., Barrow, G. I., Miller, D. C., Turk, F., and Turk, S. (1973). Mites in the personal environment and their role in skin disorders. *British Journal of Dermatology* **89**, 401–409.

Higby, W. E. (1946). A new canary plague (Cytoditidae). *All Pets Magazine* December, 8–9.

Hirst, S. (1919). Studies on Acari. No.1. The Genus *Demodex*, Owen. British Museum of Natural History, London.

Hirst, S. (1922). "Mites Injurious to Domestic Animals." Economics Series No. 13. British Museum of Natural History, London.

Hoban, P. A. (1990). A review of desert bighorn sheep in the San Andres Mountains, New Mexico. *Desert Bighorn Council Transactions* **34**, 14–22.

Hogsette, J. A., Butler, J. F., Miller, W. V., and Hall, R. D. (1988). "Annotated Bibliography of the Northern Fowl Mite, *Ornithonyssus sylviarum* (Canestrini & Fanzago), (Acari: Macronyssidae)." Miscellaneous Publication No. 76. Entomological Society of America, Lanham, MD.

Hong, C. S., Park, H. S., and Oh, S. H. (1987). *Dermatophagoides farinae*, an important allergenic substance in buckwheat-husk pillows. *Yonsei Medical Journal* **28**, 274–281.

Houck, M. A., Ed. (1994). "Mites: Ecological and Evolutionary Analyses of Life-History Patterns." Chapman & Hall, London.

Huebner, R. J., Jellison, W. L., and Pomerantz, C. (1946). Rickettsialpox—a newly recognized rickettsial disease. IV. Isolation of a rickettsia apparently identical with the causative agent of rickettsialpox from *Allodermanyssus sanguineus*, a rodent mite. *Public Health Reports* **61**, 1677–1682.

Hughes, A. M. (1976). "The Mites of Stored Food and Houses," 2nd ed. Ministry of Agriculture, Fisheries and Food, Technical Bulletin No. 9. H. M. Stationery Office, London.

Hull, W. B. (1970). Respiratory mite parasites in nonhuman primates. *Laboratory Animal Care* **20**, 402–406.

Johnson, L. S., and Albrecht, D. J. (1993). Effects of haematophagous ectoparasites on nestling house wrens, *Troglodytes aedon*: who pays the cost of parasitism? *Oikos* **66**, 255–262.

Kambarage, D. M. (1992). Sarcoptic mange infestation in goats. *Bulletin of Animal Health and Production in Africa* **40**, 239–244.

Kass, E. M., Szaniawski, W. K., Levy, H., Leach, J., Srinivasan, K., and Rives, C. (1994). Rickettsialpox in a New York City hospital, 1980–1989. *New England Journal of Medicine* **331**, 1612–1617.

Kaufmann, J. (1996). "Parasitic Infections of Domestic Animals: A Diagnostic Manual." Birkhauser Verlag, Basel.

Kawamura., A., Jr., Tanaka, H., and Tamura, A., Eds. (1995). "Tsutsugamushi Disease." Univ. of Tokyo Press, Tokyo.

Kazacos, E. A., Kazacos, K. R., and Demaree, H. A., Jr. (1983). Notoedric mange in two fox squirrels. *Journal of the American Veterinary Medical Association* **183**, 1281–1282.

Keh, B. (1974). Dermatitis caused by the bat mite *Chiroptonyssus robustipes* (Ewing) in California. *Journal of Medical Entomology* **11**, 498.

Kethley, J. (1982). Acariformes. In "Synopsis and Classification of Living Organisms" (S. P. Parker, Ed.), Vol. 2, p. 117. McGraw-Hill, New York.

Kilpio, O., and Pirila, V. (1952). A new tyroglyphid mite causing dermatitis. *Acta Dermato-Venereologica* **32**, 197–200.

Kim, K. C., Haas, V. L., and Keyes, M. C. (1980). Populations, microhabitat preference and effects of infestations of two species of *Orthohalarachne* (Halarachnidae: Acarina) in the northern fur seal. *Journal of Wildlife Diseases* **16**, 45–51.

Kirk, H. (1949). Demodectic mange. *Veterinary Record* **61**, 394.

Kirkwood, A. C. (1980). Effect of *Psoroptes ovis* on the weight of sheep. *Veterinary Record* **107**, 469–470.

Klompen, J. S. H. (1989). Ontogeny of *Rhyncoptes grabberi*, n. sp. (Acari: Astigmata: Rhyncoptidae) associated with *Macaca mulatta*. *Journal of Medical Entomology* **26**, 81–87.

Klompen, J. S. H. (1992). "Phylogenetic Relationships in the Mite family Sarcoptidae (Acari: Astigmata)." Miscellaneous Publication No. 180. Museum of Zoology, University of Michigan, Ann Arbor.

Koutz, F. R., Chamberlain, D. M., and Cole, C. R. (1954). *Pneumonyssus caninum* in the nasal cavity and paranasal sinuses. *Journal of the American Veterinary Medical Association* **122**, 106.

Krantz, G. W. (1978). "A Manual of Acarology," 2nd ed. Oregon State Univ. Book Store, Corvallis.

Krinsky, W. L. (1983). Does epizootic lymphocytic choriomeningitis prime the pump for epidemic rickettsialpox? *Review of Infectious Diseases* **5**, 1118–1119.

Kumar, D., Raisinghani, P. M., and Manohar, G. S. (1992). Sarcoptic mange in camels: a review. In "Proceedings of the First International Camel Conference, Dubai" (W. R. Allen, A. J. Higgins, I. G. Mayhew, D. H. Snow, and J. F. Wade, Eds.), pp. 79–82. R. W. Publ., Newmarket, UK.

Kummel, B. A., Estes, S. A., and Arlian, L. G. (1980). *Trixacarus caviae* infestation of guinea pigs. *Journal of the American Veterinary Medical Association* **177**, 903–908.

Lackman, D. B. (1963). A review of information on rickettsialpox in the United States. *Clinical Pediatrics* **2**, 296–301.

Lange, R. E., Sandoval, A. V., and Meleney, W. P. (1980). Psoroptic scabies in bighorn sheep (*Ovis canadensis mexicana*) in New Mexico. *Journal of Wildlife Diseases* **16**, 77–82.

Lawrence, R. F. (1952). A new parasitic mite from the nasal cavities of the South American toad *Bufo regularis* Reuss. *Proceedings of the Zoological Society of London* **121**, 747–752.

Lee, K. J., Lang, C. M., Hughes, H. C., and Hartshorn, R. D. (1981). Psorergatic mange (Acari: Psorergatidae) of the stumptail macaque (*Macaca arctoides*). *Laboratory Animal Science* **31**, 77–79.

Lefer, L. G., and R. P. Rosier. (1988). Presence of a mite in the female genital tract: some comments (Pyroglyphidae, *Dermatophagoides*). *International Journal of Acarology* **14**, 91–92.

Le Fichoux, Y., Rack, G., Motte, P., Dellamonica, P., and Marty, P. (1980). Dermatite prurigineuse due a *Pyemotes zwoelferi* Krczal, 1963, a propos de plusieurs cas dans les alpes-maritimes. *Acta Tropica* **37**, 83–89.

Li, C., and Li, L. (1990). Human pulmonary acariasis in Anhui Province: an epidemiological survey (in Chinese). *Chinese Journal of Parasitology and Parasitic Diseases* **8**, 41–44.

Lowenstine, L. J., Carpenter, J. L., and OConnor, B. M. (1979). Trombiculosis in a cat. *Journal of the American Veterinary Medical Association* **175**, 289–292.

Ma, E. P., and Wang, D. S. (1992). Tarsonemid mites. In "Researches of Acarology in China" (X. B. Chen and E. P. Ma, Eds.), pp. 34–37. Chongqing Publ., Chongqing.

Malhotra, S. K., and Capoor, V. N. (1982). A new species of *Avitellina* Gough (1911) from Garhwal Hills with a revised key to species of subgenus *Avitellina* Raina (1975). *Proceedings of the Indian Academy of Parasitology* **3**, 12–16.

Martinez Maranon, R., and Hoffman, A. (1976). Tres casos de infestacion del intestino humano por ácaros en el sur de Veracruz. *Revista de Investigacion en Salud Publica* **36**, 187–201.

Matthes, H. F. (1994). Investigations of pathogenesis of cattle demodicosis: sites of predilection, habitat, and dynamics of demodectic nodules. *Veterinary Parasitology* **53**, 283–291.

Mazet, J. A. K., Boyce, W. M., Mellies, J., Gardner, I. A., Clark, R. K., and Jessup, D. A. (1992). Exposure to *Psoroptes* sp. mites is common among bighorn sheep (*Ovis canadensis*) populations in California. *Journal of Wildlife Diseases* **28**, 542–547.

McGarry, J. W. (1993). Identification of *Cheyletiella* eggs in dog feces. *Veterinary Record* **132**, 359–360.

Mellanby, K. (1972). "Scabies," 2nd ed. Classey, Hampton, UK.

Meng, Y. C., Zhuge, H. X., Lan, M. Y., and Zhon, H. F. (1991). Experimental study on transmission of hemorrhagic fever with renal syndrome virus by mites, *Ornithonyssus bacoti* (Hirst). In: "Proceedings of the VIII International Congress of Acarology, Ceske Budejovice, Czechoslovakia, (1990)." Vol. 2, pp. 35–39. SPB Academic Publishing br, The Hague, Netherlands.

Miller, W. H. (1984). Diseases of domestic animals. In "Mammalian Diseases and Arachnids" (W. B. Nutting, Ed.), Vol. 2, Chapt. 6, pp. 115–126. CRC Press, Boca Raton, FL.

Mills, O. H., Jr., and Kligman, A. M. (1983). The follicular biopsy. *Dermatologica* **167**, 57–63.

Möller, A. P. (1990). Effects of parasitism by a haematophagous mite on reproduction in the barn swallow. *Ecology* **71**, 2345–2357.

Morgan, K. L. (1992). Parasitic otitis in sheep associated with *Psoroptes* infestation: a clinical and epidemiological study. *Veterinary Record* **130**, 530–532.

Mulla, M., and Medina, M. S., Eds. (1980). "Domestic Acari of Colombia: Bionomics, Ecology, and Distribution of Allergenic Mites, Their Role in Allergenic Diseases" (bilingual English/Spanish). Colciencias, Bogota.

Nadchatram, M. (1970). A review of intranasal chiggers with descriptions of twelve species from east New Guinea (Acarina: Trombiculidae). *Journal of Medical Entomology* **7**, 1–29.

Naltsas, S., Hodge, S. J., Gataky, G. J., Jr., and Owen, L. G. (1980). Eczematous dermatitis caused by *Dermanyssus americanus*. *Cutis* **25**, 429–431.

Narsapur, V. S. (1988). Pathogenesis and biology of anoplocephaline cestodes of domestic animals. *Annales de Recherches Veterinaires* **19**, 1–17.

Nutting, W. B. (1976a). Hair follicle mites (Acari: Demodicidae) of man. *International Journal of Dermatology* **15**, 79–98.

Nutting, W. B. (1976b). Hair follicle mites (*Demodex* spp.) of medical and veterinary concern. *Cornell Veterinarian* **66**, 214–231.

Nutting, W. B. (1976c). Pathogenesis associated with hair follicle mites (Acari: Demodicidae). *Acarologia* 17, 493–506.

Nutting, W. B., Ed. (1984). "Mammalian Diseases and Arachnids." Vol. 1, Pathogen Biology and Clinical Management. Vol. 2, Medico-Veterinary Laboratory, and Wildlife Diseases, and Control. CRC Press, Boca Raton, FL.

Nutting, W. B., Firda, K. E., and Desch, C. E., Jr. (1989). Topology and histopathology of hair follicle mites (Demodicidae) of man. In "Progress in Acarology" (G. P. ChannaBassavana and C. A. Viraktamath, Eds.), Vol. 1, pp. 113–121. Oxford & IBH Publ., New Delhi.

Oberem, P. T., and Malan, F. S. (1984). A new cause of cattle mange in South Africa: *Psorergates bos* Johnston. *Journal of the South African Veterinary Association* 55, 121–122.

O'Brien, D. J., Gray, J. S., and O'Reilly, P. F. (1994). Survival and retention of infectivity of the mite *Psoroptes ovis* off the host. *Veterinary Research Communications* 18, 27–36.

OConnor, B. M. (1984). Phylogenetic relationships among higher taxa in the Acariformes, with particular reference to the Astigmata. In "Acarology VI" (D. A. Griffiths and C. E. Bowman, Eds.), Vol. 1, pp. 19–27. Ellis Horwood, Chichester.

Oehlschlaegel, G., Bayer, F., Disko, R., Fechter, H., and Mahunka, S. (1983). *Tarsonemus hominis* in Hautbindegewebe. *Hautarzt* 34, 632–634.

Onderscheka, K., Kutzer, E., and Richter, H. E. (1968). Die Raeude der Gemse und ihre Bekampfung. II. Zusammenhaenge zischen Ernaehrung und Raeude. *Zeitschrift für Jagdwissenschaft* 14, 12.

Orkin, M., Maibach, H. I., Parish, L. C., and Schwartzman, R. M., Eds. (1977). "Scabies and Pediculosis." Lippincott, Baltimore.

Otto, Q. T., and Jordaan, L. C. (1992). An orf-like condition caused by trombiculid mites on sheep in South Africa. *Onderstepoort Journal of Veterinary Research* 59, 335–336.

Ottoboni, F., di Loreto, V., Cantoni, A., Lozzia, G. C., Rota, P., Melej, R., Bagnato, A., di Loreto, V., and Domenichini, G. (1989). [Investigations into allergic diseases among raw ham workers in Langhirano and San Daniele.] In "La difesa antiparassitaria nelle industrie alimentari e la protezion degli alimenti." Atti del 4o simposio. pp. 235–241. (G. Domenichini, Ed.) Camera di commercio industria artigianato e agricoltura, Piacenza, Italy.

Parodi, A. S., Rugiero, H. R., Greenway, D. J., Mettler, N., Martinez, A., Boxaca, M., and De la Barerra, J. M. (1959). Aislamiento del virus Junin (F. H. E.) de los ácaros de la zona epidemica (*Echinolaelaps echidninus*, Berlese). *Prensa Medica Argentina* 46, 2242–2244.

Pass, D. A., and Sue, L. J. (1983). A trombiculid mite infestation of canaries. *Australian Veterinary Journal* 60, 218–219.

Pence, D. B. (1973). The nasal mites of birds from Louisiana. IX. Synopsis. *Journal of Parasitology* 59, 881–892.

Pence, D. B. (1975). "Keys, Species and Host List, and Bibliography for Nasal Mites of North American Birds (Acarina: Rhinohyssinae, Turbinoptinae, Speleognathinae, and Cytoditidae)." Museum, Texas Tech. University, Special Publication No. 8. Texas Tech. Press, Lubbock.

Pence, D. B. (1984). Diseases of laboratory animals. In "Mammalian Diseases and Arachnids" (W. B. Nutting, Ed.), Vol. 2, Chapt. 7, pp. 129–187. CRC Press, Boca Raton, FL.

Pence, D. B., Matthews, F. D., III, and Windberg, L. A. (1982). Notoedric mange in the bobcat, *Felis rufus*, from south Texas. *Journal of Wildlife Diseases* 18, 47–50.

Pence, D. B., Spalding, M. G., Bergan, J. F., Cole, R. A., Newman, S., and Gray, P. N. (1997). New records of subcutaneous mites (Acari: Hypoderatidae) in birds, with examples of potential host colonization events. *Journal of Medical Entomology* 34, 411–416.

Philip, C. B. (1948). Tsutsugamushi disease (scrub typhus) in the World War II. *Journal of Parasitology* 34, 169–191.

Philip, C. B., and Hughes, L. E. (1948). The tropical rat mite, *Liponyssus bacoti*, as an experimental vector of rickettsialpox. *American Journal of Tropical Medicine* 28, 697–705.

Philip, C. B., and Kohls, G. M. (1945). Studies on Tsutsugamushi disease (scrub typhus, mite-borne typhus) in New Guinea and adjacent islands: Tsutsugamushi disease with high endemicity on a small South Sea island. *American Journal of Hygiene* 42, 195–202.

Phillips, C. J., Jones, J. K., and Radovsky, F. J. (1969). Macronyssid mites in oral mucosa of long-nosed bats: occurrence and associated pathology (*Radfordiella*). *Science* 165, 1368–1369.

Pitariu, T., Dinulescu, N., Panaitescu, D., and Silard, R. (1978). Cholangiocholecystitis, an acute attack with acarids in B bile (in Romanian). *Revista de Igiena, Bacteriologie, Virusologie, Parazitologie, Epidemiologie, Pneumoftiziologie* 23, 189–192.

Pitariu, T. N., Popescu, I. G., and Banescu, O. (1979). Acarids of pathological significance in urine (in Romanian). *Revista de Igiena, Bacteriologie, Virusologie, Parazitologie, Epidemiologie, Pneumoftiziologie* 24, 55–59.

Pratt, H. D., and Stojanovich, C. J. (1969). Acarina: illustrated key to some common adult female mites and adult ticks. In "CDC Pictorial Keys: Arthropods, Reptiles, Birds and Mammals of Public Health Significance," pp. 26–44. US Department of Health and Human Services, Public Health Service, Centers for Disease Control and Prevention. U.S. Government Printing Office, Washington, DC.

Prisich, I. I., Dobarskaia, L. I., and Zosimova, A. G. (1986). Acariasis in children with chronic digestive system diseases (in Russian). *Meditsinskaia Parazitologiia i Parazitarnye Bolezni* 4, 50–51.

Prosl, H., Rabitsch, A., and Brabenetz, J. (1985). Zur Bedeutung der Herbstgrasmilbe—*Neotrombicula autumnalis* (Shaw 1790)—in der Veterinarmedizin: nervale Symptome bei Hunden nach massiver Infestation. *Tierarztliche Praxis* 13, 57–64.

Quoix, E., le Mao, J., Hoyet, C., and Pauli, G. (1993). Prediction of mite allergen levels by guanine measurements in house-dust samples. *Allergy (Copenhagen)* 48, 306–309.

Rafferty, D. E., and Gray, J. S. (1987). The feeding behaviour of *Psoroptes* spp. Mites on rabbits and sheep. *Journal of Parasitology* 73, 901–906.

Raina, M. K. (1975). A monograph on the genus *Avitellina* Gough, 1911 (Avitellinidae: Cestoda). *Zoologische Jahrbuecher Abteilung fuer Systematik Oekologie und Geographie der Tiere* 102, 508–552.

Raulston, G. L. (1972). Psorergatic mites in platas monkeys. *Laboratory Animal Science* 22, 107.

Rollor, E. A., III, Nettles, V. F., Davidson, W. R., and Gerrish, R. R. (1978). Otitis media caused by *Psoroptes cuniculi* in white-tailed deer. *Journal of the American Veterinary Medical Association* 173, 1242–1243.

Rufli, T., and Mumculoglu, Y. (1981). The hair follicle mites *Demodex folliculorum* and *Demodex brevis*: biology and medical importance. A review. *Dermatologica* 162, 1–11.

Samšiňák, K., Palička, P., Zitek, K., Mališ, L., and Vobrázková, E. (1976). Are the mites of the genus *Tarsonemus* really parasites of man? *Folia Parasitologica* 23, 91–93.

Samuel, W. M., Welch, D. A., and Smith, B. L. (1991). Ectoparasites from elk (*Cervus elaphus nelsoni*) from Wyoming. *Journal of Wildlife Diseases* 27, 446–451.

Schultz, H. (1975). Human infestation by *Ophionyssus natricis* snake mite. *British Journal of Dermatology* 93, 695–697.

Schuster, R., and Murphy, P. W., Eds. (1991). "The Acari: Reproduction, Development and Life-History Strategies." Chapman & Hall, London.

Scott, D. W. (1979). Canine demodicosis. *Veterinary Clinics of North America* 9, 79–92.

Seier, J. V. (1985). Psorergatic acariasis in vervet monkeys. *Laboratory Animals* 19, 236–239.

Sengbusch, H. G. (1977). Review of oribatid mite-anoplocephalan tapeworm relationships (Acari; Oribatei: Cestoda; Anoplocephalidae). In "Biology of Oribatid Mites" (D. L. Dindal, Ed.), pp. 87–102. State Univ. of New York, College of Environmental Science and Forestry, Syracuse.

Shaddock, J. W., and Pakes, S. P. (1978). Protozoal and metazoal diseases. In "Pathology of Laboratory Animals" (K. Benirschke, F. M. Garner, and T. C. Jones, Eds.), Vol. 2, p. 1587. Springer-Verlag, Berlin.

Sheldon, W. (1966). Psorergatic mange in the sooty mangabey (*Cercocebus torquates atys*) monkey. *Laboratory Animal Care* **16**, 276.

Shelley, W. B., Shelley, E. D., and Welbourn, W. C. (1985). *Polypodium* fern wreaths (Hagnaya). A new source of occupational mite dermatitis. *Journal of the American Medical Association* **253**, 3137–3138.

Skoda-Smith, S., Mullen, G. R., Oi, F., and Atkinson, T. P. (1996). Angioedema following dust mite exposure presenting as suspected food allergy (*Dermatophagoides farinae*). *American Academy of Allergy and Immunology* **97**, 228.

Smiley, R. L. (1970). A review of the family Cheyletiellidae. *Annals of the Entomological Society of America* **63**, 1056.

Smiley, R. L., and OConnor, B. M. (1980). Mange in *Macaca arctoides* (Primates: Cercopithecidae) caused by *Cosarcoptes scanloni* (Acari: Sarcoptidae) with possible human involvement and descriptions of the adult male and immature stages. *International Journal of Acarology* **6**, 283–290.

Smith, K. E., Quist, C. F., and Crum, J. M. (1997). Clinical illness in a wild turkey with *Laminosioptes cysticola* infestation of the viscera and peripheral nerves. *Avian Diseases* **41**, 484–489.

Smith, T. G. (1996). The genus *Hepatozoon* (Apicomplexa: Adeleina). *Journal of Parasitology* **82**, 565–585.

Snyder, D. E., Hamir, A. N., Hanlon, C. A., and Rupprecht, C. E. (1991). Notoedric acariasis in the porcupine (*Erethizon dorsatum*). *Journal of Wildlife Diseases* **27**, 723–726.

Soulsby, E. J. L. (1965). "Textbook of Veterinary Clinical Parasitology." Davis, Philadelphia.

Soulsby, E. J. L. (1982). "Helminths, Arthropods and Protozoa of Domesticated Animals," 7th ed. Lea & Febiger, Philadelphia.

Southcott, R. V. (1976). Arachnidism and allied syndromes in the Australian Region, Records of the Adelaide Children's Hospital, North Adelaide, South Australia, pp. 97–186.

Southcott, R. V. (1984). Diseases and arachnids in the Tropics. In "Mammalian Diseases and Arachnids" (W. B. Nutting, Ed.), Vol. 2, Chapt. 2, pp. 15–56. CRC Press, Boca Raton, FL.

Spicer, G. S. (1987). Prevalence and host-parasite list of some nasal mites from birds (Acarina: Rhinonyssidae, Speleognathidae). *Journal of Parasitology* **73**, 259–264.

Staley, E. C., Staley, E. E., and Behr, M. J. (1994). Use of permethrin as a miticide in the African hedgehog (*Atelerix albiventris*). *Veterinary and Human Toxicology* **36**, 138.

Stone, W. B., Parks, E., Weber, B. L., and Parks, F. J. (1972). Experimental transfer of sarcoptic mange from red foxes and wild canids to captive wildlife and domestic animals. *New York Fish and Game Journal* **19**, 1–11.

Strandtmann, R. W., and Wharton, G. W. (1958). "A Manual of Mesostigmatid Mites Parasitic on Vertebrates." Contribution No. 4. Institute of Acarology, University of Maryland, College Park.

Stromberg, P. C., and Guillot, F. S. (1987a). Hematology in the regressive phase of bovine psoroptic scabies. *Veterinary Pathology* **24**, 371–377.

Stromberg, P. C., and Guillot, F. S. (1987b). Bone marrow response in cattle with chronic dermatitis caused by *Psoroptes ovis*. *Veterinary Pathology* **24**, 365–370.

Stromberg, P. C., Fisher, W. F., Guillot, F. S., Pruett, J. H., Price, R. E., and Green, R. A. (1986). Systemic pathologic responses in experimental *Psoroptes ovis* infestation of Hereford calves. *American Journal of Veterinary Research* **47**, 1326–1331.

Stunkard, H. W. (1940). The morphology and life history of the cestode *Bertiella studeri*. *American Journal of Tropical Medicine* **20**, 305–333.

Sweatman, G. K. (1957). Life history, non-specificity, and review of the genus *Chorioptes*, a parasitic mite of herbivores. *Canadian Journal of Zoology* **35**, 641.

Sweatman, G. K. (1958a). Biology of *Otodectes cynotis*, the ear canker mite of carnivores. *Canadian Journal of Zoology* **36**, 849.

Sweatman, G. K. (1958b). On the life history and validity of the species in *Psoroptes*, a genus of mange mites. *Canadian Journal of Zoology* **36**, 905–929.

Sweatman, G. K. (1958c). Redescription of *Chorioptes texanus*, a parasitic mite from the ears of reindeer in the Canadian arctic. *Canadian Journal of Zoology* **36**, 525.

Sweatman, G. K. (1971). Mites and pentastomes. In "Parasitic Diseases of Wild Mammals" (J. W. Davis and R. C. Anderson, Eds.), Chapt. 1. pp. 3–64. Iowa State Univ. Press, Ames.

Sweatman, G. K. (1984). Diseases of wildlife. In "Mammalian Diseases and Arachnids" (W. B. Nutting, Ed.), Vol. 2, Chapt. 8, pp. 189–232. CRC Press, Boca Raton, FL.

Taboada, M. de F. (1954). Pulmonary acariasis in Spain. An illustrative case report. *British Medical Journal* **4859**, 437–438.

Tamura, A., Ohahsi, N., Urakami, H., and Miyamura, S. (1995). Classification of *Rickettsia tsutsugamushi* in a new genus, *Orientia* gen. nov., as *Orientia tsutsugamushi* comb. nov. *International Journal of Systematic Bacteriology* **45**, 589–591.

Tandon, N., Chatterjee, H., Gupta, S. K., and Hati, A. K. (1988). Some observations on house dust mites in relation to naso-bronchial asthma in Calcutta, India. In "Progress in Acarology" (G. P. ChannaBasavanna and C. A. Viraktamath, Eds.), Vol. 1, pp. 163–168. Brill, Leiden.

Tenorio, J. M. (1974). A new species of *Lynxacarus* (Acarina: Astigmata: Listrophoridae) from *Felis catus* in the Hawaiian Islands. *Journal of Medical Entomology* **11**, 599–604.

Theiler, M., and Downes, W. G. (1973). "The Arthropod-Borne Viruses of Vertebrates." Yale Univ. Press, New Haven, CT.

Tomalski, M. D., and Miller, L. K. (1991). Insect paralysis by baculovirus-mediated expression of a mite neurotoxin gene (*Pyemotes tritici*). *Nature* **352**, 82–85.

Traub, R., and Wisseman, C. L., Jr. (1974). The ecology of chigger-borne rickettsiosis (scrub typhus). *Journal of Medical Entomology* **11**, 237–303.

Traub, R., Hertig, M.., Lawrence, W. H., and Harris, T. T. (1954). Potential vectors and reservoirs of hemorrhagic fever in Korea. *American Journal of Hygiene* **59**, 291.

Traver, J. R. (1951). Unusual scalp dermatitis in humans caused by the mite *Dermatophagoides* (Acarina: Epidermoptidae). *Proceedings of the Entomological Society of Washington* **53**, 1.

van Bronswijk, J. E, and De Kreek, E. J. (1976). *Cheyletiella* (Acari: Cheyletiellidae) of dog, cat and domesticated rabbit, a review. *Journal of Medical Entomology* **13**, 315–327.

van Bronswijk, J. E., and Sinha, R. N. (1971). Pyroglyphid mites (Acari) and house dust allergy: a review. *Journal of Allergy* **47**, 31–52.

van der Hammen, L. (1972). A revised classification of the mites (Arachnidea, Acarida) with diagnoses, a key, and notes on phylogeny. *Zoologische Mededelingen* **47**, 273–292.

Wharton, G. W., Jr. (1976). House dust mites. *Journal of Medical Entomology* **12**, 577–621.

Wharton, G. W., Jr., and Fuller, H. S. (1952). "A Manual of the Chiggers: The Biology, Classification, Distribution, and Importance to Man of the Larvae of the Family Trombiculidae (Acarina)." Memoirs of the Entomological Society of Washington, No. 4. Entomological Society of Washington, Washington, DC.

Whitaker, J. O., Jr. (1982). "Ectoparasites of Mammals of Indiana." Indiana Academy of Science, Monograph 4. Indiana Academy of Science, Indianapolis.

Wiedman, F. D. (1916). *Cytoleichus penrosei*, a new arachnoid parasite found in the diseased lungs of a prairie dog, *Cynomys ludovicianus* (Pneumocoptidae). *Journal of Parasitology* **3**, 82–89.

Williams, J. F., and Williams, C. S. (1978). Psoroptic ear mites in dairy goats. *Journal of the American Veterinary Medical Association* **173**, 1582–1583.

Williams, J. F., and Williams, C. S. (1982). Demodicosis in dairy goats. *Journal of the American Veterinary Medical Association* **180**, 168–169.

Wilson-Hanson, S., and Prescott, C. W. (1985). Trombidiosis in cats. *Australian Veterinary Journal* **62**, 202–203.

Wisniewski, H. M., Sigurdarson, S., Rubenstein, R., Kascsak, R. J., and Carp, R. I. (1996). Mites as vectors of scrapie. *Lancet* **347**, 1114.

Woolley, T. A. (1988). "Acarology: Mites and Human Welfare." Wiley, New York.

Wraith, D. G., Cunnington, A. M., and Seymour, W. M. (1979). The role and allergenic importance of storage mites in house dust and other environments. *Clinical Allergy* **9**, 545–561.

Wright, F. C., and Glaze, R. L. (1988). Blackbuck antelope (*Antilope cervicapra*), a new host for *Psoroptes cuniculi* (Acari: Psoroptidae). *Journal of Wildlife Diseases* **24**, 168–169.

Wright, F. C., Guillot, F. S., and Meleney, W. P. (1981). Transmission of psoroptic mites from bighorn sheep (*Ovis canadensis mexicana*) to domestic sheep, cattle and rabbits. *Journal of Wildlife Diseases* **17**, 381–386.

Wright, F. C., Riner, J. C., and Guillot, F. S. (1983). Cross-mating studies with *Psoroptes ovis* (Hering) and *Psoroptes cuniculi* Delafond (Acarina: Psoroptidae). *Journal of Parasitology* **69**, 696–700.

Wright, F. C., Riner, J. C., and Fisher, W. F. (1984). Comparison of lengths of outer opisthosomal setae of male psoroptic mites collected from various hosts. *Journal of Parasitology* **70**, 141–143.

Yashikawa, M., Hanaoka, Y., Yamada, Y., *et al.* (1983). Experimental proof of itching papules caused by *Cheyletus malaccensis* Oudemans. *Annual Report of Tokyo Metropolitan Research Laboratory of Public Health* **34**, 264–276.

Yeatts, J. W. G. (1994). Rabbit mite infestation (*Psoroptes cuniculi*). *Veterinary Record* **134**, 359–360.

Yeruham, I., Rosen, S., and Hadani, A. (1986). Sheep demodicosis (*Demodex ovis* Railliet, 1895) in Israel. *Revue d'Elevage et de Medecine Veterinaire des Pays Tropicaux* **39**, 363–365.

Yunker, C. E. (1973). Mites. In "Parasites of Laboratory Animals" (R. J. Flynn, Ed.), Chapt. 15, pp. 425–492. Iowa State Univ. Press, Ames.

Yunker, C. E., Binninger, C. E., Keirans, J. E., Beecham, J., and Schlegel, M. (1980). Clinical mange of the black bear, *Ursus americanus*, associated with *Ursicoptes americanus* (Acari: Audycoptidae). *Journal of Wildlife Diseases* **16**, 347–356.

Zahler, M., Hendrikx, W. M. L., Essig, A., Rinder, H., and Gothe, R. (2000). Species of the genus *Psoroptes* (Acari: Psoroptidae): a taxonomic consideration. *Experimental & Applied Acarology* **24**, 213–225.

Zenoble, R. D., and Greve, J. H. (1980). Sarcoptid mite infestation in a colony of guinea pigs. *Journal of the American Veterinary Medical Association* **177**, 903–908.

Zuchowska, E. (1991). Swierzb ssakow w ogrodach zoologiczynch [Scabies in zoo mammals]. *Wiadomosci Parazytologiczne* **37**, 123–125.

# 24

# TICKS (*Ixodida*)

DANIEL E. SONENSHINE, ROBERT S. LANE, AND WILLIAM L. NICHOLSON

*Ticks are* notorious as vectors of human and other animal disease agents. They transmit a greater variety of infectious organisms than any other group of bloodsucking arthropods and, worldwide, are second only to mosquitoes in terms of their public health and veterinary importance. Ticks transmit numerous protozoan, viral, bacterial (including rickettsial), and fungal pathogens. The study of ticks has contributed greatly to our ability to understand and control the spread of infectious diseases.

The first documentation of the transmission of a disease agent by an arthropod was Smith and Kilbourne's report in 1893 that Texas cattle fever, caused by the protozoan *Babesia bigemina*, was transmitted by the cattle tick *Boophilus microplus*. This epochal report soon led to a new era of discovery of the role of insects, mites, and ticks in the transmission of some of the deadliest diseases of humankind. Within a few years, scientists showed that the lethal East Coast fever (ECF) of African cattle was caused by the tick-borne protozoan *Theileria parva* (1904); that relapsing fever was caused by the tick-borne spirochete *Borrelia duttoni* (1905); that Rocky Mountain spotted fever (RMSF) was caused by the tick-borne rickettsia *Rickettsia rickettsii* (1906); and that certain types of encephalitis, e.g., Russian spring-summer encephalitis (RSSE), were caused by tick-borne viruses (1932). This process of discovery is continuing. The most recent examples are the findings that ticks are the vectors of the etiologic agents of Lyme disease and the human ehrlichioses. Since its recognition as a distinct entity in Connecticut in 1977, Lyme disease is now the most important vector-borne disease of humans in the United States, Europe, and Asia. Additionally, ticks can cause severe or sometimes fatal illnesses because of proteins injected with their saliva. They also can be injurious to domestic livestock and wildlife because of the blood loss that occurs when large numbers of ticks feed. The wounds that they produce may create sites for secondary infections and diminish the value of livestock by damaging their hides.

TABLE I
Families and Genera of Ticks

| Family | Subfamily (subgroup) | Genera |
|---|---|---|
| Ixodidae | Ixodinae (Prostriata) | *Ixodes* |
| | Amblyomminae (Metastriata) | *Amblyomma, Aponomma* |
| | Haemaphysalinae (Metastriata) | *Haemaphysalis* |
| | Hyalomminae (Metastriata) | *Hyalomma* |
| | Rhipicephalinae (Metastriata) | *Dermacentor, Cosmiomma, Boophilus, Margaropus Nosomma Anomalohimilaya Rhipicentor Rhipicephalus* |
| Argasidae | Argasinae | *Argas* |
| | Ornithodorinae | *Ornithodoros* |
| | Otobinae | *Otobius* |
| | Antricolinae | *Antricola* |
| | Nothoaspinae | *Nothoaspis* |
| Nuttalliellidae | | *Nuttalliella* |

Modified from Keirans, 1992.

## TAXONOMY

Ticks constitute the suborder Ixodida of the order Parasitiformes and are exclusively parasitic. The Ixodida contains three families: the Ixodidae, Argasidae, and Nuttalliellidae (Table I). The family Ixodidae is subdivided into the Prostriata, representing the single genus *Ixodes*, and the Metastriata, which includes the remaining 13 genera. There are approximately 650 species in this family, or approximately 80% of all the tick species described (Table I). According to Keirans (1992), the Argasidae contain 5 genera and approximately 172 species, whereas Klompen and Oliver (1993) recognize only 4 genera. The Nuttalliellidae is a monospecific family, represented by only 1 species, *Nuttalliella namaqua*.

The taxonomy of the Ixodida has been addressed by numerous authors, arranged by zoogeographic region: (1) *Afrotropical*, Arthur (1965), Hoogstraal (1956), and Matthysse and Colbo (1987); (2) *Australian*, Roberts (1970); (3) *Nearctic*, Clifford *et al.* (1961), Cooley and Kohls (1944, 1945), Cooney and Hays (1972), Durden and Keirans (1996), Furman and Loomis (1984), Keirans and Clifford (1978), Keirans and Litwak (1989), Sonenshine (1979), Yunker *et al.* (1986), and Webb *et al.* (1990); (4) *Neotropical*, Fairchild *et al.* (1966), Jones *et al.* (1972); (5) *Oriental*, Kohls (1950), Kohls (1957), and Yamaguti *et al.* (1971); and (6) *Palearctic*, Arthur (1963), Balashov (1972), Filippova (1966, 1967), and Hillyard (1996). Some general references include

Clifford *et al.* (1973), Strickland *et al.* (1976), Savory (1977), Hoogstraal (1985), Woolley (1988), and Keirans (1992).

The following provides descriptions of some of the more important tick genera.

## FAMILY IXODIDAE

### Genus *Ixodes*

This is the largest tick genus, with an estimated 235 species. Members of this genus are known as the *Prostriata*. They are easily recognized by their distinctive anal groove, which encircles the anus anteriorly. Males have sclerotized ventral plates, which are absent in males of other genera. The genus is worldwide in distribution, including Antarctica. Four species are particularly important as vectors of microbial agents to humans: the *black-legged tick* (*Ixodes scapularis*) in eastern North America; the *castor bean tick*, or *sheep tick* (*I. ricinus*) in Europe and western Asia; the *taiga tick* (*I. persulcatus*) in northeastern Europe and northern Asia; and the *western black-legged tick* (*I. pacificus*) in the far western United States.

### Genus *Dermacentor*

This is one of the most important genera of metastriate ticks, with 30 species. The basis capituli appears rectangular when viewed dorsally. A pair of medially directed spurs occurs on the first pair of coxae. The palps are short and thick. The scutum is almost always ornamented. Most *Dermacentor* spp. are three-host ticks that feed on diverse mammals. Adults attack medium-sized or large mammals, whereas the immatures feed on small mammals and lagomorphs. *Dermacentor* species are found mostly in Europe, Asia, and North America. In North America, important species are the *American dog tick* (*Dermacentor variabilis*), the *Rocky Mountain wood tick* (*D. andersoni*), the *Pacific Coast tick* (*D. occidentalis*), and the *winter tick* (*D. albipictus*).

### Genus *Boophilus*

Ticks of the genus *Boophilus* are small and lack ornamentation. The genus contains five species. The basis capituli is short and broad, with rounded lateral margins. These ticks are one-host parasites of ungulates. *Boophilus* ticks are found in most regions of the world. Among the important species are the *cattle tick* (*Boophilus annulatus*) and the *tropical fever tick, southern cattle tick* (*B. microplus*).

### Genus *Rhipicephalus*

Ticks of the genus *Rhipicephalus* are easily recognized by the hexagonal shape of the basis capituli when viewed dorsally. Important species include the *brown dog tick* (*Rhipicephalus sanguineus*) and the *brown ear tick* (*R. appendiculatus*). *Rhipicephalus* ticks mainly parasitize mainly mammals and rarely (as immature stages only) parasitize birds or reptiles. Representative species are found throughout the world. *R. sanguineus* is cosmopolitan in distribution. The genus contains 75 described species.

## Genus *Haemaphysalis*

This is the second largest genus, which is recognized by the pronounced lateral projection of palpal segment 2 in most species (including all North American species), which extends well beyond the basis capituli. These small ticks lack eyes. *Haemaphysalis* species parasitize birds and mammals in most regions of the world. An important species is the *rabbit tick Haemaphysalis leporispalustris*. The genus contains approximately 155 species.

## Genus *Hyalomma*

This is a relatively small genus of 30 species of medium-sized to large ticks. They are characterized by their elongated palps, which are at least twice as long as wide. The distinct eyes are located in sockets adjacent to the postero-lateral edges of the scutum. *Hyalomma* ticks are unornamented. Most species live in xeric environments where they parasitize small and medium-sized wild mammals and livestock. Some species also parasitize birds and reptiles. The distribution of *Hyalomma* species is limited to the Old World, primarily in arid or semiarid habitats. An important subspecies is *Hyalomma marginatum marginatum*. Other important species are *H. truncatum* in Africa and *H. asiaticum* in central Asia.

## Genus *Amblyomma*

The adults of most species in this genus are medium or large in size. The palps are long, with segment 2 at least twice as long as segment 3. The scutum is usually ornamented with varying-colored iridescent patterns. Eyes are present but are not situated in sockets. Virtually all terrestrial vertebrates serve as hosts, although amphibians are rarely attacked. The distribution is worldwide, primarily in humid tropical or subtropical regions. Examples of important species include the *Gulf coast tick* (*Amblyomma maculatum*) and the *lone star tick* (*A. americanum*) in North America; the *tropical bont tick* (*A. variegatum*) in Africa and on some islands in the Caribbean Sea; and the *bont tick* (*A. hebraeum*) in Africa. The genus contains about 102 species.

The remaining genera of the Ixodidae contain relatively few species, only one of which (*Anocentor*) is known to be important in pathogen transmission.

These include *Aponomma, Anocentor, Cosmiomma, Nosoma, Anomalohimalaya, Rhipicentor,* and *Margaropus*.

## FAMILY ARGASIDAE

### Genus *Argas*

*Argas* ticks have a flattened body margin, a lateral sutural line, and a leathery, folded cuticle. The many small integumental folds usually have a button-like appearance, each with a pit on its top. Most species parasitize bats or birds. The genus is worldwide in distribution, mostly in xeric environments or dry caves in otherwise humid environments. Examples of important species are the *fowl tick* (*Argas persicus*) and the *pigeon tick* (*A. reflexus*). About 58 species have been described.

### Genus *Ornithodoros*

Nymphs and adults have a leathery cuticle with innumerable tiny wrinkles (mammillae) and a rounded body margin; they lack a lateral, sutural line. Mammillae are smaller and more numerous than those found in *Argas*. The host range is diverse and includes reptiles, birds, and mammals. The genus is worldwide in distribution. Examples of important species include the *African tampan* (*Ornithodoros moubata*) and the *cave tick* (*O. tholozani*). The genus contains approximately 101 species.

### Genus *Otobius*

The integument of the nymphs is spinose, whereas that of the adults is granulated. There are just two nymphal instars. The adults do not feed and the hypostome is vestigial. *Otobius* ticks are found in North America, Africa, and Asia. The genus contains two species: the *spinose ear tick* (*Otobius megnini*) and *O. lagophilus*.

### Genus *Antricola*

These ticks possess a tuberculated cuticle. The females have a distinctive, scooplike hypostome; the hypostome is vestigial in the males. All are parasites of New World bats. Thus far, none has been implicated in the transmission of microbial disease agents. There are about 10 species.

### Genus *Nothoaspis*

The monospecific genus *Nothoaspis* is similar to *Antricola*, but the anterior dorsal surface bears a smooth shield-like structure, the pseudoscutum. Klompen and Oliver (1993) consider the genera *Antricola* and *Nothoaspis* invalid and place them and almost all other bat-associated argasids into the genus *Carios*.

## FAMILY NUTTALLIELLIDAE

The only known species in this family, *N. namaqua*, occurs in southern Africa. It shares features with both the Argasidae and the Ixodidae but also has several unique morphological traits. This tick has ball and socket joints that articulate the leg segments, a small, dorsal pseudoscutum, and a highly wrinkled cuticle with numerous pits and elevated rosettes. It has been collected from the nests of rock hyraxes and swallows in South Africa and Tanzania. We will not consider it further, because it is rare and of no known medical or veterinary importance.

## MORPHOLOGY

### EXTERNAL ANATOMY

The major external regions of ticks are the *capitulum* (gnathosoma), *idiosoma*, and the *legs* (Figs. 24.1 to 24.3). The capitulum consists of the *basis capituli*, which articulates with the body, the segmented *palps*, the *chelicerae*, and the toothed *hypostome*. The capitulum of ixodid ticks is located at the anterior end of the body. Females bear paired clusters of pores, the *porose areas*, located dorsally on the *basis capituli*. The porose areas secrete antioxidants that inhibit degradation of the waxy compounds in the secretions of *Gené's organ*. The *chelicerae* are located on the dorsal aspect of the capitulum. Their shafts, surrounded by spinose sheaths, lie between the palps and often extend even farther anteriorly than the palps. Each chelicera bears two digits distally. The larger, medial digit can be moved laterally; the smaller outer digit resides in a cavity of the medial digit and moves with it. Both digits have sharp *denticles*. The chelicerae are used to cut host tissues during attachment. The hypostome is a prominent, ventrally located structure that bears rows of recurved teeth on its ventral surface; teeth are absent in some nonfeeding males. A narrow food canal is located on the middorsal surface. The palps consist of four distinct segments. In nymphs and adults of most ixodid species, the small terminal segment is recessed in a cavity in segment 3 and bears numerous fine setae at its tip.

The capitulum of argasids is similar (Fig. 24.4). However, it is situated just below an anteriorly protruding body extension, or *hood*, and is not visible dorsally in nymphs or adults. The four palpal segments are about equal in size. Small flaps, the cheeks, occur alongside the capitulum in many species and can be folded to cover the delicate mouthparts.

The idiosoma is the body exclusive of the capitulum. It is divided into two parts: the anterior *podosoma*, which bears the legs and the genital pore, and the posterior *opisthosoma*, the region behind the coxae that bears the spiracles and the anal aperture. The cuticle is relatively tough, with *sclerotized plates (sclerites)* in certain locations. It serves as the site of muscle attachment and protects the animal from desiccation and injury. The cuticle bears numerous sensory setae as well as various pores representing the openings of dermal glands or sensilla.

The legs are jointed and articulate with the body via the *coxae*. Larvae are easily recognized by the presence of only three pairs of legs, whereas nymphs and adults have four pairs of legs. The structure of the legs is similar in the Ixodidae and Argasidae. Each leg is divided into six segments, the *coxa, trochanter, femur, patella (genu), tibia,* and *tarsus*. The coxae are inserted ventrally and allow limited rotation in the anteroventral and dorsoventral planes. The other segments can be flexed, so that the legs can be either folded against the ventral body surface for protection or extended for walking. A pair of claws and a pad-like *pulvillus* are present on each tarsus of most species. The pulvillus is absent in argasid nymphs and adults. An odor-detecting sensory apparatus, *Haller's organ*, is evident on the dorsal surface of the tarsus of leg I in all stages. This organ consists of an anterior pit and a posterior capsule. Gustatory, thermosensory and mechanosensory functions also have been associated with this organ. Variations in the structure of Haller's organ are useful for distinguishing genera and species.

### Ixodidae

Ixodid ticks, also called hard ticks, are illustrated in Figs. 24.1 to 24.3. Females have a hard cuticular plate or scutum on the anterior half of the dorsal body surface (Fig. 24.1A). In males, the scutum occupies virtually the entire dorsal surface (Fig. 24.2A). Elsewhere, the cuticle contains tiny surface folds, which give it a fingerprint-like appearance when viewed at high magnification. The body of the female expands enormously during feeding as new cuticle is synthesized to accommodate the blood meal. In males, however, the larger scutum limits expansion. The scutum bears setae and tiny pores termed *sensilla auriformia*. The latter are believed to serve as proprioceptive organs. When present, a simple eye occurs along each postero-lateral margin of the scutum.

Immediately posterior to the scutum in the females are paired *foveal pores* (absent in *Ixodes*) from which the volatile sex pheromone, 2,6-dichlorophenol, is emitted. The dorsal body surface posterior to the scutum, the *alloscutum*, has innumerable fine folds. In females, a paired protrusible organ, *Gené's organ*, lies in the dorsal foramen between the scutum and the capitulum (capitular foramen). The ends of this organ protrude during oviposition

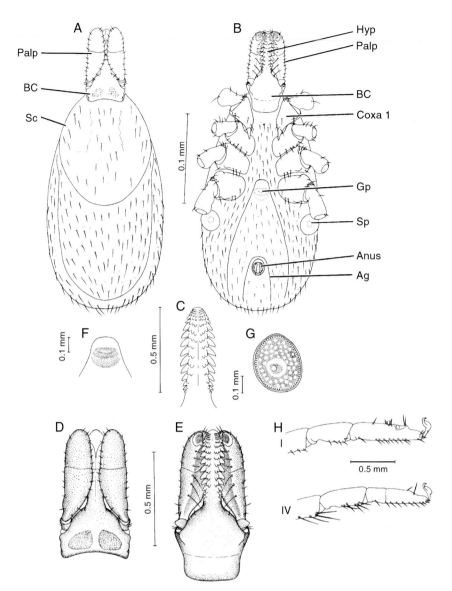

**FIGURE 24.1** External morphology of representative female ixodid tick (*Ixodes pacificus*). (A) Dorsal view; (B) ventral view; (C) hypostome; (D) capitulum, dorsal view; (E) capitulum, ventral view; (F) spiracular plate; (G) genital pore; (H) legs I and IV. Ag, anal groove; BC, basis capituli; Gp, genital pore; Hyp, hypostome; Sc, scutum; Sp, spiracle.

and apply wax to each egg as it is deposited. In *Ixodes* males, hard sclerotized plates cover the ventral body surface (Fig. 24.2B). In females, the genital pore is a U- or V-shaped opening, with prominent marginal folds (Fig. 24.1B), but in males it is covered by a movable plate (Fig. 24.2B). Other ventral structures include paired *spiracular plates* behind coxae 4 (each containing a small *ostium* that opens to the respiratory system) and the anal aperture, located near the posterior margin. The entire body is covered by numerous setae and the porelike sensilla auriformia. Larvae possess few setae but their number and

relative placement provide valuable taxonomic characters for generic and subgeneric differentiation (Fig. 24.5).

### Argasidae

The major external body features of argasid ticks, also known as soft ticks, are illustrated in Fig. 24.4. The body margins are rounded in most species. In *Argas,* however, they are flattened and covered by small marginal discs. Eyes, when present, occur on folds lateral to the coxae. A tiny coxal pore, the opening of the duct from the paired

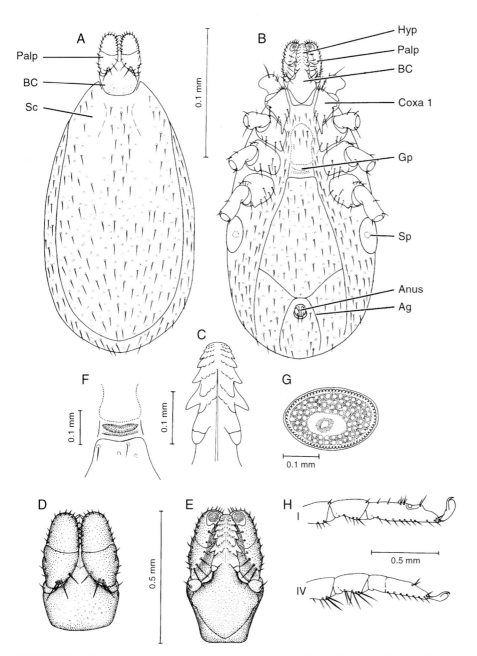

**FIGURE 24.2**  External morphology of representative male ixodid tick (*Ixodes pacificus*). (A) Dorsal view; (B) ventral view; (C) hypostome; (D) capitulum, dorsal view; (E) capitulum, ventral view; (F) genital pore; (G) spiracular plate; (H) legs I and IV. Ag, anal groove; BC, basis capituli; Gp, genital pore; Hyp, hypostome; Sc, scutum; Sp, spiracle.

coxal glands, occurs bilaterally between the coxae of legs I and II. The spiracular plates, located between coxae 3 and 4, are relatively small and inconspicuous. In females, the genital pore appears as a horizontal slit surrounded by a prominent fold. In males, the pore is subtriangular or suboval, without a genital apron. There are no foveal pores. The body features of the larvae are illustrated in Fig. 24.6.

## INTERNAL ANATOMY

The internal organs of a typical tick are illustrated in Fig. 24.7. The organs are bathed in a circulating fluid, the *hemolymph*. The hemolymph is a watery medium rich in salts, amino acids, soluble proteins, and other dissolved substances. In addition, it contains several types of *hemocytes*, the most prominent of which are the phagocytes

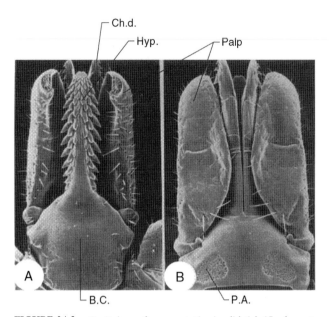

**FIGURE 24.3** Capitulum of representative ixodid tick (*Ixodes scapularis*), scanning electron micrographs. (A) Ventral view; (B) dorsal view. BC, basis capituli; Ch.d., cheliceral digit; Hyp, hypostome; P.A., porose area. From Biology of Ticks, Vol. 1 by Daniel E. Sonenshine-1991 by Oxford University Press, Inc. used by permission of Oxford University Press, Inc.

(plasmatocytes and granulocytes). A simple heart, situated middorsally, filters and circulates this vital body fluid. Muscles extend from the dorsal and ventral cuticular surfaces to the inner surfaces of the coxae, chelicerae, and other structures.

The most prominent internal organ is the *midgut*, a large saclike structure with numerous lateral diverticuli. The shape of the midgut depends upon the state of engorgement. In unfed ticks, the diverticuli are narrow, tubelike structures. In fed ticks, the diverticuli enlarge and obscure most of the other organs as they fill with blood. Branches of the tracheal system ramify over the surfaces of the diverticuli and surround the other internal organs. Ticks respire through these innumerable tiny air tubes, which open to the exterior via the paired spiracles.

Paired salivary glands are situated anterolaterally. These large glands, which resemble clusters of grapes, are connected via the salivary ducts to the mouthparts. Their salivary secretions empty into the *salivarium* located between the chelicerae and the hypostome. Tick saliva contains pharmacologically active compounds that facilitate attachment and suppress host inflammatory responses.

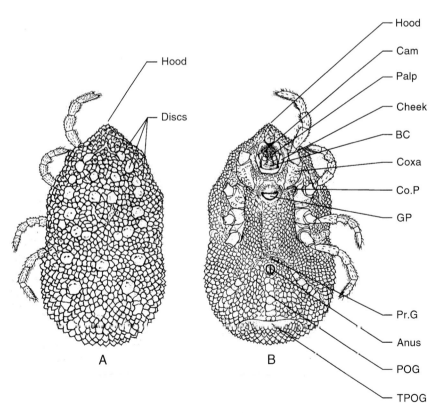

**FIGURE 24.4** External morphology of a generalized argasid tick (*Ornithodorus*). (A) Dorsal view; (B) ventral view. BC, basis capituli; Cam, camerostome; Co.P, coxal pore; GP, genital pore; Pr.G, preanal groove; POG, postanal groove; TPOG, transverse postanal groove.

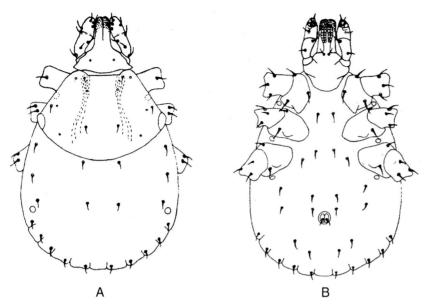

**FIGURE 24.5**    Larva of representative ixodid tick (*Dermacentor variabilis*), with legs beyond coxae and trochanters not shown. (A) Dorsal view; (B) ventral view. (From Clifford *et al.*, 1961; with permission of the Entomological Society of America.)

The salivary glands eliminate excess water from the blood meal. In ixodid ticks, most water in the blood meal is extracted by specialized salivary gland cells and excreted into the host as the tick feeds.

Other prominent internal structures are the reproductive organs. In males, these include the *testes*, the tubular *vasa deferentia*, the *seminal vesicle*, and the *ejaculatory duct* that is connected to the genital pore. The ejaculatory duct is obscured by the large, multilobed *accessory gland* that secretes the components of the *spermatophore*. In females, the reproductive organs include the *ovary*, *paired oviducts, uterus, vagina*, and the *seminal receptacle*.

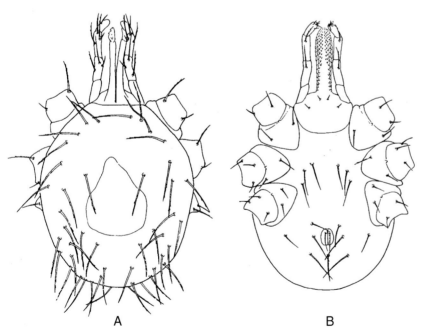

**FIGURE 24.6**    Larva of generalized argasid tick, with legs beyond coxae and trochanters not shown. (A) Dorsal view; (B) ventral view.

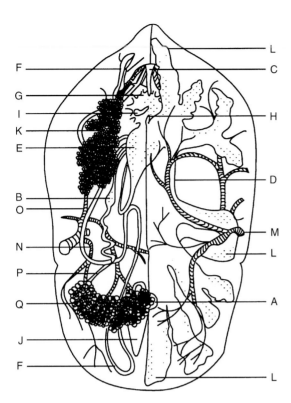

**FIGURE 24.7** Internal anatomy of typical argasid tick, female; midgut shown on right side, with midgut removed on left side to reveal underlying organs. (A) anus; (B) ampulla of oviduct; (C) pharynx; (D) median region of midgut; (E) uterus; (F) Malpighian tubule; (G) brain (synganglion); (H) esophagus; (I) accessory gland; (J) rectal sac; (K) salivary gland; (L) midgut diverticulum; (M, N, O) different regions of tracheal trunks; (P) oviduct; (Q) ovary.

The ovary is small and inconspicuous in unfed ticks but expands enormously during feeding and especially after mating. In gravid females, the ovary is distended with large, amber-colored eggs.

In argasid and ixodid ticks, excretion is accomplished by the *Malpighian tubules,* a pair of long, coiled structures that empty into the *rectal sac.* Nitrogenous wastes are excreted in the form of *guanine.* In argasid ticks, paired *coxal glands* adjacent to the coxae of leg I extract excess water and salts accumulated during feeding and excrete this watery waste via the *coxal pores.* Each gland consists of a membranous sac that serves as a filtration chamber and a coiled tubule that selectively reabsorbs small, soluble molecules and ions. Relapsing fever spirochetes may be transmitted to vertebrate hosts via the *coxal fluid* of infected ticks.

The central nervous system in ticks is fused to form the *synganglion* located antero-ventrally above the genital pore. The synganglion, which regulates the function of the structures described above, is the fused central nervous system. Large pedal nerves extend from the synganglion to the legs; smaller nerves innervate the palps, chelicerae, cuticular sensilla, and internal organs.

## LIFE HISTORY

The life cycle includes four stages: the egg, larva, nymph, and adult. Ixodid ticks have only one nymphal instar, whereas argasid ticks have two or more nymphal instars. All ticks feed on blood during some or all stages in their life cycle. Larvae attack hosts, feed, detach, and develop in sheltered microenvironments where they molt to nymphs. Nymphs seek hosts, feed, drop, and molt to adults (except in argasid ticks, which molt into later nymphal instars). Adult ticks seek hosts, feed, and, in the case of engorged females, drop off to lay their eggs.

In contrast to most other hematophagous arthropods, ticks can be remarkably long-lived. Many can survive for one or more years without feeding. Their life cycles vary greatly, with the greatest differences evident between the Ixodidae and Argasidae.

### IXODID LIFE CYCLES

Immature and adult ticks each take a blood meal, except for the nonfeeding males of some species. Following contact with the host, a tick uses its chelicerae to puncture the skin and uses its hypostome to securely anchor itself. Attachment is reinforced by secretion of cementing substances with the saliva into and around the wound site. Females feed only once. Following mating, females suck blood rapidly (24–48 hr) and swell enormously. Replete, mated females drop from their hosts, find a sheltered location, and subsequently oviposit hundreds to thousands of eggs. For *D. variabilis,* the average egg production is 5,380 (Sonenshine and Tignor, 1969). For *H. impeltatum,* the reported average is nearly 10,700 eggs per female (Logan *et al.,* 1989). The greatest number ever recorded was produced by an *Amblyomma nuttalli* female that laid close to 23,000 eggs (Arthur, 1962). The eggs are deposited in a single, continual mass over many days or weeks. The female dies upon completion of egg laying.

Males swell only slightly during feeding. They usually remain on their hosts, feed repeatedly, and inseminate several females. Mating usually occurs on the host. Certain species of *Ixodes,* however, mate on their hosts, in nests, or in vegetation. Many *Ixodes* males have vestigial hypostomes, and these species invariably mate off the host. Except for *Ixodes* species, males and females require a blood meal to stimulate oogenesis and spermatogenesis. More than 90% of the life cycle is spent off the host.

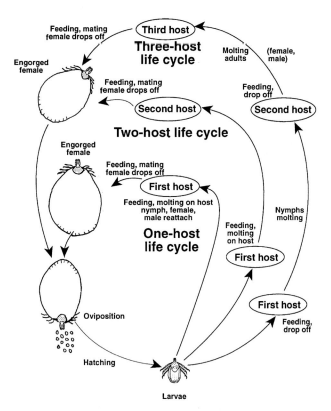

**FIGURE 24.8**   Three basic life cycles of ixodid ticks: (1) One-host ticks (inner circle) in which the larva, nymph, and adult all attach to, and develop on, a single host (e.g., *Boophilus annulatus*). (2) Two-host ticks (middle circle) in which larva and nymph feed on one host and adult attaches to and feeds on a second host (e.g., *Hyalomma dromedarii*). (3) Three-host ticks (outer circle) in which larva, nymph, and adult each parasitize a different host (typical of most ixodid ticks). Most argasid ticks have a multihost life cycle involving more than three hosts, with several nymphal instars, each potentially feeding on a different host.

Molting usually occurs in some sheltered microhabitat such as soil, leaf litter, or host nests. After molting, nymphal and adult ticks must seek another host and feed. When host seeking and feeding occur in all three parasitic stages, the pattern is termed a *three-host life cycle* (Fig. 24.8). This is characteristic of more than 90% of ixodid species.

In tropical climates with frequent rainfall, developmental times are relatively short, and several generations may occur each year. In regions with alternating dry and rainy seasons, the life cycle is longer because ticks cease host-seeking during the driest period. In colder temperate or subarctic regions, development is slower, and ticks commonly undergo diapause during the coldest months. As a result, the life cycle may take two or more years. An example of a diapausing species is *D. variabilis*. Larvae feed on mice or other small mammals, mostly in spring. Fed larvae drop off and molt to nymphs that again attack small mammals. Fed nymphs drop off and molt within a

few weeks. If the adults feed and reproduce in the same year, the entire life cycle can be completed within several months. Thus, under favorable conditions in nature, the typical three-host life cycle can be completed in less than 1 year. In the laboratory, it can be completed even sooner. However, adverse environmental conditions can prolong the life cycle to two or more years. The life cycle of *I. ricinus* may require up to 4 years in the northern parts of its range in Europe (Hoogstraal, 1985). In Ireland, the life cycle takes 3 years, with each stage requiring approximately 1 year before developing to the next (Gray, 1991).

A few ixodid species exhibit a *two-host* or *one-host life cycle* (Fig. 24.8). For example, in the two-host camel tick (*H. dromedarii*), fed larvae molt on their hosts, and the unfed nymphs reattach soon after emergence. Following engorgement, the nymphs drop off the host, molt, and feed as adults on a second host. In the one-host cattle tick *B. annulatus* and in other species of *Boophilus*, all stages feed and molt on the same host. Mating also occurs on this host. Replete, fertilized females drop off the host and oviposit in soil.

## ARGASID LIFE CYCLES

In contrast to the ixodids, most argasids have two or more nymphal instars in their life cycle, each of which must consume a blood meal. This pattern is termed the multihost life cycle. Molting occurs off the host in cracks, crevices, or beneath debris in or near the nest. Argasid females take repeated small blood meals and lay small batches of eggs (typically < 500 eggs per batch) after each feeding. These are termed *multiple gonotrophic cycles*. The interval between feedings is typically several months. As many as six gonotrophic cycles have been reported in some species. Mating usually occurs off the host. Because of the multiple nymphal instars that may number six or even seven in some species, argasid ticks often live for many years. In addition, these ticks are highly resistant to starvation that can extend their longevity even further. In some species that feed on migratory bats or birds, diapause serves to delay oviposition or development during the periods when hosts are absent.

The larvae of most *Ornithodoros* species that parasitize bats and birds remain attached to their hosts for many days, just as ixodid ticks do. Following the larval blood meal, they molt twice without additional feeding. Thereafter, the life cycle is similar to the typical argasid pattern. Another species with an unusual life cycle is *Otobius megnini*. This tick exhibits a high degree of host- and body-site specificity that regulates its feeding and development. Females do not feed and are autogenous (i.e., oviposit without feeding).

# BEHAVIOR AND ECOLOGY

Most ticks are nonnidicolous species living in forests, savannahs, second-growth areas of scrub and brush, and grassy meadows. Others remain buried in sand or sandy soils, under stones, in crevices, or in the litter, duff, and rotting vegetation at the floor of woods and grasslands. In contrast, almost all argasids and some ixodids, especially species of the genus *Ixodes,* are *nidicolous.* They live in the nests, burrows, caves, or other shelters used by their hosts.

Nonnidicolous ticks are active during certain periods of the year when climatic conditions favor development and reproduction. During such periods they engage in *host-seeking behavior.* In temperate and subpolar regions, seasonal activity is regulated by ambient temperature, changing photoperiod, and incident solar energy. In tropical regions, where day length or temperature vary only slightly throughout the year, tick activity is often controlled by the transition from the dry to the rainy season. Host-seeking ticks exhibit at least two strategies for locating potential targets for their blood meals. *Ambush* ticks climb onto weeds, grasses, bushes, or other leafy vegetation to wait for passing hosts. When stimulated by the presence of a host, they extend their forelegs anterolaterally (*questing behavior*) and cling to the hair, feathers, or clothing of the passing host. *Hunter* ticks emerge from their refuges in the soil, sand, or duff when excited by host odors and run rapidly across the ground to attack hosts.

After contacting a potential vertebrate host, the tick must perceive appropriate host-recognition cues that enable it to determine whether to attach and feed or to drop off and resume host-seeking. Odors, radiant heat, visual images, or vibrations stimulate the tick and enable it to recognize its prospective host. Odors are probably the most important stimuli. Electrophysiological studies have shown that larvae of *B. microplus* respond to odors from extracts of cattle skin but not to dry air. Human breath also elicits a response, but not as vigorously as that caused by the cattle extracts. Among the more important attractants emitted by hosts are carbon dioxide in animal breath and ammonia in urine and other animal wastes. Other odors that attract ticks are butyric acid and lactic acid, which occur commonly in sweat and other body fluids. Small increases in radiant heat excite ticks and act synergistically with host odors. Visual cues may be important, especially in certain hunter ticks that can discriminate dark shapes against the bright background of the sky. Host-seeking ticks of many species respond to shadows, resulting in extension of their legs to facilitate contact. Other stimuli that elicit questing behavior include vibrations, sound, and tactile cues. Vibrating the grass stems on which ticks are perched can elicit questing behavior almost immediately. *B. microplus* larvae respond to sounds in the 80–800 Hz range, typical of the frequencies produced by feeding cattle, whereas the sounds produced by barking dogs reportedly attract *R. sanguineus* (Waladde and Rice, 1982). Tactile stimuli perceived when ticks contact their hosts, in combination with short-range attractants such as heat and odor, help to determine the selection of suitable feeding sites. Ticks will not attach to a host unless the appropriate stimuli are received in a particular sequence.

Feeding behavior, even on preferred hosts, is not a uniform process. In ixodids, a tick may crawl about the host for several hours in search of a suitable feeding site. Shortly after they attach, many ixodid ticks secrete *cement* during the first 1–2 days to secure themselves at the wound site. Ixodid ticks feed gradually because first they must create new cuticle to accommodate the massive blood meal. Typical *attachment periods* range from as few as 2 days for larvae to as long as 13 days for females. When feeding is completed, the weight of blood and other fluids consumed ranges from 11 to 17 times the tick's prefeeding body weight in ixodid larvae and from 60 to 120 times the tick's prefeeding body weight in ixodid females. Measurements of blood volume consumed range from 0.7 ml to as high as 8.9 ml per female in some ixodid species (Balashov, 1972). Nymphal and adult argasid ticks attach for only brief periods. This can be as little as 35 to as much as 70 min for adults. These ticks do not secrete cement during attachment; instead, they attach solely with the mouthparts, especially the hypostome. Argasid ticks swell to the extent that their cuticles can stretch. New cuticle is not secreted during feeding as it is in ixodid ticks.

The timing of drop-off from the host offers important ecological advantages. For nonnidicolous ticks, such drop-off rhythms are synchronized with host behavioral patterns. This tends to disperse fed ticks in optimal habitats where they can develop and reproduce. Photoperiod appears to be the dominant exogenous factor affecting drop-off patterns. The daily light:dark cycle induces a regular rhythm of feeding and dropping off. This effect, termed *photoperiodic entrainment,* is partially reversible. In a series of elegant experiments with *Haemaphysalis leporispalustris,* George (1971) showed the existence of an endogenous drop-off rhythm that is entrained by the *scotophase,* or dark period. The rhythm was affected only partially by changing the photoperiodic regime and was maintained even when hosts were held in constant darkness. Detachment may occur while the hosts are inactive in their nests or burrows or, alternatively, it may be coordinated with the period of maximum host activity. In *D. variabilis,* fed larvae and nymphs drop soon after night begins. In contrast, immatures of *H. leporispalustris* drop off during daylight hours when their lagomorph hosts are confined in their forms or warrens.

Mating occurs during feeding or off the host. In the metastriate Ixodidae, unfed adults are sexually immature and require a blood meal to stimulate gametogenesis. Mating is usually regulated by sex pheromones and follows a complex, hierarchical pattern of responses. Feeding females secrete a volatile *sex attractant pheromone,* usually *2,6-dichlorophenol,* that excites males feeding nearby on the same host. The males detach and seek the females that they recognize by means of the *mounting sex pheromone* (a mixture of cholesteryl esters) on the female's body surface. The male climbs onto the dorsum of the female, then moves to the ventral side and searches for the gonopore. Once a male locates the gonopore, it probes the opening with its chelicerae and inserts a *spermatophore* into the female's vulva. Spermatophores, containing spermatozoa, are produced in the large accessory gland of the sexually mature male during the mating process. At this time, the spermatophore emerges from the male's genital pore, whereupon the male seizes it with his mouthparts and inserts it into the female's vulva.

In prostriate Ixodidae and in argasid ticks, the adults become sexually mature soon after the nymphal molt. These ticks usually mate in the nests or in vegetation, although ticks of some prostriate species also may also mate on the host. There is evidence that mating in argasid ticks is regulated by one or more sex pheromones, although no specific compound has been identified to date.

A preovipositional period of up to several weeks precedes egg laying in ixodid ticks. During oviposition, the cuticle of the vulva (vestibular vagina) softens and evaginates as the eggs pass through it, thereby serving as an ovipositor. The emerging eggs are waxed by *Gené's organ.* The process of oviposition continues for several weeks. Typically, about 50–60% of the female's body weight at the time of drop-off is converted to eggs. The number of eggs deposited is directly proportional to the size of the engorged female. At the completion of oviposition, the spent female dies. Thus, there is only a single gonotrophic cycle among ixodids.

In the Argasidae, mated females commence oviposition soon after feeding but deposit clutches containing only a few hundred eggs. Following oviposition, the females remain active and seek hosts again. These ticks feed and lay eggs after each meal. They do not need to mate again. The number of gonotrophic cycles varies but rarely exceeds six.

All ticks exhibit varying degrees of host specificity. More than 85% of argasid and ixodid ticks exhibit relatively strict host specificity (Hoogstraal and Aeschlimann, 1982). However, the evolutionary significance of this phenomenon is uncertain. More recent studies using molecular methods suggest that most of the existing tick–host association patterns may be explained as artifacts of biogeography and ecological specificity, as well as

incomplete sampling (Klompen *et al.,* 1996). Regardless of how host specificity evolved, it is clear that many species are specialists. Examples include certain species of argasids (e.g., *Nothoaspis* and *Antricola*) that infest only bats, and the ixodids *D. albipictus, B. annulatus,* and *B. microplus,* which feed only on large ruminants. Many of the nidicolous ticks are highly host-specific. Examples include *I. marxi,* which feeds almost exclusively on squirrels, *A. tuberculatum,* which, as adults, parasitizes the gopher tortoise, and *Argas arboreus,* which attacks herons. At the opposite extreme are ticks that are opportunistic species with catholic feeding habits. Examples include *I. ricinus* and *I. scapularis.* Larvae and nymphs of these species feed readily on lizards, birds, small mammals, and larger hosts such as sheep and humans. Adults feed on larger mammals, especially ungulates, and also attack humans. Over 300 species of vertebrates have been recorded as hosts for *I. ricinus,* and over 120 have been recorded for *I. scapularis.*

Host specificity is influenced by evolutionary history, ecological and physiologic factors, and the ability of the ticks to avoid host rejection. Many species belonging to the less specialized and phylogenetically primitive genus *Amblyomma* and all species of *Aponomma* feed on reptiles or primitive mammals. Ticks adapted to a specific habitat type (e.g., grassland) encounter only those vertebrates adapted to the same habitat. Questing height also is important. Ticks questing on or near the ground are exposed mostly to small animals, while those questing higher in the vegetation are more likely to encounter larger animals.

The extent to which different hosts are utilized depends upon host behavior and opportunities for contact, such as foraging range, time of day and time spent foraging, habitats visited, and other factors. White-footed mice, which forage extensively on the ground, acquire numerous *I. scapularis* immatures, whereas flying squirrels, which spend little time on the ground, are rarely infested. A similar relationship exists among migratory birds. In the United States, ground-feeding birds that forage in habitats shared with cottontail rabbits are heavily infested with immatures of the rabbit tick *H. leporispalustris.* Birds that forage above ground in bushes or trees rarely encounter these ticks. Acceptance of a vertebrate animal also is dependent on physiological factors and the ability of the ticks to recognize it as a host.

Host utilization may be influenced by the ability of ticks to evade or suppress host homeostatic systems and avoid rejection (Ribeiro, 1987). This was first reported in a landmark paper by Trager (1939), who noted that the feeding success of *D. variabilis* on guinea pigs declined with frequent re-exposures. The guinea pig is a South American rodent and an unnatural host for this North American tick. When fed on its natural hosts (e.g., white-footed mice), *D. variabilis* experiences little if any rejection. Similarly, *I. scapularis* saliva contains pharmacologically active compounds that suppress host mediators of edema and inflammation, such as anaphylatoxins, bradykinin, and other kinins. Other salivary components prevent release of host inflammatory agents from leukocytes while enhancing vasodilation, which brings more blood to the mouthparts. *I. scapularis* apparently lacks compounds to suppress histamine. Because histamine-induced edema does not occur in the white-footed mouse, this does not deter tick feeding on this host. However, histamine-containing basophils are very abundant in guinea pigs, and cutaneous basophil hypersensitivity develops rapidly even after a single feeding by these ticks (Ribeiro, 1989). *I. scapularis* saliva also contains an enzyme that destroys complement, thereby facilitating the survival of pathogens such as *Borrelia burgdorferi* ingested during blood feeding (Valenzuela *et al.*, 2000).

Ticks occur in many terrestrial habitats ranging from cool, arboreal northern forests to hot, arid deserts. Each species, however, has become adapted to specific types of habitats where generally it is found in greatest abundance. Typical habitat associations of nonnidicolous ixodid ticks include forests, meadows and other clearings, grasslands, savannahs, and semidesert or desert areas. At one end of the spectrum are species that have very limited resistance to desiccation and occur in cool, moist forests (e.g., *I. scapularis, I. ricinus*). In the middle are the majority of species that can survive at least brief periods of desiccation during host seeking or development (e.g., *D. variabilis, Amblyomma maculatum,* and *A. americanum*). At the other end of the spectrum are the desiccation-tolerant species adapted to survive in arid steppes, semideserts, and other xeric environments (e.g., *Hyalomma asiaticum*).

Water balance is a critical determinant of a tick's ability to wait for hosts. Ticks may quest for weeks or months while waiting for a host. When they begin to desiccate, they retreat to more sheltered, humid microenvironments, such as the rotting vegetation in a meadow or damp leaf litter on the forest floor. They secrete a hygroscopic salivary secretion onto their mouthparts that collects atmospheric water (direct sorption). After repeated cycles of secretion and drinking the condensed water, the rehydrated ticks are able to resume host seeking. Some ticks are able to remain in the questing position for many days without rehydration, while others must return to their humid microenvironments each day.

*I. scapularis* is an example of a tick with very limited *tolerance to desiccation*. Consequently, it is most abundant in dense, humid, forest habitat or in dense shrub-dominant habitats adjacent to large rivers, bays, or the Atlantic Ocean. Another desiccation-intolerant species is

*I. ricinus.* This tick is widespread in the British Isles, Europe, and western Asia, where it frequents woodlands, meadows, pastures, and ecotones.

*D. variabilis* is an example of a species exhibiting greater tolerance to desiccation. It flourishes in the ecotone between secondary-growth deciduous forests and lush, grassy meadows, as well as along secondary roads and trails in forested habitats. The dense ecotonal vegetation provides shade, increased moisture, protection from intense solar radiation, and food plants that support the tick's mammalian hosts. This type of environment is ideal for the immature stages of *D. variabilis*. Adults, with their greater resistance to desiccation and greater mobility, venture farther afield to quest in sunlit meadows or along roads and trails.

The camel tick *H. dromedarii* is an example of a desiccation-tolerant species. This desert tick is common in the steppes and semidesert habitats in large areas of North Africa and the Middle East. Larvae and nymphs generally live in rodent burrows. Adults bury themselves in sand and duff near their hosts, especially around caravansaries and similar locations where camels and other livestock are kept.

Nidicolous ticks living in or near the nests of their hosts are adapted to highly specialized environments. Normally the temperature and relative humidity in a burrow, cave, or similar type of shelter are more uniform throughout the year than in the external macroenvironment. The higher relative humidity in such microenvironments is due in part to the presence of hosts, their wastes, and the plant materials they use to construct or line their nests. Nidicolous ticks exhibit behavioral patterns that restrict their distribution to these sheltered locations. They avoid bright sunlight and low humidity, the type of conditions prevalent at the entrances of burrows or caves. Confined within these cryptic, restricted locations, nidicolous ticks become active when hosts are present. However, when hosts are absent, they may wait for up to several years for hosts to return, or until they die of starvation.

*Seasonal activity* refers to the period of the year when ticks actively seek hosts. For example, *D. variabilis* larvae emerge from winter *diapause* in spring to feed on small mammals, especially mice and voles. Activity accelerates rapidly as increasing numbers of larvae emerge from overwintering sites to attack hosts, culminating in the seasonal peak within a few weeks. Thereafter, activity continues unabated but larval abundance declines as more individuals find hosts, desiccate, or die of starvation. Nymphal and adult ticks also feed during the warm spring and early summer months. In the southern parts of its range, overwintering *D. variabilis* adults emerge early and soon overlap with those that develop from nymphs fed in the spring. Thus, the seasonal peak for adults occurs in early summer. As a result, most females oviposit in July and August and the newly hatched larvae enter diapause as day length diminishes. In the southeastern United States, the entire life cycle is completed in one year. Occasionally, a small secondary peak of *D. variabilis* larval activity occurs in the fall. In the northern part of its range, however, tick activity is delayed due to cooler spring temperatures and shorter day lengths. As a result, although larvae and nymphs feed in the late spring and summer, adults emerge too late in the summer to commence questing activity. These adults undergo diapause and emerge the following spring. This pattern of feeding and diapause results in a 2-year life cycle.

*I. scapularis* also exhibits distinct seasonal activity periods. In the northern part of its range, larval activity does not occur until middle or late summer and the nymphs that molt from the fed larvae diapause until the following spring. Nymphs that feed in the spring molt in summer, but the young adults delay host-seeking until fall. This pattern results in a 2-year life cycle. In the southernmost part of its range, *I. scapularis* activity occurs earlier in the year. These southern populations may complete their life cycle in just 1 year.

A few tick species are active during the cooler months of the year, especially fall and winter. Larvae of the *winter tick* (*D. albipictus*), a one-host tick that feeds on horses, deer, elk, moose, and other large ungulates, commence host-seeking activity in late summer or early fall. Larvae and nymphs feed and molt on the same hosts and the resulting adults reattach, feed, and mate. Replete females drop off the host and oviposit in the soil. In the northernmost parts of its range, adults usually do not appear until late winter, with peak occurrence in April. Subsequent oviposition and hatching occur in late spring, but larvae diapause, presumably in response to increasing daylight, and do not commence host seeking until after an extended period of declining photoperiod. Development proceeds faster farther south. On stanchioned bovines held in stalls at Kerrville, TX, in the southern United States, the entire process of feeding, molting, and production of engorged females was completed in 21–36 days (Drummond *et al.*, 1969). Engorged females that drop off their hosts in winter do not oviposit until the following spring.

In tropical regions, where day length is nearly uniform throughout the year and where there is no prolonged dry season, the seasonal activity of many tick species (e.g., *R. appendiculatus*) is often influenced by the distribution of rainfall. In colder regions in southern Africa, *R. appendiculatus* diapauses during the dry season.

In contrast, most argasid ticks do not exhibit patterns of seasonal activity. This is especially true for ticks infesting the nests or burrows of nonmigratory hosts such as rodents and carnivores. However, nidicolous ticks that parasitize migratory birds and bats tend to delay oviposition so that hatching occurs at about the time the hosts return.

*Diapause* is an important behavior that enables ticks to survive adverse environmental conditions and conserve energy until conditions improve. Diapausing ticks become inactive, reduce their metabolic rates, and do not feed on hosts even when given the opportunity. Newly emerged larvae, freshly molted nymphs, and adults of many species diapause before seeking hosts, particularly if they emerge during periods of declining photoperiod. This is termed host-seeking diapause. As noted above, diapause enables *D. variabilis* larvae and adults to survive the cold winters that occur throughout most of the tick's range. It also determines the length of the life cycle: 1 year in the south but 2 years in the northern part of its range. Diapause delays activity of *I. scapularis* nymphs at more northern latitudes so that they do not commence host seeking until spring or early summer. It also may delay adult activity that usually begins in fall, often several months after molting.

Another type of diapause is morphogenetic diapause, in which development or oviposition is delayed. In *D. marginatus*, oviposition rather than hatching is delayed. Thus, females that feed in spring or early summer lay eggs immediately, but those that feed in late summer or early fall oviposit the following spring. A remarkable example of morphogenetic diapause occurs in certain argasid ticks that inhabit the nests of birds or the roosts of bats. Females of the bat tick *Ornithodoros kelleyi* oviposit immediately after feeding in spring. However, those that feed in fall delay oviposition until the following spring. Because the bats migrate to cold caves or caverns far from the tick's normal habitats, this ovipositional delay avoids the risk that larvae will emerge at a time when all hosts are absent.

# TICK SPECIES OF MEDICAL-VETERINARY IMPORTANCE

The following ticks are important as household pests, species that transmit disease agents to humans, and species that are injurious to livestock or transmit disease agents to animals. The more important tick-borne diseases of humans and other animals are listed in Tables II and III.

The *brown dog tick* (*R. sanguineus*) is a common household pest throughout most of the world. Its primary host is the dog; all life stages feed on these animals. However, in many areas bordering the Mediterranean Sea, western Asia, and Africa, this tick also feeds readily on a wide range of wildlife (especially small mammals) and also

TABLE II

Representative Tick-Borne Diseases of Public Health Importance and Associated Characteristics

| Disease | Causative agent | Primary tick vector species | Affected host(s) |
|---------|-----------------|------------------------------|------------------|
| Human babesiosis | *Babesia microti, B. divergens, B. major* | *Ixodes scapularis, I. ricinus* | Humans, mice, cattle |
| Tick-borne encephalitis | *Flavivirus*[a] | *I. ricinus, I. persulcatus* | Rodents, insectivores, carnivores, humans, etc. |
| Kyasanur Forest disease | *Flavivirus*[a] | *Haemaphysalis spinigera* | Monkeys, small mammals, carnivores, birds, cattle, humans |
| Powassan encephalitis | *Flavivirus*[b] | *Ixodes, Dermacentor,* and *Haemaphysalis* spp. | Rodents, hares, carnivores |
| Colorado tick fever | *Coltivirus*[b] | *Dermacentor andersoni* | Rodents, carnivores, humans domestic animals |
| Crimean-Congo hemorrhagic fever | *Nairovirus*[c] | *Hyalomma m. marginatum, H. m. rufipes,* others | Hares, hedgehogs, small mammals, humans |
| Rocky Mountain spotted fever | *Rickettsia rickettsii* | *Dermacentor variabilis, D. andersoni,* others | Small mammals, carnivores, rabbits, other mammals |
| Boutonneuse fever[d] | *R. conorii* | *Rhipicephalus, D. marginatus, D. reticulatus,* others | Small mammals, hedgehogs, dogs |
| Human monocytic ehrlichiosis | *Ehrlichia chaffeensis* | *Amblyomma americanum, D. variabilis* | Humans, deer |
| Human ehrlichiosis | *E. ewingii* | *A. americanum* | Dogs, humans |
| Human granulocytic ehrlichiosis | *E. phagocytophila* | *I. scapularis, I. pacificus, I. ricinus* | Rodents, humans, dogs |
| Q fever | *Coxiella burnetii* | Many tick species | Large domestic livestock, humans |
| Lyme disease | *Borrelia burgdorferi, B. afzelii, B. garinii, B. bissettii* | *I. scapularis, I. ricinus, I. pacificus, I. persulcatus,* others | Small mammals, some birds |
| Tick-borne relapsing fever | *Borrelia* spp. | *Ornithodoros* spp. | Various mammals |
| Tularemia | *Francisella tularensis* | *Haemaphysalis leporispalustris,* others | Lagomorphs, rodents, carnivores |
| Tick paralysis | Tick proteins | *I. holocyclus, I. rubicundus, D. variabilis, D. andersoni* | Cattle, sheep, dogs humans, other mammals, others |
| Tick-bite allergies | Tick proteins | *Argas reflexus, Ornithodoros coriaceus, I. pacificus,* etc. | Humans |

[a] Family Flaviviridae.
[b] Family Reoviridae.
[c] Family Bunyaviridae.
[d] Also known as Mediterranean spotted fever.

attacks humans. It often infests houses, especially when dogs are kept indoors, which can produce considerable distress when the owners encounter thousands of these ticks. Seasonal activity peaks in summer, although activity peaks can occur throughout the year when ticks inhabit heated homes. This species is the primary vector of *Rickettsia conorii*, which causes *Boutonneuse fever* in many Mediterranean countries.

The *brown ear tick* (*Rhipicephalus appendiculatus*) is a major pest of livestock in southern and southeastern Africa. Hosts include most domestic ruminants and many wildlife species. This tick attaches predominantly in and around the ears of its hosts. *R. appendiculatus* is the vector of ECF, a protozoan disease that afflicts ruminant livestock within its range.

The genus *Haemaphysalis* contains numerous species that attack mammals and birds. In North America, an important species is the *rabbit tick* (*Haemaphysalis leporispalustris*). This tick occurs throughout most of North America. Larvae and nymphs attack ground-feeding birds as well as lagomorphs, while adults feed only on lagomorphs. Larvae and nymphs are active in the late

**TABLE III**
Representative Tick-borne Diseases of Veterinary Importance

| Disease | Causative agent | Primary tick vector species | Affected host(s) |
|---|---|---|---|
| Bovine babesiosis | *Babesia bigemina* | *Boophilus annulatus, B. microplus,* others | Cattle |
| East Coast fever | *Theileria parva* | *Rhipicephalus appendiculatus* | Cattle, buffalo |
| Tropical theileriosis | *T. annulata* | *Hyalomma anatolicum* | Cattle, horses |
| Feline cytauxzoonosis | *Cytauxzoon felis* | *Dermacentor variabilis* (?) | Cats |
| Louping ill | *Flavivirus*[a] | *Ixodes ricinus* | Sheep |
| African swine fever | *Iridovirus* | *Ornithodoros moubata porcinus, O. erraticus* | Domestic pigs, wild boars, warthogs |
| Tick-borne fever | *Ehrlichia phagocytophila* | *I. ricinus* | Domestic and wild ruminants |
| Canine ehrlichiosis | *E. canis, E. ewingii, E. phagocytophila* | *R. sanguineus, I. ricinus Amblyomma americanum,* others | Dogs |
| Q fever | *Coxiella burnetii* | *D. andersoni,* others | Most domestic animals |
| Heartwater | *Cowdria ruminantium* | *A. hebraeum, A. variegatum,* others | Ruminants |
| Anaplasmosis | *Anaplasma marginale, A. centrale, A. ovis* | *D. andersoni, D. occidentalis, R. sanguineus,* others | Cattle, sheep, other ruminants |
| Borrelioses | *Borrelia burgdorferi* | *I. scapularis, I. ricinus I. pacificus, I. persulcatus* | Dogs, cats, cattle, horses |
| Avian spirochetosis | *B. anserina* | *Argas persicus* | Turkeys, chickens, other birds |
| Epizootic bovine abortion | Unknown, possibly *B. coriaceae* | *O. coriaceus* | Cattle, deer |
| Tularemia | *Francisella tularensis* | *D. andersoni* | Sheep, horses, rabbits, game birds |
| Tick paralysis | Tick proteins | *I. rubicundus, R. evertsi evertsi, D. andersoni, A. walkerae* | Sheep, cattle, goats, dogs, other mammals, chickens |
| Tick toxicoses | Tick proteins | *O. savigny, O. lahorensis, A. persicus* | Cattle, sheep, birds |
| Sweating sickness | Tick proteins | *Hyalomma truncatum* | Cattle, sheep, other ruminants, dogs |

[a] Family Flaviviridae.

summer and fall, while adults feed in the spring. This species contributes to the maintenance of RMSF among wildlife. In India, an important species of this genus is *H. spinigera,* which occurs in dense forest habitat. Larvae feed on small mammals and ground-feeding birds, but nymphs and adults attack larger animals, including monkeys, cattle, and even humans. Larvae are active during October and November, nymphs from November to June, and adults mostly in July and August. This tick is the vector of the virus which causes *Kyasanur Forest disease.*

The *American dog tick* (*D. variabilis*) is a major pest of people and domestic animals throughout much of the eastern and south central United States as well as some areas of southeastern Canada. Tick populations generally decline west of the Mississippi River basin, although *D. variabilis* may be locally abundant in some parts of the midwestern and far western United States. Larvae and nymphs feed on small mammals, but adults attack

dogs, other medium sized mammals, livestock, and humans. Larvae and nymphs are active in late winter and spring, while adults are most abundant in late spring and early summer. This species is the major vector of the agent of RMSF, caused by *Rickettsia rickettsii,* in the eastern United States. It also transmits the agents of *tularemia* and *anaplasmosis.* In western North America, the closely related *Rocky Mountain wood tick* (*D. andersoni*) is an important pest attacking humans, livestock, and wildlife. Adults and nymphs of this tick attack almost any medium- or large-sized mammal, whereas larvae attack small mammals. Adults and nymphs are active in late spring and early summer, while larvae are most abundant in the summer. *D. andersoni* is the primary vector of *R. rickettsii* and *Colorado tick fever* (CTF) virus in this region. It also transmits *Anaplasma marginale,* a rickettsia which causes *anaplasmosis* in domestic ruminants. In the Pacific Northwest, *D. andersoni* is an important cause of *tick paralysis.*

The *blacklegged tick* (*I. scapularis*) is widespread throughout large areas of the eastern, south central, and midwestern United States. The immature stages usually feed on small mammals, lizards, and birds, while adults are most common on white-tailed deer. All stages of *I. scapularis* will bite humans. Nymphal ticks, the stage most likely to transmit *Lyme disease* spirochetes to people, are active in late spring and early summer. Adults are active in the fall and early spring. Larvae are most abundant in the summer. *I. scapularis* is the primary vector of the *Lyme disease* spirochete *B. burgdorferi;* the protozoan *Babesia microti,* which causes *human babesiosis;* and *Ehrlichia phagocytophila,* the agent of *human granulocytic ehrlichiosis.*

In Europe, the *castor bean tick,* or *sheep tick* (*I. ricinus*), is a major pest of livestock and humans. This tick ranges from Ireland and Britain across continental Europe to Iran and southward to the Mediterranean Sea. In Britain and Ireland, it is commonly found in poorly maintained, overgrown sheep pastures that contain dense mats of moist, rotting vegetation ideal for tick development and survival. On the European continent, *I. ricinus* abounds in mixed hardwood/pine forests and shrubs but rarely in grassy meadows. Larvae and nymphs attack mostly small mammals, insectivores, birds, and lizards. Adults are found most commonly on sheep, other domestic ruminants, and deer. However, this tick may attack virtually any vertebrate, including humans. Seasonal activity varies greatly in different regions throughout the tick's range. *I. ricinus* transmits the agents of *Lyme disease* which, in Europe, include *Borrelia burgdorferi, B. garinii,* and *B. afzelius.* In addition, *I. ricinus* is the major vector of the virus which causes *Tick-borne encephalitis* (TBE) and of the rickettsia, *E. phagocytophila.* In Ireland, Britain, and some areas of western Europe, *I. ricinus* also transmits the virus which causes *louping ill* in sheep and also transmits the bacterium, *Staphylococcus aureus,* which causes *tick pyaemia* in sheep.

In Australia, an important species is the *Australian paralysis tick* (*I. holocyclus*). This tick is found along the eastern coast of Queensland and Victoria provinces. It feeds on most wild mammals, domestic animals, and humans. *I. holocyclus* is notorious as the cause of *tick paralysis* in Australia. In contrast to other diseases caused by an infectious microbe, tick paralysis is caused by a proteinaceous material, holocyclotoxin, secreted in the tick's saliva. Even the bite of a single tick may be sufficient to cause a fatal paralysis.

In Africa and the former Soviet Union, an important species is *Hyalomma marginatum marginatum.* This tick occurs in the Crimea and adjacent areas of the former USSR and in North Africa. Larvae and nymphs attack hares, hedgehogs, and birds. Adults attack larger mammals, including domestic ruminants and humans. This tick is one of the most important vectors of *Crimean-Congo hemorrhagic fever* virus.

The *lone star tick* (*Amblyomma americanum*) is one of the most notorious tick pest species in the United States. It is found along the Atlantic coast from New York to Florida and west into Texas and Oklahoma. *A. americanum* larvae, nymphs, and adults readily attack humans and companion animals, as well as livestock and wildlife. Virtually any mammal or ground-feeding bird may be infested. It is often abundant in areas with large populations of deer, which serve as the primary hosts for the adult ticks. In the southeastern United States, nymphs and adults emerge from their winter diapause and commence host-seeking activity in late spring. Larvae generally appear in late summer. Seasonal activity may be delayed farther north. *A. americanum* has been implicated as a vector of the agent of *human monocytic ehrlichiosis, E. chaffeensis.* It is also suspected of transmitting *E. ewingii* to humans. Another important species in the United States is the *Gulf Coast tick* (*A. maculatum*), which is found in the southeastern and south central United States and Mexico. Larvae and nymphs attack a wide range of birds and mammals, but adults feed largely on ruminants. These ticks feed mainly on the head and ears. *A. maculatum* can cause severe injury to the skin of cattle and other livestock, often rendering the hides useless from the bites of these ticks or from secondary infections.

In Africa, the *bont tick* (*A. hebraeum*) and the *tropical bont tick* (*A. variegatum*) attack livestock as well as wild ruminants. In addition, *A. variegatum* larvae and nymphs will feed on ground-feeding birds, including herons and other migratory birds. *A. hebraeum* is restricted to southern Africa, but *A. variegatum* ranges throughout most of sub-Saharan Africa, Madagascar, and several islands in the Caribbean. These ticks are the major vectors of the rickettsia (*Cowdria ruminantium*) that causes *heartwater* in ruminants.

The *cattle fever tick* (*Boophilus annulatus*) ranges throughout large areas of North Africa, southern Europe, and western Asia and parts of North America, Central America, and South America. Only intensive surveillance has prevented its reintroduction into the United States from tick-infested herds in Mexico, where it was formerly widespread. This one-host tick feeds almost exclusively on cattle, but it also infests white-tailed deer. *B. annulatus* is active throughout the year in the tropics. It a major pest of cattle, causing reduced weight gains and milk production in heavily infested cattle. This tick is best known for its role in the transmission of the protozoan *Babesia bigemina,* which causes *Texas cattle fever.*

Among the Argasidae, the *fowl tick* (*Argas persicus*) is an important parasite of poultry in the Old World. All stages feed on these birds. Populations of this tick can reach enormous numbers in poultry barns and can cause high mortality due to exsanguination. This tick is the vector of the rickettsia *Aegyptianella pullorum*, which causes *fowl disease* in domestic fowl. In the Mediterranean region, it is a vector of *Borrelia anserina*, the agent of *fowl spirochetosis*, an important poultry disease. In the New World, the fowl tick exists as a complex of three species—*Argas radiatus*, *A. sanchezi*, and *A. miniatus*—in addition to the introduced but now established *A. persicus*.

The genus *Ornithodoros* includes several species that live in poorly maintained homes or shelters, where they hide in cracks and crevices of walls, ceilings, and attics. In the western United States, *O. hermsi* often infests mountain cabins and other dwellings. Although rodents that infest dwellings are the principal hosts of *O. hermsi*, humans may be attacked when they enter such dwellings if rodents have been killed or driven out. This tick is notorious as a vector of the *relapsing fever* spirochete *B. hermsii*.

In eastern Africa, the human-biting *African tampan* (*O. moubata*) coexists with people and animals in mud huts, where the tick hides in the walls. This species is the major vector of the *relapsing fever* spirochete, *B. duttoni*.

Another important argasid in western North America is the *spinose ear tick* (*Otobius megnini*). It frequently infests livestock, especially cattle and horses, and most domestic ruminants. *O. megnini* also attacks wild ruminants, especially deer, antelope, and mountain sheep. The larvae and second stage nymphs feed, but not the adults. The ticks feed in the ears, causing injury to the auditory canal and secondary infections.

## PUBLIC HEALTH IMPORTANCE

Ticks are of public health significance mainly because of the animal disease agents transmitted by them, which include an array of bacterial, viral, and protozoan disease agents (Harwood and James, 1979; Sonenshine, 1993). They also are important because their attachments may cause various kinds of dermatoses or skin disorders, such as inflammation, pain, and swelling. Rarely, they invade the auditory canal, producing a condition known as *otoacariasis*. Certain species of ticks may cause a flaccid, ascending, and sometimes fatal paralysis known as *tick paralysis*. Individuals bitten repeatedly by some ticks may develop *allergic* or even *anaphylactic reactions* (Van Wye et al., 1991).

Among the biological factors that contribute to the high vector potential of ticks are their persistent blood-sucking habit, longevity, high reproductive potential, relative freedom from natural enemies, and highly sclerotized bodies that protect them from environmental stresses. Most of these factors are discussed in the preceding biology section. Further, the slow feeding behavior of ixodid ticks permits wide dispersal and increases the likelihood of acquiring pathogens during attachment to a host. Transovarial transmission of microbial disease agents occurs with variable efficiency in many vector ticks, and this phenomenon contributes to the maintenance and spread of some tick-borne agents.

Several other biological attributes of ticks that enhance their vector potential can be added to the foregoing list. First, pharmacologically active substances present in the saliva of ticks may promote feeding success and aid transmission of microbial agents (Ribeiro, 1987). For example, the saliva of *I. scapularis* has antiedema, antihemostatic, and immunosuppressive properties. Second, ticks imbibe large quantities of blood during each feeding period. Indeed, certain species may increase their body weight by 100-fold or more. This is actually an underestimate of the amount ingested because feeding ticks concentrate the blood meal by secreting copious amounts of host-derived fluid back into the host. Third, ticks take multiple blood meals during their lifetimes. Those individuals that attain adulthood and that successfully feed as adults feed three (ixodids) or more (argasids) times. Fourth, following acquisition of an infective blood meal, many larval or nymphal vector ticks pass microorganisms transstadially.

It should be noted that ticks are far more efficient than insects in maintaining microbial agents in their bodies (Hoogstraal, 1980). In ticks, most internal tissues change gradually during development and therefore transstadial survival of pathogens occurs commonly. In holometabolous insects, however, the extensive internal changes that occur during molting seem to have a harmful effect on most microorganisms that cause human disease.

As reviewed recently (Lane, 1994; Nuttall and Labuda, 1994), ticks transmit microbes by several routes, including salivary secretions (e.g., Lyme disease spirochete, CTF virus, the agent of heartwater, and spotted fever group rickettsiae), coxal fluids (certain species of relapsing fever spirochetes), regurgitation (e.g., possibly the spirochetes that cause Lyme disease), and feces (Q fever organisms). A novel type of transmission, *saliva-activated transmission*, occurs in the case of some tick-borne arboviruses (Jones *et al.*, 1992). In this model, one or more proteins secreted in tick saliva potentiate virus transmission. Moreover, this phenomenon seems to be the mechanism underlying *nonviremic transmission*, whereby arboviruses are transmitted between infected and uninfected ticks feeding simultaneously on a vertebrate host having no or very low levels of viremia (Nuttall and Jones, 1991; Nuttall and Labuda, 1994). Transmission between co-feeding infected and uninfected ticks, which also has been demonstrated for the Lyme disease spirochete, *B. burgdorferi*, is important epidemiologically for two reasons (Randoph *et al.*, 1996). First, some vertebrates that do not develop systemic infections still can serve as competent hosts for infecting vector ticks, and, second, it adds yet another transmission route for certain tick-borne pathogens. Although some tick-borne agents may be transmitted by two routes (e.g., transmission of certain relapsing fever spirochetes via coxal fluid secretions and by saliva), only one route is usually of enzootic or epizootic (epidemiologic) significance.

In this section, we review some of the more important tick-borne diseases of public health concern, which are summarized in Table II. The causative agent, clinical manifestations, ecology, and epidemiology of each disease are discussed. Diagnosis, treatment, and disease prevention are largely omitted. For further information about these diseases, readers should consult Sonenshine (1993) and the selected references provided below.

## HUMAN BABESIOIS

Human babesiosis is an infectious disease caused by several species of protozoans in the genus *Babesia*. They, along with their close relatives, *Theileria* spp., also are known as *piroplasms*. *Babesia* spp. resemble malarial parasites (*Plasmodium* spp.) and other blood-infecting protozoans, especially as regards their developmental cycles. More information about these important parasites can be found in Ristic (1988).

Human babesiosis occurs principally in the eastern United States, most commonly on islands off the coast of New England and on Long Island, NY, where the etiologic agent is *Babesia microti*. A few cases have been reported from the western United States and Europe. The disease in Europe is believed to be caused by *B. microti* and *B. divergens*. In the western United States and Missouri, the causative agent is an undescribed species of *Babesia*.

In humans, *Babesia* species produce a malarial-like disease without the periodicity that often accompanies the human malarias. Following an incubation period of 1 to 4 weeks, the clinical course varies according to the etiologic agent. Splenectomized persons infected with *B. divergens* or *B. microti*, or elderly persons infected with *B. microti*, tend to develop severe or sometimes fatal illnesses. Signs and symptoms at onset include fever, chills, profuse sweating, headache, and generalized muscle aches. Joint pain, nausea, vomiting, and prostration may occur. Parasitemia and the resultant clinical course may persist for up to several months, with severe anemia, jaundice, and hemoglobinuria.

With few exceptions, *Babesia* spp. develop entirely in circulating red blood cells. Sporozoites are introduced via the saliva of a *Babesia*-infected tick during feeding. Once they gain entry into the bloodstream, most parasites develop asexually within red blood cells. Occasionally, *Babesia* spp. invade lymphocytes, and only subsequent generations develop in the erythrocytes. Within the host cell, the parasites develop into *trophozoites* termed *meronts*, which multiply asexually by binary fission to produce *merozoites*. Some of the merozoites escape from the disintegrating host cells to invade other erythrocytes and continue the cycle. Other merozoites develop into gametocytes called *piroplasms* after entering previously uninfected erythrocytes. The gametocytes remain in an arrested state of development until they are ingested by a feeding tick.

In many individuals, babesiosis is a mild, self-limiting disease that requires only supportive therapy. In persons experiencing severe, fulminating infections, quinine and clindamycin in combination are the drugs of choice.

When *Babesia*-infected blood is ingested by ticks, the gametocytes commence development, but the asexual stages are destroyed. The gametocytes escape from the dying host cells and transform into gamete-forming cells called gamonts, which develop structures (rays and spines) that are subsequently used to penetrate cells. Following gametic fusion, the resulting zygotes invade the tick's digestive epithelium, develop into motile *kinetes*, and migrate to other internal organs. *B. bigemina* and some other *Babesia* spp. invade the tick's ovaries and are *transmitted transovarially* to the next generation. However, *B. microti* is not transmitted transovarially by *I. scapularis*. Instead, immature ticks are infected while feeding on a parasitemic host, and the parasites invade the salivary glands, multiply, and are *passed transstadially*. In the northeastern and upper midwestern United States, *B. microti* is maintained in a transmission cycle involving *I. scapularis* and the white-footed mouse. Meadow voles also are competent reservoir hosts. Most people who acquire the infection are bitten by nymphal ticks. In the western United States and Missouri, the primary tick vector and reservoir host(s) of the undescribed *Babesia* agents have not been identified. In Europe, human cases usually stem from infection with *B. divergens*, which is transmitted by *I. ricinus*. In Britain, *I. trianguliceps* can transmit the infection (*B. microti*) among small mammals (Randolph, 1991, 1994, 1995).

## TICK-BORNE ENCEPHALITIS/COMPLEX

First described as RSSE in 1932 from the far-eastern region of the former Soviet Union, TBE was recognized after World War II in central Europe, where it was termed *central-European encephalitis* (CEE). RSSE and CEE are now considered a single entity, TBE. TBE is one of at least a dozen related but distinguishable serotypes of tick-borne flaviviruses (family Flaviviridae) that constitute the *TBE complex,* which includes such viruses as *louping ill, Kyasanur Forest disease, Omsk hemorrhagic fever,* and *Powassan encephalitis.* Each of these viruses produces a clinically distinctive disease.

TBE is currently endemic over wide areas of Europe and northern Asia. Illness in humans is accompanied by high, often biphasic, fever and headache, followed soon afterward by inflammation of the brain (*encephalitis*) and meninges (*meningitis*). Some patients develop muscle weakness or paralysis, especially in the right shoulder muscles. In the Far East, case fatality rates are relatively high (up to 54%), whereas the disease in Europe is considerably milder.

In Europe and the Far East, the most important vectors of the virus are *I. ricinus* and *I. persulcatus.* Other tick species that have been found infected naturally (*I. arboricola, I. hexagonus, I. trianguliceps*) may amplify viral infection. Although most mammals are susceptible to TBE virus, rodents (e.g., the bank vole and the yellownecked field mouse) and insectivores are believed to be the chief reservoir hosts.

*Powassan encephalitis,* named after the town in Ontario, Canada, where it was first recognized, has been found in a number of scattered localities in the United States, Canada, and the former Soviet Union. In its acute phase, this disease is marked by encephalitis, severe headache, and fever. Nausea, labored breathing, and neurologic disorders, including partial paralysis, occur frequently. Recovered patients may suffer permanent nerve damage. The case fatality rate for the 22 cases that have been reported from North America was 23% (Calisher, 1994).

Tick vectors of Powassan encephalitis virus belong to the genera *Ixodes, Dermacentor,* and *Haemaphysalis.* In the United States, isolates of the virus have been obtained from *I. cookei* and *I. marxi* in the east and *D. andersoni* and *I. spinipalpis* in the West. *I. cookei* feeds on various wild and domestic animals and on humans. The marmot (woodchuck) is an important host of *I. cookei* and an excellent reservoir of the virus. Similarly, the snowshoe hare amplifies populations of vector ticks and the virus. The virus has been isolated twice from naturally infected foxes, a red squirrel, a white-footed mouse, and a spotted skunk, but the reservoir competence of these species remains to be determined. Antibodies to the virus have been detected in 38 wild and 5 domestic mammalian species. *D. andersoni* is the most important vector in the western United States and Canada. In the former Soviet Union, the virus has been isolated from *Haemaphysalis neumanni, I. persulcatus,* and *D. silvarum* and from mosquitoes.

## COLORADO TICK FEVER

CTF is caused by a *Coltivirus* in the family Reoviridae. Two recognized and nine probable serotypes of coltiviruses are known (Calisher, 1994). The other recognized serotype, *Eyach virus,* has been isolated from *I. ricinus* and *I. ventalloi* in Germany and France. The distribution of CTF approximates that of its primary tick vector, *D. andersoni,* in western North America. CTF virus has been isolated from ticks, humans, or both from parts of the United States and Canada; most human cases have been reported from Colorado.

Symptoms usually appear within 4 days following the attachment of an infected tick. The disease is characterized by a biphasic fever, headache, muscle aches, and joint pain. Some patients experience eye pain, intolerance of light, chills, sore throat, and nausea. The virus develops in most internal organs and may spread to the brain or bone marrow. Although CTF is sometimes depicted as a mild febrile illness, convalescence may be prolonged and some patients take several weeks to recover. Case fatality rates are very low, usually less than 0.2%.

In endemic regions, people engaged in outdoor activities in mountainous or highland areas from about 4,000 feet to over 10,000 feet are at risk of exposure to virus-infected ticks. In Rocky Mountain National Park, Colorado, natural foci occur on south-facing slopes covered with open stands of pine and shrubs on dry, rocky surfaces. Cases are reported from March to November, but most occur in the spring and early summer when adult and nymphal ticks are active.

Besides *D. andersoni,* CTF virus has been isolated from *D. albipictus, D. occidentalis, D. parumapertus, H. leporispalustris, I. sculptus, I. spinipalpis,* and *O. lagophilus.* Larvae and nymphs of *D. andersoni* feed on small mammals, especially ground squirrels, mice, and rabbits. Nymphs, which quest higher in vegetation than do larvae, also attack larger mammals, such as small carnivores and humans. Important hosts of the immatures include golden-mantled ground squirrels, deer mice, bushy-tailed woodrats, chipmunks, and rabbits. Adults parasitize larger mammals, such as porcupines, elk, deer, antelope, carnivores, and humans. Competent *reservoir hosts* include the golden-mantled ground squirrel, least chipmunk, deer mouse, bushy-tailed woodrat, and porcupine.

## ROCKY MOUNTAIN SPOTTED FEVER

This disease was first recognized in the Bitterroot Valley of western Montana in 1872. RMSF is widely distributed throughout most of the United States and, to a lesser extent, in Canada and South America. In the

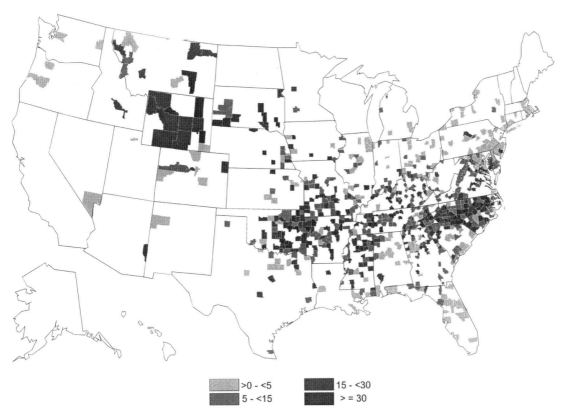

**FIGURE 24.9**    Distribution of human cases of Rocky Mountain spotted fever in the United States, shown as annual incidence (cases per million population), 1993–1996. (Courtesy of Viral and Rickettsial Zoonoses Branch, Centers for Disease Control and Prevention.)

United States, the disease was recognized only in the west until the 1930s, when cases were detected for the first time in the east. Since 1985, about 600 to 800 cases of RMSF have been reported yearly, with an annual national incidence ranging from 0.24 to 0.32 per 100,000 population. Most cases now occur east of the Mississippi River in the south central and southeastern states, especially along the Atlantic coast (Fig. 24.9). Cases tend to occur in foci in rural areas and suburban communities near major population centers. In the southeastern states, the seasonal peak of reported cases typically occurs in July, coincident with or shortly after the period of peak abundance of adult *D. variabilis* (Fig. 24.10). In the northeastern states, the peak is usually in May or early June, though a bimodal pattern may occur in this region.

RMSF is caused by a *rickettsia* of the *spotted fever group, R. rickettsii,* which multiplies freely in the cytoplasm and occasionally in the nuclei of host cells. It is a severe disease of the circulatory system, with significant mortality in untreated or inappropriately treated cases. Rickettsiae multiply in the epithelial linings of the capillaries, smooth muscle of arterioles, and other blood vessels. After an incubation period of about 7 days, patients develop fever, intense headaches, joint pain, muscle aches, nausea, and

other symptoms. A characteristic *maculopapular rash* occurs in most patients several days after onset of symptoms. It consists of countless tiny, pink or reddish spots, some of which may coalesce. The rash first appears on the hands and feet, gradually spreads to cover the entire body, and may persist for a week or longer. This particular pattern of progression is an important clinical feature of RMSF and helps to distinguish it from rashes produced by other vector-borne disease agents, such as epidemic typhus and allergic reactions. Severe cases may culminate in delirium or coma. Death can occur at any time during the acute clinical phase as a result of renal failure, clotting within blood vessels, shock, or encephalitis. Currently, 2–5% of RMSF patients in the United States die despite the availability of effective antibiotic therapy. Even treated patients may die.

*R. rickettsii* is transmitted by the bite of ixodid ticks. In the United States and Canada, *D. variabilis* and *D. andersoni* are the primary *vectors*. The elegant, pioneering works of Dr. Howard T. Ricketts at the turn of the 20th century elucidated the role of *D. andersoni* (initially reported incorrectly as *D. occidentalis*) and its vertebrate hosts in the transmission cycle of *R. rickettsii*. Ricketts detected the agent in wild-caught ticks and demonstrated

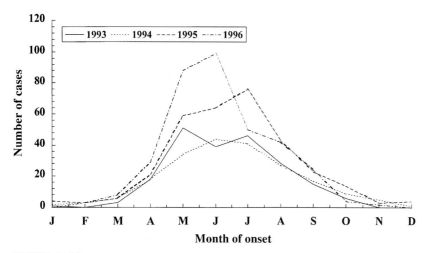

**FIGURE 24.10**   Seasonal occurrence of human cases of Rocky Mountain spotted in the United States, 1993–1996, based on number of cases by month of onset of symptoms. (Courtesy of Viral and Rickettsial Zoonoses Branch, Centers for Disease Control and Prevention.)

experimentally that *D. andersoni* could transmit it to susceptible laboratory and wild rodents by the bite. Further, he showed that *R. rickettsii* is passed transstadially and transmitted transovarially within populations of ticks. His contributions laid the foundation for subsequent studies of tick-borne zoonotic agents in the United States and abroad.

*D. variabilis* is abundant throughout eastern North America, but it has a much more limited distribution in the west, where it is not known to transmit *R. rickettsii.* In contrast, *D. andersoni* is restricted to western North America. The immature stages of both tick species feed on rodents and other small mammals, while the adults attack larger mammals, including people.

Other tick species that help maintain *R. rickettsii* in nature include *H. leporispalustris,* which feeds on birds and rabbits, and *I. texanus,* which feeds on raccoons. The *lone star tick, Amblyomma americanum,* has been found infected naturally with *spotted fever group rickettsiae* that are believed to be nonpathogenic for humans. Although suspected as a vector of RMSF rickettsiae, especially in endemic areas outside the distribution of *D. variabilis,* it is not considered to be a primary vector of *R. rickettsii.*

Larvae and nymphs maintain the infection from year to year and infect susceptible rodents when the ticks emerge to feed in the spring. Infected ticks must remain attached for at least 10 hr before transmission can occur; this is known as the *reactivation phenomenon.* The delay in transmission is due to the fact that *R. rickettsii* seems to be in an avirulent state in unfed ticks. The rickettsiae become virulent only after prolonged attachment of the tick to its host or following ingestion of blood by ticks.

Most humans who contract RMSF are infected by the bite of adult ticks in late spring or summer, although

nymphs occasionally transmit the infection. Only about 1–3% of adult ticks in most foci are infected with *R. rickettsii.* Ticks can be assayed for evidence of rickettsial infection by examination of their hemolymph using immunofluorescence assays. However, precise estimates of tick-infection prevalences with *R. rickettsii* are complicated by the potential presence of nonpathogenic spotted fever group rickettsiae, such as *R. montana, R. bellii, and R. rhipicephali.* Newer tests that employ species-specific monoclonal antibodies or polymerase chain reaction assays should provide a more reliable means for determining tick infection prevalences.

Isolations of *R. rickettsii* have been made from numerous small- and medium-sized wild mammals. Species that have been implicated as natural reservoirs include meadow voles and deer mice. In these animals, there are few if any obvious signs of clinical disease during infection.

The period when rickettsiae are present in the blood of *reservoir hosts* is usually brief, often less than 1 week. Ticks feeding on infected animals may acquire rickettsiae, which produce generalized infections in tick tissues. In western North America, people normally become infected when they enter tick-infested habitats while engaged in outdoor activities in rural areas. In eastern North America, humans acquire their infections both in rural and peridomestic settings because dogs, which are significant hosts of adult *D. variabilis,* carry infected ticks into the home environment.

## BOUTONEUSE FEVER

Boutonneuse fever, also known as *Indian tick typhus, Kenya tick typhus, Crimean tick typhus, Marseilles fever,*

*Mediterranean spotted fever,* and *Mediterranean tick fever,* shares many features with RMSF. However, the causative agent, *R. conorii,* does not occur in the Americas. It has an extensive range in southern Africa, India, central Asia, the Middle East, Europe, and North Africa. A similar disease, *South African tick typhus,* is believed to be caused by the closely related *R. africae.*

Patients with Boutonneuse fever develop fever, chills, severe headaches, and a rash. In addition, a button-like ("boutonneuse") ulcer called an *eschar* or *tache noir* usually forms at or near the site of tick attachment. The disease is generally milder than RMSF and most patients recover without antibiotic treatment. However, strains vary in virulence and one that occurs in Israel has caused severe illness and several deaths. In temperate regions, cases of Boutonneuse fever are most common in late spring and summer, coincident with the seasonal activity of the primary tick vectors.

*R. conorii* is transmitted by several species of ixodid ticks in seven genera (*Amblyomma, Boophilus, Dermacentor, Haemaphysalis, Hyalomma, Ixodes,* and *Rhipicephalus*). In Europe, *D. reticulatus, D. marginatus,* and *I. ricinus* are important vectors. *Rhipicephalus sanguineus* is the principal vector in southern Europe, the Middle East, and North Africa, especially in countries bordering the Mediterranean Sea. Larvae and nymphs of *R. sanguineus* feed on small mammals, especially rodents and hedgehogs, whereas the adults feed mainly on larger mammals, including humans. Lagomorphs, rodents, and possibly birds can serve as *reservoir hosts.* Dogs are susceptible to infection and transport vector ticks into and around human domiciles. Development of *Rickettsia conorii* within populations of ticks is similar to that of *R. rickettsii* in *D. andersoni* and *D. variabilis.*

## HUMAN EHRLICHIOSIS

Ehrlichiae are obligate intracellular organisms that invade the cells of the vertebrate hematopoietic system. Several species are important to human and veterinary health. The various species of the genus *Ehrlichia* grow within cytoplasmic vacuoles of monocytes, granulocytes, lymphocytes, or platelets. Upon infection of a cell, the ehrlichiae divide by binary fission to form microcolonies, known as *morulae.* Cells may contain one or many morulae. Although they are not commonly detected in routine examination of stained peripheral blood smears, the presence of morulae is helpful in presumptive diagnosis (Fig. 24.11).

Human ehrlichiosis was first described in Japan in the 1950s. The etiologic agent, *E. sennetsu,* is transmitted to humans by an unknown mechanism, but ingestion of infected fish parasites is suspected (Rikihisa, 1991). In the United States, ehrlichiosis was first recognized as a febrile illness following a tick bite. The disease is now known to

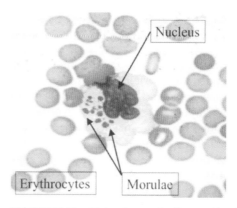

**FIGURE 24.11** Light micrograph of human blood, showing a granulocyte infected with morulae of *Ehrlichia phagocytophila,* the causative agent of human granulocytic ehrlichiosis. (Courtesy of Viral and Rickettsial Zoonoses Branch, Centers for Disease Control and Prevention.)

be caused by one of three species: *E. chaffeensis, E. phagocytophila,* or *E. ewingii.*

*E. chaffeensis* is primarily found in the southeastern and south central United States. This species invades the monocytic leukocytes (Fig. 24.12). The disease, called *human monocytic ehrlichiosis,* manifests as an acute illness with high fever, severe headaches, aching muscles and joints, and other nonspecific signs and symptoms. A rash is not common but may occur in about 20–30% of patients and usually does not involve the palms or soles. The disease is usually mild, although severe cases can occur and may result in death. Over 750 cases of *E. chaffeensis* infection have been reported since its recognition (McQuiston *et al.,* 1999).

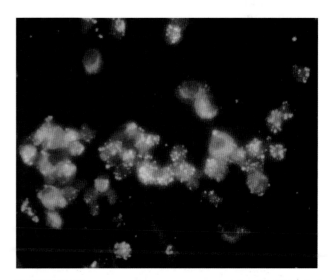

**FIGURE 24.12** Immunofluorescence assay showing *Ehrlichia chaffeensis,* causative agent of human monocytic ehrlichiosis, in canine monocytic cells. Ehrlichiae and morulae are seen as tiny, yellow–green coccoid bodies. (Courtesy of Viral and Rickettsial Zoonoses Branch, Centers for Disease Control and Prevention.)

The primary *vector* for *E. chaffeensis* is *A. americanum* (Anderson *et al.*, 1992, 1993). This tick species feeds readily on white-tailed deer, which serve as one reservoir for the ehrlichiae. Other wild and domestic animals have been identified as potential reservoirs based on serologic, cultural, and molecular studies.

In 1994, a new ehrlichial pathogen was identified in Wisconsin and Minnesota. The infected patients were found outside the known geographic range for *E. chaffeensis* and the agent invaded the granulocytes, so the disease was provisionally named *human granulocytic ehrlichiosis*. Genetic studies have revealed that the organism is *E. phagocytophila (E. equi)*, known previously only as a veterinary pathogen.

This pathogen has a wide distribution in the temperate regions of the world and is found primarily in the northeastern, upper midwestern, and far western United States. The primary vector in the eastern United States is *I. scapularis*, while *I. pacificus* transmits the pathogen in northern California. Novel cycles of infection involving woodrats (*Neotoma* spp.) have been identified in the western United States, whereby the pathogen is maintained by *I. spinipalpis* among *Neotoma fuscipes* in California and among *N. mexicana* in Colorado (Nicholson *et al.*, 1999; Zeidner *et al.*, 2000). Long-lasting bloodstream infections in woodrats provide infectious blood meals to feeding ticks. In Europe, the pathogen appears to be transmitted to humans by the bites of *I. ricinus* (Petrovec *et al.*, 1999).

*E. ewingii* was recently detected in the granulocytes of human patients from Missouri in 1999 (Buller *et al.*, 1999). Since then, additional cases have been reported, primarily in immunocompromised patients. The etiologic agent has been described from several southern states where it was known to be a cause of illness in dogs. Experimental evidence has shown adult *A. americanum* to be an efficient vector, and recent studies have detected the organism in this tick species (Wolf *et al.*, 2000). The reservoir for the pathogen is not known but is suspected to be wild or domestic canines.

## Q FEVER

First recognized among livestock handlers in Australia in 1935, Q fever is now known to occur on four other continents (Europe, Asia, Africa, and North America) and is probably worldwide in distribution. The etiologic agent, *Coxiella burnetii*, is a bacterium that develops in the phagolysosomes of the cytoplasm of susceptible cells. *C. burnetii* can survive for months or years outside host cells under environmental conditions that are lethal to other rickettsiae. It can survive in dried tick feces, dried or frozen tissues, soil, and water.

After an incubation period of about 20 days, Q fever is characterized by sudden onset of fever, chills, sweats, diarrhea, sore throat, painful sensitivity to light, muscle pain, and headache. Fever may persist for 2 weeks and show a biphasic pattern. Fatigue, enlargement of the liver, and inflammation of the lungs accompanied by a mild cough and chest pain occur frequently. A rash is usually absent; when present, it appears on the trunk and shoulders. Q fever may become chronic, in which case it causes inflammation of the lining of the heart and its valves. The case fatality rate is less than 1% in acute cases but may rise to 30% in chronic cases.

Transmission by ticks was first reported in 1938. Both argasid and ixodid ticks have been found infected naturally with *C. burnetii*. Subadult ticks infected while feeding on a rickettsemic host develop a generalized infection in their tissues. Following the transstadial molt, nymphs or adults transmit *C. burnetii* by the bite, and females can pass the organism *transovarially*. Argasid ticks also can disseminate the organism via *infectious coxal fluids*. Notably, *C. burnetii* can survive in contaminated tick feces for as long as 6 years, which facilitates spread to humans and domestic animals.

*C. burnetii* is maintained in enzootic cycles involving domestic animals (e.g., sheep, cattle, goats), wildlife, and their associated ticks. For example, a parasitic cycle exists among Australian kangaroos (*Macropus major* and *M. minor*), the marsupial bandicoot (*Isoodon torosus*), and their associated host-specific ticks. Transmission to cattle and humans occurs when the wild mammals also are parasitized by the nonspecific *Ixodes holocyclus*. In mammals, infection is usually asymptomatic, but abortions sometimes occur. Small mammals (e.g., *Apodemus*, *Microtus*, *Clethrionomys*, *Arvicola*, and *Pitymys*) living in and around agricultural communities may link the domestic and feral cycles. These animals develop high rickettsemias and shed the organism in their feces for weeks after becoming infected. Dogs, cats, birds, and reptiles also are susceptible to infection and may play a role in maintaining the infection in natural habitats.

Although ticks are important in maintaining the pathogen horizontally and vertically in enzootic cycles, they rarely transmit the infection to humans by the bite. Instead, persons who handle infected animals or their products, or materials contaminated by tick feces, are at increased risk of acquiring *C. burnetii*. Tick excreta are an important source of infection because they are often highly contaminated and easily aerosolized. However, aerosols emanating from afterbirth membranes and associated fluids, blood, urine, feces, nasopharyngeal discharges, and milk containing high concentrations of the organism constitute the most common means for spreading the infection. As these materials dry, *C. burnetii* can be spread in aerosolized dust and debris present in animal stalls, barns, store-rooms, and similar facilities. The most common site of Q fever epidemics is on farms or in farming communities, usually when domestic animals are being handled, such as during wool-shearing, lambing, calving, and slaughtering. Milk and milk products are a particularly important means of disseminating *C. burnetii* to humans; the organism may survive in contaminated milk and butter for up to 3 months.

## LYME DISEASE

Lyme disease, also known as *Lyme borreliosis, erythema chronicum migrans, Bannwarth's syndrome, tick-borne meningopolyneuritis,* and other names, is a tick-borne *spirochetosis* caused by at least four species of closely related bacteria, *Borrelia burgdorferi* sensu stricto (s.s.), *B. afzelii, B. bissettii,* and *B. garinii*. Six additional members of the *B. burgdorferi* sensu lato (s.l.) complex have been described thus far, none of which appears to infect humans. First recognized as a new form of *inflammatory arthritis* in the environs of Old Lyme, CT, in the mid-1970s, Lyme disease or related disorders have been reported from most states in the United States and in Canada and various European and Asian countries. Suspected human cases of the disease have been reported from Australia, South America, and Africa, although *B. burgdorferi* s.l. has not been isolated from vertebrates or ticks from these continents. In the United States, more than 128,000 cases have been reported since 1982, when systematic national surveillance was initiated. Lyme disease is the most common vector-borne disease in the United States, accounting for more 95% of all such illnesses, with an average annual incidence of almost 5 per 100,000 population (Dennis, 1998). A total of 16,273 cases were reported in 1999. It is also the most common vector-borne infection afflicting humans in temperate regions of the world. Although the etiologic agent was not discovered until 1981 (Burgdorfer *et al.*, 1982), human cases have been documented in the medical literature since the early 20th century (Afzelius, 1921).

*B. burgdorferi* s.s. normally develops extracellularly in *Ixodes* ticks (Fig. 24.13), although spirochetes have been found occasionally in oocytes of the ovaries and in secretory cells of the salivary glands (Burgdorfer *et al.*, 1989). When injected into humans by a feeding tick, spirochetes multiply and disseminate in the skin. Gradually, they invade the blood-stream and may spread throughout the body, often localizing in the bursae of the large joints, in the heart, and in the nervous system. Clinical signs and symptoms usually appear within 1 to 2 weeks following the bite of an infectious tick. Most cases occur during the late spring or summer, coincident with the seasonal activity of the nymphal stages of the primary vectors

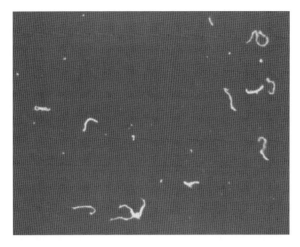

**FIGURE 24.13** Lyme disease spirochete, *Borrelia burgdorferi*, in a smear prepared from midgut of adult tick (*Ixodes pacificus*). (Photo by R. S. Lane.)

(see below). *I. persulcatus* is an exception; the adult stage of this Eurasian tick seems to be the primary life stage that transmits spirochetes to humans.

Early-stage Lyme disease is characterized by nonspecific constitutional ("flu-like") symptoms and an erythematous skin rash, *erythema migrans* (EM), which is present in 60–80% of patients. EM is a slowly expanding, usually circular or elliptical but sometimes triangular lesion that often exhibits bright red outer margins and partial central clearing (Fig. 24.14). Most patients have one EM lesion at the site of tick attachment, but 25–50% may develop multiple satellite lesions. The rash should not be confused with typical tick-bite lesions, especially when the mouthparts break off in the wound.

Untreated patients may manifest no further signs or symptoms of illness, or they may go on to develop

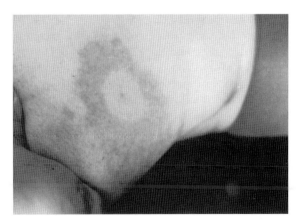

**FIGURE 24.14** Erythema migrans (EM) skin lesion in human, common in early stage of Lyme disease. (Courtesy of Ross Ritter, Potter Valley, CA.)

*late stage Lyme disease* within 1 month to several months. Late manifestations include, either alone or in combination, *cardiac, neurologic, arthritic,* or *further dermatologic abnormalities.*

Over 40 species of ixodid ticks and 1 species of argasid tick have been found infected naturally with *B. burgdorferi* or closely related spirochetes. In most endemic foci, however, a member of the *I. ricinus complex* serves as the primary vector to people. Thus, spirochetes are transmitted to humans by *I. scapularis* (formerly *I. dammini* in part; Oliver *et al.,* 1993) and *I. pacificus* in eastern and western North America, respectively; by *I. ricinus* in Europe and western Asia; and by *I. persulcatus* in eastern Europe and Asia. Other *Ixodes* ticks that seldom bite humans and do not belong to the *I. ricinus* complex, such as *I. spinipalpis* (*I. neotomae*) in the western United States, *I. dentatus* in the eastern United States, and *I. ovatus* in Japan, may serve as efficient *enzootic* (maintenance) *vectors* of *B. burgdorferi* or related spirochetes. Ticks in other genera occasionally serve as secondary vectors to people. For example, *A. americanum* in New Jersey, United States, has been implicated even though several experimental studies have shown that it is an incompetent vector of certain isolates of *B. burgdorferi.*

In *Ixodes* immatures, development of *B. burgdorferi* begins with ingestion of an infectious blood meal. Spirochetes accumulate in the midgut diverticula, where they multiply by binary fission. Following the transstadial molt and refeeding on another host, spirochetes escape from the midgut, enter the hemocoel, and migrate to the salivary glands (Zung *et al.,* 1989). In some ticks, borreliae spread to other organs as well. Thus, spirochetes are maintained within populations of vector ticks by *transstadial passage* and by replenishment as noninfected ticks feed on infectious hosts. *Transovarial transmission* has been documented for each of the primary vectors, but it appears to be an inefficient process for perpetuating and distributing certain borrelial *genospecies* (e.g., *B. burgdorferi* s.s. by *I. pacificus* and *I. scapularis*).

In addition to ticks, spirochetes have been detected in mosquitoes, deer flies, and horse flies in the northeastern United States, and anecdotal accounts in this region and in Europe suggest that some individuals may acquire spirochetal infections following the bites of these bloodsucking insects. Although the overall role of insects in the ecology of *B. burgdorferi* and related spirochetes appears to be minimal, further investigation is warranted.

Wherever the ecology of *B. burgdorferi* has been studied intensively, one or more species of rodents or insectivores have been implicated as primary *reservoir hosts.* Reservoirs are those vertebrates that are commonly infected with the spirochete, that maintain spirochetes within their tissues for prolonged periods, if not for life, that have a significant amount of contact with vector ticks,

**FIGURE 24.15**   Dusky-footed wood rat (*Neotoma fuscipes*), a reservoir host of the Lyme disease agent *Borrelia burgdorferi* sensu lato in the far western United States. (Courtesy of R. J. Keiffer, Hopland, CA.)

and that readily infect ticks that feed on them. Competent reservoirs include the white-footed mouse, dusky-footed woodrat (Fig. 24.15), California kangaroo rat, and several other rodents in North America and such animals as the common shrew, bank vole, wood mouse, yellow-necked field mouse, hares, and pheasants in Europe. In some geographic regions, certain species of birds play a significant enzootiologic role by providing populations of immature ticks with blood meals, by infecting vector ticks with spirochetes, and by transporting infected ticks considerable distances, thereby establishing new foci of infection. On the other hand, certain species of vertebrates that are excellent hosts of vector ticks, such as white-tailed deer and western fence lizards (Fig. 24.16) in North America, are *incompetent reservoirs* and therefore reduce tick infection prevalences and the risk of human exposure to spirochetes. In the eastern United States, white-tailed deer are

**FIGURE 24.16** Western fence lizard (*Sceloporus occidentalis*), a primary host of *Ixodes pacificus* larvae and nymphs. (Courtesy of Jack Kelly Clark, University of California, Davis.)

transmission of a zoonotic disease agent to humans and other animals. More recently, the mechanism responsible for the borreliacidal activity in preimmune sera from the western fence lizard and the southern alligator lizard was demonstrated to reside in the alternative complement pathway. In Europe, complement-mediated borreliacidal effects observed for specific combinations of vertebrate–host serum and different genospecies of *B. burgdorferi* s.l. generally coincided with what was known about the reservoir competence of the mammalian and avian species evaluated (Kurtenbach *et al.,* 1998).

In the United States, Lyme disease is most prevalent in the northeastern states, especially in New York, Pennsylvania, New Jersey, and southern New England (Fig. 24.17). New York alone has reported more than 4000 cases per year in recent years, representing approximately 25–30% of all reported cases in the United States. Other major regional foci occur in the upper Midwest, especially in Wisconsin and Minnesota, and in northern California. In the northeastern United States, people living in close proximity to forests or in suburban communities having a mosaic patchwork of wooded areas and homes have the highest risk of exposure to spirochete-infected ticks. White-tailed deer thrive in these habitats and, consequently, *I. scapularis* abounds. Moreover, infection rates in *I. scapularis* nymphs and adults are high. In one study in New York, 30% of nymphal and 50% of adult ticks were infected with *B. burgdorferi* (Maupin *et al.,* 1991).

In northern California, cases are most likely to occur in rural areas and wild lands, where *I. pacificus* is abundant (Fig. 24.18). However, infection prevalences in that

important as *amplifying hosts* of *I. scapularis,* especially the adults, even though they are unable to infect ticks.

Besides being an incompetent host for *B. burgdorferi* in the far western United States, the western fence lizard was found to contain a heat-labile, *spirochete-killing (borreliacidal) factor* in its blood (Lane and Quistad, 1998). This factor destroys spirochetes in the midgut diverticula of infected nymphs, while they feed on lizards with the result, after the transstadial molt, that the adult ticks are spirochete-free. Eliminating spirochetal infections from vector ticks such as *I. pacificus* reduces the force of

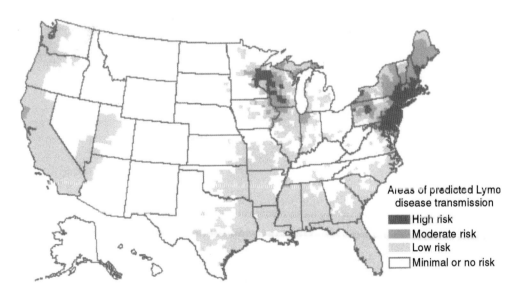

**FIGURE 24.17** Approximate distribution of four categories of predicted risk of humans acquiring Lyme disease in the United States. (Courtesy of Centers for Disease Control and Prevention.)

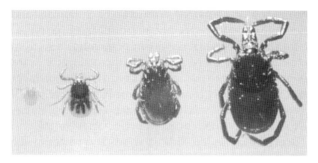

**FIGURE 24.18** Relative sizes of larva, nymph, and adult male and female (left to right) of the western blacklegged tick (*Ixodes pacificus*). (Courtesy of former Scientific Photographic Laboratory, University of California, Berkeley.)

tick (typically 1−2% in adults and 2−15% in nymphs) are generally much lower than they are in northern populations of *I. scapularis,* and the risk of infection to humans is correspondingly lower. Cases occur throughout the year but are most often seen in spring and early summer, when the nymphs reach peak densities. By contrast, *I. pacificus* adults are largely active in fall and winter, when temperatures are cool and humidities are high (Fig. 24.19).

In Europe, Lyme disease is most prevalent in northern countries such as Sweden, Germany, Austria, and Switzerland. Relatively few cases have been reported

**FIGURE 24.19** A host-seeking (questing) female of the western blacklegged tick (*Ixodes pacificus*). (Photo by R. S. Lane, University of California, Berkeley.)

from Britain or Ireland, although *I. ricinus,* the primary European vector, is widespread in these islands. The reason for the paucity of human cases in Great Britain is not clearly understood.

Collectively, several ecological and epidemiological factors have contributed to the current epidemic of Lyme disease in the northeastern United States and Europe. They are the occurrence of an abundant, efficient tick vector in these regions, the presence of numerous natural hosts for the immature and adult stages of the vectors, high rates of spirochetal infection in reservoir populations, and close proximity of susceptible human populations to populations of vector ticks.

## TICK-BORNE RELAPSING FEVER

Relapsing fever has been known since ancient times. Early descriptions confused tick-borne and louse-borne relapsing fever. Transmission by ticks was not recognized until the pioneering work of Dutton and Todd (1905), who detected spirochetes in *Ornithodoros moubata* from east Africa. Tick-borne relapsing fever (*tick-borne spirochetosis, endemic relapsing fever*) is caused by about 14 species of *Borrelia,* each of which is typically associated with 1 species of *Ornithodoros* tick. Six additional species of borreliae associated with individual *Ornithodoros* ticks are of unknown or little human health significance. Thus, the high *vector specificity* of the relapsing fever spirochetes stands in marked contrast to the lower vector specificity of the Lyme disease spirochetes, *B. burgdorferi* s.l.

The 14 relapsing fever spirochetes of known public health importance and their associated vector ticks are *B. duttonii/O. moubata* in central, eastern, and southern Africa; *B. hispanica/O. erraticus* (large variety) in North Africa and southwestern Europe; *B. crocidurae, B. merionesi, B. microti* ,and *B. dipodilli/O. erraticus* (small variety) in Africa, the Near East, and Central Asia; *B. persica/O. tholozani* from northeast Africa and Asia; *B. caucasica/O. verrucosus* from the Caucasus Mountains of the former Soviet Union to Iraq; *B. latyschewii/O. tartakovskyi* from central Asia and Iran; *B. hermsii/O. hermsi, B. parkeri/O. parkeri,* and *B. turicatae/O. turicata* from the western or southwestern United States; *B. mazzottii/O. talaje* from the southern United States, Mexico, and Central and South America; and *B. venezuelensis/O. rudis* from Central and South America (Burgdorfer and Schwan, 1991).

Onset of the disease in humans is characterized by fever, chills, and a throbbing headache, usually without a pronounced rash or an ulcer at the bite site. The episode of fever usually lasts about 3 to 5 days, during which time spirochetes are present in the peripheral blood. Symptoms recur after several days without fever. This alternating cycle of febrile and afebrile periods may be repeated two or more times, which extends the illness for several weeks. Other signs or symptoms that often occur are muscle aches, joint pain, abdominal pain, nausea, vomiting, diarrhea, and a petechial rash on the trunk.

The relapsing nature of the disease is explained by the *antigenic variability* of the borreliae. That is, some spirochetes are able to alter their protein composition, probably through transposition of the genes encoding them. Consequently, new populations of spirochetes emerge in an infected host and multiply before the host can mount an effective antibody response against them.

In their tick vectors, borreliae ingested with blood disseminate to the internal organs, including the salivary glands. Spirochetes are passed transstadially so that once infected, ticks remain so for life. This, together with the long life span of *Ornithodoros* ticks, enhances the likelihood that relapsing fever spirochetes will persist in tick-infested habitats for prolonged periods. *Transovarial transmission* of spirochetes has been demonstrated in *O. moubata, O. erraticus, O. tholozani, O. tartakovskyi, O. verrucosus, O. turicata,* and *O. hermsi,* but not in *O. parkeri, O. talaje,* or *O. rudis* (Burgdorfer and Schwan, 1991). Borreliae also invade the coxal glands of some *Ornithodoros* spp. and can be transmitted to hosts via *coxal fluid* excreted during or soon after the blood meal.

Primary *reservoir hosts* of nearly all borreliae transmitted by *Ornithodoros* ticks are rodents. Two notable exceptions are *Borrelia duttonii* and *B. coriaceae.* Humans are the reservoir host of *B. duttonii,* the cause of *East African tick-borne* or *endemic relapsing fever.* Columbian black-tailed deer and possibly other subspecies of mule deer are apparently reservoirs of *B. coriaceae,* the putative agent of *epizootic bovine abortion* (EBA) in the far western United States. Limited clinical and serologic evidence suggests that *B. coriaceae* may occasionally infect humans in California.

Epidemiologically, tick-borne relapsing fever is a highly focal disease. In the western United States and Mexico, *O. turicata* inhabits caves and the burrows and nests of rodents, terrapins, and snakes; it also infests homes, slaughterhouses, and pigsties, where it is likely to encounter humans. Another North American vector is *O. hermsi,* which infests cavities in dead trees, fallen logs, log cabins, and even homes in mountainous areas, where it feeds on various rodents. People become infected when they are bitten indoors by *O. hermsi* at night, especially if rodent hosts have been driven out or eliminated through house-cleaning or rodent-trapping or -poisoning activities. In 1973, the largest outbreak of relapsing fever ever recorded in North America was attributed to infection with *Borrelia*-infected *O. hermsi* ticks. This episode involved 62 persons who slept overnight in rustic cabins at the North Rim, Grand Canyon National Park, Arizona (Boyer *et al.,* 1977).

Another vector of relapsing fever spirochetes in western North America is *O. parkeri,* which typically occurs at lower elevations than *O. hermsi.* However, *B. parkeri* rarely infects humans because *O. parkeri* ticks inhabit the burrows of rodents and therefore seldom have an opportunity to feed upon people.

*O. erraticus,* a parasite of rats, mice, gerbils, and other small mammals, is an important vector in North Africa, whereas *O. tholozani* is the primary vector in the Middle East, the Balkans, and the southern part of the former Soviet Union. *O. tholozani,* an indiscriminate feeder, is common in human shelters, caves, rocky overhangs, and other situations where livestock are housed. In southern Africa, *O. moubata* infests mud and thatch huts used by native peoples, surviving in cracks and crevices in the walls, bed posts, and other locations. While feeding, infected ticks excrete spirochetes in coxal fluid onto the open feeding wound. In such situations, ticks and humans maintain longterm, intimate contact. Recent changes in home construction, including use of modern building materials and larger homes divided into different rooms, have greatly reduced tick infestations and, concomitantly, the incidence of relapsing fever.

## TULAREMIA

Tularemia was first identified as a distinct clinical entity in Japan in 1837, where it was attributed to infected hares. It was first recognized in the western United States in 1911 by McCoy, who described it as "a plague-like disease of rodents." The organism was detected in tissues of the California ground squirrel. In the United States, it has long been associated with hunters who acquire the infection while skinning wild rabbits (hence the colloquial name, "rabbit fever") and with people who handle infected livestock, especially sheep. Workers in the former Soviet Union recognized this disease among trappers handling "water rats" and established that it was caused by the same bacterium. For excellent overviews of tularemia in North America, see Jellison (1974) and Bell (1988).

The causative agent, *Francisella tularensis,* is a pleomorphic, gram-negative, aerobic bacterium. It exists in different forms, termed *biovars.* A biovar is an infrasubspecific rank that is applied to strains having special biochemical or physiological properties. Two biovars of *F. tularensis* occur in North America: a highly virulent form associated with rabbits, sheep, and ticks (biovar *tularensis*), and an apparently waterborne, less virulent form associated with beavers, muskrats, and voles (biovar *palaearctica*). Biovar *tularensis* is known only from North America and is fatal in 5–7% of untreated patients; biovar *palaearctica* occurs in Asia, Europe, and North America and is rarely fatal to humans. A third biovar, *mediaasiatica*, from Central Asia, has been proposed.

*F. tularensis* is transmitted by many blood-feeding arthropods besides ticks, including deer flies, mosquitoes, and fleas, and by handling infected animals, inhaling contaminated dust, or ingesting contaminated water or insufficiently cooked meat. Patients infected with *F. tularensis* experience fever, headache, and nausea, which are usually accompanied by development of an ulcerated lesion at the site of inoculation. Other clinical manifestations include enlargement of regional lymph nodes, pneumonia, and occasionally a rash. Although seven clinical types have been recognized, the ulceroglandular form accounts for about 80% of cases (Jellison, 1974). The pneumonic form is particularly prone to produce severe illness; if left untreated, it may persist for 2 to 3 months and become chronic thereafter.

Many species of ixodid ticks have been found infected naturally with *F. tularensis*. In North America alone, this agent has been detected in or isolated from at least 13 species of ixodids in 4 genera (1 *Amblyomma* sp., 5 *Dermacentor* spp., 2 *Haemaphysalis* spp., and 5 *Ixodes* spp.). Ticks are considered by some workers to be reservoirs of *F. tularensis* or to be at least part of a multi-host reservoir system along with their primary vertebrate hosts. Transstadial passage of the agent occurs in vector ticks, but earlier claims that transovarial transmission also occurs have not been confirmed.

Although tularemia is mainly an infection of wild lagomorphs and rodents, natural infections have been reported in numerous species of wild and domestic mammals and birds. Fish, frogs, and toads have been found infected occasionally.

Epizootics are spread mainly by water or ticks. Waterborne epizootics occur in western Siberia, where lakes and other fresh-water habitats are contaminated by infected dead and dying muskrats and microtine rodents (voles). Fur trappers and others who handle carcasses of infected water rats or muskrats, or water contaminated by them, have contracted the disease in large numbers. *I. apronophorus* transmits *F. tularensis* among water rats, furthering the spread of the disease along shorelines.

In parts of North America, *F. tularensis* is acquired by hunters as they skin freshly killed rabbits. Thus, lagomorphs were identified as the source of infection in over 80% of the cases of tularemia acquired in California between 1927 and 1951; of these, jack-rabbits were implicated in 71% of the cases where a distinction was made between jackrabbits and cottontail rabbits. On the other hand, ticks are much more significant than lagomorphs as a source of human infection in the south central United States, especially in Arkansas and Missouri, where the lone star tick, *A. americanum*, is the primary vector (Jellison, 1974).

## TICK PARALYSIS

Tick paralysis is a host reaction to compounds secreted in the saliva of feeding ticks. This malady has been reported from North America, Europe, Asia, South Africa, and eastern Australia. It was first reported in Australia in 1824 by William Howell, who described a tick "which buries itself in the flesh and would in the end destroy either man or beast if not removed in time" (cited in Harwood and James, 1979).

The affliction is characterized by a progressive, *flaccid, ascending paralysis*. In humans, it usually begins in the legs with muscle weakness and loss of motor coordination and sensation. Paralysis gradually extends to the trunk, with loss of coordination in the abdominal muscles, back muscles, and eventually the inter-costal muscles of the chest. Paralysis of the latter may lead to death from respiratory failure. During advanced stages, the patient may be unable to sit up or move the arms and legs, and chewing, swallowing, and speaking may become difficult. The condition progresses rapidly, and death may ensue within 24 to 48 hr after the onset of symptoms. In North America, the case fatality rate is about 10% in the Pacific Northwest, and most of those who die are children (Gregson, 1973). In Washington state alone, 33 cases were reported between 1946 and 1996, mostly in children less than 8 years old (Dworkin *et al.*, 1999). In most cases, symptoms abate within hours following detection and removal of the offending tick or ticks, and recovery may be complete within 48 hr. If paralysis has progressed too far, complete recovery may take up to 6 weeks. In contrast, paralysis induced by the Australian tick, *I. holocyclus*, may worsen after tick removal and full recovery may take up to several weeks.

The nature of the toxin(s) causing tick paralysis has not been determined for most species of ticks. Intensive research on the salivary components of *I. holocylcus* has revealed the existence of a protein, termed *holocyclotoxin*, that produces paralytic symptoms. A different salivary gland protein has been implicated as the toxin in female *D. andersoni*.

Typically, only female ticks can induce tick paralysis, but nymphs of the *Karoo paralysis tick I. rubicundus* can cause paralysis in laboratory rabbits, and larvae of the soft tick *Argas walkerae* produce a toxic protein that can paralyze chickens. Female ticks must be attached to the host for several days (usually about 4 to 6) before they begin secreting the toxins. For an excellent review of the mechanisms of pathology among the tick paralyses, see Gothe *et al.* (1979).

Forty-six species of ticks in 10 genera reportedly cause tick paralysis in humans, other mammals, and birds. Worldwide, the ticks of greatest concern are *I. holocyclus*, *I. rubicundus*, *D. andersoni*, and *D. variabilis*.

In North America, tick paralysis in humans and domesticated animals has been reported most often from the Pacific Northwest in the United States and British Columbia in Canada. Cases in humans and other animals occur in spring and early summer, coincident with the activity period of adult *D. andersoni*. In the eastern United States, cases in dogs and humans are caused by *D. variabilis*, whereas in California this tick has been associated only with paralysis in dogs.

In contrast to paralysis produced by *Dermacentor* species or *I. holocyclus*, where feeding by one female is sufficient to cause this condition, severity of the disease in South Africa is directly related to the number of attached *I. rubicundus*.

In England and France, human paralysis has been attributed infrequently to *I. hexagonus*. In California, several mild cases have been ascribed to *I. pacificus* prior to the 1950s, but these earlier observations were not well documented and no new cases have been reported since then.

## TICK-BITE ALLERGIES

The bites of many species of ticks can cause host reactions other than paralysis. These range from minor, localized, *inflammatory reactions* that subside soon after tick removal to severe, *systemic reactions* involving skin rash, fever, nausea, vomiting, diarrhea, shock, and death. Severe toxic or allergic reactions may follow the bites of the soft ticks *A. brumpti, A. reflexus, O. coriaceus, O. moubata,* and *O. turicata*. In Europe, severe reactions and even loss of consciousness have occurred following attacks by the pigeon tick, *A. reflexus,* which infests buildings where pigeons roost. This tick is particularly prone to bite people if the birds have been driven away. In the far western United States, reports that bites of the *pajaroello tick* (*O. coriaceus*) are more feared than those of rattlesnakes seem exaggerated, but severe allergic reactions characterized by edema, pain, erythema, tissue necrosis, ulceration, and prolonged healing have been documented.

*I. pacificus* sometimes causes severe allergic reactions and, in rare cases, *anaphylactic shock* in persons previously sensitized to its bite (Van Wye *et al.,* 1991). Likewise, individuals bitten by *I. holocyclus* may experience anaphylactic reactions involving tick-specific immunoglobulin E.

# VETERINARY IMPORTANCE

Ticks are of veterinary concern mostly because of the many microbial disease agents they transmit to livestock, companion animals, and wildlife. These include arboviruses, bacteria, rickettsiae, and protozoans. Ticks also are important because of the debilitating and sometimes fatal host reactions produced in domesticated livestock and companion animals as a result of the feeding activities of certain species (tick toxicosis, tick paralysis). Moreover, wildlife as well as livestock may suffer from exsanguination, leading to anemia and death. In Oklahoma and Texas, for example, significant mortality in white-tailed deer fawns has been associated with heavy infestations of *Amblyomma americanum*. Other species routinely or incidentally invade the auditory canal of bovines or other mammals, a condition known as otoacariasis, which may be accompanied by serious secondary bacterial infections. Livestock and poultry that are heavily infested with ticks may experience economically significant reductions in body weight, egg production, or general unthriftiness. Wounds caused by tick bites may attract myiasis-producing flies, which in turn may reduce the value of the hide, fleece, or carcass. The veterinary impact of ticks has been reviewed by Mahoney (1977), Williams *et al.* (1985), Lancaster and Meisch (1986), McDade (1990), Norval *et al.* (1992), and others.

As in the preceding section, no attempt was made to cover all tick-borne diseases of veterinary importance; instead, representative diseases of major concern have been included. Further, tick-borne diseases affecting wildlife are largely omitted because they are beyond the scope of this chapter.

## BABESIOSIS

In cattle, the disease produced by *Babesia* species, especially *B. bovis, B. divergens,* and *B. bigemina,* is severe. Animals develop high temperatures, cease feeding, and become dehydrated. As babesiosis progresses, the animals become lethargic and eventually lapse into a coma and sometimes die. Considerable variation in severity has been noted in different geographical regions, which is attributable to differences in virulence among the parasites as well as variation in susceptibility of diverse cattle breeds. Nursing animals are protected by passive immunity acquired from antibodies in the colostrum of immune cows. Babesiosis also affects other domestic animals and numerous species of wildlife.

The most important vectors of *bovine babesiosis* are species of the genus *Boophilus*, especially *Boophilus microplus* and *B. annulatus*. In some areas of Africa, *B. decoloratus* also is an important vector. Although eradicated from the United States in the early decades of the 20th century, *B. annulatus* and *B. microplus* still occur in Mexico and stringent controls are maintained to prevent their reintroduction. Moreover, white-tailed deer are potential hosts for *B. annulatus* and could serve as a wild reservoir for this vector tick. Relict populations of *B. annulatus* are believed to exist in some areas of Florida (Sonenshine, 1993).

Dogs are often victims of *canine babesiosis,* caused primarily by *B. canis.* The most important vector is the dog tick, *Rhipicephalus sanguineus,* although *D. marginatus* and *D. reticulatus* may be important locally as vectors. Babesiosis also affects other animals, such as horses and deer.

## EAST COAST FEVER

ECF is a disease of cattle and domestic buffalo caused by *T. parva* (family Theileridae), an intracellular protozoan parasite. The disease, which has been known from East Africa since the 19th century, is widespread in central and southern Africa as well.

An estimated 25 million cattle are at risk for acquiring ECF. Infected animals develop enlarged lymph glands, become listless, stop feeding, and cough frequently. Fever occurs a few days after onset of symptoms, followed by diarrhea and mucous discharges from the eyes and nose. Mortality may exceed 90% in adult animals but is usually much less in calves. *T. parva* causes a more severe illness in some cattle, termed *corridor disease,* which is caused by organisms derived from wild buffalo and transmitted to cattle by ticks. In other cattle, it produces a much milder illness known as *January disease.* Earlier workers considered these two clinical entities to be caused by separate subspecies of *T. parva,* but recent molecular studies do not support this classification. The two conditions are now believed to represent different expressions of the same disease.

Other *Theileria* spp. cause disease in livestock and wildlife in different regions of the world. One of the most important is *T. annulata,* the etiologic agent of *tropical theileriosis,* a somewhat milder disease than ECF. Tropical theileriosis is transmitted by species of *Hyalomma,* especially *Hyalomma anatolicum anatolicum.* It is distributed widely throughout North Africa, southern Europe, the Middle East, and elsewhere in central Asia, India, and China. Domestic cats are prone to suffer from a theilerial disease caused by *T. felis (Cytauxzoon felis).* In the United States, this disease, known as *feline cytauxzoonosis,* is often fatal in untreated animals. *D. variabilis* is believed to be the primary vector. Another host of *T. felis* is the bobcat, in which it apparently does not cause illness.

The life cycles of *Theileria* species are generally similar to those of *Babesia* species, with some notable differences. Following ingestion of host blood by larval or nymphal ticks, gamete formation and fertilization occur in the tick's midgut, after which the parasites invade the midgut epithelium. The resulting *kinetes* migrate exclusively to the salivary glands so that transmission occurs solely by the bite, not transovarially. After the transstadial molt, the resultant nymphal or adult ticks may transmit the infection to another host.

In vertebrates, the parasites undergo multiplication (*schizogony*) within lymphoid tissue or lymphocytes instead of in the erythrocytes, as do most *Babesia* species. The resulting schizonts proliferate in lymphoid cells that have been transformed, the so-called *macroschizonts.* Lymphocytes with large schizonts, commonly known as *Koch's blue bodies,* appear a few days after onset of symptoms. Later, lymphocytes infected with microschizonts appear. The numerous *merozoites* released from these cells invade erythrocytes, where they develop into *piroplasms* that will later form *gametes* when ingested by feeding ticks.

The primary vector of *T. parva* is *R. appendiculatus,* a tick whose geographical distribution coincides largely with that of ECF throughout much of eastern and southern Africa. Corridor disease occurs outside the known range of *R. appendiculatus* but within that of another competent vector, *R. zambesiensis.* The occurrence of ECF coincides with the distribution of its primary wild *reservoir,* the cape buffalo and, to a lesser extent, with that of secondary reservoirs such as the waterbuck. Movement of cattle has played a major role in the periodic outbreaks of ECF during the 20th century. Outbreaks tend to occur when highly susceptible breeds (e.g., taurine breeds) are introduced into areas endemic for *T. parva.*

## LOUPING ILL

Although known to sheep herders in Scotland for centuries, louping ill was not recognized as a distinct clinical entity until 1913, and its viral etiology was not established until 1931. It has long been regarded as a disease of sheep, but it occasionally infects cattle, horses, pigs, and humans, often with severe or fatal consequences. Louping ill is caused by a flavivirus that is antigenically similar to other members of the *TBE complex* and is the only member of the complex present in the British Isles. A disease similar to or identical with louping ill occurs sporadically in several other European countries.

Infected sheep lose their appetites and become feverish. About the 5th day after onset, the fever rises and the animals become uncoordinated and develop tremors. Seriously ill animals walk with an awkward, erratic, "louping" gait, hence the name of the disease. Many sheep die shortly after locomotor signs appear, but others develop a chronic condition that may persist for several weeks. Mortality may reach 100% in some susceptible herds. Recovered animals often show signs of permanent *neurologic damage*.

Historically, the disease has been most prevalent in areas of unimproved pastures and moorlands where *I. ricinus* is abundant. The virus is passed transstadially in ticks but not transovarially. Field and laboratory studies suggest that the vector competence of *I. ricinus* for louping ill virus is not high. For example, in enzootic areas of northern Britain, only 0.1 to 0.4% of ticks are infected with the virus.

Although *I. ricinus* has a multitude of vertebrate hosts, tick populations inhabiting sheep rangeland are supported principally by this animal. Moreover, sheep exposed to louping ill virus develop high viremias and are infective to ticks for several days. Other vertebrates, both domestic and wild, that may contribute to the maintenance of the virus are cattle, goats, and several species of ground-inhabiting birds, such as red grouse, willow grouse, and ptarmigan. These species serve as hosts to *I. ricinus* and occasionally develop viremias high enough to infect feeding ticks.

## AFRICAN SWINE FEVER

This disease was first recognized in Kenya in 1921 as a catastrophic illness that killed 99% of infected pigs. Since then, sporadic epidemics of African swine fever have been reported from many African countries south of the Sahara, in Europe, and in the western hemisphere in Cuba, Haiti, the Dominican Republic, and Brazil.

The disease is caused by a large *icosahedral DNA virus* (family Iridoviridae) that attacks cells of the reticuloendothelial system, especially monocytes. The host range of the virus is limited solely to domestic pigs, European wild boars, warthogs, and bush pigs, all species of the family Suidae. In its acute form, animals develop fever about 3 days post-infection, the fever persists for 3 or 4 days, the temperature drops, and death ensues. In its subacute form, an irregular fever lasts for 3 or 4 weeks, whereupon the animals either recover or die. In its chronic form, animals may survive for long periods before succumbing to a secondary illness, most commonly pneumonia. Chronically infected animals usually experience stunted growth and emaciation and serve as long-term reservoirs of the virus.

The primary *vectors* in Africa and indeed the only proven natural vectors are *O. moubata porcinus* and *O. erraticus*. In the New World, *O. coriaceus*, *O. turicata*, and other *Ornithodoros* species have been demonstrated to be competent experimental vectors. The occurrence of several competent vectors in this region has raised concerns about the potential spread of African swine fever virus into North America. Besides tick transmission,

susceptible pigs may acquire the infection by direct contact with, or by ingesting the secretions or excretions of, infected animals. This mode of transmission, if unchecked, can result in rapid dissemination of the virus among herds of domestic swine.

## EHRLICHIOSIS

The diseases caused by members of the genus *Ehrlichia* have long been known in veterinary medicine. These pathogens infect the leukocytes and platelets of livestock, companion animals, and wildlife. The ehrlichiae grow as distinct microcolonies, or *morulae*, within the cytoplasm of host cells. The disease manifestations caused by these agents can range from asymptomatic to fatal.

*Canine ehrlichiosis* due to *E. canis* was first recognized in 1935. The cosmopolitan distribution of this pathogen corresponds with that of its primary vector, *R. sanguineus*. In the United States, serosurveys of civilian and military dogs have revealed that the disease is present in most states. Dogs infected with *E. canis* develop fever, conjunctivitis, and swelling of various tissues. The disease causes a reduction in the blood cell number and has been referred to as canine tropical pancytopenia. Infected animals stop eating, lose weight, and frequently appear depressed. Acute infection is often followed by a debilitating chronic phase accompanied by anemia and sometimes hemorrhagic bleeding. The German shepherd breed is particularly susceptible to acute, severe illness. These animals suffer from low white blood cell counts and damage to the lymph glands, bone marrow, and spleen. Animals with severe infection usually die without antibiotic treatment. In other breeds, the disease is often milder and ranges from asymptomatic to chronically symptomatic.

Transstadial transmission of *E. canis* occurs in *R. sanguineus*, and transstadially infected nymphal or adult ticks can transmit *E. canis* to susceptible dogs. Transovarial transmission occurs rarely, if at all (Groves *et al.*, 1975). Dogs apparently serve as the primary reservoir of *E. canis* because inapparent infections can persist for over 5 years, and the agent is continually present in chronically infected dogs (Groves *et al.*, 1975).

Infection by *E. ewingii* causes a disease in dogs known as *canine granulocytic ehrlichiosis*. The pathogen grows within the neutrophils of infected animals and was once erroneously thought to be an atypical form of *E. canis*. The infection is usually mild and may manifest as *polyarthritis*. Although its actual distribution may be wider, *E. ewingii* is primarily found in the southern United States. The agent is passed from nymphal to adult *A. americanum*, and transstadially infected adults can transmit the infection to susceptible dogs (Anziani *et al.*, 1990). Naturally infected lone star ticks have been identified recently in North Carolina (Wolf *et al.*, 2000).

*E. platys* infects canine platelets and causes their numbers to fluctuate over time. This infection is rather mild and is often diagnosed as *canine cyclic thrombocytopenia.* The tick vector(s) of this species have not been determined. The species has been identified in the southern United States, Greece, Taiwan, and Japan.

In Great Britain and elsewhere in Europe, *I. ricinus* transmits the causative agent of *tick-borne fever, E. phagocytophila,* to sheep, cattle, and goats. Weight loss, reduced milk production, and abortion can occur in infected animals. The organism infects granulocytes and induces a marked immunosuppression that may predispose animals to secondary infections and reduce their antibody response to vaccination. In the United States and Europe, *E. phagocytophila* may cause a mild illness in dogs. As has been noted for the other *Ehrlichia* species, *E. phagocytophila* appears to be passed transstadially, but not transovarially, in the tick vector. A number of other ehrlichial species have been detected in ruminants in various parts of the world but will be omitted from this discussion.

## Q FEVER

*Coxiella burnetii,* the rickettsia that causes Q fever, is widespread in populations of domestic livestock and wildlife, but infection in mammals other than humans is typically benign. For that reason, the disease is treated in the public health section. As noted above, however, *C. burnetii* is a veterinary pathogen because it occasionally induces abortion in pregnant cows and ewes. Since infected animals excrete large numbers of *C. burnetii* in their waste and in birthing tissues and fluids, they pose a significant health risk for animal handlers.

## HEARTWATER

Heartwater is an ehrlichial disease of large ungulates (livestock and game) that is caused by *Cowdria ruminantium.* The disease occurs primarily in sub-Saharan Africa, especially in eastern and southern regions, and in Madagascar and several Caribbean islands.

Infected cattle develop fever, become disoriented, and show signs of motor disorder, especially abnormal walking and muscle twitching. Within a few days, animals appear listless and exhibit loss of appetite despite constant chewing. As the illness progresses, they develop convulsions and die soon afterwards. Dead and dying animals show a massive accumulation of fluids in the pericardium, the membrane surrounding the heart, and edema in the lungs, brain, and other organs. Some animals survive and become immune. Calves are protected by maternal antibodies transferred in milk and, when infected, develop a mild illness or none at all. The disease is quite severe in sheep and goats, although the motor symptoms are usually not as prominent as they are in cattle.

The agent of heartwater is transmitted by at least seven species of *Amblyomma* ticks, all of which have indiscriminate feeding habits. In view of its enormous geographic range and adaptability to varying climatic conditions, *A. variegatum* is the primary vector in most enzootic areas of Africa and the Caribbean. *A. hebraeum,* which is usually found in cooler climates, is an important vector in southern Africa. Other *Amblyomma* species are important locally as vectors to livestock and wild ungulates. *C. ruminantium* is maintained transstadially within populations of ticks, and the infection is usually transmitted by the bite of females. Ticks remain infected for long periods, possibly for life. Although transovarial transmission has been reported, it appears to be an inefficient process for maintaining *C. ruminantium.*

In semiarid regions, livestock and wildlife congregate around the few water sources remaining after the rainy season. Thickets and other dense vegetation provide them with shelter against predators as well as habitat for roosting birds. Larval and nymphal ticks feed readily on such hosts, and the high humidity present in sheltered environments enhances tick survival. Consequently, large populations of vector ticks often occur in and around such foci, which promotes disease transmission. After a focus has persisted for many years, a stable level of transmission known as *endemic stability* may ensue. Introduction of exotic livestock species that lack immunity to heartwater disrupts this stability and often leads to epidemics of the disease. Similarly, rapid resurgences of *Cowdria*-infected tick populations following drought also can lead to devastating epidemics. Moreover, endemic stability can be disrupted by excessive use of pesticides, which destroys the natural herd immunity that results from constant, low-level challenge by small numbers of infected ticks.

## ANAPLASMOSIS

Anaplasmosis was first described in South Africa in 1910 by Theiler, who identified and named the primary agent 1 year later. This parasite, *Anaplasma marginale,* and two related species, *A. centrale* and *A. ovis,* infect red blood cells of cattle and sheep throughout much of the world. Anaplasmosis is now considered one of the most important diseases of livestock. *A. marginale* is a pleomorphic, coccoid rickettsia that occurs in membrane-bound inclusions called colonies in the cytoplasm of infected erythrocytes.

Onset is abrupt following a lengthy incubation period of about 3–6 weeks. Common clinical manifestations include fever of several days' duration, labored breathing, frequent urination, constipation, and a hemolytic anemia. Animals usually recover when infected with strains of mild virulence, but 30 to 50% of those infected with highly virulent strains may die. Moreover, the severity of the disease and case fatality rates increase with age. Cattle that recover from acute anaplasmosis maintain a persistent, low-level parasitemia which protects them from reinfection and contributes to their importance as reservoirs.

Approximately 20 species of ixodid ticks serve as vectors of *Anaplasma* spp. (Kocan *et al.*, 1992). In North America, the primary vectors are *D. andersoni, D. occidentalis,* and *D. variabilis;* in Europe and Africa, they are *R. sanguineus, R. simus, R. bursa, B. microplus, B. annulatus, B. decoloratus,* and *I. ricinus.* Blood-sucking flies in the family Tabanidae (horse flies and deer flies) have been implicated as mechanical vectors. *A. marginale* undergoes a complex developmental cycle in ticks involving five morphological forms. Details of the life cycle have been elucidated in *D. andersoni* and are presumably representative of the parasite's development in its other tick vectors (Kocan *et al.*, 1989). Further, the genetic and morphologic characteristics of *Anaplasma* species and their development within ticks are similar to those of *C. ruminantium* and *Ehrlichia* species in their tick vectors. The classification of these organisms is being revised to reflect these relationships.

*A. marginale* is passed transstadially, but not transovarially, within ticks. Female *D. andersoni* infected as nymphs begin transmitting the infection by the 6th day of feeding on a susceptible host, whereas male ticks that acquire infection as adults can transmit the pathogen within 24 hr. Male ticks are of considerable enzootiologic importance because they feed repeatedly, they readily transfer between hosts that are in close contact, and therefore they are capable of transmitting *A. marginale* to multiple hosts.

Cases occur frequently on farms located adjacent to tick-infested woodlands. In the eastern United States, the presence of white-tailed deer is considered a risk factor because this cervid is an important host of vector ticks. However, white-tailed deer are not competent reservoirs of *A. marginale* and therefore cannot serve as a source of infection for noninfected ticks. In the western United States, not only are mule deer primary hosts of vector ticks, but they are also competent *reservoirs* of *A. marginale.* The two regions having the highest prevalence of anaplasmosis in the United States are the Southeast and the Pacific Northwest. Overall losses attributable to the disease in the United States were estimated at US$300 million in 1986.

## BORRELIOSES

Borrelioses are diseases of birds and mammals caused by spirochetes in the genus *Borrelia.* Important tick-borne borrelioses include *avian spirochetosis, Lyme disease, tick spirochetosis,* and possibly EBA. *Avian spirochetosis* is a highly fatal disease of turkeys, pheasants, geese, doves, chickens, and canaries in Europe, Africa, Siberia, Australia, Indonesia, and India and North, Central, and South America. It causes severe losses to the poultry industry in certain countries. Infected birds develop high fever and diarrhea and become cyanotic. Birds that survive develop a long-lasting immunity. *Argas persicus* and related ticks (subgenus *Persicargas*) transmit the etiologic agent, *Borrelia anserina,* via infectious tick feces. *Transovarial transmission* occurs and ticks can remain infective for 6 months or longer.

Dogs, cats, cattle, horses, and possibly sheep can be infected with the etiologic agent of *Lyme disease, B. burgdorferi* s.l. This disease or related disorders have been reported from numerous countries on five continents: Africa, Asia, Europe, and North and South America. Earlier claims that *B. burgdorferi* is present in Australia have not been confirmed by recent field and laboratory studies, nor have isolates been obtained in Africa or South America. Populations of domestic dogs living in areas highly endemic for *B. burgdorferi* can have seroprevalence rates as high as 76%, but relatively few seropositive animals manifest overt clinical signs. Indeed, 24–53% of apparently healthy dogs in some areas have significant antibody titers to *B. burgdorferi.* In hyperendemic foci of the northeastern and far western United States, the most commonly observed clinical manifestations among dogs are lameness, inappetence, fever, and fatigue.

Lyme disease has been reported in cattle and horses in the northeastern and upper mid-western United States. In the Delaware River valley, serum antibodies against *B. burgdorferi* have been detected in about 10% of the horses tested and in 60% of the animals from one farm. Arthritis, edema, and dermatitis have been observed in some of these animals.

*Tick spirochetosis* is a benign disease of cattle, sheep, and horses that occurs in South Africa and Australia (Burgdorfer and Schwan, 1991). Infected *Boophilus* ticks also have been identified along the United States/Mexico border. Infected animals experience one or two attacks of fever, loss of appetite, weight loss, anemia, and weakness. The causative agent, *B. theileri,* is transmitted by *Rhipicephalus* spp. and probably other ixodid ticks.

EBA is a major disease of rangeland cattle in the far western United States, particularly in California. In this state alone, it has been estimated that annual calf losses due to EBA range from 5 to 10% at a cost of US$5 to $15 million. Circumstantial evidence suggested that a spirochete, *B. coriaceae,* detected in all three stages of the soft tick *O. coriaceus* may be causally related to EBA (Lane et al., 1985). This spirochete is passed transstadially and transovarially and also is excreted in coxal fluid once replete ticks have detached from a host.

## TULAREMIA

Tularemia, caused by the bacterium *F. tularensis,* is primarily a disease of rodents and lagomorphs (rabbits and hares) in the Northern Hemisphere. Although best known for its public health importance (see above), *F. tularensis* also is a veterinary pathogen that can cause devastating epizootics in domestic sheep. Reliable reports of epizootics in North America date back to 1923, when serious losses occurred in eastern Montana and southern Idaho (Jellison, 1974). These outbreaks shared several features: animals were put on rangeland enzootic for *F. tularensis* in early spring, they grazed in sagebrush areas where *D. andersoni* ticks were abundant, and they became heavily infested with ticks. *D. andersoni* is a competent vector of virulent strains of *F. tularensis.* As many as 50% of a herd may become sick and 10% may die within a few days. Diseased animals that survive such outbreaks may lose weight and condition during their illnesses. Sheep usually are quite resistant to infection with *F. tularensis,* but reduced vitality of animals after a long winter, shortage of feed, and exposure to early spring storms and heavy infestations of ticks predispose flocks to epizootics. Over 14,000 cases of tularemia in sheep were recorded in the United States between 1923 and 1945. In contrast, only about 40 cases had been reported in sheep from Canada by 1945.

Epizootics of tularemia among sheep constitute a risk factor for humans. Jellison and Kohls (1955) presented records of 189 human cases of tularemia associated with the sheep industry in the United States. Of these, 66 cases occurred among sheep shearers. Other individuals found to be at risk were sheep owners and herders and housewives.

Horses infected with *F. tularensis* can become ill or die. The agent was isolated from two foals that died of an acute illness in Montana in 1958. Three other horses (one mare and two foals) from the same locality became ill but recovered and had high agglutination titers to *F. tularensis.* All five horses were infested abundantly with ticks. Notably, one of the veterinarians involved in the study became ill with tularemia.

In North America, upland game birds are sometimes infected with *F. tularensis.* The agent has been isolated from ruffed grouse, sharp-tailed grouse, sage grouse, bobwhite quail, and pheasant. An epizootic in sage grouse was reported from central Montana in which the birds were heavily parasitized by the ixodid tick *Haemaphysalis chordeilis.* Likewise, severe epizootics in domestic turkeys infested with the same tick have been reported. The significance of these sporadic outbreaks is that they occasionally expose hunters and veterinarians to infection with *F. tularensis.* For example, among 375 cases of tularemia reported by the Minnesota Department of Health between 1926 and 1946, the source of infection in 15 cases was attributed to birds as follows: pheasant, 10; partridge, 2; grouse, 2; and pheasant or partridge, 1 (Jellison, 1974).

## TICK PARALYSIS

The first reports of tick paralysis in livestock originated in Australia in 1890 and in British Columbia, Canada, in 1912. This condition is most common in livestock and pets and causes injury or death to thousands of animals each year. Tick paralysis has been reported from many countries in Europe, Asia, Africa, and North America.

In South Africa, *I. rubicundus* is responsible for annual losses totaling tens of thousands of sheep and up to 9% of game animals in some areas. Other animals affected include goats, cattle, and species of wild antelope. Induction of paralysis by the *Karoo paralysis tick I. rubicundus* is directly related to the number of ticks feeding on a host. Stock losses between 1983 and 1986 averaged nearly 29,000 animals per year of which 91% were sheep. Stock farmers regard tick paralysis as one of the most important problems affecting their operations. The disease occurs in hill rangeland or mountainous terrain covered with a "Karoo" type of vegetation, which is grassy areas interspersed with shrubs or trees. Although *I. rubicundus* parasitizes many wild mammals, only antelopes are known to develop paralysis.

Another tick that paralyzes sheep and goats in South Africa, although not as severely or as often as *I. rubicundus,* is *R. evertsi evertsi.* This species has been recognized as a cause of tick paralysis since 1900. The induction, duration, and severity of the paralysis is related to the number of female ticks that have engorged to body weights of 15 to 21 mg.

In North America, three species of *Dermacentor* ticks cause paralysis in companion animals and livestock. In the eastern and western United States, *D. variabilis* is a common cause of tick paralysis in domestic dogs. In the Sierra Nevada foothills of northern California, for example, an average of six cases was seen in two veterinary practices during a 1-year investigation (Lane et al., 1984). Dogs were infested with a mean of 32 ticks; 98% were *D. variabilis* adults. Another tick from the same region, *D. occidentalis,* is responsible for occasional cases of tick paralysis in cattle, ponies, and deer, but not in dogs.

The most important species of paralysis-inducing tick in North America is *D. andersoni,* which was responsible for paralyzing more than 3800 sheep and cattle in the Pacific Northwest (British Columbia, Washington, Idaho, and Montana) between 1900 and the early 1970s. Individual outbreaks have involved up to 320 animals, with cases occurring most frequently from April to June when adult *D. andersoni* activity is greatest. In Australia, *I. holocyclus* induces paralysis in dogs and humans. This tick inhabits a narrow zone along the eastern coast of Queensland and Victoria. Drugs administered along with hyperimmune serum to dogs with advanced paralysis have improved their chances for full recovery. Among several drugs tested, phenoxybenzamine hydrochloride, an *α* adrenergic blocking agent, has been found to be most effective.

In Europe and Asia, tick paralysis is scattered widely. In Macedonia and Bulgaria, paralysis in sheep, goats, and cattle has been attributed to *Hyalomma punctata,* and in Crete and the former Yugoslavia and the former Soviet Union, livestock are sometimes paralyzed by the bites of *I. ricinus.*

Birds are susceptible to tick paralyses also. Larvae of *A. walkerae* induce paralysis in chickens in South Africa. A toxic fraction isolated from replete larvae of this argasid tick consists of two proteins having ''membranophilic'' properties and molecular masses of 32 and 60 kDa; extracts containing these proteins induced paralysis in 1-day-old chicks. In the southeastern United States, a number of species of wild birds, especially passeriforms, are paralyzed by *I. brunneus* females.

## TICK TOXICOSES

Toxic reactions have been associated with the bites of certain species of ticks, notably argasids. In Africa, cattle bitten by *O. savignyi* may die of toxicosis in just 1 day. Sheep attacked by *O. lahorensis* in eastern Europe and in the southern region of the former Soviet Union may tremble, gnash their teeth, exude frothy saliva, experience paralysis, and sometimes perish. In Europe, *A. persicus* can cause leg weakness in ducks and geese; this condition resembles a true toxicosis and not a paralysis. A toxic illness that affects cattle, goats, other livestock, and even dogs in large areas of central, eastern, and southern Africa, India, and Sri Lanka is known as *sweating sickness*. Wild hosts include eland, antelope, and zebra. The illness, which begins four or more days after tick attachment, is characterized by fever, loss of appetite, sweating, lachrymation, and salivation, but no paralysis. Approximately 75% of afflicted animals die. The active principle in the African tick, *H. truncatum,* is a salivary gland protein.

## PREVENTION AND CONTROL

Historically, control of ticks and tick-borne diseases almost always was accomplished with pesticides, called *acaricides,* to kill the ticks and drugs to kill the infectious agents. The cattle tick *Boophilus annulatus* was eradicated in the United States by dipping cattle in pesticide solutions, thereby eliminating the deadly Texas cattle fever. Quarantine, pasture rotation, and elimination of deer also were used in the effort to eradicate this vector. Graham and Hourrigan (1977) estimated that reintroduction of cattle ticks (*Boophilus* spp.) could cost the United States cattle industry more than US$1 billion/year to achieve eradication again. Costs to the worldwide cattle industry were estimated at more than US$7 billion. Damage to other livestock and valuable wildlife is much more difficult to estimate. Losses due to human illnesses have never been calculated.

Treatment with acaricides provides the most widely used means to control or prevent tick attacks. Promising alternatives, such as *vaccines* or *pheromone/acaricidal treatments,* are being investigated but none is available for broad-scale use. For further information, the reader should consult Drummond *et al.* (1988).

### PERSONAL PROTECTION

Preventive measures are the most effective means for protecting people who enter tick-infested habitats. People should wear boots, socks, long trousers, and light-colored clothing. Trousers should be tucked into the boots, socks drawn over trousers, and the socks taped to form a tight seal. The clothing should be treated with a *repellent* or *acaricide.* Exposed skin also should be treated with repellents or acaricides suitable for use on humans. Each person should conduct *self-examinations* for ticks during and after exposure to tick-infested areas. Early *removal of attached ticks* is important in minimizing the risk of contracting tick-borne diseases. Ticks should be removed by grasping the capitulum as close to the skin as possible with a pair of fine forceps and gently pulling the tick with a slow, steady force until its mouthparts release their hold. Turning or twisting the tick should be avoided to prevent the hypostome breaking off in the wound.

The most widely used personal protectant is the repellent *N*, *N*-diethyl-m-toluamide (DEET), available as a lotion or a spray. Applications should be repeated as needed to maintain maximum protection but should be applied cautiously on children to avoid adverse reactions that occasionally follow overuse. In addition, permethrin is effective when applied to clothing before entering tick-infested habitats. However, permethrin should not be applied to bare skin.

## ACARICIDES

Acaricides are pesticides used to kill ticks and mites. Acaricides include *chlorinated hydrocarbons* (e.g., dichlorodiphenyltrichloroethane; DDT), *organophosphorus compounds* (e.g., Diazinon), *carbamates* (e.g., carbaryl), *pyrethroids* (e.g., permethrin, flumethrin), *formamidines*, and *avermectins*. The pyrethroids are among the safest and most effective pesticides and are now widely used for tick control.

One way to kill ticks on host animals is to dip livestock and pets in a pesticide bath. When used for cattle, this is termed a cattle dip. *Dipping* is not always effective. Often, ticks hidden in sheltered locations (e.g., between the toes or in the ears) are missed and survive to lay eggs and re-establish the pest population. Acaricides also can be delivered as sprays, using high-pressure sprayers to provide a mist that can reach every part of the animal's body. They can also be delivered as *pour-ons* or *spot-ons*. These are formulations in which the acaricide is mixed with surfactants to spread the liquid over the animal's hair coat. Finally, they may be applied as *dusts*, in which acaricides are mixed with talc and deposited directly onto the animal's fur. The familiar "flea powders" for pets, which are effective against ticks as well as fleas, and the dust bags used for treating cattle are examples of acaricidal dusts.

To achieve long-lasting efficacy, acaricides can be incorporated into plastic or other suitable matrices that provide a slow release of the toxicant over a period of weeks or months. *Plastic collars*, such as the familiar flea and tick collars, are widely used for control of ticks on cats and dogs. Similarly, acaricide-impregnated plastic ear tags are widely used for control of ear-infesting ticks (*Gulf Coast tick* and *spinose ear tick*) on cattle and other large domestic animals. However, they are much less effective for control of ticks that attach around the groin, udder, and other parts of the hindquarters of these animals. *Systemic acaricides* offer another means of providing long-lasting and effective tick control. In this case, the toxicant is introduced into the host's blood to kill ticks as they feed on the treated animals. Unfortunately, most acaricides are too toxic to administer to animals systemically. An exception is *ivermectin*, which can provide excellent control of certain ticks on cattle for 2–3 months (Drummond *et al.,* 1988).

The development of *acaricide resistance* by ticks is a continuing concern. Ticks have been found to be resistant to cyclodiene insecticides, chlorinated hydrocarbons, organophosphorus insecticides, and pyrethroids. Resistance to these acaricides is not universal. Resistance may occur in one or two species, while other species in the same locality remain acaricide-susceptible. Recently, cattle ticks in central Queensland were found resistant to all of the pyrethroids currently in use. Resistance of cattle ticks to pyrethroids and organophosphorus compounds has also been found in Mexico and poses concerns regarding the possible re-establishment of these ticks into the United States. Continued research is necessary to discover and develop new pesticide products to overcome resistance to compounds already in use.

## PHEROMONE-ASSISTED CONTROL

The difficulties and high cost of tick control on animals have stimulated interest in alternatives to the conventional methods described above. Such alternatives help to reduce the use of acaricides. Research with tick pheromones suggests that combinations of pheromones and acaricides can be significantly more effective for controlling ticks than the acaricide alone, because ticks are unlikely to develop resistance to their own pheromones. A pheromone/acaricide combination applied to a single spot on cattle can be effective in killing the *Gulf Coast tick*. Another promising device is the "tick decoy" in which the sex pheromone 2,6-dichlorophenol and an acaricide are impregnated into plastic beads, on the surface of which "mounting" sex pheromone is smeared (Hamilton and Sonenshine, 1989). Male ticks are attracted to decoys on the animal's hair coat and killed. This also disrupts mating activity, so that any surviving females cannot lay viable eggs. For the livestock-parasitizing *bont ticks* (*Amblyomma hebraeum* and *A. variegatum*) a tail-tag decoy was developed that uses a mixture of tick-specific phenols to attract the ticks to specific sites on cattle and kills them when they attach nearby. Field trials with the tail-tags have demonstrated excellent efficacy for up to 3 months (Norval *et al.,* 1996).

## PASSIVE TREATMENT

Another way to apply acaricides to animals is by means of self-treating devices. Animals seeking food or nesting materials visit these devices and acquire an acaricide, spreading it over their fur and skin to kill ticks. An example is the biodegradable cardboard tube containing a *permethrin-impregnated cotton ball* (Mather *et al.,* 1987). Mice collect the cotton for nesting material, thereby spreading the pesticide among nest mates. Such tubes have been effective in reducing populations of *I. scapularis* and the occurrence of Lyme disease in some localities, especially in residential communities; however, they have not been effective in other situations. Another example is the *self-treating tick applicator* for controlling blacklegged ticks (*I. scapularis*) on white-tailed deer (Sonenshine *et al.*, 1996). Animals become coated with oil containing an acaricide as they remove food from the applicator. A similar technique was used in Zimbabwe for treating wild ungulates (Duncan and Monks, 1992).

## HORMONE-ASSISTED CONTROL

Hormones and other *insect growth regulators* such as methoprene also have been used to disrupt tick development in laboratory experiments. *Analogs* or *mimics of ecdysteroids* and juvenile hormone are effective in killing ticks by delaying their development, disrupting oviposition, or killing the larvae when they hatch from eggs deposited by treated females. However, these compounds do not appear to be uniformly effective against all types of ticks.

## VACCINES

Anti-tick vaccines also have been investigated. In Australia, studies are in progress to develop a vaccine for control of the cattle tick *B. microplus*. A recombinant antigen vaccine has been tested and there are plans for its commercial development. Although it is possible that antigen-resistant strains of cattle ticks may ultimately appear, large-scale vaccination of cattle herds with the recombinant vaccine offers a promising alternative to acaricides.

## MANAGEMENT

Management practices provide another means of reducing tick numbers. Many acaricides can be applied directly to vegetation. However, because ticks commonly occur in microhabitats covered by vegetation, leaf litter, soil, and other natural materials, or in the nests, burrows, and other cavities used by their hosts, they often do not come into direct contact with these toxicants. Therefore, to be effective, the acaricides must reach the ticks as vapors or by contact when the ticks move about while seeking hosts. Public opposition to treatment of natural habitats with pesticides has made it unpopular to use this form of tick control except for the most compelling reasons. In recent years, acaricidal treatment of natural areas has been limited largely to military bases or selected recreational areas. Alternatives include *habitat modifications,* such as *burning* or *clearing vegetation,* or *host removal* (e.g., removal of deer by hunting or deer exclusion fences). Burning or clearing vegetation removes the dense cover under which

ticks shelter, thereby reducing ground-level humidity as well as exposing them to intense ultraviolet radiation and heat. Such changes can make the habitat unsuitable for tick survival. Management practices integrating cattle-breed resistance, rotational grazing, and selected use of acaricides have proved successful in controlling populations of *B. microplus* in Australia.

*Tick surveys* often are conducted to determine whether or not tick control is warranted and, if so, when it should be implemented. The most common method for sampling ticks is the use of a *flag* or *drag* cloth pulled or dragged through the vegetation. Ticks collected on the cloth are counted as the number of a given species per unit of distance dragged (e.g., 100 m) or the number collected per hour of dragging. Although absolute measures of tick population densities can not be obtained in this manner, the relative abundance of ticks in different areas sampled can be determined. An alternative to dragging is the use of *carbon dioxide traps.* Carbon dioxide gas from a block of dry ice or from a compressed-gas cylinder is the tick attractant. Ticks adhering to or crawling around the trap are counted after a few hours of operation. When more reliable estimates of tick abundance are required, a mark-and-recapture method can be used. With this method, the numbers of ticks recaptured from a previous sample are compared with the number of unmarked ticks to obtain an estimate of the entire tick population in the area studied.

For tick-infested cattle, horses, mules, and other livestock, a time-tested method is the *scratching technique,* whereby livestock inspectors pass their hands over different regions of the animal's body to detect attached ticks. A similar technique is used in combination with *visual inspection* to examine wild animals. For example, investigators can be assigned to deer check stations during hunting season to count all ticks on hunter-killed animals. Another technique for sampling ticks is to trap small- and medium-sized wild animals and hold them over trays filled with water or alcohol to catch fed ticks as they detach.

## ERADICATION

In a few cases, tick eradication may be practicable. An example is the Cattle Tick Fever Eradication Program that was initiated in the southern United States in 1907. This program led to the eradication of *B. annulatus*, *B. microplus*, and Texas Cattle Fever from the United States by 1960. Attempts to eradicate *B. microplus* in other areas (US Virgin Islands, Argentina, Uruguay, Australia, and Papua New Guinea) have been unsuccessful, despite reductions of over 99% of the tick populations in some localities. Reinvasion by ticks and their rapid repopulation of areas in the eradication zone are major contributing factors. Attempts to eradicate *A. variegatum,* a vector of the agent of heartwater and other livestock diseases, from islands in the Caribbean similarly have been unsuccessful. Enzootic stability, rather than eradication, is the preferred method of controlling tick-borne diseases of livestock in several African countries (Norval *et al.,* 1992). Occasionally, however, eradication has been achieved in the case of exotic species which were recognized soon after their introduction to a new area. An example includes the eradication of *R. evertsi* soon after it was introduced into a wild animal compound in Florida. Eradication is easiest if exotic species are identified as soon as possible after their introduction. This was addressed following the discovery of *A. marmoreum* and *A. sparsum,* both vectors of heartwater, on nine different premises in Florida (Burridge *et al.,* 2000).

# REFERENCES AND FURTHER READING

Afzelius, A. (1921). Erythema chronicum migrans. *Acta Dermato-Venereologica* **2,** 120–125.

Anderson, B. E., Sumner, J. W., Dawson, J. E., Tzianabos, T., Greene, C. R., Olson, J. G., Fishbein, D. B., Olsen-Rasmussen, M., Holloway, B. P., and Azad, A. F. (1992). Detection of the etiologic agent of human ehrlichiosis by polymerase chain reaction. *Journal of Clinical Microbiology* **30,** 775–780.

Anderson, B. E., Sims, K. G., Olson, J. G., Childs, J. E., Piesman, J. F., Happ, C. M., Maupin, G. O., and Johnson, B. J. B. (1993). *Amblyomma americanum:* a potential vector of human ehrlichiosis. *American Journal of Tropical Medicine and Hygiene* **49,** 239–244.

Anziani, O. S., Ewing, S. A., and Barker, R. W. (1990). Experimental transmission of a granulocytic form of the tribe Ehrlichieae by *Dermacentor variabilis* and *Amblyomma americanum* to dogs. *American Journal of Veterinary Research* **51,** 929–931.

Arthur, D. R. (1962). "Ticks and Disease." Pergamon, Oxford.

Arthur, D. R. (1963). "British Ticks." Butterworths, London.

Arthur, D. R. (1965). "Ticks of the Genus *Ixodes* in Africa." Athlone Press, London.

Balashov, Y. S. (1972). Bloodsucking ticks—(Ixodoidea)—vectors of disease of man and animals (English translation). *Miscellaneous Publications of the Entomological Society of America* **8,** 163–376.

Bell, J. F. (1988). Tularemia. In "CRC Handbook Series in Zoonoses. Section A. Bacterial, Rickettsial and Mycotic Disease" (J. H. Steele, Ed.), Vol. 2, pp. 161–193. CRC Press, Boca Raton, FL.

Boyer, K. M., Munford, R. S., Maupin, G. O., Pattison, C. P., Fox, M. D., Barnes, A. M., Jones, W. L., and Maynard, J. E. (1977). Tick-borne relapsing fever: an interstate outbreak originating at Grand Canyon National Park. *American Journal of Epidemiology* **105,** 469–479.

Buller, R. S., Arens, M., Hmiel, S. P., Paddock, C. D., Summer, J. W., Rikihisa, Y., Unver, A., Gaudreault-Keener, M., Manian, F. A., Liddell, A. M., Schmulewitz, N., and Storch, G. A. (1999). *Ehrlichia ewingii,* a newly recognized agent of human ehrlichiosis. *New England Journal of Medicine* **341,** 148–155.

Burgdorfer, W. (1977). Tick-borne diseases in the United States: Rocky Mountain spotted fever and Colorado tick fever. *Acta Tropica* **34,** 103–126.

Burgdorfer, W., and Schwan, T. G. (1991). Borrelia. In "Manual of Clinical Microbiology" (A. Balows, W. J. Hausler, Jr., K. L. Herrman, H. D. Isenberg, and H. J. Shadomy, Eds.), 5th ed., pp. 560–566. American Society of Microbiologists, Washington, DC.

Burgdorfer, W., Barbour, A. G., Hayes, S. F., Benach, J. L., Grunwaldt, E., and Davis, J. P. (1982). Lyme disease—a tick-borne spirochetosis? *Science* **216,** 1317–1319.

Burgdorfer, W., Hayes, S. F., and Corwin, D. (1989). Pathophysiology of the Lyme disease spirochete, *Borrelia burgdorferi,* in ixodid ticks. *Reviews of Infectious Diseases* **11**(Suppl. 6), S1442–S1450.

Burger, D. B., Crause, J. C., Spickett, A. M., and Neitz, A. W. H. (1991). A comparative study of proteins present in sweating-sickness-inducing and non-inducing strains of *Hyalomma truncatum* ticks. *Experimental & Applied Acarology* **13,** 59–63.

Burridge, M. J., Simmons, L. A., and Allan, S. A. (2000). Introduction of potential heartwater vectors and other exotic ticks in Florida on imported reptiles. *Journal of Parasitology* **86,** 700–704.

Calisher, C. H. (1994). Medically important arboviruses of the United States and Canada. *Clinical Microbiology Reviews* **7,** 89–116.

Camus, E., and Barre, N. (1995). Vector situation of tick-borne diseases in the Caribbean Islands. *Veterinary Parasitology* **57,** 167–176.

Clifford, C. M., Anastos, A., and Elbl, A. (1961). The larval ixodid ticks of the eastern United States. *Miscellaneous Publications of the Entomological Society of America* **2,** 213–237.

Clifford, C. M., Sonenshine, D. E., Keirans, J. E., and Kohls, G. M. (1973). Systematics of the subfamily Ixodinae (Acarina: Ixodidae). I. The subgenera of *Ixodes. Annals of the Entomological Society of America* **66,** 489–500.

Cooley, R. A., and Kohls, G. M. (1944). "The Argasidae of North America, Central America and Cuba." American Midland Naturalist, Monograph No. 1. Univ. of Notre Dame Press, Notre Dame.

Cooley, R. A., and Kohls, G. M. (1945). "The Genus *Ixodes* in North America." National Institute of Health Bulletin No. 184. US Govt. Printing Office, Washington, DC.

Cooney, J. C., and Hays, K. L. (1972). "The Ticks of Alabama (Ixodidae: Acarina)." Alabama Agriculture Experiment Station, Bulletin No. 426, Auburn University, Auburn, AL.

Dennis, D. T. (1998). Epidemiology, ecology and prevention of Lyme disease. In "Lyme Disease" (D. W. Rahn and J. Evans, Eds.), pp. 7–34. American College of Physicians, Philadelphia.

Drummond, R. O., Whetstone, T. M., Ernst, S. E., and Gladney, W. J. (1969). Biology and colonization of the winter tick in the laboratory. *Journal of Economic Entomology* **62,** 235–238.

Drummond, R. O., George, J. E., and Kunz, S. E. (1988). "Control of Arthropod Pests of Livestock: A Review of Technology." CRC Press, Boca Raton, FL.

Duncan, I. M., and Monks, N. (1992). Tick control on eland (*Taurotragus oryx*) and Buffalo (*Syncerus caffer*). *Journal of the South African Veterinary Association* **63,** 7–10.

Durden, L. A., and Keirans, J. E. (1996). "Nymphs of the Genus *Ixodes* (Acari: Ixodidae) of the United States: Taxonomy, Identification Key, Distribution, Hosts and Medical-Veterinary Importance." Thomas Say Foundation Monographs, Entomological Society of America, Lanham, MD.

Dutton, J. E., and Todd, J. L. (1905). The nature of tick fever in the eastern part of the Congo Free State, with notes on the distribution and bionomics of the tick. *British Medical Journal* **2,** 1259–1260.

Dworkin, M. S., Shoemaker, P. C., and Anderson, D. E. (1999). Tick paralysis: 33 human cases in Washington state, 1946–1996. *Clinical Infectious Diseases* **29,** 1435–1439.

Fairchild, G. B., Kohls, G. M., and Tipton, V. J. (1966). The ticks of Panama (Acarina: Ixodoidea). In "Ectoparasites of Panama" (R. L. Wenzel and V. J. Tipton, Eds.), pp. 167–219. Field Museum of Natural History, Chicago.

Filippova, N. A. (1966). Argasid ticks (Argasidae) (in Russian). Fauna SSSR, Paukoobraznye 4(3).

Filippova, N. A. (1967). Ixodid ticks of the subfamily Ixodinae (in Russian). Fauna SSSR, Paukoobraznye. 4(4).

Furman, D. P., and Loomis, E. C. (1984). The ticks of California (Acari: Ixodida). *Bulletin of the California Insect Survey* **25**, 1–239.

George, J. E. (1971). Drop-off rhythms of engorged rabbit ticks, *Haemaphysalis leporispalustris* (Packard, 1896) (Acari: Ixodidae). *Journal of Medical Entomology* **8**, 461–479.

Gothe, R., Kunze, K., and Hoogstraal, H. (1979). The mechanisms of pathogenicity in the tick paralyses. *Journal of Medical Entomology* **16**, 357–369.

Graham, O. H., and Hourrigan, J. L. (1977). Eradication programs for the arthropod parasites of livestock. *Journal of Medical Entomology* **13**, 629–658.

Gray, J. S. (1991). The development and seasonal activity of the tick *Ixodes ricinus:* a vector of Lyme borreliosis. *Review of Medical and Veterinary Entomology* **79**, 323–333.

Gregson, J. D. (1973). Tick paralysis: an appraisal of natural and experimental data. Monograph No. 9. Canada Department of Agriculture, Ottawa.

Groves, M. G., Dennis, G. L., Amyx, H. L., and Huxsoll, D. L. (1975). Transmission *of Ehrlichia canis* to dogs by ticks (*Rhipicephalus sanguineus*). *American Journal of Veterinary Research* **36**, 937–940.

Hamilton, J. G. C., and Sonenshine, D. E., inventors; Center for Innovative Technology, assignee. (1989). "Methods and Apparatus for Controlling Arthropod Populations." US patent 4,884,361.

Harwood, R. F., and James, M. T. (1979). "Entomology in Human and Animal Health," 7th ed. Macmillan Co., New York.

Hillyard, P. D. (1996). "Ticks of Northwest Europe." Natural History Museum, London.

Hoogstraal, H. (1956). "African Ixodoidea. I. Ticks of the Sudan (with Special Reference to Equatoria Province and with Preliminary Reviews of the Genera *Boophilus, Margaropus,* and *Hyalomma*)." Department of the Navy, Bureau of Medicine and Surgery, Washington, DC.

Hoogstraal, H. (1980). The roles of fleas and ticks in the epidemiology of human diseases. In "Proceedings of the International Conference on Fleas" (R. Traub and H. Starcke, Eds.), pp. 241–244. Balkema, Rotterdam.

Hoogstraal, H. (1985). Argasid and nuttallielid ticks as parasites and vectors. *Advances in Parasitology* **24**, 135–238.

Hoogstraal, H., and Aeschlimann, A. (1982). Tick host specificity. *Bulletin de la Societe Entomologique Suisse* **55**, 5–32.

Jellison, W. L. (1974). "Tularemia in North America 1930–1974." University of Montana Foundation, Missoula, MT.

Jellison, W. L., and G. M. Kohls. (1955). Tularemia in sheep and sheep industry workers in western United States. *Public Health Monographs* **28**, 1–17.

Jones, E. K., Clifford, C. M., Keirans, J. E., and Kohls, G. M. (1972). The ticks of Venezuela (Acarina: Ixodoidea) with a key to the species of *Amblyomma* in the Western Hemisphere. *Brigham Young University Science Bulletin Biological Series* **174**, 1–40.

Jones, L. D., Hodgson, E., Williams, T., Higgs, S., and Nuttall, P. A. (1992). Saliva-activated transmission (SAT) of Thogoto virus: relationship with vector potential of different haematophagous arthropods. *Medical and Veterinary Entomology* **6**, 261–265.

Keirans, J. E. (1992). Systematics of the Ixodidae (Argasidae, Ixodidae, Nuttalliellidae); an overview and some problems. In "Tick Vector Biology: Medical and Veterinary Aspects" (B. Fivaz, T. Petney, and I. Horak, Eds.), pp. 1–21. Springer-Verlag, Berlin.

Keirans J. E., and Clifford, C. M. (1978). The genus *Ixodes* in the United States: a scanning electron microscope study and key to the adults. *Journal of Medical Entomology* Suppl. 2.1–149.

Keirans, J. E., and Litwak, T. R. (1989). Pictorial keys to the adults of hard ticks, family Ixodidae (Ixodida: Ixodoidea), east of the Mississippi River. *Journal of Medical Entomology* **26**, 435–448.

Klompen, J. H. S., and Oliver, J. H., Jr. (1993). Systematic relationships in the soft ticks (Acari: Ixodida: Argasidae). *Systematic Entomology* **18**, 313–331.

Klompen, J. H. S., Black, W. C., IV, Keirans, J. E., and Oliver, J. H., Jr. (1996). Evolution of ticks. *Annu. Rev. Entomol.* **41**, 141–161.

Kocan K. M., Stiller, D., Goff, W. L., Edwards, W., Wickwire, K. B., Stick, W., Yellin, T. N., Ewing, S. A., Palmer, G. H., Barron, S. J., Hair, J. A., and McGuire, T. C. (1989). A review of the developmental cycle of *Anaplasma marginale* in *Dermacentor spp.* ticks. In "Proceedings of the 8th National Veterinary Hemoparasite Disease Conference, St. Louis, MO." pp. 129–148.

Kocan, K. M., Goff, W. L., Stiller, D., Edwards, W., Ewing, S. A., McGuire, T. C., Hair, J. A., and Barron, S. J. (1992). Development of *Anaplasma marginale* in salivary glands of male *Dermacentor andersoni. American Journal of Veterinary Research* **54**, 107–112.

Kohls, G. M. (1950). "Ticks (Ixodoidea) of the Philippines." National Institutes of Health Bulletin No. 192. US Govt. Printing Office, Washington, DC.

Kohls, G. M. (1957). "Malaysian parasites XVIII. Ticks (Ixodoidea) of Borneo and Malaya." Studies from the Institute for Medical Research, Federation of Malaya, No. 28. pp. 65–94. Govt. Press, Kuala Lampur.

Kurtenbach, K., Sewell, H.-S., Ogden, N. H., Randolph, S. E., and Nuttall, P. A. (1998). Serum complement sensitivity as a key factor in Lyme disease ecology. *Infect. Immun.* **66**, 1248–1251.

Lancaster, J. L., and Meisch, M. V. (1986). "Arthropods in Livestock and Poultry Production." Ellis Horwood, Chichester.

Lane, R. S. (1994). Competence of ticks as vectors of microbial agents with an emphasis on *Borrelia burgdorferi.* In "Ecological Dynamics of Tick-Borne Zoonoses" (D. E. Sonenshine and T. N. Mather, Eds.), pp. 45–67. Oxford Univ. Press, New York.

Lane, R. S., and Quistad, G. B. (1998). Borreliacidal factor in the blood of the western fence lizard (*Sceloporus occidentalis*). *Journal of Parasitology* **84**, 29–34.

Lane, R. S., Peek, J., and Donaghey, P. J. (1984). Tick (Acari: Ixodidae) paralysis in dogs from northern California: acarological and clinical findings. *Journal of Medical Entomology* **21**, 321–326.

Lane, R. S., Burgdorfer, W., Hayes, S. F., and Barbour, A. G. (1985). Isolation of a spirochete from the soft tick, *Ornithodoros coriaceus:* a possible agent of epizootic bovine abortion. *Science* **230**, 85–87.

Logan, T. M., Linthicum, K. J., Kondig, J. P., and Bailey, C. L. (1989). Biology of *Hyalomma impeltatum* (Acari: Ixodidae) under laboratory conditions. *Journal of Medical Entomology* **26**, 479–483.

Mahoney, D. F. (1977). *Babesia* of domestic animals. In "Parasitic Protozoa" (J. Kreier, Ed.), pp. 1–52. Academic Press, New York.

Mather, T. N., Ribeiro, J. M. C., and Spielman, A. (1987). Lyme disease and babesiosis: acaricide focused on potentially infected ticks. *American Journal of Tropical Medicine and Hygiene* **36**, 609–614.

Matthysse, J. G., and Colbo, M. H. (1987). "The Ixodid Ticks of Uganda." Entomological Society of America, College Park, MD.

Maupin, G. O., Fish, D., Zultowsky, J., Campos, E. G., and Piesman, J. (1991). Landscape ecology of Lyme disease in a residential area of Westchester County, New York. *American Journal of Epidemiology* **133**, 1105–1113.

McDade, J. E. (1990). Ehrlichiosis—a disease of animals and humans. *Journal of Infectious Diseases* **161**, 609–617.

McQuiston, J. H., Paddock, C. D., Holman, R. C., and Childs, J. E. (1999). The human ehrlichioses in the United States. *Emerging Infectious Diseases* **5**, 635–642.

Nicholson, W. L., Castro, M. B., Kramer, V. L., Summer, J. W., and Childs, J. E. (1999). Dusky-footed wood rats (*Neotoma fuscipes*) as reservoirs of granulocytic ehrlichiae (Rickettsiales: Ehrlichieae) in northern California. *Journal of Clinical Microbiology* **37**, 3323–3327.

Norval, R. A. I., Perry, B. D., and Young, A. S. (1992). "The Epidemiology of Theileriosis in Africa." Academic Press, London.

Norval, R. A. I., Sonenshine, D. E., Allan, S. A., and Burridge, M. J. (1996). Efficacy of pheromone-acaricide impregnated tail-tag decoys for control of bont ticks, *Amblyomma hebraeum* on cattle in Zimbabwe. *Experimental & Applied Acarology* **20**, 31–46.

Nuttall, P. A., and Jones, L. D. (1991). Non-viraemic tick-borne virus transmission: mechanism and significance. In "Modern Acarology, Vol. 2. Proceedings, 8th International Congress of Acarology" (F. Dusbabek and V. Bukva, Eds.), pp. 3–6. Academia Prague and SPB Academic Publ., The Hague.

Nuttall, P. A., and Labuda, M. (1994). Tick-borne encephalitis subgroup complex. In "Ecological Dynamics of Tick-Borne Zoonoses" (D. E. Sonenshine and T. N. Mather, Eds.), pp. 351–391. Oxford Univ. Press, New York.

Oliver, J. H., Jr., Owsley, M. R., Hutcheson, H. J., James, A. M., Chen, C., Irby, W. S., Dotson, E. M., and McClain, D. K. (1993). Conspecificity of the ticks *Ixodes scapularis* and *I. dammini* (Acari: Ixodidae). *Journal of Medical Entomology* **30**, 54–63.

Petrovec, M., Sumner, J. W., Nicholson, W. L., Childs, J. E., Strle, F., Barlic, J., Lotric-Furlan, S., and Avsic-Zupanc, T. (1999). Identity of ehrlichial DNA sequences derived from *Ixodes ricinus* ticks with those obtained from patients with human granulocytic ehrlichiosis in Slovenia. *Journal of Clinical Microbiology* **37**, 209–210.

Randolph, S. E. (1991). The effect of *Babesia microti* on feeding and survival in the tick vector *Ixodes trianguliceps*. *Parasitology* **102**, 9–16.

Randolph, S. E. (1994). Density-dependent acquired resistance to ticks in natural hosts, independent of concurrent infection with *Babesia microti*. *Parasitology* **108**, 413–419.

Randolph, S. E. (1995). Quantifying parameters in the transmission of *Babesia microti* by the tick *Ixodes trianguliceps* amongst voles (*Clethrionomys glareolus*). *Parasitology* **110**, 287–295.

Randolph, S. E., Gern, L., and Nuttall, P. A. (1996). Co-feeding ticks: epidemiological significance for tick-borne pathogen transmission. *Parasitology Today* **12**, 472–479.

Ribeiro, J. M. C. (1987). Role of saliva in blood feeding by arthropods. *Annu. Rev. Entomol.* **32**, 463–478.

Ribeiro, J. M. C. (1989). Role of saliva in tick/host interactions. *Experimental & Applied Acarology* **7**, 15–20.

Rikihisa, Y. (1991). The tribe Ehrlichieae and ehrlichial diseases. *Clinical Microbiology Reviews* **4**, 286–308.

Ristic, M. (1988). "Babesiosis of Domestic Animals and Man." CRC Press, Boca Raton, FL.

Roberts, F. H. S. (1970). "Australian Ticks." Commonwealth Scientific and Industrial Organization, Melbourne.

Savory, T. (1977). "Arachnida." Academic Press, New York.

Sonenshine, D. E. (1979). "Ticks of Virginia (Acari: Metastigmata)." Research Division Bulletin 139. Insects of Virginia No. 13. Virginia Polytechnic Institute and State Univ., Blacksburg.

Sonenshine, D. E. (1991). "Biology of Ticks." Vol. 1. Oxford Univ. Press, New York.

Sonenshine, D. E. (1993). "Biology of Ticks." Vol. 2. Oxford Univ. Press, New York.

Sonenshine, D. E., and Tignor, J. A. (1969). Oviposition and hatching in two species of ticks in relation to moisture deficit. *Annals of the Entomological Society of America* **62**, 628–640.

Sonenshine, D. E., Allan, S. A., Norval, R. A. I., and Burridge, M. J. (1996). A self-medicating applicator for control of ticks on deer. *Medical and Veterinary Entomology* **10**, 149–154.

Strickland, R. K., Gerrish, R. R., Hourrigan, J. L., and Schubert, G. O. (1976). "Ticks of Veterinary Importance." Agriculture Handbook. No. 485. Animal and Plant Health Inspection Service, US Department of Agriculture, Washington, DC.

Trager, W. (1939). Acquired immunity to ticks. *Journal of Parasitology* **25**, 57–81.

Valenzuela, J. G., Charlab, R., Mather, T. N., and Ribeiro, J. M. C. (2000). Purification, cloning and expression of a novel salivary anticomplement protein from the tick, *Ixodes scapularis*. *J. Biol. Chem.* **275**, 18717–18723.

Van Wye, J. E., Hsu, Y.-P., Terr, A. I., Lane, R. S., and Moss, R. B. (1991). Anaphylaxis from a tick bite. *New England Journal of Medicine* **324**, 777–778.

Waladde, S. M., and Rice, M. J. (1982). The sensory basis of tick feeding behavior. In "Physiology of Ticks" (F. D. Obenchain and R. Galun, Eds.), pp. 71–118. Pergamon Press, Oxford.

Webb, J. P., Jr., Bennett, S. G., and Challet, G. L. (1990). The larval ticks of the genus *Ixodes* Latreille (Acari: Ixodidae) of California. *Bulletin of the Society of Vector Ecology* **15**, 73–124.

Williams, R. E., Hall, R. D., Broce, A. B., and Scholl, P. J. (1985). "Livestock Entomology." Wiley, New York.

Wolf, L., McPherson, T., Harrison, B., Engber, B., Anderson, A., and Whitt, P. (2000). Prevalence of *Ehrlichia ewingii* in *Amblyomma americanum* in North Carolina. *Journal of Clinical Microbiology* **38**, 2795.

Woolley, T. A. (1988). "Acarology: Mites and Human Welfare." Wiley, New York.

Yamaguti, N., Tipton, V. J., Keegan, H. L., and Toshioka, S. (1971). Ticks of Japan, Korea and the Ryukyu Islands. *Brigham Young University Science Bulletin Biological Series* **15**(1), 1–226.

Yunker, C. E., Keirans, J. E., Clifford, C. M., and Easton, E. R. (1986). *Dermacentor* ticks (Acari: Ixodoidea: Ixodidae) of the New World: a scanning electron microscope atlas. *Proceedings of the Entomological Society of Washington* **88**, 609–627.

Zeidner, N. S., Burkot, T. R., Massung, R., Nicholson, W. L., Dolan, M. C., Rutherford, J. S., Biggerstaff, B. J., and Maupin, G. O. (2000). Transmission of the agent of human granulocytic ehrlichiosis by *Ixodes spinipalpis* ticks: evidence of an enzootic cycle of dual infection with *Borrelia burgdorferi* in northern Colorado. *Journal of Infectious Diseases* **182**, 616–619.

Zung, J. L., Lewengrub, S., and Rudzinska, M. A. (1989). Fine structural evidence for the penetration of the Lyme disease spirochete *Borrelia burgdorferi* through the gut and salivary tissues of *Ixodes dammini*. *Canadian Journal of Zoology* **67**, 1737–1748.

# TAXONOMIC INDEX

# SUBJECT INDEX

## A

Aadvarks, Glossinidae 308
Acalyptratae, 131
Acanthocephala, beetles as intermediate
    hosts 91, 98
    primates 42
Acari, *see* Mites
Acariasis, enteric acariasis 471
    esophageal acariasis 500
    gastric acariasis 500
    internal acariasis 453, 471–472,
        500–501
    nasal acariasis 503
    oral acariasis 500
    psoroptic otoacariasis 501
    pulmonary acariasis 471, 503
    respiratory acariasis 500, 503
    urinary acariasis 471
Acaricides, 553–554
Acarology, 8
Acarophobia, 6, 12, 453, 474–475
Acetic acid, 407
Acetylcholine, 389
Acronictinae, 374
Aculeate, 383
Adenotrophic vivipary, Hippoboscoidea
    352
Adephaga, 89
Adrenaline, *see* Epinephrine
African eyeworm, 271–272
African horsesickness, 163, 173,
    177–178
African sleeping sickness, 303, 309–312
African swine fever, 62, 549
African trypanosomiasis, 1, 3
Agas, 163

Agramonte, Aristides 231
Agu, 240
Ague, 240
Albendazole, 256
Aleppo Boil, 157
Allergies and allergic reactions, *see also*
    Hypersensitivity
    bed bugs 83
    beetles 89–91
    black flies 192, 193
    body lice 55
    Chironomidae 136
    cockroaches 38, 40–41
    Dermestidae 89
    desensitization therapy 406
    equine allergic dermatitis 173, 181
    flea bites 113, 121
    flea exuviae 114
    general 4, 11
    honey bees 390, 403
    Hymenoptera 403–406
    inhalational allergies 378
    mites 453, 460, 461, 468–471, 500,
        502
    moths 378
    silk-induced 378
    tick bites 547
Alphaviruses, 224, 225–229, 249
Altamont virus, 224
Amblycera, 46
American trypanosomiasis, *see* Chagas
    disease
American screwworm, *see* New World
    screwworm
American cockroach, 32–34, 38;
    pathogen transmission 42

American dog tick, 518, 532
Ammotrechidae, 425
Amphibians, biting midges 168
    chiggers 458, 459–460
    flesh flies 334
    hepatozoonosis 507
    myiasis 331, 334
    respiratory mites 505
    toad blow flies 331
    Triatominae 72
Amplification, epidemiological 17
Amplification cycles, 21
Amplifying host, 17, 26
Anal papillae, 208
Analgesics, 7
Anaphylactic shock, 198, 405; *see also*
    Anaphylaxis
Anaphylaxis, bed bug bites 83
    Hymenoptera stings 404–406
    tick bites 547
    tsetse fly bites 312
Anaplasmosis 532, 550–551
    bovine 62, 550–551
    ovine 358
    Tabanidae 263, 274
Anautogeny, black flies 190, 268
    Diptera 133
    mosquitoes 214–215
    muscid flies 286
Ankle bands, 346
Annoyance, by arthropods 4, 11
Annual transmission potential, 248
Anopheliinae, 204
Anoplocephalidae, 508–510
Anoplura, 45–65
Anteaters, trypanosomiasis 77